MECHANICS OF JOINTED AND FAULTED ROCK – MJFR-3

PROCEEDINGS OF THE THIRD INTERNATIONAL CONFERENCE ON MECHANICS OF JOINTED AND FAULTED ROCK – MJFR-3 / VIENNA / AUSTRIA / 6-9 APRIL 1998

Mechanics of Jointed and Faulted Rock

Edited by

HANS-PETER ROSSMANITH
Institute of Mechanics, Vienna University of Technology, Austria

Taylor & Francis
Taylor & Francis Group
LONDON AND NEW YORK

The texts of the various papers in this volume were set individually by typists under the supervision of each of the authors concerned.

Published by
2 Park Square, Milton Park, Abingdon, Oxon, OX14 4RN
270 Madison Ave, New York NY 10016

Transferred to Digital Printing 2007

ISBN 90 5410 955 6

Publisher's Note
The publisher has gone to great lengths to ensure the quality of this reprint but points out that some imperfections in the original may be apparent

Mechanics of Jointed and Faulted Rock, Rossmanith (ed.) © 1998 Taylor & Francis, ISBN 90 5410 955 6

Table of contents

Faults

Joints

Mining and underground construction

Slope stability

Hydromechanics

Mechanics of Jointed and Faulted Rock, Rossmanith (ed.) © 1998 *Taylor & Francis, ISBN 90 5410 955 6*

Preface

The Mechanics of Jointed and Faulted Rock (MJFR) conferences have become the traditional meetings of experts in this highly interdisciplinary field of research. As an international conference MJFR-3 is designed for presentation and publication of technical contributions in the broad subject of the mechanics of jointed and faulted rock.

Geotechnology in general and rock mechanics in particular have been constantly transformed by various factors. New perspectives are appearing at an ever accelerating rate and some of the pertinent decisive factors that are of major influence are:

- Cost reduction at increasing power of computation in computer hardware and software;
- Environmental awareness and concern and induced pressure on construction and mining industry;
- Use of advanced technology and techniques in all fields of geotechnical engineering;
- Awareness and concern about safety in large scale mining and excavation operations;
- Increased understanding of geotechnical materials and their behavior;
- Fierce market competition in science and engineering between the excellently educated and highly motivated researchers from the former Eastern European countries and their Western colleagues requires fast, reliable, economically competitive solutions to modern geotechnical problems;
- Manageability and solution of large scale problems involving complex interaction of various interdisciplinary fields of research.

The MJFR-3 conference again provides a forum for the presentation of new research results and discussion for a wide group of experts from all fields of the geosciences, rock mechanics, tectono-mechanics, geophysics, mining engineering, petroleum engineering, earthquake engineering, rock dynamics, tunneling etc.

The main objective of this conference is to present the state-of-the-art in those areas of engineering and science associated with the mechanics and physics of jointed and faulted rock. Special attention is given to the interface between theoretical concepts and modelling and practical applications.

Highlights at this conference will be contributions offering solutions to complex problems in underground constructions and mining, dam engineering and foundation, 3D-modelling, time-dependent behavior, dynamic loading and complex interaction between jointing and fluid flow, chemical degradation, porous rock and seismicity.

The third international conference MJFR-3 was again organized within the Vienna University of Technology by the Institute of Mechanics in cooperation with the Austrian Society of Geomechanics (ISRM Austrian National Group). More than 300 engineers and scientists have expressed their strong interest in this conference and have contributed more than 100 contributions of which a selection has been included in this proceedings volume.

Invited plenary lectures highlight the main themes of the conference and include:

- Mining below 3000m and challenges for the South African gold mining industry. G.R.Gurtunca, RSA;
- Fractures: Flow and geometry, L.J.Pyrak-Nolte, USA;
- Strain-induced damage of rocks, S.Sakurai et al., Japan;
- Seismic wave propagation in fractured rock, L.Myer, USA;
- Poroelastic applications, J.-C.Rogiers et al., USA;
- The contribution of Historical Earthquake Research to the analysis of seismic hazard and the tectonic stress field, R.Gutdeutsch, Austria;
- Quantitative tectonofractography – An appraisal, D.Bahat, Israel,
- Catastrophic sliding over a fault caused by accumulation of dilation zones, A.Dyskin et al., Australia.

It is hoped that the new results and experience exchanged at this conference will be taken into consideration and put to work by scientists and engineers who are concerned about the mechanics of jointed and faulted rock and interested in solving geotechnical problems in the future.

The Organizing Committee wishes to thank the authors for the timely submission of the original manuscripts and would like to acknowledge the help, efforts and advice offered by many colleagues.

The editor would like to express his sincere thanks to Dr Natalia Kouzniak, Dr Koji Uenishi, DI Rudolf Knasmillner and Claus Böswarth for their help in preparing the conference as well as this proceedings volume.

Vienna, April 1998
H.P.Rossmanith
(Chairman/Organizer)

Plenary lectures

Mechanics of Jointed and Faulted Rock, Rossmanith (ed.) © 1998 Taylor & Francis, ISBN 90 5410 955 6

Mining below 3000m and challenges for the South African gold mining industry

R.G.Gürtunca
Division of Mining Technology, CSIR, Johannesburg, South Africa

ABSTRACT: Potential new mining projects at ultra-depth are identified and some of the mining challenges at these depths are discussed. A new research programme, called DEEPMINE, is described and some of the important research projects are highlighted. Finally, the future of the South African gold mining industry is briefly discussed in terms of the development of new technology and people.

1 INTRODUCTION

Gold production in South Africa has decreased from 1200 tons in 1970 to 495 tons in 1996. There are three major reasons for the reduction in gold output: low productivity, high cost coupled with a low gold price and depletion of high grade gold reserves. This has made ultra-deep mining, between three and five kilometres underground, more attractive. A number of projects have been initiated as shown in Figure 1.

Current mining operations exceed depths of three kilometres and mines are anticipated to reach depths in excess of five kilometres, in some instances. For example, Anglogold has disclosed that, besides deepening Western Deep Levels to depths of up to five kilometres , the group plans the establishment of two mines near Carletonville, with depths similar to Western Deep Levels. JCI's South Deep mine will access 60 million ounces of gold, significant amounts of which are at ultra-depth. Gold Fields' Driefontein mine also envisages mining between three and four kilometres depth. AVGOLD too, is far advanced in its planning to develop the northern extension of the Free State goldfield. In addition, exploration has identified other deep level ore reserves, which could be developed in the future, such as Potchefstroom Gap and the extensions of ERPM and Durban Deep mines as shown in Figure 1.

Figure 2 shows the current and predicted mining depths on the gold mines in South Africa. At present, only 5% of production occurs below three kilometres and it is estimated (Willis, 1997) that some 40% of total South African production will be sourced from below these depths by 2015, assuming a favourable economic environment.

However, the South African gold mining industry faces a number of challenges to be able to mine economically and safely at ultra-depth. These challenges are:

i) rockbursts induced by high rock stresses (i.e. between 80 to 130 MPa vertical virgin stress),

ii) high virgin rock temperatures of 70-80 degrees Celsius at 5 km depth,

iii) the effect on human physiology, of the environment at these depths, and

iv) maximising productivity and minimising working costs under these conditions.

In this paper the various problems associated with mining at ultra-depth are discussed and the establishment of a new research programme called DEEPMINE is briefly described. This has been an initiative between the Council for Scientific and Industrial Research (CSIR), the University of the Witwatersrand in conjunction with the Foundation for Research and Development (FRD), and the South African gold mining industry.

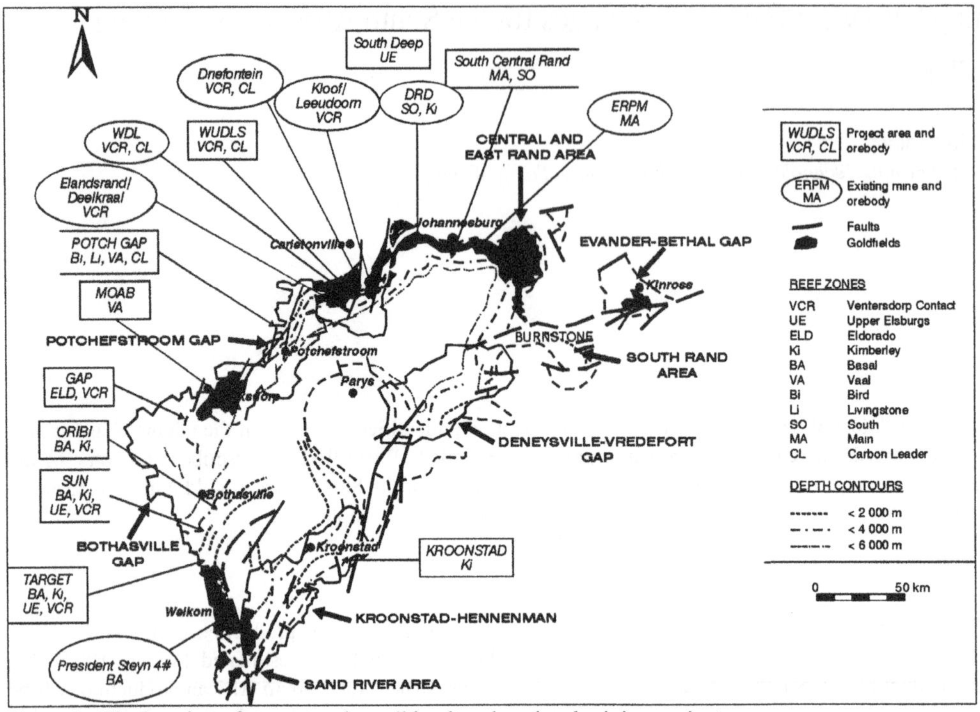

Figure 1. The location of current and possible ultra-deep level mining projects.

2 ROCK ENGINEERING CHALLENGES

Rockfalls and rockbursts account for more than a quarter of the total injuries in the mining industry and more than half of all fatalities, at present. In terms of mining at ultra-depth, the rock mass environment could be described as follows from a rock engineering point of view.

First of all, the virgin stresses at depths between three and five kilometres will be higher. The vertical component of the virgin stress tends to increase according to the mass of the overburden, that is, at about 27 MPa per 1000 m of depth below surface. Therefore it is predicted that the vertical virgin stress will vary between 80-120 MPa at ultra-depth. The horizontal virgin stresses are subject to considerable variation, but tend to increase according to Hor.Stress=10 MPa + 10 MPa /km of depth. Figure 3 shows typical virgin stress measurements carried out in South Africa. As shown in the figure, the k ratio (i.e. the ratio between horizontal stress and vertical stress) might become less than 0,5 at ultra-depth, and the lower values of the k ratio can increase the likely incidence of seismicity. However, this point should be verified by measuring virgin stresses at those depths. The expected increase in seismicity could occur in the vicinity of geological discontinuities due to low levels of clamping forces.

Figure 4 shows the relationship between mining depth and the ratio of rockbursts over rockfall fatalities. The figure clearly displays that the ratio of rockburst over rockfall fatalities increases significantly as the depth of mining increases. This is also an indication of higher seismicity levels as mining becomes deeper.

It is also expected that the degree of fracturing around deep level stopes will be much greater. Figure 5 shows the results of computer modelling which simulates fracturing around a stope at 2,5 and 5 km depths respectively. As expected, the rock fracturing becomes more extensive and particularly shear type fractures become more dominant. This means that the support of the face area would become more crucial in terms of keeping the fractured rock in place during violent seismic events. Therefore, it is very important for rock engineers to understand how the rock fractures and behaves during normal closure and during seismic events.

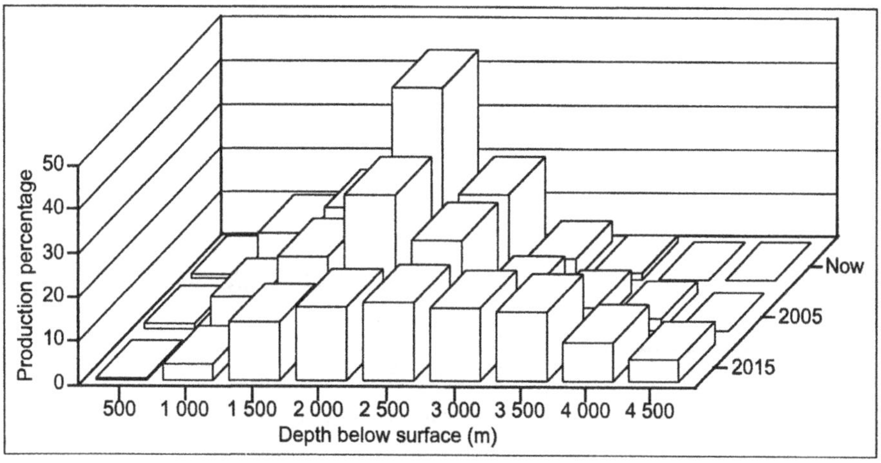

Figure 2. Current and predicted mining depths.

If one understands the interaction between intensively fractured rock mass and the seismic waves which travel through this medium it is possible to design more effective support systems to reduce rockburst damage in ultra-deep level stopes.

Seismic monitoring is extremely important for ultra-deep level mining. Measuring location and magnitude of seismic events could help in understanding how the rock mass behaves and explains the instabilities of geological discontinuities. Reliable seismic monitoring equipment has already been developed but the interpretation of seismic data to achieve a reliable seismic risk management system has not yet been successful.

It is also envisaged that more mechanised systems will be used in the future and this requires the development of new face area support systems for mechanised mining. These support systems should be effective in reducing rockburst damage and be able to move at the higher face advance rates of mechanised mining. Short roof bolts together with a membrane type of support system could be an option for use in the face area and along gullies.

The support of access ways, mainly tunnels, would also be a major challenge. If a mining method is used where tunnels are pre-developed then the support of these tunnels could become very difficult unless tunnels are excavated deeper in the footwall. Shotcrete and membrane support systems are gaining more acceptance for use as tunnel support.

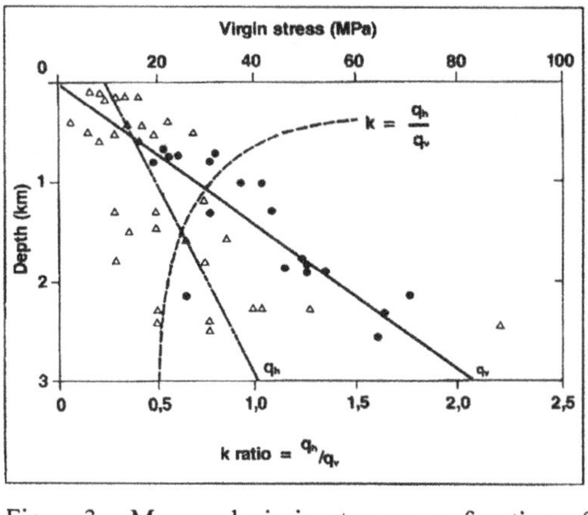

Figure 3. Measured virgin stresses as function of depth.

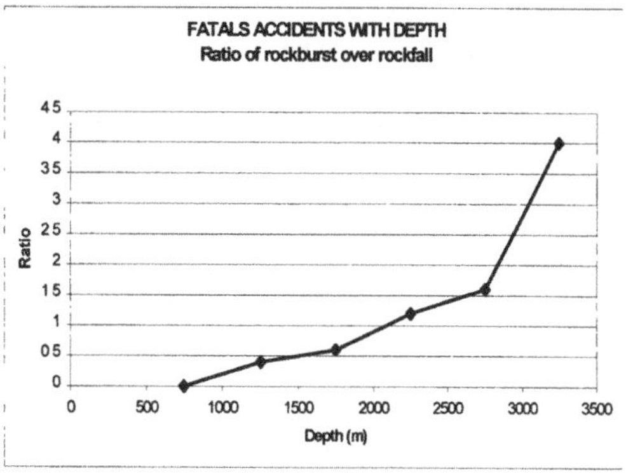

Figure 4. The relationship between depth of mining and the ratio of rockburst over rockfall fatalities.

All the factors discussed above should be integrated into a mining method where productivity is maximised and the occurrence of large seismic events is minimised. There are a number of ideas on new mining methods but these methods should be evaluated and tested against productivity and safety principals.

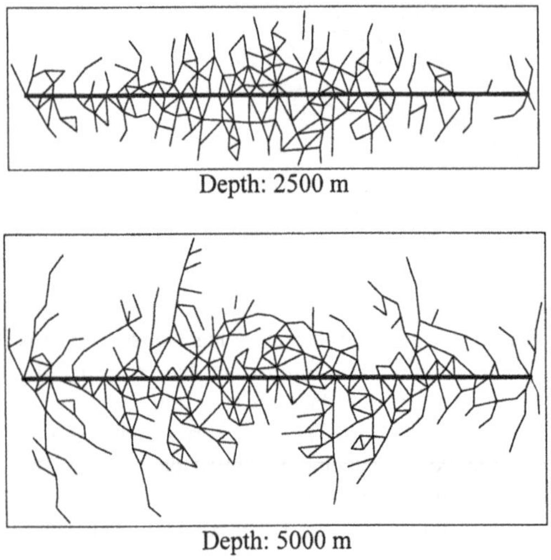

Figure 5. Effect of depth on the fracturing surrounding a tabular stope of span 20m.

3 UNDERGROUND VENTILATION AND COOLING

It is well established that the profound effect of depth on the underground environment is almost exclusively a consequence of the increased amount of low grade heat added to the ventilation air. In the context of South African gold mining, the average depth of mining has increased steadily over the past few decades. Simultaneously, the total installed refrigeration capacity on gold mines has grown in an apparently exponential manner. It is very probable that these trends will continue into the future and hence it is becoming ever more urgent that optimal solutions to the problems of deep mining be sought.

The temperature of the underground rock mass in its undisturbed state increases with depth in an approximately linear relationship. At a given depth, the temperature of the undisturbed rock is referred to as the virgin rock temperature, or VRT. The rate at

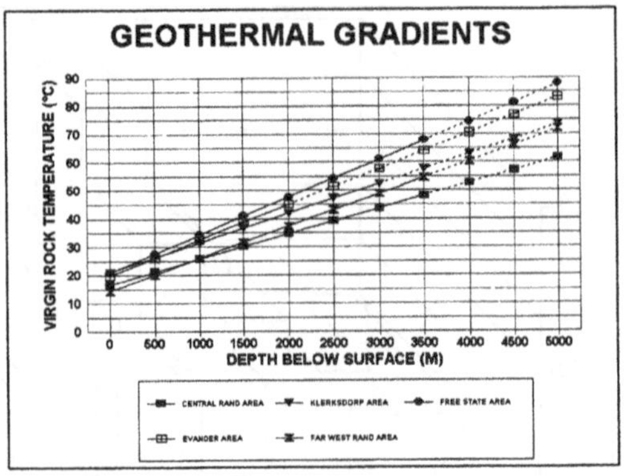

Figure 6. VRT versus depth for some SA mining areas.

which the temperature increases with depth is termed the geothermal gradient. This quantity varies considerably with location, and typical South African values are shown in Figure 6. The figure shows that VRTs are likely to reach above 70^0 C at ultra-depth.

Research in the past has indicated that human productivity decreases by as much as 10% for each degree centigrade increase in wet bulb temperature above a value of $27{,}5^0$ C. Additionally, the physiological discomfort of wet bulb temperatures above this value diminishes alertness, and a sharp increase in accident rates has been noted where this temperature is exceeded. The wet bulb temperature to a large extent determines the effective cooling power of the air with respect to the human body. An increase in wet bulb temperature inhibits the evaporative cooling by which the human body rejects metabolic heat.

Current refrigeration practices are at present adequate for the task of removing superfluous heat, but it is unlikely that they will remain effective at greater depth. The considerable capital and running costs associated with refrigeration plants will very probably necessitate a reassessment of these practices, since the cooling costs appear to grow significantly with depth as shown in Figure 7. The cost of cooling of a base case mine at 2 km depth is assumed to be unity. The cost factor indicates, for example, that the cost at 3 km is roughly six times as high. If perpetuated in their present guise, these practices will soon comprise a severe limiting factor on profitability. Specifically the practice of bulk air cooling will become progressively less efficient at increased depths owing to the greater temperature

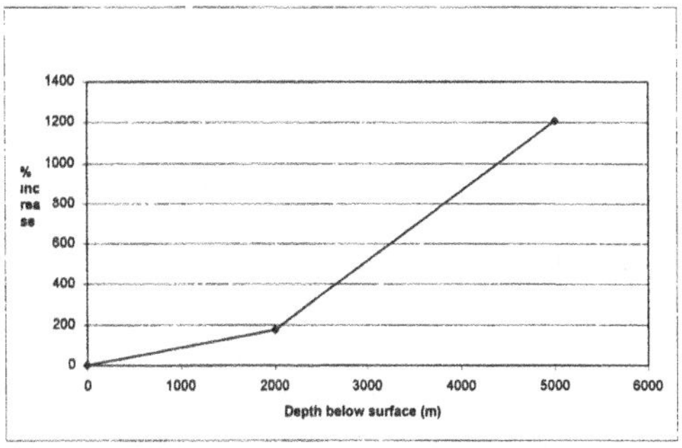

Figure 7. Cooling costs versus mining depth.

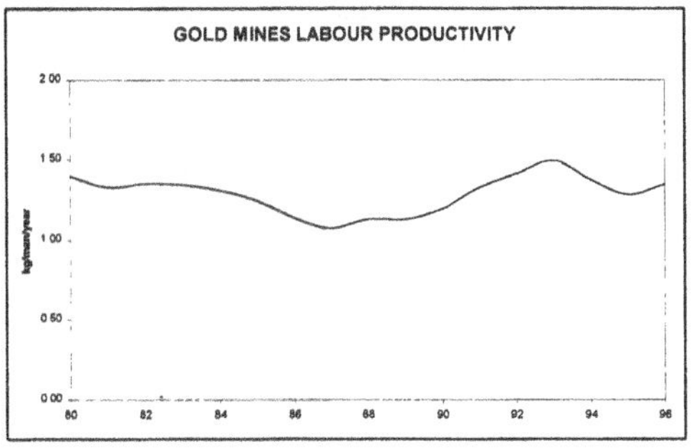

Figure 8. Change in labour productivity on South African gold mines.

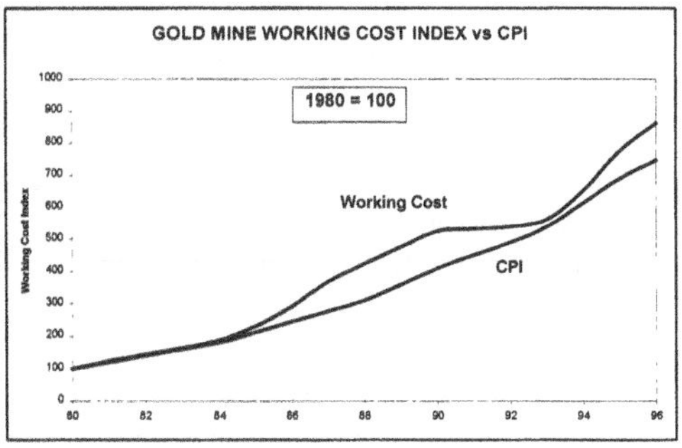

Figure 9. Change of working cost with respect to consumer price index.

differences, between the rock and the cooled air, which increase the rate of heat pick-up. In extreme cases, it would become necessary to cool the air several times before it reaches the working areas.

4 PRODUCTIVITY AND COST CHALLENGES

Productivity at the gold mines is relatively low due to the labour intensive nature of underground gold operations. Figure 8 shows the change in labour productivity since 1980. As shown in the figure, the productivity rates have not improved significantly over the years and this problem is exacerbated by greater increases in working costs with respect to the consumer price index, as displayed in Figure 9. The negative influence of these two factors (i.e. productivity and working cost) has reduced the profitability levels of South African gold mines. Obviously, the low gold price is also a major contributing factor to the reduction in profit margin of the gold mines.

Although the future of the gold mining industry is under serious threat in terms of mining at present depths and ultra-deep level mining, the mining industry has been introducing a number of measures. One of the major initiatives has been the formation of a new research programme called 'DEEPMINE' to address the problems of mining at ultra-depth mainly in terms of productivity and safety.

5 DEEPMINE RESEARCH PROGRAMME

A collaborative research programmed called "DEEPMINE" is being established at the time this paper was written. DEEPMINE was conceived by the CSIR Division of Mining Technology (Miningtek), the gold mining industry and the University of the Witwatersrand together with the Foundation for Research and Development (FRD). The objective of DEEPMINE is the development of technology for mining economically and safely between 3000 and 5000 metres. The project involves undertaking research and development into technologies which will enable profitable ultra-deep mining to take place in an environment which is at least as safe and healthy for employees as the top one-third of the industry, at present, under acceptable levels of financial risk. The vision of the project is to provide solutions and technologies, which will enable future mining operations to achieve concentrated mining with high productivity, high output and low resources.

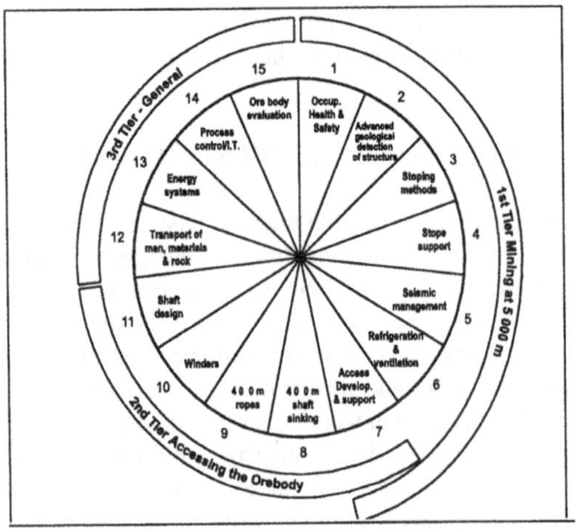

Figure 10. DEEPMINE technology wheel.

A technology element wheel which was originally constructed (Diering, 1997) to identify research needs is illustrated in Figure 10. The figure shows that the technology elements are divided into three tiers. The first tier represents the technology elements which are critical to mining at ultra-depth. The second tier elements are those required for accessing the ultra-deep mineral resources and the third tier represents general technology issues which are not necessarily unique to deep level mining.

The research needs are to be developed with respect to two scenarios. The scenario I projects would involve the optimisation and application of present technology to ultra-deep level mining or the development of new technologies which could be implemented within three to five years. The scenario II projects are for long-term future technologies such as the development of an underground mechanisation system, which could take 10-15 years of research and development.

Although the DEEPMINE research programme will produce many new technologies, some of those unique to ultra-deep mining are to be highlighted here. One of the projects would deal with barotrauma caused by moving people up and down 3-4 km long shafts. The physiological effects of increased barometric pressure on human exposure to airborne pollutants are also to be investigated. Since it is planned to have higher face advance rates and production, it is crucial to know all the geological features ahead of mining by integrating geophysical and geological data sets. There will be significant research investment in the development of new geophysical techniques to provide this information.

Some of the present mining methods would be tested for suitability to mine at 4-5 km depth. There may be a requirement for new methods or for modifying the existing ones.

One of the challenges would be to determine the effects of face advance rates on seismicity. This means that the time dependent behaviour of the rock mass should be better understood. There will be a number of projects in the rock engineering field but the most important one, which will give input to most of the projects in the DEEPMINE programme, is the understanding of the underground environment at ultra-depth. This obviously involves the understanding of rock mass behaviour at these depths under static and dynamic conditions. Seismic management will become a crucial issue because if the faces experience continuos rockburst damage then it could become very difficult to achieve high face advance rates and more importantly the safety of workers become a serious problem. New rockburst resistant support systems, which could be used in mechanised mining environment are also required.

Refrigeration and cooling is a major problem at these depths. At present, thirty per cent of the working cost of a gold mine is the refrigeration and cooling cost and these costs are likely to become higher at ultra-depth. This requires innovative alternatives to current ventilation and cooling techniques.

If a new shaft is to be sunk to 4 km depth it would take about 10 to 12 years before a mine can realise a positive cash flow. Obviously, this is a very long time to see a return on a more than \$1 billion investment. It is crucial that shaft sinking and development of access ways have to be done rapidly to reduce the 12 years by half.

The use of shotcrete and membrane support on tunnel walls would also be researched and applied. Conventional triangular strand ropes are almost certainly not going to be suitable. Instead, ropes which are torsionally neutral will be required, such as the non-spin ‘fishback’ construction used for kibble and stage ropes. What is not known is how these ropes would behave in a conventional hoisting situation, and what the expected life would be (Diering, 1997).

One of the new technologies, which could become available in three to five years time is the hydraulic transportation of broken ore from the gold face to surface. This technology could help to make a major change in mining layout design and to reduce costs significantly.

In summary, two hundred and ten projects have been identified in the research programme and these projects will be carried out during the next five to seven years. It is expected that the programme will produce about fifty MSc and PhD theses and is due to start 1 April 1998.

6 DISCUSSION

In this section of the paper the vision for future ultra-deep level gold mines in South Africa is discussed. It is expected that the gold mines will start reducing the number of employees, however, they will employ more qualified people and this will lead to higher productivity levels. It is envisaged that the gold mines will start applying more new technology and there will be incremental changes and improvements in the short to medium term. It is believed that the DEEPMINE programme will help the gold mines to provide these new technologies in the short term. However, it should be noted that one of the problems facing the industry is the effective implementation of new technologies. The DEEPMINE programme, while developing technologies, will also train new undergraduate and post-graduate engineers and scientists who understand these new technologies. These graduates will be more open minded to accept new technologies and implement them while on the gold mines.

However, future gold mines will be significantly different from the present ones. The incremental changes will improve productivity and reduce costs to a certain extent but the real challenge is to mechanise gold production and increase productivity levels at least ten times. It is believed that mines in the future (i.e. 15-20 years time) will use machines to break the rock face which will be operated remotely by highly-skilled people in air-conditioned control rooms positioned in tunnels. As indicated previously in this paper, the rockburst and cooling problems are maximised at stope faces where the gold reef is excavated. At present, the workers spend most of their time at the face where they need protection from rockbursts and heat problems. However, if machines are used in place of humans to break the rock and transport it away from the face area then, in the case of rockbursts, the human casualties could be eliminated and since the machines could work at high temperatures, the cooling requirements at the face area could be minimised. In the case of machine malfunctions and breakdowns, the technicians could use special suits to attend the problems at the face area. If there were to be longer term disruption due to installation or removal of machines then those isolated areas could be provided with cool air. There would be no need to cool the whole mine, as is done at present

This situation would change the number and education level of people working underground. Approximately 200 people could produce the same amount of gold produced by five thousand people today. People working underground would be mostly engineers and technicians instead of unskilled workers.

A second strategy which should be implemented, is the development of a local manufacturing industry, which could produce the necessary machines and equipment for the gold mines in South Africa. This would create new jobs where the present labour force could be re-employed with some necessary retraining.

I personally believe that although there have been significant changes in the market place in terms of mergers and acquisitions of mining companies, the key to the survival of the gold mines beyond the year 2000 is the development and growth of their people and the use of new technologies.

7 CONCLUSIONS

In this paper, the potential new mining projects at ultra-depth have been identified and some of the mining challenges at those depths have been discussed. These mining challenges are, rockburst problems, cooling and ventilation of underground workings and improved productivity.

A new research collaborative programme, called DEEPMINE has been described to find solutions for the challenges discussed. This research programme will develop new technologies and train under and post-graduate students during the next five to seven years.

It is believed that South African gold mines will go through incremental changes over the next three to five years but in the long-term, the gold mines should be mechanised to achieve significant increases in productivity and to improve safety to enable mining at ultra-depth.

REFERENCES

Schweitzer, J.K. and Johnson, R.A. 1997. Geotechnical classification of deep and ultra-deep Witwatersrand mining areas, South Africa. Mineralium Deposita 32: pp. 335 -348.

Willis, R.P.H. 1997. Towards an integrated system for deep level mining using new technology. *Proc. 4th Int. Symp. on Mine Mechanisation and Automation,* Brisbane, Vol. 2, pp. A 9.-.14.

An Industry Guide to Methods of Ameliorating the Hazards of Rockfalls and Rockbursts, 1988 edition. COMRO.

Diering, D.H. 1997. Ultra-deep level mining - future requirements. Journal of South Afr. Inst. Mining and Metallurgy. Vol. 97, No 6. pp. 249 - 255.

Mechanics of Jointed and Faulted Rock, Rossmanith (ed.) © *1998 Taylor & Francis, ISBN 90 5410 955 6*

Fractures: Flow and geometry

L.J. Pyrak-Nolte
Department of Physics, Purdue University, West Lafayette, Ind., USA

ABSTRACT: An analysis of the interrelationship between fracture specific stiffness and fluid flow through a fracture was performed on data from ten different rock cores each containing a single fracture. The data were from rock cores from different geographic locations and of different sizes and appear to follow the same sigmodial stiffness-fluid flow curve that shows a nine-order-of-magnitude decrease in flow with a three order of magnitude increase in fracture specific stiffness.

1 INTRODUCTION

Figure 1 shows four direct relationships between the mechanical and hydraulic properties of a single fracture subjected to normal stress: (1) fluid flow through the fracture depends on the aperture distribution of the fracture; (2) flow through a fracture depends on the contact area of the fracture; (3) fracture specific stiffness depends on the amount and spatial distribution of the contacts; and (4) fracture specific stiffness depends on the aperture distribution of the fracture. Because fluid flow and fracture specific stiffness depend the geometry of the fracture through the size and spatial distribution of aperture and contact area, fluid flow through the fracture and fracture specific stiffness must be implicitly related.

To uncover this implicit relationship, it is first necessary to quantitatively establish the direct relationships illustrated by Figure 1. The first direct relationship between fluid flow and fracture aperture was established from experimental, theoretical, and numerical investigations of fluid flow through a fracture. Lomize (1951) found that laminar flow between two glass plates depends on the cube of the separation (aperture) between the plates. This relationship is the Reynolds equation for viscous flow between parallel plates and is often referred to as the cubic law. The applicability of the cubic law to flow through fractures has been explored by many investigators both experimentally and analytically (Snow, 1965; Iwai, 1976; Gangi, 1978; Witherspoon et al., 1980; Engelder & Scholz, 1981; Raven & Gale, 1985; Pyrak-Nolte et al., 1987; Gale, 1987; Tsang & Tsang, 1987; Pyrak-Nolte et al., 1988; Hakami, 1988; Zimmerman et al., 1991; Iwano et al., 1995; Durham & Bonner, 1995; etc.). Deviations from cubic law behavior have been attributed to irreducible flow (Pyrak-Nolte et al., 1987), surface roughness (Kranz et al., 1979; Raven & Gale, 1985; Brown, 1987), tortuosity (Tsang 1984; Cook, 1992), and a non-linear relationship between the hydraulic aperture and mechanical displacement (Pyrak-Nolte et al, 1988). While deviations from cubic law have been observed overall fluid flow through a fracture will depend on the aperture of the

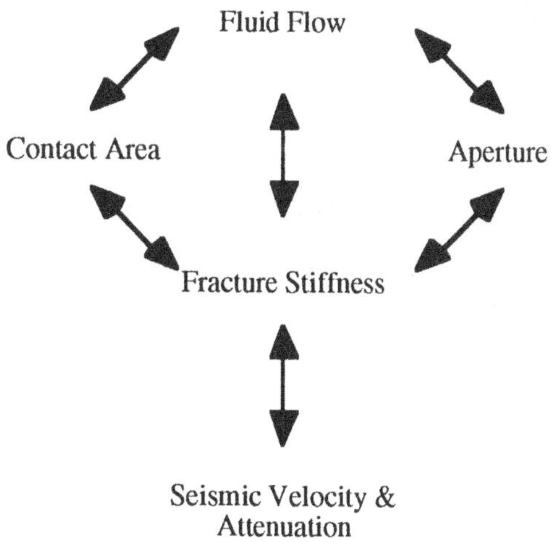

Figure 1. Fracture stiffness and fluid flow through a fracture are implicitly interrelated through the geometry of the fracture.

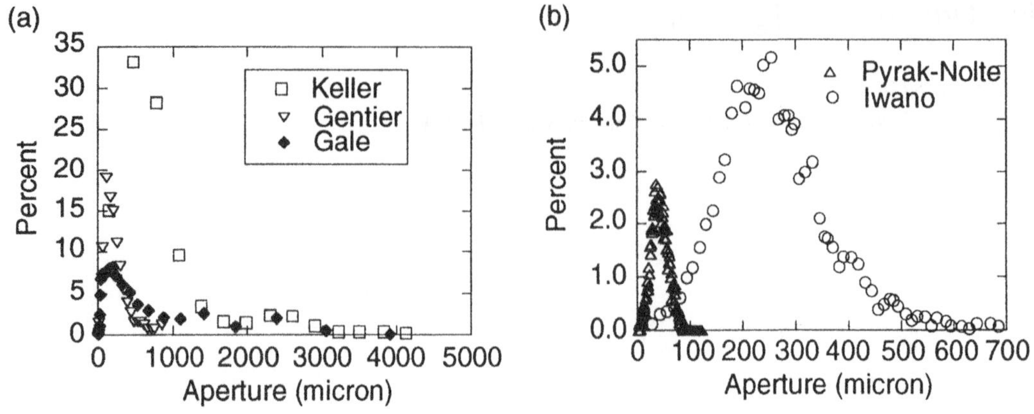

Figure 2a&b. Measured aperture distributions for single fractures from sample C of Keller (1997), sample H1 of Gale (1987), natural tensile fracture from Gentier et al. (1990), natural fracture from Iwano & Einstein (1995), and data from a fracture in a coal core from Pyrak-Nolte (unpublished). Histogram is expressed in percent for comparison.

fracture. Much of the recent work on understanding the relationship between fluid flow and fracture aperture has focused on imaging and quantifying the aperture distributions observed in single fractures (Figure 2). These measurements were made using a variety of techniques including x-ray tomography (Pyrak-Nolte et al., 1997; Keller, 1997), surface topography (Durham & Bonner, 1995; Durham, 1997; Iwano & Einstein, 1995), and void casting/injection (Gale, 1987; Gentier et al., 1989; Hakami, 1990 & 1995). The data in Figure 2 show that for a wide variety of fractures the aperture size distribution for single fractures can be either gaussian or log-normal.

Only a few investigations have studied the second direct relationship that relates fluid flow to contact area because measurement of the contact area between two rough surfaces in contact is difficult. Methods to measure contact area as a function of normal loading have involved pressure-sensitive paper (Duncan and Hancock, 1966), deformable films (Iwai, 1976; Bandis et al., 1983) and photomicroscopy of Wood's metal injected fractures (Pyrak-Nolte et al. 1987). Pyrak-Nolte et al. (1987) observed that fluid flow through the fracture decreased as contact area increased. As stress on the fracture increased, the apertures of the fracture were closed, thereby increasing the contact area between the two surfaces and decreasing the connectivity of the void space.

The third and fourth direct relationships that relate fracture specific stiffness to contact area, and fracture specific stiffness to displacement, have been studied by several investigators through measurements of asperities or surface roughness, and theoretical and numerical analysis of asperity deformations (Goodman, 1976; Swan, 1983; Bandis et al., 1983; Barton et al, 1985; Brown & Scholz, 1985 & 1986; Cook, 1992). Fracture specific stiffness is defined as the ratio of the increment of stress to the increment of displacement caused by the deformation of the void space in the fracture. As stress on the fracture increases, the contact area between the two fracture surfaces also increases raising the stiffness of the fracture. From elasticity methods and experimental data, fracture specific stiffness depends on the elastic properties of the rock and depends critically on the amount and distribution of contact area in a fracture that arises from two rough surfaces in contact(Brown & Scholz, 1985; Hopkins et al. 1987 & 1990). Kendall & Tabor (1971) showed experimentally and Hopkins et al. (1987, 1990) have shown numerically that interfaces with the same amount of contact area but different spatial distributions of the contact area will have different stiffnesses. Greater separation between points of contacts results in a more compliant fracture or interface. In addition, Bandis et al. (1983) and Pyrak-Nolte et al. (1987) noted that more compliant fractures tended to have larger apertures.

Sufficient experimental and theoretical evidence exists to support the four direct relationships illustrated in Figure 1. Because of these direct relationships, fluid flow through the fracture and the fracture specific stiffness are implicitly related through the geometry of the fracture. Pyrak-Nolte (1995 & 1997) presented experimental evidence to support the interrelationship between fracture specific stiffness and fluid flow through a fracture, i.e., a fracture with a high specific stiffness will support less fluid flow than a more compliant

Table 1. Sample name, source of data reference, sample dimensions, rock type, and type of flow measurement for each sample used in the analysis.

Sample Name	Reference	Sample Dimensions (length, diameter)	Rock Type	Flow Measurement
H1	Gale (1987)	0.338 m, 0.159 m	Granite, URL, Manitoba	Radial
STR2	Gale (1987)	0.483 m, 0.152 m	Stripa Granite	Radial
Sample 1	Raven & Gale (1985)	0.7 m, 0.100 m	Charcoal Black Granite, Coldsprings, Minnesota	Radial
Sample 2	Raven & Gale (1985)	0.7 m, 0.150 m	Charcoal Black Granite, Coldsprings, Minnesota	Radial
Sample 3	Raven & Gale (1985)	0.7 m, 0.193 m	Charcoal Black Granite, Coldsprings, Minnesota	Radial
Sample 4	Raven & Gale (1985)	0.7 m, 0.294 m	Charcoal Black Granite, Coldsprings, Minnesota	Radial
E30	Pyrak-Nolte et al. (1987)	0.07 m, 0.052 m	Stripa Granite	Diametric
E32	Pyrak-Nolte et al. (1987)	0.07 m, 0.052 m	Stripa Granite	Diametric
E35	Pyrak-Nolte et al. (1987)	0.07 m, 0.052 m	Stripa Granite	Diametric
Granite	Witherspoon et al. (1980)	length = 0.207 m, width = 0.121 m, height = 0.155 m	Granite	Straight Flow

fracture. This is an important inter-relationship because measurements of seismic velocity and attenuation (Pyrak-Nolte et al., 1990 & 1992) can be used to determine remotely the specific stiffness of a fracture in a rock mass. If this relationship holds, seismic measurements of fracture specific stiffness can provide a tool for predicting the hydraulic properties of a fractured rock mass. It is this implicit relationship between fracture specific stiffness and fluid flow that will be further examined in this paper using previously published data for ten different rock cores.

2 DATA AND SAMPLES

The hydraulic and mechanical data used in this paper have been obtained from Witherspoon et al. (1980), Raven & Gale (1985), Gale (1987), and Pyrak-Nolte (1987). Table 1 lists the sample name, source of the data, rock type, sample dimensions, and type of flow measurement made. All the studies used an applied normal load only. The values of fluid flow and fracture displacement were obtained from published data and fracture specific stiffness was determined numerically from the inverse of the slope of the displacement-stress curves for each sample. In previous work on the interrelationships among fracture properties, Pyrak-Nolte (1995) used values of fluid flow that were corrected for irreducible (Pyrak-Nolte et al., 1988) or residual flow. In this paper, no correction is made for irreducible flow.

2.1 *Mechanical and Hydraulic Properties as a Function of Normal Loading*

Traditionally, investigators have examined the mechanical and hydraulic properties of a single fracture as a function of stress and then developed relationships between stress and flow or stress and displacement. Figure 3 shows the flow per unit

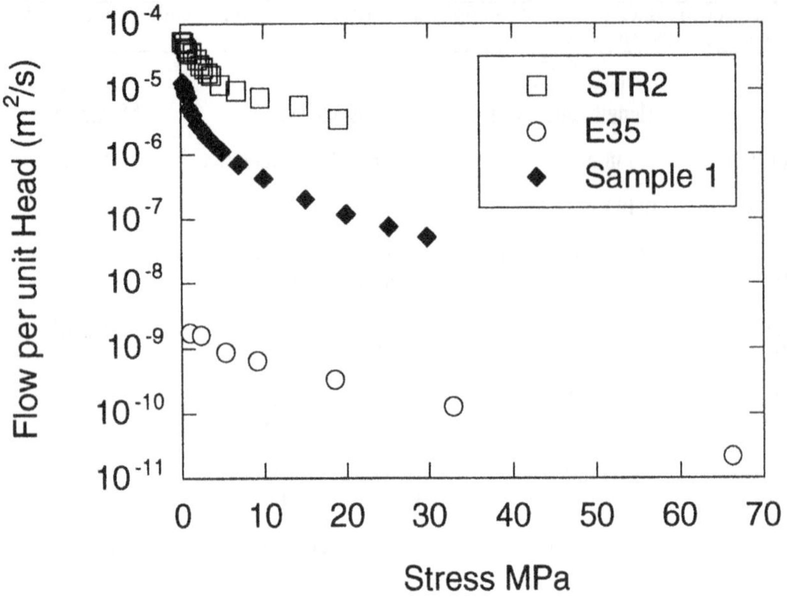

Figure 3. Flow per unit head as a function of stress for three different samples each containing a single fracture.

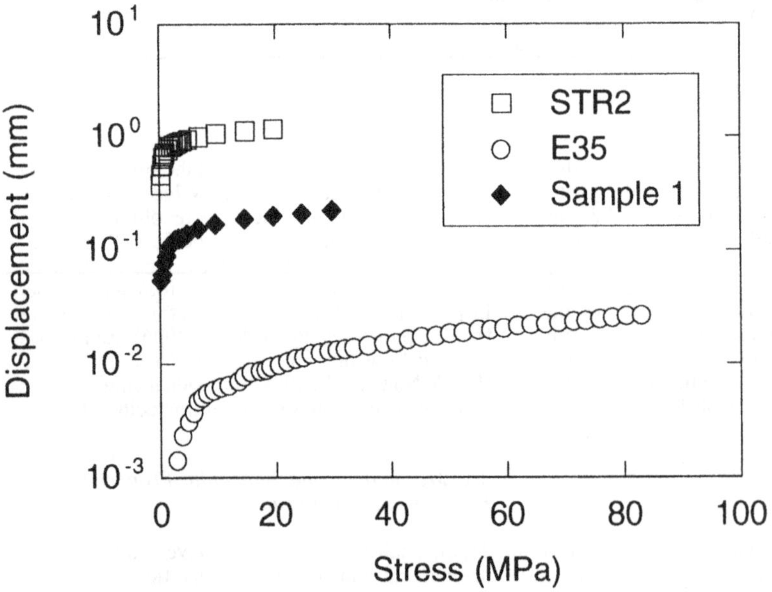

Figure 4. Displacement as a function of applied normal stress for three different samples each containing a single fracture.

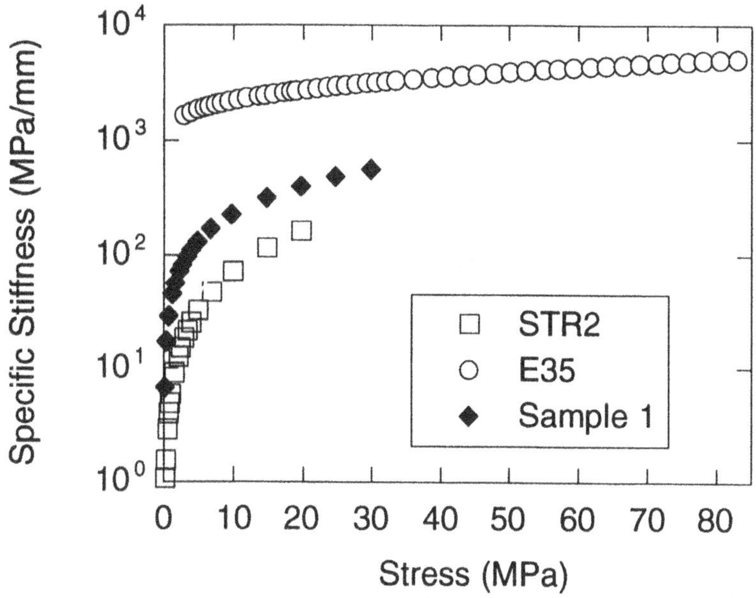

Figure 5. Fracture specific stiffness as a function of applied normal stress for three different samples each containing a single fracture.

head as a function of normal stress for fracture samples STR2, Sample 1, and E35 for data taken during the first loading cycle. All of the fractures exhibited a decrease in the amount of fluid flow with increasing normal stress on the fracture. As stress is applied the apertures are reduced in size and the contact area increases thus reducing the flow. While the three samples exhibit the same qualitative trends, the amount of flow among the fracture samples vary from one to several orders of magnitude. Similarly, the displacement (Figure 4) and the fracture specific stiffness (figure 5) for fracture samples STR2, Sample 1, and E35 exhibit the same qualitative trends with increasing applied normal load but vary in the amount of displacement by hundreds of microns. As stress is applied to the fracture, the displacement increases as apertures are closed by the applied normal load and the fracture specific stiffness also increases as the contact area increases with an increase in stress. By examining the data as a function of stress, all three fractures appear to behave very differently and any interrelationship among the fracture properties is obscured. This arises because stress is not the link between the hydraulic and mechanical properties of a fracture. The link between these properties is the fracture geometry and how it deforms under stress. For instance, by comparing the mechanical deformation data and fracture specific stiffness, it is observed that the fracture with smallest displacement (E35) is the stiffest of the three fractures. In addition, the most compliant fracture (STR2) supported the most fluid flow.

3 ANALYSIS OF THE INTERRELATIONSHIP BETWEEN FRACTURE SPECIFIC STIFFNESS AND FLUID FLOW

In this section, we will examine the interrelationship between fracture specific stiffness and fluid flow per unit head for all ten samples. Figure 6 contains the fluid flow - fracture specific stiffness interrelationship observed for 10 samples ranging over several orders of magnitude in the amount of flow and the fracture specific stiffness, and ranging from 0.052 m to 0.295 m in size. All of the data shown in Figure 6 is from the first loading cycle and the fluid flow measurements are not corrected for irreducible flow. There appears to be two types of relationship between fracture specific stiffness. The first relationship is exhibited by the data from samples STR2, Sample 1, Sample 2, Sample 3, E30, E32, E35 which tend to fall on a sigmodial curve that shows a nine-order-of-magnitude decrease in flow with a three order of magnitude increase in fracture specific stiffness. The second relationship illustrated by fracture samples H1, Sample 5, and Granite, shows that the flow is essentially independent of stiffness or stress. The

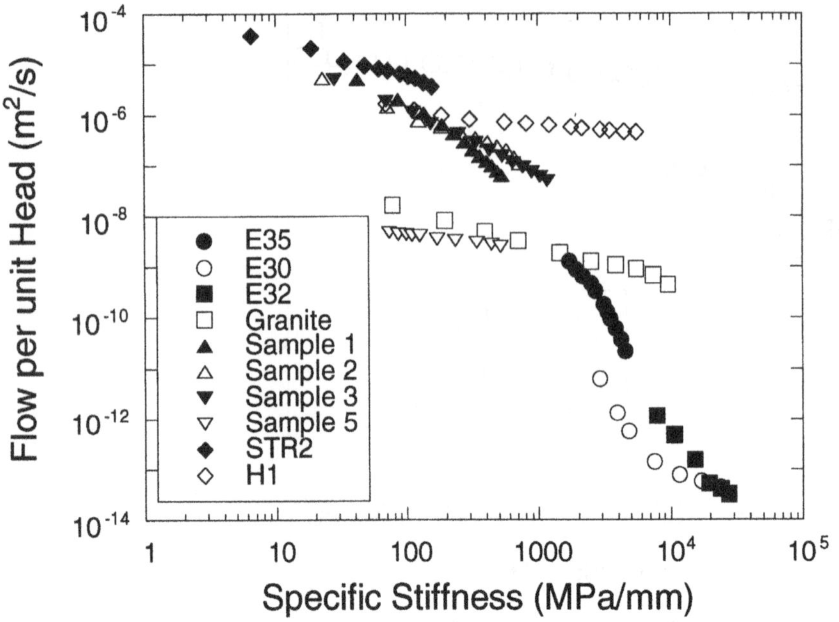

Figure 6. Observed fluid flow - fracture specific stiffness inter-relationship for ten samples each containing a single fracture.

the same behavior, i.e., the same approximate decrease in flow with increasing fracture specific stiffness. However, by the 2nd loading cycle, the samples no longer exhibit the same inter-relationship. The data for the three samples show similar trends (slopes), but the magnitude of the flow for a given stiffness varies by an order of magnitude. This may suggest that frictional effects during unloading and re-loading may be a function of scale size or that the inter-relationship may break down after repeated cycling. More experimental data are needed to explore scale and loading affects. Also, it is important to note that Sample 5 data from Raven & Gale (1985) did not exhibit the same interrelationship as Sample 1, 2, and 3 even for the first loading cycle (Figure 6).

4 SUMMARY

While it is clear that a single fracture will exhibit a relationship between flow and stiffness, there are stiffness independence suggests that there is a critical path (Nolte et al., 1989) that is insensitive to increases in contact area or closure, i.e., an irreducible flow.

One question concerning the inter-relationship between fluid flow and fracture specific stiffness is whether this relationship scales with sample size? The data from Raven & Gale (1985) can be used to examine the effect of sample size on the interrelationship between fluid flow and fracture specific stiffness because the sample cores were all extracted from the same fracture in the same rock. Figures 7 and 8 compare the fracture specific stiffness - fluid flow interrelationship for samples of different sizes subjected to the same loading history. For the first loading cycle, the three samples exhibit several broader questions. First, how does this relationship depend on the fracture geometry. And second, are there subsets of fractures that will all obey the same relationship. The answer to the first question is almost certainly yes, because of the well-established relationships between flow and aperture and between flow and contact area. The answer to the second question is partly addressed here. Data from several samples appear to follow the same sigmodial stiffness-fluid flow curve even though the rock cores were from different geographic locations and were different sizes. However, if the hydraulic path (i.e. the critical path or path of least resistance) is unaltered or only slightly altered by the application of a normal load, fluid flow through the fracture will be insensitive to changes in fracture specific stiffness.

Several questions remain, such as: Will rocks from the same tectonic setting fall on the same fracture-specific stiffness -fluid flow curve, or will

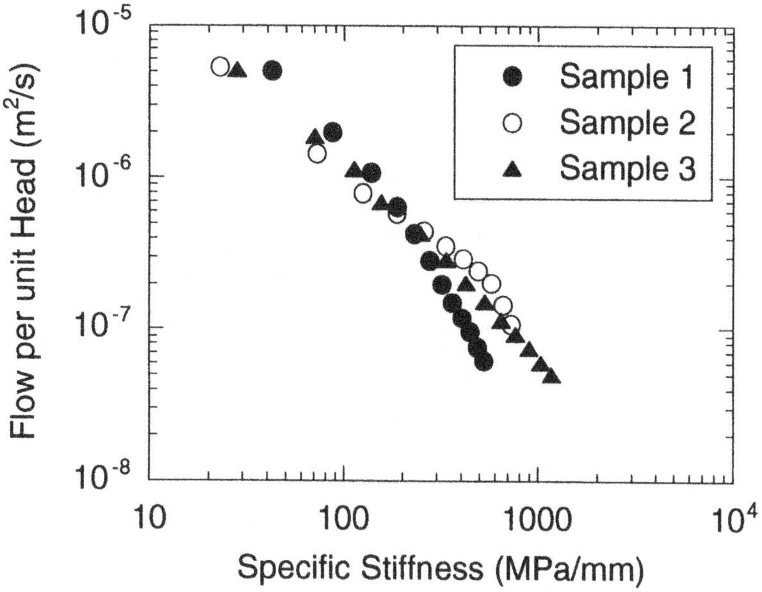

Figure 7. Comparison of fluid flow -specific stiffness relationship for samples taken from the same fracture but with different dimensions. Data are from the first loading cycle.

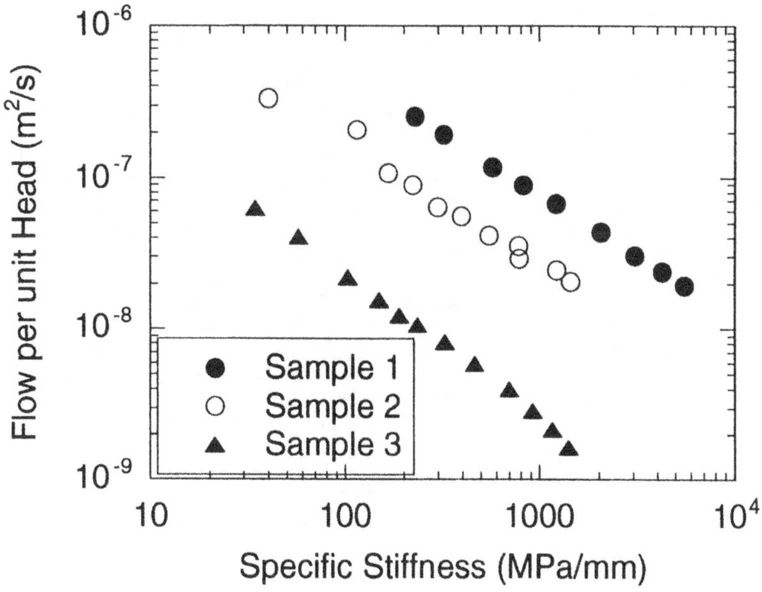

Figure 8. Comparison of the fluid flow - fracture specific relationship for data from the 2nd loading cycle for three samples with different dimensions.

there be different universality classes based on rock type or stress history? Will repeated stress cycling alter the relationship? Will the relationship be scalable to the field scale? These questions cannot be addressed without more experimental results or without experimentally validated numerical codes to simulate fluid flow and fracture deformation.

5 ACKNOWLEDGMENTS

The author wishes to thank the National Science Foundation - Young Investigator Award from the Division of Earth Sciences (NSF/94 58373-001) for support of this research.

6 REFERENCES

Bandis, S. C., A. C. Luden, and N. R. Barton. 1983. Fundamentals of rock joint deformation. *Int. J. Rock Mech. Min Sci. Geomech. Abstr.* 20(6): 249-268.

Barton, N.R., Bandis, S. and K. Bakhtar. 1985. Strength, deformation and conductivity coupling of rock joints. *Int. J. Rock Mech. Min Sci. Geomech. Abstr.* 22(3): 121-140.

Brown, S. R. and C. H. Scholz. 1985. Closure of random surfaces in contact. *J. Geophys. Res.*, 90:B7: 5531-5545.

Brown, S. R., and C. H. Scholz. 1986. Closure of rock joints. *J. Geophys. Res*, 91:4939.

Brown, S.R. 1987. Fluid flow through rock joints: The effect of surface roughness. *J. Geophysical Res.* 92:1337-1347.

Duncan, N. and K.E. Hancock. 1966. The concept of contact stress in assessment of the behavior of rock masses as structural foundations. *Proc. First Congress Int. Soc., Rock Mech.* 2:487-492. Lisbon.

Durham, W.B. and B.P. Bonner. 1995. Closure and fluid flow in discrete fractures. In L.R. Myer, N.G.W. Cook, R.E. Goodman & C.F. Tsang (eds) *Fractured and Jointed Rock Masses.* 441:446. Rotterdam: Balkema.

Durham, W.B. 1997. Laboratory observations of the hydraulic behavior of a permeable fracture from 3800 m depth in the KTB pilot hole. *J. Geophys. Res.* 102:B9:18405-18416.

Engelder, T. and C.H. Scholz. 1981. Fluid flow along very smooth joints at effective pressure up to 200 megapascals. *Mechanical Behavior of Crustal Rocks. Amer. Geophys. Union Mongraph* 24:147.

Gale, J. E. 1987. Comparison of coupled fracture deformation and fluid models with direct measurements of fracture pore structure and stress-flow properties. In I.W. Farmer, J.J.K. Daemen, C.S. Desai, C.E. Glass & S.P. Neuman. (eds) *Rock Mechanics: Proceedings of the 28th US Symposium.* Tucson, Arizona. Rotterdam. 1213-1222. Rotterdam: Balkema.

Gangi, A. F. 1978. Variation of whole- and fractured-porous-rock permeability with confining pressure. *Int. J. Rock Mech. Min. Sci. & Geomech. Abstr.* 15: 249-257.

Gentier, S., Billaux, D. and L. van Vliet. 1989. Laboratory testing of the voids in a fracture. *Rock Mechanics and Rock Engineering.* 22:149-157.

Gentier, S. 1990. Morphological analysis of a natural fracture. Selected papers from the 28th Int. Geological Congress. Washington, D.C. July 9-19, 1989. *Hydrogeology.* 1:315-326.

Goodman, R.E. 1976. *Methods of geological engineering in discontinuous rock.* p172. New York: West Publishing.

Hakami, E. 1988. *Water flow in single rock joints.* Licentiate. Thesis. Lulea University of Technology. Sweden.

Hakami, E. 1995. Joint aperture measurements - An experimental technique. In L.R. Myer, N.G.W. Cook, R.E. Goodman & C.F. Tsang (eds) *Fractured and Jointed Rock Masses.* 453:456. Rotterdam: Balkema.

Hopkins, D. L., Cook, N.G.W. & L.R. Myer. 1987. Fracture stiffness and aperture as a function of applied stress and contact geometry. In I.W. Farmer, J.J.K. Daemen, C.S. Desai, C.E. Glass & S.P. Neuman. (eds) *Rock Mechanics: Proceedings of the 28th US Symposium.* Tucson, Arizona. 673-680. Rotterdam. : Balkema

Hopkins, D. L., Cook, N.G.W. and L.R. Myer. 1990. Normal joint stiffness as a function of spatial geometry and surface roughness. *International Symposium on Rock Joints.* 203-210. Rotterdam: Balkema.

Iwai, K. 1976. *Fundamental studies of fluid flow in a single fracture.* Ph.D. Thesis. 208pp. Univ. of California.

Iwano, M. and H. H. Einstein. 1995. Laboratory experiments on geometric and hydromechanical characteristics of three different fractures in granodiorite. In T. Fujui (ed.) *Proceedings of the Eight International Congress on Rock Mechanics.* 2:743:750. Rotterdam: Balkema.

Keller, A.A. 1997. High resolution cat imaging of fractures in consolidated materials. In K. Kim (ed.) *Proceddings for the 36th U.S. Rock Mechanics Symposium and ISRM International* Symposium. 1:97:106. New York: Columbia University.

Kendall, K. and D. Tabor. 1971. An ultrasonic study of the area of contact between stationary and sliding surfaces. *Proc. Royal Soc. London, Series A.* 323: 321-340.

Kranz, R.L., Frankel, A.D., Engelder, T. and C.H. Scholz. 1979. The permeability of whole and jointed barre granite. *Int. J. Rock Mech. Min. Sci. & Geomech. Abstr.* 16: 225-234.

Lomize, G.M. 1951. Water flow through jointed rock (in Russian). Gosenergoizdat. Moscow.

Nolte, D.D., Pyrak-Nolte, L.J., and N.G.W. Cook. 1989. Fractal Geometry of the Flow Paths in Natural Fractures and the Approach to Percolation. *Pure and Applied Geophysics.* 131:1/2:111.

Pyrak-Nolte, L.J., Myer, L.R., Cook, N.G.W., and P.A. Witherspoon. 1987. Hydraulic and mechanical properties of natural fractures in low permeability rock. In G. Herget & S. Vongpaisal. (eds) *Proceedings of the Sixth International Congress on Rock Mechanics.* Montreal, Canada. August 1987. 225-231.Rotterdam:Balkema.

Pyrak-Nolte, L.J., Cook, N.G.W., and D.D. Nolte. 1988. Fluid Percolation through Single Fractures. *Geophysical Research Letters.* 15:11:1247-1250.

Pyrak-Nolte, L.J., Myer, L.R., and N.G.W. Cook. 1990. Transmission of Seismic Waves across Natural Fractures. *Journal of Geophysical Research.* 95:B6:8617-8638.

Pyrak-Nolte, L.J., Xu, J., and G.M. Haley. 1992. Elastic Interface Waves Propagating in a Fracture. *Physical Review Letters.* 68:24:3650-3653.

Pyrak-Nolte, L.J. 1995. Interrelationship between the hydraulic and seismic properties of fractures. In L.R. Myer, N.G.W. Cook, R.E. Goodman & C.F. Tsang (eds) *Fractured and Jointed Rock Masses.* 111-117. Rotterdam: Balkema.

Pyrak-Nolte, L.J. 1997. The seismic response of fractures and the interrelations among fracture properties. *Int. J. Rock Mech. Min. Sci. & Geomech. Abstr.* 33:8: 787-802.

Pyrak-Nolte, L. J., Montemagno, C.D. and D.D. Nolte. 1997. Volumetric imaging of aperture distributions in connected fracture networks. *Geophys. Res. Letters.* 24:18:2343-2346.

Raven, K.G. and J.E. Gale. 1985. Water flow in a natural rock fracture as a function of stress and sample size. *Int. J. Rock Mech. Min. Sci. & Geomech. Abstr.* 22:4: 251-261.

Snow, D.T. 1965. *A parallel plate model of fractured permeable media.* Ph.D. Thesis. 331pp. Univ. of California.

Swan, G. 1983. Determination of stiffness and other joint properties from roughness measurements. *Rock Mech. Rock Eng.* 16:19-38.

Tsang, Y.W. 1984. The effect of tortuosity on fluid flow through a single fracture. *Water Resources Res.* 20:9:1209-1215.

Tsang, Y.W. and C.F. Tsang. 1987. Channel model of flow through fractured media. *Water Resources Res.* 23:467.

Witherspoon, P.A., Wang, J.S., Iwai, K. and J.E. Gale. 1980. Validity of cubic law for fluid flow in a deformable rock fracture. *Water Resources Res.* 16:6:1016-1024.

Zimmerman, R.W., Kumar S. and G. S. Bodvarsson. 1991. Lubrication theory analysis of the permeability of rough-walled fractures. *Int. J. Rock Mech. Min. Sci. & Geomech. Abstr.* 28:4:325-331.

Mechanics of Jointed and Faulted Rock, Rossmanith (ed.) © *1998 Taylor & Francis, ISBN 90 5410 955 6*

Strain-induced damage of rocks

S. Sakurai, A. Hiraoka & K. Hori
Kobe University, Japan

ABSTRACT: This paper deals with a constitutive equation, which can represent damage induced in rocks by maximum shear strain. In the constitutive equation, the anisotropic damage parameter and dilatancy coefficient are introduced so that the normality rule commonly used in elasto-plastic constitutive equation is not needed. The validity of the proposed constitutive equation is proved by using sands tested in laboratory. The anisotropic damage parameter and the dilatancy coefficient are determined for different type of rocks to discuss the damage process of rocks during the increase of maximum shear strain.

1 INTRODUCTION

The mechanical characteristics of rocks are extremely complex and entirely depend on confining pressure. It is well known that the deformational behavior of rocks are no more isotropic as they are compressed under certain levels of stress. This behavior is known as a stain-induced anisotropy. This anisotropy may be caused by damages occurring in a rock mass along the direction of the maximum shear strain. In other words, it may occur due to strain-induced damage.

In order to formulate a constitutive equation, the normality rule, based on the theory of plasticity is usually adopted. However, the normality rule was originally developed for ideally plastic materials like steel. Thus, if we adopt the normality rule for materials like rocks, we need to use a non-associated flow rule, which requires the so-called plastic potential function. However, it is extremely difficult to determine this function even in the laboratory, and almost impossible in in-situ. To overcome this difficulty in adopting the normality rule, the "anisotropic damage parameter" is introduced. The dilatancy effect is also represented by introducing "dilatancy coefficients." Both the anisotropic damage parameter and the dilatancy coefficient can be determined easily in the laboratory.

In this paper, the constitutive equation introducing the anisotropic damage parameter and the dilatancy coefficients is described, and the results of laboratory tests are used to determine the anisotropic damage parameter of several different rocks, as well as their dilatancy coefficient.

2 STRESS-STRAIN RELATIONSHIP

To determine the mechanical characteristics of geomaterials, a cyclical specimen is generally prepared and tested under uniaxial or triaxial state of stress. The global and local coordinate systems are assumed as shown in Fig. 1. Both are expressed in the cylindrical coordinate systems. It is assumed that the stress-strain relationship is expressed in the local coordinate system as follows:

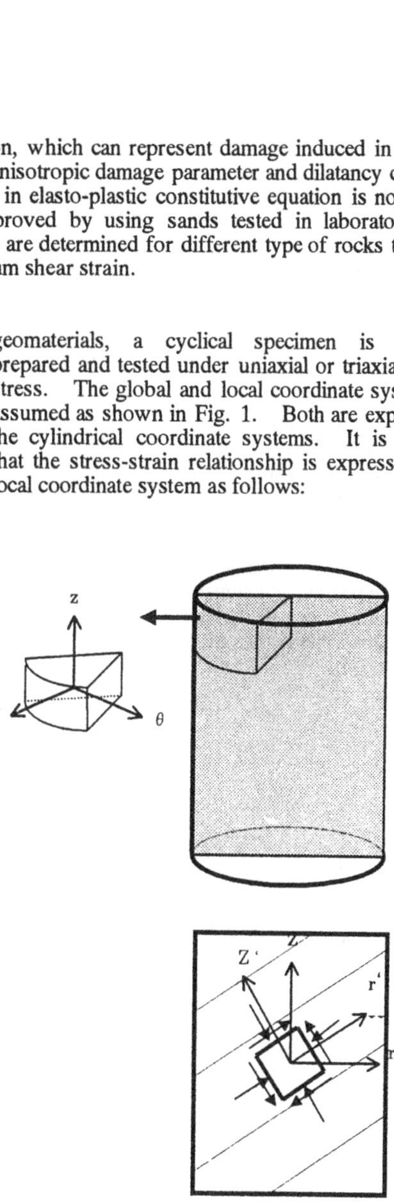

Fig. 1 Global and local coordinate systems for a cylindrical specimen

$$\{\varepsilon'\} = [c']\{\sigma'\} \quad (1)$$

In more detailed,

$$\begin{Bmatrix} \varepsilon_{r'} \\ \varepsilon_{z'} \\ \varepsilon_{\theta'} \\ \gamma_{r'z'} \end{Bmatrix} = \frac{1}{E}\begin{bmatrix} 1 & -\nu & -\nu & c \\ -\nu & 1 & -\nu & c \\ -\nu & -\nu & 1 & c \\ c & c & c & 1/m \end{bmatrix}\begin{Bmatrix} \sigma_{r'} \\ \sigma_{z'} \\ \sigma_{\theta'} \\ \tau_{r'z'} \end{Bmatrix} \quad (2)$$

where E : Young's modulus
ν : Poisson's ratio
$m = \dfrac{1}{2(1+\nu)} - d$
d : Anisotropic damage parameter
c : dilatancy coefficient

If the anisotropic damage parameter d is zero, Eq. (2) becomes the same as the one for isotropic elastic materials. When the material is compressed under an applied load and deformed beyond a certain level, the yielding point, then damages start to occur. The anisotropic damage parameter d then increases from zero as the damage increases, but it never goes beyond the value $1/2(1+\nu)$. Therefor, the value m decreases from $1/2(1+\nu)$ to zero.

The physical meaning of the value m is the ratio of shear modulus to Young's modulus, that is $m = G/E$. Therefore, it is assumed that shear modulus decreases when the damage starts to occur, while Young's modulus never changes. In other words, the shear modulus decreases most in the direction of sliding planes along which the damage is occurring.

The stress-strain relationship expressed in the global coordinates can be derived form Eq. (1) by the common transformation practice as follows:

$$\{\varepsilon\} = [C]\{\sigma\} \quad (3)$$

$$\text{where} \quad [C] = [T][C'][T]^T \quad (4)$$

$$[T] = \begin{bmatrix} \cos^2\theta & \sin^2\theta & 0 & \cos\theta\sin\theta \\ \sin^2\theta & \cos^2\theta & 0 & -\cos\theta\sin\theta \\ 0 & 0 & 1 & 0 \\ -2\cos\theta\sin\theta & 2\cos\theta\sin\theta & 0 & \cos^2\theta - \sin^2\theta \end{bmatrix} \quad (5)$$

θ : Angle between r' and r axies. When the Mohr-Coulomb failure criterion validates, this angle is expressed as,

$$\theta = 45^\circ + \frac{\phi}{2} \qquad (\phi : \text{Internal friction angle})$$

When the cylindrical specimen is compressed, two conjugate sets of sliding planes may theoretically occur. Therefore, the effects of these two sets of sliding planes must be taken into account in the formulation of the constitutive equation given in Eq.(3).

3 VALIDITY OF THE PROPOSED STRESS-STRAIN RELATIONSHIP

The validity of Eq. (3) has been substantiated in such a way that the anisotropic damage parameter as well as the dilatancy coefficient of sands are determined by the two different laboratory tests, the tensional test and the triaxial test. The results are then compared with each other.

3.1 *Torsion Tests*

A hollow cylindrical specimen (outer diameter: 7cm, inner diameter: 3cm, height: 14cm) was twisted under confining pressure applied from the outer and inner surface of the cylinder. The specimen was tested in the unsaturated condition. During the torsion tests axial force was applied by loading and unloading, while shear strain was kept constant. The axial loading and unloading tests was done several times at the different levels of shear strain.

In order to investigate the influence of applying axial loading and unloading on the results of torsion tests, two different tests were carried out. One was a torsion test carried out continuously without applying the additional axial force, and the other test applied axial loading and unloading at certain level of shear strain during the torsion test. The shear stress and shear strain relationship obtained by the two different ways are shown in Fig. 2 to be almost identical. This means that there is no influence of applying the axial force on the shear stress and shear strain relationship.

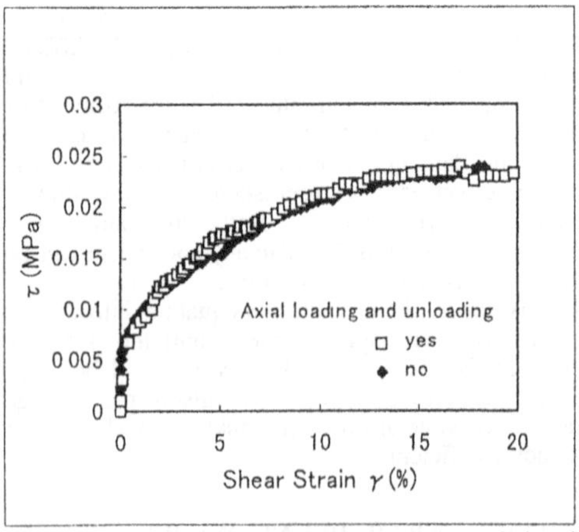

Fig.2 Shear stress and shear strain relationship obtained by two different ways.

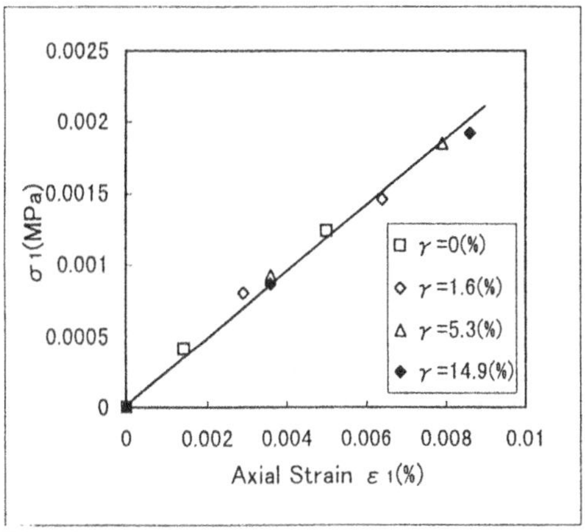

Fig. 3 Axial stress and axial strain relationship obtained under different shear strain levels.

The axial stress and axial strain relationship obtained under different shear strain levels is shown in Fig. 3. It can be seen from this figure that the axial stress and axial strain are almost identical to each other. This means that the axial stress and axial strain relationship is independent of shear strain, so that Young's modulus is constant all the time. However, it is easily understood from Fig. 1 that shear modulus defined as the slope of the shear stress and shear strain curve decreases with increase of shear strain. In this paper the secant shear modulus is used.

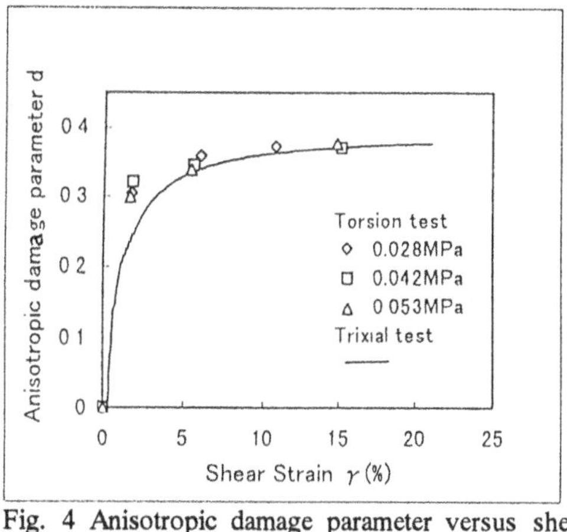

Fig. 4 Anisotropic damage parameter versus shear strain

The value *m* defined as the ratio of the secant shear modulus to Young's modulus can be calculated. The anisotropic damage parameter *d* is then calculated and plotted in Fig. 4. It is seen from this figure that the anisotropic damage parameter *d* starts to increase drastically, and converge to the value of $1/2(1+\nu)$. This means that damage starts to occur in the specimen even at a very small strain level, and it increases with the increase of shear strain. The dilatancy coefficient *c* is also obtained and plotted in Fig. 5. This figure indicates that the dilatancy coefficient also increases drastically as shear strain increases. This implies that the material shows the transition phenomena from isotropic characteristics to anisotropic ones, depending on the magnitude of shear strain. It is thus understood that the proposed constitutive equation introducing the anisotropic damage parameter together with the dilatancy coefficient can represent the transition phenomena from the isotropic to anisotropic characteristics of materials. In addition, this anisotropy tends to increase with increase of shear strain. Thus, this is called as a strain-induced anisotropy.

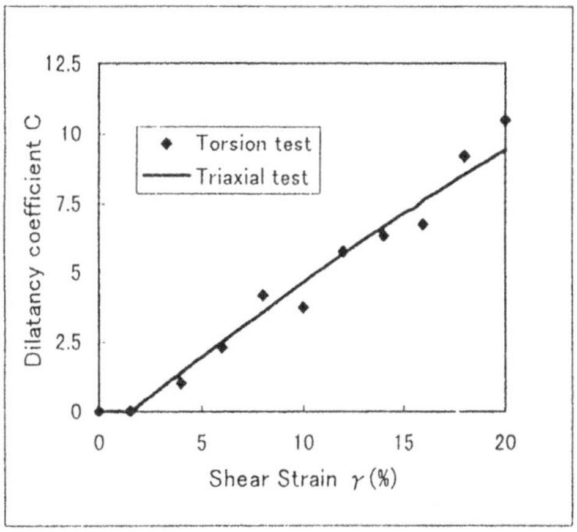

Fig. 5 Dilatancy coefficient versus shear strain.

3.2 *Triaxial Tests*

The conventional triaxial tests were conducted on a cylindrical specimen (diameter: 5cm, height: 10cm) of the same sands as those used for the torsion tests. Damage may have occurred in the specimen when the applied load reached a certain level. It is obvious from considering the Mohr-Coulomb failure criterion that the damage started to occur along the planes diagonal to the axial loading direction by the angle of $45°-\phi/2$, and reached the final collapse along one of the planes, which is sometimes called shear band. Considering this damage process, shear modulus defined in the direction of the shear band was determined by back analysis. The dilatancy coefficient were also back-calculated under the condition that Young's modulus was kept constant. As determining the shear modulus and Young's modulus, the ratio of them yielded the value of *m*. The anisotropic damage parameter *d* was then calculated and plotted to the shear strain as shown in Fig. 4. It is understood from this figure that there is a slight difference in the results between the torsion and triaxial tests. However, the difference may not be so serious for engineering practice. The dilatancy coefficient back-calculated from the results of triaxial tests is also given in Fig. 5. It is seen from this figure that the results obtained from torsion and triaxial tests are nearly identical to each other.

It is noted that the anisotropic damage parameter and the dilatancy coefficient introduced in the proposed constitutive equation were determined independently by the two different tests. The results of the tests are nearly identical to each other. This implies that the constitutive equation can be used for representing the behaviour of the materials in more general conditions. It should be emphasized that this constitutive equation has a great advantage in that it does not require any conventional plasticity theory such as normality rule, plastic potential function, etc.. Instead, the two parameters *d* and *c* are introduced to simulate both the shear and volumetric deformations of geomaterials.

4 DAMAGE OF ROCKS

The anisotropic damage parameter and the dilatancy coefficient of rocks are determined from the results of triaxial tests. The triaxial test results used in this study are those obtained by different researchers on several different types of rock.

4.1 *Tuff*

The specimen of tuff (diameter: 5cm, height: 10cm) were tested under a natural and dry condition. The axial stress and axial strain relationship is shown in Fig.6, while the relationship between volumetric strain and axial strain is shown in Fig.7.

Young's modulus and Poisson's ratio were determined from the tangential lines near the origin of the curve, as 3700MPa and 0.2, respectively. It is obvious from these two figures that non-linear behaviour appears as axial strain increases. This is simply because damage started to occur in the specimen. The anisotropic damage parameter was then determined by back analysis of the triaxial test results, and shown in Fig.8. The dilatancy coefficients were also determined as shown in Fig.9. In this back analysis the internal friction angle was assumed to be 30°.

It is seen from Fig.8 and 9 that the damage starts to occur when the maximum shear strain reaches a certain level.

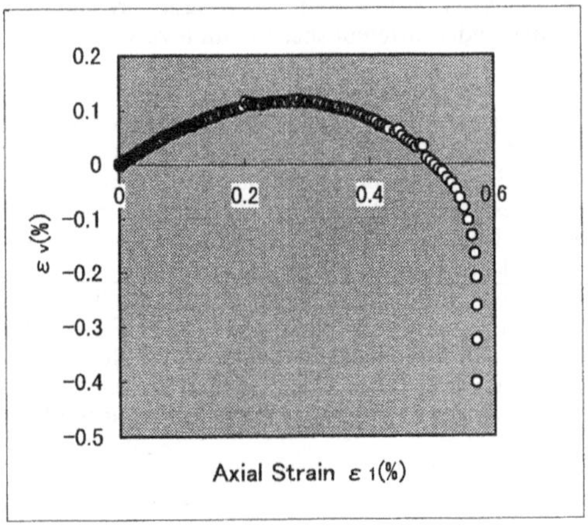

Fig. 7 Volumetric strain versus axial strain of tuff.

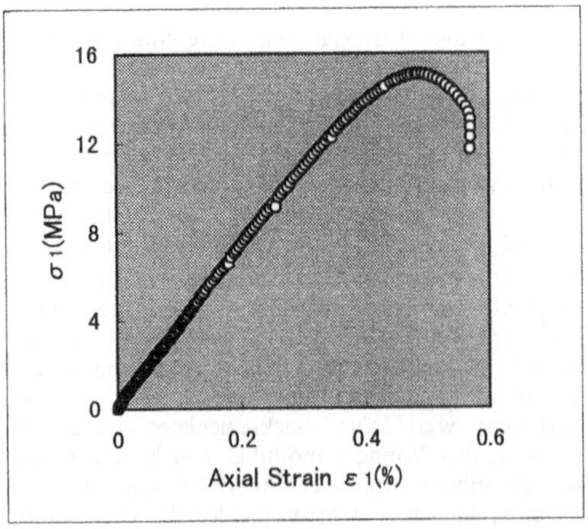

Fig. 6 Axial stress and axial strain relationship of tuff.

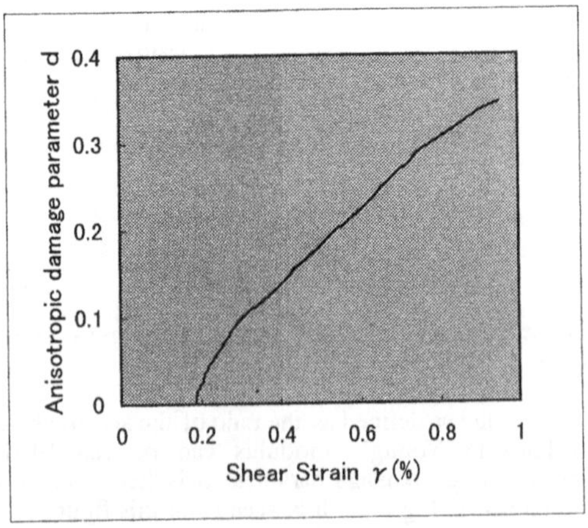

Fig. 8 Anisotropic damage parameter of tuff.

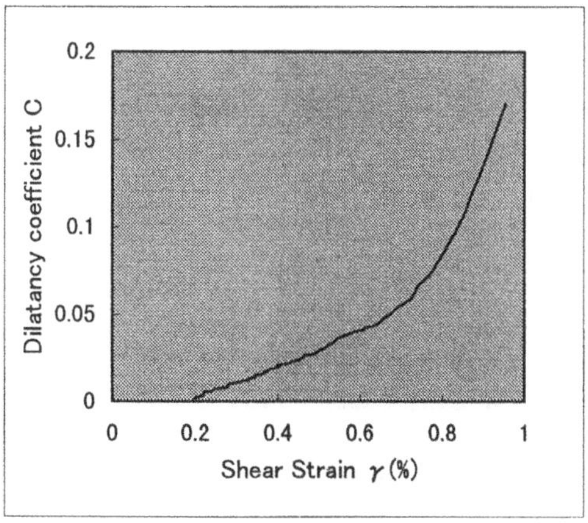

Fig. 9 Dilatancy coefficient of tuff.

4.2 *Andesite*

The specimen of andesite is 3.5cm in diameter and 7cm in height. The porosity of the rock was 15.2%. The triaxial tests were carried out under a dry condition by applying different confining pressure. Assuming the Mohr-Coulomb failure criterion, the internal friction angle was obtained as 45°. The stress and strain relationship in the axial direction is shown in Fig.10, for different confining pressures. The volumetric strain versus axial strain is shown in Fig.11. The anisotropic damage parameter and the dilatancy coefficient were then determined, and shown in Fig.12 and Fig.13, respectively.

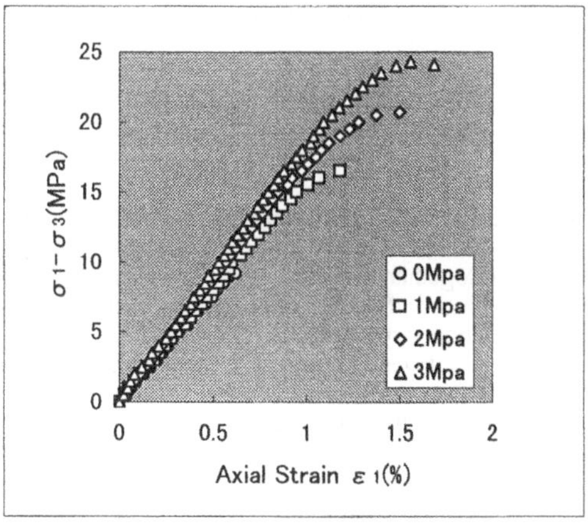

Fig.10 Axial stress and axial strain relationship of andesite.

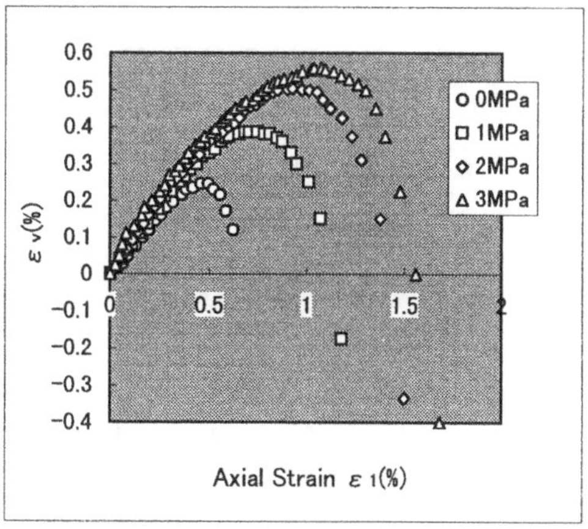

Fig.11 Volumetric strain versus axial strain of andesite.

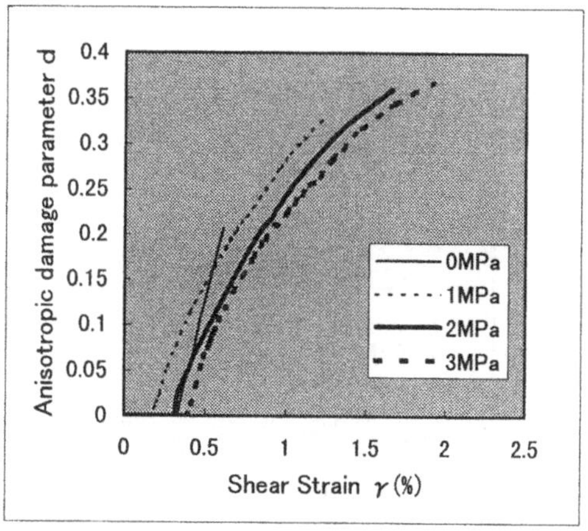

Fig. 12 Anisotropic damage parameter of andesite.

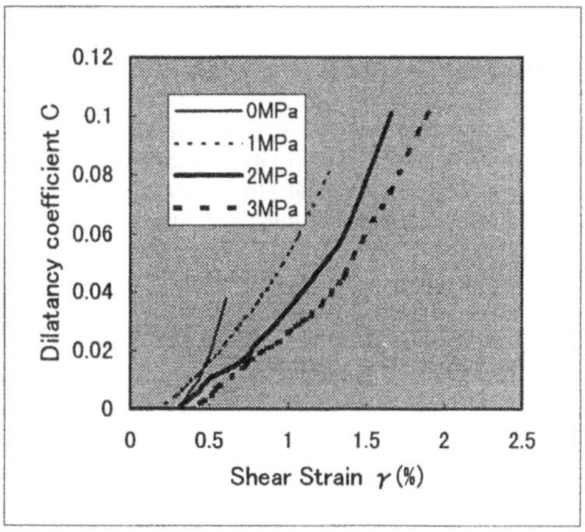

Fig.13 Dilatancy coefficient of andesite.

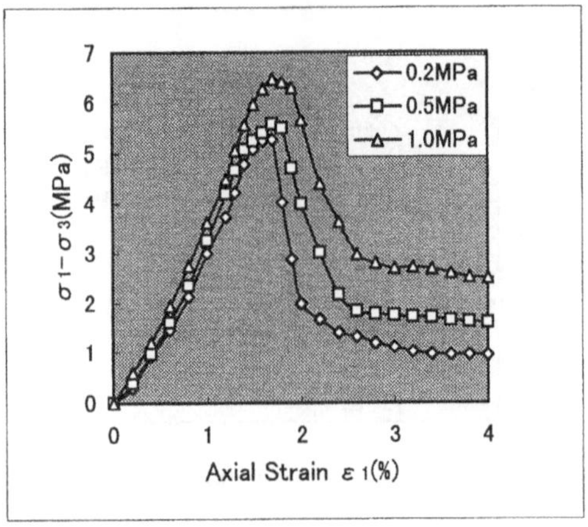

Fig.14 Axial stress and axial strain relationship of mudstone.

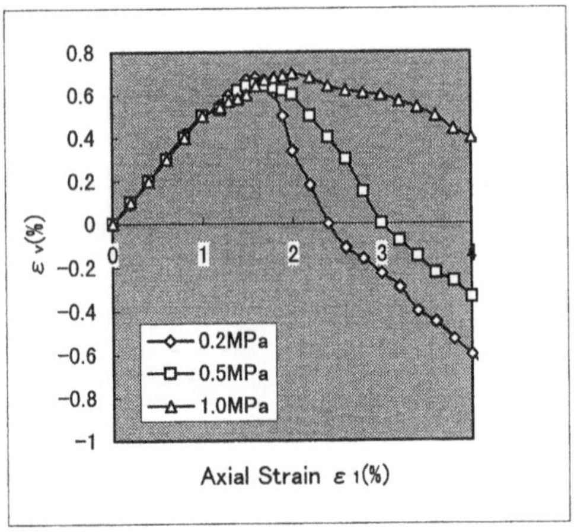

Fig.15 Volumetric strain versus axial strain of mudstone.

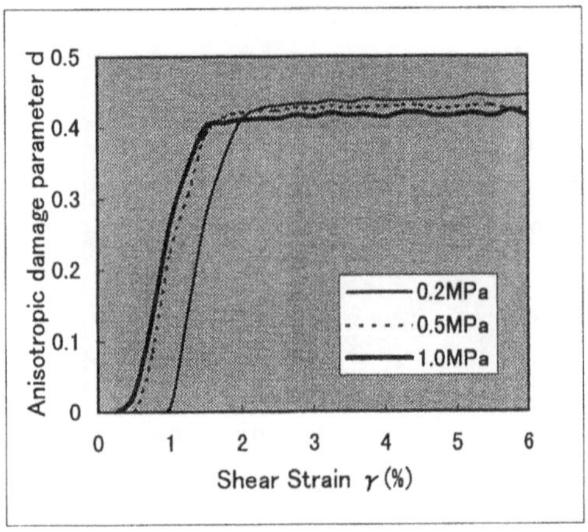

Fig.16 Anisotropic damage parameter of mudstone.

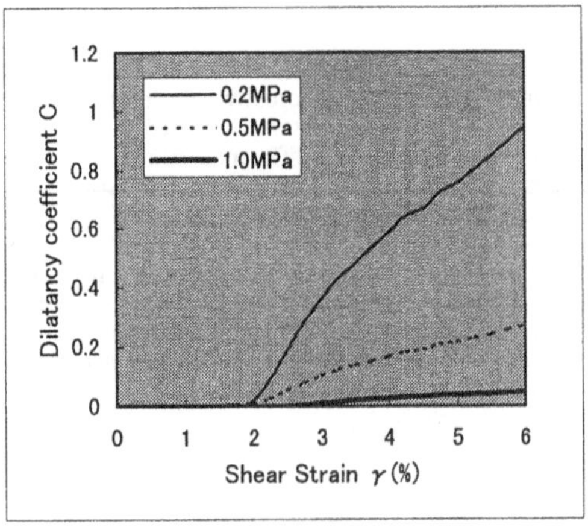

Fig.17 Dilatancy coefficient of mudstone.

4.3 *Mudstone*

The mudstone was tested under a natural moisture content of 25%. The porosity of the rock specimen tested was 0.68. The axial stress-strain relationship obtained for different confining pressure are shown in Fig.14. The volumetric strain versus axial strain is shown in Fig.15. The results shown in these two figures were used to determine the anisotropic damage parameter and the dilatancy coefficient, as shown in Fig.16 and Fig.17, respectively.

5 CONCLUSION

The constitutive equation described here has been derived by assuming damages occur in rocks along the direction of the maximum shear strain. In order to represent the damages, the anisotropic damage parameter together with dilatancy coefficients have been introduced, both of which can be determined easily by laboratory tests.

It is advantageous that the constitutive equation described here does not require any conventional theory of plasticity, such as normality rule, plastic potential function, etc.. The adequacy of the constitutive equation has been demonstrated by showing that the anisotropic damage parameter and the dilatancy coefficient determined by torsion tests and triaxial tests show the good agreement each other.

It is concluded that the anisotropic damage parameter is nearly independent from confining pressures, while the dilatancy coefficient entirely depends on confining pressure.

REFERENCES

Sakurai, S. N. Deeswasmongkol, and S. Shinji, Back Analysis of determining material characteristics in cut slopes, Proc. Int. Sympo. Eng. in Complex Rock Formations, Beijin, 1986.

Sakurai, S. Monitoring the stability of cut slopes, Proc. 2nd Int. Sympo. Mine Planning and Equipment Selection, Calgary, Balkema, 1990, pp269-274.

Sakurai, S., Monitoring of slope stability by means of GPS, Proc. 9th Int. Sympo. Deformation Measurements, Hong Kong, 1996.

Mechanics of Jointed and Faulted Rock, Rossmanith (ed.) © *1998 Taylor & Francis, ISBN 90 5410 955 6*

Seismic wave propagation in fractured rock

Larry Myer
Earth Sciences Division, Ernest Orlando Lawrence Berkeley National Laboratory, Calif., USA

ABSTRACT: The seismic displacement discontinuity model supposes that displacements are locally discontinuous in the plane of a fracture. Normal and shear displacements may be coupled and viscous loss may occur in addition. The consequences of these assumptions on theoretically predicted wavefields are many, including frequency dependent amplitude and velocity changes in transmitted and reflected waves, generations of converted waves and waves trapped as interface waves on the fracture and guided waves between parallel fractures. A number of these characteristics have been observed in laboratory experiments on artificial and natural fractures. Field observations are more limited but have provided evidence of the validity of the model at practical engineering scales.

INTRODUCTION

Rocks contain discontinuities at all scales from grain boundaries and cracks at the microscale to joints and fractures at the macroscale to faults at the mesoscale. It has also been recognized for a long time that the effect of these discontinuities, if averaged over some volume, is a reduction in the stiffness of the rock. From the theory of elasticity for a propagating seismic wave, this reduction in stiffness translates into a reduction in wave velocity. If discontinuities are preferentially oriented in one direction the average stiffness will be reduced more in one direction than another, resulting in an effective anisotropic medium. One of the important characteristics of wave propagation in anisotropic media is that a shear wave propagating at an oblique angle to an axis of symmetry will be split into components traveling at different velocities related to the degree of anisotropy. Thus, shear wave splitting is often taken as diagnostic of fracture orientation and density. These concepts constitute the most common approach in seismic geophysics to modeling wave propagation in fractured rock (e.g. O'Connell and Budiansky 1974, Crampin 1981, Hudson 1981, Thomsen 1996 and others).

A fundamental assumption in this approach is that an appropriate sized volume can be defined over which the effects of the discontinuities can be averaged. This volume is normally defined by wavelength, and it is argued intuitively that if discontinuities are closely spaced compared to wavelength then it is appropriate to use average properties. It is also intuitive to believe that average properties are not meaningful if the wavelength is comparable to, or less than, the average spacing. In this case, the discontinuity must be explicitly modeled. With the advent of piezoelectric and other sources in the kilohertz range, the latter condition is frequently encountered in engineering practice.

This paper will focus on the properties of the seismic wavefield when discontinuities are modeled explicitly. The result is a number of effects which are not predicted by an averaging approach, and can be used to better characterize fractured rock.

THE DISPLACEMENT DISCONTINUITY MODEL

The reduction in stiffness of rock containing a discontinuity arises from excess deformation localized at the discontinuity. Figure 1a is a schematic illustration of the uniaxial loading of a sample containing a discontinuity. Below is a sketch of the average displacement, measured relative to the bottom rigid support. Displacement increases linearly until, at the position of the discontinuity, a local, step increase in displacement occurs. This step increase in displacement, referred to as a displacement discontinuity, is in excess to that which would occur if the sample were intact, as shown schematically in Figure 1b. Though the displacement field is discontinuous, stresses, on average, are continuous. As is well known,

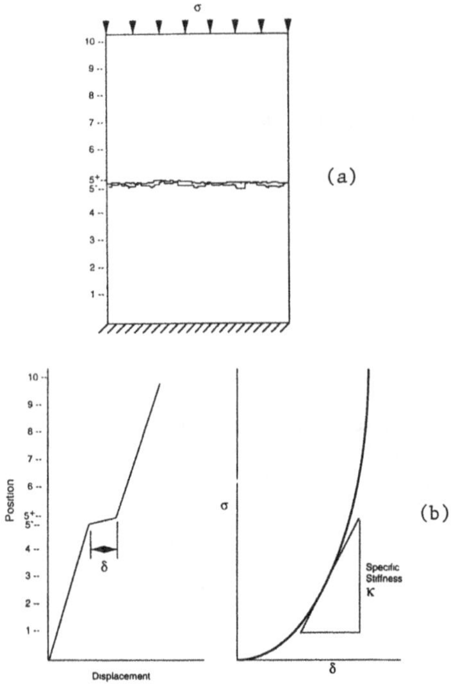

Figure 1. Definition of displacement discontinuity in rock. a) Loading of a sample with a single fracture; b) Left, displacement as function of position, bottom as reference; Right, displacement discontinuity as a function of far field stress.

(Goodman 1976, Bandis et al. 1983, Pyrak-Nolte, et al. 1987) the magnitude of the displacement discontinuity is a nonlinear function of stress. At low stresses, there are few areas of contact between the surfaces of a discontinuity. As stress increases, so do areas of contact so that the displacement discontinuity decreases. The tangent slope to the stress displacement discontinuity curve is referred to as the specific stiffness of the discontinuity.

Generalizing these ideas for dynamic loading under shear as well as normal stress results in the seismic displacement discontinuity boundary conditions (Figure 2):

$$
\begin{aligned}
u_z^I - u_z^{II} &= \tau_{zz} / \kappa_z, \tau_{zz}^I = \tau_{zz}^{II} \\
u_x^I - u_x^{II} &= \tau_{zx} / \kappa_x, \tau_{zx}^I = \tau_{zx}^{II} \\
u_y^I - u_y^{II} &= \tau_{zy} / \kappa_y, \tau_{zy}^I = \tau_{zy}^{II}
\end{aligned}
\tag{1}
$$

where

u = particle displacement ($u^I - u^{II} = \delta$)

τ = stress

κ = specific stiffness

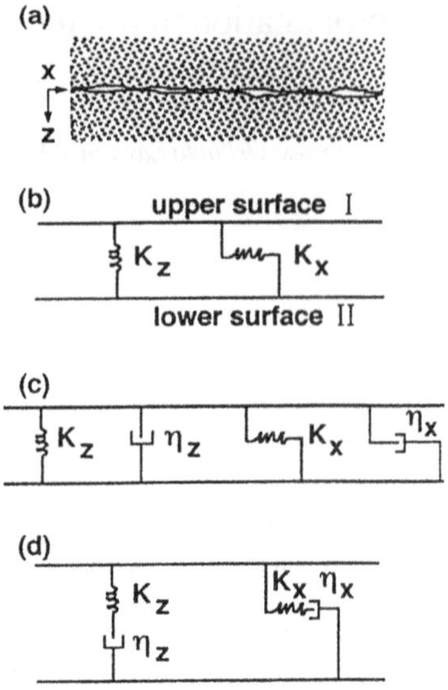

Figure 2. a) Discontinuity; b) Elastic boundary conditions; c) Kelvin rheologic model; d) Maxwell rheologic model.

I,II = superscripts referring to media above and below discontinuity.

These boundary conditions apply over all scales from micro to meso with the restriction that the wavelength is long compared to the spacing between the areas of contact in a discontinuity as well as the size of these areas.

Another assumption in Eq. (1) is that the force displacement relationship in the normal and shear directions are independent. A situation in which this might not be the case is one where the two opposing surfaces of a discontinuity are very rough and loaded in shear. This would result in nonhomogeneous loading on the surfaces and coupling between the normal and shear force displacement relationship. In this case, the boundary conditions for a P- or S_v- wave become

$$
\begin{aligned}
(u_z^I - u_z^{II})\kappa_{zz} + (u_x^I - u_x^{II})\kappa_{zx} &= \tau_{zz}, \tau_{zx}^I = \tau_{zx}^{II} \\
(u_z^I - u_z^{II})\kappa_{zx} + (u_x^I - u_x^{II})\kappa_{xx} &= \tau_{zx}, \tau_{zz}^I = \tau_{zz}^{II}
\end{aligned}
\tag{2}
$$

Solutions of the elastodynamic equations for plane waves incident upon a discontinuity modeled by Eq. (1) have been presented by a number of authors (Schoenberg 1980, Pyrak-Nolte et al. 1990, Gu et al. 1996a, and others). The solution including cross coupling terms (Eq. 2) is discussed in Nakagawa 1998 and Myer et al. 1998. Figure 3 presents results for the case of normal incidence in which $|T|$ is the magnitude of the transmission

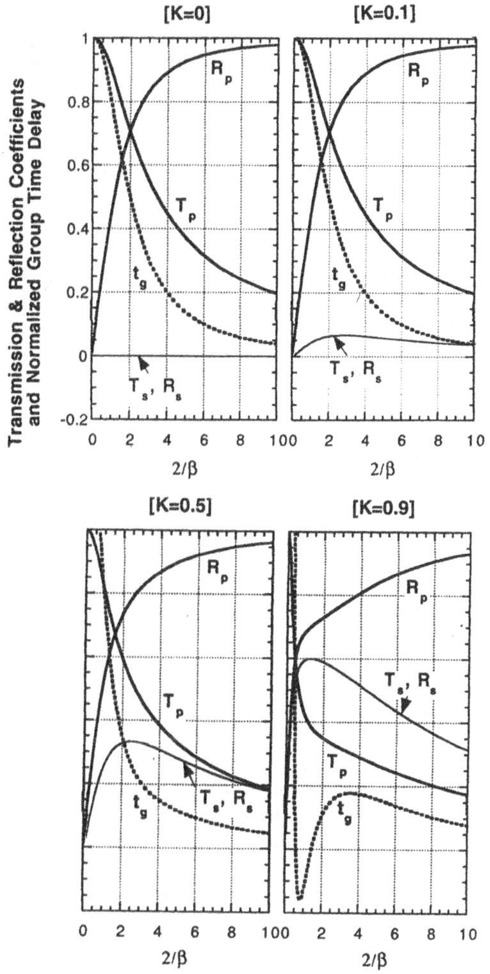

Figure 3. Magnitude of transmission and reflection coefficients and group time delay for P-wave normally incident upon an elastic interface with cross coupled stiffness.

coefficient, $|R|$ is the magnitude of the reflection coefficient and t_g is a normalized group time delay for the transmitted wave. The parameter

$$\beta \equiv \frac{z\kappa_{zz}/\omega}{z_p}$$

where

ω = angular frequency
$z_p = \rho c_p$ for P-waves
$c_p = \sqrt{\lambda + 2\mu / p}$
ρ = density
λ, μ = Lame's constants for media on either side of discontinuity

Plots are shown for a range of the values of the ratio $K = \kappa_{zy}/\kappa_{zz}$. For $K = 0$ the cross coupling stiffness terms are zero, corresponding to the solution for conditions given by Eq. (1).

Results in Figure 3 show that the seismic displacement discontinuity model predicts that a single joint or fracture should cause a frequency dependent change in the amplitude and velocity of a propagating seismic wave. High frequency components of a wave are preferentially reflected. Because the discontinuity is elastic, as described by Eqs. (1) or (2), there is no energy dissipation; energy is either transmitted or reflected. The curve for t_g implies that high frequency components are slowed less than low frequency components.

The effect of nonzero cross coupling terms is to generate converted waves. For the case of a normally incident P-wave as shown in Figure 3, transmitted and reflected S-waves are formed with amplitudes which are frequency and stiffness dependent. The negative values of t_g in one panel imply that, under some conditions, the P-wave velocity could be increased.

Limiting conditions for the displacement discontinuity model occur for specific stiffness approaching zero or infinity. For $\kappa \to 0$, the solution reverts to that for a plane wave incident upon a free boundary, and for $\kappa \to \infty$ the solution is equivalent to those of classical seismology in which both displacements and stresses are continuous at the boundaries between layers.

The elastic boundary conditions do not account for energy dissipation which could occur due to partial saturation or clay in the discontinuity. Two very simple rheological models which also have practical significance are the Kelvin and Maxwell models. For Kelvin rheology Eqs. (1) become:

$$(u_z^I - u_z^{II})(\kappa_z - i\omega\eta_z) = \tau_{zz},\ \tau_{zz}^I = \tau_{zz}^{II}$$
$$(u_x^I - u_x^{II})(\kappa_x - i\omega\eta_x) = \tau_{zx},\ \tau_{zx}^I = \tau_{zx}^{II}$$
$$(u_y^I - u_y^{II})(\kappa_y - i\omega\eta_y) = \tau_{zy},\ \tau_{zy}^I = \tau_{zy}^{II} \tag{3}$$

and for Maxwell rheology:

$$(u_z^I - u_z^{II}) = \tau_{zz}\left(\frac{1}{\kappa_z} + \frac{i}{\omega\eta_z}\right),\ \tau_{zz}^I = \tau_{zz}^{II}$$
$$(u_x^I - u_x^{II}) = \tau_{zx}\left(\frac{1}{\kappa_x} + \frac{i}{\omega\eta_x}\right),\ \tau_{zx}^I = \tau_{zx}^{II}$$
$$(u_y^I - u_y^{II}) = \tau_{zy}\left(\frac{1}{\kappa_y} + \frac{i}{\omega\eta_y}\right),\ \tau_{zy}^I = \tau_{zy}^{II} \tag{4}$$

where
K = specific viscosity (units of viscosity per length).

The magnitude of the transmission and reflection coefficient for a normally incident wave are plotted in Figure 4. The solution becomes equivalent to that of the elastic interface when the specific viscosity becomes large (e.g. $z/2\eta = 0.01$). When the specific viscosity is small (e.g. $z/2\eta = 5$), the solution is dominated by the viscous element. Thus, both $|T|$ and $|R|$ become frequency independent. Since energy is lost by viscous dissipation at the discontinuity, $|R|^2 + |T|^2 \neq 1$.

For oblique angles of incidence of P- or S_v plane wave, the displacement discontinuity model predicts that converted waves will be generated even though the material properties on either side of the discontinuity are the same. Snell's law defines the relationship between the angles of incidence, reflection, and refraction. For S_v incident waves, a critical angle exists above which transmitted and reflected waves are no longer real valued. Transmissions and reflection coefficients for oblique angles of incidence have been developed by several authors (Schoenberg 1980, Pyrak-Nolte et al. 1990, Gu et al. 1996a and others).

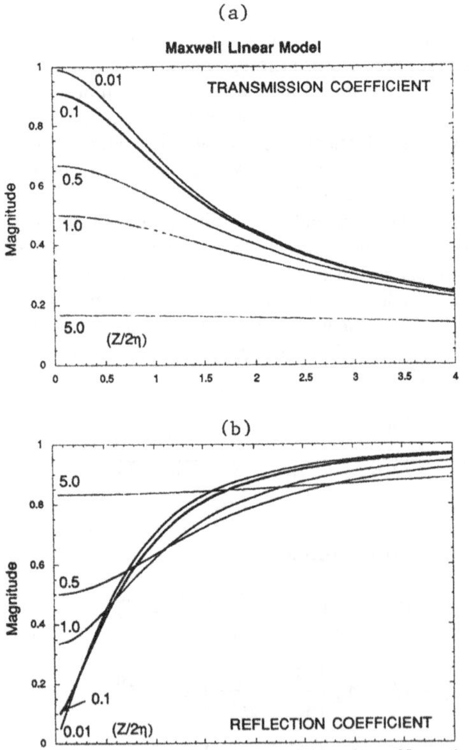

Figure 4. a) Magnitude of the transmission coefficient for normal incidence assuming Maxwell rheologic model for boundary conditions; b) Magnitude of the reflection coefficient.

It is well known that energy will propagate along a free surface as a Rayleigh wave or, under some conditions, along an interface between materials of contrasting properties (Stoneley 1924). It is now recognized that a discontinuity such as a fracture will support an interface wave even though the materials on either side of the discontinuity have the same properties. Solutions for an inhomogeneous plane wave propagating along a discontinuity described by the elastic displacement discontinuity model have been obtained by Pyrak-Nolte and Cook 1987, Gu et al. 1996b, Nihei et al. 1995 and others. Two interface waves are predicted with phase velocities which lie between the Rayleigh wave and shear body wave velocities and are dependent upon the specific stiffness of the discontinuity. The particle motion of these waves is elliptic and exponentially decays in amplitude with distance from the discontinuity. These waves are therefore known as generalized Rayleigh waves.Boundary element simulations for point sources located on or near a discontinuity (Gu et al. 1996b) showed that an interface wave traveling at a velocity close to the P-body wave should exist in addition to the generalized Rayleigh waves.

Figure 5 presents waveforms for a cylindrical source located on an elastic discontinuity of varying specific stiffness. At very high stiffness, only a P-wave is present, representing the response of the medium without a discontinuity. As specific stiffness decreases a generalized Rayleigh wave and a P-interface wave develop.

Finally, Nihei et al. 1994 and Nihei et al. 1998 have shown that energy can be trapped as guided waves between two fractures even though there are no material property contrasts. A family of dispersive trapped wave modes is predicted for each value of specific stiffness. The modes have x-component particle notion which is either symmetric or antisymmetric with respect to the center of the waveguide. Figure 6 shows particle motions for the lowest order antisymmetric mode for two values of specific stiffness of the discontinuities. For low values of specific stiffness, most energy is trapped between the discontinuities. As specific stiffness increases, more energy couples across the discontinuities indicating a transition to the bulk body wave.

Neither interface waves nor guided waves are predicted if the classical approach of replacing a fractured medium by effective anisotropic properties is followed. If such waves exist in practice, they could be diagnostic of fracturing at scales of interest in many engineering projects. Observations of these waves at laboratory and field scale is discussed below.

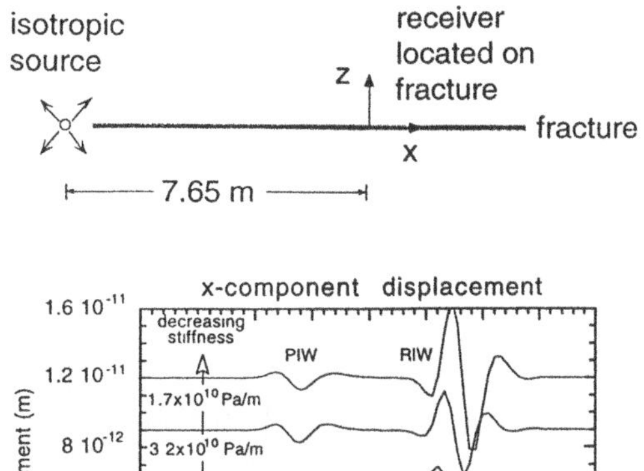

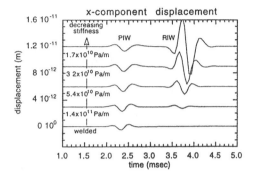

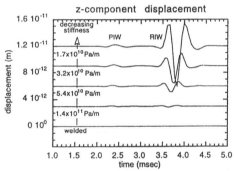

Figure 5. x and z components of particle displacements for different values of fracture specific stiffness from boundary element simulation.

LABORATORY MEASUREMENTS

The displacement discontinuity model for normal incidence waves has been validated in laboratory measurements on idealized fractures formed by lead foil strips sandwiched between two steel cylinders placed end to end (Myer et al. 1985). A reference measurement was made in which a solid disk of the foil was placed between the cylinders. Strip thickness (0.03 mm), width (1.0 mm), and spacing (variable from 1 mm to 7 mm) were small compared to wavelength (23 mm for P-wave at center frequency). The strips were placed in parallel, effectively forming a coplanar array of cracks (Figure 7). For such a geometry analytic expressions can be used to calculate the stiffness and magnitude of the displacement discontinuity, δ, (Tada et al. 1973, Myer et al. 1995 and others).

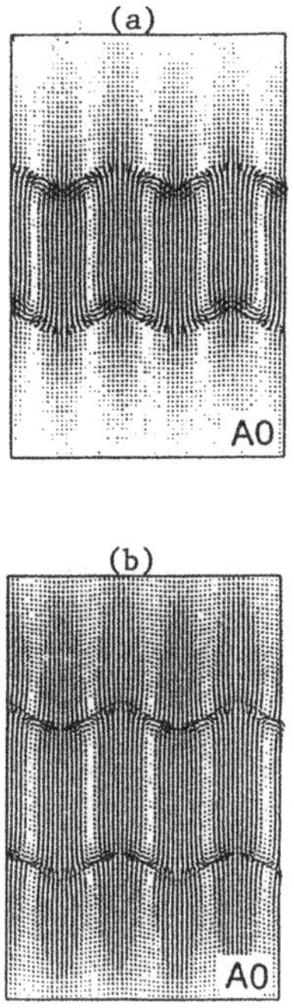

Figure 6. Predicted mode shapes for the lowest order antisymmetric fracture channel wave for a frequency of 1 kHz, P-wave velocity of 3118 m/s, S-wave velocity of 1800 m/s; a) specific stiffness of $1 \times 10^9 P_a/m$; b) specific stiffness of $1 \times 10^{10} P_a/m$.

The "predicted" curves in Figure 7 were therefore generated without "adjustable" variables. The magnitude of the transmission coefficient, $|T|$, as a function of frequency, is uniquely determined from the calculated stiffness. The product of $|T|$ and the measured spectral amplitudes for the reference test yields the predicted spectral amplitudes. It should be noted that previous reference to these lead foil strip tests (Myer et al. 1985 and Myer et al. 1995) presented results for s-wave transmission which were erroneously labeled as P-wave results. Subsequent tests were performed in which lead strips were used to form two parallel idealized fractures. Results showed that, for widely spaced fractures, the magnitude of the transmission coefficient for a wave propagating across a set of parallel fractures is $|T|^N$ where $|T|$ is the value for a single fracture and N is the number of fractures (Myer et al. 1995).

The displacement discontinuity model has been used to successfully simulate laboratory P- and S-wave transmission measurements on specimens containing single natural fractures (Pyrak-Nolte et al. 1990). The frequency dependent reduction in amplitude of waves transmitted across the fractures varied with the amount of normal stress applied to

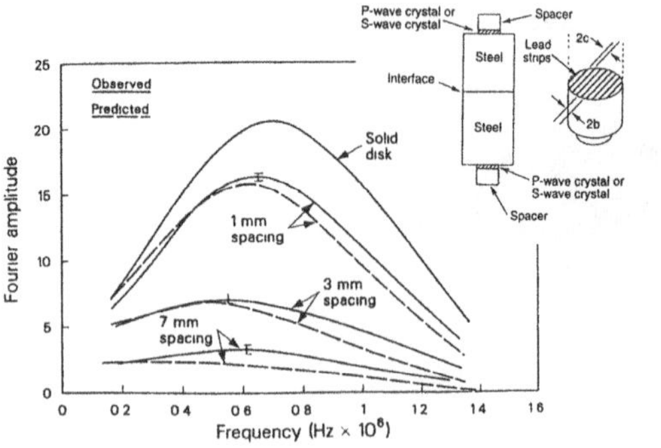

Figure 7. Amplitude spectra obtained from experiment compared with those predicted using "solid disk" spectrum as reference.

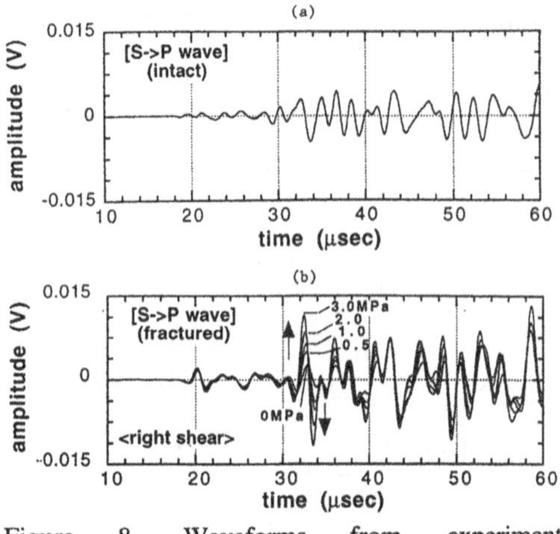

Figure 8. Waveforms from experiment demonstrating stiffness crosscoupling; a) P-wave receiver receiver response for intact sample; b) P-wave receiver response for fractured sample with shear stress applied to generate first motion indicated by arrows.

the fractures, reflecting the effect of changes in specific stiffness as illustrated in Figure 1. Good agreement between observed and modeled results was achieved by using values of fracture stiffness of the same order of magnitude, though somewhat higher than those measured under pseudostatic loading conditions. Elastic boundary conditions in the displacement discontinuity model well simulated P- and S-wave propagation across dry natural fractures and P-wave propagation across water saturated fractures. S-wave propagation across water saturated fractures was better simulated by assuming rheologic boundary conditions (Pyrak-Nolte et al. 1990).

Evidence for the existence of converted waves formed due to cross-coupling terms in the stiffness matrix has been provided by laboratory tests by Nakagawa 1998. A piezoelastic S-wave source was placed at one end of a cylindrical granite sample and a piezoelastic P-wave receiver at the opposite end. The received waveform for the intact specimen as shown in Figure 8a constituted a reference signal. The sample was then fractured under Brazilian loading creating a fracture perpendicular to the axis of the sample. A shear load was imposed on the fracture and the transmission experiment repeated. Results (Figure 8b) show the arrival of a converted P-wave which grows in amplitude as shear stress on the fracture increases. The increasing shear stress created an asymmetric loading on asperities of contact in the fracture plane and, consequently, cross coupled stiffness terms.

Fluids of differing chemistries and clay coatings of differing mineralogy are ubiquitous in fractured rock systems. The presence of these additional components not only changes the stiffness of a fracture but also can result in dissipation of energy. Suárez-Rivera 1992 and Suárez-Rivera et al. 1992 explored the effects of clay coatings on propagation of S-waves across fractures and found that even thin clay coatings can cause significant attenuation if the clays absorb pore fluids, or, in civil engineering terms, are prone to swelling behavior. He further found that the displacement discontinuity model incorporating rheologic boundary conditions well simulated the result of laboratory experiments. Figure 9 shows the results of the transmission and reflection of an S-wave by a thin Na-montmorillonite layer containing 27% water by weight. The clay layer was about 5μm thick and was sandwiched between quartz disks. The reference spectrum refers to the case when no clay was present for the transmitted wave and a free surface was present for the reflected wave. Measurements were made at different levels of load applied normal to the quartz disks confining the layer. Experimental results were well simulated assuming a Maxwell rheologic model in which single values of specific stiffness and specific viscosity were used to fit the transmitted, reflected and viscous loss spectral data at each normal load.

Laboratory confirmation of the existence and properties of fracture interface waves has been carried out by several investigators. Pyrak-Nolte et al. (1992) placed two aluminum blocks together, forming an interface or idealized fracture between them. Both slow and fast interface waves were observed. It was also shown, in accordance with theory, that no interface wave is present for source shear motion in the plane of the fracture (S_h source). In similar tests on rock fracture, Rayleigh-type interface waves have been observed by Ekern et al.

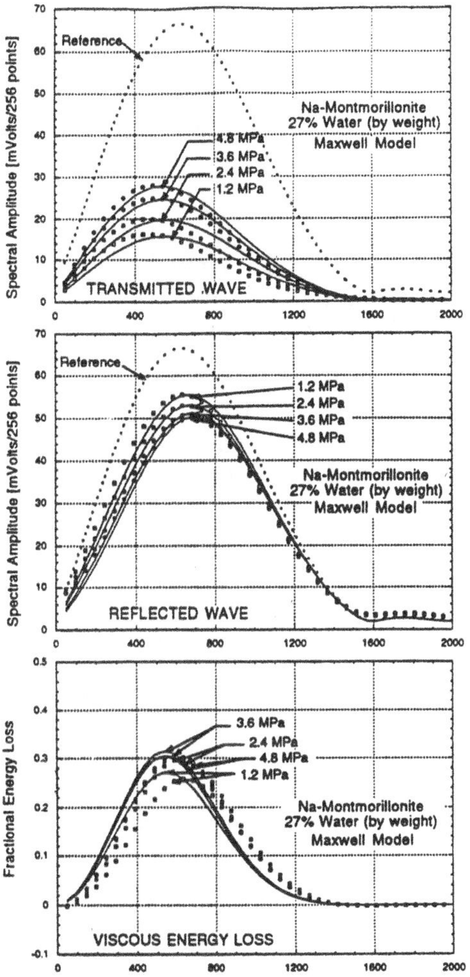

Figure 9. Comparison of observed and predicted amplitude spectra and energy loss for shear waves propagated across a thin layer of clay for different values of normal stress on the clay layer. Predicted values based on Maxwell model and the indicated reference spectra.

1995 and Roy and Pyrak-Nolte 1995. A compressional-mode interface wave was observed by Roy and Pyrak-Nolte 1997. Using a sheet of Plexiglas containing a single idealized fracture, Fan et al. (1996) confirmed particle motion patterns and decay in amplitude away from a discontinuity as predicted by theory.

Nihei et al. (1998) have recently demonstrated at laboratory scale the existence of trapped waves in a channel formed by two fractures. Two parallel fractures were generated in a 30.5 cm square by 1.05 cm thick slab of marble. A piezoelectric shear source was placed on the edge of the slab and located midway between the fractures. Particle motions were mapped on the surface of the slab and are shown in Figure 10 a,b for two different levels of normal stress across the fractures. The particle motions are consistent with the lowest order antisymmetric mode (see Figure 6). At the higher stress, Figure 10b, larger amplitudes were observed outside the channel formed by the fractures. At the higher stress, the specific stiffness of the fractures was higher, allowing coupling of energy outside the channel, in accordance with theory.

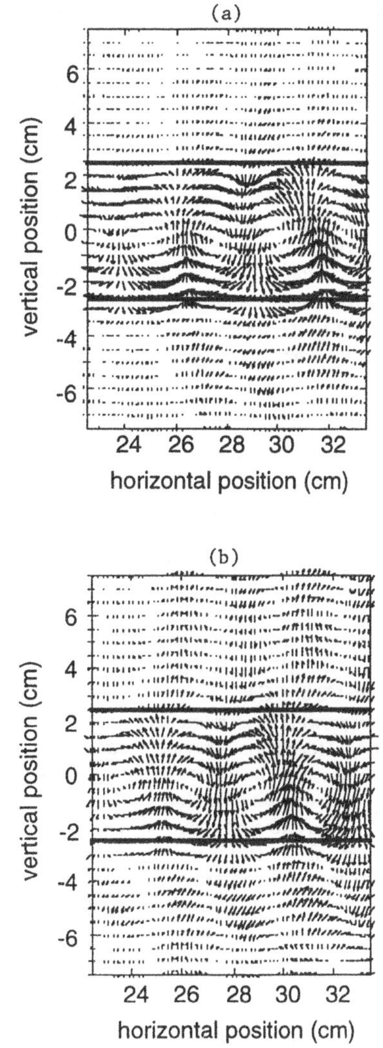

Figure 10. Measured particle motions in a fractured marble plate; a) Low normal stress across fractures; b) Higher normal stress

FIELD OBSERVATIONS

A small scale cross-well test performed in basalt provided some of the first evidence for applicability of the displacement discontinuity model at field scale (King et al. 1986, Myer et al. 1995). Measurements were made between horizontal drill holes, labeled C1, C2, C3, C4, as shown in Figure 11a, spaced 3m apart in the wall of a drift above the water table. The primary fracture set as shown in the figure was formed by the basalt columns which ranged in thickness from about 0.2m to 0.4m. Crosswell measurements were made with a 1.0m source and receiver spacing. The effect of propagating across the column-forming fractures is shown in Figure 11b, which presents first arriving P-wave pulses for different borehole pairs. Pulses which crossed column-forming fractures were slowed and attenuated compared to those propagating along a column, qualitatively in accordance with predictions of the displacement discontinuity model.

Results of quantitative modeling of four of the C2-C4 crosswell raypaths are shown in Figure 11c. The modeling procedures consisted of first selecting a reference pulse from the data set for the C1-C2 measurements. After transforming into the frequency domain and correcting for path length and intrinsic attenuation in the rock mass, this spectrum was multiplied by the transmission coefficient assuming elastic boundary conditions in the displacement discontinuity model. The result was inverse transformed to the time domain for comparison with the observed pulses. The number of fractures at each raypath location was held constant (at 3) while fracture stiffness was changed to obtain the best fit with observations. The stiffness varied between 1.0×10^{11} and 5.0×10^{11} mPa/m units for the results shown. It is seen that the simple elastic model captures quite well the amplitude reduction, dispersion and time delay experienced by the pulses which crossed the fractures between C2 and C4.

More recent crosswell measurements in a shallow fractured limestone formation provide evidence of the existence of fracture interface waves at field scale. A comprehensive set of high resolution (1 to 10kHz) crosswell measurements have been obtained between five vertical wells (Majer et al. 1997). Majer et al. also found that a single, strata-bound vertical fracture with strike as shown in Figure 12a intersected well GW-5. For source positions in GW-5 an arrival corresponding to an interface wave would be expected in GW-2, which is also located on, or immediately adjacent to, the fracture. The arrival in GW-2 interpreted as the interface wave is marked in Figure 12b. The delay time for this arrival is also consistent with results of a theoretical

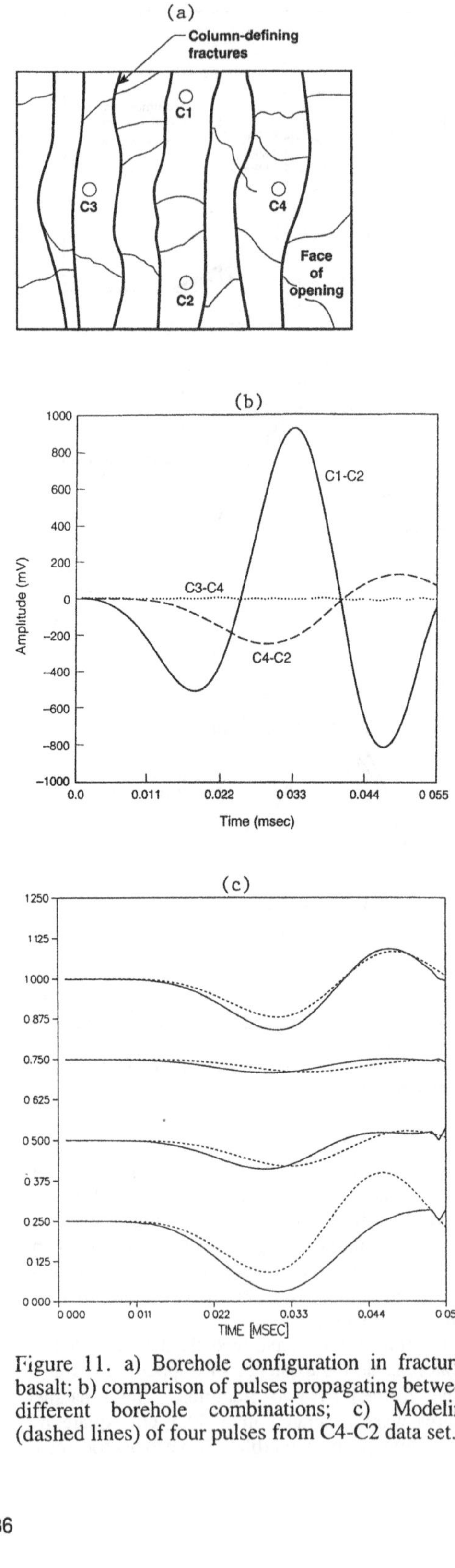

Figure 11. a) Borehole configuration in fractured basalt; b) comparison of pulses propagating between different borehole combinations; c) Modeling (dashed lines) of four pulses from C4-C2 data set.

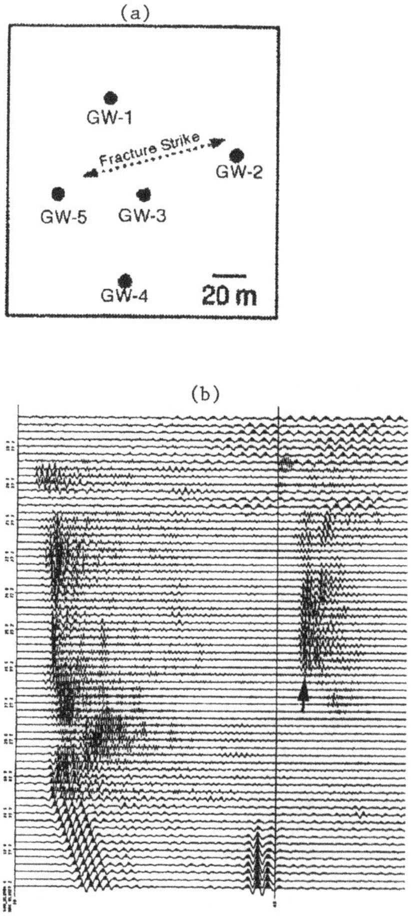

Figure 12. a) Plan view of well locations; b) GW-5 to GW-2 crosswell results; arrow indicating arrival of interface wave; c) GW-3 to GW-4 crosswell results.

model (see Figure 5) constructed with source characteristics and rock properties relevant to the in-situ test.

CONCLUSIONS

When the wavelengths are comparable to or less than the average spacing of discontinuities it is no longer appropriate to average their effects on wave propagation. The displacement discontinuity model has been shown to capture many aspects of the physics associated with the interaction of seismic waves with individual discontinuities. Theoretical studies have defined the transmission and reflection characteristics of plane waves propagating across elastic and rheologic discontinuities. Propagation parallel to discontinuities results in energy being trapped as interface and guided waves. Such waves are not predicted by effective media approaches.

Laboratory studies have validated many aspects of the displacement discontinuity model through measurements of transmission and reflection of plane waves incident on elastic and rheologic discontinuities. Laboratory measurements also established the existence and properties of interface and guided waves which had not previously been redognized.

Field observations and applications of the displacement discontinuity model are few. The model provides potential for improved characterization of fractured rock, but implementation will require further development of interpretational tools.

ACKNOWLEDGMENTS

I would like to acknowledge Professor Neville Cook, past and present graduate students and my colleagues, particularly Dr. Kurt Nihei, without whom the concepts in this paper would not have progressed. I also acknowledge Dr. John Queen and others at CONOCO for their collaborative contribution to in-situ testing of these concepts. I thank Dr. Nihei for numerical results shown in Figure 5 and Dr. John Peterson for results shown in Figure 11. The work was carried out under U.S. Department of Energy Contract No. DEA03-76SF00098. Support was provided by the Director, DOE Office of Energy Research, Office of Basic Energy Science, the Director, DOE Office of Fossil Energy, Office of Oil, Gas and Shale Technologies, and the Gas Research Institute.

REFERENCES

Bandis, S.C., A.L. Lumsden & N.R. Barton 1983. Fundamentals of rock joint deformation. *Int. J. of Rock Mechanics and Mining Sciences & Geomechanical Abstracts* 20(6):249-268.

Crampin, S. 1981. A review of wave motion in anisotropic and cracked elastic media. *Wave Motion* 3:343-391.

Ekern, A., R. Suárez-Rivera & A. Hansen 1995. Investigation of interface wave propagation along planar fractures in sedimentary rocks. In J. Daemen & R. Schultz (eds), *Rock Mechanics, Proceedings of the 35th U.S. Symposium* 161-167, Balkema.

Fan, J., B. Gu, K.T. Nihei, N.G.W. Cook & L.R. Myer 1996. Experimental and numerical investigation of fracture interface waves. In M. Aubertin, R. Hassani, & H. Mitri (eds), *Rock Mechanics, Tools and Techniques, Proceedings*

of 2nd North American Rock Mechanics Symposium 1:845-851.
Goodman, R.E. 1976. *Methods of Geological Engineering Discontinuous Rock.* St. Paul: West Publishing Co.
Gu, B., R. Suárez-Rivera, K. Nihei & L.R. Myer 1996a. Incidence of plane waves upon a fracture. *J. Geophys. Res.* 101(B11):25337-25346.
Gu, B., K. Nihei, L. Myer & L. Pyrak-Nolte 1996b. Fracture interface waves. *J. Geophys. Res.* 101(B1):827-835.
Hudson, J.A. 1981. Wave speeds and attenuation of elastic waves in material containing cracks. *Geophys. J.R. Astron Soc.* 64(1):133-150.
King, J.S., L.R. Myer & J.J. Rezowalli 1986. Experimental studies of elastic-wave propagation in a columnar-pointed rock mass. *Geophysical Prospecting* 34(8):1185-1199.
Majer, E.L., J.E. Peterson, T. Daley, B. Kaelin, L. Myer, J. Queen, P. D'Onfro & W. Rizer 1997. Fracture detection using crosswell and single well surveys. *Geophysics* 62(2):495-504.
Myer, L.R., D. Hopkins & N.G.W. Cook 1985. Effects of contact area of an interface on acoustic wave transmission. In *Research and Engineering Application in Rock Mechanics, Proceedings of the 26th U.S. Symposium on Rock Mechanics* 549-556. Rotterdam: Balkema.
Myer, L.R., D. Hopkins, J. Peterson & N. Cook 1995. Seismic wave propagation across multiple fractures. In Myer, Cook, Goodman and Tsang (eds), *Fractured and Jointed Rock Masses*:105-110. Rotterdam: Balkema.
Myer, L.R., K.T. Nihei & S. Nakagawa 1998. Dynamic properties of interfaces. H.P. Rossmanith (ed), *Damage and Failure of Interfaces*. Balkema (in press).
Nakagawa, S. 1998. Acoustic resonance characteristics of rock and concrete including fracture, *Ph.D. Thesis*, University of California, Berkeley.
Nihei, K.T., L.R. Myer, N.G.W. Cook & W. Yi 1994. Effects of non-welded interfaces on guided SH-waves. *Geophysical Research Letters* 21(9):745-748.
Nihei, K.T., B. Gu, L.R. Myer, L.J. Pyrak-Nolte & N.G.W. Cook 1995. Elastic interface wave propagation along a fracture. *Proc. 8th Int. Cong. Rock Mech., Tokyo, 25-29 Sept 1995*:Rotterdam: Balkema.
Nihei, K.T., W. Yi, L.R. Myer & N.G.W. Cook 1998. Fracture channel waves. *J. of Geophysical Research*, submitted.
O'Connell, R.J. & B. Budiansky 1974. Seismic velocities in dry and saturated cracked solids. *J. Geophys. Res.* 79:5412-5426.
Pyrak-Nolte, L.J. & N.G.W. Cook 1987. Elastic interface waves along a fracture. *Geophys. Res. Lett.* 14:1107-1110.
Pyrak-Nolte, L.J., L.R. Myer, N.G.W. Cook & R.A. Witherspoon 1987. Hydraulic and mechanical properties of natural fractures in low permeability rock. *Proceedings of 6th International Congress of Rock Mechanics*: I:225-232, Rotterdam:Balkema.
Pyrak-Nolte, L.J., L.R. Myer & N.G.W. Cook 1990. Transmission of seismic waves across single natural fractures. *J. Geophys. Res.* 95:8617-8638.
Pyrak-Nolte, L.J., J. Xu & G. Haley 1992. Elastic interface waves propagating in a fracture. *Physical Review Letters* 69(24):3650-3653.
Roy, S. & L.J. Pyrak-Nolte 1995. Interface waves propagating along tensile fractures in dolomite. *Geophysical Research Letters* 22(20):2773-2776.
Roy, S. & L.J. Pyrak-Nolte 1997. Observation of a distinct compressional-mode interface wave on a single fracture. *Geophysical Research Letters* 24(2):173-176.
Schoenberg, M. 1980. Elastic wave behavior across linear slip interfaces. *J. Acoust. Soc. Am.* 68(5):1516-1521.
Stoneley, R. 1924. Elastic waves at the surface of separation of two solids. *Proceedings of the Royal Society, London*. 106:416-428.
Suárez-Rivera, R. 1992. The influence of thin clay layers containing liquids on the Propagation of shear waves. *Ph.D. Thesis*, University of California, Berkeley.
Tada, H.P., P.C. Paris & G.K. Irwin 1973. *The stress analysis of cracks handbook.*Helbertown, PA: Del Research Corp.
Thomsen, L. 1986. Weak elastic anisotropy. *Geophysics* 51(10):1954-1966.

Mechanics of Jointed and Faulted Rock, Rossmanith (ed.) © 1998 Taylor & Francis, ISBN 90 5410 955 6

Poroelasticity applications

J-C. Roegiers, L. Cui & M. Bai
Rock Mechanics Institute, University of Oklahoma, Norman, Okla., USA

ABSTRACT: Poroelasticity is the theory governing mechanics of fluid saturated porous elastic media as it takes into consideration the interaction between solid deformation and pore fluid flow. This paper describes poroelasticity modeling in rock mechanics as well as some poroelasticity applications in the petroleum industry.

1 INTRODUCTION

Rock in underground environments is generally porous and permeated with fluids. The mechanical responses of the solid matrix and the pore fluid of a saturated po-rous body subjected to external loading are usually coupled together. This coupling effect may result in much higher pore pressures than predicted by the conventional (uncoupled) theory. Therefore, engineering activities such as excavation and fluid withdrawal can trigger a number of more critical phenomena. The theory that rigorously couples the solid and fluid deformation-diffusion mechanisms, known as poroelasticity, was pioneered by Biot (1941) and has recently received wide attention in the rock mechanics community and petroleum industry.

This paper describes recent research on modeling saturated porous media for various environments and introduces poroelasticity applications in the petroleum industry.

2 FUNDAMENTALS

Various presentations of the theory of poroelasticity have been adopted by various authors (Biot 1941, Rice & Cleary 1976, Detournay & Cheng 1993). The notation system used in Detournay & Cheng (1993) is adopted herein. Basic equations of poroelasticity for isotropic materials are first described below.

- Stress-restrain relationship

$$\sigma_{ij} = 2G\epsilon_{ij} + \frac{2G\nu}{1-2\nu}\epsilon\delta_{ij} - \alpha p\delta_{ij} \qquad (1)$$

In the above, σ_{ij} is the total stress tensor (positive for tension); ϵ_{ij} is the strain tensor of the porous body; $\epsilon = \epsilon_{ii}$ is the volumetric strain; G is the shear modulus; ν is the drained Poisson's ratio; and α is the Biot's effective stress coefficient which can be expressed by (Nur & Byerlee 1971):

$$\alpha = 1 - \frac{K}{K_s} \qquad (2)$$

in which K and K_s are the bulk moduli of the drained porous body and the solid grain, respectively.

- Volumetric variation relation

$$p = M(\zeta - \alpha\epsilon) \qquad (3)$$

where ζ is the volumetric variation of pore fluid content in the porous body, and M is the Biot's modulus which can be expressed by (Brown & Korringa 1975):

$$\frac{1}{M} = \frac{\phi}{K_f} + \frac{\alpha - \phi}{K_s} \qquad (4)$$

in which ϕ is the porosity, and K_f is the bulk modulus of the pore fluid.

- Equilibrium equation

$$\sigma_{ij,j} = b_i \qquad (5)$$

where b_i is the body force vector.

- Continuity equation

$$\frac{\partial \zeta}{\partial t} + q_{i,i} = 0 \qquad (6)$$

where q is the pore fluid discharge vector.

- Darcy's law

$$q_i = -\frac{k}{\mu} p_{,i} \tag{7}$$

where k is the intrinsic permeability, and μ is the viscosity of the pore fluid.

In the above, the materials have been assumed linear, isotropic, and elastic; the quasi-static state and the absence of a fluid sink or source have also been assumed.

3 MATERIAL ANISOTROPY

Rocks generally exhibit some degree of anisotropy, i.e., the material properties are dependent on the directions. Constitutive equations for anisotropic poroelastic media were developed by Biot (1955). However, until lately, further studies of anisotropic poroelasticity were not conducted. Carroll (1979) and Thompson & Willis (1991) studied constitutive relations; Aoki et al. (1993) conducted measurements of anisotropic poroelastic constants; and, recently, Abousleiman et al. (1996) and Cui et al. (1996) investigated material anisotropy effects on several poroelastic problems.

In anisotropic poroelasticity, relations (1), (3), and (7) are no longer valid. Instead, their corresponding equations are, respectively, expressed by:

$$\sigma_{ij} = D_{ijkl}\epsilon_{kl} - \alpha_{ij} p \tag{8}$$

$$p = M(\zeta - \alpha_{ij}\epsilon_{ij}) \tag{9}$$

$$q_i = -\frac{k_{ij}}{\mu} p_{,j} \tag{10}$$

where D_{ijkl} is the anisotropic drained elastic modulus tensor, α_{ij} is the tensor of anisotropic Biot's effective stress coefficients, and k_{ij} is the anisotropic permeability tensor.

The number of independent material properties for an anisotropic poroelastic body is greater than for a corresponding anisotropic elastic body. For a most general anisotropic case, 28 independent material properties are required for poroelasticity *vs.* 21 for elasticity; for an orthotropic case, 13 for poroelasticity *vs.* 9 for elasticity; and for a transversely isotropic case, 8 for poroelasticity *vs.* 5 for elasticity. Obviously, the measurements for anisotropic poroelastic properties are much more difficult, since more limited conditions are required.

Using a micromechanics approach, relations among various material properties of anisotropic poroelastic media were obtained (Thompson & Willis 1991, Cheng 1997). Applying these relations, some anisotropic poroelastic properties necessary for the constitutive equations may be evaluated from other measurable quantities. For example, with the assumption of microhomogeneity and microisotropy of the solid matrix, α_{ij} and M of an anisotropic porous material can be expressed by (Cheng 1997):

$$\alpha_{ij} = \delta_{ij} - \frac{D_{ijkk}}{3K_s} \tag{11}$$

$$M = \frac{K_s}{\left(1 - \frac{D_{iijj}}{9K_s}\right) - \phi\left(1 - \frac{K_s}{K_f}\right)}. \tag{12}$$

Unlike isotropic poroelasticity, it is seen from the above constitutive equations that the coupling between the pore pressure and stress/strain depends on the direction; and that the shear deformation of the porous body influences the pore pressure distribution. Until recently, in many applications this anisotropic behavior has been ignored by assuming the tensor of Biot's effective stress coefficients to be isotropic (i.e., $\alpha_{ij} = \alpha\delta_{ij}$). This assumption is not appropriate for most anisotropic rocks. The measurements of a transversely isotropic shale (Aoki et al. 1993) demonstrated that Biot's effective stress coefficients are anisotropic. For a transversely isotropic rock, Cui et al. (1996) showed that using $\alpha_{ij} = \alpha\delta_{ij}$ may cause significant errors (see Figure 1).

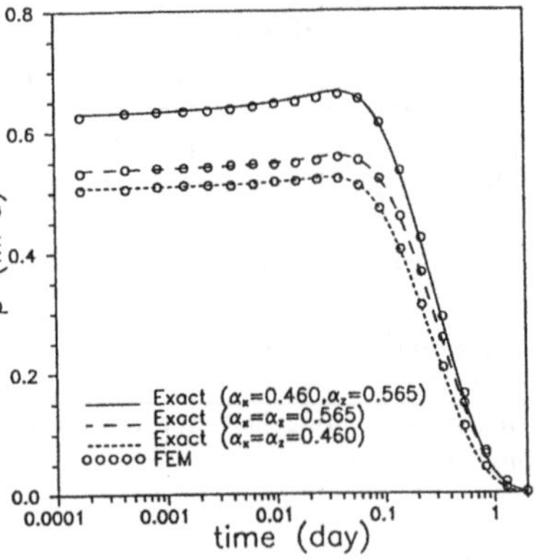

Figure 1: For a 2-D anisotropic consolidation (Mandel's problem), pore pressure histories varying with assumptions of α_{ij}. $\alpha_x = \alpha_y = 0.460$ and $\alpha_z = 0.565$ are exact values of the coefficients; $\alpha_x = \alpha_y = \alpha_z = 0.460$ and $\alpha_x = \alpha_y = \alpha_z = 0.565$ are the isotropy assumptions for these coefficients.

4 THERMAL EFFECT

The interactive responses of fluid flow and heat

transfer in deformable porous media are demonstrated in a triangular-fashion interplay, where the changes in either solid body strain, pore pressure, or material temperature result in the corresponding disequilibrium of momentum, mass and energy. For a linear, quasi-steady elastic system, a fully-coupled thermoporoelastic formulation (Elsworth 1993, Bai & Roegiers 1994a, Jing et al. 1995) may be described in the sequences of conservations of the momentum, mass and energy as (Bai et al. 1996):

$$Gu_{i,jj} + \frac{G}{1-2\nu}u_{k,ki} = \alpha p_{,i} + \beta T_{,i} \quad (13)$$

$$\frac{k}{\mu}p_{,kk} = \alpha\dot{\varepsilon}_{kk} - \alpha_h\dot{T} + \alpha^*\dot{p} + Q_f \quad (14)$$

$$K^*T_{,kk} = -\frac{k\,s}{\mu}p_{,k}T_{,k} + \beta T_0\dot{\varepsilon}_{kk} + s^*\dot{T} + Q_h \quad (15)$$

where a superscripted dot identifies the derivative with respect to time t, u_i is the displacement vector of the porous body, T is the temperature variation, β is the thermal expansion factor, $\alpha^* = 1/M$, α_h is the thermal expansion coefficient, K^* is the thermal conductivity, s is the intrinsic heat capacity for fluid, s^* is the lumped intrinsic heat capacity, T_0 is the reference temperature, and Q_f and Q_h are the internal as well as external fluid and heat sources.

In equations (13), (14) and (15), in addition to the traditional quasi-steady mechanical equilibrium, transient fluid flow and heat transport, coupling terms are shown as the first two terms on the right-hand side of each equation. They represent, respectively: $\alpha p_{,i}$: modification of the effective stress due to the influence of fluid pressure; $\beta T_{,i}$: adjustment of total strain as a result of thermal expansion; $\alpha\dot{\varepsilon}_{kk}$: mass rate changes expelled by the solid volumetric strain; $\alpha_h\dot{T}$: rate variation as a result of thermal expansion (or contraction) of fluid and solid; $ksp_{,k}T_{,k}/\mu$: disequilibrium of energy transport due to forced thermal convection in fluid flow; and, $\beta T_0\dot{\varepsilon}_{kk}$: energy lost as a result of solid elastic deformation in the form of thermal expansion or contraction.

5 FLOW IN FRACTURED MEDIA

The response of fractured-porous rock masses is significantly different from an intact porous media because of their unique flow characteristics along with its unique mechanical response to internal and external applied loads. The governing equations representing flow through fractured-porous media were initially proposed by Barenblatt et al. (1960). The mathematical model was further developed as a potential reservoir simulator by Warren & Root (1963). Recent advances in the dual-porosity modeling extend the traditional dual-porosity approach to encompass the coupled processes (Bai et al. 1993). Research efforts have also been focused on identifying local influences such as advective flow (Bai & Roegiers 1994b) and nonlinear flow near a well (Bai et al. 1994) in a dual-porosity medium. Accompanying numerical advances include the development of a three-dimensional finite element model capable of evaluating coupled flow-deformation problems in dual-porosity media (Bai et al. 1995a).

In the dual-porosity poromechanical formulation, the governing equations for the solid and fluid phases can be written as (Wilson & Aifantis 1982, Bai et al. 1995b):

$$Gu_{i,jj} + \frac{G}{1-2\nu}u_{k,ki} - \sum_{m=1}^{2}\alpha_m p_{m,i} = 0 \quad (16)$$

$$\frac{k_m}{\mu}p_{m,kk} - \alpha_m\dot{\epsilon}_{kk} - \alpha_m^*\dot{p}_m \pm \Gamma(\Delta p) = 0 \quad (17)$$

where m=1 and 2, represent the matrix and fractures, respectively; α_m, α_m^*, k_m and p_m are the corresponding quantities of α, α^*, k and p in the matrix and fractures, respectively; Γ is the transfer coefficient, Δp is the pore pressure difference between fractures and matrix blocks.

6 NUMERICAL MODELING

The mathematical structure of poroelasticity is more complicated than its counterpart in the conventional theory of elasticity because of the time-dependency of the problem and the coupling of the pore fluid effect. Usually, numerical techniques must be employed to solve most engineering problems. Three popular numerical methods for continuum mechanics, the finite element method, the finite difference method and the boundary element method, are usually used for poroelasticity problems. Among them, the finite element method is the most popular.

Application of the finite element method to poroelasticity was pioneered by Sandhu & Wilson (1969) for the case of incompressible fluid and solid constituents for soil mechanics applications. Since then, this work has been extended to the cases of compressible constituents (Ghaboussi & Wilson 1973), dynamic problems (Zienkiewicz & Shimoi 1984), nonlinear and thermally coupled problems (Lewis & Schrefler 1987), etc. The procedure to establish the finite element equations of poroelasticity has been formalized by Zienkiewicz & Taylor (1989). Some commercial codes such as ABAQUS (Hibbit et al. 1989) are also capable of dealing with poroelasticity. Recently, aiming at applications in the petroleum industry and in mining engineering, a more general anisotropic constitutive model with anisotropic Biot's effective stress coefficients was formulated in finite element equations of poroelasticity (Cui et al. 1996); the finite element formulation for generalized plane strain poroelastic problems (Cui et al. 1997b) was derived; and a three-dimensional finite element model for dual-porosity media (Bai et al. 1995a) was also developed.

The finite element equations for poroelasticity in incremental form can be expressed as (Cui et al. 1996):

$$\begin{bmatrix} \mathbf{K} & \mathbf{G} \\ \mathbf{G}^{\mathrm{T}} & \mathbf{L}+\Theta\Delta t\mathbf{H} \end{bmatrix} \begin{Bmatrix} \Delta\hat{\mathbf{u}} \\ \Delta\hat{\mathbf{p}} \end{Bmatrix} = \begin{Bmatrix} \Delta\mathbf{F}_1 \\ \Delta\mathbf{F}_2 - \Delta t\mathbf{H}\hat{p}_n \end{Bmatrix} \tag{18}$$

where,

$$\mathbf{K} = \int_\Omega \mathbf{B}_u^{\mathrm{T}}\mathbf{D}\mathbf{B}_u d\Omega \tag{19}$$

$$\mathbf{G} = -\int_\Omega \mathbf{B}_u^{\mathrm{T}}\boldsymbol{\alpha}\mathbf{N}_p d\Omega \tag{20}$$

$$\mathbf{L} = -\int_\Omega \frac{1}{M}\mathbf{N}_p^{\mathrm{T}}\mathbf{N}_p d\Omega \tag{21}$$

$$\mathbf{H} = -\int_\Omega \mathbf{B}_p^{\mathrm{T}}\boldsymbol{\kappa}\mathbf{B}_p d\Omega \tag{22}$$

$$\Delta\mathbf{F}_1 = \int_\Gamma \mathbf{N}_u^{\mathrm{T}}\Delta\mathbf{t} d\Gamma \tag{23}$$

$$\Delta\mathbf{F}_2 = \int_\Gamma \Delta Q\mathbf{N}_p^{\mathrm{T}} d\Gamma\,. \tag{24}$$

In the above, $\Delta = ()_{n+1} - ()_n$ in which n indicates the nth time step; $\hat{\mathbf{u}}$ is the column vector of nodal values of displacements; $\hat{\mathbf{p}}$ is the column vector of nodal values of the pore pressure; Θ is the weighing coefficient for a time stepping scheme; $\mathbf{B}_u$ is the strain matrix calculated from the shape function matrix $\mathbf{N}_u$ for displacements; $\mathbf{B}_p$ is the gradient matrix calculated from the shape function matrix $\mathbf{N}_p$ for the pore pressure; $\mathbf{D}$ is the elastic modulus matrix; $\boldsymbol{\alpha}$ is the matrix of Biot's effective stress coefficients; and, $\boldsymbol{\kappa}$ is the permeability matrix.

In addition to the finite element method, other numerical methods have also been applied to poroelastic problems. The boundary element method has been implemented in poroelasticity using Laplace transform (Cheng & Detournay 1988) and time stepping technique (Dargush & Banerjee 1989). Recently, the boundary element method was applied to hydraulic fracturing (Detournay & Cheng 1991, Boone & Ingraffea 1990). Finite difference method is another popular numerical method used in continuum mechanics. Recently, this method was also applied to solve nonlinear poroelastic problems such as numerical modeling of hydraulic fracturing (Yuan 1996).

7 APPLICATIONS IN THE PETROLEUM INDUSTRY

7.1 *Borehole Stability*

In the petroleum industry borehole failure is a continuing problem that results in huge financial losses. Conventional methods for stability analysis of boreholes are based on Kirsch's elastic solution (Bradley 1979, Fjær et al. 1992). However, many borehole failure phenomena such as time-dependency (Fjær et al. 1992) can hardly be predicted using this approach. Since boreholes are usually drilled through fluid saturated formations, the stress concentration around the borehole results in a disturbance of the pore pressure field; and the variation of the pore pressure strongly influences the stability. Therefore, poroelasticity is a proper theory to model borehole stability in fluid saturated formations.

Detournay & Cheng (1988) derived a poroelastic analytical solution for vertical boreholes under plane strain conditions. Later, Cui et al. (1997a) developed a poroelastic model for inclined boreholes (i.e., the borehole axis is deviated from principal directions of the *in-situ* stress tensor.) Poroelastic models showed that effective stresses and pore pressures around the borehole wall are time-dependent (e.g., see Figure 2).

In addition, the analysis also showed that the pore pressure around the borehole could be much higher or lower than both formation and well pressures (Cui et al., in prep.) This is quite differ-

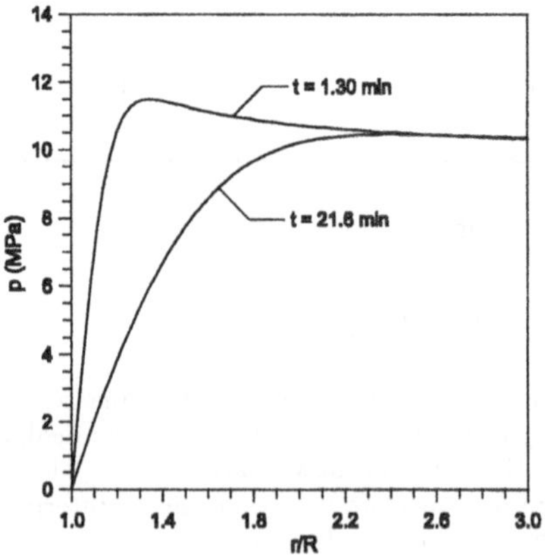

Figure 2: For an inclined borehole, the pore pressure around the borehole in the direction close to the minimum horizontal *in-situ* stress. r= the radial distance; and R= the borehole radius.

ent from the assumption in conventional models that the pore pressure around the borehole wall is the same as the formation pore pressure (impermeable model) or the well pressure (permeable model). All these indicated that the predictions on the borehole stability based on the poroelastic models would be quite different from the ones based on the elastic models. Figure 3 presents critical mud pressures varying with the borehole inclination. The significant quantitative differences between elastic and poroelastic models are clearly seen in the figure; and the time-dependent failure was implied in poroelastic predictions. Besides, the poroelastic analyses also showed that failure might initiate inside the borehole wall instead of right at the borehole wall. It was also found that both shear failure and tensile failure (spalling) may occur inside the borehole wall in the early stages after the excavation (Cui et al. 1995). For example, Figure 4 indicates the possibility of shear failure initiating inside the borehole wall.

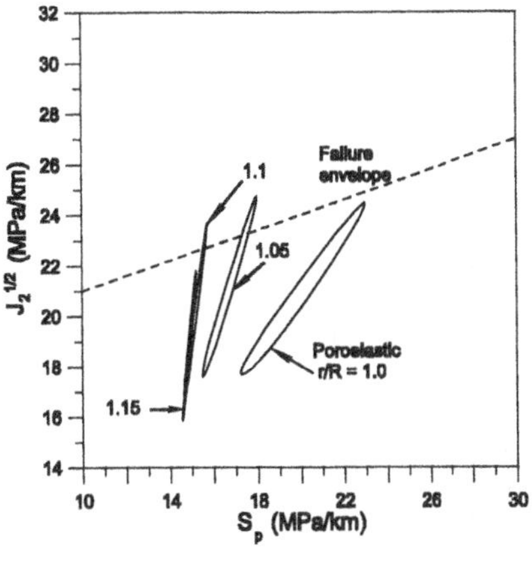

Figure 4: Stress clouds at a small time shifting in the stress space with the change of the radial distance. The failure envelope represents the Drucker-Prager criterion.

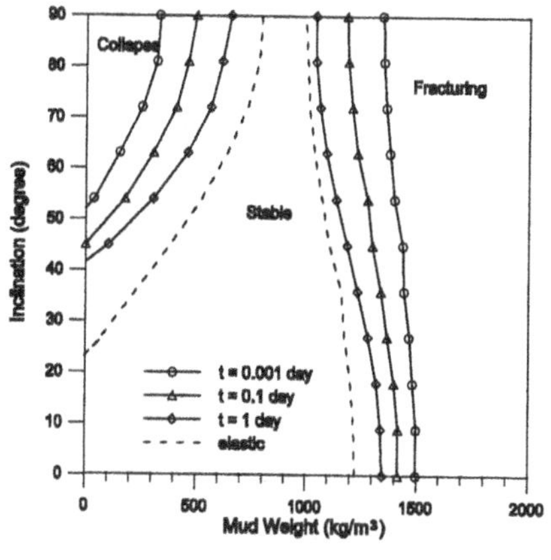

Figure 3: Safe range of mud weight varying with the borehole inclination for an inclined borehole.

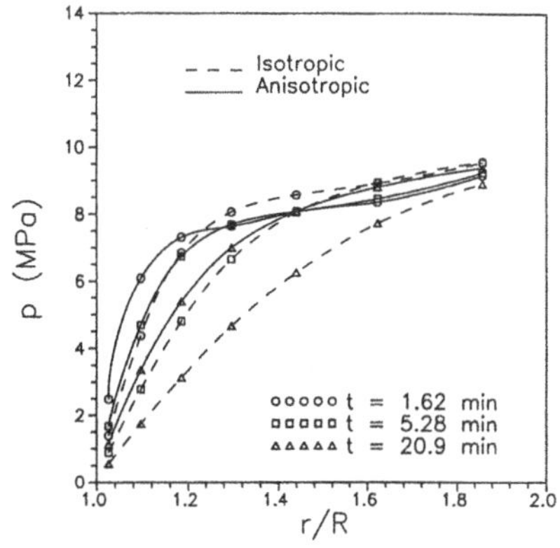

Figure 5: Pore pressure distributions around borehole varying with time.

Other poroelastic effects on the borehole stability in fluid saturated formations have also been investigated. For example, the material anisotropy effect on the borehole stability was studied recently by Abousleiman et al. (1995) and Cui et al. (1996). Poroelastic analyses showed that the permeability anisotropy influences the borehole stability significantly (Cui et al. 1996). For an inclined borehole drilled through a transversely isotropic formation with a high degree of permeability anisotropy but a low degree of material anisotropy, Figure 5 shows that pore pressure distributions around the borehole for anisotropic and isotropic cases are quite different, indicating that the permeability anisotropy is not a negligible factor for the borehole stability.

7.2 *Hydraulic Fracturing*

Hydraulic fracturing is widely used in the petroleum industry to enhance the production of hydrocarbons and to measure *in-situ* stresses at great depths. Many models have been developed to simulate hydraulic fracturing (Mendelsohn 1984). However, most proposed models ignored the coupled poroelastic effect which is induced by leakoff of the fracturing fluid in the formation and stress variations around the fracture.

Recently, many authors proposed models for hydraulic fracturing with consideration of poroelasticity (Detournay et al. 1990, Detournay & Cheng 1991, Boone & Ingraffea 1990). Poroelastic analyses of hydraulic fracturing seemed to conclude that the pore pressure around the fracture and the dimension of the fracture might be quite different from the ones predicted using traditional approaches.

For a PKN model, by taking into account the poroelastic effect, Detournay et al. (1990) demonstrated that the fracturing pressure had increased significantly compared to the prediction without poroelasticity (see Figure 6). Similar phenomena were also observed for other poroelastic models such as the penny-shaped fracture (Yuan 1996). For a plane strain poroelastic model of hydraulic fracturing Detournay & Cheng (1991) observed that the fracture volume as well as the stress intensity factor at the fracture tips are strongly influenced by the poroelastic effect and these quantities are also time-dependent (e.g., see Figure 7).

In conclusion, the poroelastic effect should be included in the hydraulic models for porous formations. Especially, when hydraulic fracturing is used for the determination of the state of formations such as *in-situ* stresses, leakoff coefficient, etc., poroelastic effects must be considered for a correct interpretation of these parameters.

7.3 *Naturally-Fractured Reservoirs*

For the typical naturally-fractured reservoirs, undesirable consequences during production may be reflected in: (a) difficulties in controlling initial production rate as a result of rapid fracture flow; (b) interrupted production due to insufficient oil yield from the matrix blocks into the conductive fractures after a period of flow to the well; and, (c) fluctuating fluid flow in the form of variable pressure profiles as a result of nonuniform distribution of percolating fracture clusters.

Assuming the subscripts 1 and 2 represent matrix and fractures, respectively, the following ratios are defined as: (a) permeability ratio $r_0 = k_1/k_2$; (b) compressibility ratio $r_1 = K_f/(K_n d)$ where K_f and K_n are the bulk moduli of fluid and fractures, d is the fracture spacing; (c) $r_2 = K_f/K_s$ where K_s is the bulk modulus of solid grain; and, (d) $r_3 = K_f/E$ where E is the Young's modulus. For a fixed permeability ratio, r_0, and compressibility ratio, r_3, pressure profiles are sensitive to variation of the compressibility ratios r_1 and r_2, as illustrated in Figure 8 (Bai et al. 1995b). It is obvious that the pressure change is delayed as r_1 increases and r_2 decreases. This behavior resembles an equivalent homogeneous reservoir with soft fractures and stiff solid grains. However, it is important to note that the depleting pressure magnitude at early periods appears greater than the reservoir pressure as depicted in Figure 8. This poromechanical impact cannot be observed if a conventional flow model is used alone.

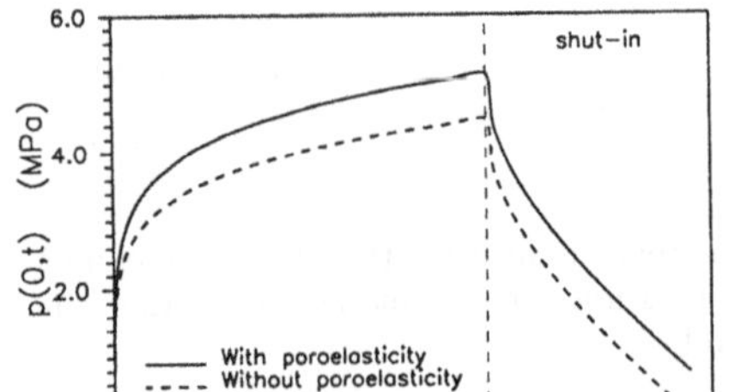

Figure 6: Fracturing pore pressure varying with time for a PKN model (Detournay & Cheng 1990).

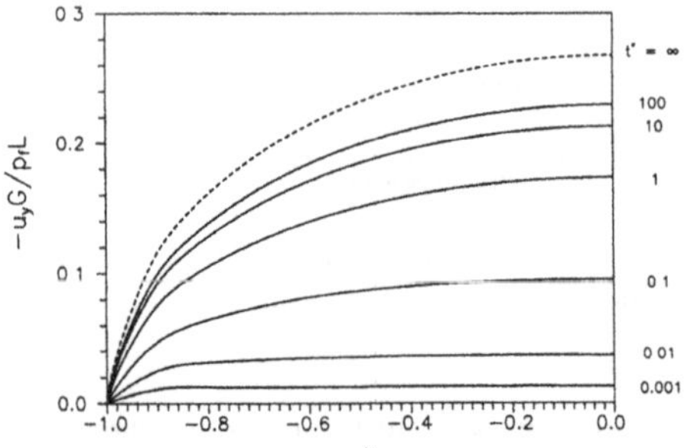

Figure 7: Normalized fracture profile for mode 2 of a Griffith fracture (Detournay & Cheng 1991). p_f= fluid pressure applied at the fracture wall; L= the fracture length; u_y= the normal displacement of the fracture wall; and x= the coordinate along the fracture.

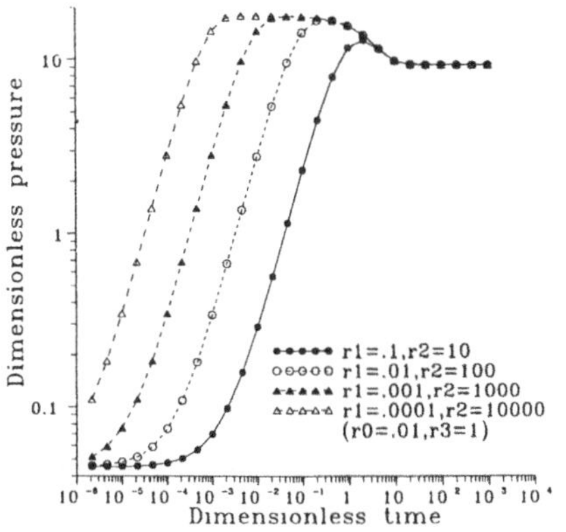

Figure 8: Pressure *vs.* various compressibility ratios.

8 SUMMARY

The coupling effect between stress/strain and the pore fluid pressure of fluid saturated media may be very significant in some engineering problems. This effect can be dealt with appropriately using the theory of poroelasticity. This paper reviewed recent work on the modeling of fluid saturated porous rocks using poroelasticity for various situations such as isotropic and anisotropic materials, thermal stress problems, as well as fractured media. The applications of poroelasticity to the petroleum industry were also described. The numerical examples showed that erroneous results could be obtained by ignoring the poroelastic effect.

ACKNOWLEDGMENT

This work is supported by a NSF grant to the Rock Mechanics Research Center, the Oklahoma Center for the Advancement of Science and Technology, and the O.U. Rock Mechanics Consortium.

REFERENCES

Abousleiman, Y., L. Cui, A.H-D. Cheng & J-C. Roegiers 1995. Poroelastic solution of an inclined borehole in a transversely isotropic medium. *Proc. 35th U.S. Symp. on Rock Mechanics.* J.J.K. Daemen & R.A. Schultz (eds). 313-318.

Abousleiman, Y., A.H-D. Cheng, L. Cui, E. Detournay & J-C. Roegiers 1996. Mandel's problem revisited. *Géotechnique.* 46 (2): 187-195.

Aoki, T., C.P. Tan & W.E. Bamford 1993. Effects of elastic and strength anisotropy on borehole failures in saturated rocks. *Int. J. Rock Mech. Min. Sci. & Geomech. Abstr.* 30: 1031-1034.

Bai, M., Y. Abousleiman & J-C. Roegiers 1996. Some thoughts on thermoporoelastic coupling. *Proc. 11th Engineering Mechanics Conference, ASCE.* Ft. Lauderdale, FL. 48-51.

Bai, M., D. Elsworth & J-C. Roegiers 1993. Modeling of naturally fractured reservoirs using deformation-dependent flow mechanisms. *Int. J. Rock Mech. Min. Sci. and Geomech. Abstr.* 30(7): 1185-1191.

Bai, M., D. Elsworth, J-C. Roegiers & F. Meng 1995a. A three-dimensional dual-porosity poroelastic model. *Proc. of CCMRI - International Mining Tech'95.* Beijing, China. 184-202.

Bai, M., Q. Ma & J-C. Roegiers 1994. A nonlinear dual-porosity model. *J. Appl. Math. Modelling.* 18: 602-610.

Bai, M. & J-C. Roegiers 1994a. Fluid flow and heat flow in deformable fractured porous media. *Int. J. Engng. Sci.* 32(10): 1615-1633.

Bai, M. & J-C. Roegiers 1994b. On the correlation of nonlinear flow and linear transport with application to dual-porosity modeling. *J. Petroleum Sci. Engng.* 11: 63-72.

Bai, M., J-C. Roegiers & D. Elsworth 1995b. Poromechanical response of fractured-porous rock masses. *J. Petroleum Sci. Engng.* 13: 155-168.

Barenblatt, G.I., I.P. Zheltov & N. Kochina 1960. Basic concepts in the theory of seepage of homogeneous liquids in fissured rocks. *Prikl. Mat. Mekh.* 24(5): 852-864.

Biot, M.A. 1941. General theory of three dimensional consolidation. *J. Appl. Phys.* 12: 155-164.

Biot, M.A. 1955. Theory of elasticity and consolidation of a porous anisotropic solid. *J. Appl. Phys.* 26: 182-185.

Boone, T.J. & A.R. Ingraffea 1990. Numerical procedure for simulation of hydraulically driven fracture propagation in poroelastic media. *Int. J. Numer. Anal. Meth. Geomech.* 14: 27-47.

Bradley, W.B. 1979. Failure of inclined boreholes. *J. Energy Resour. Tech., ASME.* 101: 232-239.

Brown, R.J.S. & J. Korringa 1975. On the dependence of the elastic properties of a porous rock on the compressibility of the fluid. *Geophys.* 40(4): 608-616.

Carroll, M.M. 1979. An effective stress law for anisotropic elastic deformation. *J. Geophys. Res.* 84: 7510-7512.

Cheng, A.H-D. & E. Detournay 1988. A direct boundary element method for plane strain poroelasticity. *Int. J. Numer. Anal. Meth. Geomech.* 12: 551-572.

Cheng, A.H-D. 1997. Material coefficient of anisotropic poroelasticity. *Int. J. Rock Mech. Min. Sci.* 34: 199-205.

Cui, L., Y. Abousleiman, A.H-D. Cheng, J-C. Roegiers & D. Leshchinsky 1995. Stability analysis of an inclined borehole in an isotropic poroelastic medium. *Proc. 35th U.S. Symp. on Rock Mechanics.* J.J.K. Daemen & R.A. Schultz (eds). 307-312.

Cui, L., Y. Abousleiman, A.H-D. Cheng, V.N. Kaliakin & J-C. Roegiers 1996. Finite element analysis of anisotropic poroelasticity: a generalized Mandel's problem and an inclined borehole problem. *Int. J. Numer. Anal. Methods Geomech.* 20: 381-401.

Cui, L., A.H-D. Cheng & Y. Abousleiman 1997a. Poroelastic solution for an inclined borehole. *J. Appl. Mech., ASME.* 64: 32-38.

Cui, L., V.N. Kaliakin, Y. Abousleiman & A.H-D. Cheng 1997b. Finite element formulation and application of poroelastic generalized plane strain problems. *Int. J. Rock Mech. Min. Sci.* 34(6): 953-962.

Cui, L., S. Ekbote, Y. Abousleiman, M. Zaman & J-C. Roegiers, in prep. Borehole stability analyses in fluid saturated formations with non-penetrating walls. Accepted by *3rd North American Rock Mechanics Symposium.* Cancun, Quintana Roo, Mexico. June 3-5, 1998

Dargush, G.F. & P.K. Banerjee 1989. A time domain boundary element method for poroelasticity. *Int. J. Numer. Meth. Eng.* 28: 2423-2449.

Detournay, E. & A.H-D. Cheng 1988. Poroelastic response of a borehole in a non-hydrostatic stress field. *Int. J. Rock Mech. Min. Sci. Geomech. Abstr.* 25: 171-182.

Detournay, E., A.H-D. Cheng & J.D. McLennan 1990. A poroelastic PKN hydraulic fracture model based on an explicit moving mesh algorithm. *J. Energy Res. Tech., ASME.* 112: 224-230.

Detournay, E. & A.H-D. Cheng 1991. Plane strain analysis of a stationary hydraulic fracture in a poroelastic medium. *Int. J. Solids Struct.* 37: 1645-1662.

Detournay, E. & A.H-D. Cheng 1993. Fundamentals of Poroelasticity. *Comprehensive rock engineering: principles, practice & projects. Vol. II, Analysis and design method.* C. Fairhurst (eds.) 113-171. Pergamon Press.

Elsworth, D. 1993. Computational methods in fluid flow. *Comprehensive Rock Engineering.* J.A. Hudson (ed.) 2: 173-189. Pergamon Press.

Fjær, E., R.M. Holt, P. Horsrud, A.M. Raaen & R. Risnes, 1992. *Petroleum related rock mechanics*, Elsevier Science.

Ghaboussi, J. & E.L. Wilson 1973. Flow of compressible fluid in porous elastic media. *Int. J. Num. Meth. Eng.* 5: 419-442.

Hibbit, Karlsson & Sorensen, Inc. 1989. *ABAQUS Theory Manual, Version 4.8.*

Jing, L., C.-F. Tsang & O. Stephansson 1995. DECOVALEX - An international co-operative research project on mathematical models of coupled THM processes for safety analysis of radioactive waste repositories. *Int. J. Rock Mech. Min. Sci. & Geomech. Abstr.* 32(5): 389-398.

Lewis, R.W. & B.A. Schrefler 1987. *The finite element method in the deformation and consolidation of porous media.* John Wiley & Sons.

Mendelsohn, D. 1984. A review of hydraulic fracture modeling–Part I: General concepts, 2D models, motivation for 3D modeling. *J. Energy Res. Tech., ASME.* 106: 369-376.

Nur, A. & J.D. Byerlee 1971. An exact effective stress law for elastic deformation of rock with fluids. *J. Geophys. Res.* 76(26): 6414-6419.

Rice, J.R. & M.P. Cleary 1976. Some basic stress diffusion solutions for fluid-saturated elastic porous media with compressible constituents. *Reviews of Geophysics and Space Physics.* 14(4): 227-241.

Sandhu, R.S. & E.L. Wilson 1969. Finite-element analysis of seepage in elastic media. *J. Eng. Mech. Div. ASCE.* 95: 641-652.

Thompson, M. & J.R. Willis 1991. A reformulation of the equations of anisotropic poroelasticity. *J. Appl. Mech., ASME.* 58: 612-616.

Warren, J.E. & P.J. Root 1963. The behavior of naturally fractured reservoirs. *J. Soc. Pet. Eng.* 3: 245-255.

Wilson R. K. & E. C. Aifantis 1982. On the theory of consolidation with double porosity. *Int. J. Engng. Sci.* 20: 1009-1035.

Yuan, Y. 1996. *Simulation of penny-shaped hydraulic fracturing in porous media.* Ph.D. dissertation. University of Oklahoma.

Zienkiewicz, O.C. & T. Shiomi 1984. Dynamic behaviour of saturated porous media; the generalized Biot formulation and its numerical solution. *Int. J. Numer. Anal. Methods Geomech.* 8: 71-96.

Zienkiewicz, O.C. & R.L. Taylor 1989. *The finite element method. 4th ed.* McGraw-Hill.

Mechanics of Jointed and Faulted Rock, Rossmanith (ed.) © 1998 Taylor & Francis, ISBN 90 5410 955 6

The contribution of Historical Earthquake Research to the analysis of seismic hazard and the tectonic stress field

R.Gutdeutsch
Institute of Meteorology and Geophysics, University of Vienna, Austria

ABSTRACT: In view of the fast increase of population in many countries the analysis of seismic hazard has become more and more important even for regions with low seismicity. Therefore the demand for a more reliable earthquake hazard estimation requires to extend the observation time of seismicity to earlier centuries. Knowledge about earthquakes of the "pre-instrumetal time" before 1900 comes from contemporary documentation, the sources, mainly texts, but depictions or photos can serve as helpful tools. Methods of historical earthquake research are visualized by the presentation of 3 historical events.

1 INTRODUCTION

In Central Europe great earthquakes are rare. It may happen, that the design of planned critical technical structures as NPP´s or waste deposits needs more detailed data than can be gained from the present empirical knowledge of the site. Even when the probability of earthquake occurence is low, the increasing construction activity of technical structures forms a *risk* to the environment. Therefore the demand for a more reliable earthquake hazard estimation requires to extend the observation time of seismicity to earlier centuries. Since 1900 earthquakes are recorded by seismometers. Their records allow quantitative statements as amplitude and frequency content of ground motion and herefrom important parameters as magnitudes M_L, maximum intensity I_0 and focal solutions of events. Earlier earthquakes have been handed down by many kinds of written texts, mainly by narrative sources, but also by invoices or lists of the damage caused by the historical earthquake. In rare cases artistic or photographic depictions of the damage by earthquakes are still available.

The processing of historical messages in order to gain seismological parameters is the main emphasis of Historical Earthquake Research. Seismologists and historians co-operate to find answers to questions such as follows:

1) How strong was the historical earthquake and which was its regionally specific mechanism?

2) What is the recurrence rate of damaging earthquakes in certain regions and is it constant?

3) What seismic hazard follows from the investigation of historical earthquakes for a site?

As a matter of fact Central Europe was the place of rather active seismity in the course of centuries. Figure 1 shows a map of epicenters with maximum intensities greater than VI. Open circles indicate epicenters before 1900. In this paper I will stress three examples: (1) the earthquake of Neulengbach 1590, I_0 = IX in the Viennese Wood, (2) the Komarno earthquake 1763, I_0 = VIII or VIII-IX next to the Danube on the border between Slovakia and Hungary and (3) the Friuli earthquake 1348, I_0 = X, which is wellknown as the "Great Villach earthquake" (In this paper intensities are given in Roman letters and correspond to the European Macroseismic Scale (EMS92, see Grünthal 1993))

2 THE BASIS PROBLEM: INCOMPLETE AND INHOMOGENEOUS HISTORICAL EARTHQUAKE DATA IN CATALOGUES

Statistical conclusions can be drawn if the used data set is *complete* and *homogeneous*. At any rate we have to test the data, whether this condition is satisfied. This is true not only for physical measurements. It is true for historical earthquake data as well. We will visualize this problem by the example of earthquake catalogues.

When we would like to learn more about an historical earthquake, at first we have to ask the respective national earthquake catalogues. Most national earthquake catalogues go back to pre-

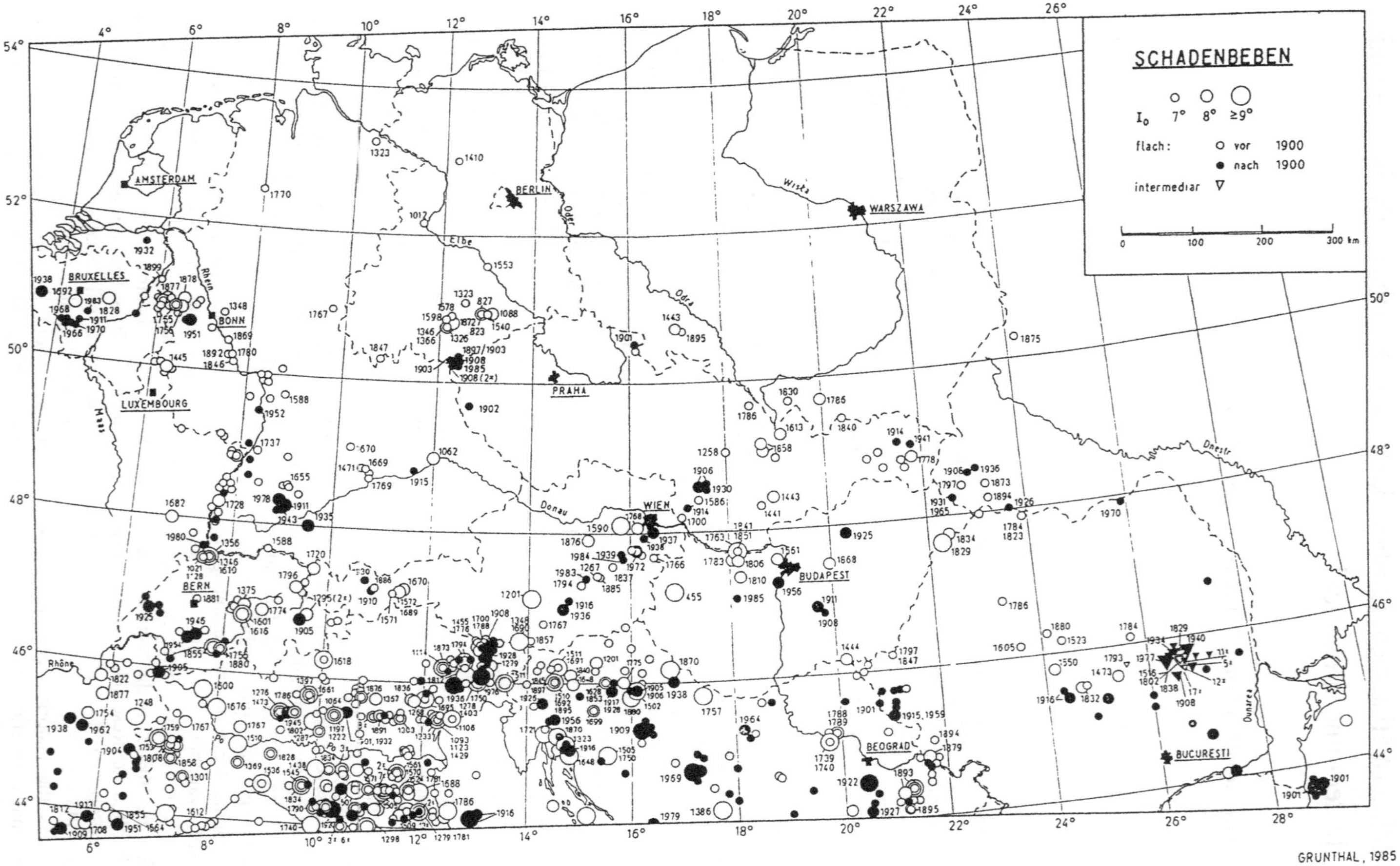

Figure 1. Epicenters of known damaging earthquakes in Central Europe sensu latu (after Grünthal 1988).

instrumental times. They give lists of the events with their epicentral co-ordinates, focal time, maximum intensity and in many cases the magnitude.

The authors of earlier catalogues found it sufficiant to add a reference list. This list included every published text, as known sofar, independently of its importance. For instance, texts written more than 100 years after the earthquake (literature) and contemporary sources, in some cases reports written by eye-witnesses, have been used in the same way, although their qualities as a earthquake message are different. The used data material was not *homogeneous*. An improvement has been achieved by the CFT (Boschi et al 1995) of Italy, but the CFT includes the strongest earthquakes only.

Catalogues provide an appropriate tool of testing the completeness criterion. The Italian earthquake catalogue (Postpischl et al 1985) realizes the inhomogenity of data but leaves the question open how to treat this problem. Figure 2 shows the numbers N of events per 50 years for earthquakes with maximum intensities VIII, XI an X. The general increase of N with time between 1000 and 1800 p.C. asymptotically reaches a constant value approximately at 1800 and probably indicates the incompleteness of data - provided that the seismicity was constant during the centuries. Since 1800 the earthqakes with $I_0 \geq$ VIII have been documented completely. In earlier times, for example, in 1200 only 1/10 of them have been noticed by contemporaries! One has to keep in mind that the contemporaries were not familiar with systematic inquiries of building damage and effects which is necessary here. This kind of systematic research work has been slowly developed during the centuries. This fact visualizes the problem of any statistical statement from historical earthquake data. The more a comprehensive searching for contemporary sources is necessary.

3 METHODS OF HISTORICAL EARTHQUAKE RESEARCH

Historical earthquake research aims to a precise knowledge what actually has happened during the earthquake and hence of its numerical parameters. The long term objectives consist of statistically derived hazard maps. It makes use of working methods of historians. Their conclusions are drawn from contemporary sources (flow chart 1).

Historians have developed special methods to find messages of earthquakes which can be expected and possibly be found. It resembles a fisherman´s work, who throws his net out, where he expects the best fishes. Anyway, the mostly time consuming part of his work consists in the systematic searching for sources.

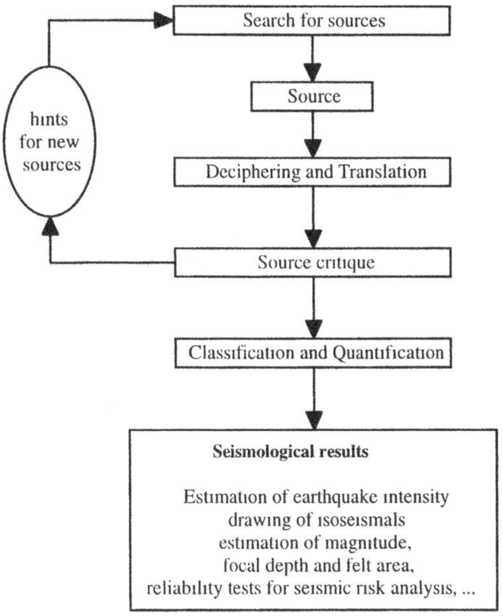

Flow chart 1. Methods of Historical Earthquake Research

A contemporary text refering to the earthquake, sofar not known in the scientific public can be of fundamental importance, as it can considerably change scientific opinions about that event. One can imagine, that such sudden findings sometimes were celibrated as a great discovery. A contemporary text, once found has to be deciphered and translated. This important work requires professional training. The next step, the *source critique* is the equivalent of the historian´s working method to the error calculation of natural scientist. It has to be taken into account, that verbal expressions of a language change with time. Therefore contemporary depictions, in fortunate circumstances, can help. The background of the contemporary writer or painter, his interests and education is necessary to know. The next step, the *quantification* forms the most problematic step of the procedure. It forms the link between historical and the seismological working methods. The lack of completeness mentioned in the discussion of catalogues is present in the original texts as well.

Figure 3 presents an example. It refers to an earthquake in the Early Enlightement time, happened in September 28th, 1590 in Lower Austria (see figure 1). The text is part of a handwritten report in German by the clerk of Koenigstetten to the archbishop of Passau. It describes the damage caused by the earthquake in villages and towns in the district of Tullnerfeld/Danube. This document refers to the damage caused in the village of Zwentendorf, ca. 20 km distant from the suspected

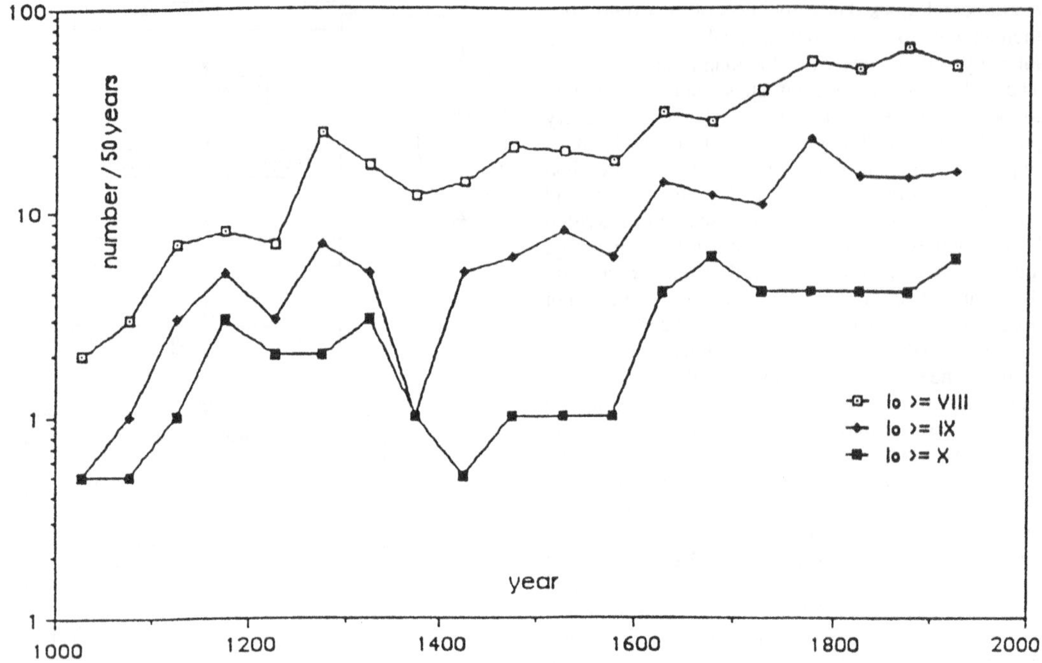

Figure 2. Numbers N of earthquakes per 50 years in Italy (after Postpischl et al 1985).

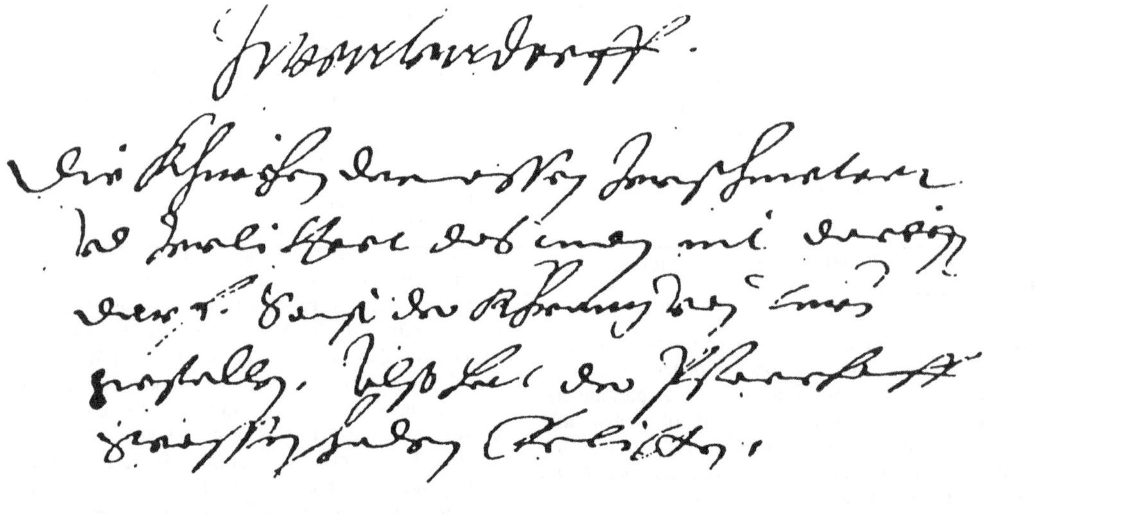

Figure 3. Handwritten report of damage caused to the church of Zwentendorf during the earthquake September 28th, 1590 in Lower Austria , (after Gutdeutsch et al. 1988)

epicenter. It says that the church has been damaged considerably so it could not be used any more and that the upper rim of its tower has dropped down on the parsonage where it caused considerable devastation.

4 QUANTIFICATION - THE LINK BETWEEN HISTORICAL AND SEISMOLOGICAL WORKING METHODS

This example visualizes the difficuly of intensity estimations of historical earthquakes well: Today we use the modern intensity scale EMS92(Grünthal 1993). Figure 4 shows an excerpt of it. The EMS92 is an ordinale scale which associates a Roman number designing the intensity with earthquake effects. In principle it is not allowed to interpolate between intensities. For instance, if *damage class* 3 has been done to *few* buildings of *vulnerability class B* we associate intensity VII for this site. *Few* means up to 10% (maybe 20%) of all buildings and vulnerability class B means "simple stone houses".

As a matter of fact, the message in figure 3 would not allow an intensity estimation for several reasons:

1. The historical building does not exist anymore. Therefore we do not know the vulnerabilitiy class. But, in our case we have good reason to assume vulnerability class B. Figure 4 shows that the vulnerability class has been defined with a given uncertainty. This uncertainty follows from the specific masonary quality of the building. We do not have such information for the church of Zwentendorf. Our general experience is, that historical notices about the quality of masonary of buildings are rare enough.
2. We have to take into account, that this was the only message from Zwentendorf. We do not know, whether or not more buildings have suffered damage here. A comprehensive and extensive search for damanged houses has not been carried out. Therefore we cannot decide between "few", "many" and "most". In this case we can make suspections only. We suspect, that more comparable damage occured in Zwentendorf, because similar damages have been reported from all villages in that area. But how sure is this suspection? By no means these considerations are theoretical games. – They reached a very practical importance in 1978:

This Neulengbach earthquake from 1590 played an important role during the public debate and the plebicit in Austria 1978 which decided against the running of the NPP in Zwentendorf. Later - together with other arguments - it lead to the decision against the construction of nuclear power plants in Austria. It can be seen as an irony of fate, that during the debate, we did not know this document yet (figure 3). It has been detected nine years later, in 1986. Therefore the experts had to rely on estimates only using empirical equations of intensity decaý with distance and the geology of this area. Consequently the public debate run with a poor basis of background knowledge. It was a great luck that the early estimation of intensity VII-VIII for Zwentendorf was conservative and correct in the range of uncertainty.

5 THE ARTISTIC AND TECHNICAL DEPICTION - A HELP OF HISTORICAL TEXT INTERPRETATIONS

In 1763, June 28th, a destructive earthquake happened next to the town of Komárno/Danube. The event occured in a time of highly developed govermental organisation, which characterized the cultural prosperity during the reign of emporer Maria Theresia. Many sources have been handed down which make an approximate estimation of the meizoseismic zone and the epicenter possible. Zsiros et al(1988) associate this shock with Io = IX, while Karník and Szeidowitz 1990 suppose Io = VIII-IX only. This discrepancy led to a political controverse about the water power plant in Nagymaros at the Danube. One written source (in Latin) report dramatic destruction in Komarno as

"..... The Jesuit church, the future parish church....collapsed with the two towers nearly completely, walls and roof are standing but damaged..."

What does "collapsed nearly completely" mean? Did the towers break at the level of the tower clocks? Did the roof of the tower or the complete tower including the walls collapse? Obviously, these sentences are unsufficiant for a decision of the damage class according to the EMS92. The damage class determines the intensity. Here the scientific and artistic depictions can help. Furtunately two contemporary oil paintings and one technical drawing of Komárno, months or weeks after the earthquake, have been handed over (Broucek et al 1990) and reproduced in figure 5. The upper two ones are partial views of oil paintings. Note that the damage of the Jesuit church and college has been represented differently by the painters respectively the draughtsman. But they support the estimate of intensity VIII-IX rather than IX.

Thus, the contemporary depiction can be of help for the interpretation, but the help of limited importance.

3 MACROSEISMIC INTENSITY SCALE

3.1 Classifications used in the EMS

3.1.1 Differentiation of structures (buildings) into vulnerability classes

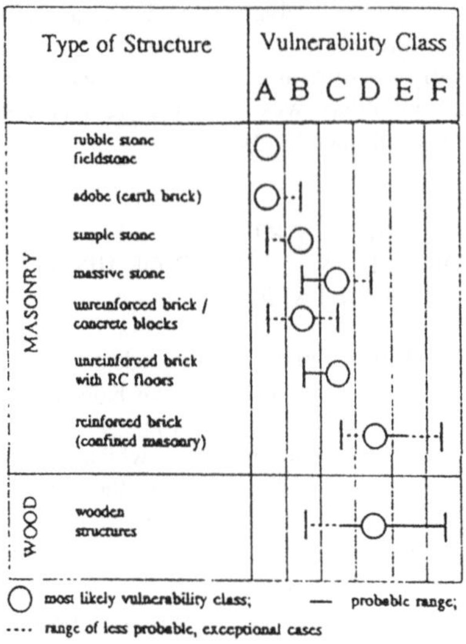

Type of Structure
Vulnerability Class
A B C D E F
MASONRY
rubble stone fieldstone
adobe (earth brick)
simple stone
massive stone
unreinforced brick / concrete blocks
unreinforced brick with RC floors
reinforced brick (confined masonry)
WOOD
wooden structures
most likely vulnerability class;
probable range;
range of less probable, exceptional cases

3.1.2 Definition of quantity

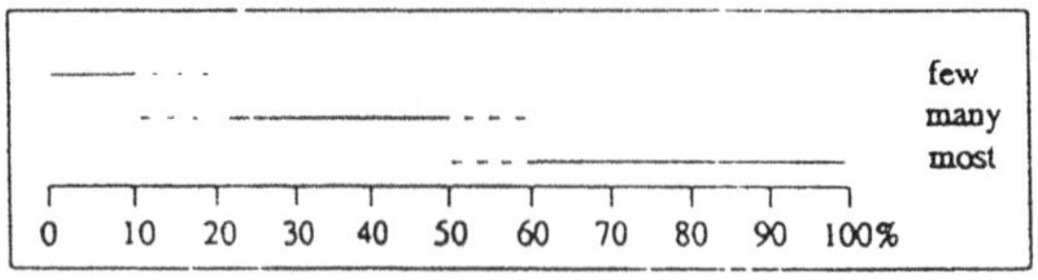

3.1.3 Classification of damage

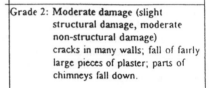

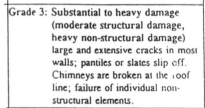

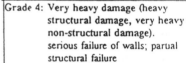

Grade 1:	**Negligible to slight damage (no structural damage)** . hair-line cracks in very few walls; fall of small pieces of plaster only. Fall of loose stones from upper parts of buildings in very few cases only.
Grade 2:	**Moderate damage (slight structural damage, moderate non-structural damage)** cracks in many walls; fall of fairly large pieces of plaster; parts of chimneys fall down.
Grade 3:	**Substantial to heavy damage (moderate structural damage, heavy non-structural damage)** large and extensive cracks in most walls; pantiles or slates slip off. Chimneys are broken at the roof line; failure of individual non-structural elements.
Grade 4:	**Very heavy damage (heavy structural damage, very heavy non-structural damage).** serious failure of walls; partial structural failure
Grade 5:	**Destruction (very heavy structural damage)** total or near total collapse

VII. Damaging

a) Most people are frightened and try to run outdoors. Many find it difficult to stand, especially on upper floors.
b) Furniture is shifted and top-heavy furniture may be overturned. Objects fall from shelves in large numbers. Water splashes from containers, tanks and pools
c) Many buildings of vulnerability class B and a few of class C suffer damage of grade 2 Many buildings of class A and a few of class B suffer damage of grade 3, a few buildings of class A suffer damage of grade 4 Damage is particularly noticeable in the upper parts of buildings

Figure 4. Excerpt of the EMS92 (after Grünthal 1993), vulnerability classes, type of structures, damage classes and description of intensity VII.

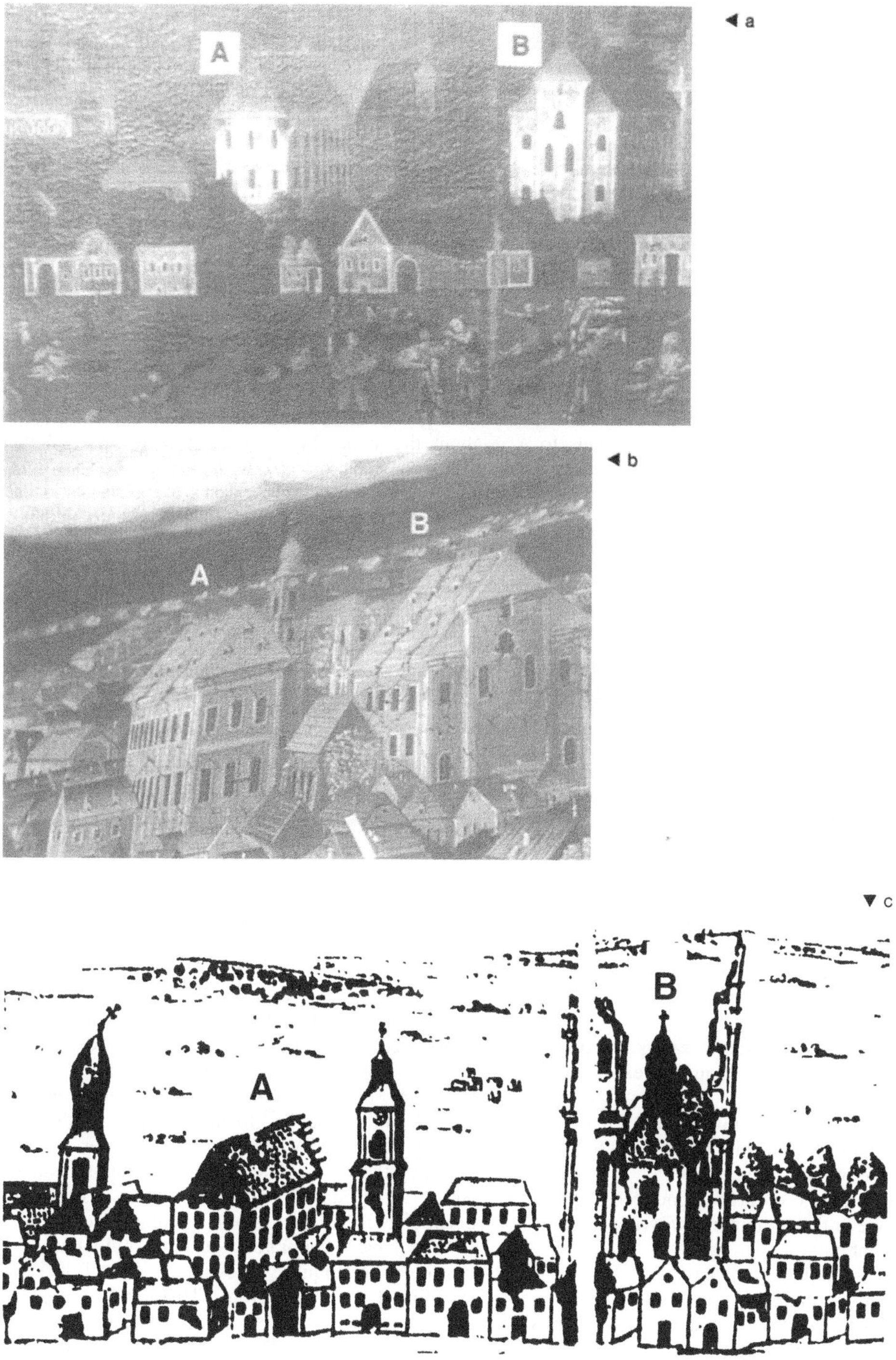

Figure 5. Depictions of the Komárno earthquake of 1763 (after Broucek et al 1992). Legend: a) Enlargement of Friedel´s painting of Komàrno. b) Enlargement of the votiv painting of Komàrno. c) Enlargement of Kastner´s drawing of Komàrno. A: Jesuit college; B: Jesuit Church.

6 AN ATTACK TO THE PROBLEM OF INCOMPLETE AND INHOMOGENEOUS HISTORICAL EARTHQUAKE DATA - THE FRIULI EARTHQUAKE OF JANUARY 25th, 1348

We see, the seismologist has to cope with the uncertainty caused by incomplete and inhomogeneous historical data which often are hard to interprete. Evidently, when the data by themselves are incomplete, we have no chance to improve this fate - which is typical for historical research. But we can arrange an inquiry of searching contemporary sources as completely, exhausively and comprehensively as possible. Herewith the problem is shifted to another level: from the data themselves to the data collection method. This means, that the inquiry – not necessarily the data - satisfies the demand of completeness. This concept leads to a deeper insight of uncertainty and helps a lot for the interpretation.

I will explain the procedure by another example, the South Alpine earthquake of January 25th, 1348. Older papers call it "The Great Villach Earthquake". Hammerl (1992) in her thesis found that the epicenter was not in Carinthia rather than in North Italy, in the Friuli county. It caused lot of damage and killed many people in Friuli and probably in Villach as well. Hammerl expected this historical earthquake as being not too different from the wellknown Friuli earthquake of May 6th, 1976, I_0 = X, focal depth 5 km (This focal depth has been taken in agreement with the determination given by Gutdeutsch and Hammerl (1998)). This event has been thoroughly studied by Prochaskova et al (1978) and many others. Hammerl used the area including the IV degree isoseismal (figure 6) of the 1976 event where she expected contemporary sources. In this area she carried out a comprehensive search for serial sources, which are annales of monasteries mainly. The messages from these sites haven been grouped into 6 categories:

- -1 A contemporary source exists, but does not mention the event during the expected time span.
- 0 The source refers to the event, but does not say whether this notice refers to the place of the source.
- 1 The earthquake has been noticed at the place of the contemporary writer
- 2 Slight building damage.
- 3 Severe building damage.
- 4 Destruction of buildings.

In a later paper (Gutdeutsch et al 1998) categories 1 to 4 have been associated with an *à priori-probability* distributions pi (i = 1,2,3) with intensities according to Rüttener's method (1995). For instance category 1 means: p_1 = 20 % for intensity III, p_2 = 40 % for intensity IV, p_3 = 40 % for intensity V. This method takes into account the expected uncertainty of historical data and their influence to the intensity estimation. It avoids interpolation.

7 SCALING OF HISTORICAL EARTHQUAKES BY A WELLKNOWN EARTHQUAKE OF THE 20th CENTURY - THE MODEL EARTHQUAKE

On this basis we later developed the method of *scaling* historical earthquakes by comparison with an earthquake of the 20th century (Gutdeutsch and Lenhardt 1996, Gutdeutsch and Hammerl 1998). This model or comparison earthquake should be of the same order of magnitude as the historical earthquake and the epicenters of both should be about the same. In the present case we used the isoseismals of the 1976 event as the basis of constructing the model earthquake (figure 7). A least square fit has been carried out in order to find the epicenter and focal depth with the minimum RMS of the intensity distribution. This method makes use of the empirical relation between geometrical spreading factor and the focal depth of seismic waves in the Eastern Alps (Franke und Gutdeutsch, 1974). The epicenter turned out as the same as the epicenter of the 1976 event but its focal depth of 15 km was greater. I would like to emphazise, that the mean error of the spatial intensity distribution was plus-minus one to two degrees EMS92! It visualizes the uncertainty due to the quantification of narrative texts of the 14th century into intensities. This result makes us more modest in our expectations.

8 INTERNATIONAL ORGANISATIONS

The increasing importance of Historical Earthquake Research reflects the activity of international organisations. The European Community has established programs as the project "Review of Historical Seismicity", This project dealt with the European seismicity in the time window from the second half of the 17th to the first half of the 18th century. The IAEA (International Atomic Energy Agency), the IUGG (International Union of Geodesy and Geophysics) and the ESC (European Seismological Commission), a commission of the IUGG have elaborated general guidelines of Historical earthquake Research. The ESC has established a working group "Historical Earthquake data" with seismologists and historians from 15 european countries. This working group has the following tasks: 1. elaboration of guidelines and recommendations, 2. revision of European earthquake catalogues, for historical earthquake investigation, 3. publication of monographs, case

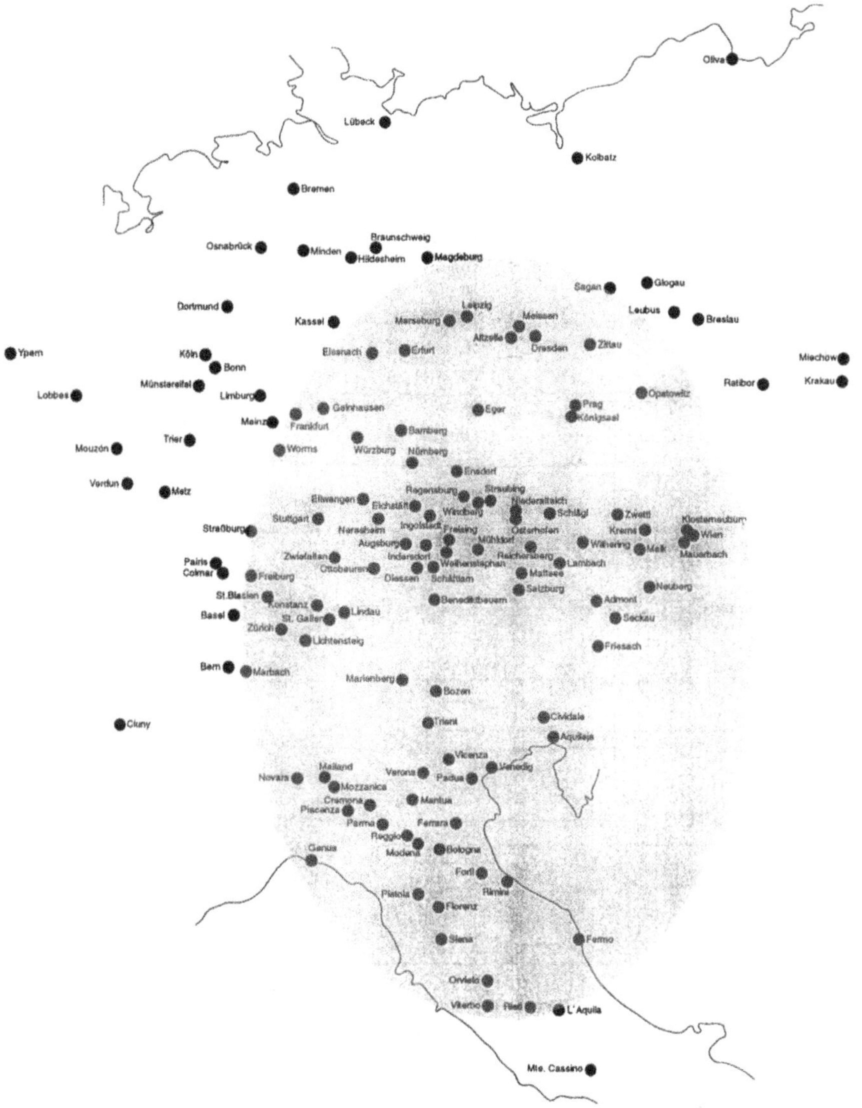

Figure 6. Places of exhaustiv searching for contemporary serial sources of the 1348 earthquake (after Hammerl 1992). The shaded area includes the IV isoseismal of the Friuli event of May 6th, 1976.

histories of historical earthquakes and proceedings related to scientific conferences, 4. organisation of symposia and workshops on Historical Earthquake Research, 5. collection and distribution of important data of Historical Earthquake Research and making them available to investigators, 6. establishment of a data bank of Historical Earthquakes.

9 CONCLUSION

For sure, historical earthquake research is an important tool for estimates of seismological parameter of pre-instrumental earthquakes. But the formulation of criteria of homogenity and completeness of data as well as their appropriate quantification into physical parameters represents still a great challenge. We have to acknowledge, that a natural boundary of the "resolution power" of focal parameters of historical earthquakes exists.

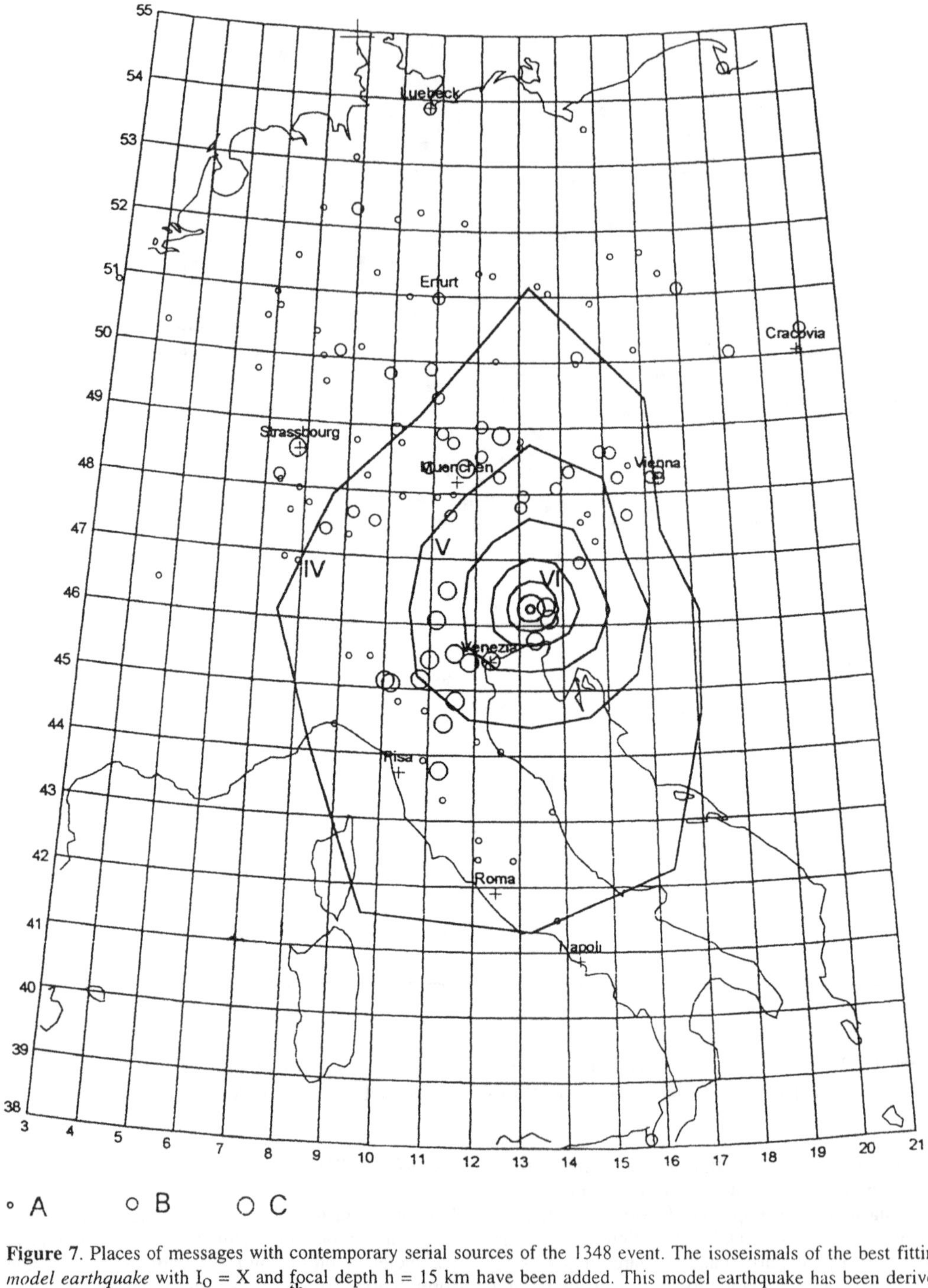

Figure 7. Places of messages with contemporary serial sources of the 1348 event. The isoseismals of the best fitting *model earthquake* with I_O = X and focal depth h = 15 km have been added. This model earthquake has been derived from the Friuli earthquake of May 6th with I_O = X and h = 5 km.

Legend:

A: The contemporary source does not mention the 1348 earthquake.

B: The source mentions the 1348 earthquake not refering to the place, where it has been noticed.

C: The contemporary source makes sure, that the earthquake has been noticed at that place.

10 REFERENCES

Boschi et al: Catalogo die forti terremoti in Italia, dal 461 a.C. at 1980 (1996), CFT.

Broucek, I., Eisinger, U., Farkas, V., Gutdeutsch, R., Hammerl, Ch. & Szeidowitz, V.: Reconstruction of building damage caused by the 1763 earthquake in Komárno/Danube from contemporary depictions of the same site and from respective texts. Proceedings of the ESC General Assembly Barcelona, September 1990.

Eisinger, U., Gutdeutsch, R., & Hammerl, Ch.: Historical Earthquake Research - an example of interdisciprinary Cooperation between Geophysicists and Historians, in Historical Earthquakes in Central Europe, Monography V1, 1992, Abhandlungen der Geologischen Bundesanstalt 48/1992, Vienna 1992.

Grünthal, G. (Editor): EMS92 European Macroseismic Scale 1992 (updated MSK-Scale 1992), European Seismological Commission, Luxembourg 1993.

Gutdeutsch, R. & Hammerl, Ch.: Naturkatastrophen in der Historischen Forschung - am Beispiel des Neulengbacher Bebens von 1590, pp. 52 - 69, Österreichische Gesellschaft für Geschichte der Naturwissenschaften, Mitt. Jg. 8, 1-4, 1988.

Gutdeutsch, R. & Hammerl, Ch.: The record threshold of historical earthquakes, submitted for publication in Journal of Seismology, 1998.

Franke, A. & Gutdeutsch, R.: Makroseismische Abschätzungen von Herdparametern österreichischer Erdbeben aus den Jahren 1905 - 1973. In: Journal of Geophysics Vol. 40, 173-188, 1974.

Gutdeutsch, R. & Lenhardt, W.: Re-interpretation of the South Alpine earthquake of January 25th 1348, extended abstract, XXV. General Assembly of the - European Seismological Commission, Reyjavik 1996, pp. 634-638.

Hammerl, Ch.: Das Erdbeben vom 25. Jänner 1348, Dissertation, University of Vienna, 1992.

Kárník, V. & Szeidowitz, V.: Komárno 1763. Paper, presented on the 3rd workshop meeting "Historical Earthquakes in Central Europe" Liblice/CSFR, April 1990.

Postpischl, D. (Editor): Catalogo dei terremoti Italiani dall'anno 1000 al 1980 - Consiglio Nazionale delle Richerche Progretto Finalizzato Geodinamica, Bologna (CNR) 1985.

Procháskovä, D. (Editor): Atlas of Isoseismal maps, Central and Eastern Europe, KAPG, Working group 4.3 Geophysical Institute of the Czechoslovak Academy of Sciences, Prague 1978.

Rüttener, E.: Earthquake Hazard Evaluation for Switzerland, Schweizerischer Erdbebendienst 1995.

Zsíros, T., Mónus, P. & Tóth, L.: Hungarian Earthquake Catalogue 456 - 1986, Seismological Observatory geodetic and Geophysical Research Institute Hungarian Academy of Sciences, Budapest (1988).

Mechanics of Jointed and Faulted Rock, Rossmanith (ed.) © 1998 Taylor & Francis, ISBN 90 5410 955 6

Quantitative tectonofractography – An appraisal

Dov Bahat
Department of Geological and Environmental Sciences & the Deichmann Rock Mechanics Laboratory of the Negev, Ben Gurion University of the Negev, Beer Sheva, Israel

ABSTRACT: Quantitative-fractoraphy concerns the methods which enable the determination of the tensile failure-stress on fracture surfaces of engineering materials. The mirror radius on the fracture surface, which is a key parameter for calculation and other important morphological parameters are briefly discussed. Fractographic methods are applied to geological exposures of considerably larger dimensions, with the objective of estimating fracture paleostresses. The fracture mechanics formulae which are applied to joints cutting granitic rocks and sandstone, and some estimated paleostress results are presented. The assessment of the results, further tectonophysical implications, and the merits of quantitative tectonofractography are considered. Finally, some future challenges are outlined.

1 INTRODUCTION

Fractography concerns the analysis of fracture surface morphology (or fracture markings) and related features, and their causes and mechanisms in engineering materials.Woodworth (1895, 1896) set the stage for the science of fractography by astute observations of the morphology of joints in geologic exposures. He also noted the essential morphologic similarities between fracture markings in glass samples and rock outcrops. Following the spectacular experimental fracture results by De Freminville (1914), Preston (1931) recognized that a fracture surface constitutes specific morphological features in a logical order. The realization that the stippled perimeter of the smooth mirror surface defines its boundary (Fig.1) (Smekal 1936, 1940; Terao 1953; Shand 1954) enabled the estimation of fracture stress by fractography. The estimation of fracture stress has become the most elaborate topic of quantitative fractography, and this topic will be at the centre of this study.

Tectonofractography is a new branch of tectonics and tectonophysics, which applies fractographical analysis to rock fractures and to regional fracture systems (joint sets) with the objectives of identifying the tectonic processes that produced the fractures, and determining the mechanical conditions involved (Bahat 1991). Fractographic results from glass experimentation and particularly from polycrystalline ceramics and metals are most useful in discussing analogies to fracture observations in much larger natural exposures, which obey the same mechanical formulae.

This study consists of two parts. An appraisal of quantitative tectonofractography will follow a summary of the principles of quantitative fractography.

2 THE QUANTITATIVE MIRROR PLANE

Fractures in glass and ceramic bodies and in rocks propagate from preexisting flaws. Those which are responsible for failure are the severest. In polycrystalline non-metallic materials they are usually taken to be the longest and straightest grain boundaries (Friedman & Logan 1970). Such a grain boundary is designated as the initial flaw length $2c_i$, being distinct from the critical flaw length $2c_{cr}$ (Figs.1, 2). A fracture mirror plane extends from the critical flaw to three differently defined radii, r_m (the mirror-mist boundary); r (the mist-hackle boundary); and r_b (the initiation of macroscopic crack branching) (Mecholsky & Freiman 1979). Many investigators (e.g. Levengood 1958) have obtained the semi-empirical formula:

$$\sigma_f \cdot r_j^{1/2} = A_j \quad (1)$$

where σ_f is the fracture stress, r_j is the distance to a particular boundary, and A_j is the corresponding mirror constant. An important aspect of this equation is its validity for tensile, flexural and biaxial (tension) loading, and it is useful in the study of mixed modes, which renders it applicable to the analysis of shaped bodies under complicated stress states (Mecholsky & Rice 1984). Naturally, there are some exceptions and these include cases of large stress gradients and materials of very low or very high strengths.

Mecholsky et al. (1976) show that the elastic modulus, E, is proportional to the mirror constant A and probably to the fracture energy γ, but that the latter is highly dependent on local microstructure. Good correlations between E and A were also found by Kirchner et al. (1976) for several hot pressed and reaction bonded ceramics. Kirchner & Gruver (1973) verified the relationship $\sigma_f r^{1/2}$ = constant, for alumina ceramics at room temperature and at elevated temperatures.

Kirchner & Gruver (1974) used fracture stress and measurements of mirror radius to estimate the reduction of residual surface stress resulting from annealing of flint glass, and to estimate the residual stresses induced in steatite and silicon-carbide by quenching. Conway & Mecholsky (1989) use crack branching data for measuring near-surface residual stresses in tempered glass. Kirchner et al (1975) used specific features of the fracture mirror to interpret impact fractures in brittle materials.

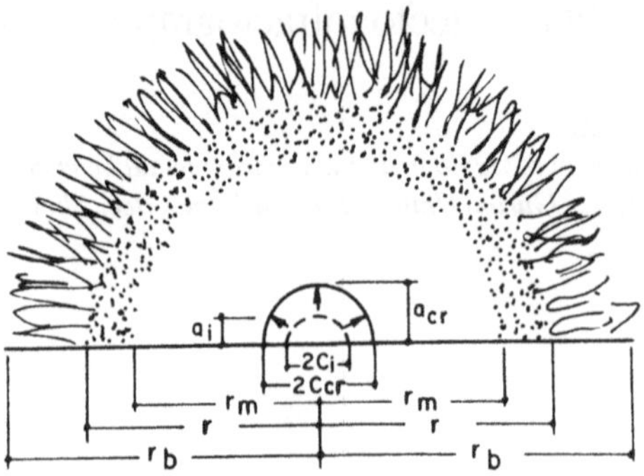

Figure 1. Fracture surface showing idealized initial flaw length $2c_i$ and depth a_i, and critical flaw length $2c_{cr}$ and depth a_{cr}, as well as the three mirror radii, r_m, r and r_b (modified from Mecholsky Jr. & Freiman 1979).

3 BASIC PARAMETERS OF THE MIRROR PLANE

3. 1 The shape of the mirror Plane

The mist, hackle and branching boundaries in isotropic materials are circular when formed under uniform tension (Fig. 1). The mirror boundary may be seriously affected by stress gradients, secondary defects, edge effects, the ratio of stress to specimen size, and asymmetric mirrors which occasionally occur (e. g. Leonard 1979; Bahat 1991, Fig. 2. 33).

In tempered (toughened) glass the mirror in the fracture surface is semi-elliptical (Shand 1967) when the fracture is slow or the fracture stress is low, but becomes more circular with increasing the rate of deflection or at high fracture stress (Hagan et al. 1979). Hagan et al. observed that in annealed glass the radius r_s of the mirror on the tensile surface is always less than the radius r_b of the mirror extending into the bulk of the sample (where r_b/r_s lies between one and two) but in thermally toughened glass this ratio is less than or equal to unity. The combined effects of the stress gradient and the surface edge on the shape of the mirror zone in glass were treated by Kirchner & Kirchner (1979) and by Bahat et al. (1982).

3. 2 Flaw size and shape

The sharp transition between the flaw, $2c_{cr}$, and the mirror plane (Figs. 1, 2) designates the loci of K_{Ic} stress conditions and indicates the initiation of the catastrophic stage of failure, namely, an abrupt transition to rapid fracture propagation.

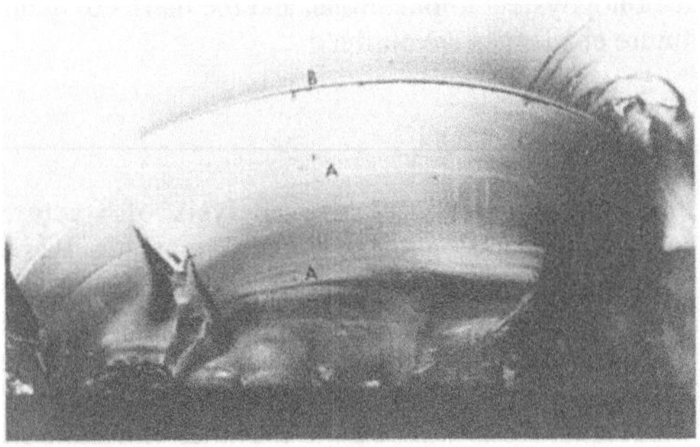

Figure 2. Early growths of initial flaw A and radial scars. B- flaw to mirror transition curve at the boundary of the critical flaw. Length of critical flaw diameter is 0.9 mm (from Bahat et al. 1982).

Krohn & Hasselman (1971) showed the relation of flaw size to mirror in the fracture of glass, and Bansal (1977) demonstrated the importance of the overall flaw shape in the fracture behavior, which led to a modified Griffith equation (Irwin 1962, Bansal 1976):

$$\sigma_f = Z/Y(2E\gamma)^{1/2}/c_{cr}^{1/2} = (Z/Y)(K_{Ic}/c_{cr}^{1/2}) \quad (2)$$

where E and γ are the Young's modulus and fracture surface energy, respectively, σ_f is the fracture stress at the origin, Z and Y are geometric constants, c_{cr} is the shorter radius of the flaw at the onset of catastrophic fracture (the flaw depth in the case of a surface flaw and the half depth in the case of a subsurface flaw) and $K_{Ic} = (2\gamma E)^{1/2}$. The combination of eqs. (1) and (2) gives (Bansal & Duckworth (1977):

$$(Z/Y)(r/c_{cr})^{1/2}=A/K_{Ic} \quad (3)$$

Krohn & Hasselman (1971) also suggested that the ratio between A (a mirror parameter) and K_{Ic} (a parameter associated with the flaw perimeter) is an absolute constant independent of type of material. Bansal (1977) found that this constant is 2. 32 for soda-lime glass.

3. 3 Mirror to flaw size ratio

Mecholsky et al. (1974, 1976) found that the outer mirror r, to flaw size ratio is shown to vary about a value of 13:1. Thus the mirror constants are used to predict critical flaw sizes in glass and ceramic materials. Rice (1979) observed smaller mirror to flaw-size ratios in polycrystalline bodies compared with those in glass and offered an explanation (Rice 1984), based on the assumption that mirror sizes are linearly related to fracture energies. Accordingly, mirror-to-flaw-size (r_m/c_{cr}, Fig. 1) ratios should be proportional to the ratio of fracture energy for crack propagation (γ_p) before branching to that at the initiation of mist (γ_m):

$$(r_m/c_{cr})(\gamma_p/\gamma_m)=\text{constant} \quad (4)$$

For glasses $\gamma_p/\gamma_m = 1/2$ and $r_m/c_{cr} = 12\text{-}13$. γ_p/γ_m for polycrystals varies from 3/4 to 20/21 because nucleation of mist involves low γ at grain boundaries and this gives $r_m/c_{cr} \sim 6$ to 8, for polycrystals, in good agreement with experimental observations (Rice 1984). It follows that due to the readiness of mist formation in polycrystals since γ is reduced, the onset of hackles is closer to the fracture origin. This trend is intensified with the increase of grain size. A poor distinction between c_i and c_{cr} (Fig. 1) is possibly an important factor contributing to the above discrepancies in the ratio of mirror to flaw size.

Flaws deviating from a single plane may lead to different parts of the flaw acting essentially independently (Rice 1984). A single flaw would act as two flaws with each generating its own portion of a mirror. Extreme distortions of the mirror are associated with pores. These effects seem to increase with the decrease of the average angle of the flaw relative to the tensile axis, namely with the increase of shearing and the reduction of relative tension on the flaw.

3. 4 Terminology

The terminology of fractography is problematic, particularly in its quantitative part. On this issue (starting from fractography in glass), Frechette (1984) and Rice (1984) represent substantially different approaches. The evolution of these differences and their implications are discussed by Bahat (1991, p.118). The approach adapted in the present paper is that of Rice (1984).

3. 5 Applied fractography

Fractography is widely used in the analyses of fracture processes in glass, ceramics, single-crystal, polymeric and metallic engineering materials. The level of sophistication reached in the utilization of fractography of glass bottles probably surpasses that in other engineering fields. Quite often the results of such analyses are brought up in court in product liability law suits. Hertzberg (1976, Fig. 14. 9) presents an intriguing fractographic analysis of a fractured metal.

4 TECTONOFRACTOGRAPHY

This chapter relates to several aspects of quantitative tectonofractography. It concerns: a. the fracture mechanics formulae that were applied to joints, b. estimations of paleostresses, c. assessments of the results, d. tectonophysical implication, e. the merits of quantitative tectonofractography, and, f. future challenges.

4. 1 Fracture mechanics formulae

Bahat & Rabinovitch (1988) applied principles of quantitative fractography derived from ceramics in the analysis of a tectonic joint in granite from Sinai, using two methods. First, they calculated σ_f by "the stress intensity method" on the basis of the empirical relationship (Randall 1966):

$$\sigma_f=K_{Ic}(Q/1.2\pi c_{cr})^{1/2} \quad (5)$$

where K_{Ic} is the critical stress intensity factor (the fracture toughness), Q is the modifying geometrical factor and c_{cr} is the radius of the critical flaw (section 3. 2 above). They derived the radius of the critical flaw c_{cr} from r/c_{cr}= constant (section 3. 3). This ratio has been theoretically postulated (Anthony & Congleton 1968), and empirically shown (Bansal 1977; Mecholsky & Freiman 1979) to be constant for various glass and ceramic materials where it varies from 13. 9 to 16. 7 (Mecholsky & Freiman 1979). Therefore Bahat & Rabinovitch assumed its relevance to rocks.

The second method, which is termed "the surface energy method", directly uses the radius r together with Griffith's equation modified for bifurcation by Congleton & Petch (1967). For the calculation of σ_f, they applied the expression (section 2):

$$\sigma_f=2G(E\gamma/\pi r)^{1/2} \quad (6)$$

where E is Young's modulus, γ is surface energy, r is the radius length of the mirror plane and G is the enhancement factor which varies from $2\sqrt{2}$ to $5\sqrt{2}$. The component $\sqrt{2}$ of the G factor relates to the position of the Griffith crack with respect to the tip of the parent fracture (almost insignificant in this case); and the change in the range 2 - 5, is a function of the strain rate ahead of the parent fracture. An increase in the fracture velocity would favour the upper end of this range (Congleton & Petch 1967).

4. 2 Estimation of paleostresses

The above mentioned methods were applied to the analyses of fractures on which the mirror boundaries (Fig.1) could be clearly measured and calculated. Such fractures were sampled from granite from East Sinai, Egypt (Bahat & Rabinovitch, 1988), from the Navajo sandsone at Zion National Park, Utah, and from granodiorite at the SW cliff of Half Dome in Yosemite Valley, California (Bahat et al. 1995). Although the procedures of both trigonometric determination of the mirror radius and the fracture mechanics calculations slightly varied from location to location the basic principles were maintained.

The data on two fracture surfaces are presented below, from the track to the Hidden Canyon and from the Red Arch. Both are well known locations at Zion National Park (Bahat et al. 1995). First, the stress intensity method was applied to the Hidden Canyon mirror (Fig. 3a). Values of K_{Ic} vary considerably for different sandstones, reflecting rock properties and experimental techniques performed by different investigators. Since we did not have data on the fracture toughness of the Navajo sandstone we averaged the results for sandstones qualified by Atkinson & Meredith (1987, Table 11.3) as group c (five results of presumed high reliability) and obtained K_{Ic} = 1. 030± 0.84 MPam$^{1/2}$. We then applied the two extremes 0.190 MPam$^{1/2}$ and 1.870 MPam$^{1/2}$ in the calculation. The Q factor ranges in value from 1. 0 for a long shallow flaw to 2. 46 for a semi-circular flaw (Randall, 1966). Since we did not know the shape of the flaw, we used both values in the calculation. We derived the radius of the critical flaw c_{cr} from r/c_{cr}= constant. This is an acceptable constant for fracture mirrors in wide varieties of glasses and ceramics (see below). We used for this ratio the value r/c_{cr}=15. 3, and obtained c_{cr}=0.093m. There is no need to calculate the two extremes which vary only insignificantly from each other when applied to eq. 5. The results for σ_f range from 0.321 MPa to 4.953 MPa.

Applying the surface energy method to the Hidden Canyon mirror, we calculated σ_f for the two extreme values of G, 2.8284 and 7.0711. We used E= 1. 67x 10^{10} Pa, the mean of 8 measurements of E values for the Navajo sandstone (Hatheway & Kiersch, 1982). We averaged six surface energy results obtained by different methods for quartz (Atkinson & Meredith,1987). This average is γ = 2. 54 Jm^{-2}. We also used the value γ =88 Jm^{-2}, obtained for Tennessee sandstone (Atkinson & Meredith,1987). The results for σ_f range from 0.550 MPa to 8.089 MPa.

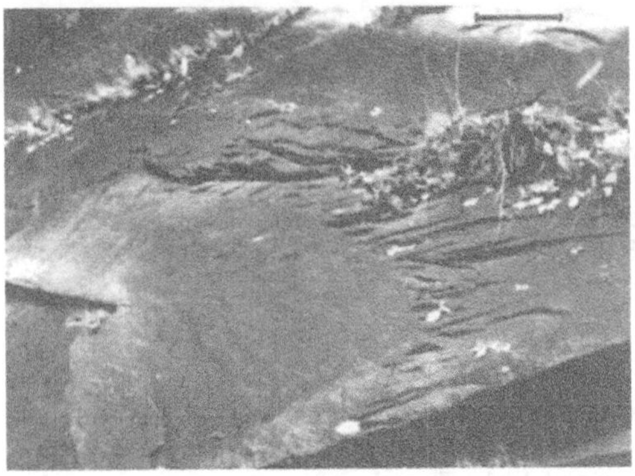

Figure 3a. Mirror planes from Zion National Park, Utah, USA. (from Bahat et al. 1995). a. Along the trail to the "Hidden Canyon". Scale bar is 30 cm (left column).

Figure 3b. In the Red Arch a rim of 10 m uniform width at lower part of arch (between two arrows).

For calculating the paleostress on the Red Arch mirror (Fig. 3b) we applied the same procedure used for calculating σ_f for the Hidden Canyon fracture. Accordingly, r was calculated to be 31. 9 m and C_{cr}= 2. 08 m. The σ_f values obtained by the stress intensity method range from 0. 060 MPa to 1. 047 MPa and the σ_f values obtained by the surface energy method range from 0. 116 MPa to 1. 713 MPa. The maximum and minimum results obtained by the two methods for the two investigated joints are summarized in Table 1.

4. 3 Assessments of the results

The adaptation of fractographic-quantitative methods from material-science to large field outcrops involves many unknown variables (Bahat 1979) so that a selection of approximate values is required. Accordingly, estimates of σ_f ranges (rather than precise values) should be considered (Segall & Pollard 1983).

The stress intensity method is not directly related to the fracture geometry since c_{cr} must be determined empirically from the radius r, which may cause an increased error. The surface energy method, on the other hand, directly uses the radius r together with Griffith's equation modified for bifurcation by Congleton & Petch (1967). The latter straightforward method in which geometry and stress are directly related, considerably reduces the error.

It would be appropriate to compare the new results to experimental data. Robinson (1970) gives four results for tensile strength of Navajo sandstone cores derived by uniaxial tension tests. For two cores parallel to bedding results were 1. 2 MPa and 3. 0 MPa, and for two cores normal to bedding 0. 5 MPa and 1. 0 MPa values were obtained. It appears that the tensile strength is a function of the core orientation, with a mean of 2.1 MPa for cores parallel to bedding, and 0.75 MPa for cores taken normal to bedding. Probably, fracture propagation had higher resistance across bedding in the cores cut parallel to bedding, in analogy to increase in strength when fracture propagates across acicular micro structure in chain silicate ceramics (Beall et al. 1986).

In assessing the fracture stress results, two opposing contributions to rock strength should be considered. The investigated joints propagated normal to bedding, hence, closer results to the mean 2.1 MPa should be expected. On the other hand, slower fracture propagation in nature compared with rapid fracture in the laboratory, often results in lower strength values (Mould & Southwick 1959; Scholz 1972).

For the Red Arch fracture surface, the stress intensity and the surface energy methods yield fracture stresses ranging from 0.1 to 1.0 MPa and from 0.1 to 1.7 MPa, respectively. The results by both methods overlap, and they are close to the lower strength range obtained by Robinson (1970). These results are considerably below the σ_f values derived for the Hidden Canyon fracture surface. The critical flaw for the Red Arch fracture surface is much larger than the critical flaw for the Hidden Canyon fracture surface (2.08m and 0.09m, respectively). For this reason, the fracture stress is much lower at the Red Arch fracture surface. This corresponds to previous observations in ceramics. A review of the relationships of fracture stress, flaw size and shape, and mirror to flaw size ratios in glass and ceramics is given in (Bahat 1991, p.108).

The morphological definition of c_{cr} is often quite clear in glass (Figs. 1 and 2), but grain boundaries and pores in polycrystalline materials (ceramics and rocks) reduce the clarity of fracture features.In Fig. 4

Table1. Estimation of fracture paleostress σ_f on joints cutting sandstone in Zion National Park*

The stress intensity method

The Hidden Canyon

K_{Ic} ($MPam^{1/2}$)	Q	c_{cr} (m)	σ_f (MPa)
0.190	1.000	0.093	0.321
1.870	2.460	0.093	4.953

The Red Arch

K_{Ic} ($MPam^{1/2}$)	Q	c_{cr} (m)	σ_f (MPa)
0.190	1.000	2.080	0.060
1.870	2.460	2.080	1.047

The surface energy method

The Hidden Canyon

G-factor	E ($x10^{10}$Pa)	γ (Jm-2)	r (m)	σ_f (MPa)
7.0711	1.670	88.000	1.43	8.089
2.8284	1.670	2.540	1.43	0.550

The Red Arch

G-factor	E ($x10^{10}$Pa)	γ (Jm-2)	r (m)	σ_f (MPa)
7.0711	1.670	88.000	31.90	1.713
2.8284	1.670	2.540	31.90	0.116

*See the meaning of terms in text.

the boundaries of the critical flaw are not as clear as in Fig. 2, and this may reduce the confidence in the calculation by the stress intensity method (see more on the prediction of critical flaw sizes in Mecholsky et al. (1974, 1976)). In the present paper c_{cr} was derived by the r/c_{cr} ratio in order to be applied in eq. 5. by the stress intensity method. Nevertheless, the overlap between the results by the two methods suggests that, the treatment of the r/c_{cr} ratio by the stress intensity method is reasonable.

Figure. 4. Fracture initiation in granite at the scar near A, which is the critical flaw. Curve B marks the circular perimeter of the hackle boundary. Distance from A to B, r = 360 mm (from Bahat & Rabinovitch 1988).

4. 4 Tectonophysical implication

The Red Arch mirror mostly fractured under fatigue conditions (characterized by slow fracture propagation). However, unstable conditions (rapid propagation) could have been attained at late stages of the fracture (Fig. 3b), in analogy with the experimental results by Hertzberg (1976, Fig.14.9). Hertzberg shows a mirror plane surrounded by a rim of en echelons, and the fractographic calculations indicate that the fatigue crack probably approached unstable conditions during its late stages. The general relationship $K= \sigma (c \pi)^{1/2}$ requires that the stress σ (or the fracture stress σ_f) is in inverse proportion to the square root of the fracture-length c (or mirror radius), where K is the stress intensity factor (Sih et al. 1962). Accordingly, under a given stress (a given overburden pressure), the stress intensity increases in proportion to the squre root of the fracture-length, and beyond a certain length the fracture becomes unstable. This provides a theoretical basis for a joint to grow slowly up to a certain size as a smooth surface, beyond which new stress intensity conditions would alter the fracture surface morphology (Bahat 1991, p. 243).

Joints generated by critical conditions which assume great fracture velocities (up to about a third of seismic velocity in a given rock) are rare, due to their common suppressive frictional stress caused by compression in the crust. However, there are geological conditions that may provide the local required stresses which enable rapid fracture, as occasionally revealed by fracture mirrors rimed with twisted en echelon segments (e.g. Figs. 3a, 5a). This suggestion is supported by previous observations of unstable rockbursts in the Chelmsford granite quarry (Holzhausen 1989) .

4.5 The merits of the method

Tectonofractography in general, and quantitative tectonofractography in particular have several distinct merits.

a. Fractography can be a deterministic tool in determining the fracture sequence in rocks. The determination of the sequence of joints cutting each other is most difficult if fractography is not taken into account. However, it is clear that a young fracture cuts fracture markings of a previous joint (Bankwitz 1966, Fig. 9).

b. The ubiquity of fracture markings in rocks makes fractography accessible and useful.

c. The quantitative methods stem from the solid theory of fracture mechanics.

d. There is a vast information on experimental resuls that may be compared to geological data.

e. Quantitative tectonofractography is applicable in a wide size-range of rock exposures.

f. This seems to be the only method which enables a direct estimate of fracture paleostress which can be performed by a single measurement, because geometry and stress are directly related.

4. 6 Future challenges

The frontiers of tectonofractography follow the innovations of fractography. Accordingly, future developments in the former may be influenced by elaboration in the latter. This refers particularly to: a. estimation of the critical flaw size, b. residual stresses, c. the rate of fracture propagation, d. distinct behavior of en echelon segmentation versus hackling along the en echelon-rim which surrounds the mirror plane, and e. fractal applications.

Estimation of the critical flaw size. There are several techniques for estimating c_{cr}. One of them suggests, that when σ_f and K_{Ic} are known, a very good approximation of c_{cr} is given by (Mecholsky 1991):

$$\sigma_f = K_{Ic}/[1.24(c)^{1/2}] \qquad (7)$$

where c is given as $c=(a \cdot b)^{1/2}$; and a and b represent the semi-minor and semi-major of the semi-elliptical flaw, respectively. This estimate may have a number of applications. For instance, understanding the rules that determine joint spacing depends on flaw size parametrization (Gross 1993).

Residual stresses. The presence of residual stress and the magnitude of the residual stress can be obtained by a study of the shape of the fracture mirror boundary (Conway, Jr. and Mecholsky Jr. 1989; Srinivasan et al. 1996). Possibly, the study of residual strain/stress in rocks (Friedman and Logan 1970) may gain from fractography.

Rate of fracture propagation. There is fractographic evidence that slow crack growth was manifested by isolated cracks and often by an intergranular fracture mode before the transgranular fast fracture became predominant (Breder et al. 1996).

En echelon segmentation versus hackling along the hackle-rim. Although diagrams generally present hackle rims of uniform widths, photographs from both engineering materials and geological exposures quite often reveal deviations from such widths. There are two distinct rims of radial cracks which bound mirror planes on exfoliation surfaces which cut sandstone in Zion National Park (Bahat et al. 1995).

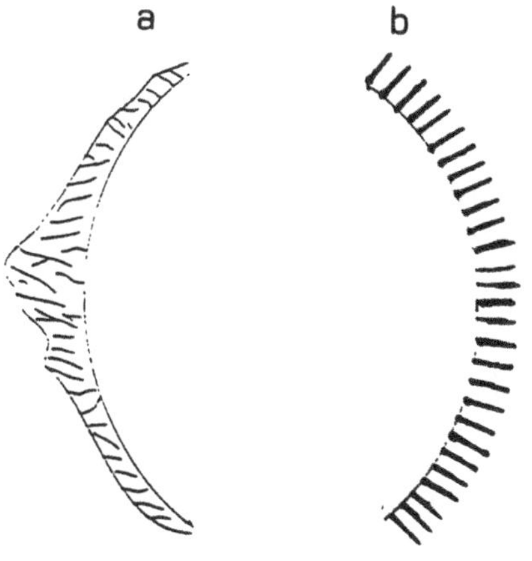

Figure 5. Diagram of partial mirrors with: a. A rim of hackles of non uniform width, and b. A rim of en echelon segments of uniform width.

One of them consists of somewhat hackled en echelon segments which display different lengths, resulting in a non uniform width of the rim (Figs. 3a, 5a). The other one consists of en echelon segments of equal lengths which are not hackled, resulting in a uniform rim width (Figs.3b, 5b). Inspection of 11 photographs of complete or almost complete mirrors from both engineering materials and geological exposures (Bahat 1991) reveals the ratio of 8/3 in favor of rims with non uniform widths.

Understanding the full mechanical significance of the "spectrum" between rims of hackles and rims of en echelon segments around mirrors appears to be quite a challenge. This understanding would improve the quantitative determination / estimation in both, fractography and tectonofractography.

Fractal applications. Fractal investigation of fracture, in various scales, may also gain from quantitative fractography (Mecholsky 1991).

5 CONCLUSIONS

1. Fracture mechanics formulae which enable the calculation of fracture stresses for engineering materials can be adapted successfully to estimate paleostresses on joints from geological exposures.
2. Two formule that were based on different mechanical parameters were applied in the calculation, analyzing two joints of distinct properties.
3. The overlap between the results by the two methods reaffirms their credibility, particularly, in reference to the problematic size determination of the critical flaw.
4. Considering the merits of the existing results, it is encouraging to pursuit additional challenges of quantitative tectonofractography.

ACKNOWLEDGMENT

Photographs 3a and 3b were taken in cooperation with Ken Grossenbacher and Kenzi Karasaki.

REFERENCES

Atkinson, B. K. & P. G. Meredith 1987. The theory of subcritical crack growth with applications to minerals and rocks.In B. K.Atkinson (ed.), *Fracture mechanics of rock.Academic Press*:: 111-166, New York: Academic Press.

Bahat, D. 1979. Theoretical considerations on mechanical parameters of joint surfaces based on studies on ceramics.*Geol Mag*. 116: 81-92.

Bahat, D. 1991. *Tectonofractography*. Heidelberg: Springer-Verlag.

Bahat, D. & A. Rabinovitch 1988. Paleostress determination in a rock by fractographic method.*J. Struct. Geol.* 10: 193-199.

Bahat, D., G. Leonard & A. Rabinovitch 1982. Analysis of symetric fracture mirrors in glass bottles. *Int. J. Fract.* 18: 29-38.

Bahat, D., K. Grossenbacher & K. Karasaki 1995. Investigation of Exfoliation Joints in Navajo Sandstone at the Zion National Park and in Granite at the Yosemite National Park by Tectonofractographic Techniques. *LBL-36971; UC-400*, 67 pp.

Bankwitz, P. 1966. Uber klufte.2 die bildung der kluftflache und eine systematik ihrer strukturen.*Geologie* 15: 896-941

Bansal, G. K. 1976. Effect of flaw shape on strength of ceramics.*J. Am. Ceram.*. Soc. 59: 87-88.

Bansal, G. K.1977. On fracture mirror formation in glass and polycrystalline ceramics. *Phil. Mag.* 35: 935-944

Bansal, G.K. & W. H. Duckworth 1977. Fracture stress as related to flaw and fracture mirror sizes. *J. Am. Ceram. Soc.* 60: 304-309.

Breder, K., T. J. Mroz, A. A. Wereszczak & V. J. Tennery 1996. Utilization of fractography in the evaluation of high temperature dynamic fatigue experiments. In J. R. Varner, V. D. Frechette & G. D. Quinn (eds.) *Fractography of glasses and ceramics III*, The American Ceramic Society, Ceramic transactions, 64: 353-366.

Congleton, J. & N. J. Petch 1967. Crack branching. *Phil. Mag.* 16: 749-760.

Conway, J. C. Jr & J. J. Mecholsky Jr 1989. Use of crack branching data for measuring near-surface residual stresses in tempered glass.*J. Am. Ceram. Soc.* 72: 1584-1587.

De Freminville, M. Ch. 1914. Recherches sur la fragilite - l'eclatement. *Rev. Met* . 11: 971-1056.

Frechette, C. D. 1984. Markings on crack surfaces of brittle materials: a suggested unified nomenclature. In J. J. Mecholsky Jr. & S. R. Powell Jr.(eds.), *Fractography of ceramic and metal failures:* Special Technical Publication, 827: 104-109. Philadelphia: ASTM.

Friedman, M. & J. M. Logan 1970. The influence of residual elastic strain on the orientation of experimental fractures in three quartzose sandstones. *J. Geophys. Res.* 75: 387-405.

Gross, M. R. 1993. The origin and spacing of cross joints: examples from the Monterey Formation, Santa Barbara Coastline, California. *J. Struct. Geol.* 15: 737-751.

Hagan, J. T., M. V. Swain & J. E. Field 1979. Fracture-strength studies on annealed and tempered glasses under dynamic conditions. *Phil. Mag.* 39: 743-756.

Hatheway, A. W. & G. A. Kiersch 1982. Engineering properties of rock. In R. S. Carmichael (ed.), *Handbook of physical properties of rocks* II: 289-331, Boca Raton, Florida: CRC Press.

Hertzberg, R. W. 1976. *Deformation and fracture mechanics of engineering materials*. New York: John Wiley & Sons.

Holzhausen, G. R. 1989. Origin of sheet structure,1. Morphology and boundary conditions. *Engineering Geology* 27: 225-278.

Irwin, G. R. 1962. Crack-extension force for a part-through crack in a plate.*J. Appl. Mech.* 29: 651-654.

Kirchner, H. P. & R. M. Gruver 1973. Fracture mirrors in alumina ceramics.*Phil. Mag.* 30: 1433-1446.

Kirchner, H. P. & R. M. Gruver 1974. Fracture mirrors in polycrystalline ceramics and glass. In R. C. Bradt, D. P. H. Hasselman & e. F. Lange (eds.), Fracture mechanics of ceramics, Vol. 1: 309-320, New York, Plenum Press.

Kirchner, H. P. & J. W. Kirchner 1979. Fracture mechanics of fracture mirrors. *J. Am. Ceram. Soc.* 62: 198-202.

Kirchner, H. P., R. M. Gruver & W. A. Sotter 1975. Use of fracture mirrors to interpret impact fractures in brittle materials. *J. Am. Ceram. Soc.* 58: 188-191.

Kirchner, H. P., R. M. Gruver RM & W. A. Sotter 1976. Fracture stress-mirror size relations for polycrystalline ceramics. *Phil. Mag.* 33: 775-780.

Krohn, D. A. & D. Hasselman 1971. Relation of flow size to mirror in the fracture of glass. J. Am. Ceram. Soc. 54: 411.

Levengood, W. C. 1958. Effect of origin flaw characteristics on glass strength.*J. Appl. Phys.* 29: 820-826.

Mecholsky, Jr., J. J. & S. W. Freiman 1979. Determination of fracture mechanics parameters through fractographic analysis of ceramics.In S. W. Freiman (ed.)*Fracture mechanics applied to brittle materials* : 678: 136-150. Philadelphia: ASTM.

Mecholsky Jr., J. J., S. W. Freiman & R. W. Rice 1976. Fracture surface analysis of ceramics. *J. Mater. Sci.* 11: 1310-1319.

Mecholsky Jr., J. J., R. W. Rice & S. W. Freiman 1974. Prediction of fracture energy and flaw size in glasses from measurments of mirror size. *J. Am. Ceram. Soc.* 57: 440-443.

Mecholsky Jr., J. J. & R. W. Rice 1984. Fractographic analysis of biaxial failure in ceramics. In J. J. Mecholsky Jr, &S. R. Powell Jr. (eds), *Fractography of ceramic and metal failures:* ASTM, STP, 827: 185-193. Philadelphia: ASTM.

Mecholsky, Jr., J.J. 1991. Quantitative fractography: an assessment. In V. D. Frechette & J. R. Varner (eds.), *Fractography of glasses and ceramics II*: Ceramic transactions, 17: 413-451. Westerville, Ohio: The American Ceramic Society.

Mould, R. E. & R. D. Southwick 1959. Strength and statistic fatigue and abraded glass controlled ambient conditions: 2 Effect of various abrassions and the universal fatigue curve. *J. Am. Ceram. Soc.* 42: 582-592.

Preston, F. W. 1931. The propagation of fissures in glass and other bodies with special reference to the split-wave front. *J. Am. Ceram. Soc.* 14: 419-427.

Randall, P. N. 1966. Plain strain crack toughness testing of high strength metallic materials. ASTM, STP, 410: 88-129.Philadelphia: ASTM.

Rice, R. W. 1979. The difference in mirror-to-flaw size ratios between dense glasses and polycrystals. *J. Am. Ceram. Soc.* 62: 533-535.

Rice, R.W.1984.Ceramic fracture features, observations, mechanisms and uses.In J. J. Mecholsky, Jr. & S. R. Powell, Jr. (eds.), *Fractography of ceramic and metal failures:* ASTM, STP, 827:1-5. Philadelphia: ASTM.

Robinson, E. S. 1970. Mechanical disintegration of the Navajo Sandstone in Zion canyon, Utah. *Geol. Soc. Amer. Bull.* 81: 2799-2806.

Scholz, C. H. 1972. Static fatigue of quartz. *J. Geophys. Res.* 77: 2104-2114.

Segall, P. & D. D. Pollard 1983. Joint formation in granitic rock of the Sierra Nevada. *Geol. Soc. Am. Bull.* 94: 563-575.

Shand, E. B. 1954. Experimental study of fracture of glass : 2 Experimental data. *J. Am. Ceram. Soc.* 37: 559-572.

Shand, E. B. 1967. Breaking stresses of glass determined from fracture surfaces. *Glass Ind.* 47: 190-194.

Smekal, A.1936. Die Festigkeitseigenschaften sproder korper. *Natur. Wiss.* 15: 106-188.

Smekal, A. 1940. Ultraschalldispersion und Bruchgeschwindigkeit. Phys. Zeit. 41: 475-480.

Srinivasan, G.V., J. Gibson, S. K. Lau, A. A. Wereszczak & M. K. Ferber 1996. The origin of strength limiting defects in a toughened SiC (HEXOLOY SX-SiC). In J. R. Varner, V. D. Frechette & G.D. Quinn (eds), *Fractography of glasses and ceramics III:* Ceramic transactions, 64: 181-192. Westerville, Ohio: The American Ceramic Society.

Terao, N. 1953. Sur une relation entre la ressistance a la rupture et le foyer d'eclatement du verre . *Jour Phys Soc Japan* 8: 545-549.

Woodworth, J. B. 1895. Some features of joints.*SCI* 2: 903-904.

Woodworth, J. B. 1896. On the fracture system of joints ,with remarks on certain great fractures. Boston Soc. Nat . Hist. Proc. 27: 63-184.

Yukawa, S., D. P. Timo & A. Rubio 1969. Fracture design practices for rotating equipment. In H. Liebowitz (ed.) *Fracture*: 65-157, New York: Academic Press.

Mechanics of Jointed and Faulted Rock, Rossmanith (ed.) © 1998 Taylor & Francis, ISBN 90 5410 955 6

Catastrophic sliding over a fault caused by accumulation of dilation zones

A.V. Dyskin, A.N. Galybin & B.H. Brady
The University of Western Australia, Nedlands, W.A., Australia

ABSTRACT: A mechanism of rockburst caused by a sudden sliding over a fault repeatedly subjected to local sort-time tensile loads (eg, by tensile phases of incident waves caused by blasting operation or by neighbour seismic events) is considered. It is assumed that, normally, friction between rough faces of the fault prevents their sliding. The action of each local tensile load however opens a part of the fault enabling local sliding and formation, due to the interaction between rough faces, of a dilation zone. As the dilation zones accumulate their interaction enhance this process, which eventually makes the average size of the dilation zones equal to the average distance between the centres of the zones. This will cause their coalescence and, hence, the formation of a macro-zone of sliding. This may be interpreted as a catastrophic sliding which is a major seismic event. A model is proposed which allows a prediction of the catastrophic sliding and the evaluation of the magnitude of the seismic event. Two simple methods of computing the interaction between randomly located dilation zones are used: the self-consistent method and the approximation of periodic locations. It is shown that both methods give similar results.

1. INTRODUCTION

Catastrophic sliding over a fault triggered by mining operations is one of the sources of mine tremors (seismic events) and can initiate a rockburst (eg, Gibowicz & Kijko, 1994).

Initially, faults in earth's crust may stay in natural equilibrium when possible sliding is prevented by high friction. However, over time, under the action of mining operations, especially if the fault is repeatedly subjected to cyclic loading, the equilibrium may be violated, resulting in a sudden (unstable) sliding over a part of the fault. If this part is large enough, the sliding will manifest itself as a strong seismic event that can cause severe damage to neighbour underground structures.

One of the mechanisms of catastrophic sliding can be associated with the roughness of the fault faces. Interaction of the rough faces results in friction with the friction angle mainly determined by the average inclination of asperities (eg. Goodman, 1989). When the fault is repeatedly subjected to local sort-time tensile loads (eg, by tensile phases of incident waves caused by blasting operations or by neighbour seismic events), the local tensile stresses may open the corresponding parts of the fault thus eliminating friction and enabling sliding. Due to the interaction between rough faces, this sliding will also create a dilation zone which, in its own term, will impose additional tensile stresses on the fault. These tensile stresses will help opening new parts of the fault thus facilitating the creation of new sliding and, hence, new dilation zones. As the dilation zones accumulate, their average size becomes equal to the average distance between the centres of the zones. The subsequent coalescence of the zones will produce catastrophic sliding and, as a result, a seismic event.

The magnitude of the event and, therefore, a possibility of initiating a rockburst will depend on the dimensions of the large scale sliding which are determined by the characteristic size of nonuniformity of either fault friction (roughness) or the initial stress state.

This paper presents a model of the above mechanism, which paves a way to predict the catastrophic sliding and to estimate the magnitude of the associated seismic event.

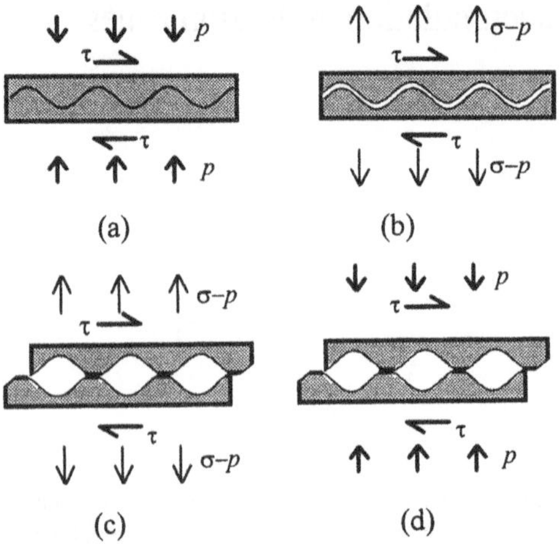

Figure 1. Action of instantaneous tensile wave of magnitude σ on fault with rough faces: (a) shear stresses cannot overcome friction; (b) strong tensile wave opens the fault, temporarily removing friction; (c) in the absence of friction shear stress moves the fault faces; (d) friction between contacted parts prevents the sheared faces from relaxing.

2. FORMATION OF SLIDING AND DILATION ZONES

2.1 *Formation of a single dilation zone*

The formation of a single dilation zone can be associated with the interaction between rough faces of the fault, Figure 1. Suppose that initially sliding over the fault is not possible, Figure 1a, ie. $|\tau|<|p|\tan(\phi)+c$, where τ and p are shear and normal components of the initial stress state respectively acting on the fault plane, ϕ is the macroscopic friction angle which is mainly determined by the average angle of the asperity slopes (eg, Goodman, 1989), c is the macroscopic cohesion.

Consider the situation when the fault is locally hit by a tensile wave with a magnitude $s>p$. Then the affected part will be opened, Figure 1b, and friction between the fault surfaces will be removed over this part. This will enable the shear stress to produce tangential displacements until the asperities contact each other, Figure 1c. Further displacement will be prevented by friction. After the wave is gone friction will also prevent the faces from returning back, Figure 1d.

It should be noted that the condition $s>p$ is not essential. Even if $s<p$ the dilation zone may still be formed if $|\tau|<|p-s|\tan(\phi)+c$, ie. s is high enough to enable sliding. The the dilation will be produced due to the interaction of the rough surfaces.

The affected part of the fault will subsequently act as an opened and sheared crack, Figure 2a. Since the fault cannot withstand singular tensile stresses created at the crack tips (the fault tensile strength is assumed to be negligible), the initial crack will propagate until the closing action of the ambient compression, p, equilibrates the opening action of the initial crack. This can be expressed by the condition of zero fracture toughness, $K_I=0$, Figure 2b.

2.2 *Model of dilation zone*

It is assumed that the local tension acts sufficiently long to enable shearing displacement of the fault faces to be greater than the average asperity length. Then, after the tensile stress is gone the created initial dilation zone will have the uniform macroscopic opening equal to the average asperity height, $2h$ (greater opening is not possible due to the closing action of the ambient compression, Dyskin & Galybin, 1997). Therefore the initial dilation zone acts as a wedge of thickness $2h$ and the length $2b$. Mathematically, this can be represented by a couple of opposite edge dislocations at points $\pm b$ with

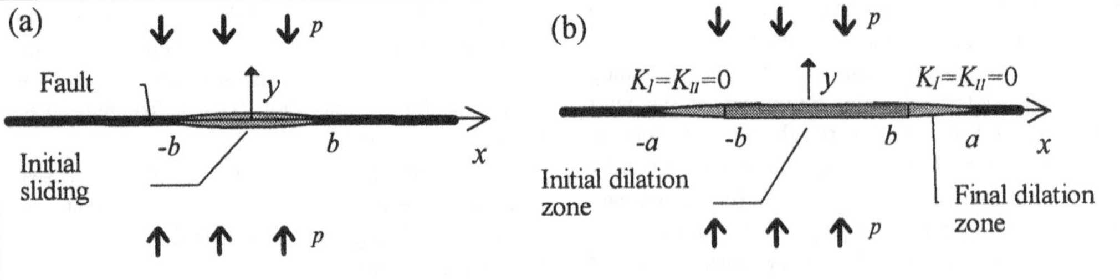

Figure 2. Formation and propagation of the dilation zone triggered by a tensile wave: (a) initial shear and dilation crack; (b) final dilation zone formed by propagating crack.

Burgers vector $2h$ inserted into the fault, Figure 2b.

Outside the initial dilation zone there might still be some opening and, therefore, shearing proportional to the opening with the factor tanϕ, Dyskin & Galybin (1997). The reason for the proportionality is in the fact that after opening the corresponding parts of the fault faces will get out of contact, Figure 3a, and then the acting shear stress will produce tangential displacement until the asperities are in the contact again, Figure 3b.

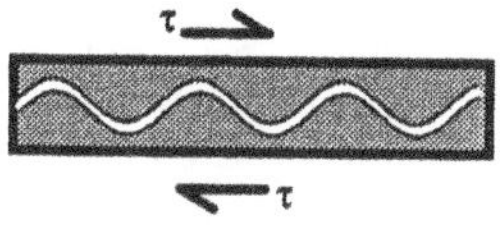

(a)

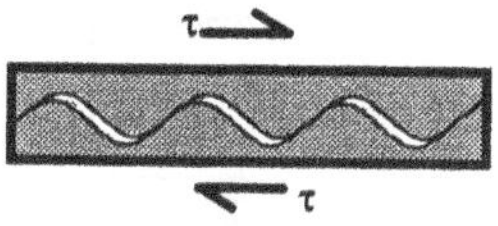

(b)

Figure 3. On the proportionality between normal and tangential displacements: (a) normal displacement removes contact and, hence, friction between the parts of the fault faces; (b) the shear stress produces a proportional tangential displacement restoring the contact.

The length of this final dilation zone, $2a$, can be determined by the condition K_I=0 which results from the assumption of zero tensile resistance of the fault.

It should be also noted that faults usually have some resistance to shearing. However due to pre-opening and the proportionality between tangential and normal displacements, K_{II}=0 as well, Figure 2b. This means that the tips of the shearing zone will coincide with the tips of the dilation zone. Hereafter only the dilation zone will be considered. This is possible because elastic problems for rectilinear cracks situated on a line allow decomposition into the problems for pure normal and pure shear cracks.

The rock mass is assumed to be homogeneous, elastic and isotropic.

The corresponding quasistatic model can be formulated as follows:

$$\sigma_y(x) = 0, \quad b < |x| < a$$
$$\left[u_y'(x)\right] = 0, \quad |x| < b \qquad (1)$$

Here square brackets around a function denote jump of the value of the function across the crack contour.

Thus, we have arrived at a problem for a pure Mode I crack opened by the pair of dislocations and equilibrated by the ambient pressure, p.

In order to solve the plane strain elastic problem with boundary conditions (1) it is convenient to introduce a new unknown function proportional to the derivative of the jump in normal displacements:

$$Q(x) = \frac{E}{4(1-\nu^2)}\left[u_y'(x)\right] \qquad (2)$$

where ν is Poisson's ratio, E is Young's modulus.

Using (2) one can express the additional normal stresses created by the crack on the crack line as follows (Muskhelishvili, 1953)

$$\sigma_y(x) = \frac{1}{\pi}\int_{-a}^{a}\frac{Q(t)}{t-x}dt \qquad (3)$$

By taking into account that $Q(x)=-Q(-x)$ and $Q(x)=0$ on $(-b,b)$ formula (3) can be rewritten:

$$\sigma_y(x) = \frac{2}{\pi}\int_{b}^{a}\frac{tQ(t)}{t^2-x^2}dt \qquad (4)$$

The boundary conditions (1) lead to the following simple singular integral equation

$$\frac{2}{\pi}\int_{b}^{a}\frac{tQ(t)}{t^2-x^2}dt = p, \quad b < |x| < a \qquad (5)$$

Here $p>0$ is the magnitude of compression at infinity.

Solution of (5) satisfying the condition K_I=0 and the condition of single-valuedness of displacement is

$$Q(t) = -p\sqrt{\frac{a^2-t^2}{t^2-b^2}}\,\mathrm{sgn}(t)H\left[\left(a^2-t^2\right)\left(t^2-b^2\right)\right] \qquad (6)$$

where $H(t)$ is the step function: $H(t)$ =1 if t>0, otherwise $H(t)$ =0.

Substituting (6) into (4), integrating and adding the stresses at infinity one obtains the following expression for the normal stress created by the dilation zone

$$\sigma_y(x) = \begin{cases} p & b \le |x| \le a \\ p\left[1 - \sqrt{\dfrac{x^2 - a^2}{x^2 - b^2}}\right] & \text{otherwise} \end{cases} \tag{7}$$

If b is known, a can be determined by using the condition that the opening of the central part, $(-b,b)$ is equal to $2h$:

$$\int_{-a}^{-b} \left[u_y'(t)\right] dt = 2h \tag{8}$$

After substituting (2) and (6) into (8) one gets the following equation for the determination of the unknown length of the final dilation zone

$$\mathrm{K}(k) - \mathrm{E}(k) = \frac{1}{2(1-\nu^2)} \frac{E}{p} \frac{h}{a}, \quad k = \sqrt{1 - \frac{b^2}{a^2}} \tag{9}$$

where $\mathrm{K}(k)$ and $\mathrm{E}(k)$ are complete elliptic integrals of the first and second kind respectively.

From here

$$a = b / x_*(\lambda) \tag{10}$$

where x_* is the root of equation

$$\mathrm{K}\left(\sqrt{1 - x_*^2}\right) - \mathrm{E}\left(\sqrt{1 - x_*^2}\right) = \lambda x_* \tag{11}$$

and

$$\lambda = \frac{1}{2(1-\nu^2)} \frac{E}{p} \frac{h}{b} \tag{12}$$

Figure 4 shows the solution of equation (11) as a function of dimensionless parameter λ.

For small values of x_* the asymptotics for the elliptic integrals can be used (Jahnke, et al., 1960)

$$\mathrm{K}(k) = \Lambda + o(\Lambda), \ \mathrm{E}(k) = 1 + o(\Lambda), \ \Lambda = \ln\left(\frac{4}{x_*}\right) \tag{13}$$

Now (11) will assume the following asymptotic form

$$\ln\left(\frac{4}{x_*}\right) = \lambda x_* \tag{14}$$

It is seen that $x_* \to 0$ and $a \to \infty$ as $\lambda \to \infty$ ($p \to 0$).

It should be noted now that the action of the dilation zone and, subsequently, the final length are determined by the asperity height rather than by friction.

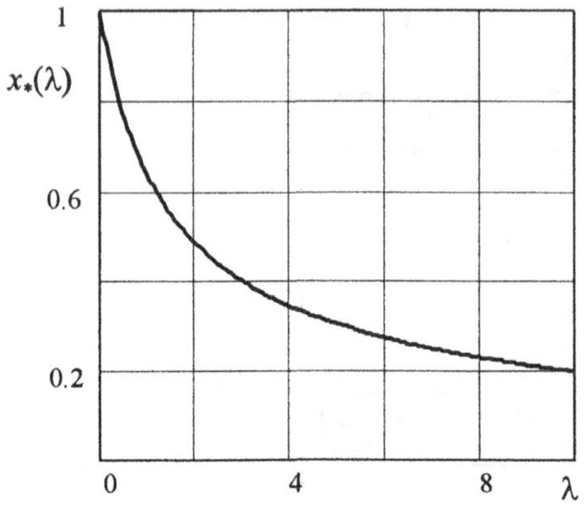

Figure 4. Solution of equation (11).

3. ACCUMULATION OF DILATION ZONES. CATASTROPHIC SLIDING

Suppose now that new waves come, creating initial dilation zones at random locations. Let, in average, the sizes of the initial dilation zones be the same, $2b$. Then, the presence of already accumulated and developed dilation zones will produce additional tensile stresses with an average magnitude $\langle\sigma\rangle$ which can be found as follows.

If the dilation zones do not interact then the average stress produced by them is

$$\langle\sigma\rangle = N\left[\int_{-\infty}^{-a} \sigma_y(x)dx + \int_{a}^{\infty} \sigma_y(x)dx\right] \tag{15}$$

where N is the number of dilation zones per unit length of the fault.

Due to equilibrium, the improper integrals in (15) can be reduced to the integral over the contact (initial dilation zone)

$$\langle\sigma\rangle - p = -2N\int_0^b [\sigma_y(x) - p]dx \qquad (16)$$

which, after substituting expression (7) for stresses, gives

$$<\sigma>= pNa\mathrm{E}\left(\frac{b}{a}\right) + p \qquad (17)$$

The exact solution of the problem for interacting dilation zones situated at random locations is complicated. Nevertheless there are two approximations which are simple enough to allow closed form solutions. One of them, a simple self-consistent method will be considered in subsection 3.1. Another one the approximation of periodically situated zones will be considered in subsection 3.2.

3.1 Self-consistent method

The interaction can affect both the local value of compression, p, and the final length of dilation zones. The simplest way to take this into account is to replace p with p-$<\sigma>$ in (12) and (17); this is a method of effective field similar to the one used by Kachanov (1993). This gives the following equation for the determination of the average stress $<\sigma>$:

$$<\sigma>= (p-<\sigma>)\left[1 + Nb\frac{\mathrm{E}(x_*(\lambda(p-<\sigma>)))}{x_*(\lambda(p-<\sigma>))}\right] \qquad (18)$$
$$\lambda(p-<\sigma>) = \frac{1}{2(1-\nu^2)}\frac{E}{p-<\sigma>}\frac{h}{b}$$

Equation (18) can be rewritten as an equation for unknown x_*=b/a:

$$\frac{4x_* + \omega\mathrm{E}(x_*)}{\mathrm{K}\left(\sqrt{1-x_*^2}\right) - \mathrm{E}\left(\sqrt{1-x_*^2}\right)} = \frac{2}{\lambda} \qquad (19)$$

where $\omega=2Nb$ is the dimensionless concentration of initial dilation zones, N is the number of the zones per unit length of the fault and λ is given by (12).

In average, the final length of the dilation zone, $2a$, cannot exceed the average distance, N^{-1}, between the dilation zones. This means that there is a critical length, $2a_{cr}=N^{-1}$, at which the dilation zones will coalesce into a macro-zone of sliding and dilation. This can be interpreted as catastrophic sliding.

The corresponding critical concentration of the initial dilation zones satisfies the following equations

$$\omega_{cr}^{-1}=a_{cr}/b \text{ or } \omega_{cr}=x_*. \qquad (20)$$

After substituting the second equation of (20) into (19) one has an equation to find the critical concentration:

$$\frac{[4 + \mathrm{E}(x_*)]\omega_{cr}}{\mathrm{K}\left(\sqrt{1-\omega_{cr}^2}\right) - \mathrm{E}\left(\sqrt{1-\omega_{cr}^2}\right)} = \frac{2}{\lambda} \qquad (21)$$

The solid line in Figure 5 shows the solution of this equation. The analysis of (21) also shows that $\omega_{cr}\to 0$ as $\lambda\to\infty$ $(p\to 0)$ and $\omega_{cr}\to\infty$ as $\lambda\to 0$ $(p\to\infty)$.

3.2 Periodic array of dilation zones

If the dilation zones are situated periodically, then equation (5) should be replaced with the following one (eg, Savruk, 1981)

$$\frac{1}{d}\int_{-a}^{a} Q(t)\cot\left[\frac{\pi(t-x)}{d}\right]dt = p \qquad (22)$$

where d is the distance between centers of the neighbour cracks.

Using substitutions $\eta=\cos(2\pi t/d)$ and $\xi=\cos(2\pi x/d)$ one can transform (22) into the problem of inversion of the singular integral of the Cauchy-type. Then, the solution of (22) will be

$$Q(t) = \begin{cases} p\sqrt{\dfrac{\cos\frac{2\pi a}{d} - \cos\frac{2\pi t}{d}}{\cos\frac{2\pi t}{d} - \cos\frac{2\pi b}{d}}}\,\mathrm{sgn}(t), & |b| < t < |a| \\ 0 & \text{otherwise} \end{cases} \qquad (23)$$

To determine the dilation zone length one should use equation (8). Then, for the critical concentration of initial dilation zones $\omega_{cr}=2b/d$ one has the following equation

$$\lambda = \int_1^{\frac{1}{\omega}} \sqrt{\frac{\sin\left[\frac{\pi}{4}(1+4\omega\eta)\right] - \sin\left[\frac{\pi}{4}(1-4\omega\eta)\right]}{\sin\left[\frac{\pi}{4}\omega(1+\eta)\right] - \sin\left[\frac{\pi}{4}\omega(\eta-1)\right]}}\,d\eta \qquad (24)$$

The dependence ω_{cr} against λ is shown in Figure 5 by the dashed line.

It is seen that both approximations give results not very much different from each other. This suggests that the simplest approximation based on the self-consistent method can be used to analyse this type of phenomena in a more complicated 3-D case.

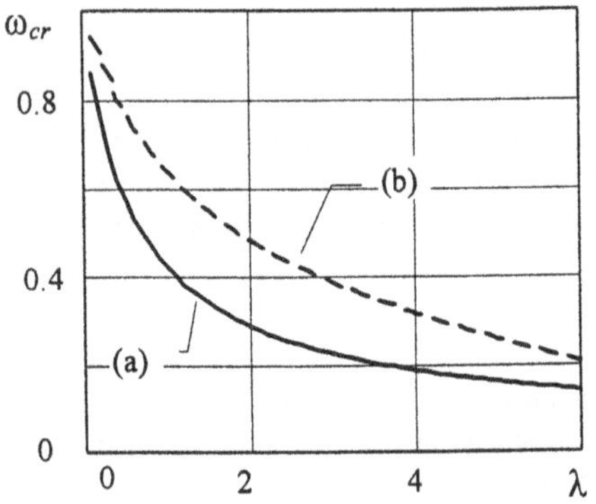

Figure 5. Critical concentration of initial dilation zones corresponding the catastrophic sliding:
(a) - approximation by self-consistent method;
(b) - periodic approximation.

3.3 Magnitude of seismic event associated with catastrophic sliding

In order to evaluate the magnitude of the seismic event let us assume that L is the characteristic macroscopic length of the formed sliding zone (it can be the length of a part of the fault affected by the waves or a characteristic length of non-uniformity of friction properties or the initial stress state). Then the energy release associated only with sliding will be $U \sim (\pi/2)\tau^2 L^2/E$ per unit length in the direction perpendicular to the drawings in Figure 2. For moderate values of the parameters τ=1 MPa, E=10 GPa and the sliding zone length, L, and width of 1 km one has $U \sim 10^{11}$ J. This approximately corresponds to the magnitude of 4 on Richter's scale (eg, Kasahara, 1981).

Also, if L is the length of the part of the fault that is affected by the waves, the number of waves before the catastrophic sliding is $M_{cr}=\omega_{cr}L/b$. When the waves come with the rate ν, the time before the catastrophic sliding will be $T_{cr}=\nu\omega_{cr}L/b$.

4. CONCLUSIONS

A model is proposed that considers the formation of local dilation zones (open cracks) in faults with rough surfaces initially hold by friction. It is demonstrated that such zones can propagate further by opening first and then by shearing. The accumulation and interaction of these zones lead to an increase in their average length until it reaches the average distance between their centres. This creates conditions for the coalescence of the dilation zones and the formation of one large zone where (catastrophic) sliding is possible. The critical concentration of the initial dilation zones at which the initiation of the catastrophic sliding is possible depends on a parameter comprising the ratio of ambient compression to the rock mass Young's modulus and the ratio of the initial dilation zone length to the height of the roughness of the fault faces.

ACKNOWLEDGMENT. The authors acknowledge the support of ARC large grant A89600726 (1996-1998).

5. REFERENCES

Dyskin, A.V. & A.N. Galybin 1997. The equilibrium of dilating shear cracks with rough surfaces. *Advances in Fracture Research, Proc. 9th Intern. Conf. on Fracture*, 4, 2251-2258.

Gibowicz, S.J. & A. Kijko 1994. *An introduction to Mining Seismology*. San Diego, New York, Boston, London, Sydney, Tokyo, Toronto: Academic Press.

Goodman, R.E. 1989. *Introduction to Rock Mechanics*. John Wiley & Sons.

Jahnke, E., F. Emde & F. Lösch 1960. *Tables of Higher Functions*. New York: McGraw-Hill

Kachanov, M.1993. On the effective moduli of solids with cavities and cracks. *Intern. Journal of Fracture* **59**, R17-R21.

Muskhelishvili, N.I. 1953. *Some Basic Problems of the Mathematical Theory of Elasticity*. P. Noordhoff Ltd, Groningen, the Netherlands.

Kasahara, K. 1981. *Earthquake Mechanics*. Cambridge: Cambridge University Press.

Savruk, M.P. 1981. *Two-Dimensional Problems of Elasticity for Bodies with Cracks*. Kiev, Naukova Dumka (in Russian).

Geology

Mechanics of Jointed and Faulted Rock, Rossmanith (ed.) © *1998 Taylor & Francis, ISBN 90 5410 955 6*

Mapping concealed faults in 3-D along the Wasatch Fault zone

A.K. Benson & S.T. Hash
Department of Geology, Brigham Young University, Provo, Utah, USA

ABSTRACT: Close-order gravity and magnetic surveys were conducted to locate concealed faults in Mapleton, Utah County, Utah, a developing city along the Wasatch Front, eighty kilometers south of Salt Lake City, Utah. Regional trends were calculated using trend analysis, and residual data were modeled using self-consistent iterative modeling. Ambiguities in the subsurface models were reduced by (1)incorporating data from previous geophysical surveys, surface mapping, and well data; (2)integrating the gravity and magnetic data; and (3)correlating the modeled cross sections. The modeled profiles and integrated 3-D interpretation indicate that there are four major fault blocks beneath the Mapleton area. These faults are potential earthquake threats, especially for possible surface rupture.

1 INTRODUCTION

Eighty percent of the population of Utah, or more than 1.6 million people, live along the Wasatch Front (Gori and Hays,1992). Though no large earthquakes (M_S>7.0) have occurred along the Wasatch Front since Utah was settled, geologic and geomorphic evidence indicate a series of episodic events throughout the past 10-15 million years (Hamblin, 1976). Clearly, there is a potential threat to lives and property in this area. Mitigating casualties and damages caused by earthquakes in Utah requires the interest and concern of legislators, planners, engineers, and developers. Continued research is needed to add to the awareness of potential earthquake hazards and to increase our understanding of how to mitigate their effects.

This research integrates close-order gravity and magnetic surveys to locate buried faults in the Mapleton area. Mapleton is a rapidly-growing community of 4,500 people located along the Wasatch Front, about 80 km south of Salt Lake City, Utah (Fig. 1). This area is located within the Basin and Range Province, where E-W to ESE-WNW minimum horizontal principal stresses during the past 15-20 million years has resulted in a broad zone of extensional normal faulting (Zoback, 1983). Previous research using gravity and magnetic techniques (Cook and Berg, 1961; Mabey, 1992; Zoback, 1983; D.A. Davis, 1983) have covered relatively large areas (1,500 to 3,500 km^2 areas). This research in Mapleton will concentrate on a smaller area of approximately 30 km^2 (Fig. 1), and geophysical stations are relatively closely spaced, with approximately 50% closer station spacing than any previous regional survey.

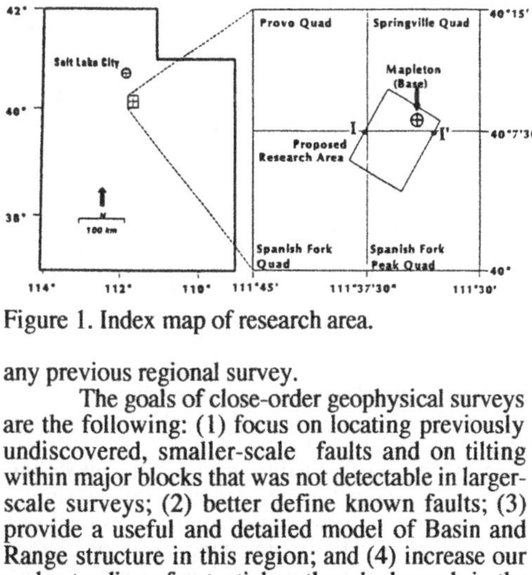

Figure 1. Index map of research area.

The goals of close-order geophysical surveys are the following: (1) focus on locating previously undiscovered, smaller-scale faults and on tilting within major blocks that was not detectable in larger-scale surveys; (2) better define known faults; (3) provide a useful and detailed model of Basin and Range structure in this region; and (4) increase our understanding of potential earthquake hazards in the Mapleton area and provide some details useful for zoning, building, and emergency planning.

2 GEOPHYSICAL DATA

A Worden gravimeter, with a maximum precision of 0.01 mGals, was used to collect the gravity data, and a Geometrics proton precession magnetometer, with a maximum precision of 1 nT, was used to collect magnetic field data. Gravity and magnetic data were collected at 190 stations within the Mapleton research area.

All gravity and magnetic data stations were measured using a looping technique, with the duration of each loop less than 3 hours. All loops originated and terminated at a base station established in the NW1/4 of the NW1/4 of section 14 T.8 S., R.3 E. (Figure 1). Gravity field data were processed using conventional procedures, which included drift, latitude, free-air, Bouguer, terrain, and regional corrections (Telford et al., 1990). A maximum error of 0.03 mGal was found at any repeated station over a period of three months.

At least three magnetic field readings were taken at each data station. The average of these three readings was recorded, unless the variation exceeded 5 nT. Stations exceeding a 5 nT variation between readings were either rechecked at a later date, or not used. Magnetic data were corrected for temporal variations using the same looping technique applied to the gravity data. Corrections for latitude and longitude variations were made by interpolating between IGRF (International Geomagnetic Reference Field) values provided by the USGS. Magnetic data are not included for some stations due to local cultural interference. Gravity and magnetic residuals were calculated using the trend-surface analysis option of GRIDZO™.

3 DATA MODELING AND INTERPRETATIONS

Nine profiles of data (designated A-A' through I-I') were chosen for modeling, based on distribution of survey locations, trends within the contoured residual data, and the geology and topography in the area. In this paper, we will specifically examine geological models for profile I-I' (Figure 1), and the modeling results for all of the profiles will be integrated into the final 3-D interpretation. GM-SYS™, a modeling program developed by Northwest Geophysical Associates, Inc. was used to model the residual gravity and magnetic data. Since there are many different density distributions that can be modeled to match a residual anomaly, steps must be taken to establish the validity of a model. The "best-fit" models resulted from an iterative modeling process to achieve low percentage error (maximum deviation of 2%) between the observed residual and modeled values, while maintaining close compliance with modeling constraints. Since the gravity data were considered to be the primary set of geophysical data, with the magnetic data serving as a secondary set for the Mapleton research area, one of the more important modeling constraints involved density values.

A systematic decrease in valley fill density towards the center of the basin was modeled. The higher densities modeled towards the basin edge are characteristic of coarse, poorly sorted fluvial and alluvial sediments found near the mountain fronts. The relatively lower densities modeled towards the center of the basin are characteristic of the finer, well sorted, lucustrine sediments found there. Mabey (1992) emphasizes the point that using varying densities in the modeling process is very important for accurately modeling the location and dip of normal faults.

Another very important constraint focuses on the orientation of the faulted blocks. As described by Stewart (1977), two general models of Basin and Range structure have been proposed: horst and graben, and tilted-block with listric faults. Regional gravity data (Mabey, 1992), as well as physiographic and structural features (Stewart, 1980), show predominantly east-dipping structural tilt along the Wasatch Front. These tilt patterns would favor the tilted block model, since according to Stewart (1980), the horst and graben model probably would not create this amount of tilting. Conjugate faulting occurs in the area, with local horst and graben structures resulting, but the dominant structural model for the research area would favor the tilted-block, listric fault model.

The surface geology, available well data, and previously collected geophysical data were also utilized as modeling constraints. Of particular value were geophysical surveys by Benson and Mustoe (1995), and D.A. Davis (1983). These surveys provided both local and regional geophysical modeling, as well as density and susceptibility values. Likewise, surface mapping of faults by Machette (1992) and Davis and Gurgel (1983) were valuable for locating exposed faults and for comparing modeled structures with surface morphology. A list of drill holes compiled by Case (1985) provided some minimum depths to bedrock in the Mapleton area, as well as providing lithology data to a depth of 3,962 meters from the Gulf test well northwest of Spanish Fork, Utah.

Fault locations, depths to fault blocks, and densities were compared at profile intersections during the modeling process. The models for all profiles were modified to be self-consistent with each other, as well as correlate with the topography and geology observed in the field and on maps. Through iterative modeling, a model of each profile was generated which best complies with the previously-discussed modeling constraints. Locations of profile intersections are marked by the letter corresponding to the intersecting profile (e.g. "C" for Profile C-C', as in Figure 2).

Fault planes or zones in the magnetic models were modeled as prisms of magnetic mineralization, with magnetic susceptibilities of 0.008 (cgs). This value is based on previous work in the area by Benson and Mustoe (1995). Susceptibilities of 0.00005 (cgs) for alluvium and 0.001 (cgs) used for faulted blocks are based on values measured in the field using a Kappameter (Model KT-5) susceptibility meter.

By assuming deposition of magnetic material(s) along the fault planes, or within the fault zones, reasonable fits to the residual magnetic data were obtained for all the magnetic profiles using similar fault locations and depths to bedrock as determined from the gravity modeling. The magnetic

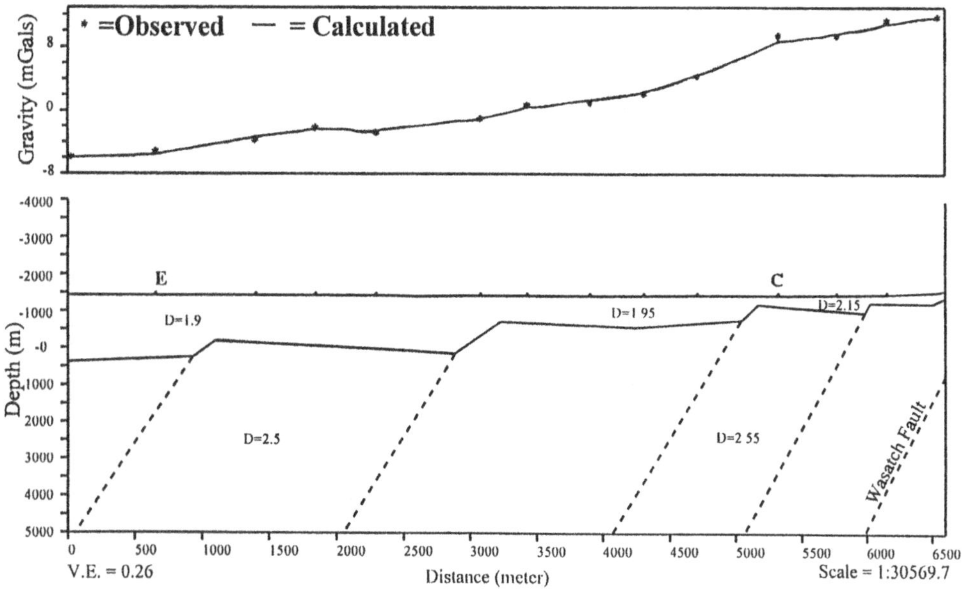

Figure 2. Gravity model of profile I-I'. Location of the Wasatch Fault is inferred by surface mapping.

response to faults in this area could be related to secondary mineralization along the fault planes, or deposition of limonite, hematite, or iron oxide along the more permeable fault planes during ground-water movement (Benson and Mustoe, 1995). Possible sources of this magnetic material include (1) Precambrian Mineral Fork Tillite, a glacial deposit composed primarily of gravel, but with some igneous clasts; and/or (2) Lower Cambrian Tintic Quartzite, a marine transgressive series of quartzite which contains some interbedded shales in its upper 150 meters, and a chloritzed basalt flow 300 meters above its base (Davis, 1983; Hintze, 1988).

Profile I-I' is 6.5 km long and is positioned in an east-west direction 2 km south of Maple Canyon (Figure 1). I-I' has a total residual gravity variation of 13.72 mGals and a total residual magnetic variation of 295 nT.

The gravity model for profile I-I' (Figure 2) shows a very good fit between the residual and modeled data. Profile I-I'(Figure 1) is intersected by Profiles E-E', and C-C'. The Wasatch Fault is labeled, with the location inferred by previous mapping (Machette, 1992). Four major fault blocks are modeled beneath the Mapleton area, bounded by west-dipping normal faults.

In this model, the density of valley alluvium ranges from 1.9 g/cc at the west end of the profile, to 2.15 g/cc in the sediments at the east end of the profile. Depths to faulted blocks range from less than 100 meters near the mouth of Maple Canyon, to 1,800 meters at the west end of the profile.

The magnetic model for I-I' (Figure 3) shows a good fit between the residual magnetic data and the modeled data. Fault locations on the magnetic model are comparable to the gravity model, though depths to the faulted blocks are 100 to 200 meters deeper at the east end of the magnetic model, indicating that magnetic mineralization occurs deeper along the fault plane than where density contrasts occur.

Since it is not possible to obtain a unique solution in the modeling process, alternative models, which vary from previously mentioned constraints, were considered for all the profiles. The "best-fit" models that matched all of the constraints most closely were selected to represent the fault locations and the orientations of the faulted blocks.

4 INTEGRATED 3-D INTERPRETATION

A surface map was generated (Figure 4) which shows the connected fault traces obtained from the best-fit models for all the profiles projected to the surface. These traces correspond to surface positions where there is a good likelihood that surface rupturing could occur. The eastern-most fault trace corresponds to mapped traces of the Wasatch Fault (Machette, 1992; Davis and Gurgel, 1983). The segmenting seen in the western-most fault trace may be related to the sharp concavity of the southwest bend of the Wasatch fault zone in this area. Approximate locations of the four major fault blocks beneath the Mapleton area are mapped.

Combining information about the depths and connected fault traces from the "best-fit" models, a three-dimensional subsurface model of the faulted blocks was generated (Figure 5). The four major

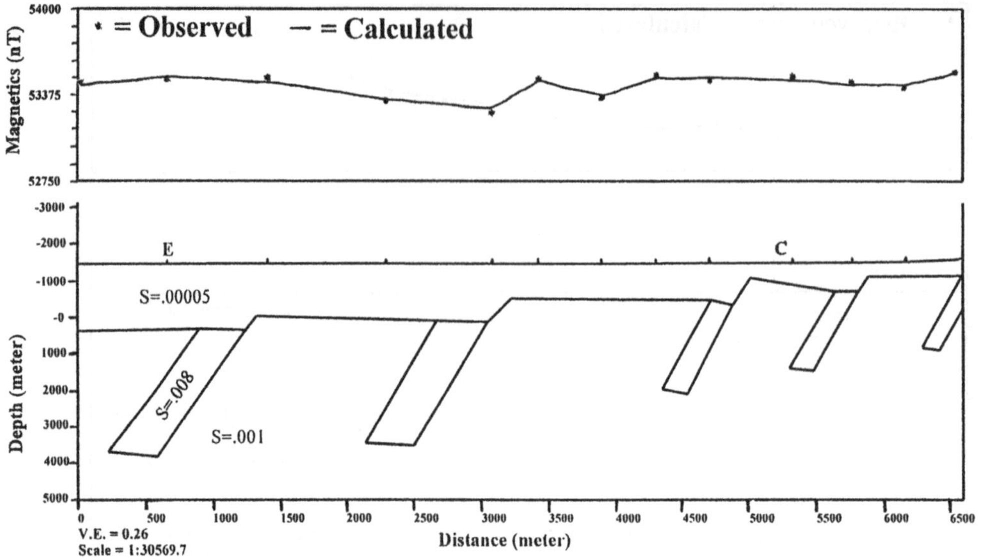

Figure 3. Magnetic model of profile I-I'.

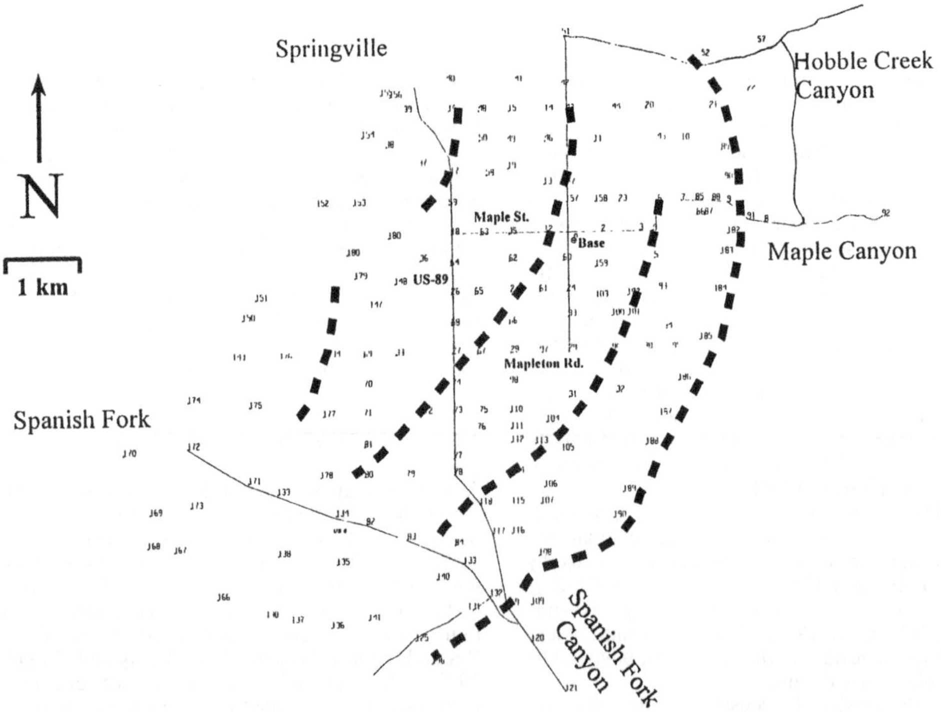

Figure 4. Surface map of connected fault traces obtained from the "best-fit" models. The trace to the far east is the Wasatch Fault mapped by Machette (1992).

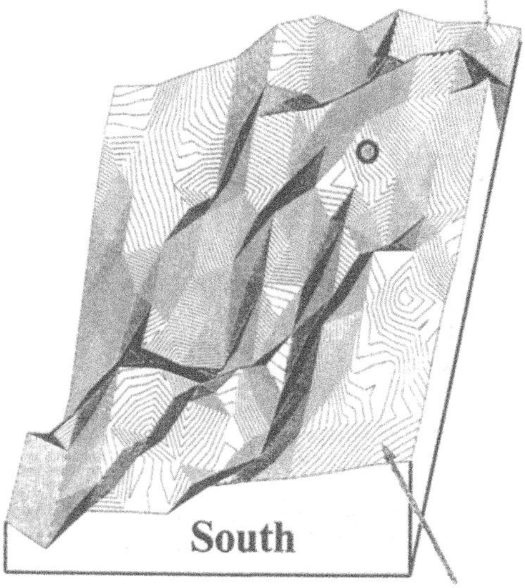

Figure 5. Subsurface 3-d model of fault locations and depths to faulted blocks obtained from the "best-fit" models. Circle is base station location.

fault blocks, with some segmentation are mapped. This segmentation may help compensate for the bend in the Wasatch fault zone in this area. The mouths of Spanish Fork Canyon and Hobble Creek Canyon, as well as the Mapleton base station are labeled for reference. The Wasatch Fault Zone is located just east of this three-dimensional model.

5 CONCLUSIONS

Based on close compliance with the following modeling constraints: (1) data from previous geophysical surveys, surface mapping, and well data; (2) tilted-blocks bounded by listric faults; (3) decreasing density towards the basin center; (4) integrating the gravity and magnetic data; and (5) correlating the modeled cross sections, best-fit models were generated for close-order gravity and magnetic data collected in the Mpaleton, Utah, U.S.A. area.

Fault locations and depths to fault blocks found in the modeling process for a series of strategically-located profiles were combined to construct a three-dimensional subsurface map, as well as a surface map of possible fault locations. These maps indicate that there are four major fault blocks beneath the Mapleton area. These fault blocks are bounded by a series of west-dipping synthetic, listric faults. Depths to the fault blocks range from less than one hundred meters in east Mapleton to nearly 2,000 meters in west Mapleton.

The results of this research support the concern of many geologists regarding potential earthquake hazards along the Wasatch Front. Concealed faults in the Mapleton area should be considered a potential earthquake threat, especially for possible surface rupture. Since it is technically and economically difficult to build structures that will withstand 2-3 meters of offset through their foundation, the most reasonable hazard reduction plan is to avoid building over these major faults.

REFERENCES

Benson, A.K.& N.B. Mustoe 1995. Analyzing shallow faulting at a site in the Wasatch fault zone, Utah, USA, by integrating seismic, gravity, magnetic, and trench data. *Engineering Geology,* 40: 139-156.

Case, W.F. 1985. Significant drill holes of Wasatch front valleys. *Utah Geol. and Min. Surv. Open File Rep.,* 82: 1-78.

Cook, K.L. & J.W. Berg 1961. Regional gravity survey along the Central and Southern Wasatch Front, Utah, *U.S Geol. Surv., Prof. Pap.,* 316-E: 75-89 (scale 1:500,000).

Davis, D.A. 1983. Gravity survey of Utah and Goshen Valley and adjacent areas. Unpubl. M.Sc. thesis, University of Utah, Salt Lake City, Utah, 141 pp.

Davis, F.D. & K.D. Gurgel 1983. Geologic Map of of the southern Wasatch Front, Utah. *Utah Geol. and Min. Surv.,* Map 55-A (scale 1:100,000).

Gori, P.L. & W.W. Hays (eds.) 1992. Introduction: Assessment of regional earthquake hazards and risk along the Wasatch Front, Utah. *U.S. Geol. Surv., Prof. Pap.,* 1500: 1-7.

Hamblin, W.K. 1976. Patterns of displacement along the Wasatch fault. *Geology,* 4: 619-622.

Hintze, L.F. 1988. Geologic History of Utah. *Brigham Young University Geol. Studies,* Special Publication 7: 1-199.

Mabey, D.R. 1992. Subsurface geology along the Wasatch front. In P.L. Gori & W.W. Hays (eds.) Assessment of regional earthquake hazards and risk along the Wasatch Front, *Utah. U.S. Geol. Surv., Prof. Pap.,* 1500-C: 1-16.

Machette, M.N. 1992. Surficial geologic map of the Wasatch fault zone, eastern part of Utah Valley, Utah County and parts of Salt Lake and Juab Counties, Utah. *U.S. Geol. Surv.,* Map I-2095 (scale 1:50000).

Stewart J.H. 1977. Basin-range structure in western North America: A review. In R.B. Smith & G.P. Eaton (eds.) Cenozoic tectonics and regional geophysics of the western Cordillera. *Geol. Soc. of America,* Memoir 152: 1-31.

Stewart, J.H. 1980. Regional tilt patterns of late Cenozoic basin-range fault blocks, western United States. *Geol. Soc. of America Bull.,* Part I, 91: 460-464.

Telford, W.M., L.P. Geldart & R.E. Sheriff 1990. *Applied Geophysics*, Second Editon. London: Cambridge University Press.
Zoback, M.L. 1983. Structure and Cenozoic tectonics along the Wasatch fault zone, Utah. In Miller, D.M., Todd, V.R., and Howard, K.A., Editors, Tectonics and stratigraphy of the eastern Great Basin. *Geol. Soc. of America*, Memoir 157: 3-27.

Mechanics of Jointed and Faulted Rock, Rossmanith (ed.) © 1998 Taylor & Francis, ISBN 90 5410 955 6

Stress perturbation near two fault zones in Himalayas-field measurements and numerical simulation

S. Sengupta, D. Joseph, C. Nagaraj & A. Kar
National Institute of Rock Mechanics, Kolar Gold Fields, Karnataka, India

ABSTRACT: Near a fault the orientation of maximum horizontal stress (σ_H) seems to rotate relative to the regional stress (σ_R) orientation in a consistent manner. The maximum horizontal stress direction is reported to have been rotated even 90 degrees to the regional stress. In this paper, the state of stress near two fault zones located in two hydroelectric projects in North Western Himalayas is discussed. In each case the stress rotation has been noticed distinctly. To understand factors responsible for the rotations, the faults are simulated using a two dimensional Distinct Element Method incorporating Mohr-Coulomb model and a parametric study has been conducted. The parametric study has revealed that a major geological structure could considerably influence the stress setup around it. It has also been inferred that external parameters like angle (α) formed by geological structure with regional stress orientation (σ_R) and ratio of boundary stress magnitude (K_0) are more responsible for stress rotation as compared to the internal parameters like cohesion (c), angle of internal friction (ϕ), bulk and shear moduli (K, G) and joint stiffness (K_s, K_n). Thus any major civil engineering structures falling within this perturbed stress due to the geological structure may have to be designed accordingly.

1 INTRODUCTION

The distribution of in-situ stresses in terms of magnitude and orientation effects the geometry, shape, dimensioning, excavation sequence and orientation of underground excavations. The in-situ stress magnitude and orientation are found to be controlled by the major geological structures like fold, fault and intrusive. Though considerable advances have been made in the measurement of in-situ stress over the past 25 years, the interpretation of these measurements with respect to geological structures have received significant attention only over the past 10 years. From different construction sites all over the world a large number of observations implicating the rotation of stress orientation as much as by 90° near geological structures are reported. A number of investigators have also postulated various theories to explain the phenomena.

Anderson (1951) cited reorientation of principal stress trajectories due to primary fault formation and postulated that reorientation of the stress trajectories takes place only at the ends of the principal fault. Ez (1962) measured stresses around a network of strike slip faults in the principal anticline of the Donetz Coal Basin, and found reorientation of regional stress by almost 90° around the longitudinal faults in the core of the anticline.

Zoback (1992) presumed that a strength contrast between a low frictional resistance fault inside a strong crust is responsible for rotation of σ_H in the San Andreas Fault in Central California.

We have encountered the stress rotational phenomena at Baspa and Dulhasti Hydroelectric Projects in Himalayas where faults at the vicinity of the measurement sites seem to have rotated the σ_H direction by 87° and 24° respectively. We tried to understand the problem by parametric study using numerical modelling (2D-UDEC code) incorporating the rock and fault properties along with the stress ratio and angle between the maximum horizontal stress and the fault as input parameters. The model calculations indicate that the elastic properties contrast, friction angle of fault contact, angle between geological structure, regional horizontal stress, and the ratios of the magnitudes of the regional stresses are the prime factors responsible for rotation phenomena of the horizontal stresses.

Table 1. Direction of σ_R by focal mechanism

Latitude	Longitude	Azimuth	Reference
32°30′N	78°E	267°	Choudhury and Srivastava (1976)

2 DESCRIPTION OF THE PROJECT SITES

2.1 *Baspa Hydroelectric Project, Himachal Pradesh (In Himalayas)*

2.1.1 *Introduction*

The Baspa Hydroelectric Project is located in Greater Himalayas in Kinnaur district of Himachal Pradesh. The project envisages construction of a barrage at Kuppa village near Sangla across the river Baspa, a tributary of river Satluj. The head water will be diverted into a 7.77 km head race tunnel, into the underground power house complex of 92 m x 19 m x 41 m (height) at Karcham through one number of steel lined pressure shaft to generate 300 MW of power.

2.1.2 *Regional stress province*

The project area is located at latitude 31°30′N and longitude 78°15′E. The focal mechanism solution for this part of the Himalayas postulated from the Kinnaur earthquake of 1975 which probably was caused by N-S trending Kaurik-Chango fault gives a direction of maximum compression (σ_R) given in Table 1.

2.1.3 *Local perturbed stress province*

Two major shear zones with appreciable movements are the nearest geological structures to the testing sites. They strike N25°E, dipping 45° towards east. The width of the wider shear zone is around 5 m. The perturbed stress was measured by hydrofracture method inside five boreholes at the vicinity of the fault zone. The stress tensors calculated are shown in Table 2.

Thus the following observations are made:
1) The direction of the regional maximum horizontal compression σ_R by focal mechanism = 267°
2) The direction of the perturbed maximum horizontal compression σ_H near the fault = 180°
3) The total rotation of the maximum horizontal compression = 087°

Figure 1 shows the regional and local perturbed maximum horizontal stresses.

Table 2. The stress tensors calculated by GENSIM

Major In-situ stress parameters:	
Minimum Horizontal Stress (σ_h)	5.40 MPa
Maximum Horizontal Stress (σ_H)	10.80 MPa
Vertical Stress (σ_V) *	9.71 MPa
Direction of σ_H	180°

* The vertical stress corresponds to an overburden o 360 m and density of rock 2.75 gm/cm^3

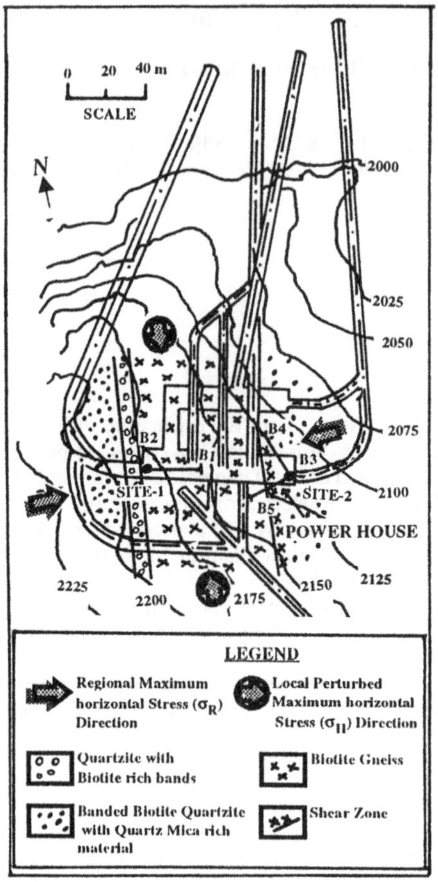

Figure 1. Plan showing the orientations of the regional and local perturbed maximum horizontal stresses, Baspa Hydroelectric Project.

2.2 *Dulhasti Hydroelectric Project, Jammu and Kashmir (In Himalayas)*

2.2.1 *Introduction*

Dulhasti Hydroelectric Project is located at around 70 km west of Srinagar, in Jammu and Kashmir. The project envisages construction of a 20 m high barrage across river Chenab. The power station will house four units of turbines of 120 MW each to have a total generating capacity of 480 MW.

2.2.2 *Regional stress province*

The project area is located at latitude 33°30′N and longitude 75°E in the North Western Himalayas.

The nearest focal mechanism solution to the project site with P. axis azimuth (σ_R azimuth) is given in Table 3.

Table 3. Direction of σ_R by focal mechanism

Latitude	Longitude	Azimuth	Reference
33°42′N	75°30′E	046°	Chandra (1978)

2.2.3 *Local perturbed stress province*

A major fault/shear zone of 10 m to 20 m in width with orientation N50^{0}E dipping 70^{0}N and with appreciable movement along thrust fault is the closest major discontinuity to the testing area. The shear zone is only 20 m south of the testing site seems to have perturbed the stress orientation. The perturbed stress was measured by hydrofracture method inside three boreholes.

Table 4. The stress tensors calculated by GENSIM

Major in-situ stress parameters:	
Minimum Horizontal Stress (σ_h)	5.40 MPa
Maximum Horizontal Stress (σ_H)	10.40 MPa
Vertical Stress (σ_V) **	7.15 MPa
Direction of σ_H	070°

** The vertical stress corresponds to an overburden of 270 m and density of rock 2.75 gm/cm^3

Thus the following observations are made:
1) The direction of the regional maximum horizontal compression σ_R by hydrofracture method = 046°
2) The direction of the perturbed maximum horizontal compression σ_H near the fault = 070°
3) The total rotation of the maximum horizontal compression inside the fault = 024°

Figure 2 shows the regional and local perturbed maximum horizontal stresses.

3 NUMERICAL ANALYSES

A two-dimensional DEM programme, UDEC, simulates both the case histories. In the analyses the following assumptions are made:
1) Plane strain conditions applied in the model.
2) The intact rock blocks are assumed to behave as non-linear elastic/plastic material.
3) Joint area contacts as elastic/plastic with Coulomb slip failure.
4) Measured maximum compression direction and magnitude are assumed to be the perturbed stress direction and magnitude in and around the discontinuity.
5) The fault is modelled as low modulus material respectively relative to their surrounding material.
6) Regional stress directions are assumed to be the boundary stress directions.

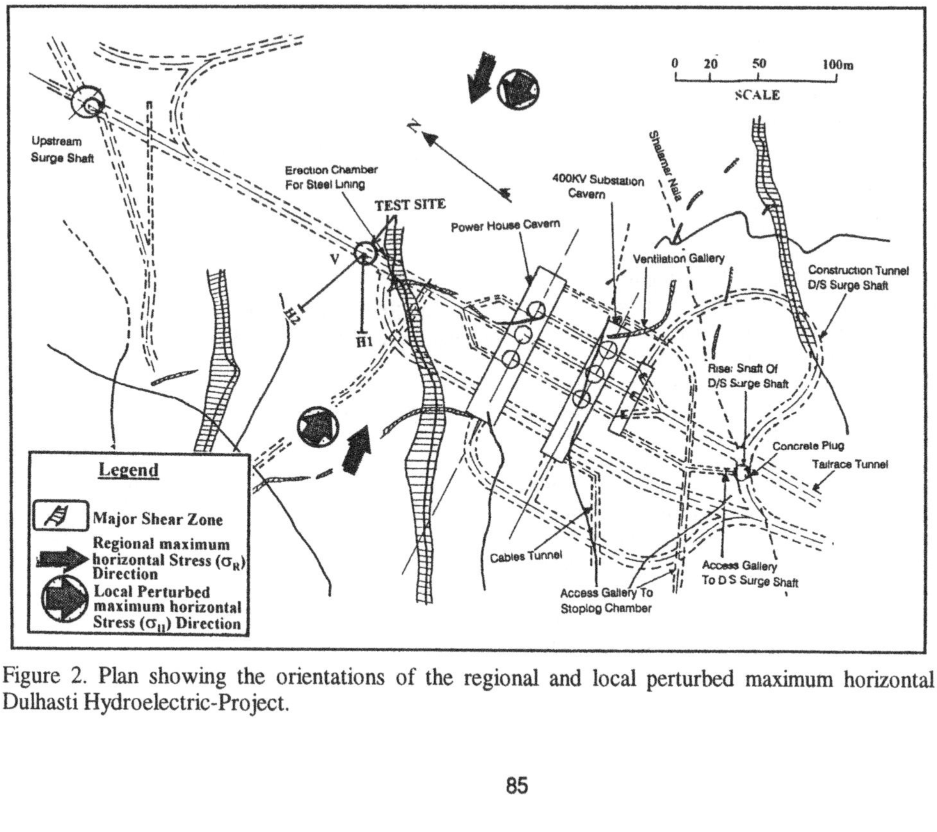

Figure 2. Plan showing the orientations of the regional and local perturbed maximum horizontal stresses, Dulhasti Hydroelectric-Project.

7) Magnitudes of the boundary stresses (unknown) are manipulated systematically so as to achieve the horizontal stress magnitudes around the discontinuity close to the measured stresses.

3.1 *Geometry and boundary conditions*

The two dimensional computational model is used which consists of a block with an intersecting fault. It is subjected to a bilateral tectonic stress field. The configuration, the dimensions and the initial boundary conditions for different cases are taken in accordance with the field conditions.

3.2 *Modelling strategy*

Different models are tried by varying the following parameters:

1) Bulk moduli (K) and shear moduli (G) of both intact rock and the discontinuity.
2) Cohesion (c) and friction angle (ϕ) of intact rock, discontinuity and the discontinuity contact.
3) Normal stiffness (K_n) and shear stiffness (K_s) of the discontinuity contact.
4) Angle (α) between the boundary stress direction and the strike of the discontinuity.
5) Ratio of boundary stress magnitudes (K_0).

3.3 *Simulation results*

The parameters used in the analysis can be grouped as follows:

1) External parameters like α, K_0, which are independent to the material properties of the rock and geological discontinuity and
2) Internal parameters which are governed by the material properties of geological structure in concern and the surrounding rock. Parameters like c, ϕ, shear and bulk moduli of intact rock and discontinuity (G, K), joint stiffness (K_s, K_n), fall in this category.

The material properties for the two projects are found to be almost identical along with the perturbed stress ratios. But there are marked differences in angle (α) between boundary stress direction and the geological structure for the two projects (For Baspa Project α = 60° and for Dulhasti Project α = 4°). The α values are kept in accordance with the field conditions. The magnitudes of the regional stresses (σ_R/σ_r = ratio between maximum and minimum regional horizontal stresses) are varied systematically to achieve the stress magnitudes as close as possible to the measured stress magnitudes.

The fault is treated as a low modulus material with definite boundary surrounded with high modulus isotropic material. Along the fault contacts with zero cohesion, only slippage is allowed but not block

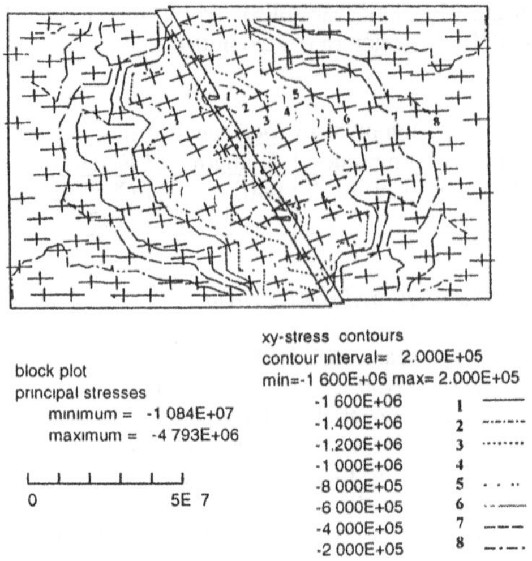

Figure 3. Effects of the fault on horizontal stress orientations for Baspa Hydroelectric Project.

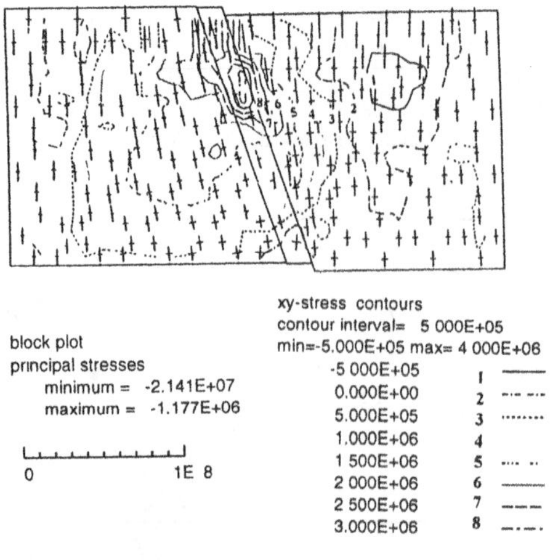

Figure 4. Effects of the fault on horizontal stress orientations for Dulhasti Hydroelectric Project

rotation as in that case the angle between regional stress and the geological structure would not have represented the true field condition. Both the actual location (outside the fault) in the field and the measured stress magnitudes are considered while selecting the monitoring segment in the model to study the rotation of the stress direction in simulated condition.

3.3.1 *Effects of fault on stress trajectory*

Figure 3 shows stress tensors along with the shear stress contours in and around the fault model.

The modelling studies reveal a rotation of σ_H around the fault equal to 24° in the monitoring segment as against 87° measured in the field in case of Baspa Hydroelectric Project (Figure 3).

Regarding the magnitudes, the simulated σ_H (11.20 MPa) and σ_h (4.50 MPa) are comparable with the field (10.80 MPa and 5.40 MPa).

The modelling studies reveal a rotation of σ_H around the fault equal to 13° in the monitoring segment as against 24° measured in the field in case of Dulhasti Hydroelectric Project (Figure 4).

Regarding the magnitudes, the simulated σ_H (10.70 MPa) and σ_h (5.40 MPa) are comparable with the field (10.40 MPa and 5.40 MPa).

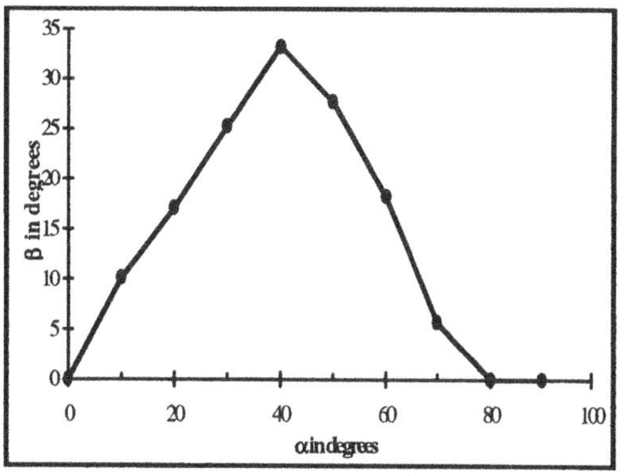

Figure 5. The effect of angle (α) between horizontal stress direction and the strike of the geological structure on the degree of horizontal stress deflection (β).

4 PARAMETRIC STUDY

To check the factors influencing the stress deflections a systematic parametric study is carried out. Shear and bulk moduli values (G, K), angle of friction (ϕ), cohesion (c), boundary stress ratio (K_0), the angle (α) between regional stress and the geological structures are taken in to consideration along with other factors (Sengupta et al., 1997).

The results of the parametric study are as follows:

1) A low friction angle to the fault contact deflects the stress direction by a higher degree.

2) A higher stress ratio between two horizontal stresses deflects the stress more.

3) There is significant deflection of stresses due the variations in elastic moduli between geological structure and the surrounding rocks.

4) The maximum deflection of stress orientation is noted when the angle between regional stress and the geological structures is around 40°.

From the parametric study it can be concluded that three parameters viz. angle (ϕ) of friction of the discontinuity contact, boundary stress ratio (K_0) and angle (α) formed by the discontinuity plane with the maximum horizontal stress are having influence on stress rotation (β). External parameters like α and K_0 which do not depend on the material properties of either discontinuity or surrounding rock, are more responsible for stress rotation than internal parameters like c, ϕ, bulk and shear moduli (K, G), joint stiffness (K_s, K_n) which depend on material properties.

5 DISCUSSION AND CONCLUSIONS

The difference of β evaluated by field data and numerical methods may be attributed to the isotropic conditions assumed for both country rock and geological structure in numerical analysis. This is in marked contrast to the actual field condition where rock as well as the geological structure is highly anisotropic. Zhang et al. (1994) has shown that the rotation of stresses in an anisotropic material (where the moduli in X and Y directions are different) is very high. Nevertheless the numerical analyses have given scope to understand the role of various parameters responsible for stress rotation.

Figure 5 shows the effect of (angle between regional horizontal stress direction and geological structure (α) on stress rotation (β), whereas Figure 6

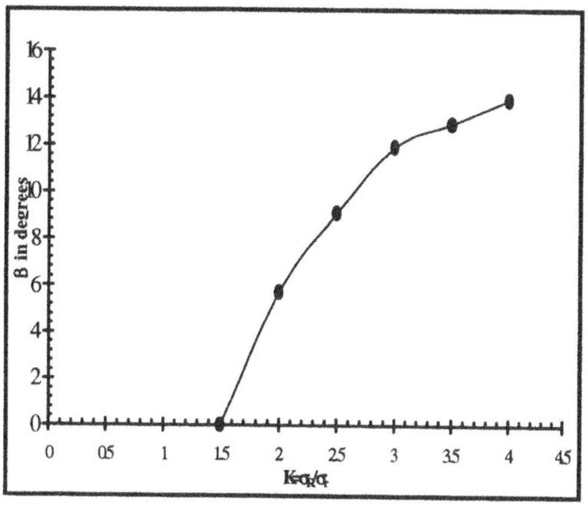

Figure 6 shows the effect stress ratio (K_0) on the degree of horizontal stress deflection (β) due to the fault/shear zone.

shows the effect of stress ratio (K_0) on stress rotation. These plots are obtained by parametric study.
The above studies confirm that a major geological structure can considerably influence the stress setup around it. Any major structures designed within this perturbed stress due to the geological structure may have to be designed accordingly. Thus this paper gives an idea of the application of rock stress measurement in mining and civil engineering projects.

ACKNOWLEDGEMENTS

We are thankful to the Director National Institute of Rock Mechanics (NIRM), KGF, India, for permission to publish this paper. Mr.N.Reddy of NIRM is also thankfully acknowledged for his extensive support during numerical modelling. We are also thankful to the authorities and all technical staffs of Baspa and Dulhasti Hydroelectric Projects for rendering help during field measurements.

REFERENCES

Anderson, E.M. 1951. The dynamics of faulting and dyke formation with applications to Britain. Oliver and Boyd, London, Edinburgh.

Chandra, U. 1978. Seismicity, earthquake mechanisms and tectonics along the Himalayan mountain range and vicinity. *Phy. Earth and Planet. Int.* 16, 109-131.

Chaudhury, H.M. and Srivastava, H.N. 1976. Seismicity and focal mechanisms of some recent earthquakes in northeast India. *Annali di Geophysica.* 29(1-2), 41-56.

Ez, V.V. 1962. Influence of Hercynian folding on the structure of the Caledonian structural stage in the Karatau mountain ridge and storeyed nature of folding. *In: Skladchatye deformatisii zemmoi kory, ikh tipy mekhanism o brazovaniya, Moscow.*

Sengupta, S., Joseph, D., Nagaraj, C. & Kar, A. 1997. Influence of a fault and a dyke on stress distribution at two project sites in India. *Proc. Int. Symp. on Rock Stress, Kumamoto, Japan.* pp. 397-402.

Zhang, Y.Z., Dusseault, M.B., Yasir, N.A. 1994. Effects of rock anisotropy and heterogeneity on stress distribution at selected sites in North America. *J. Eng. Geol.* 37, 181-197.

Zoback, M.L. 1992. First and Second Order Patterns of Stress in the Lithosphere, The World Stress Map Project, *J. Geophys. , Res.* 97, 11.703-11.728.

Mechanics of Jointed and Faulted Rock, Rossmanith (ed.) © 1998 Taylor & Francis, ISBN 90 5410 955 6

On interaction of arbitrary oriented faults systems

A.S. Bykovtsev
Regional Academy of National Sciences, Tashkent, Uzbekistan

D.B. Kramarovsky
Institute of Seismology, Tashkent, Uzbekistan

D.I. Bardzokas
National Technical University of Athens, Greece

ABSTRACT: Interaction and conditions for motion of an arbitrary oriented ruptures system are studied by the methods of mathematical simulation. The system consists of arbitrary number of ruptures with arbitrary oriented vectors of final dislocation on them and is in the field of external stresses. The use of a number of a fracture mechanics methods allows to obtain exact analytical solutions which make it possible to predict behaviour of ruptures system. Equations describing the conditions under which the ruptures can grow were obtained in the form of functions of the ruptures orientation, their length, fracture type on them (shear, tensile, mixed type of fracture), values of the external stresses, physical and mechanical parameters of surroundings. Analyses of the results obtained for different values of the above mentioned parameters and for different number of interacting ruptures simulating different geodynamic situation is given. The results obtained can be used for the long-term prediction of a possible earthquake location.

1 INTRODUCTION

Studying the deformation and cracking processes in the Earth's crust, besides the general cognitive purpose related to the geodynamic processes research, acquires a great significance in connection with solving a number of particular practical problems. Thus, for instance, it is well known that preparation of a tectonic earthquake source proceeds by way of disruptive zone formation in the source region, its formation being a unification and merging of small ruptures into one or several somewhat larger ruptures. Analysing the stress state in the vicinity of large tectonic fracture ends can provide useful information on probable places of the future disastrous events - the so-called long-term prediction of the earthquakes, and also substantially improve the existing methods for a detailed seismic subdivision of the earth. The statement of quantitative relationships between the basic parameters of rupture processes and definition of other characteristics of the geophysical fields will allow us to better understand the nature of geodynamic processes occurring in the earth's interior.

It is necessary first of all to analyse conditions of interaction between separate arbitrary oriented ruptures.

2 FORMULATION OF THE TASK

Let N arbitrary oriented dislocation ruptures with the values of displacement vectors of final dislocations b^j (b_1, b_2, b_3)=const (j=1...N) are given in an homogeneous isotropic elastic medium. Angles between the ruptures and x-axis of Oxy base coordinate system define as α_j (Fig. 1). Coordinates of the initial and final tips of the ruptures are x_1^j, y_1^j and x_2^j, y_2^j correspondingly in Oxy coordinate system. Let lengths of these ruptures define as l_j.

The main goal of solution to the problem is to determine displacements and stresses generated by these rupture systems, and formulate criteria, conditions and directions of their further growth.

Earlier the problem was considered [1,2] for ruptures situated along one straight line, and Cartesian coordinates were oriented in such a way that the x-axis coincides with the rupture line.

Exact analytical solutions for determination of the displacements fields and components of the stresses tensor generated by such rupture systems in coordinate system connected with them were obtained.

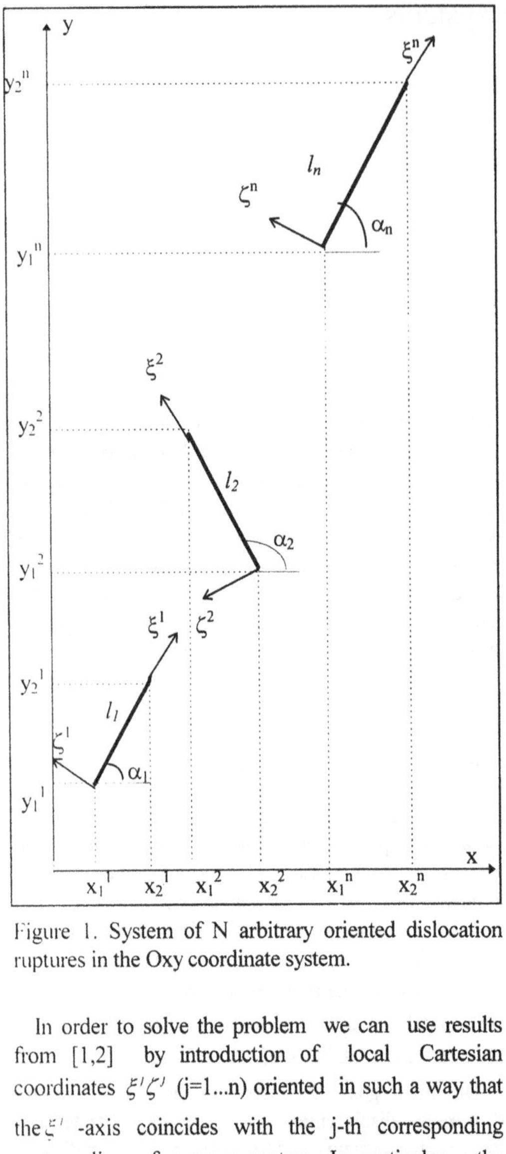

Figure 1. System of N arbitrary oriented dislocation ruptures in the Oxy coordinate system.

In order to solve the problem we can use results from [1,2] by introduction of local Cartesian coordinates $\xi^j\zeta^j$ (j=1...n) oriented in such a way that the ξ^j -axis coincides with the j-th corresponding rupture line for every rupture. In particular, the transformation of Oxy coordinate system to the $\xi^j\zeta^j$ systems can be given as follows:

$$\xi^j = (x - x_1^j)\cos\alpha_j + (y - y_1^j)\sin\alpha_j$$
$$\zeta^j = -(x - x_1^j)\sin\alpha_j + (y - y_1^j)\cos\alpha_j \qquad (1)$$

With such a variables transformation coordinates of the j-th rupture tips in the $\xi^j\zeta^j$ coordinate system connected with the j-th rupture are 0 and l_j. Field of stresses generated by the rupture in this coordinate system can be represented using complex functions w_i by the ratios [1,2]:

2.1 Mode I:

$$2Gu_\xi = \frac{\chi - 1}{2}\operatorname{Re} w_2 - \zeta \operatorname{Im} W_2;$$

$$2Gu_\zeta = \frac{\chi + 1}{2}\operatorname{Im} w_2 - \zeta \operatorname{Re} W_2;$$

$$\sigma_{\xi\xi} = \operatorname{Re} W_2 - \zeta \operatorname{Im} W_2';$$

$$\sigma_{\zeta\zeta} = \operatorname{Re} W_2 + \zeta \operatorname{Im} W_2';$$

$$\sigma_{\xi\zeta} = -\zeta \operatorname{Im} W_2' \qquad (2)$$

2.2 *Mode II:*

$$2Gu_\xi = \frac{\chi + 1}{2}\operatorname{Im} w_1 + \zeta \operatorname{Re} W_1$$

$$2Gu_\zeta = \frac{1 - \chi}{2}\operatorname{Re} w_1 - \zeta \operatorname{Im} W_1$$

$$\sigma_{\xi\xi} = 2\operatorname{Im} W_1 + \zeta \operatorname{Re} W_1'$$

$$\sigma_{\zeta\zeta} = -\zeta \operatorname{Re} W_1'$$

$$\sigma_{\xi\zeta} = \operatorname{Re} W_1 - \zeta \operatorname{Im} W_1' \qquad (3)$$

2.3 *Mode III*:

$$u_z = \operatorname{Im} w_3;$$

$$\sigma_{\xi z} = G \operatorname{Im} W_3; \qquad \sigma_{\zeta z} = G \operatorname{Re} W_3; \qquad (4)$$

where:

w'(z) = W_i(z), (i=1,2,3); $z = \xi + i\zeta$;

$\chi = 3 - 4\nu$ for plane deformation;

$\chi = (3 - 4\nu)/(1 + \nu)$ for generalised plane stress state;

G is the modulus of rigidity;

ν is the Poisson's ratio;

u_i are components of the displacements vector;

σ_{ij} are the components of the stresses tensor.

To simplify the notation we omit index j for ξ,ζ,z. variables.

We have for the function w_i [1,2]

$$w_i = A_i^j Ln\frac{z-l_j}{z} \qquad (5)$$

where

$$A_i^j = \frac{2Gb_i^j}{\pi(\chi+1)}, \quad (i=1,2,), \qquad A_3^j = \frac{b_3^j}{2\pi}$$

Substituting (5) into (2)-(4) it is possible to determine both components of the displacements vector and components of the stresses tensor generated by the j-th rupture in the coordinate system connected with the rupture. Therefore, the stresses can be represented as follows:

$$\sigma = (\sigma(\xi - l_j,\zeta) - \sigma(\xi,\zeta)),$$

In particular, in the vicinity of the rupture tips components of the stresses tensor $\sigma(\xi,\zeta)$ are as follows:

$$\sigma_{\xi\xi} = r^{-1}[-A_1^j(2+\cos 2\phi)\sin\phi + A_2^j \cos\phi\cos 2\phi];$$

$$\sigma_{\zeta\zeta} = r^{-1}[A_1^j \sin\phi\cos 2\phi + A_2^j(2-\cos 2\phi)\cos\phi];$$

$$\sigma_{\xi\zeta} = r^{-1}\cos 2\phi[A_1^j\cos\phi + A_2^j\sin\phi];$$

$$\sigma_{\xi z} = -r^{-1}A_3^j\sin\phi; \qquad \sigma_{\zeta z} = r^{-1}A_3^j\cos\phi;$$

$$r = \sqrt{\xi^2+\zeta^2}; \qquad \phi = arctg(\zeta/\xi) \qquad (6)$$

Stresses generated by every other rupture in coordinate system connected with the rupture are determined in the same way.

In order to obtain general fields of displacements and stresses for the system consisting of N-ruptures it is better to choose some coordinate system (for example, Oxy) and rewrite the ratios obtained. Transforming from $\xi^j\zeta^j$ local coordinate systems to Oxy main system we obtain the following ratios for components of the stresses tensor [3]:

$$\sigma_{xx} = \sigma_{\xi\xi}\cos^2\alpha_j + \sigma_{\zeta\zeta}\sin^2\alpha_j - \sigma_{\xi\zeta}\sin 2\alpha_j;$$

$$\sigma_{yy} = \sigma_{\xi\xi}\sin^2\alpha_j + \sigma_{\zeta\zeta}\cos^2\alpha_j + \sigma_{\xi\zeta}\sin 2\alpha_j;$$

$$\sigma_{xy} = \frac{1}{2}(\sigma_{\xi\xi} - \sigma_{\zeta\zeta})\sin 2\alpha_j + \sigma_{\xi\zeta}\cos 2\alpha_j;$$

$$\sigma_{xz} = \sigma_{\xi z}\cos\alpha_j - \sigma_{\zeta z}\sin\alpha_j;$$

$$\sigma_{yz} = \sigma_{\zeta z}\cos\alpha_j + \sigma_{\xi z}\sin\alpha_j \qquad (7)$$

Components of the general field of stresses are the sum of components of the stresses generated by every rupture and determined by the given ratios.

For the prediction of earthquake sequences in zones of active tectonic faults it is necessary to formulate criteria of the beginning of rupture tips growth and determine direction of their moving.

Let us determine conditions under which the ruptures can grow. We can use method of invariant Γ-integrals developed in [4] for continuum with arbitrary geological and electromagnetic properties. In the present case let us assume that external tectonic fields of stresses σ_{ij}° are applied at infinity. Besides, every rupture generates stresses in the vicinity of all other ruptures tips which can be determined by (2)-(5) with use of transformation of coordinates.

Let σ_{ij}^{1} be the summary stresses in the vicinity of tips of the j-th rupture generated by all other ruptures. Then we get the following ratios for Γ-integrals in every tip of the dislocation ruptures:

$$\Gamma_{\xi}^{j} = b_1^j(\sigma_{\xi\zeta}^{\circ} + \sigma_{\xi\zeta}^{1} - \frac{A_1^j}{l_j}) + b_2^j(\sigma_{\zeta\zeta}^{\circ} + \sigma_{\zeta\zeta}^{1} - \frac{A_2^j}{l_j}) + b_3^j(\sigma_{\zeta z}^{\circ} + \sigma_{\zeta z}^{1} - \frac{A_3^j}{l_j}$$

$$\Gamma_{\zeta}^{j} = -b_1^j(\sigma_{\xi\xi}^{\circ} + \sigma_{\xi\xi}^{1} - \frac{A_2^j}{l_j}) - b_2^j(\sigma_{\xi\zeta}^{\circ} + \sigma_{\xi\zeta}^{1} - \frac{A_1^j}{l_j}) - b_3^j(\sigma_{\xi z}^{\circ} + \sigma_{\xi z}^{1} \qquad (8)$$

Index j means that we consider the j-th rupture; $\xi^j\zeta^j$ is the coordinate system connected with the j-th rupture. In (8) notation we omit index j at $\xi\ \zeta$ variables.

According to the general theory of motion of the displacement discontinuity surfaces in solids there is a certain critical value Γ_c characterising the beginning of motion of the rupture surface.

Rupture growth begins when Γ-integral value becomes equal to some value Γ_c Then criteria of the beginning of motion of the rupture surface in vicinity of tectonic faults might be represented as follows:

Rupture tip is immovable when $|\Gamma| < \Gamma_c$;

Rupture tip is movable when $|\Gamma| = \Gamma_c$;

where:

$$\Gamma = \Gamma_{\xi} \cos\Theta + \Gamma_{\zeta} \sin\Theta, \Theta = arctg(\Gamma_{\xi} / \Gamma_{\xi}) \qquad (9)$$

In general case the value Γ_c should be determined experimentally and considered as a function of the rupture tips location, rupture plane orientation and velocity of the rupture motion.

Direction of the rupture tip growth is expressed by the ratio:

$$\Theta^{j} = arctg\,(\Gamma_{\zeta}^{j} / \Gamma_{\xi}^{j}) \qquad (10)$$

3 RESULTS

Computational results are represented by the Figures 2, 3 as isolines of the maximum tangent stresses constructed with use of the solution obtained for two (Fig. 2) and three (Fig.3) ruptures.

For numerical calculation there were chosen the following parameters:

-ruptures length - 100 m;

-components of final dislocation on every rupture:

$b_1 = 1$ m, $b_2 = b_3 = 0$.

The following dimensionless values of stresses (σ/G, G is the modulus of rigidity) correspond to the isolines denoted on the figures by digits from 1 to 5:

1-0.00011; 2-0.00033; 3-0.00055; 4-0.00094;

5-0.00381 for two ruptures on Figure 2;

1-0.00025; 2-0.00047; 3-0.00071; 4-0.00134;

5-0.00544 for three ruptures on Figure 3.

Analyses of the results represented by Figures 2, 3 show that the isolines of stresses are situated along the ruptures. The stresses reach maximum values in the vicinity of the rupture, and the stresses considerably decrease moving away from them. Stresses generated by the system of three ruptures are greater than those for two ruptures. Note that Figures 2, 3 show stresses generated by the rupture only when the external field of stresses is not applied. In case when some external stresses are applied the general field of stresses should be presented as a sum of the given external stresses and stresses generated by the ruptures.

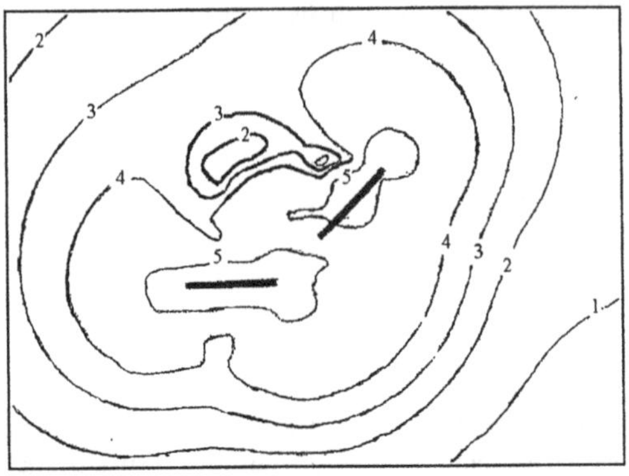

Figure 2. Isolines of the maximum tangent stresses generated by the two ruptures.

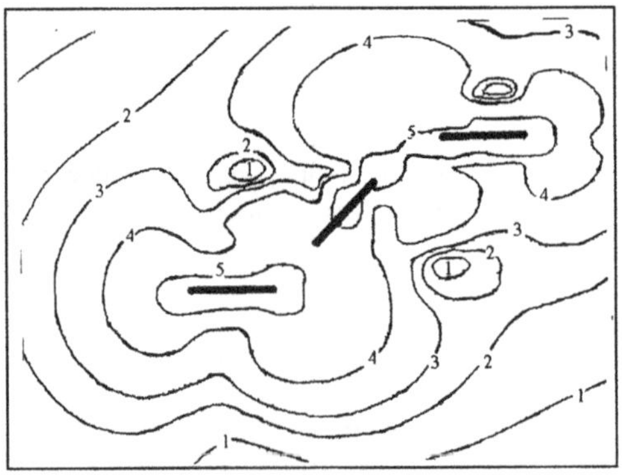

Figure 3. Isolines of the maximum tangent stresses generated by the three ruptures.

Let us investigate in more detail the case of interaction of two ruptures of shear type, (i.e. $b_1=1$, $b_2 = b_3 = 0$) arbitrary oriented to each other. This model very often takes place in theoretical seismology and tectonics when modelling process of occurrence of a new earthquake source zone. Using this model we can predict consequences of earthquake occurrence in vicinity of active tectonic faults.

Change of Γ-integral values for system consisting of two shear type ruptures arbitrary oriented to each other is shown on Figures 4,5. Curvilinears 1-4 in Figures

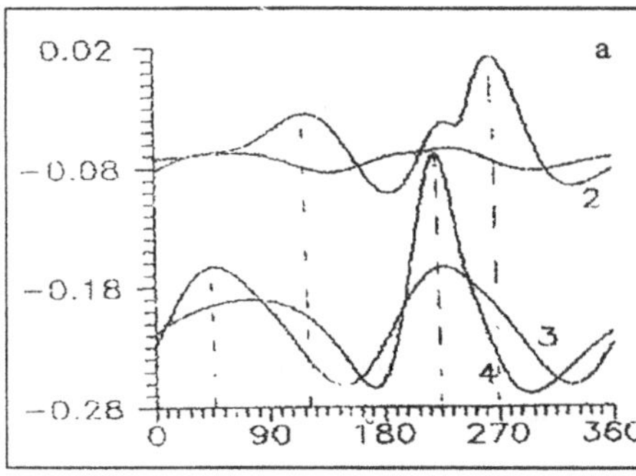

Figure 4a. Change of the general vector of **Γ**-integrals when rotating the lesser rupture around its fixed tip (3) for configuration represented on Figure 6a.

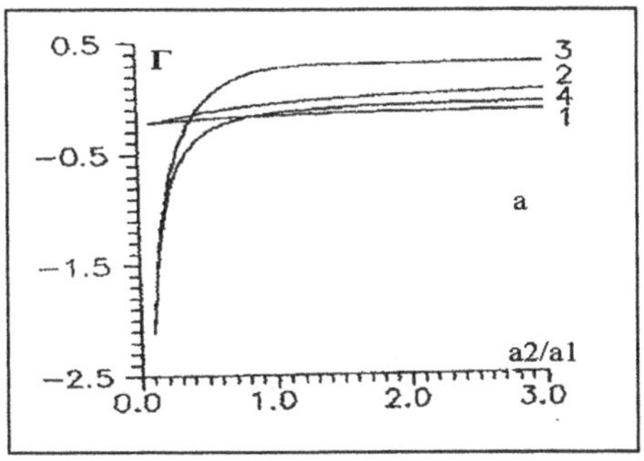

Figure 5a. Change of the general vector of **Γ**-integrals when changing length a_2 of the second rupture represented on Figure 7a.

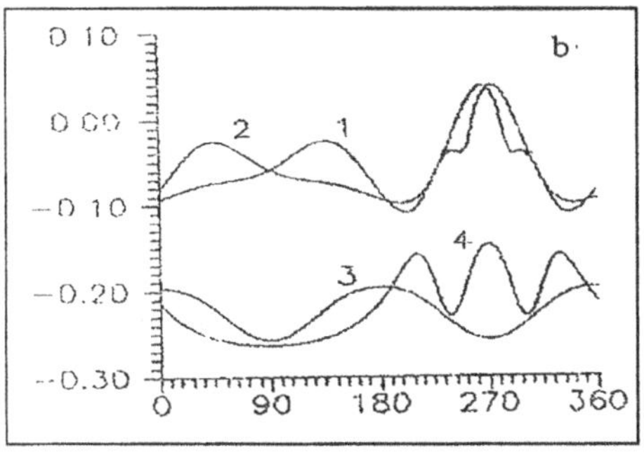

Figure 4b. Change of the general vector of **Γ**-integrals when rotating the lesser rupture around its fixed tip (3) for configuration represented on Figure 6a.

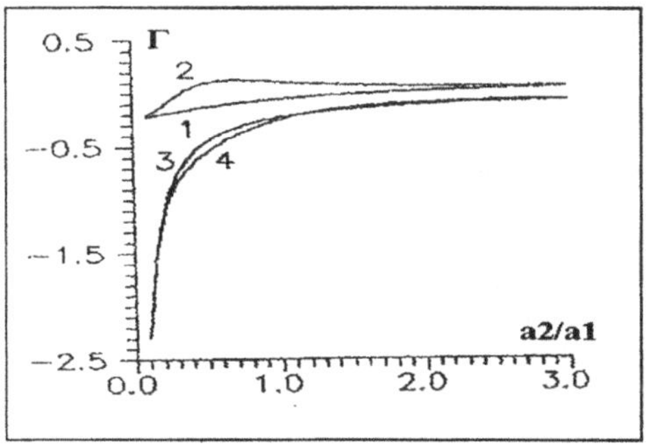

Figure 5b. Change of the general vector of **Γ**-integrals when changing length a_2 of the second rupture represented on Figure 7b.

4a, 4b correspond to **Γ**-integral values obtained for rupture tips shown as points (1)-(4) on Figures 6a6b on condition that greater rupture is immovable and lesser rupture is rotating around its fixed tip (3).

Figure 4a corresponds to rupture location shown on with Figure 6a on condition that lesser rupture is located far from the greater one. Figure 4b corresponds to rupture location shown on with Figure 6b on condition that lesser rupture is located over the greater one.

The abscess axis on Figure 4 shows the angle between the rotated rupture and the x-axis. The ordinate axis shows the Γ/G values (G is the modulus of rigidity) for four tips of two ruptures (digits (1)-(4) on Figure 6).

Figure 4 shows that **Γ**-integrals reach maximum values in tips of the greater rupture. Therefore greater rupture is more predisposed to movement than the lesser one. **Γ**-integrals reach maximum values when distance between two ruptures in the process of one rupture rotation becomes minimum.

Figure 4a analysis indicates that ruptures configuration shown on Figure 6a is characterized by 2 maximum for every rupture tip. These maximum might be reached at different angles being characterized rupture location to each to other. For example, in vicinity of rupture tip shown as (2) on Figure 6a maximum **Γ**-integral might be reached when lesser rupture is oriented at the angle of $130^0 \pm \varepsilon$ and $270^0 \pm \varepsilon$. In vicinity of rupture tip shown as (4) on Figure 6a maximum **Γ**-integral might be reached when lesser rupture is oriented at the angle of $45^0 \pm \varepsilon$ and $225^0 \pm \varepsilon$.

Figure 4b analysis indicates that ruptures configuration shown on Figure 6b is characterized by 2 maximum for **Γ**-integral curvilinears 1-2 computed in (1), (2) tips. Curvilinears 3-4 have another characteristics. We can see that ruptures in vicinity of digit (3) are more stable for all angle values in diapason of $90^0 \pm \varepsilon$ and $270^0 \pm \varepsilon$. Rupture behavior in vicinity of digit (4) is characterized by more stability when angle is variable from 0^0 to 180^0. If angle changes from 180^0 to 360^0 **Γ**-integral values reach three local maximum in which rupture

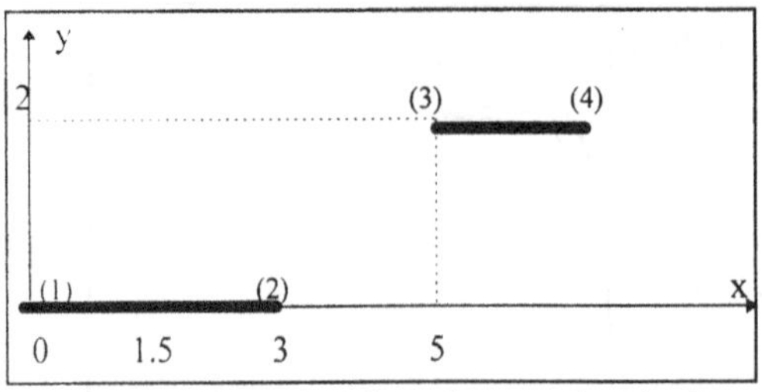

Figure 6a. Ruptures location when rotating the lesser rupture around its fixed tip (3). The lesser rupture is located far from the greater one.

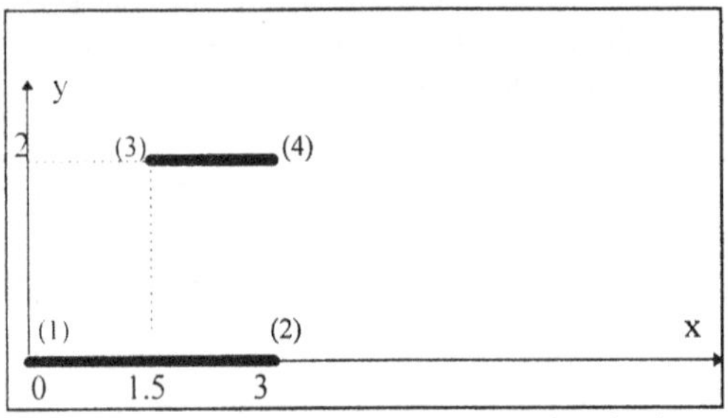

Figure 6b. Ruptures location when rotating the lesser rupture around its fixed tip (3). The lesser rupture is located over the greater one

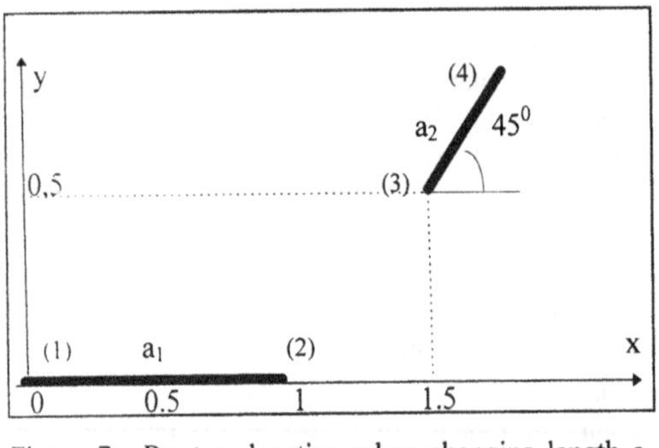

Figure 7a. Rupture location when changing length a_2. The lesser rupture is located far from the greater one.

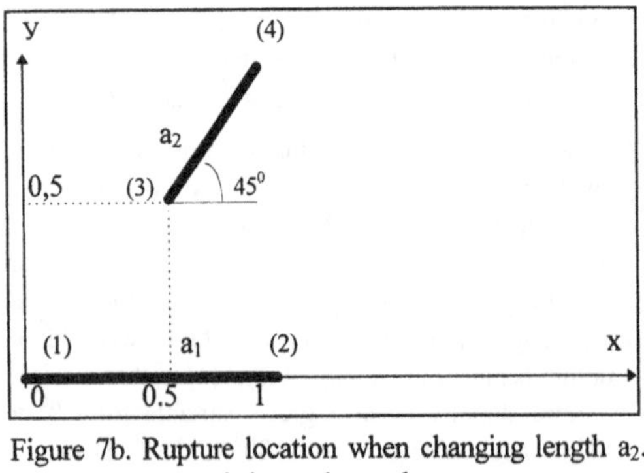

Figure 7b. Rupture location when changing length a_2. The lesser rupture is located over the greater one.

location is immovable and two local minimum which characterize the most stable position of 4th tip of lesser rupture.

For Figures 5a, b construction a system of two ruptures is considered. Location and orientation of one rupture are fixed and length of the other rupture changes (Figures 7a, b respectively). The absciss axis in Figure 5 shows values of ratio of the rupture lengths, the ordinate axis shows the Γ/G values. Curvilinears 1-4 correspond to four tips of two ruptures (digits (1)-(4) on Figure 7).

Figure 5 shows that the **Γ**-integral values depend on orientation of the rupture to each other.

4 CONCLUSIONS

From analysis of Figure 5a we can note that when one of the ruptures is not located above the other rupture (see Figure 7a) digit (3) of the greater rupture is more predisposed to movement. After digit (3) the most predisposed to the movement is digit (2), and then digit (4). As analysis shows the most stable in this system is digit (1). In case rupture is located above the other rupture (see Figure 7b) analysis of Figure 5b shows that digit (2) of the greater rupture is more predisposed to movement. After digit (2) the most predisposed to the movement is digit (1), then digit (3), and the most stable is digit (4).

REFERENCES

Bykovtsev, A.S. 1983. On the conditions of growth of two collinear dislocation ruptures. *Applied Mathematics and Mechanics (PMM),* v. 47.

Bykovtsev, A.S., Semenova, J.S. 1995. On interaction of a dislocation ruptures system // *Problems of Mechanics*. No 3-4.

Cherepanov, G.P. 1974. *Mechanics of Brittle Fracture.* Moscow: Nauka,.

Muskhelishvili, N.I. 1966 *Some Basic Problems of the Mathematical Theory of Elasticity*. Moscow: Nauka.

Mechanics of Jointed and Faulted Rock, Rossmanith (ed.) © 1998 Taylor & Francis, ISBN 90 5410 955 6

Morphotectonic indications for the opening of Davis Strait

Adrian E. Scheidegger
Section of Geophysics, Technical University, Vienna, Austria

ABSTRACT: General geological observations indicate that the southern parts of Greenland and of Baffin Island (northern Canada) are separating from each other in consequence of an ongoing opening of Davis Strait. We present additional evidence for this view from the standpoint of morphotectonics: tectonic features expressing themselves in the morphology of the area. In particular, the orientation structure of rock-joints and of the fiords in the two areas is such that both types of features can be interpreted as predesigned by the shear in the neotectonic stress field.

1 INTRODUCTION

Davis Strait is the body of water that separates the southern parts of Greenland and those of Baffin Island (Fig.1); it connects Baffin Bay with the Labrador Sea. Its origin is generally assumed to be the result of an initial rift and subsequent drift between Greenland and Eastern Canada: The system of channels that now separates the Arctic islands is evidently a drowned valley system modified by Pleistocene glaciation. The larger channels, many straight and arcuate-walled, are controlled by graben or rift-valley structures: such a structure also separates West Greenland from Baffin Island and formed Davis Strait and Baffin Bay; it is presumed to have been an offshoot of the Mid-Atlantic Ridge (Andrews et al. 1972). The timing of the initial rift has been estimated from Late Paleozoic to Early Tertiary, but since the paleomagnetic evidence is scarce, there is a considerable amount of uncertainty (Sharma and Athavale, 1975). This paper aims at presenting morphotectonic evidence in support of the above view on the genesis of Davis Strait.

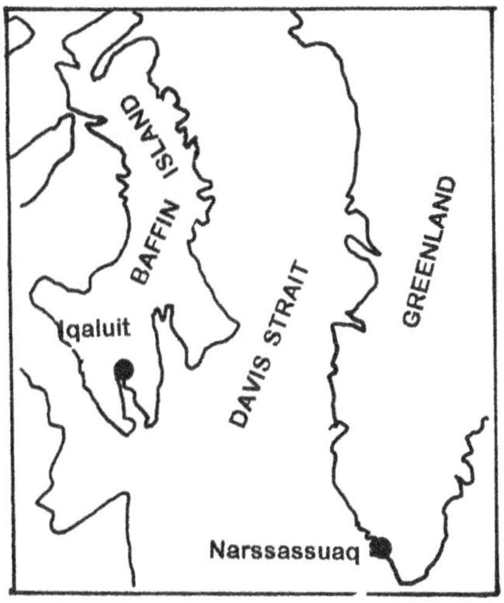

Fig.1. The North American Arctic showing the relevant parts of Greenland and Baffin Island

2 BACKGROUND

The morphotectonics of an area can best be assessed by making studies of the orientation patterns of joints and topographic features.

For joints, the orientation data are treated by the statistical method of Kohlbeck and Scheidegger (1977): In order to determine the prevalent directions in joint systems, it is assumed that the polar axes representing joints are a priori distributed according to so-called 'Dimroth (1963) - Watson (1966)' probability distributions, which, on a sphere, correspond to Gaussian distributions on a plane. Applied to a practical problem, the procedure consists in the determination of the 'best-fitting' parameters of the theoretical distribution(s) by minimizing the mean-square deviation sum of predicted versus measured values by an iterative procedure. For a joint system representing two conjugate joint sets (i.e. forming a grid), two a priori Dimroth-Watson distributions are assumed and the best-fitting parameters (representing joint set 1 and joint set 2) are determined by computer; a similar procedure applies for three assumed joints sets.

The calculations are performed using polar or dip-axes, but for geological purposes the strike/trend values or strike/trend-rose diagrams are more convenient. Azimuths are given in degrees N>E. The bisectrices of two conjugate joint sets are the principal stress directions of the stress field (P= compression; T= tension) that caused them. This kind of "parametric" statistics is not free of a-priori assumptions, viz. of the assumption at of the *existence of a set number* of Dimroth-Watson distributions (and therefore of the existence of a set number of parameters for describing the latter). Non-parametric statistics can, on occasion, be more useful: the directions of the (absolute) maxima on a direction-rose are independent of any a-priori assumptions.

For other topographic features, any sets of linear "elements" (in our case: fiords) are treated analogously to joints, applying the Kohlbeck-Scheidegger (1977) evaluations to the polar axes of such elements, assuming a virtual "dip" of the linears of 90°. Again, the actual calculations are performed with the polar axes, but the for the final representations, strikes or trends are given. In this vein, this paper undertakes to investigate joint orientations on the margins of Davis Strait and relate them to the neotectonic environment.

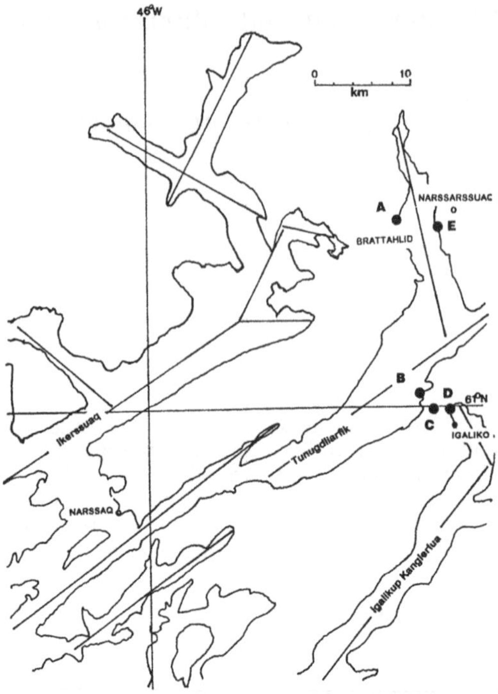

Fig.2. Detailed map of the Greenland-region around Narssarssuaq with measurement locations (A-E) shown

3 GREENLAND

3.1 General Situation

Greenland, with an area of 2,186,000 sqkm, is the largest island in the Arctic archipelago of North America; it stretches from 60°N to 84°N latitude and 10°W to 70°W longitude. Some 80% of its area are covered by the "inland ice" and peripheral glaciers, leaving only 383,600 sqkm of accessible coastal land surface (Bondam 1975). For the present study, the South of Greenland was visited; headquarters was the town of Narssarssuaq at latitude 61°10'N and longitude 45°30'W. Outings were made to Brattahlid, Igaliko and the inland ice cap to the NE of Narssarssuaq. Fig.2 will serve to identify these localities.

3.2 Geology

Very little is known about the rocks underneath the 80% of Greenland that is covered by ice. Most of the knowlege of Greenland geology derives from a study of its coasts and extrapolations therefrom. Thus, the greater part of Greenland seems to be an extension of the Precambrian Shield of North America along its northern and eastern margins. This shield area is bounded by Paleozoic fold belts. The Precambrian basement consists of three major units: on either side of the central Archean block, two younger mobile belts of similar age have been recognized (Bondam 1975; Langel and Thorning 1982).

The central (Archean) block extends on the West coast (which mainly concerns us in this paper) from 61°45'N to 66°30'N. The dominant rocks are granulite or amphibolite facies gneisses of grano-dioritic to quartz-dioritic composition with several successions of metasedimentary and metavolcanic rocks. To the N of Qaqortoq [formerly Julianehaab] (60°34'N), i.e. in our area, the gneisses and supracrustals are cut off by an extensive areas of granite and ganodiorite which was emplaced during several different periods beginning about 2600 Ma ago. The active period ended with the intrusion of post-kinematic alkaline rocks (norite, diorite, granite), ending about 1780 Ma ago.

At their southern margin, the gneisses are unconformably overlain by a thick sequence of virtually unmetamorphosed sedimentary and volcanic rock suites, with ages ranging from 1850 to 1740 Ma. After the plutonic events around 1700 Ma ago, southern Greenland was deeply eroded before the beginning 1200 Ma ago of extensive basaltic volcanism, forming the Gardar volcanic province, accompanied by faulting and dyke intrusion. The southern area is known as the Ketilidian Mobile Belt.

North of the Archean rocks lies the second, Nagssugtoqidian mobile belt, with the deformation taking place in several episodes between 1700 and 2700 Ma ago. This deformation is related to strongly horizontal movements in shear belts.

Finally, sedimentation of both, continental and marine character continued into the Mesozoic; major tectonic activity occurred with N-S rifting and NW-SE fractures and faults (geological information after Bondam 1975 and after Langel & Thorning 1982).

3.3 Joint Orientations

Joint orientations were measured at 5 locations in the general vicinity of Narssarssuaq (Locs. A-E in Fig.2):

Loc.A: Brattahlid. Crossing the Tunugdliarfik [formerly Eiriks Fiord] from Narssarssuaq, the ancient site of Brattahlid, now the Inuit (Eskimo) establishment of Qagssiarssuk, was reached; the outcropping rocks were mainly gneisses.

Loc.B: Itileq. The Tunugdliarfik was followed by boat toward its mouth as far as the landing place "B" on the map in Fig.2. Rock outcrops were again gneisses.

Loc.C: Kongevejen. From Itileq, a footpath (Kongevejen, "King's Path") leads across the isthmus between the Tunugdliarfik and the Igalikup Kangerdlua [formerly Einars (Igaliko) Fiord]. On the isthmus, an outcrop showed the usual Precambrian rocks.

Loc.D: Igaliko. The Kongevejen ends at the Inuit town of Igaliko where outcops of schists are in evidence.

Loc.E: Narssarssuaq. Finally, the road from the hotel to the airport of Narssarssuaq skirts many outcrops of gneisses.

The measurements of the locations A-E (all) were then evaluated according to the ususal Kohlbeck-Scheidegger (1977) method. The results are shown in Table 1; Fig.4a gives the corresponding strike-rose diagram.

3.4 Morphology

Quaternary glaciation lead to the formation of the Greenland Ice Cap 2-4 Ma ago. Advances to the coast took place during the ice ages; since the last ice age there has been an isostatic uplift of some 150m. The local morphology has been greatly afffected by glacial action, which resulted in the formation of large fiords which show a grid-like pattern. In order to confirm this visual impression, we have digitized the median lines of the fiords shown in Fig.2 in steps of 2.5 km and treated the orientation data by the Kohlbeck-Scheidegger method. The results of this procedure is shown on Table 1; Fig. 4c shows the trend-rose diagram. It is seen that the values of the fiord-directions are very close (within 4°) to those of the joint-directions: The fiords as well as the joints strike NW-SE and NE-SW and the

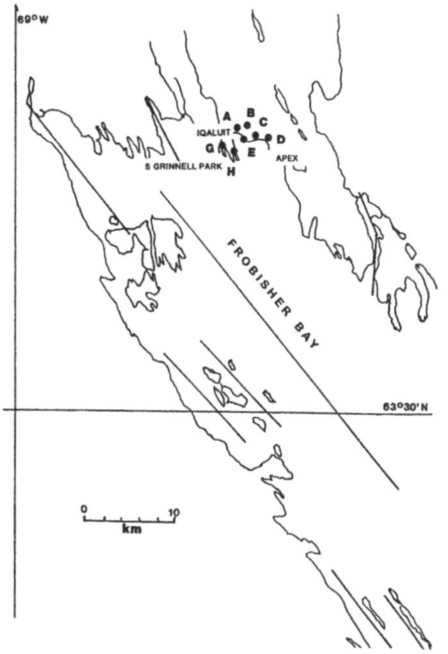

Fig.3. Detailed map of the Baffin-Island-region around Iqaluit with measurement locations (A-H) shown

Table 1: Joint-strike and fiord-trend evaluations for Baffin Island and Greenland

Loc.	No.	Max.1	Max.2	Ang.	P	T
GREENLAND						
Joints	102	140±12	47±05	88	3	93
Fiords	91	136±10	49±02	87	3	93
BAFFIN ISLAND						
Joints						
2 distr.	133	132±00	51±07	81	91	2
(1,2) ¢ 3		136±10	43±10	87	180	90
Fiords						
1 distr	46	140±03				

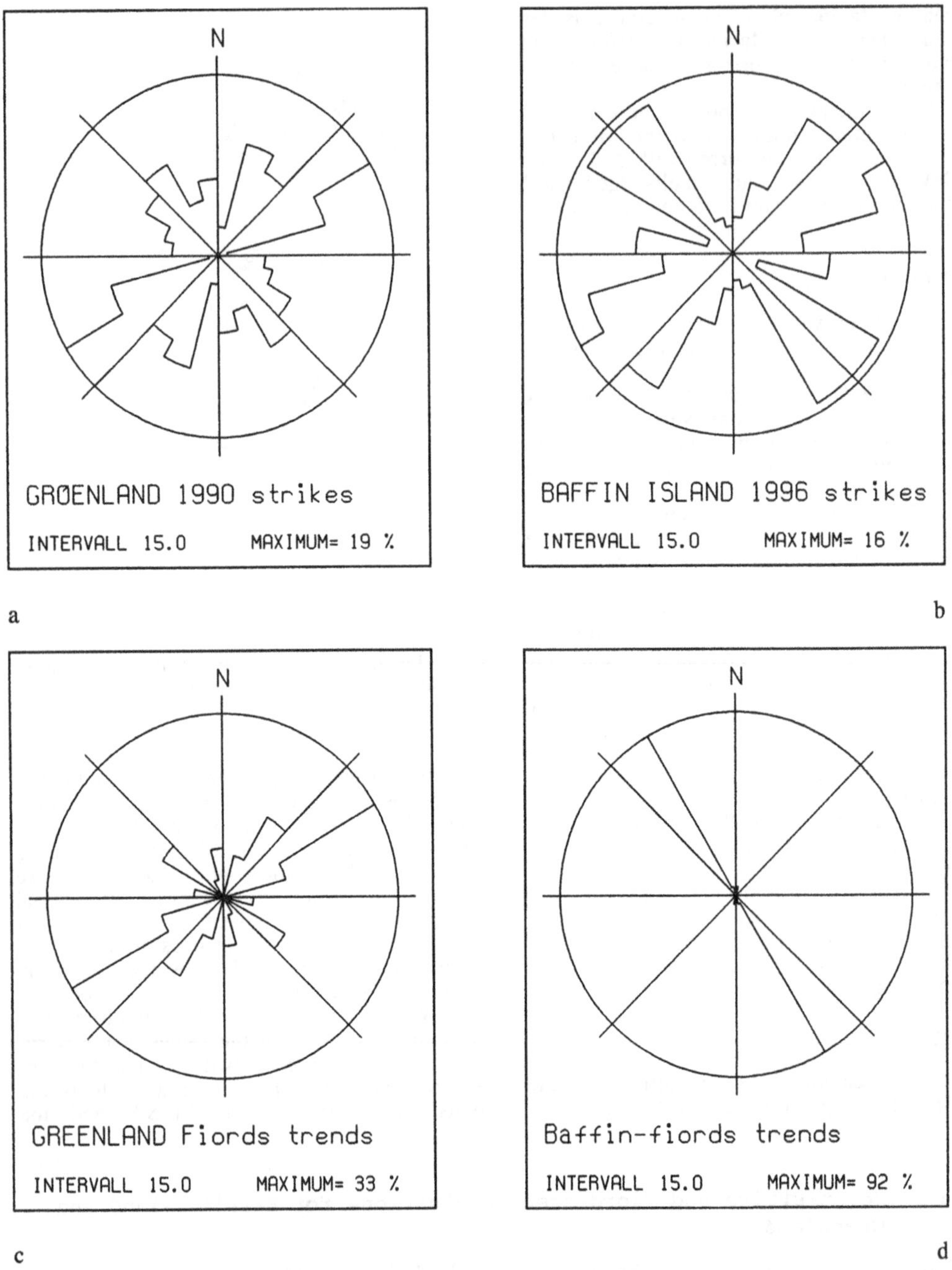

Fig.4. Strike/trend diagrams (a,b) of joints and (c,d) of fiords as indicated

corresponding principal stress directions trend NS and EW, where EW is the "tension", as corresponds to an "opening" of Davis Strait.

4 BAFFIN ISLAND

4.1 General Situation

Baffin Island is the south-easternmost island of the Canadian Arctic Archipelago. It comprises an area of 474,000 sqkm, has high (>2000 m) glaciated mountains in the North and a plateau (with tundra vegetation) in the South. For the present study, the area around its capital, Iqaluit (formerly Frobisher Bay), was visited; it lies on Frobisher Bay at a latitude of 63°45'N and a longitude of 68°31'W, about 2000 km north of Montréal. Studies were made in the closer environment of Frobisher Bay: Sylvia Grinnell Park, Apex and the Hills around (cf. Fig.3).

4.2 Geology

In a general way, the oldest rocks of the Canadian Arctic Islands lie to the SE, and the age of exposed rocks decreases towards the NW. The Canadian Shield of Precambrian metamorphic rocks forms a basement complex along the southern edges of the archipelago: This includes the eastern and southern parts of Baffin Island which belong to the Churchill Province of the Shield, consisting of Archean and early Proterozoic sedimentary and plutonic rocks, mainly metamorphosed and converted to granitic gneisses during the Hudsonian orogeny 1.65-1.85 Gy ago (Christie, 1975).

4.3 Joint Orientations

Joint orientations were measurements made at 7 locations around Iqaluit (cf. Fig.3), all in the banded granitic gneisses of the Churchill Province of the Canadian Shield:

Loc.A in the center of Iqaluit, near the Arctic Ventures store

Loc.B behind the Nunatta College building

Loc.C at the lookout near the Canadian Fisheries Building

Loc.D on the Apex Road

Loc.E at the cemetary

Loc.G at the lookout of Sylvia Grinnell Park

Loc.H near the picknic shelter of Sylvia Grinnell Park.

The data were evaluated according to the method of Kohlbeck and Scheidegger (1977). Fig.4b shows the corresponding strike-rose diagram. We have summarized the main results in Table 1, together with those of W Greenland.

4.4 Morphology

The morphology around Iqaluit is characteristic of the Canadian Shield under arctic conditions. The most impressive feature around Iqaluit are the fiords including the fiord-like Frobisher Bay: The tidal range is very high (ca. 20m sea-level change); thus, at low tide, the bay is practically dry; at high tide it is filled with water. Of much interest is also Sylvia Grinnell Park: here, the typical puddles and tundra-covered hummocks of the Canadian Shield under arctic conditions are prominent. The Sylvia Grinnell River forms two water-falls before it empties through Koojenessee Inlet into the sea. The fiord/river systems have a characteristic parallel-like pattern. It is evident from the map in Fig.3 that one of the joint sets found strikes along the trend of the local fiords, including Frobisher Bay itself. In order to confirm this impression, we have digitized the median lines of the fiords/bays shown in Fig.3 in steps of 2.5 km and made an orientation study by the Kohlbeck-Scheidegger (1977) method (assuming only 1 a priori distribution). The result is also shown in Table 1; Fig 4d shows the corresponding trend-rose diagram: the one significant maximum lies within 4° of one of the joint maxima. There is therefore every indication that these fiords have been tectonically predesigned.

5 COMPARISON ACROSS DAVIS STRAIT

One sees at once that the results for Greenland and Baffin Island are practically identical, except that, for 2 distributions in Baffin Island, the directions of P and T are reversed. This, however, is a common possible error inasmuch as the identification of which of the principal stress directions is the compression direction, depends by Mohr's criterion on the angle between conjugate joint sets and is therefore quite uncertain. If the calculation is made for Baffin Island for 3 assumed joint sets, and the third set is disregarded, then the P and T directions are reversed again, i.e. they correspond to those found in Greenland. This would also exactly correspond to the view that there is ongoing rifting between Greenland and Baffin Island across Davis Strait, i.e. a stress relief in the E-W direction; this corresponds to a T direction also trending E-W. This rift would be the result of a relative rotation of Greenland and Baffin Island with the North Pole as center.

REFERENCES

Andrews, J.T., Buennell, G.K., Wray, J.L., & J.D.Ives 1972. An early Tertiary outcrop in north-central Baffin Island, N.W.T., Canada: environment and significance. Canad. J. Earth Sci. 9:233-238

Bondam, J. 1975. Greenland, *in* Encyclopedia of World Regional Geology, Pt.1: Western Hemisphere; ed. R.W.Fairbridge. Stroudsburg, Pa., Dowden, Hutchinson and Ross, p.292-300

Christie, R.L. 1975. Canada - Arctic Archipelago. Encyclop. World Regional Geol. (ed. R. W. Fairbridge, Dowden, Hutchinson and Ross, Stroudsburg Pa.) 1:145-156

Dimroth, E. 1963. Fortschritte der Gefügestatistik. Neues Jahrbuch der Mineralogie, Monatshefte 1963:186-192.

Kohlbeck, F.K. & A.E. Scheidegger. On the theory of the evaluation of joint-orientation measurements. Rock Mechanics 9: 9-25

Langel, R.A. & L.Thorning 1982. A satellite magnetic anomaly map of Greenland. Geophys. J. Roy.Astr,Soc., 71:599-612
Sharma, F.V. & R.N. Athavale 1975. Paleomagnetic evidence relating to the Cenozoic drift of Greenland. Tectonophysics 29: 209-221
Sörensen, K. 1983. Growth and dynamics of the Nordre Strömfjord shear zone. J.Geophys.Res. 88(B4): 3419-3473
Watson, G. S., 1966: The statistics of orientation data. J.Geology 74:786-797.

Mechanics of Jointed and Faulted Rock, Rossmanith (ed.) © 1998 Taylor & Francis, ISBN 90 5410 955 6

Neotectonic synthesis of the mediterranean Languedoc region, Southern France

F.A. Audemard M.
FUNVISIS, Caracas, Venezuela

ABSTRACT: Paradoxically to the very low seismic activity of the mediterranean Languedoc region in southern France, many references of neotectonic indices in this region are reported in the literature. The critical evaluation of them has allowed to produce a first synthetic and comprehensive compilation of such young, potentially-active tectonic structures, aiming at detecting future research fields. This compilation has revealed that this region has undergone two successive tectonic regimes in Neogene times, shifting from a NW-SE extension to a N-S compression, sometimes between the middle Miocene and the early Pliocene. This shift and the present compressional regime need to be related to the collisional stage between Africa and Western Europe since the late Pliocene. This neotectonic (Quaternary geology) approach is the only reliable means of assessing seismic hazard in regions of very low seismic activity, where most of the few epicentral locations do not clearly correlate with suspected-active major faults.

1 INTRODUCTION

The mediterranean Languedoc region in southern France is characterized by a very low contemporary seismicity -only a dozen of earthquakes have been felt in the last 250 years-, but several researchers have been paradoxically very much interested in the Neogene and/or Quaternary tectonic indices of this region since the 1940's. Due to its very low seismic activity, the seismic hazard assessment of this region can basically be approached by means of classical geomorphological and geological studies in order to identify the active tectonic features and characterize their seismic potential. In this paper, a first synthetic compilation of such young (Neogene and Quaternary), potentially active tectonic features is presented, based on the critical evaluation of many references of neotectonic indices existing in the french literature.

2 STUDY AREA

This study focuses in a region of southern France, located between the Massif Central and the Mediterranean sea, comprising the Languedoc "garrigues" and low plains, the eastern termination of the Eastern Pyrenees, the Montaigne Noire, the Saint-Chinian range and the Lodève basin.

3 STRUCTURAL FRAMEWORK

Two large lithological units with different mechanical behaviours can be distinguished in the Languedoc region: a Paleozoic basement deformed by the Hercynian and late-Hercynian orogeny and an unconformable Mesozoic sedimentary cover that in turn is locally overlain by Tertiary sediments. This Mesozoic cover is affected by the Pyrenean orogeny during the early Tertiary (Paleocene-Eocene), producing NE-SW folds -such as the Montpellier fold and the Saint-Chinian high-, and reactivating left-laterally pre-existing NE-SW faults such as: the Cévennes, Matelles-Corconne and Nîmes faults. Then, this region is affected by an Oligocene NW-SE trending extension (Arthaud et al. 1980-81, Etchecopar et al. 1981, Taha 1986, Obid 1987) responsible for reactivation as normal faults of those NE-SW striking faults that bound newly-formed half-grabens.

Therefore, it is necessary to keep in mind that the basement of this region has already undergone a polyphasic tectonic evolution when assessing its Neogene history. In turn, this Neogene evolution of the Languedoc region is also polyphasic since: a)- several authors (Schwobthaler & Vogt 1955, Ellenberger 1961, Bonnet 1974) describe NE-SW faults cutting the Miocene transgressive deposits; and Combes (1981) and Phan Trong (1989) confirm the occurrence of such Miocene NW-SE extension west of Montpellier by means of microtectonic

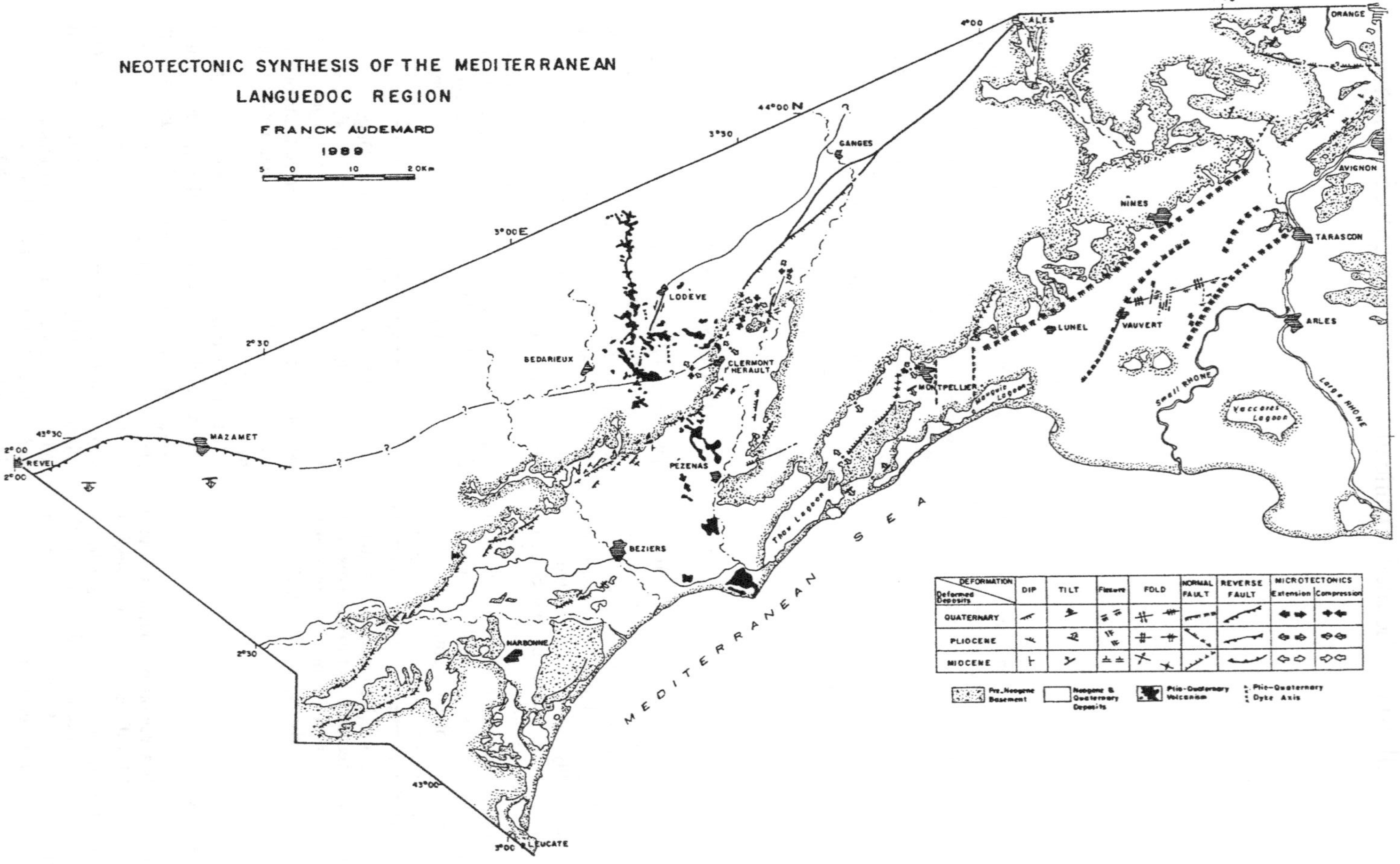

FIGURE 1 Neotectonic features of the Languedoc region, Southern France (after Audemard, 1989)

analysis; and b)- Bousquet & Philip (1976) and Philip (1987) establish that southern France is undergoing a N-S compression and/or E-W extension nowadays, which Tapponnier (1977) considers as a straightforward consequence of the locking of the western Mediterranean subduction, implying a collisional stage between Africa and Europe during the Plio-Quaternary.

4 NEOTECTONIC INDICES

The critical evaluation of the relevant neotectonic references regarding the Languedoc region has allowed to gather the structures in three major trends (Fig. 1): a)- a very conspicuous NE-SW striking fault set, such as the Cévennes and Nîmes faults and the Cévenole flexure between Lieuran Cabrières and Fabrezan (SW of Clermont l'Herault), b)- an E-W trending set, comprising the Mazamet reverse fault along the northern edge of the Montaigne Noire and the Vauvert (Costières) anticline located south of Nîmes, and c)- a N-S trending set of structures that comprises the Plio-Quaternary fissural volcanism and several small normal faults, all located in the surrounds of Montpellier, such as: Clamouse, La Mosson, Lez and Saint Aunès-Carnon faults. Because of lack of space, this paper only discusses one structure per trend (for more details, refer to Audemard 1989).

4.1 *The Nîmes fault*

The Nîmes fault is one of the major features of the post-Pyrenean orogeny in the Languedoc region. This NE-SW striking fault can be followed between the cities of Orange and Montpellier in the study area (Fig. 1), but it extends beyond the Rhone river towards the northeast.

In landspace, the Nîmes fault is revealed by the linear aspect of the southern edge of the Cretaceous "garrigues" between Remoulins and near Sommières (Bonnet 1962). Along some sectors (in the Uzés region), the fault shows a conspicuous SE-looking scarp that puts in contact the Cretaceous sedimentary rocks with the Pliocene deposits. However, this scarp does not seem of tectonic origin because it has been deeply attacked by *lithophaga*, implying that it corresponded to a Pliocene sea cliff. Moreover, this scarp is drowned by Quaternary deposits further north, meaning that the fault should be in a more southern position if it happens to be active in this sector. Nevertheless, several evidences of tectonic activity of this Neogene feature, as a down-to-the-south "normal" fault, are reported in the literature at several localities: a)- the Miocene deposits are tilted towards the fault (to the NW) close to Sernhac (Schwobthaler & Vogt 1955); b)- a road cut of the Lédenon-Sernhac road (D223) exposes the fault cutting the late Pliocene (Astian) which is tilted 40°N (Bonnet 1974, BRGM's Geologic map of Nîmes -scale 1:50,000-); c)- the Astian-Plaisancian contact is at 25 m in elevation at the Nîmes Hospital whereas it is 10 m deep in the Milhaud-4 well (in the Vistrenque), implying a 40 m throw between the two blocks (Bonnet 1974); d)- 40 m of vertical throw of the Villafranchian –early Pleistocene- conglomerates between the elevated patches located at the towns of Mus, Ballargues and Mas de Belle-Vue and the equivalent deposits observed in the subsurface of the Lunel plain (Bonnet 1962); and e)- Schowbthaler & Vogt (1955) mention that the Rhodanian conglomerates are capping the "garrigues" at 200 m in elevation whereas they are never above 140 m high in the Costières du Gard.

From the above-mentioned evidences, there is no doubt that the Nîmes fault has been active along part of its trace since the Miocene with a persistent down-to-the-south movement. However, if this region is undergoing a N-S compression today, as proposed by Bousquet & Philip (1976), Philip (1987) and Audemard (1989), this fault should present left-lateral and reverse components. A trench excavated across the eastern segment of this fault at Courthézon –out of the study area– (Chateauneuf-du-Pape area, east of the Rhone river), has proved the occurrence of south-vergent reverse slip –consistent with the down-to-the-south component observed in the study area– on this fault during the late Pleistocene (Carbon et al. 1993, Grellet et al. 1993).

4.2 *The Montaigne Noire and Mazamet fault*

Many authors have focused their attention on the Montaigne Noire and its northern edge, and their most relevant neotectonic observations are: a)- the present asymmetric cross profile of the Montaigne Noire relief: a south gentle slope (6 to 8%) against a north steep one (David 1920, Brum Ferreira 1975); b)- the presence of faceted spurs dipping 30°N along some sectors of the northern edge of this relief (David 1920); c)- a sharp change in topography and vegetation, associated with several natural water springs at the foot of the northern edge of the Montaigne Noire (Ellenberger 1938); d)- the rather straight trace of the northern edge that suggests the presence of a steeply-dipping underlying structure; e)- the south steep dip of this fault was observed at the Aupillac gully by Ellenberger (1938) who interpreted this northern boundary as a reverse fault (Mazamet fault), f)- the contact between the Paleozoic rocks of the Montaigne Noire on the south and the Tertiary rocks of the Castres basin on the north is mainly masked by Quaternary colluvial and/or alluvial deposits, but several authors report a fault breccia overlain by the gneissic rocks of the Montaigne Noire at several sites (Ellenberger 1938, Birot et al. 1968, Mouline et al. ?); which is Quaternary in age (Mindel or Riss), based on palynological content (Birot et al. 1968, Mouline et al. ?); g)- the arcuate shape (concave to the south) of the Mazamet fault trace along the northern edge of the Montaigne Noire (Fig. 1) is another argument in favor of a south dipping, north-vergent reverse fault, thus sharing the hypothesis proposed by Ellenberger (1938), Brum Ferreira, 1975 and Mouline et al. (?) that the Montaigne Noire is thrusted northwards onto the Castres basin. Such thrusting is Quaternary in age (based on the palynological age of breccia) and may be responsible for the v-shaped cross-profile of some valleys on the northern slope of the Montaigne Noire. The Quaternary activity of this structure requires a N-S compression.

4.3 *The Fissural volcanism in Languedoc*

The Languedoc landscape is dominated by volcanic plateaus (Fig. 1). This volcanism is characterized by: a)- volcanic centers that are roughly aligned N-S (Gastaud et al. 1983); b)- dyke swarms within the Lodève basin that present two main trends: N150-170° and N0-030°; orientations that coincide with the post-Miocene maximum horizontal stress directions obtained by Combes (1981) in the Arboras-Clermont l'Herault region; c)- an age variable in space, but that is always within the late Pliocene-Quaternary time span (Gillot 1972, Ildefonse 1972, Gastaud et al. 1983, among others). Besides, small conjugated reverse faults are reported by Combes (1981) in the volcano-sedimentary tuff cropping out at Rièges, that are generated by a compression oriented between N100° and 190°.

If dyke opening propagates in the direction of the maximum stress, as proposed by Anderson (1951), it may be concluded that this region has been undergoing a roughly N-S compression since the late Pliocene.

5 RESULTS

The critical evaluation of many references of neotectonic indices existing in the literature has made possible a first synthetic compilation of such young –Neogene and Quaternary-, potentially active tectonic features. The main results from this critical compilation are:

5.1 *Consistency of microtectonic data*

Although the small number of stress tensors calculated from microtectonic analyses of brittle deformations in Neogene or Quaternary sedimentary units of this region –either by other researchers or by the author-, two regionally-coherent stress tensors arise:

1. an older tensor characterized by a minimum horizontal stress oriented NW-SE (extensional regime);
2. a more recent tensor characterized by a N-S trending maximum horizontal stress (compressional regime).

Relative chronology of stress fields is established by direct observation of cross-cutting relationship between different fault striation generations on faut planes.

5.2 *Coherence of regional-scale structures*

Two separate sets of structures are observed:

1. a set of NE-SW striking normal faults or flexures, bounding the Neogene Languedoc basin on the north, such as the Cévennes and Nîmes faults and the Cévenole flexure between Lieuran Cabrières and Fabrezan (SW of Clermont l'Herault);
2. another group of structures gathered in two orthogonal orientations:

- E-W trending thrust faults (Mazamet fault on the northern edge of Montaigne Noire) and folds (Costières-du-Gard anticline) and
- N-S trending extensional features: late Pliocene-Quaternary fissural volcanism of the Languedoc, and normal faults such as: Clamouse, la Mosson, Lez and Saint-Aunès-Carnon.

5.3 *Accordance between microtectonic data and related large-scale structures*

There is a perfect agreement between calculated stress tensors from microtectonic data collected in this region and spatial configuration of large-scale tectonic features. This allows us to postulate that:

1. a Miocene (or slightly younger) extensional regime, characterized by a NW-SE trending minimum horizontal stress, is responsible for the activity of the set of NE-SW striking normal faults that bound the Neogene Languedoc basin mainly on the north;
2. the prior regime is later replaced by a compressional one characterized by a N-S trending maximum horizontal stress and/or an E-W trending minimum horizontal stress. This stress tensor is still active, as suggested by the recent age of the fissural volcanism of L'Escandorgue, and is responsible for the abovementioned volcanism and the potential activity of N-S striking normal faults (Clamouse, Mosson, Lez and Saint-Aunès-Carnon) and E-W trending compressional features such as the Mazamet fault and Costières-du-Gard (Vauvert) anticline.

5.4 *Age of shifting from extensional to compressional regime*

In the southernmost Languedoc region, the shift from one regime to the other is not well constrained but it might have happened between middle Miocene and early Pliocene. This regime change should be related to the beginning of the collision between northern Africa and western Europe (Tapponnier 1977).

5.5 *Reactivation of pre-existing tectonic features*

It is highly probable that some faults generated or activated during the extensional regime could have been reactivated during the later compressional regime if their orientation favored it. For instance, the NE-SW striking Nîmes fault, that moved as a down-to-the-south normal fault during the Miocene extensional regime, may have been reactivated as a left-lateral reverse fault during the present stress field. Paleoseismic trenching performed by a french team across the easternmost segment of this fault at Courthézon, east of the Rhone river, has revealed that this fault has moved as a north-upthrown low-angle reverse fault in the late Pleistocene. Such slip is in perfect agreement with the present N-S maximum horizontal stress proposed for Southern France and other regions of the Northern Mediterranean realm.

RECOMMENDATIONS

This region, although it has interested many researchers from the beginning of this century, still remains badly understood in regard of its Quaternary tectonic history. Most likely-active tectonic features in this region require more detailed and comprehensive studies, among which we could mention: seismic profiling across faults and folds, paleoseismic trenching across faults such as the Mazamet and Nîmes faults, likewise that the one already dug at Courthézon (Chateauneuf-du-Pape area, east of the Rhone river); in combination with a thorough microtectonic evaluation of the entire region. In this regard, a big step forward was made with Rebaï (1988)'s work.

ACKNOWLEDGEMENTS

This work was carried out at the former laboratory of Structural Geology of the Université des Sciences et Techniques du Languedoc (U.S.T.L.), now Université de Montpellier II (Montpellier, France), in 1989 as part of my postgraduate studies. Therefore, I would like to thank Dr. Maurice Mattauer for receiving me at his laboratory. This study was carried out under the guidance of Dr. Jean-Claude Bousquet and Dr. Hervé Philip, whom I am very much indebted. Besides, my thanks also go to the latter one and to Dr. Philippe Combes for making most references available.

REFERENCES

Anderson, E. 1951. *The Dynamics of faulting and dyke formation with application in Britain.* 2° ed., Edimburg: Oliver & Boyd. 206 pp.

Arthaud, F., M. Ogier & M. Séguret. 1980-81. Géologie et géophysique du golfe du Lion et de sa bordure nord. *Bull. BRGM.* 1(3):175-193.

Audemard, F.A. 1989. Néotectonique du Languedoc mediterranéen: examen critique et synthèse des données existantes. *Mémoire D.E.A Université des Sciences et Techniques du Languedoc, Montpellier, France.* 60 pp + appendices.

Birot, P. et al. 1968. Néotectonique sur le versant Nord-Ouest de la Montaigne Noire. *C. R. Acad. Sci. Paris.* 267(D): 1815-1816.

Bonnet, A. 1962. Note sur la liaison entre les tectoniques superficielles et profonde de la Camargue. *Bull. Serv. Carte Géol.* 59(269): 251-259

Bonnet, A. 1974. Stratigraphie et tectonique du Plio-Quaternaire du Languedoc oriental. I les Costières du Gard et la Vistrenque. *Bull. Soc. Etn. Sc. Nat. Nîmes.* 54: 7-34.

Bousquet, J.C. & H. Philip. 1976. Observations microtectoniques sur la compression nord-sud quater-naire des Cordillères bétiques orientales (Espagne meridionale-Arc de Gibraltar). *Bull. Soc. Géol. France.* 18(3): 711-724.

Brum Ferreira, A. 1975. Le relief du versant sud de la Montaigne Noire. *Rev. Géogr. des Pyrénées et du Sud-Ouest.* 46(1): 27-54.

Carbon, D., Ph. Combes, M. Cushing & Th. Granier. 1993. Enregistrement d'un paléoséisme dans les sédiments du Pleistocène supérieur dans la vallée du Rhone: essais de quantification de la déformation. *Géol. Alpine.* 69:33-48.

Combes, P. 1981. Néotectonique de la Basse Vallée de l'Herault (Rive droite) et de la faille des Cévennes. *Mémoire D.E.A Université des Sciences et Techniques du Languedoc, Montpellier, France.*

David, A. 1920. Le relief de la Montaigne Noire. *Ann. Géogr.* 29:241-260.

Ellenberger, F. 1938. Problèmes de tectonique et de morphologie tertiaires. Gresigne et Montaigne Noire. *Bull. Soc. Hist. Nat. Toulouse.* 52:327-364.

Ellenberger, F. 1961 Age pliocène probable des limons jaunes a galets du narbonnais occidental ("mollasses de Thézan", etc) et jeux de failles tardifs. *C. R. Som. Soc. Géol. France.* 7:183-184.

Etchecopar, A., G. Vasseur & M. Daignières. 1981. An inverse problem in microtectonics for the determination of stress tensors from fault striation analysis. *J. Struct. Geol.* 3:51-65.

Gastaud, J., R. Campredon & G. Féraud. 1983. Les systèmes filoniens des Causses et du Bas Languedoc (Sud de la France): géochronologie, relation avec les paléocontraintes. *Bull. Soc. Géol. France.* 25(5): 737-746.

Gillot, P.Y. 1972. Chronometrie de quelques intrusions volcaniques du Sud du plateau du Larzac. *C. R. Acad. Sci. Paris*. 274(D): 2855-2858.

Grellet, B., Ph. Combes, Th. Granier, H. Philip & B. Mohammadioun. 1993. Sismotectonique de la France metropolitaine dans son cadre géologique et géophysique. *Mém. Soc. Géol. France,* 2Vol., 164:1-75 + appendices.

Ildefonse, J.P. 1972. Mise en evidence de la transition paléomagnetique Gauss-Matuyama dans les formations volcaniques de L'Escandorgue, Herault, France. *Earth & Planetary Sci. Letters*. 14: 249-254.

Mouline et al. ? . Le rebord NE de la Montaigne Noire dans la région de Revel.

Obid, A. 1987. Determination des directions de compression et distension par la microtectonique cassante, dans la région de Montpellier. *These Doctorat Université des Sciences et Techniques du Languedoc, Montpellier, France*. 121 pp.

Phan Trong, T. 1989. Superposition et perturbation du champ de contraintes et modelisation numérique. *Thèse Doctorat Institut Physique du Globe–Paris VII*. 295 pp.

Philip, H. 1987. Plio-Quaternary evolution of the stress field in Mediterranean zones of subduction and collision. *Annales Geophysicae*. 5B(3): 301-320.

Rebaï, S. 1988. Le champ de contrainte actuel en Europe et dans les régions méditerranéenes. *Mémoire D.E.A Université des Sciences et Techniques du Languedoc, Montpellier, France*.

Schwobthaler, J.P. & H. Vogt. 1955. Aspects de la morphogénèse plio-quaternaire dans le Bas-Rhône occidental. *Bull. Soc. Lang. Géogr*. 26(1): 13-59 and 26(2): 67-126.

Taha, M. 1986. Apport de la microtectonique cassante aux problèmes des trajectoires des contraintes et de leurs perturbations (Exemple du Nord de Montpellier). *These Doctorat d'Etat Université des Sciences et Techniques du Languedoc, Montpellier, France*. 155 pp.

Tapponnier, P. 1977. Evolution tectonique du système alpin en Méditerranée: poinçnnement et écrassement rigide-plastique. *Bull. Soc. Géol. France*. 19(3): 437-460.

Mechanics of Jointed and Faulted Rock, Rossmanith (ed.) © *1998 Taylor & Francis, ISBN 90 5410 955 6*

Lithological and structural models of the Gideå Study Site, Ångermanland Province, Sweden

Sven Follin, Lars Mærsk Hansen & Jan Hermanson
Golder Associates, Solna, Sweden

Abstract
During the period 1981-1983, SKB (Swedish Nuclear Fuel and Waste Co.) performed surface and borehole investigations at the Gideå site, North-east Sweden as part of a site selection program for a spent nuclear fuel repository. This study presents structural and lithological models to be used as input for hydraulic modelling. A three dimensional lithology model demonstrates that paragneiss is intermingled by elongated bodies of orthogneiss. The gneiss complex is intersected by steeply dipping dolerites. Rock contacts between dolerites and gneisses are not extensively fractured, implying no hydraulic pathways along dolerite dykes. There is no correlation of statistical significance between log k and rock type. Lithological sequence does not change within repository depth, whereas the hydraulic conductivity is higher down to 200 m of depth than at repository depth (500 m). Twelve fracture zones have been identified in the area, many of which are highly water conductive, and strike NNE to ENE, parallel with the main regional stress.

1 SITE LOCATION

Gideå site is located in NE Sweden, some 600 km North of Stockholm (Figure 1). The site is characterised by a low relief, 100 - 130 m a.s.l. Typical characteristics of the area are well exposed rock, with flat top surfaces, eroded by glaciation and with some steep edges. Swamps and moraine deposits with coniferous wood comprise the Quaternary cover.

Figure 1 Location of the Gideå study site.

2 OVERVIEW OF REGIONAL GEOLOGY

The regional geology has been described by Lundqvist et al. (1990). The rocks in the region are Precambrian comprising veined paragneiss, migmate, intrusive prim-orogenic granitoids (orthogneiss), pegmatite, and dolerite dykes and sills. The age and structural relationships of the different rock types and geological events are outlined in Table 1.

2.1 Lithology

The site is located in a formation of greywacke with subordinate schist, phyllite and slate in different stages of metamorphsm. Some rocks show well preserved sedimentary structures. Thermal metamorphism, due to intrusion can also be observed. Due to regional tectonic processes, a large amount of the sedimentary rocks have been altered into paragneiss of different stages of metamorphism, from a common gneissic foliation, to veined gneiss and migmatite. Most of the study area lies in rocks of latter two types. The gneissic and schistose nature of the rock is due to a regional folding process. In the region, the fold axis is horizontal or gently dipping, oriented more or less East-West.

Table 1 Main geological events, rocks and age relationships of the region

Age, million years	Volcanic and sedimentary rocks	Plutonic rocks	Veins, dykes and sills	Geological events
400	Ordovician graywacke, limestone, shale		Possible dolerite	Caledonian thrusting. Possible marine intrusions
500	Cambrian sandstone and shale			Sedimentation in the Caledonian and in the Bothnian Sea
600		Alnö Massif	Carbonatitic dykes	Intrusion and volcanic activity at Alnön. Fracturing
	Tillite, quartzite, varved slate (Vendian)			Glaciation Sedimentation
800	Sandstone (Rifeikan)			
1200			Postjotnian dolerites	Intrusion of large volumes of dolerite dykes and sills
	Sandstone, shale, conglomerate (Jotnian)			Fracturing. Sedimentation
1600		Anorogenic intr: Granitoids, Anorthosite, Gabbro	Porphyrite, Porphyry, granite veins	Formation of massifs, dykes and veins, thermal metamorphism
1700		Postorogenic granitoids	Turingen Dolerite	Postorogenic intrusions. Thermal metamorphism. Fracturing
1800		Serorogenic granite and pegmatite	Dolerite	Serorogenic intrusion, Fracturing
1850		Bothnian superficial rocks	Amphibolites	Regional metamorphism, migmatisation and folding
	Sandstones, Rhyolite	Primorogenic Granitoids, Diorite and Gabbro	Amphibolite and Greenstone dykes	Primorogenic intrusions Sedimentation and volcanism
1900				

Table 2. Lithological units of the Gideå Site

Rock type	Characteristics	% of core material	Formation and age, million years
Gneissose metasediments	Layered graywacke gneiss with no veins	90	Sedimentation and volcanism, 1900
Veined gneiss	Fine to medium grained layers of graywacke gneiss varved with veins of granitic material		Folding, metamorphism and migmatisation, 1850
Migmatite	Graywacke or veined gneiss in an irregular mixture with granitic material		
Granite veins	Gneissose granodiorite as layers along with the regional foliation	6	Primorogenic intrusion, 1850
Pegmatite	Small massifs and vertical narrow dykes	4	Serorogenic intrusion, 1800
Dolerite	Vertical dykes with lengths of up to 2 km and with a thickness of up to several metres	2	Postjotnian intrusion of dykes and sills, 1200

Granitic and granodioritic dykes intruded during the Svecokarelian orogeny (1800 Ma). Some granite intrusions in the study area are of prim-orogenic age, and follow the foliation of the migmatite. Others are of ser- or post-orogenic age. In all these rocks, there are pegmatite veins of different ages, often difficult to determine.

A main characteristic of the region is the presence of post-Jotnian dolerite, which has intruded in the other rocks, 500 Ma years after the granite intrusions and the folding of the rock complex. The dolerites form flat lying sheets, up to several hundred metres thick, or vertical dykes, up to 15 metres thick.

2.2 Rock stresses and tectonic zones

Regional rock stresses have been calculated from focal mechanism or measured by means of

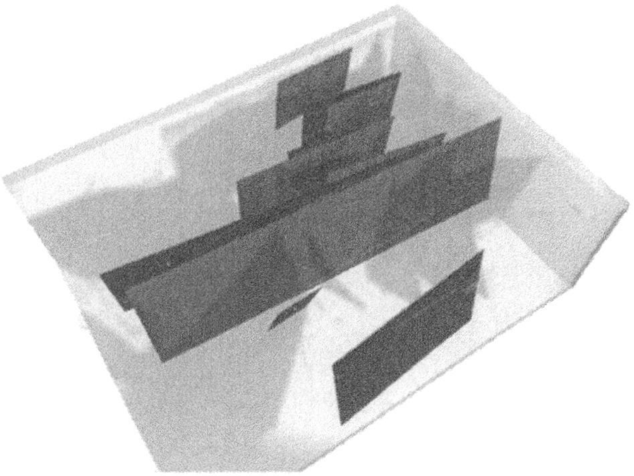

Figure 2.
Three-dimensional model of dolerite bodies

Figure 3.
Three-dimensional model of foliation, fold structures and orthogneiss bodies

hydraulic fracturing, at a number of locations from the site, to the Norwegian coast in the NW. The maximum regional horizontal stresses strike NNE to ENE (Müller et al. 1992), in agreement with stress measurements on site (Ahlbom et al. 1991). Some regional lineaments strike in this direction, which thus may form the present tensile fracture set corresponding to the stress field. Predominant lineaments, however, strike NW-SE and East-West, the latter parallel with most of the dolerite dykes. This indicates an East-West trending post-Jotnian palaeostress. North-South striking post-Jotnian normal faults exist, pre- and postdating the dolerite intrusions. Earthquakes are known to have occurred in the 1920's (Sahlström 1930) and in 1983 (Kulhánek & Wahlström 1985), the latter of a magnitude of 4.1, indicating that the area is still tectonically active to a minor extent. The epicenter was located at 40 km of depth.

3 SITE INVESTIGATIONS

3.1 Methods

Surface investigations include geological mapping along with seismic, magnetic, electric and electromagnetic methods. Magnetic survey shows a number of E-W striking anomalies interpreted as dolerite dykes, some of which also appear in rock exposures. Seismic velocity anomalies, electric and electro-magnetic survey show lineaments representing fracture zones in the rock mass.

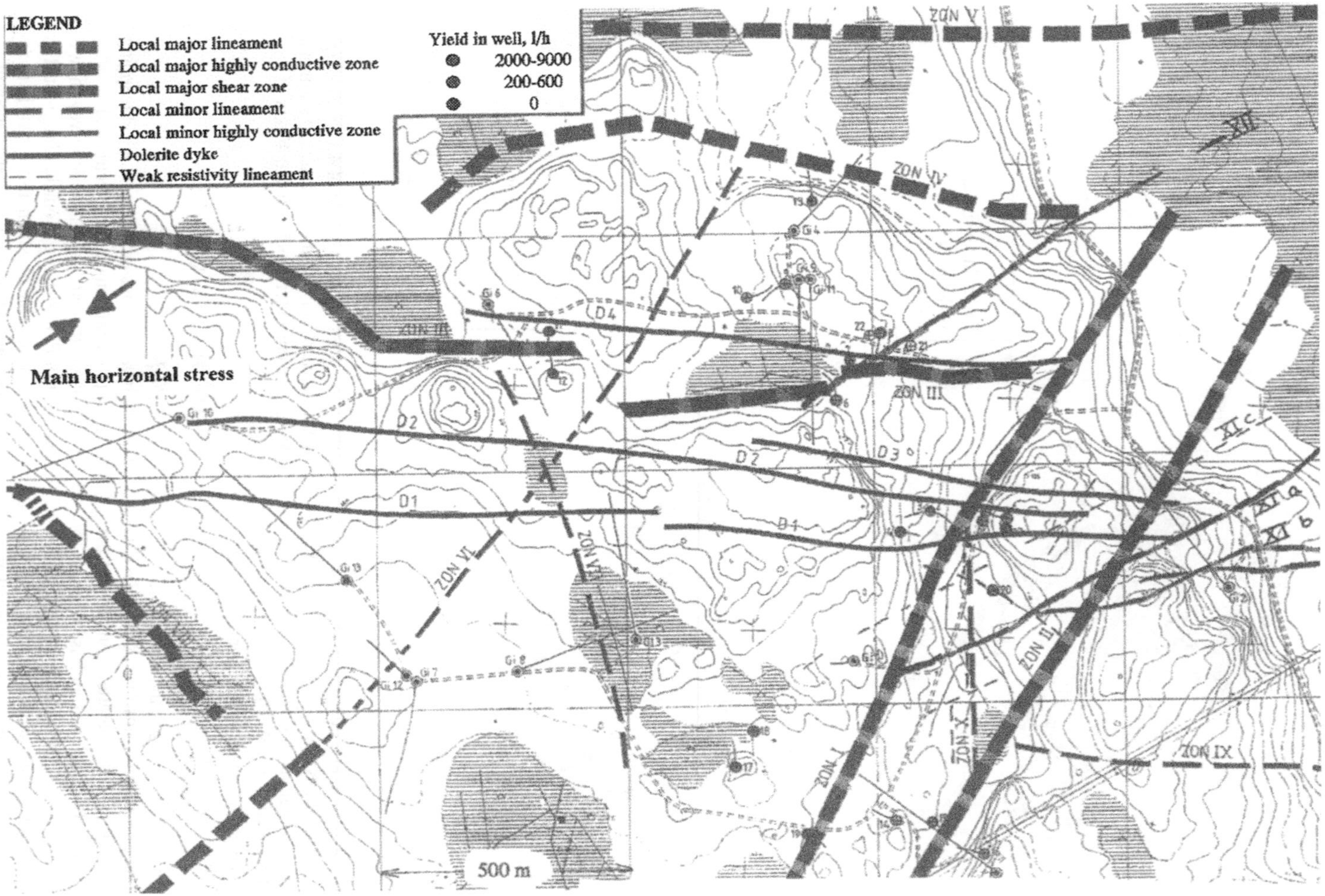

Figure 4. Map of dolerites and interpreted fracture zones in the Gideå Study Area

Figure 5. Three -dimensional model of fracture zones

Subsurface investigations comprise percussion and core drilling and geophysical measurements in in holes. Percussion drilling was carried out in order to investigate dip directions and water occurrence in fracture zones, while the purpose of coring was been to investigate rock type, joint frequency, fill minerals, and hydraulic conductivity.

3.2 results

The general results from the subsurface investigation can be summarised as follows;

- paragneiss-migmatites dominate
- orthogneiss occur frequently and in variable thickness throughout the site
- a number of steeply dipping dolerites can be correlated with surface investigations
- existing core logging does not distinguish well between granitic migmatite, orthogneiss and granitic dykes
- ten zones can be distinguished with variable degree of reliability
- difficulties to distinguish zones from surrounding bedrock due to little difference in fracture frequency
- conservative interpretation exclude subhorizontal zones

4 GEOLOGICAL SITE MODEL

4.1 Lithological units

Six main rock types exist in the area, as outlined in Table 2. Three are paragneisses of the same origin, in the corelogs called migmatite. The foliation of the gneiss is of secondary nature, with no remainders of any original bedding. The orthogneiss is of prim-orogenic age and form bodies along the fold axes. It cannot easily be traced between the drill holes, due to the folding and the wide spacing of the holes. The extent and thickness of the orthogneiss bodies is limited, due to its metamorphic appearance.

One larger steep, NS trending pegmatite dyke intersects the area. The pegmatite is competent in relation to the migmatite and is usually easy to see on outcrops. It is not extensively fractured and is not believed to be a hydraulic pathway, but not considered an impermeable boundary.

Based on borehole intercepts, the dolerites seem generally planar. The width varies from 25 m to only a few meters. No sub-horizontal dolerites have been identified with certainty. Any subhorizontal dykes must be limited in extent as no dolerite can be extrapolated through more than two boreholes. Further, a vertical hole, drilled to intersect subhorizontal dolerites, failed.

The dolerites often feature columnar joints, filled with calcite. The dolerites are not believed to be any major hydraulic pathways, nor do they seem correlated to the fracture zones. They are likely to act as impermeable hydraulic boundaries against transport in the orthogneiss bodies, but not

Table 3. Statistics of $\log_{10}(K_{25m})$ of the 25 packer tests at Gideå. Data are gathered according to the lithological logs of boreholes KG01-KGI08 and KGI11 at Gideå. By occurrence it is meant here that the considered rock type is present, either alone or as one of 2-4 possible rock types.

All data	Migmatite	Pegmatite	Granite	Dolerite
Number of intervals	200	109	61	37
Average $\log_{10}(K_{25m})$	-9.62	-9.68	-9.45	-9.69
Median $\log_{10}(K_{25m})$	-10.26	-10.35	-9.85	-10.38
STD $\log_{10}(K_{25m})$	1.69	1.68	1.41	1.66
95% confidence interval for the $\log_{10}$ average, lower/upper	-9.86/-9.38	-10.01/-9.36	-9.81/-9.09	-10.24/-9.15
$K_{25m} > 10^{-11}$ m/s				
Number of intervals	132	69	53	26
Average $\log_{10}(K_{25m})$	-8.82	-8.82	-9.19	-8.20
Median $\log_{10}(K_{25m})$	-9.07	-9.07	-9.46	-8.80
STD $\log_{10}(K_{25m})$	1.57	1.55	1.34	2.82
95% confidence interval for the $\log_{10}$ average, lower/upper	-9.10/-8.55	-9.19/-8.45	-9.56/-8.83	-9.30/-7.09

against transport in the fracture zones.

Figures 2 and 3, respectively, demonstrate the dolerite dykes and the folded nature of the rock mass.

4.2 Fracture zones

Tectonic structures are shown in Figure 4. Water conductive zones, I, II, XI and XII, all strike NNE to ENE , parallel to the maximum horizontal stress at the site and in the region.

Figure 5 shows an overview of the fracture zone model. The input to the model is electric, electro-magnetic, seismic refraction and other geophysical anomaly interpretation, data from percussion drilling and data from core drillings, including drill hole geophysics and hydraulic tests.

4.3 Hydraulic properties of lithological units

Regression analyses of $\log_{10}(K_{25m})$ for all core lengths greater than 0 m render flat regression lines, hence low degrees of determination ($r^2 \approx 0$). In summary, the approach used in this study suggests no correlation between $\log_{10}(K_{25m})$ and rock type nor the number of rock contacts.

The lack of correlation is quantified in statistical terms in Table 3. Regardless of whether all data or only data above the lower measurement limit are considered, the 95% confidence intervals for the different means show great overlaps, thus suggesting insignificant differences between the samples.

REFERENCES

Ahlbom K, Andersson J E, Nordqvist R, Ljunggren C, Tirén S & Voss C, 1991. Gideå Study Site. Scope of activities and main results. SKB Tech Rep 91-51. SKB, Stockholm.

Lundqvist T, Gee D, Karis L, Kumpulainen R, Kresten P, 1990. Beskrivning till berggrundskartan över Västernorrlands Län (in Swedish with an English Summary). SGU Ba 31. Geological Survey of Sweden, Uppsala.

Kulhánek, O., Wahlström, R. 1985. Macroseismic observations in Sweden 1980-1983. SGU C 808.

Müller B, Zoback M L, Fucha K, Mastin L, Gregersen S, Pavoni N, Stephansson O & Ljunggren C, 1992. Regional Patterns of Tectonic Stress in Europe. J. Geoph Res. 97, B, pp 11783-11803.

Mechanics of Jointed and Faulted Rock, Rossmanith (ed.) © *1998 Taylor & Francis, ISBN 90 5410 955 6*

Controls of range-bounding faults adjacent to the Yellowstone hotspot, USA

D. R. Lageson
Department of Earth Sciences, Montana State University, Mont., USA

ABSTRACT: Four categories of structural control have been identified that help explain the geometry and map distribution of late Cenozoic, range-bounding normal faults adjacent to the Yellowstone hotspot. Type I range-bounding faults are those that are influenced by the fabric (planar anisotropy) of adjacent rocks; Type II are those inherited from fault zones associated with the Middle Proterozoic Belt Basin; Type III are those that reactivate frontal and/or lateral ramps in older contractional structures; and Type IV are those defined by the limbs of older fault-propagation folds. Combinations of two or more of these types are common. These observations suggest that the distribution of late Cenozoic normal faults across the northern Rocky Mountain region generally lacks a direct correlation to an ideal stress/strain field resulting from mantle hotspot doming. Rather, structural inheritance from features that predate formation of the hotspot seems to play an important role in determining the geometry and orientation of these young normal faults.

1 INTRODUCTION

1.1 Regional tectonic setting

The contemporary structural architecture of the central and northern Rocky Mountains surrounding the Yellowstone hotspot, northwestern United States, is the result of an exceedingly long and complex tectonic history that has spanned over 2.5 billion years. From youngest to oldest, the region lies at the juncture of seven major tectono-magmatic provinces: 1) the Quaternary Yellowstone hotspot and its Neogene track, the eastern Snake River Plain; 2) the eastern margin of mid- to late Cenozoic crustal extension of the Basin and Range province; 3) the Eocene Absaroka volcanic field and associated extensional faulting; 4) the Late Cretaceous to early Tertiary, contractional, Laramide foreland province; 5) the mid-Cretaceous to early Tertiary foreland fold and thrust belt (Sevier orogenic belt); 6) the Middle Proterozoic Belt Basin; and 7) the western and northwestern(?) margins of the Archean Wyoming province. Additionally, the region is flanked to the west by accreted terranes of the Pacific-margin tectonostratigraphic collage, located west of the $^{87}Sr/^{86}Sr$ = 0.706 line, and is traversed by several major lineaments or zones of well-documented recurrent tectonic activity.

1.2 The Contemporary Yellowstone Hotspot

The surface manifestation of the Yellowstone hotspot currently resides in the northwest corner of Wyoming and adjacent parts of Montana and Idaho. The hotspot is marked by voluminous Quaternary silicic volcanism, high topography (~2,500 m), high levels of regional seismicity, a long wavelength (~1,000 km-wide) geoid anomaly, a very low Bouguer gravity anomaly (-250 mGal), active uplift and subsidence within Yellowstone's youngest (Lava Creek) caldera, widespread geothermal features, and most notably, extremely high heat flow (~2,000 mWm^{-2}, ~30 times continental average) (e.g. Smith & Braile 1984, 1994; Pierce & Morgan 1992). As North America moved relatively southwestward over the Yellowstone hotspot during the past 16-17 Ma, a record of crustal subsidence, basaltic volcanism, and progressively older silicic volcanism was left in the hotspot's wake as the eastern Snake River Plain (Armstrong et al. 1975; Christiansen 1984; Smith & Braile 1984; Pierce & Morgan 1992). These geological and geophysical characteristics have been integrated by many modern workers into a deep mantle, plume-driven hotspot model for the origin of Yellowstone (e.g. Pierce & Morgan 1992). However, recent tomographic images of the upper mantle velocity structure beneath the track of the

Yellowstone hotspot fail to discriminate between a deep-seated versus upper mantle source for the thermal plume that drives the Yellowstone tectonomagmatic system (Saltzer & Humphreys 1997). The ultimate depth and source of the Yellowstone thermal plume remains a mystery and the subject of much interpretation. Despite the obvious appeal of the deep mantle-derived plume theory as a comprehensive explanation for many of the present and past features of the Yellowstone hotspot, alternative hypotheses have been presented in the literature based on crustal-scale tectonic features (e.g. Eaton et al. 1975; Christiansen & McKee 1978; Smith & Braile 1994). For example, the Yellowstone-Snake River Plain system corresponds to a prominent embayment in the $^{87}Sr/^{86}Sr = 0.706$ line and is roughly parallel to several northeast-trending tectonic zones and geophysical anomalies (Christiansen & McKee 1978). These relationships may indicate that the track of the hotspot through time was partly controlled by crustal-scale zones of weakness, although the lack of a major northeast-trending tectonic or geophysical lineament immediately northeast of the present-day Yellowstone hotspot makes this point difficult to evaluate (Pierce & Morgan 1992). Perhaps more important, there seems to be a direct correlation between the temporal and spatial positions of the hotspot relative to diachronous crustal extension across the intermountain west; i.e. the present and past positions of the hotspot correspond to the eastward-propagating margin of Neogene crustal extension (Pierce & Morgan 1992). Although the contemporary heat flow and history of voluminous volcanic eruptions from Yellowstone may require a mantle-derived thermal plume (source depth unknown), the position of the hotspot through time is suggestive of a crustal-scale propagating crack-tip that is accommodating, and keeping pace with, Basin-and-Range extension (Hamilton 1989). Also, the possibility that the track of the hotspot may be at least partly influenced by crustal-scale features, and not entirely dictated by deep mantle processes, has yet to be entirely ruled out.

Regardless of the origin of the hotspot, it is clear that the Yellowstone hotspot is a significant thermal, density, and topographic perturbation of the crust. Pierce & Morgan (1992, p.19) described the topographic anomaly surrounding Yellowstone as a "crescent of high terrain" (so-called "bow wave" of the hotspot; Anders et al. 1989) about 350 km wide and standing 0.5-1.0 km higher than surrounding areas, superimposed on a roughly circular geoid anomaly (Milbert 1991). Complications exist where this crescent overlaps previously existing, basement-cored uplifts and basins of Laramide and Sevier vintage, such that not all of the high terrain is solely accounted for by heat-induced hotspot uplift. Pierce & Morgan (1992) have subdivided this crescent-shaped arc into four major belts of neotectonic faulting, with the zone of active Holocene faulting occurring along the central axis of the crescent. Closest to the Snake River Plain is a zone of waning to inactive fault activity, whereas the outer perimeter of the crescent is undergoing waxing fault activity. The asymmetry of these neotectonic belts across the Snake River Plain (belt IV, inactive faults, occurs only on the south side of the SRP) may be indicative of different rates, styles, or orientations of extension across the track of the Yellowstone hot-spot.

2 LATE CENOZOIC RANGE-BOUNDING FAULTS

In the region surrounding the Quaternary Yellowstone caldera system, range-bounding normal faults fall into four general categories based on structural control. Type I range-front faults are geometrically controlled by the dip of metamorphic foliation in adjacent basement-cored fault blocks (typically gneissic banding and schistosity within the proximal footwall block, but also including ductile shear zones within foliation planes). A good example is the Teton normal fault in western Wyoming (Figure 1) where the average strike/dip of footwall foliation along the Teton range-front is 354°, 40° E (Figure 2), sub-parallel to the average northerly trend of Quaternary fault scarps (Gilbert et al. 1983; Smith et al. 1993). Other examples of Type I range-fronts in southwest Montana include the Deep Creek normal fault bordering the Paradise Valley along the northwest flank of the Beartooth Range, and the northern Gallatin Range front fault adjacent to the southern Gallatin Valley. Also, it cannot be ruled out that bedrock fabrics may control the geometry of extensional faults in areas where the Precambrian basement is deeply buried.

Type II range-front fault systems are those controlled by Middle Proterozoic fault zones, such as the southwest Montana transverse fault zone (Perry line) in the Gallatin Valley, the west flank of the Big Belt and Bridger Mountains (Townsend line), and other lineaments and/or fault zones having origins dating back to the formation of the Belt Basin, ~1.4-1.1 Ga (Winston 1986).

Figure 1. Oblique aerial photograph of Jackson Hole and the Teton Range, western Wyoming. Snake River forms the braided floodplain at lower left. The Grand Teton, highest peak in the range, rises to an elevation of 4,197 m. View to the northwest.

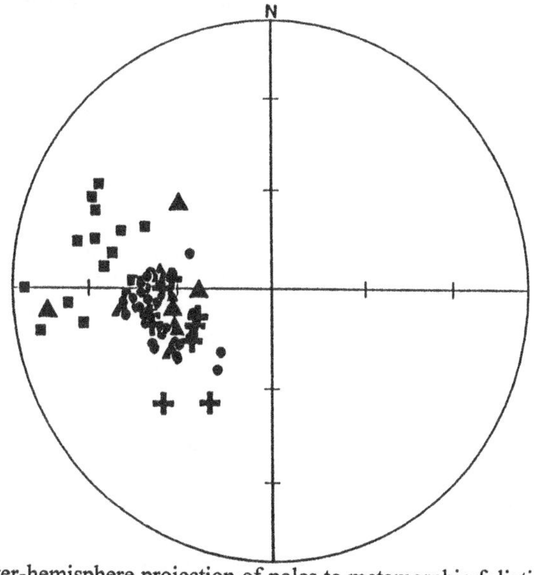

Figure 2. Equal area, lower-hemisphere projection of poles to metamorphic foliation (n=68) in the proximal footwall block of the Teton normal fault. Average strike/dip of foliation is 354°, 40°E, subparallel to trend of range-front fault scarps. Data sets: triangles = Avalanche Canyon; squares = base of Mt. Teewinot; dots = Garnet Canyon; crosses = Cascade Canyon.

Type III range-front fault zones are those controlled by the dip of frontal and/or lateral-ramps in either the basement-involved Laramide foreland province (IIIa) or the Sevier fold and thrust belt (IIIb). Examples of Type IIIa faults in southwest Montana include the Bridger Range (Lageson 1989), Madison Range (Kellogg et al. 1995), Snowcrest Range and southern Pioneer Mountains, and in western Wyoming the overlap of the southern Teton and Gros Ventre Ranges (Lageson 1992). Examples of Type IIIb normal faults in northwest Montana include those bordering the Swan Range, Mission Mountains, western Glacier National Park, and the Big Belt Mountains. In western Wyoming the Hoback and Grand Valley listric normal faults are classic examples of Type IIIb faults (Royse et al.

1975). Also, the Wasatch normal fault in north-central Utah has reactivated a large frontal ramp in the hinterland of the Sevier orogenic belt where basement rocks were stacked into an antiformal duplex culmination, the crest of which has collapsed along the ramp interface to form the modern range-bounding fault (Yonkee 1992).

Type IV range-front fault zones are those defined by the dip of Phanerozoic cover strata, initially created as the steep limbs of fault-propagation folds of Laramide or Sevier vintage. Examples in southwest Montana include the west flank of the Tobacco Root Mountains and normal faults on the east flank of the Ruby Mountains. The above classification scheme is entirely empirical and based on field observations and published maps.

In some cases, range-bounding normal faults are controlled by combinations of pre-existing structural features. For example, the normal fault along the west flank of the Bridger Range in southwest Montana soles into a basement-rooted Laramide frontal ramp (Type IIIa), and is also parallel to a major Middle Proterozoic lineament (Type II). Similarly, the orientation of the southern Teton normal fault (Figure 1) is controlled by basement foliation in its footwall (Type I) and by the position of a lateral ramp along the Cache Creek thrust (Type IIIa).

3 DISCUSSION

Rocky Mountain geologists have debated the topic of "reactivation" of preexisting structures during subsequent tectonic events for decades. Although it is generally accepted that reactivation of frontal thrust ramps in the Sevier orogen by late Cenozoic normal faults has occurred throughout the eastern Basin and Range province (e.g. cross sections by Royse 1993), other forms of reactivation and/or structural control of geologically young normal faults are not as widely understood or appreciated. Exceptions to the "rule of reactivation" can always be found in the field and specific kinematic correlations between younger and older structures are sometimes difficult to document. With respect to Laramide structures, Hamilton (1988, p. 31) stated that "This concept [reactivation] is unsupported by field evidence; no major structures have been demonstrated to have the predicted multistage histories." However, this point only underscores the need for detailed structural and stratigraphic analyses when proposing a multistage slip history for a given fault. For Type I normal faults, a case for structural control can be made statistically based on stereographic analysis of footwall fabrics (Figure 2) coupled with other structural data regarding fault geometry. For Type II normal faults, one must be able to document a reversal in the age of stratigraphic juxtaposition across the fault, requiring opposite senses of hanging wall displacement through geologic time. For example, younger-over-older stratigraphic relations followed up-dip by older-over-younger relations indicate reactivation of a former normal fault as a contractional fault (Lageson 1989). For Type III and IV normal faults, reactivation and/or structural control may be confirmed through a combination of reflection seismic profiles, map relations, or drilling.

The distribution and geometry of late Cenozoic range-bounding normal faults adjacent to the Yellowstone hotspot are not consistent with an isotropic stress/strain field associated with thermal doming of a homogeneous crust. Unlike Hawaiian-type hotspots where tensional stress and extensional strain are accommodated by radial rift zones and annular fault systems, the deformational field surrounding the Yellowstone hotspot is strongly influenced by preexisting structures and country-rock fabrics. The distribution of extensional structures in the northern intermountain west also differs from that of continental-based RRR-triple junctions where extensional strain is often symmetrically disposed around a central thermal "high." The examples cited in this paper emphasize the point that areas with a history of long-lived tectonic activity and instability are characterized by recurrent motion on pre-existing zones of weakness, regardless of the orientation of the contemporary stress field. Therefore, the geometry and distribution of late Cenozoic range-bounding normal faults may be explained by their past structural history and/or the influence of bedrock fabric, in combination with their orientation relative to the east-west tensional stress regime of the Basin and Range province and the thermally-driven domal perturbation of the Yellowstone hotspot.

The northern Rocky Mountains suggest that a "least work" principle has operated over several major tectonic events, whereby it has been mechanically easier to reactivate older fault zones (regardless of their orientation) and adjust kinematically to new stress fields, than it is to create entirely new faults within the crust. Therefore, it is very important to understand the progressive strain history and complexities of the heterogeneous crust in order to make sense of the array of Neogene-Quaternary normal faults in the region.

Such an understanding will also provide better insight to the controlling factors of fault segmentation and characteristic rupture lengths, which will assist in seismic risk assessment of large range-bounding faults. Lastly, this approach will lead to a better understanding of how fracture patterns and bedrock fabrics within and adjacent to fault zones may control fluid migration (e.g. Type I faults) and, in general, to a more complete understanding of the overall kinematic history of a given fault zone.

4 CONCLUSIONS

Within the setting of complex tectonic influences adjacent to the Yellowstone hotspot, the structural geometry and distribution of late Cenozoic normal faults is not simply the result of a homogeneous and isotropic crust responding to thermal doming. Rather, the northern Rocky Mountain region is dominated by fault zones, lineaments, shear zones, basement fabrics and province boundaries that have experienced multiple episodes of reactivation and reuse through geologic time, often with an opposite sense of displacement relative to older tectonic events. Even the so-called "bow-wave" of neotectonic displacements and uplift associated with the Yellowstone hotspot has taken advantage of preexisting structural weaknesses, resulting in a non-symmetrical pattern of crustal displacement adjacent to the hotspot. Furthermore, the hotspot track itself may be controlled by preexisting discontinuities in the crust as suggested by earlier workers, although there is by no means consensus on this point.

ACKNOWLEDGEMENTS

The author would like to thank Todd Feeley and Jim Schmitt (both at Montana State University), Karl Kellogg and Kenneth Pierce (both at U.S. Geological Survey, Denver), and Dave Adams (Jackson, Wyoming) for their helpful comments and reviews of this paper. Robert B. Smith (University of Utah) provided informative comments regarding the source depth of the Yellowstone plume and led the author to several geophysical papers. Any observational or interpretation errors presented herein are the author's responsibility.

REFERENCES

Anders, M.H., J.W. Geissman, L. Piety & T. Sullivan 1989. Parabolic distribution of circum-eastern Snake River Plain seismicity and latest Quaternary faulting: migratory pattern and association with the Yellowstone hotspot. *Jour. Geophys. Res.* 94: 1589-1621.

Armstrong, R.L., W.P. Leeman & H.E. Malde 1975. K-Ar dating, Quaternary and Neogene volcanic rocks of the Snake River Plain, Idaho. *Am. Jour. Sci.* 275: 225-251.

Christiansen, R.L. 1984. Yellowstone magmatic evolution: its bearing on understanding large-volume explosive volcanism. In *Explosive volcanism: inception, evolution, and hazards.* National Research Council Studies in Geophysics, National Academy Press: 84-95. Washington, D.C.

Christiansen, R.L. & E.H. McKee 1978. Late Cenozoic volcanic and tectonic evolution of the Great Basin and Columbia Intermontane regions. In R.B. Smith & G.P. Eaton (eds), *Cenozoic tectonics and regional geophysics of the western Cordillera.* Geol. Soc. Am. Memoir 152: 283-311.

Eaton, G.P. & 7 others 1975. Magma beneath Yellowstone National Park. *Science* 188:787-798.

Gilbert, J.D., D. Ostennaa & C. Wood 1983. Seismotectonic study, Jackson Lake Dam and reservoir, Minidoka Project, Idaho-Wyoming. *U.S. Bureau Reclamation Seismotectonic Report 83-8*: 122 p.

Hamilton, W.B. 1988. Laramide crustal shortening. In C.J. Schmidt & W.J. Perry Jr. (eds), *Interaction of the Rocky Mountain foreland and the Cordilleran thrust belt.* Geol. Soc. Am. Mem. 171: 27-39.

Hamilton, W.B. 1989. Crustal geologic processes of the United States. In L.C. Pakiser & W.D. Mooney (eds), Geophysical framework of the continental United States. *The Geol. Soc. of Am. Memoir 172*: 743-781.

Kellogg, K.S., C.J. Schmidt & S.W. Young, 1995. Basement and cover-rock deformation during Laramide contraction in the northern Madison Range (Montana) and its influence on Cenozoic basin formation. *Am. Assoc. Petrol. Geol. Bull.* 79: 1117-1137.

Lageson, D.R. 1989. Reactivation of a Proterozoic continental margin, Bridger Range, southwestern Montana. In D.E. French & R.F. Grabb (eds), *Geologic resources of Montana Volume I.* Montana Geol. Soc. 1989 Field Conf. Guidebook: Montana Centennial Edition: 279-298.

Lageson, D.R. 1992. Possible Laramide influence on the Teton normal fault, western Wyoming. In P.K. Link, M.A. Kuntz & L.B. Platt (eds), *Regional geology of eastern Idaho and western Wyoming.* Geol. Soc. Am. Memoir 179: 183-196.

Milbert, D.G. 1991. GEOID90: a high-resolution geoid for the United States. *EOS, Trans. Am. Geophys. Union* 72: 49.

Pierce, K.L., & L.A. Morgan 1992. The track of the Yellowstone hot spot: volcanism, faulting, and uplift. In P.K. Link, M.A. Kuntz & L.B. Platt (eds), *Regional geology of eastern Idaho and western Wyoming.* Geol. Soc. Am. Memoir 179: 1-53.

Royse, F. Jr. 1993. An overview of the geologic structure of the thrust belt in Wyoming, northern Utah, and eastern Idaho. In A.W. Snoke, J.R. Steidtmann & S.M. Roberts (eds), *Geology of Wyoming.* The Geol. Surv. of Wyo. Mem. 5: 273-311.

Royse, F. Jr, M.A. Warner & D.L. Reese 1975. Thrust belt structural geometry and related stratigraphic problems, Wyoming-Idaho-northern Utah. In D.W. Bolyard (ed), *Symposium on deep drilling frontiers in the central Rocky Mountains.* Rocky Mountain Association of Geologists: 41-54. Denver, Colorado.

Saltzer, R.L. & E.D. Humphreys 1997. Upper mantle P wave velocity structure of the eastern Snake River Plain and its relationship to geodynamic models of the region. *Jour. Geophys. Res.* 102, B6: 11, 829-11, 841.

Smith, R.B. & L.W. Braile 1984. Crustal structure and evolution of an explosive silicic volcanic system at Yellowstone National Park. In *Explosive volcanism: inception, evolution, and hazards.* National Research Council Studies in Geophysics, National Academy Press: 96-109. Washington D.C.

Smith, R.B. & L.W. Braile 1994. The Yellowstone hotspot. *Jour. Volc. and Geothermal Res.* 61: 121-187.

Smith, R.B., J.O.D. Byrd & D.D. Susong 1993. The Teton fault, Wyoming: seismotectonics, Quaternary history, and earthquake hazards. In A.W. Snoke, J.R. Steidtmann & S.M. Roberts (eds), *Geology of Wyoming.* The Geol. Surv. of Wyo. Mem. 5: 629-667.

Winston, D. 1986. Middle Proterozoic tectonics of the Belt Basin, western Montana and northern Idaho. In S.M. Roberts (ed), *Belt Supergroup: a guide to Proterozic rocks of western Montana and adjacent areas.* Montana Bur. Mines & Geol. Spec. Pub. 94: 245-257.

Yonkee, W.A. 1992. Basement-cover relations, Sevier orogenic belt, northern Utah. *The Geol. Soc. Am. Bull.* 104: 280-302.

Mechanics of Jointed and Faulted Rock, Rossmanith (ed.) © 1998 Taylor & Francis, ISBN 90 5410 955 6

Recent tectonic stress field research in Shanxi graben system

Fangquan Li & Meijian An
Institute of Crustal Dynamics, SSB, Beijing, People's Republic of China

ABSTRACT: By comprehension of the focal mechanism solutions of earthquake and the data of the measurements of the in situ stress , the tectonic stress field in Shanxi region was summarized, which indicated that the stress state in this region was so different from that of the surrounding regions. The next, by fitting of the measured data, the boundary force which influenced the distribution of the stress field in this region has been studied using the inversion method. The result of the inversion showed the distribution of the stress field in this region and it's influence factors clearly.

1 INTRODUCTION

Several years ago, Shanxi Graben system was studied mainly and in detail, in China. First, by the use of the method of seismogeology, geophysics and seismological observation. The next, because of the need of engineering, many of the in situ stress measurements have been made in Shanxi. All of the observations and the study not only made us know this region comprehensively, but also the detail observing data were gotten out for the deep study of geodynamics, this means that all of above gave out a good condition of study of geodynamics.

In the numerical analysis of the tectonic stress field, the finite element inversion was often used. In the numerical analysis, the genetic finite element inversion is one kind of effective finite element inversion methods. which seeks all elements by genetic algorithm (An, M. Y. Shi & F. Li, 1997,1998). Doing genetic algorithm, not only the model boundary force of the linear and nonlinear finite element inversion can be done, but also the inversion of the property of the model material can be done.

In this paper, first, Shanxi recent tectonic stress field was comprehensively analyzed, it was based on this that the inversion algorithm was done to the observed data by the genetic finite element inversion, hope to know the boundary effect between Shanxi recent tectonic stress field and land-block deeply.

2 THE ACTIVE TECTONICS AND THE RECENT TECTONIC STRESS FIELD IN SHANXI REGION

2.1 *The simple description of the active tectonics*

In this region, the Cenozoic era tectonics developed so much, there are some fault basins, faults and folds in the new strata (Zhang B. et al., 1989, Editorial Board for the Lithospheric Dynamics Atlas of China, 1991). Shanxi graben system and its around grabens stretched to en echelon arrangement-as the character S. There are some great faults on the one or two sides of the faulted basin, which means that the whole of the basin is graben and semi-graben basin coming from the control of fault to fault depression. And the graben usually was shaped in Miocene epoch-Pliocene epoch. In this region, the important character of the tectonic action in the Cenozoic era is the new action of old faults and the generation of new faults, which showed up with the development of fault depression of fault basin. According to the tectonic frame and the activity characteristics of the basin of Cenozoic era in the middle part of this region, comprehensively considering, it was recognized that in this region, in Cenozoic era and it's period, the tectonic action of faults and fault depressions was caused by the control of the region stress field, which was tensile in the direction of NW-NNW and compressed in the direction of NE-NEE mainly.

The recent active tectonics in this region basically succeeded to the main active characteristics since the Cenozoic era, and the faults and the fault-block controlled by it were taken as the leading factor, the activity characteristics of which was successive and nonhomogeneous.

2.2 *The data of the recent tectonic stress field*

First, the focal mechanism solutions in Shanxi region, the results of the solutions of the fault plane of earthquakes were made as many as more than 30, the magnitudes of earthquakes were bigger than $M_L \geq 4.0$, the results of the comprehensive solution of the fault plane of small earthquakes which were observed by the 15 seismic observation stations, the earthquakes occurred nearby the stations, and the results were taken from the comprehensive solutions of fault plane of small earthquakes observed by many stations of Lingfeng telemetered seismic net (Liu W. et al., 1993). In figure 1, the thin line indicates the orientation of P axes of the fault plane solutions of earthquakes, the magnitudes of which were bigger than $M_L \geq 4.0$ (sum of data, 29), and the orientation of P axes of solutions of the comprehensive fault plane of small earthquakes (sum of data, 7).

Next, it is about the in situ stress measurements. since the 1980s, because of the need of the engineering of water conservancy which is going to draw the water of Yellow River to Shanxi, more than ten in situ stress measurements were made in this region (overcoring and hydrofracturing).

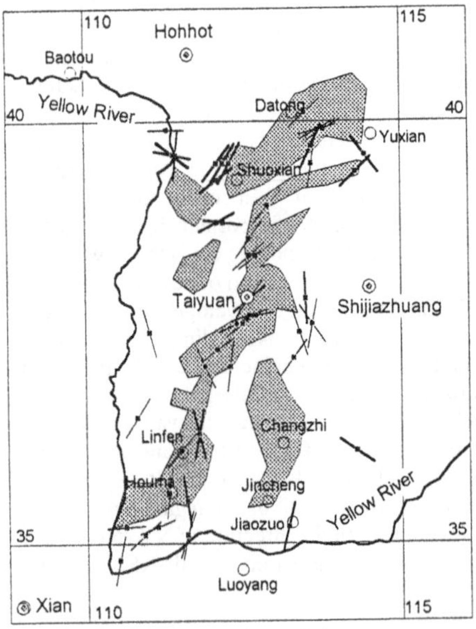

Figure 1. The distribution of the recent tectonic stress field.

Especially, several years ago, the measurements of by hydrofracturing were made in deep (Li, F. et al., 1984, 1992). And these results of the in situ stress measurements are the very important basic data for the studying recent tectonic stress field in Shanxi graben. In figure 1, the thick line is the indication of the orientation of the horizontal principal maximum stress measured in situ, the sum of the data is 16.

In figure 1, the polygon with shadow means the fault basins (according to Wang, Y. et al., 1989). The thin line indicates the horizontal orientation of P axis of stress field from earthquake. And the thick line shows the direction of the horizontal maximum principal stress measured in situ.

2.3 *Recognition of the recent tectonic stress field*

From the data of the in situ stress measurements, it was discovered that there is a clear different between the tectonic stress field in Shanxi graben and that of in the north of China, east of Taihang mountain. In the east of Taihang mountain, the direction of the horizontal maximum principal stress is near EW, and in the west of Taihang mountain, in the fault basins the direction of the horizontal maximum principal stress is almost in the direction of NNE and NNW, and the scatter is not even (Li, F. & G. Liu, 1986). The nonhomogeneousity of the scatter can be found in the data of the focal mechanism solutions. Li, Q. thought that the land-block in the north of China is acted on by an uniform crustal stress field with the better unanimous, and the direction of the axis of the maximum principal compressive stress is in the direction of NNE-SWW, the axis of maximum tensile principal stress is in the direction of NNW-SSE, all of them are near horizontal. But it was also recognized that in Shanxi earthquake zone, the crustal stress field has it's complicity in some degree, not the same as in another region which has the better uniformity. Xue, H. et at. also thought that the axes of principal tensile stresses in Shanxi fault basins were near the horizontal though, but the orientation, the upward angle of the axes of the principal stresses are bigger than that in the north of China, it's characteristic is basically different from that in the north of China. With more and more in situ stress measurements were made in this region and the focal mechanism solutions were studied more detaily, the nonuniformity was discovered step by step.

The next, if the data of the in situ stress measurements and the data of the focal mechanism solutions were checked only, both of them have better unanimity (Li, F. 1992). Therefore, in one side this means that the two kinds of data can be checked by each other, in another side it means that the two kinds of methods can be compensated by each other.

2.4 *The relationship between tectonic stress field and the active structure in Shanxi*

From figure 1, according to the data of observation of stress field and the distribution of the fault basins, there is a relativity between the orientation of the horizontal maximum principal stress and the distribution of the fault basins, especially in the inside of the fault basins, the orientation of the horizontal maximum principal stress and the long axis of the fault basin are parallel. Considering the fault basins to be made in Miocene epoch-Pliocene epoch, and the inheritance of the tectonic action in this region, it was thought to be reasonable that the distribution of the recent tectonic stress field and the structure basin has the unanimous.

3 THE FITTING INVERSION OF THE RECENT TECTONIC STRESS FIELD IN SHANXI

In order to study the tectonic stress field in Shanxi deeply, the method of the genetic finite element inversion was used for the fitting inversion of the tectonic stress field in Shanxi. Because the recent tectonic stress field in Shanxi is controlled by horizontal tectonic force mainly (Editorial Board for the Lithospheric Dynamics Atlas of China, 1991, Wang, Y. 1994), therefore in the computation, the two models were used, one was two dimensional linear elasticity, the other was elastic-plasticity.

3.1 *The model and parameters of the inversion of the genetic finite element*

Shanxi was the major body of the studying region for the inversion, the north boundary of the region is to Huhehaote-Baotou fault basin and the north boundary of Datong-Weixian fault depression zone, the east boundary is the Mt. Taihang fault, the south boundary is to the south boundary of Lingfen fault basin, the western boundary is to longitude of 110 degree.

The studying region was divided into 221 quadrilateral elements, and there are 252 nodes in total (Figure 2). According to the tectonic property of the rock formation, the materials in the region was divided into two kinds of properties, the fault basin was material number two, the others were the number one (Figure 2). The Poisson ratio of material number one is 0.25, the Young's modules is 1.1×10^{11} Pa (Dziewonski, 1981). By the use of the criterion of Mohr-Coulomb (Owen and Hinton 1980), the inner frictional angle is 36°.

In figure 2 the elements with thick line are with material number two, the other elements are with material number one.

According to the tectonic structure condition in Shanxi and the surrounding regions, the western

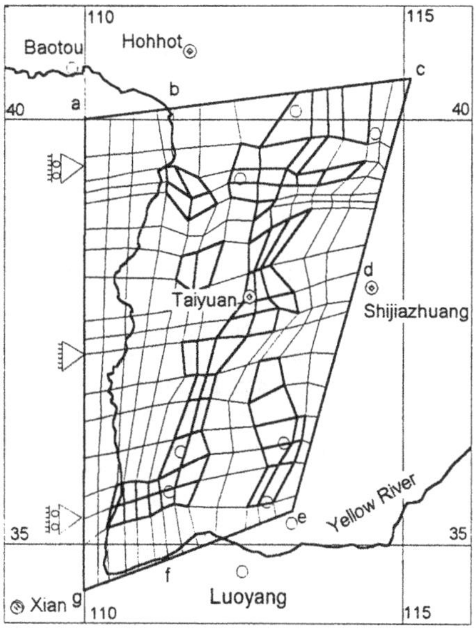

Figure 2. The finite element model.

boundary (ag) was defined as the restrictive boundary for the direction of E-W, and the center node was the fixed point. The other 6 boundaries were the boundaries of boundary force, they were ab, bc, cd, de, ef and fg respectively. In the inversion, it was supposed that all boundaries had same boundary stress. The three parameters from the inversion of stress boundary were the maximum principal stress, minimum principal stress, it's ratio and the orientation angle of the maximum principal stress. The start seeking intervals of the three inversion parameters for all the stress boundaries were same, they were 0.0~100.0, 0.0~1.0, and -90.0° ~90.0° respectively.

In the process of optimization for the genetic algorithm, the mean square deviation of the observing value and the computational value of the direction of the principal stress was used as the optimizing objective function D_d. In the process of the computation, the dynamic genetic algorithm was used (An, M. & Y. Shi 1996), and the parameters of the genetic algorithm were selected as follows: the groups of species were 16; the fixed variable probability was 0.02. The fixed commuting probability was 0.90. The bits of the binary code for every parameter were 15, that means that the precision of the every parameter was 0.00003 $\triangle$, $\triangle$ was the length of the single interval.

3.2 *The computation and the results of inversion of the genetic finite element algorithm*

First, the boundary force was inversed by the use of

linear elastic model. The property of the material iteration times for every cycle was I_{max}=50, then the seeking interval was optimized for one time. While the seeking interval was optimizing, the new interval was not restricted, that means the model was the unknown restrictive interval (An, M. & Y. Shi 1996). After the ideal interval was found, in last seeking cycle the iteration was $I_{last,max}$=100, and the last optimization model was obtained. The optimization solution of the angle fitting mean square deviation was D_d=31.5° from the computation. According the optimizing interval and result of the rough seeking computation, the start seeking interval was contracted and the precision computation was going on. In the precision computation, N=3 I_{max}=40. the seeking interval was optimized for one time, and the model was supposed to be as an unknown restrictive interval. $I_{last,max}$=100. The fitting mean square deviation of the optimizing model was gotten from the computation, it was D_d=31.1°, the mean fitting remain error was 24°. The scatter of the vector of the boundary force is in figure 3, which was from the computation of the principal stress that came from the precision computation seeking.

According the scatter of the principal stress in figure 3, it is clear that the recent stress field obtained from computation is related closely to distribution of the fault basin and other structures and so on in Shanxi, that means the direction of the horizontal maximum principal stress is parallel to the long axis of the fault basin. Looking the scatter of the boundary force, the effect of the inversion boundary force which influences on Shanxi, and the velocity difference of the recent movement which comes from Shanxi region moving relatively to the around land-block (Ding, G & Y. Lu 1989), there is a better corresponding between these two events. The computation boundary force, and according to seismological data, this region is controlled by the compression in the direction of NE--NEE (Editorial Board for the Lithospheric Dynamics Atlas of China ,1991), the two events are consistent.

In Figure 3, the arrow indicates the direction of the boundary force, the length of the arrow is directly proportional to the square root of the absolute value of the boundary force. The line means the orientation of the computation principal stress.

The next, while the boundary force was inverted, the Poisson's ratio and the Young's modules of the material number two was inverted too, from the inversion, the material number two was more easy deformation than the material number one. In the inversion, for the different tectonic unit, the corresponding material property was used, so that it was good for the fitting model. But from the result of the fitting inversion, the material property inside

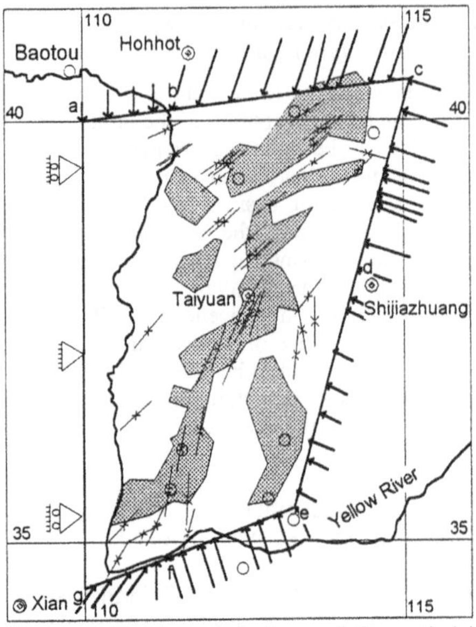

Figure 3. The boundary force of the optimizing model and the distribution of the computation principal stress.

the region was the second factor for the influence on the distribution of the tectonic stress field.

The third, the inversion of the model was done by the use of the plastic-elasticity model, but the result from this model was almost same as that from the elastic model.

In summary, according to the result of the inversion of the plane stress model, the effect of the boundary force was the main factor which influenced on the recent tectonic stress field in Shanxi, the material and the property inside the region were the second factor which influenced on the distribution of the tectonic stress field.

4 SUMMARY

Analyzing the tectonic stress field in Shanxi, it was recognized that the tectonic stress field in Shanxi is so different from that around the region. Through the fitting inversion, it is recognized that the effect of the boundary force between the blocks is the main determinative factor for the recent tectonic stress field in Shanxi, and it is clear that the computation stress field is consistent with the scatter of the fault basin. Because the observing data of the recent stress field in Shanxi is closely related to the distribution of the fault basin, therefore it was thought that the effect of the mutual boundary force of the lithospheric plates is also the important factor which influences the distribution of the active tectonic unit.

After recognizing the effect of the boundary force and the distribution of the tectonic stress field in Shanxi region, it is the studying object for future that the numerical studying is used in the whole north of China, so that the complexity of influencing tectonic stress field in this region and the effect mechanism of the earth dynamics should be more known.

REFERENCE

Dziewonski, A.M. & D.L. Anderson 1981. Preliminary reference earth model. *Phys. Earth Planet. Int.* 25: 297-356.

Owen, D.R.J. & E. Hinton 1980. *Finite elements in plasticity Theory and practice*. Pineridge Press Ltd.

An, M. & Y. Shi 1976. Dynamic genetic algorithms inversion of earthquake focal mechanism from P waves first motions. *Earthquake Research in China*. 12(4):394-402.

An, M., Y. Shi & F. Li 1997. Preliminary researches for genetic algorithms-finite element method (GA-FEM) inversion with ideal model, *Collected Works of Crustal Structure and Crustal Stress*. 10. Beijing: Seismological Press. (in Chinese and in Press).

An, M., Y. Shi & F. LI 1998. GA-FEM application on the inversion controlling the tectonic stress field of China. *Acta Seismol. Sin.* 20: (in press).

Ding, G.& Y. Lu 1989. Recent motions of intraplate blocks. *Lithospheric Dynamics Atlas of China*: 21. Beijing: China Cartographic Publishing House.

Li, F. 1992. Comparison of the results obtained from different techniques such as overcoring, hydrofracturing, breakouts and focal mechanism solutions. *Acta Seismol. Sin.* 5(4):709-717.

Li, F. & G. Liu 1986. The present state of stress field in China and the related problems. *Acta Seismol. Sin.* 8(2): 156-171(in Chinese).

Li, F., Y. Li, E. Wang, Q. Zhai, S. Bi, J. Zhang, P. Liu, Q. Wei & S. Zhao 1983. *Experiments of in situ stress measurements using stress relief and hydraulic fracturing techniques*. In M.D. Zoback & B.C. Haimson (ed.). *Hydraulic fracturing stress measurements*: 130-134. Washington, D.C.: National Academy Press.

Li, F., Q. Zhai, S. Bi, P. Liu, J. Zhang & S. Zhao 1986. In situ stress measurement by hydraulic fracturing and preliminary results. *Acta Seismol. Sin.* 8(4): 107-114.

Li, Q. 1980. General features of the stress field in the crustal of North China. *Acta Geophys. Sin.* 23(4):376-388(in Chinese)

Liu, W., X. Zhao, W. An & K. Zhang 1993. Crustal Stress field in Shanxi. *Earthquake Research in Shanxi*. (3): 3-11(in Chinese).

Wang, Y. Active faults of Ordos area lithospheric dynamics. *Study and Application on recent Geodynamics*: 222-227. Beijing: Seismological Press. (in Chinese).

Wang, Y., Q. Deng & S. Zhu 1989. Lithospheric dynamics of North China. *Lithospheric Dynamics Atlas of China*: 59. Beijing: China Cartographic Publishing House.

Xue H. & J. Yan 1984. The contemporary crustal stress field around the Ordos block. *Acta Geophys. Sin.* 27(2): 144-152 (in Chinese).

Zhang, B., K. Yu & S. Jia 1989. Seismotectonics of Shanxi province. *Lithospheric Dynamics Atlas of China*: 31. Beijing: China Cartographic Publishing House.

Editorial Board for the Lithospheric Dynamics Atlas of China 1991. *Lithospheric Dynamics of China*: 279-288. Beijing: Seismological Press. (in Chinese).

Mechanics of Jointed and Faulted Rock, Rossmanith (ed.) © *1998 Taylor & Francis, ISBN 90 5410 955 6*

History of folding in the Magura nappe, Outer Carpathians, Poland

A. K. Tokarski & A. Świerczewska
Institute of Geological Sciences, Polish Academy of Sciences, Kraków, Poland

ABSTRACT: This paper traces the development of folding within a portion of the Krynica slice which is the innermost tectonic slice of the Magura nappe in Poland. Analysis of map-scale structures has been combined with studies of small-scale structures, mineral veins and diagenetic maturity. Five groups of small-scale structures: hydroplastic faults, deformation bands, joints, sandstone dykes and brittle faults have been studied. Results indicate that map-scale folding and thrusting as well as cross-fold jointing occurred in Paleocene–Eocene times during deposition of the strata involved. The folding and thrusting were completed when these strata were still poorly indurated.

The folding, thrusting and cross-fold jointing occurred simultaneously due to numerous permutations of the principal stress axes. The bulk of calcite mineralization was introduced into the strata involved after completion of folding. Still later, the entirely indurated strata were cut by brittle faults.

1. INTRODUCTION

The object of this study was to trace in detail the process of folding within the Western Outer Carpathians. A portion of the Magura nappe has been chosen as the study region, because this is the only area within the Western Outer Carpathians where the timing of folding and process of induration of the strata involved during the folding are relatively well known (Tokarski et al. 1995; Świerczewska & Tokarski, in press). Our specific aims have been: (1) to date folding-related thrusting, faulting and jointing; (2) to date calcite mineralization and; (3) to establish the stress regime during the folding.

The small-scale structures, calcite mineralization and clay minerals have been used as tools to asses the degree of induration of the strata involved during folding. Field observations have been combined with microscopic studies, including cathodoluminescence, and X-Ray analysis.

2. REGIONAL SETTING

The Western Outer Carpathians (Figure 1) are a north-verging fold-and-thrust belt composed mostly of Lower Cretaceous through Lower Miocene flysch. The belt comprises several nappes. In the Polish segment of the belt the Magura nappe is the innermost of these nappes. This nappe is subdivided by north- verging reverse faults into four slices, the innermost of which is the Krynica slice which has been the locus of our research (Figure 2). The results of our recent studies on deformation bands (Tokarski et al. 1995; Świerczewska & Tokarski, in press) show that folding within the Magura nappe started at least during Eocene times, earlier than thought hitherto (e.g. Roca et al. 1995 and references therein).

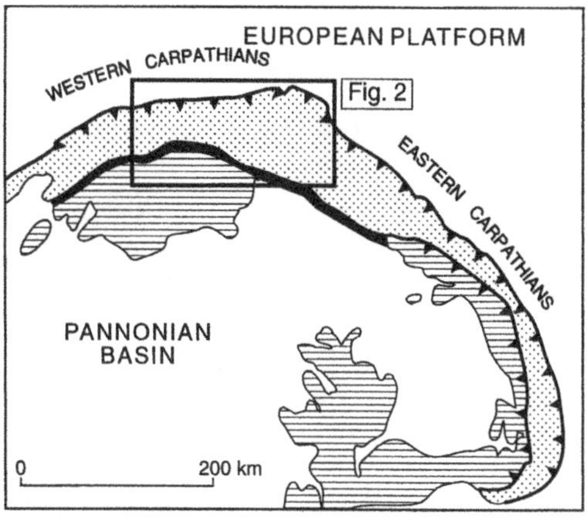

Figure 1. Sketch map of Carpathian arc showing Inner Carpathians (ruled), Pieniny Klippen Belt (black), Outer Carpathians (stippled) and location of Fig. 2

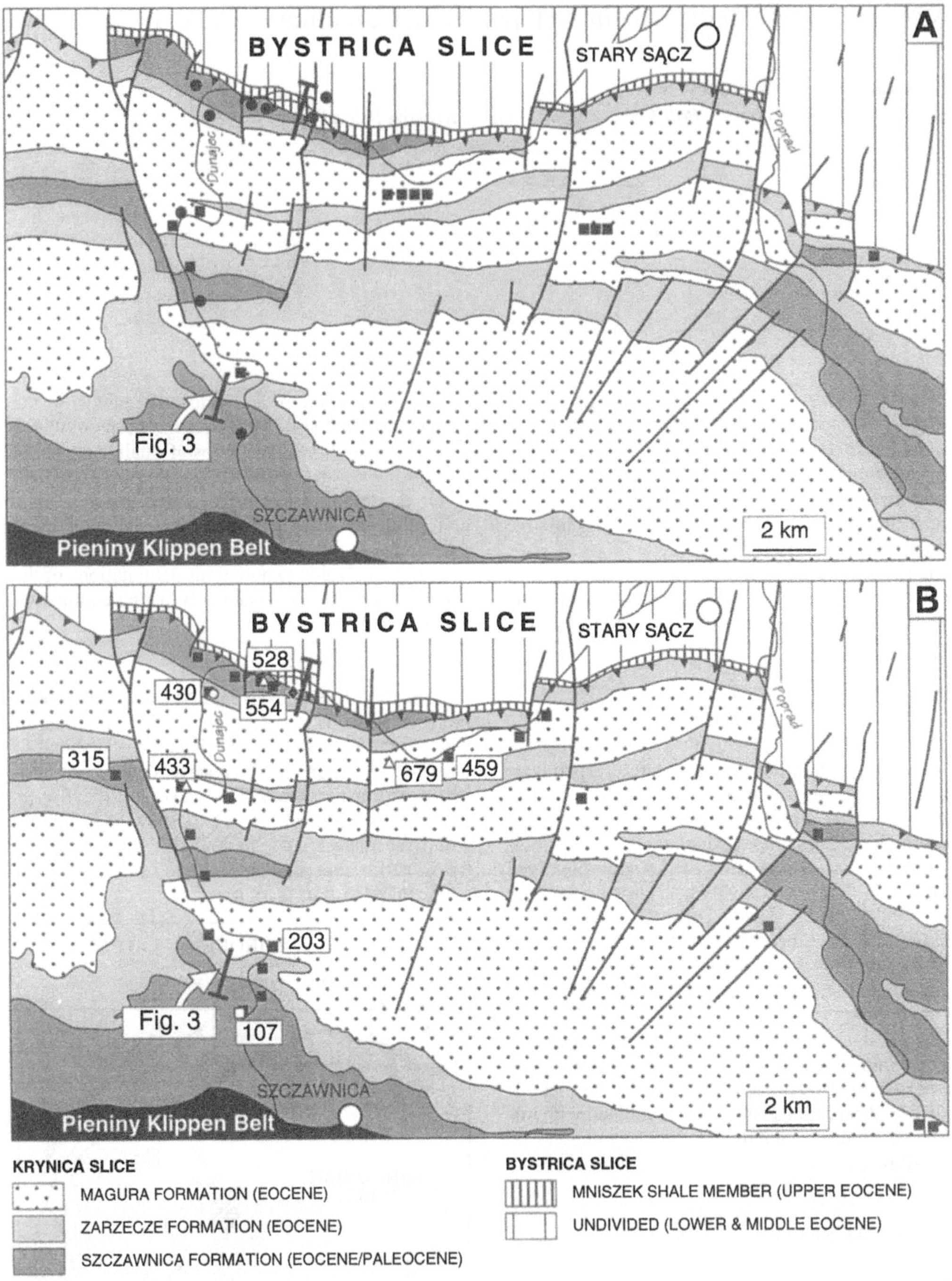

Figure 2. Geological map of study area showing locations of studied and/or sampled exposures and location of Fig. 3. Geology after Burtan et al. (1981) modified. A – Deformation bands (squares) and clay minerals (dots); B – Joints (black squares), hydroplastic thrusts (black diamond), sandstone dykes (white dot), calcite mineralization (white square) and brittle faults (white triangles). Numbers refer to exposures discussed in the text

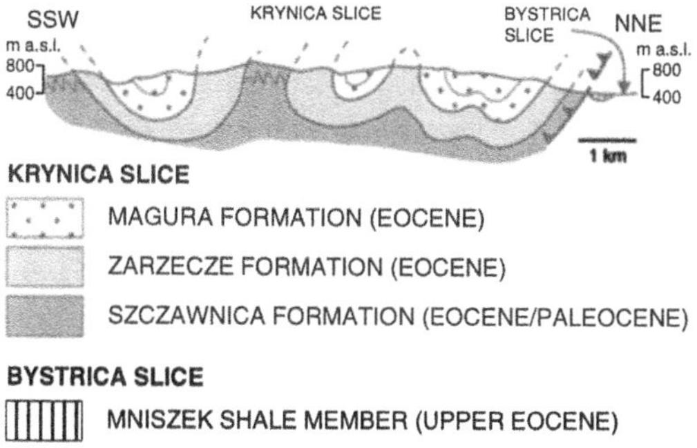

Figure 3. Cross-section of the study area. For location see Fig. 2

3. MAP-SCALE STRUCTURES

In the study area (Figure 2), the Krynica slice comprises Paleocene through Eocene strata. To the south, the Krynica slice contacts with the Pieniny Klippen Belt along subvertical faults, whereas towards the north the slice is thrusted over the Bystrica slice of the Magura nappe along the Krynica overthrust (Figure 3). In the study area, the Krynica slice is deformed in open folds, except locally where the Szczawnica Formation is tightly folded. The axes of the map-scale folds trend W-E to NW-SE (Figure 2).

In the hanging wall of the Krynica overthrust, the strata which abut the thrust surface, range in age from the Paleocene up to Middle Eocene. In contrast, in the footwall of the overthrust, the thrust surface contacts mostly with the Upper Eocene Mniszek Shale Member, which spans the youngest strata in the southern part of the Bystrica slice. The Mniszek Shale Member in the discussed area contains intercalations of olistostrome. This olistostrome appears to be not affected by the map-scale folding.

4. SMALL-SCALE STRUCTURES

Five groups of small-scale structures have been studied: hydroplastic faults, deformation bands, joints, sandstone dykes and brittle faults.

4.1 *Hydroplastic faults*

In the hanging wall of the Krynica overthrust, close to the thrust surface, the Paleocene sandstones of the Szczawnica Formation (Figure 2B – station 554) are cut by small-scale thrusts which are parallel to the map-scale overthrust (Figure 4A). This points to a genetic link between the small-scale thrusts and the Krynica overthrust. In thin sections (Figure 5), these small-scale thrusts display features of hydroplastic faults (Petit & Laville 1987). The thrusts show undulating surfaces which do not crosscut quartz grains. In the host rock close to the thrusts, the grain size is considerably reduced and some mica flakes are oriented parallel to the thrust and oblique to the bedding. Signs of feldspar cataclasis are common in the host rock close to the small-scale thrusts, but we have not observed there any traces of quartz cataclasis and calcite cataclasis.

4.2 *Deformation bands*

Deformation bands (DB) have been observed in thick-bedded sandstones at 12 stations within the Eocene strata of the Zarzecze and Magura formations (Figure 2A). The majority of the observed DB strike parallel to the strike of the host strata (fold-parallel). These fold-parallel DB display exclusively reverse offsets. These features suggest that the DB are related to the map-scale folding.

Seventy DB have been studied in thin-sections (Świerczewska & Tokarski, in press). These DB present full spectrum from DB with no cataclasis to DB with strong feldspar cataclasis and with some quartz cataclasis. The cross-cutting relationships indicate that the cataclastic DB post-date DB showing no cataclasis. The relationship of the DB to sedimentary fea-

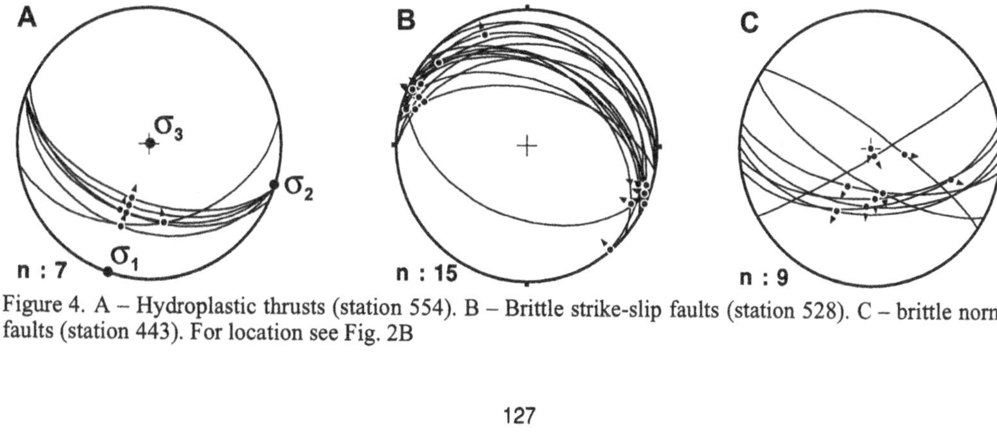

Figure 4. A – Hydroplastic thrusts (station 554). B – Brittle strike-slip faults (station 528). C – brittle normal faults (station 443). For location see Fig. 2B

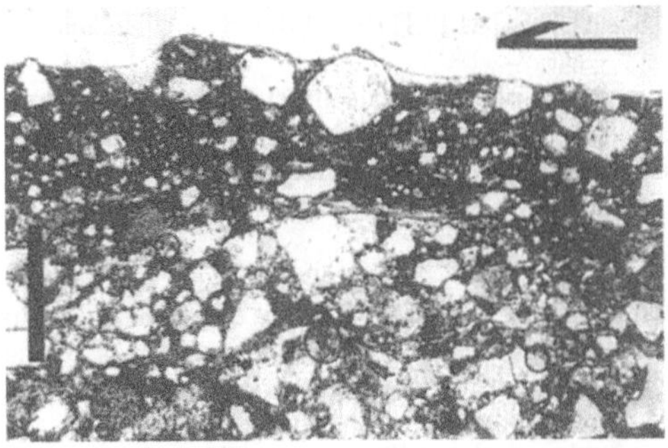

Figure 5. Crossed-polars photomicrograph of hydroplastic thrust (station 554). Note undulating fault surface and reduction of the grain size close to this surface. Arrow indicates sense of movement of lacking fault limb. Scale bar is 0.5 mm long. For location see Fig. 2B

tures indicates that the earliest DB were formed during deposition of the strata involved (Świerczewska & Tokarski, in press). The geometrical relationship of DB to the map-scale folds indicates that DB with minor cataclasis were formed in places after the folding was completed (Tokarski et al. 1995). All DB contain calcite cement, but we have not observed calcite cataclasis in any of the examined DB.

4.3 *Joints*

Joints have been studied in sandstones at 23 stations within the whole discussed stratal sequence (Figure 2B). At particular stations there occur 1-5 sets of joints. These joints comprise three sets of cross-fold joints and two sets of fold-parallel joints. This study has been focused on the cross-fold joints. These joints comprise single set of T joints striking subperpendicular to the map-scale fold axes and two conjugated sets of D_1 and D_2 joints striking at high angles to the map-scale fold axes. All cross-fold joints are oriented perpendicular to stratification. Cross-fold joints were observed in all stations: D_1 and D_2 joints (Figure 6A) at 19 stations, T joints (Figure 6B) at 16 stations and all three sets (Figure 6C) at 13 stations. The observed acute angle between D_1 and D_2 joints is 19–80° (mean value 50°). The deviation between strike of the T joints and strike of the acute bisector between the D_1 and D_2 joins is 0–12° (mean value 6°). No stratigraphic control on the development of the joints has been observed.

4.4 *Sandstone dykes*

At station 430 (Figure 2B) a net of sandstone dykes stick out from the top surface of a sandstone bed which is 7 m thick. Particular dykes are up to 85 cm long, up to 10 cm thick and they stick out up to 4 cm above the sandstone bed. The dykes are oriented at high angles to stratification. Orientations of traces of 139 dykes on bedding surface have been measured. Plot of these traces (Figure 7A) is very similar to the plot of traces of cross-fold joints on this surface (Figure 7B).

4.5 *Brittle faults*

The strata of the Krynica slice are cut by numerous small-scale reverse, strike-slip and normal brittle faults (Figure 4B, C). Cross-cutting relationships show that the faults post-date deformation bands (Świerczewska & Tokarski, in press). The faults are commonly related to calcite mineralization. In thin-sections (Figure 8), these faults display planar clear-cut surfaces cutting quartz grains. Close to fault surfaces, quartz grains commonly reveal fractures which dip in the direction of movement of the opposite fault limb. Cataclasis of quartz, feldspar and calcite is very common close to the discussed small-scale faults.

5. CALCITE MINERALIZATION

Mineralization of joints in sandstone has been studied at one station (Figure 2B - station 107) (Świerczewska et al. 1997). Numerous joints are filled there by calcite veins which were formed in four successive stages (1-

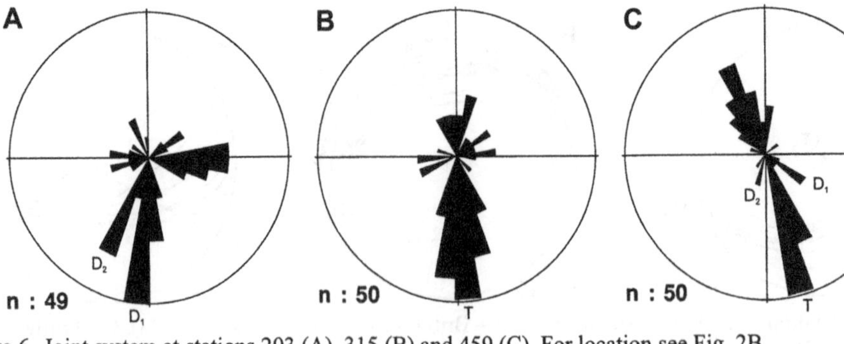

Figure 6. Joint system at stations 203 (A), 315 (B) and 459 (C). For location see Fig. 2B

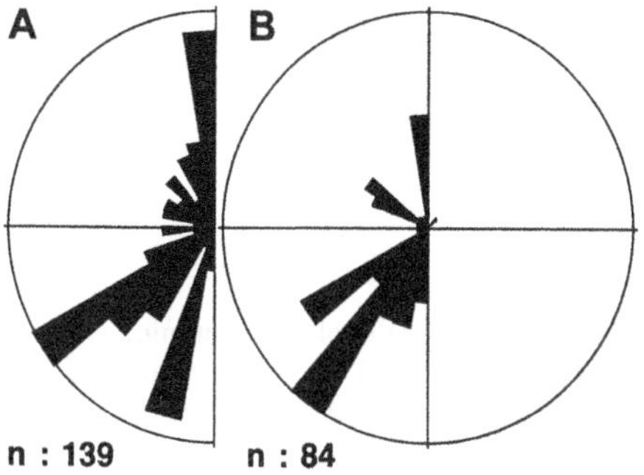

Figure 7. A – Sandstone dykes and B – joints at station 430. For location see Fig. 2B

4). The mineralization is largely of post-folding age except for the stage (1) which possibly occurred when the strata were still in horizontal position. During this stage veins of columnar calcite filling D_1 and D_2 joints were formed. Orientation of calcite fibre shows that the veins were formed due to fold-parallel extension.

6. DIAGENETIC MATURITY

The diagenetic grade has been studied tracing transformation of smectite to illite (Środoń, 1984). Samples have been taken at seven sites within the Krynica slice and at two sites in the footwall of the Krynica overthrust within the Bystrica slice (Figure 2A). The content of smectite varies between 15-35% which corresponds to diagenetic grade up to advanced diagenesis. The diagenetic grade neither increases towards the bottom of the studied stratal sequence nor changes across the Krynica overthrust.

7. DISCUSSION

7.1 *Thrusting*

The small-scale hydroplastic thrusts in the hanging wall of the Krynica overthrust (Figure 5) were formed when the Paleocene host strata were poorly indurated, i.e. possibly during Paleocene times. We believe that this may constrain the lower limit of the time of thrusting on the Krynica overthrust. In our interpretation, the origin of the Upper Eocene olistostrome in the footwall of the Krynica overthrust can be related to a relief generated by overthrusting. The olistostrome is unfolded that may indicate that it was emplaced after cessation of thrusting. Thence, the Late Eocene age of the olistostrome may constrain the upper time limit of the thrusting. The lack of change in diagenetic maturity of clay minerals across the overthrust confirms an early age of the thrusting.

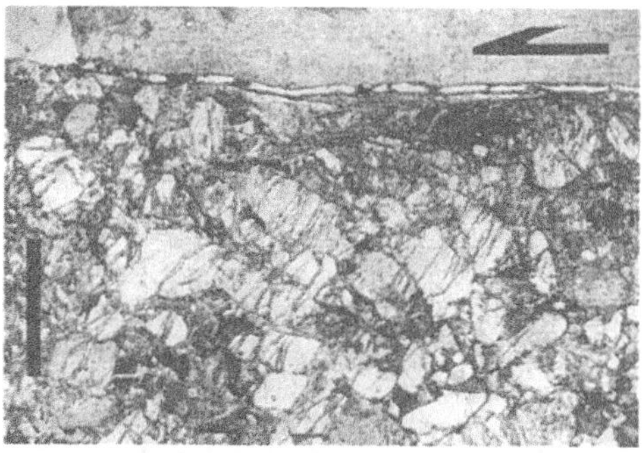

Figure 8. Plane light photomicrograph of brittle normal fault (station 679). Note planar clean-cut fault surface and fractured quartz grains close to this surface. Arrow indicates sense of movement of lacking fault limb. Scale bar is 0.5 mm long. For location see Fig. 2 B

7.2 *Folding*

The analysis of the deformation bands has shown that they were formed during and after map-scale folding. The folding started et least during deposition of the Eocene strata and was completed when these strata were still poorly indurated, most probably during Eocene times. The constant diagenetic grade of the studied stratal sequence suggests that the illitization postdated folding. This confirms early age of the folding.

7.3 *Jointing*

The subperpendicular alignment of both T joints and the acute bisector between the D_1 and D_2 joints to the map-scale fold axes indicates a relationship between the cross-fold joints and the map-scale folds. In the Outer Carpathians, cross-fold joints are early features formed when the host strata were still in horizontal position (Mastella et al. 1997). Moreover, in the studied area, at least the Eocene portion of the discussed stratal sequence was folded syndepositionally. It appears therefore, that the cross-fold joints recorded stress regime which occurred during deposition of the strata involved. This very early age of the cross-fold joints is supported by the similar orientations of the joints and sandstone dykes (early features) at the station 430 (Figure 7). The similar development of the cross-fold joints in the whole discussed stratal sequence indicates that the overall stress arrangement remained constant during Paleocene to Eocene times in the studied area.

7.4 *Induration and diagenetic maturity*

The results of analysis of deformation bands and hydroplastic faults indicate that the map-scale folding and thrusting along the Krynica overthrust occurred when the strata involved were poorly indurated. This is confirmed by results of studies of the diagenetic grade indicating that the diagenetic maturity of the studied stratal sequence was accomplished only after completion of the map-scale folding and thrusting.

7.5 *Calcite mineralization*

No calcite cataclasis has been found in the studied deformation bands. This indicates that calcite mineralization was introduced into the host strata mostly after completion of the map-scale folding. This is confirmed by the lack of signs of calcite cataclasis in the hydroplastic small-scale thrusts. The bulk of calcite mineralization filling joints is of a post-folding age as well.

7.6 *Faulting*

The cross-cutting relationships between the faults and deformation bands (Świerczewska & Tokarski, in press) indicate that the faults post-dated the map-scale folds. Moreover, the microscopic features of the small-scale brittle faults (Figure 8) indicate that the faults were formed after calcite cementation of the strata involved and after complete induration of these strata.

7.7 *Stress permutations*

It follows from the foregoing discussion that during deposition of the studied stratal sequence folding, thrusting, jointing, dyking and veining (phase 1 of mineralization) all occurred simultaneously. This seems to be hardly possible as the particular deformations were formed in different stress regimes: folding and thrusting in the regime of reverse faults, jointing in the regime of strike-slip faults, while dyking and veining in the extensional regime. This apparent paradox can be explained by the occurrence of numerous permutations of the intermediate principal stress (σ_2) from horizontal (folding and thrusting) to vertical (jointing) position as well as by fluctuations in the magnitude of minimum principal stress (σ_3). Stress permutations during single tectonic events have been documented in numerous studies worldwide (e.g. Angelier 1977; Angelier et al. 1990).

8. CONCLUSIONS

1. Within the studied portion of the Magura nappe the map-scale folding and thrusting as well as cross-fold jointing occurred in Paleocene–Eocene times during deposition of the strata involved. 2. The folding and thrusting were completed when the strata involved were still poorly indurated. 3. The folding, thrusting and jointing occurred simultaneously due to numerous stress permutations. 4. The bulk of calcite mineralization was introduced after completion of folding. 5. Still later, the completely indurated strata were cut by brittle faults.

REFERENCES

Angelier, J. 1977. Sur l'évolution tectonique depuis le Miocène supérieur d'un arc insulaire Méditerranéen: l'arc Égeén. Rev. Géogr. Phys. Géol. Dynam., 19: 271-94.

Angelier, J., F. Bergerat, H.T. Chu W.S. Juang & C.Y Lu. 1990. Paleostress analysis as a key to margin extension : the Penghu Islands, South China Sea. Tectonophysics, 183: 161-176.

Burtan, J., J. Golonka, N. Oszczypko, Z. Paul & A. Ślączka. 1981. Geological Map of Poland, 1:200000, Sheet Nowy Sącz. Warszawa.

Mastella, L., W. Zuchiewicz, A. K. Tokarski, J. Rubinkiewicz, P. Leonowicz & R. Szczęsny. 1997. Application of joint analysis for paleostress reconstructions in structurally complicated settings: Case study from Silesian nappe, Outer Carpathians (Poland). Przegląd Geol., 45: 1064-1066.

Petit J.-P. & E. Laville. 1987. Morphology and microstructures of hydroplastic slickensides in sandstone. In: M.E. Jones & R.M.F. Preston (eds), Deformation of sediments and sedimentary rocks, Geol. Soc. Special Publ., 29: 107-121.

Roca, E., G. Bessereau, E. Jawor, M. Kotarba & F. Roure. 1995. Pre-Neogene evolution of the Western Carpathians: Constrains from the Bochnia-Tatra Mountains section (Polish Western Carpathians). Tectonics, 14: 855-873.

Środoń, J. 1984. X-Ray powder diffraction identification of illitic materials. Clays and Clay Minerals, 32: 337-349.

Świerczewska, A. & A.K. Tokarski. In press. Deformation bands and history of folding in the Magura nappe, Western Outer Carpathians (Poland). Tectonophysics.

Świerczewska, A.,V. Hurai & A.K. Tokarski. 1997. Structural control on mineralization of joints: case study from Paleogene flysh, Outer Carpathians (Poland). Przegląd Geol., 45: 1107-1108.

Tokarski, A.K., A. Świerczewska & M. Banaś. 1995. Deformation bands and early folding in Lower Eocene flysch sandstone, Outer Carpathians, Poland. In: H.-P. Rossmanith (ed.), Mechanics of jointed and faulted rock: 323-327. Balkema.

Faults

Mechanics of Jointed and Faulted Rock, Rossmanith (ed.) © 1998 Taylor & Francis, ISBN 90 5410 955 6

Extensional joints and faults: A 3D mechanical model for quantifying their ratio – Part 1: Theory

Riccardo Caputo
Dipartimento di Scienze della Terra, Università G. d'Annunzio, Chieti Scalo, Italy
Giovanni Santarato
Istituto di Mineralogia, Università di Ferrara, Corso Ercole I d'Este, Italy

ABSTRACT: Due to the brittle behaviour of rocks during deformation in surficial conditions, numerous fractures develop, both extension joints and faults. The ratio between the number of extension joints and the number of faults (*j/f*) which form in a rock mass has been investigated. In order to simulate the stress evolution and the consequent generation of fractures during time, a mechanical model is proposed, while a specific computer program has been prepared. The model takes into account i) the stochastic distribution of the tensile strength of rocks, ii) the temporal and local stress variations caused by continuous fracturing events and iii) the sealing effect occurring along fractures with time. Several geological parameters, such as the tectonic stress rates and their ratio, the depth (*i.e.* the confining pressure) and the role of the pore-fluid pressure have been also considered. In the companion paper (this volume), numerical calculations and comparison with field examples are discussed.

1 INTRODUCTION

Most geologists agree that joints (most of which are extension fractures) are commonly very much more abundant than faults. The proportion between the two kinds of structures depends on many factors, among which the tectonic setting, but the ratio between the number of extension joints and the number of faults (here called the *j/f* ratio) is commonly very high.

The principal aim of the present research is to investigate the evolution of a rock mass during a brittle deformational event. For this purpose a mechanical model is proposed to describe the occurrence of both extension joints and faults. A dedicated computer program, JF3, has been written in FORTRAN77. In particular, one of the major goals of the project is to quantify the ratio between the two kinds of fractures as a function of the diverse parameters which govern the deformation of rocks. In the present paper, the analytical model and the computer program are described and discussed.

2 THE MODEL

2.1 *Genetic partition of the stress tensor*

Any stress field can be represented at any point by a unique stress tensor, $\mathbf{T}_{total}$. However, stress fields originate from several sources, including thermal, overburden, pore fluid, tectonic and diagenetic processes, all independently pervading with different magnitudes, rock volumes of different size. Thus, the stress tensor at any point can be visualised as the sum of several tensors each representing a different genetic component:

$$\mathbf{T}_{total} = \mathbf{T}_{thermal} + \mathbf{T}_{grav} + \mathbf{T}_{fluids} + \mathbf{T}_{tectonic} + ..$$

Sometimes one component prevails over the others and the stress field can be easily characterised. More often, however, the stress field is composite and the magnitude of each individual genetic contribution is difficult, if not impossible, to recognise. Each genetic component varies in space and time, with particular gradients and rates, respectively, in a different way from the other components. This implies that the evolution of the total stress tensor at any point in a rock mass can be very complex.

Moreover, we emphasise that fractures do not care about the origin of the stress field components when they form and do not yield any information about the different genetic stress components at the time of their formation. Rocks behave according to the total and unique *in situ* stress state. Nevertheless, this stress state is undoubtedly the result of the overlapping influences of several individual genetic components and from this intrinsic behaviour the basic assumptions for the mechanical model we propose in the followings have been developed.

When dealing with a stress field, the volume of the rock within which it is more or less uniform also needs to be considered. For example, stress trajectories are commonly smoothly curving on the scale of continents (*e.g.* Zoback & Zoback 1991). On the other hand, from rock mechanics it is well known that even a microscopic fracture strongly perturbs the local stress field (*e.g.* Jaeger & Cook 1979, Atkinson 1987). At an intermediate scale, structural geologists studying outcrops within which there are numerous fractures, are often perplexed which approach to follow. A conclusion that we reach from observation of natural fractures and consideration of the variability of the *in situ* stress field with depth and location is that before fractures initiate and propagate the stress field is uniformly oriented, but between fracture propagation episodes it is much less regularly oriented.

2.2 *Assumptions*

The basic assumptions and the mathematical notations used for the present research, mainly follow the work of Caputo (1995), where a mechanical model of extensional brittle deformation has been set up by using a simplified approach of Hooke's law.

Figure 1. Nearly horizontal Liassic limestones and mudstones cut by abundant extension joints. West of Lilstock, Somerset.

Also in the present note, we consider a volume of rock of outcrop scale (of the order of 10^3 m^3; Fig. 1), and a stress tensor whose principal components are σ_x, σ_y and σ_z parallel to the E-W, N-S and vertical directions, respectively (symbols are listed in Table 1). The convention of positive values of stress being compressive is also followed.In order to define and apply a stress tensor to the given volume of rock and especially in order to tentatively quantify the process, it is useful to analyse in detail the possible genetic components as above mentioned. If this volume is supposed to be at a given depth, z, then the 'gravitational' component of the stress field, here considered to be the confining pressure, p_c, is given by ρgz, where ρ is the mean density of the overlying rocks and g is the gravity. This notation is realistic one on the basis of the results of *in situ* measurements made in deep bore-holes (*e.g.* Ranalli & Chandler, 1975). Another important genetic component is contributed by the pore-fluid pressure, p_f. In shallow crustal conditions, say, less than 1000 m, the value of p_f normally ranges between hydrostatic and lithostatic, that is, $0.4 < \lambda_e < 1$ (*e.g.* Mouchet and Mitchell, 1989).

Two different 'tectonic' genetic components have been considered. The first is a far-field strike-slip regime within which $\sigma_x=\sigma_3$, $\sigma_y=\sigma_1$ and $\sigma_z=\sigma_2$, while the second is generated by a far-field extensional regime within which $\sigma_x=\sigma_3$, $\sigma_y=\sigma_2$ and $\sigma_z=\sigma_1$. In the former case, the gravitational genetic component can be considered to be constant in time while, if the tectonic 'engine' being considered is constant, σ_x and σ_y will be functions of time. In the latter case, deposition is assumed to occur due to extension and subsidence (*viz.* p_c increasing with time), while the tectonic genetic component is characterised by a tensile x-axis ($\sigma_x<0$). In both regimes, the total stress tensor will continuously vary as a function of time due to the applied stress rates being, in the former case, $\underline{\sigma}_x<0$ and $\underline{\sigma}_y>0$ and, in the latter case, $\underline{\sigma}_x<0$ and $\underline{p}_c>0$.

As a consequence, failure conditions can be locally achieved. Therefore, all the components of the local stress tensor experience a significant drop in magnitude either positive or negative as a result of the sudden release of elastic energy. After the failure event, the same stress rates which progressively 'charged' the rock mass and generated the failure conditions, and which were supposed to be constant in time, firstly, continue to affect the local stress field as well as the whole rock-mass; secondly, generate continuous variations of the tensor components; and, thirdly, potentially restore the failure conditions (Caputo, 1995).

There is abundant field evidence that the growth of both extension fractures (*e.g.* dilational veins) and shear fractures (*i.e.* faults) involves cyclical processes. Specific discussion on the cyclic behaviour of the stress tensor within rock volumes during brittle deformational events is in Caputo and Hancock (in press). For the purpose of the model here presented, we assume that: (a) both extension and shear fracturing are cyclic processes and (b) that all fractures seal after a given time thus allowing the far-field stresses to 'work' again on the fractured rock volume.

2.3 *Analytical approach*

Following the mechanical proposed by Caputo (1995) for the study of orthogonal joint systems and according to the different genetic components of the stress tensor as above discussed, the final tensors which apply to the considered rock volume, for the strike-slip regime (left) and the tensional regime (right), respectively, are the following:

$$\begin{array}{ll} \sigma_x = t \cdot \underline{\sigma}_x + p_c - p_f & \sigma_x = t \cdot \underline{\sigma}_x + t \cdot \underline{p}_c - p_f \\ \sigma_y = t \cdot \underline{\sigma}_y + p_c - p_f & \sigma_y = t \cdot \underline{p}_c - p_f \\ \sigma_z = p_c - p_f & \sigma_z = t \cdot \underline{p}_c - p_f \end{array} \tag{1}$$

With increasing time, the stress tensor will evolve and the stress ellipsoid will change its shape. Both tectonic regimes we took into account are characterised by $\underline{\sigma}_x < 0$ (*i.e.* tension). As a consequence, at a given time (t_f), σ_x equals the tensile strength of the rock, T, and an extension joint forms perpendicular to the x-direction. When failure occurs, σ_x undergoes a positive stress drop. Indeed, within a volume near the fracture, the absolute value is strongly reduced from an original negative (*i.e.* tensile) value (Pollard & Segall, 1987; Caputo, 1995). An expression of this positive stress drop was reported after the 1983 (M=7.3) Borah Peak earthquake which was on a normal fault. Accompanying the earthquake there was violent fountaning of ground water from fissures striking parallel to the rupture trace (Wood *et al.*, 1985). This process confirms the occurrence of a sudden post-failure increase (*i.e.* positive stress drop) of σ_3 (Sibson, 1991). However, after the catastrophic failure event, the confining pressure and the pore fluid pressure are fully restored, but not the tectonic stress component. Thus, the stress tensor will be temporarily and locally as follows

$$\begin{array}{ll} \sigma_x \approx p_c - p_f & \sigma_x \approx t \cdot \underline{p}_c - p_f \\ \sigma_y = t \cdot \underline{\sigma}_y + p_c - p_f & \sigma_y = t \cdot \underline{p}_c - p_f \\ \sigma_z = p_c - p_f & \sigma_z = t \cdot \underline{p}_c - p_f \end{array} \tag{2}$$

representing the local (*i.e.* referred to the volume element) and temporal stress tensor which is evidently different from that occurring before fracturing. Until the fracture is open, the local stress field is perturbed because σ_x remains nil on the surface and a strong gradient exists in the x-direction of the stress tensor (Pollard & Segall, 1987; Caputo, 1995). In contrast, the orthogonal component of the stress rate continues to act at all scales. Only once cementation has sealed the fracture, at time t_c, will the tectonic component of the stress rate in the x-direction operate again in the near-field volume of the fracture.

As concerns faulting, because the far-field tectonic component of the strike-slip regime here considered is characterised by $\underline{\sigma}_x < 0$ and $\underline{\sigma}_y > 0$, the differential stress ($\sigma_y - \sigma_x$) increases with time. Consequently, shear fractures can also locally occur. Two different mechanical criteria were applied. Firstly, was checked if a critical stress difference, σ_c, is achieved (Ranalli & Yin, 1990), that is, when

$$\sigma_1 - \sigma_3 \geq \sigma_c \tag{3}$$

and, secondly, was applied the three-dimensional extension of the Griffith criterion (Murrell, 1963)

$$(\sigma_z - \sigma_x)^2 + (\sigma_x - \sigma_y)^2 + (\sigma_y - \sigma_z)^2 = \\ = 24 \cdot T \cdot (\sigma_x + \sigma_y + \sigma_z) \tag{4}$$

Likewise when jointing occurs, also during faulting the stress tensor is perturbed; the new local stress conditions not mimicing the far-field tectonic regime. As a first approximation, it is assumed that after each faulting event the stress drops to a constant value which is a function of the distance from the fault plane. For simplicity, the stress drop is distributed throughout an ellipsoid-shaped volume around the fault.

3 THE PROGRAM

3.1 *Flow-chart of the program*

Numerical simulations of the complex processes occurring during brittle deformation were performed by considering a cube of rock and dividing it into N^3 elementary cells. As a first approximation at the working scale of the model, the stress state throughout each cell can be represented by a unique stress state calculated *via* equations [1] at its centre, with coordinates i, j and k. Obviously, the smaller the side of the cell, the more realistic and accurate is the result. With this assumption, the stress field calculated for each cell should be considered as a rough average for the whole cell.

The whole procedure can be schematically represented by a flow-chart, with several operations performed for each elementary cell, summarised as follows. Firstly, the tensile strength, T_{ijk}, is randomly attributed according to a Weibull distribution law characteristic of all natural rocks (Weibull, 1951). This operation (oper. 1) takes into account situations in real rock masses where hetereogenities always exist. It is well known from tensile tests on identical materials and under the same mechanical conditions, that the stress required for the failure of the different specimens is statistically determined, and the probability density function of failure at any stress σ (*i.e.* tensile strength) is given by the function $m \cdot \sigma^{m-1} \cdot \exp(-\sigma^m)$, where *m* is the Weibull modulus.

Secondly, the constant genetic components of the stress field are applied (oper. 2), followed by the components which are functions of time (oper. 3) starting at t=0.

Thirdly, according to equations [1], the stress tensor for each elementary cell is calculated for discrete time increments, t_s (oper. 4).

Fourthly, the existence of local extensional failure conditions is checked, that is if $\sigma_{x,ijk} \leq T_{ijk}$ (oper. 5). Everytime tensional failure conditions are achieved, the program enters a subroutine (oper. 6) that generates an extension joint within the elementary cell and recalculates the stress tensor according to equations [2]. Moreover, in order to simulate the perturbed stress field surrounding a fracture in a layered rock mass (Caputo, 1995; Caputo and Hancock, in press) also for the contiguous cells a partial stress drop has been assumed and the local stress tensor is recalculated accordingly.

The next step is to check whether the conditions for shear failure occur (oper. 7) according to the above mechanical criteria (equations [3] and [4]). If failure conditions are verified, the program enters another subroutine (oper. 8) generating the fault and recalculating the stress tensor for the elementary cells along the fracture plane as well as for several other cells within an ellipsoid-shape volume around the fault. The induced stress drop in the surrounding cells decreases with distance from the fault and falls to zero at a maximum distance of a fixed number of elementary cells.

Eventually, the last step of the loop seals joints and faults if a given time, t_c, has gone by (oper. 9). Thus, the stress tensor of a cell previously affected by an open fracture can again freely grow as a result of the influence of the tectonically generated stress rates. Then, the program repeats operations 4 to 9 and the whole loop continues for a total time, t_t.

3.2 *Computer-dependent parameters*

Computer simulations of a continuous process in space and time, as the one here considered, must undergo two main constraints: the amplitude of the core memory of the computing machine and the velocity of the math processor. This implies, first, the necessity of limiting calculations to a finite volume and especially to a reduced number of elementary cells, N^3; second, to impose a proper finite time step, t_s, to the program. Time step and the total cell number have both direct influence on the required computing time and their optimisation tests are discussed in the following.

Moreover, the process must run long enough in order to avoid that the initial transient of the function perturbs the results. In our specific problem, this occurs when the *j/f* ratio assumes a constant trend.

In order to perform realistic numerical simulations, the size of the considered rock volume must contain the largest structures we analyse (*i.e.* faults). In particular, due to the specific aims of the research, that is an attempt of quantifying a brittle deformational process, this size should be big enough to contain a large number of both structures. According to common field observations in weakly deformed sedimentary terrains, a minimum volume is represented by a 10 m sided cube. In order to subdivide this cube into elementary cells, the smaller structures (*i.e.* extension joints) should be considered. In natural examples of layered rocks, most joints do not affect more than a single bed and, because in our model joints affect one elementary cell, the size of the cell must be proportional to the bed thickness.

Therefore, as a first step of the computing process, it was important to check if and how this computer constraint affects the final results. After a number of tests varying N^3, from 30^3 to 100^3, and keeping constant the remaining parameters, we found that a total cells number $N^3=50^3$ seems to be a reasonable minimum value being a good compromise between stability of the algorithm and the required machine time and memory consumption. This value thus implies an average bed thickness of 20 cm which is comparable with field observations of jointed rocks (Fig. 1).

As concerns the time step, t_s, the closure time, t_c, is necessarily the reference value and therefore the former parameter must be shorter, or at most equal, to the latter one ($t_s \leq t_c$). We verified that the algorithm was stable even at the maximum admitted value, that is $t_s = t_c$, while possible ranges of the closure time are discussed in the following section.

At the beginning of the process, the contribute of the time-dependent genetic stress components is nil, and the system is progressively charged with time. Because in natural rocks tensile failure conditions are much more easy to obtain than shearing, during this initial period only extensional joints occur before the first faults are produced. Necessarily the *j/f* ratio starts from an infinite value, but rapidly diminishes with time and stability of the function is commonly reached for $t_t \geq 10^{12}$ sec, which corresponds to about 30 ka (Fig. 2). According to the assumed values of t_s, stability of the *j/f* function occurs after several thousands loops, commonly in the range 10^3-10^4. From a geological point of view, it should be noted that the commonly accepted period of activity for a brittle tectonic phase is 2-3 orders of magnitude longe. Accordingly, this transient period of the model represents a very short time span of the whole deformational event thus being not influent for the final result.

Furthermore, possible boundary effects have been

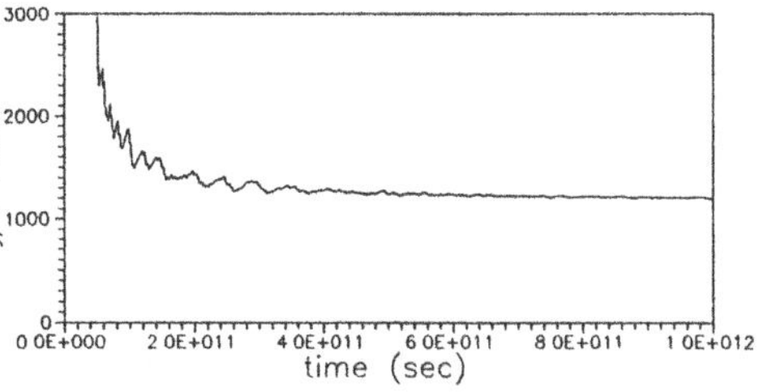

Figure 2. Example of the *j/f* ratio as a function of time. After an initial transient period, the stability of the function is reached. T = $-2\cdot10^6$ Pa; ρ = 2600 kg/m3; depth = 200 m; $\sigma_x = -5\cdot10^{-4}$ Pa/sec, $\sigma_y = 2\cdot10^{-4}$ Pa/sec; $\lambda_e = 0.83$; $t_t = 1\cdot10^{12}$ sec; $t_s = 2\cdot10^8$ sec; $t_c = 2\cdot10^8$ sec.

considered. In fact, according to the model when a failure event occurs, a perturbation of the stress field is induced in several elementary cells surrounding the fractured one(s). Therefore, the cyclic behaviour of the stress tensor within a generic cell (Caputo, 1995; Caputo and Hancock, in press) can be also induced by that of the surrounding cells. For this reason, in a finite cubic model, all boundary cells can be affected by nearby-induced stress-drops from their internal side only. If we consider that the number and distribution of the perturbed cells is different in the two fracture mechanisms, it is clear that these boundary cells give a spurious contribution to the *j/f* ratio. In order to detect this boundary effect, several tests have been performed by varying the size of the matrix border. Within this outer shell border, the complete computing procedure was performed, fractures could freely occur and produce their effects on the nearby cells, but they were not counted and did not contribute to the *j/f* ratio. Results show that with a matrix border of 5 cells all around, the obtained values are quite stable.

3.3 *Model parameters*

In this section realistic values of the many parameters involved in the model are introduced into the above equations in order to numerically simulate the process.

First of all, as concerns the tensile strength, the choice was critical due to the large range of possible values and the strong influence of this parameter in the model. Values between 0.5 and 2 MPa have been assumed corresponding to fairly indurated sedimentary rocks.

As previously discussed, the random Weibull distribution is the best suited to describe the local variability of the tensile strength of the elementary cells. For testing, we assigned different values to the modulus *m* of the exponential Weibull function (*m* = 4, 8 and 16). The histograms relative to the three distributions are plotted in Fig. 3. The choice of *m* has a strong influence on the behaviour of the *j/f* ratio. Indeed, according to the four dimensional approach of the model, two elementary cells with different values of the tensile strength necessarily require different time spans to reach failure conditions. Therefore, due to the Weibull distribution of the tensile strength and the cyclic behaviour of the stress tensor for each cell, at the beginning of the process the *j/f* ratio has a periodic trend. The width and the amplitude of this initial period is a direct function of the Weibull modulus, *m*. The amplitude is greatest and the period is shortest with the highest Weibull modulus (*m* = 16), while the transient period during which the oscillations of the function occur, is shortest with a lower modulus (*m* = 4). This feature can be easily interpreted. In fact, if all cells have similar values of the tensile strength (*i.e.* a sharp distribution, high *m* modulus), at the beginning of the process and for several cycles, failure events occur almost simultaneously giving a peak in the *j/f* curve. In the opposite case, with a spread distribution (*m* = 4), the more fragile cells (*i.e.* lower tensile strength) have a stress cyclicity soon overlapping with that of the more resistant cells (*i.e.* higher tensile strength) and therefore the oscillations of the *j/f* curve quickly disappear. Following the work of Brzesowsky (1995), we tentatively assume *m* = 8.

Due to the stochastic initial conditions imposed to the model, we tested whether the choice of the random file affects the numerical simulations and eventually how it does. For this purpose, the

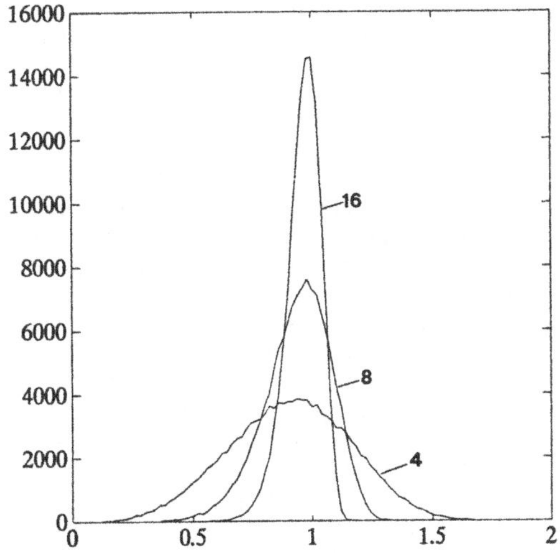

Figure 3. The randomly generated data sets showing a Weibull distribution as obtained with modulus *m* = 4, 8 and 16. Values in abscissa are normalised to 1 and these values have been used for initialising the tensile strength matrix.

program was run several times, keeping the parameters constant, but using different random files generated with the same Weibull modulus. Though the overall Weibull statistical distribution for all files is identical, computer-generated local variations can cause the variability of the results of about ±10%. Consequently, all numerical results here obtained must be regarded as a rough quantitative estimate of the investigated phenomenon and the comparison of different results can be solely performed when the initial random file is the same.

No data are available concerning the closure time, t_c. Too many geometric, physical and chemical parameters would be necessary to quantify the time required for sealing of a fracture. However, a tentative value in the range of 5 to 100 years has been assumed (Eugenio Scandale written communication, 1995). Indeed, the geological meaning of this parameter is relevant, because it determines the length of the transient at the beginning of the simulated tectonic process.

Also more geological parameters such as depth, pore-fluid pressure, subsidence and the tectonic stresses have been varied. Variations of each of these parameters also heavily affect the final as discussed in the companion paper (this volume).

4 CONCLUDING REMARKS

The proposed model should be considered as a first attempt of quantifying extension and shear fractures (and especially their ratio) occurring in a rock volume during a brittle deformational event. The complexity of the natural phenomenon and the large number of involved parameters, forced us to simplify the mechanical approach to the problem. However, the choice of applying the linear elastic theory to the model, seems justified by the following considerations. Although, the linear elastic fracture mechanics approach would be more correct to describe in detail single fracturing events (Kenneth Cruikshank, written communication, 1995), Caputo (1995) has shown that also the simplified approach here applied provides similar analytical solutions. But, above all, due to the large volumes of rock here considered (*viz.* number of elementary cells) and because the present research is an attempt of describing the macroscale evolution during brittle deformational events, only the mesoscale effects of the fracturing process are of interest here. In line with this premise is, for example, the choice of using the tensile strength, T, instead of the fracture toughness to determine extensional failure events. Indeed, several laboratory tests and studies on the effects of the size of specimens, confirm that the rock behaviour under tensile stress conditions is sufficiently well described by the former parameter (*e.g.* Jaeger and Cook, 1979).

The characteristics of the present research can be summarised as follows. i) the stochastic distribution of the tensile strength (T), and probably of other mechanical parameters, generates a complex behaviour of the rock-mass during brittle deformation; ii) the principal parameters affecting the *j/f* ratio have been identified as the tensile strength, T; the confining pressure (*viz.* depth); the ratio between the two horizontal stress rates (Σ) and the ratio between the pore-fluid pressure and the confining pressure (λ_e); iii) within the assumed ranges, these parameters can generate variations of the *j/f* ratio of several orders of magnitude; iv) in particular tectonic conditions such as the strike-slip regime (at great depth and/or with low Σ values), shear fractures can be the only forming structures and no extension joints can occur, therefore nullifying the *j/f* ratio, whereas in the tensional regime the opposite can occur thus genereting an infinite *j/f* ratio.

ACKNOWLEDGEMENTS

The Palaeostress Methodologies Project (resp. P.L. Hancock) supported by the Peri-Tethys Consortium provided funds for fieldwork in England and Wales. Computer facilities and granted computing time (grant # t1pfezt1, resp. G. Santarato) supported by Cineca (Bologna, Italy) are warmly acknowledged.

REFERENCES

[see complete reference list in the companion paper]

Table 1: List of symbols

x, y, z = orthogonal co-ordinate axes system
σ_x, σ_y, σ_z = principal stresses in x,y,z directions
σ_1, σ_2, σ_3 = maximum, intermediate and minimum principal stresses, respectively

rock properties: T = tensile strength
ρ = rock density

genetic stress components:
$\underline{\sigma}_x$, $\underline{\sigma}_y$ = tectonic stress rates in the x- and y-directions, respectively
$\Sigma = \underline{\sigma}_x / \underline{\sigma}_y$ (stress rates ratio)
p_f = pore-fluid pressure
p_c = confining pressure
$\lambda_e = p_f / p_c$

t = time
t_t = total time of program running
t_f = time at failure
t_c = time at sealing
j = number of extensional failure events
f = number of shear failure events

Mechanics of Jointed and Faulted Rock, Rossmanith (ed.) © 1998 Taylor & Francis, ISBN 90 5410 955 6

Extensional joints and faults: A 3D mechanical model for quantifying their ratio – Part 2: Applications

Riccardo Caputo
Dipartimento di Scienze della Terra, Università G. d'Annunzio, Chieti Scalo, Italy
Giovanni Santarato
Istituto di Mineralogia, Università di Ferrara, Corso Ercole I d'Este, Italy

ABSTRACT: By applying the model proposed and discussed in the companion paper (Caputo & Santarato, this volume), numerous calculations have been performed for better understanding the role played by single mechanical and 'geological' parameters and the possible mutual influence. *j/f* ratios have been also estimated from natural examples of faulted and jointed rocks. According to the geological and tectonic conditions of these examples, the observed values are in good agreement with those obtained from the numerical simulations. Possible further applications of the model and the use of the program are also discussed.

1 NUMERICAL SIMULATIONS

Numerous computer simulations have been carried out by using the JF3 program proposed by Caputo & Santarato (this volume; in the following referred to as C&S). At first, the several mechanical and 'geological' parameters have been varied one by one in order to understand the contribution of each one. In a second phase, parameters have been changed two by two, to control their possible inter-dependence. In the following, the effects of the different parameters are briefly discussed due to the limited space. During computation, unless differently specified the values of the other parameters are kept constant (as indicated between brackets in Tab. 1) and referred to as 'standard values'.

1.1 *Strike-slip regime*

Depth is obviously a crucial parameter because it directly controls the confining pressure, p_c, and the pore pressure, p_f. Accordingly, jointing can be greatly enhanced in shallow conditions or inhibited at depth. All diagrams clearly show this pattern.

The parameter with the strongest influence on the *j/f* function is undoubtedly the ratio between the two orthogonal stress-rates, Σ ($-\underline{\sigma}_x/\underline{\sigma}_y$). By varying Σ of one order of magnitude (Fig. 1a), that is from 0.4 ($\underline{\sigma}_x = -2 \cdot 10^{-4}$, $\underline{\sigma}_y = 5 \cdot 10^{-4}$ Pa/sec) to 5 ($\underline{\sigma}_x = -10^{-5}$, $\underline{\sigma}_y = 2 \cdot 10^{-4}$ Pa/sec), at 200 m depth the *j/f* ratio varies of two orders of magnitude, that is from about 40 to 2800. It is noteworthy to say that this variation is linear and strongly depends on depth.

Although in a minor way, also the ratio λ_e (p_f/p_c) directly affects the number of jointing events. By changing λ_e from 0.46 to 0.96, the *j/f* ratio generally increases being also proportional to depth. However, the effects of depth variations on the *j/f* ratio are greatly reduced with very λ_e (*i.e.* high pore-fluid pressures). Also the effects of other parameters such as the tensile strength seem to be attenuated when p_f/p_c is relatively high.

Concerning the tensile strength, by increasing the value, the *j/f* ratio generally also increases (Fig. 2b). Indeed, if the number of extension fractures slightly decreases, the number of shearing events is progressively greatly increased. The results also show that when the tensile strength is low and the confining pressure (*viz.* depth) is sufficiently high, or alternatively Σ is relatively low, jointing can be completely inhibited and shearing is the only fracture mechanism operating in the rock mass. At this regard, it is noteworthy to say that although we precautionally considered two different failure criteria for shear fracturing, according to the results the only mechanism capable of generating faults seems to be the modified Griffith criterion (eq. [4] in C&S). Therefore, in order to better investigate this behaviour of the model, that is the variation of the *j/f* ratio as induced by the cumulative effects of tensile strength and depth or of tensile strength and Σ, we calculated for both extensional and shearing fracturing mechanisms, the time required to load an elementary cell at failure conditions as caused by the two assumed orthogonal stress rates. We thus inverted equations [1a] and [4] (in C&S), respectively, to obtain the failure time, t_f, as a function of the tensile strength, T. The results are represented in Fig. 2, where the two curves have been calculated for different confining pressures (*viz.*

depth, Fig. 2a), as well as for different Σ (Fig. 2b). The computed failure times are intended as the minimum times required for failure within an elementary cell, if no stress drops are induced by nearby cells.

When, for a given tensile strength the shearing failure time is lower than the jointing failure time (Fig. 2), no extension fracturing can occur in that elementary cell. Indeed, according to the model, as a consequence of faulting the induced stress drop is assumed in both orthogonal directions (x and y) and thus both stress components are temporarily nullified. Therefore, next loading conditions will be once more favourable to faulting. In contrast, when the tensile strength generates joints more easily than faults (*i.e.* shorter failure time), the opposite does not occur and shear fractures can also form. In fact, because the primary stress drop induced by jointing is limited to the x-direction, extensional fracturing can only delay but not definitely impede the occurrence of faulting conditions.

In practice, the model is more complex due to the continuous possible stress variations of the elementary cells as induced by the nearby ones (*viz.* induced stress drops; Caputo, 1995) and especially due to the stochastic Weibull initial distribution of

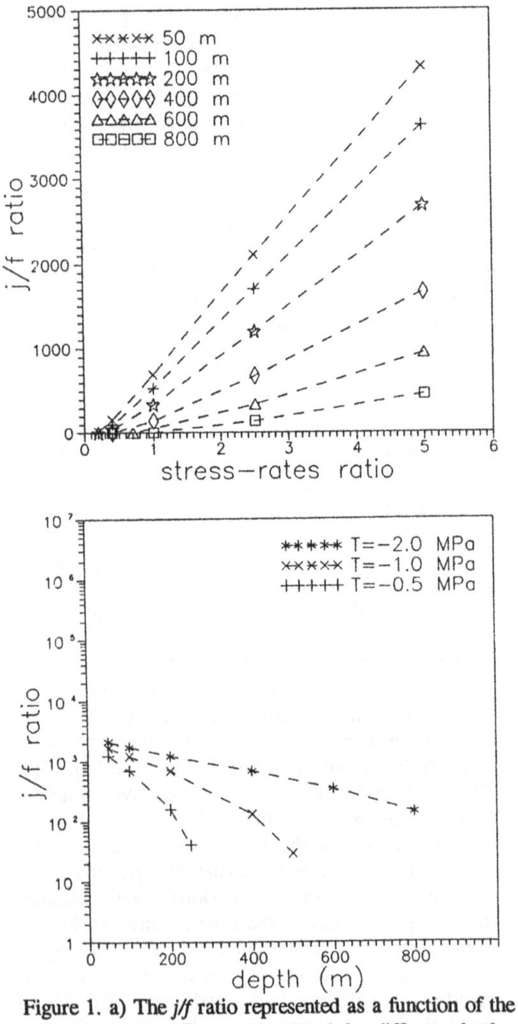

Figure 1. a) The *j/f* ratio represented as a function of the stress-rates ratio, Σ, and calculated for different depths. b). the *j/* ratio represented as a function of depth and calculated for different tensile strengths, T. Other parameters as in Table 1.

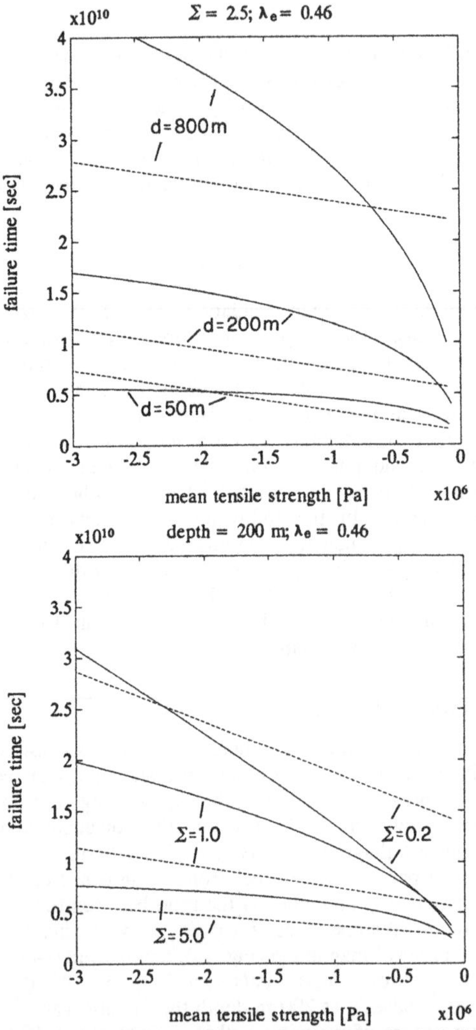

Figure 2. The failure time, t_f, for jointing (dashed lines) and faulting as a function of the tensile strength, T, obtained by inversion of equations [1a] and [4] of C&S, and calculated for different depths (a) and Σ (b). Other parameters as in Table 1.

the tensile strengths. In fact, when a mean tensile strength of -1.0 MPa is assumed, the real possible range of values is from about -0.35 to -1.25 Mpa (Fig. 3 in C&S). Accordingly, the failure times can vary significantly from cell to cell, and some of the cells can stand below a critical strength, where the shearing failure time is less than the jointing failure time, while others are well above. The former cells can generate only faults, the latter both kinds of structures.

As a further consequence of this behaviour, we can explain the occurrence of extensional joints even if their failure time, as computed for the mean tensile strength, is longer than for shearing events. For

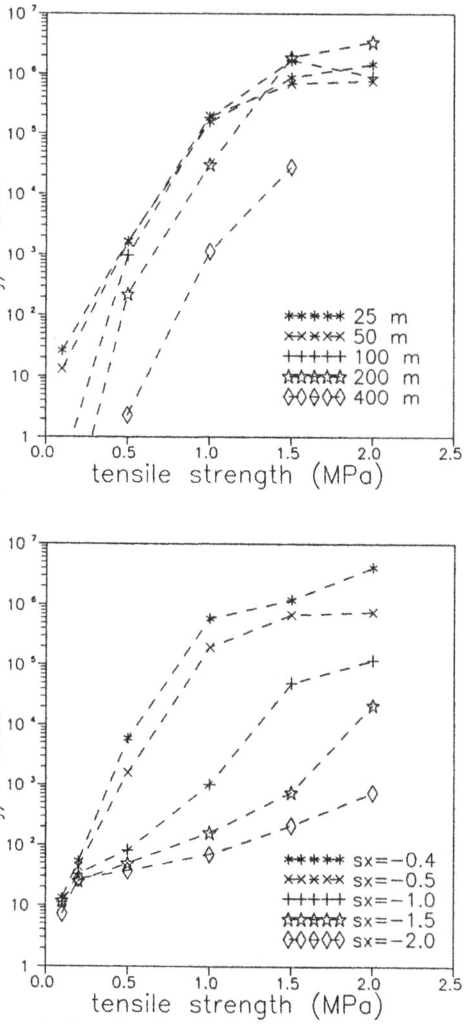

Figure 3. The *j/f* ratio represented as a function of the tensile strenght, T, and calculated (a) for different depths and (b) for different stress-rates, $\underline{\sigma}_x$ (10^{-4} Pa/sec). Other parameters as in Table 1.

example, the real range of values for T = -2.0 MPa is from about -0.7 to -2.5 MPa. Therefore, with $\Sigma = 0.4$ (Fig. 2b) some of the cells can still 'produce' extensional joints thus providing a *j/f* ratio not nil.

1.2 *Tensional regime*

The model seems to be less stable and regular than in the transcurrent regime. According to the different stress conditions, this behaviour could be due to the fact that the *j/f* ratio is commonly much higher, while shearing events are generally few. Therefore, local effects induced by the particular random initial conditions can bias the results of the numerical simulations as observed when testing different matrixes with the same Weibull distribution (see companion paper). Moreover, the tensional regime is characterised by variations of the *j/f* ratio much larger than within the transcurrent regime,

The most influencing parameters are the tensile strength, T, and the stress-rate, $\underline{\sigma}_x$. By increasing the tensile strength of one order of magnitude (Fig. 3a), the *j/f* ratio increases of several orders of magnitude. This variation is also influenced by other parameters such as the pore-fluid pressure and depth. In the former case, when assuming high λ_e values, the *j/f* ratio decreases as it occurs at increasing depths of initial burial.

As concerns the stress-rate, it also strongly influences the computed *j/f* ratio (Fig. 3b). For example, by increasing $\underline{\sigma}_x$ of a factor 5 (from -0.4 to $-2 \cdot 10^{-4}$ Pa/sec), the *j/f* ratio generally decreases from one to four orders of magnitude, according to the chosen tensile strenght (Fig. 3b). Also in this case, the variation is a function of the pore-fluid pressure showing a behaviour similar to the tensile strenght. In contrast, within the assumed range (25-400 m) the influence of depth has no relation with that of the stress-rate.

2 COMPARISON WITH FIELD EXAMPLES

In this section natural examples of faulted and jointed rocks have been considered for comparison with the proposed model.

In order to estimate the *j/f* ratio from natural examples, we need first to calculate the number of extensional joints and faults affecting volumes of rock as large as possible. Outcrop conditions generally limit the exposure size and in practice a statistical approach must be followed. Well layered and subhorizontal sedimentary sequences are best suited for this purpose also because they commonly indicate fairly deformed rocks. In brief, the following empirical procedure has been applied. Several vertical sections orthogonal to the main joint set are selected. Alternatively, in case a system of two coeval orthogonal joint sets is present, a section

diagonal to both sets must be selected. All systematic joints are counted layer by layer along a fixed long section, say 10 m. These values are then averaged.
To calculate the number of joints per stratum within a given area of a layer we need to observe different outcrops of the same sequence, but parallel to layering (*i.e.* horizontal). In this way it is possible to observe the lateral extension of the single fractures. In particular, the whole number of distinct joints affecting a given surface, say $10 \cdot 10\ m^2$, can be compared with the number of joints crossing randomly chosen sections (but always vertical ones) and an empirical ratio thus inferred. Indeed, this correction factor is necessary to estimate the number of joints per stratum. Therefore, by multiplying the latter number by the number of strata in a 10 m thick sequence, the number of extensional joints affecting a 10-m-sided cube of rock is obtained.

By repeating this procedure for as more cases as possible, and for volumes of different size, a good average, suitable for the purposes of the present research, can be obtained.

During the field work, particular care must be paid to observe whether extension joints represent just one or more extensional fracturing events. Indeed, several natural fractures show clear evidence of being the result of multiple fracturing events as documented by their vein infilling (Ramsay, 1980).

Figure 4. Abundant orthogonal joints but only two normal faults cutting Late Pleistocene sediments in Southern Thessaly, Greece. The exposure is about 5 m high.

In the proposed model, extensional fracturing events are counted and therefore, attention must be payed to check which kind of joints prevails in the rock mass and eventually to estimate the average number of events that each vein represents. As concern the number of faults, a similar procedure can be followed. However, due to the commonly minor diffusion of shear fractures with respect to extension joints and the greater dimensions of the shear planes, some correction factor must be introduced. Finally, all the vaules obtained from the different samples are normalised for $1\ m^3$ and the *j/f* ratio is calculated.

The first field example is from Thessaly, Central Greece (Fig. 4). Late Pleistocene slightly indurated red clastic sediments were deformed by a single extensional tectonic phase within which $\sigma_z = \sigma_1$, $\underline{\sigma}_x < 0$ (Caputo 1991). The tensile strength was probably low, burial depth was at most some tens of meters, while the pore-fluid pressure was certainly nearly hydrostatic. The result was the formation of a system of two orthogonal and vertical joint sets, and a set of normal faults striking parallel to one of the two joint sets. Cross-cutting and abutting relationships and especially the very young age of the deposits, clearly indicate that all the three sets of fracture planes (*i.e.* two joint sets and a fault set) are geologically coeval (Caputo, 1991). The measured average number of extension joints per cube meter is about $1.2 \cdot 10^3$ *j*/m^3 while the inferred number of faults is 0.02 *f*/m^3. Consequently, the estimated *j/f* ratio is $6 \cdot 10^4$.

The second field example comes from the Bristol Channel basin where well bedded, 10-50 cm-thick layers of Liassic marine micritic limestones, intercalated with mudstones, are spectacularly exposed. These rocks are affected by several tectonic structures. However, Roberts (1974) suggests that two orthogonal sets of pervasive fractures, trending roughly 280° and 190°, are possibly the earliest formed tectonic structures in the region. New field observations demonstrate that also numerous normal faults are coeval with those extensional features and possibly related to a rifting stage. Also in this case, $\sigma_z = \sigma_1$, $\underline{\sigma}_x < 0$. We have no data about T, but due to the longer hystory of these sediments and the certainly greater depth of burial, say some hundred meters, compaction had probably partly lithified these sediments. Accordingly, the pore-fluid pressure was probably very high possibly due to the contribution of different factors such as the alternating waterproof marly layers and a high sedimentation rate, which caused compaction without complete seepage and consequent overpressures.

At the locality illustrated in Fig. 5, the faults, although relatively abundant are, unlike the joints, not periodic (*i.e.* evenly spaced). Some of the faults are spaced about 100 m apart, while others are more closely spaced (1-10 m) within fault zones about 100 m or more apart. In Fig. 6 is represented the spatial

Figure 5. Gently dipping Liassic limestones and mudstones cut by a few normal faults and abundant orthogonal extension joints. West of Lilstock, Somerset. The structural setting shown this picture is represented by cube A in Fig. 6.

relationships depicted in Fig. 1 of C&S and Fig. 5, respectively. In Fig. 1 of C&S, the more common situation in the Liassic rocks of the Bristol Channel basin is shown; that is within a 10-m-sided volume of rock there are abundant joints but no faults.

In both cases, as well as in several other nearby outcrops, the measured number of extension joints is between 0.8 and $1.5 \cdot 10^2$ *j*/m^3. In contrast, within faulted rock volumes, like that represented in Fig. 5, the number of faults is about 0.24 *f*/m^3 and thus the calculated *j/f* ratio is $3.3\text{-}6.3 \cdot 10^2$. At localities unaffected by faulting (Fig. 1 in C&S) represented by cube B in Fig. 6, the *j/f* ratio would be infinite as a consequence of the complete lack of faults. Again, being conservative we estimate that for a setting such as that shown in Fig. 6 only about 10% of the 10-m-sided cubes will be crossed by at least some mesoscale faults. By introducing this correction factor, we tentatively estimate that *j/f* ratio will be probably one order larger and about $5 \cdot 10^3$.

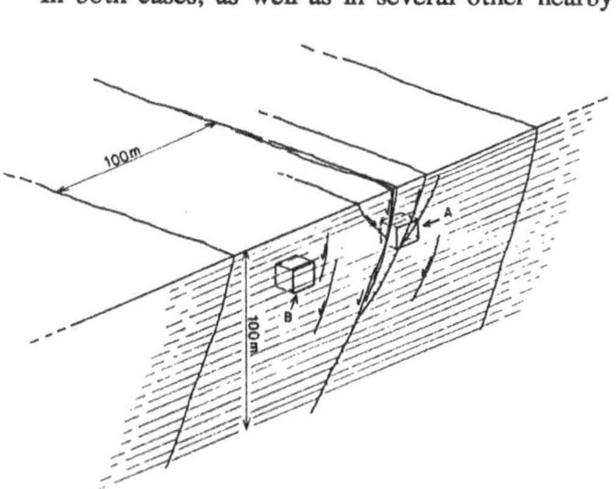

Figure 6. Block diagram representing the total volume of rock being considered for the Bristol Channel basin example. The small cubes (A and B referred to in the text) represent two typical volumes analysed as field examples and shown in Fig. 1 of C&S and in Fig. 5.

3 CONCLUDING REMARKS

As stated in C&S, the proposed model should be considered as a first attempt of quantifying extension and shear fractures (and especially their ratio) occurring in a rock volume during a brittle deformational event.

The numerical values of the *j/f* ratio obtained from the field examples are comparable with those obtained during the numerical simulations. Indeed, although exact values of the tensile strength are not

Table 1: Values used for the numerical simulation

mechanical properties of the rock:

$T = -0.5$ to $-2.0 \cdot 10^6$ Pa [$-2.0 \cdot 10^6$]

$\rho = 2.6 \cdot 10^3$ kg·m^{-3}

critical stress difference: $\sigma_c = 130$ MPa

$\sigma_c = 80$ MPa

depths: $z = 25$ to 800 m [200 m and 50 m]

pore-fluid pressure: $p_f = 0.46$ to $0.96 \cdot p_c$ [$0.46 \cdot p_c$]

assumed sealing time: $t_c = 2$ to $30 \cdot 10^8$ sec [$2 \cdot 10^8$]

tectonic genetic components:

transcurrent regime

$\underline{\sigma}_x = -10$ to $-2 \cdot 10^{-4}$ Pa·sec^{-1} [$-5 \cdot 10^{-4}$]

$\underline{\sigma}_y = 2$ to $10 \cdot 10^{-4}$ Pa·sec^{-1} [$2 \cdot 10^{-4}$]

tensional regime

$\underline{\sigma}_x = -10$ to $-5 \cdot 10^{-4}$ Pa·sec^{-1} [$-5 \cdot 10^{-4}$]

$\underline{\sigma}_y = 0$

[in brackets are 'standard values' as defined in text]

available, the very shallow depth inferred from the geological and structural setting of the Thessaly case study, in contrast to the deeper setting of the Bristol Channel example, agrees well with the general trends of the function (Fig. 3a), also considering the different role played by the pore-fluid pressure. According to these estimates, possible values of T, for the two examples, are 0.5-1 MPa and >1 MPa, respectively. Further consistency between numerical values obtained from field examples and computer simulations occurs if we consider the magnitude of the stress-rate. In fact, if Thessaly region is weakly deformed (*viz.* low $\underline{\sigma}_x$), the Bristol Channel Basin suffered a rifting stage (*viz.* high $\underline{\sigma}_x$) confirming the difference in tensile strength of the two affected rocks (Fig. 3b).

As a final comment, we would like to stress the potential applications of the present model and its computer program, *via* an 'inversion' approach. Indeed, the possibility to infer and distinguish the single genetic components of a palaeostress field, such as stress-rates and pore-fluid pressure, is certainly of great interest. It is especially appealing the relative simplicity to collect the required basic data set, that is the number of extension joints, of faults and their ratio per unit volume. In fact, other physical parameters, such as the tensile strength, can be potentially measured in the laboratory, thus further reducing the mathematical ambiguity. While depth, and consequently the confining pressure, can be estimated according to stratigraphy. Another practical application of the present research is the possibility to estimate the secondary porosity of a rock-mass (*i.e.* contemporaneously open fractures) induced by a given tectonic stress regime at depth.

REFERENCES

Atkinson, B.K. (ed) 1987. *Fracture mechanics of rock*. Academic Press, London.

Brzesowsky R. 1995. Micromechanics of sand grain failure and sand compaction. *Geologica Ultraiectina*, **133**, pp. 180.

Caputo, R. 1991. A comparison between joints and faults as brittle structures used for evaluating the stress field. *Ann. Tectonicae*, **5**, 1, 74-84.

Caputo, R. 1995. Evolution of orthogonal sets of coeval extension joints. *Terra Nova*, **7**, 5, 479-490.

Caputo, R. & Hancock, P.L. (in press). *Crack-jump* mechanism of microvein formation and its implications for stress cyclicity during extension fracturing. *J. Geodynamics.*

Jaeger, J.C. & Cook, N.G.W. 1979. *Fundamentals of rock mechanics*. Chapman and Hall, London.

Mouchet, J.P. & Mitchell, A. 1989. *Abnormal pressures while drilling*. Manuels techniques elf aquitaine, **2**, Elf Aquitaine, Boussens, France.

Murrell, S.A.F. 1963. A criterion for brittle fracture of rocks and concrete under triaxial stress and the effect of pore pressure on the criterion. *Proc. Fifth Rock Mechanics Symposium*, University of Minnesota. In: *Rock Mechanics* (edited by Fairhurst, C.). Pergamon, Oxford, 563-577.

Pollard, D.D. & Segall, P. 1987. Theoretical displacements and stresses near fractures in rock: with applications to faults, joints, veins, dikes, and solution surfaces. In: *Fracture mechanics of rocks* (edited by Atkinson, B.K.). Academic Press, 277-349.

Ramsay, J.G. 1980. The crack-seal mechanism of rock deformation. *Nature*, **284**, 135-139.

Ranalli, G. & Chandler, T.E. 1975. The stress field in the Upper crust as determined from in situ measurements. *Geol. Rund.*, **64**, 2, 653-674.

Ranalli, G. & Yin, Z.-M. 1990. Critical stress difference and orientation of faults in rocks with strength anisotropies: the two-dimensional case. *J. struct. Geol.*, **12**, 8, 1067-1071.

Roberts, J.C. 1974. Jointing and minor tectonics of the Vale of Glamorgan between Ogmore-by-Sea and Lavernock Point, South Wales. *Geol. J.*, 9, 653-674.

Sibson, R.H. 1991. Loading of faults to failure. *Bull. Seismol. Soc. Am.*, **81**, 6, 2493-2497.

Weibull, W. 1951. A statistical distribution function of wide applicability. *J. Appl. Mech.*, **18**, 293-297.

Wood, S.H., Wurts, C., Ballenger, N., Shaleen, M. & Totorica, D. 1985. The Borah Peak, Idaho, earthquake of October 28, 1983: hydrologic effects. *Earthquake Spectra*, **2**, 127-150.

Zoback, M.D. & Zoback, M.L. 1991. Tectonic stress field on North America and relative plate motions. In: *The geology of North America, Decade Map Vol. 1, Neotectonics of North America* (edited by Slemmons, D.B., Engdahl, E.R., Zoback, M.D. & Blackwekk, D.D.). Geol. Soc. Am., Boulder, Colorado, 339-366.

Mechanics of Jointed and Faulted Rock, Rossmanith (ed.) © 1998 Taylor & Francis, ISBN 90 5410 955 6

From joint and fault quantification to simulation of 3-D conceptual fracture models

C.Castaing, B.Bourgine, J.P.Chilès & A.Genter
BRGM, Orléans, France

ABSTRACT: The behaviour of fractured reservoirs depends on the geometry of small-scale fractures that have to be represented by 3D models. Well data are the most important source of information about small-scale fractures. However well data are not sufficient to define the 3D fracture geometry and so cannot be used alone to predict the connectivity and size of unfractured blocks. It is proposed, therefore, that characteristic end-member fracture geometries defined from outcrops should be used as an additional source of information. Thus, well data can be used to choose between various fracture templates, which then provide the required information about the size and connectivities of the fractures. The method was applied to Triassic sandstones intersected by a deep well in the Rhine graben, France. A 3D conceptual model was simulated, which combines (1) the fracture orientation and spacing distributions defined in the well, and (2) the fracture size distribution derived from a field analogue exposed in Saudi Arabia.

1 INTRODUCTION

The main difficulty in developing stochastic fracture models for subsurface rock volumes is that our knowledge of sub-seismic fractures is limited to cores and borehole images, which only provide 1D data. Well data are not sufficient to define 3D fracture systems, and cannot therefore be used alone to infer important parameters, such as fracture size and connectivity. The method proposed for modelling 3D fracture systems from well data is based on the input of additionnal 2D/3D information, which is taken from characteristic end-member fracture geometries derived from outcrop that are used as templates. The method involves three steps: (1) a geological and statistical fracture description of the well; (2) selection of a relevant fracture template; and (3) simulation of a 3D conceptual near-well fracture model (Figure 1). The selection between various templates will depend on the well fracture data themselves, and on the geological setting of the site. The simulation of a 3D conceptual model will be done by linking the selected fracture template to the quantitative parameters measured on boreholes, i.e. number of fracture sets, and orientation, spacing and thickness of fractures. The dimensions and shapes of fractures will be given by the fracture template. This method was applied to fractures in the Triassic Buntsandstein Sandstone intersected by Well EPS1, in the Rhine graben, France. The selected fracture template comes from a well-exposed sandstone formation in Saudi-Arabia, whose fracture systematics was determined on the ground and by means of low-level aerial photographs.

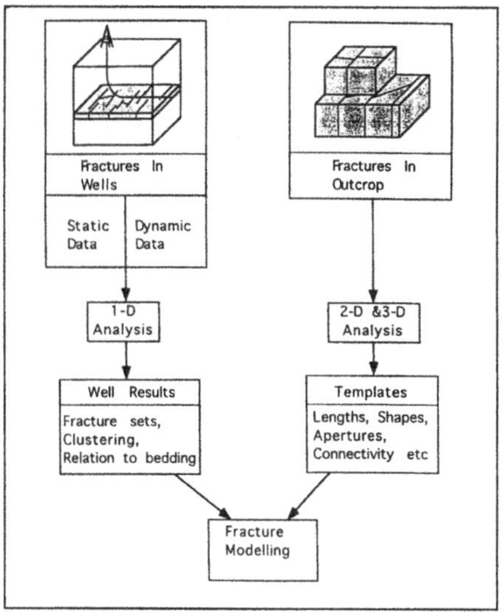

Figure 1: Organogram showing procedure for fracture modelling.

2 FRACTURE DESCRIPTION OF THE WELL

2.1 Geological Setting

The drilled formation is the Triassic Buntsandstein Sandstone intersected by Well EPS-1 as part of the Soultz geothermal programme (Baria et al. 1995). The site is located within the Upper Rhine Valley (France) and forms a local horst structure within the Rhine graben, bounded by north-trending normal faults dipping either west or east. A core provides a continuous section of the Buntsandstein Sandstone. The continental sandstone succession was intersected over approximately 400 m (1009 to 1417 m). At 1417 m it lies unconformably on the granitic basement and at 1009 m it is stratigraphically overlain by the transgressive Muschelkalk lagoonal-marine succession. Fractures were studied directly from core analysis and oriented by comparison with borehole imaging techniques.

2.2 Fracture analysis

The fracture population discussed in this paper only includes naturally formed fractures as observed on cores, whose structural analysis has defined two fracture types, i.e. steeply-dipping filled-joints and normal faults (Genter & Traineau 1996). The core data provide the depth, strike, dip, spacing, and thickness for 350 fractures intersected between 1009 and 1417 m. Three fracture sets, oriented N 0° 75°W, N 170° 75°E (major sets) and N 120° 75°N (minor set), have been determined (Figure 2). For reasons of clarity, only the two major sets will be taken into account in the following, and considered as forming a near-vertical N 175°-trending single set, dipping steeply east or west, and comprising 126 fractures.

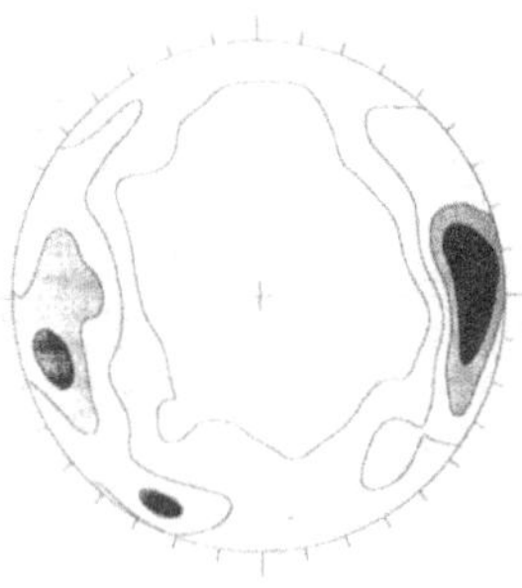

Figure 2: Schmidt projection (lower-hemisphere) of fracture poles observed on Well EPS1 cores. The different contours indicate 10%, 30%, 50% and 90% of the whole fracture population.

The filled-joints are small-scale Mode I fractures filled by hydrothermal deposits. They group into clusters alternating with poorly-fractured zones (Figure 3). The degree of clustering can be estimated by the coefficient of variation, which is the standard deviation over the mean (Cox & Lewis 1966). Values greater than 1 indicate positive clustering, while values less than 1 indicate anticlustering. A value of 1 is consistent with a random Poisson process giving an exponential spacing distribution. In Well EPS1 the spacings give values of about 1.70 indicating high clustering. The faults correspond to Mode II reactivation of joints within the clusters, as shown by striation marks indicating normal faulting.

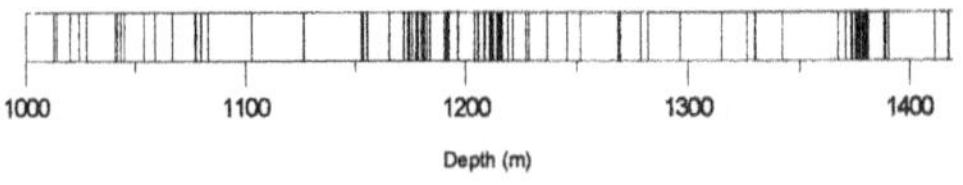

Figure 3: Log of fracture intersections of the two major fracture sets along Well EPS1.

The cumulative fracture density versus depth enables a better determination of the number and size of the fracture clusters, which forms three main clusters centred around 1045, 1185 and 1380 m depth (Figure 4).

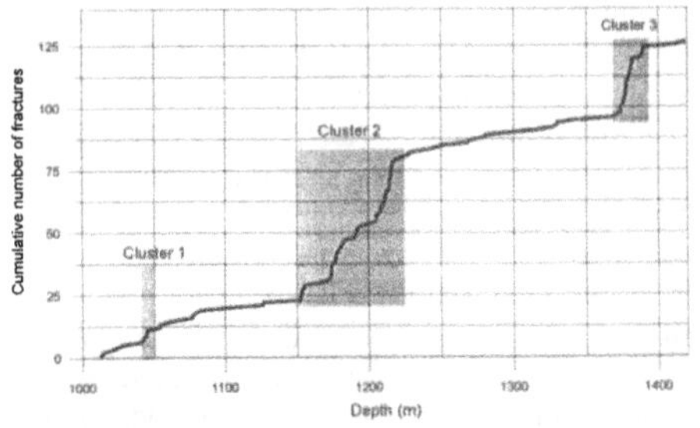

Figure 4: Cumulative number of fractures versus depth showing three main fracture clusters.

The cumulative distribution of fracture spacings derived from the two major sets taken together, and corrected for the angles between the well and the fracture planes, follows approximately a gamma-law distribution (Figure 5a). Within the clusters, the spacing distribution is also a gamma law, whereas the distribution is exponential between the clusters (Figure 5b).

(a)

(b)

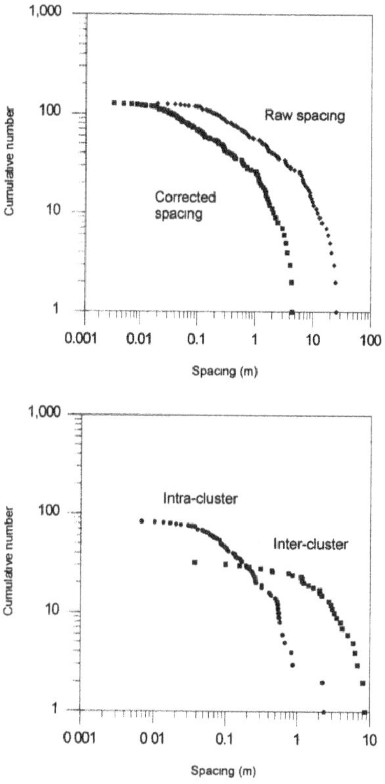

Figure 5: Cumulative distributions of fracture spacings in Well EPS1, considering the whole sampling line (a), and the clustered and non-clustered zones separately (b).

In conclusion, structural data recorded from wells can provide a good geometrical characterisation of the orientation and spatial distribution of the fractures that intercept the well (Genter et al. 1997). But cores do not provide any reliable information on the dimensions, shape, and connectivity of fractures.

3 SELECTION OF A RELEVANT FRACTURE TEMPLATE

3.1 Characteristic end-member fracture geometries

In the framework of the European Joule project (Aarseth et al. 1997), large fracture databases have been collected, and a series of end-member fracture geometries has been recently defined. They can be grouped into five main models: (1) non-clustered stratabound (fractures regularly spaced passing through one bed only and abutting against bedding planes); (2) non-clustered nested (fractures regularly spaced passing through one or several beds and abutting against bedding planes); (3) clustered non-stratabound (fractures not regularly spaced and not bound within any beds); (4) clustered stratabound; and (5) clustered nested models. Models (1), (2), and (3) are presented in Figure 6. These models have been designed to require only a limited number on input parameters, and can be combined to enable simulation of relatively complex fracture systems.

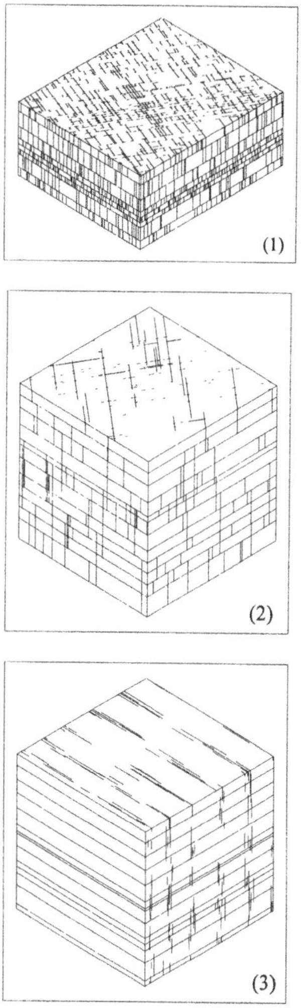

Figure 6: Examples of non-clustered stratabound (1), non-clustered nested (2), and clustered non-stratabound models (3).

3.2 Characteristics of the selected fracture template

Among the different well-exposed sites studied in the framework of the European Joule project, fracture geometries were systematically defined in some different sandstone formations, using ground fracture mapping and low-level aerial photographs (Aarseth et al., 1997). The Saq Sandstone studied in Saudi Arabia appears to be the most relevant fracture template for the Buntsandstein Sandstone (Castaing et al. 1996). Like the Buntsandstein Sandstone, the Saq Sandstone forms a massive continental formation of about 660 m thick submitted to a comparable extensional tectonic regime. Thus, the fracture length distribution defined in the Saq Sandstone was used as an additional source of information for simulating a 3D conceptual fracture model in the environment of Well EPS1 (Figure 7).

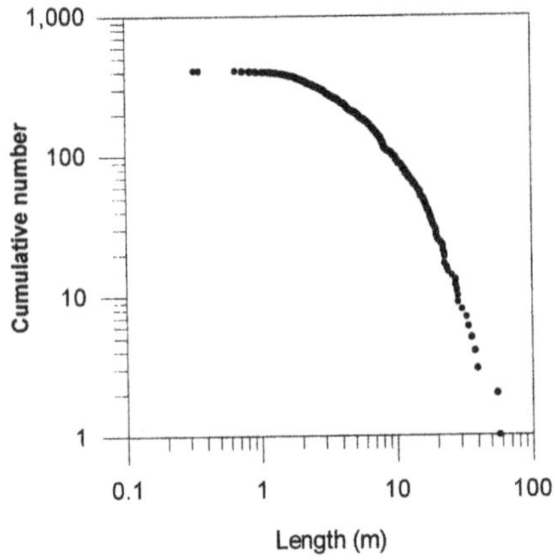

Figure 7: Cumulative distribution of fracture length in the Saq Sandstone.

4 SIMULATION OF A 3D CONCEPTUAL NEAR-WELL FRACTURE MODEL

The fact that the Buntsandstein Sandstone is a homogeneous rock formation with no well-expressed sedimentary beds and shows a clustered fracture spacing, enables the selection of a clustered non-stratabound model for representing its 3D organization. The simulation of the conceptual near-well fracture model combines (1) the fracture orientation and spacing distributions defined in Well EPS1, and (2) the fracture length distribution defined in the Saq Sandstone. The fracture spacing analysis of Well EPS1 has shown that fractures group into clusters, but there are also single fractures more randomly distributed between the clusters (Figures 3, 4). For this reason, the stochastic simulation process was divided into two independent steps comprising (1) the simulation of the fracture clusters, and (2) the simulation of the single fractures. The input parameters for the simulation are presented in Table 1.

Table 1. Input parameters for simulating the 3D near-well fracture model.

Input parameters	Clusters and fractures within clusters	Single fractures between clusters
Azimuth	N 175° (± 5°)	N 175° (± 5°)
Dip	70°W (± 5°)	70°W & 70°E (±5°)
Cluster centres	1045, 1185 & 1380 m depth	n.a.
Cluster length	Intersect the whole simulated volume	n.a.
Cluster width	1.5, 22 & 7 m	n.a.
Fracture density	4.2, 2.5 & 4.2 fracture/m	0.35 fracture/m
Fracture spacing distribution	Gamma law simulated with an exponential distribution	Exponential distribution
Fracture length distribution	Gamma law	Gamma law
Fracture shape	Rectangle	Rectangle
Fracture shape ratio	2	2

4.1 Simulation of fracture clusters

For the simulation of the near-well model, the location of the centres of the three clusters was conditioned on the well data. As the clusters are the locus of large normal faults, it was decided that each cluster would intersect the entire simulated volume. The number of fractures to be generated in each cluster was determined in order to reproduce the fracture linear density measured on cores. As the probability of intersecting one fracture is linked to the area of the fracture, the calculation of the number of fractures took into account the width of the clusters as well as the size and shape of the fractures. The distance between each fracture and the centre of the cluster was determined using an uniform distribution ranging over [-w/2; +w/2], w being the cluster width. This led to an exponential spacing distribution within each cluster, which approximates the distribution defined on cores (Figure 5b).

4.2 Simulation of single fractures

Between the clusters, the centres of single fractures were then generated using a 3D Poisson process, in order to reproduce their real exponential spacing distribution (Figure 4b). Both fractures, within and between clusters, were generated (1) assuming a rectangular shape with a shape ratio (length over height) of 2, and (2) according to the length distribution taken from the Saq Sandstone dataset.

The results of the simulation are visualized in Figure 8, where it is possible to recognize the prolongation of the three fracture clusters and the single fractures distributed within the whole volume. A fictitious well drilled within the simulated 3D fracture system is visualized in Figure 9. It gives a log of fracture intersections that is in good agreement with the fracture intersections of Well EPS1 (Figure 10).

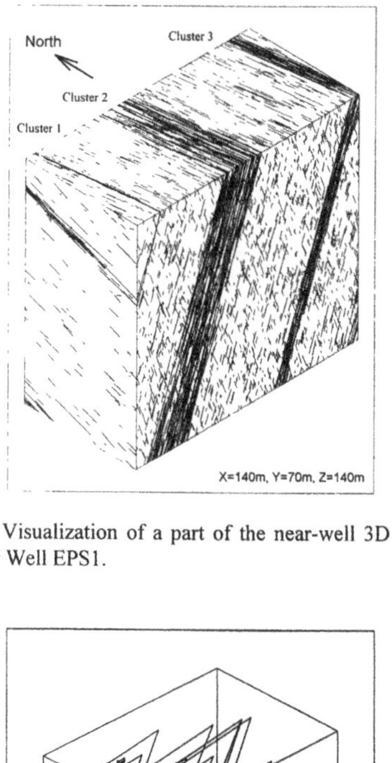

Figure 8: Visualization of a part of the near-well 3D fracture model for Well EPS1.

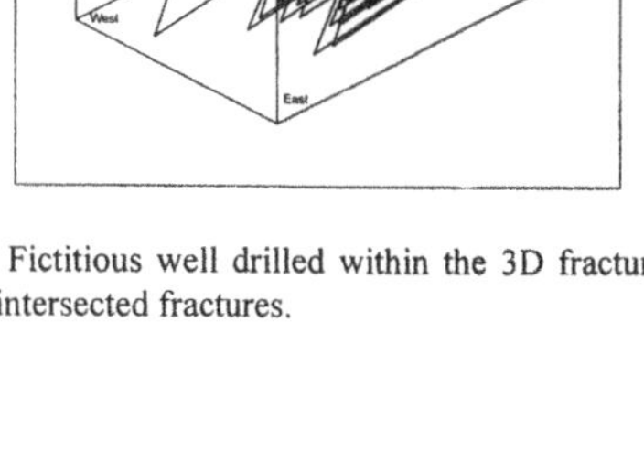

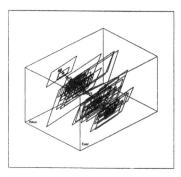

Figure 9: Fictitious well drilled within the 3D fracture model showing intersected fractures.

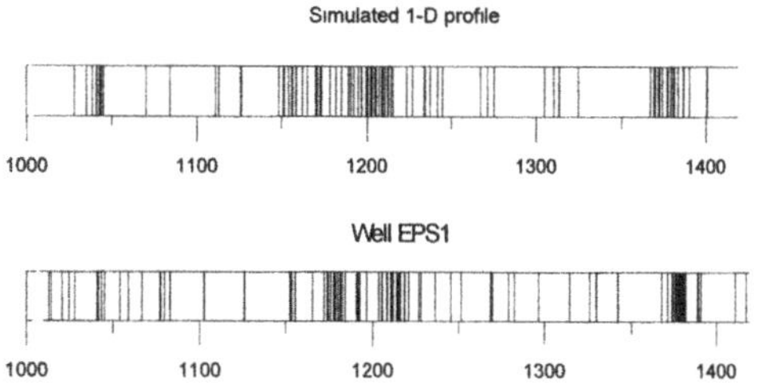

Figure 10: Comparison between logs of fracture intersections obtained from the fictitious well and Well EPS1, corrected for the angles between wells and fracture planes.

5 CONCLUSIONS

An attempt to simulate 3D fracture systems from 1D well data has been proposed in this paper. The well data, combined with general knowledge of the geological setting, were used to choose between various fracture templates, which then provided the required information to simulate the fracture systems in 3D. Upscaling of the near-well fracture model is now under progress, considering that the median cluster intersected by Well EPS1 represents a large-scale normal fault evidenced by seismics.

REFERENCES

Aarseth E.S., B. Bourgine, C. Castaing, J.P. Chilès, N.P. Christensen, M. Eeles, E. Fillion, A. Genter, P.A. Gillespie, E. Hakansson, K.Z. Joergensen, H.F. Lindgaard, L. Madsen, N. Odling, C. Olsen, J. Reffstrup, R. Trice, J.J Walsh & J. Watterson (1997). Interim guide to fracture interpretation and flow modelling in fractured reservoirs, *European Commission Report EUR 17116*, Joule II.

Baria R., J. Garnish, J. Baumgärtner, A. Gérard & R. Jung 1995. Recent developments in the European HDR research programme at Soultz-sous-Forêts (France). *Proceedings of World Geothermal Congress*, Florence, Italy, 18–31 May 1995, 2631–2637.

Castaing C., M.A. Halawani, F. Gervais, J.P. Chilès, A. Genter, B. Bourgine, G. Ouillon, J.M. Brosse, P. Martin, A. Genna & D. Janjou 1996. Scaling relationships in intraplate fracture systems related to Red Sea rifting. *Tectonophysics*, 261: 291–314.

Cox D.R. & P.W. Lewis 1966. *Statistical analysis of series of events*, London, Methuen.

Genter A. & H Traineau 1996. Analysis of macroscopic fractures in granite in the HDR geothermal EPS-1 well, Soultz-sous-Forêts (France). *Journal of Volcanology and Geothermal Research*, 72: 121–141.

Genter A., C. Castaing & P. Martin 1997. Evaluation de la fracturation des réservoirs par forages: comparaison entre les données de carottes et d'imagerie de paroi. *Revue de l'Institut Français du Pétrole*, Janvier-Février 1997, 52: 49–64.

Mechanics of Jointed and Faulted Rock, Rossmanith (ed.) © 1998 Taylor & Francis, ISBN 90 5410 955 6

Influence of density, pressure and mean grain diameter on shear band localization in granular materials

E. Bauer
Institute of Mechanics, Technical University Graz, Austria

J. Tejchman
Technical University of Gdansk, Poland

ABSTRACT: The focus of the present paper is to investigate the influence of density, pressure and mean grain diameter on shear banding in dry granular materials based on a hypoplastic constitutive model. With a unified description of the interaction between pressure level and density the hypoplastic model can be applied to a wider range of pressures and densities. A bifurcation analysis is carried out to study the onset of strain localization and the shear band orientation under plane-strain compression. The post-bifurcation behavior, i.e. the shear band evolution, shear band thickness and distribution of density, is investigated with finite element calculations using a polar hypoplastic model.

1 INTRODUCTION

Predictions of shear band localization are closely related to the constitutive relation and the quantity of the constitutive constants. In order to investigate the stress and density sensitive behavior of granular materials a unified description of the interaction between stress and density is necessary. To this end a special hypoplastic constitutive model is proposed to describe the three-dimensional stress-strain behavior of cohesionless granular materials. Originally, the hypoplastic concept was developed for a non-polar continuum, e.g. Kolymbas, 1987, Wu et al., 1996, Gudehus, 1996, Bauer, 1996. However, the investigation of shear banding based on a non-polar continuum model is restricted to the prediction of the onset of possible shear bands, e.g. Kolymbas, 1981, Chambon et al., 1990, Wu and Sikora, 1991. The post-bifurcation behavior within shear zones is characterized by dilatancy, strain softening and polar effects, i.e. pronounced grain rotations and couple stresses occur (Mühlhaus and Vardoulakis, 1987, Tejchman, 1989, Oda, 1993). Recently, a non-polar hypoplastic constitutive law according to Gudehus and Bauer has been extended by quantities which are characteristics of a polar continuum. The extended model takes into account Cosserat rotations, couple stresses and an internal length. It can be shown that a realistic prediction of the shear zone thickness is possible with a constitutive relation which includes the mean grain diameter as an internal length (Tejchman, 1997). In this case, finite element results converge to a finite size of the shear zone thickness and boundary value problems are mathematically well-posed (de Borst et al., 1992).

2 POLAR HYPOPLASTIC MODEL

The proposed polar hypoplastic model for the case of plane strain has the following general form (Tejchman and Bauer, 1996, Tejchman 1997):

$$\overset{\circ}{\sigma}_{ij} = f_s \left[L_{ij}(\hat{\sigma}_{kl}, \hat{m}_k, D^c_{kl}, k_k d_{50}) + \right.$$
$$\left. + f_d N_{ij}(\hat{\sigma}_{ij}) \sqrt{D^c_{kl} D^c_{kl} + k_k k_k d^2_{50}} \right], \quad (1)$$

$$\overset{\circ}{m}_i / d_{50} = f_s \left[L^c_i(\hat{\sigma}_{kl}, \hat{m}_k, D^c_{kl}, k_k d_{50}) + \right.$$
$$\left. + f_d N^c_i(\hat{m}_i) \sqrt{D^c_{kl} D^c_{kl} + k_k k_k d^2_{50}} \right], \quad (2)$$

wherein $\overset{\circ}{\sigma}_{ij}$ is the Jaumann stress rate, $\overset{\circ}{m}_i$ - the Jaumann rate of couple stresses, d_{50} - the mean grain diameter, $\hat{\sigma}_{ij} = \sigma_{ij}/\sigma_{kk}$ - the normalized stress, $\hat{m}_i = m_i/(\sigma_{kk}\, d_{50})$- the normalized couple stress, f_s - the stiffness factor, f_d - the density factor, D^c_{ij} - the rate of deformation, k_i - the rate of curvature, and L_{ij}, N_{ij}, L^c_i and N^c_i - the tensor- and vector-valued functions, respectively. The Jaumann stress rate and the Jaumann couple stress rate are defined as:

$$\overset{\circ}{\sigma}_{ij} = \dot{\sigma}_{ij} - W_{ik}\,\sigma_{kj} + \sigma_{ik}\,W_{kj}\ , \tag{3}$$

$$\overset{\circ}{m}_i/d_{50} = \dot{m}_i - [\,W_{ik}\, m_k - m_k\, W_{ki}\,]/2\ , \tag{4}$$

wherein $\dot{\sigma}_{ij}$ and $\dot{m}_i$ are the components of the time derivative of the Cauchy stress tensor and the Cauchy couple stress vector, respectively. The rate of deformation D^c_{ij} and the rate of curvature k_i are defined by the following relations:

$$D^c_{ij} = D_{ij} + W_{ij} - W^c_{ij}\ \ , \ \ k_i = w^c_{,i}\ , \tag{5}$$

$$D_{ij} = [v_{i,j} + v_{j,i}]/2\ \ , \ \ W_{ij} = [v_{i,j} - v_{j,i}]/2\ . \tag{6}$$

Herein D_{ij} is the classical (non-polar) stretching tensor, W_{ij} denotes the classical (non-polar) spin tensor, W^c_{ij} is the rate of Cosserat rotation with $W^c_{kk} = 0$, $W^c_{21} = -W^c_{12} = w^c$ for plane strain, and $v_{i,j}$ denotes the gradient of velocity.
For the tensor-valued functions L_{ij}, N_{ij},L^c_i and N^c_i in Eq.(1) and Eq.(2) the following expressions are used:

$$L_{ij} = \hat{a}^2\, D^c_{ij} + \hat{\sigma}_{ij}\,(\hat{\sigma}_{kl}\, D^c_{kl} + \hat{m}_k\, k_k\, d_{50})\ , \tag{7}$$

$$L^c_i = k_i\, d_{50} + \hat{m}_i\,(\hat{\sigma}_{kl}\, D^c_{kl} + \hat{m}_k\, k_k\, d_{50})\ , \tag{8}$$

$$N_{ij} = \hat{a}\,(\hat{\sigma}_{ij} + \hat{\sigma}^*_{ij})\ \ , \ \ N^c_i = a^c\,\hat{m}_i\ , \tag{9}$$

where

$$\hat{a}^{-1} = c_1 + c_2\,\sqrt{\hat{\sigma}^*_{kl}\,\hat{\sigma}^*_{kl}}\left[1 - \frac{\sqrt{6}\,\hat{\sigma}^*_{kl}\,\hat{\sigma}^*_{lm}\,\hat{\sigma}^*_{mk}}{[\hat{\sigma}^*_{kl}\,\hat{\sigma}^*_{lk}]^{3/2}}\right] , \tag{10}$$

with

$$c_1 = \sqrt{\frac{3}{8}}\,\frac{(3-\sin\varphi)}{\sin\varphi}\ \ , \ \ c_2 = \frac{3}{8}\,\frac{(3+\sin\varphi)}{\sin\varphi}\ . \tag{11}$$

Herein $\hat{\sigma}^*_{ij}$ denotes the deviatoric part of $\hat{\sigma}_{ij}$ and φ is the critical angle of internal friction. The dimensionless polar constant a^c controls the influence of the polar quantities on the material behavior.
The scalar functions f_s and f_d take into account the influence of the current density and pressure. The stiffness factor f_s is proportional to the granular hardness h_s and depends on the stress level σ_{kk}, i.e.

$$f_s = \frac{h_s}{n\, h_i}\,\frac{(1+e_i)}{e}\left(-\frac{\sigma_{kk}}{h_s}\right)^{1-n}\ , \tag{12}$$

$$\text{with}\ \ h_i = \frac{1}{c_1^2} + \frac{1}{3} - \left(\frac{e_{i0}-e_{d0}}{e_{c0}-e_{d0}}\right)^{\alpha}\frac{1}{c_1\sqrt{3}}\ . \tag{13}$$

Herein e is the current void ratio, e_i is the void ratio of maximum looening and n is a constant. The density factor f_d represents a relation between the current void ratio e, the void ratio e_d of maximum densification and the critical void ratio e_c, i.e.

$$f_d = \left(\frac{e-e_d}{e_c-e_d}\right)^{\alpha}\ , \tag{14}$$

where α is a positive constant. The current void ratio is updated during calculations by the following evolution equation:

$$\dot{e} = (1+e)\,D_{kk}\ . \tag{15}$$

The values of e_i, e_d and e_c are assumed to decrease with the pressure level according to the following compression law:

$$\frac{e_i}{e_{i0}} = \frac{e_d}{e_{d0}} = \frac{e_c}{e_{c0}} = \exp[-(\sigma_{kk}/h_s)^n] \tag{16}$$

wherein e_{i0}, e_{d0} and e_{c0} are the values of e_i, e_d, and e_c for pressure equal to zero, respectively.

For $d_{50} \to 0$ or purely coaxial homogeneous deformations there are no polar effects, i.e. $m_i = \dot{m}_i = 0$, and $W_{ij} = W^c_{ij}$, and the polar hypoplastic model reduces to the non-polar one (Gudehus, 1996, Bauer, 1996):

$$\overset{\circ}{\sigma}_{ij} = f_s\left[\,L_{ij}(\hat{\sigma}_{kl}, D_{kl}) + f_d\, N_{ij}(\hat{\sigma}_{ij})\,\sqrt{D_{kl}\, D_{kl}}\,\right] , \tag{17}$$

with

$$L_{ij} = \hat{a}^2\, D_{ij} + \hat{\sigma}_{ij}\,(\hat{\sigma}_{kl}\, D_{kl})\ \ , \ \ N_{ij} = \hat{a}\,(\hat{\sigma}_{ij} + \hat{\sigma}^*_{ij})\ .$$

For the present calculations the following material constants for a medium sand are used: $e_{i0} = 1.3$, $e_{d0} = 0.51$, $e_{c0} = 0.82$, $\varphi = 30°$, $h_s = 190$ MPa, $\alpha = 0.3$, $n = 0.5$, $d_{50} = 0.5$ mm, $a^c = 1.0$.

3 ANALYSIS OF SHEAR BAND BIFURCATION

Basaed on the non-polar hypoplastic model (17) the possibility of a shear band bifurcation will be studied for an initially homogeneous specimen which is homogeneously strained up to a certain state (σ_{ij}, e) under plane strain compression. It will be examind as to whether – besides a continuous velocity field and a homogeneous velocity gradient field, – the field equations also allow an alternative non-uniform deformation for which the velocity gradient has an jump across a discontinuity plane, i.e.

$$[\![v_{i,j}]\!] = g_i n_j \neq 0 \,. \tag{18}$$

The jump of the velocity gradient in (18) is represented by the dyadic product of vectors **g** and **n** with regard to the local coordinate system $\bar{x}_i$ in Figure 1. $\mathbf{n} = [0, 1, 0]^T$ denotes the unit normal to the discontinuity plane and $\mathbf{g} = [g_1, g_2, 0]^T$ can be interpreted as a vector that represents the orientation of the shear band evolution. Then the jump of the stretching and the spin tensor are related to the jump of the velocity gradient (18), viz. $[\![D_{ij}]\!] = [g_i n_j + n_i g_j]/2$ and $[\![W_{ij}]\!] = [g_i n_j - n_i g_j]/2$.

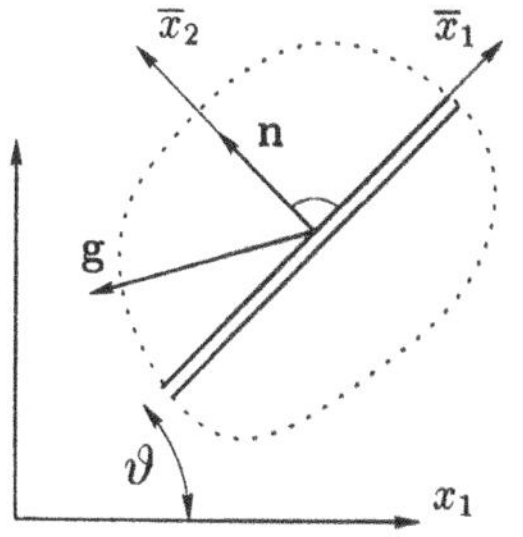

Figure 1: Orientation of the shear band

The stress components in the coordinate systems x_i and $\bar{x}_i$ are related by the transformation

$$\bar{\sigma}_{ij} = Q_{ik} Q_{jl} \sigma_{kl} \tag{19}$$

with

$$[Q] = \begin{bmatrix} \cos\vartheta & \sin\vartheta & 0 \\ -\sin\vartheta & \cos\vartheta & 0 \\ 0 & 0 & 1 \end{bmatrix} .$$

Herein the transformation matrix $[Q]$ depends on the shear band inclination ϑ, which is the angle between the direction of the minimum principal stress, i.e. the global direction x_1, and the local direction $\bar{x}_1$. For continuing equilibrium, the jump of the traction rate accros the discontinuity must be zero (Rice and Rudnicki, 1980), i.e.

$$[\![\dot{\bar{\sigma}}_{ij}]\!] n_j = 0 \,. \tag{20}$$

The jump of the stress rate in Eq. (20) can be related to the Jaumann stress rate according to

$$[\![\dot{\bar{\sigma}}_{ij}]\!] = [\![\overset{c}{\bar{\sigma}}_{ij}]\!] + [\![W_{ik}]\!] \bar{\sigma}_{kj} - \bar{\sigma}_{ik} [\![W_{kj}]\!] \,. \tag{21}$$

With respect to the response of the hypoplastic model, i.e. $\overset{c}{\bar{\sigma}}_{ij}$, the following fundamental equation for shear band analysis can be obtained:

$$[f_s \, L_{ij}(\bar{\sigma}_{kl}, [\![D_{kl}]\!]) + f_s \, f_d \, N_{ij}(\bar{\sigma}_{kl}) [\![\sqrt{D_{kl} D_{kl}}]\!] + \\ + [\![W_{ik}]\!] \bar{\sigma}_{kj} - \bar{\sigma}_{ik} [\![W_{kj}]\!]] n_j = 0 \,. \tag{22}$$

It can be noted that the influence of different incremental stiffnesses due to a different velocity gradient outside and inside the shear band can be taken into account by the same constitutive relation because there is no need in hypoplasticity to differentiate between separate constitutive relations for loading and unloading. Between the jump of the magnitudes of the stretchings on both sides of the discontinuity, i.e. $\lambda = [\![\sqrt{D_{ij} D_{ij}}]\!]$ and the amount of $[\![D_{ij}]\!]$, i.e $\beta = \sqrt{[\![D_{ij}]\!][\![D_{ij}]\!]}$, the inequality $\lambda \leq \beta$ is valid independent of the amount of the stretching outside and inside of the shear band. It can been shown that the smallest possible bifurcation stress ratio will be obtained for $\beta = \lambda$ (Bauer, 1998). Because Eq. (22) is positively homogeneous of first order with respect to g_i the amount of $[\![D_{ij}]\!]$ is arbitrary, e.g. it can be set to $\beta = 1 \; s^{-1}$ so that $\beta = \lambda = 1$. With respect to $\beta = \sqrt{g_1^2/2 + g_2^2} = \lambda = 1$ and Eq. (22) the following relation for the shear band inclination angle ϑ can be obtained:

$$b_4 \, w^4 + b_3 \, w^3 + b_2 \, w^2 + b_1 \, w + b_0 = 0 \,. \tag{23}$$

The coefficients $b_0 \ldots b_4$ in Eq. (23) are functions of the current state $(\bar{\sigma}_{ij}, e)$. The possibility of shear band bifurcations can now be examined. For real solutions to Eq. (23) shear bands appear in pairs $\pm\vartheta_t$, because of $w = \cos^2\vartheta = \cos^2(\kappa\pi \mp \vartheta)$ with $\kappa = 0, 1$. For plane strain compression tests the influence of the minimum principal stress σ_1 and the void ratio on the shear band inclination ϑ is shown in Figure 2.

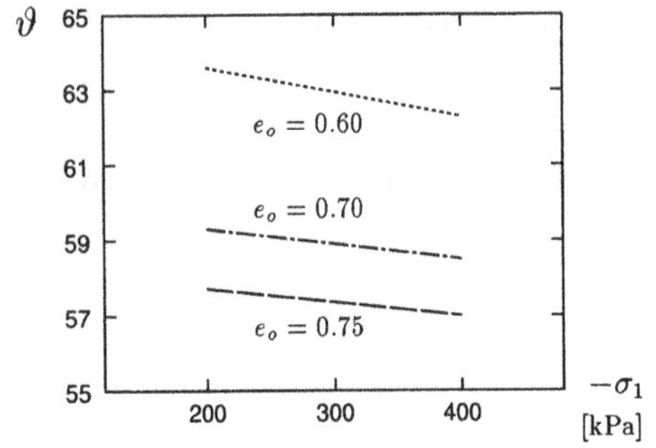

Figure 2: Shear band orientation ϑ versus horizontal stress

For the same stress σ_1 the angle ϑ increases with a decrease of the void ratio but for an increase of σ_1 the shear band inclination decreases. A similar dependence of the shear band inclination on pressure and density has been observed in experiments, e.g. Desrues et al. (1989) and Yoshida et al. (1993).

4 NUMERICAL SIMULATIONS OF PLANE STRAIN COMPRESSION

In order to trace the evolution of shear bands a finite element calculation is carried out based on the polar-hypoplastic model. The analysis is performed with a specimen 14 cm high and 2 cm wide. The finite element mesh includes 1440 triangular elements with linear shape functions for displacements and the Cosserat rotation. The size of the elements is choosen no larger than 5 times the mean grain diameter, so that the calculated shear band thickness does not depend upon the mesh refinement. The test is controlled by an increase of the vertical displacement at the top side. The top and bottom are smooth. To initiate the starting point for shear band localization a slightly enlarged void ratio is introduced for a single element in the middle of the left side of the specimen.

Figure 3 presents the effect of the initial void ratio e_o, the confining pressure σ_0 and the mean grain diameter d_{50} on the calculated normalized load-displacement curve. All curves increase first, show a pronounced peak, decrease and reach almost the same residual state. The angle of internal friction in the residual state, i.e. $\psi = 34°$ for $d_{50} = 0.5$ mm, is higher than the assumed critical friction angle, i.e. $\varphi = 30°$, which is defined for triaxial compression. The lower the initial void ratio e_o and the confining pressure σ_0, the higher the maximum vertical force P. With an increase of the initial void ratio, the confining pressure, and the mean grain diameter, the vertical displace-

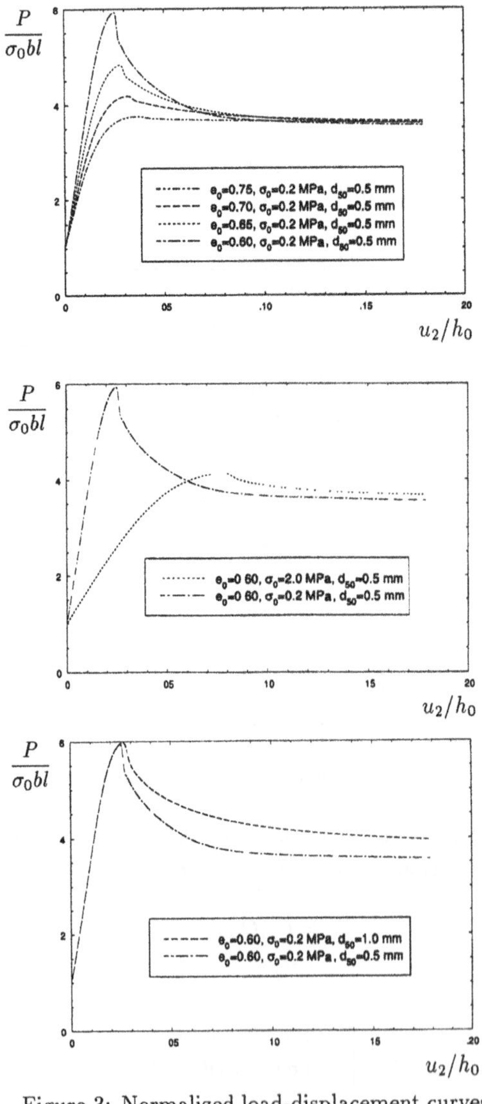

Figure 3: Normalized load-displacement curves

ment u_2 related to the maximum force P becomes larger. The results are in agreement with plane strain compression tests by Vardoulakis (1980) and Yoshida et al (1993).

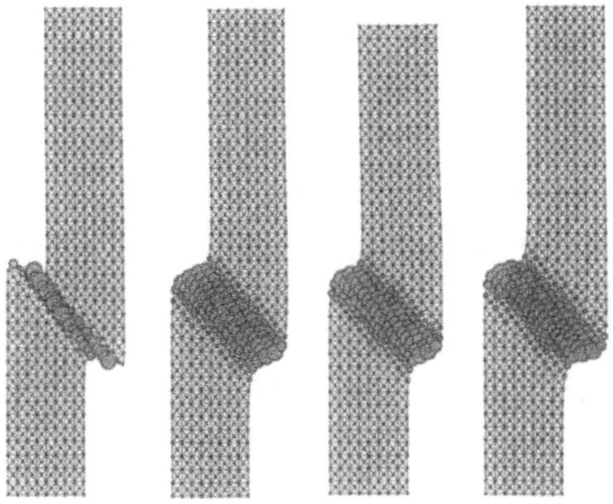

(a) (b) (c) (d)

Figure 4: Calculated deformations and distribution of the Cosserat rotations (marked by circles) for: (a) $e_0 = 0.60$, $\sigma_0 = 0.2$ MPa, $d_{50} = 0.5$ mm,
(b) $e_0 = 0.70$, $\sigma_0 = 0.2$ MPa, $d_{50} = 0.5$ mm,
(c) $e_0 = 0.60$, $\sigma_0 = 2.0$ MPa, $d_{50} = 0.5$ mm,
(d) $e_0 = 0.60$, $\sigma_0 = 0.2$ MPa, $d_{50} = 1.0$ mm

Figure. 4 shows the deformed meshes and calculated distribution of the Cosserat rotation for various initial states. The shear band is characterized both by a concentration of the Cosserat rotation and an increase of the void ratio. The Cosserat rotation is only noticeable in the shear zone. The void ratio changes across the shear zone. The largest void ratio is in the middle of the zone, which reaches the critical value for large shearing (Tejchman, 1997). The thickness of the shear zone rises with an increase of the initial void ratio, confining pressure and mean grain diameter. An increase of the thickness of a shear zone with increasing e_0, σ_0 and d_{50} can be explained due to the fact that with an increase of these factors, the rate of material softening decreases (Figure 3). The thickness is about $t_s \approx 11 \times d_{50}$ ($e_0 = 0.60$), $t_s \approx 25 \times d_{50}$ ($e_0 = 0.70$) for $\sigma_0 = 0.2$ MPa and $d_{50} = 0.5$ mm. For $e_0 = 0.6$, $d_{50} = 0.5$ mm and a larger confining pressure, $\sigma_0 = 2.0$ MPa, $t_s \approx 23 \times d_{50}$. A mean grain diameter of $d_{50} = 1.0$ mm results in $t_s \approx 12 \times d_{50}$ for $e_0 = 0.6$ and $\sigma_0 = 0.2$ MPa.

The behavior of stresses and couple stresses in the middle of the shear zone versus the vertical displacement of the specimen is shown in Figure 5. The non-symmetry of the stress tensor, i.e. $\sigma_{12} \neq \sigma_{21}$, and the couple stresses take place immediately at the onset of the shear zone formation. The shear zone is created before the peak stress occurs. This result confirms the solution of the bifurcation analysis (Bauer, 1998). With advanced vertical compression all stresses and couple stresses tend towards a residual state.

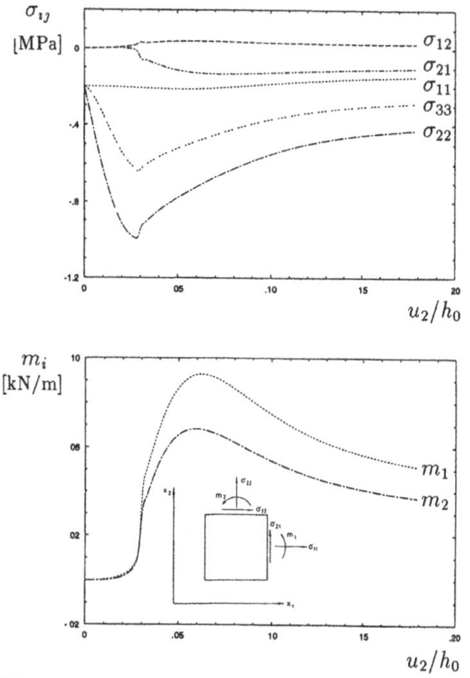

Figure 5: Stresses σ_{ij} and couple stresses m_i versus normalized vertical displacement

5 CONCLUSIONS

A special hypoplastic constitutive model is proposed, which covers a broad range of densities and pressures using only one set of constitutive constants. For initially homogeneous plane strain compressions an analysis of shear band bifurcation is carried out. The inclination of shear bands with regard to the direction of the minor principal stress decreases with an increase of the initial void ratio and the pressure level. The post-bifurcation behavior is traced using a finite element calculation based on a polar hypoplastic constitutive law. It follows from the calculations that:

i.) The polar approach is effective as a regularization method when shear localization is dominant. The Cosserat rotations and couple stresses are only noticeable in shear zones. The Cosserat rotation and the increasing void ratio are suitable indicators for shear zones.

ii.) The thickness of shear zones rises with an increase of the initial void ratio, the pressure level and the mean grain diameter.

REFERENCES

Bauer, E. (1996): Calibration of a comprehensive hypoplastic model for granular materials. *Soils and Foundations Vol. 36, No. 1, 13–26.*

Bauer, E. (1998): The dependence of shear banding on pressure and density in hypoplasticity. *Proc. of the 4th Workshop on Localization and Bifurcation Theory for Soils and Rocks, Gifu, 1997, Publication in preparation.*

Borst, R. de, Mühlhaus, H. B., Pamin, J., Sluys, L. Y. (1992): Computational modelling of localization of deformation. *Proc. 3rd Int. Conf. Comp. Plast., ed. by D. R. Owen, E. Onate and E. Hinton, Pineridge Press, 483-508.*

Chambon, R., Charlier, R., Desrues, J., Hammad, W.(1990): A rate type constitutive law including explicit localization: development and implementation in a finite element code. *2nd Eur. Spec. Conf. on Num. Meth. in Geotechn. Engn., Santander.*

Desrues, J., Hammad, W. (1989): Shear banding dependency on mean stress level in sand. *Proc. of the Int. Workshop on Numerical Methods for Localization and Bifurcation of Granular Bodies, Gdansk, Poland, pp. 57–67.*

Gudehus, G. (1996): A comprehensive constitutive equation for granular materials. *Soils and Foundations Vol. 36, No. 1, 1–12.*

Kolymbas, D. (1987): A novel constitutive law for soils. *2nd Int. Conf. on Constitutive Laws for Engineering Methods, Tucson, Arizona.*

Kolymbas, D. (1981): Bifurcation analysis for sand samples with non-linear constitutive equation. *Ingenieur-Archive, 50, pp. 131–140*

Mühlhaus, H.-B., Vardoulakis, I. (1987): The thickness of shear bands in granular material. *Geotechnique 37, pp. 271–283.*

Oda, M. (1993): Micro-fabric and couple stress in shear bands of granular materials. *Powders and Grains, ed. by C. Thornton, Balkema, 161-167.*

Rice, J., Rudnicki, J. W. (1980): A note on some features on the theory of localization of deformation. *Int. J. Solids Structures, 16, pp. 597–605.*

Tejchman, J. (1989): Scherzonenbildung und Verspannungseffekte in Granulaten unter Berücksichtigung von Korndrehungen. *Publication of the Institute of Soil and Rock Mechanics, Karlsruhe University, No. 117.*

Tejchman, J., Bauer, E. (1996): Numerical simulation of shear band formation with a polar hypoplastic constitutive model. *Computers and Geotechnics, Vol. 19, No. 3, pp. 221–244.*

Tejchman, J. (1997): Shear localization and autogeneous dynamic effects in granular bodies. *Publication of the Institute of Soil and Rock Mechanics, Karlsruhe University, No. 140.*

Vardoulakis, I. (1980): Shear band inclination and shear modulus in biaxial tests. *Int. J. Num. Anal. Meth. Geomech., 4, pp. 103–119.*

Wu, W., Sikora, Z. (1991): Localized bifurcation in hypoplasticity. *Int. J. Engng. Sci. Vol. 29, No. 2, pp. 195–201.*

Wu, W., Bauer, E., Kolymbas, D. (1996): Hypoplastic constitutive model with critical state for granular materials. *Mechanics of Materials 23, 45–69.*

Yoshida, T., Tatsuoka, F., Siddiquee, M.S.A., Kamegai, Y., Park, C.S. (1993): Shear banding in sands observed in plane strain compression. *Proc. of the 3th Int. Workshop on Localisation and Bifurcation Theory for Soils and Rocks, Grenoble, France, 1994 Balkema, p 170.*

Joints

Mechanics of Jointed and Faulted Rock, Rossmanith (ed.) © 1998 Taylor & Francis, ISBN 90 5410 955 6

Mechanical behaviour of Dionysos marble smooth joints: I. Experiments

G. Armand & M. Boulon
Laboratory of Soils, Solids and Structures, University Joseph Fourier, Grenoble, France

C. Papadopoulos, M. E. Basanou & I. P. Vardoulakis
Laboratory of Testing and Materials, NTUA, Athens, Greece

ABSTRACT: In order to model later on the mechanical behaviour and the stability of ancient monuments submitted to seismic excitations, a basic study of the frictional properties of the contact between elements of masonry has been undertaken. Experimental data about the contact properties of smooth Dionysos marble joints are presented. Three main influences are investigated: the normal stress level, the shear velocity, and the cyclic degradation through cyclic tests at constant normal force. In part II (Numerical modeling) of this work (companion paper) a Perzyna type viscoplastic constitutive model for frictional interfaces is proposed.

1 INTRODUCTION

The preservation of old monuments is a great problem for architects. They have to study the mechanical behaviour and the stability of ancient monuments submitted to different critical loadings. Particularly if the behaviour of marble drums of columns, like those of the Parthenon, is studied, the behaviour in the dynamic range has to be carefully investigated. The determination of mechanical behaviour of joints under dynamic loading is of the highest importance for the characterisation of masonry stone assemblies. Zambas (1988) has shown that the dry joint mechanism plays a dominant role on the stability of the structural system for the famous articulated-jointed monuments of classical Greek architecture. The full history of the joint response contributes to the overall system behaviour and must therefore be adequately understood and quantified to allow numerical modelling (Papadopoulos et al(1998)).
Many of previous laboratory studies have mainly focused on rock joint behaviour under static or quasi-static loads. But other authors like Bandis et al (1983), Barbero et al (1996), Kana et al (1996), investigated rock joint behaviour under dynamic loading. The authors present here the results of a systematic testing program, designed for displaying the phenomena related to the mechanical behaviour of Dionysos smooth marble joints under dynamic and cyclic loading. The main attention has been devoted to investigate the influence of loading rate and history on joint response and on typical rock joint properties such as friction coefficient and dilatancy.

2 DIRECT SHEAR TESTING MACHINE

The tests were performed on a new servo-controlled direct shear machine (BCR3D). This device has been designed for testing mechanical behaviour and hydraulic conductivity of rock and concrete joints. The concept of this shear box described by Boulon (1995) is to prevent the rotation of the joint walls. It should be noticed that the shearing movement in one horizontal direction is produced by two symmetrical movements which leads to two main advantages: first the normal force is centered on the active part of the sample (no rotation), and secondly this feature allows for magnifying the shearing velocity in a ratio two. The guidance system is very complex due to the set of degrees of freedom necessary on each axis, and also regarding the requirement of minimising the dry friction within the machine. There exists one pseudo-static and one dynamic version of the machine, using two different sets of electro mechanical jacks, with brushless motors.

2.1 *Configuration for dynamic tests*

To realise the 2 D dynamic tests the machine was equiped with three mechanical jacks (one for normal direction and two for the tangential (horizontal) direction. Nominal loads up to 2500 daN can be applied. The range of shear velocity is between 0.001 m/s and 0.5 m/s . The maximum allowed displacement is 0.07 m in each horizontal opposite direction. The maximum sample size could be 0.10 m * 0.10 m for both the upper and the lower walls of the sample. More over the system has been designed to be as rigid as possible.

2.2 *Metrology, data acquisition, motor control*

The BCR3D is equiped with one LVDT and one load cell for each motor. The data acquisition is carried out digitally, using a high frequency data acquisition card. The motors used are brushless motors, which movements are controlled separately by PID filters programmed by a special software (MOOG).

3 EXPERIMENTAL PROGRAM

Direct shear tests were conducted on Dionysos marble joints.

3.1 *Dionysos marble*

The Dionysos marble is excavated from Dionysos quarries near Pentellikon mountain in Greece. This type of marble is composed by 98 % of calcite , 0.5 % of muscovite, 0.3 % of serecite, 0.2 % of quartz and 0.1 % of chlorite. Its grain size is around 0.4 mm, its density around 2717 kg/m3 and its absorption coefficient by weight around 0.11 %. The crystals are polygonic and mainly of equals size. It is of white colour with a few thin parallel ash-green vains containing locally silver areas due to the existence of chlorite and muscovite.

The most important physical and mechanical properties of Dionysos marble are as follows:

Uniaxial compression strength: 80 MPa

Uniaxial tension strength: 7.5 MPa

Young's modulus of elasticity: 75 GPa

Poisson's ratio: 0.33

3.2 *Sample preparation*

The smooth joints of Dionysos marble consisted of two half blocks of marble. The half blocks have been obtained by sawing and the contact faces by polishing. The height of these blocks was 5.50 cm and their sheared cross-section was 6.55 cm* 6.55 cm. After this stage the two walls were adjusted by special screws and grouted separately in their internal steel shear box. The grouting mixture of cement and water was in ratio 2:1 and the final mortar strength was about 8 MPa. Special care has been given so that the joint plane was centered between the two boxes and was parallel to the shear direction. For every sample, it was ensured that the average joint plan was aligned in a position parallel to the shear direction within 1° or 2°. After the mortar's strengthening the steel boxes were placed in the shear machine.

3.3 *Test program*

The experimental program related to smooth Dionysos marble joints is described in table 1. The parameters of these tests have been chosen in order to investigate the effect of normal stress (from 3 kN to 10 kN), the effect of the shearing velocity (from 0.01 m/s to 0.1 m/s), and also the cyclic degradation. In a typical test, the normal load was first applied up to its prescribed value. Then by servo-controlled mode a shear displacement was applied to the upper and to the lower half boxes, while the normal load was held constant. Each part of the sample moved from the initial mated position to a preset maximum shear displacement, followed by a reversal shear movement, in which each part, after passing the initial position, reached a preset position in the opposite direction. Then in a complete cycle each part return in its initial position. An harmonic sinusoidal excitation has been selected for the shear displacement since it allows the most convenient basis for the study of the results. Each sample was submitted to ten cycles in terms of shear displacement.

During these tests in which the average normal force was held constant and the shear displacement history was prescribed, the resulting shear force versus shear displacement and normal displacement versus shear displacement were recorded. The normal force versus normal displacement was also recorded.

4 TEST RESULTS

The typical loading program in terms of prescribed shear displacement or velocity is presented at the first figure (fig 1). It shows that the real displacement history is not far from the prescribed harmonic sinusoidal excitation. But when the sense

Table 1: Set of direct shear tests on smooth Dionysos marble joints.

Test number	1	2	3	4	5	6	7	8	9	11 12 (*)
Normal force (kN)	3.	5	10	3	5	10	3	5	10	10
Velocity (m/s)	0.01	0.01	0 01	0 05	0.05	0.05	0 1	0 1	0.1	5e-5
Frequency (Hz)	0 15	0.15	0.15	0.5	0 5	0 5	1	1	1	
Max. Displ. (cm)	±1 7	±1.7	±1.7	±2.5	±2.5	±2.5	±2.5	±2.5	±2.5	±0.8
Numb. of cycles	10	10	10	10	10	10	10	10	10	0

(* tests 11 and 12 were carried out at prescribed normal stress, and at low shear velocity : 0.05 mm/s)

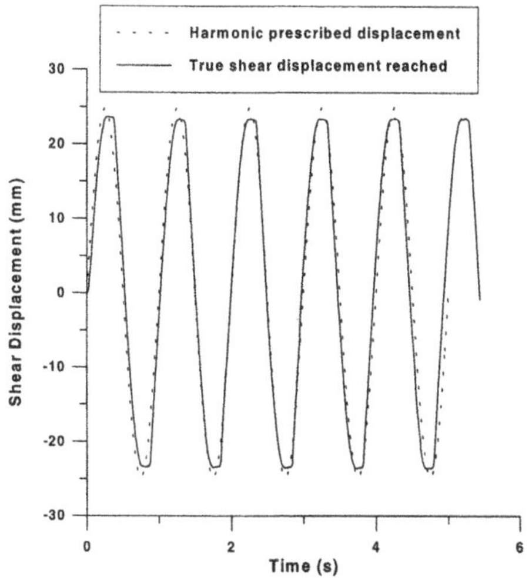

Figure 1. Typical loading program of smooth Dionysos marble joint in terms of history of shear displacement (Test nr 9, normal force 10 kN, frequency 1 Hz, max. displacement 2.5 cm, 5 cycles)

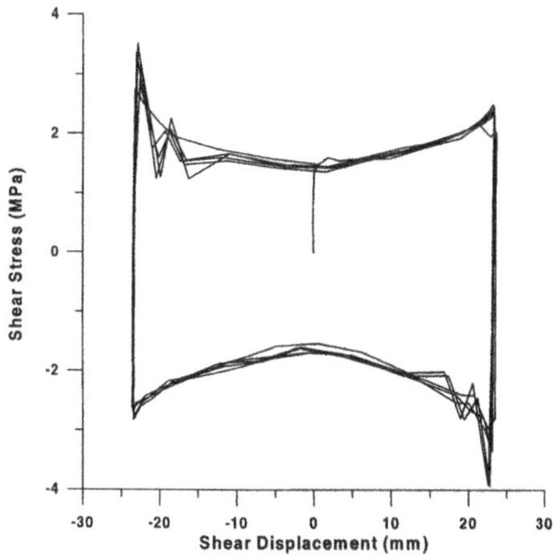

Figure 2. Typical response of smooth Dionysos marble in terms of shear stress (Test nr 9, normal force 10 kN, frequency 1 Hz, max. displacement 2.5 cm, 5 cycles).

of shearing changes, there is a little lateness of the shear relative displacement due to the mechanical play of the jacks. Figure 2 shows the typical response of a joint in terms of shear stress versus shear displacement, and exhibits significant peaks at each change of sense of shearing, even after several cycles. The response in terms of normal relative displacement appears at figure 3 and shows dilatancy and contraction within the first cycle, but a general contractive trend after several cycles. In addition it seems that there is a general slope (normal versus

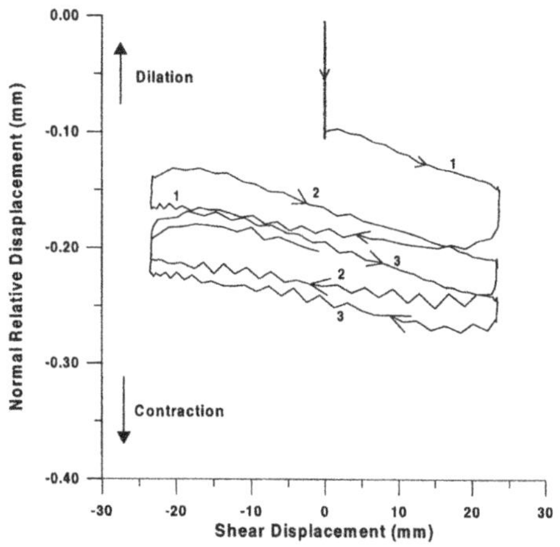

Figure 3. Typical response of smooth Dionysos marble in terms of normal relative displacement (Test nr 9, normal force 10 kN, frequency 1 Hz, max. displacement 2.5 cm, 3 cycles).

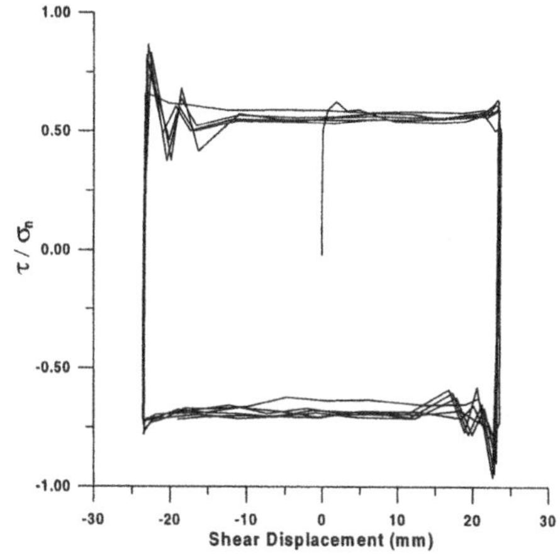

Figure 4. Typical response of smooth Dionysos marble in terms of friction coefficient (Test nr 9, normal force 10 kN, frequency 1 Hz, max. displacement 2.5 cm, 5 cycles).

shear relative displacement). It could also be noticed that the magnitude of the normal relative displacement is extremely small, according to the very small size of the asperities of the polished joint. A typical curve of friction coefficient versus shear displacement is given in figure 4. For the increasing displacement starting from zero the friction coefficient increases to a corresponding limit at which slippage occurs. This shear stress limit is maintained until motion stops at maximum displacement, where the friction coefficient reverses and continues proceeding around a loop, once for each cycle of the harmonic sinusoidal displacement.

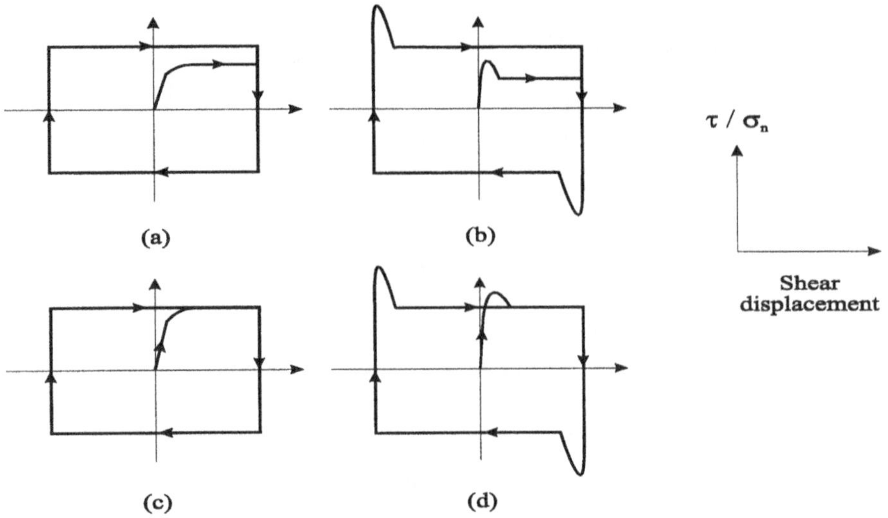

Figure 5. Tests at constant normal load on smooth Dionysos marble joints. Typical friction coefficient versus shear displacement curves.

(a) low normal load and low velocity
(b) low normal load and high velocity
(c) high normal load and low velocity
(d) high normal load and high velocity

A small dissymetry of the friction coefficient is readable.

5 DISCUSSION

A complete analysis of the set of shear tests shows that both the level of normal stress and the level of shear velocity have a significant effect on the shape of mobilisation of friction, on the friction coefficient, and on the ratio of dilatancy-contraction of the joint. Typical curves of friction coefficient versus shear displacement are summarised in figure 5. It could be seen (fig 5 a, b) that for the lower values of normal load, the magnitude of the friction coefficient, during sliding, increases from an initial minimum value to a higher value at the second cycle and remains rather unchanged for the next cycles. Huang et al (1993) observed the same phenomena in some low normal load pseudo-static tests. The magnitude of the shear stress during sliding gradually increases from a minimum value during the first cycle of shear displacement to a higher one, more constant as cycling continued. The explanation of this phenomenon is, after Huang, that debris produced by the initial cycle impede shear displacement in the subsequent cycles, thus requiring a higher level of shear stress. The increase of normal load seems to reduce or to eliminate the difference between initial and residual values of friction coefficient (fig 5c, d). The effect of the increase in velocity is easy to see and the comparison between fig 5a and fig 5b or between fig 5c and fig 5d exhibits the presence of a peak of shear stress at each change of the sense of shearing. In addition the relative amplitude of this peak (related to the residual shear stress value) is an increasing function of the velocity and a decreasing function of the normal stress level, as it can be seen in figure 6. It follows that the faster is the application of the rate of displacement, the greater is the additional shear resistance at each change of sense of shearing. A part of the explanation of this peak of shear stress could be, as it is shown in figure 7, that the shear velocity and the shear acceleration are not exactly the prescribed ones. In fact when the change of shearing occurs the mechanical plays induce that the shear velocity and the shear acceleration are equal to zero. After that an important acceleration is needed to reach the prescribed displacement. The peak of shear stress (or the peak of the friction coefficient) occurs together with this peak of acceleration. The mechanical plays ever exist even if jacks have been build with minimum mechanical plays. But if the rock walls temporarily stick together the shear relative displacement vanishes too. Then the momentary cessation of relative displacement at each peak is probably due to both effects : on one hand the existence of mechanical plays and on the

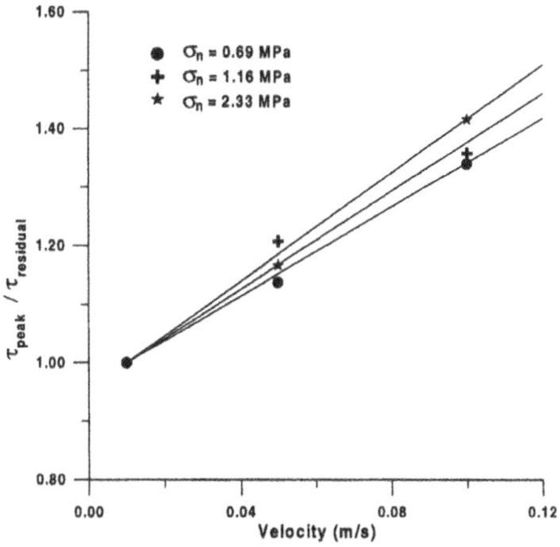

Figure 6. Tests at constant normal load on smooth Dionysos marble joints. Relative amplitude of the peak of shear stress as a function of the relative velocity and of the normal stress level.

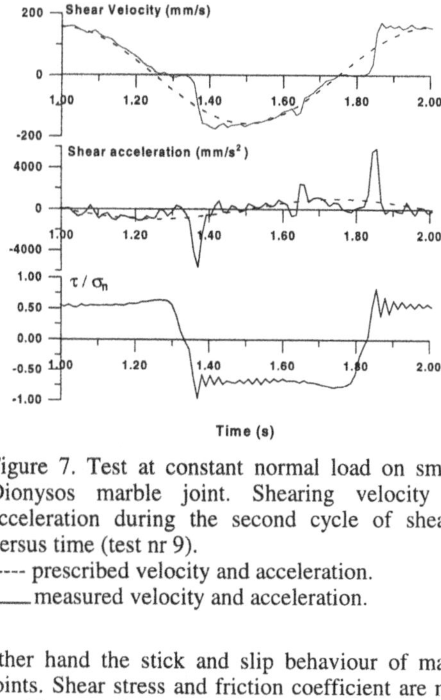

Figure 7. Test at constant normal load on smooth Dionysos marble joint. Shearing velocity and acceleration during the second cycle of shearing versus time (test nr 9).

----- prescribed velocity and acceleration.

——— measured velocity and acceleration.

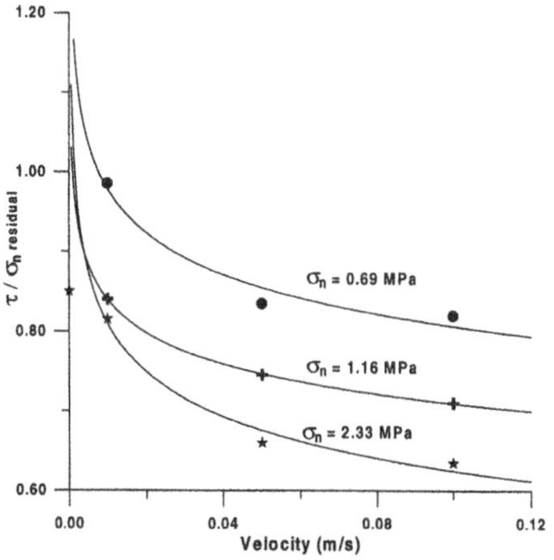

Figure 8. Tests at constant normal load on smooth Dionysos marble joints. Residual friction coefficient as a function of the relative velocity and of the normal stress level.

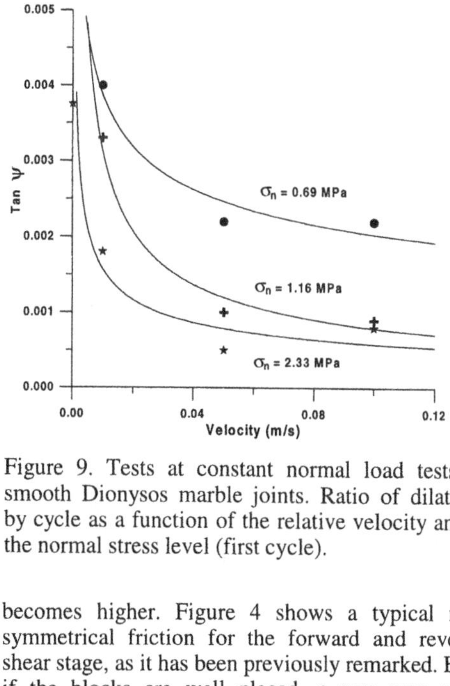

Figure 9. Tests at constant normal load tests on smooth Dionysos marble joints. Ratio of dilatancy by cycle as a function of the relative velocity and of the normal stress level (first cycle).

other hand the stick and slip behaviour of marble joints. Shear stress and friction coefficient are more sensitive to the rate of acceleration than to the relative shear displacement. The evolution of the mean value of the residual friction coefficient seems itself to be a non linear function of the normal stress level and of the relative shear velocity (figure 8). The data were fitted by power-form least square curves. A softening of the residual strength is observed and the magnitude of the friction coefficient decreases as the compression stress becomes higher. Figure 4 shows a typical non-symmetrical friction for the forward and reversal shear stage, as it has been previously remarked. Even if the blocks are well placed, a non zero angle between the shear plane imposed by the machine and the joint plane can subsist. This assumption is reinforced by the general slope (normal versus shear relative displacement) noticed in figure 3. So a correction can be used to take into account this non zero angle (called α). If we assume that the friction coefficient has to be symmetric, we find α obeying to equation 1.

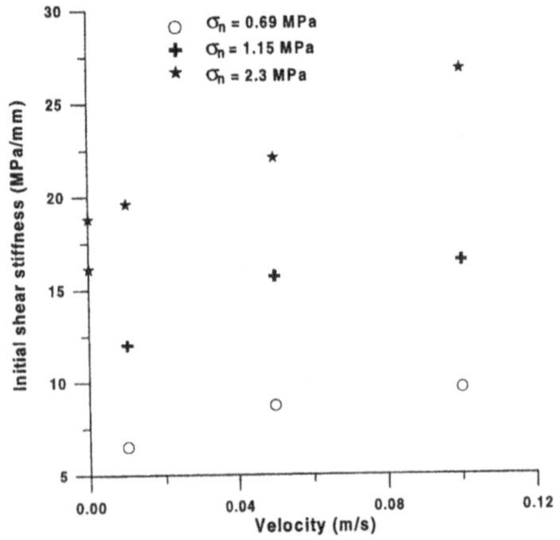

Figure 10 : Tests at constant normal load on smooth Dionysos marble joints. Initial shear stiffness as a function of the relative velocity and the normal stress level (first loading).

$$tg(2\alpha) = \sigma_{n0}*(\tau_m^+ + \tau_m^-)/(\sigma_{n0}^2 - \tau_m^+ * \tau_m^-) \qquad (1)$$

with σ_{n0} normal stress

τ_m shear stress

τ is taken > 0 in the sense of the first shearing and < 0 in the opposite sense.

The geometrical angle defined by that way varies from -3.6 ° to 6.1 ° within tests nr 1 to 12. The authors have estimated that the maximum errors is 2°, so the non symmetrical friction for forward and reverse shearing is not only due to a bad position of the joint. An anisotropy of the friction seems to develop after the first part of the first cycle of shearing. This other way of interpretation has been developed in the companion paper (Papadopoulous et al 1998). For the moment no attempt has been made to relate this phenomenon to the initial asperity condition and joint roughness.

The dilatancy effect related to a cycle, being calculated as the ratio of normal relative displacement to the shear displacement is presented in figure 9. The influencing factors are also the normal stress level and the relative rate of shearing. The stiffness parameters of the Dionysos smooth marble joint have been calculated. The normal stiffness is about 14.8 ± 8.1 MPa/mm. The initial shear stiffnesses are summarised in figure 10. These stiffnesses are estimated with a low precision, nevertheless a velocity and a stress effects seem to be present. The same effect can be seen on the shear stiffnesses measured on each cycle, but the values of the stiffness do not change with the number of cycles.

6 CONCLUSIONS

As the scope of this study is the Dionysos marble joint behaviour with a prescribed smoothness (the one between marble drums in a temple column or marble units in a masonry wall), no attempt has been made to relate the above results with various joint roughness patterns and topographies. The realisation of good dynamic tests is not easy. The test presented here are not perfect but they allow some assumptions and conclusions. The increase of normal load eliminates the difference between the initial and residual value of friction coefficient, arisen from a possible exasperation of the surface at first scraping. The residual friction resistance softens with the increasing loading velocity. Normal load plays a significant role in the dynamic behaviour as it appears to intensify the friction resistance. A high velocity shearing creates an offset peak at the beginning of each inversion of sense of shearing. In fact, this peak seem to be correlated with peaks of acceleration. Dilatant-contractant cycles are observed during shearing loops, but their magnitude is extremely small. The general behaviour during many cycles is contractant.

ACKNOWLEDGMENTS

The authors would like to thank the EU through its European program Environment (EV5V-CT93-0300).

REFERENCES

Bandis S., Lunsden A. c. and Barton N., (1983), Fundamentals of rock joint deformation, IJRMMS, 13, 255-279.

Barbero m., Barla G. and Zaninetti A., (1996), dynamic shear strength of rock joints subjected to impulse loading, IJRMMS, 33, 141-151.

Boulon M., (1995), A 3D direct shear device for testing themechanical behaviour and the hydraulic conductivity of rock joints, Proc. MJFR-2 Conf., Vienna, Austria, Balkema pub, 407-413.

Huang X., Haimson B. C., Plesha M. E., Qiu X., (1993), An investigation of the mechanics of rock joints- Part I. Laboratory investigation, IJRMMS, 30, 257-269.

Kana D. D., Fox D. J. and Hsiung S. M., 1996, Interlock/Friction model for dynamic shear response in natural jointed rock, IJRMMS, 33, 371-386.

Papadopoulos C., Basanou M., Armand G.,Vardoulakis I., Boulon M., 1998, Mechanical behaviour of Dionysos marble smooth joints: II. Constitutive modeling, MJFR-3 Conf., Vienna, Austria.

Zambas C., (1988) Principles for the structural restoration of the Acropolis Monuments, The Enggineering Geology of ancient Works, Monuments and Historical Sites, P. Marinos and G. Koukis Eds, Balkema pub, Vol. 3, 1813-1819.

Mechanics of Jointed and Faulted Rock, Rossmanith (ed.) © 1998 Taylor & Francis, ISBN 90 5410 955 6

Mechanical behaviour of Dionysos marble smooth joints under cyclic loading: II.Constitutive modeling

Ch.Th. Papadopoulos, M.E. Basanou & I.P. Vardoulakis
NTUA, Laboratory of Testing and Materials, Athens, Greece

M. Boulon & G. Armand
University Joseph Fourier, Laboratory 3S, Grenoble, France

ABSTRACT: In part I of this work (Mechanical behavior of Dionysos marble smooth joints under cyclic loading : I. Experiments), experiments on Dionysos marble for the determination of the mechanical properties of smooth, precut interfaces, were presented. These experiments have shown clearly normal stress and velocity dependence of friction coefficient values. In order to model mathematically such phenomena a Perzyna type viscoplastic constitutive model for frictional interfaces is proposed. In accordance with the experimental results, the internal non-linear dynamics of interface degradation are explored by adding a rate-dependence on the viscoplastic behavior.

1 INTRODUCTION

Determination of mechanical behavior of rock joints under dynamic loading is of outmost importance for characterization of jointed rock masses, formulation of constitutive models for rock interfaces and evaluation of blocky structures response under seismic loading. Specially for masonry stone assemblies, such as the famous articulated-jointed monuments of classical Greek architecture, the dry joint mechanism plays a dominant role on the structural system stability (Zambas 1988). The full history of joint response contributes to the overall system behavior and therefore must be adequately understood and quantified for meaningful numerical reasons.

Dynamic theories on frictional sliding of rock surfaces have been developed by Dietrich (1979) in order to describe the earth fault sliding and earthquake generation mechanisms. Rice and Tse (1986) analyze cyclic slip motion on basis of advanced frictional rate and state dependent constitutive models.

Most of previous laboratory studies have mainly focused on rock joint behavior under static or quasi-static loads. This paper presents the results of a systematic testing program, designed to uncover and display important phenomena in mechanical behavior of smooth marble joints under dynamic cyclic loading. Most of attention was devoted to investigate the influence of loading rate and history in joint response and in typical rock joint properties such as friction coefficient and dilatancy. The tests performed have shown clearly normal stress and velocity dependence of friction coefficient values. The inability of standard viscoplastic constitutive models to reproduce this shearing velocity softening of friction coefficient has been justified by noting that standard models assume that the process of interface degradation is quasi-static. In view of the experimental results this assumption can not be sustained. Accordingly, in this paper the internal non-linear dynamics of interface degradation are explored by adding a rate dependence on the viscoplastic behavior.

2 EXPERIMENTAL RESULTS

The tests were performed on a new servo-controlled direct shear machine (BCR 3D).The device was designed and fabricated at the Laboratory 3S of I.M.G. for the purpose of testing mechanical behavior and hydraulic conductivity of rock or concrete joints (Boulon 1995).

Direct shear tests were conducted on Dionysos Marble joints. Dionysos-Pentelikon Marble has been used as natural building stone for the construction of Parthenon (Korres 1988). The sheared surface of each sample half was polished so that the real smoothness, of an interface between marble drums in a column or marble units in a wall, was approximated as good as possible. Smoothing was realized through progressive finery grinding.

Dionysos Marble is excavated from Dionysos quarries near Pentelikon mountain. This type of marble is composed by 98% of calcite, 0.5% of muscovite, 0.3% of serecite, 0.2% of quartz and 0.1% chlorite. Its grain size varies around 0.4 mm, its density around 2717 kg/m^3 and its absorption coefficient by weight around 0.11%. Its crystalls are polygonics and mainly of equal size. It is of white color with a few thin parallel ash-green vains containing locally silver areas due to the existence of chlorite and muscivite. The experimental program comprised dynamic shear tests with different ranges of normal forces and velocities on nine marble samples. An harmonic sinusoidal excitation was selected for the tests, since it allows the most convenient basis for the study of the results. Each sample was submitted to a 10-cycle shear displacement. In the following we summarize the conclusions given in the first part of this work (Armand et al. 1997)

Typical curves of friction coefficient vs. shear displacement are summarized in Figure 1. For an increasing displacement starting at zero, the friction coefficient increases to a corresponding limit in which slippage occurs. This shear stress limit is maintained until motion stops at maximum displacement, where the friction coefficient reverses and continues proceeding around a loop, once for each cycle of the harmonic sinusoidal displacement. In the same Figure is also specified the terminology, that is adopted for the rest of the text.

The curves in Figs 1a & 1b that correspond to the lower values of normal load show that the magnitude of friction coefficient, during sliding, increases from an initial minimum value at the first cycle, to a higher value at the second cycle and remains rather unchanged for the next cycles. An explanation for this phenomenon, that it also observed in some low normal stress quasi- static experiments by Huang et al. (1993), could be that, any debris produced by the initial cycle serves to impede shear displacement in the subsequent cycles, thus requiring a higher shear stress magnitude. Perhaps this behavior is accompanied with a creation of a permanent anomalous surface structure, compared with the intact one, during the first cycle shearing. To a similar explanation could be attributed the typical feature of non- symmetrical friction coefficient values for the forward and reversal shear stage in the same cycle. After the first cycle, an anisotropy of the friction resistance for the forward and reversal shear stage seems to develop on the joint surface. For the moment no attempt has been made to relate this phenomenon to the initial asperity condition and rock joint roughness. Such an attempt is currently in progress.

In Figs 1c & 1d, that correspond to the higher values of normal load, the value of friction coefficient seems to remain less or more about the same level during the cycles. The increase of normal stress appears to eliminate the difference between initial and residual values of friction coefficient and also obtuses the anisotropy observed for the forward and reversal shear stage.

The mean values of the residual friction coefficient (for both forward and reversal shear stage in 10 cycles) are plotted in Figure 2. vs. the shearing velocity and for different levels of normal stress. From the plots a softening of the residual friction strength with increasing loading velocity is observed. The normal stress plays a significant role on the dynamic response of the interface. The magnitude of friction coefficient decreases as the compression stress becomes higher.

Another very important feature is the offset peak of the friction coefficient loop that occurs at the starting part of each cycle particularly in high velocity tests (Figs 1b & 1d). This behavior seems to be independent of the loading history (number of cycles preceded) and is more pronounced for lower normal loads.

The ratios of mean values of peak-residual friction coefficient to residual friction coefficient

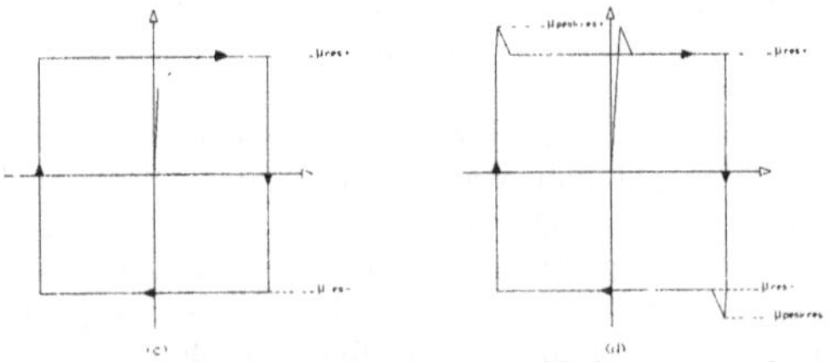

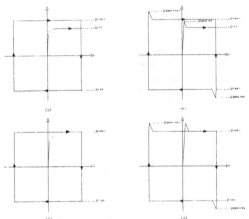

Figure 1. Typical friction coefficient vs. shear displacement curves.
(a) low normal load, low velocity
(b) low normal load, high velocity
(c) high normal load, low velocity
(d) high normal load, high velocity

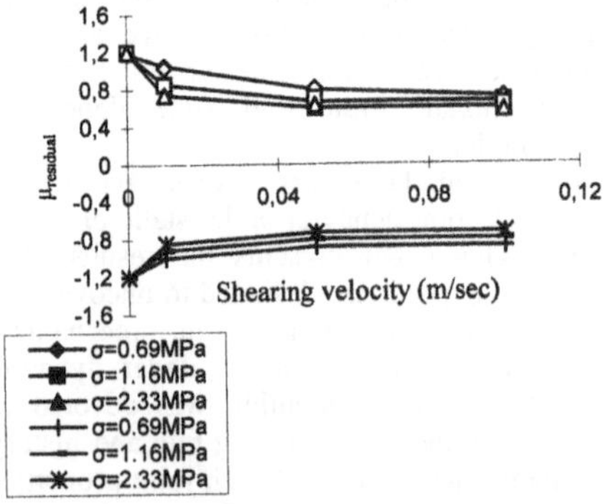

Figure 2. Mean values of μ_{res} (friction coefficient) vs. shearing velocity for reverse and forward cycles.

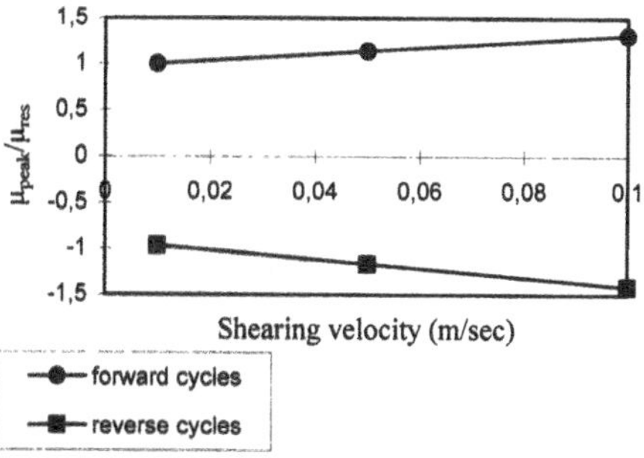

Figure 3. Ratios of mean values of peak and residual friction coefficient respectively vs. shearing velocity.

$\mu_{peak,res.}/\ \mu_{res.}$ (for both forward and reversal shear stages in 10 cycles) vs. shearing velocity is plotted in Figure 3.

The data points were fitted by linear least square lines. The ratio $\mu_{peak,res.}/\ \mu_{res.}$ increases gradually from unity (low velocity) to values greater than 1 as velocity increases. It follows that the faster is the rate of displacement application, the greater is the additional shear resistance at the beginning of each loop. This extra shear resistance decreases over a certain shear distance making a 'bump' appear in friction coefficient vs. tangential displacement curve i.e. an apparent peak point. Furthermore the peaks are somewhat clipped, which indicates that there is a momentary cessation of relative displacement at each peak.

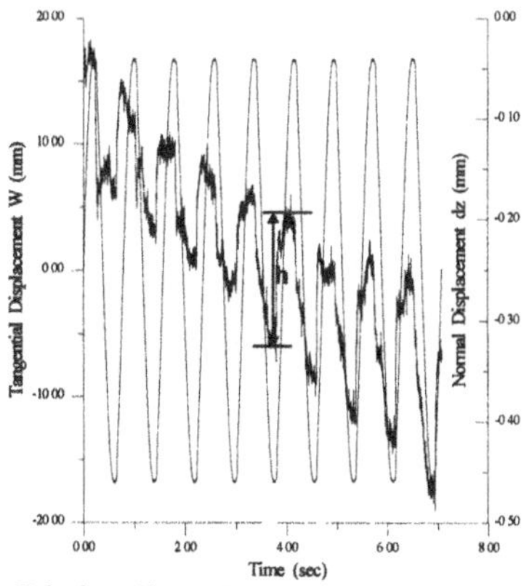

Figure 4. Friction Shear Tests on Dionysos Marble. Test #3#, normal force:10 kN, velocity: 0.01m/sec, frequency:0.15Hz.

Thus, the rocks momentarily stick together, and then break away from each other to reverse the direction of motion. This stick-slip behavior is followed by an oscillatory weakening of the friction coefficient.

A characteristic curve of normal displacement vs. time is shown in Figure 4.

A degradation of normal displacement magnitude during time with dilatant-contractant cycles that correspond to the shearing cycles is observed. A gradual decrease of normal displacement is displayed, that indicates an apparent surface damage during cyclic shearing. Dilatant-contractant behavior could be attributed to the surface anisotropy created under the first scraping between the joint's parts, or even to the migration of asperity debris during joint sliding. The resulting values of dilatancy angle are considered extremely small and without any significance for the constitutive interface model.

3 A SIMPLE VISCOPLASTIC VELOCITY SOFTENING MODEL

The observed velocity softening behavior can not be reproduced by means of a standard viscoplastic constitutive model. Within the frame of viscoplastic theory, the experimental observations may be adequately described mathematically by introducing a strain - rate dependence of friction coefficient, Rice and Tse (1986) and di Prisco et al.(1997). The proposed constitutive model has the simplest possible mathematical structure, which is necessary to reproduce qualitative features of the experimental observations.

An interface behavior involves a displacement pattern of u and v components in local x and y co-ordinate directions under a normal σ and a shear stress τ, shown in Figure 5.

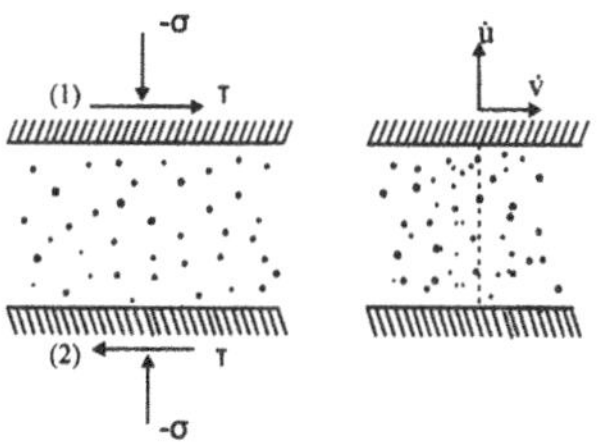

Figure 5. Stress and displacement components on interface

The velocity vector is decomposed into a reversible (elastic) part and an irreversible (viscoplastic) part :

$$s = \begin{Bmatrix} \dot{u} \\ \dot{v} \end{Bmatrix} = \begin{Bmatrix} \dot{u}^{e} \\ \dot{v}^{e} \end{Bmatrix} + \begin{Bmatrix} \dot{u}^{vp} \\ \dot{v}^{vp} \end{Bmatrix} \qquad (1)$$

where the overdot denotes time rate. The stress vector is written as :

$$\sigma = \begin{Bmatrix} \tau \\ \sigma \end{Bmatrix} \tag{2}$$

The elastic part of velocity vector is linked to stress - rate vector as follows :

$$\begin{Bmatrix} \dot{u}^e \\ \dot{v}^e \end{Bmatrix} = \begin{bmatrix} k_s & 0 \\ 0 & k_n \end{bmatrix} \begin{Bmatrix} \dot{\tau} \\ \dot{\sigma} \end{Bmatrix} \tag{3}$$

where subscripts n and s denote normal and shear components, and k_n , k_s are the elastic normal and shear stiffness of the interface. As the proposed model focuses on the cyclic behavior of the joint under large scale displacement, the elastic response is ignored and only the viscoplastic one is considered.

By following Perzyna's original approach, the viscoplastic velocity vector is defined by a viscoplastic flow rule of the form :

$$\dot{s}^{vp} = \gamma\, \Phi(F) \frac{\partial Q}{\partial \sigma} \tag{4}$$

In this equation the viscoplastic potential Q gives the direction of $\dot{s}^{vp}$, the viscous nucleus is a scalar function of the yield function F and γ is a material constant, known as the fluidity parameter with dimension of inverse time.

Adopting an anisotropic (for forward and reversal shear motion) Mohr - Coulomb plasticity model, the yield surface F (Figure 6) is defined as:

$$F_{1,2} = \tau + (r \pm k)\sigma_n \tag{5}$$

with

$$r = \frac{1}{2}(\mu_1 - \mu_2)$$
$$k = \frac{1}{2}(\mu_1 + \mu_2) \tag{6}$$

where μ_1 , μ_2 is friction coefficient in forward and reversal direction correspondingly. If $\mu_1 = \mu_2$, then $r = 0$ and $k = \mu_1 = \mu_2 = \mu$.

The plastic potential is defined as :

$$Q_{1,2} = \tau + \tan(\psi_{1,2})\sigma_n + \text{const.} \tag{7}$$

If dilatancy is negligible, as it results from the experimental observations, then :

$$Q_1 = Q_2 = \tau + const\,, \frac{\partial Q}{\partial \sigma} = \begin{Bmatrix} \frac{\partial Q}{\partial \sigma} \\ \frac{\partial Q}{\partial \tau} \end{Bmatrix} = \begin{Bmatrix} 0 \\ 1 \end{Bmatrix} \tag{8}$$

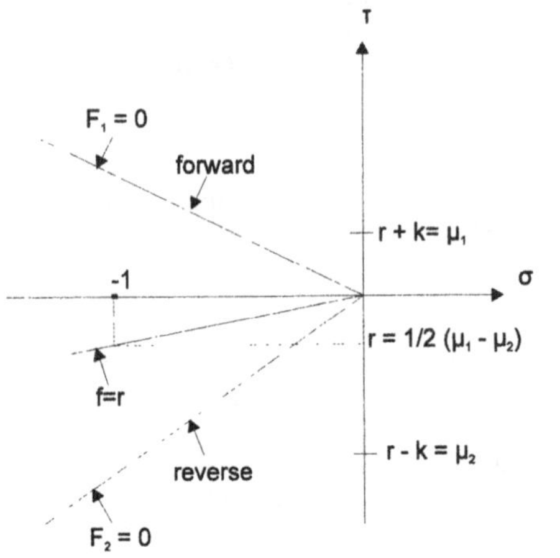

Figure 6. Yield surface representation

The viscous nucleus $\Phi(F)$ describes the evolution of the viscoplastic strain vector. Using the Foeppl - Macauley brackets, a simple representation of viscous nucleus is :

$$\Phi(F) = \prec \frac{F}{-\sigma} \succ = \begin{cases} -F/\sigma & \text{if } F \rangle 0 \\ 0 & \text{if } F \le 0 \end{cases} \tag{9}$$

or

$$\Phi(F) = \begin{cases} f + \mu_{1,2} \text{ if } f_{1,2} \ge 0 \text{ with } f = \dfrac{F}{-\sigma} \\ 0 \quad \text{else} \end{cases} \tag{10}$$

The definition of viscous nucleus $\Phi(F)$ implies that the viscoplastic velocity is not nil, if and only if, the current state of stress lies outside the yield locus.

Material constant γ introduces the physical time dimension into the problem. One can observe that for the limit $\gamma \to \infty$, the constitutive model becomes plastic and its dependency on the time vanishes. Through γ a material time scaling factor $t_c = 1/\gamma$, is introduced. Transformation $t' = t/t_c$ rescales absolute time to the material's internal one. According to this time rescaling, material time derivatives transform as,

$$\frac{d}{dt} = \gamma \frac{d}{dt'} = \frac{1}{t_c} \frac{d}{dt'} \tag{11}$$

In the following, we will assume that time is always rescaled by the material time scaling factor t_c. Similarly, material time differentiation will be in terms of the dimensionless time factor t'.

The friction coefficient evolution during shearing displacement is assumed to follow the constitutive equation :

$$\dot{\mu}_{1,2} = \alpha_{1,2}(\mu_{1,2-res} - \mu)\dot{u} \tag{12}$$

where $\mu_{1,2\text{-res.}}$ is the residual value of $\mu_{1,2}$ and $a_{1,2}$ is a coefficient that controls the $\mu_{1,2}$ decrease.

Solving the differential equation for a single constant velocity test, we obtain the following expression :

$$\mu_{1,2} = \mu_{1,2-res} + (\mu_{1,2-peak} - \mu_{1,2-res})e^{\alpha_{1,2}u} \tag{13}$$

where $\mu_{1,2\text{-peak}}$ is the peak value of $\mu_{1,2}$ in the beginning of a shearing motion.

The velocity softening of friction coefficient observed in experiments (Figure 2) is described by the following equation :

$$\frac{\mu_{1,2-res} - \mu^{\infty}_{1,2-res}}{\mu^{0}_{1,2-res} - \mu^{\infty}_{1,2-res}} = \\ = \exp\left\{ -\frac{\beta_{1,2}}{\mu^{0}_{1,2-res} - \mu^{\infty}_{1,2-res}} \dot{u} \right\} \tag{14}$$

where $\mu_{1,2\text{-res}}$ is an extrapolation value of $\mu_{1,2}$ for $\dot{u}$ = 0

$\mu_{1,2\text{-res}}$ is the limit value of $\mu_{1,2}$ for $\dot{u} \rightarrow \infty$ and

$\beta_{1,2}$ is a coefficient that controls the $\mu_{1,2\text{-res}}$ decrease.

The dependence of the limit value of friction coefficient for $\dot{u} \rightarrow \infty$ on the normal stress is assumed to follow a linear law :

$$\mu^{\infty}_{1,2-res} = A_{1,2} + B_{1,2}(-\sigma) \tag{15}$$

A linear law is assumed also for the relation between the ratio of peak to residual values of friction coefficient and the velocity :

$$\frac{\mu_{1,2-peak}}{\mu_{1,2-res}} = \Gamma_{1,2} + \Delta_{1,2}\dot{u} \tag{16}$$

4 CALIBRATION OF MODEL PARAMETERS

Two sets of seven parameters are required to define the anisotropic viscoplastic joint behavior. One set counts for the forward shearing motion, the other on the reversal. All the parameters are evaluated from the experimental results. The procedure for the determination of these parameters is presented below

- Parameters $A_{1,2}$, $B_{1,2}$

Parameters A_1, A_2 and B_1, B_2 are evaluated from the linear least square fit of $\mu_{1\text{-res}}$ vs. σ_n and of $\mu_{2\text{-res}}$ vs. σ curves correspondingly.

- Parameters $\Gamma_{1,2}$, $\Delta_{1,2}$

Parameters Γ_1, Γ_2 and Δ_1, Δ_2 are evaluated from the linear least square fit of $\mu_{1\text{-peak}}/\mu_{1\text{-res}}$ vs. $\dot{u}$ and of $\mu_{2\text{-peak}}/\mu_{1\text{-res}}$ vs. $\dot{u}$ curves correspondingly.

- Velocity softening parameters $\beta_{1,2}$ and interpolation value $\mu_{1,2\text{-res}}$

Parameters β_1, β_2 are determined from the $\mu_{1\text{-res}}$ vs. udot.. and of $\mu_{2\text{-res}}$ vs. $\dot{u}$ curves correspondingly. The interpolation value is the same for both forward and reversal cycles $\mu_{1\text{-res.}} = \mu_{2\text{-res.}} = 1.20$.

- Parameters a_1, a_2

Parameters a_1, a_2 are determined from the μ_1 vs. u and of μ_2 vs. u curves correspondingly.

The values of all parameters are summarized in Table 1.

Table 1. Parameter values

Forward cycles		Reversal cycles	
Parameter	Value	Parameter	Value
A_1	0.74	A_2	0.91
B_1	-0.074	B_2	-0.085
Γ_1	0.97	Γ_2	0.92
Δ_1	3.44	Δ_2	4.96
β_1	43.5	β_2	92.2
α_1	1.20	α_2	1.20
$\mu_{1\text{-res}}$	0.98	$\mu_{2\text{-res}}$	0.83

The fluidity parameter γ is obtained from the ratio between model parameters α and β.

γ [sec^{-1}] = α [m^{-1}] / β [sec/m], here γ_1=0.0275, γ_2=0.013.

5 MODEL VALIDATION

The constitutive model was validated by backanalysing the laboratory tests. The results were obtained by using the parameters presented in Table 1. Figure 7 show the comparison between backanalysis and test data for the test conducted at σ =2.33MPa, $\dot{u}$ =0.05m/sec for both forward and reversal cycles. Analogous comparison is shown in Figure 8 for the test conducted at σ = 2.33MPa, $\dot{u}$ = 0.1m/sec. Comparisons prove that the proposed constitutive model provides satisfactory description of the viscoplastic behavior of the interface.

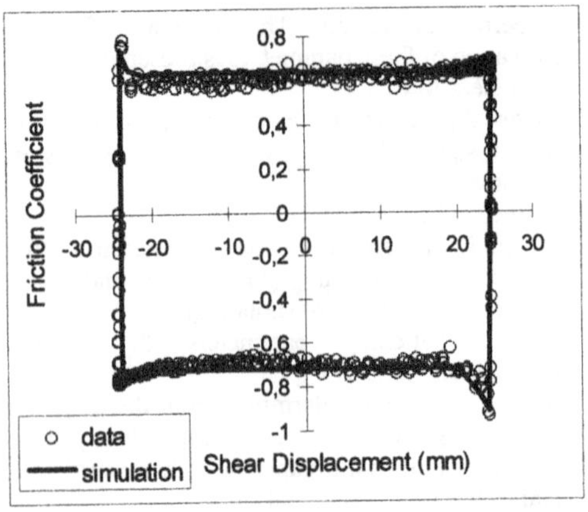

Figure 7. Experimental data and model prediction for test conducted at σ=2.33MPa, u̇ =0.05m/sec.

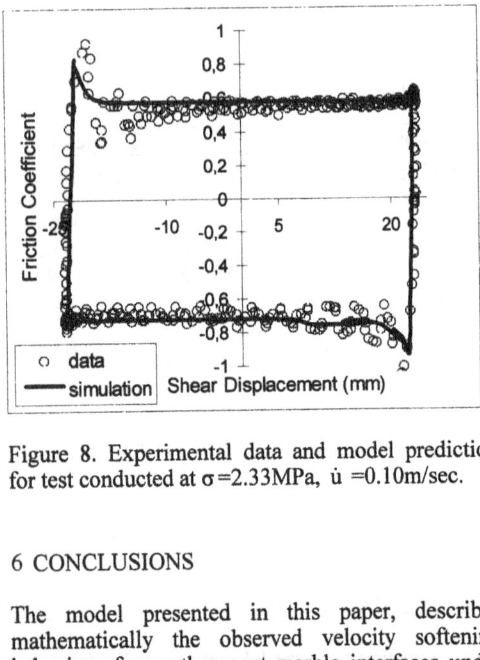

Figure 8. Experimental data and model prediction for test conducted at σ=2.33MPa, u̇ =0.10m/sec.

6 CONCLUSIONS

The model presented in this paper, describes mathematically the observed velocity softening behavior of smooth precut marble interfaces under dynamic cyclic loading. The non-linear dynamics of interface degradation are explored by adding a rate dependence on the viscoplastic behavior. The structure of the model is conceptually simple and involves the minimum number of parameters. Elastic response, as well as dilatancy are neglected here as it seems natural in joints under large number of cyclic displacements. The observed oscillatory weakening of the friction coefficient after peak in the beginning of each shearing cycle is not modeled here, but it is a phenomenon that requires further study.

The model parameters are determined from the experimental results. It has been demonstrated by backanalysis that the model describes satisfactorily the viscoplastic behavior of the joint. Simulations illustrate the versatility and the consistency of the model.

7 AKNOWLEDGEMENTS

The Authors would like to thank the EU through its European program Environment (EV5V-CT93-0300).

8 REFERENCES

Zambas C. 1988. Principles for the structural restoration of the Acropolis Monuments, *The Engineering Geology of Ancient Works, Monuments and Historical Sites*, P. Marinos and G. Koukis Eds., A. A. Balkema, Vol. 3, 1813-1818.

Dietrich H. J. 1979. Modeling of rock friction : 1. Experimental results and constitutive equations, *J. Geophys. Res.* 84, 2161- 2168.

Rice R. J. and Tse, T. S. 1986. Dynamic motion of a single degree of freedom system following a rate and state dependent friction law, *J. Geophys. Res.* 91, 521- 530.

Boulon M. 1995. A 3D direct shear device for testing the mechanical behavior and the hydraulic conductivity of rock joints, *Proc. of the Second International Conference on the Mechanics of Jointed and Faulted Rock*, Vienna, Austria, A.A Balkema, 407- 413.

Korres E. 1988. The geological factor in ancient Greek architecture, *The Engineering Geology of Ancient Works, Monuments and Historical Sites*, P. Marinos and G. Koukis Eds., A. A. Balkema, Vol. 3, 1779-1793.

Armand G., Boulon M., Papadopoulos Ch., Basanou M., Vardoulakis I. 1997. Mechanical behavior of Dionysos marble smooth joints under cyclic loading : I. Experiments, *MJFR-3 Conf.*, Vienna, Austria.

Huang X., Haimson B. C., Plesha M. E. and Qiu X. 1993. An investigation of the mechanics of rock joints- Part I. Laboratory investigation, *Int. J. Rock Mech. Min. Sci. & Geomech. Abstr.* 30, 257-269.

di Prisco C., Imposimato S., Vardoulakis I. 1997. Strain rate softening of loose sand, *submitted for publication.*

Perzyna P. 1966. Fundamental problems in viscoplasticity advances in Applied Mechanics, *Academic Press,* Vol. 9, 243-377.

Mechanics of Jointed and Faulted Rock, Rossmanith (ed.) © *1998 Taylor & Francis, ISBN 90 5410 955 6*

A static fatigue constitutive law for joints in weak rock

Mark K. Larson
Spokane Research Laboratory, National Institute for Occupational Safety and Health, Wash., USA

ABSTRACT: A constitutive model for rock interfaces based on the principles of static fatigue is proposed. The model is applicable to weak rock or joints with cohesion. Friction and dilation are decreased as plastic work increases. At equilibrium, the creep time increases by one time-step interval, and interface cohesion at each node decreases according to a power function. A constant-load, direct-shear simulation showed that the model is capable of mimicking constant-rate and tertiary creep and predicting failure. Tests on core from a coal mine showed constant-rate and, sometimes, tertiary creep, but also weak transitory creep. The cohesion deterioration function may need modification to simulate primary-phase creep. If model parameters are known for a site, the model can predict the onset of tertiary creep and failure. Such predictions can help engineers make better entry design and support decisions, which will reduce the likelihood of roof falls and increase safety for underground miners.

1 INTRODUCTION

Researchers from the Spokane Research Laboratory of the National Institute for Occupational Safety and Health (NIOSH) are developing numerical modeling tools to forecast time-dependent deformation around coal mine entries. Conditions such as weak, layered strata sometimes cause deformation rates to increase so that the roof eventually fails, possibly resulting in injuries and fatalities. Therefore, efforts have been focused on developing models that simulate time-dependent deformation at joints and other planes of weakness. Such models may be applied to other problems, such as slope stability, tunneling, and mining problems involving weak joints.

Most time-dependent interface models are of the viscous-creep type. The rate of slip at some location along an interface is considered to be a continuous (nonlinear) function of the shear stress. Often, therefore, friction, dilation, and cohesion are weakened by displacement and are coupled with the creep model, as Fakhimi (1992) did in his discontinuum model.

In this paper, the principles of static fatigue, or stress corrosion, are applied to cohesion along planes of weakness or joints. The goal is to describe time-dependent mechanisms involving reduction of cohesion coupled with weakening of friction and dilation caused by displacement. This idea seems reasonable because micromechanisms of static fatigue and creep are similar. For example, microcracking is the principal underlying process causing rock creep and static fatigue. Also, both static fatigue and creep rates are affected by temperature and the presence of corrosive agents, such as water. Finally, preliminary modeling of a coal mine site suggested that reductions in joint interface cohesion play a significant role in deformation of strata over time.

2 STATIC FATIGUE

2.1 *Points from the literature*

Wawersik (1974) tested intact and jointed rock specimens (Westerly Granite and Navaho Sandstone) in uniaxial and triaxial compression. Creep of the rock was observed, particularly when water was present. Very little creep was seen in a specimen with a clean, interlocked joint. Based on these tests, Wawersik concluded that creep of jointed rock follows the same general behavior as that of intact material, but with larger amounts of shearing strain for jointed rock.

Charles (1958) conducted investigations of static fatigue in soda-lime glass. He assumed that surface flaws grow by corrosive interactions between water vapor and components of the glass and that the rate of this reaction is determined by local stress conditions and temperature, pressure, and composition of the atmosphere. To simulate subcritical crack growth in a sliding crack model, Kemeny (1991) used the empirical Charles power law,

$$\dot{c} = c_0 \, e^{\left(\frac{-H}{RT}\right)} K_I^{\,n}, \tag{1}$$

where $\dot{c}$ is crack velocity, H is activation enthalpy, R is gas constant, T is absolute temperature, K_I is mode I stress intensity factor, and c_0 and n are constants.

Several tests and studies of the behavior of intact, brittle, crystalline rock under static fatigue have been conducted. Scholz (1972) and Martin (1972) investigated static fatigue and crack growth in quartz. Each proposed, based on their experiments, that static fatigue might be described by

$$t_f = t_0 \, P^{-\alpha} \, e^{\left(\frac{E}{RT} - K\sigma\right)}, \tag{2}$$

where t_f is time to failure, P is partial pressure of water in cracks, E is activation energy for the fatigue process, σ is differential stress, R is gas constant, T is absolute temperature, and t_0, α, and K are empirical constants.

Lajtai et al. (1987) tested specimens of granite in direct contact with water and observed a decrease in strength. Lajtai and Schmidtke (1986) tested specimens of granite and anorthosite in which specimen strength decreased to about 60% of its dry, short-term strength in tests lasting from a few minutes to 17 days. The conclusions from these tests were that (1) creep occurs principally because of microcracking; (2) crack tip strength, long-term strength, and, sometimes, short-term strength are decreased by increasing temperature or by the presence of moisture, while creep strain rate and slow crack velocity are increased; (3) creep rate is dependent on the applied stress-to-strength ratio; (4) in creep tests, crack growth is limited by the rate at which corrosive agents can decrease crack tip strength; and (5) other internal surfaces, such as grain boundaries, pores, and preexisting cracks, also promote crack development.

Price (1964) studied time-dependent strain of coal-measure rocks (Pennant Sandstone, calcareous sandstone, sandstone, siltstone, and nodular, muddy limestone) using bending and compressive creep tests. He could describe the time-deflection results of the bending tests with a Bingham-Voigt model. For heterogeneous rock types, expansion of specimens under constant compressive load can be explained in terms of the release of pockets of residual strain energy formed during the rock's geologic history. He suggested that cohesion among components breaks down during a creep test, thus releasing localized concentrations of strain energy.

2.2 *Mathematical formulation of model*

From the previous subsection, modeling a single plane of weakness with an interface that initially has non-zero cohesion and deteriorates over time according to some function appears reasonable. From equations 1 and 2, dependent parameters might include pore pressure, temperature, activation energy for the fracturing process, and stress level or stress intensity. Although the former three parameters may be significant, for ease of demonstration in this paper, only a measure of stress level will be used as a dependent factor.

Displacement-weakening friction and dilation are likely to be important processes. Therefore, a form of these processes is included in this formulation. What follows is a mathematical description of the model.

Incremental displacements normal and parallel to the joint are represented by

$$\Delta u_n = \Delta u_n^e + \Delta u_n^p \tag{3}$$

and

$$\Delta u_s = \Delta u_s^e + \Delta u_s^p, \tag{4}$$

where Δu_n and Δu_s are relative normal and shear increments of displacement, respectively, and the superscripts e and p refer to the elastic and plastic components of displacement.

The normal (σ_n) and shear (σ_s) stress increments are

$$\Delta\sigma_n = K_n \, \Delta u_n^e \tag{5}$$

and

$$\Delta\sigma_s = K_s \, \Delta u_s^e, \tag{6}$$

where K_n and K_s are normal and shear stiffnesses.

The yield function is

$$f = |\sigma_s| + \sigma_n \tan(\phi) - C \qquad (7)$$

and the plastic potential function is

$$g = |\sigma_s| + \sigma_n \tan(\psi). \qquad (8)$$

C, ϕ, and ψ are mobilized cohesion, friction angle, and dilatation angle, respectively.[1] The plastic components of the relative displacements follow the flow rules

$$\Delta u_n^p = \lambda \frac{\partial g}{\partial \sigma_n} = \lambda \tan(\psi) \qquad (9)$$

and

$$\Delta u_s^p = \lambda \frac{\partial g}{\partial \sigma_s} = \lambda\, sgn(\sigma_s). \qquad (10)$$

where λ is a scalar multiplier (not a constant) determined by requiring that the new stress point be located on the shear yield surface.

Incremental plastic viscous work is

$$\Delta W_p = \sigma_s \Delta u_s^p - \sigma_n \Delta u_n^p. \qquad (11)$$

ϕ and ψ are based on plastic work, such as

$$\phi = \phi_{re} + (\phi_p - \phi_{re})\, e^{-\alpha W_p} \qquad (12)$$

and

$$\psi = \psi_p\, e^{\left[-\left(\frac{W_p}{W_{sl}}\right)^2\right]}, \qquad (13)$$

where ϕ_{re} is residual friction angle, ϕ_p is peak friction angle, ψ_p is peak dilation angle, and W_{sl} is an empirical constant.

Cohesion is represented as a combination of residual and time-dependent components, so that

$$C(t) = C_{re} + C_{SF}(t,S), \qquad (14)$$

where C_{re} is the residual (i.e., constant minimum value) cohesion, $C_{SF}(t,S)$ is the static fatigue component of cohesion that deteriorates over time, t is time, and S is a safety factor. Various functions are possible to represent the time-dependent change in C_{SF}. One structure for such a function could be represented incrementally according to

$$\Delta C_{SF} = -h(S, C_{SF})\, \Delta t, \qquad (15)$$

where $h(S, C_{SF})$ is an empirical function appropriate for the joint or plane of weakness. For demonstration purposes, in this paper the author has chosen the function

$$h(S, C_{SF}) = C_{SF}\, S^{h_1}, \qquad (16)$$

where

$$S = \frac{-\sigma_n \tan(\phi) + C}{|\sigma_s|}, \qquad (17)$$

and h_1 is an empirical constant. If S is calculated to be less than 1.0, it is set equal to 1.0.

It is recognized that even if the function, h, can successfully simulate deterioration of cohesion for a constant set of environmental condition, a change in environmental conditions may make any parameter of the function variable or may make the function invalid.

3 CODING OF THE MODEL

The finite-difference code, FLAC, was selected as a vehicle to code the model because of the embedded computer language, FISH, and access to FLAC's internal data array. Several FISH functions were written to overwrite the interface calculation task in the main cycle loop. This is accomplished in three steps. First, interface interaction forces and out-of-balance forces are saved in memory before FLAC executes the interface calculation step. Second, the new interface forces are calculated and saved in place of those forces calculated in the regular interface calculation step. Third, these forces are added to the previously saved out-of-balance forces. Friction and dilation are updated as the solution steps toward equilibrium. At equilibrium, a creep time is increased by the time-step interval, and the cohesion is updated.

A purely elastic case was selected for a test. The case was run with and without the FISH functions. Any differences between the equilibrium states were

[1] The sign convention for stress is positive for tension.

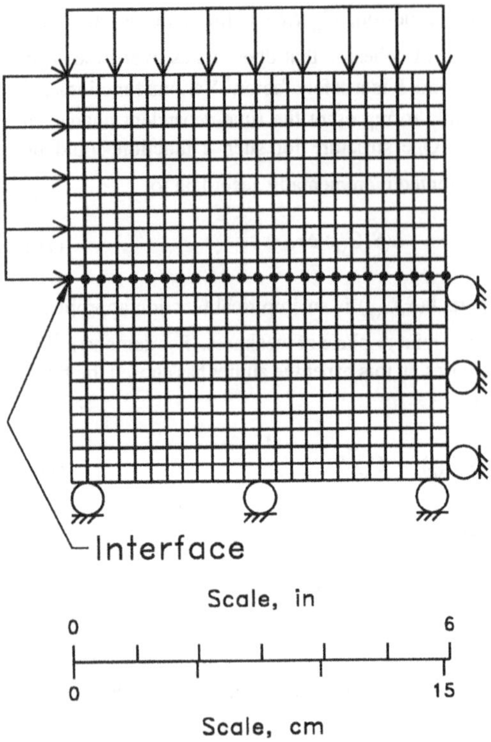

Figure 1.—Direct-shear model grid and boundary condition locations

small, and further experiments showed them to be within the computational precision of FLAC.

A purely elastic case was selected for a test. The case was run with and without the FISH functions. Any differences between the equilibrium states were small, and further experiments showed them to be within the computational precision of FLAC.

4 MODEL DEMONSTRATION

A direct-shear test was simulated in a preliminary demonstration of the model. Figure 1 shows the model grid and boundary conditions. A normal stress of 3.45 MPa (500 psi) was applied for all tests. Friction and dilation angles were initially at 30° and 10°, respectively. Initial cohesion was 1.38 MPa (200 psi), so that initial shear strength was 3.37 MPa (489 psi). Constant shear stresses of 2.09, 2.41, 2.59, 2.76, 2.93, 3.10, and 3.28 MPa (300, 350, 375, 400, 425, 450, and 475 psi) were applied.

Figure 2 is a plot showing x-direction displacements of three points over creep time up to a maximum of 4000 hours. These three points are on the upper half of the interface, that is, the left-most, center, and right-most grid points. Within 4000 hours, tertiary creep and failure set in to all but the two cases with the lowest shear stress. The figure shows some early transitory creep, but not like a classic decaying

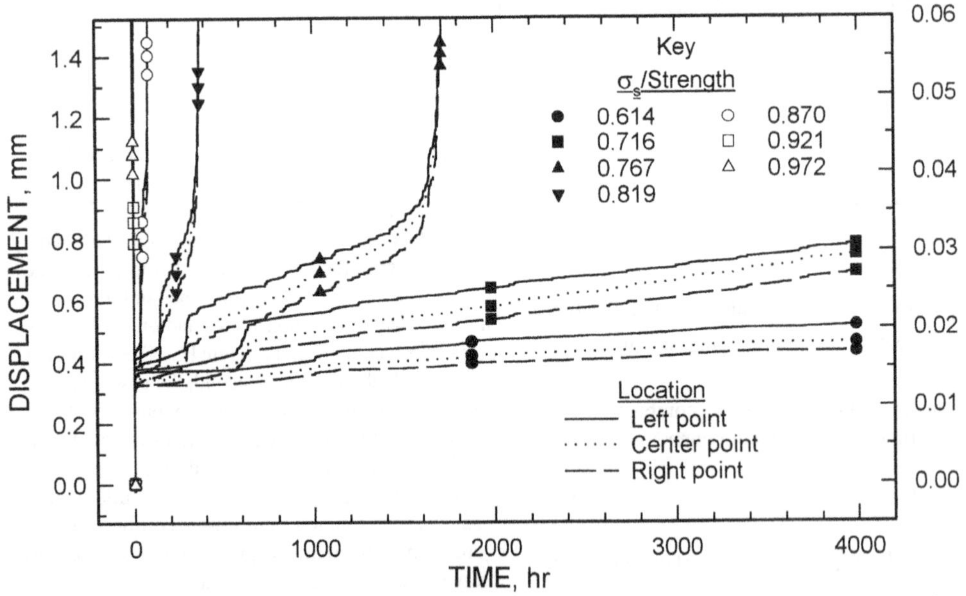

Figure 2.—X-direction displacement versus creep time as calculated by FLAC model with static fatigue FISH functions for various initial shear stress-to-shear strength ratios. Symbols are data points selected to distinguish between sets of lines.

primary-phase creep. Constant-rate creep seems to predominate until tertiary creep takes over. Perhaps primary creep may be simulated by choosing another function for h, possibly in which time, instead of just time step is a direct variable.

An interesting result comes from an examination of the progression of cohesion deterioration in the case where the σ_s-to-strength ratio is 0.716. Figure 3 plots cohesion by location along the interface (left to right) after 4000 hours of creep time. If cohesion deterioration represents the degree of microcracking, then it is evident that microcracking was prevalent on the left side of the interface but did not progress completely across the interface or plane of weakness.

Another observation was that if the σ_s-to-strength ratio is low enough, failure will never occur. No case was run that showed no deterioration, but it is not hard to see that, for practical purposes, this statement is true.

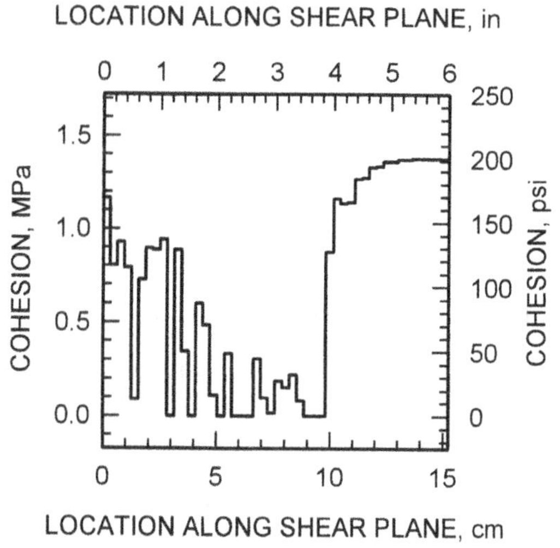

Figure 3.—Cohesion along interface at 4000 hr for σ_s/strength = 0.716

5 COMPARISON WITH LABORATORY TESTS

Larson and others (1995) and Larson and Maleki (1995) reported measurements of roof deformation over time at two sites in a western U.S. coal mine. The roof consists of a carbonaceous mudstone having weak planes throughout along bedding. At site 1, primary-phase creep was prevalent, whereas secondary-phase creep dominated at site 2. Several 152-mm-diameter (6-in-diameter) cores were taken from the roof near site 1. Several direct-shear tests were conducted at the laboratory. The shearing plane was not selected according to any observed plane of weakness, but according to convenience of sample length to the specimen frame.

Figures 4 and 5 are samples of the results of these tests. Figures 4*A* and 5*A* show the history of loading, where a negative value is an applied compressive stress. Figures 4*B* and 5*B* show the history of relative displacement measured on different sides of the specimen (near and far) and an average of the two. Secondary creep was present and predominated in both cases. Primary creep is shown in Figure 5b and may be shown in Figure 4b, but lack of measurement points makes it difficult to confirm. In the case of higher shear stress, tertiary creep and failure resulted just before 550 sec. In the other case, failure had not occurred at 48 hours.

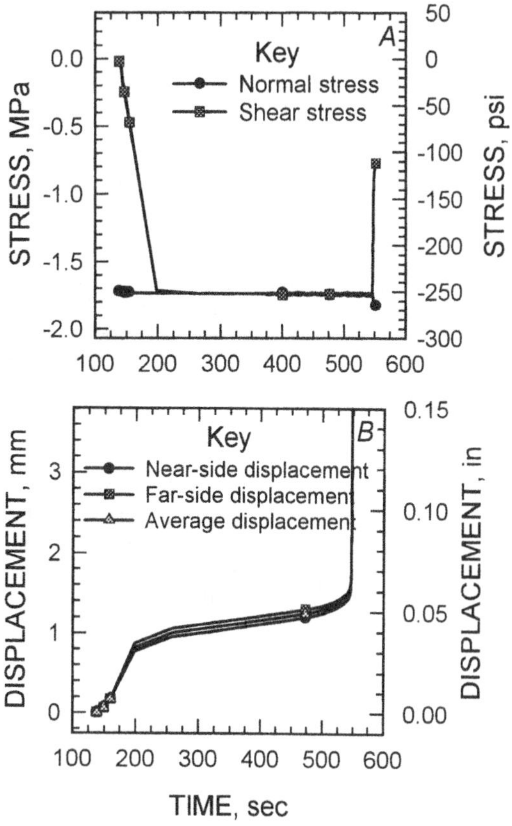

Figure 4.—Results of direct-shear tests on mudstone sample 02. Symbols are data points selected to distinguish between lines. *A*, Normal stress and shear stress over time; *B*, average and relative shear displacement of points on near and far sides of specimen over time.

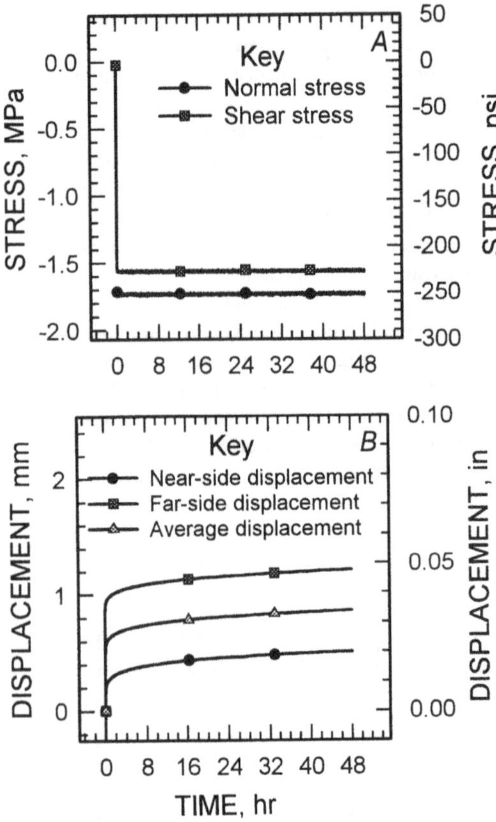

Figure 5.—Results of direct-shear tests on mudstone sample 03. Symbols are data points selected to distinguish between lines. *A*, Normal stress and shear stress over time; *B*, average and relative shear displacement of points on near and far sides of specimen over time.

The deformational curves from the mudstone samples have similar characteristics to the curves from the numerical model. The failure of the model to simulate transitory creep may be remedied with the choice of another cohesion deterioration function, but may not be necessary when predicting the time of failure unless transitory creep is judged to be an important part of the mechanics.

Further tests are planned in which deformation of coal mine roof and slopes will be simulated. Comparisons with real measurements will provide a good measure of the ability of the model to predict time of failure.

6 CONCLUSIONS

The mathematical formulation of cohesion reduction coupled with friction and dilation weakening has been successfully coded with FISH functions in FLAC. Preliminary direct-shear tests show the model appears to simulate progression of microcracking along a plane of weakness. Comparison of these test results with laboratory direct-shear tests of core taken from a coal mine roof shows similar creep curve characteristics. Secondary creep can be reasonably simulated, but primary creep simulation would likely require adjustment of the cohesion deterioration function.

Further tests must be executed to evaluate the capability of the model to simulate deformation and failure over time in the field.

ACKNOWLEDGMENTS

The author is grateful to Jeffrey K. Whyatt, mining engineer, Spokane Research Laboratory, for many useful discusions in the development of the model. Cheng Ho Lee, formerly with Itasca Consulting Group, provide initial guidance in writing FISH functions. Christine Detournay, of Itasca Consulting Group, provided information that aided in debugging the functions. Karen Walker, a student at Gonzaga University, Spokane, WA, assisted with the laboratory tests and data reduction.

REFERENCES

Charles, R. J. 1958. Static fatigue of glass. *J. Applied Phys.* 29 (11):1549-1560.

Fakhimi, A. A. 1992. *The Influence of Time Dependent Deformation of Rock on the Stability of Underground Excavations.* Ph. D. dissertation, Univ. Minnesota, 186 pp.

Kemeny, J. M. 1991. A model for non-linear rock deformation under compression due to sub-critical crack growth. *Int. J. Rock Mech. Mine. Sci. & Geomech. Abstr.* 28(6): 459-467.

Lajtai, E. Z., and R. H. Schmidtke1986. Delayed failure in rock loaded in uniaxial compression. *Rock Mech. Rock Eng.* 19(1):11-25.

Lajtai, E. Z., R. H. Schmidtke, and L. P. Bielus 1987. The effect of water on the time-dependent deformation and fracture of a granite. *Int. J. Rock Mech. Min. Sci. & Geomech. Abstr.* 24(4):247-255.

Larson, M. K., C. L. Stewart, M. A. Stevenson, M. E. King, and S. P. Signer 1995. A case study of a deformation mechanism around a two-entry gate road system involving probable time-dependent behavior. In S. S. Peng (ed.), *Proceedings of the Fourteenth International Conference on Ground Control in*

Mining, pp. 295-304. Morgantown, WV: West Virginia Univ.

Larson, M. K., and H. Maleki 1996. Geotechnical factors influencing a time-dependent deformation mechanism around an entry in a dipping seam. In L. Ozdemir, K. Hanna, K. Y. Haramy, and S. Peng (eds.), *Proceedings, 15th International Conference on Ground Control in Mining*, pp. 699-710. Golden, CO: Colorado School of Mines.

Martin, R. J., III, 1972. Time-dependent crack growth in quartz and its application to the creep of rocks. *J. Geophys. Res.* 77(8):1406-1419.

Price, N. J. 1964. A study of time-strain behavior of coal-measure rocks. *Int. J. Rock Mech. Min. Sci. & Geomech. Abstr.* 1(2):277-303.

Scholz, C. H. 1972. Static Fatigue of Quartz. *J. Geophys. Res.* 77(11):2104-2114.

Wawersik, W. R. 1974. Time-dependent behavior of rock in compression. In *Advances in Rock Mechanics: Reports of Current Research, Themes 1-2. Proceedings of the Third Congress of the International Society for Rock Mechanics*, Vol. II, Part A, pp. 357-363. Washington, D.C.: National Academy of Sciences.

Mechanics of Jointed and Faulted Rock, Rossmanith (ed.) © 1998 Taylor & Francis, ISBN 90 5410 955 6

A constitutive model for the mechanical behaviour of bolted rock joint

F. Pellet & M. Boulon
Laboratoire 3S, IMG, Université Joseph Fourier, Grenoble, France

ABSTRACT : A new equation for the modelling of the shearing process of a bolted rock joint is presented in this paper. The complete curve of the bolt contribution to the shear strength as a function of the displacement along the joint is described by a power function. The determination of the required parameters needs to know the bolt contributions and the corresponding displacements at both the yield and the failure limits of the bolt. These parameters are computed based on an analytical model developped for a sliding along the joint. The main characteristic of this model is the account of the interaction of the axial force and the shear force mobilised in the bolt in both the elastic and plastic stages. Moreover, the large plastic displacements of the bolt occurring during the loading process are taken into account. The comparison of numerical versus experimental data gives reliable results.

1 INTRODUCTION

The reinforcement of rock masses by untensioned rock bolts is widely used to ensure stability and to restrain deformation of rock engineering structures (Aydan, 1989). A lot of experimental programs were performed in order to describe in a qualitative manner the behaviour of bolted rock joint. The mechanical action of the bolt is, however, still difficult to assess, especially for bolts installed in fractured rock masses (Holmberg, 1991).

The present study proposes a new constitutive model for the mechanical behaviour of bolted rock joints subjected to shearing. It allows the computation of both the strength and the deformability properties of a bolted rock joint, when the failure mechanism involves sliding displacements along the rock joints.

2 SCOPE OF THE MODEL

This model was developped by Pellet (1994). In this section only the main computation steps are reminded.

For a complete description of this model the reader may refered to Pellet & Egger (1996)

2.1 *Description of the deformational process of the bolt*

The analytical description of the behaviour of the bolted rock joint takes into account the interaction phenomena developped between the bolt, the grout and the surrounding medium. The objective is to establish the relation between the force, R_O, acting in the bolt at the joint level and the corresponding displacement, U_O, on the joint during the shearing process.

When the bolted rock joint is subjected to a shear displacement, the bolt deflects, and at the same time, the host material (i.e., grout or rock) provides a reaction. The bolt is thus axially and transversally loaded by a set of forces composed of an axial force, a shear force and a bending moment.

Based on experimental studies (Spang & Egger, 1990), the deformed shape of the bolts exibits two singular points. The first one is the intersection between the joint and the bolt (point O) where the curvature of the deformed shape of the bolt is zero. The second one is the point of the maximum curvature (point A) where the bending moment is maximum and the shear force is zero (Figure 1).

Experimental studies also show that the yield limit of the bolt corresponds to the appearance of two plastic hinges located in the point of maximum bending moment. Thus the bolt yields by the combination of a bending moment and an axial force in point A. Beyond the yield limit, the bolt behaves plastically until it breaks down in point O under the action of the axial and shear forces.

From the beginning of the loading process, the host material supplies a reaction on a bolt's length which increases until the bolt yields. Even though at the early stage of the loading process the host material may behaves elastically, a rigid perfectly plastic behaviour is assumed. This assumption is justified because the elastic portion of the material behaviour is extremely small in comparison to the plastic portion.

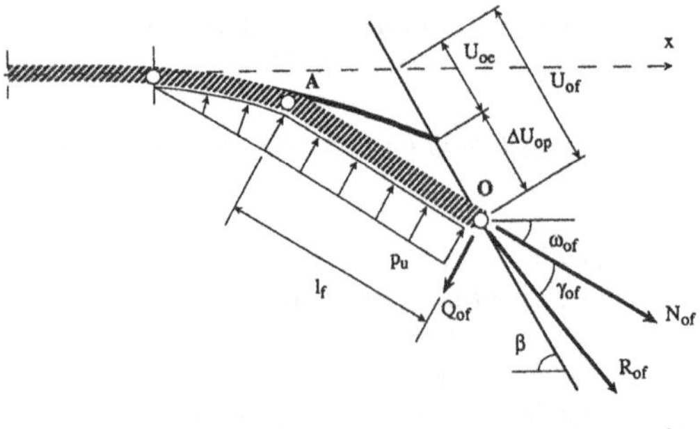

Figure 1 : Loading system and deformed shape of a bolt

2.2 *Yield limit of the bolt*

The shear force acting in point O which leads to the yield limit in point A (i.e., upper fibber of the bolt section reach yield stress of the bolt material) is obtained by the following equation :

$$Q_{oe} = \frac{1}{2}\sqrt{p_u\, D_b\left(\frac{\pi\, D_b^2\, \sigma_{el}}{4} - N_{oe}\right)}$$

where,

Q_{oe} : shear force in O at the yield stress

N_{oe} : axial force in O at the yield stress

p_u : rock or grout reaction

D_b : bolt diameter

σ_{el} : yield stress of the bolt material

2.3 *Failure criterion of the bolt*

For the failure of the bolt material in point O, the Tresca criterion is used. The maximum force acting in the bolt at point O is expressed as a combination of the shear force and of the axial force. The interaction formula is expressed here as a function of the shear force :

$$Q_{of} = \frac{\pi\, D_b^2}{8}\,\sigma_{ec}\sqrt{1 - 16\left(\frac{N_{of}}{\pi\, D_b^2\, \sigma_{ec}}\right)^2}$$

where,

Q_{of} : shear force in O at the failure

N_{of} : axial force in O at the failure

σ_{ec} : failure stress of the bolt

These two relations are represented in Figure 2. The elastic limit is parabolic, whereas the failure criterion is elliptical.

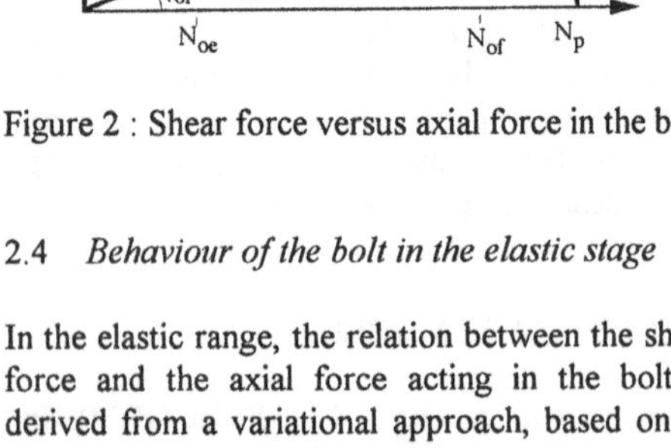

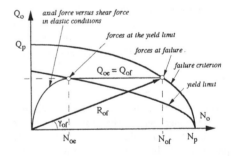

Figure 2 : Shear force versus axial force in the bolt

2.4 *Behaviour of the bolt in the elastic stage*

In the elastic range, the relation between the shear force and the axial force acting in the bolt is derived from a variational approach, based on an energetic concept (Pellet, 1994).

$$Q_o = \sqrt[3]{\frac{3\ N_o\ p_u^2\ \pi^3\ tg\beta\ D_b^2}{1024\ b}}$$

This relation is drawn in Figure 2. The determination of the forces at the yield limit leads to the following third order equation :

$$Q_{oe}^3 + Q_{oe}^2\left(\frac{3\ p_u\ \pi^3\ D_b\ tg\beta}{256\ b}\right) - \left(\frac{3\ p_u^2\ \pi^4\ D_b^4\ tg\beta\ \sigma_{el}}{4096\ b}\right) = 0$$

The solution of this equation furnishes the shear force and the axial force in the bolt. Then, if the forces which act in the bolt are known, it is possible to compute the displacement, U_{oe}, and the rotation, ω_{oe}, of the bolt, which are represented in Figure 2.

$$U_{oe} = \frac{8192\ Q_{oe}^4\ b}{E\ \pi^4\ D_b^4\ p_u^3\ \sin\beta}$$

$$\omega_{oe} = -\frac{2048\ Q_{oe}^3\ b}{E\ p_u^2\ \pi^3\ D_b^4}$$

2.5 *Behaviour of the bolt in the plastic stage*

In the plastic stage, the bending stiffness of the bolt drops due to the appearance of the plastic hinges. Therefore the part O-A of the bolt behaves as a truss, and only the axial force grows. The shear force in the bolt remains constant ($Q = Q_{oe}$ at point O), and the shape of the bolt between the hinges can be considered linear.

The determination of the forces acting in the bolt at failure is simply achieved by solving the following equations, which are drawn in Figure 2 :

$$Q_{of} = Q_{oe}$$

$$N_{of} = \frac{\pi\ D_b^2}{4}\ \sigma_{ec}\sqrt{1 - 64\left(\frac{Q_{oe}}{\pi\ D_b^2\ \sigma_{ec}}\right)^2}$$

As the deformed shape of the bolt between points A and O is linear, the computation of the displacement and the rotation at the end of the bolt can be done by the use of the large displacements formulation. It is shown by simple geometrical consideration that the following expression gives a good approximation of the increment of plastic rotation, $\Delta\omega_{op}$:

$$\Delta\omega_{op} = \arccos\left[\frac{l_e}{l_f}\sin^2\beta \pm \sqrt{\cos^2\beta\left(1 - \left(\frac{l_e}{l_f}\right)^2\sin^2\beta\right)}\right]$$

where, l_e, is the distance between bolt extremity (point O) and the location of the maximum bending moment (point A) when the yield limit is reached and l_f is the length of the part O-A at failure. ε_f is the bolt strain at failure.

The solution of the above equation gives the increment of plastic rotation and then allows for the calculation of the plastic displacement, ΔU_{op}:

$$\Delta U_{op} = \frac{Q_{oe}\ \sin\Delta\omega_{op}}{p_u\ \sin(\beta - \Delta\omega_{op})}$$

As the bolt strain at failure is equal to the sum of the elastic and the plastic strains, the total rotation, ω_{of}, and the total displacement, U_{of}, of the end of the bolt at failure are respectively computed by the followings formulas :

$$\omega_{of} = \omega_{oe} + \Delta\omega_{op}$$

$$U_{of} = U_{oe} + \Delta U_{op}$$

2.6 *Bolt contribution to the joint shear strength*

At each step of the loading process, both the axial and the shear forces acting at the end of the bolt as well as the rotation and the displacement are known.

Thus by combining forces and using the Mohr-Coulomb criterion for joint strength, it is possible to compute the bolt contribution, T_b, to the total reinforced rock joint strength :

$$T_b = R_{ot} + R_{on} \, tg \, \phi_j$$

where,

R_{on} : force normal to the joint
R_{ot} : force tangential to the joint
ϕ_j : joint friction angle

3 THE NEW CONSTITUTIVE MODEL

Based on this analytical formulation, a new constitutive model called BJM for Bolted Joint Model was developped. The bolt contribution versus joint displacement is described by the following power function.

$$T_b = a \, U_j^{\,n}$$

The two parameters a and n are determined using the values of the displacements and the bolt contributions at both the yield limit and the failure limit. It leads to the following expressions :

$$n = \frac{\log(T_{bf} - T_{be})}{\log(U_{jf} - U_{je})} \qquad a = \frac{T_{bf}}{U_{jf}^{\,n}}$$

4 EVALUATION OF THE THEORY

In order to evaluate the reliability of the constitutive model, a comparison between the numerical results and some experimental data was achieved. The details of these tests were reported by Pellet et al., 1995.

4.1 *Direct shear tests reported by Spang (1988)*

These tests were performed on samples of concrete and sandstone having the following mechanical characteristics.

- sandstone strength $\sigma_c = 10$ MPa
- concrete strength $\sigma_c = 40$ MPa
- grout strength $\sigma_{cg} = 36$ MPa
- bolt diameter $D_b = 8$ mm
- yield limit of steel $\sigma_{ec} = 670$ MPa
- strain at failure $\varepsilon_f = 13.5$ %

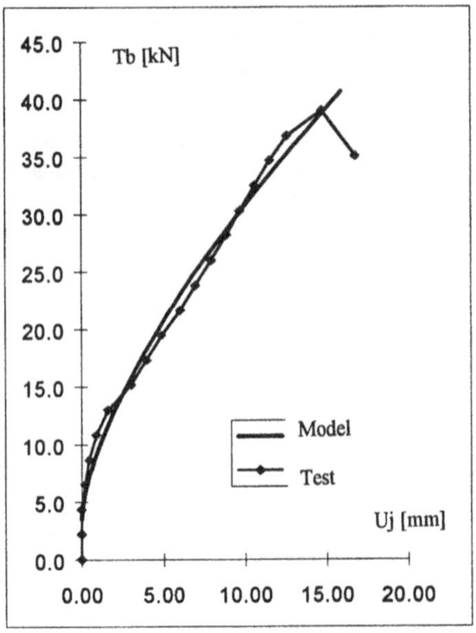

Figure 3 : Bolt contribution versus joint displacement for test carried out by Spang

- joint friction angle $\phi_j = 32$ degrees

The analytical results shown in Figure 3 are close to the experimental data for both the prediction of the bolt contribution and the joint displacement.

4.2 *Tests reported by Pellet (1994)*

Tests were carried out with a high capacity press on large models built in limestone. The models were reinforced by steel bars grouted with an epoxy resin. The mechanical properties of the materials were the following.

- strength of the rock $\sigma_c = 150$ MPa
- strength of the grout $\sigma_{cg} = 60$ MPa
- bolt diameter $D_b = 3$ mm
- yield limit of steel $\sigma_{ec} = 600$ MPa
- strain at failure $\varepsilon_f = 20$ %
- joint friction angle $\phi_j = 25$ degrees

In this case also a good agreement is obtained between analytical and experimental results shown in Figure 4.

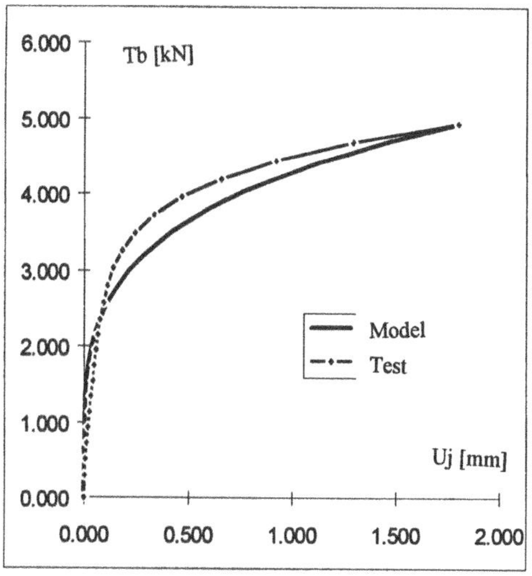

Figure 4 : Bolt contribution versus joint displacement for test carried out by Pellet 1994

5 CONCLUSION

The analytical model developped for this work allows one the prediction of the complete load-displacement curve for a bolted rock joint subjected to shearing. The maximum bolt contribution to the shear strength of the bolted rock joint, as well as the maximum displacement on the joint can be obtained.

The comparison of the analytical predictions with the tests results obtained on large scale models or by direct shear tests gives good results.

Based on this analytical development, it is possible to perform the stability analysis of the reinforcement system for practical application of rock structures such as rock slopes and underground cavities.

REFERENCES

Aydan Ö., The stabilisation of rock engineering structures by rockbolts, Ph.D. Thesis, Nagoya University, Japan, 1989

Holmberg M., The mechanical behaviour of untensionned grouted rock bolts, Ph. D. Thesis, Royal Institute of Technology, Stockholm, Sweden, 1991

Pellet F., Egger P., Analytical model for the mechanical behaviour of bolted rock joints subjected to shearing, Rock Mechanics and Rock Engineering, Springer Verlag, vol 29, no 2, pp 73–97, 1996

Pellet F., Egger P., Ferrero A.M., Contribution of fully bonded rock bolts to the shear strength of joints : Analytical and experimental evaluation, Proc. 2nd Int. Conf. on Mechanics of Jointed and Faulted Rock, MJFR-2, Vienna, 1995

Pellet F., Strength and deformability of jointed rock masses reinforced by rockbolts, English Translation of the Doctoral Thesis 1169, Swiss Federal Institute of Technology, Lausanne, Switzerland, 1994

Spang K., Egger P., Action of fully-grouted bolts in jointed rock and factors of influence, Rock Mech. and Rock Eng., Vol 23, pp 201-229, 1990

Mechanics of Jointed and Faulted Rock, Rossmanith (ed.) © *1998 Taylor & Francis, ISBN 90 5410 955 6*

3-D-laser measurements and representation of roughness of rock fractures

F. Lanaro, L. Jing & O. Stephansson
Division of Engineering Geology, Department of Civil and Environmental Engineering, Royal Institute of Technology, Stockholm, Sweden

ABSTRACT: 3-D-laser digitisation is applied for measuring the roughness of rock fracture surfaces and reconstructing the geometry of joint aperture. The suggested representation of roughness, which combines the Re-scaled Range Analysis technique for fractals with the concept of 3-D co-latitude, provides information about the angularity, anisotropy and scale-dependency. The method is applied to a schistosity joint of a hard schist.

1. INTRODUCTION

One of the most challenging problems in rock mechanics is the characterisation of roughness of rock joint surfaces. Roughness is difficult to quantify and, even when quantified, the result strongly depends on the chosen method of representation.

Due to technological limitations, many studies have concentrated on two-dimensional characteristics of roughness, by studying cross-section profiles. Although their results have indicated important properties of the joint surfaces (Barton & Choubey 1977; Brown & Scholz 1985; Power & Tullis 1991; Odling 1994), few authors have shown that joint morphology must be handled in a purely three-dimensional manner, considering both anisotropy and volumetric features (e.g. aperture) (Liu & Sterling 1990; Muralha 1992; Riss et al. 1995).

Three-dimensional-laser-topography digitisation extends the applicability of two-dimensional methods for roughness characterisation, with vast extension to the density of the collected data, the accuracy of the measurements and the repeatability of the results. In addition, it allows the surfaces, or the joint, to be considered as a whole and not as a set of discrete and independent cross-sections. This is extremely important for the understanding of the mechanical, hydraulic and thermal behaviour of natural rock fractures.

In the present paper, results of three-dimensional-laser scanning of both walls of a rock joint are presented, and significant statistical parameters are evaluated and analysed. An attempt to combine the Re-scaled Range Analysis technique (Feder 1988; Schmittbuhl et al. 1995) with the angular description of the asperities by 3-D co-latitude (Riss et. al. 1995) is introduced. The influence of the sampling dimension on the standard deviation of the height of the asperities with respect to the local interpolation plane is investigated. Moreover, the distribution of the slope of the local interpolation planes is studied, taking into account its spatial orientation and the influence of the sampling dimension.

2. DESCRIPTION OF THE EQUIPMENT

The 3-D-laser-digitising system consists of a laser sensor mounted on a co-ordinate-measuring machine (CMM) with three axes (Fig. 1). The scanner sensor, manufactured by Kréon® Industrie (France), uses a technique of fine triangulation by means of a light source and the reflected light rays from the sample. The laser projects a light strip 25 mm wide on the surface of the object and defines a "laser plane". The images of the object cut by the laser plane are captured by two CCD (Charge Coupled Device) cameras. Since the position of the cameras with respect to the laser source is known, it is possible to project the images back into the plane by mathematical transformations.

The scanner picks 600 points along the laser strip, 50 μm apart from each other; it has an accuracy of ±20 μm and a resolution of 10 μm. It can scan up to 15,000 points per sec, depending on the selected scanning step size.

The system is able to digitise samples with an

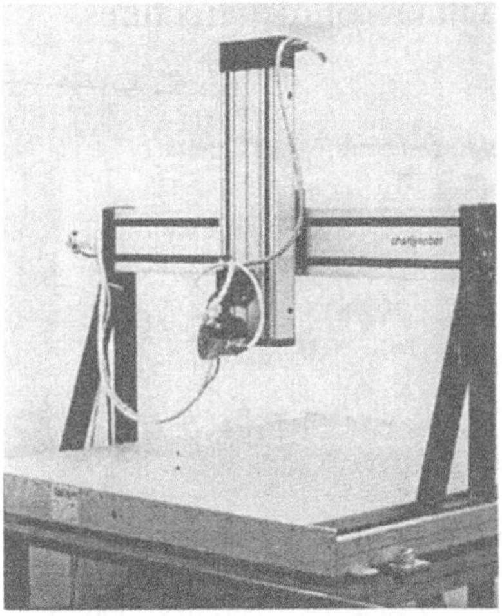

Figure 1. Overview of the 3-D-laser-scanner equipment.

overall dimension of 1000×1040×420 mm and locate the position of the scanner head with the accuracy of ±50 μm. The scanner moves and must maintain a distance of between 50 and 100 mm from the object during scanning. The sensor path can be programmed before hand, based on the shape of the object. The laser sensor can be rotated in steps of 90° in case of shadows or overhangs. Since the scanner sensor is always inclined by 22.5° with respect to the vertical axis, vertical surfaces can also be easily scanned.

An electronic control unit connects the laser sensor, the machine controller for the CMM and a computer, and processes the signal in real time. The collected data consist of 3-D co-ordinates constituting large clouds of points. The origin of the data point co-ordinates is chosen by the operator. Those co-ordinates are stored as ASCHII or binary format files and handled with reverse engineering packages like Surfacer© by Imageware (USA).

The advantage of this 3-D-laser digitising system is that it can digitise small details, it does not enlarge the shape of the scanned object and it is accurate in measuring sharp angles due to the density of the collected data. Moreover, the scanning process is fully automatic with programmed scanning paths, so that large samples can be scanned without inspection of the operator, as required for conventional profile scanners.

3. MEASURING TECHNIQUE

The high accuracy of the 3-D laser scanner makes it possible to analyse small objects with short distances

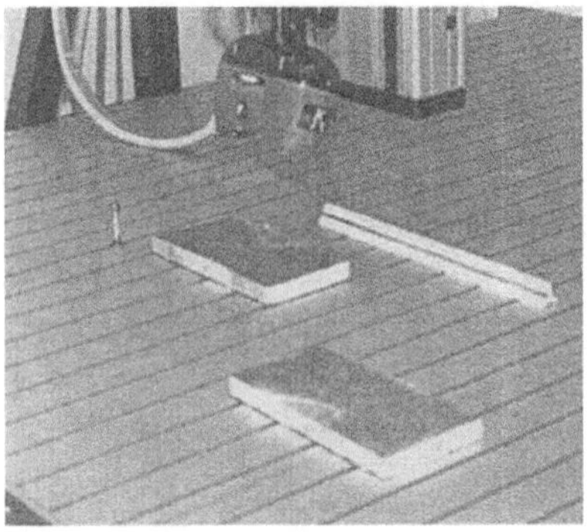

Figure 2. A joint sample of schist.

between them; which is the case for measurement of the aperture distribution and the contact area between the walls of rock joints.

Before measurement, a number of calibrated steel spheres of 7 mm in diameter are fixed onto the two blocks constituting the joint sample. They should lay close to each other so that they remain contemporary visible by the scanner sensor. There must be at least three spheres on each of the two blocks so that the spatial position with respect to each other can be determined.

In PHASE I the closed joint sample is placed on the scanning table and the reference spheres are scanned first. In PHASE II, the joint is opened and the surfaces of the two walls and reference spheres are individually scanned (Fig. 2).

To reconstruct the geometry of the closed joint, the clouds of points of the two surfaces obtained in PHASE II are rotated so that the reference spheres coincide with the position that they had in PHASE I. This operation is called "registration".

4. ROUGHNESS REPRESENTATION

Natural joint surfaces are often considered as geo-statistical non-stationary entities by many authors (Sayes & Thomas 1978; Berry & Hannay 1978) because of their statistical properties generally vary due to the effects of different or non-contemporary geological generating processes. These features strongly depend on the scale of observation. It is complicated to reproduce the surfaces by superimposition of Euclidean geometrical forms.

By comparing power spectra of the topography of many different natural surfaces, such as micro-fractures, rock joints, faults and structural contacts, in a first approximation all of them have spectral density function G of the form:

$$G(\omega) = C\omega^{-a} \quad (1)$$

where ω is the frequency.

The power exponent a for non-overhanging surface profiles ranges between 2 and 3, and C is a proportionality constant. Since the spectrum has a power law with negative exponent, the long wavelengths have higher amplitude and contribute more to the overall roughness than do short wavelengths (Brown 1995).

The variance of the asperity height is related to the integral of the power spectrum of the surface in the frequency domain. The relationship (Chatfield 1975) follows from the expression of the auto-covariance function. Its maximum value corresponds to the variance σ^2 of the data set:

$$\sigma^2 = \int_{\omega_{min}}^{\omega_{max}} G(\omega)d\omega \quad (2)$$

In this equation ω_{min} and ω_{max} are the highest and lowest frequencies of the power spectrum of the surface topography. The power spectrum quantifies how the overall variance of a series of elevation measurements varies corresponding to the harmonic components in which it can be decomposed.

However, caution must be taken when applying spectral analysis to non-stationary surfaces because the techniques have been developed primarily for stationary data (Houng 1989; Fox 1985). When non-stationarity is observed, the data have to be prepared, divided into segments which are on the first order stationary. Boundaries have to be placed where the statistical properties vary too rapidly (Malinverno 1995).

Fractal geometry (Mandelbrot 1983), on the other hand, provides another framework for describing scale invariant features of recursive nature and diagonal symmetry of non-stationary objects.

Schmittbuhl et al. (1995) found that all the approaches they used to evaluate the Hurst exponent for natural joint surfaces in granite and gneiss gave a self-affine fractal structure, despite the anisotropy of some of their samples. Other studies emphasise a self-affine fractal dimension independent of the fracture mode or the material (Sayes & Thomas 1978).

Self-similar and self-affine fractals are related to the power spectra density function. The power law form of the power spectrum, on the other hand, does not assure that the spectrum belongs to a fractal function because fractals are characterised by only random phase spectra (Brown 1995). For a self-affine fractal the following exponential relation holds:

$$\sigma^2(h) = Ah^{2H} \quad (3)$$

where σ^2 is the variance of the residual of the asperity profile referred to the interval h, once the trend is removed, and A is a coefficient of proportionality. In the Re-scaled Range Analysis technique, the Hurst exponent H is determined by the slope of the linear interpolation of σ^2 versus h in a log-log plot (Schmittbuhl 1995). The advantage of this technique over spectral analysis is that it is simpler to apply and the resulting plots contain less noise (Odling 1994).

A way to describe the angularity of the surface (Riss et al. 1995) is to build a mesh of triangles, squares or hexagons based on the scanned points, and calculate the co-latitude angle α between their normal and the normal of the surface mean plane.

The co-latitude and the variance σ^2 of the residual of the asperity height, once the elevations of mean plane of the fracture have been reduced from the original surface elevation, could characterise the morphology of the fracture globally.

5. NUMERICAL ALGORITHM OF 3-D CO-LATITUDE (α,θ)

The Re-scaled Range Analysis technique is applied in this study for the purpose of roughness representation of the walls of a rock joint. The interpolation plane over the topography of the sample is first calculated. The sample is then sub-divided into small areas by drawing a grid of planes orthogonal to the interpolation plane with a pattern of rectangles or squares, whose edge is the window size h. The portion of sample surface included in a singular cell is extracted and a new local interpolation plane is determined with the least squares method. The distances between the local interpolation plane and the points on the joint surface are taken to be the asperity height of the rough surface. The statistics of their distribution are calculated as the standard deviation σ and the bridge distance d between the farthest point from the plane, one per each side. Additionally the normals of the local interpolation plane are determined.

The two systems of orthogonal planes have traces on the global interpolation plane. The lines are used as reference directions on the plane. Orthogonal planes of an angle θ with respect to the reference direction are created and intersect the local interpolation planes. The angle between the global interpolation plane and these lines is the apparent co-latitude (slope) α of the plane, and it can have positive and negative values (Fig. 3).

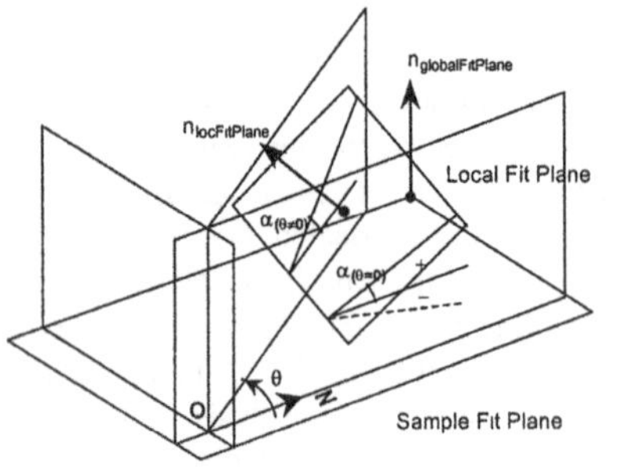

Figure 3. Definition and description of the parameters used by the numerical algorithm.

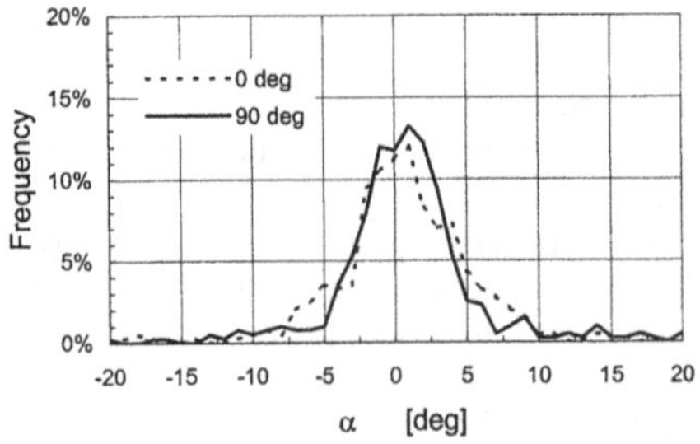

Figure 4. Frequency distribution of the slope α for a window size h of 2 mm and azimuth θ of 0° and 90°.

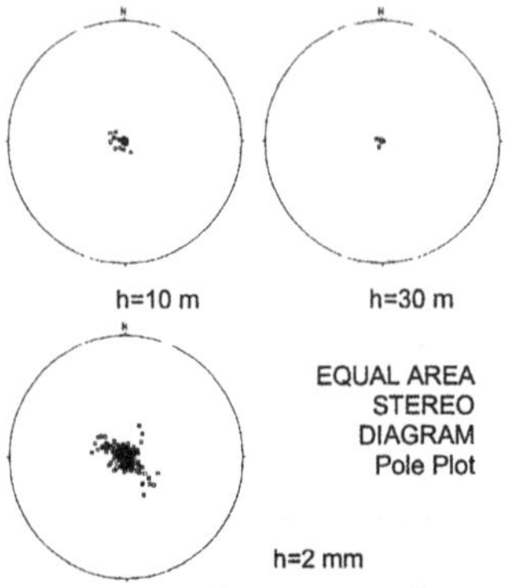

Figure 5. Equal-area-stereo diagrams of the normal to the local interpolation planes for different window sizes.

6. RESULTS OF AN EXAMPLE JOINT

The analysed sample is an artificial joint in Serv schist from the Offerdal area (Central Sweden) of the size of 200×100 mm. The topography of the two walls of the sample consists on clouds of 505,854 and 501,305 points respectively, arranged on an irregular grid of 0.2 mm side interval. The following results are those of the lower wall of the measured joint.

A series of window sizes were used (2, 5, 10, 20, 30, 40 mm). For each of them the distribution of slope α for different azimuth θ (0°~90°) was calculated.

The histograms of frequency distribution of the slope α, for an assigned azimuth, are not perfectly symmetric with respect to the nil slope. They systematically appear to be widely spread when considered along the 0° azimuth (Fig. 4). This characteristic remains for larger window sizes, but in this case most of the values happen to fall in the interval -1° and 3°. It appears that the dispersion of the distributions increases with the decrease of window size, testifying that smaller sampling cells capture better the real shape of the rough surface.

The data of slope angle α are also plotted in an equal-area-stereo diagram, where the downward normals of the interpolation planes for different window sizes are represented (Fig. 5). The smaller the window size is, the wider is the range of slopes and a preferential orientation is evident.

By plotting the standard deviation σ of the asperity height reduced by the elevations of the local interpolation plane, versus the window size, the log-log diagram in Figure 6 is obtained. The results show that the standard deviation is scale-dependent and increases with the window size. If the most frequent value of σ is chosen from the distribution for a given window size, then a clear-cut increase is observed for h equal to 20 mm. Assuming self-affinity of the surface, an Hurst exponent of 0.529 is calculated for h smaller than 20 mm. For larger values of h new and higher frequent values of σ appear. This can be due to the fact that the number of data decreases and the algorithm detect a larger feature of roughness which for small window sizes was removed by taking out the local planar trend.

After registration of the data containing the topography of the walls of the joint, the aperture distribution was obtained by calculating the distance between the facing points of the two surfaces (Fig. 7).

7. DISCUSSION AND CONCLUDING REMARKS

The 3-D-laser measuring technique offers a variety of applications for roughness representation and calculation. It provides rapid scanning with dense and highly accurate data sets. The surface topography can be considered as a 3-D entity and its spatial features can be revealed. Furthermore, the proposed technique for reconstructing the geometry

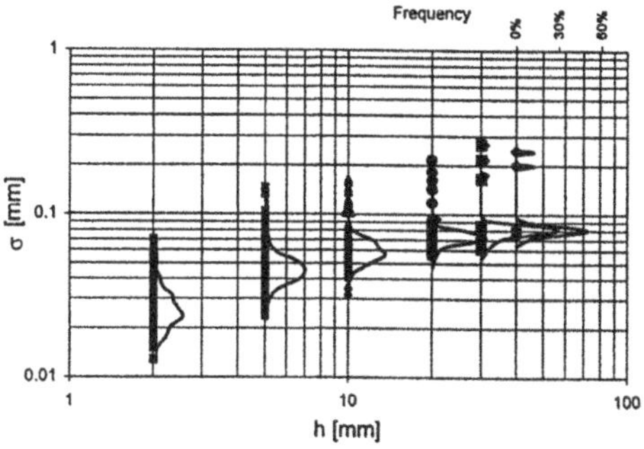

Figure 6. Standard deviation σ of the reduced asperity height versus window size *h*.

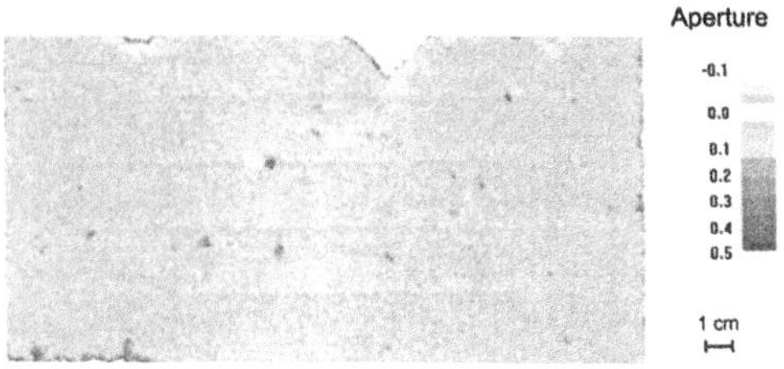

Figure 7. Aperture distribution for the joint sample in Figure 1.

of the aperture distribution between the surfaces of a rock joint appears to be a major advancement. With this technique, it is possible to study the void distribution of a rock joint, the contact area, and their evolution when one side of the joint is dislocated with respect to the other.

Adding the azimuth θ to the representation of roughness (Riss et al. 1995), allows to have different co-latitude values according to either forward or backward directions are considered. This is of great importance when dealing with strongly anisotropic surfaces, where the direction of evaluation of the slope cannot be neglected. Moreover, the co-latitude distribution very much depends on the density of the scanned data over which it is calculated. This study demonstrates that the co-latitude or the slope α, can be defined for a certain data spacing or window size, but like all the other roughness descriptors, varies with the scale of sampling window.

The proposed method is able to identify stationarity of a surface, in the sense that nil co-latitude indicates constancy of the mean, and constant standard deviation of the reduced aspertity heigth implyes independence from scale.

8. ACKNOWLEDGEMENTS

This work has been supported by grant E-AD/EG 03447-359 from the Swedish Natural Science Research Council.

9. REFERENCIES

Barton, N., & Choubey, V., 1977. The shear strength of rock joints in theory and practice. *Rock Mechanics*, Vol. 10: 1-54.

Berry, M.V., & Hannay, J.H., 1978. Topography of random surfaces, *Nature*, 273: 573.

Brown, S.R., 1995. Measuring the dimension of self-affine fractals: example of rough surfaces, *Fractals in the Earth sciences*: 77-87. Barton and La Pointe (eds). New York: Plenum Press.

Brown, S.R. & Scholz, C.H., 1985. Broad bandwidth study of the topography of natural rock surfaces, *J. of Geophys. Res.*, Vol. 91: 12575-12582.

Chatfield, C., 1975. The analysis of time series: Theory and practice: 263. New York: John Wiley.

Feder, J., 1988. *Fractals*. New York: Plenum Press.

Fox, C.G., 1989. Empirical derived relationships between fractal dimension and power law from frequency spectra, *Pure and Applied Geophysics*, Vol. 23: 1-48.

Houng, S.E., 1989. On the use of spectral methods for the determination of fractal dimension, *Geophys. Res. Lett.*, Vol. 16: 673-676.

Liu, H., & Sterling, R.I., 1990. Statistical description of the surface roughness of rock joints, *Proc. of the 31st U.S. Symp. on Rock Mechanics, Golden, CO, US*: 277-284. Hustrulid and Johnson (eds).

Muralha, J., 1992. Fractal dimension of joint roughness surface. *Fractured and Jointed Rock Masses*: 205-212. Myer, Cook, Goodman & Tsang (eds). Rotterdam: Balkema.

Malinverno, A., 1995. Fractals and ocean floor topography: a review and a model, *Fractals in the Earth sciences*: 107-130. Barton and La Pointe (eds). New York: Plenum Press.

Mandelbrot, B.B., 1983. The fractal geometry of nature: 468. New York: W.H. Freeman.

Odling, N.E., 1994. Natural fracture profiles, fractal dimension and joint roughness coefficients, *Rock Mech. and Rock Eng.*, Vol. 27 (3): 135-153.

Power, W.L., Tullis, T.E., 1991. Euclidean and fractal models for the description of rock surface roughness, *J. of Geophys. Res.*, Vol. 96: 415-424.

Riss, J., Gentier, S., Archambault, G., Flamand, R., Sirieix, C., 1995. Irregular joint shear behaviour on the basis of 3D modelling of their morphology: morphology description end 3D modelling, *Proc. 2nd Int. Conf. Mech. of Jointed and Faulted Rock, MJFR-2, Vienna, Austria*; Rossmanith (ed.), Vol. 2: 157-162.

Sayles, R.S., and Thomas, T.R., 1978. Surface topography as a non stationary process, *Nature*, 271: 431-434.

Schmittbuhl, J., Schmitt, F., Scholz, C., 1995. Scaling invariance of crack surfaces, *J. of Geophys. Res.*, Vol. 100: 5953-5973.

Mechanics of Jointed and Faulted Rock, Rossmanith (ed.) © 1998 Taylor & Francis, ISBN 90 5410 955 6

What's 'rough'? Modelling the vague meaning of joint roughness classifiers by means of empirical fuzzy sets

Thomas Raab & Qian Liu
Institute of Engineering Geology and Applied Mineralogy, Technical University Graz, Austria

ABSTRACT: We will argue, that the meaning of linguistic geological data is inherently vague in nature. Based on this premise, we will present some results of a demoscopic pilot study on 55 international experts in the fields of engineering geology and rock mechanics. The study was carried out in order to assess empirical fuzzy membership functions for the classifiers VERY SMOOTH, SMOOTH, ROUGH, and VERY ROUGH pertaining to the parameter "joint roughness". In contrast to probability theory which can only deal with purely statistical (i.e., syntactic) information measures, possibility theory formulated in terms of fuzzy set theory offers a powerful tool for the treatment of semantic imprecision of information. In general, the assessment of realistic membership functions for the linguistic entities in question is the first and crucial step with regard to practical applications of fuzzy set theory. Technically speaking, the transformation of psychological probabilistic data from our pilot study into possibilistic fuzzy sets becomes possible by means of the probability-possibility consistency principle. Dealing with semantic vagueness is necessary for geological purposes, when one considers the influences of the observing embodied subject´s perception and cognition on data structures. In general, the mathematical modelling of linguistic variables should provide a viable interface between geological data sampling and computational models used in geology and rock mechanics.

1 INTRODUCTION AND MOTIVATION

Geological pattern recognition includes the perception and the classification of geological entities, i.e. objects or properties of these objects. In contrast to the analytical and experimental approach of the exact and the engineering sciences, geology has to advance more by observation, description, and classification than by experiment and calculation, at least at the first stages of research (Hagner 1963, Frodeman 1995). Yet, classifications are almost always vague and uncertain which has led some to the conclusion that so-called qualitative data are wholly subjective. However, it is impossible to discover a quality apart from a quantity and vice-versa. As the quantity changes the quality changes with it (Birch 1951, Li & Yen 1995). This relation can be easily demonstrated by means of the ancient baldhead paradox. From the two logically proper premises that a man with $n \in \mathbb{N}$ hairs is "baldheaded" and so is a man with $n + 1$ hairs, we can deduce by iteration that plainly every man is "baldheaded". Obviously, the meaning of the quality "baldheaded" can not be treated in terms of Cantorian set theory. Unfortunately, this is also the case with most geological categories. As, for instance, the quantitative mineral composision of a rock material varies, the qualitative discriptions (e.g. color, quartz-content etc.) and the lithological categorization varies with it.

How to represent this semantic vagueness properly in order to provide a realistic interface between geological data sampling and computational models is the topic of this paper. Such a procedure should in part be able to quantify the vague outcomes of a "geologic frame of mind" (Hagner 1963) into an engineering one. In section 2 we will try to provide a sound argumentative basis for the necessity for dealing with semantic vagueness in geology. Section 3 deals with two mathematical frameworks for dealing with vagueness and uncertainty in general, probability theory and possibility theory, and their relation to each other as expressed in the probability-possibility consistency principle. This will also enable us to transform probabilistic psychological data from a demoscopic expert study into possibilistic fuzzy sets. In section 4 we present the method of this empirical study on 55 specialists from the fields of engineering geology and rock mechanics, which was carried out in order to obtain empirical membership functions for linguistic classifiers for the parameter "discontinuity roughness". Section 5 presents an explicit algorithm for yielding fuzzy membership function from probabilistic histograms and the resulting membership functions.

2 THE INTRINSIC IMPRECISION OF GEOLOGICAL JUDGMENTS

Following the positivistic semantical account (Frege 1966), each geological concept's reference is supposed to be expressible by means of Cantorian set theory (Lakoff 1987). There should always be observational conditions necessary and sufficient for the correct geological categorization of patterns in the field and/or in the laboratory. In the case of geology correct means that (a) the observed geological entity should be either or not a member of the set referring to the class or classifier in question, (b) the definition should unambiguously specify the necessary and sufficient observables for this classification. From a cognitive science perspective we have to add (c) that these observables can be practically realized by an observer and/or experimentation. It is easy to see that the positivistic account draws on classical binary logic, more precisely the law of the excluded middle by Aristotle (Klir and Folger 1988).

However, when taking the influences of an embodied observing subject into account it turns out that this positivistic ideals are an illusionary claim (Lakoff 1987). The aim at an "ideal geological language" which would enable a mapping of percepts onto clear-cut and well-defined categories is not only an illusion but also highly counter-intuitive.

Roughly speaking, there are three reasons implying that geological classification cannot be subject to positivistic semantics. Consider, for instance, the lithological categorization of a rock material at an outcrop and/or in the laboratory. Assume, moreover, that the material can be referred to as an "feldspar-rich sandstone", which is defined as a sandstone containing more than 25% and less than 50% feldspar (Füchtbauer & Müller 1980). An definitely yes-or-no classification, is not possible for the following reasons:

(a) *The pragmatic objection:* It is not unlikely that the feldspar content varies even within one outcrop so that it is higher or lower 25 % every few centimeters. The abstraction of this objective variability already yields fuzzy classifications.

(b) *The psychophysical judgment imprecision objection:* The given exact definition is not strictly applicable in the field, because one simply cannot estimate (i.e. perceive and classify) a mineral content in such an exact manner. The psychophysical judgment imprecision of geological parameter estimates in situ has been demonstrated by Raab & Brosch (1996).

(c) *The formal objection:* Definitions based on crisp thresholds are themsevels an idealization, because they require infinitely accurate observations/measurements. Theoretically speaking, an infinitely precise measurement cannot even be obtained from and/or represented after laboratory measurements.

From these objections we infer that geological categories are generally prototype-structured, that is, they refer to best exemplars and gradually less good exemplars (Rosch 1973). The formalism of fuzzy set theory as a basis of possibility theory provides a restricted and approximate practical tool for representing this prototype structure of semantic vagueness (Zadeh 1982).

However, the empirical data that we obtained by the demoscopic pilot study are probabilistic themselves. In order to use these data for a fuzzy membership estimation we thus need a theory and a formalism which relate probability and possibility.

3 PROBABILITY AND POSSIBILITY

Since the introduction of the theory of fuzzy sets (Zadeh 1965), one of the major developments in this field was the establishment of a theory of possibility (Zadeh 1978). The theory of possibility concerns, in contrast to the theory of probability which deals exclusively with statistical measures of purely syntactic information, its meaning. This provides a more viable method for dealing with judgment under uncertainty, for instance, for controlling and prognosis based on geo-information systems (e.g. Liu et al. 1997).

Like the probability distribution of a random variable, a fuzzy variable is associated with a possibility distribution x and it is defined to be numerically equal to the membership function x of the fuzzy restriction F (Zadeh 1978):

$$\Pi_x = \mu_F \tag{3.1}$$

A general method for the measurement of membership functions μ_F based on methods of mathematical psychology (e.g. Norwich & Turksen 1984, Hersh & Caramazza 1976), has not been proposed so far (Zimmermann 1991). For engineering practice, however, the most efficient way to establish membership functions seems to be the use the relationship between probability and possibility. By elaborating Zadeh's general definition of a possibility measure, the following relation between possibility and probability was established by Dubois & Prade (1982, 1983).

Let $\Pi(A)$ be the possibility measure of an event A ($A \subset \mathbf{X}$), then

$$\Pi(A) = 1 - \textit{impossibility of } A \qquad (3.2)$$

and the impossibility of event A is the necessity of its opposite event $N(\bar{A})$, so we get:

$$\Pi(A) = 1 - N(\bar{A}) \qquad (3.3)$$

Denoting $\pi_i = \Pi(\{x_i\})$, we have another form of $\Pi(A)$

$$\Pi(A) = \max_{x_i \in A} \pi_i \qquad (3.4)$$

According to Dubois & Prade (1983), the degree of necessity of event A ($\subseteq$X) is the extra amount of probability of elementary events in A over the amount of probability assigned to the most frequent elementary event outside A, i.e.

$$N(A) = \sum_{x_j \in A} \max(p_j - \max_{x_k \notin A} p_k, 0) \qquad (3.5)$$

If $A=A_i$

$$N(A_i) = \sum_{j=1}^{i} (p_j - p_{j+1}), \quad i = 1,\ldots,n \qquad (3.6)$$

For $N(\bar{A})$ in (3.3)

$$\begin{aligned} N(\bar{A}) &= N(X - \{x_i\}) \\ &= N(A_{i-1}) \qquad (3.7) \\ &= \sum_{j=1}^{i-1} (p_j - p_i) \quad \textit{for} \quad i>1 \end{aligned}$$

Combine (3.3), (3.4) and (3.7), then π_i's, the possibility distributions of event A, are easily obtained from its p_i's:

$$\begin{aligned} \pi_i &= 1 - N(X - \{x_i\}) \\ &= 1 - N(A_{i-1}) \qquad (3.8) \\ &= 1 - \sum_{j=1}^{i-1} (p_j - p_i) \quad \textit{for} \quad i>1 \end{aligned}$$

and $\pi_1=1$, taking into account the constraint

$$\sum_{i=1}^{n} p_i = 1 \qquad (3.9)$$

we get

$$\pi_i = i \cdot p_i + \sum_{j=i+1}^{n} p_j \quad i = 1,\ldots,n \qquad (3.10)$$

Hence, the relationships between necessity, probability and possibility is:

$$\forall A: \quad N(A) \leq P(A) \leq \Pi(A) \qquad (3.11)$$

Relation (3.11) is referred to as the probability-possibility consistency principle (Zadeh 1978, Dubois & Prade 1988). It suggests that "what is possible may not be probable and what is improbable need not be impossible" (Zadeh 1978). This thought might intuitively be familiar to geologists and rock engineers.

It is clear from (3.10) that the possibility distribution and the probability density function have the same shape. This bijective relation between probability and possibility can be used for estimating membership functions. If X is the set of patterns and $x \in X$ and f(x) is any function (or discrete value) representing a smoothed histogram, then the possibility distribution $\Pi(D)$ ($D \subset X$) is

$$\Pi(D) = \frac{sup_{x \in D} f(x)}{sup\, f(x)} \qquad (3.12)$$

and in the sense of Zadeh (1978), the membership function of D is

$$\mu_D(x) = \pi(x) \qquad (3.13)$$

4 THE DEMOSCOPIC METHOD

4.1 *The questionnaire and its methodical problems*

We conducted a questionnaire survey on 55 professionals with expertise in the fields of engineering geology, rock mechanics or related fields. The total number of mailed questionnaires was 470. They were sent to persons, who were selected from the current membership listings of the International Society of Rock Mechanics (ISRM) and the International Association of Engineering Geologists (IAEG). The total backflow of 55 questionnaires represents a percentage of ca. 12%.

For the data sampling, we designed the questionnaire in order to obtain the subjective thresholds between the linguistic classifiers for five engineering geological parameters (uniaxial compressive strength, discontinuity spacing, discontinuity roughness, trace length, discontinuity aperture). The subjective ranges of each classifier had to be marked as continuous bars on logarithmic bar-scales (Fig. 4-1). The applicability of the bar-scale method for research on geological perceptual judgments has been demonstrated in a pilot study on the psychophysics of geological in situ parameter estimation (Raab & Brosch 1996). Moreover, the questionnaire contained questions concerning information about the experts' professional background, the years of experience, the geographical area, where expertise has been built up, and a comment possibility.

3. Roughness of Discontinuities

This section pertains to the parameter *"Roughness of a Discontinuity (scale: 10 cm approx.)"*.
Please mark your subjective guesses of the ranges for the linguistic description of this parameter. You may use the Typical Roughness Profile Chart given below. The classifiers from left to are:

① very smooth (slickensided) | ② smooth | ③ rough | ④ very rough

Roughness of Discontinuity (JRC)

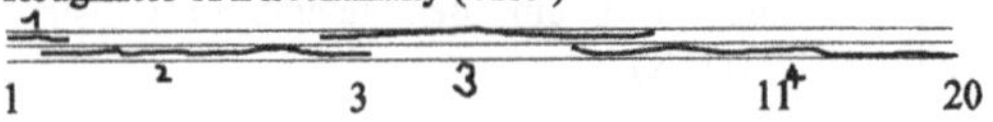

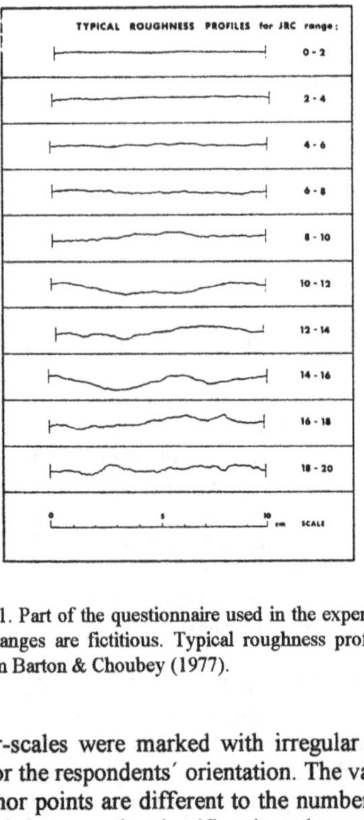

Figure 4-1. Part of the questionnaire used in the expert survey. Marked ranges are fictitious. Typical roughness profile chart taken from Barton & Choubey (1977).

The bar-scales were marked with irregular anchor points for the respondents' orientation. The values of the anchor points are different to the number of the thresholds between the classifiers in order to warrant that there would be no accumulation of subjective thresholds at the anchors (avoidance of "anchoring effects"). After having received several remarks that the subjective threshold sample statistics would only yield the "mean international standard" in use, we decided to use only the data pertaining to perceptual judgments that are not cognitively influenced by existing classification standards (e.g. Brown 1981). Unfortunately, only the parameter "joint roughness" fulfilled this requirement. As far as we know, there are no international standards in use that map linguistic classifiers of roughness onto the JRC scale. Moreover, only with regard to roughness we used perceptual figures. This methodical flaw certainly limits general conclusions based on the survey. In the questionnaire design, the following points had to be taken into account from a methodical point of view:

(a) Psychophysics. Psychophysics is the study on the relation between the intensity of a stimulus, the roughness of joint surfaces in our case, and the intensity its subjective tactile or visual perception. Based on extensive empirical studies Stevens (1975) postulates a general power law function as the psychophysical law: the subjective perceptual magnitude ψ grows as a power function of the stimulus magnitude ϕ.

$$\psi = k\phi^{\beta} \tag{4.1}$$

with k being a constant depending on the units of measurement and β serving as the characteristic variable for differing for each sensory modality. For tactile roughness, β was determined as 1.5 (Stevens & Harris 1962, Stevens 1975). The psychophysical law is the reason, why we used logarithmic bar-scales in the design of the questionnaire for the expert study. This was done in order to adjust the scale of the answering space to the perceptual magnitude, thereby also warranting a adjusted data variance for the different stimulus magnitudes.

(b) Barton's JRC scale. The Joint Roughness Coefficient (JRC; Barton 1973, Barton & Choubey 1977) serves as the objective physical scale onto which the linguistic roughness classifiers are mapped. It is a dimension-less parameter describing the roughness of discontinuity surfaces at the cm-range. Roughness at this scale provides the friction of a discontinuity in absence of a filling and dilatation (Johnson & De Graff 1988). The JRC was determined by empirical testing and can be expressed by the following relation (Barton & Choubey 1977):

$$JRC = \frac{\arctan\left(\frac{\tau}{\sigma_n}\right) - \phi_b}{\log_{10}\left(\frac{JCS}{\sigma_n}\right)} \tag{4.2}$$

with τ being the peak shear strength, σ_n the effective normal stress, JCS the joint wall compressive strength, and ϕ_b the basic friction angle which is often around 30°. Note that for completely smooth planar surfaces the JRC equals zero. The strength of Barton's relation is that it does not only serve to fit and extrapolate experimental data, but also to predict it. It involves three constants, viz. JRC, JCS, and ϕ_b. Having a quantification of linguistic classifiers for the two constants JRC and JCS, and fuzzy numbers for ϕ_b

thus seems a desirable goal with regard to the calculation of a vague quantification of the peak shear strength τ of unfilled joints from descriptive linguistic data. On the concept of fuzzy numbers, cf. Klir & Folger (1988).

(c) Linguistic classifiers for joint roughness. In order to enable an easy implementation of classifiers by means of fuzzy membership functions, the universe of discourse of the variables has to be a continuum. For this reason the scheme of VERY SMOOTH-SMOOTH-ROUGH-VERY ROUGH seemed adequate for our purposes (cf. Goodman 1976).

Since the subjective thresholds between the classifiers were assessed in a fuzzy way, that is, allowing some overlap on the bars, we then measured each single of the extreme points of the overlaps. For statistical analysis of the normality of the sample thresholds, we then used the mean of these upper and lower extreme points. For the approximation of the fuzzy membership functions, we had to transform the continuous bars into discrete data by means of partitions which were adjusted to the logarithmicity of the scales (numerically smaller partitions at the lower part of the JRC interval and larger partitions at the upper part, equ. (5.10)).

4.2 *Testing of possible scaling effects*

If the hypothesis H_0 that the sample threshold data for each threshold do not significantly differ from a standard normal distribution cannot be rejected, then we can assume that the logarithmicity of the scale did not significantly bias the subjective threshold estimates. The fitting of the frequency histograms to standard normal pdfs was performed by χ^2-tests. Table 4-1 shows the number of classes used (class.), the χ^2 values, the degrees of freedom (d.f.) and p-values for each threshold sample data. For all three sample threshold data the hypothesis H_0 was confirmed. Thus, we infer that the data are randomly distributed and further statistical analyses and data transformation is valid.

Table 4-1. χ^2-tests for sample threshold data [JRC], n = 48, H_0 = data fit normal pdf

threshold	class.	χ^2	d.f.	p	H_0
VERY SMOOTH / SMOOTH	8	1.10	1	0.295	+
SMOOTH / ROUGH	10	2.39	3	0.496	+
ROUGH / VERY ROUGH	12	1.94	2	0.379	+

Table 4-2 shows the sample statistics of the thresholds as mean, standard deviation (st. dev.), minimum and maximum value (min., max.). One statistical outlier was identified by means of a David-Hartley-Pearson test (Sachs, 1984). Comparing the means to the approximately corresponding roughness profiles from Barton (Fig. 4-1), one sees that they are fairly in accordance with intuition.

Table 4-2. Descriptive statistics for the sample threshold data [JRC], n = 48

threshold	mean	st.dev	min.	max.
VERY SMOOTH / SMOOTH	2.72	0.91	1.0	4.5
SMOOTH / ROUGH	5.94	2.04	1.5	10.65
ROUGH / VERY ROUGH	12.06	2.51	5.0	20.0 (outlier)

5 ASSESSMENT OF EMPIRICAL FUZZY SETS BY RATIONAL FUNCTION APPROXIMATION

5.1 *Algorithm*

The algorithm we will present in the following was developed by Devi & Sarma (1985). The function f(x) in (3.12) can be approximately expressed by a rational function

$$\hat{f}(x) = \frac{\sum_{k=0}^{m} a_k y^k}{\sum_{j=0}^{p} b_j y^j} \tag{5.1}$$

As in common engineering practice as well as in our survey membership functions are scaled on a continuous range, i.e. the function is unimodal, the zeroth degree numerator is accurate enough in (5.1):

$$\hat{f}(x) = \frac{1}{\sum_{j=0}^{p} \theta_j y^j} \tag{5.2}$$

where y is some function of x, and for an unimodal form of $\hat{f}(x)$ it is controlled by the mean (a) and the standard deviation (b) of the specific histogram, i.e.

$$y = \left| \frac{x-a}{b} \right|^q \tag{5.3}$$

where q (≥ 1) is an integer. By using a least squares fit the linear equation solving θ can be obtained:

$$\hat{Z}\,\theta = Z \tag{5.4}$$

where

$$\hat{Z} = \begin{pmatrix} \hat{z}^0 & \hat{z}^1 & \dots & \hat{z}^p \\ \hat{z}^1 & \hat{z}^2 & \dots & \hat{z}^{p+1} \\ \vdots & \vdots & \dots & \vdots \\ \hat{z}^p & \hat{z}^{p+1} & \dots & \hat{z}^{2p} \end{pmatrix} \tag{5.5}$$

with

$$\hat{z}^k = \sum_{j=1}^{l} f_j^2(x) y_j^k \tag{5.6}$$

l is the number of class intervals in the specific probability histogram and

$$Z = (z^0 \ z^1 \ \dots \ z^p)^t \tag{5.7}$$

with

$$\hat{z}^k = \sum_{j=1}^{l} f_j(x) y_j^k \tag{5.8}$$

and

$$\theta = (\theta_0 \ \theta_1 \ \dots \ \theta_p)^t \tag{5.9}$$

By using (5.4), (5.9), (5.2), (3.12) and (3.13) the estimated membership function is obtained. For our calculation processes a small computer program was developed where the user can vary the values for q and p in order to obtain the optimal fitting.

5.2 *Gained membership functions*

For the rational function approximations, we had to make the continuous bars representing the respondents' estimated ranges of the roughness classifiers discrete. This was done by partitioning them by an integer P yielded through:

$$P = INT\left(\frac{5\,(t_i - t_{i-1})}{t_{i-1}}\right) \tag{5.10}$$

where t_i is the upper mean threshold and t_{i-1} is the lower mean threshold of the specific bar. With this procedure we obtained the following sample sizes: $n_{VERY\ SMOOTH}=466$, $n_{SMOOTH}=350$, $n_{ROUGH}=348$, and $n_{VERY\ ROUGH}=235$. With an interval sizes of 0.8 JRC starting at 0.9 JRC we get the dot histograms depicted in Figure 5-1.

Using the algorithm of section 5.1 we get the membership functions with the following parameters: VERY SMOOTH (l=4, p=5, q=1), SMOOTH (l=13, p=5, q=2), ROUGH (l=21, p=11, q=1), VERY ROUGH (l=19, p=8, q=1). The functions are depicted in Figure 5-2. Note that the function for VERY ROUGH is distorted as an effect of the logarithmicity of the sampling scale. Furthermore, the JRC scale is open. So we almost forced the respondents to rate JRC=20 as VERY ROUGH, thereby forcing μ(20) for VERY ROUGH yielding 1. Thus, the function has generally to be regarded questionable. The blip in the function graph at about JRC=18 may also be a hint to a subjective discontinuity between the perception of the last two profiles in the typical roughness profile chart from Barton & Choubey (1977, comp. Fig. 4-1)

According to the obtained functions, the prototypical JRC values for a SMOOTH and a ROUGH joint surface is about 3.8 and 8.5, respectively (comp. Fig. 4-1).

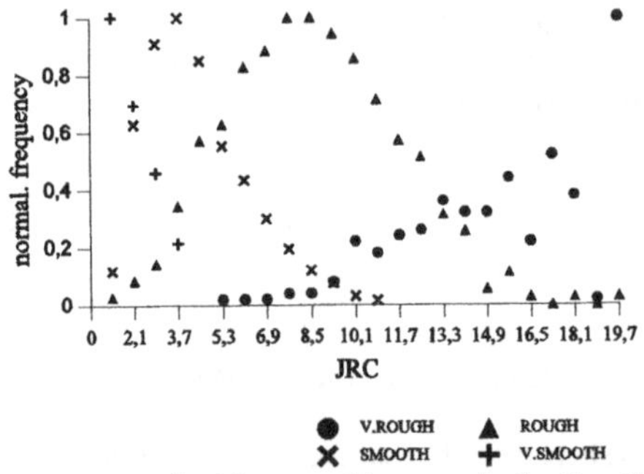

Figure 5-1. Normalized frequency histograms for the discretized data of VERY SMOOTH, SMOOTH, ROUGH, VERY ROUGH.

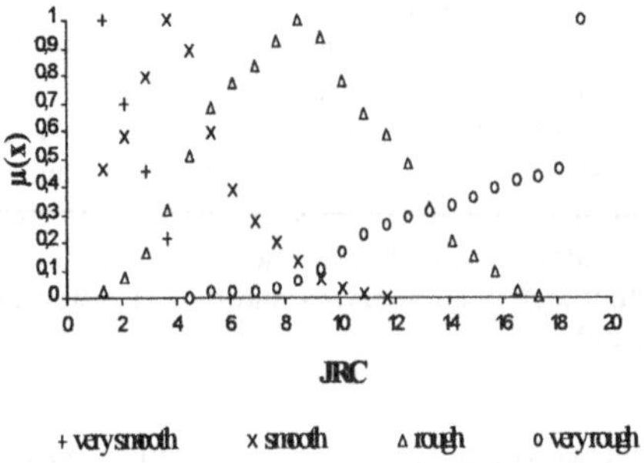

Figure 5-2. Obtained empirical membership functions for the linguistic classifiers VERY SMOOTH, SMOOTH, ROUGH, VERY ROUGH.

6 CONCLUSIONS

By presenting some theoretical considerations on the classification of geological entities, we argued that geological linguistic categories have an intrinsically vague meaning. As the meaning of information cannot be dealt with by probability theory, we postulated that the adequate mathematical treatment of semantic vagueness is possibility theory, of which the mathematical basis is the theory of fuzzy sets. The relation between the theory of probability and the theory of possibility was outlined. It provides the basis for the use of probabilistic data from psychological studies for the generation of membership functions.

An empirical demoscopic method for the assessment of fuzzy membership functions for linguistic classifiers of joint roughness was elaborated and used in an demoscopic expert survey. The method is generally sensitive, because several specific influences on the data sampling have to be considered. However, we believe that it is generalizable. Note that the gained functions are a representation of the average and international conventional vagueness of roughness classifiers. Their relation to functions that would be obtained by testing single individuals is not clear. In practice the decision whether to use functions based on sample or individual testing depends on the problem setting and the amount of individual expertise that should be incorporated. If this amount is great, the presented method is not advisable and functions should be assessed by testing relevant individuals in order to incorporate some of their expertise.

ACKNOWLEDGMENTS

The first author gratefully acknowledges a Ph.D. grant of the Austrian Academy of Sciences, Vienna, and a grant for short-term research at UC-Berkeley from the Austrian Federal Ministry of Science and Transport. He wants to thank Judith Fischer, Prof. M.F. Peschl (University Vienna) and Prof. F.J. Brosch (TU Graz) for advice. Prof. H.-P. Rossmanith (TU Vienna) and Prof. E. Raab (University Graz) provided help in the design of the questionnaire. Both authors are grateful to the experts who participated at our survey and gave invaluable comments: 12 anonymous persons, plus (without title) P.M. Acevedo (C), J. Antikainen (SF), A.K. Benson (USA), F. Blaha (CZ), J. Braybrooke (AUS), M. Cai (PRC), G. Clarke (AUS), J. Corominas (E), D.M. Cruden (CAN), M.G. Culshaw (UK), L.G. de Vallejo (E), A. den Outer (NL), M.B. Dusseault (CAN), P.R. Evans (AUS), T. Francis (RSA), R.K. Goel (IN), R.E. Goodman (USA), A.W. Hathaway (USA), J.N. Hutchinson (UK), M. Jessen (NZL), E.C. Kalkani (GR), Y. Kanaori (J), L.K. Kauranne (SF), J.R. Keaton (USA), G. Koukis (GR), I. Koumantakis (GR), F. Loset (N), R.J. Maharaj (TRI), G. Melidoro (I), C.L. Moore (AUS), G. Mostyn (AUS), A. Niini (SF), M. Romana (E), M.S. Rosenbaum (UK), A. Rustan (S), B.H. Sadagah (SA), O.N. Sikazwe (ZAM), V.K. Singh (IN), A. Smallwood (UK), M. Sorriso-Valvo (I), O. Stephansson (S), P. Vuorela (SF), W. Wittke (D).

REFERENCES

Barton, N. 1973. Review of a new shear-strength criterion for rock joints. *Eng.Geol.* 7: 287-332.

Barton, N. & V. Choubey 1977. The shear strength of rock joints in theory and practice. *Rock Mech.* 10: 1-54.

Birch, L.C. 1951. Concept of nature. *Am.Scientist* 39: 294-302.

Brown, E.T. (ed.) 1981. *Rock Characterization, Testing and Monitoring. ISRM Suggested Methods.* Oxford: Pergamon.

Devi, B.B. & V.V.S. Sarma 1985. Estimation of fuzzy memberships from histograms. *Inf.Sci.* 35: 43-59.

Dubois, D. & H. Prade 1982. On several represenations of an uncertain body of evidence. In: M.M. Gupta & E. Sanchez(eds.). *Fuzzy Information and Decision Processes.* Amsterdam: North -Holland: 167-181.

Dubois, D. & H. Prade 1983. Unfair coins and necessity measures: towards a possibilistic interpretation of histograms. *Fuzzy Sets Systems* 10: 15-20.

Dubois, D. & H. Prade 1988. *Possibility Theory.* New York: Plenum.

Frege, G. 1966. On sense and reference. In: P. Geach & M. Black (eds.). *Translations from Philosophical Writings of Gottlob Frege.* Oxford: Basil Blackwell.

Frodeman, R.1995. Geological reasoning: geology as an interpretive and historical science. *Geol.Soc.Am.Bull.* 107/8: 960-968.

Füchtbauer, H. & G.Müller 1980. *Sedimente und Sedimentgesteine.* Stuttgart: Schweizerbart.

Goodman, R.E. 1976. *Methods of Geological Engineering.* St. Paul: West.

Hagner, A.F. 1963. Philosophical aspects of the geological sciences. In: C.C. Albritton (ed.). *The Fabric of Geology.* San Francisco: Freeman: 233-241.

Hersh, H.M. & A. Caramazza 1978. A fuzzy set approach to modifiers and vagueness in natural language. *J. Exp. Psych.: Gen.* 105/3: 254-276.

Johnson, R.B. & J.V. De Graff 1988. *Principles of Engineering Geology.* New York: Wiley.

Klir, G.J. & T.A. Folger 1988. *Fuzzy Sets, Uncertainty, and Information.* Eaglewood Cliffs: Prentice-Hall.

Lakoff, G. 1987. *Women, Fire, and Dangerous Things. What Categories Reveal about the Mind.* Chicago: University of Chicago Press.

Li, H.X. & V.C. Yen 1995. *Fuzzy Sets and Fuzzy Decision making.* Boca Raton: CRC Press.

Liu, Q., F.J. Brosch, K. Klima, G. Riedmüller & W. Schubert 1997. Evaluation of data during tunneling by using an expert system. In: J. Golser & W. Schubert (eds.). *Tunnels for People. World Tunnel Congress 1997.* Rotterdam: Balkema.

Norwich, A.M. & I.B. Turksen 1984. A model for the measurement of membership and the consequences of its empirical implementation. *Fuzzy Sets Systems* 12: 1-25.

Raab, T. & F.J. Brosch 1996. Uncertainty, subjectivity, experience: a comparative pilot study. *Eng.Geol.* 44: 149-167.

Rosch, E. 1973. Natural categories. *Cogn.Psych.* 4: 328-350.

Stevens, S.S. 1975. *Psychophysics.* New York: Wiley.

Stevens, S.S. & J.R.Harris 1962. The scaling of subjective roughness and smoothness. *J.Exp.Psych.* 64/5: 489-494.

Zadeh, L.A. 1965. Fuzzy sets. *Inf.Contr.* 8: 338-353.

Zadeh, L.A. 1978. Fuzzy sets as a basis for a theory of possibility. *Fuzzy Sets Systems* 1: 3-28.

Zadeh, L.A. 1982. A note on prototype theory and fuzzy sets. *Cognition* 12: 291-297.

Zimmermann, H.-J. 1991. *Fuzzy Set Theory and its Applications.* Boston: Kluwer.

Mechanics of Jointed and Faulted Rock, Rossmanith (ed.)© 1998 Taylor & Francis, ISBN 90 5410 955 6

Some remarks about an in situ shear vane technique for claystone

A.Clerici
Earth Sciences Department, Milan University, Italy
D.Martello
Italy

ABSTRACT: The paper proposes a simple and rapid test to measure in situ the drained shear strength of claystones. The technique derives from the shear vane technique commonly used in soil mechanics. The tool is similar to the traditional shear vane, but modified in such a way to let it usables in harder materials. The proposed test doesn't give the sharp value of share strength but, nevertheless, gives its order of magnitude from which the friction angle of the rock mass can be derived. Therefore it can be used in many practical problems involving claystone to avoid forming a doubtful estimate of this parameter of the rock mass.

1 INTRODUCTION

In engineering geology there is always necessity to have numerical information about the resistance l.s. of the material, i.e. soil and/or rock mass, involved from the work in project.

As known, both in the field of the mechanics of hard rock and of the soil mechanics, laboratory tests and in situ methods allow to measure the mechanical behaviour of the materials.

The results obtained in laboratory on soil samples don't need normally further elaboration, while those obtained on rock material are to be corrected somehow before they can be referred to the rock mass that constitutes, in every problem of rock mechanics, the physical object that is necessary to know.

The execution of direct tests on the rock mass is much more complex and onerous: as far as it concerns rock mass with a rigid behaviour, the problem is essentially of cost and operative. The situation is different in laboratory and in situ tests for weak rocks, for which specific tests and operational guides and standards are very few.

The term "weak rock" includes all rocks that, because of the low resistance of the material (soft rock), or severe alteration (weathered rock) and/or the presence of a great number of joints (intensely fractured rock), have mechanical characteristics that put them in an intermediate position between those of soil and those of hard rock. Soils with an elevated degree of over-consolidation also belong to this category (Koichi, Hayashi & Nishimatsu, 1981; Clerici, 1992).

A common feature of all these materials is that they are very sensitive to sampling operations; moreover, in these materials the preparation of specimens to submit to laboratory tests is often impossible. This is one of the reason of the shortage of the number of procedures describing tests that can be performed in an equipped laboratory.

This leads to the disappointing situation for which, really in the weaker and more complex rocks, the engineering geologist must confine necessary his opinion about the mechanical behaviour of the rock mass to subjective and qualitative descriptions and therefore formulate hypothesis based on his own experience more than on measured data.

The purpose of this study is to propose a proper procedure to measure in situ the drained shear resistance of a class of weak rock, namely the claystone, through a simple and rapid test.

This technique derives from the shear vane technique commonly used in soil mechanics to measure the same parameter - see also B.L.Wiid (1981), the Standard ASTM D 2573 and the ISRM Suggested Methods (1981).

2 THE TOOL AND THE EQUIPEMENT

The shape of the tool used for the tests is similar to the traditional shear vane, partially modified in such a way to let it usable in a relatively hard material and therefore suitable to bear great stresses in comparison with those applied while operating in soils.

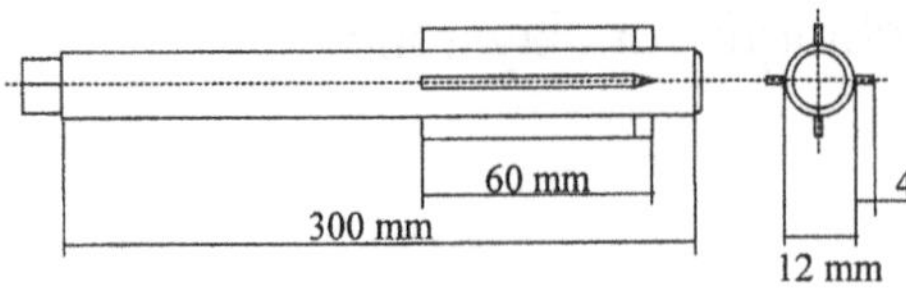

Figure 1. Tool dimensions. Not in scale.

As shown in fig. 1, the main difference is connected with the dimensions, that are greater than the traditional ones, and the shape.

Not being available on the market, the utensil has been built on purpose using a steel of good resistance, moulded as to have four thickset and strong rectangular vanes and an inner cylindrical rod (that doesn't exist in the version for soil) of 1.2 centimetres of diameter.

The dimensions have been stated on the basis of repeated tests. The vanes have length of 6 cm, width of 0.4 cm and thickness of 0.1 cm.

The inner rod, 25 cm long, allows tests at different depth and has a termination on which a dynamometer is attached.

The essential tools for the execution of the in situ test are the following:

a battery drill that guarantees a good work autonomy;

a series of twists to drill the rock with different diameters up to the dimension of the inner rod;

a 80 N*m dynamometer;

a 9 N*m dynamometer;

a series of bushes to connect and block the dynamometers to the inner rod;

a hammer;

a form to record the measurements.

The «direct reading» dynamometers are provided with a quadrant where the applied torque moment is pointed out by a needle-memory which facilitates the reading of the peak force.

3 TEST PROCEDURE

Before all, it is necessary to perform a hole in the rock using the series of twists drill by which it is possible to ream the hole progressively. This way of proceeding allows to cause the minimum damage to the rock mass. The final diameter of the hole will be equal, or larger of a fraction of millimetre, than that of the inner rod, such as to limit the damage in the rock while forcing the tool in the hole.

The phase of perforation, besides being essential to prepare the rock mass to the execution of the test, gives also some information on discontinuities spacing, on the homogeneity of the material in depth and on the possible presence of anomalies that would make of scarce significance the result of the test.

The maximum depth where the tool allows to perform the test is limited within 25 cm from the surface. This is clearly a limitation; on the other hand we can notice that, usually, the superficial part of the rock mass is the weakest one and that it is also the part that we are used to describe, for example when we refer to a geomechanics classification.

The results of the test are therefore conservative, but they reflect the strength of what we can observe directly. Furthermore our intention was to keep the tool the most possible light and portable.

Once reached the diameter and the stated depth, the tool is struck with a hammer up to the depth of test.

In this phase it is necessary to check the sharp direction of the tool in deepening with respect to the hole: deviations, inflexions, rotations or anomalous movements of the tool invalidate the test result.

Hereinafter we have shown 123 results. Some operative difficulties and uncertainties have suggested to discard 17 tests; this means that 87 % of tests were executed successfully.

When the tool is positioned correctly at the right depth, the 80 N*m dynamometer is connected and a torsion is applied to the whole system up to the failure of the rock. The rate of the stress application must be the most possible constant and near at 6°/min.

After 10 turns completed, the residual torque moment of the re-moulded material can be measured using the 9 N*m dynamometer which gives more precise readings for the low values.

The test is concluded with the extraction of the tool from the rock mass and with the examination, when possible, of the material fixed between the vanes. This can show the plane of failure and give further indications to the interpretation of the results.

If the test is performed at a limited depth (10 - 15 cm), it can be repeated deepening the same hole as to arrive to the undisturbed material.

The measured values, the geology of the rock mass and the test conditions, i.e. the depth and the inclination of the hole in comparison with the stratification or the scistosity planes, are recorded on a form.

Undoubtedly, the test cannot give the sharp value of shear strength, nevertheless it give its order of magnitude.

The test is feasible also in laboratory, when the preparation of specimens extruded from a sampler is deemed difficult or impossible. In this case, the sampler is fixed in an extruder, with an end locked by the thrust piston and wound by metallic clamps that wrap and stiffen it. The piston and the clamps revent movements of the specimen inside the sampler and radial deformations of it during the test.

The test gives the value of the shear resistance that can be confirmed later, in case the preparation of specimens to submit to lab tests of conventional type is possible.

Table 1

Rock mass	Comparison method
1) Riva di Solto (BG) claystone	RMR
2) Riva di Solto (BS) claystone	Back-analisys, RMR
Montoggio (GE) claystone: 3) Unit A 4) Unit B 5) Unit C	Lab direct shear test
6) Ellero (TO) clay and silt	Lab direct shear test

Table 2. Riva di Solto claystone (Rota Imagna).

N°	Z [mm]	T [N*m]	T_r [N*m]	τ [MPa]	t_r [MPa]
1	14	44	1.2	0.7164	0.020
2	23	47.5	1	0.7733	0.016
3	12	70	2	1.1397	0.033
4	14	38	1.1	0.6187	0.018
5	23	74	1.6	1.2048	0.026
6	14	46	1.2	0.7489	0.020
7	23	56	1.1	0.9117	0.018
8	14	45	1.3	0.7326	0.021
9	23	43	1.3	0.7001	0.021
10	14	40	1	0.6512	0.016
11	23	49	1.1	0.7978	0.018
12	10	34	1	0.8129	0.024
13	23	52	1.7	0.8466	0.028
14	10	62	1.5	1.4824	0.036
15	12	74	1.5	1.2048	0.024
16	14	62		1.0094	
17	14	67		1.0908	
18	12	69		1.1234	
19	14	43		0.7001	
20	14	74		1.2048	
21	14	52		0.8466	
22	23	78		1.2699	
23	14	30		0.4884	
24	23	65		1.0583	
25	14	61		0.9931	
26	14	66		1.0748	
27	14	63		1.0257	
28	14	60		0.9769	
29	23	68		1.1071	
30	14	40		0.6512	
31	14	43	0.85	0.7001	0.014
32	23	61	1.2	0.9931	0.020
33	14	41	0.8	0.6675	0.013
34	23	60	1.3	0.9767	0.021
35	14	39	1	0.6350	0.016
36	23	58	0.9	0.9443	0.015
37	14	45	0.95	0.7326	0.015
38	23	68	1.1	1.1071	0.018
39	14	42.5	1	0.6919	0.016
40	23	66	1.5	1.0745	0.024
41	14	44	0.9	0.7164	0.015
42	23	53.5	1.3	0.8710	0.021
43	14	46	0.85	0.7489	0.014
44	14	51	1.5	0.8303	0.024
45	23	74	2.1	1.2048	0.034
46	14	42	1	0.6838	0.016
47	23	73	1.8	1.1885	0.029
48	14	60	1.5	0.9769	0.024
49	23	77	1.9	1.2536	0.031
50	14	53.5	1.2	0.8710	0.020
51	23	74.5	1.5	1.2129	0.024
52	14	61	1.7	0.9931	0.028
53	23	71.5	1.75	1.1641	0.028
54	14	41.5	1.1	0.6757	0.018
55	23	59.5	1.1	0.9687	0.018
56	14	54	1	0.8792	0.016
57	14	49.5	0.9	0.8059	0.015
58	23	64.5	1.35	1.0501	0.022
59	14	32	1.25	0.5210	0.020
60	23	45.5	15	0.7408	0.024
61	23	66.5	1.8	1.0827	0.029
62	14	50.5	1.8	1.0827	0.029
63	23	68	1.9	1.1071	0.031
64	14	43	1.55	0.7001	0.025
65	23	64	1.7	1.0420	0.028

Table 3. Riva di Solto claystone (Lumezzane).

N°	Z [mm]	T [N*m]	T_r [N*m]	t [MPa]	t_r [MPa]
1	20	21	1.8	0.3256	0.0293
2	20	32	1.2	0.5210	0.0195
3	20	14	1.5	0.2279	0.0244
4	20	13	1.2	0.2117	0.0195
5	20	22	1.5	0.3582	0.0244
6	15	22	1.4	0.3585	0.0227
7	15	18	1.3	0.2931	0.0211
8	20	23	1.3	0.3745	0.0211
9	15	12	1.3	0.1954	0.0211
10	15	30	1.4	0.4884	0.0227
11	15	19	1.2	0.3093	0.0195
12	15	22	1	0.3582	0.0162
13	15	18	1.2	0.4304	0.0286
14	20	30	1.6	0.4884	0.0260
15	20	25	1.6	0.5977	0.0382
16	20	23	1.8	0.3745	0.0293
17	20	29	1.8	0.4721	0.0293

Table 4. Montoggio claystone, unit A.

N°	CAMPIONE	T [Nm]	Tr [Nm]	t [MPa]	t_r [MPa]
1	SISC3	6.4	0.3	0.1042	0.0211
2	SISC3	6.2	0.3	0.1009	0.0211
3	SI3C3	5.8	0.2	0.0944	0.0032
4	SI3C3	8	0.4	0.1302	0.0032
5	SP5C5	6	0.5	0.0977	0.0081
6	SP5C5	8	0.5	0.1302	0.0081
7	SI6C1	5	0.4	0.0814	0.0244
8	SI6C1	5	0.4	0.0814	0.0244
9	SI6C1	4	0.35	0.0651	0.0056
10	SI6C1	4	0.4	0.0733	0.0244
11	SI6C1	5	0.5	0.0814	0.0081
12	SI6C1	5.5	0.5	0.0895	0.0081

Table 5. Montoggio claystone, unit B.

N°	CAMPIONE	T [Nm]	Tr [Nm]	t [MPa]	t_r [MPa]
1	SISC3	10.8	0.2	0.1758	0.0032
2	SI3C3	17	1.7	0.2768	0.0276
3	SI3C3	21	1.8	0.3419	0.0293
4	SP5C5	15	0.7	0.2442	0.0293
5	SP5C5	14	0.4	0.2279	0.0244

4 METHODOLOGY OF STUDY

To verify the reliability of the results of the tests, the same have been compared with the results obtained with different traditional methods, particularly with

Table 6. Montoggio claystone, unit C.

N°	CAMPIONE	T [Nm]	Tr [Nm]	t [MPa]	t_r [MPa]
1	SP5C6	58	1.8	0.9443	0.0293
2	SP5C6	60	1.8	0.9769	0.0293
3	SP5C6	30	1.3	0.4884	0.0211
4	SP5C6	54	1.5	0.8792	0.0244
5	SI6C1	52	1.3	0.8466	0.0211

Table 7. Ellero claystone.

N	T [N*m]	Tr [N*m]	τ_{Wiid} [MPa]	τ_r [MPa]
1	72	7	1.3263	0.114
2	44.5	3.8	0.8197	0.062
3	52	7	0.9579	0.114
4	58.5	8	1.0776	0.130
5	74	7.8	1.3632	0.127
6	48	6.2	0.8842	0.101
7	48.5	7.8	0.8934	0.127
8	42	7.2	0.7737	0.117
9	47.5	7.8	0.8750	0.127
10	62	6.4	1.1421	0.104
11	70	9	1.2895	0.147
12	68	4.8	1.2526	0.078
13	66	5.6	1.2158	0.091
14	62	6	1.1421	0.098
15	55	5.2	1.0132	0.085
16	64	5.8	1.1790	0.094
17	52	6.4	0.9579	0.104
18	45	6	0.8290	0.098
19	68	8.8	1.2526	0.143

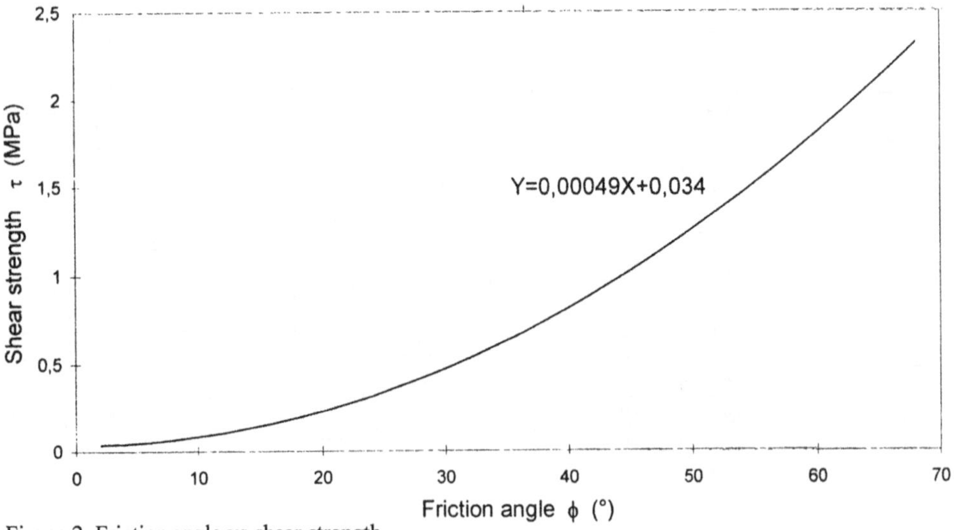

Figure 2. Friction angle vs shear strength.

laboratory direct shear tests on material sampled in boreholes, with the results of back-analysis of an active landslide and with the parameters derived from the RMR classification of rock masses (Bieniawski, 1973).

Table 1 shows the six groups of samples considered in this study.

The drained shear resistance has been calculated by means of the following equations:

a) the one used by Wiid (1981):

$$\tau = \frac{2T}{\pi d^2 L} \tag{1}$$

b) the general relationship proposed by Osterberg (1956) and Ladd (1977):

$$\tau = \frac{T}{\frac{\pi D^3}{2} \times \left(\frac{H}{D} + \frac{a}{2}\right)} \tag{2}$$

c) the relationship considered in the ASTM D 2573 standard for mechanical shear vane:

$$\tau = \frac{T}{\pi D^3 + 0.37\left(2D^3 - d^3\right)} \tag{3}$$

where: T is the peak torque moment in N*m; d is the diameter of the inner rod in m, D is the external diameter of the vanes in m.

The comparison of the results calculated using the different equations has underlined little differences, however within 10 %. From now on we consider the results obtained especially with the last expression.

5 RESULTS OF THE TESTS

On the Riva di Solto claystone outcropping in Rota Imagna (BG), 65 measure have been performed with the following characteristics; depth of the tests ranging between 14 cm and 23 cm; varying inclination in comparison with stratification. The rock is jointed black claystone.The results are in table 2.

On the Riva di Solto claystone outcropping in Lumezzane (BS) 17 measures have been performed, as shown in table 3. The rock is black claystone, intensely jointed and folded. Depth of the tests ranges between 14 cm and 23 cm.

On the Montoggio claystone (GE) 22 measurements have been carried out in laboratory, on undisturbed samples collected in boreholes.

The rock material is very compact dark-black grey claystone with thin clayey layers. Three technical units can be distinguished:

Unit A: predominantly clay;
Unit B: claystone and clay;
Unit C: predominantly claystone.

Results are in table 4 -5 - 6.

On silts and clays outcropping in Ellero (TO) 19 measures have been performed in laboratory; tests were carried out on specimen sampled in boreholes.

The material was over-consolidated, dark grey silt and clay: as a matter of fact this is the sole group of tests not performed on weak rock and, as hereafter pointed out, the results cannot be computed with the same procedure.

The results are listed in table 7.

6 ELABORATION OF DATA

The comparison between the values of shear strength obtained from the tool and the values of friction angle acquired through direct laboratory tests allows to draw the diagram shown in figure 2.

The equation of the curve is the following:

$$\phi = \sqrt{\frac{\tau - 0.034}{0.000494}} \tag{4}$$

where ϕ is the friction angle and τ is the shear strength.

On the same diagram we have also plotted the results obtained with different methods (back-analysis, RMR Classification): these results fit quite well the proposed curve.

Finally, we have plotted the results of the tests carried out on Ellero silt an clay: in this case it is clear that they don't fit the curve.

It is worth underlining that the measured shear strength is referred, in consideration of the low block size index, to the rock mass as a whole, and not to the intact rock material nor to the discontinuities.

Furthermore it is important to point out that the rapidity of the execution of the proposed test, the intense jointness of the rock material (and consequently its high permeability) makes the test a drained test.

This mean that its results can be matched with the ones obtained in laboratory direct tests.

7 CONSIDERATIONS

The 123 tests executed do not allow to make definitive remarks on the significance of the proposed methodology.

On the other hand, the tests carried out on claystones fit quite well the curve hereabove computed.

Many other tests on many other claystone formations would be necessary, but the proposed relationship appears to give a representative value of the shear strength of the rock mass from which the peak friction angle can be derived.

The test is neither time consuming nor costly and allows to overcome the difficulties of sampling and preparing the specimens for laboratory tests. Therefore it can be used in many practical problems involving claystones to avoid forming a doubtful estimate of this basic parameter of the rock mass.

ACKNOWLEDGEMENTS

This research was supported by a grant from the italian GNDCI (Gruppo Nazionale Difesa Catastrofi Idrogeologiche, Unit 2.4 - Local Coordinator Prof. P. Massiotta.

REFERENCES

Akai, K., Hayashi, M. & Nishimatsu, Y. 1981. *Proceeding of the International Symposium on Weak Rock. Soft, fractured and weathered rock.* Tokio: Balkema.

Clerici, A. 1992. Engineering geological characterization of weak rocks: classification, sampling and testing. In J.A.Hudson (ed) *ISRM Symposium: Eurock '92*: 179-184. British Geotechnical Society, London.

Wiid, B.L. A shear vane technique to measure the in situ strength of weak rock. In Akai K., Hayashi M & Nishimatsu Y., *Proceedings of the International symposium on weak rock, Tokio, 21 24 September 1981:* 485 - 490.

ASTM - D2573 - 72 (Reapproved 1978). *Standard method for "field vane shear test in cohesive soil".*

ISRM, 1981. Suggested Method for in situ determination of shear strengths using a torsional shear test. *ISRM Suggested Methods rock caracterization testing and monitoring part II:* 137 - 140. Ed E.T. Brown Pergamon Press.

Bieniawski, Z.T. 1973. Engineering classification of jointed rock masses. *Trans. S. Afr. Inst. Civ.Eng. 15: 334 - 335.*

Mechanics of Jointed and Faulted Rock, Rossmanith (ed.) © 1998 Taylor & Francis, ISBN 90 5410 955 6

Description of a fracture morphology in regard of its behaviour during shearing

J. Riss
CDGA, Université Bordeaux, Talence, France

S. Gentier
BRGM DR, Orléans, France

R. Flamand & G. Archambault
CERM UQAC Boulevard de l'Université Chicoutimi, Qué., Canada

ABSTRACT: The understanding of the mechanical behaviour of rock joints during shear displacement needs an accurate description of the morphology of both walls of the fracture. This paper is a contribution to the very detailed description of the fracture asperities distribution. Asperities are any features of whatever size contributing to the roughness of the walls of the fracture. Assuming fracture surfaces are looking like a geomorphologic landscape, asperities are considered to be local or regional rough patches lying between catchment basins. Image analysis and specially mathematical morphology associated with geostatistical methods are very useful and efficient tools in order to characterise the roughness of a fracture.

1 INTRODUCTION

The hydromechanical stability analysis, the design of various workings in jointed rock masses (rock slopes, underground openings, dam foundations) and the fluid circulation analysis require the estimation of compressibility, deformability and shear strength characteristics of the individual joints and joints sets. « Although major progresses have been made in laboratory testing of, and model development for, rock joints in the past, there still remain several uncertainties about joint behaviour » (Stephansson & Jing 1995). First of all is the roughness : these authors wrote : « the characterisation of roughness is the dominant factor affecting all other properties » of rock joints, but « unfortunately, an adequate and unique mathematical representation of roughness has not be yet realised and remains a great challenge ».

Assuming fracture surfaces are looking like a geomorphologic landscape, the roughness of a fracture is considered to be made of local or regional rough patches lying between catchment basins. Since the aim of this paper is to correlate these morphological features to the damaged areas developed on the surfaces during shearing, we have first collected data $z = f(x,y)$ about a natural fracture in order to infer a geostatistical 3D reconstruction of the morphology and, next, shearing tests have been performed by working on replicas identical to the original fracture. At the end of each shearing test, grey level images of both the upper and the lower walls were recorded by means of a black and white camera so that we could characterise the location, the size and the shape of the damaged areas developed on the sliding surfaces.

2 GEOMETRY OF THE FRACTURE

The work has been carried out on a sample that has been cored across a natural fracture within the well studied Guéret granite of France (Gentier 1986). A set of profiles were recorded in 4 various directions (Fig. 1) giving rise to 7563 (lower surface) and 7556 (upper surface) co-ordinates $\{x,y,z\}$. Based first on the calculus of the experimental variograms of elevations z and first derivatives z', and next on the design of well fitted theoretical models, a 3D reconstruction (conditional simulation) of the upper and lower surfaces is available for the characterisation of the asperities (Chilès & Gentier 1993, Riss et al. 1997). No major differences were detected while modelling lower and upper surfaces. Figure 2 shows the topography of the lower wall.

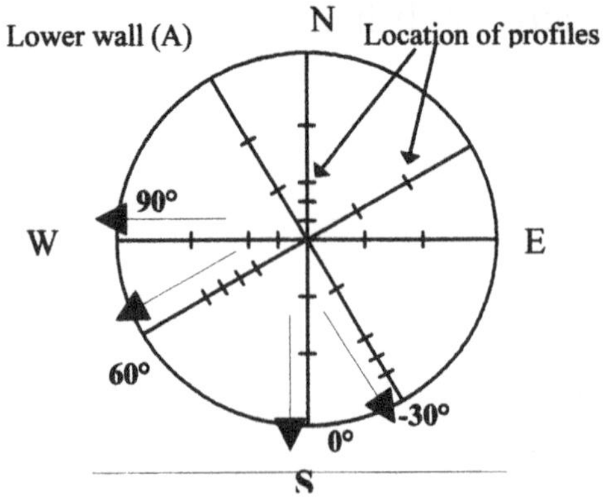

Fig. 1. Diagram showing onto the projection of the lower surface of the fracture the 4 shearing directions (-30°, 0°, 60°, 90°). Profiles were recorded parallel to these directions. Geographical coordinates N, E, S, W will be used for facilities.

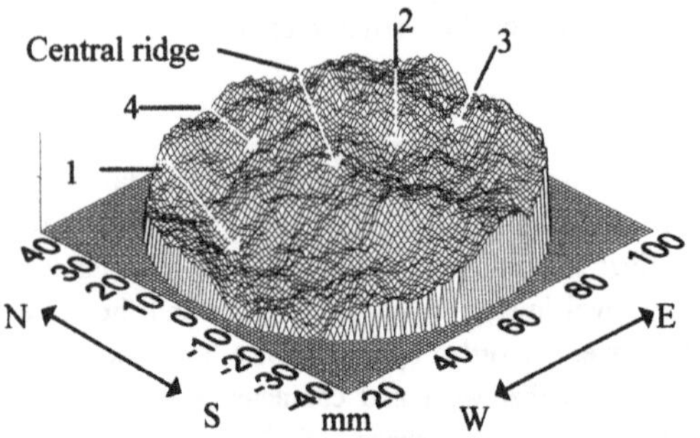

Fig. 2. The morphology of the lower wall of the fracture inferred from a geostatistical conditional simulation ; difference of altitude between the highest and lowest points equals 8.9 mm, dip direction is EW, strike direction is NS. Numbers 1, 2, 3, 4 refer to morphological features (see below).

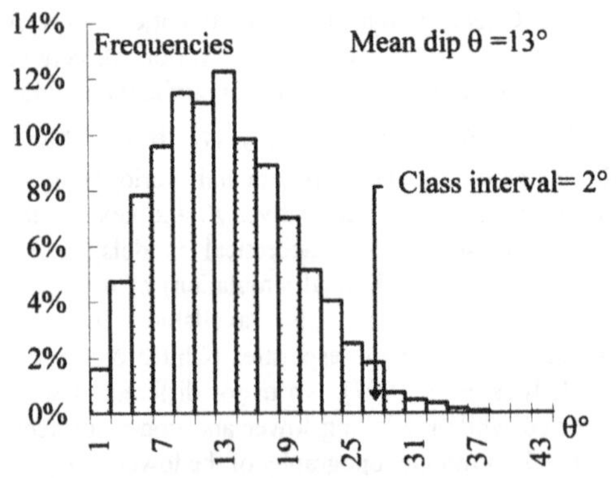

Fig. 3. Dip distribution : values along the horizontal axis are the centres of the class angle.

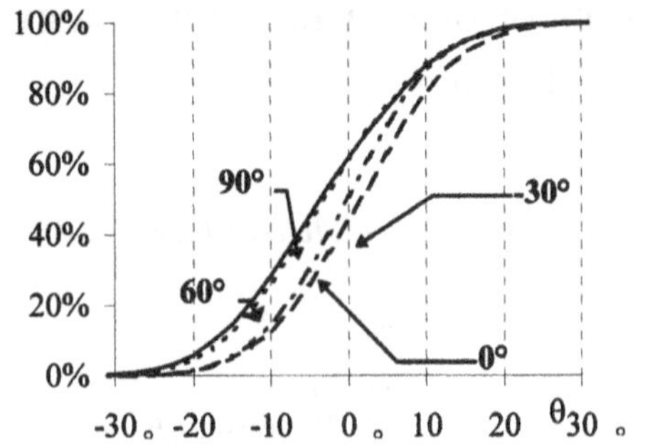

Fig. 4. Distribution of the apparent dips along the 4 shearing directions. Negative angles are associated with parts of the fracture facing up the upper wall during shearing, the positive for positive angles (see Fig. 5).

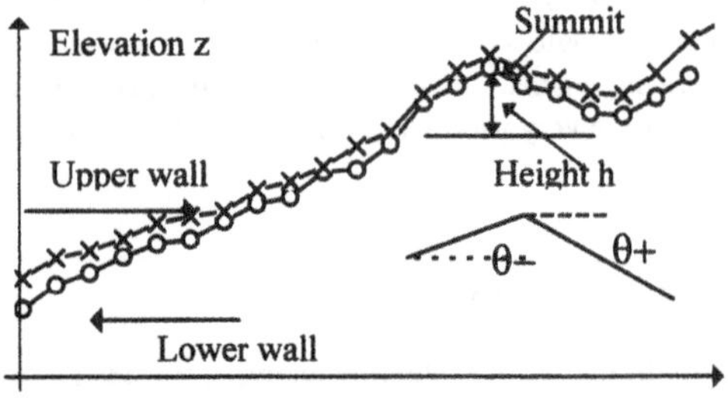

Fig. 5. Definition of positive and negative apparent dips. Arrows show the shear direction of both upper and lower walls of the fracture.

Based on the observation of the lower surface, the main geometrical characteristics of the fracture are :

1. a general trend to dip (5°) westwards deduced from a linear regression $z = f(x,y)$ ($r = 0.90$) with a residual variance $s^2_{z,xy} = 0.86\ mm^2$; this result is derived from the experimental data {x, y, z},

2. a mean dip (13°) calculated from the individual dips at each grid node of the 3D conditional simulation (Fig. 3),

3. the cumulative distribution of apparent dips (Fig . 4) are shifted towards the left for the shearing directions quite parallel to the dip direction (90° and 60°). The mean dips are respectively {5°5, 5°6, 8°4, 8°8} and, {-8°6, -7°5, -7°9, -8°1} for the 4 directions (-30°, 0°, 60°, 90°). These values are inferred from the 3D geostatistical simulation,

4. the topographical features (Fig. 2) contributing to the roughness of the fracture are mainly composed of a central ridge running north south (//0°) through the sample associated to four main valleys and adjacent crest lines ; on its western side there is a main valley looking southwards (1) and on the

eastern side there are two perched up valleys looking southwards (2) and (3). An adjacent valley (4) is looking mainly westwards. These features associated with the dipping trend of the fracture are the centimetric contribution to the roughness of the fracture ; in a more detailed analysis there are millimetric basins and ranges mainly distributed on the east part of the fracture as can be seen figure 2. The geometrical characteristics of the upper wall of the fracture are similar to those of the lower wall.
In the next paragraph, because of the anisotropy of the morphology of the fracture, it will be shown that the mechanical behaviour during shearing is highly dependant of the shearing direction.

3. MECHANICAL RESULTS

Shearing tests have been performed under various applied normal stresses σ_N (7, 14, 21 MPa) with a series of identical cement mortar replicas of the natural fracture; the shear velocity was 0.5 mm/mn. Shear displacements were interrupted for each normal stress and chosen in the following sequence : Δu=0.35 mm, 0.50 mm, 2.00 mm, 5.00 mm and four different directions θ of shearing have been used (0°, -30°, 60°, 90°). It must not be forgotten that a replica was only once used for a specific shearing test (given σ_N, Δu and shearing direction).

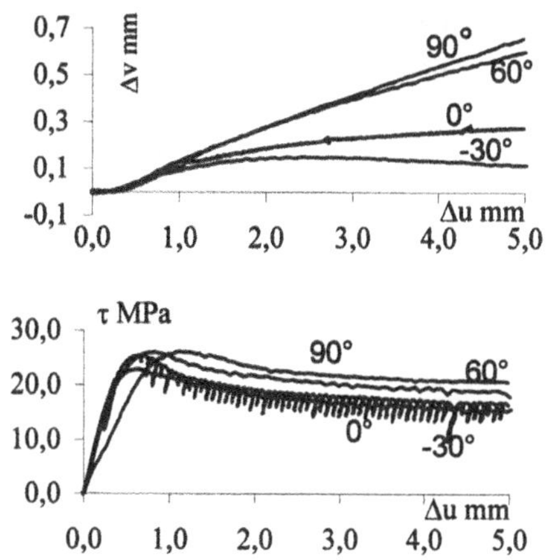

Fig. 6. Vertical displacement (Δv) and shear stress (τ) vs. shear displacement (Δu) of the joint with σ_N= 21 MPa and 4 directions of shearing (-30°, 0°, 60° and 90°).

Figure 6 shows the mechanical curves for σ_N=21 MPa ; others curves for σ_N=7 or 14 MPa are also available. They show similar phenomenon. All things being equal, when the shearing direction is parallel to the dip direction of the fracture (cases 90° and 60°opposite to case 0° and -30°):

1. the peak shear stress occurs at greater Δu and at higher shear stress τ,
2. the residual shear stress is higher and,
3. the decreasing of the dilatancy (Δv vs. Δu) is different.

Since the mechanical behaviour is different from one direction to the other, we will show the consequence of the shearing on the morphology. In order to show the damaged areas occurring during sliding, grey level images of both the upper and the lower walls were recorded by means of a black and white camera, at the end of the shearing tests. Thus we can characterise the location, the size and the shape of these damaged areas.

4. DAMAGED AREAS

4.1 *Grey level Images*

Figure 7 shows some of the recorded grey level images of the lower part of the fracture. The more or less white zones of these images are the damaged areas (gouge material or crushed material).

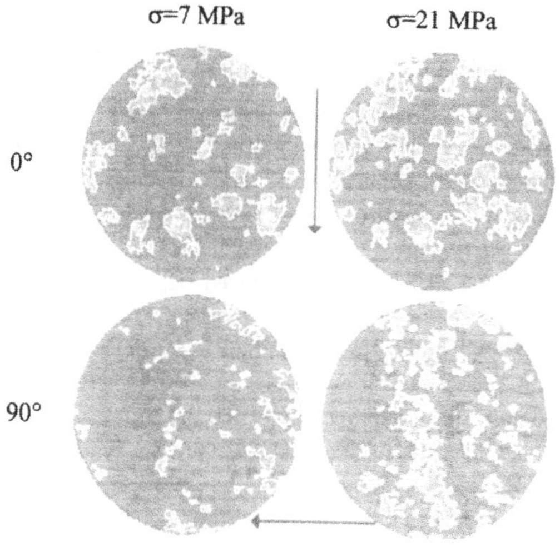

Fig. 7. Some grey level images of the lower surface where the damaged zones are outlined. Arrows show the shearing direction for the lower wall.

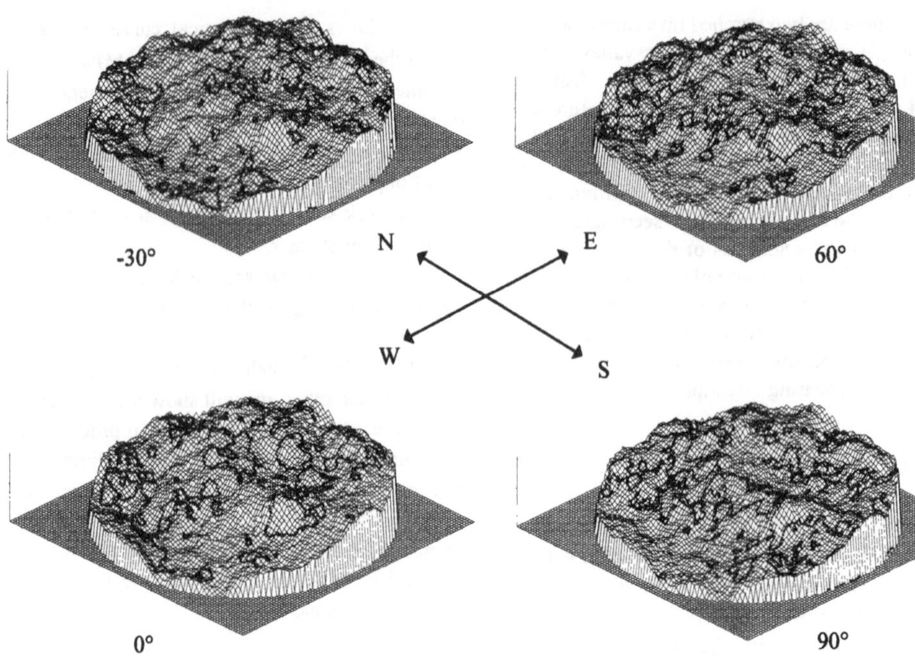

Fig.8. Damaged zones superimposed onto the topography ($\sigma_N = 21$ MPa , $\Delta u = 5$ mm). On the left, shearing directions (-30°, 0°) are quite parallel to the strike direction (N S) of the fracture ; on the right the shearing directions (60°, 90°) are quite parallel to the dip direction of the fracture (E W).

Using a specific algorithm we have developed (Riss et al. 1996), damaged areas can be segmented and black and white images were generated so that we can calculate the co-ordinates of their boundaries. The analysis of these images suggests some conclusions. The damaged area are distributed over the joint surfaces without concentration at some specific locations. These damaged areas increase with increasing applied normal stress mainly by dilation and connection of them. The trend of the linked damaged zones, regardless of their size, is in an orientation almost perpendicular to the shear direction. It may also be noticed (Riss et al. 1996) that corresponding damaged areas exist on each joint wall and their relative position when both surfaces are superposed is available. Since a numerical 3D simulation of the topography and the co-ordinates of the boundaries of the damaged zones are available, it will be very interesting to superimpose the boundaries onto the 3D map of the fracture.

4.2 *Superimposition of the damaged areas onto the topography*

Figure 8 shows the locations, the size and the shape of the damaged zones in regard of the various directions of shearing. All the previous conclusions

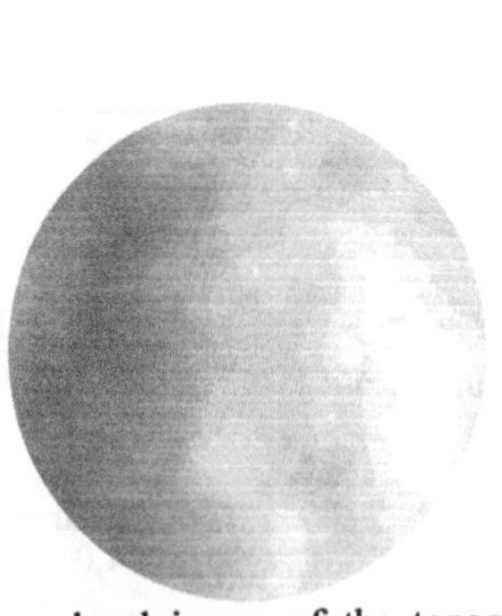

Fig. 9. Grey level image of the topography of the lower wall of the fracture. $f(z) = az+b$ (z values are those inferred at the grid nodes by the geostatistical simulation, $f(z) \in [5\ 250]$, pixel size is $0.5 mm^2$ and a grey level represents 33µm.

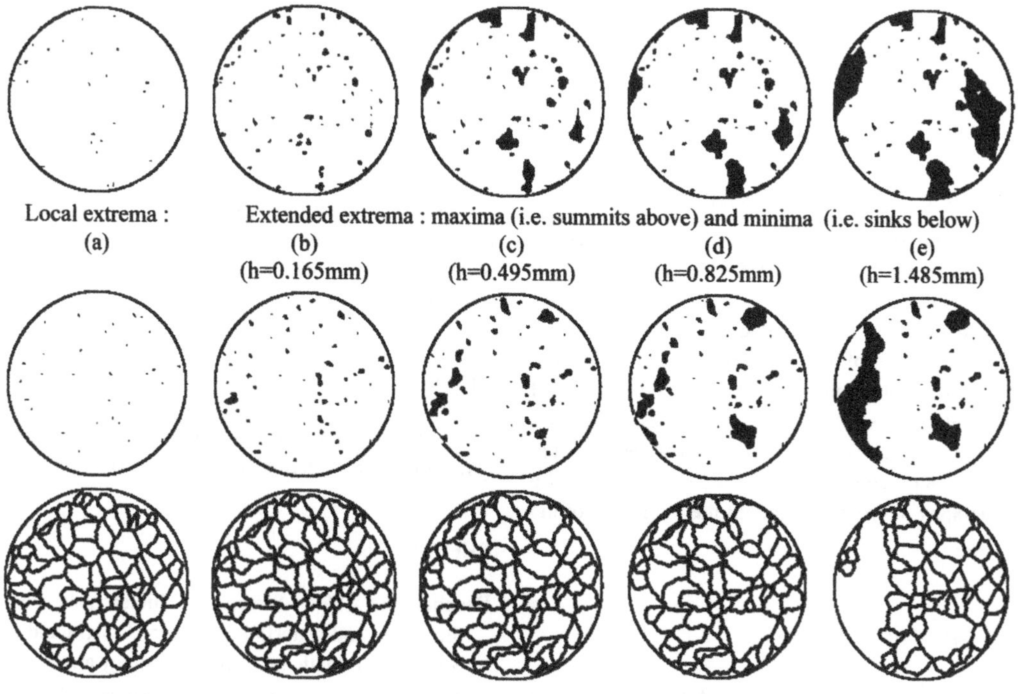

Fig. 10. Maps of the local extrema (i.e. asperities) of the grey tone function with increasing heights h. Last rows shows the crest lines and the watershed basins of the surface associated with the previous minima.

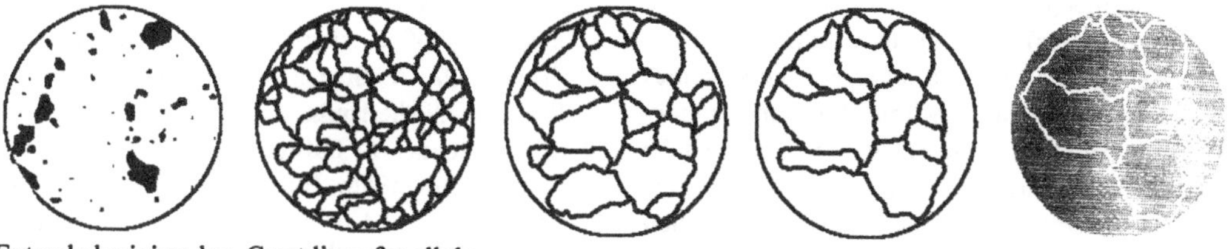

Fig. 11. Crest lines since minima considered to be irrelevant are deleted : i.e. various scales of observation.

remain true but it must be added that all the parts of the surface that have been damaged are mainly the hillsides facing up the opposite side of the fracture (here, the upper wall) whatever their elevation z. When the shearing direction is quite parallel to the dip direction of the fracture, it can be seen that the main topographical features (central ridge and east sides of valleys n°2 and 3) are concerned ; in other words all hillsides or ridges oriented from north to south and dipping westwards will be damaged. On the opposite, when the shear direction is parallel to the strike direction of the fracture the hillsides, ridges or cuesta dipping southwards will be damaged. The main conclusion is not only about the orientation of the damaged zones (that is quite obvious) but about the height (h) of the asperities above the local mean planes.

Since the degradation of the topography during shear process concerns the asperities (local summits, peaks or ridges), we will try to characterise these asperities whatever their altitude by taking into account their size, sharpness and elevation over local smoothed mean planes. Image analysis and mathematical morphology (Coster & Chermant 1985, Serra 1981) are very efficient methods to manage such a problem.

5. MAPS OF THE FRACTURE ASPERITIES

5.1 *Grey level Image of the topography*

Image analysis methods are very useful to analyse the roughness of a non planar surface e.g. the surface of the walls of a fracture. In order to operate with such methods we have created a grey level image of the lower wall of the fracture assuming that the levels of the grey tone function f(z) are proportional to the elevations (z) of the fracture. Figure 9 shows such an image. We will now present the minima, maxima and crest lines or rough patches of the topography in order to highlight the roughness of the surface.

5.2 *Maps of the asperities*

Let us assume that the summits of the lower surface are the sinks of the upper surface and vice versa. Then, figure 10 (first and second rows) shows where the asperities of the fracture are located, depending of their heights (h) but whatever their altitudes. First of all, figure 10 (a) shows the local summits or sinks and next, figure 10 (b, c, d, e) shows the asperities with their surroundings delimited by an arbitrary increasing (from left to right) height h. Let now h be equal to a arbitray value, then, the size of the surrounding area of a summit gives information about the sharpness of the asperity : the larger the area is, the smoother the asperity is. The change in the shape gives information on the anisotropy of the asperity. Supposing that sharp asperities will be split away during shearing and sliding will occur along smooth asperities we can predict what parts of the fracture will be damaged during shearing.

Figure 10 (last row) shows the design of the dividing lines (crest lines) of the watershed basins assuming they were computed with the previous extended minima. Then, by such a mean we can simulate a more and more simple representation of the topography, on one hand because of the coalescence of the minima with the increasing of h consequently by increasing the size of the homogeneous zones and on the other hand by eliminating some of the minima considered to be irrelevant from a mechanical point of view (Fig. 11). From the shearing point of view each watershed basin can be assumed to have an homogeneous roughness.

CONCLUSION

At the moment the mathematical representation of the roughness remains a great challenge but added to the present work further developments are in progress in order to solve the problem.

ACKNOWLEDGEMENT

The present research was supported by the BRGM (France) and the University Bordeaux I (France).

REFERENCES

Beucher, S. 1994. Watershed, hierarchical segmentation and waterfall algorithm. J. Serra and P. Soille (eds), Mathematical Morphology and its Application to Image Processing, Kluwer Academic Publishers : 69-76.

Coster, M. & J.L. Chermant 1989 . Précis d'analyse d'images. Presses du CNRS, France: p. 560.

Chilès, J.P. & S. Gentier 1993. Geostatistical modelling of a single fracture. Geostatistics Troia'-1992. A. Soarez (ed), Kluwer Academic Publisher, Netherlands.

Riss, J., S Gentier, K. Laffréchine, R. Flamand & G. Archambault 1996. Binary Images of Sheared Rock Joints : Characterisation of Damaged Zones. M. M. M ; 1996 ; 7 ;521-526.

Riss, J., S. Gentier, G. Archambault & R. Flamand 1997 Sheared rock joints : dependence of damaged zones on morphological anisotropy. Int J Rock Mech & Min Sci 1997 ; 34 (3–4) : Paper No. 258.

Serra, J. 1982. Image Analysis and Mathematical Morphology. Academic Press : p. 610.

Stephansson, O. & L. Jing 1995. Testing and modeling of rock joints. Mechanics of Jointed and faulted Rock, Rossmanith (ed.) : 37-47. Rotterdam : Balkema.

Mechanics of Jointed and Faulted Rock, Rossmanith (ed.) © *1998 Taylor & Francis, ISBN 90 5410 955 6*

Four joint genetic groups and their distinct characteristics

Dov Bahat
Department of Geological and Environmental Sciences, Ben Gurion University of the Negev, Beer Sheva, Israel

ABSTRACT: Joints in sedimentary rocks are divided into four, the burial, the syntectonic, the uplift and the post-uplift groups. Joints from these groups differ in their various properties. It is demonstrated how sets in specific outcrops can be assigned to one of the four joint groups, and how each group is correlated with specific fractographic features and certain joint charcteristics that play important roles in engineering and environmental projects. In a joint system certain sets may propagate unidimensionally and others may propagate bidimentionally. A wrong appreciation of the joint propagation mode may result in erroneous analyses of joint spacing. Burial and syntectonic joints may develop at much greater depths than uplift and post uplift joints. Joint aperturesmay vary considerably between burial and uplift single-layer joints. Uplift joints may provide better water drainage than syntectonic joints.

1 INTRODUCTION

Many geological, hydrological, engineering and environmental technologies are profoundly influenced by the properties of joints in the bedrock. Joints control the pathways for subsurface fluid flow and, for ore-forming fluids, geothermal fluids, oil, gas and pollutants. The design and building of large structures, such as highways and nuclear-waste repositories depend on the thorough cognizance of the fracture properties of the rock mass. The influence of joints on the transport and mechanical properties of the rock depends on a series of joint characteristics. Joint characteristics include the fracture origin, mode of propagation in relation to the boundaries of the rock layer, size, spacing, ratio of size/spacing, orientation, persistence, depth below the ground surface, interaction with other fractures, interaction with fluids, aperture, infilling, surface roughness and wall strength.

Although there are various classifications of joints based on geometric properties (e. g. Price 1966) investigators were reluctant to suggest classifications based on genetic criteria, presumably due to the complexities of jointing. One important genetic classification was offered by Engelder (1985). He suggested a general scheme of joint classification in clastic rocks from the Devonian sedimentary basins in the Appalachian Plateau, USA. His scheme consisted of four categories: The tectonic, the hydraulic, the unloading and the release joint types.

The present paper relates to joint characterization that has been investigated in chalks from various tectonic environments so that by concentrating on a single rock the problem of "material properties" was eliminated.This enabled one to ellucidate and classify several groups of joints which vary from each other and reflect distinct fracture mechanisms. Examples are drawn from the Negev and Judean deserts in southern Israel.

Tectonofractography concerns the analysis of various morphological features of the fracture surface, their tectonophysical interpretation and their correlation with various joint characteristics (Bahat, 1991). This technique was widely used in the classification. Investigation of joints in other rocks revealed many resemblences to the fracture properties and grouping initially observed in chalk, hence, supporting the above classification. Accordigly, joints in sedimentary rocks are divided into four genetic groups, the burial, the syntectonic, the uplift and the post-uplift joints (Bahat 1991; Engelder et al. 1993; Bankwitz & Bankwitz 1994).

The objective of this paper is, to demonstrate a few representative cases (specific outcrops) where field relationships enable to assign joints to one of the four joint groups, to correlate this group with specific fractographic features; and, to show correlations between particular joint groups and their distinct joint characteristics, as well as their applications.

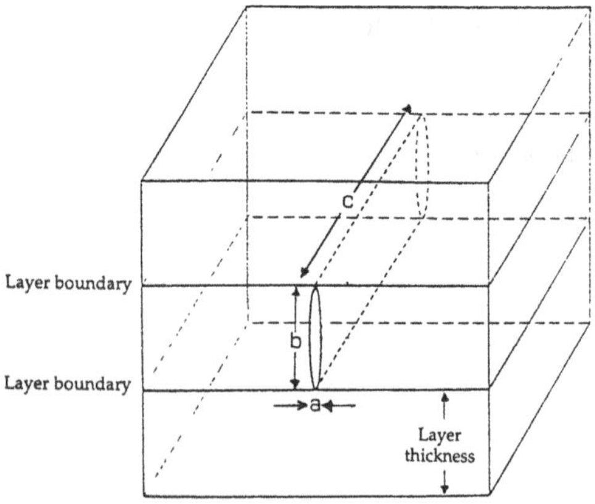

Figure 1. Dimensions of a single-layer joint. a. Thickness (aperture), b.Width (height) generally equivalent to layer thickness, and, c. length (modified from Wu & Pollard 1995).

2 FOUR JOINT GROUPS

2. 1 Burial joints

The burial phase is the historical stage that includes sedimentation, downwarping and diagenesis, all preceding the phases of syntectonic deformation and uplift. The outcrop which is used as an example consists of slightly folded Lower Eocene (Mor formation) thin chalk layers (40 cm to 90 cm thick) which alternate with beds of chert nodules (up to 10 cm thick) (Bahat & Grossman 1988, station 20). The cross-fold joints (which occur sub-normal to the fold axis) that cut the chalks of the Mor formation arrest at the boundary of the chalk layer (Fig.1) with the chert beds (Fig. 2a). The fault that displaces these joints contains nodules and fragments of chert along its trace. The solidification of chert is associated with the diagenetic stage of the chalk (Knauth & Lowe 1978). Chert does not occur in the overlying Middle Eocene (Horsha Formation). Therefore, both, the cross-fold jointis and the fault occurred before the sedimentation of the Middle Eocene. Hence, they were formed during downwarping, by a process controlled by the tectonic regime that prevailed in the Lower Eocene. The common orthogonality of the cross-fold joints with the strike-parallel joints (which occur sub-parallel to the fold axis), and their alternate fracture in the same beds of the Mor Formation implies that both sets are genetically linked (Bahat 1989).The general term given to joints that cut a single layer and do not cross its boundaries is, single-layer joints (S. L.) (Fig. 1).

Opening (aperture) is < 0.1 mm in most burial joints and there is almost no secondary mineralization. Spacing in S. L. ranges between 12 and 30 cm. The ratio of joint height (Fig.1) T to mean spacing S is T/S. T/S ranges between 1. 2 and 3. 1 in S. L. (Bahat 1988a).

Generally, the burial joints that cut the Mor Formation in two sets orthogonal to each other (328° and 059°) propagate bidimentionally (radially, both laterally and vertically). Two other joint sets which strike 309° and 344° propagate unidimensionally (only laterally). The bidimentional propagation is demonstrated by concentric undulations (rib markings) on surfaces of orthogonal cross-fold joints and strike-parallel joints. In Fig. 2a two undulations at centre start in the middle of the layer at a surface of a previous joint (which is normal to picture); and one undulation at right starts at the upper layer boundary. The unidimensional propagation is shown by horizontal plumes which parallel bedding in sets 309° and 344° (Fig. 2b) that do not form orthogonal relationship.

2. 2 Syntectonic joints

Syntectonic joints include those fracture phenomena which are associated with intense deformation. Typically, these joints are linked with strongly folded layers and/or adjacent faults. Generally, they temporally form between the burial and uplift phases, and display fracture surface morphologies which differ from those in the other two groups.

Syntectonic joints are well exposed in Santonian chalks (Menuch Formation) near 'Arad, in the Ye'elim Valley outcrop (Bahat 1991) where the Menuha chalks are overlain by endemically-folded chert of the (Campanian) Mishash Formation. Joint spacing decreases from 20-40 cm to less than 1 cm in thick chalk layers (≥ 3 m) that occur close to intensely folded and faulted chalk layers and chert beds (Fig. 2c). There is an up to about 20° deviation from verticality of some joints cutting the chalk close to the fold, implying more than one generation of joints.

Steinitz (1974) regards the genesis of the Mishash chert as a diagenetic process, involving increase of volume and associated intraformational contortion and dislocation. Thus, if the joints were formed in association with the chert folding, they were created through the Late Campanian-Maastrichtian-Paleocene, during the most intense tectonic period in the region (Bentor & Vroman 1960; de Sitter 1962). That is, jointing in the Menuch chalks occurred before the earlier uplifts of the Late Eocene (Benjamini 1984).

Opening in syntectonic joints may vary between ~ 1 mm and ≥ 1 cm. The syntectonic joints are enriched with secondary mineralization. Fillings consist of a 1 cm thick dark, coarse-grained plates rich in gypsum, covered by a 1 mm layer of predominantly white fine-grained calcite with traces of aragonite, celestine, quartz and montmorillonite. T/S may increase

Figure 2a. Series of four concentric undulations on 327° joints cutting a Lower Eocene chalk layer about 0.5 m thick, which is bounded between chert beds about 0.07 m thick (marked by two horizontal arrows).

Figure 2b. A bilateral asymmetric plume 127 cm long, is confined within an elliptical arrest line in a chalk layer bounded between chert beds.

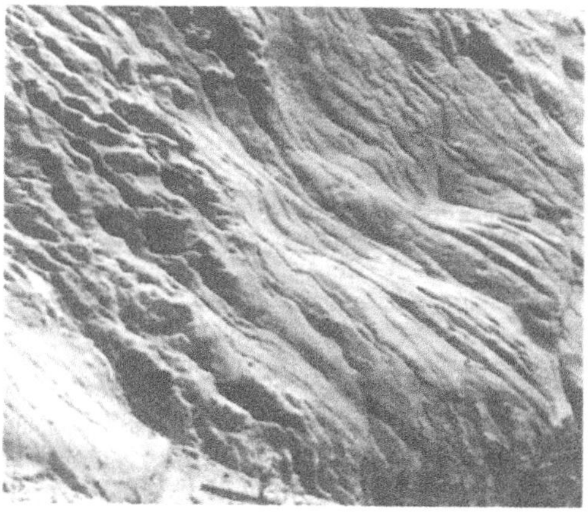

Figure 2c. Joints in a chalk layer more than 3 m thick in the Mishash Formation (Campanian) below strongly folded chert and chalk beds. Joint spacing is down to 10 mm. Note pen for scale.

Figure 2d. Concentric circular rib markings in Santonian chalk near 'Arad, southern Israel.

Figure 2e. Multi layer joints in a fracture zone cutting an outcrop 10 m thick. Scale bar is 30 cm.

Figure 2f. Some nine undulations convexing downward on a joint that cuts Middle Eocene chalks at Haelah Valley.

drastically with reduced distance from faults and intense folds. In the Ye'elim Valley outcrop T/S ranges between 10 and ≥ 300.

Generally, in thin layers S. L. initiate both at layer boundaries (Fig. 2a), where boundary effects become predominant, and from within the layer. However, the thicker the layer becomes, the less influental are boundaries, and joint initiation preferentially occurs around the middle of the layer (Fig. 2d).

2. 3 *Uplift joints*

Whereas S. L. are relatively small (height ≤ 1m) multi-layer joints (M. L.) are large (height > 5m). They penetrate through layer boundaries and cut outcrops consisting of many layers of both chalk and chert. These joints are straight and continous, without kinks at layer boundaries. M. L. appear in two styles. First, individual M. L. occur in large spacings (5-15 m) and are termed simple M. L. (S.M. L.). M. L. of the second style occur as fracture zones consisting of adjacent two to five joints which are closely spaced (1-5 cm) and occasionally are associated with considerable openings (≤ 1cm); these are termed fracture zone M. L. (F. Z. M. L.). Figure 2e shows alternating chalk layers and chert beds along a road cut in Lower Eocene chalks near Beer Sheva. S. L. (partly erroded by machines) arrest at chert beds. At centre are multi layer joints in a fracture zone cutting an outcrop 10 m thick. The wide F. Z. M. L. which cross chert beds are spaced only about 1 cm apart. Accordingly, T/S for S.M. L. is about 1. 9, but it may range between 10 and ≥ 100 for F. Z. M. L. (Bahat 1988a).

In correspondence with the general rule that joint spacing increases with layer thickness (Price 1966), the large height and spacing of S.M. L. indicate that they developed after the completion of the deposition of many layers in the formation, and the latter responded to fracture as a single "layer". Their late development, their penetration through chert, and their occasional openings imply fracture under intense tensile conditions. The F. Z. M. L. suggest fracture under intense pore pressures, not above the water table. It is interpreted that the M. L. were formed in association with the uplifts from Late Eocene (Benjamini 1984) through the Neogene. They are termed uplift joints.

Uplift joints rarely show fracture markings characteristic to burial or syntectonic joints. The following fracture markings are typical to uplift joints:

1) Rib markings on uplift joints are generally large (meters to tens of meters) compared to smaller rib markings on burial joints (centimeters to meters).

2) Typically, fracture markings initiate at upper parts of joints (Fig. 2f), not in association with layer boundaries.

3) Undulations on uplift joints generally convex downward, indicating vertical downward propagation rather than upward (Fig. 2f).

4) Overprinting of fracture markings may occur when uplift joints propagate along pre-existing joints (Bahat 1991).

2. 4. *Post Uplift Joints*

Post uplift joints, or exfoliation joints, occur in sets which form above the ground surface under controlled geomorphological conditions. They often mimic surfaces which have been formed by previous tectonic or erosional processes.

Post-uplift joints occur north of Beer Sheva (Bahat 1991, p. 301). They cross the local Lower Eocene chalks at 55° to 75° angles and their attitudes correspond to the local topography. Characterristically, they occur in composite fractures consisting of several individual joints that initiate separately from a common centre, and propagate approximately on the same plane to different directions. The composite fractures cut several layers. These joints display surface morphologies quite different from the fracture markings of most joints from the other three groups, suggesting different fracture mechanisms. Post uplift joints consisting of composite fractures were also chracterized in sandstone and in granite (Bahat et a. 1995).

3 JOINT CHARACTERIZATION

The four joint groups form under unlike geological setups while being exposed to different fracture conditions. Consequently, they attain distinct spatial and fractographic properties, as well as diverse joint characteristics. Four features which are implied from the above observations are exemplified below in connection with some applied problems.

First, it is important to establish, do joints propagate unidimesionally or bidimensionally. A wrong appreciation of the joint propagation mode may result in erroneous analyses of joint spacing (Wu & Pollard 1995). Roberts (1961) observed that flaggy beds normally have well developed plumes. Syme Gash (1971) suggests that a running fracture is dependent on the energy-to-thickness ratio: if the stratum or test plate is within the critical thickness, a running fracture with straight plume will result. But, if the stratum is too thick, fan-like plumes will develop.

A summary of a large number of fractographies which disclose the modes of joint propagation indicates that burial joints generally propagate in thin layers both radially (concentric undulations, Fig. 2a) and in one direction, parallel to layer boundaries (lateral plumes, Fig. 2b). Syntectonic joints often cut

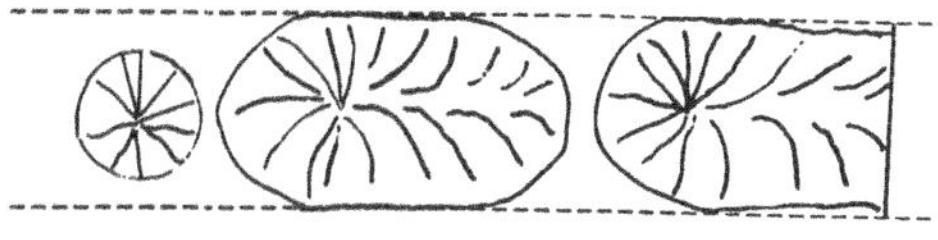

Figure 3. Three styles of joint propagation in a layer defined in Fig.1.

thick layers in concentric manners. Upift joints characteristicaly propagate vertically downward; and post uplift joints may propagate radially by the coalescence of earlier smaller joints (Bahat et al. 1995).

Bahat (1991a, p. 157) suggested a general scheme of joint propagation under various constraints. This scheme is summarized in Fig. 3. It shows that, the unconstrained crack propagates radially (Fig. 3a). On further growth, layer boundaries provide limits to crack width (Fig.1), Therefore, fracture length is unconstrained, and the joint proceeds to grow laterally (Fig. 3b), but may be arrested by an existing surface (Fig. 3c). Different fractures in a given layer that have the same width may have a considerable range of lengths (Bahat, 1988b). The implication is that there is no scale dependence between joint width and joint length, in contrast to scale dependence that was observed between fracture length and fracture thickness (aperture) (Toriumi & Hara 1995; Walmann et al. 1996).

Second, the maximum depth of a given fracture network is of great interest to the environmentalist. It is anticipated that networks which consist of burial joints that form by extension under conditions of increasing loading and intensifying effective stresses could extend to considerable depths (several km) (Secor 1965). Syntectonic joints show high T/S values,that reflect fracture under intense stresses, suggesting fracture under conditions of drastic pore pressures. Surfaces of these joints are decorated by rythmic concentric undulations which are centered in the middle of the layer. This geometry corresponds well with the model of periodic rise and fall of pore pressure during joint formation (Secor 1969). The secondary mineralization also testifies to intense flow of solutions along the joints. These joints should exist at depths much greater than fracture systems dominated by uplift that develop close to the ground surface (depths of several hundreds of meters), or post-uplift joints that develop above it.

Third, new criteria for the distinction between burial S. L. and uplift S. L. were recently defined (study in preparation). Burial S. L. of a given set generally display a uniform aperture, very often quite small (< 1mm) (Bahat, 1991, p.257). On the other hand, apertures of adjacent uplift S. L. of the same set may vary considerably (Fig. 4). Clearly, water

Figure 4. A set of S. L. uplift joints, showing several clossed incipient joints next to one with an aperture of several cm. Scale bar is 20 cm.

flow through these two fracture types may differe very significantly.

Fourth, joint intensity (given by the joint area per unit rock volume), is an important parameter for the construction engineer. A formation dominated by syntectonic joints will have a greater joint intensity than a formation cut by S. M. L., due to their differences in T/S values. However, to the environmentalist this should not necessarily suggest a corresponding better water drainage by the syntectonic joints, because they often have smaller apertures and they contain secondary mineralization; so that their "effective aperture" could be smaller compared to that of S. M. L. Furtheremore, uplift joints populated by many F. Z. M. L. would average large T/S values that could intensify liquid flow.

REFERENCES

Bahat, D.1988a. Early single-layer and late multi-layer joints in the Lower Eocene chalks near Beer Sheva, Israel. *Annal. Tecton* . 2: 3-11.

Bahat D.1988b. Fractographic determination of joint length distribution in chalk.*Rock Mech. Rock Engin.* 21: 79-94.

Bahat, D.1989. Fracture stresses at shallow depths during burial. *Tectonophysics* 169: 59-65.

Bahat, D.1991.Tectonofractography, Heidelberg: Springer-Verlag.

Bahat D. & N. F. Grossmann.1988. Regional jointing and paleostresses in Eocene chalks around Beer Sheva. *Israel J. Earth Sci.*. 37: 181-191.

Bahat, D., K. Grossenbacher & K. Karasaki 1995. Investigation of exfoliation joints in Navajo sandstone at the

Zion National Park and in granite at the Yosemite National Park by tectonofractographic techniques. *Lawrence Berkeley Laboratory* -36971; UC-400.

Bankwitz, P. & E. Bankwitz 1994. Event related jointing in rocks on Bornholm island (Denmark). *Z. geol. Wiss.* 22, 97-114.

Benjamini, C. 1984. Stratigraphy of the Eocene of the 'Arava Valley (Eastern and Southern Negev, Southern Israel). *Isr. J. Earth Sciences* . 33: 167-177.

Bentor, Y. K. and Vroman, A. 1960. The geological map of Israel, scale 1: 250,000, sheet 16, Mt. Sodom. 2nd edition, with explanatory notes, 117 pp., Government Printer, Jerusalem.

de Sitter, L. U. 1962. Structural development of the Arabian shield in Palestine. Geol. Mijnbouw 41: 116-124.

Engelder T (1985) Loading Paths to joint propagation during cycle: an example of the Appalachian Plateau,USA.J Struct Geol 7: 459-476.

Engelder, T., Fischer, M. P. and Gross, M. R. (1993) Geological aspects of fracture mechanics: *Geological Socciety of America, A short course manual note, Annual Meeting, Boston Mass.* 281 p.

Knauth, L. P. & D. R. Lowe 1978. Oxygen isotope geochemistry of cherts from the Onverwacht group (3.4 billion rears, Transvaal, South Africa, with implications for secular variations in the isotopic composition of cherts. *Earth Planet. Sci. Lett.*. 41: 209-222.

Price NJ (1966) Fault and joint development in brittle and semi-brittle rock,Pergamon Press,New York,pp 110-161.

Roberts J. C. 1961. Feather-fracture and the mechanics of rock jointing .*Am J Sci* . 259: 481-492.

Secor Jr, D. 1965. Role of fluid pressure in jointing. *Am. J. Sci.* 263: 633-646.

Secor Jr, D. 1969. Mechanics of natural extension fracturing at depth in the
earth's crust. In: *Research in tectonics* ,*Geol Surv Canada Paper*, pp 3-47.

Steinitz, G.1974. The deformational structures in the Senonian bedded cherts of Israel.*Ph.D.,The Hebrew University*,p 126

Syme Gash, P. J. 1971.Surface features relating tobrittle fracture.*Tectonophysics* 12: 349-391.

Toriumi, M. & E. Hara 1995. Crack geometries and deformation by the crack-seal mechanism in the Sambagawa metamorphic belt.*Tectonophysics* 245: 249-261.

Walmann, T., A. Malthe-Sorenssen, J. Feder, T. Jossang, T. & P.Meakin 1996. Scaling relations for the lengths and widths of fractures. *Phys. Rev. Lett.* 77: 5393-5396.

Wu, H. & D. D.Pollard 1995. An experimental study of the relationship between joint spacing and layer thickness.*J.Struct. Geol.* 17: 887-905.

Mechanics of Jointed and Faulted Rock, Rossmanith (ed.) © *1998 Taylor & Francis, ISBN 90 5410 955 6*

Mechanical behaviour of natural joints of granodiorite under high normal stress

G. Armand & M. Boulon
Laboratory of Soils, Solids and Structures, University Joseph Fourier, Grenoble, France
N. Hoteit
Direction Scientifque, Service Géomécanique, ANDRA, Châtenay-Malabry, France
S. Cannic
Institute Dolomieu, University Joseph Fourier, Grenoble, France

ABSTRACT: The authors have studied the mechanical behaviour of natural joints in granodiorite. These joints have been extracted from conventional cores and they are naturally cemented and filled by thin or thick layers of calcite. They have been tested by pseudo-static cyclic direct shear tests fully computer controlled. Despite the fact that these tests are 2D, they have been performed using a 3D direct shear box allowing to measure displacements and forces in two orthogonal directions of the plane of the joint. After presenting the experimental data, a synthesis will be presented in the framework of constitutive parameters, which is a first step for modelling the mechanical behaviour of jointed media.

1 INTRODUCTION

It is well known that the properties of rock masses in the upper part of the earth crust are strongly governed by the presence of rock discontinuities like cracks, joints, faults, etc. As long as life-time of underground storage constructions is studied, the behaviour of discontinuities becomes a field of interest requering both laboratory and in situ tests. Direct shear tests are commonly used to determine the behaviour of joints. The shear strength of clean joints has been discussed by many authors like Patton (1966), Barton (1976), Bandis et al (1981), Hutson and Dowding (1990).

But filled joints have been studied much less systematically despite they are widespread in the nature. Barton (1974) emphasised the role of clay fillings, and other authors like Papaliangas et al (1993) studied models of sandstone discontinuities filled with different thickness of dry pulverised fuel ash. Indraratna and Haque (1997) sheared regular saw-tooth joints with bentonite as an infill material between the joint sides. They showed that an increase of the fill thickness significantly reduces the shear strength and reduces the residual strength less markedly. The current study by the authors is an attempt for investigating the shear behaviour of real infilled joints of calcite in granodiorite masses.

2 DESCRIPTION OF ROCK MATERIAL

Direct shear tests were conducted on infilled joints of calcite in granodiorite.

2.1 *Sample preparation*

Joints have been extracted from conventional cores (9 cm of diameter) excavated in the area of Poitiers located in the center of France. The cores have been artificially maintained during sawing in order to prevent any relative displacement of the joint walls. The sample of rock is parallelepiped and the size of the joints is approximately 4 cm* 4 cm.

Screws were used for maintaining the sample in the internal boxes in order to adjust the average plane of the joint parallel to the plane of shearing. To allow the preparation of this closed joint, a removable brace was disposed between the upper and the lower internal boxes during grouting. The mortar used for grouting has an uniaxial compressive strength of about 50 MPa after five days.

2.2 *Mineralogical description*

The core samples correspond to granodioritic rocks. This kind of rock exhibits an homogeneous petrographic character. It shows porphyric texture and is characterised by the following igneous assemblage: quartz ± feldspar (Labrador) ± biotite ± amphibole ± Fe-Ti oxide. Weathering and/or low-grade metamorphism are responsible for the replacement of biotite and amphibole by chlorite, and feldspar by sericite.

The joints studied have been cut across the available cores. Two types of joints have been tested. The first type called thin joint, is characterised by a hydrothermal infill, of thickness smaller than 0.3 mm, and is composed by carbonate (calcite) chlorite

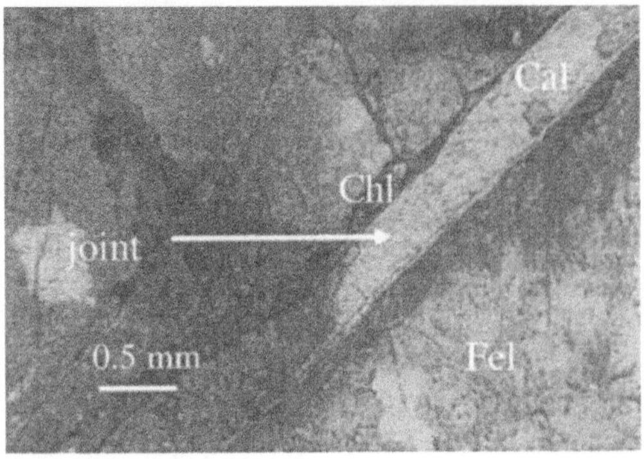

Photo 1. Thin section of a infill joint of calcite.
(Cal = Calcite, Chl = Chlorite, Fel = felspar)

and serecite. The second type, called thick joint has the same mineralogical composition, but is thicker (0.3 to 2 mm) (photo 1). It should be noticed that a third type exists, very thin (lower than 0.1 mm) and showing no observable infill. This third type has not been tested.

2.3 *Mechanical properties*

The most important physical and mechanical properties of granodiorite are as follows.
Uniaxial compressive strength: 100 to 120 MPa
Uniaxial tensile strength: 4 to 14 MPa
Young's modulus of elasticity: 68 to 70 GPa
Poisson's ratio: from 0.25 to 0.3
The joint roughness has not been measured before the test because joints were closed. The back calculated value of the JRC is about 8.1 to 14.3.

3 DIRECT SHEAR TESTING MACHINE AND TEST PROGRAM

The test were performed on a new 3D servo-controlled direct shear machine (BCR 3D). The concept of the BCR3D has been developed by Boulon (1995), and is exposed in a companion paper (Armand et al 1997). The principal characteristic of this apparatus is that the shearing is due to two symmetric movements of the two part of the joint, and so the normal force remains centered on the active part of the joint at any time. The shear velocity is prescribed using brushless motors and a PID computer control. By that way the precision of the shear velocity is better than 1 % even in stage of sudden rupture as displayed by the experiments.

Despite the fact that these tests are 2D , they have been performed with a 3D direct shear box allowing to measure displacements and forces in the third direction (direction of the joint orthogonal to the shearing direction). Many different tests can be conducted with this apparatus. Both tests at prescribed normal stress and at constant normal stiffness were carried out. The experimental program consists of 21 tests whose main data are reported in table 1. Each test has been conducted at a rate of shear deformation of 0.05 mm/s, at ambient temperature, the maximal shear displacement beeing equal to 10 mm. There was no prescribed displacement in the orthogonal direction subjected to a very low stiffness.

Table 1. Experimental program of direct shear tests on granodiorite joints

Number of tests	Initial Stress (MPa)	Normal Stiffness (MPa/mm)
3	5	0
4	10	0
2	15	0
6	20	0
2	5	70
2	5	120
1	10	120
1	15	120

4 TEST RESULTS

A high frequency data acquisition card has been used for recording the experience data during the tests. Data were recorded every 12 millisecond, leading to a very large number of measurement for each test. Then a special procedure of smoothing has been adopted. It was based on the moving average technique and the size of the local window of smoothing could grow from 1 to 51 points taken into account for averaging. So peak of shear stress isn't cut by this smoothing technique if the local windows size has been suitably choosen. Since it is impossible to represent full details of each individual test in this paper, only a few typical tests at prescribed normal stress test and at constant normal stiffness with thin and thick joints of calcite will be considered. Every sample came from the same core, but joints are called « thin » when the thickness is under 0.5 mm and the other are called « thick ». Figure 1 shows the evolution of the shear stress versus shear displacement during a test at prescribed normal stress on a thin joint. First the shear stress reaches a peak, then the joint breaks . Figures 2 shows the evolution of the shear stress on a thick joint, exhibiting a less important peak of shear stress. Figure 3 shows the evolution of the normal relative displacement versus the tangential relative displacement during the first test. First there is a slight compression due to normal loading. When the shearing displacement starts really a new closure appears. After that a dilatant behaviour takes place up to the breaking of the joint, and after that

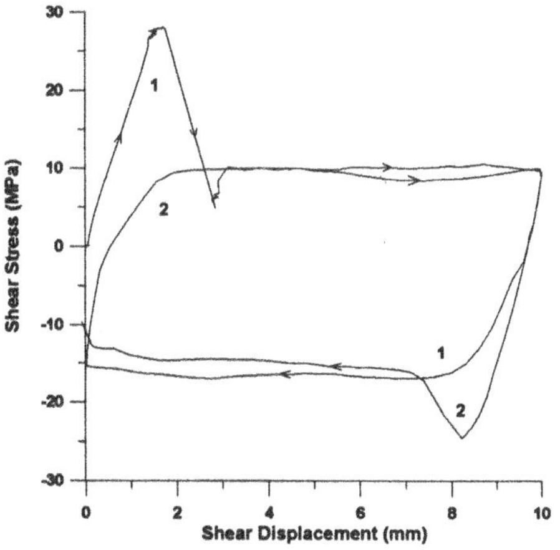

Figure 1. Typical response of a thin joint in granodiorite in terms of shear stress (Test nr 9 at prescribed normal stress, initial normal stress: 20 Mpa). Cycles nr 1 and 2.

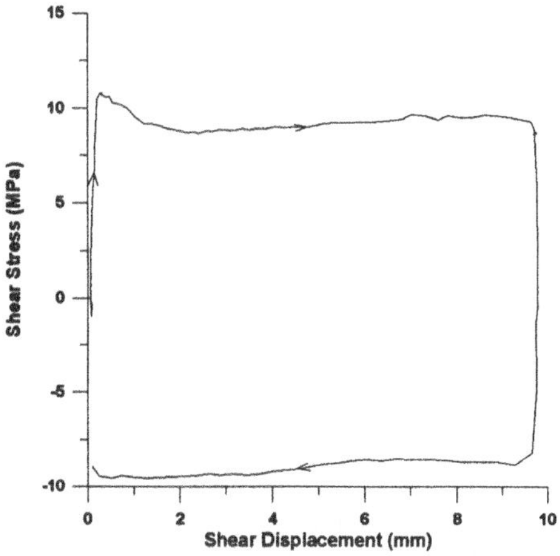

Figure 2. Typical response of a thick joint in granodiorite in terms of shear stress (Test nr 25 at prescribed normal stress, initial normal stress: 20 MPa)

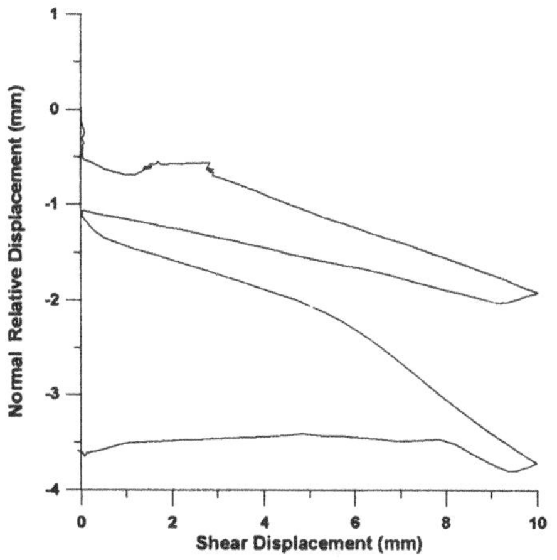

Figure 3. Typical response of a thin joint in granodiorite in terms of normal relative displacement (Test nr 9 at prescribed normal stress, initial normal stress: 20 MPa)

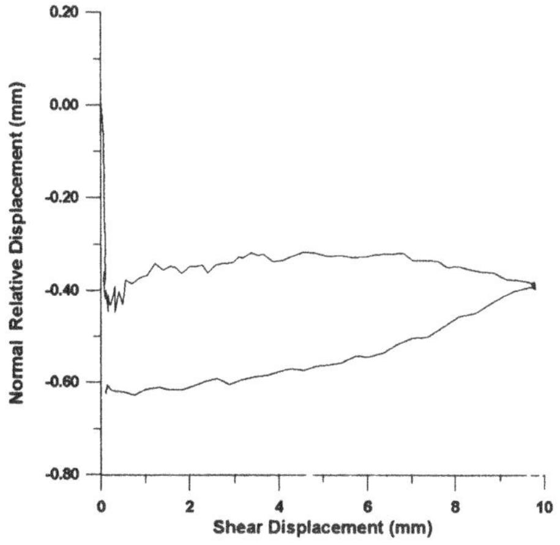

Figure 4. Typical response of a thick joint in granodiorite in terms of normal relative displacement (Test nr 25 at prescribed normal stress, initial normal stress 20 MPa)

contraction develops. For the thick joint, in figure 4, the same stages are repeated but the dilatancy phase continues until 8 mm of shearing. The rate of contraction corresponding to the end of the cycle is smaller than the rate of contraction for thin joints. For tests at constant normal stiffness, figure 5 shows the stress path for a thin joint in which the normal and shear stress grow highly (from 5 MPa to 22 MPa for the normal stress and from 0 to 25 MPa for the shear stress). As soon as the joint breaks both the normal and the shear stress decrease. Figure 6 shows the stress path during a test at constant normal stiffness with a thick joint. There is no real peak. The normal stress and shear stress increase and follow a Mohr-Coulomb line. Their maximum level is lower than for the thin joint (Normal stress : 14 MPa/22 MPa ; Shear stress : 8 MPa/25 MPa). Figure 7 shows the behaviour of a joint in the third direction. It is the shear displacement orthogonal to the prescribed shear displacement for the first cycle, related to four tests at prescribed normal stress (20 MPa). The forces were also recorded and were negligible (less than 0.4 MPa) because displacement were nearly free. As it could be seen in some cases, the orthogonal displacement is important (0.8 mm) for a cycle and in an other case less than 0.15

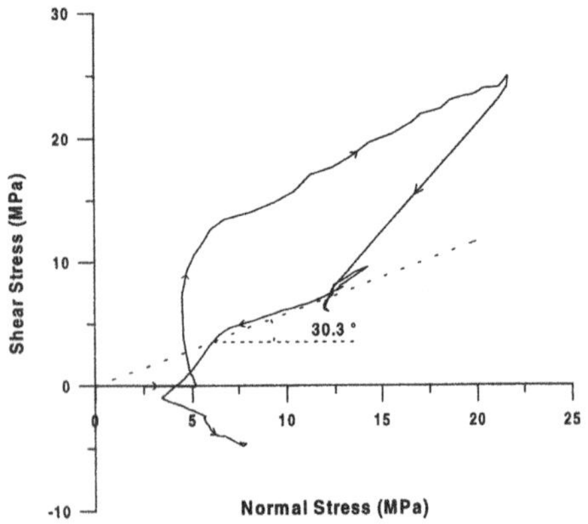

Figure 5. Typical response of a thin joint in granodiorite in terms of stress path (Test nr 34 at constant normal stiffness, initial normal stress 5 MPa, prescribed normal stiffness 70 MPa/mm)

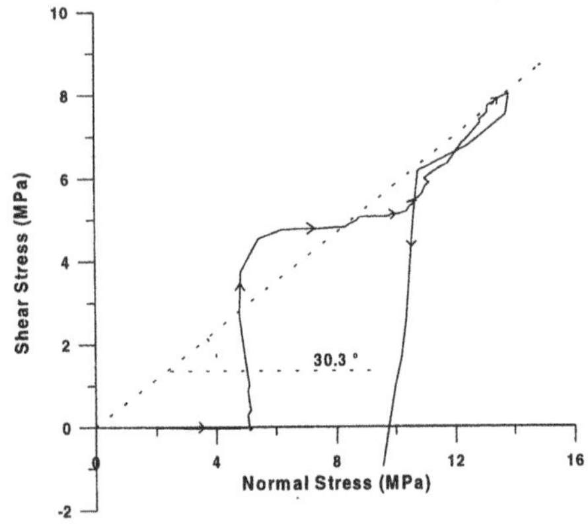

Figure 6. Typical response of a thick joint in granodiorite in terms of stress path (Test nr 24 at constant normal stiffness, initial normal stress 5 MPa, prescribed normal stiffness 120 MPa/mm)

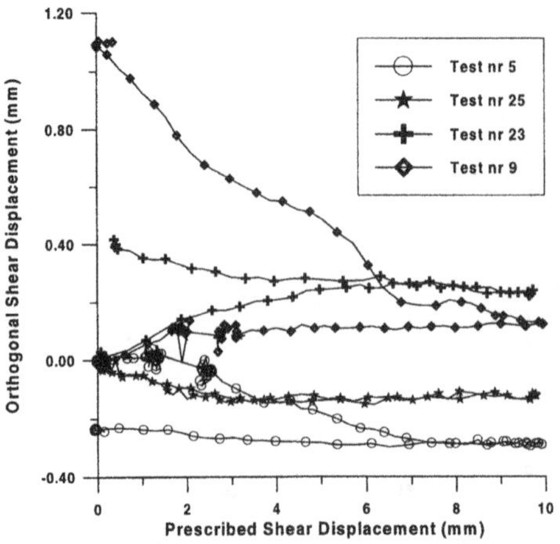

Figure 7. Typical response of joint in terms of orthogonal shear displacement versus prescribed shear displacement during tests at prescribed normal stress (20 MPa).

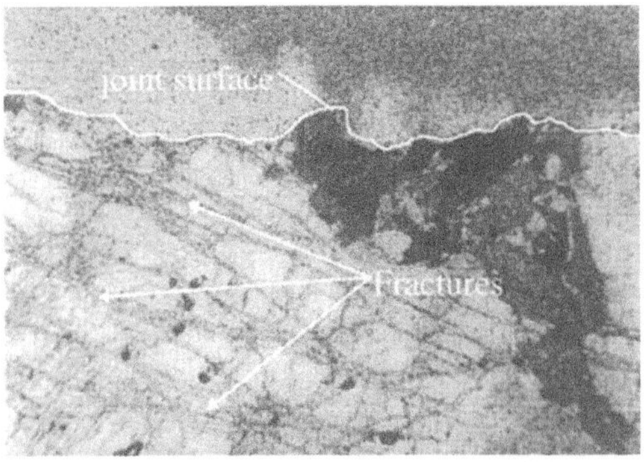

Photo 2. Thin section of an infilled joint after shearing under 20 Mpa, showing the development of a secondary fractures-system.

mm. These different behaviours show that the orthogonal shear displacement is due to the morphology of the joint, to the inclination of the average plane in orthogonal direction with the shear plane and to the joint roughness.

On the other hand, thin sections were made in order to understand the shear mechanism within the joint itself and in the volume of rock close to the joint. Structures cracks have been observed in two planes. The first one is oriented perpendicularly to the sheared joint and parallel to the shearing direction, whereas the second one is orthogonal to the sheared joint and to the shearing direction. The first effect of the mechanical loading is the reactivation of pre-existing joints. This reactivation corresponds to the displacement along the thin joints and do not depend upon their orientation. The second effect corresponds to the extraction of grains of quartz and feldspar. This extraction induces the formation of cavities along the wall of the sheared joint, while the extracted grains are crushed and form a cataclastic zone. These features could be related to the friction between the footwall and the hanging wall of the joint. Finally, the shearing induces the development of a secondary system of fractures. It has been only observed in the plane oriented parallely to the shearing direction. This system is composed of sub-parallel fractures, which make an angle of 10° to 20° with the sheared joint (Photo 2). Such induced structures are located at the vicinity of the sheared joint. However, the thickness of the wall, affected by the shearing, increases

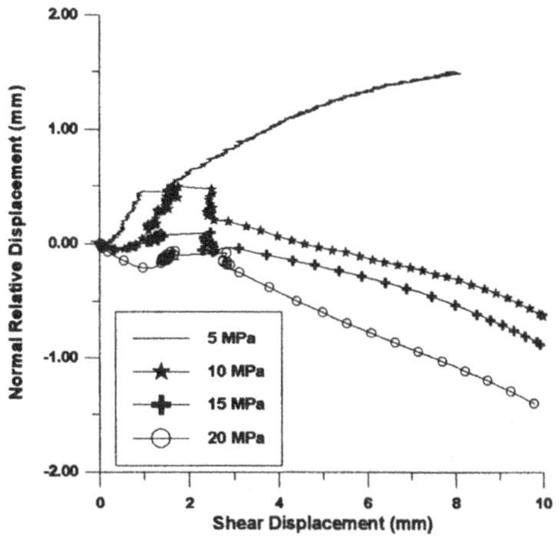

Figure 8. Direct shear tests at prescribed normal stress on thin joints in granodiorite. Normal relative displacement versus tangential relative displacement.

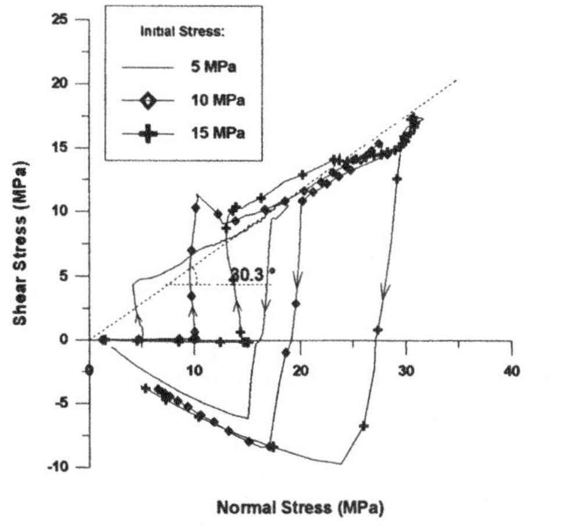

Figure 10. Direct shear test at constant normal stiffness on thick joint in granodiorite, stress path (Normal stiffness 120 MPa/mm).

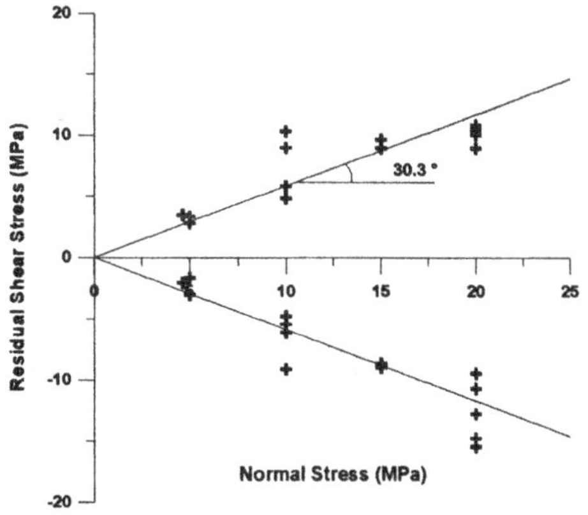

Figure 9. Direct shear test at prescribed normal stress on joints in granodiorite. Residual value of shear stress as a function of the normal stress.

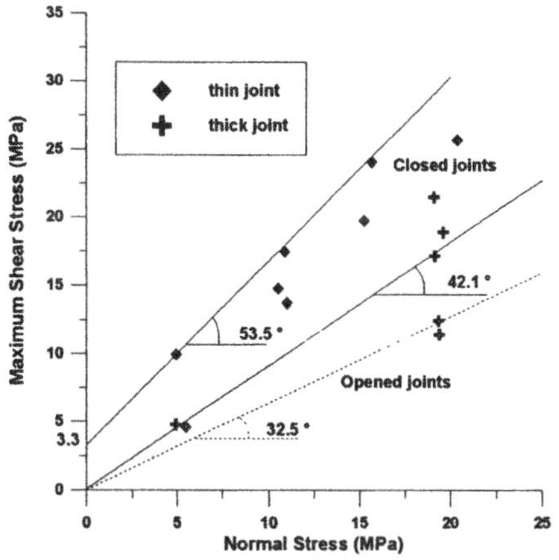

Figure 11. Direct shear test at prescribed normal stress on joints in granodiorite. Peak of shear stress as a function of the normal stress.

with the intensity of the normal stress. For example, 5 mm of each rock wall have been affected by the reactivation of joints for a normal stress of 10 MPa, while about 10 mm of each wall are affected under 20 MPa.

5 DISCUSSION

Typical curves are presented in section 4, but the effect of normal stress is not synthetised. In fact with a normal stress of 5 MPa acting under thin joints, the curve of shear stress have the same shape, but the curves of relative normal displacement are different as it appears in figure 8. The dilatancy occurs along all the first part of the cycle even after the rupture of the joint. This dilatancy variation is well known and has been reported by many authors. The whole sample follows the mechanism of matched joint with irregular surfaces described by Archambault et al (1997). The first phase is an elastic mobilisation of shear stress by friction, with a new closure of the joint. After there is a mobilisation phase to peak shear with an increasing dilatancy. Then the joint is broken and the shear stress collapses to the residual shear stress. If the normal stress is 5 MPa, dilatancy occurs during the residual phase of stable sliding, but if the normal stress is greater then contractancy occurs during the end of the test. Archambault says that the dilatancy rate must be zero or near zero for low and very low normal load and tends to disappear with shear displacement. Here contractancy appears because, as a thin section shows it, there is a degradation of the joint surface but also a degradation of the wall masse in an important zone around the joint. An other aspect is that the size of the joint is small and the shear displacement is 25% of the wall length, then material is squeezing out during shearing. For thick joints, the same things appears but the contact area of the two sides of the joint is less important due to the calcite thickness, then the disturbed zone around the joint is less developed and the contractancy rate becomes smaller. As it has been described in the literature for infilled joints, the fill thickness of calcite reduces significantly the shear strength and the initial dilatancy. For the residual strength, this influence is not measurable as it can be seen in figure 9, in which all residual values of strength are plotted. The residual friction angle defined by a Mohr Coulomb model is about 30.3°. It' s near the residual angle given by Barton (31° to 35°) for granite joints. With tests at constant normal stiffness on thick joint, the residual value is the same (figure 10). For the peak of shear strength, figure 11 shows the peak of stress for every joints tested. Some joint which are called open joint, were broken prior the shearing, then the peak of shear stress is less important. It is difficult to define an average value of the peak of shear stress versus normal stress, so a zone is defined delimited by two Coulomb lines. The first line shows a small initial cohesion of 0.1 MPa and a friction angle of 42.1°. The second line has an initial cohesion of 3.3 MPa with a friction angle of 53.5 °. Thin joints are near the second line and thick joint are near the first line. But the effective joint thickness has probably to be defined taking into account the presence of the fractured zones near the joint itself. The stiffness parameters of the joints (Normal and shear stiffness) are reported in table 2. It should be noticed that the normal stiffnesses of the thin and thick joints are not very different. On the opposite, there is a big contrast between the shear stiffnesses.

Table 2. Stiffness parameters of the joints

Type of Joint	Normal stress (MPa)	Normal stiffness (MPa/mm)	Shear stiffness (MPa/mm)
thin	5	14.3±2.5	8.9
	10		16.9
	15		21.6
	20		23.2
thick	5	37.7±2.3	80.7
	10		227
	15		250.2
	20		286.3

CONCLUSIONS

The tests reported here were conducted on real infilled joints of calcite in granodiorite, so some aspects like infill thickness cannot be well defined. On the other hand the roughness of the joint has not been measured and its influence couldn't be well understood. But the analysis of thin sections allowed a better understanding of mechanism occurring due to the shearing, like extraction of quartz and feldspar, the grains crushing , reactivation of pre-existing joints, and the development of a secondary system of fractures. All this events are more or less due to the stress level. The infilled thickness plays an important role in the behaviour of infilled joints. The shear strength at peak decreases with the thickness, and the dilatancy decreases to. For this series of tests the peak of strength varies in the range {(cohesion: 0.1 MPa, friction angle 42.1°) - (cohesion 3.3 MPa, friction angle 53.5°)} and the residual friction angle is about 30.3°±2.6°.

REFERENCES

Armand G., Boulon M., Papadopoulos C., Vardoulakis I. P., 1998. Mechanical behaviour of Dionysos marble smooth joints : I Experiments, Proc. of the MJFR-3 Conf., Wien, Austria

Archambault G., Gentier S. 1997. Riss J., Flamand R., The evolution of void spaces (permeability) in relation with rock shear behaviour, IJRMMS, 34:3-4, paper No. 014.

Bandis S.C., Lumsden A. C. and Barton N. R. 1981. Experimental studies of scale effects on the shear behaviour of rock joints, 18, 1-21.

Barton N. R. 1976. The shear strength of rock joints. IJRMMS, 13, 255-279.

Barton N. R. 1974. A rewiew of the shear strength of filled discontinuities in rock. Norwegian Geotechnical Institute Pub. No 105, pp. 38.

Boulon M. 1995. A 3D direct shear device for testing the mechanical behaviour and the hydraulic conductivity of rock joints. Proc. of the MJFR-2 Conf., Wien, Austria, Rossmanith Ed., Balkema,pp. 407-413.

Hutson R. W., Dowding C. H. 1990. Joint asperity degradation during cyclic shear, IJRMMS, 27, No2. pp 109-119.

Indraratna B., Haque A. 1997. Experimental study of shear behaviour of rock joints under constant normal stiffness conditions, IJRMMS, 34:3-4, paper No. 141.

Papaliangas T., Hencher S. R., Lumsden A. C., Manolopoulous S. 1993. The effect of frictional Fill Thickness on the shear strength of rock discontinuities, IJRMMS, Vol 30,No 2, pp 81-91.

Patton F.D. 1966. Multiple modes of shear failure in rock and related materials. Ph.D Thesis, university of Illinois, 282 pages.

Mechanics of Jointed and Faulted Rock, Rossmanith (ed.) © 1998 Taylor & Francis, ISBN 90 5410 955 6

The effect of shear displacement on the void geometry of aperture distribution of a rock fracture

In Wook Yeo & Robert W. Zimmerman
Department of Earth Resources Engineering, Imperial College, London, UK
Michael H. de Freitas
Department of Civil Engineering, Imperial College, London, UK

ABSTRACT: Aperture replicas of a natural sandstone fracture were made at 0 mm, 1 mm, and 2 mm shear displacements. Both the number of contact points and the fractional contact area decrease with increasing shear displacement. As shear displacement increased, the ratio of standard deviation to mean aperture increased slightly, and the aperture became more closely correlated in the direction parallel to the roughness ridges than in the direction of shear.

1 INTRODUCTION

The hydraulic and mechanical properties of a rock fracture are to a great extent controlled by the geometry of the void space - in particular, by the distribution of apertures and contact areas of a rock fracture. The void geometry, and hence the hydraulic conductivity of the rock fracture, is in turn strongly influenced by the displacement history of the rock fracture. Experiments have shown that a few millimetres of shear displacement can cause the hydraulic conductivity to increase by two orders of magnitude (Olsson & Brown, 1993), and make the fracture become an anisotropic medium to fluid flow (Gentier et al., 1997).

Only a few researchers have studied the distribution of contact areas and apertures in a rock fracture under a normal stress (Pyrak-Nolte et al., 1990, Hakami, 1995). Furthermore, due to shear displacement, laboratory investigations of the changes in void geometry, and consequently anisotropic hydraulic properties, have not thus far been made. For a better understanding of the relationship between void space geometry, applied stress, and hydraulic conductivity, complete knowledge of the details of the fracture void geometry is required.

2 EXPERIMENTAL APPARATUS

Two epoxy resin replicas were made and fitted together to form matched copies of the natural rock fracture. To make epoxy resin replicas of a rock fracture, red sandstone of Permian age was chosen, whose surface could be seen to have anisotropic roughness. The ridges of roughness were used to define the x direction. The experimental device was designed to conduct radial and unidirectional flow tests on a rock fracture that is subject to normal loading and shear displacement. A 20 cm × 20 cm fracture specimen is placed into a shear box. Holes are drilled through the top half of the fracture replica, terminating at the fracture, so as to monitor hydraulic head on the fracture (Yeo et al., 1996).

A loading platform supported on four long legs is placed on the upper half of the fracture; it is designed to apply normal load to the fracture by simply putting weights on the platform. Four side plates were fixed to the lower shear box by screws, to precisely locate the position of the upper shear box and to ensure the exact relative location of the rock surfaces, and shear boxes, during aperture measurements and flow tests at 0 mm, 1 mm and 2 mm shear displacement. Exact shear displacement of the upper shear box relative to the lower box was required, as this displacement had to be reproduced exactly on a number of occasions. The amount of shear is controlled by precisely-dimensioned metal gauge blocks inserted between the side plate and lower shear box, and between the side plate and upper shear box on opposite sides, ensuring zero, 1 mm and 2 mm shear displacement without rotation, when required (refer to Yeo, 1997 for details). The present paper focuses on the measurements of void geometry due to shear displacement.

3 MEASUREMENT PROCEDURE

The casting method was adopted for measuring the aperture in this experiment. Silcoset, which is a liquid silicone rubber that cures to form a resilient, solid silicone rubber with the addition of curing agents, was used to make a replica of the void geometry of the rock fracture. The aperture replicas were made under a constant normal load of 10 kPa, with three different shear displacements: 0 mm, 1 mm and 2 mm. Shear displacements were applied along the *y* direction, which is normal to the ridges of roughness. To make aperture replicas, the shear boxes were first positioned accurately using the side plates to ensure 0 mm, 1 mm and 2 mm shear displacements as required: the lower specimen was in its lower shear box at this time. Silcoset was then poured on the surface of the lower half of the fracture, the upper half of the rock fracture was placed on the lower surface, and the loading platform was then placed on the upper half of the fracture to apply a normal load of 10 kPa. Monitoring holes and drainage holes were open to allow excess Silcoset to flow out of the fracture to prevent air bubbles from being entrapped. The Silcoset solidified in 24 hours. After separating the rock fracture, the solidified Silcoset replica was removed.

In the aperture measurement, it is imperative that the upper and lower surfaces of the rock fracture have the same exact relative position throughout the aperture measurements and flow tests. In order to investigate the reproducibility of this relative positioning and of the aperture replicas themselves, duplicate casts were made for each shear displacement: 0 mm, 1 mm, and 2 mm.

Aperture replicas were scanned to determine the void geometry, the contact area and its spatial distribution. Each aperture replica was sliced at every 5 mm in the *y* direction, except regions close to monitoring holes, into forty strips. Aperture is defined as the vertical thickness at each known coordinate. Apertures were measured at every 5 mm interval along these sections as a function of location, using the microscope of the "MINILOAD hardness tester" made by Leitz Wetzlar. A micrometer with increments of 1 μm is affixed to the measuring stage, allowing the aperture to be measured to within ± 1 μm. These measurements were carried out at 1600 known *x* and *y* coordinates for each shear condition.

4 REPRODUCIBILITY

The reproducibility of the replicas was measured by comparing their weights of duplicate aperture replicas. Table 1 shows the weights of aperture replicas for different shear displacements. Weight comparisons between duplicate aperture replicas indicate that the replicas were reproduced within less than 2% error, and some of this error was due to the precision with which the surplus Silcoset (which existed at the sides and in the monitoring holes used for drainage of Silcoset) was cut away. Weights were converted into mean apertures using the known density of Silcoset, 1.19×10^3 kg/m^3. Mean aperture shows very good agreement with that measured using the microscope, indicating that the mean aperture of a rock fracture can be easily and accurately measured using aperture replicas made with Silcoset.

Table 1. Reproducibility of aperture replicas as estimated by comparing weights (columns 2, 3); comparison of mean aperture calculated based on the density of the Silcoset with that measured by microscope (columns 4, 5).

Shear Displa.	Sample	Weight (g)	Mean aperture (μm) by weight	Mean aperture (μm) by microscope
0 mm	1	28.54	599.58	
	5	29.05	610.29	606.55
1 mm	4	33.87	711.56	711.96
	8	34.17	717.79	
2 mm	6	47.91	1006.51	
	7	48.84	1026.05	1024.16

5 APERTURE DISTRIBUTION

Fig. 1 shows scanned photographs of aperture replicas. Because the translucency of Silcoset depends on thickness, contact points are clearly distinguished as black spots, and lighter regions correspond to larger aperture. With increasing shear displacement, the aperture replicas become lighter, indicating that the aperture increases due to dilation of the fracture. The lighter regions, i.e., larger aperture regions, can be presumed to be the paths preferentially taken by the fluid, because the fluid takes the paths of least resistance. The aperture replicas show that at zero shear displacement, larger and smaller apertures are evenly distributed over the fracture, indicating that the fracture is an isotropic medium. The fracture becomes an anisotropic hydraulic medium with increasing shear displacement because valleys having smaller apertures exist at the middle of the fracture, and also at the northern edge along the direction normal to the shear displacement. The number of contact points and fractional contact area decreases, although the fractional contact area is

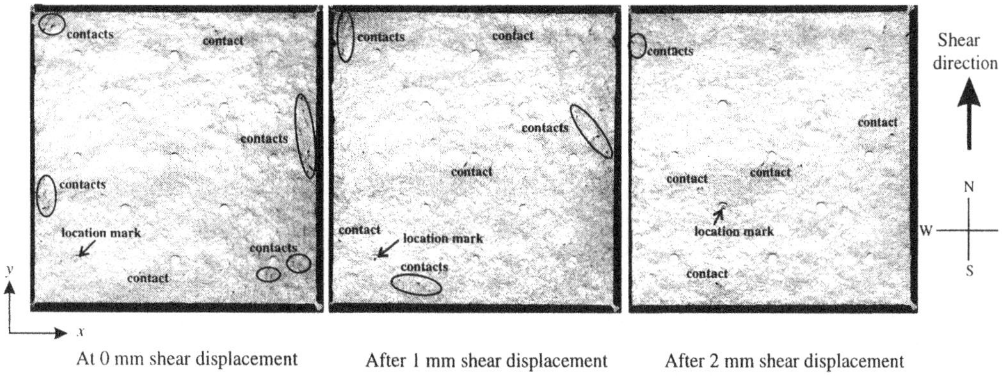

Fig. 1. Aperture replicas showing void geometry of the fracture under different shear displacements. Lighter regions correspond to larger aperture; contact regions are indicated by the small black spots.

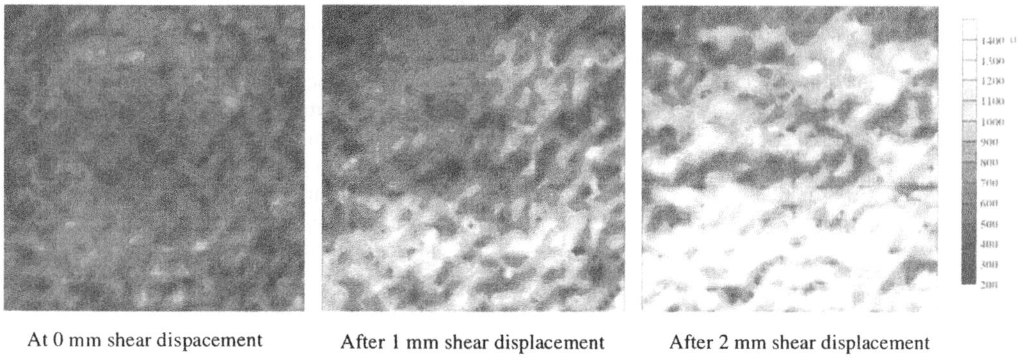

Fig. 2. Distributions of apertures measured at different shear displacements.

too small to quantify, as shear displacement increases. Considering that the aperture replicas were made at a very low normal stress of 10 kPa, it is not surprising that only a few contact points support the fracture (Fig. 1). Iwai (1976) showed that the fractional area of the fracture plane that is occupied by contact was 0.001 at 260 kPa of effective normal stress, and increased to the range of 0.1 – 0.2 at 20 MPa.

Aperture replicas of samples 5, 4, and 7 were selected to measure apertures for 0 mm, 1 mm and 2 mm shear displacement, respectively. Fig. 2 shows aperture distributions for each shear displacement, created by the microscopically measured apertures. Before shear displacement, larger apertures are distributed in the northern and southern regions of the fracture trending in the x direction, and along the y direction at the east side of the fracture. Smaller apertures are located at the western and eastern edges, which corresponds to the regions having contact points in Fig. 1. With increasing shear displacement, larger aperture regions are located at the southern part of the fracture. As observed in Fig. 1, the smaller apertures form the valleys, which trend in the x direction. The fracture becomes inclined towards the north. The fact that there is a relatively large contact point at the southern part (Fig. 1) refutes the possibility that the upper half of the fracture has tilted when shear-displaced. These results suggest that the fracture becomes more anisotropic with regard to fluid flow with increasing shear displacement. With respect to the concept of a "representative elementary volume", a homogeneous medium does not exist in this fracture after 1 mm and 2 mm shear displacements (possibly not even at zero shear displacement).

Mean aperture and standard deviation increase with increasing shear displacement (Table 2), as does the ratio of standard deviation to mean aperture. This ratio shows a greater increase from 0 mm to 1

Table 2. Statistics of apertures measured at 0 mm, 1 mm and 2 mm shear displacement.

Shear displacement	0 mm	1 mm	2 mm
mean, <e> (μm)	607	712	1024
Stan. dev., s (μm)	160	221	332
s/<e>	0.26	0.31	0.32
Max. aperture (μm)	1217	1496	1083
Min. aperture (μm)	59	53	84

mm shear displacement than from 1 mm to 2 mm shear displacement. On the other hand, mean aperture increases sharply when the fracture is shear-displaced from 1 mm and to 2 mm. The frequency histograms are generally well-fitted by a normal distribution (Fig. 3).

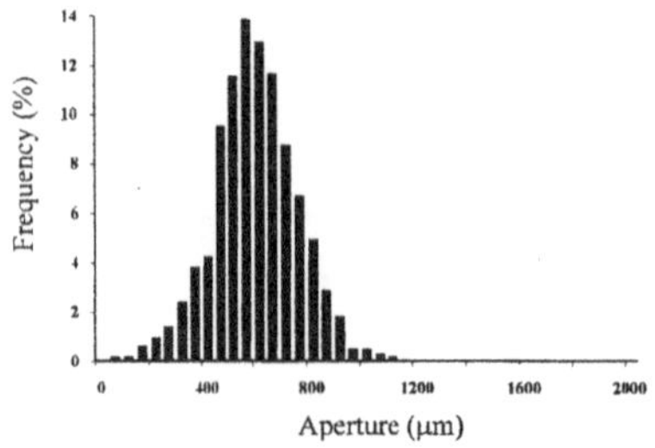

At 0 mm shear displacement

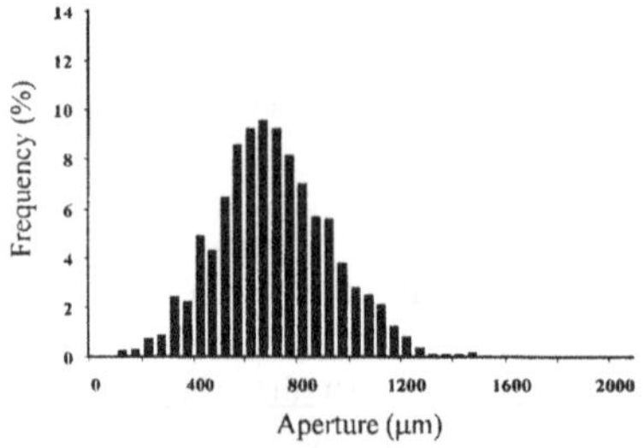

After 1 mm shear displacement

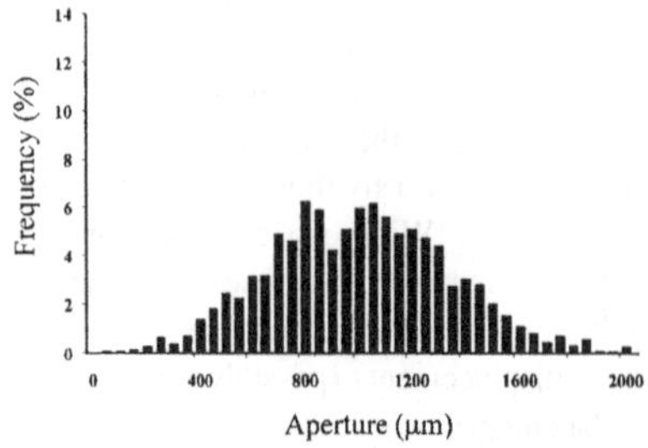

After 2 mm shear displacement

Fig. 3. Frequency histograms of apertures measured at 0 mm, 1 mm and 2 mm shear displacements.

6 SPATIAL CORRELATION STRUCTURE

Semivariograms have been used to study the spatial correlation structure (Iwano & Einstein, 1993, Hakami, 1995). Semivariance $\gamma(l)$ is calculated using the following equation (Clark, 1979):

$$\gamma(l) = \frac{1}{2n}\sum_{i=1}^{n}[e(x_i) - e(x_{i+l})]^2, \qquad (1)$$

where l is the lag distance, n is the number of observation pairs, $e(x_i)$ is the aperture taken at location x_i, and $e(x_{i+l})$ is the aperture taken l intervals away. If the lag distance is small, the compared apertures tend to be very similar, consequently leading to small semivariance values. As the apertures being compared move further apart, they become less correlated to each other, and their difference becomes larger, which results in a larger semivariance value. At some lag distance the apertures being compared are too far apart to be correlated to each other, and their squared differences become equal in magnitude to the variance of apertures. The sill is defined as a flat level that the semivariogram develops when the semivariance no longer increases. Ideally, the sill is equal to aperture variance. The lag distance at which the semivariance approaches the sill is referred to as the "range", or the "spatial correlation length", and defines a neighbourhood within which all apertures are closely correlated to each other.

Semivariance was calculated both omnidirectionally, and in the x and y directions, for each shear displacement using the computer package VARIOWIN (Pannatier, 1993), so as to investigate the directional structures of spatial correlation (Fig. 4). At 0 and 1 mm shear displacements, semivariograms do not develop clear sills for either the omnidirectional case or the y direction, resulting in difficulties calculating ranges. Even though the flat sills are weakly developed at the x direction, the high "nugget effect" (where the semivariogram does not go through the origin but rather has a nonzero value) obstructs the accurate determination of ranges. The ratio of the value of the nugget effect to the flat sill is over 0.6, suggesting that, although modelled by a semivariogram, more than 60% of the aperture correlation structure is unpredictable (Clark, 1979).

At 2 mm shear displacement, the flat sills are seen to develop for all directions. Although the flat plateau is lower than the variance of apertures (shown by the dotted horizontal line in the figure), it was used as a sill to model the semivariograms. The semivariogram for the omnidirectional case can be fitted by a spherical model with a sill of 82500 μm^2 and a range of 16 mm. In the x and y directions, the

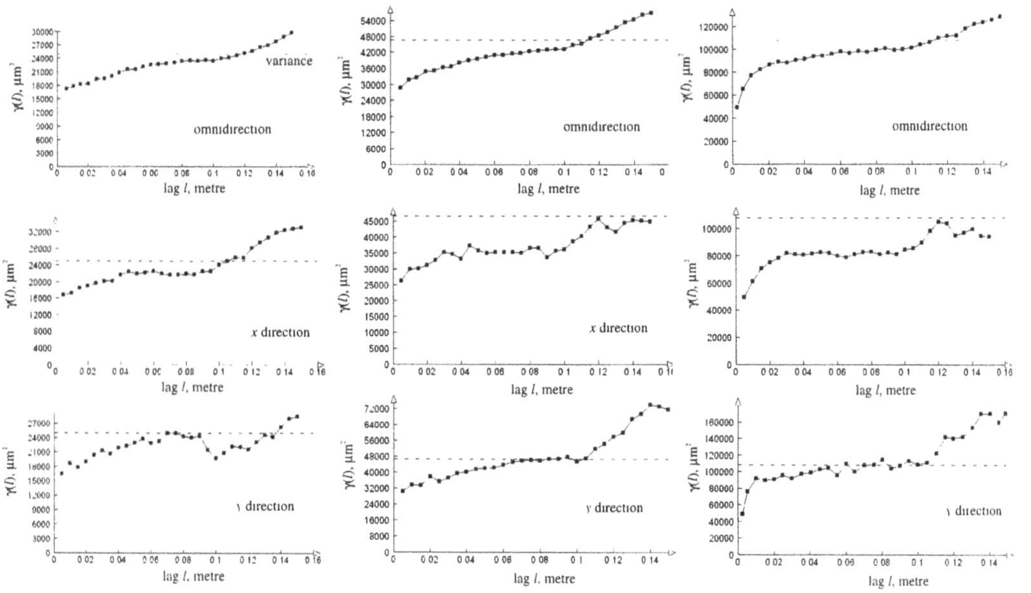

At 0 mm shear displacement | After 1 mm shear displacement | After 2 mm shear displacement

Fig. 4. Semivariograms for spatial correlation structure of apertures. Dotted horizontal line represents the variance of the apertures. x axis is lag l (m), y axis is γ (μm^2).

semivariogram can be well-described by a spherical model, with a 92400 μm^2 sill and 24 mm range, and by a spherical model with a 97900 μm^2 sill and 12 mm range, respectively. These different values of range show that the aperture is much more closely correlated in the x direction than in the y direction, which is confirmed by Figs. 1 and 2 which also show that the apertures are aligned in the x direction at 2 mm shear displacement.

Because accurate ranges cannot be obtained in both the x and y directions at 0 mm and 1 mm shear displacements, and in an ideal case, the variance is considered to be equal to the sill, the effect of shear displacement on the anisotropy of the spatial correlation structure was studied using the relationship between semivariograms and the variance of apertures. At zero shear displacement, semivariograms are not clearly distinguished from one another in either the omnidirection, or in the x and y directions, indicating that the fracture has a similar spatial correlation structure in all directions. As shear displacement increases, semivariograms in the x direction lie lower compared with the variance of apertures than those in the y direction, suggesting that the aperture distribution become more closely correlated in the x direction than in the y direction, with increasing shear displacement.

7 CONCLUSIONS

The aperture replicas were reproduced to within less than 2% error. Mean aperture calculated by weighing the aperture replica showed very good agreement with that measured by microscope. Aperture replicas using Silcoset visually provided information on the paths taken by the water as it flows through the fracture. As shear displacement increased, the number of contact points and fractional contact area decreased. Aperture measurement showed that the ratio of standard deviation to mean aperture increased slightly from 0.26 to 0.32 with increasing shear displacement. Frequency histograms were well-fitted by a normal distribution.

Spatial correlation structure studies using the semivariogram indicated that as shear displacement increased in the direction normal to the roughness ridges aligned along the x direction, the apertures became more closely correlated to each other in the x direction rather than in the y direction. Both the semivariogram study and the scanned photographs of the aperture replicas implied that the fracture would become an anisotropic hydraulic medium when sheared.

The work reported in this paper is part of a larger research programme aimed at studying the effect of shear displacement on aperture distribution and permeability. The measured apertures were used as input data for numerical modelling of flow through the fracture, using the Reynolds equation. Radial and unidirectional flow tests were conducted at 0 mm, 1 mm and 2 mm shear displacement. The results of the flow tests were compared to the predictions of the numerical simulations. For more details on this, refer to Yeo (1997).

ACKNOWLEDGEMENTS

This work was supported by the Ministry of Education of the Republic of Korea, by a UK Overseas Research Students Award, and by a grant from the Natural Environment Research Council, UK (Grant GR9/02289). This support is gratefully acknowledged.

REFERENCES

Clark, I. 1976. *Practical Geostatistics*, Elsevier Applied Science Publishers, London.

Gentier, S., Lamontagne, E., Archambault, G. & Riss, J. 1997. Anisotropy of flow in a fracture undergoing shear and its realtionship to the direction of shearing and injection pressure. *Int. J. Rock Mech.* 34:3-4, Paper No. 094.

Hakami, E. 1995. *Aperture Distribution of Rock Fractures*. PhD thesis. Royal Institute of Technology, Stockholm, Sweden.

Iwai, K. 1976. *Fundamental Studies of Fluid Flow through a Single Fracture*. PhD thesis, University of California, Berkeley.

Iwano, H. & Einstein, H. H. 1993. Stochastic analysis of surface roughness, aperture and flow in a single fracture. in *Proc. Int. Symp. EUROCK '93*, L. Ribeiro e Sousa and N. F. Grossmann (eds.), Balkema, Rotterdam, 135-141.

Olsson, W. A. & Brown, S. R. 1993. Hydromechanical response of a fracture undergoing compression and shear. *Int. J. Rock Mech.* 30, 845-851.

Pannatier, Y. 1993. MS-Windows programs for exploratory variography and variogram modelling in 2D, in *Statistics of spatial processes: theory and applications*, Bari, Italy, Sep. 27-30, 1993, 225-231.

Pyrak-Nolte, L. J., Nolte, D. D., Myer, L. R., & Cook, N. G. W. 1990. Fluid flow through single fractures, in *Proc. Int. Symp. Rock Joints*, N. Barton and O. Stephansson (eds.), Balkema, Rotterdam, 405-412.

Yeo, I. W., Zimmerman, R. W. & de Freitas, M. H. 1996. Design and analysis of an experimental apparatus to measure directional permeabilities of a rock fracture under normal and shear loading. in *Proc. Int. Symp. EUROCK '96*, G. Barla (ed.), Balkema, Rotterdam, 1223-1227.

Yeo, I. W. 1997. *Anisotropic Hydraulic Properties of a Rock Fracture under Normal and Shear Loading*, PhD thesis. University of London (Imperial College), London, UK.

Mechanics of Jointed and Faulted Rock, Rossmanith (ed.) © *1998 Taylor & Francis, ISBN 90 5410 955 6*

Development of shear testing equipment to investigate the creep of discontinuities in hard rock

U.W.Vogler, D.F.Malan & K.Drescher
Division of Mining Technology, CSIR, Johannesburg, South Africa

ABSTRACT: This paper describes the development of shear equipment to test the creep of discontinuities in hard rock. The shear box was constructed from an existing soil shear box which allows for easy and cost-effective replication of the equipment. As the creep rate of some discontinuities in hard rock is small, the design included a novel arrangement of six displacement transducers that allowed for measurements directly on the rock surface. This prevented the data from being obscured by creep components of the epoxy used to mount the samples. Finally, the developed data acquisition unit and testing methodology are discussed and typical creep results are illustrated.

1 INTRODUCTION

Knowledge of the mechanical properties of rocks is essential to understand and simulate the behaviour of underground excavations. The long-term stability of these excavations is determined by the time-dependent behaviour of the rock mass. Vast quantities of laboratory data for the creep of intact rock is available (Lama & Vutukuri 1978). The closure rate of deep excavations in hard rock is however much larger than can be expected for the creep of intact rock and significant time-dependent movements can be observed within a period of hours or days (Malan 1997, Malan et al. 1997). This is an indication that the time-dependent behaviour of rock on an engineering scale is dominated by other mechanisms. Many workers (Tan & Kang 1980, Bowden & Curran 1984) suggested that creep deformations surrounding excavations may be governed by the rheological properties of discontinuities. Unfortunately very limited research has been done on the shear creep behaviour of discontinuities. (In this paper the shear creep of discontinuities is defined as the time-dependent relative displacement of opposing discontinuity surfaces caused by a shear stress smaller than the shear strength of the discontinuity.) One factor contributing to this lack of knowledge is the unavailability of suitable testing equipment. Some workers (Wawersik 1974, Solberg et al. 1978, Schwartz & Kolluru 1984) investigated the shear creep of discontinuities by using inclined discontinuities in rock specimens loaded in triaxial or uniaxial testing equipment. A drawback of this method is that the normal and shear loads applied to the discontinuities cannot be adjusted independently. Although some laboratories are equipped for direct shear tests on rock, not all of the available shear equipment is suitable for creep testing, as constant shear and normal stresses need to be applied over long time periods. Bowden & Curran (1984) used a specially designed bi-directional shear machine to investigate the creep of joints in shale. A hydraulic system with accumulators was used to maintain constant loads over long periods.

This paper describes the development of cost-effective shear equipment with an innovative displacement transducer design suitable for measuring the creep of discontinuities in hard rock.

2 TESTING REQUIREMENTS

A study of the requirements for measuring the long-term creep of discontinuities in hard rock led to the following needs.

A normal load has to be applied to the discontinuity which remains constant for long periods. For the sliding of discontinuities surrounding underground excavations, it is known that the normal stress is not

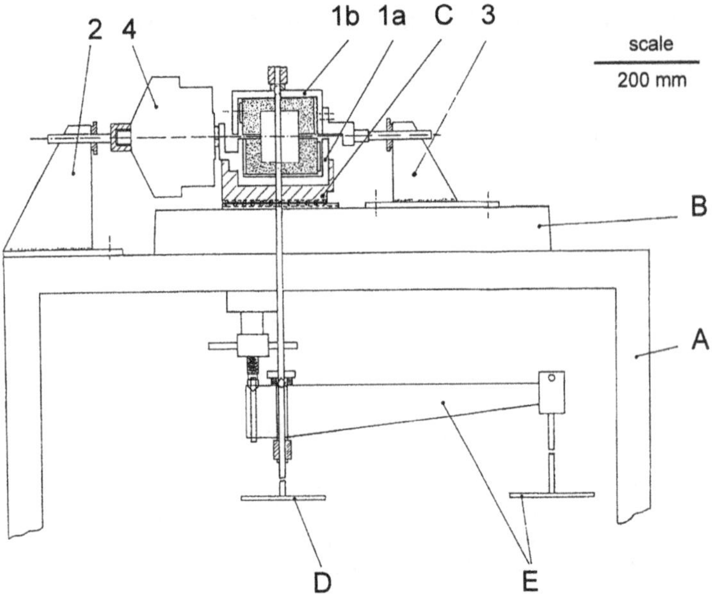

Figure 1. Development of the shear creep equipment from the Wykeham Farrance soil shear box. The existing components are A) Stand, B) Support block on the stand, C) Lower half of the soil specimen mounting box, D) Direct (1:1) normal loading device, E) Lever (10:1) normal loading device. The modifications are 1a) Lower half of box for mounting the specimen, 1b) Upper half of box for mounting the specimen, 2) Bracket with adjusting screw and support for the air cylinder, 3) Bracket with adjusting screw for positioning the upper box, 4) Air cylinder/piston to apply shear load.

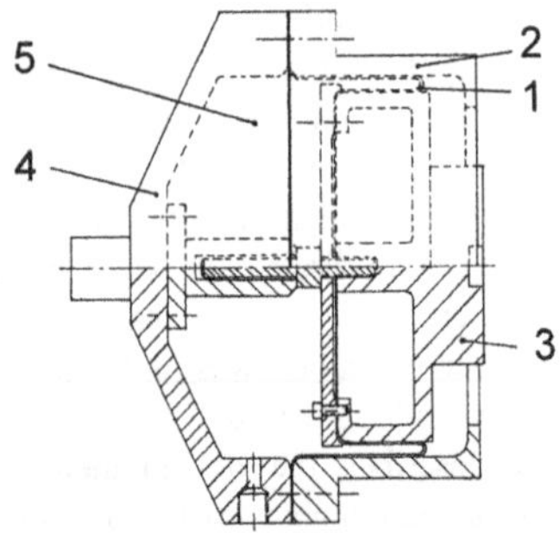

Figure 2. Details of the developed air cylinder. 1) Flexible "Bellofram" rubber seal of 150 mm diameter, 2) Bellofram cylinder. This cylinder is partially closed at the right hand side, limiting the movement of the piston to the right, 3) Bellofram piston, 4) Cylinder head with connection for air supply, guide for the piston and stop to limit the movement of the piston to the left, 5) Compressed air space.

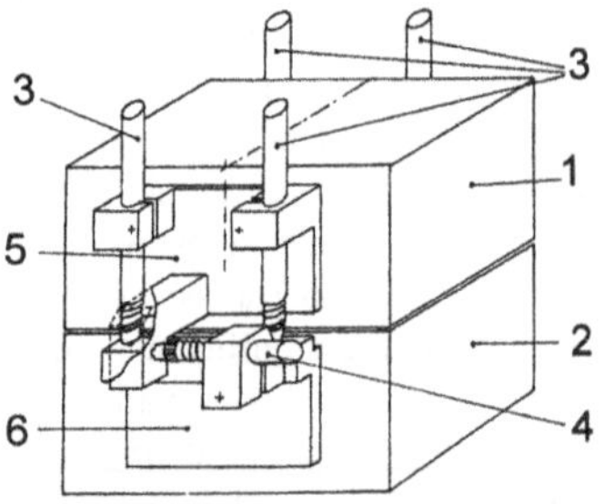

Figure 3. Attachment of the LVDTs to the discontinuity specimen. 1) Top half of the shear specimen. This half of the specimen is held stationary by the bracket numbered as 3 in Figure 1, 2) Bottom half of the shear specimen with direction of movement from left to right, 3) LVDT's for measuring vertical displacements, 4) LVDT's for measuring horizontal displacements, 5) Upper fixing plate. Note that these attachments serve as fixtures for the vertical LVDT's and provide targets for the horizontal LVDT's, 6) Lower fixing plates. These attachments serve as fixtures for the horizontal LVDT's and provide targets for the vertical LVDT's. Note that the direction of shear displacement is such that when the sample fails, the horizontal LVDT's move away from the targets and are not damaged.

constant, but varies due to the combined effect of dilation and the stiffness of the surrounding rock. Therefore, the emphasis has recently been placed on developing shear equipment with a constant normal stiffness (Ooi & Carter 1987, Mouchaorab & Benmokrane 1994). As little is known about the shear creep of discontinuities, it was decided to use a constant normal load for the initial testing phase as it was more cost-effective and easier to implement. A further requirement of the normal load is that it should not be affected by power failures as a reduction in this load can lead to premature failure of the sample. The shear load should remain constant for long periods and should preferably also not be affected by power failures. A reduction in shear load would however only lead to a loss of creep data and not to failure of the sample.

The shear and normal displacements of the discontinuity must be accurately measured. The displacements were expected to be relatively small during the initial creep stages, but large just before failure of the specimen. Preliminary studies using another shear box showed that for some discontinuities in hard rock, the creep rate can be very small. This poses the danger that the discontinuity creep data may be obscured by creep components from the equipment itself. Traditionally, shear specimens are mounted by casting them into a mould having a geometry suitable to fit the sample holder of the testing equipment. A mounting medium of epoxy or cement is used. It was feared that this mounting medium could exhibit significant creep when subjected to stress over long periods. A very important additional requirement for testing of discontinuities in hard rock is that the measuring transducers should be mounted directly onto the rock specimen. To enable this, the specimen needs to be easily accessible.

3 EQUIPMENT DEVELOPMENT

3.1 Shear box

After a survey of the available standard shear testing equipment, it was found that an existing 'Wykeham Farrance' shear box, normally used for shear tests on soils, could be modified to meet all the requirements. This also proved to be a cost-effective solution, as it could be modified with limited funds in a short period of time. The deadweight method of applying normal load was retained as it is an ideal method to maintain a constant load over a long period of time. To apply the constant shear load, the existing shear driving mechanism was replaced with a pneumatic system. This consisted of a specially developed air cylinder/piston and a system of regulating valves. Although the use of compressed air resulted in a system with a very low stiffness, this did not present any problems as only the pre-failure creep behaviour was of interest. The complete system is illustrated in Figures 1 and the details of the air cylinder in Figure 2.

3.2 Displacement transducers

The upper and lower halves of the box were designed to allow unrestricted access to the two sides of the specimen. This enabled the mounting of four vertical LVDT's (linear variable differential transformers) and two horizontal LVDT's directly onto the sample. This is illustrated in Figure 3. For simplicity a prismatic test specimen with a sawcut discontinuity is shown. Cores or irregular specimens can, however, easily be cast into a prismatic mould. The only requirements are that the two faces on which the LVDT's are fixed are flat and protruding from the mould. The maximum shear area which can be accommodated is 125 mm × 100 mm. The open design has the disadvantage that the specimen cannot be tested with the discontinuity submerged in water. It is nevertheless possible to test the specimens 'wet' by filling the bottom tray with water and wrapping a thin strip of water absorbent cloth (which should also hang in the water) around the discontinuity. This method proved successful as described in Malan (1997).

The LVDT mountings are cemented to the surface of the rock. As there is no shear stress applied to these attachments, they will not creep relative to the rock surface. From the six LVDT's, the complete spatial movement (including all rotations) can be resolved. As the range of the horizontal LVDT's was quite small, another horizontal LVDT with a larger range was used outside the shear box to measure any large scale horizontal movement. The horizontal and vertical LVDT's used were models GTX500Z and GTX1000 from RDP Electronics Ltd. Figure 4 illustrates the completed shear box and Figure 5 the LVDT's attached to the sample.

Figure 4. The shear box developed for the testing of discontinuities in hard rock.

Figure 5. Details of the LVDT's attached to the rock sample.

3.3 Data acquisition system

To record the creep data, a Hewlett Packard 3497A datalogger, desktop computer and a printer (as backup) are used. The logging system and the conditioning units for the LVDT's and the transducer for monitoring the air pressure are powered by an uninterrupted power supply. This will keep the entire system running for 4 to 6 hours after a power failure. This proved invaluable for creep testing as a number of power failures did occur during the testing program described in Malan et al. (1998).

The software was developed so that the data for a particular test is stored as an electronic file and also sent to the printer as a hard copy. This prevents data loss in the case of long duration power failures or computer crashes.

During testing, the data acquisition unit is triggered to log a data point by either of the following two methods. One of the horizontal LVDT's is monitored and when the recorded displacement is larger than a user-specified threshold value, the logging is triggered. If the displacement is less than the threshold value, the logging will be triggered after a certain time period has elapsed. Both the threshold value and time period can be changed interactively while the test is running.

In the software, the data is not immediately converted to the appropriate displacement or pressure units, but stored as direct voltage measurements. Converting the data to the appropriate units is only done afterwards. This has the advantage that any of the measuring devices can be exchanged without having to modify the software. Furthermore, any of the transducers can be re-calibrated and the data re-evaluated without difficulty.

3.4 Measurement accuracy

With the high sensitivity of the datalogger and the LVDT's, great accuracy can be obtained. Tests illustrated that the displacement can be measured to an accuracy of 0.002 mm. The measuring equipment is slightly temperature sensitive and the value above is for the equipment housed in a climate controlled laboratory where the temperature variation was not more than 2°C over a 24 hr period.

At a typical air pressure of 92 kPa used during one of the tests, this pressure, and therefore the shear load, remains constant to within 0.8 % over a 48h period. The normal force remains perfectly constant due to the deadweight loading arrangement.

4 TYPICAL CREEP RESULTS

The equipment described above was used in a testing program described in Malan (1997) and in the accompanying paper Malan et al. (1998).

A typical test is conducted by mounting the specimen and applying the appropriate normal stress by choosing the correct deadweight for normal loading. A shear stress below the expected shear strength is applied. This is increased after a typical period of 24 or 48 hours. This cycle is repeated until the sample fails. Note that the normal load remains constant throughout the test. The increase in shear load is achieved by manually adjusting a regulator valve to obtain the required air pressure in the cylinder. Note that this is done very carefully over a period of several seconds to prevent sudden loading and premature failure of the sample.

As an illustration of the capabilities of the equipment, Figures 6 and 7 are included. These results are for a sawcut discontinuity in quartzite with artificial gouge infilling (crushed quartzite) of 2 mm thickness. The normal load was 1 MPa and the shear load 0.48 MPa (no previous loading stages). Note the low steady-state creep phase in the secondary phase illustrating the need for sensitive equipment.

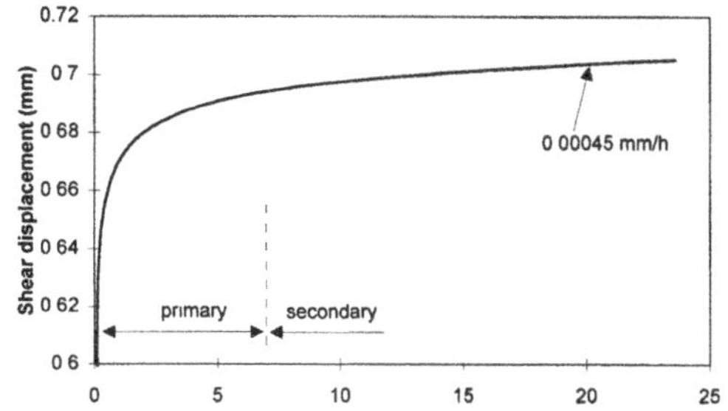

Figure 6. Typical shear creep behaviour of a discontinuity with gouge infilling. This result is the average value of the two horizontal LVDT's. Note that the range of the y-axis is chosen from 0.6 to 0.72 mm to illustrate the typical creep profile, but the origin of the data is at 0 mm.

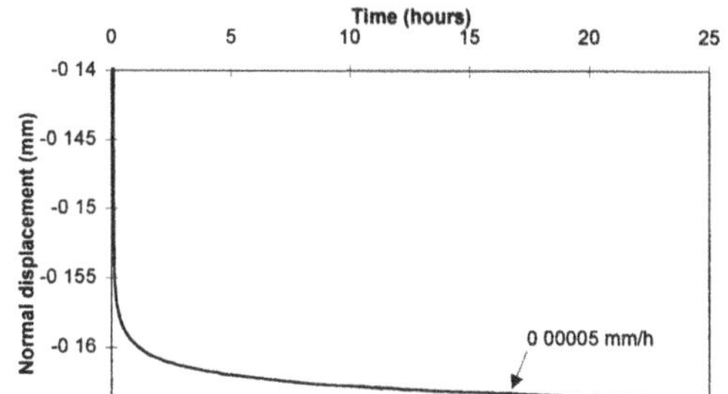

Figure 7. Normal displacement accompanying the results in Figure 6. This result is the average of the four vertical LVDT's. The negative values indicate compaction.

5 CONCLUSION

This paper describes the development of shear testing equipment to investigate the creep behaviour of discontinuities in hard rock. The equipment was developed by modifying an existing soil mechanics shear box. This is a cost-effective solution which allows for easy replication of the equipment. As the shear creep rate can be low for some specimens, it is important to measure the creep movements directly on the rock surface and thereby avoid obscuring the data by creep components from the equipment or the epoxy material used to mount the specimens. The shear box was designed to have easy access to the sample allowing the displacement transducers to be mounted on the rock itself using special attachments. Design of the data acquisition system for long-term shear testing is also important to prevent data loss during power failures or unnecessarily large data files due to high sampling rates. The techniques used for this design are also discussed in the paper. Typical discontinuity creep results obtained with the equipment illustrate the success of the design methodology.

ACKNOWLEDGEMENTS

This work forms part of the research program of Rock Engineering, CSIR Mining Technology and was funded by the Safety in Mines Research Advisory Committee (SIMRAC). D.F. Malan is working towards a Ph.D. degree at the University of the Witwatersrand and the equipment described here was developed for that study. The authors would like to thank Dr J. Napier, Dr R. Durrheim and Mr A. Haile for reviewing the manuscript and making several useful comments.

REFERENCES

Bowden, R.K. & Curran, J.H. 1984. Time-dependent behaviour of joints in shale. In: C.H. Dowding & M.M. Singh (eds), *Proc. 25th Symp. Rock Mech.*: 320-327.

Lama, R.D. & Vutukuri, V.S. 1978. Handbook on mechanical properties of rocks - Testing techniques and results - Volume III: 209-320 Trans Tech Publications, Germany.

Malan, D.F. 1997. Identification and modelling of time-dependent behaviour of deep level excavations in hard rock. *Draft PhD thesis*, University of the Witwatersrand, Johannesburg, South Africa.

Malan, D.F., Vogler, U.W. & Drescher, K. 1997. Time-dependent behaviour of hard rock in deep level gold mines. *J. S. Afr. Inst. Min. Metall.* 97:135-147.

Malan, D.F., Drescher, K. & Vogler, U.W. 1998. Shear creep of discontinuities in hard rock surrounding deep excavations. *Ibid.*

Mouchaorab, K.S. & Benmokrane, B. 1994. A new combined servo-controlled loading frame/direct-shear apparatus for the study of concrete or rock joint behavior under different boundary and loading conditions. *Geotechnical Testing Journal* 17:233-242.

Ooi, L.H. & Carter, J.P. 1987. A constant normal stiffness direct shear device for static and cyclic loading. *Geotechnical Testing Journal* 10:3-12.

Schwartz, C.W. & Kolluru, S. 1984. The influence of stress level on the creep of unfilled rock joints. In: C.H. Dowding & M.M. Singh (eds), *Proc. 25th Symp. Rock Mech.*: 333-340.

Solberg, P.H., Lockner, D.A., Summers, R.S., Weeks, J.D. & Byerlee, J.D. 1978. Experimental fault creep under constant differential stress and high confining pressure. In: Y.S. Kim (ed.), *19th US Symp. Rock Mech.*: 118-121.

Tan, T. and Kang, W. 1980. Locked in stresses, creep and dilatancy of rocks, and constitutive equations. *Rock Mech.* 13:5-22.

Wawersik, W.R. 1974. Time-dependent behaviour of rock in compression. In: Advances in Rock Mechanics. In: G.B. Wallace (ed.), *Proc. 3rd Congr. Int. Soc. Rock Mech.*: 357-363.

Mechanics of Jointed and Faulted Rock, Rossmanith (ed.) © 1998 Taylor & Francis, ISBN 90 5410 955 6

A formulation of shear strength for rock joints included two or three different asperities

H. Kusumi & K. Nishida
Department of Civil Engineering, Kansai University, Suita, Osaka, Japan

K. Teraoka
Department of Geological Research, Sanyo Co. Ltd, Nara, Japan

ABSTRACT: We carried out the experimental study in order to investigate the influence of the surface roughness of rock joints on the shear strength of those. We made some artificial plaster specimens which have three different types of surface profiles, i.e. regular triangular profile, irregular triangular profile and JRC profiles. All of these specimens are applied on the direct shear test. On the other hand, before direct shear test, in order to investigate the method of quantitative estimation of irregular joint profile, the measurement and the analysis of joint profile for each specimen using laser profilometer have conducted.
As the results, we proposed the shear strength criteria which is expressed by the power function, and it is recognized that this criteria is applied to various rock joints.

1. INTRODUCTION

In this study, we carried out the experimental investigation to make clear the relationships between the surface roughness of rock joints and the shear strength. The plaster specimens which have some different joint profiles were used for the experiments.
First of all, from the results of direct shear tests for the joints with regular teeth, we proposed a simple shear strength criterion which considers the surface roughness. And we examined the application of the proposed criterion to the joints with irregular triangular teeth. Furthermore, from the results of direct shear tests for the three types of JRC profiles, the application of it to the irregular joints was examined.

2. SPECIMEN

The plaster was chosen for the material of specimens, because numerous specimens of similar quality can be easily produced. The specimens made from the material which mixed plaster, fine grain sand and water, these weight are in the ratio 1:1:0.6. After pouring this material into the mold, these specimens are curing for two weeks under 20° temperature and 20% less humidity condition. Table 1 shows some properties of each plaster specimen using the experiments. Figure 1 shows the profile of specimen with triangular teeth. In this figure, from left side, this specimen has three asperities with angle i_a, i_b and i_c and each asperity angle is chosen from three angles 15°, 20° and 25°. In this study, these specimens are termed the irregular teeth specimen, and expressed as $IPi_a\text{-}i_b\text{-}i_c$. The specimens with the angle $i_a=i_b=i_c=i$ are termed the regular teeth specimen, and expressed as RPi. The asperity height (h) of each specimen is constant(4.47mm). The shapes of plaster

Table 1 Some properties of plaster specimen.

Unit Weight (kN/m³)	Uniaxial Compressive Strength σ_c (MPa)	Tensile Strength by Brazilian Method σ_t (MPa)
15.8	20.9	2.5

Table 2 Length of each specimen.

Specimen	Length (mm)
RP15	100.0
RP20	73.62
RP25	57.46
IP15-20-15	91.22
IP15-20-20	82.42
IP15-25-15	85.83
IP15-25-25	71.65
IP20-25-20	68.23
IP20-25-25	62.85
IP15-25-20	77.63

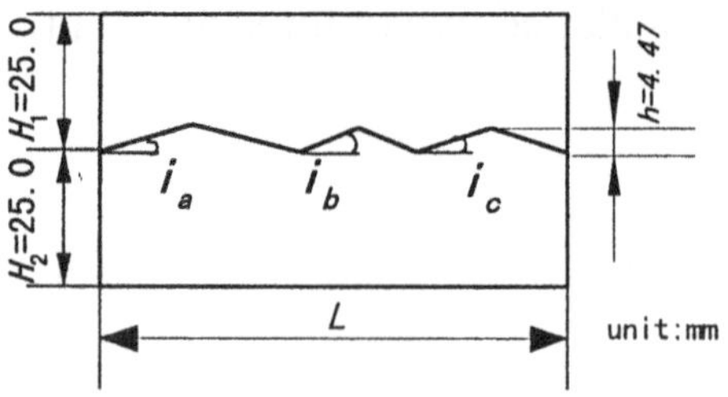

Figure 1 Shape of specimen with teeth discontinuity.

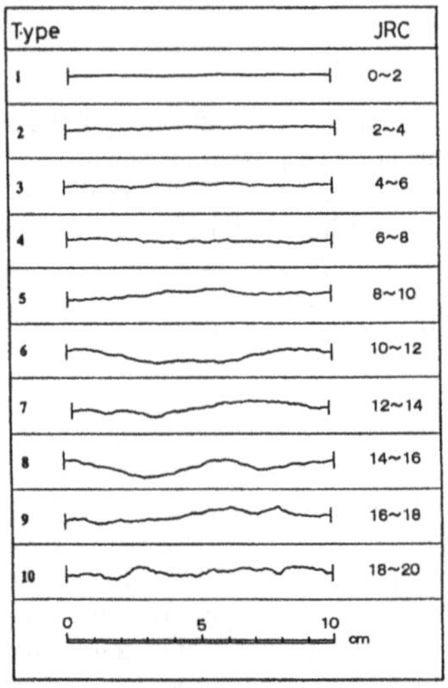

Figure 2 Typical range of JRC values associated with each profile.

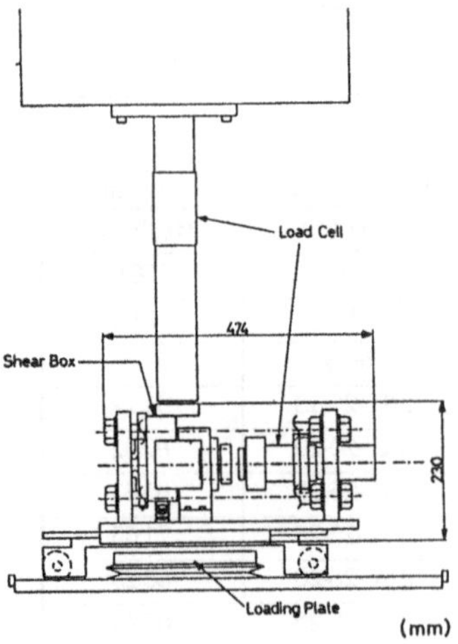

Figure 3 Testing apparatus.

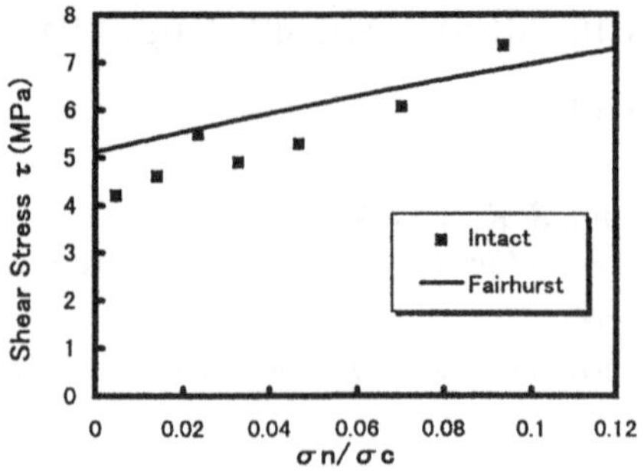

Figure 4 Shear strength for intact specimens.

specimens are rectangular prisms 50mm wide, 50mm high and length depends on the specimen types. **Table 2** shows the length of each specimen. **Figure 2** shows the typical range of JRC values proposed by Barton. Three profiles were chosen from Barton's JRC profiles, which are JRC 2-4, JRC 10-12 and JRC 18-20, and the specimens which have these joint profiles in the shear direction were produced. These specimens have no roughness in the direction perpendicular to the shear direction. The plaster material of JRC specimens are the same as the teeth specimens. In this study, these specimens are termed JRC profile specimen, and expressed as *J2-4*, *J10-12* and *J18-20*. The shapes of JRC profile specimens are rectangular prisms 50mm wide, 50mm high and 100mm long. The intact and the saw-cut specimen were also prepared, and these specimen's material is the same to the other specimens.

3. TESTING MACHINE AND PROCEDURE

Figure 3 shows the direct shear testing machine. The shear box consists of moving box (left hand) and fixed box (right hand), and two boxes separating on shear zone. The inner box exists in each shear box, and after the specimen puts into inner box, shear test can be started. The shear and normal load are applied on vertical and horizontal direction respectively. The direct shear test is done by the strain control method under constant normal stress condition, the shear speed being 0.1mm/min for every specimen. For teeth specimens, applied normal stress is seven steps between 0.1 and 2.0MPa, and for JRC profile specimens, five steps between the same range. Prior to direct shear test, the measurement of the joint profile is conducted for JRC profile specimens. Laser profilometer used for the measurement has a resolution of 2μm. Placing a specimen on the X-Y stage, the measurement of the joint profile is made by fixed laser profilometer moving the X-Y stage along the shear direction. Each joint profile was measured at 1.0mm spacing, and the data was recorded on a personal computer as X-Y coordinate data.

4. FORMATION OF SHEAR STRENGTH FOR ROCK JOINTS

4.1 *Estimation of shear strength criterion for regular teeth specimen*

In this section, we tried to propose a new shear strength criterion considering the joint roughness, in which the parameters could be easily determined. **Figure 4** shows the relationship between peak shear strength and normal stress for intact specimens. A solid line in this figure denotes the peak shear strength curve of rock substance obtained by Fairhurst's equation shown as **Equation(1)**. In this equation, it is expressed that the compressive stress is plus and the tensile stress is minus.

$$\tau = \sigma_c \frac{m-1}{n}\left(1+n\frac{\sigma_n}{\sigma_c}\right)^{1/2} \quad (1)$$

$$n = \sigma_c / (-\sigma_t) \qquad m = (n+1)^{1/2}$$

In **Equation(1)**, σ_c is the uniaxial compressive strength of solid rock, and σ_t is the tensile strength of solid rock. **Figure 5** shows the same relationships for regular teeth specimens and saw-cut specimens. For discontinuous specimen, it is well-known that peak shear strength line does not show a straight line but a non-linear curve raised above. So, a power law shown in **Equation(2)** was proposed to express the behavior that described above.

$$\tau_p = a \cdot \sigma_n^{\ b} \quad (2)$$

In **Equation(2)**, a and b are coefficients. The physical meanings of coefficients a and b are not interpreted clearly, and the property of joint formed rock is not considered. Therefore, in this study, **Equation(2)** was transformed into **Equation(3)** shown below.

$$\tau_p = a_1 \cdot \left(\frac{\sigma_n}{\sigma_c}\right)^{b_1} \quad (3)$$

In **Equation(3)**, a_1 and b_1 are coefficients. Coefficient a_1 represents the peak shear strength under the stress condition that the normal stress corresponds to the uniaxial compressive strength of solid rock. Therefore, a_1 can be expressed by the shear strength of rock substance under the stress condition described above, if the shear strength of rock joint and rock substance almost agree under that condition. So, in **Figure 4**, Fairhurst's shear strength curve shows the good agreement with the shear strength of intact specimens, that a_1 is expressed by **Equation(4)** which is rearranged **Equation(1)** by substituting $\sigma_n = \sigma_c$ into.

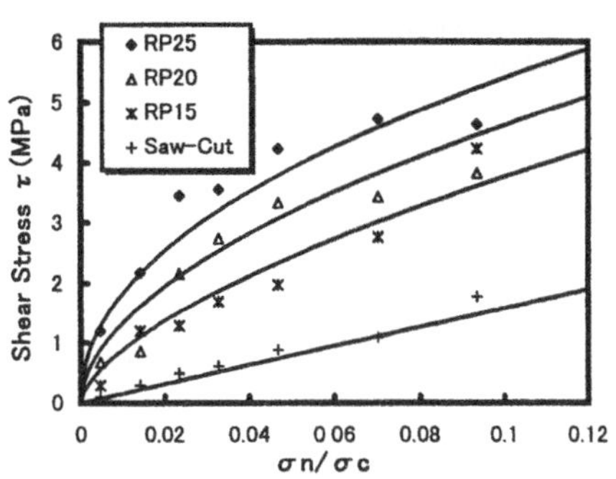

Figure 5 Shear strength for regular teeth specimens.

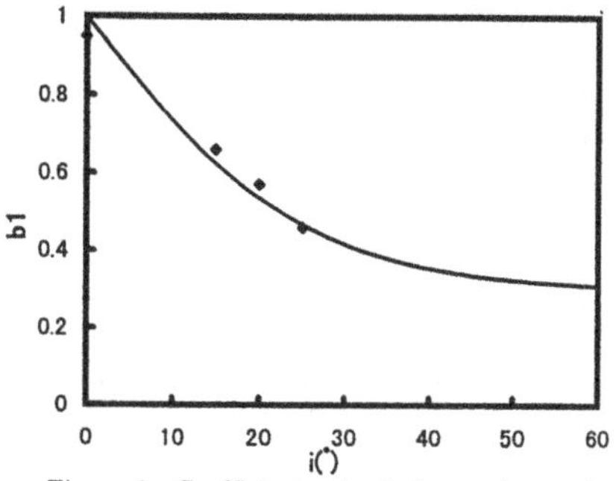

Figure 6 Coefficient value b_1 for regular teeth specimens.

$$a_1 = \sigma_t \left\{\left(\frac{\sigma_c}{-\sigma_t}+1\right)^{1/2} - 1\right\} + \sigma_c \quad (4)$$

Figure 6 shows b_1 values for each type of specimen which is obtained by **Equation(3)** and **Equation(4)**. In this figure, the dotted line shows b_1 value for intact specimen. Essentially, **Equation(3)** can not applied on intact specimen, because it has the cohesive force. Nevertheless comparing with the other regular teeth specimens, b_1 value of intact specimen was obtained. In this figure, b_1 value decrease with the increase of the asperity angle i. Because the shear strength of discontinuous specimen dose not exceed that of intact specimens, it can be considered that b_1 value of the regular teeth specimen approaches that of intact specimen. Here, b_1 value can be expressed as following equation.

$$b_1 = 1 - 0.7 \cdot \tanh(0.04 \cdot i) \quad \textbf{(5)}$$

In **Figure 5**, solid lines show the shear strength curve for each type of specimen obtained by **Equation(3)**, **(4)** and **(5)**. In this figure, it is apparent that the shear strength of each type of specimen approximately agree with the shear strength curve obtained by equation described above.

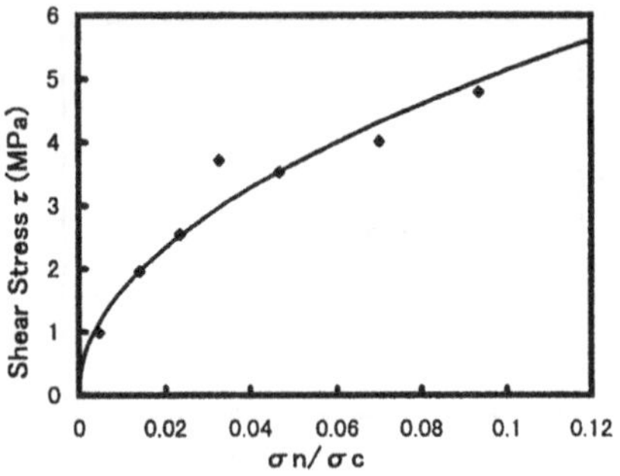

Figure 7 Shear strength for IP20-25-25 specimens.

4.2 *Estimation of shear strength criterion for irregular teeth specimen*

Prior to the estimation of shear strength for irregular teeth specimen, it is necessary to estimate the roughness of joint profiles. In this study, the joint roughness is expressed as the initial dilation angle i_0, and it tried to be applied on **Equation(6)** proposed by Dight and Chiu.

$$i_0 = i_{ave} + k_i \cdot SD_i \quad \textbf{(6)}$$

In **Equation(6)**, i is a inclination angle between adjacent datum, i_{ave} is a mean value of i calculated for each profile, and SD_i is a standard deviation of i for each profile. k_i is a coefficient, and unit for triangular teeth profile. For example, i_{ave} is zero, k_i is unit and i_0 becomes SD_i for irregular teeth specimens. **Figure 7** shows the relationship between the peak shear strength and the normal stress for *IP20-25-25* specimens. In this figure, a solid line shows the peak shear strength curve obtained by **Equation(3)**,**(4)**,**(5)** and **(6)**. In this case, i_0 value obtained by **Equation(6)** is 23.2 °. This figure shows that experimental values agree well with the calculated curve. Therefore, for irregular teeth specimen, it is recognized that the shear strength criterion of power formula which includes the estimation the joint profile by initial dilation angle can accurately represents the shear strength of these.

4.3 *Estimation of shear strength criterion for JRC profile specimen*

With respect to JRC profile specimen, it is necessary to estimate the roughness of joint profile quantitatively, to express the peak shear strength. A quantitative estimation method for irregular rock joints has been proposed by authors as follows.
The estimation of first order roughness of joint profile is necessary, because the second order roughness already

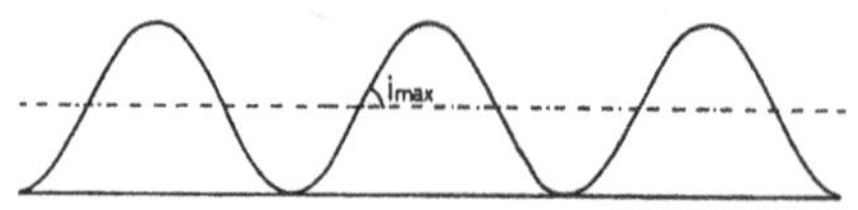

Figure 8 An example of sinusoidal profile.

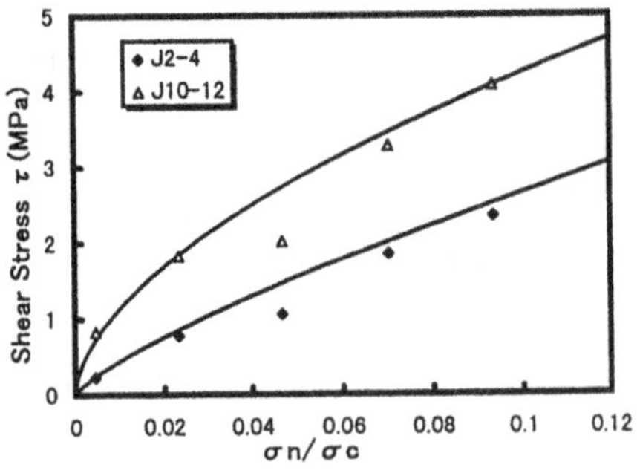

Figure 9 Shear strength for JRC profile specimens.

includes the estimation of the friction angle of joint surface. It is convenient to use a regular profile which simply models for the irregular profile in order to estimate the irregular joint roughness. A sinusoidal profile shown in **Figure 8** has the variation of inclination similar to the natural joints, so it can be expected that this profile is a good model for the natural joints. Here, it is assumed that the maximum angle of inclination for a sinusoidal profile can be represented by the first order roughness of joint surface. The results of analysis for some sinusoidal profiles show that i_{max}/SD_i is approximately $\sqrt{2}$, and hence i_{max} can be expressed by following equation.

$$i_{max} = i_{ave} + \sqrt{2} \cdot SD_i \quad \textbf{(7)}$$

The results of direct shear tests under very low

normal stress indicate that initial dilation angle can be expressed by i_{max} in **Equation(7)**. Therefore, the initial dilation angle i_0 is substituted into i_{max} in **Equation(7)** and **Equation(7)** becomes as follows.

$$i_0 = i_{ave} + \sqrt{2} \cdot SD_i \qquad \textbf{(8)}$$

In the present study, **Equation(8)** was adopted to express the joint roughness. **Figure 9** shows the relationships between peak shear strength and normal stress for *J2-4* specimen and *J10-12* specimen. Solid lines in this figure are the calculated peak shear strength curves for each type of specimen obtained by **Equation(3),(4),(5)** and **(8)**. The initial dilation angle obtained by **Equation(8)** for *J2-4* specimen and *J10-12* is 8.3 ° and 17.7 ° respectively. From this figure, the curved lines show a similar tendency with measured values. Therefore, it is recognized that the quantitative estimation method using **Equation(8)** and the shear strength criterion of power formula proposed by this study are successful for JRC profile specimens.

5 .CONCLUSION

In this study, from the results of direct shear tests for regular teeth specimen, we proposed a simple shear strength criterion of power formula which includes the the estimation of the surface roughness. Furthermore, we examined the application of the proposed criterion to the irregular teeth specimen and JRC profile specimen. The results of this study are as follows.

1) For irregular teeth specimen, it is recognized that the shear strength criterion of power formula which make a application with the quantitative value of the joint profile estimated by initial dilation angle can accurately represents the shear strength of these.

2) For JRC profile specimen, it is recognized that the peak shear strength can be expressed by the shear strength criterion of power formula using the initial dilation angle. Therefore, it is confirmed that the shear strength criterion proposed in this study is applicable for irregular joints similar to actual rock joints.

REFERENCES

Barton, N. and Choubey, V. (1977). The shear strength of rock joints in theory and practice. Rock Mechanics, 10:1-2, 1-54.

Dight, P.M. and Chiu, H.K. (1981). Prediction of shear behaviour of joints using profiles. International Journal of Rock Mechanics and Mining Science & Geomechanics Abstract, 18, 369-386.

Fairhurst, C. (1964). On the validity of Brazilian test for brittle materials. International Journal of Rock Mechanics and Mining Science, 1, 535-546.

Ladanyi, B. and Archambault, G. (1970). Simulation of shear behaviour of a jointed rock mass. *11th Symposium of Rock Mechanics, AIME,* 105-125.

Hobbs, D. W. (1970). The behavior of broken rock under triaxial compression, International Journal of Rock Mechanics and Mining Science , 7, 125-148.

Murrell, S. A. F. (1965). The effect of triaxial stress systems on the strength of rocks at atmospheric temperatures, Geophysical journal, 10, 231-281.

Dynamics

Mechanics of Jointed and Faulted Rock, Rossmanith (ed.) © 1998 Taylor & Francis, ISBN 90 5410 955 6

Velocity amplification considered as a phenomenon of elastic energy release due to softening

A.M. Linkov
Institute for Problems of Mechanical Engineering & All-Russian Institute for Rock Mechanics and Mining Surveying (VNIMI), St-Petersburg, Russia

R.J. Durrheim
CSIR, Division of Mining Technology, Johannesburg, South Africa

ABSTRACT: A new theoretical explanation is given to the phenomenon of particle velocity amplification, which manifests itself in a significant (4-10 fold) increase near and at the surface of an excavation compared to the particle velocity of a wave inside solid rock. A mechanism of energy release is proposed which emphasizes the role of softening at interacting surfaces of cracks or/and blocks. As softening occurs in a limit state, the effect appears only under sufficiently high stresses. In particular, the effect can be expected, and appears in fact, at cracks near openings in hard rock in deep mines.

1 INTRODUCTION

The amplification of wave motion on the walls of excavations in the deep, hard rock gold mines of South Africa has been recently observed and reported: peak velocity and acceleration parameters at the surface of an excavation indicate a 4 to 10 fold increase when compared to measurements within the solid rock (Durrheim et al. 1996). The observed amplification is considerably greater than the two-fold amplification expected at a free surface. The effect is, additionally, although indirectly, confirmed by observations of wall-rock velocities which sometimes are of an order of 10 m/s (Ortlepp 1993). The author of the cited paper distinguishes between source mechanism and damage mechanism; he writes (p. 104): "This dichotomy will highlight paradoxes such as, importantly, how the relatively low PPV generated in the rock mass does not reconcile with the high velocity displacements of the wall rock." From this we may see a clear recognition of the differences between parameters of an incident wave and those appearing near the surface of an excavation. McGarr (1996) also states that the velocity of the wall rock is "substantially greater than ground velocities associated with the primary seismic events". Discussing a quantitative relation between these velocities, he writes: "Whereas several types of evidence suggest that slip across a fault at the source of an event generates nearby particle velocities of at most, several m/s, numerous observations, in nearby damaged tunnels, for instance, imply wall-rock velocities of the order of 10 m/s and greater".

Three different mechanisms have been suggested to explain the source of energy for this phenomenon: (i) resonance, which is discussed, with reservations, by Durrheim et al. (1996); (ii) trapping of energy within a channel (Spottiswoode, pers. comm. 1997); and (iii) energy release due to slab buckling (McGarr, 1996). All three hypotheses invoke changes in the structure of rock near an excavation due to growth of cracks induced by rock pressure. The first and the second hypotheses consider purely geometrical changes; which implies that amplification may occur even in unloaded rock, given that the crack geometry exists. The third hypothesis, in contrast, requires the supply of elastic energy from highly stressed rock. Hence, it implies that the amplification will not appear in unloaded rock. Resonance is not considered to be a likely mechanism as there is no sufficiently long periodic excitation (Spottiswoode, pers. comm. 1997). The "channel" hypothesis seems more plausible. While this hypothesis cannot explain the fact that the amplification occurs both at surfaces with nearby cracks parallel to them and at surfaces with nearby cracks perpendicular to them, 'channelling' may provide its input into amplification. In the extreme case, where a small seismic event triggers a greater event (a rockburst in particular), the main source of energy supply for velocity amplification is the elastic energy stored in rock. Consequently, the third hypothesis appears the most attractive.

McGarr (1996) has explicitly expressed the energy hypothesis, attributing high wall-rock velocities to the buckling of rock slabs stressed near to their compressive strength. Note, however, that buckling is but one of a variety of possible mechanisms for instability with energy excess. The buckling mechanism places emphasis on the aspect of a geometrical nonlinearity appearing in slab flexure. However, there exists another type of nonlinearity *viz.* a physical nonlinearity. The latter is of major importance for compressed rock (Linkov 1994 and 1995). Taking this as the point of departure, we suggest another mechanism of energy release which places emphasis on a physical, rather than geometrical, nonlinearity of highly compressed rock. This nonlinearity is termed softening (or weakening) and has been shown to be responsible for dynamic phenomena in mines (Cook 1965; Salamon 1970; Linkov 1994 and 1995).

In conclusion of this brief review, note that the pioneering papers by Ortlepp (1993) and McGarr (1996), although concerned with wave amplification, do not address it in a direct way. They do not trace amplification of the same seismic wave coming from a remote source to an excavation. In this paper, we attempt to treat the problem explicitly by directly studying the amplification of the same wave. Our aim is to consider wave amplification as a phenomenon of energy release due to softening.

2 FIELD MEASUREMENTS

Measurement of the strong ground motion produced by a rockburst is subject to two main difficulties. Firstly, it is necessary to predict when and where a seismic event is likely to happen in order to record the near-source ground motion within a reasonable time period. Measurements were made at a preconditioning research site on Blyvooruitzicht Gold Mine (Kullmann et al. 1996) where a stabilizing pillar was mined at a depth of 1900 m below surface. The Carbon Leader Reef was largely mined out during the 1970's, except for 40 m broad strike-parallel stabilizing pillars. Total closure of the old stope had since occurred, relieving the stress sufficiently for the stabilizing pillar to be extracted. Nevertheless, the rockburst hazard was considered to be high, and preconditioning was implemented in order to reduce the hazard of face bursts. A microseismic system was used to monitor the seismicity associated with the preconditioning and mining of the pillar, and to determine the source parameters of the seismic events. A sampling rate of 10 kHz was used. The geophones were grouted in boreholes drilled 10 m or more into the solid rock. All seismic events which occurred within the network, and had local magnitudes greater than -1.8, were detected. A location accuracy of ±5 m was verified through the location of blasts. The ground motions produced by preconditioning blasts and any subsequent seismic events were recorded by the microseismic system.

Secondly, special instruments are required to make measurements within a stope where mining is taking place. A robust ground motion monitor was developed by the CSIR to perform this function. The CSIR Ground Motion Monitors were configured to trigger at accelerations greater than 1 g. Accelerographs were installed in a footwall drive, crosscut and stope at the preconditioning reserach site. During a two month period, 147 seismic events were recorded by both the microseismic system and at least one Ground Motion Monitor (Durrheim et al. 1996). The local magnitudes ranged from -1.9 to 1.8.

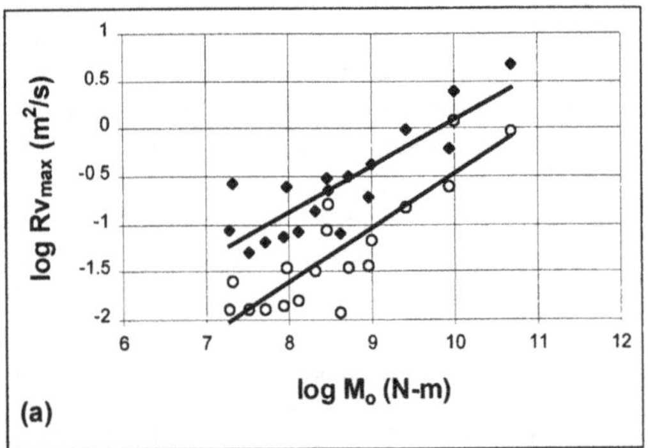

Figure 1. Measurement of the peak velocity parameter Rv_{max} at Blyvooruitzicht Gold Mine. Measurements made in the solid rock are indicated by open circles. CSIR Ground Motion Monitor measurements made on the wall of the footwall drive are indicated by solid diamonds. R is the hypocentral distance.

An estimate of the site effect was made by comparing measurements of the peak velocity parameter (Rv_{max} where R is the hypocentral distance) and peak acceleration parameter (ρRa_{max} where ρ is the density) made in the solid rock with measurements made on the wall of the excavation.

The distance between the hypocentres and Ground Motion Monitors ranged from 20 m to 50 m. When compared to the recordings within the solid rock, an amplification of v_{max} on the wall of the excavation by 4- to 10-fold was indicated (Figure 1).

3 WAVE AMPLIFICATION AT SOFTENING CONTACTS

3.1 *Amplification of a reflected wave*

The essence of wave amplification due to softening can be explained by considering the simplest scheme of a plane shear incident wave. A displacement wave $w=f(x - ct)$ arrives at a moment $t=0$ at the plane boundary $x=0$ (Figure 2(a)). The displacements generate shear stresses τ. These displacements and stresses are added to the displacement u_0 and stresses σ_0 of the initial state; the latter is supposed to be given. The total displacements u and shear stresses σ_τ are the sums

$$u = u_0 + w\,;\; \sigma_\tau = \sigma_0 + \tau.$$

The contact interaction at the boundary $x=0$ occurs via specific "springs" which exhibit softening with the softening modulus M_c after reaching the limit shear strength τ_m corresponding to the shear displacement u_m in a reversible process with the contact modulus E_c (Figure 3). The analytical expression for the interaction if given by the formula

$$-\sigma_\tau(0,t) = \begin{cases} E_c u(0,t) & u<u_m \\ \tau_m - M_c\,[u(0,t) - u_m] & u_m<u<u_* \\ \tau_* & u_*<u \end{cases} \quad (1)$$

The "minus" sign at $-\sigma_\tau(0,t)$ accounts for the fact that the normal to the surface $x=0$ is taken in the direction of the x axis; the value of the shear strength τ_m depends on the normal stress in the media; assuming the normal stress is fixed, the shear strength τ_m is also fixed.

Inside the medium $(x < 0)$, the total shear stress σ_τ is related to the total displacement, u, by Hooke's law $\sigma_\tau = G\partial u/\partial x$ where G is the shear modulus of the medium. Due to the linear elasticity of the medium the same law relates the values in the wave:

$$\tau(x,t) = G\,\partial w/\partial x \quad (2)$$

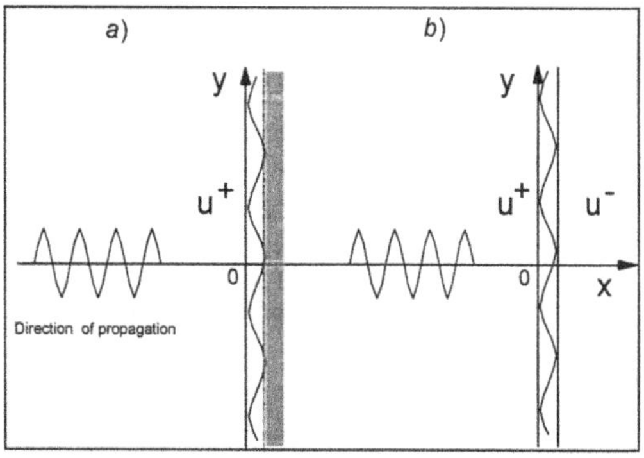

Figure 2. Shear wave impinging on a softening contact.
(a) Contact between soft and very stiff media.
(b) Contact between two similar half-planes.

The initial (static) displacement u_0 at the boundary $x=0$ may correspond to either the ascending (point A) or the descending (point B) portion of the curve of contact interaction shown in Figure 3. In the latter case the contact is in a limit state. Note that in this state there is an alternative: further contact deformation may proceed as either softening (path BC) or elastic unloading (path BD), depending on the direction of displacements in the incident wave. We shall first discuss the case when displacements, u, in the wave have the same direction as initial displacement u_0, i. e. the case when unloading does not occur. Then from (1) it follows for the displacement and shear stress in the wave at the contact:

$$\tau(0,t) = -\,k\,w(0,t) \quad (3)$$

$$\text{where } k = \begin{cases} E_c & u<u_m \\ -M_c & u_m<u<u_* \end{cases}$$

The mathematical formulation of the problem is as follows. We need to solve the 1-D wave equation:

$$\partial^2 w/\partial t^2 - c\,\partial^2 w/\partial x^2 = 0 \quad (4)$$

under
(i) the initial condition $w=f(x,0)$ $(x<0)$,
(ii) the boundary condition (3) at $x=0$ $(t>0)$, and
(iii) the dependence

$$\tau(0,t) = G\,\partial w/\partial x \quad (t>0) \quad (5)$$

expressing Hooke's law (2) for points on the boundary $x=0$. The general solution of (4) is represented by the sum of the incident $f(x,t)$ and reflected $g(t)$ waves $w(x,t) = f(x-ct) + g(x+ct)$. Then (3) and (5) yield the ordinary differential equation for the reflected wave:

$$(1/c)dg(0+ct)/dt + (k/E)g(0+ct) = (1/c)df(0-ct)/dt - (k/E)f(0-ct)$$

Its solution is

$$g(0+ct) = f(0-ct) - 2\omega_0 \int_0^t exp[-\omega_0(t-\tau)]\, f(0-c\tau)d\tau \quad (6)$$

where $\omega_0 = ck/G$. (7)

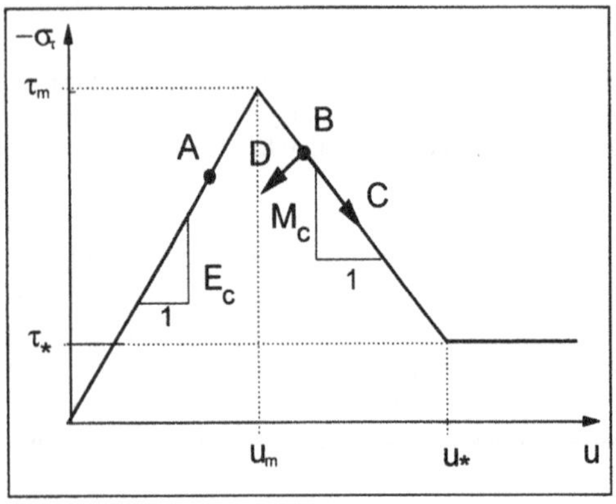

Figure 3. Curve showing the contact interaction with softening modulus.

The modulus of the inverse value of ω_0 is the characteristic time of the contact interaction:

$$T_c = |G/(ck)| \quad (8)$$

The latter can be also written in terms of the attenuation $Z = G/c$ as:

$$T_c = |Z/k| \quad (9)$$

The modulus is used for softening, because the coefficient k is negative ($k = -M_c$). The solution (6) accounts for both the continuous increase of $f(0)$ from the zero value, and the case when it is presented by a step function at $t=0$. In the latter case (6) yields:

$$g(0+ct) = -1 + 2exp(-\omega_0 t) \quad (10)$$

From (10) we may see that for elastic interaction (k, $\omega_0 > 0$) the reflected wave is equal to +1 at small times ($t \ll T_c$), which corresponds to the reflection from a free surface. Opposite to that, at large times ($t \gg T_c$), $g(0+ct)$ tends to -1, which corresponds to the reflection from a rigid boundary. Both these conclusions are obvious from the physical point of view.

For softening interaction (k, $\omega_0 < 0$) the picture changes dramatically because the exponent in (10) increases with time. The velocity and acceleration also grow with the same positive power. Their direction coincides with that of the incident wave. Thus we have exponential amplification of displacements, velocities and acceleration due to softening.

Observing the solution (6) for an arbitrary incident wave $f(x-ct)$, we conclude that this effect is general if the contact reaction is softening ($\omega_0 < 0$). In particular, for a harmonic wave with frequency ω, and incident wave function $f(\xi) = exp(i\omega\xi)$ we have from (6):

$$g(0+ct) = [2/(1+\omega/\omega_0)]\, exp(-\omega_0 t) - [(1-i\omega/\omega_0)/(1+i\omega/\omega_0)]\, exp(i\omega t)$$

The first term in the r. h. s. grows infinitely and suppresses the second (harmonic) term in a case of softening ($\omega_0 < 0$). We see that stationary reflected harmonic waves are impossible for contact interaction in a regime of softening. Note also that the last formula is true only for active softening, i. e. for positive phase of velocity in the incident wave. After it, we have unloading and the result of integration in (6) changes.

3.2 *Amplification of reflected and transmitted waves*

The scheme of Figure 2(a) has served to reveal the effect in the simplest form. It corresponds to a wave coming from a compliant medium at the contact with a very stiff medium. Meanwhile, the observed "amplifying" influence of softening is quite general. It can also be easily revealed for the case when a contact generates both a reflected and transmitted wave. In particular, it appears when a plane incident shear wave arrives at a contact between two similar half-planes (Figure 2(b)). This corresponds to the interaction of a wave with a crack or fault in a limit state. Shear traction τ stays continuous through the contact:

$$\tau^+(0,t) = \tau^-(0,t) = \tau(0,t) \quad (11)$$

and the contact condition (3) now takes the form:

$$\tau(0,t) = -k\,\Delta w(0,t) \quad (12)$$

Herein, $\Delta w(0,t) = w^+ - w^-$ is the displacement discontinuity at the contact $x=0$; the superscript "plus" marks the value from the side of the incoming incident wave; the superscript "minus" refers to the opposite side of a crack or fault. The equations (2), (4) and (5) and initial conditions are the same as in the previous case when applied to the left or to the right of the contact.

For the scheme Figure 2(b) we have a transient wave $h(x-ct)$ in addition to the incident $f(x-ct)$ and reflected $g(x-ct)$ waves. Hence, the general solution of (4) is presented as:

$$w^+(x,t) = f(x-ct) + g(x+ct), \quad w^-(x,t) = h(x-ct) \qquad (13)$$

Substitution of (13) into (2), (11) and (12) gives a system of ordinary differential equations with respect to functions $g(0+ct)$ and $h(0-ct)$:

$$dg(0+ct)/dt + \omega_0\, g(0+ct) - \omega_0\, h(0-ct) = df(0-ct)/dt - \omega_0 f(0-ct)$$
$$dh(0-ct)/dt + \omega_0\, h(0-ct) - \omega_0\, g(0+ct) = \omega_0 f(0-ct) \qquad (14)$$

where ω_0 is given by (7). From (14) we find for the reflected wave:

$$g(0+ct) = f(0-ct) - 2\omega_0 \int_0^t \exp[-2\omega_0(t-\tau)]\, f(0-c\tau)d\tau \quad (15)$$

which differs from (6) only by the multiplier 2 in the exponent in the integrand. This multiplier accounts for the fact that the scheme of Figure 2(b) includes two similar surfaces of contact interaction, whereas the scheme of Figure 2(a) contains only one surface. For the transmitted wave, it follows from (14), (15):

$$h(0-ct) = 2\omega_0 \int_0^t \exp[-2\omega_0(t-\tau)]\, f(0-c\tau)d\tau \qquad (16)$$

If the incident wave is given by the step function, (15) and (16) yield:

$$g(0+ct) = \exp(-2\omega_0 t); \quad h(0-ct) = 1 - \exp(-2\omega_0 t) \qquad (17)$$

From these expressions we see that for elastic interaction $(k,\ \omega_0>0)$, the instant reaction $(t<<1/\omega_0)$ of a contact is that of a free surface, while the long-term reaction $(t>>1/\omega_0)$ is that of an ideal contact. In other words, high frequency incident waves are mostly reflected while low frequency waves do not feel the contact. The same is obvious from the solution for the harmonic incident wave function $f(\xi) = \exp(i\omega\xi)$. In this case

$$g(0+ct) = [1/(1+1/2i\omega/\omega_0)]\exp(-2\omega_0 t) + [1/2i\omega/\omega_0/(1+1/2i\omega/\omega_0)\exp(i\omega t)$$

and

$$h(0-ct) = [1/(1+1/2i\omega/\omega_0)][\exp(i\omega t) - \exp(-2\omega_0 t)] \qquad (18)$$

In the case of elastic contact $(k,\ w_0>0)$ the terms containing $\exp(-2\omega_0 t)$ decay and we have a well-studied stationary regime (e. g. Pyrak-Nolte 1996). In case of a softening contact $(k,\ \omega_0<0)$ all the expressions (15)-(18) contain a positive exponent and grow infinitely with time. We have amplification of the displacement, velocity and acceleration. Obviously, a stationary regime is impossible.

4 DISCUSSION

In the previous section we have seen that amplification due to softening has a characteristic time $T_c = -1/\omega_0$, or substituting (7) and (8):

$$T_c = G/(cM_c) = Z/M_c \qquad (19)$$

where G is the elastic shear modulus; M_c is the contact softening modulus; c is the velocity of shear wave propagation. For estimations of T_c in hard elastic rock, take $2G = 7.5\text{x}10^4$ MPa, c = $5\text{x}10^3$ m/s, and $M_c = M/d$ where M is the shear softening modulus of the material interacting at the contact and d is the average contact thickness. Assume that M is of an order of $2G$. Then $T_c = 10^{-4}\, d$ where d is taken in meters, T_c is in seconds. For $d < 10^{-3}$ m, the characteristic time T_c is less then 10^{-7} s. This means that amplification may be pronounced even for very short impulses with a duration of microseconds. In numerical calculations it will appear as a new seismic event generated by the incident wave. Note also the interesting fact that, as it follows from (17) and (18), displacements in the amplified transient wave have a direction opposite to that in the incident wave. This may serve to check the proposed mechanism of amplification.

So far we have considered the simplest case of a plane incident wave. Hence, our study was restricted to wavelengths less than the size of the surfaces in contact, and in particular, the size of cracks. For cracks with sizes of 0.2 to 1 m, this corresponds to rather high frequencies exceeding 5 kHz.

Meanwhile, the effect of amplification, observed in mines, occurs at frequencies of about 0.5 kHz, which is an order less than the given estimate.

The case of 'low' frequencies is more difficult for analytical treatment. Nevertheless, from physical considerations, it may be seen that amplification will occur at these frequencies as well. Indeed, one may use a quasi-static approximation to reveal energy excess from instability due to contact softening (Linkov 1994 and 1995). This instability appears as exponential growth of the particle velocity (Linkov 1997). In other words, in a quasi-static approximation we again have exponential amplification caused by elastic energy release due to softening. Having a set of cracks with interacting surfaces at or near a limit state, one can expect their input into amplification. This action certainly produces changes in waveforms when compared with those in the incident wave far from the cracked, highly stressed zone. It should also be mentioned that amplification resulting from crack propagation, induced by the incident wave, is a particular case of the mechanism suggested.

5 CONCLUSIONS

The conclusions of this study can be summarised as follows.

1. Wave amplification may be attributed to energy release from rock under stresses high enough to generate cracks at a state close to instability; the amplification occurs due to physical nonlinearity at interacting softening surfaces.

2. Wave amplification appears as an exponential growth of displacements, velocities and accelerations with a characteristic time of the order Z/M_C where Z is the rock attenuation; M_C is the contact modulus of softening.

3. Stationary waves do not exist for the regime of amplification.

4. The triggering of seismic events, rockbursts in particular, by an incident wave or in apparent "static" conditions, may be an extreme case of amplification.

Further investigation of the proposed mechanism is necessary. It may be carried out by means of numerical simulation of waves impinging on cracks close to an unstable state. This will provide synthetic data on the magnitude of amplification and changes in waveforms.

ACKNOWLEDGEMENTS

This work was carried out while Prof. Linkov was a visiting scientist at CSIR: Miningtek. It was funded by the Safety in Mines Research Advisory Committee (SIMRAC) as part of project GAP201. The authors would like to thank Dr S. Spottiswoode for valuable discussions and Drs J. Napier and A. Milev for reviewing the manuscript.

REFERENCES

Cook, N. G.W. 1965. A note on rockbursts considered as a problem of stability. *J. South African Inst. Mining and Metallurgy*, 65: 437-446.

Durrheim, R.J., D.H. Kullmann, R.D. Stewart, & M. Grodner 1996. Seismic excitation of the rock mass surrounding an excavation in highly stressed ground. In M. Aubertin, F. Hassani and H. Mitri (eds), *Proc. 2nd North American Rock Mechanics Symposium:* 389-394. Rotterdam: Balkema.

Kullmann, D.H., R.D. Stewart & M. Grodner 1996. A pillar preconditioning experiment on a deep-level South African gold mine. In M. Aubertin, F. Hassani and H. Mitri (eds), *Proc. 2nd North American Rock Mechanics Symposium:* 375-380. Rotterdam: Balkema.

Linkov, A.M. 1994. *Dynamic phenomena in mines and the problem of stability*. Lisboa, Cedex: Int. Soc. for Rock Mech, P-1799.

Linkov, A. M. 1995. Key-note address: Equilibrium and stability of rock masses. In T. Fujii (ed.), *Proc. 8th Int. Congress on Rock Mechanics.* Rotterdam: Balkema.

Linkov, A. M. 1997. Key-note address: New geomechanical approaches to develop quantitative seismicity. In S.J. Gibowicz & S. Lasocki (eds), *Proc. 4th Int. Symp. on Rockbursts and Seismicity in Mines:* 151-166. Rotterdam: Balkema.

McGarr, A. 1996. A mechanism for high wall-rock velocities in rockbursts. In S. Talebi (ed.), *Proc. Workshop on Induced Seismicity (2nd North American Rock Mechanics Symposium)*, p. 9.

Ortlepp W. D. 1993. High ground displacement velocities associated with rockburst damage. In P. Young (ed.), *Proc. 3rd Int. Symp. on Rockbursts and Seismicity in Mines:* 101-106. Rotterdam, Balkema.

Pyrak-Nolte, L.J. 1996. The seismic response of fractures and the interrelations among fracture properties. *Int. J. Rock. Mech. Min. Sci. & Geomech. Abstr.* 33: 787-802.

Salamon, M.D.G. 1970. Stability, instability and design of pillar working. *Int. J. Rock. Mech. Min. Sci.* 7: 613-631.

Spottiswoode, S.M. 1993. Seismic attenuation in deep-level mines. In R.P Young (ed.), *Proc. 3rd Int. Symp. on Rockbursts and Seismicity in Mines:* 409-414. Rotterdam, Balkema.

Mechanics of Jointed and Faulted Rock, Rossmanith (ed.) © *1998 Taylor & Francis, ISBN 90 5410 955 6*

Rayleigh pulse-induced fault slip: Model and application

K. Uenishi, H. P. Rossmanith & R. E. Knasmillner
Institute of Mechanics, Vienna University of Technology, Austria

ABSTRACT: Rayleigh (R-) pulse interaction with a partially contacting strike-slip fault is experimentally as well as numerically investigated. This study is intended to offer an improved understanding of earthquake rupture mechanisms. The fault is subjected to static pre-stresses. Using dynamic photoelasticity in conjunction with high speed cinematography, the evolution of the R-pulse interaction is recorded by means of isochromatic fringe patterns (contours of maximum in-plane shear stress). It is shown that fault instability (slip) can be triggered by a pulse which propagates along the fault interface at Rayleigh wave speed. Numerically, a finite difference wave propagation simulator SWIFD is used for a quantitative analysis of the problem. Dynamic rupture in laterally heterogeneous structures is discussed by considering the effect of acoustic impedance mismatch on the wave patterns. The results indicate that upon fault rupture, head waves, which carry a relatively large amount of concentrated energy, can be generated which propagate off the fault contact region. Such head waves can cause concentrated wave-induced damage in a particular region located inside an adjacent acoustically softer area. This type of damage concentration was observed in Kobe, Japan, on the occasion of the 1995 Hanshin earthquake.

KEYWORDS: Contact mechanics, Dynamic faulting, Dynamic photoelasticity, Earthquake rupture mechanism, Finite difference method, Hanshin (Kobe) earthquake, Rayleigh wave.

1 INTRODUCTION

Earthquakes and rockbursts belong to the class of catastrophic phenomena which pose great threat to mankind and they have been associated with mysticism for a long time. Scientifically, seismological observations made possible by the worldwide installation of a standardized seismic network suggest the idea that shallow earthquakes are caused by fault instabilities and it is currently accepted that dynamic faulting is the origin of the majority of seismic events (e.g. Scheidegger 1982, Mandl 1988).

The classical approach to fault rupture mechanisms assumes rupture initiation to occur at one point on the fault and to spread across the fault surface. Slip is assumed to extend across the entire fault zone until arrest of the rupture occurs. However, Heaton (1990) showed, based on seismic inversions, that local rise times for fault slip are much shorter than would be expected by the classical approach. As a model for the short rise time observed, he suggested that self-healing pulses of slip occur in earthquake ruptures (Turcotte 1997).

Based on the results obtained by laboratory frictional experiments, fundamentally the same fault rupture mechanism has been proposed by Brune et al. (Brune et al. 1993, Anooshehpoor & Brune 1994, Brune 1996). Their idea is that dynamic rupture on a fault is triggered by a pulse with a separational section propagating along the fault, or in other words, by ripples which propagate along the fault and slip occurs while the compressive stress is reduced (Mora & Place 1994). It is suggested that the excitation of Rayleigh waves on a rupture surface can lead to pulses of separation (Turcotte 1997).

These mechanisms, however, have not been confirmed in a conclusive manner: mathematically, rupture pulses with interface (fault) separation were studied by Comninou and Dundurs (1977, 1978a,b). They showed that a pulse involving separation can stably propagate along an interface when two elastic media are compressed and simultaneously sheared, though the validity of their solution was questioned by Freund (1978) from an energy point of view (Anooshehpoor & Brune 1994).

Numerically, Day (1991) indicated that, along a statically pre-stressed interface between similar materials, pulses of separation decay rapidly rather than propagate in a self-sustaining manner. However, Mora and Place (1994) showed, again numerically, that self-healing pulses can be sustained if interface surface roughness is present. Andrews and Ben-Zion (1997) conducted 2D numerical simulations of dynamic rupture along a planar material interface governed by simple friction, showing self-sustaining propagation of slip pulse and spontaneous break up of the propagating pulse to a number of smaller pulses.

In the present study, the basic mechanisms of dynamic fault rupture (instability) caused by a Rayleigh (R-) pulse will be investigated (Rossmanith & Uenishi 1997, Uenishi et al. 1997a,b). First, the results of laboratory model experiments utilizing dynamic photoelasticity will be presented. Second, the problem will be numerically simulated using the SWIFD finite difference simulator (Rossmanith & Uenishi 1996). Finally, the results obtained by the model investigation will be discussed in detail and applied to explain the damage concentration caused by the 1995 Hanshin (Kobe), Japan, earthquake.

2 EXPERIMENTAL INVESTIGATIONS

2.1 *Experimental set-up*

A Cranz-Schardin type multiple spark gap camera is used to record a sequence of isochromatic fringe patterns of the interaction between a R-pulse and a statically pre-loaded, non-welded partially contacting interface. The isochromatic fringe patterns are produced by a light-field circular polariscope with nearly monochromatic light.

The experimental model is schematically shown in Figure 1. The model consists of two plates of Araldite B which are statically pre-loaded in compression. The dimensions of the plates are selected so as to prevent reflected waves from impinging upon the region of contact and altering the results. The upper surface of the plate 2 shows a very blunt double wedge cut such that only the central section of the surface would initially be in contact with the plate 1.

The contact region is not glued (non-welded) and during the interaction process the contact can recede and be enforced depending on the relative position of the R-pulse with respect to the contact region. The dynamic disturbance is generated in plate 1 by detonating a small amount of explosive.

2.2 *Results*

Figure 2 shows two snapshots of isochromatic fringe patterns of the dynamic wave interaction process. The R-pulse propagates in the upper plate 1 from left to right and interacts with the contact region. The time scale t indicates the time elapsed from the instance of maximum stress amplification at the lhs edge of the contact region.

Figure 2(1) pertains to the event where the incident R-pulse impinges upon the lhs edge of the contact region. It is clearly seen that the strength of the stress singularity, expressed by the fringe order about the lhs edge, is larger than that about the rhs edge where, in this phase, the dynamic effect is still negligible. This stress amplification about the lhs edge is due to the particle motion on the free edge of the upper plate 1 where the leading part of the incident R-pulse induces a back- and downward movement of the particles. The surface particles which are already in contact jointly move towards the lower plate 2, and thus increase the stresses about the lhs edge.

Theoretically, the R-pulse does not exist in the contact region and there must be other kinds of generalized interface wave disturbances which carry the energy across and along the contact region. As the incident (generalized) R-pulse approaches the rhs edge of the contact region, seen in Figure 2(2), partial wave energy transmission occurs across the interface into the lower plate 2. Due to the separational movements of surface particles in the trailing part of the R-pulse, the corresponding fringes in the lower plate 2 are missing (transient interface separation).

3 NUMERICAL SIMULATIONS

The dynamic R-pulse interaction problem is numerically investigated by using the SWIFD finite difference wave propagation simulator (Rossmanith & Uenishi 1996). For the contact problem conditions of plane stress are assumed to prevail. In addition, the R-pulse is assumed to interact with a contact region characterized by a Mohr-Coulomb friction criterion. The interface is pre-stressed and the tensile strength of the interface is set to a very low level.

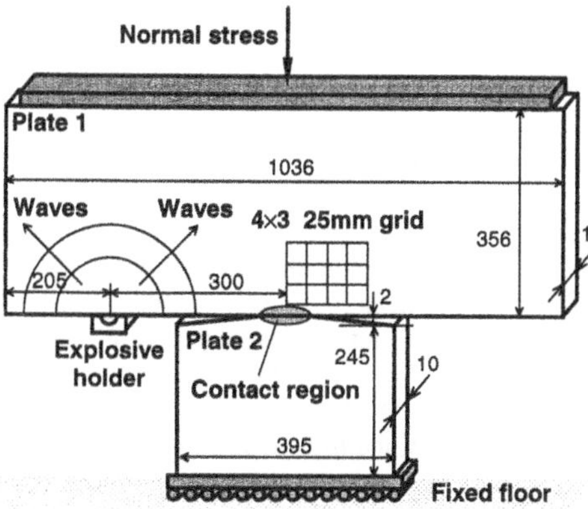

Figure 1. The experimental model set-up for Rayleigh pulse interaction investigation [all lengths in mm].

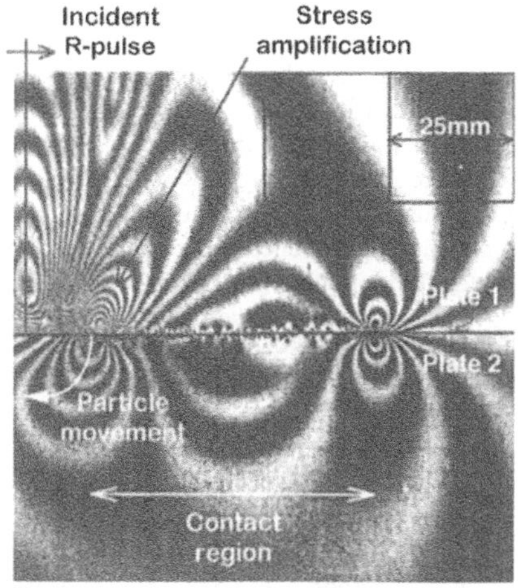

(1) $t = 0\mu s$

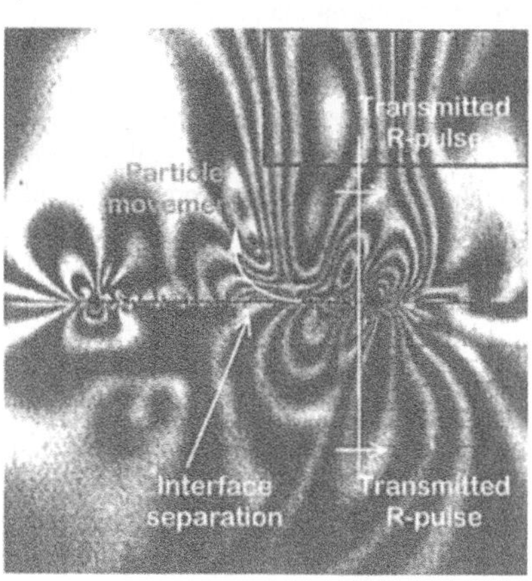

(2) $t = 53\mu s$

Figure 2. Experimentally obtained snapshots of isochromatic fringe patterns.

3.1 *Pulse energy partition*

It is informative to evaluate the relative amount of energy transmitted across and reflected at the contact area during the interaction process. A method to calculate the R-pulse energy from isochromatic fringe patterns or displacement data has been developed to evaluate the pulse energy partition. Results (Uenishi et al. 1997b) indicate that for a pulse length comparable to the contact length, more than one half of the energy initially contained in the incident R-pulse is radiated into the far-field in the form of bulk waves, and only a small portion of the total energy is transmitted along the free surface in the form of a new R-pulse. The energy carried by the reflected R-pulses is negligibly small compared with that contained in the transmitted R-pulses.

3.2 *Influence of the acoustic impedance mismatch*

During the course of numerical investigations, pulse interaction with a contact region between dissimilar materials was also studied. It turned out, that the acoustic impedance mismatch of the two contacting materials plays a crucial role in wave transmission and reflection at the interface. Figure 3 shows a numerically generated isochromatic fringe pattern pertaining to the case where the incident R-pulse speed $(c_R)_1$ lies between the shear (S-) and longitudinal (P-) wave speeds of the lower material 2, $(c_S)_2 < (c_R)_1 < (c_P)_2$. The snapshot shown was taken at 60μs after the maximum stress amplification at the lhs edge of the contact region. Since the slip pulse propagates at the Rayleigh wave speed of the acoustically harder material 1, $(c_R)_1$, the pulse energy is transferred along the contact region at transonic speed with respect to the acoustically softer material 2. A shear-type (S-) head wave is generated which propagates from the contact region into material 2.

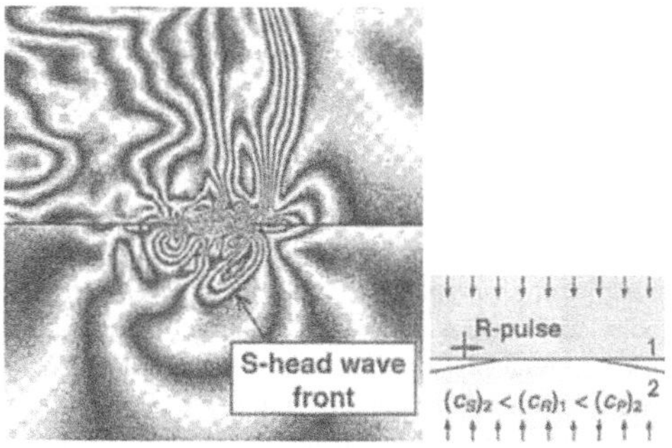

Figure 3. Numerically generated isochromatic fringe pattern pertaining to the case where the contact region lies between dissimilar materials: $(c_S)_2 < (c_R)_1 < (c_P)_2$.

4 DISCUSSION

4.1 *Fracture along an interface vs. fracture in a monolithic medium*

Figure 2 has revealed that the initiation of interface instability is controlled by the incident R-pulse.

The relation between the Rayleigh wave (pulse) speed c_R and the shear wave speed c_S is approximately given by (Viktorov 1962):

$$c_R / c_S \approx (0.87 + 1.12\nu^*) / (1 + \nu^*), \qquad (1)$$

where ν^* is generalized Poisson's ratio [$\nu^* = \nu$ (Poisson's ratio) for plane strain; and $\nu^* = \nu/(1+\nu)$ for plane stress conditions]. For example, for rock with ν^*=0.25, c_R/c_S = 0.92, which indicates that a slip pulse propagates at a speed of about 90% of the shear wave speed of the material. This relatively high rupture propagation speed seems contradictory to the previous theoretical predictions.

In the past, rupture associated with earthquakes was studied in the framework of dynamic fracture mechanics (crack propagation) in a monolithic material. Broberg (1960) showed that the speed of a remotely loaded traction-free crack in a homogeneous material cannot exceed the Rayleigh wave speed c_R of that material, and that an infinite amount of energy would have to be fed into the crack tip in order to maintain further crack extension at speed c_R if the stress intensity factor was non-zero (e.g. Freund 1990). This requirement obviously makes it energetically impossible for a crack in a homogeneous solid to exceed the material's Rayleigh wave speed (Lambros & Rosakis 1995a).

Recently, dynamic interface fracture mechanics has attracted the attention of researchers in fracture mechanics. Due to the intrinsic difficulties associated with this problem, however, only a few theoretical studies have been performed and there is a strong disagreement as to the theoretically predicted value of the terminal velocity for propagating interface cracks. Willis (1973) suggested that the terminal speed of an interface crack should be slightly larger than the Rayleigh wave speed of the more compliant of the two constituents, while Atkinson (1977) claimed that the terminal speed should be the lower Rayleigh wave speed of the two materials (Lambros & Rosakis 1995a). Weertman (1980) studied an edge dislocation moving along a material interface at a constant velocity and suggested that a dynamic reduction in compressive normal traction may allow a slip pulse to propagate in a self-sustaining manner near the shear wave speed of the softer material along an interface governed by a constant coefficient of friction (Andrews & Ben-Zion 1997).

More recently, using the experimental technique of shearography, Rosakis et al. (Lambros & Rosakis 1995a,b, Liu et al. 1995) have investigated dynamic interfacial delamination in dissimilar media. They showed experimentally as well as analytically that, under certain loading conditions, the speed of a propagating interface crack in a bimaterial system may exceed the Rayleigh and the shear wave speeds of the softer material (Lambros & Rosakis 1995a,b, Liu et al. 1995). This result, consistent with the high propagation speed of rupture pulse observed in this study, indicates the fundamental difference between interface fracture and fracture in a monolithic medium.

4.2 *Influence of fault topography*

Real faults are not perfectly planar and their topography shows roughness on all scales, containing jogs or steps. In modeling earthquakes, one consequence of this complexity is the introduction of the concept of asperities. Seismological asperities, or the regions of very high slip, are important in earthquake hazard analysis because the failure of asperities radiates most of the high-frequency seismic energy into the far-field (Lay & Wallace 1995).

In this study, it has been shown that a relatively large amount of pulse energy is radiated into the far-field in the form of bulk waves from a slipping contact region. This observation suggests that, when scaled up, the contact region can be considered as an asperity on a geological fault. Hence, the energy partition pattern obtained by the model investigation is important in the evaluation of the influence of asperity rupture.

4.3 *The 1995 Hanshin (Kobe), Japan, earthquake*

On 17 January 1995, at 5:46 a.m. local time, an earthquake of moment magnitude 6.9 struck the region of Kobe and Osaka (Hanshin region) in the west-central part of mainland Japan. Seismic inversion (Wald 1995, 1996) indicates that the rupture was initiated at a shallow depth on a fault system running through the city of Kobe and propagated bilaterally: along the Suma/Suwayama faults toward the city of Kobe and along the Nojima fault [Fig.4(1)]. Strong ground motion lasted for some 20 seconds and caused considerable damage within a radius of 100km from the epicenter, but most severely affected were Kobe and its neighboring cities.

One of the mystifying phenomena observed in Kobe is the emergence of the strip of the most severely damaged zone [the Japan Meteorological Agency Intensity 7 zone indicated in dark gray in Figure 4(1)]. This strip, about 20km long, is found close, but not parallel, to the Suma/Suwayama faults. Damage due to liquefaction was scarcely observed inside this strip, and it is suggested that the damage was directly due to the seismic waves.

A 2D model including a plane normal to the fault plane is regarded as appropriate for an approximate first order analysis of the rupture mechanisms of the Hanshin earthquake, because the Suma/Suwayama faults dip steeply, nearly 90°, and a near-source, SH directivity pulse from strike-slip faulting was recorded (Toki et al. 1995) in the city of Kobe.

Beneath the lhs edge [in Fig.4(1)] of the strip, a region of relatively large slip (asperity) was found, and arrival of a concentrated shear disturbance as well as the resulting large (particle) velocity were recorded in central Kobe (Wald 1995, 1996). This indicates that the rupture-induced shear wave was of a head wave type.

As described above, an asperity on a geological fault corresponds to a contact region in the model. In Figure 3, where an interface is located between dissimilar materials [$(c_S)_2 < (c_R)_1 < (c_P)_2$], a S-head wave is observed in the numerical simulation. In the model, material 1 fits the acoustically harder region in the foothills of the Rokko Mountains, where soils are very shallow or rock outcroppings are found and the damage tended to be relatively minor. Material 2 corresponds to the acoustically softer region (where soft alluvial soils primarily prevail) which includes the "damage strip".

During dynamic interaction, each particle in the materials experiences a history of velocity and acceleration. The maximum values of these quantities, the peak particle velocity (PPV) and the peak particle acceleration (PPA), are very important and often used practical design parameters in many applications such as blasting in mines and quarries, and in engineering seismology.

Figure 4 shows the contours of high PPV [Fig.4(2)] and PPA [Fig.4(3)] obtained by the numerical simulation [$(c_S)_2 < (c_R)_1 < (c_P)_2$]. It is interesting to note that the high PPV (or PPA) region is found in a narrow band, similar to the shape of the "damage strip" in Kobe [Fig.4(1)]. The angle between the fault (interface) and the high PPV (PPA) region is influenced by the S-head wave generated during the dynamic interaction (Fig.3). For the 2D simulation, this angle is approximately 45°. For the Hanshin earthquake, the angle between the fault and the "damage strip" is 18°. The difference in these angles is presumably because of the local geological as well as the 3D effect on the real fault rupture process. However, this result shows that a simple 2D model may be able to offer the information about the seismic rupture and the ensuing dynamic wave phenomena, although a more sophisticated study would have to be based on a 3D analysis.

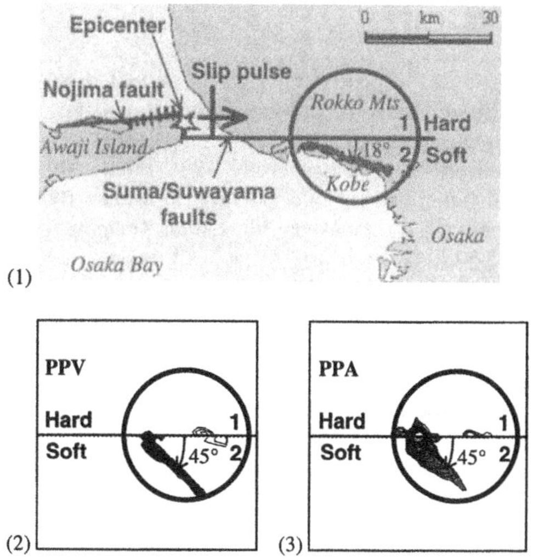

Figure 4. Comparison of Hanshin (Kobe) earthquake and numerical simulation for the case $(c_S)_2 < (c_R)_1 < (c_P)_2$.
(1) The Japan Meteorological Agency Intensity 7 (the most severely damaged) zone (dark gray region) associated with the Hanshin earthquake; and contours of (2) high PPV (peak particle velocity) and (3) high PPA (peak particle acceleration) obtained from the numerical simulation.

5 CONCLUSIONS

The purpose of this study was to obtain an improved understanding of Rayleigh (R-) pulse interaction with a partially contacting, statically pre-stressed fault. The observed phenomena of dynamic fault slip can be attributed to the particle motion in a R-pulse. The existence of an interface pulse with a separational zone propagating along a fault in the Earth crust under very high compressive stress conditions, as suggested by Brune et al. (Brune et al. 1993, Anooshehpoor & Brune 1994, Brune 1996) and Andrews and Ben-Zion (1997), may resolve the long standing paradoxes associated with earthquakes: the origin of short rise times in earthquake slip; dynamic contributions to spatio-temporal slip complexities; the anomalous P-wave radiation; the heat flow paradox; and the low static shear stress levels on geological faults.

ACKNOWLEDGMENTS

This work was financially supported by the Austrian National Science Foundation (FWF) through Research Project No P10326-GEO.

REFERENCES

Andrews, D.J. & Ben-Zion, Y. 1997. Wrinkle-like slip pulse on a fault between different materials. *J. Geophys. Res.* 102: 553-571.

Anooshehpoor, A. & Brune, J.N. 1994. Frictional heat generation and seismic radiation in a foam rubber model of earthquakes. *Pure Appl. Geophy.* 143: 735-747.

Atkinson, C. 1977. Dynamic crack problems in dissimilar media. In G.C. Sih (ed.), *Mechanics of Fracture Vol.4*: 213-248, Leyden: Noordhoff.

Broberg K.B. 1960. The propagation of a Griffith crack. *Ark. Fys.* 18: 159.

Brune, J.N. 1996. Particle motions in a physical model of shallow angle thrust faulting. *Proc. Indian Acad. Sci. (Earth Planet. Sci.)* 105: L197-L206.

Brune, J.N., Brown, S. & Johnson, P.A. 1993. Rupture mechanism and interface separation in foam rubber models of earthquakes: a possible solution to the heat flow paradox and the paradox of large overthrusts. *Tectonophysics* 218: 59-67.

Comninou, M. & Dundurs, J. 1977. Elastic interface waves involving separation. *J. Appl. Mech.* 44: 222-226.

Comninou, M. & Dundurs, J. 1978a. Can two solids slide without slipping? *Int. J. Solids Structures* 14: 251-260.

Comninou, M. & Dundurs, J. 1978b. Elastic interface waves and sliding between two solids. *J. Appl. Mech.* 45: 325-330.

Day, S.M. 1991. Numerical simulation of fault propagation with interface separation (abstract). *EOS Trans. AGU* 72: 486

Freund, L.B. 1978. Discussion: elastic interface waves involving separation. *J. Appl. Mech.* 45: 226-228.

Freund, L.B. 1990. *Dynamic fracture mechanics.* Cambridge: Cambridge University Press.

Heaton, T.H. 1990. Evidence for and implications of self-healing pulses of slip in earthquake rupture. *Phys. Earth Planet. Int.* 64: 1-20.

Lambros, J. & Rosakis, A.J. 1995a. Shear dominated transonic interfacial crack growth in a bimaterial – I. Experimental observations. *J. Mech. Phys. Solids* 43: 169-188.

Lambros, J. & Rosakis, A.J. 1995b. Development of a dynamic decohesion criterion for subsonic fracture of the interface between two dissimilar materials. *Proc. R. Soc. Lond.* A451: 711-736.

Lay, T. & Wallace, T.C. 1995. *Modern global seismology.* San Diego: Academic Press.

Liu, C., Huang, Y. & Rosakis, A.J. 1995. Shear dominated transonic interfacial crack growth in a bimaterial – II. Asymptotic fields and favorable velocity regimes. *J. Mech. Phys. Solids* 43: 189-206.

Mandl, G. 1988. *Mechanics of tectonic faulting: models and basic concepts.* Amsterdam: Elsevier.

Mora, P. & Place, D. 1994. Simulation of the frictional stick-slip instability. *Pure Appl. Geophys.* 143: 61-87.

Rossmanith, H.P. & Uenishi, K. 1996. PC software assisted teaching and learning of dynamic fracture and wave propagation phenomena. In H.P. Rossmanith (ed.), *Teaching and Education in Fracture and Fatigue*: 253-262. London: E & FN Spon.

Rossmanith, H.P. & Uenishi, K. 1997. Fault dynamics – dynamic triggering of fault slip. In K. Sugawara and Y. Obara (ed.), *Proceedings of the International Symposium on Rock Stress*: 27-34, Rotterdam: A.A. Balkema.

Scheidegger, A.E. 1982. *Principles of geodynamics.* Berlin: Springer-Verlag.

Toki, K., Irikura, K. & Kagawa, T. 1995. Strong motion data recorded in the source area of the Hyogo-Ken-Nanbu earthquake, January 17, 1995, Japan. *J. Nat. Disas. Sci.* 16: 23-30.

Turcotte, D.L. 1997. Earthquakes, fracture, complexity. In J.R. Willis (ed.), *Proceedings of the IUTAM Symposium on Nonlinear Analysis of Fracture*: 163-175, Dortrecht: Kluwer Academic Publishers.

Uenishi, K., Rossmanith, H.P. & Knasmillner, R.E. 1997a. Interaction of Rayleigh pulse with non-uniformly contacting interfaces. In *Proceedings of the International Conference on Materials and Mechanics '97*: 95-100, Tokyo: The Japan Society of Mechanical Engineers.

Uenishi, K., Rossmanith, H.P., Knasmillner, R.E. & Böswarth, C. 1997b. Rayleigh pulse interaction with partially contacting dissimilar interfaces. To appear in H.P. Rossmanith (ed.), *Proceedings of the First International Conference on Damage and Failure of Interfaces*, Rotterdam: A.A. Balkema.

Viktorov, I.A. 1962. Rayleigh waves in the ultrasonic range. *Soviet Physics – Acoustics* 8: 118-129.

Wald, D.J. 1995. A preliminary dislocation model for the 1995 Kobe (Hyogo-Ken Nanbu), Japan, earthquake determined from strong motion and teleseismic waveforms. *Seism. Res. Let.* 66: 22-28.

Wald, D.J. 1996. Slip history of the 1995 Kobe, Japan, earthquake determined from strong motion, teleseismic, and geodetic data. *J. Phys. Earth* 44: 489-503.

Weertman, J. 1980. Unstable slippage across a fault that separates elastic media of different elastic constants. *J. Geophys. Res.* 85: 1455-1461.

Willis, J.R. 1973. Self-similar problems in elastodynamics. *Phil. Trans. R. Soc. Lond.* 274: 435-491.

Mechanics of Jointed and Faulted Rock, Rossmanith (ed.) © 1998 Taylor & Francis, ISBN 90 5410 955 6

Modeling of characteristics of waves propagating in arbitrary anisotropic media

M.N.Luneva
Institute of Tectonics and Geophysics, FEB RAS, Khabarovsk, Russia

ABSTRACT: The results of computation of main characteristics (slowness surface, ray surface, phase velocity and polarization vector) of three body seismic waves propagating in a homogeneous anisotropic elastic medium of arbitrary symmetry are presented. The study was directed to the correct solution of eigenvalues-eigenvectors problem and shear wave recognition in numerical algorithms for arbitrary direction of propagation in anisotropic media. The effects of wave propagation and their connection with the symmetry elements were investigated in anisotropic media for all basic types of symmetry.

1 INTRODUCTION

In fact, the study of anisotropic properties of the Earth's interior materials plays an important role in seismic and seismological exploration. The origin of the Earth's material anisotropy can be different, either intrinsic (minerals) or induced by the existing or palaeo stress field (cracking, fracturing) or formed under certain conditions (deposits), etc. Uniform homogeneous elastic materials are divided into 8 systems of symmetry, which are described definitely by 8 types of elastic constant tensors including isotropic symmetry (Fedorov F.I. 1968), or into 32 groups of symmetry in accordance with a combination of the basic symmetry elements such as the center (C), plane (P) and axis (L_k, k=2, 3, 4, 6) of symmetry. Each system or even group in the same system of symmetry has specific features.

Wave propagation in anisotropic media differs significantly from that in isotropic media and produces particularly distinctive effects on shear wave behaviour. Although the general theory of wave propagation in anisotropic media is well-known (e.g. Crampin 1984, Petrashen 1980, 1984), there exist some impediments concerning 3-D wavefield modeling in homogeneous elastic anisotropic layered media. One of them under this study is the solution of eigenvalues-eigenvectors problem with the correct recognition of body wave types in numerical algorithms for arbitrary anisotropic media (Petrashen 1980, Fryer & Frazer 1987).

This paper presents a brief description of some aspects in the computing of main characteristics of body waves propagating in arbitrary anisotropic media (slowness, ray and phase velocities, polarization vector). To study the wave effects and to test the numerical algorithm the wave characteristics for all the systems of symmetry were constructed and investigated. The values of elastic constants were taken from experimental mineralogical data (Clark 1966). Under theoretical and numerical study the criteria for body wave recognition in arbitrary anisotropic media are proposed.

2 BASIC THEORY

The mathematical background for the modeling of body wave characteristics for arbitrary anisotropic media is based on the zero approximation of the asymptotic ray theory (Petrashen 1980, 1984).

Let us consider a homogeneous arbitrary anisotropic elastic medium in the Cartesian coordinate system $\Omega\ (x_1,x_2,x_3)$. The relationship between stress τ and strain ε satisfies the Hook's law

$$\tau_{ik} = \sum_{l,q=1}^{3} C_{iklq}\varepsilon_{lq} = \sum_{l,q=1}^{3} C_{iklq}\frac{\partial u_q}{\partial x_l}, \tag{1}$$

where $\mathbf{u}$ is the displacement vector. The properties of anisotropic medium are defined by the four-order symmetric tensor of elastic constants C with components $C_{iklq} = C_{\alpha\beta}$ and density ρ. Let a harmonic plane wave propagate in the direction of unit vector $\mathbf{n}$ with a displacement vector $\mathbf{u}$

$$\mathbf{u} = U\mathbf{e}\cdot\exp[i\omega(t-\mathbf{n}\cdot\mathbf{r}/V)], \tag{2}$$

where U is the displacement amplitude, $\mathbf{e}$ is the unit polarization vector, $\mathbf{r}$ is the radius-vector, V is the phase velocity in the direction $\mathbf{n}$, ω is the circular frequency and t is the time.

The elastodynamic equation of elastic anisotropic medium in terms of the components of the displacement vector $\mathbf{u}$ for lack of body forces gives the system of three differential equations

$$\sum_{k,l,q=1}^{3}\frac{\partial}{\partial x_k}[C_{iklq}\frac{\partial u_q}{\partial x_l}] = \rho\frac{\partial^2 u_i}{\partial t^2}, \quad i=1,2,3. \tag{3}$$

After substitution of eq. (2) into eq. (3) one obtains the system of algebraic equations with the unknown values of polarization vector components e_i and phase velocity V

$$\sum_{k,l,q=1}^{3}\lambda_{iklq}n_k n_l e_q - V^2 e_i = 0, i=1,2,3, \lambda_{iklq} = C_{iklq}/\rho. \tag{4}$$

The condition for the solving of system (4), called Christoffel equations, requires that

$$\left|\Lambda_{iq} - V^2\delta_{iq}\right| = \left|\sum_{k,l=1}^{3}\lambda_{iklq}n_k n_l - V^2\delta_{iq}\right| = 0, \tag{5}$$

$$\Lambda_{iq} = \sum_{k,l=1}^{3}\lambda_{iklq}n_k n_l, \quad i,q=1,2,3. \tag{6}$$

The two-order matrix Λ_{iq} is real, symmetric and positively defined for any values of $n_i \neq 0$. In fact, equations (4) and (5) present the eigenvalue-eigenvector problem, where eigenvalues correspond to three phase velocities V_r^2 (r =1, 2, 3) and eigenvectors correspond to three mutually orthogonal unit polarization vectors $\mathbf{e}^r$. The solutions demonstrate that in anisotropic medium, there are three plane body waves in every direction, one of them is the quasi-compressional wave (qP) and the other two are the quasi-shear waves (qS_1, qS_2). Also, eq. (5) defines three slowness surfaces $\mathbf{p}^r = \mathbf{n}/V_r$ as a function of propagation direction.

A plane wave of r-type can be also characterized by the ray velocity (or group velocity) ξ^r, which corresponds to the energy velocity vector, and its components can be written

$$\xi_k^r = \sum_{i,l,q=1}^{3}\lambda_{iklq}p_l^r e_q^r e_k^r, \quad k=1,2,3. \tag{7}$$

Phase velocity V_r and ray velocity ξ^r, as a rule, are not equal to each other neither in value nor in direction.

3 EIGENVALUES AND EIGENVECTORS

Let us construct the body plane wave parameters for the propagation direction $\mathbf{n}$ in arbitrarily chosen plane Π crossing the center point O of the coordinate system Ω. Introduce the local system of coordinates Ω° with the corresponding orthogonal basic vectors $\mathbf{x}_1^\circ, \mathbf{x}_2^\circ, \mathbf{x}_3^\circ$ and the center at point O. Two coordinate vectors $\mathbf{x}_1^\circ$, $\mathbf{x}_3^\circ$ belong to the plane Π and vector $\mathbf{x}_2^\circ$ is normal to Π. Then, the propagation direction vector can be written

$$\mathbf{n} = n_1\mathbf{x}_1^\circ + n_3\mathbf{x}_3^\circ, \quad n_2 = 0. \tag{8}$$

For every direction $\mathbf{n}$ construct three mutually orthogonal unit vectors $\mathbf{g}_1$, $\mathbf{g}_2$, $\mathbf{g}_3$, where $\mathbf{g}_1 = \mathbf{n}$, $\mathbf{g}_3$ is normal to $\mathbf{n}$ and belongs to plane Π, and $\mathbf{g}_2$ is normal to Π. These unit vectors correspond to three unit polarization vectors of P, SH, SV wave types constructed as for isotropic media.

The tensor of elastic constants has to be transformed correspondingly from C into C° for the local system Ω° using the tensor transformation law. In this case, the Christoffel matrix Λ_{iq} is simplified considering that $\lambda_{iklq} = C^\circ_{iklq}/\rho$. The Christoffel equations (5) may be reduced to a cubic equation with eigenvalues $\chi = V^2$

$$\chi^3 + a_1\chi^2 + a_2\chi + a_3 = 0, \tag{9}$$

where a_1, a_2, a_3 are the real coefficients. The analytical solutions of this equation are known and will be

$$\chi_1 = u + d - \frac{a_1}{3},$$

$$\chi_2 = -\frac{(u+d)}{2} - \frac{a_1}{3} + i(u-d)\frac{\sqrt{3}}{2},$$

$$\chi_3 = -\frac{(u+d)}{2} - \frac{a_1}{3} - i(u-d)\frac{\sqrt{3}}{2}, \tag{10}$$

where $u = (q + \sqrt{p^3+q^2})^{1/3}, d = (q - \sqrt{p^3+q^2})^{1/3}$,

$p = (3a_2 - a_1^2)/9, \quad q = -(a_3 - a_2 a_1/3 + a_1^3/27)/2.$

Three roots obtained define three phase velocities $V^r = (\chi_r)^{1/2}$ which are real, positive and correspond strictly to qP, qS_1 and qS_2 waves except certain cases. For the travelling waves, the relationship between roots is defined to be $\chi_1 > \chi_3 \geq \chi_2$.

Analysis of the Christoffel matrix Λ_{iq} gives three definite forms (11) which lead to three ways of finding eigenvectors associated with the direction of wave propagation and symmetry elements of the media. The first form Λ'_{iq} with the equality of any two elements defines shear wave singularities, where their slowness surfaces touch each other. It occurs when the direction of propagation $\mathbf{n}$ is parallel to the highest order

$$\Lambda^1_{iq}=\begin{pmatrix}\Lambda_{11}&0&0\\0&\Lambda_{22}&0\\0&0&\Lambda_{33}\end{pmatrix},\quad \Lambda^2_{iq}=\begin{pmatrix}\Lambda_{11}&0&\Lambda_{13}\\0&\Lambda_{22}&0\\\Lambda_{31}&0&\Lambda_{33}\end{pmatrix},$$

$$\Lambda^3_{iq}=\begin{pmatrix}\Lambda_{11}&\Lambda_{12}&\Lambda_{13}\\\Lambda_{21}&\Lambda_{22}&\Lambda_{23}\\\Lambda_{31}&\Lambda_{32}&\Lambda_{33}\end{pmatrix}. \qquad (11)$$

symmetry axis L_k (Fedorov 1968). In this case, the pure compressional and shear waves propagate along **n**. The polarization vector of P-wave will be $\mathbf{e}^1=\mathbf{n}$. The shear wave eigenvectors will be perpendicular to **n** and can be chosen arbitrarily, so we get them to be equal to $\mathbf{e}^2=\mathbf{g}_2$ and $\mathbf{e}^3=\mathbf{g}_3$ that corresponds to pure SH and SV polarization. For the symmetry axis of even order (k = 2, 4, 6), the slowness surfaces or phase velocity functions of shear waves come into tangential contact at the point where their phase and ray velocities are equal: $V_2=V_3$, $\xi^2=\xi^3$. There is an exception for the trigonal system with the three-fold axis where the contact between the shear wave slowness surfaces is neither the tangential one nor intersection point and that corresponds to $V_2=V_3$, whereas $|\xi^2|=|\xi^3|$. In case there is no equality between elements in matrix Λ^1_{iq}, three pure waves propagate, their slowness surfaces being separated.

The second form Λ^2_{iq} associates with the existence of a single pure shear wave. According to Fedorov (1968) and numerical computations for anisotropic media including all the systems of symmetry, it takes place when the plane Π coincides with the symmetry plane P or the plane Π is perpendicular to the symmetry axis of even order L_k. To provide non-trivial solutions the system of equations (4) must be reduced to two equations taking into consideration that $(\Lambda_{22}-V_r^2)e_2^r=0$. The latter means the propagation of shear wave with pure polarization of SH type along the $\mathbf{x}_2^\circ$-axis and normally oriented to the plane P. When $\Lambda_{22}=V_2^2$, qS_1 wave has polarization $\mathbf{e}^2=\mathbf{g}_2$. When $\Lambda_{22}=V_3^2$, we replace the roots of eq. (10): $V_2=(\chi_3)^{1/2}, V_3=(\chi_2)^{1/2}$. The last situation means that there exist intersection points of shear wave slowness surfaces in this plane where $V_2=V_3$. The correlation of phase velocities changes between intersection points and can be either $V_1>V_2\geq V_3$ or $V_1>V_3\geq V_2$. The polarization vectors of qP and qS_2 waves lie in the plane P and rotate synchronously.

When the Christoffel matrix has the third form Λ^3_{iq} (11), there is no problem to compute eigenvectors and the correlation of phase velocities is stable and expressed by $V_1>V_3>V_2$. Note, in this case, the ray velocity and polarization vectors of three body waves may vary significantly in space.

Note, the idenification of shear waves as qS_1 or qS_2 depends on the local system orientation like definition of SV and SH waves for isotropic case.

4 NUMERICAL EXAMPLES

In this section, we display some numerical examples of three body wave characteristics computed in anisotropic media . The results are presented by the plots of slowness surface, phase velocity, ray surface, angle shift and angle of ray inclination for the chosen planes with variation of propagation direction **n** in the range from 0° to 360°. The orientation of planes Π in space is defined by two angles Θ and Ψ, which represent the angles between the plane P and axes $\mathbf{x}_3$ and $\mathbf{x}_1$, relatively. The angle shift is defined as the space angle of inclination α^r between polarization vector $\mathbf{e}^r$ and the corresponding vector $\mathbf{g}_r$. The ray surface is plotted as the projection of ray velocity vectors on plane Π. The angle β^r defines the angle between the ray (or ray velocity vector) and plane Π.

4.1 Cubic system Fluorite

Figure 1 shows wave characteristics in four differently oriented planes in anisotropic media of the highest symmetry.

a) Plane Π_1: $\Theta=0°$, $\Psi=0°$. Plane Π_1 belongs to the vertical symmetry plane P and includes 2 axes L_4 and 2 axes L_2. Two symmetry planes, which are oriented at 45° from the vertical axis and coincide with the axis L_2, cross normally Π_1.

The tangential contacts between shear wave slowness surfaces are observed at 4 points (γ = 0°, 90°, 180°, 270°) parallel to the axes L_4. Along these singular directions three body waves have polarization of pure compressional and shear waves and $V_2=V_3$. Because $\Pi_1\subset P$, the qS_1-wave has pure polarization of SH type. Phase velocity of qS_1-wave is constant for all directions. Polarization vectors of qP and qS_2 waves rotate synchronously in plane Π_1. The correlation of phase velocities is stable $V_1>V_3\geq V_2$, and ray velocity vectors belong to the plane.

Two particular directions associate with the axes L_2. It is known (Fedorov 1968), that pure P- wave propagates along any symmetry axis, but behaviour of shear waves may be different. In this case, two waves are also pure because one of them is already defined above as SH -wave.

b) Plane Π_2: $\Theta=0°$, $\Psi=15°$. Plane Π_2 includes vertical four-fold axis L_4.

Three pure waves propagate along the singular directions parallel to L_4. Except these directions, the

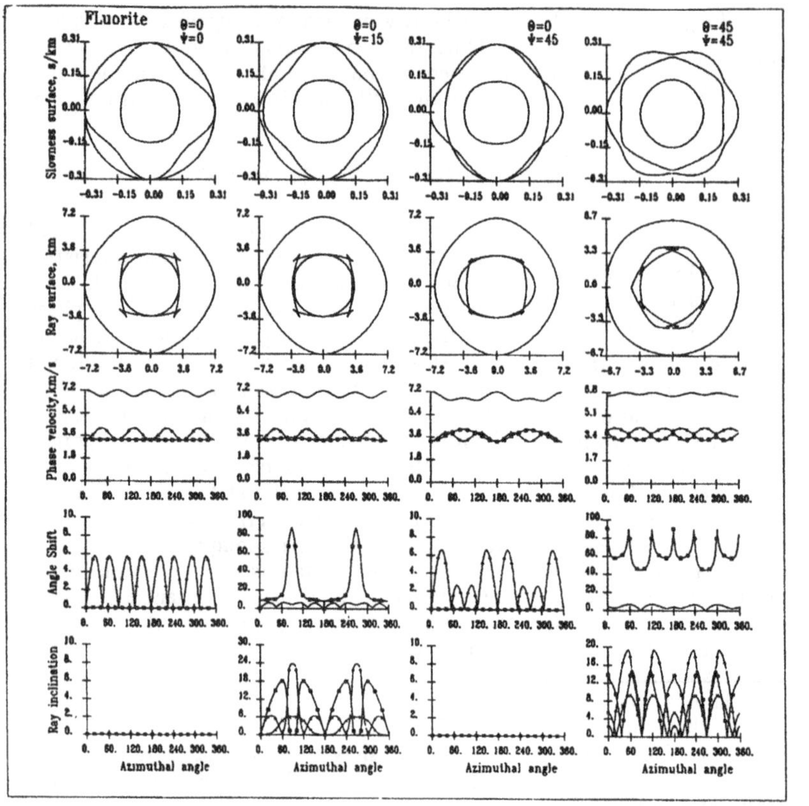

Group of symmetry: $3L_4 4L_3 6L_2 9PC$

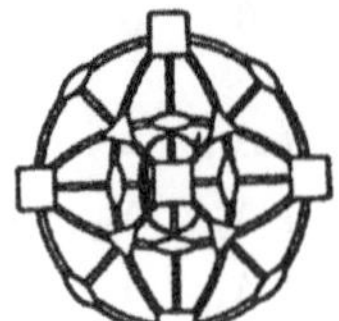

Fig. 1 Behaviour of the slowness surface, ray surface, phase velocity, angle shift and angle of ray inclination of three body waves propagating in cubic fluorite in differently oriented planes. Solid line, squares and triangles mark qP, qS_1, qS_2 waves, respectively.

phase velocity of qS_2-wave is always higher than that of qS_1-wave. The quasi-shear wave polarization vectors vary synchronously in space, and the angle shift changes in the range of 90°. The ray velocity vectors (or rays) lie beyond the plane.

c) Plane Π_3: $\Theta = 0°$, $\Psi = 45°$. Plane Π_3 belongs to the inclined symmetry plane P and includes vertical axis L_4, horizontal axis L_2 and 2 axes L_3 declined from the vertical axis at 54°44'.

The singular directions parallel to the axis L_4 with the pure wave polarization are observed at 2 points. Because $\Pi_3 \subset P$, qS_1-wave has pure polarization of *SH* type. Two particular directions with three pure wave propagation along the axes L_3 define 4 points of intersection at the curves (γ = 54°44', 125°16', 234°44', 305°16'). Other particular directions are parallel to L_2 and give as well three pure waves at points γ = 90°, 270°, where $V_2 > V_3$. Polarization vectors of qP and qS_2 waves rotate synchronically in the plane.

d) Plane Π_4: $\Theta = 45°$, $\Psi = 45°$. Plane Π_4 does not associate with any symmetry element, but locates near the symmetry axis L_3.

In this plane, there is no particular direction of wave propagation. The phase velocity correlation is constant ($V_1 > V_3 > V_2$). The quasi-shear wave polarization vectors vary synchronically in space. Slowness and ray curves tend to the hexagonal symmetry. Note that in the plane oriented normally to L_3, the curves have distinct hexagonal symmetry.

4.2 Trigonal system Calcite

Figure 2 displays specific properties associated with the vertical three-fold axis. The horizontal coordinate axis is oriented along the axis L_2.

a) Plane Π_1: $\Theta = 90°$, $\Psi = 0°$. The plane is perpendicular to the vertical axis L_3 and includes 3 horizontal axes L_2. Three vertical symmetry planes cross Π_1 between axes L_2.

Wave curves have hexagonal symmetry in horizontal plane with separation of slowness and ray surfaces of the waves. Both shear waves have cusps at ray surface. There are 12 particular directions. Six of them belong to the lines of

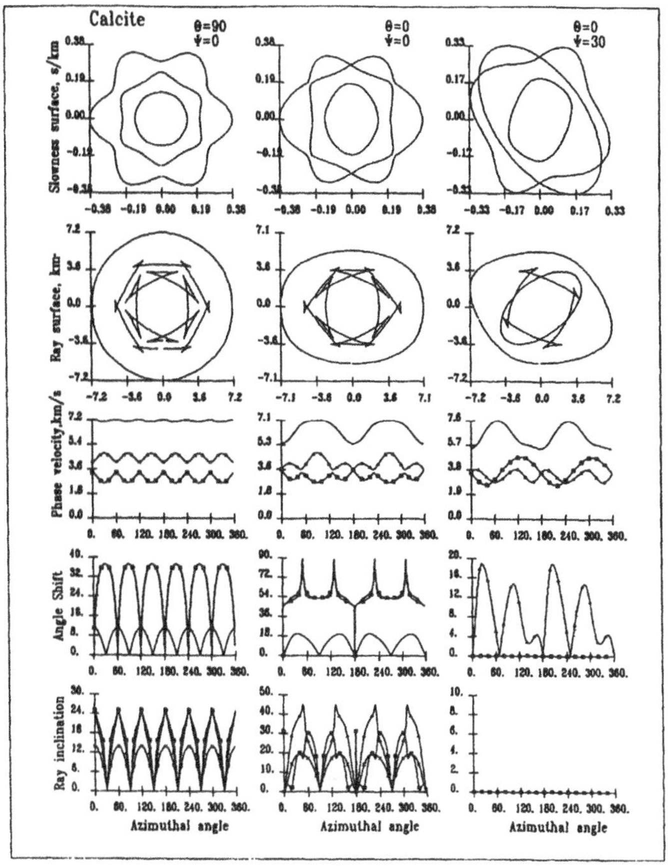

Fig. 2 As in Fig.1, but for trigonal calcite.

Group of symmetry: $L_3 3L_2 3PC$

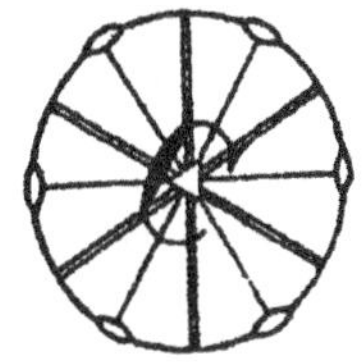

intersection of symmetry planes P with Π_1. Along these directions ($\gamma = 0°, 60°, ..., 300°$), qS_2-wave has pure polarization of SV type, and the displacements of the other waves belong to the vertical symmetry planes. The other 6 directions are parallel to the axes L_2. Along these directions, pure P-waves propagate.

b) Plane Π_2*:* $\Theta = 0°$, $\Psi = 0°$. The plane includes the vertical axis L_3 and horizontal axis L_2.

Three pure waves propagate along the vertical axis L_3 with $V_2 = V_3$, $|\xi^2| = |\xi^3|$. Along the horizontal axis L_2, the pure P-wave propagates. The ray surfaces of shear waves have a complicated form, and inclination of rays reaches the value of about 45°. The correlation of phase velocities of waves is described as $V_1 > V_3 \geq V_2$.

c) Plane Π_3*:* $\Theta = 0°$, $\Psi = 30°$. Π_3 belongs to the vertical symmetry plane and includes vertical axis L_3.

In this plane, three pure waves propagate along the vertical axis L_3 also. Because $\Pi_3 \subset P$, the qS_1-wave has pure polarization of SH type. The correlation of shear wave phase velocities changes at 2 points of intersection.

4.3 Monoclinic system Hornblende

Monoclinic medium is characterized by 3 symmetry elements only. The horizontal axis is normal to the two-fold axis and belongs to the symmetry plane. So, when the directions of wave propagation belong to P, qS_1-wave has pure polarization of SH type (Fig. 3; $\Theta = 90°$, $\Psi = 0°$) and in the directions parallel to L_2 the pure P-wave will propagate. In any other planes, it is observed a significant variation of wave parameters in space, but with constant relationship of phase velocities: $V_1 > V_3 > V_2$ (Fig. 3).

CONCLUSIONS

Theoretical and numerical studies show that numerical solution of eigenvalue-eigenvector problem can be correctly and easily performed in the local coordinate system for the chosen planes.

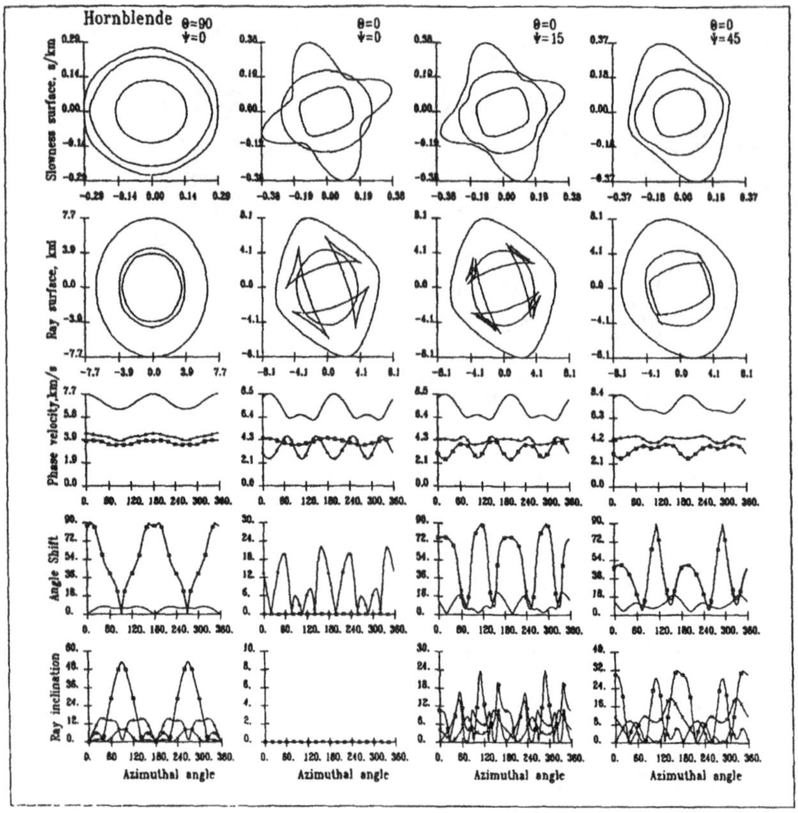

Group of symmetry:
L_2PC

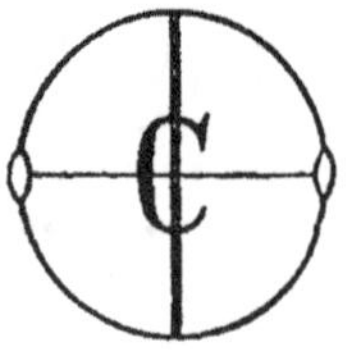

Fig. 3 As in Fig.1, but for monoclinic hornblende.

It gives also advantages in the computing of wavefield in 3-D layered media. Wave properties depend on the combination of symmetry elements of medium and on directions also. The wave effects can be formulated for different directions as follows:

1) Singular directions correspond to the propagation of three pure waves with the shear wave slowness surface touching. It occurs when the direction of wave propagation is parallel to the symmetry axis of the highest order.

2) Particular directions correspond to the directions along which some waves propagate with pure polarization. If the wave direction belongs to the symmetry plane or is perpendicular to the symmetry axis of even order, there exists one wave with pure *SH* polarization. In this case, intersection points may exist between shear wave surfaces. The directions parallel to the symmetry axis are always associated with the pure *P*-wave propagation. Combination of symmetry elements in certain directions may give propagation of some waves with pure polarization.

3) If the direction of wave propagation does not associate with symmetry elements, there is no contact or intersection point between waves, and the phase velocity relationship is stable ($V_1 > V_3 > V_2$), polarization and ray velocity vectors may vary significantly in space.

REFERENCES

Clark S.P(ed). 1966. Handbook of physical constants. *Yale University. New Haven. Connecticut. Geol. Soc. of America, Inc. Memoir 97.*

Crampin S. 1984. An introduction to wave propagation in anisotropic media. *Geophys. J. R. astr. Soc.*. 76: 17-28.

Petrashen G.I. 1980. Wave propagation in elastic anisotropic media. *Leningrad: Nauka (in Russian)*.

Petrashen G.I (ed). 1984. Bulk wave propagation and methods of wave field computation in elastic anisotropic media. *Leningrad: Nauka (in Russian)*.

Fedorov F.I. 1968. Theory of elastic waves in crystals. *New York: Plenum Press.*

Fryer G.J. & L.N. Frazer 1987. Seismic waves in stratified anisotropic media - II. Elastodynamic eigensolutions for some anisotropic system. *Geophys. J. R. astr. Soc.. 91: 73-101.*

Mechanics of Jointed and Faulted Rock, Rossmanith (ed.) © 1998 Taylor & Francis, ISBN 90 5410 955 6

Time-dependent elastic properties in the earth and their causes

G.H.R. Bokelmann
Institute of Geophysics, Bochum University, Germany

ABSTRACT: A technique for accurate measurement of time-dependent changes in seismic propagation velocities are presented The focus of this technique is to monitor changes in elastic stresses acting on the subsurface down to a depth of several kilometers. In addition, motion of fluids can also be detected by this technique. The method utilizes elastic waves, which are generated continuously by large imperfectly balanced machines running synchronously with the power network. We make use of the solid Earth tides as test signals, since they are associated with known stresses in the Earth's interior. That test signal suggests that we are indeed observing effects related to changes within the Earth. In that sense, the method should be superior to surface deformation measurements, in which the changes at depth are frequently masked by near-surface artefacts

1 INTRODUCTION

Despite the known complexity of the earthquake source process, a simple fact is that earthquakes occur precisely at the time when stresses in the subsurface reach the critical value. This suggests that methodologies for measuring changes of stress near potential earthquake locations have great potential for the tantalizing question of seismic hazard estimation.

This is complicated, however, by the fact that 1) most quakes are initiated far from the Earth's surface, at depths between 5 and 20 km, and 2) that the Earth's crust appears to be in a state of near-criticality (Crampin 1994) indicating that stress changes leading to quake initiation are likely to be relatively small compared with the absolute level of stress.

Current drilling technology (hydraulic fracturing) allows measurement of stress levels only with large uncertainties easily exceeding 10%. These uncertainties are several orders of magnitude larger than the temporal stress changes of interest.

On the other hand, seismic waves can propagate through the crust down to the seismogenic zone. Importantly, the velocities of such elastic waves depend on the stress level. This dependence may be greatly enhanced, if distributed fluid-filled cracks are present throughout the medium, since the cracks can then be deformed more strongly than a homogeneous rock matrix. Since velocities are derived from measurement of arrival times (or phase), we obtain optimum accuracy, if we remove the tradeoff with path length by sampling the same ray path at different times. This is guaranteed if both source and receiver are fixed in space. In this paper we demonstrate that we can obtain such sources for free. In fact, waves radiated by vibrations of certain large machines can be used to continuously monitor changes in the velocity of crustal waves. Observations discussed below suggest that these changes are in fact related to stress changes within the crust. An added feature of velocity monitoring is that nature supplies us with an ideal test signal, the solid Earth tides, which impose known stress changes on the crust. Observed solid Earth tides thus serve as a confirmation and more importantly for calibrating observed velocity changes with respect to stress changes.

2 OBSERVATIONS

2.1 *Narrowband signals in the noise*

The typical spectrum of a long time window recorded by a short-period seismic sensor (1Hz eigenfrequency) is displayed in Figure 1. The velocity-proportional instrument represents one of 25 stations of the GERESS array (Harjes 1991) in the Bohemian Forest of Bavaria. Such spectra typically show a strong maximum near 0.2 Hz (microseismic noise), the instrumentally-caused decay to lower frequencies and the smooth decay to higher frequencies. In addition we find a number of sharp peaks. The strongest peak is located near 2.083 Hz which agrees closely with one of the possible rotation frequencies of power-network synchronous machines, which is f = 50/24Hz = 2.083Hz (Rentzsch 1968). We will focus on that signal in this paper. The observation of the signal is not new. At that frequency, a signal had in fact been observed even before the advent of digital data (Plesinger & Wielandt 1974 and references therein).

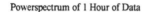

Powerspectrum of 1 Hour of Data

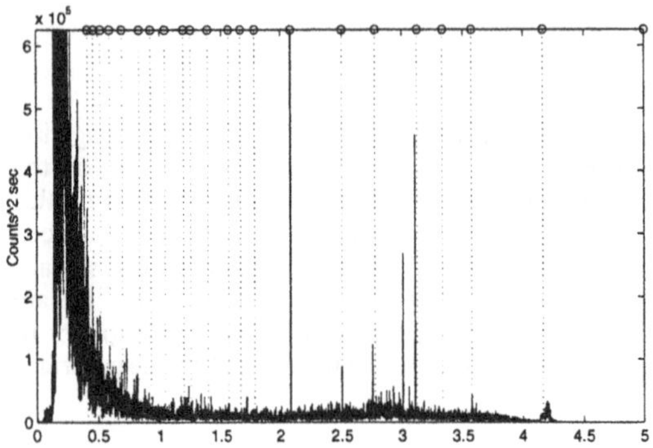

Figure 1. Averaged spectrum of a 1-hr time window compared with allowed rotation frequencies for machines running synchronously with the 50Hz power network (circles at top and dotted lines). The x-axis shows frequency in Hz

The display of frequency versus time (Baisch 1997; Bokelmann 1997) showed that the frequency of the signal at 50/24 Hz is actually remarkably stable.

Several of the other observed peaks are also at machine frequencies, several are not.

2.2 *Further inferences about the signal*

From the analysis of array data of the GERESS array Baisch (1997) found that the wave propagates with an apparent velocity of about v_0= 4 km/sec. This value corresponds to the apparent velocity of the dominant phases of regional seismograms observed at the same location, suggesting that the energy of this narrowband signal propagates as a regional phase, most likely of Lg type.

The wave was shown to arrive from a direction of 348° at GERESS. Together with a direction estimate from a second array in Bavaria the source was constrained to lie near the German-Czech border.

The distance to which the signal can be detected is constrained differently: In the data analyzed previously, there was a time-window of 11 hours duration, in which the signal was apparently switched off. Analysis of other stations distributed over Germany also showed the gap suggesting 1) that the waves generated by the source near the German-Czech border propagate over several hundred kilometers and 2) that there is a single source which dominates the coherent wavefield at that frequency in the area near the German-Czech border.

3 TEMPORAL VARIATIONS

Temporal changes of the phase velocity of the wave can be determined from the difference $\Delta\phi(t)$ of the phase at two stations:

$$\Delta v(t) = -\frac{v_0^2}{2\pi f x}\Delta\phi(t) + const \qquad (1)$$

Our test station pair has a separation of 3.8 km. The projection onto the signal propagation direction has a length of x=3.3 km. For simplicity we display in Figure 2 the measured phase differences themselves instead of the velocity changes as a function of time, for the frequency f = 50/24 Hz.

The gaps in the 17-day time series represent contaminated time windows due to transient events as earthquake waves. They were elimininated from the time series

There is an indication of a slight trend in the phase differences. That may be the effect of a long-time change in the stress acting on the subsurface. While that is the kind of signal we will eventually be most interested in, we focus in this paper on whether the time series in Figure 2 contains our stress test signal, the solid Earth tides.

Periodic signals as the tides are best analyzed by a spectrum analysis of the time series. In this case, such

Phase Differences for 17 Days

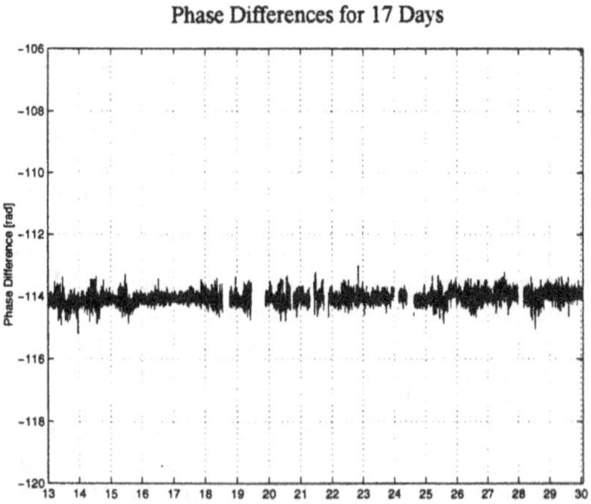

Figure 2. Phase difference as a function of time. The number gives the day within march of 1997.

an analysis is complicated due to the gaps though. We have to either resort to interpolation techniques or to spectrum estimation operating on non-equidistant-sampled time series as the Lomb-Scargle-Periodogram (Press & Rybicki 1989). Here we take the first approach and find interpolated values for the gap after severe median-filtering. The resulting spectrum using a subwindow length of 5.7 days is shown in Figure 3.

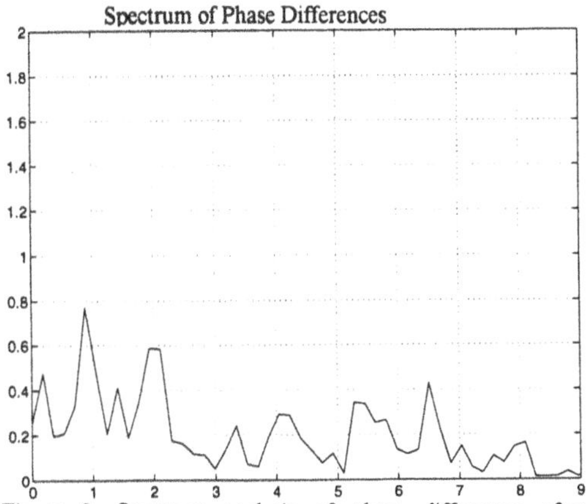

Figure 3 Spectrum analysis of phase differences after interpolation and removing the mean (5.7-day subwindow length) The x-axis is in counts/day.

We obtain two dominant peaks indicating periodicities of 1/day and 2/day. Are these periodicities caused by the solid Earth tides? For comparison we predict the deformation due to tidal stresses (Wilhelm & Zürn 1984) for the vicinity of the GERESS array. Figure 4 displays predicted horizontal deformations for a time window of 6 days and the associated spectra. Since we do not know beforehand, which tidal deformation component we are observing, we display two horizontal components of deformation, parallel to the propagation direction of the wave and perpendicular to it.

Predicted tidal deformations show periodicities of 1/day and 2/day, just like the observed phase differences of the elastic wave. This suggests that tidal stresses and associated deformations cause the periodic changes in the propagation velocity of the wave. Judging from the two components of the tidal deformations, the phase differences are apparently more sensitive to the component of stress or deformation orthogonal to the propagation direction than parallel. This is consistent with what we expect for a shear wave as Lg.

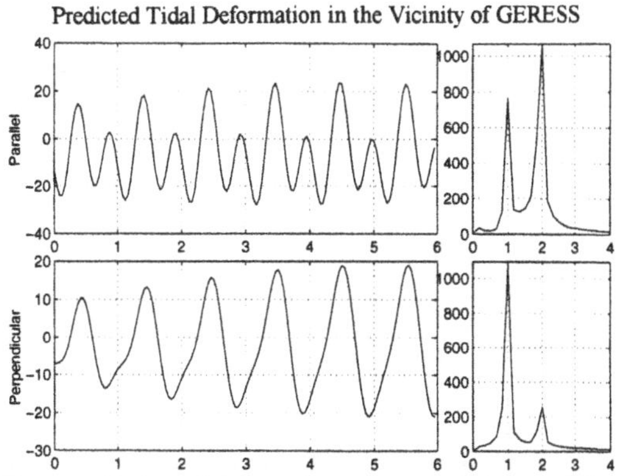

Figure 4. Horizontal deformation (nanostrains) due to solid Earth tides for a time window of 6 days (left). On the right the associated spectra (after removing the mean) are shown in counts/day. The upper half is for the deformation component parallel to the wave propagation direction, the lower half perpendicular to it.

4 CONCLUSIONS

We have illustrated a methodology for continuously monitoring changes of seismic velocity within the Earth's crust. Since the wave propagation mechanism is that of a regional phase, most likely of Lg-type, the velocity changes should be interpreted as changes of a crustal average velocity with time.

The observation of the tidal test signal suggests that the velocity variations are related to stresses and associated deformation within the crust. The trend in the phase differences we see may therefore be due to changes in the stress level within the crust. From the tidal test signal it appears that we can resolve stress changes as small as the tides, which are at most on the order of 10^{-1} bar. In comparison, stresses released by earthquakes are 1 to 3 orders of magnitude larger. Thus, the proposed measurement tool should indeed be rather useful for observing the tectonic stress buildup within the crust.

The temporal change of stress is one of the two critical factors involved in the occurence of earthquakes. The other factor is the value of the critical stress. It is known that this value depends on the amount of pore fluids in the material (Nur & Booker 1972). Interestingly, the amount of fluid also influences the wave velocity of elastic waves. Thus the two crucial players in triggering earthquakes may in principle both show up in observations of velocity changes. In the case of the above data example however, substantial water flow in large regions of the subsurface appears unlikely as a possible cause of the trend in Figure 2 due to the short time scale of only 17 days.

The wave we observe propagates within Central Europe. Most likely. similar conditions exist in many other locations on the Earth. Using waves generated by suitable machines (power plants etc.) supplies us with a tool for simple monitoring of the state of the crust at low cost. However, the result of long time studies must be awaited to see whether these signals can be used for very long time windows, that is for monitoring over time spans of years.

To discuss various methods for monitoring changes in the elastic properties of the crust, a special session was recently organized at the American Geophysical Union Fall Meeting 1997 in San Francisco (*Temporal Variations of Elastic Properties of the Earth's Crust;* Conveners: G. Bokelmann & R. Aster). A number of approaches based on data from natural seismic events and from explosions deliberately set up to measure changes in wave velocities with time were presented. The method for continuous measurement using machine signals, which was presented here, may be particularly interesting due to its continuous coverage, low cost and simplicity. All available methods should be tested in a long-time experiment possibly within the context of large-scale monitoring efforts of certain endangered urban areas

ACKNOWLEDGEMENTS.

Help of Stefan Baisch in data analysis is gratefully acknowledged. I also wish to thank Walter Zürn for valuable discussions The project described here has been funded in part by the Deutsche Forschungsgemeinschaft.

REFERENCES

Baisch, S. 1997. Untersuchung schmalbandiger Signale im GERESS-Rauschspektrum. *Diplomarbeit, Ruhr-Universität Bochum.*

Bokelmann, G.H.R. 1997. Messung zeitlicher Variationen von Wellengeschwindigkeiten in der Erdkruste. *Mitt. der Deut. Geophys. Ges.* 3: 6-12.

Crampin, S. 1994. The fracture criticality of crustal rocks. *Geophys. J. Int.* 118: 428-438.

Harjes, H.-P. 1991. Design and siting of a new regional array in Central Europe, *Bull. Seism. Soc. Am.* 80: 1801-1817.

Nur, A., Booker, J.R. 1972. Aftershocks caused by fluid flow? *Science*. 175: 885-887.

Plesinger, A., Wielandt, E. 1974. Seismic noise at 2 Hz in Europe, *J. Geophys.* 40: 131-136.

Press, H., Rybicki, G.B. 1989. Fast algorithm for spectral analysis of unevenly sampled data. *Astrophys. J.* 338: 277-280.

Rentzsch, H. 1968. *Handbuch für Elektromotoren*. Brown Bovery & Cie AG, Mannheim.

Wilhelm, H., Zürn, W. 1984. Tides of the Earth, in: Landolt-Börnstein. *Zahlenwerte und Funktionen aus Naturwissenschaft und Technik*, Neue Serie, Group V, Volume 2, Geophysics of the Solid Earth, the Moon and the Planets, Springer-Verlag, Berlin.

Mechanics of Jointed and Faulted Rock, Rossmanith (ed.) © 1998 Taylor & Francis, ISBN 90 5410 955 6

Electrical resistivity of jointed rock and its correlation to elastic wave velocity

I. Sekine & H. Nishimaki
Toda Corporation, Tokyo, Japan

A. Saito
Mitsui Mineral Development Engineering Co., Ltd, Tokyo, Japan

R. Yoshinaka
Saitama University, Japan

ABSTRACT: In order to interpret resistivity profiles derived from electrical and electromagnetic surveys, it is necessary to study the influence of joints on rock resistivity. In this paper, the resistivity of a rock mass is expressed by the resistivity model in which intact rock resistivity and fracture resistivity are regarded in parallel, and by introducing a coefficient α that varies with the state of fracture, we generalized this model for a rock mass that includes various fractures. To confirm this model experimentally, the results of calculation were compared with the experimental results. When α was 0.3, the calculation results agreed well with the experimental results. Lastly, as elastic wave velocity is useful for rock mass evaluation, we derived an equation that indicated the correlation of resistivity and elastic wave velocity for fractured specimens. We confirmed that this equation was applicable under confining pressure.

1 INTRODUCTION

Clarification of the geological structure in problems such as underground excavation is important, to maintain safety and low cost during excavation works. Geophysical investigation, by methods such as electromagnetics and resistivity tomography, has recently been applied to construction projects such as underground excavation. In order to interpret resistivity profiles derived from these methods, two points need to be considered. The first is to study the influence of fractures on rock mass resistivity. There is general agreement that resistivity decreases as fracturing increases. But the manner in which rock mass resistivity decreases with the state of fracture in the rock mass is unclear. The second point, which is more important, is to investigate the correlation between rock mass resistivity and geoengineering properties. The value of elastic wave velocity is frequently used for evaluating rock mass. If we can understand the correlation between resistivity and elastic wave velocity, we can obtained data to evaluate the rock mass indirectly by resistivity measurements. From field measurements, however, it is difficult to derive quantitative data related to these points, therefore we conducted laboratory tests.

In this paper, we propose a resistivity model of rock mass including many fractures. To confirm this model experimentally, the results of calculation were compared with the experimental results of specimens including many fractures. By using this model, we showed the resistivity of fractured rock mass for various conditions. We then determined the correlation of resistivity and elastic wave velocity for fractured specimens. We also studied the influence of confining pressure on this correlation.

2 ELECTRICAL RESISTIVITY OF ROCK MASS

2.1 *Resistivity model of rock mass*

We considered that the resistivity of a rock mass containing many fractures is given by the resistivity model in which intact rock resistivity and fracture resistivity are regarded in parallel. The current flows through intact rock, and crosses various fractures. However, the influence of such fractures can be disregarded, because the influence of the fractures crossing a current flow is small (Sekine et al., 1997).

This model is expressed by the following equation:

$$\frac{1}{\rho}=\frac{\alpha \cdot t \cdot N}{\rho_J}+\frac{1}{\rho_I} \tag{1}$$

where, ρ is the resistivity of rock mass including fractures, ρ_J is the resistivity of fractures filled with water or clay, ρ_I is the intact rock resistivity, α is the coefficient that varies with the state of

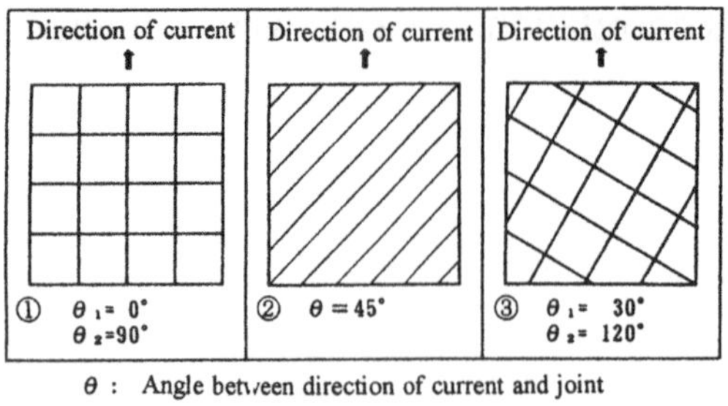

Figure 1. Rock mass model including two series of continuous fractures.

fracture, t is the average of apparent thickness of fracture opening along the research line, and N is the frequency of fracture.

This resistivity model was applied to the rock mass model including two series of continuous fractures. This rock mass model is shown in Figure 1. The results calculated by equation (1) gave a value of 0.5 for model ①, 0.71 for model ②, and 0.68 for model ③.

2.2 *Resistivity measurement of fractured rock*

To investigate the resistivity of rock mass containing many fractures, we made rock mass models including fractures by heating from granite and diorite. The specimens were 5 cm diameter and 10 cm long. These specimens were heated in an electrical muffle furnace. By changing the maximum heated temperatures in the range from 450 ℃ to 1,000 ℃, we were able to prepare specimens of various effective porosities. Before heating, the initial porosity of these specimens was about 0.7%. After heating, the maximum porosity of the specimens was about 8%. These specimens contained many open cracks.

These specimens were saturated under various water resistivities. The resistivity of the specimens was measured by the four-electrode method (Parkhomenko 1967). A schematic of the measurement apparatus with electrodes attached to a sample is shown in Figure 2. The resistivity of the sample ρ may be calculated using following formula:

$$\rho = \pi d^2 V / 4LI \qquad (2)$$

where, d is diameter of the sample, V is voltage across the potential electrodes, L is the separation between the upper and lower potential electrodes, and I is current.

2.3 *Experiment results*

By regarding the resistivity of these specimens as rock mass resistivity, the results of calculation by equation (1) were compared with the experimental results. The relationship between rock resistivity and effective porosity for various pore water resistivities is shown in Figure 3. From this figure, the resistivity of fractured specimen was divided by that of the intact rock specimen, as shown in Figure 4. By regarding these experiment results as those of a fractured rock mass, we confirmed the above-mentioned model. Specifically, we regarded the increase in effective porosity from that of intact rock to be due to the fractured part. By considering that the resistivity of pore water is equal to that of fracture, we calculated the resistivities of fractured specimens by equation (1). These results were added to Figure 4.

When α of equation (1) was 0.3, the

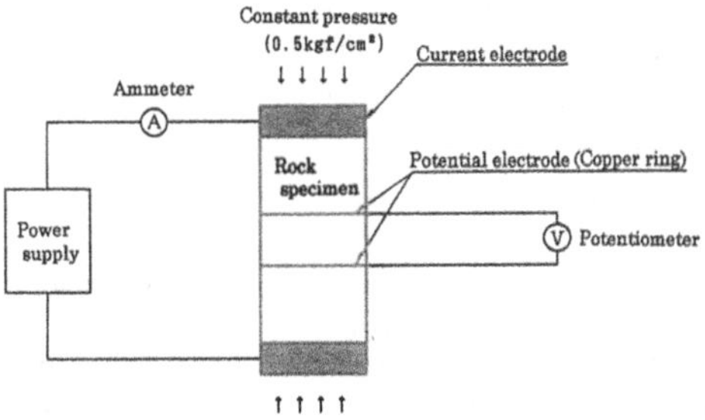

Figure 2. A schematic of the measurement apparatus.

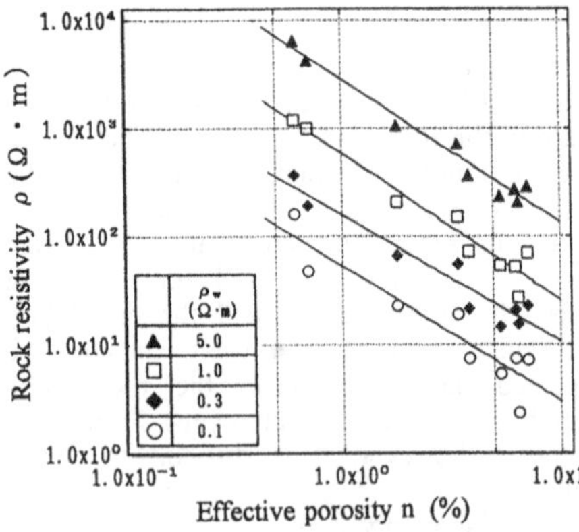

Figure 3. The relationship between rock resistivity and effective porosity.

calculation results agreed well with the experimental results. This value (α = 0.3) was smaller than that of the model containing two series of continuous fractures shown in Figure 1. In order to study the cause of this discrepancy, the fractured specimens were observed with a polarizing microscope (Photograph 1). These observations showed that the fractures of these heated specimens were bent and discontinuous, and also pockets of pores were included in these specimens. These factors made α small. Therefore it was considered that the value of α (= 0.3) obtained from these specimens was due to the state of fractures.

From equation (1), the value of ρ / ρ_i was calculated from the various values of α, t· N, ρ_i / ρ_j. The results of calculation are summarized in

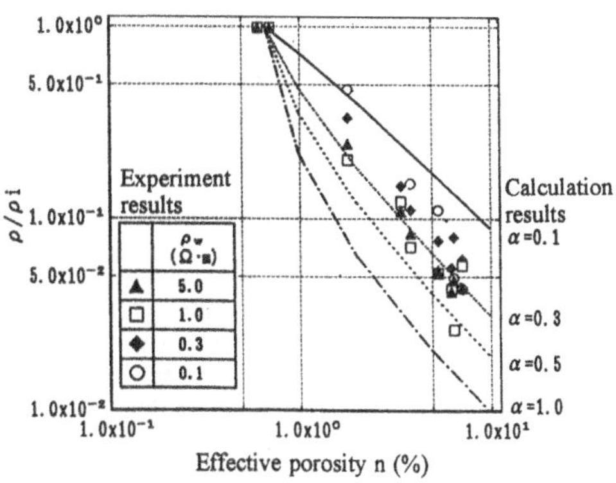

Figure 4. Comparison of the experimental results with calculated results.

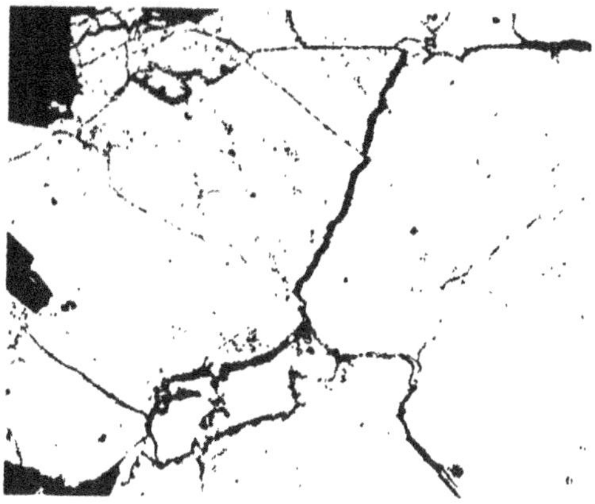

Photograph 1. Crack observation of heated granite.

Figure 5. As this figure indicates, the resistivity of rock mass composed of high resistivity rock decreases as fracturing increases. However, the resistivity of rock mass composed of low resistivity rock hardly decreases with fracturing increased. From Figure 5, we can estimate the resistivity of rock mass containing fractures. Until now, it has been qualitatively known that rock resistivity decreases as fracturing increases. But it was not understood quantitatively how rock resistivity decreases with it. Figure 5 shows quantitatively how rock resistivity decreases as fracturing increases.

3 RELATIONSHIP BETWEEN RESISTIVITY AND ELASTIC WAVE VELOCITY

In Figure 5, the rock resistivity that indicated high resistivity such a granite decreases largely by a fracture, because a large difference between a intact rock resistivity and a fracture resistivity is recognized. Therefore, the relation between a resistivity and elastic wave velocity is studied for the rock that shows a high resistivity. Elastic wave velocity, effective porosity and resistivity of fractured specimens were measured. These specimens were made by heating from granite and diorite with the same method that showed it to 2.2 clauses. The relationship between elastic wave velocity and effective porosity is shown in Figure 6. The white circles (○) are the experimental data derived from granite specimens and black circles (●) are those of diorite. The double circles indicate experimental data for intact unheated rocks. As it is well known that the elastic wave velocity of rock is given by the *time average equation*, the results calculated by this equation were added to

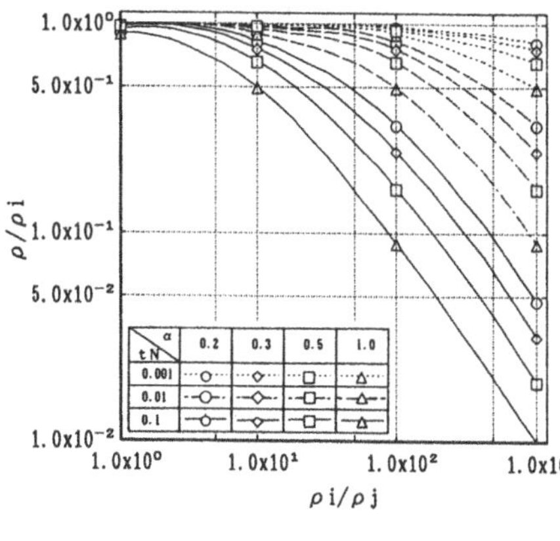

Figure 5. Calculation results of jointed rock mass.

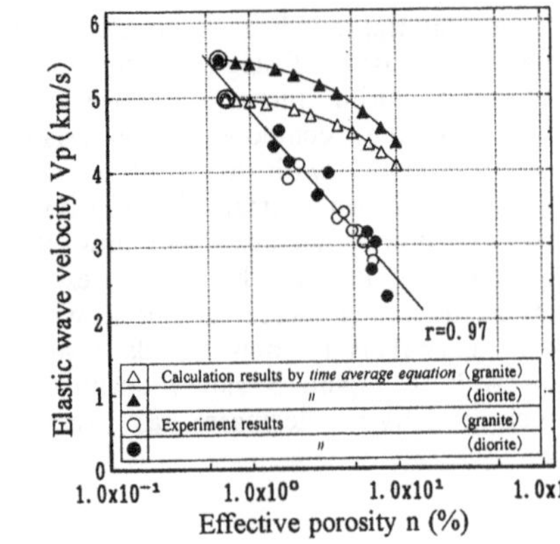

Figure 6. Relationship between elastic wave velocity and effective porosity.

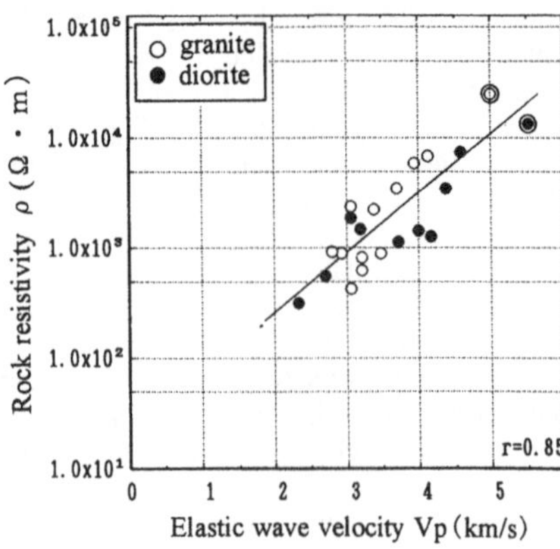

Figure 8. Relationship between rock resistivities and elastic wave velocity.

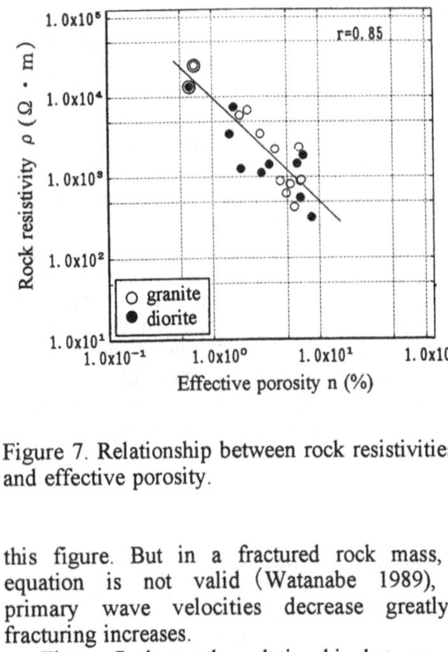

Figure 7. Relationship between rock resistivities and effective porosity.

this figure. But in a fractured rock mass, this equation is not valid (Watanabe 1989), and primary wave velocities decrease greatly as fracturing increases.

Figure 7 shows the relationship between rock resistivity and effective porosity. The resistivity measurements were carried out with pore water with a resistivity of 80 Ω·m. This pore water resistivity is often observed in Japanese mountainous regions. Resistivities were found to be strongly related to effective porosity as was shown by Archie (1942).

Therefore resistivity and elastic wave velocity change together as fracturing increases. As Figure 8 indicates, resistivities were found to be strongly related to elastic wave velocities, and the correlation between resistivity ρ and primary wave velocity Vp can be mathematically expressed as follows.

$$Vp = \frac{Vp_0}{\alpha} \cdot \ln(\rho / \beta) \qquad (3)$$

where, Vp_0 is unit elastic wave velocity, and α, β are coefficients. β was obtained from the data of Figure 8 ($\beta = 40.7\ \Omega \cdot m$). By using this equation, we can convert resistivity into elastic wave velocity that is often used in the evaluation of the rock mass.

4 EFFECT OF CONFINING PRESSURE

4.1 *Experiment method*

When applying equation (3) to practical problems such as the evaluation of rock mass for tunneling, the effect of confining pressure becomes a serious problem. In order to resolve this problem, we measured resistivities and elastic wave velocities of fractured rock under confining pressure. The effect of confining pressure on elastic wave velocity is being studied by many researchers, for example, the research of Christensen et al. (1985). The effect of confining pressure on rock resistivity was investigated by Brace et al. (1968), Lockner et al. (1985) and Parkhomenko (1967), etc., but the influence of confining pressure on the relationship between resistivity and elastic wave velocity for the fractured rock is not well-known.

The resistivity measurement apparatus under confining pressure is shown in Figure 9. The rock specimen was isolated from the pressure cell by an isolator inserted between them. The elastic wave velocity was measured inside the other pressure cell that incorporated an ultrasonic sensor into the pedestal. The maximum confining pressure was 5 MN/m^2, which is roughly equivalent to an earth covering of 200 m. Pressure was applied in drainage condition.

The fractured rock specimens were made from intact granite by heating. The effective porosity of intact rock was 0.65%. The maximum effective porosity of heated granite was 5.19%. The diameter of these specimens was about 5 cm, and the length was about 10 cm.

4.2 *Experiment results*

The effect of confining pressure on elastic wave velocity is shown in Figure 10. The elastic wave velocities of fractured specimens increase as confining pressure increases. Figure 11 shows the influence of confining pressure on the resistivities. The resistivities increase as confining pressure increases.

Based on the experiment results indicated in Figure 11 and 12, the relationship between elastic wave velocity and resistivity under confining pressure is shown in Figure 12. This figure agrees well with Figure 8. These results indicate that equation (2) obtained from experimental data of

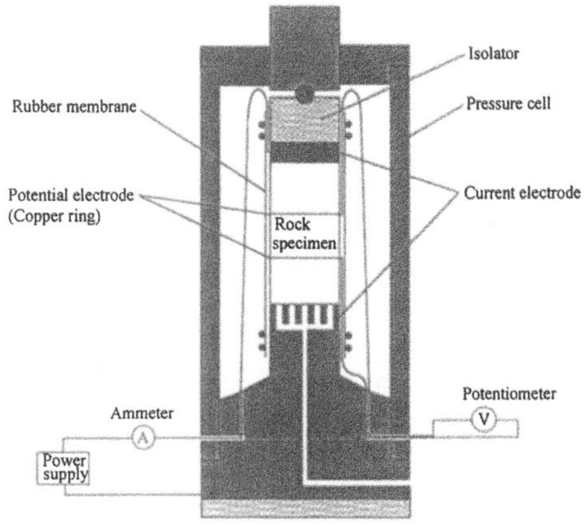

Figure 9. View of the resistivity measurement apparatus under confining pressure.

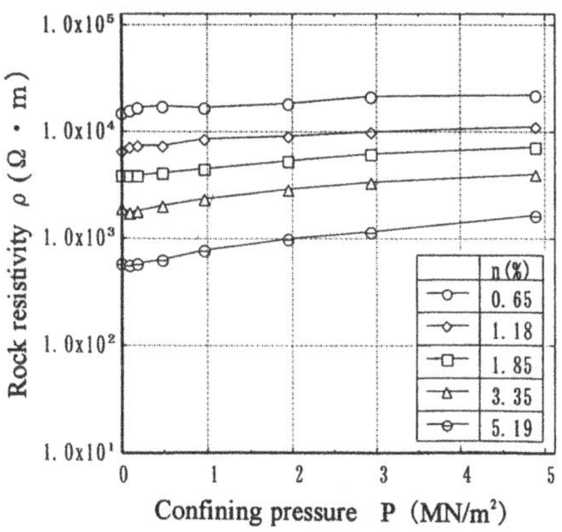

Figure 11. Effect of confining pressure on rock resistivity.

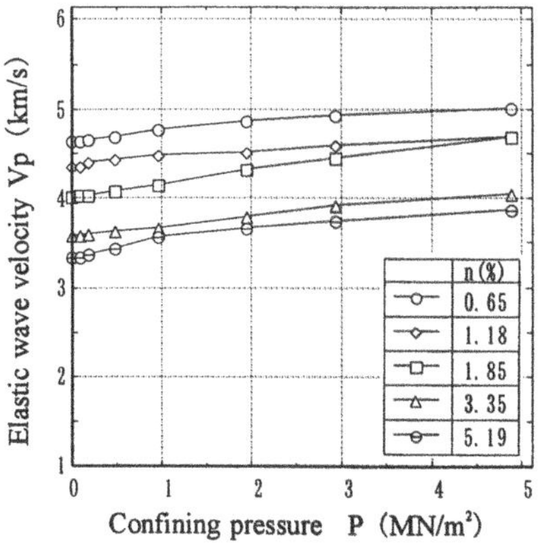

Figure 10. Effect of confining pressure on elastic wave velocity.

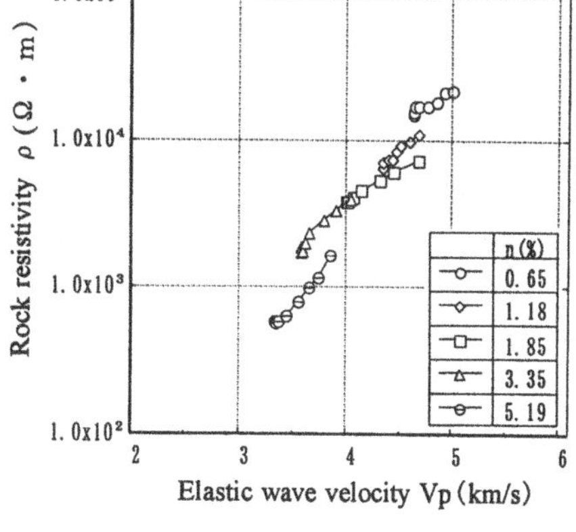

Figure 12. Influence of confining pressure on rock properties.

Figure 8 is generally applicable under confining pressure. Accordingly, resistivity measurement is thus useful for the evaluation of jointed rock.

5 CONCLUSIONS

These results lead to the following conclusions.

(1) The resistivity model of rock mass showed that intact rock resistivity and fracture resistivity are in parallel. By introducing a coefficient α that varies with the state of fracture, we generalized this model for a rock mass that includes various fractures. When α was 0.3, the calculation results of this model agreed well with the experimental results of fractured specimens.

(2) Resistivity and elastic wave velocity change as fracturing increases. Resistivity was found to be strongly related to elastic wave velocity. An equation expressing the correlation between resistivity and primary wave velocity was presented.

(3) Resistivity and elastic wave velocity increase as confining pressure increases. The above-mentioned equation was applicable under confining pressure.

REFERENCES

Archie, G. E. 1942. The electrical resistivity log as an aid in determining some reservoir characteristics, *Trans. A. I. M. E.*, Vol. 146: 54-62

Brace, W. F. 1968. Further studies of the effects of pressure on electrical resistivity of rocks, *Journal of geophysical research*, Vol. 73, No. 16: 5407-5421

Christensen, N. I. 1985. The influence of pore pressure and confining pressure on dynamic elastic properties of Berea sandstone, *Geophysics*, Vol. 50: 207-213.

Keller, G. V. 1966. *Electrical methods in geophysical prospecting*. Pergamon Press: 33-39

Lockner, D. A. & Byerlee, J. D. 1985. Complex resistivity measurements of confined rock, *Journal of geophysical research*, Vol. 90, No. B9: 7837-7847

Parkhomenko, E. I. (Translated by Keller, G. V.) 1967. *Electrical properties of rocks*, New York: Plenum Press.

Sekine, I., Nishimaki, H., Ishigaki, K., Hara, T. & Saito, A. 1997. Influence of fracture and its filling materials on rock resistivity, *Jour. Japan Soc. Eng. Geol.*, Vol. 38, No. 4: 213-223.

Watanabe, T., Sassa, K., Ashida, Y. & Kishimoto, M. 1989. Effects of low velocity zone consisting of multiple thin layers on a P wave, *Butsuri-Tansa (Geophys. Explor.)*, Vol. 42, No. 2: 75-81

Mechanics of Jointed and Faulted Rock, Rossmanith (ed.) © 1998 *Taylor & Francis, ISBN 90 5410 955 6*

Mechanical anisotropy of granites: Evidence from comparative experimental study and fabric analysis

R. Přikryl
Institute of Geochemistry, Mineralogy and Mineral Resources, Faculty of Science, Charles University, Prague, Czech Republic

Z. Pros & T. Lokajíček
Geophysical Institute, Academy of Sciences of the Czech Republic, Prague, Czech Republic

ABSTRACT: The experimental study of mechanical properties of granites proved remarkable anisotropy. The anisotropy of apparently isotropic rocks is explained by the preferred orientation of rock fabric elements. The magnitude of mechanical anisotropy fairly correlates with the symmetry and magnitude of rock fabric.

1 INTRODUCTION

Most of igneous rocks including granites are considered homogeneous and isotropic material when examining mechanical behaviour. The testing of mechanical properties is then derived based on the assumption that microcracks and rock forming minerals are arranged randomly without any obvious preferred orientation.

The mechanical behaviour of rocks fits badly into this assumption because they possess ordered microcrack systems and/or mineral preferred orientation. Such explanation can be found in experimental studies on mechanical behaviour of metamorphic rocks (e.g. Donath 1961; Gottschalk et al. 1990 among others) but also of some igneous rocks (Rodrigues 1966; Douglas & Voight 1969). On the other hand, only few researchers made an attempt to explain anisotropic character of rock mechanical properties by detailed analysis of rock fabric (e.g. Douglas & Voight 1969; Peng & Johnson 1972).

This paper presents contribution to the laboratory investigation of mechanical anisotropy of apparently isotropic intact rocks of granitic composition. The observed anisotropy of mechanical behaviour is explained with the help of detailed fabric analysis.

2 EXPERIMENTAL MATERIAL

Granites under study belong to the variety of igneous rocks used as building stones in the Czech Republic. The samples come from several operating quarries in the Moldanubian pluton in southern part of the Bohemian Massif and from the Smrčiny and Karlovy Vary plutons outcropping in the western part of the Czech Republic. The structural orientation of each block was carefully marked.

The rocks under study cover grain-size range from fine-grained to low coarse-grained rocks. At the same time, they differ in the density of microdefects. According to their mineral composition, they are true granites exhibiting no macroscopically visible preferred orientation of fabric elements. An orthogneiss was also included to the study as a rock possessing macroscopically well-expressed anisotropic fabric.

3 MECHANICAL PROPERTIES

3.1 *Experimental*

The uniaxial compressive stress was applied on prisms of 80 x 40 x 40 mm. The specimens were sampled from each rock type in three mutually perpendicular directions according to the hardway, grain and rift plane orientations (X, Y, Z direction).

For each specimen, foil-type gauges or LVDTs were mounted to measure axial and radial strains. The tests were carried out on "soft" machine at constant stress rate to the failure of the specimens. The tangent Young's modulus and uniaxial compressive strength were computed according to the procedure suggested by Douglass and Voigth (1969).

3.2 *Stress-strain behaviour and Young's moduli*

The rocks under study were grouped in several classes according to their stress-strain behaviour.

Only one granite exhibited linear elastic behaviour from the very beginning of the loading (Fig. 1A). The negligible difference in deformation for different loading directions makes it possible to consider this rock type quasiisotropic material.

Second group of granites under study proved linear elastic behaviour only in two directions whereas the third direction gives a remarkable initial inelastic yielding (Fig. 1B). Plastic behaviour of this loading direction progressively changes to an elastic one with increasing axial stress. This most compliant direction is oriented perpendicularly to the dominant set of microcracks.

Most of granites under study exhibited initial non-linear stress-strain behaviour in all directions that is progressively replaced by a linear one at a certain stress level (Fig. 1C). The remarkable deformational anisotropy for principal loading directions is caused by the intensity of microcracking.

The rock of anisotropic fabric - orthogneiss - proved also the highest anisotropy of stress-strain behaviour (Fig. 1D). The most compliant direction is oriented perpendicularly to the foliation plane. The deformation in the direction of lineation proved almost linear stress-strain behaviour.

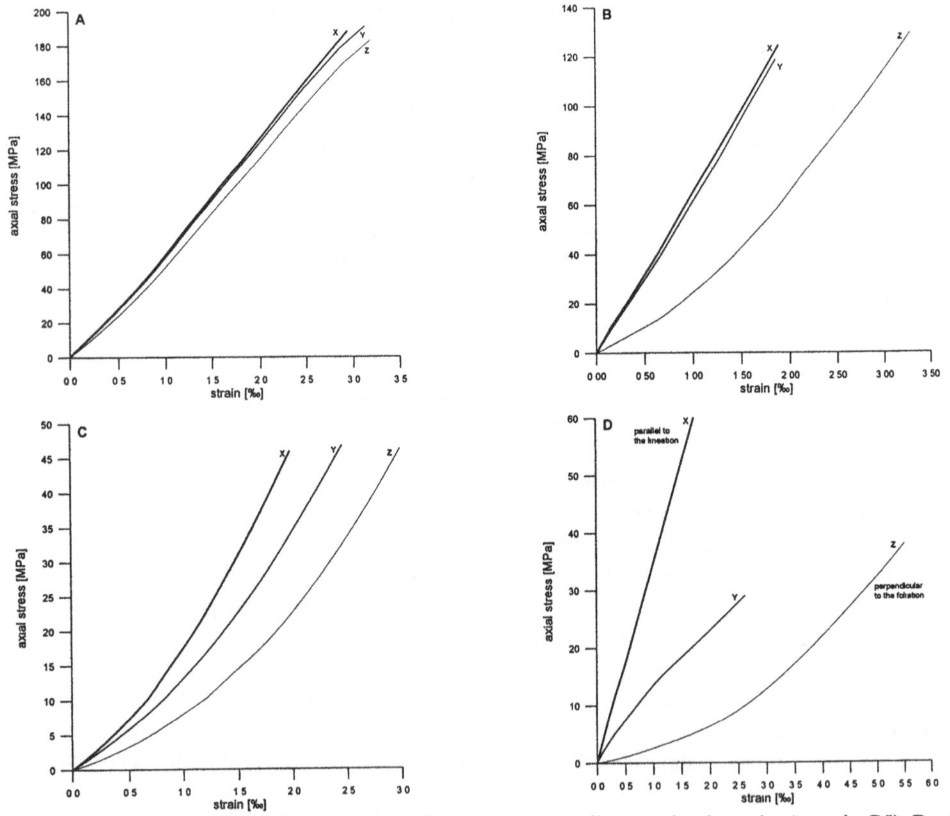

Figure 1. Stress-strain behaviour of rocks under study. A - medium-grained granite (sample G5), B - fine-grained granite (sample RP1), C - coarse-grained granite (sample G9), D - orthogneiss (sample RP3).

The anisotropic character of deformational properties is also documented by the values of tangent Young's modulus and their anisotropy that are summarized in Table 1.

Table 1. Tangent Young's modulus (E_T) and anisotropy of rocks under study.

sample	E_T max [GPa]	E_T min [GPa]	anisotropy
granite G5	65.9	62.1	1.06
granite RP1	66.3	46.8	1.42
granite G9	34.0	23.3	1.46
orthogneiss RP3	34.8	12.5	2.78

3.3 *Uniaxial compressive strength*

Some rocks under study proved remarkable anisotropy of uniaxial compressive strength as documented in Table 2. The highest strength was observed in X-direction that coincides with the pole to the hardway plane in granites and with the lineation in orthogneiss. This fact as well as observed anisotropy are further explained by fabric analysis in next sections.

Table 2. Uniaxial compressive strength (UCS) of rocks studied.

sample	UCS max [MPa]	UCS min [MPa]	anisotropy
granite G5	193.2	180.6	1.07
granite RP1	269.3	209.2	1.29
granite G9	112.1	95.8	1.17
orthogneiss RP3	115.4	52.6	2.19

4 ROCK FABRIC CHARACTERIZATION

4.1 *General*

Fabric analysis aims to define individual elements of rock fabric including texture (crystallographic preferred orientation), structure (shape preferred orientation and all geometric aspects of grains) and defects (presence and orientation of microcracks).

In general, rock fabric can be studied directly in optical microscope by various methods (e.g. image analysis, U-stage study). The second possibility is to examine fabric by some indirect methods allowing quantification of defects (e.g. differential strain analysis) or whole rock fabric.

In this study, the ultrasonic testing was selected as a method for the whole rock fabric characterization. The optical study was employed as a complementary method for the precise interpretation of the results of ultrasonic testing.

4.2 *Ultrasonic testing*

The velocity of elastic waves of rocks belongs to the important parameters in geophysics, petrophysics and rock engineering. One of the methods is capable not only to determine elastic properties but also to characterize rock fabric.

The method for ultrasonic laboratory testing of rocks developed at the Geophysical Institute (Prague, Czech Republic) enables to measure P-wave velocities on spherical samples in many directions under hydrostatic pressure conditions. The method and apparatus are described in detail by Pros & Vaněk (1960), Pros & Podroužková (1974), Pros (1977) and Pros et al. (1998). The geometry of the specimen makes it possible to obtain velocity data in sufficiently dense net and thus to have knowledge on spatial distribution of velocities. Moreover, the apparatus is designed for the measurement of elastic wave velocities at the hydrostatic pressure conditions up to 400 MPa.

The increasing hydrostatic pressure closes microcracks in specimen. The observation of the changes of P-wave velocity distribution at atmospheric pressure and at higher hydrostatic pressure conditions thus enables to separate anisotropy related to the presence of microcracks and anisotropy caused by the anisotropic fabric of rock-forming minerals.

The velocity data are displayed spatially in the form of contour diagrams (lower hemisphere equal area projection) where isolines present limits of the same velocity. The rocks under study show several patterns of P-wave velocity spatial distribution (Figure 2). The first one is almost isotropic pattern of sample G5 as observed at all confining pressure levels except weak transversally isotropic pattern at atmospheric pressure. This pattern suggests the lack of preferred orientation of any fabric element. The second observed pattern is well-developed transversal isotropic pattern of P-wave velocity isolines that occurs in sample RP1 due to the presence of one dominant set of microcracks. This fabric symmetry sustains even at high confinement and could be explained by weak magmatic foliation defined by the mica texture (preferred orientation of mica basal planes). The third pattern - pseudoorthorhombic - is represented by sample G9. The pattern observed at atmospheric pressure rapidly disappears and is not observed at higher confinement. This suggests the major contribution of three sets of microcracks to the rock fabric. The last P-wave velocity isolines pattern is well-pronounced pseudoorthorhombic pattern as recorded for all confining pressure levels in orthogneiss sample RP3. This pattern is explained by the influence of shape and crystallographic preferred orientation of micas and by strong shape preferred orientation of other rock forming minerals (quartz and feldspars). The density of isolines is lower in high confinement domain due to the closure of cleavage cracks in micas and grain boundary cracks.

4.3 *Fabric analysis using optical microscopy*

Fabric analysis in optical microscope aims to give an precise explanation to observed P-wave velocity patterns and to quantify some structural elements such as shape preferred orientation of minerals.

The study of rock texture (quartz c-axes, poles to the basal planes (001) of micas, plagioclase polysynthetic twinning (010)) and defects (poles to the planes of intragranular and multigranular microcracks) was carried out with a conventional universal stage mounted on a polarising microscope. The data were plotted out as contour diagrams.

The quantitative analysis of grain geometries (microstructures - grain shape, size etc.) and rock modal composition was performed by an image measurement system (Přikryl 1998) employing SIGMASCAN software. The whole process consists of source image acquisition from thin section, pre-processing of an image, image digitizing, measurement of individual parameters, and data analysis and output. This complex analysis made it possible to quantify in detail shapes and shape preferred orientation of grains, grain size distribution and modal composition. Shape preferred orientation of major rock forming minerals seems to be crucial for the interpretation of mechanical anisotropy of rocks under study.

5 DISCUSSION OF RESULTS

The stress-strain behaviour of granitic rocks subjected to the uniaxial compressive stress can be interpreted in terms of the symmetry of rock fabric as derived from non-destructive ultrasonic testing.

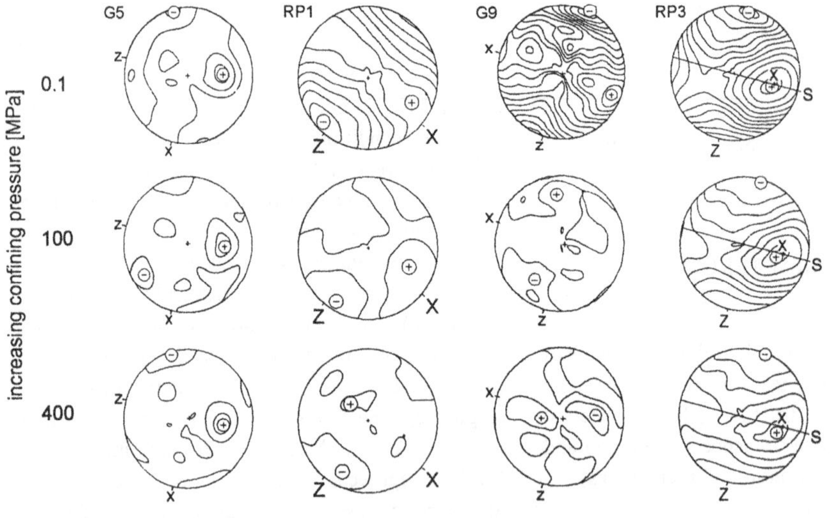

Figure 2. Spatial distribution of P-wave velocities.

The granites under study can be divided in three main groups. The first one - called quasiisotropic - involves rocks with quasilinear stress-strain behaviour and low anisotropy of Young's modulus. The granites of transversally isotropic symmetry of rock fabric form the second group. Those rocks possess quasilinear and isotropic stress-strain behaviour in two directions and remarkably non-linear behaviour in the third direction. This behaviour can be explained by the effect of present one major set of microcracks as well as by the preferred shape orientation of minerals (Brace 1965). The symmetry of rock fabric was confirmed by the study of spatial distribution of P-wave velocities and by optical study. Last group involves rocks of orthorhombic fabric symmetry. Granites show remarkable non-linear anisotropic stress-strain behaviour due to the presence of microcracks (e.g. Walsh 1993) whilst the mechanical anisotropy of orthogneiss is explained by rock macrofabric (e.g. Tremmel and Widmann 1970; Dobereiner et al. 1993).

All rocks under study exhibited certain degree of strength anisotropy. The uniaxial compressive strength of granites under study can differ as much as 30 % between rift and hardway planes. This fact is explained by the shape preferred orientation of rock forming minerals (Figure 3). The lowest

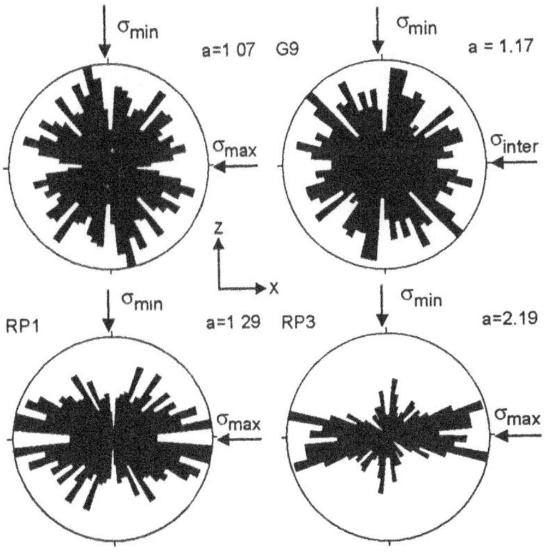

Figure 3. Comparison of mineral shape preferred orientation and directions of uniaxial compressive strength and strength anisotropy (a).

strength anisotropy was observed in the rock without shape preferred orientation of minerals. The highest strength anisotropy in orthogneiss is explained by simultaneous influence of shape preferred orientation of minerals and rock macrofabric.

The comparison of the P-wave velocity extremes distribution and deformability of rocks under uniaxial compression shows that the direction of maximum P-wave velocity correlates with the least deformable direction and vice versa. It is well known fact from numerous geophysical studies that there exists tight control of minimum P-wave velocity direction and dominant crack system (e.g. Simmons et al. 1975; Babuška et al. 1977). The coincidence of the position of minimum P-wave velocity and the most compliant direction in granites could be thus easily attributed to the present microcracks.

Study of P-wave velocity spatial distribution and behaviour of rock spheres under confining pressure presents useful tool for non-destructive study of rock fabric. The anisotropy of P-wave velocity pattern observed at atmospheric pressure and low confining pressure is caused dominantly by the presence of microcracks. This is in agreement with the optical study of microcrack distribution. The closure of microcracks at higher hydrostatic pressures revealed presence of low texture anisotropy in granites and higher texture anisotropy in orthogneiss.

The precise study of fabric of building stones on microscale is very important. Quantitative analyses of rock microstructures with the help of image analysis, textures as well as microdefects should be inseparable part of the interpretation of rock mechanical properties. The image analysis is an excellent tool allowing geologists accurate description of rock microstructure, i.e. modal composition, grain size, shape orientation of grains etc. Only the precise knowledge of rock fabric can correctly explain the rock mechanical properties.

6 CONCLUSIONS

The experimental study of mechanical behaviour of granitic rocks under uniaxial compression revealed following aspects:

- stress-strain behaviour under uniaxial compression is largely influenced by the presence of microcracks;

- the anisotropy of stress-strain behaviour is governed by the density of different sets of microcracks but can be also explained by contemporaneous influence of shape preferred orientation of minerals;
- the anisotropy of uniaxial compressive strength is high in rocks where the shape preferred orientation of major rock forming minerals is well developed (orthogneiss and some granites) but is low in rocks where this fabric feature is missing.

The results of this study suggest that most of granites (and other igneous rocks as well) cannot be considered neither continuum neither isotropic materials. This fact should be reflected not only in the laboratory testing of those rocks but also during planning of large engineering underground structures.

REFERENCES

Babuška, V., Pros, Z., & W.Franke 1977. Effect of fabric and cracks on the elastic anisotropy in granodiorite. *Publ. Inst. Geophys. Pol. Acad., Sc.* 117: 179-186.

Brace, W.F. 1965. Some new measurements of linear compressibility of rocks. *J. Geophys. Res.* 70(2): 391-398.

Dobereiner, L., Durville, J.L., & J.Restituito 1993. Weathering of the Massaic gneiss (Massif Central, France). *Bulletin of International Association of Engineering Geologists* 47: 79-96.

Donath, F.A. 1961. Experimental study of shear failure in anisotropic rocks. *Geol. Soc. Am. Bull.* 72: 985-990.

Douglas, P.M. & B.Voight 1969. Anisotropy of granites: a reflection of microscopic fabric. *Géotechnique* 19: 376-398.

Gottschalk, R.R., Kronenberg, A.K., Russell, J.E. & J.Handin 1990. Mechanical anisotropy of gneiss: failure criterion and textural sources of directional behaviour. *J. Geophys. Res.* 95: 21613-21634.

Peng, S. & A.M.Johnson 1972. Crack growth and faulting in cylindrical specimens of Chelmsford granite. *Int. J. Rock Mech. Min. Sci.* 9: 37-86.

Pros, Z. 1977. Studies of anisotropy in elastic properties of rocks at uniform pressures on spherical samples. In M.P.Volarovich & H.Stiller (eds), *High Pressure and Temperature Studies of Physical Properties of Rocks and Minerals:* 56-67. Kiev: Naukova Dumka. (In Russian.)

Pros, Z. & J.Vaněk 1960. Experimental study of a pulse method for measuring elastic parameters of rocks on samples. *Studia geoph. et geod.* 4: 338-349.

Pros, Z. & Z.Podroužková 1974. Apparatus for investigating the elastic anisotropy on spherical rock samples at high pressures. *Veröff. Zentralinst. Physik der Erde* 22: 42-47.

Pros, Z., Lokajíček, T., & K.Klíma 1998. Laboratory study of elastic anisotropy on rock samples. *Pure Appl. Geophys.* 151 (in print).

Přikryl, R. 1998. *The effect of rock fabric on some mechanical properties of rocks: an example of granites.* Doctoral thesis, Prague: Charles University.

Rodrigues, F.P. 1966. Anisotropy of granites. *Proc. 1st Congress of the International Society of Rock Mechanics, Lisbon, September 25 - October 1, 1966,* Volume I, Theme 3: 721-731.

Simmons, G., Todd, T., & W.S.Baldridge, 1975. Toward a quantitative relationship between elastic properties and cracks in low porosity rocks. *American Journal of Science* 275: 318-345.

Tremmel, E., & R.Widmann 1970. Das Verformungsverhalten von Gneis. *Proc. 2nd Congress of the International Society for Rock Mechanics, Beograd, September 21-26, 1970,* Vol. I, Theme 2: 9 pp.

Walsh, J.B. 1993. The influence of microstructure on rock deformation. In J.A.Hudson (ed.), *Comprehensive rocks engineering. Principles, practice & projects.* 243-254. Oxford: Pergamon Press.

Mechanics of Jointed and Faulted Rock, Rossmanith (ed.) © *1998 Taylor & Francis, ISBN 90 5410 955 6*

Dynamic modelling of a fault and dyke in the vicinity of mining

T. Dede
CSIR Division of Mining Technology, Johannesburg, South Africa

ABSTRACT: In order to gain an understanding of fault and dyke behaviour, a series of numerical models have been investigated. The dynamic finite difference program *WAVE* (Hildyard et al. 1995) is used for the modelling of a stope in a deep mine. The slip along a vertical discontinuity, situated within the positive excess shear stress lobe of the stope, initiates seismic waves. The stope-discontinuity geometry is modelled in both two and three dimensions. The fault is modelled by reducing the fault cohesion instantaneously to zero, or by applying a slip-weakening law to the fault interface. Peak particle velocity values at the stope were stored in *WAVE* to compare the different mining scenarios. Results from using *WAVE* in three dimensions show that if the angle between the face direction and the strike of the fault or dyke is kept at more than 35°, then it decreases the possible seismic event due to slip on the discontinuity.

1 INTRODUCTION

In the mining industry, modelling of the elastodynamic behaviour of the rockmass is being undertaken to gain a fundamental understanding in a number of areas, including: dynamic fault slip and rockburst mechanisms; the interaction of seismic waves with tabular stopes and geological structures such as dykes, and determining geometric factors which influence the magnitude of motions in stopes; the influence of local and regional support (such as backfill) on dynamic motions; and the interaction of seismic waves with the stope fracture zones.

The elastodynamic finite-difference program *WAVE* (Hildyard et al. 1995) is used in this study. Numerical studies using *WAVE* have shown that particle velocities generated by slip on a simulated fault are in broad agreement with established distance-velocity correlations in three dimensions. *WAVE* has proven to be a useful tool for analysing elostodynamic problems in mining. This is firstly because it is exclusively orientated toward mining applications (and hence it is easy to apply stope and crack elements in the model). Secondly, due to its efficiency, it has a fast turn-around time on simple problems (which is good for experimentation) and it allows order of magnitude larger meshes than other available mesh-based dynamic codes, which is important for large-scale three-dimensional models (Napier et al. 1995).

WAVE is used to study various mining cases to obtain the effect of:

1. dyke stiffness,
2. bracket pillar width,
3. stope span,
4. stope dip,
5. angle of approach to a discontinuity.

This work describes the numerical two- and three-dimensional *WAVE* model used in the investigation, and presents the comparisons of results based on the above variables. Finally, conclusions are drawn as to the use of a dynamic code in gaining an understanding of fault and dyke behaviour.

2 THE *WAVE* MODELS

2.1 *Two-dimensional stope-dyke model*

In the first model, the effect of stiffness of a dyke on a seismic event is studied. A horizontal stope of 152 m span is subjected to a stress field of 80 MPa in the vertical and 40 MPa in the horizontal direction. The stress field interacts with the stope and generates the zero excess shear stress (ESS) contour (Napier 1987, Ryder 1988) depicted in Figure 1. To simulate the slip in this model, a fault is placed next to the dyke. The ESS can be defined as the difference between the prevailing shear stress prior to slip and the dynamic strength of the fault plane. Thus,

$$ESS = |\tau| - \mu\sigma_n,$$

where $\mu=\tan\phi$, ϕ is the friction angle, τ is the shear stress, and σ_n is the normal stress. In this model, the friction angle on the fault-dyke interface is assumed to be 30°. A vertical dyke is situated ahead of the stope and, with no cohesion on the fault-dyke interface, the section of the fault contained within the zero ESS contour will rupture.

WAVE is a finite-difference code that is able to model wave propagation in a two- or three-dimensional elastic medium. Second order interlaced finite-difference equations are used on an orthogonal grid with uniform grid spacing. *WAVE* has the ability to model dislocations that can represent faults or tabular stopes (Daehnke 1995).

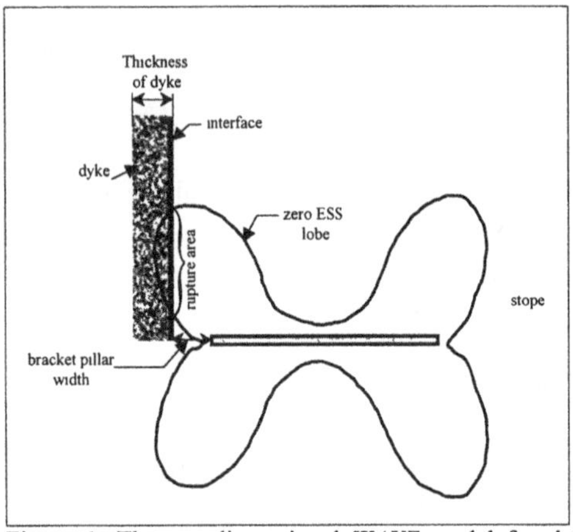

Figure 1. The two-dimensional *WAVE* model for dyke cases.

The *WAVE* analyses were completed in two steps: In the first step, the discontinuity is locked such that no slip can occur, the external stress field is applied, and *WAVE* is cycled until the stope reaches an equilibrium closure when the maximum velocities in the finite-difference mesh are small. This produces the static solution.

In the second step, the discontinuity cohesion is reduced and the fault ruptures. The interface cohesion is either set to zero instantaneously, resulting in a sudden discontinuity rupture along the length of the discontinuity situated within the zero ESS lobe, or a slip-weakening law is applied to the interface. In the latter case, the cohesion along the whole discontinuity is reduced so that slip just commences at a point on the discontinuity. As slip occurs at this point the cohesion is reduced linearly (Figure 2), and the shear stress is increasingly transferred to an adjacent grid point until they slip. Thus, by the application of a slip-weakening law to the interface, slip occurs progressively which is considered to be more realistic than an instantaneous rupture model.

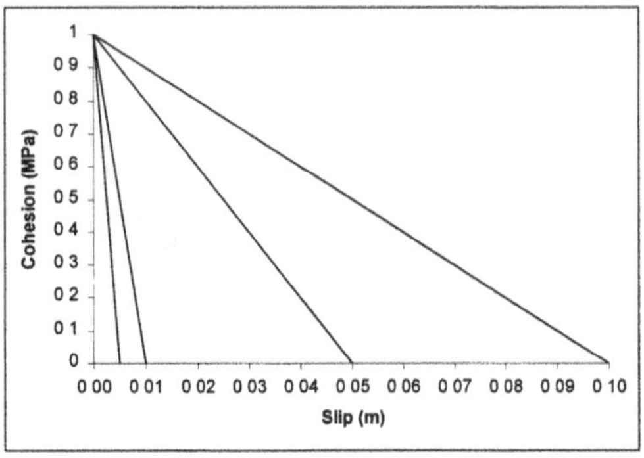

Figure 2. The *WAVE* slip-weakening relationship for four slip-weakening rates.

Table 1. Physical parameters used in the *WAVE* models.

Young's Modulus	78 GPa
Poisson's Ratio	0.21
Shear stiffness of discontinuity	1e10 GPa/m
Normal stiffness of discontinuity	1e12 GPa/m

The physical parameters used in the models are given in Table 1. The sequence of snapshots in Figure 3 illustrates the propagation of particle velocity contours, which are initiated by the slip at the dyke interface. Different dyke stiffnesses were implemented in the model. Different stiffnesses of a dyke produce different maximum peak particle velocities in the analyses. The history of the peak particle velocities in the stope were stored, and the maximum velocities were used for comparison. Note that in this study the maximum particle velocities are only used in a relative sense to the stope for the same mining configurations. As the Young's modulus of the dyke is increased, the maximum peak particle velocity increases (Figure 4). This is also true for the Poisson's ratio of the dyke, but the increase in the maximum peak particle velocity is not significant as in the case of Young's Modulus (Figure 5).

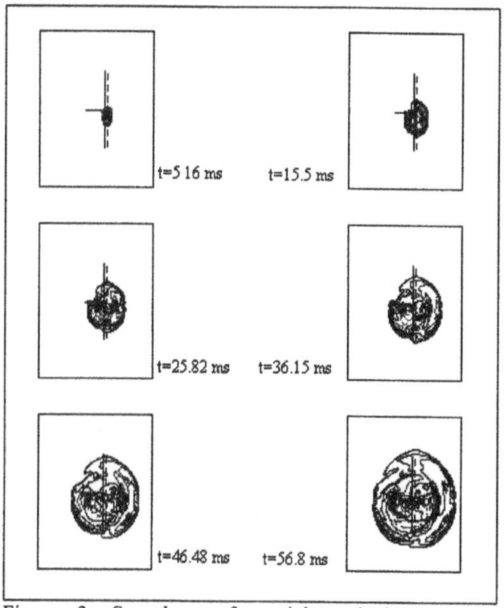

Figure 3. Snapshots of particle velocity contours showing the wave propagation.

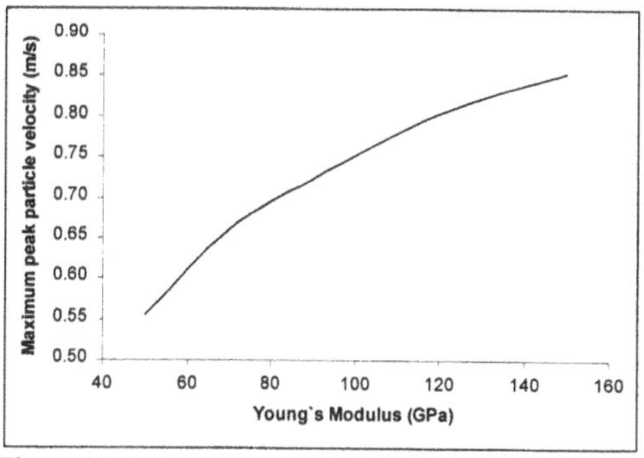

Figure 4. Maximum peak particle velocity throughout the stope for various stiffnesses of dyke.

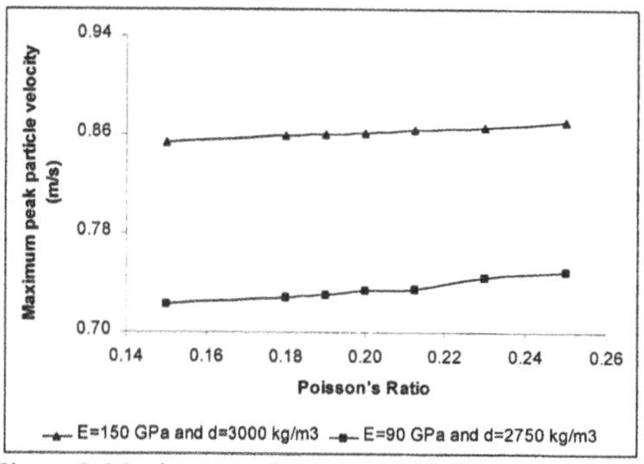

Figure 5. Maximum peak particle velocity throughout the stope for various Poisson's ratios of dyke.

In subsequent studies, various slip-weakening values were introduced into the model. Slip weakening displacement (swd) is the amount of slip required for the cohesion of the interface to become zero. Figure 6 illustrates the variation in peak particle velocity with slip weakening displacement. As the swd value is decreased (i.e. more sudden fault rupture), the maximum peak particle velocity in the stope is increased (Figure 7).

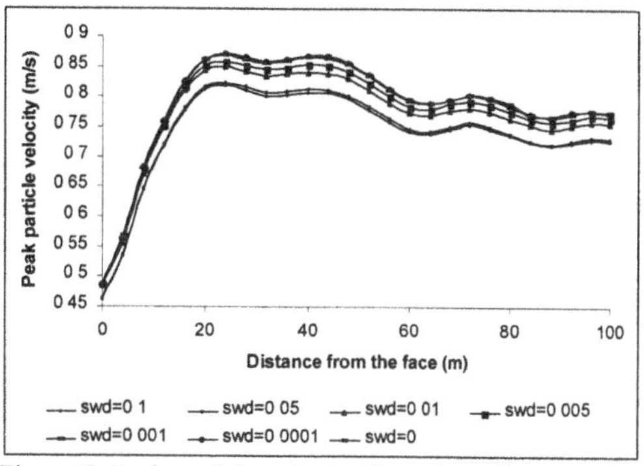

Figure 6. Peak particle velocity throughout the stope for various slip weakening displacement values (swd).

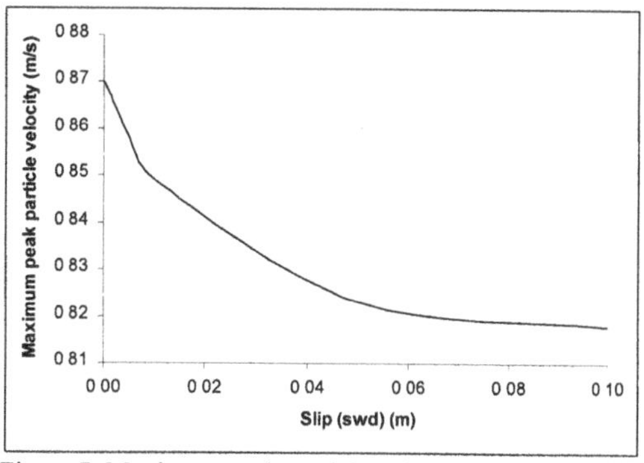

Figure 7. Maximum peak particle velocity throughout the stope for various slip weakening values (swd).

Table 2. Variables used in the *WAVE* model.

SPAN (m)	PILLAR WIDTH (m)	STOPE DIP (°)	DEPTH (m)
52	4	0	2000
76	8	4.5	2200
100	12	10.7	2400
124	16	15.5	2600
152	20	20 4	2800
200	24	20 4	3000
224	28		
248			

2.2 Two-dimensional stope-fault model

In the second *WAVE* model, the effects of bracket pillar width, stope span, stope dip and depth of stope on the maximum peak particle velocity in the stope have been analysed. The mining geometry is similar to the previous model which is shown in Figure 1. Different mining scenarios have been analysed which are described in Table 2.

The effect of the above parameters can be summarized as:

Bracket pillar width: A horizontal stope is subjected to a stress field, at a depth of 3000 m below surface, in which bracket pillar widths have been varied from 4 m to 28 m for different stope spans. Decreasing the bracket pillar width increases the maximum peak particle velocity in the stope. This is verified for various stope spans (Figure 8) and is predicted in the static analysis of bracket pillars (Dede & Handley 1997).

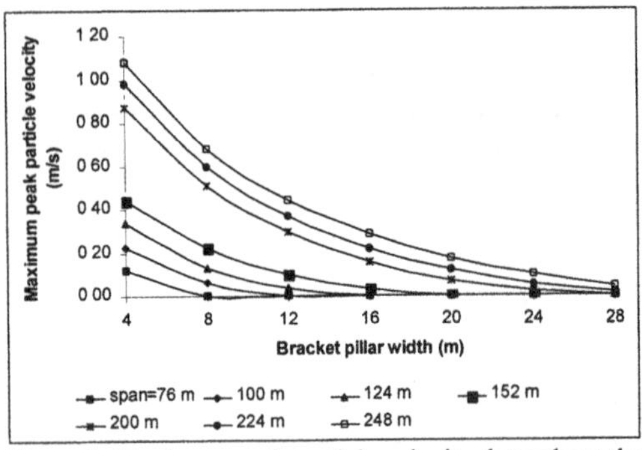

Figure 8. Maximum peak particle velocity throughout the stope for various bracket pillar widths at different stope spans.

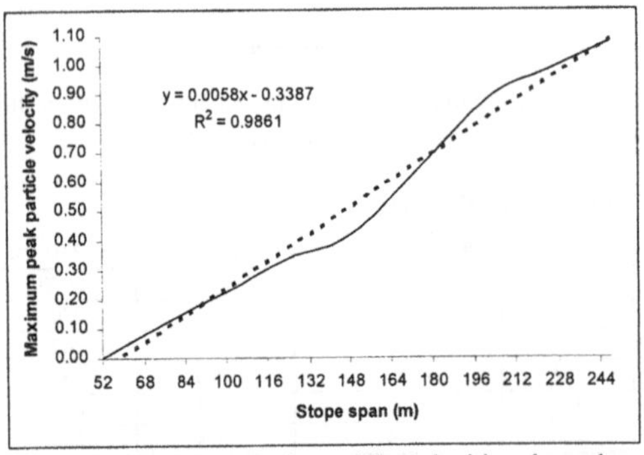

Figure 9. Maximum peak particle velocities throughout the stope for various stope spans at a bracket pillar width of 4 m.

Stope span: The same mining conditions are used for analyzing the effect of span on the peak particle velocity in the stope. Figure 9 illustrates the results in which increasing the stope span increases the peak particle velocities. There is a linear relationship between stope span and the peak particle velocities.

Stope dip: From the *WAVE* formulation point of view, the stope and the discontinuity are required to be either perpendicular or parallel to each other. In order to solve this difficulty and to gain some understanding about an inclined stope behaviour, a "stepped stope" configuration has been implemented in the model which is shown in Figure 10. The static results were compared to *DIGS* (Napier 1990) and give similar values. Therefore, various stope dip cases were analysed (Figure 11). There is an approximately linear relationship between stope dip and the maximum peak particle velocity in the stope for dips up to 20°.

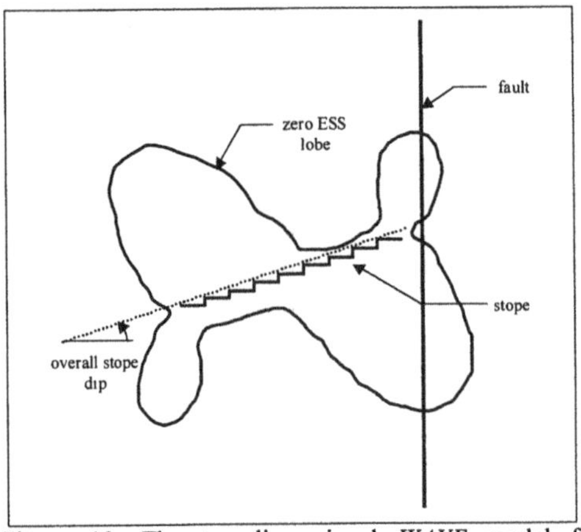

Figure 10. The two-dimensional *WAVE* model for "stepped stope" configuration.

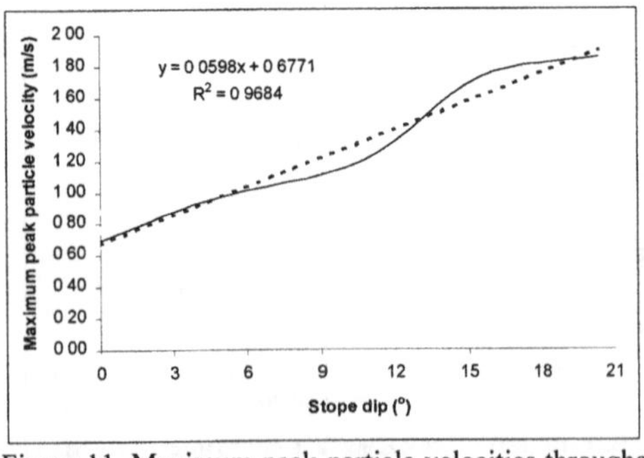

Figure 11. Maximum peak particle velocities throughout the stope for various stope dips at a bracket pillar width and stope span of 4 m and 152 m respectively.

2.3 Three-dimensional stope-fault model

General mining experience for approaching a fault or dyke indicates that shear failure can be reduced if the shear stresses caused by mining are not aligned parallel to the structure. This can be accomplished by maintaining an angle of 30° or more between the face orientation and the strike of the fault or dyke (COMRO 1988). This criteria has not been checked by any dynamic numerical program before.

A horizontal stope of 75 m span is subjected to a stress field of 70 MPa in the vertical and 35 MPa in the horizontal direction. Note that the approaching angle is the overall angle of all faces as shown in Figure 12. This angle was varied from 0° to 71° to assess the effect on the stope, while keeping the amount of mined out area constant. In the first step, the discontinuity is locked such that no slip can occur, the external stress field is applied, and *WAVE* is cycled until the stope reaches an equilibrium closure to calculate the static stresses. Increasing the approaching angle decreases the excess shear stress (ESS) on the fault (Figure 13).

In the second step, the discontinuity cohesion is reduced and the fault ruptures. The slip-weakening value of 0.0001 m is applied to the interface. At this stage, the results of dynamic analysis can be analysed in three parts:

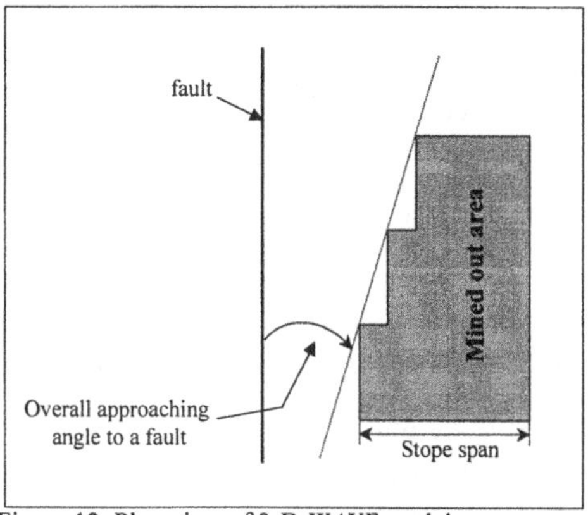

Figure 12. Plan view of 3-D *WAVE* model.

Slip on the fault: Increasing the angle of approach decreases the maximum amount of slip on the fault which can be estimated from *ESS* on the fault obtained from static results (Figure 14).

Peak particle velocity: The maximum peak particle velocity was stored during the *WAVE* analysis both in the stope and on the fault. The maximum peak particle velocity in the stope shows a similar trend as that of on the fault (Figure 15). An increase in the angle of approach decreases the maximum peak particle velocity in the stope and on the fault as well.

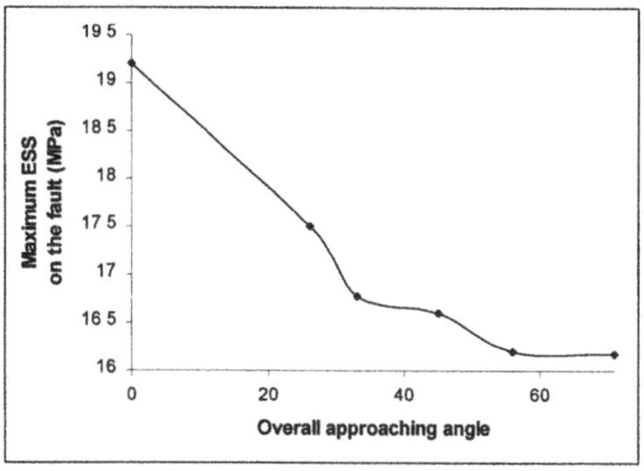

Figure 13. The maximum ESS for different approaching angles.

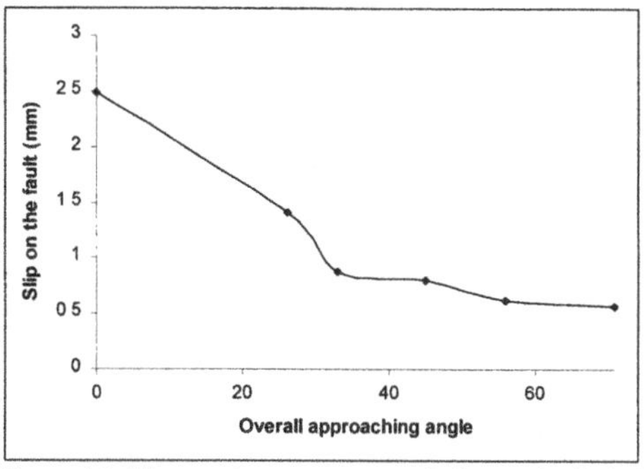

Figure 14. The maximum amount of slip on the fault for different approaching angles.

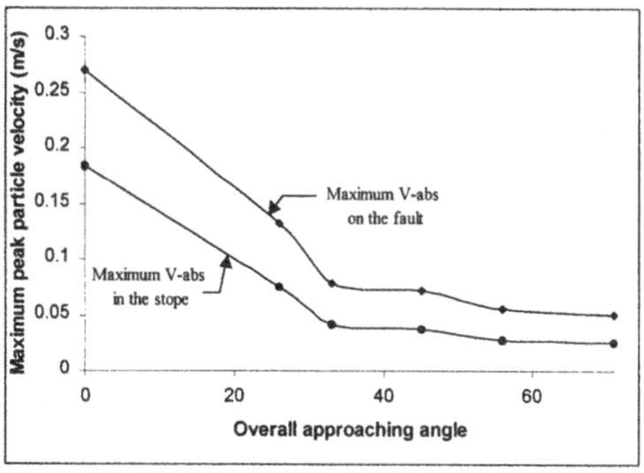

Figure 15. The maximum peak particle velocity in the stope and on the fault for different approaching angles.

Convergence velocity: The maximum relative normal velocity (convergence velocity) in the stope was stored during the *WAVE* analysis. It shows a similar trend as in the previous cases in which the most favourable angle can be set as 35° or greater for approaching a discontinuity (Figure 16 and 17).

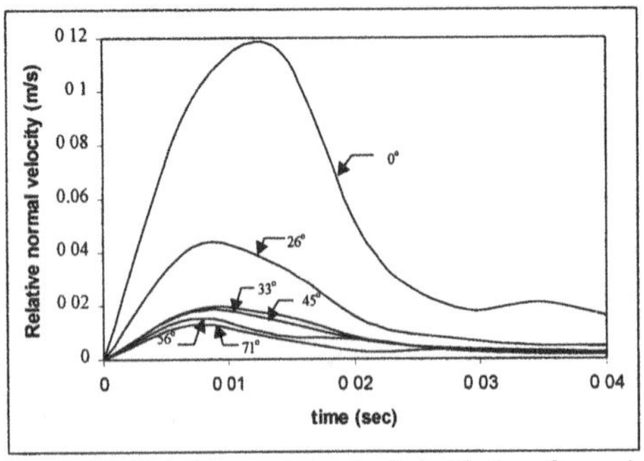

Figure 16. Relative normal velocity with time for various approaching angles to a discontinuity.

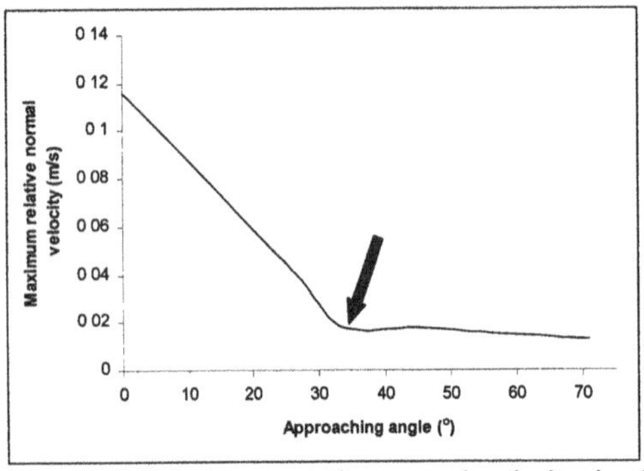

Figure 17. Maximum relative normal velocity in the stope for various approaching angles to a discontinuity.

3 CONCLUSIONS

A number of parametric studies have been conducted on two- and three-dimensional stope models. This has extended work carried out by Dede and Handley (1997) on the static analysis of bracket pillars to a dynamic analysis where the velocities in the stope have been used as a criterion for evaluating mining geometries. Numerical modelling is an essential tool for the identification of potentially hazardous stope areas prone to excessive velocities. *WAVE* is an efficient numerical tool for the modelling of fault slip and wave propagation. The importance of this work is to show that there may be underlying mechanisms which are not directly predicted by a static analysis. Different mining scenarios can be compared and the most favorable one selected.

It has been shown that the optimum angle of approach between a longwall mining face and a discontinuity should not be less than 35° if mining takes place on one side of the discontinuity.

4 ACKNOWLEDGEMENTS

The author wishes to thank SIMRAC (Safety in Mines Research Advisory Committee) for funding the project under GAP223. The CSIR Division of Mining Technology is also gratefully acknowledged for the support provided for the writing and publishing of this paper. Finally the author wishes to thank Dr. Napier, Dr. Durrheim and Mr. Hildyard for their guidance and input in this paper.

5 REFERENCES

COMRO 1988. An Industry Guide to Methods of Ameliorating the hazards of rockfalls and rockbursts, 1988 edition, *Chamber of Mines Research Organization.*

Daehnke, A. 1995. Comparisons between *WAVE* and seismological data. *J. S. Afr. Inst. Min. and Metall.*, pp. 197-206, Sept. 1995.

Dede, T. & Handley, M.F. 1997. Bracket pillar design charts. *SIMRAC Interim Project Report (GAP223)*, Jan. 1997.

Hildyard, M.W., Daehnke, A. & Cundall, P.A. 1995. *WAVE*: A computer program for investigating elastodynamic issues in mining. *Proc. 35th U.S. Symp. On Rock Mech.* 1995 pp. 519-524.

Napier, J.A.L. 1987. The application of excess shear stress to the design of mine layouts. *J.A. Afr. Inst. Min. Metall.*, vol. 87, no. 12. Dec. 1987. pp. 397-405.

Napier, J.A.L. 1990. Modelling of fracturing near deep level gold mine excavations using a displacement discontinuity approach. *Mechanics of Jointed and Fractured Rock*, In H.P. Rossmanith (ed.). Balkema, Rotterdam, pp. 709-715.

Napier, J.A.L, Hildyard, M.W., Kuijpers, J.S., Daehnke, A., Sellers, E.J., Malan, D.F., Siebrits, E., Ozbay, M.U., Dede, T. & Turner, P.A. 1995. Develop a quantitative understanding of rockmass behaviour near excavations in deep mines. *SIMRAC Final Project Report (GAP029)*, Dec. 1995.

Ryder, J.A. 1988. Excess shear stress in the assessment of geologically hazardous situations *J. S. Afr. Inst. Min. Metall.* vol 88, no. 1. Jan. 1988. pp. 27-39.

3D Modelling

Mechanics of Jointed and Faulted Rock, Rossmanith (ed.) © *1998 Taylor & Francis, ISBN 90 5410 955 6*

Three dimensional modelling of seismicity in deep level mines

J.A.L. Napier
CSIR Division of Mining Technology, Johannesburg, South Africa

ABSTRACT: A method is proposed for the analysis of time dependent stress transfer and seismic recurrence processes in deep level mines. An essential feature of the method is to represent large scale failure as a hierarchical process on many length scales. A particular scheme for generating sets of cracks which are equivalent to existing finer scale fracture sets, in a given spatial region, is outlined. The scheme for equivalent crack allocation is tested for two particular cases in which specific sets of cracks are assigned in a given region and appears to work well. An example to illustrate the potential analysis of time dependent energy release associated with a simplified mining problem is given.

1 INTRODUCTION

For mine design purposes it is very desirable to represent stress transfer and seismic recurrence effects as a definite mechanistic (though not necessarily deterministic) process which can be analysed numerically in routine mine design studies. One of the major challenges in reaching this goal is the ability to model large scale damage, recognizing that fracturing occurs on a multiplicity of length scales. The essence of the problem, therefore, is not merely to implement efficient numerical solution procedures for large systems of equations but, in addition, to be able to represent the self-organizing fracture coalescence processes on many scales. A number of attempts to address these goals have been made using renormalization solution concepts (Madden 1983, Allegre *et. al.* 1995, Main 1995) based on concepts of critical phenomena in physics (Wilson 1979). This paper presents some initial steps towards the formulation of a general solution scheme which incorporates time dependent stress relaxation effects in a three dimensional model and which also allows hierarchical failure processes to be modelled by approximating the inelastic strain in a given region as a special set of appropriately aligned crack elements. Failure is represented by means of discontinuity elements at multiple hierarchical levels, enabling fracture coalescence and clustering to be manifested at all scales. The formulation of the time dependent stress relaxation model is first discussed, followed by a suggested scheme for the allocation of the equivalent crack elements.

2 FRACTURE REPRESENTATION

Cracks can be modelled conveniently by means of small strain dislocations or so-called "displacement discontinuities" (Crouch and Starfield 1983). The displacement discontinuity vector is defined at point Q of a crack surface by the jump in the displacement vector $u_i\,(Q)$ across the surface:

$$D_i(Q) = u_i^-(Q) - u_i^+(Q), \tag{1}$$

where the component i ranges from 1 to 3 and the superscripts + and - designate the sides of the crack corresponding respectively to the positive and negative directions of a defined normal $n_i(Q)$. In the case of a discontinuity across a flat surface A, it is convenient to define a local coordinate axis system where the discontinuity falls in the x-y plane and the z-axis is perpendicular to the discontinuity.

If the discontinuity vector $D = [D_x, D_y, D_z]^T$ is constant over the surface A, the stress tensor components τ_{ij} at a point P in the local axis system are given by

$$T = KD \tag{2}$$

where $T = [\tau_{xx}, \tau_{yy}, \tau_{zz}, \tau_{yz}, \tau_{zx}, \tau_{xy}]^T$ and where the components of the influence matrix K are expressed in terms of the derivatives of the potential integral I over the surface. Specifically,

$$K =$$

$$\begin{bmatrix} 2I_{,xz} - zI_{,xxx} & 2\nu I_{,yz} - zI_{,xxy} & -I_{,xx} - 2\nu I_{,yy} - zI_{,xxz} \\ 2\nu I_{,xz} - zI_{,xyy} & 2I_{,yz} - zI_{,yyy} & -I_{,yy} - 2\nu I_{,xx} - zI_{,yyz} \\ -zI_{,xzz} & -zI_{,yzz} & I_{,zz} - zI_{,zzz} \\ -\nu I_{,xy} - zI_{,xyz} & I_{,zz} + \nu I_{,xx} - zI_{,yyz} & -zI_{,yzz} \\ I_{,zz} + \nu I_{,yy} - zI_{,xxz} & -\nu I_{,xy} - zI_{,xyz} & -zI_{,xzz} \\ aI_{,yz} - zI_{,xxy} & aI_{,xz} - zI_{,xyy} & -bI_{,xy} - zI_{,xyz} \end{bmatrix} \tag{3}$$

and

$$I(P) = \frac{G}{4\pi(1-\nu)} \int_A \frac{dS_Q}{r} \ , \tag{4}$$

where G is the shear modulus, ν is Poisson's ratio, $a = (1-\nu)$, $b = (1-2\nu)$ and r is the distance between the field point P and the integration point Q. When the surface A is a polygon, I can be evaluated in closed form. Caution must be exercised in evaluating the derivatives of I when the field point P falls on the extension of any of the sides of the polygon in the x-y plane. When P is chosen at a specific point P_0 inside the polygon A, the local traction vector components τ_{xz}, τ_{yz}, τ_{zz} at point P_0 correspond to the fifth, fourth and third rows of equation (3). The total traction vector at P_0 can be written in the condensed form

$$\begin{bmatrix} T_x \\ T_y \\ T_z \end{bmatrix} = \begin{bmatrix} K_{xx} & K_{xy} & 0 \\ K_{yx} & K_{yy} & 0 \\ 0 & 0 & K_{zz} \end{bmatrix} \begin{bmatrix} D_x \\ D_y \\ D_z \end{bmatrix} + \begin{bmatrix} E_x \\ E_y \\ E_z \end{bmatrix} \tag{5}$$

where $[E_x \ \ E_y \ \ E_z]^T$ is the "external" influence due to the primitive stress field and other mobilized discontinuities. Equation (5) can be used as the nucleus of an iterative scheme to solve the mutual interaction between an assembly of discontinuity elements A^i (Ryder and Napier 1985). This scheme can be extended to allow for the time dependent relaxation of a set of discontinuity elements if it is postulated that the rate of shear slip is proportional to the net shear stress acting at point P_0. In particular, assume that

$$dD_s / dt = \kappa(\tau - \tau_r)^\beta \ , \tag{6}$$

where $D_s = \sqrt{D_x^2 + D_y^2}$ is the total slip and κ and β are parameters controlling the slip rate. The resultant shear stress τ and shear resistance τ_r are given by

$$\tau = \sqrt{T_x^2 + T_y^2} \qquad \text{and} \tag{7}$$

$$\tau_r = S_0 - \mu T_z \tag{8}$$

respectively, where S_0 is a residual cohesion and μ is the coefficient of friction controlling the resistance to slip. (T_z is assumed to be negative when compressive). Equation (6) can be written in discrete form as

$$\Delta D_s = \kappa(\tau - \tau_r)^\beta \Delta t \tag{9}$$

where Δt is a suitably small time step. The slip components are chosen to be in a direction $[D_x^0, D_y^0]^T$ which tends to reduce the shear stress to zero. Specifically, once ΔD_s is determined from equation (9), D_x and D_y are given by

$$\begin{bmatrix} D_x \\ D_y \end{bmatrix} = \gamma \begin{bmatrix} D_x^0 \\ D_y^0 \end{bmatrix} \tag{10}$$

$$\text{where} \quad \begin{bmatrix} D_x^0 \\ D_y^0 \end{bmatrix} = -\begin{bmatrix} K_{xx} & K_{xy} \\ K_{yx} & K_{yy} \end{bmatrix}^{-1} \begin{bmatrix} E_x \\ E_y \end{bmatrix} \tag{11}$$

$$\text{and} \quad \gamma = \Delta D_s / \sqrt{(D_x^0)^2 + (D_y^0)^2} \tag{12}$$

3 HIERARCHICAL SOLUTION STRATEGY

Consider a problem comprising N displacement discontinuity elements. In order to evaluate the total traction vector T^i at point P_0^i of element i it is necessary to compute the sum

$$T^i = \sum_{j=1}^{N} K^{ij} D^j + V^i \tag{13}$$

where K^{ij} represents the influence at receiving point P_0' due to sending element j as given by equation (3) with a suitable stress rotation to the local receiving element coordinate system. V^i is the primitive stress state at point P_0'. Designate the computational effort required to compute one of the entries in the sum of equation (13) as a "basic influence computation". Consequently, the total number of influence computations required for the evaluation of the traction vector at all receiving elements is equal to

$$I_B = N^2. \tag{14}$$

In order to reduce this computational effort, it is assumed that the problem space is partitioned into a number of cubic cells such that each cell forms one of eight equal sized members of a "parent" cell and itself contains eight "daughter" cells (an "octree" structure). Suppose that the side length of each finest level cell is L_F, then the size L_n of a cell at a given level n is given by

$$L_n = 2^{n-1} L_F \quad ; \quad n \geq 1 \tag{15}$$

The representative cell volume increases by a factor of eight for each unit increase in n ($L_{n+1}^3 = 8L_n^3$). It is further assumed that at the finest level the cells are located in fixed spatial positions and have unique address indices (i,j,k). In particular, the cell (1,1,1) is assumed to occupy the cube $0 \leq x < L_F$, $0 \leq y < L_F$, $0 \leq z < L_F$. The index address of any cell at any level n can be determined uniquely by the following recurrence relations.

$$i_{n+1} = \lfloor i_n + 1 \rfloor / 2 \ \text{ if } \ i_n \geq 0$$
$$i_{n+1} = -\lfloor |i_n| \rfloor / 2 \ \text{ if } \ i_n < 0. \tag{16}$$

It should be noted that a crude scheme, similar to the hierarchical process described here, termed "lumping", was used by Ryder and Napier 1985 for the solution of large scale tabular mining problems on two dimensional discontinuity surfaces.

Consider now a solution strategy where the stress influences at all elements in a finest level cell (level 1) are computed directly from "sending" elements within a specified inner region surrounding the cell. Outside this inner region and inside a specified outer region, the stress influence is computed at R designated interpolation points surrounding the receiving cell by using a fixed number S of equivalent discontinuity elements which are constructed in each sending cell. Let the overall problem dimension be D and suppose that the average element density is ρ. The total number of elements N is therefore given by

$$N = \rho D^3. \tag{17}$$

The number of finest level cells is

$$N_F = (D / L_F)^3. \tag{18}$$

Figure 1 shows the disposition of a receiving cell A surrounded by an inner "ring" of fine cells whose width is defined as a multiple g of the cell size, L_F, and an outer "ring" which is a multiple of the parent cell size, $2L_F$. It should be noted that the inner ring falls on the grid system of cell A and the outer ring falls on the grid system of the parent cell A_P. The sizes of the inner and outer rings, h and H respectively, are determined by the cell gap parameter g by the following expressions.

$$h = (2g+1)L_F \quad ; \quad H = 2h \tag{19}$$

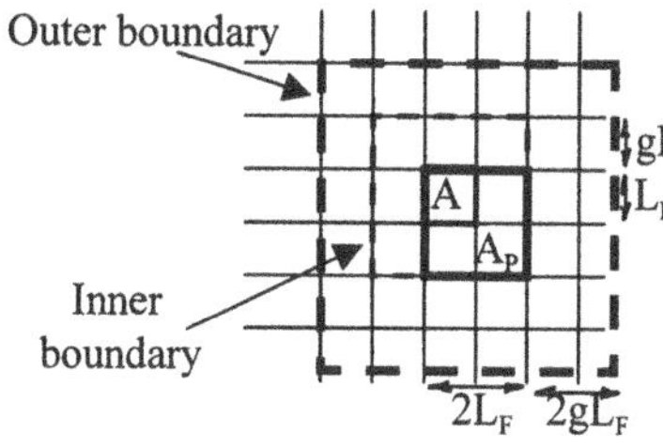

Figure 1. Two dimensional section of a fine level cell surrounded by an inner and an outer boundary determined by a cell gap value of $g = 1$.

Hence, the total number of direct influences on elements within the inner rings of all finest level cells is given by

I_D = (Number of elements in each fine cell)
X (Number of elements in inner ring)
X (Number of fine cells)

Using the parameter definitions this can be expressed as

$I_D = (\rho L_F^3)(\rho h^3)N_F$, whence,

$$I_D = (2g+1)^3 N^2 (L_F / D)^3 \quad (20)$$

The total number of indirect influences from cells in the region between the inner and outer rings is given by

I_F = (Number of influences per sending cell)
X (Number of sending cells)
X (Number of receiving points per cell)
X (Number of receiving cells)

This can be shown to be equal to

$$I_F = 7SR(2g+1)^3 (D / L_F)^3 \quad (21)$$

The total number of influences, $I_D + I_F$, is therefore

$$I_T = (2g+1)^3 [N^2 (L_F / D)^3 + 7SR(D / L_F)^3] \quad (22)$$

It can be seen from equation (22) that there is a tradeoff between the number of direct influences, which decreases as L_F is decreased, and the number of cell influences which increases as L_F is decreased. The optimum choice of D/L_F is given by

$$(D / L_F)^* = (N^2 / 7SR)^{1/6} \quad (23)$$

and the corresponding minimum number of influence computations is given by

$$I_T^* = 2(2g+1)^3 \sqrt{7SR}\, N \quad (24)$$

The most significant point of equation (24) is that the total computational effort is proportional to N rather than to N^2. (This conclusion is still valid if all the remaining coarse level cells, above the first level, are considered). Furthermore, the breakeven number of elements, N_c, for the hierarchical scheme to be more efficient than direct evaluation of influences, is such that $I_T^* < N^2$. Consequently, the hierarchical scheme is more efficient if

$$N > N_c = 2(2g+1)^3 \sqrt{7SR} \quad (25)$$

In the particular case when $g=1$, $S=3$ and $R=27$, N_C=1286.

4 EQUIVALENT SENDING ELEMENT STRAINS

A crucial aspect of achieving the computational performance demonstrated for the hierarchical solution scheme, is that the number, S, of equivalent discontinuities used in each sending cell should be fixed. It is also desirable for the chosen set of equivalent cracks to provide a physical representation of the coalesced fracture structure which arises from the fine level elements in a given cell. It is proposed that the equivalent crack set can be constructed using three orthogonal, rectangular shaped, discontinuity elements centred on the approximate centre of gravity of the mobilized fine level elements in the cell. The orientations of the three representative cracks are assumed to be aligned with the principal directions of the crack density tensor in the designated volume. Following Kachanov 1993, define the components of the crack density tensor α_{rs} by

$$\alpha_{rs} = (1/V)\sum_k (A_k / \pi)^{3/2} n_r^k n_s^k \quad (26)$$

where the sum is over all elements k within the volume V. A_k is the area of the kth crack and n_i^k are the components of the normal to the kth crack. Let the principal direction vectors of the crack density tensor be designated by X^1, X^2, X^3 respectively. The average strain due to all cracks in the volume V is

$$<\varepsilon_{ij}> = \frac{1}{2V}\int_A [D_i(Q)n_j(Q) + D_j(Q)n_i(Q)]dS_Q$$

or, in approximate form

$$<\varepsilon_{ij}> \approx \frac{1}{2V}\sum_k A_k [D_i^k n_j^k + D_j^k n_i^k] \quad (27)$$

where k is again summed over the crack elements falling in V and D_i^k, n_i^k designate the components of the discontinuity vector and normal vector for the kth crack, rotated to the principal axis system of the crack density tensor. Assume now that for the three equivalent cracks,

$$\tfrac{1}{2}\sum_{k=1}^{3} A_k [\hat{D}_i^k X_j^k + \hat{D}_j^k X_i^k] = <\varepsilon_{ij}> \quad (28)$$

Equation (28) represents a system of six equations (due to the symmetry of the strain tensor) in the nine unknown displacement discontinuity components $\hat{D}_i^k$ of the equivalent cracks. If it is assumed that $\hat{D}_i^k = \hat{D}_k^i$, equation (28) is satisfied provided

$$\hat{D}_i^j = 2 < \varepsilon_{ij} > /(A_i + A_j) \tag{29}$$

The areas A_i of each equivalent crack are assumed to be given by the sum of the projections of all the cracks in the representative volume onto the planes normal to each principal direction of the crack density tensor. The effectiveness of this scheme for the assignment of equivalent crack sets was tested by considering two cases. In the first case a square plane of side 100m is covered by a set of 50 random Delaunay triangular elements. The plane is assumed to be parallel to the x-axis of the global coordinate system and inclined at 70 degrees to the x-y plane. The crack elements on the plane are allowed to slip in response to a compressive stress field of 600 MPa applied in the z-direction. The resultant stress field is evaluated at particular points along a line in the y-direction, starting at the point x=50, y=50, z=50. The values of two selected components τ_{yy} and τ_{yz} are plotted as solid lines in Figure 2. The values corresponding to the equivalent crack set are plotted as broken lines in Figure 2. If it is assumed that the original set of 50 crack elements is located in a cell occupying the region with diagonally opposite vertices (-50,-50,-50) to (50,50,50) then $L_F = 100$ and the effective position of the inner boundary of the first level influence cell is at $y = 150$. It can be seen from Figure 2 that the agreement between the equivalent set of cracks and the actual stress components is excellent for values of y beyond the inner boundary.

In the second case, the vertices of the triangles used in the first test were perturbed randomly up to ±10m in the normal direction to the nominal plane to form an irregular surface. The system was solved again with the external field stress of 600 MPa parallel to the z-axis. The values of the two stress components are compared to the equivalent crack set in Figure 3, again showing excellent agreement beyond the inner boundary. The good agreement also holds if the exact and equivalent discontinuity schemes are compared for the component u_y of the displacement vector as shown Figure 4, for both the smooth and irregular surfaces.

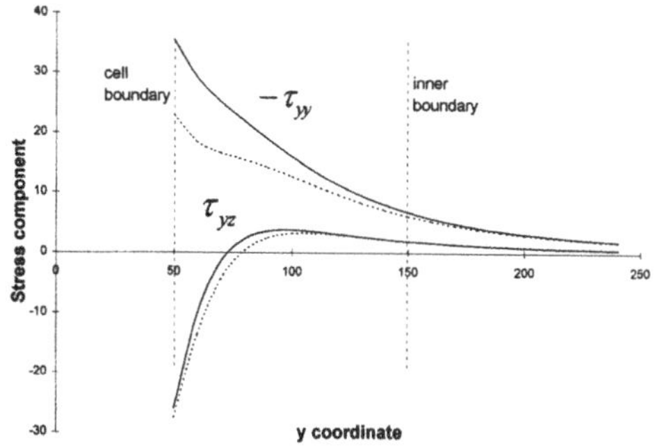

Figure 2. Comparison of stresses, generated by a set of 50 crack elements (solid lines) on a plane, to the stresses due to an equivalent single crack (broken lines).

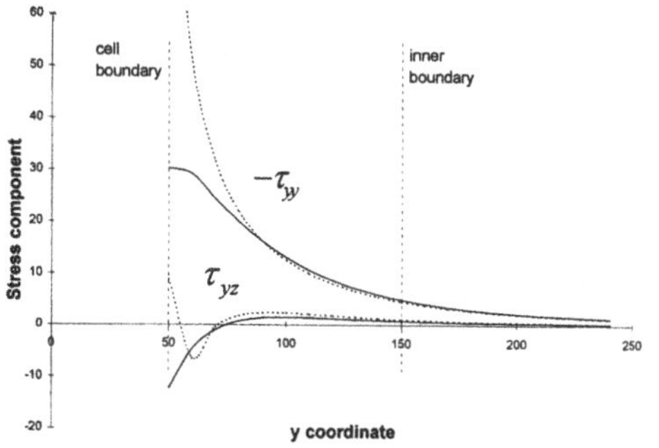

Figure 3. Comparison of stresses generated by a set of 50 crack elements, irregularly perturbed about a nominal plane (solid lines), to the stresses due to an equivalent set of three cracks (broken lines).

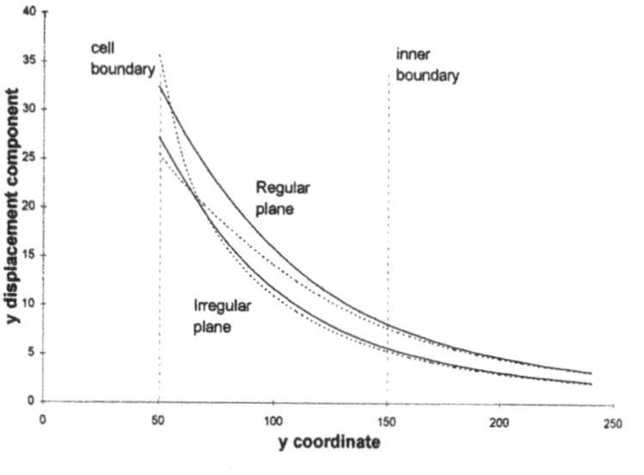

Figure 4. Comparison between the exact (solid lines) and equivalent discontinuity sets (broken lines) for the displacement component u_y.

In the case of the irregular surface, the principal directions of the crack density tensor are computed to be

$$X^1 = \begin{bmatrix} .025 \\ .942 \\ -.335 \end{bmatrix}, \quad X^2 = \begin{bmatrix} -.986 \\ -.032 \\ -.165 \end{bmatrix}, \quad X^3 = \begin{bmatrix} -.167 \\ .334 \\ .928 \end{bmatrix}.$$

The areas and local discontinuity vector components for the equivalent set of discontinuities are shown in Table 1.

Table 1. Equivalent crack areas and local discontinuity vector components for the equivalent crack set approximating the irregular crack surface.

Element number, k	Equivalent area (m^2)	DDx (mm)	DDy (mm)	DDz (mm)
1	10002	13.5	-114.2	-12.1
2	3692	-16.1	13.5	-9.9
3	2560	-114.2	-16.1	16.7

5 SEISMIC ENERGY RELEASE CYCLES

The hierarchical solution scheme described here, together with the slip relaxation model outlined in section 2, can be applied to the analysis of time dependent seismic energy release near deep level mining excavations. The implementation of this scheme is incorporated in a computer program called 3DIGS (Three dimensional Discontinuity Interaction and Growth Simulation). Full scale trials of the scheme have not been completed. An indication of the seismic energy release cycles which are generated when a remnant pillar is extracted incrementally adjacent to a fault plane are plotted in Figure 5 showing an increasing trend of seismic energy release as the pillar size is reduced. Important future investigations will address the mechanisms of stress accumulation on fault planes and the processes of stress transfer and "diffusion" in the rock mass as a function of the mining rate.

6 CONCLUSIONS

A method for treating the viscoplastic relaxation of three dimensional crack assemblies has been introduced. A hierarchical scheme for the solution of large scale assemblies of cracks is proposed and is demonstrated to have a solution time that is proportional to the number of crack elements. A method for the assignment of equivalent crack sets in a given region is also described. This provides a first step in representing fracture clustering and localization processes on multiple length scales. The proposed solution method opens some exciting possibilities for the analysis of multi-scale, three dimensional damage modelling and time dependent stress transfer processes in deep level mines.

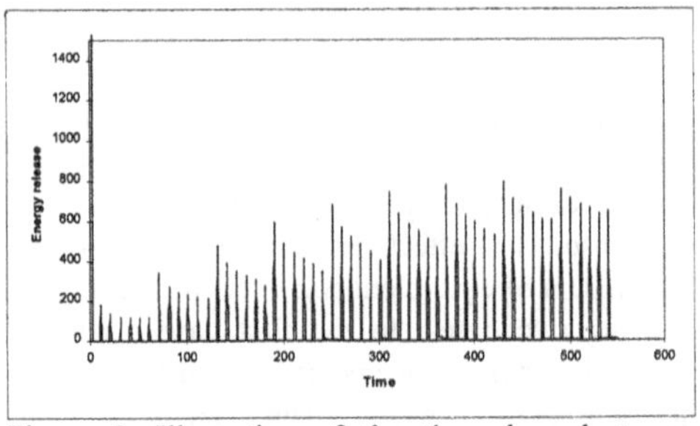

Figure 5. Illustration of the time dependent energy release cycles accompanying the mining of a remnant pillar adjacent to a fault plane.

ACKNOWLEDGEMENTS

This work forms part of the research program of the CSIR Division of Mining Technology and was funded by the Safety in Mines Research Advisory Committee of South Africa (SIMRAC). I am grateful to Mr T Dede for his assistance in preparing the diagrams.

REFERENCES

Allegre, C.J., Le Mouel, J.L., Chau, H.D. and Norteau, C. 1995. Scaling organization of fracture tectonics (SOFT) and earthquake mechanism. *Phys. Earth Planet. Inter.*, 92:215-233.

Crouch, S.L. and Starfield, A.M. 1983. *Boundary Element Methods in Solid Mechanics*, George Allen & Unwin, London.

Kachanov, M. 1993. Elastic solids with many cracks and related problems. In: J.W. Hutchinson and T.Y. Wu (Eds.), *Advances in Applied Mechanics*, 30:259-445.

Madden, T.R. 1983. Microcrack connectivity in rocks: A renormalization group approach to the critical phenomena of conduction and failure in crystalline rocks. *J. Geophys. Res.*, 88:585-592.

Main, I.G. 1995. Earthquakes and critical phenomena: Implications for probabilistic seismic hazard analysis. *Bull. Seismol. Soc. Am.*, 85:1299-1308.

Ryder, J.A. and Napier, J.A.L. 1985. Error analysis and design of a large-scale tabular mining stress analyser. In *Proceedings 5th international conference on numerical methods in geomechanics*. Nagoya:1549-1555.

Wilson, K.G. 1979. Problems in physics with many scales of length. *Scientific American*, 241:140-157.

Mechanics of Jointed and Faulted Rock, Rossmanith (ed.) © *1998 Taylor & Francis, ISBN 90 5410 955 6*

An energy-based damage model of jointed rockmass

G. Swoboda
University of Innsbruck, Austria
Q. Yang
Department of Hydraulic Engineering, Tsinghua University, Beijing, People's Republic of China
X. P. Shen
University of Innsbruck, Austria & Department of Engineering Mechanics, Tsinghua University, Beijing, People's Republic of China

ABSTRACT: In this paper, an anisotropic elasto-plastic damage model in strain space has been established to describe the behaviour of geomaterials under compression dominated stress fields. The research work focuses on rate-independent and small-deformation behaviour during isothermal processes. The salient features of geomaterials is fully defined in the starting points: (1) an equivalent state based on which the formulation is symmetric with respect to the sign of the stress/strain and (2) a general free energy function including the mechanisms of damage, plasticity and damage residual effects. The constitutive relations are developed in the well-established continuum mechanics. The prime results include: (1) a one-parameter damage dependent elasticity tensor deduced based on tensorial algebra and thermodynamic requirements; (2) a fourth-order damage characteristic tensor which determines anisotropic damaging, deduced in the framework of Rice's "normality structure"; (3) a procedure for the determination of the damage tensor according to the geological data is described here. Finally, some numerical results are presented.

1 INTRODUCTION

Geomaterials have complex mechanical behaviour, such as stress-induced anisotropy, hysteresis, dilatancy, irreversible strain and strongly path dependent stress-strain relations, which is generally associated with the existence of a great deal of micro- and meso-cracks and their propagation. Continuum damage mechanics, which employs some continuum variables to describe the micro-defects, has been an appealing framework for modeling geomaterials, see e.g., Ju (1989)), Dragon (1993), Kawamoto *et al.* (1988), Ortiz (1985), Stumvoll and Swoboda 1993. In order to fit the complicated behaviour of geomaterials, various starting points and assumptions are adopted in these models, and their formulations are generally sophisticated and have few common points, which have hindered these models from going into practical applications. The key point is that few models can take account of geomaterial features on an appropriate theoretic basis.

In this paper, we try to put all salient features of geomaterials only into starting points: the damage variable, equivalent state and free energy function. The constitutive relations, e.g., damage elasticity and damage evolution laws, will be developed in the well-established continuum mechanics without concerning the special features of geomaterials.

In this paper, a one-parameter damage dependent elasticity tensor is formulated by tensorial algebra and thermodynamic requirements.

The *equivalent state* is termed here as the state for which the formulation is symmetric with respect to the sign of the stress/strain. In this paper, an equivalent state which represents the current open cracks is developed.

In this paper, the damage evolution law is formulated in the conjugate force space based on the irreversible thermodynamics. An explicit free energy function including the mechanisms of damage, plasticity and damage residual effects is established; an analytic damage characteristic tensor is deduced in Rice's (1971) *normality structure* which can introduce the micromechanics into the thermodynamic framework.

2 DAMAGE TENSOR

The defects in geomaterials consist of microcavities and microcracks. It is reasonable to assume that the governing defects in geomaterials are microcracks. For each set of cracks, the damage tensor can be definited as:

$$\boldsymbol{\Omega} = \omega \boldsymbol{n}\boldsymbol{n} \tag{1}$$

or

$$\Omega_{ij} = \omega n_i n_j, \quad i = 1,3 \tag{2}$$

where $\boldsymbol{n} = [l, m, n]^T$ is the direction vector of damage tensor Ω of the crack set; ω is the characteristic damage parameter, the separation factor, which can also be obtained through statistical data.

2.1 *Equivalent state*

Geomaterials have much more complicated mechanical behaviour, which is generally associated with frictional sliding on crack surfaces. Sliding causes dilation by opening the crack at asperities and by inducing local tensile cracking at some angle to the crack. Frictional crack models have been investigated by many authors, see e.g., Nemat-Nasser and Obata (1988). For local tensile cracking, as shown in Fig. 1(a), the stress intensity factor of Mode I at crack tips can be computed with very good accuracy by considering an equivalent crack of length $2l$, subjected to a pair of collinear concentrated forces, τ, as well as the applied overall stress, as shown in Fig. 1(b). In this representation, τ denotes the resultant force transmitted across the preexisting crack.

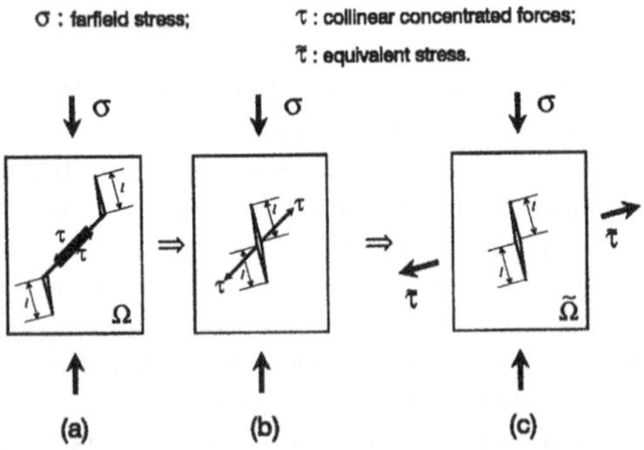

Figure 1: Schematic frictional crack equivalence

Macroscopically, the states (a) and (c) are characterized by Table 1. In this paper, the state (c) is defined as the equivalent state; $\tilde{\boldsymbol{\Omega}}$ is termed as the *effective damage tensor*, which corresponds to the equivalent cracks; $\tilde{\boldsymbol{\sigma}} = \boldsymbol{\sigma} + \tilde{\boldsymbol{\tau}}$ is called the *effective stress*, where $\tilde{\boldsymbol{\tau}}$ is the macroscopic representation of τ and can be called the *equivalent stress.*

Table 1: Macroscopic characterization of real and equivalent states

	Real state (a)	Equivalent state (c)
Damage tensor	$\boldsymbol{\Omega}$	$\tilde{\boldsymbol{\Omega}}$
Stress	$\boldsymbol{\sigma}$	$\tilde{\boldsymbol{\sigma}} = \boldsymbol{\sigma} + \tilde{\boldsymbol{\tau}}$
Elastic strain	$\boldsymbol{\varepsilon}^e$	$\tilde{\boldsymbol{\varepsilon}}^e = \tilde{\mathbf{C}} : \tilde{\boldsymbol{\sigma}}$
Elasticity tensor /Compliance	$\mathbf{D}/\mathbf{C}$	$\tilde{\mathbf{D}}/\tilde{\mathbf{C}}$
Elastic complementary energy	w	$\tilde{w} = \frac{1}{2}\tilde{\boldsymbol{\sigma}} : \tilde{\mathbf{C}} : \tilde{\boldsymbol{\sigma}}$

Equivalence of the stress intensity factors between (a) and (c) in the frictional model macroscopically implies $dw = d\tilde{w}$ during infinitesimal crack or damage propagation. Based on the equivalence, we assume $w = \tilde{w}$. If further assuming that $\tilde{\boldsymbol{\tau}}$ is independent on $\boldsymbol{\sigma}$, = i.e., $\frac{\partial \tilde{\boldsymbol{\tau}}}{\partial \boldsymbol{\sigma}} = \mathbf{0}$, then

$$\boldsymbol{\varepsilon}^e = \frac{\partial w}{\partial \boldsymbol{\sigma}} = \frac{\partial \tilde{=} w}{\partial \boldsymbol{\sigma}} = \tilde{\mathbf{C}} : \tilde{\boldsymbol{\sigma}} = \tilde{\boldsymbol{\varepsilon}}^e \tag{3}$$

Thus, the real state is equivalent to the equivalent state in terms of the elastic strain and the elastic strain/complementary energy although generally $\mathbf{D} \neq \tilde{\mathbf{D}}$. Eqn. (3) can be rearranged as

$$\boldsymbol{\sigma} = \tilde{\mathbf{D}}(\tilde{\boldsymbol{\Omega}}) : \boldsymbol{\varepsilon}^e - \tilde{\boldsymbol{\tau}} \tag{4}$$

The explicit expression of $\tilde{\mathbf{D}}(\tilde{\boldsymbol{\Omega}})$ will be developed late. Since τ acts on the equivalent crack as shown in Fig. 1(b), it is reasonable to assume that the equivalent stress $\tilde{\boldsymbol{\tau}}$ is *parallel* to $\tilde{\boldsymbol{\Omega}}$, i.e.

$$\tilde{\boldsymbol{\tau}} = g\tilde{\boldsymbol{\Omega}} \tag{5}$$

where $g \geq 0$ is assumed to be a material constant. Thus, $\tilde{\boldsymbol{\tau}}$ is independent of $\boldsymbol{\sigma}$, which is consistent with eqn. (3) but contrary to the fictitious stress introduced by Krajcinovic *et al.* (1991). Such a fictitious equivalent stress was first proposed by Dragon *et al.* (1993) from a different viewpoint.

2.2 *Effective damage tensor*

In principle, the effective damage tensor $\tilde{\boldsymbol{\Omega}}$ can be computed directly using eqn. (1) with respect to the equivalent cracks. It is difficult, however, to determine the equivalent crack. Here, by decomposing one original crack vector in the principal strain coordinate system, it is assumed that each equivalent crack is composed of only these components in tensile principal strain directions. Thus, the effective damage tensor can be calculated by the transformation

$$\tilde{\mathbf{\Omega}} = \mathbf{P}^{+} : \mathbf{\Omega} \tag{6}$$

where $\mathbf{P}^{+}$ is just the so-called "positive projection tensor" (Ortiz, 1985; Ju, 1989),

$$P^{+}_{ijkl} = Q^{+}_{ik} Q^{+}_{jl}, \qquad \text{where} \qquad \mathbf{Q}^{+} = \sum_{i=1}^{3} \hat{H}(\varepsilon_i)\mathbf{p}_i\mathbf{p}_i \tag{7}$$

where Q^{+}_{ij} is the positive (tensile) spectral projection tensor; $\mathbf{p}_i$ and ε_i are the i-th principal normal vector and principal strain, respectively, of strain $\boldsymbol{\varepsilon}$; $\hat{H}(\cdot)$ is the Heaviside function

$$\hat{H}(\varepsilon_i) = \begin{cases} 1 & \text{if } \varepsilon_i > 0 \\ 0 & \text{otherwise} \end{cases} \tag{8}$$

The effective damage tensor is symmetric, ensured by eqns (6) and (7). In some cases for geomaterials, the Heaviside function $\hat{H}$ will overestimate the difference between compression and tension. For example, in jointed rock, the joints are not perfectly contact but with some filling materials or roughness. In order to take these factors into account, the Heaviside function $\hat{H}$ in eqn. (7) can be replaced with

$$H(\varepsilon_i) = (1-h)\hat{H} + h = \begin{cases} 1 & \text{if } \varepsilon_i > 0 \\ h & \text{otherwise} \end{cases} \tag{9}$$

where h ($0 \leq h < 1$) is a material constant to reflect the properties of crack contact.

3 DAMAGE ELASTICITY

In this section, the development of an explicit expression of a damage-dependent elasticity tensor $\tilde{\mathbf{D}}(\tilde{\mathbf{\Omega}})$ is presented. Note that $\tilde{\mathbf{D}}(\tilde{\mathbf{\Omega}}) = \mathbf{D}(\mathbf{\Omega})$ if $\tilde{\mathbf{\Omega}} = \mathbf{\Omega}$. Firstly, a more general problem $\mathbf{D}(\mathbf{\Omega})$ on the real state is considered.

In general, an elasticity tensor is subject to the following general principles of continuum mechanics(Malvern, 1969): (1) *Symmetry condition* requires $D_{ijkl} = D_{jikl} = D_{ijlk} = D_{klij}$; (2) *Positive definite condition* requires the elastic potential function $W = \frac{1}{2}\varepsilon_{ij} D_{ijkl} \varepsilon_{kl}$ is positive-definite as a function of strain ε_{ij}; (3) *Material symmetry condition* requires that $\mathbf{D}(\mathbf{\Omega})$ be an isotropic tensor function.

The isotropic tensor function $D_{ijkl}(\Omega_{mn})$, which satisfies the symmetric condition, takes the form

$$\begin{aligned} D_{ijkl} &= A_1\delta_{ij}\delta_{kl} + A_2(\delta_{ik}\delta_{jl} + \delta_{il}\delta_{jk}) \\ &+ A_3(\Omega_{ij}\delta_{kl} + \Omega_{kl}\delta_{ij}) + A_4(\Omega_{ik}\delta_{jl} + \Omega_{il}\delta_{jk} + \Omega_{jk}\delta_{il} + \Omega_{jl}\delta_{ik}) \\ &+ A_5\Omega_{ij}\Omega_{kl} + A_6(\Omega_{ik}\Omega_{jl} + \Omega_{il}\Omega_{jk}) \\ &+ A_7(\Theta_{ij}\delta_{kl} + \Theta_{kl}\delta_{ij}) + A_8(\Theta_{ik}\delta_{jl} + \Theta_{il}\delta_{jk} + \Theta_{jk}\delta_{il} + \Theta_{jl}\delta_{ik}) \\ &+ A_9(\Theta_{ij}\Omega_{kl} + \Omega_{ij}\Theta_{kl}) + A_{10}(\Theta_{ik}\Omega_{jl} + \Omega_{ik}\Theta_{jl} + \Theta_{il}\Omega_{jk} + \Omega_{il}\Theta_{jk}) \\ &+ A_{11}\Theta_{ij}\Theta_{kl} + A_{12}(\Theta_{ik}\Theta_{jl} + \Theta_{il}Theta_{jk}) \end{aligned} \tag{10}$$

where $\Theta_{ij} = \Omega_{im}\Omega_{mj}$; A_1, A_2, $\cdots$, A_{12} are functions of invariants of Ω_{ij}, see e.g., Bazant (1983). All existing expressions of the damage elasticity are special cases of eqn. (10). To enforce the requirements on A_1, A_2, $\cdots$, A_{12}, like Cowin (1985), makes the expression too complicated to be used. Here, a one-parameter damage elasticity tensor is introduced

$$\mathbf{D} = \boldsymbol{\phi} \cdot \mathbf{D}^{0} \cdot \boldsymbol{\phi} \tag{11}$$

or

$$\tilde{D}_{ijkl} = \lambda\phi_{ij}\phi_{kl} + \mu(\phi_{ik}\phi_{jl} + \phi_{il}\phi_{jk}) \tag{12}$$

$$\phi_{ij} = \delta_{ij} - m\tilde{\Omega}_{ij} - (1-m)\tilde{\Omega}_{im}\tilde{\O}mega_{mj}, \qquad 0 \leq m \leq 1 \tag{13}$$

where m is assumed to be a material constant. The condition $0 \leq m \leq 1$ is needed to satisfy the positive definite requirement. Evidently, the elasticity tensor satisfies all requirements and a complete tensorial polynomial as compared with eqn. (10).

The behaviour of the damage-dependent elasticity tensor defined in eqn. (13) is illustrated by a series of numerical uniaxial tests. A sets of parallel microcracks are consisted in the cylinder. These parallel microcracks in the specimen can be characterized by a damage vector (n_i, ω), where ω is the characteristic damage value, and the normal vector n_i is determined by α, the angle between loading direction and the crack plane. The damage vector corresponds to a damage tensor $\Omega_{ij} = \omega n_i n_j$. With the increase of the damage value, the reduction of the normalized Young's modulus E/E_0 in the loading direction for different values of m is shown in Fig. 2 under the condition of $h = 1$. The results indicate that an increase of m will reduce the normalized module E/E_0.

The relation between normalized Young's modulus E/E_0 in the loading direction and the increment of damage value is shown in Fig. 3, based on the damage elasticity parameter $m = 1$. It is indicate that if no residual damage effect is considered, which means, that $h = 0$, no reduction of Young's modulus in the direction of compression. The stiffness of the material between the crack can produce such an effect and would compensate the damage effect. With an assumption of $h = 1$, the active damage tensor will be equal the nominal damage tensor for both tension and compression.

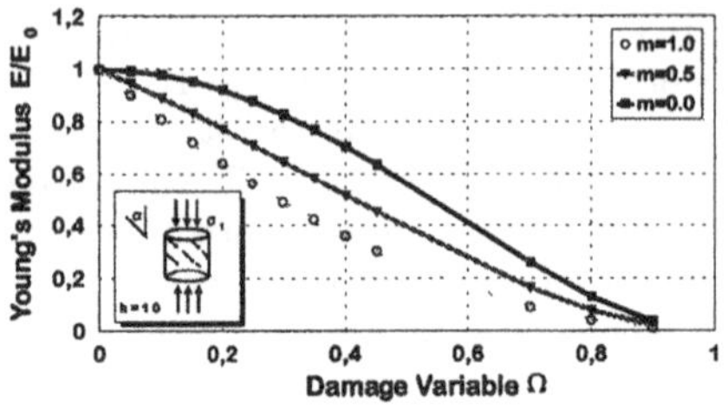

Figure 2: Effect of damage elasticity parameter m

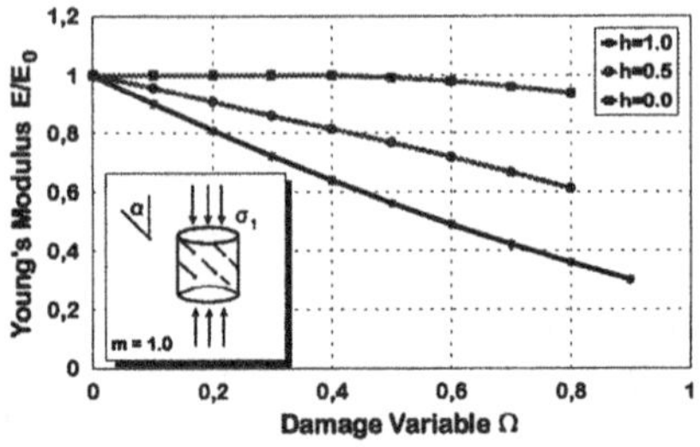

Figure 3: Effect of rock property parameter h

The relation between apparent Young's modulus E/E_0 along the loading direction and the crack angle α is shown in Fig. 4. It indicates that the horizontal component of damage tensor is of little influence on apparent Young's modulus in the vertical direction under the action of uniaxial compression.

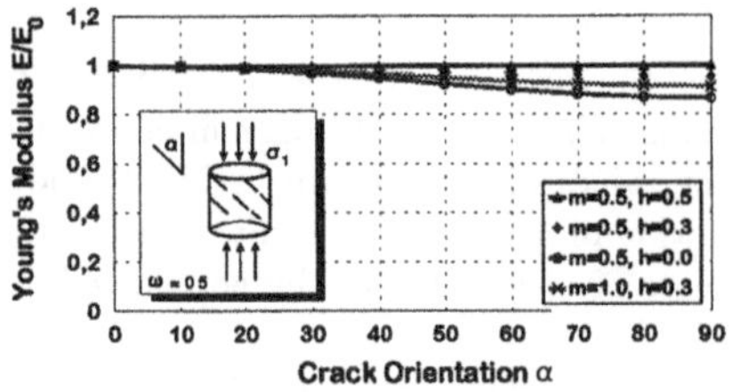

Figure 4: Effect of the crack orientation.

To study the influence of the lateral pressure on the stiffness in the vertical direction a numerical simulation with 12 20-node isoparametric elements was done. On the cylinder acts an uniform pressure σ_{top} on the top surface and lateral pressure $\sigma_{lateral}$. With this two stresses the dimensionless factor $r = \sigma_{lateral}/\sigma_{top}$ can be defined. In the example of Fig. 5, three sets of joints are assumed to be consisted in the cylinder in order to simulate the engineering situation. The initial damage tensor $\mathbf{\Omega}$ is obtained through the geological data ($\omega_1 = 0.2, \omega_2 = 0.6, \omega_3 = 0.3$). The definition will be shown in the next section.

The results for the $\sigma_z - \varepsilon_z$ diagram with different values of r are shown in Fig. 5. The reduction of the stiffness in z-direction is not very seriously influenced by the lateral pressure.

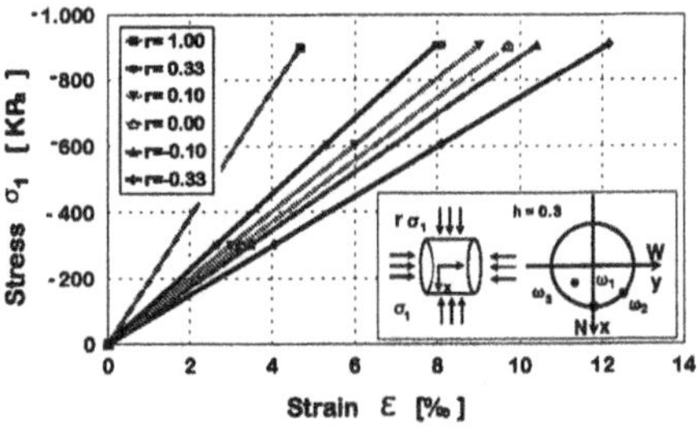

Figure 5: Influence of the lateral pressure on the stiffness in vertical direction.

3.1 *Determination of the tensor through the geological data*

The orientation of the joint set is usually identified in the field by sampling along a line, as in a borehole, or over an area, as on an outcrop surface. Statistical descriptions of the orientation data are always expressed as the orientations of normal line of the joint plane. Spherical coordinates (α, β, r) or the direction cosine (l, m, n) are usually used for this purpose. These are related to dip angle, which is definited for the angle between the horizontal and the line of maximum dip of the joint plane, β and the dip direction, which is definited for the angle between North and the horizontal projection of the line of maximum dip of joint plane measured in the clockwise direction, α, together with direction angle of the tunnel axis γ, the angle between tunnel axis and the North, as the follows:

$$\begin{cases} l = sin\beta cos[2\pi - (\alpha + \gamma)] \\ m = sin\beta sin[2\pi - (\alpha + \gamma)] \\ n = cos\beta \end{cases} \tag{14}$$

The direction vector of damage tensor Ω according to eqn. (1) and (2) is defined on the basis of the direction vector $\boldsymbol{n} = [l, m, n]^T$.

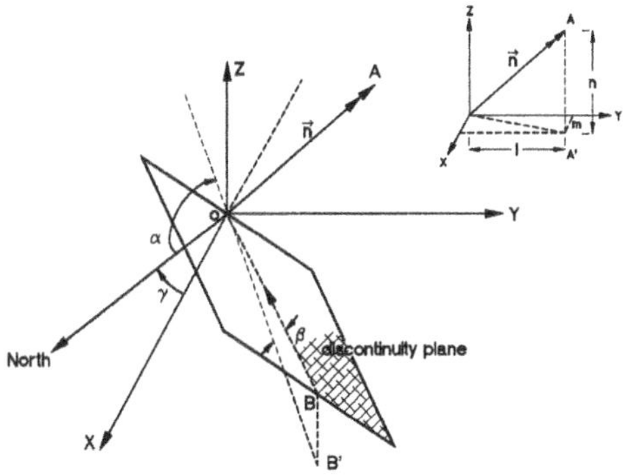

Figure 6: Relationship between orientation parameters and the spherical coordinate system.

Here an example for the calculation of damage tensor from the geological data is given. The dip direction $\alpha = 45^0$, dip angle $\beta = 45^0$, tunnel axis direction angle $\gamma = 0$, damage parameter $\omega = 0.3$ is based on the geological investigation. Through the before mentioned calculation process, the initial damage tensor would be:

$$\boldsymbol{\Omega}_0 = \begin{bmatrix} 0.075 & -0.075 & 0.106 \\ -0.075 & 0.075 & -0.106 \\ 0.106 & -0.106 & 0.15 \end{bmatrix}$$

3.2 *Global damage tensor for multiple sets of joints*

For jointed rockmass, the most common damage state is the presence of multiple sets of joints which intersecting each other. Here it is presented the techniques for determination of the global damage tensor which can represents the damage effect of the multiple joint sets system.

For each set of joints, the related damage tensor can be obtained with the techniques introduced above. For the global damage tensor, its calculation is based on the principle of energy equivalence: The energy dissipated by global damage tensor is equivalent to the sum of the energy dissipated by damage tensor related to each set of joints.

The equation for the calculation of global damage tensor is given in the following:

$$\boldsymbol{\Omega}_g = \mathbf{I} - \left[\sum_{n=1}^{N} (\mathbf{I} - \boldsymbol{\Omega}_n)^{-1} - (N-1)\mathbf{I} \right]^{-1} \quad (15)$$

where $\boldsymbol{\Omega}_g$ is the global damage tensor; $\boldsymbol{\Omega}_n$ is the damage tensor for the nth joint; N is the total number of joint sets; $\mathbf{I}$ is a second order unit tensor.

The following is the detail data for global damage tensor used in the above numerical examples of the cylinder under compression.

There are three set of joints are given in this cylinder. For the nth joint, geological data are:

$$\begin{array}{lll} \gamma = 0, & & \\ \alpha_1 = 0, & \beta_1 = \frac{\pi}{2}, & \omega_1 = 0.2 \\ \alpha_2 = -\frac{\pi}{4}, & \beta_2 = \frac{\pi}{2}, & \omega_2 = 0.6 \\ \alpha_3 = \frac{\pi}{4}, & \beta_3 = \frac{\pi}{4}, & \omega_3 = 0.3 \end{array}$$

The damage tensor respectively related to each set of joints are:

$$\boldsymbol{\Omega}_1 = \begin{bmatrix} 0.2 & 0.0 & 0.0 \\ 0.0 & 0.0 & 0.0 \\ 0.0 & 0.0 & 0.0 \end{bmatrix}$$

$$\boldsymbol{\Omega}_2 = \begin{bmatrix} 0.3 & 0.3 & 0.0 \\ 0.3 & 0.3 & 0.0 \\ 0.0 & 0.0 & 0.0 \end{bmatrix}$$

$$\boldsymbol{\Omega}_3 = \begin{bmatrix} 0.075 & -0.075 & 0.106 \\ -0.075 & 0.075 & -0.106 \\ 0.106 & -0.106 & 0.15 \end{bmatrix}$$

The global damage tensor for the three joint sets in the cylinder is:

$$\boldsymbol{\Omega} = \begin{bmatrix} 0.45946 & 0.19459 & 0.01 \\ 0.19459 & 0.38595 & -0.1 \\ 0.01 & -0.1 & 0.15243 \end{bmatrix}$$

4 FREE ENERGY FUNCTION

The specific free energy ψ, taken as the thermodynamic potential, plays a central role in describing thermodynamic processes. It determines the stress-strain relation at the current state and the conjugate forces, which are the drive forces of state changes. Here a general free energy function including the mechanisms of damage, plasticity and damage residual effects is established. Confined in isothermal and small deformation processes, it is assumed that there exists a free energy function $\psi = \psi(\boldsymbol{\varepsilon}, \boldsymbol{\varepsilon}^p, \mathbf{q}, \boldsymbol{\Omega})$, which is the function of strain tensor $\boldsymbol{\varepsilon}$, plastic strain tensor $\boldsymbol{\varepsilon}^p$, one set of suitable plastic internal variables $\mathbf{q}$, damage tensor $\boldsymbol{\Omega}$. Its explicit expression is proposed as

$$\psi(\boldsymbol{\varepsilon}, \boldsymbol{\varepsilon}^p, \mathbf{q}, \boldsymbol{\Omega}, \mathbf{p}) = \frac{1}{2}\varepsilon^e_{ij}\tilde{D}_{ijkl}\varepsilon^e_{kl} - g\tilde{\Omega}_{kl}\varepsilon_{ij} + \psi_p(\mathbf{q}) \quad (16)$$

where $\boldsymbol{\varepsilon}^e = \boldsymbol{\varepsilon} - \boldsymbol{\varepsilon}^p$ is the elastic strain tensor.

As shown in Fig. 1(b), the equivalent forces τ simply make the crack open. Its contribution to the free energy can be calculated as

$$\psi_{dp}(\tilde{\boldsymbol{\tau}}, \tilde{\boldsymbol{\Omega}}, \boldsymbol{\varepsilon}) = -\tilde{\boldsymbol{\tau}} : \boldsymbol{\varepsilon} = -g\tilde{\Omega}_{ij}\varepsilon_{ij} \tag{17}$$

where the negative sign indicates that $\tilde{\boldsymbol{\tau}}$ is not an external stress but an equivalent internal stress. ψ_{dp} is called the damage residual term because a *residual* elastic strain energy can be expected as $\psi_{dp} = -g\tilde{\Omega}_{kl}\varepsilon^p_{ij}$, after unloading, i.e., $\varepsilon^e_{ij} = 0$.

Microscopically, the residual ψ_{dp} is caused by a residual τ produced by the residual shear deformation on the main crack surfaces due to the roughness. Frictional sliding eventually causes dilatancy and splitting cracking.

5 INTERNAL DISSIPATION

The dissipative inequality takes the form

$$\tilde{\sigma}_{ij}\dot{\varepsilon}^p_{ij} + \mathbf{Q}\cdot\dot{\mathbf{q}} + Y_{ij}\dot{\Omega}_{ij} \geq 0 \tag{18}$$

where $\tilde{\sigma}_{ij}$ is the effective stress, see Table 1; $\mathbf{Q} = -\frac{\partial \psi_p}{\partial \mathbf{q}}$ is the thermodynamic forces conjugate to the plastic internal variables $\mathbf{q}$ and Y_{ij} is the thermodynamic forces conjugate to the damage internal variable Ω_{ij}

$$\begin{aligned} Y_{ij} &= -\frac{\partial \psi}{\partial \Omega_{ij}} = -\frac{\partial \psi}{\partial \tilde{\Omega}_{kl}}\frac{\partial \tilde{\Omega}_{kl}}{\partial \Omega_{ij}} = -\frac{\partial \psi}{\partial \tilde{\Omega}_{kl}}P^+_{klij} = P^+_{ijkl}\tilde{Y}_{kl} \\ \tilde{Y}_{ij} &= -\frac{\partial \psi}{\partial \tilde{\Omega}_{ij}} = \frac{1}{2}\varepsilon^e_{mn}\mathcal{D}_{mnklij}\varepsilon^e_{kl} + g\varepsilon_{ij} \end{aligned} \tag{19}$$

where, by using eqn. (13),

$$\begin{aligned} \mathcal{D}_{ijklmn} &= \lambda(\Phi_{mnij}\phi_{kl} + \Phi_{mnkl}\phi_{ij}) \\ &+\mu(\Phi_{mnik}\phi_{jl} + \Phi_{mnjl}\phi_{ik} + \Phi_{mnil}\phi_{jk} + \Phi_{mnjk}\phi_{il}) \\ \Phi_{ijkl} &= k\delta_{ik}\delta_{jl} + (1-k)(\delta_{ik}\tilde{\Omega}_{jl} + \tilde{\Omega}_{ik}\delta_{jl}) \end{aligned} \tag{20}$$

6 DAMAGE EVOLUTION LAW

The damage evolution laws developed in conjugate force space have the most general form. However, its analytic expression is very difficult to be obtained. The irreversible thermodynamics can only furnish certain limitation. In this part, the evolution law is deduced in Rice's (1971) "normality structure". which can incorporate the micromechanics into the thermodynamic framework. As pointed out by Krajcinovic *et al.* (1991), the normality structure is exactly within the framework of the self-consistent method. The results confirm the validity of the linear irreversible thermodynamics and furnish an analytic damage characteristic tensor. Due to limited space, only main conclusions are mentioned.

Consider the basic internal variables $\boldsymbol{\xi}$:

$$\boldsymbol{\xi} = \{\xi_1, \xi_2 \ldots, \xi_n\}, \qquad \xi_\alpha = \{\mathbf{n}^\alpha, r_\alpha\} \tag{21}$$

where and $\mathbf{n}^\alpha$ are the radius and normal vector of the α-th crack.

Their kinetic equations of internal variables are required to take the form

$$\dot{\xi}_\alpha = \dot{\xi}_\alpha(f_\alpha, \boldsymbol{\xi}), \qquad (\alpha = 1, 2, \ldots, n) \tag{22}$$

where $f_1, f_2, \ldots, f_n$ (collectively $\mathbf{f}$) are the thermodynamic forces acting on the internal variables,

$$f_\alpha = -V^0\frac{\partial \psi}{\partial \xi_\alpha} = -V^0\left\{\frac{\partial \psi}{\partial \mathbf{n}^\alpha}, \frac{\partial \psi}{\partial r_\alpha}\right\} \tag{23}$$

Thus, if set

$$Q = \frac{1}{V^0}\int_0^{\mathbf{f}} \dot{\xi}_\alpha(\mathbf{f}, \boldsymbol{\xi})df_\alpha \tag{24}$$

then Q is a point function of $\mathbf{f}$, i.e., $Q = Q(\mathbf{f}, \boldsymbol{\xi})$, because eqn. (22) furnishes the strictest condition to ensure the integrand to be a total differential. Therefore, the kinetic equations of internal variables can be recast as

$$\dot{\xi}_\alpha = V^0\frac{\partial Q(\mathbf{f}, \boldsymbol{\xi})}{\partial f_\alpha} \tag{25}$$

Note that the same damage state can also be characterized by the damage tensor $\boldsymbol{\Omega}$ which can be seen as the average measure of basic internal variable $\boldsymbol{\xi}$. If the potential function $Q(\mathbf{f}, \boldsymbol{\xi})$ can be rewritten as $Q(\boldsymbol{\Omega}, \mathbf{Y})$, the damage evolution law takes the form

$$\dot{\boldsymbol{\Omega}} = \frac{\partial Q(\boldsymbol{\Omega}, \mathbf{Y})}{\partial \mathbf{Y}} \tag{26}$$

If the microcracks are Griffith cracks and each crack kinks along the direction which make the system release maximum energy for the same crack increment, the kinetic equations of the basic internal variables can be deduced (Yang, 1996)

$$\dot{\xi}_\alpha = -V^0\dot{\lambda}\left\{(\mathbf{I} - \mathbf{nn})\cdot\frac{\partial \psi}{\partial \mathbf{n}^\alpha}, \left(\frac{r}{2}\right)^2\frac{\partial \psi}{\partial r_\alpha}\right\} \tag{27}$$

Finally, the damage evolution law becomes

$$\dot{\boldsymbol{\Omega}} = \dot{\lambda}_d\mathbf{J} : \mathbf{Y} \tag{28}$$

with the damage characteristic tensor $\mathbf{J}$

$$
\begin{aligned}
&\mathbf{J} = \mathbf{J}(\boldsymbol{\Omega}) = \sum_{\nu=1}^{3} \omega_\nu^2 \left(4\mathbf{T}_\nu + \tfrac{9}{4}\mathbf{N}_\nu\right) \\
&\mathbf{T}_\nu = T^\nu_{ijkl} = \tfrac{1}{4}(n_i^\nu n_k^\nu \delta_{jl} + n_i^\nu n_l^\nu \delta_{jk} + n_j^\nu n_k^\nu \delta_{il} \\
&+ n_j^\nu n_l^\nu \delta_{ik}) - n_i^\nu n_j^\nu n_k^\nu n_l^\nu \\
&\mathbf{N}_\nu = N^\nu_{ijkl} = n_i^\nu n_j^\nu n_k^\nu n_l^\nu \quad \text{(no summation for } \nu)
\end{aligned}
\tag{29}
$$

where $\mathbf{n}^\nu$ and ω_ν ($\nu = 1, 2, 3$) are the principal directions and values, respectively, of damage tensor $\boldsymbol{\Omega}$. Eqn. (29) indicates that the damage characteristic tensor can be determined uniquely by the current damage tensor.

Due to $\mathbf{Y} = \mathbf{P}^+ : \tilde{\mathbf{Y}}$, eqn. (28) indicates that the damage propagation is tensile-strain oriented.

The damage surface has the form

$$
\mathcal{F} = \mathcal{G} - \mathcal{R}, \qquad \mathcal{G}^2 = \frac{1}{2}\mathbf{Y} : \mathbf{J} : \mathbf{Y} \tag{30}
$$

where $\mathcal{G}$ is the generalized energy release rate; $\mathcal{R}$ is the damage threshold.

$$
\mathcal{R} = \max\{\mathcal{R}_0, \max \mathcal{G}\} \tag{31}
$$

where $\mathcal{R}_0$ is the damage threshold of virgin materials.

The damage multiplier is proposed as

$$
d\lambda_d = \beta \frac{d\xi}{\mathcal{G}}, \qquad d\xi = \mid d\varepsilon^e_{ij} \mid = \sqrt{d\varepsilon^e_{ij} d\varepsilon^e_{ij}} \tag{32}
$$

where $\beta > 0$ is a material constant.

The aforementioned relations are deduced for open cracks so it refers to the equivalent state. However, it can be easily proved that the same forms hold true also for the real state.

7 NUMERICAL RESULTS

The application of the theory is shown with the help of two examples. The emphasis is put on the damage behaviour and the plastic flow is neglected.

(1) *Simple supported beam* The behaviour of the model is tested with the assumption, that we have one set of joint with a dip direction: $\alpha = 270^o, \beta = 45^o$, the separation factor $\omega = 0.5$. The beam shown in Fig. 7 has an initial compressive stress $\sigma_0 = -10.0KPa$. Which means, that all joints are closed. The external loading P acted in the middle section is incremental applied. The displacement response of this loading is shown in the Fig. 7.

It shows, that at the beginning the response of the beam is elastic to the fact, that all joints are closed by the initial stress field. If the stress becomes larger, than the initial stress field the crackset will open and we get a nonlinear load-displacement response. As shown in (a) and (b) with the increasing of the load the crack start to propagate and the effective damage value becomes larger, than the initial value. The unsymetrical results shows the impact of the crack orientation.

(2) *Tunnel excavation* The excavation of a tunnel with one initial set of cracks is analyzed. Due to the initial stress field all cracks are closed. The crack orientation is: dip direction: $\alpha = 270^o, \beta = 45^o$, the separation factor $\omega = 0.5$. The results shows that the excavation activate the the initial damage in the surrounding rockmass. Which means, that in this region the cracks are opened.

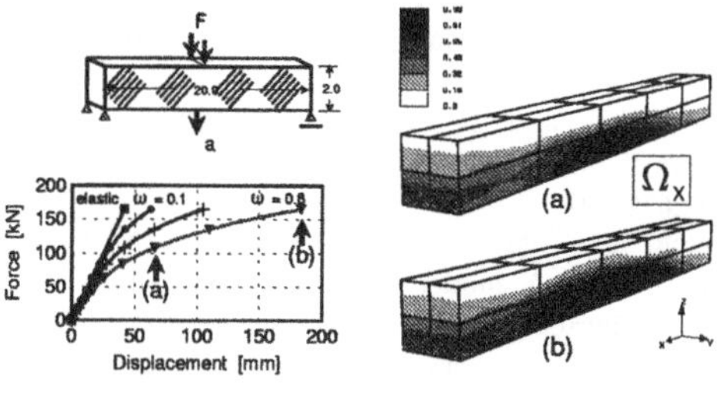

Figure 7: Simply supported beam model: active damage tensor.

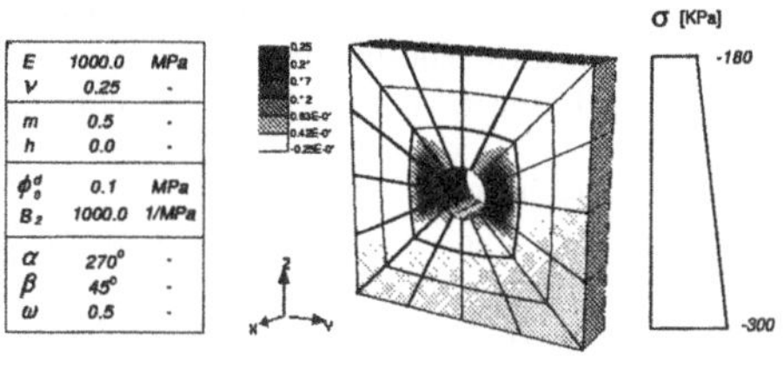

Figure 8: Tunnel model: parameter and damage distribution.

8 CONCLUSION

Continuum damage mechanics is still a very powerful tool to model geomaterials. Evidently, it is very difficult to reflect stress-induced anisotropy without a definite direct damage measure. The introduction of the equivalent state not only furnishes a basis for a compact and unitary formulation but also successfully reveals the internal deformation mechanisms of geomaterials. The normality structure is a suitable framework to establish anisotropic damage evolution laws. Numerical results show that the theory presented in this paper is suitable for solving engineering problems.

ACKNOWLEDGMENTS: The work reported here was supported by the Austrian National Science Foundation, Fonds zur Förderung der wissenschaftlichen Forschung, project S08004-TEC.

REFERENCES

Barenblatt, G. E., Zheltov, I. P. and Kochina, I. N., 1960. Basic concepts in the theory of homogeneous liquids in fractured rocks, *J. Appl. Mech.*, 24:1286-1303

Bazant, Z. P., 1983. Comment on orthotropic models for concrete and geomaterials, *J. Eng. Mech. ASCE*, 109:849-865

Cowin, S. C., 1985. The relationship between the elasticity tensor and the fabric tensor, *Mech. of Mat.*, 4:137-147

Chow, C. L. and Lu, T., J., 1989. On evolution laws of anisotropic damage, *Eng. Fracture Mech.*, 34:679-701

Dragon, A., Charlez, Ph., Pham, D. and Shao, J. F., 1993. A model of anisotropic damage by (micro)crack growth, in *Proceedings of the International Symposium on Assessment and Prevention of Failure Phenomena in Rock Engineering* (Edited by A. G. Pasamehmetoglu, *et al.*), pp. 71-78. Istanbul, Turkey, April 5-7, 1993.

Ju, J. W., 1989. On energy-based coupled elastoplastic damage theories: constitutive modelling and computational aspects, *Int. J. Solids Structures*, 25:803-833

Kawamoto, T., Ichikawa, Y. and Kyoya, T. 1988. Deformation and fracturing behaviour of discontinuous rock mass and damage mechanics theory, *Int. J. Numer. Anal. Meth. Geomech.*, 12:1-30

Krajcinovic, D., Basista, M. and Sumarac, D., 1991. Micromechanically inspired phenomenological damage model, *J. Appl. Mech.*, 58:305-316

Malvern, L. E., 1969. *Introduction to the mechanics of a continuous medium*, Prentice-Hall, Inc., Englewood Cliffs, N.J.

Nemat-Nasser, S. and Obata, M., 1988. A microcrack model of dilatancy in brittle Materials, *J. Eng. Mech. ASCE*, 55:24-35

Ortiz, M., 1985. A constitutive theory for the inelastic behaviour of concrete, *Mech. Mater.*, 4:67-93

Rice, J. R., 1971. Inelastic constitutive relations for solids: an integral variable theory and its application to metal plasticity, *J. Mech. Phys. Solids*, 19:433-455

Stumvoll, M. and Swoboda, G., 1993. Deformation behavior of ductile solids containing anisotropic damage, *J. Eng. Mech. ASCE*, 119:1331-1352

Yang, Q., 1996. *Numerical Modelling for Discontinuous Geomaterials Considering Damage Propagation and Seepage*, Ph.D Dissertation, Faculty of Architecture and Civil Engineering, University of Innsbruck, Austria, 1996.

Mechanics of Jointed and Faulted Rock, Rossmanith (ed.) © 1998 Taylor & Francis, ISBN 90 5410 955 6

3D behaviour of bolted rock joints: Experimental and numerical study

M. Kharchafi, G. Grasselli & P. Egger
Swiss Federal Institute of Technology, Lausanne, Switzerland

ABSTRACT: Fully grouted, untensioned bolts have been commonly used in rock mechanics (mines, rock fall stability, underground works) for many years. The experience accumulated on this subject gives the know-how for the bolt reinforcement calculation and execution, but it does not explain the mechanical behaviour of the bolted rock joint. The Rock Mechanics Laboratory of EPFL, Switzerland, has accumulated extensive experience over the years with experimental and numerical research on bolted rock joints. The object of this paper is to discuss and compare the results of large scale (1:1) laboratory tests of bolt-reinforced rock with those obtained by 3D FEM calculations, looking for the global behaviour of jointed rock, reinforced by bolts.

1 INTRODUCTION

The reinforcement of jointed rock masses by passive anchors is widely used because of its low cost and its proved efficacy.

However, the mechanical behaviour of passive rock bolts has not been univocally explained yet. In fact, the mechanism of resistance of the bolted system is difficult to analyse.

The bolts increase the rock mass resistance, but the value of the growth is hard to calculate, depending on the interaction of many factors like :
- mechanical properties of the rock, grout and bolt
- bolt typology
- bolt orientation
- roughness of rock joint surfaces.

As a consequence, there is no unanimously approved method for the design of the reinforcement system.

1.1 Outline of past studies

Many experimental test series were performed in order to study the mechanical behaviour of bolted rock joints. Dight (1983), Schubert (1984), Spang (1988), Egger & Zabuski (1991), Ferrero (1993), Pellet (1993), Grasselli (1995), Migliazza (1997) carried out experimental programs on different types of rock material reinforced by various elements.

Several analytical expressions have been developed to predict the behaviour of a bolted rock joint. The simplest of them considers only the axial force acting in the bolt (Bjurström 1974). By its projection on the joint plane, it is possible to calculate the reinforcement effect of the bolt. But this method does not take into account the shear force mobilised in the bolt nor the deformation of the bolt near the joint.

To take into account the deformation of the bolt during the load process more sophisticated expressions were developed, based on the theory of the beams on elastic supports formulated by the small displacements theory. Dight (1983) proposed an expression to predict the maximum force mobilised in the bolt as well as the associated displacement of the joint. The failure of the bolt is determined by the combination of axial and shear forces, and the displacement is computed taking into account the yield of the grout. Based on Dight's work, Holmberg & Stille (1992) proposed a method which gives a good prediction of the maximum bolt contribution when the bolt is inclined to the joint. Furthermore, Spang & Egger (1990) proposed empirical expressions to compute the maximum bolt contribution and the associated joint displacement. Egger & Pellet proposed numerical approaches to compute the ultimate resistance of the bolt, considering it as a semi-infinite beam loaded at one end by both an axial force and a shear force. One of the most restrictive assumption concerns the behaviour of the surrounding medium which is considered as a rigid perfectly plastic material.

1.2 Main purpose of this study

The goal of this study is to understand the influence of several parameters involved into the observed phenomena, as the number of the bolts, their inclination to the joint and their diameters.

The behaviour of bolted rock joints is then compared with an analytical model (Pellet 1993) and 3D FEM calculations carried out with Z-Soil 3D.

2 EXPERIMENTAL STUDY

2.1 Experimental set-up

To study the shear behaviour of reinforced rock joints at large scale (1 :1), a special experimental set-up was developed. As shown in Figure 1, the test uses a system of three large blocks (100x60x60 cm each), reinforced by one or two bolts for each discontinuity. The bolts are placed similarly at both joints.

Up to now, to simulate the behaviour of the bolted joint, blocks of concrete were used. The fully grouted 20 or 16 mm diameter bolts were placed into 40 or 32 mm prefabricated holes (Fig.2) at different inclinations (ß=0°, 15°, 30°, 45°).

During a test, the central block was pushed progressively down by a controlled jack (maximum force of 1000 kN).

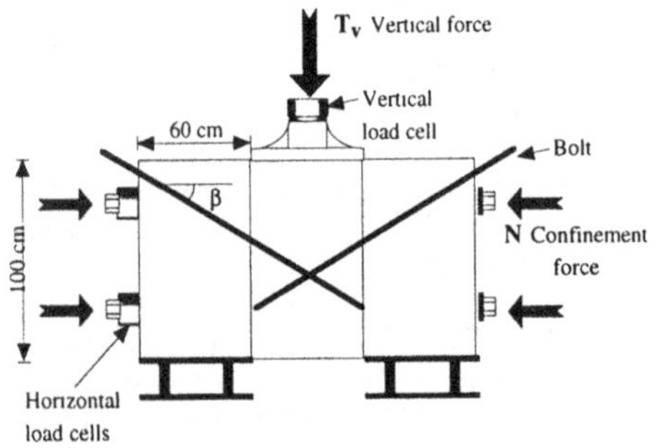

Fig.1 *Scheme of experimental set-up*

Fig.2 *Detail of the bolted zone. Transversal view :*
1= concrete
2= injection grout
3= bolt (Ø16 or Ø20)

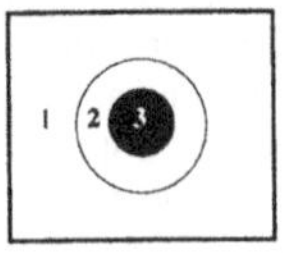

The following values were recorded:
- the vertical load applied on the central block, using a load cell (max 2000 kN), placed between the jack and the distribution plate;
- the confinement force (horizontal) using load cells (max 200 kN);
- the vertical displacement of the central block using linear transducers;
- all of the full steel bolts were equipped with five pairs of strain gauges to measure deformations on the bolt surface.

2.2 Discussion of tests results

A large number of shear tests using reinforced concrete blocks were performed at the laboratory of EPFL in order to study the mechanical behaviour of bolted rock joints.

The Figures 3-5 summarise some of the results demonstrating the effects of the parameters investigated. They show the bolt contribution T versus the vertical displacement of the central block. The bolt contribution T is computed as follows :

$$T=T_v/2-N\cdot tg\phi$$

where T_v is the vertical force applied on the central block, N is the confinement force, and $\phi=44°$ is the friction angle, directly measured by a series of non reinforced tests.

By the analysis of the experimental curves it is possible to point out a non linear trend, peculiar behaviour common to all tests, characterised by three steps:
- the first is characterised by small displacements at a large increase of load. A detachment of the bolt from the grout is already observed at this step;
- the second is characterised by large displacements corresponding to the plastification of the bolt and the progressive breaking of the grout. There is the formation of plastic hinges on the bolt ;
- the third is characterised by the nearly free deformation of the bolt until the failure.

Experimental results show that the resistance mobilised by the bolts is proportional to the total steel section, therefore two bolts provide a resistance double compared to one (Fig.3). With two bolts, a slightly larger displacement at the failure is observed than with a single one, because the bolts never mobilise 100% of the ultimate resistance at the same time.

When increasing the bolt diameter, a clear increase of rigidity of the reinforced system is observed, due to the growth of inertia of the single reinforcement (Fig. 4).

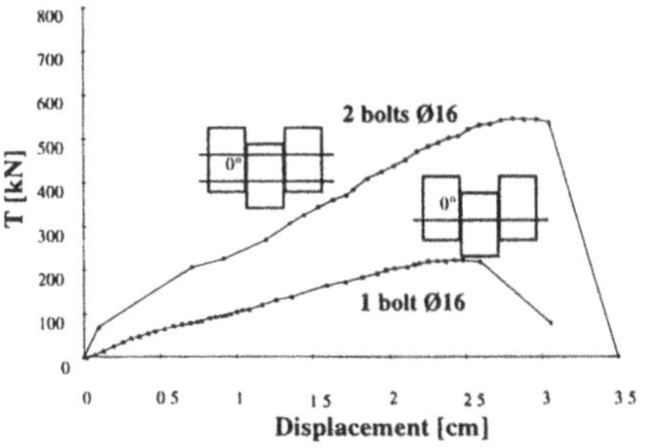

Fig.3 Effects of number of the bolts

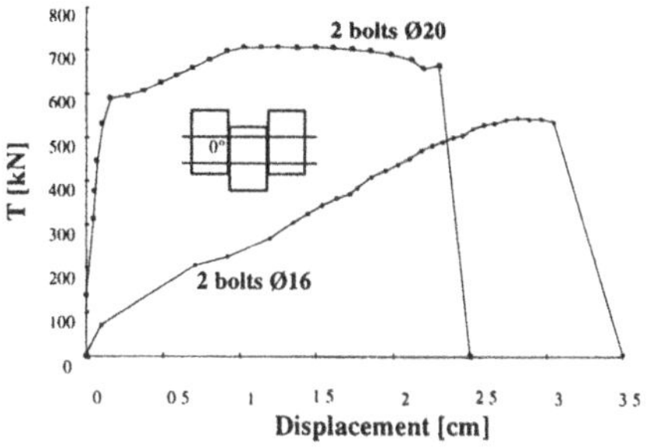

Fig.4 Effects of diameter of the bolts

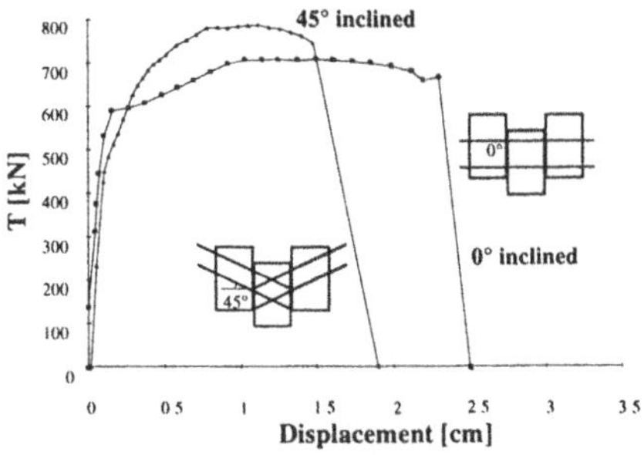

Fig.5 Effects of orientation of the bolts

The variation of the bolt inclination β affects both the maximum load mobilised by the reinforcements and the rigidity of the jointed system (Fig. 4).

The 45° inclined test shows quite the same striction and failure surface as in pure traction, and the plastic hinges do not act perfectly like boarders to stress propagation.

This confirms that the degree of tractional behaviour grows with bolt inclination and, at the same time, shear reduces. In agreement with Egger and Fernandes (1983) it is noticed that the maximum load mobilised by the bolt occurs for an initial inclination to the joint in the range of 30°-60° (Fig.5).

The distance between the hinges was measured directly on the pieces of bolts extracted from the blocks and the angle α of the bolt at failure was calculated. A linearly decreasing tendency with increasing the initial inclination angle between bolt and joint can be noticed.

Analysing the data we can predict that for an initial position angle β=80°-85° there will not be any hinge and the bolt should work only by traction. This conclusion is reasonable because the reinforcement is quite parallel to the joint.

Similar results were obtained by interpolation of scattered hinges distance data with a parabolic tendency envelope. The distance between the hinges reduces to zero for an initial angle of 80°-85°, which means that there are no hinges along the bolt.

It is worth noting that the distance between the hinges is nearly the same for 15°-45° initial bolt inclination and it takes values between one and two bolt diameters.

Analysing strain gauge data recorded during the test (Fig.6), it is possible to point out the following aspects on the behaviour of the bolt:

- antisymmetric bar deformation and displacement with respect to the intersection point between bolt and joint plane;
- the bending moment of the bar is negligible yet at small distances from the joint (8-10 cm);
- the traction load quickly decreases and disappears at 30-45 cm from the joint;
- the plastic hinges seem to work like boarders to the stress propagation and all the deformations increase between them ;
- a clear mechanism for the formation of hinges: compression on one side, and traction on the other.

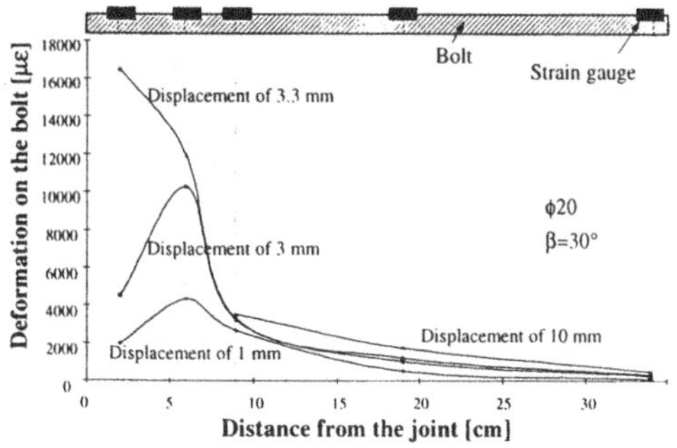

Fig.6 Deformations on the bolt surface, measured by strain gauges

2.3 Comparison with Pellet's analytical model

The test results are compared with the analytical prediction proposed by Pellet (1992). Both curves are plotted in terms of shear resistance, T_V, versus joint displacement, U_j.

Good agreement is observed between experimental and analytical results, concerning the maximum load mobilised by the bolted system.

Concerning the displacements it is noticed an overestimate of the analytical prediction of 20%.

The comparison was extended to a large number of experimental tests, carried out at the EPFL laboratory. It shows that Pellet's method gives a good prediction for computing the maximum resistance of bolted systems.

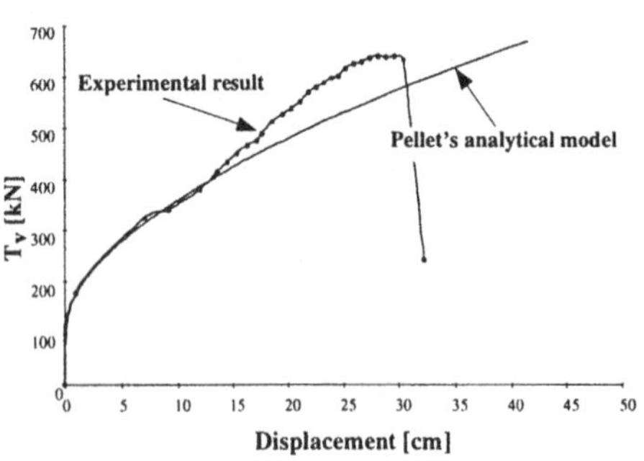

Fig.7 *Comparison between the analytical prediction proposed by Pellet and the experimental curve*

3D NUMERICAL ANALYSIS

3.1 Introduction

The main difficulties of simulating fully-grouted bolts in jointed rock are due to:
- the presence of 3 materials with very different stiffness and behaviours.
- the presence of 3 families of contacts or joints between these materials: a real joint between rock blocks and weak adherence at material limits.
- the geometry of the bolt and grout and the kind of loading; the problem is obviously 3D and cannot be treated as a 2D problem (plane strain or axisymmetric hypothesis).

In the FE simulation presented below, it was attempted to take into account all these aspects to discuss the behaviour of bolted rock joints and to compare with experimental results.

3.2 FE code and model used

For the 3D simulation, the FE code ZSOIL_3D developed within a collaboration of the Swiss Federal Institute of Technology and private companies was used (Zsoil 1997). This 3D FEM code allows the simulations of elastoplastic materials and contact joints behaviour.

3.3 Mesh used

A 3D model with a large number of elements (3388) and nodes (4057) was used. It represents a (20x20x10cm) parallelepiped surrounding the horizontal bolt (20mm diameter). The location of the discretized region, the boundary conditions and the load conditions are schematised in figure 8.

Due to the symmetry of the problem, only one half of the system is considered.

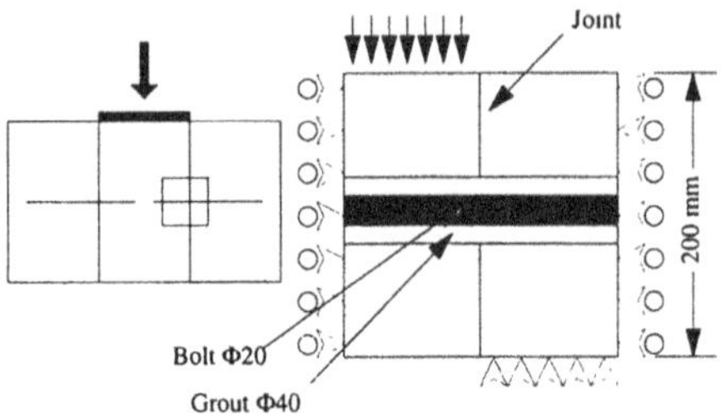

Fig.8 *3D FEM simulation data*

3.4 Characteristics of materials

The elastic characteristics of the used materials are shown below.

Elastic mat. properties	*E* [kPa]	ν
Steel	2.10^8	0.2
Grout	$1.4\ 10^7$	0.25
Concrete (Rock)	$2.4\ 10^7$	0.2

The vertical joint separating the concrete blocks and the contact surface between the bolt and the grout are simulated by elastic-perfectly plastic contact interface elements. The elastic limit is set by cohesive Mohr-Coulomb criteria and the irreversible displacements of the interface are governed by a flow rule with similar yield criteria. The adopted values for these elements are summarized below.

Contact properties	c_j [kPa]	Φ_j [deg.]	Ψ_j
joint Rock/ Rock	0	44	0
contact Steel/Grout	250	45	0
contact Grout/Rock	*Perfect contact*		

Where c_j is the cohesion, Φ_j is the friction angle and Ψ_j the dilatancy angle of the contact elements.

The surfaces of the tested concrete blocks were smooth and plane, therefore no cohesion and dilatancy were considered. A high value of cohesion is adopted for grout/steel contact because a good adherence was observed experimentally.

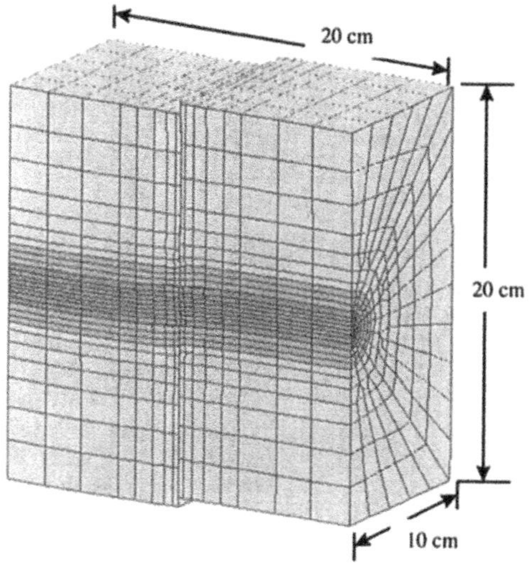

Fig.9 *3D deformed mesh*

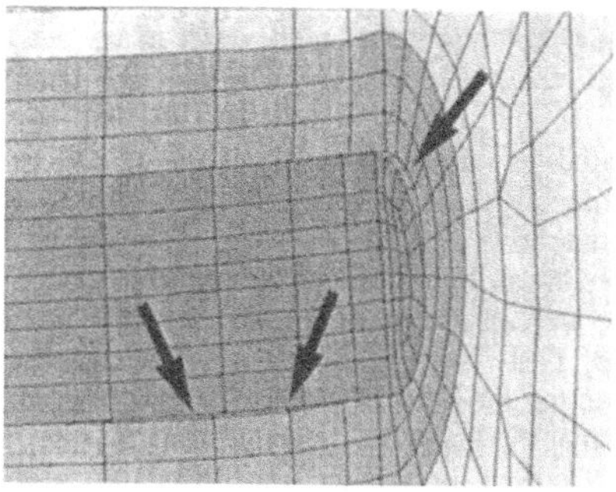

Fig.10 *Detail of the 3D deformed mesh : detachment of the bolt from the grout near the joint*

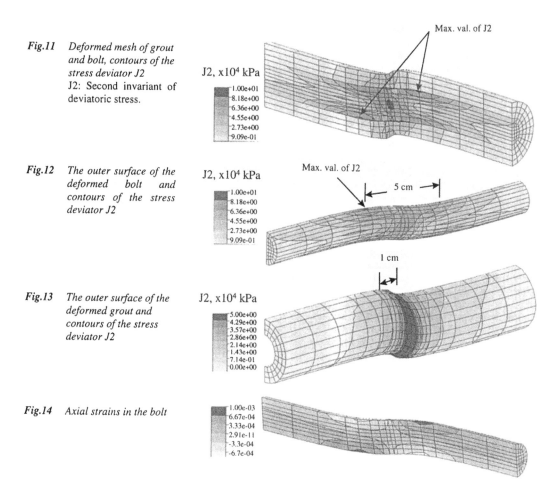

Fig.11 *Deformed mesh of grout and bolt, contours of the stress deviator J2*
J2: Second invariant of deviatoric stress.

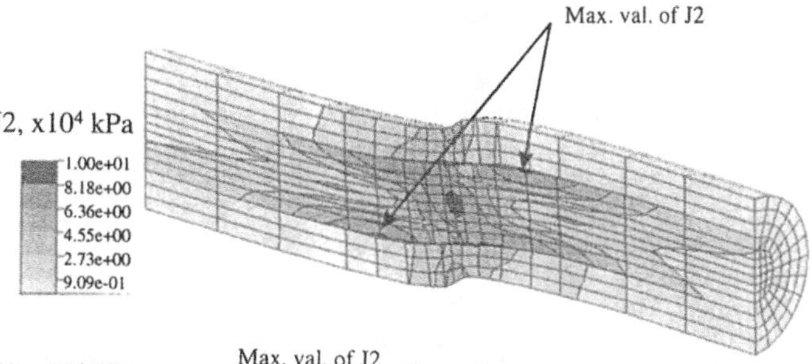

Fig.12 *The outer surface of the deformed bolt and contours of the stress deviator J2*

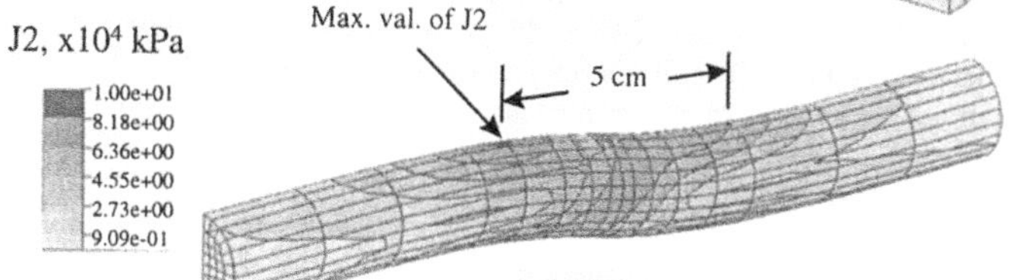

Fig.13 *The outer surface of the deformed grout and contours of the stress deviator J2*

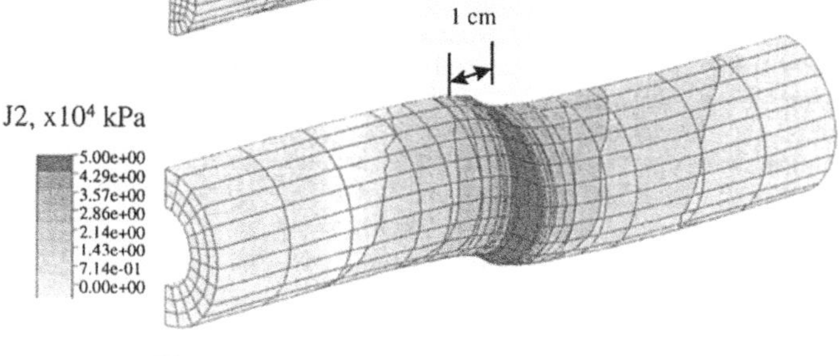

Fig.14 *Axial strains in the bolt*

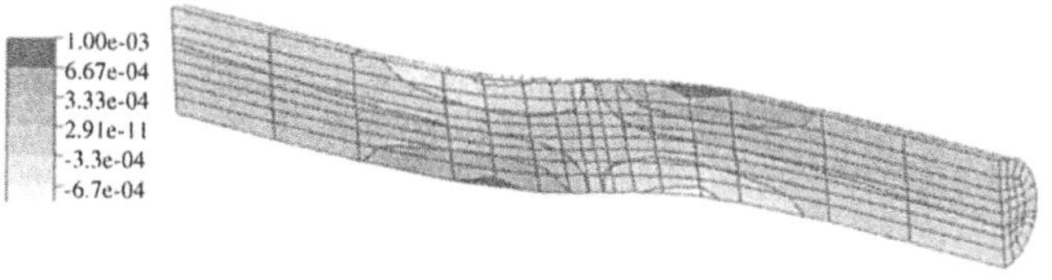

3.5 Calculation results

The 3D simulation presented was carried out for a vertical load of T_V=200 kN. The deformed mesh is shown in Figure 9. The vertical displacement calculated is 0.35 mm which is close to the experimental result obtained with a 20 mm diameter horizontal bolt

Apart from the relative displacements of the two blocks, we can observe (Fig.10) a clear detachment of the bolt from the grout. The Figure 11 representing the deformations of both grout and bolt shows that the length of detached zone is about 6 cm on each side of the vertical joint. It is comparable to the experimental observations: 40 to 60 mm.

The Figure 11 shows the contours of the deviator stress J2 (Second invariant of deviatoric stress). A large difference of the values can be observed in the two materials. The higher value of J2 is located at about 25 mm on each side of the shear plane. These values which will correspond to the development of the two hinges are due to the traction. A value less than the previous ones due to shearing is observed at the intersection between bolt and joint plane.

The Figures 12-13 show respectively the outer surface of the deformed bolt and grout. As is shown, the high values of J2 are very concentrated near the shear plane for the grout and more uniformly distributed between the maximums of J2 for the bolt.

The calculated distance between the two hinges is 50 mm, which is close to the value at failure observed in the test.

The axial strains in the bolts are shown in Figure 14. A reasonable estimation of the order of magnitude was obtained. But a direct comparison with the results is difficult due to problems with the strain gauge measurements in this test.

4 CONCLUSION

Up to now, few large scale tests are presented in the literature, because of the technical difficulties to carry them out. The experimental results obtained overcoming many difficulties linked to the large scale set-up and the bond of the strain gauges, furnish a unique and very valuable database to validate analytical and numerical analyses.

The experimental results agree with the hinges theory and Pellet's analytical solution of the problem.

The 3D FEM analysis carried out with elasto-plastic contact elements simulates very closely the behaviour of reinforced joints observed experimentally for small displacements. The development and the distance between the hinges is predicted with a good degree of accuracy.

For the study of the behaviour at larger displacements, plastification of the steel and grout has to be taken into account. This will be done by further calculations with elasto-plastic behaviour for these two materials.

REFERENCES

Bjurström, S. 1974. Shear strength of hard rock joints reinforced by grouted untensioned bolts. Proc. 3rd ISRM Cong., Denver, USA, 1194-1199

Dight, P.M. 1983. Improvements to the stability of rock walls in open pit mine. Ph.D Thesis, Monash University, Australia

Egger, P. & Fernandes, H. 1983. Nouvelle presse triaxiale - Etude de modèles discontinus boulonnés. Proc. 5th ISRM Cong., Melbourne, Australia, A171-A175

Egger, P. & Zabuski, L. 1991. Behaviour of rough bolted joints in direct shear tests. Proc. 7th ISMR Cong., Aanchen, Germany, 1285-1288

Ferrero, A.M. 1993. Resistenza al taglio di discontinuità rinforzate. Ph.D Thesis, Politecnico di Torino, Italy

Grasselli, G. 1995. Comportamento meccanico di masse rocciose discontinue rinforzate con bulloni passivi. Tesi di Laurea, Università di Parma, Italy

Holberg, M. & Stille, H. 1992. The mechanical behaviour of a single grouted bolt. Int. Symp. on Rock Support, Sudbury, Canada, 473-481

Migliazza, R. 1997. Analisi teoriche e sperimentali del comportamento a taglio di rocce discontinue bullonate. Tesi di Laurea, Università di Parma, Italy

Pellet, F. & Egger, P. 1996. Analytical model for the mechanical behaviour of bolted rock joints subjected to shearing. Rock Mech. and Rock Eng., 29 :73-97

Pellet, F. 1993. Résistance et déformabilité des massifs rocheux stratifiés par ancrages passifs. Ph.D Thesis n°1193, EPF Lausanne, Switzerland

Schubert, P. 1984. Das Tragvermögen des mörtelversetzten Ankers unter aufgezwungener Kluftverschiebung. Ph.D Thesis, Montanuniversität, Leoben, Austria

Spang, K. 1988. Beitrag zur rechnerischen Berücksichtigung vollvermörtelter Anker bei der Sicherung von Felsbauwerken in geschichtetem oder geklüftetem Gebirge. Ph.D Thesis n°740, EPFL, Lausanne, Switzerland

Spang, K. Egger, P. 1990. Action of fully-grouted bolts of jointed rock and factors of influence. Rock Mech. and Rock Eng., 23:201-229

Z_SOIL 1997, User manual. Elmepress (1985-1997)

Mechanics of Jointed and Faulted Rock, Rossmanith (ed.) © 1998 Taylor & Francis, ISBN 90 5410 955 6

Development of a 3-D block-spring model for jointed rocks

G. Li & B. Wang
Canada Centre for Mining and Energy Technology, Natural Resources Canada, Ottawa, Ont., Canada

ABSTRACT: This paper presents a three-dimensional Block-Spring Model (BSM3D) recently developed for simulation of jointed rocks. It is an extension of a two-dimensional Block-Spring Model that has been successfully applied to various rock mechanics cases. Applications of 2-D model have demonstrated that this technique is a useful tool for analyzing the behavior of jointed and faulted rocks. The extension to a 3-D model further enhanced the capability of this technique. The paper presents the algorithm and some special features of the 3-D BSM. Two examples are presented to illustrate the capabilities of the model.

1 INTRODUCTION

A two-dimensional Block-Spring Model (BSM) was developed recently for modelling the behaviour of jointed rocks (Wang and Garga, 1993). The 2-D model has been successfully applied to many mining and other rock mechanics projects and proved to be a useful tool for jointed rock analysis (Garga & Wang, 1993, Wang et al, 1995, 1997). In order to extend its capability, a three-dimensional BSM model has been developed. The development is not a simple expansion of the 2-D formulation. It is involved with much more than just adding a third coordinate into the equations. This paper summarizes the basic theory of the model and special treatment of certain three dimensional problems, e.g., simulation of contact between blocks. Two examples are also presented to illustrate the capabilities of the model.

2 BASIC ASSUMPTIONS

The 3-D Block-Spring Model is formulated based on an assumption that the rock structure is formed with rigid blocks. Four contact modes, i.e., face-to-face, edge-to-face, corner-to-face or edge-to-edge (Fig. 1), are likely to occur between blocks and considered effective in the model. Contact forces may develop depending on the relative displacements and contact stiffness between blocks. The contact forces distributed on the contact area are represented by the equivalent forces transferred to the corners of the contact area. The contact stiffness can therefore be simulated by three springs located at each corner, i.e., one normal and two shear springs (Fig. 2).

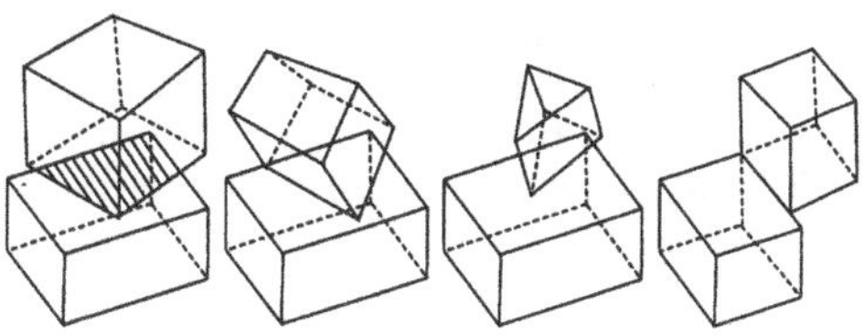

Fig. 1 Possible contact modes between blocks

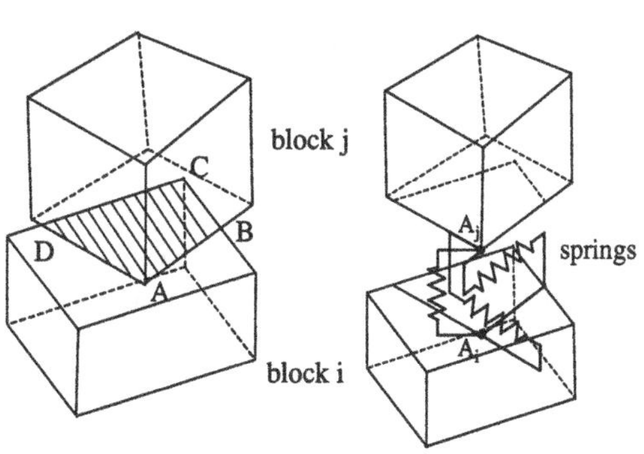

Fig. 2 Springs used to simulate contact stiffness

The positive directions of the contact forces are defined as follows:

Normal force: outward normal to the contact surface;

Shear force 1: along the first edge of the contact area, i.e., from the first corner to the second corner (from A to B in Fig.2); along the only contact edge if edge-to-face and edge-to-edge contact; or along the first edge of the contact face if corner-to-face contact;

Shear force 2: perpendicular to shear force 1.

The blocks are assumed to satisfy equilibrium conditions. In other words, the sum of all forces imposed on each block should be zero. Equilibrium equations can therefore be constructed and assembled to form a global stiffness matrix. The displacements of the blocks are obtained after solving the equilibrium equations. The contact forces are subsequently calculated from the relative displacements and the contact stiffnesses between blocks.

3 FORMULATION OF DISPLACEMENTS AND CONTACT FORCES

As the rock blocks are assumed rigid, each block has six degrees of freedom, i.e., three translations, U_{xi}, U_{yi} and U_{zi}, and three rotations, θ_{xi}, θ_{yi} and θ_{zi}, defined at the centroid of block i, as shown in Fig. 3. The displacement $\{u\}_i = \{u_{xi}, u_{yi}, u_{zi}\}^T$ at any point inside block i can be determined in terms of U_{xi}, U_{yi}, U_{zi}, θ_{xi}, θ_{yi} and θ_{zi}.

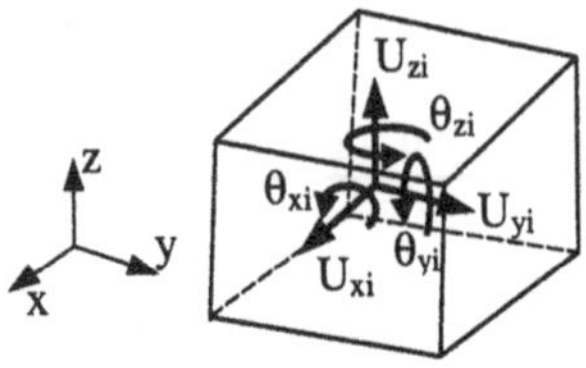

Fig. 3 Displacement of block i

Assume that point Q displaces to Q′ due to block translations and from Q′ to Q″ as a result of block rotations. Then

$$\{u\}_i = \{u'\}_i + \{u''\}_i \qquad (1)$$

where $\{u'\}_i$ and $\{u''\}_i$ are the displacements corresponding to block translations and rotations, respectively.

The translational component $\{u'\}_i$ is equal to the translations of the centroid of the block. The rotational component $\{u''\}_i$ can be evaluated from the rotation angles around each axis.

$$\{u''\} = \begin{bmatrix} 0 & \bar{z}_i & -\bar{y}_i \\ -\bar{z}_i & 0 & \bar{x}_i \\ \bar{y}_i & -\bar{x}_i & 0 \end{bmatrix} \begin{Bmatrix} \theta_{xi} \\ \theta_{yi} \\ \theta_{zi} \end{Bmatrix} \qquad (2)$$

where $\bar{x}_i$, $\bar{y}_i$ and $\bar{z}_i$ are local coordinates of point Q originated at the centroid of the block.

Equation (1) can therefore be written in the following form:

$$\{u\}_i = [B]_i \{U\}_i \qquad (3)$$

where

$$[B]_i = \begin{bmatrix} 1 & 0 & 0 & 0 & \bar{z}_i & -\bar{y}_i \\ 0 & 1 & 0 & -\bar{z}_i & 0 & \bar{x}_i \\ 0 & 0 & 1 & \bar{y}_i & -\bar{x}_i & 0 \end{bmatrix} \qquad (4)$$

$$\{U\}_i = \{U_{xi} \quad U_{yi} \quad U_{zi} \quad \theta_{xi} \quad \theta_{yi} \quad \theta_{zi}\}^T \qquad (5)$$

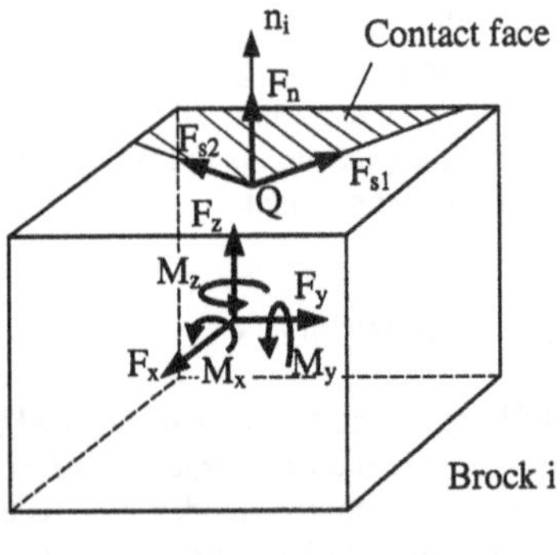

Fig. 4 Contact forces between block i and j

The contact forces are evaluated from the contact stiffnesses and the relative displacements between blocks. Suppose that point Q is a contact point between block i and j, the normal and shear components of the contact force, as shown in Fig. 4, can be determined by the following expression.

$$\{F_n\}_{ij} = [k]_{ij}\{\delta\}_{ij} \tag{6}$$

where

$$[k]_{ij} = \begin{bmatrix} k_n & 0 & 0 \\ 0 & k_{s1} & 0 \\ 0 & 0 & k_{s2} \end{bmatrix}_{ij} \tag{7}$$

$$\{F_n\}_{ij} = \{F_n \quad F_{s1} \quad F_{s2}\}^T_{ij} \tag{8}$$

$\{\delta\}_{ij}$ is the relative displacement of the two sides of contact point Q. k_n, k_{s1} and k_{s2} are normal and shear contact stiffnesses. The resultant of shear forces F_{s1} and F_{s2} is checked for shearing failure.

By expressing the relative displacement $\{\delta\}_{ij}$ in terms of the six degrees of freedom of block i and j and then transforming the contact force to the centroid of block i, the following expression is obtained:

$$\{\mathbf{F}\}_{ij} = [\mathbf{k}]_{ij}\{\mathbf{U}\}_{ij} \tag{9}$$

where

$$\{\mathbf{F}\}_{ij} = \{F_x \quad F_y \quad F_z \quad M_x \quad M_y \quad M_z\}^T_{ij} \tag{10}$$

$$[\mathbf{k}]_{ij} = [\mathbf{B}]^T_i[T]^T_i[k]_{ij}[T]_i[-\mathbf{B}_i \vdots \mathbf{B}_j] \tag{11}$$

$$[T]_i = \begin{bmatrix} \cos\alpha_1 & \cos\beta_1 & \cos\gamma_1 \\ \cos\alpha_2 & \cos\beta_2 & \cos\gamma_2 \\ \cos\alpha_3 & \cos\beta_3 & \cos\gamma_3 \end{bmatrix}_i \tag{12}$$

$[\mathbf{k}]_{ij}$ is a $[6\times12]$ stiffness matrix for blocks i and j; $\{\mathbf{F}\}_{ij}$ is a vector of forces acting on the centroid of block i which are equivalent to the contact forces; $\{\mathbf{U}\}_{ij}$ contains the displacements of blocks i and j; $[T]_i$ is a transformation matrix, in which α_k, β_k and γ_k are the angles from local coordinate axis k (k=1, 2 and 3 denotes normal and two shear directions) to global axes x, y and z.

4 EQUILIBRIUM EQUATIONS AND STRESSES WITHIN BLOCKS

The Block-Spring Model is developed based on the equilibrium conditions of all blocks. It is assumed that the blocks are in equilibrium state under a system of forces. The contact forces derived in previous section and other forces, e.g., gravity and other boundary conditions, can be assembled into the equilibrium equations. A set of global stiffness equations can subsequently be obtained as follows:

$$[\mathbf{K}]\{\mathbf{U}\} = \{\mathbf{P}\} \tag{13}$$

where $\{\mathbf{P}\}$ represents the external forces and initial stresses transformed to the centroids of the blocks; $[\mathbf{K}]$ is a global stiffness matrix in which non-zero elements occupy only a diagonal band and are symmetric; $\{\mathbf{U}\}$ is a vector of displacement that contains all the degrees of freedom of all the blocks.

The displacement vector $\{\mathbf{U}\}$ can be obtained by solving equation (13). The contact forces can be calculated subsequently using equation (6).

The average stress within each block may be calculated using a procedure presented by Drescher and De Josselin De Jong (1972). After the values of all the forces on a block are determined, the average stress $\overline{\sigma}_{ij}$ within the block may be calculated as follows:

$$\overline{\sigma}_{ij} = \frac{1}{V}\sum_{Q}\overline{x}_i(Q)F_j(Q) \tag{14}$$

where $F_j(Q)$ denotes the jth component of the force acting on point Q; and $\overline{x}_i(Q)$ is the ith local coordinate of point Q.

5 MODELING OF LARGE DISPLACEMENT

Large-scale displacement may occur along rock joints. Blocks may slide over each other as a result of shear failure and may detach upon tensile failure. An iterative procedure must be used to simulate the large displacement.

The Mohr-Coulomb failure criterion is used to examine failure conditions along the joints. If normal or shear failure occurs, the driving force is replaced with resisting force (strength) determined by the failure criterion. If that happens, the previous equilibrium state is disturbed and the blocks should undergo certain displacements to reach a new state of equilibrium. Such displacements can be determined by constructing and solving a new stiffness matrix, which forms a new cycle of iteration. Failure criteria are always applied after each iteration cycle to check the failure conditions along the joints. Such iteration continues until no failure is detected or certain blocks could not satisfy the equilibrium conditions. The coordinates of the blocks are updated after each cycle of iteration. Large displacements can therefore be modelled incrementally. Some blocks may not satisfy the equilibrium condition due to lack of support after sliding or detachment. Continued construction of equilibrium equations will result in a singular stiffness matrix with some diagonal elements appearing to be zero. These blocks can be declared unstable. However, iteration may continue upon special treatment of the unstable blocks. Temporary weak constraints may be added to the unstable blocks so that equilibrium conditions can be applied. Mathematically, zero diagonal elements in the stiffness matrix may be replaced with a small stiffness so that the matrix would become non-singular. A displacement tolerance is required in this case to control the maximum displacement during each iteration cycle.

6 ILLUSTRATIVE EXAMPLES

Example 1: Simulation of Large Displacements

This example illustrates BSM3D's capability of modelling large displacement of the blocks. Fig. 5 shows a block system with two blocks sliding on the top of other blocks. There is a 0.25m gap between blocks 4 and 5 (Fig. 5a). An external force was applied on the left-hand side of block 4 to move the block towards right. In this example, a tolerance of 0.01 m was used to control the maximum displacement of the blocks. 25 iteration steps were executed before block 4 touched block 5 (Fig. 5b). The two blocks then moved together. After another 50 iterations, the bottom centre of block 5 reached the upper-right edge of block 3 and rotational movement started to develop.

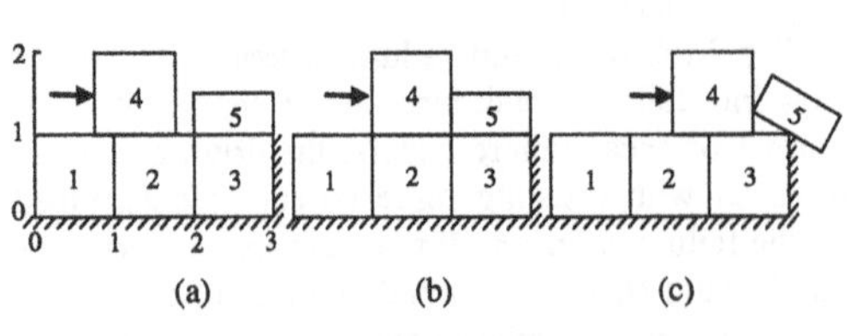

Fig. 5 Large displacement of blocks

Example 2: A Slope of Jointed Rocks

Fig. 6 shows a slope with several major structures and joints. F_1 and F_2 are two major faults, which cut the slope into three regions. The rock mass was subjected to gravity forces. The bottom and lateral boundaries of the mesh were constrained by rollers. Fig. 7 is a BSM3D mesh, in which the faults are simulated as soft layers. The material properties are listed in table 1.

Table 1. Material properties

	Argillite	Serpentinite	Discontinuity
Young's modulus (MPa)	2.1×10^4	9.65×10^3	---
Poisson's ratio	0.24	0.3	---
Unit weight (kN/m^3)	28.2	25.3	---
Cohesion (MPa)	0.70	0.20	0.05
Friction angle (deg.)	32	26	25
Normal stiffness (MPa/m)	---	---	44
Shear stiffness (MPa/m)	---	---	6.8

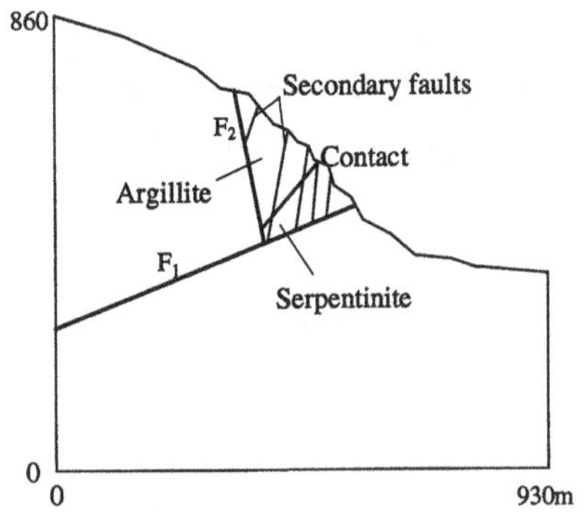

Fig. 6 Cross section of a slope

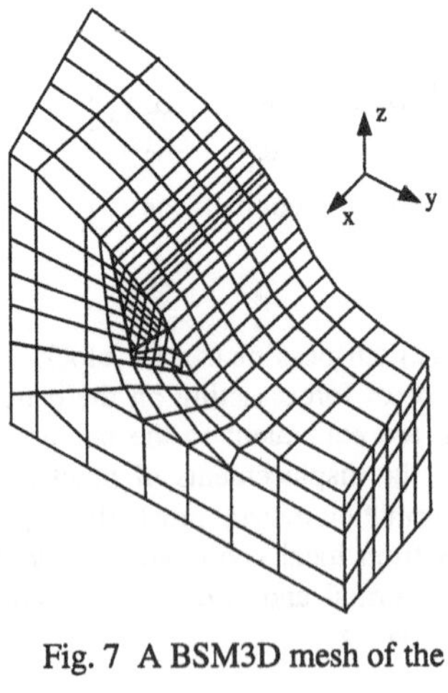

Fig. 7 A BSM3D mesh of the slope

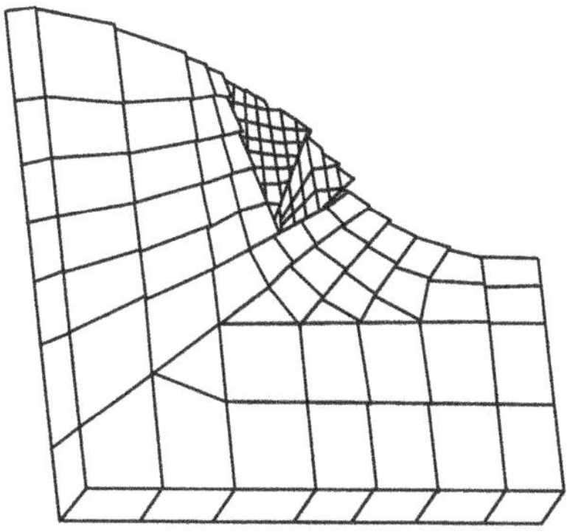

Fig. 8 Slope movement (exaggerated 80 times)

Fig. 8 shows the exaggerated displacement of the slope obtained from BSM3D. The displacements were enlarged by 80 times. It is obvious that the faults had a significant impact on the pattern of the slope movement. A maximum displacement of 0.67m occurred at the top of the wedge.

7 CONCLUTION

A three-dimensional Block-Spring Model has been developed to simulate large displacement of jointed rocks. Rock structures are simulated as a system of rigid blocks connected by springs. Equilibrium equations are used to construct a stiffness matrix so that displacements of the blocks could be determined. The contact forces are calculated from the relative displacements between blocks. The model uses the Mohr-Coulomb failure criterion to examine the failure condition along the joints. Upon failure of any contact between blocks, an iteration procedure is used so that large displacement could be modelled incrementally.

Two examples were presented to demonstrate the capability of the model on simulating large movement of the blocks as well as its capability of modelling complex rock structures.

Continued research is still being carried out to further enhance the model with user friendly pre- and post-processors and with other features such as modelling the ground support systems.

8 REFERENCES

Drescher, A. and De Josselin de Jong, G., 1972. Photoelastic verification of a mechanical model for the flow of a granular material. *Journal of the mechanics and Physics of Solids:* Vol.20, pp337-351.

Garga, V.K. and Wang, B., 1993. A numerical method for modelling large displacements of jointed rocks – Part II: Modelling of rock bolts and groundwater and applications. *Canadian Geotechnical Journal:* Vol.30, pp109-123.

Wang, B., Dunne, K., Pakalnis, R. and Vongpaisal, S., 1997. Prediction and measurement of hangingwall movements of Detour Lake Mine SLR stope. The 9th International Conference of the International Association for Computer Methods and Advances in Geomechanics, Wuhan, China, Nov. 2-7, 1997. Paper #702.

Wang, B. and Garga, V.K., 1993. A numerical method for modelling large displacements of jointed rocks – Part I: Fundamentals. *Canadian Geotechnical Journal:* Vol.30, pp96-108.

Wang, B., Yu, Y.S. and Aston, T., 1995. Stability Assessment of an Inactive Mine Using the Numerical Model. The 3rd Canadian Conference on Computer Applications in the Mineral Industry, Montréal. pp.390-399.

Mechanics of Jointed and Faulted Rock, Rossmanith (ed.) © *1998 Taylor & Francis, ISBN 90 5410 955 6*

Principles of numerical modelling of jointed rock mass

N. Pavlović
Faculty of Mining and Geology, Belgrade, Yugoslavia

ABSTRACT: Numerical methods are increasingly applied for modelling of jointed rock mass behaviour, with more or less success. However, often those who are carrying out such analysis are not familiar enough with all characteristics of geological environment and specificity of its interaction with various natural processes and human activities. Also, it is not rare that those engineers who do know well geological environment are not acquainted enough with capabilities and limitations of numerical methods. This paper attempts to contribute to better understanding of the matter in question and to closer connection of these two aspects of the same geotechnical problem.

1 INTRODUCTION

Rock mass, as a part of geological environment, has very complex characteristics that are changeable in space and time. Its behaviour in certain conditions is result of numerous, often unknown, factors that act from the very creation of rock mass until the moment when it becomes the object of engineering concern. The most of these factors is hard to include in traditional models of rock mass behaviour. Instead of complete set of factors, most often only few of them are taken into account when formulating appropriate functions of behaviour. That is the main reason why experimental data, or data obtained by observations, are in disagreement with those obtained by model analysis.

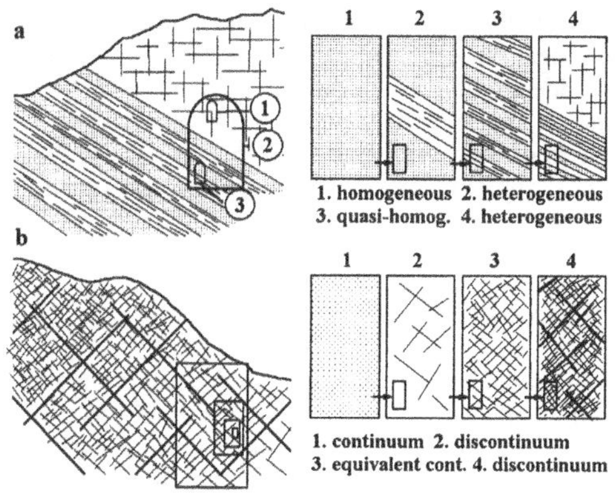

Figure 1. Influence of scale effects on changes of homogeneity (a) and mechanical behaviour (b)

Essential characteristics of geological environment are heterogeneity, anisotropy and scale effects. Scale effects especially come to a full expression when extrapolating the results of laboratory and field tests on the part of geological environment that they represent. Properties of rock mass, as well as heterogeneity and anisotropy of this properties and possible mechanical mechanisms can be qualitatively and quantitatively different in different scales. Figure 1 shows qualitative changes of heterogeneity (a) and type of mechanical behaviour (b) with changes of the size of area under investigation.

However, behaviour of rock mass depends not only on its properties but also on the other natural and human factors. Therefore, the same rock mass in different circumstances may expose different reaction. In this respect, not only the structure or excavation, but also the technology of execution of the works, can highly influence the type and intensity of interaction of geological environment and engineering activity. So, it is very important to know their possible influences and take them into account during analysis and designing.

In the previous period, when the numerical abilities where limited, the problems were solved by oversimplifications of the real problems. With the computers development, as the basic prerequisite for numerical methods application, the conditions for more objective modelling of real problems are created. With them there are possibilities for simulations of various modes of stress-strain behaviour in heterogeneous, anisotropy and discontinuous media for project schemes of arbitrary shape, and with account for various elements of technological processes involved. Such a possibility to incorporate many details in the model, however, can lead to construction of very complex models, with the idea to bring them closer to real system that they represent. That will require the more detail information, and thus the more detail investigations, and yet final results have no such effects that correspond to invested labour, time and money. The essence and art of correct and optimal modelling are to define and discriminate those fundamental aspects that represent the given problem.

2 STRUCTURE OF GEOTECHNICAL MODEL

In geotechnical studies geological environment can be treated from a different standpoints, but no matter which one is in question, in each situation exists: certain region, as a part of geological environment, certain engineering activity, and their interaction. The entireness of their individual characteristics and mutual relations represents geotechnical system.

Model of geotechnical system or geotechnical model is a complex model that consists of three basic models: 1) model of geological environment, 2) model of engineering activity and 3) model of interaction. These partial models are inseparable and only together represents geotechnical model as a whole. Between them, there are complex relations of interdependencies and conditionality (Fig. 2). In this way, characteristics of planed engineering activity, which constitute the elements of appropriate model, influence the selection of relevant parameters of the model of geologic environment. On the other hand, model of geological environment, which is highly dependent on geological conditions of the area, influence certain technical and constructive solutions, and so the vital element of this model. Their interaction, which is simulated by appropriate model, can in return have influence on both models. Inevitable components of geotechnical model are all findings and facts that are not directly presented in the model, either those are information that can not be presented in the appropriate way, or they are abstracted in the process of modelling. Although they are not included in the geotechnical model, they can have great influence on the conclusions on the problem being modelled, especially regarding the reality of results.

When formulating mathematical model of certain geotechnical system, following elements should be considered:

a) System of conditions of rock mass when certain engineering activity is performed. This system is variable in function of time and dynamics of the works. When defining this system the basic decision is choice between 2D and 3D analysis. The former is simpler, often only possible, while the latter is desirable and justified when estimated interaction is such that cannot be approximated with 2D model, and when it is practically feasible and economically acceptable.

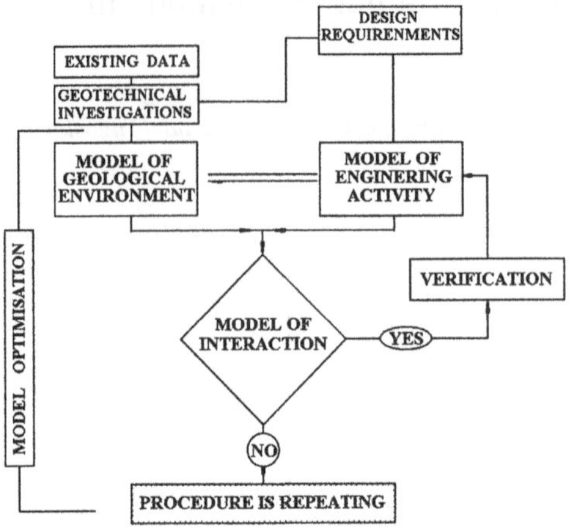

Figure 2. Structure of geotechnical model

b) System of loading: internal (virgin state of stress, groundwater pressures, dead weight, seismic forces, environmental forces etc.) and external (concentrated or regularly or irregularly distributed over the area).

c) Constitutive law: model of stress-strain behaviour of rock mass is the key element and, also the bottleneck, of any mathematical model. Adequate formulation of constitutive law means that for each quasi-homogeneous zone the appropriate type of stress-strain relation has to be formulated and its relevant parameters defined.

Considering that when processing each of the three elements there are certain amounts of approximations and idealisations, it is necessary that approximations and idealisation be balanced. In other words, one cannot process exactly and in detail the elements of system of conditions and external forces and to make greater approximations and generalisations of internal forces and constitutive laws.

3 PARAMETERS OF THE MODEL

Conclusions that result from investigations are the main source of technical data and should serve as a basis for all subsequent estimations, analyses and decisions. Obviously, it is very important for these data to be clearly defined and practically applicable. However, objective estimate of data reliability is often very hard, because there are number of subjective stages during investigations. Nevertheless, one fact cannot be denied: reliability of conclusions is never greater then the reliability of source data. Uncertainty that is inherent in the process of parameters estimates is one of the distinct characteristics of geotechnical investigations.

Natural geological environment consists of various materials, with different interrelations and influences. Besides, it is often subdivided into discrete zones of various dimensions. Even in apparently homogeneous materials, properties may vary from point to point. Although with sufficient number of measurements that variability can be relatively accurately characterised, their number is often restricted. Those facts on the variability of properties and limitations of measurements have important consequences. The main is that the rock mass should be delineated with limited number of parameters, which are defined by range of values.Also, that real ranges of variation in the area under consideration are often higher then those manifested in ranges of measured values. The aim of studying of certain parameter is not only to find out its possible values, but also to define the law relations of possible variations.

Considering that certain phenomena almost never can be completely encompassed with investigations, method of the sample is practically only way of their examination. One of the basic problems, which occur when studying population by means of the sample, is selection of sample size and location. Also, there is a question of extrapolation of the results obtained on samples on the population that it represents. This population can be a part of a medium if it is heterogeneous in respect of examined property, the whole medium if it is homogeneous, or the complete rock mass in the area if it consists of one medium that is homogeneous in respect a given property. Each quasi-homogeneous zone should be characterised by appropriate values of relevant parameters: average value, deviation from average and critical value in respects the given problem.

Extrapolation of test data on the rock mass is complex and delicate problem. It depends not only on the sample characteristics and testing conditions, but also on the rock mass properties, especially its macro-structures. Although samples may satisfy conditions in respect representativeness, size and randomness, results of testing almost never can identify with rock mass properties, mainly because:

1. It is not possible to provide absolute representativeness of samples, mainly because of great difference in size between sample and medium that it represents. Homogeneity and isotropy of rock mass greatly depend on the scale, so often they can have different qualitative and quantitative character in respect homogeneity and isotropy of the sample.

2. It is hard to simulate the real natural conditions within sample testing. This mainly relates to the size and nature of loading and duration of tests in respect the real loading and their duration in natural conditions. Besides, it is possible to occur the effects of time dependent behaviour (progressive deformations, stress relaxation etc.). Such real conditions are often hard to identify.

3. Characteristics of samples when they are testing usually do not coincide with rock mass characteristics that it has when critical interaction occurs.

Successful and objective extrapolation requires carefully planed and comprehensive approach, which consist of:

a) detail estimates of geological conditions and characterisation of rock mass with quantitative parameters, which results in definition of homogeneity and isotropy of rock mass in respect the relevant parameters and its division into quasi-homogeneous zones;

b) definition of conditions for sample selection and treatment, also the technical conditions for their testing, for each zone;

c) determination of expected unfavourable (critical) conditions in which the geotechnical system could be (especially the state of stress and underground water);

d) correlation of laboratory and field testing, also correlation with other investigations (cross-hole, seismic wave velocity on samples and in the rock mass etc.).

Respecting them, one can eliminate the risk to attribute the sample properties to the whole medium, which is not rare for those engineers that are insufficiently familiar with specific characteristics of natural geological environment.

Acquiring of reliable model parameters is on of the most complex tasks in the process of geotechnical investigations. Prognosis, even when they are carried out based on detailed investigations, involve certain degree of subjective interpretations. Depending on the complexity of natural environment, the scope of investigations and quantity and quality of collected data, also, the experience of the researcher, the corresponding degree of reliability of prediction will be accomplished. The main problem is not a fact that predicted conditions are not fully reliable, but the degree of reliability is rarely critically estimated and adequately presented. Clear estimates of reliability are necessity, because it directly influences the final formulation of the model. Probability and statistics are possible ways for treating the uncertainty. However, one should always have in mind that their laws are only approximate and do not determine the essence of the phenomenon. They only describe it, without explaining the causative relations. So, statistical results are more or less probable findings and certainly do not provide absolute objectivity, but most of all, consistency in the process of modelling. Such numerical procedures should be that part of modelling procedure that give appropriate quantitative aspect, but not its basic tools. They could not be successfully applied without clear understanding of nature and origin of data.

4 CRITERIA FOR MODEL SELECTION

Studying of stress-strain behaviour of rock mass and mechanical types of its interaction with various structures is traditionally leaned on the approaches and methods developed in other branches of technical sciences. Success of numerical modelling in other field of engineering, was a main incitement for more intensive application in geotechnics. In this respect, there were always two confronted convictions. The first, that numerical methods cannot be successfully applied because of inability to obtain sufficiently reliable quantitative data for such complex model that can with reality simulates the behaviour of rock mass. The second, that geotechnical problems should be brought to such level of understanding of mechanical mechanisms and parameters definition that models defined and successfully applied in other sciences can be implemented.

If we are thinking of possible ways of development of geotechnics in this respect, maybe should ask: weather the geotechnical problems could be placed into conventional molds? If they could not, and that is almost definite in most cases, then it is necessary to develop such approaches that will satisfy the requirements of geotechnics, and not to force the geotechnics to satisfy the existing. We must have in mind that the differences between geotechnical problems and problems in other fields are perhaps greater than similarities. So, the methods applicable in that field may not be suitable for geotechnics. Therefore, those differences must be clearly identified an accordingly tend to develop the specific

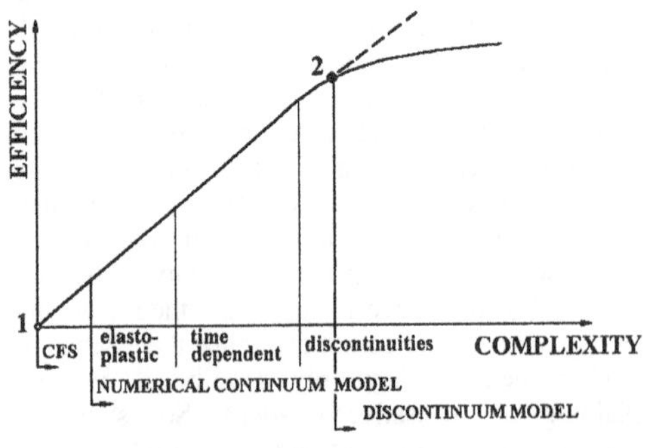

Figure 3. Qualitative relationship between complexity and efficiency of certain approach of analysis

methodology that can lead to valuable and effective results.

Every approach and each method can be developed and upgraded to a certain degree of efficiency, after which they come to the stagnation (Fig. 3). Let us take as an example the application of continuum approach for analysing the stress-deformational behaviour of discontinues rock mass. Efficiency of application, i.e. reality and reliability of the results, are getting higher with increasing the complexity of the model, starting from the simplest linear-elastic closed form solution, to the numerical models with more complex constitutive relations. However, further increasing the model complexity does not provide appropriate increasing of its efficiency. This is the moment when introducing a new approach or principle (point 2) the situation is qualitatively changed and the efficiency of the model is again compatible with its complexity.

A great number of approaches and methods exist in geotechnical theory and practice. Each of them has certain specificity, advantages and shortcomings, possible field of application and limitations. The truth is that even most sophisticated model from some reason may not be appropriate for solving the given problem. It is obvious that there is no universal method that can successfully solve different problems in different situations. Before selecting method of analysis, it is necessary to specify the basic requirements that such analysis has to fulfil in certain situation. These requirements depend on various factors, such as: type and purpose of structure, stage of designing, predicted basic type of rock mass behaviour, available fund of data and their reliability, technology of excavation and supporting etc. As for their practical utility, those methods should be simple and economic enough to justify efforts and expenses with their results. Besides, when making estimates on the capabilities and limitations of certain method, it is necessary to consider the possibilities of realisation of individual stages of modelling process.

Problems of designing and execution of complex structures in complex geological conditions can rarely be solved with one particular modelling approach. In such cases, it is necessary to consider the possibilities of application of different modelling approaches, such as the combination of photomechanical and physical modelling with numerical models. Anyway, careful and comprehensive analysis of all relevant factors, with the appreciation of relevant principles and criteria, will bring to the optimal model of a given problem.

5 SOURCES OF ERRORS IN MODELLING

In each stage of investigations and modelling, we are faced with necessity to come to certain decisions, which are often based on insufficient fund of information. In such cases, an important role has subjective factors: imagination, intuition and free engineering judgement based on overall knowledge and experience of investigator. So, there is certain risk for making wrong estimates, especially if we are not critics in respect our own attitudes.

Formulation of the model of rock mass reaction on the external influences demands certain suppositions, based on theoretical statements and/or empirical relations. Considering abstractions, idealisations and approximations that are made for simplification of the problem, the appropriate errors occur. They are the consequence of uncertain formulations of the theories that are adopted as a representative of the real physical processes being studied, formulation of boundary and initial conditions, also the determination of relevant parameters and numerical approximation.

a) Statement and formulation of the problem comprise possibility for making mistakes due to incorrect definition of basic types of interaction, as well as incorrect hypothesis on the stress-deformation behaviour of rock mass. Their correct formulation is decisive for further process of modelling, because if they are wrong all further results will be also wrong, no matter how all other elements (parameters, numerical processing etc.) are accurate.

b) Mistakes in model parameter determination may result as a consequence of two reasons: 1) errors in measuring and processing of parameters and 2) errors in selection of relevant parameters. Errors in interaction modelling and errors in parameter definition are highly interdependent. For construction of simple models, only few parameters are usually required which are relatively easy to define. Therefore, the probability for making errors due to inaccurate parameters are less, but there may be errors due to unreal formulation of rock mass actual behaviour. On the other hand, multi-parameters model can more realistically simulate the rock mass behaviour and to reduce the errors. But, here arise problems of greater number of parameters that are usually more difficult to obtain (plastic potential, law of plastic strengthening or softening, parameters of residual strengths etc.), so the errors due to incorrect parameters may occur.

c) Numerical models which are used for analysis of geotechnical problems are based on convenient idealisation of geometrical, physical and mechanical properties of rock mass. However, generally speaking, prognosis obtained by model examination will not coincide with real reaction. Basically, there are two sources of errors of such prognosis: approximation in modelling of numerical solution and inadequate knowledge on some aspects of geotechnical system. Nowadays, when the progress of computer development is evident, the second source of uncertainty is of greater importance then the first one.

6 GUIDELINES FOR MODELLING

The fact is that the mathematical modelling, when all real properties of rock mass are considered, is faced with almost insurmountable difficulties. Required degree of accuracy and reliability of predictions imposes quite large limitations in developing the appropriate model. They are by their very nature approximation of the real performance of the system. So, it is clear that they could not be absolutely correct, neither it should be their main goal. Their efficiency depends on the reality of the inputs and the degree to which the mathematical formulations really represent the original system. All that imposes the necessity to clearly understand and to "feel" the complexity of problem being analysed, that is, to recognise the rock mass reaction as fully as possible. Then it is possible to formulate the mathematical model more realistically and to perceive its results more rightly. For successful mathematical modelling, the following principles have to be honoured:

1. When constructing a model it is necessary to completely and clearly conceive the purpose of its creation and what are the expected results, i.e. on what question model should give the answer.

2. Models that are previously constructed for other similar situations can never be routinely applied for new location. It does not mean that existing models cannot be applied for solving an actual problem, but careful selection of appropriate model respecting all specificity on the given location, also the possibilities and limitations of the model. Problems in geotechnics are by their very nature and manifestations unique; the certain combination of field conditions and construction characteristics almost never repeats. Those are always new problems with specific manifestations, more or less different from other known similar conditions.

3. In the process of modelling all properties of rock mass and natural environment must be considered, even those that are not directly used for model formulation. The fact is that all properties of geologic environment can be neither quantified nor comprised in model, although many of them may have more or less influence on the environment behaviour. Besides, every researcher has certain impressions about the investigated terrain that cannot be appropriately expressed and presented. This is that thoutghful component of geotechnical model which is important in many situations that require engineering decisions, also for estimating the reality of the results of modelling.

4. When choosing or creating the constitutive laws it should be bears in mind that they reflect the essence of a phenomenon and involve parameters that can be uniquely and reliably defined. The need that constitutive law should reflect the essence of a phenomenon is a consequence of model basic

function, i.e. the fact that all phenomena, which have indirect or direct (but less important) influence on rock mass performance, could not be involved in the model. Parameters that get into the constitutive equations must be reliable enough. This means that the constitutive law should be adapted to those parameters that are reliable (having in mind the first statement of the principle), or to quantitatively define parameters that are necessary for formulation of appropriate constitutive law.

5. Model should be adapted to the fund of available data and to their reliability. This is the consequence of the fact that geotechnical models in general are constructed and analysed from the limited fund of data that are by their very nature uncertain. Therefore, for each particular case it is necessary to carefully estimate available fund of data, to conceive the possible mechanisms of rock mass reaction, and then to select that model that can give the optimal results.

6. When approximating there must not be such approximations that change basic forms of real behaviour. Approximations in geotechnical modelling are necessity that imposes the need for clear understanding of their possible consequences. For example, discontinuous rock mass exposed to a low stress field, cannot be approximated as continuum, because the basic mechanism of deformation will be qualitatively changed. But, there are possibilities for number of approximations regarding the relevant parameters (planar discontinuities, statistical data etc).

7. The level of model complexity must correspond to the stage of geotechnical modelling, the purpose of modelling and possibility of practical application. The relation between knowledge on the environment and complexity of the model should be balanced. Complex analysis with insufficient data or insufficiently reliable data could make worthless their results. In many problems, more complete picture can be obtained with simpler models, but with boundary values of data.

8. The results of model analysis should always point out its shortcomings and to contribute its improvement. Results of previous modelling always point out what are the required data and enable reinterpretation of existing. This, in return, influences the model improvement, or brings to a new idea for new types of model. So, modelling is not stationary process, but highly interactive, influencing improvement of further investigations, and in this way with new results his further qualitative and quantitative upgrading

7 CONCLUSIONS

The fact is that the numerical methods are very powerful tool for analysis and that they progress from day to day. With them one can solve a very complex problem with complex geometrical shape, in the rock mass build up of various types of rocks with complex constitutive relations, with different types of supports, excavation technology etc. However, in engineering daily practice, the numerical methods are not in use to such extent that is comparable with their capabilities. There are several reasons for that. The main one is that capability of numerical methods overcome to far the objective possibilities of reliable and quantitative defining of required input parameters.

Mathematical modelling of interaction is not a trivial task, especially a task that can be routinely realised. There exist two confronted moments: the model should reflect performance of geotechnical system as accurate as possible, on one hand, and to be as simply as possible for practical utilisation, on the other. Computer simulations are attractive and often very illustrative, with the results whose accuracy cannot be disputed if there are no mistakes in model construction. However, they are reliable as much as input data reliably represents the real behaviour of the rock mass. All that can easily create vision that the problem is successfully solved. That may be the truth, but the final assessment must be done by logical engineering decision, based on the experience and good knowledge on the essence of the problem being studied. Thus, the results of numerical modelling cannot be the substitute for inadequate understanding of the problem, but only supplement and support of logical engineering thinking. Respecting all their capabilities and advantages, one should never forget essential fact that the numerical methods are highly dependent on the reliability and degree of elaboration of the model of natural environment, as well as the fact that those methods are neither the only, nor always optimal and most acceptable.

REFERENCES

Kovari, K. 1977. The elasto - plastic analysis in the design practice of underground openings. *Finite elements in geomechnics*, John Willey & Sons ltd

Pavlović, N. 1992. *Stability of underground openings in hard rock masses*. MSc Thesis, Belgrade University; Belgrade.

Pavlović, N. 1996. *Methodology of geotechnical modelling*. PhD Thesis, Belgrade University, Belgrade.

Pavlović, N. 1996. Interaction of geological environment and engineering activity. *Int. Symp. Trends In Development of Geotechnics*. Belgrade.

Starfield, A.M. & Cundall, P.A. 1988. Towards a methodology for rock mechanics modelling. *Int. J. Rock. Mech. Min. Sci. & Geomech. Abstr. 25:* 99-106.

Mechanics of Jointed and Faulted Rock, Rossmanith (ed.) © *1998 Taylor & Francis, ISBN 90 5410 955 6*

Implementation of the Barton-Bandis criterion in a three-dimensional numerical model

J. Hadjigeorgiou & A. Ghanmi
Department of Mining and Metallurgy, Université Laval, Que., Canada

ABSTRACT: This paper reports on the potential of using numerical methods in predicting the shear behavior of rock discontinuities. For this purpose, the Barton-Bandis empirical model for joint behavior has been implemented in a three-dimensional finite difference code (FLAC3D). A series of direct shear tests under controlled normal stress were numerically undertaken on both saw-cut and natural discontinuities and compared to laboratory tests. It was demonstrated that it was possible to obtain realistic simulation of joint behavior.

1 INTRODUCTION

Rock joints play an important role in defining the strength and deformability of jointed rock masses (Bandis, 1990). The presence of joints can significantly affect the mechanical behavior of rock masses by reducing their capability to support shear and tensile loading. At the relatively low levels encountered in near surface excavations, the deformation of joints dominates the elastic deflection of the intact rock. Even under the higher levels of stress associated with heavy structures, the slippage and closure of joints constitute the major part of settlement on rock (Huang *et al.*, 1993). As joints exert such a substantial role in the static and dynamic behavior of rock masses, their formal description and experimental characterization are important in rock engineering practice. An understanding of the behaviour of discontinuities under load is essential for analysis and design in jointed rock.

Empirical models aim to capture the dominant aspects of joint behavior and to develop predictive tools. Arbitrary constants appearing in the final expressions are derived from field and laboratory observations. The objective is to reconcile the response predicted from the model with observed performance.

In this class of models, the behavior of joints is assumed to be elasto-plastic. The elastic behavior is represented by the initial elastic tangential normal and shear stiffness. The peak strength and dilatancy of rock joints is represented by a failure criterion and a flow rule, respectively. Empirical models derived in this way may not satisfy the laws of deformable body mechanics, but they have the engineering utility of providing techniques for immediate practical solutions to current engineering problems.

Analytical approaches rely on the development of constitutive relations for a joint, properly satisfying the laws of continuum mechanics. These methods formulate analytical expressions from consideration of the morphology of a joint surface (joint topography) and the micromechanics of surface deformation (dilatancy, surface damage). Input experimental data to the constitutive models are the description of the joint geometry (asperity profile) and the mechanical properties of the rock material.

Constitutive models based upon elastic-plastic theory are of interest since they have the potential for replicating joint behavior. Goodman (1974) proposed that a discontinuity surface can be represented by non-linear springs. This model, however, describes only a monotonic loading behavior. Desai *et al.* (1987) developed a finite-thickness interface model, which makes it possible to use continuum elasto-plastic theories for the interface behavior. Pleasha *et al.* (1989) proposed an elastic-plastic model assuming an exponentially decreasing of aspirity during slipping. Swan (1981) proposed a joint behavior formulation based upon an extensive literature related to metal friction and the role of roughness and waviness developed over a shear surface. Qui *et al.* (1993) have represented a model based on the characterization of joint topography using a sinusoidal asperity surface. In the past, satisfactory techniques for defining joint topography were not well established. As a result, experience with constitutive models and their implementation in numerical codes was limited.

Support for the analysis by experimental data obtained under rigorously controlled conditions becomes an essential aspect for any model validation.

2 MODELING OF JOINT BEHAVIOR

Distinct rock joints in a given rock mass are subjected to a wide variety of boundary conditions and it is almost impossible to obtain, by experiment, rock joint response for all possible boundary conditions. On the other hand, natural joints present in general three dimensional asperity profiles. Representation of such models has been studied by Gentier and Riss (1990). Empirical models of joint behavior can be implemented in most numerical codes (finite element, boundary element, finite difference, discrete element) to solve problems in which media containing a number of distinct discontinuities are subjected to external loads. The performance of numerical models in simulating problems of load deformation in rock mechanics depends on the modeling of joints behavior and their interaction with the media (Plesha *et al.*, 1989). This paper demonstrates the potential of using numerical methods in predicting the stress deformation behavior of joints. For that purpose, direct shear tests under normal constant stress have been conducted using FLAC3D, a three dimensional finite difference code (Itasca, 1996). The Barton-Bandis empirical model for joint behaviour has been implemented in FLAC3D using the Fish option. In order to compare the simulation results, a series of direct shear tests, under controlled normal stress, has been carried out on saw-cut and natural discontinuities.

3 MODEL IMPLEMENTATION

A series of empirical relations have been developed by Barton and Bandis (Bandis, 1990) to describe the effects of surface roughness on discontinuity deformation and strength. These relations, known collectively as the Barton-Bandis joint model, have been implemented into FLAC-3D.

3.1 Joint normal behavior

An hyperbolic model represented by equation (1) has been proposed to simulate the joint normal behavior under increasing normal stress. This model was derived from laboratory tests on natural discontinuities and has been extensively used in engineering analysis and design of rock structures. The normal stress, σ_n, and normal displacement, U_n, are related as follows:

$$\sigma_n = \frac{U_n k_{ni} V_m}{V_m - U_n} \tag{1}$$

where V_m is the maximum allowable joint closure (considered positive) and k_{ni} is the initial normal stiffness of the joint. The initial joint stiffness (k_{ni}), which changes with load cycle number, is calculated by:

$$k_{ni} = 0.0178\left[\frac{JCS_0}{a_{jn}}\right] + 1.748 \times JRC_0 - 7.155 \tag{2}$$

where JCS_0 is the laboratory-scale joint wall compression strength, JRC_0, the laboratory-scale roughness coefficient, and, a_{jn}, the joint aperture at zero normal stress. The maximum allowable closure (V_m) for load cycle i is given by:

$$V_{mi} = C_i\left[\frac{JCS_0}{a_{jn}}\right]^{D_i} + B_i \times JRC_0 + A_i \tag{3}$$

where A_i, B_i, C_i, D_i are constants associated with the number of load cycles.

To calculate the path for an unload cycle following a load cycle, a new V_{mi} and k_{ni} are calculated. k_{ni} is calculated from equation (2) using a new aperture value, a_{jn}, which is reduced by the irrecoverable closure (V_{irr}). V_m is recalculated from equation (3), also using a_{jn} reduced by V_{irr}. The irrecoverable closure, V_{irr}, is calculated from the following equation.

$$V_{irr} = \left[C_1 - C_2\left(\frac{JCS_0}{a_{jn}}\right)\right]\frac{u_{nl}}{100} \tag{4}$$

where u_{nl} is the maximum closure for a completed load cycle, and, C_1, C_2 are empirical constants for the current cycle.

3.2 Joint shear and dilatant behavior

A typical shear-displacement curve of a tension fracture presents, pre-peak, peak, and post peak regions. Under the peak, the shear stress-displacement relation is quasi-linear and the cuve slope is defined as the unit shear stiffness. The shear resistance of a joint is calculated using the concept of mobilized roughness. The mobilized roughness coefficient (JRC_{mob}) is a function of the joint properties: length, normal load, current shear displacement, and shear dispalcement history. To implement the shear stress model, a limiting shear stress, σ_{sl}, is calculated from the full-scale roughness coefficient (JRC_n), the joint wall compressive strength (JCS_n) and the peak shear displacement (δ_{peak}):

$$JRC_n = JRC_0\left(\frac{L_n}{L_0}\right)^{-0.02\,JRC_0} \tag{5}$$

$$JCS_n = JCS_0\left(\frac{L_n}{L_0}\right)^{-0.03\,JRC_0} \quad (6)$$

$$\delta_{peak} = \frac{L_n}{500}\left(\frac{L_n}{L_0}\right)^{0.33} \quad (7)$$

where L_0 is the laboratory-scale joint length, and L_n is the field-scale joint length. The mobilized joint roughness, JRC_{mob}, is calculated from the full-scale roughness coefficient (JRC_n):

$$JRC_{mob} = B \times JRC_n \quad (8)$$

where B is constant calculated from the ratio of current shear displacement to peak shear dispacement. The limiting shear stress (σ_{sl}) is calculated:

$$\sigma_{sl} = \sigma_n.\tan\left(JRC_{mob}.\log_{10}\left(\frac{JCS_n}{\sigma_n}\right) + \phi_r\right) \quad (9)$$

where ϕ_r is the residual friction angle. The shear stress approaches the limiting shear stress incrementally by multiplying the shear displacement increment (Δu_s) by a shear stiffness (k_s). The stiffness is defined as one of two intial linear segments of the load path, depending on shear displacement. The increment shear stress ($\Delta\sigma_s$) is calculated from the following expression:

$$\Delta\sigma_s = \Delta u_s.k_s$$

where

$$\begin{cases} k_s = \sigma_n.\tan\left(\dfrac{0.75\phi_r}{0.2\delta_{peak}}\right).L_n \ ; \ \left(\dfrac{\delta}{\delta_{peak}}\right) < 0.2 \\ k_s = \sigma_n.\tan\left(\dfrac{0.25\phi_r}{0.1\delta_{peak}}\right).L_n \ ; \ \left(\dfrac{\delta}{\delta_{peak}}\right) > 0.2 \end{cases} \quad (10)$$

4 PREDICTING JOINT BEHAVIOR

In this work, the joint behavior under constant normal stress has been investigated using numerical modeling supported by laboratory tests. These tests allow the validation of the numerical simulation and the calibration of some parameters like the shearing rate.

4.1 Experimental set-up

The experimental set-up has been reported by Lessard (1996). A series of direct shear tests were performed on different dry saw-cut basalt samples. The samples had a square shear surface of approximately 2500 mm^2 and were set in cement prior to being fitted for a portable direct shear box. Tests were performed on each sample at increasing normal stress from 1.0 to 5.0 MPa. During testing the normal load was kept constant. Tests were conducted at shearing rate of approximately 2 mm/min until a displacement of 8 mm was reached. A data acquisition system was utilized to record shear displacement, and shear and normal stress.

A series of a direct shear tests was also performed on different rough "massif sulphate" samples which had a square surface of approximately 1800 mm^2. All experiment conditions were maintained as below except the shearing rate which is reduced to 1 mm/min.

4.2 Numerical simulation

The objective of this paper is to demonstate the potential of numerical modeling in predicting joint behavior under constant normal stress. It was undertaken using the Fast Lagrangian Analysis of Continua in three dimensions (FLAC3D) program (Itasca, 1996). It is an explicit Finite Difference Solution scheme for the numerical simulation of stress and displacement in deformed media. It is based on the Lagrangian description of deformations in that the initial coordinates of particles are taken as the independent variables in defining the strain tensors. The explicit scheme simulates the propagation of a disturbance through the model by small deformation increment. Its particular advantages are the simulation facility of non linear constitutive laws without load-path sensitivity of iterative methods. Direct shear test is modelled in FLAC3D by considering two superposed blocks separed by an interface contact (Figure 1). The interface is represented as collections of triangular planes (interface elements) and points in space (interface nodes). The normal stresses, applied on the top of upper block, were maintained constant during the shearing tests, consistent with the experimental procedure. The boundary conditions involved fixing the lower block against vertical and horizontal displacements and imposing a constant shearing rate on the upper block. The shear tests were conducted in one direction (longitudinal direction) and no consideration was given on the anisotropy at this preliminary phase. The normal and shear forces developed on the joint surface were calculated at the interface nodes. The average normal and shear stresses were obtained by integrating respectively the normal and the shear forces over the joint surface area. A special subroutine "JOINT" was developed to implement the integration of the joint behavior model and to calculate the shear test parameters.

A preliminary analysis was conducted to determine the adequacy of the numerical model for study of saw-

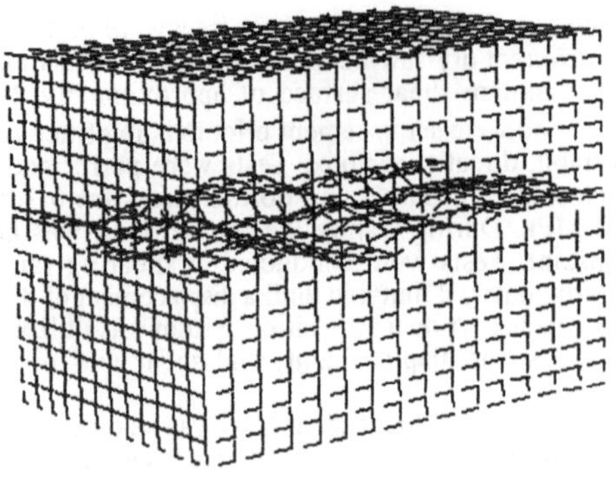

Figure1. Schematic representation of a discontinuity profile.

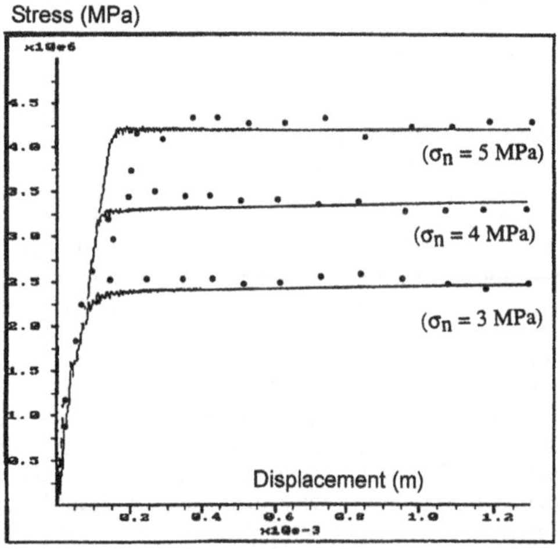

Figure 2. Stress displacement curve for saw-cut joints.

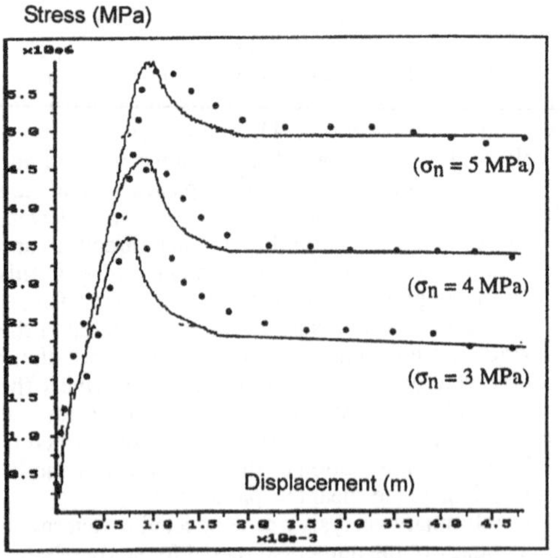

Figure 3 Stress dispacement curves for rough profile joints.

cut joint shear. The mechanical properties of both rock and joint were experimentally determined:

Rock properties (Basalt)	
Young's modulus (E)	30 GPa
Poisson's ratio (ν)	0.25
Compressive Strength (σ_c)	80 MPa
Cohesion (c)	21 MPa
Friction angle (ϕ)	35°
Joint properties (saw-cut)	
Stiffness ($k_{ni} = k_s$)	25 GPa/m
Cohesion (c)	0.0
L_0	50 mm
Friction angle (ϕ)	39°

The second objective of this paper was to examine the behavior of natural joints under conditions approximating those "in-situ" using the same procedure as the saw-cut joint. The material selected was a sulfur massif sample and the mechanical properties of both rock and joint were as below:

Rock properties (Sulfur massif)	
Young's modulus (E)	148 GPa
Poisson's ratio (ν)	0.225
Compressive Strength (σ_c)	263 MPa
Cohesion (c)	40 MPa
Friction angle (ϕ)	53°
Joint properties (rough)	
Stiffness ($k_{ni} = k_s$)	7.0 GPa/m
Cohesion (c)	0.4 MPa
L_0	45 mm
Residual friction angle (ϕ_r)	44°
JRC_0	12

5 INTERPRETATION OF RESULTS

The performance of the proposed model implemented in FLAC3D code was assessed by comparing the experimental with the predicted shear strength-displacement curves. The plots of normal and shear stresses mobilized as a function of shear displacement for the case of saw-cut joints and for different normal stresses (3, 4, and 5 MPa) were shown in Figure 2. Several observations can be made on the basis of these figures. First, the model seems to have captured the general pattern of the experimental curves. A detailed analysis indicates that the measured shear stress-displacement curves are not necessary linear in the elastic region. This behavior has particularly been observed for small shear displacements. Numerically, the relation between stress and displacements was reproduced for the different normal stresses using an average slope of k_s = 25 E9 N/m. The mobilized shear strengths were also simulated accordingly to the introduced Barton-Bandis failure criterion. The strengths were reached at approximately 0.2 mm of shear displacement. It is readily recognised that the comparison of laboratory and numerical tests would have been facilitated by employing a more sophisticated laboratory programme. This does not however detract from the essential robustness of the numerical approach. It can also be seen that the numerical model tends to smooth the shear strength-displacement curves, without paying attention to the vagarities of the experimental shear strength-displacement curves.

For the rough natural joint, the plots of normal and shear stresses mobilized as a function of shear displacement for different normal stresses (3, 4, and 5 MPa) are shown in comparison with experimental results in Figure 3. It appeared that the numerical model has correctly simulated the natural joint behavior. The general pattern of experimental curves has then been captured. In particular, the joint strengths and the shear stress-displacement realtions after peak have been simulated with reasonable precision. The relative difference between the measured and the predicted results can be attribuated to the following reasons:

- use of a constant joint stiffness can compromise numerical results when asperities are important. It has be shown that the shear stress-displacement relations present non linear curves in the elastic region. Models with variable stiffness must then be implemented to correctly reproduce the experimental curves and to improve precision;
- normal stiffness of joint was unknown and was taken equal to shear stiffness affecting joint behavior.
- the normal stresses that varied during the experimental tests have complicated the result interpretation. For a good comparison, it is then important to consider the same boundary conditions during tests.
- finally, finite difference discretization of the domain geometry was not refined and the joint profile is not well reproduced and approximated efficiently.

6 CONCLUSIONS

This paper has focused on demonstrating the possibility of predicting joint behavior using numerical modeling.

(1) The joint behavior can be predicted using a numerical code. The implemented empirical model has shown the potential of simulating the behavior of both saw-cut and rough joints under shearing tests. This can support the numerical modeling of jointed rock masses. However, since this study has been done for monotonic loading behavior, further work should be done with regards to cyclic sliding behavior.

(2) For natural joints, the contact surface presents in general a three-dimensional profile and joint behavior simulation requires then a three-dimensional numerical model. This allowed to capture the influence of the anisotropy and and the aspirity.

(3) Although difficulties noted with experimental procedure, the predicted results were representative. The model was able to perfectly reproduce every part of the shear strength-displacement curves, they were in very good agreement with the experimental results.

(4) Finally, to improve the joint behavior prdiction, finite discretization of the domain must be further refined to adequately represent the geometry of the joint profile and to minimize errors.

The practical significance of this simple exercise is to recognise that a series of numerical tests can be undertaken relatively easy once a model is set. This can greatly facilitate parametric studies where the three dimensional profile can be incorporated into the analysis as well as different orientations in shearing.

ACKNOWLEDGEMENT

The authors would like to acknowledge the financial support of the National Science and Engineering Research Council of Canada. Mr. J. S. Lessard conducted the laboratory tests.

REFERENCES

Bandis, S.C. (1990) *Mechanical proprieties of rock joints*, Barton & Stephansson (eds), Balkerna, Rotterdam, ISBN 906191 1095.

Barton, N. (1973) A review of a new shear strength criterion of rock joints. *Engineering Geology*, vol. 7, 287-332.

Desai, C. S. and Fishman K.L. (1987) Constitutive models for rocks and discontinuities (joints). *Proc. 28th U.S. Symp. on rock mechanics*, Tuscon, 609-619.

Gentier, S. and Riss, J. (1990) Quantitative description and modelling of joint morphology. *Rock Joints, Barton & Stephansson (eds) pp. 375-382*

Goodman, R.E (1974) The mechanical proprieties of joints. *Proc. 3rd Int. Congr. ISRM*, Denver, Colorado, vol. 1, Part A, 127-140.

Huang, X., Haimson, B.C., Plesha, M.E. and Qui, X. (1993) An investigation of mechanics of rock joints- Part1: Laboratory Investigation. *Int. J. Rock mech. Min. Sci. & Geom. Absts.*, vol. 30, 257-269.

Itasca Consulting Group, Inc. (1996) User manual of FLAC3D.

Lessard, J.S., (1996) Modélisation du comportement en cisaillement des discontinuités rocheuses à l'aide des réseaux neuroneaux. M. Sc. Thesis, Université Laval.

Pleasha, M.E., Ballarim, R. and Parulekar, A. (1989) A constitutive model and finite element solution procedure for dilatant contact problems. *J. Eng. Mech.* ASCE, vol. 115, 2649-2668.

Swan, G. (1981) Tribology and caracterization of rock joints. *Proc. 22nd U.S. Symp. on rock Mechanics*, Boston, USA, 402-407.
Qui, X., Pleasha, M.E., Huang, X. and Haimson, B.C. (1993) An investigation of mechanics of rock joints- PartII: Analytical Investigation. *Int. J. Rock mech. Min. Sci. & Geom. Absts.*, vol. 30, 271-287.

Fractures

Mechanics of Jointed and Faulted Rock, Rossmanith (ed.) © 1998 Taylor & Francis, ISBN 90 5410 955 6

Mechanics of rock fracturing around boreholes

Bezalel Haimson & Insun Song
Geological Engineering Program, University of Wisconsin, Madison, Wis., USA

ABSTRACT: Experiments in which borehole breakouts were produced in different rocks have revealed three major mechanisms of rock fracturing. In all tests a true triaxial stress condition was applied to cubical rock specimens. Then a central vertical borehole was drilled, which induced borehole-breakout failure, depending on the magnitudes of the far-field stresses. In crystalline granite intra- and trans-granular dilatant microcracking subparallel to the borehole wall and to the direction of the maximum horizontal stress σ_H preceded the development of breakout failure, resulting in 'V'-shaped failed zones along the σ_h spring line. In well-consolidated sandstone breakout failure took a similar 'V' final shape. However, microcracks preceding failure were mainly intergranular, and only occasionally intragranular. In a less consolidated but still competent sandstone, failure did occur at the same locations as with the other rock types but, surprisingly, the final breakout shape was a narrow linear fracture perpendicular to σ_H direction. The mechanism of this failure mode appears to be largely matrix non-dilatational disintegration leading to grain disaggregation.

1 INTRODUCTION

The occurrence of borehole breakouts, defined as failure-induced borehole cross-section elongations, is directly related to mechanical properties of individual rocks and the prevailing in situ state of stress. Extensive field evidence and laboratory experiments suggest that breakout orientation along the perimeter of vertical-boreholes is typically aligned with the direction of the minimum horizontal in situ stress σ_h (Bell and Gough, 1979; Shamir and Zoback, 1992; Haimson and Herrick, 1986; Haimson and Song, 1993; Lee and Haimson, 1993; Song and Haimson, 1997). Laboratory experiments in several limestones and granites also reveal a direct correlation between breakout geometry and in situ stress magnitudes (Haimson and Herrick, 1986; Haimson and Song, 1993; Lee and Haimson, 1993; Song and Haimson, 1997). In field situations, such as the scientific research boreholes at Cajon Pass, California, and KTB, Germany, the angular span of breakouts determined from televiewer logging combined with estimates of σ_h from hydraulic fracturing have been utilized to constrain the magnitude of the maximum horizontal in situ stress σ_H (Vernik and Zoback, 1992; Brudy et al., 1997).

The growing importance of borehole breakouts as a means of assessing in situ stress conditions in the earth crust at locations where direct measurements are either too expensive or not feasible due to hostile conditions, has brought about intensified research into the breakout-forming fracture mechanism. The ultimate goal of this research is to find more reliable means of determining in situ stresses directly from breakout observable features. Several modes of failure have been suggested as the mechanism leading to the breakout phenomenon. One is extensile cracking in the vicinity of the borehole (Zheng et al., 1989). The extensile failure model has been used to explain the episodic growth of breakouts resulting from sequential spalling of rock flakes from the borehole wall. Breakout shapes resulting from extensile cracking are deep and pointed. Another suggested failure mode is shear failure, which has been observed experimentally in a weak limestone and an artificial rock (Bandis et al., 1987 Haimson and Song, 1993). Shear failure occurs along two conjugate fractures which follow trajectories of high shear stress at and behind the borehole wall. Breakout shapes are again deep and pointed.

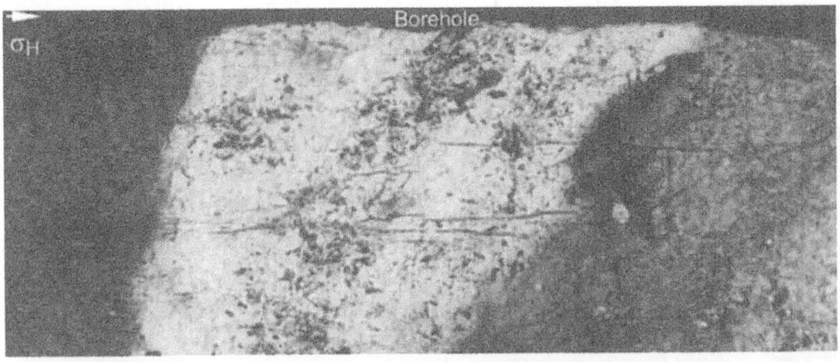

Figure 1. Intra- and trans-crystalline microcracks subparallel to borehole wall and to σ_H prior to breakout failure in granite.

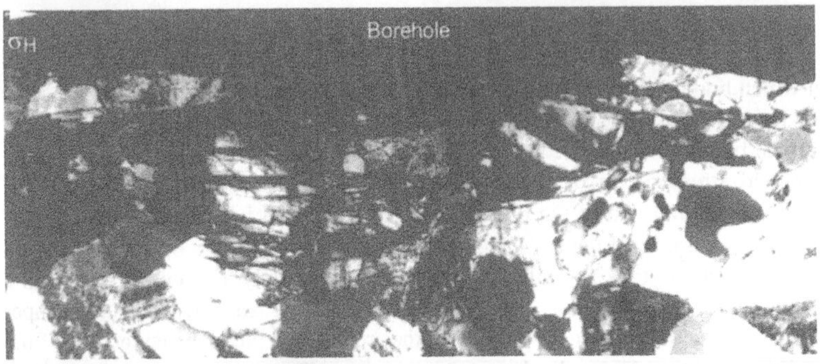

Figure 2. Increased microcracking at higher σ_H, creation of separated rock flakes, and their buckling leading to breakout initiation in granite.

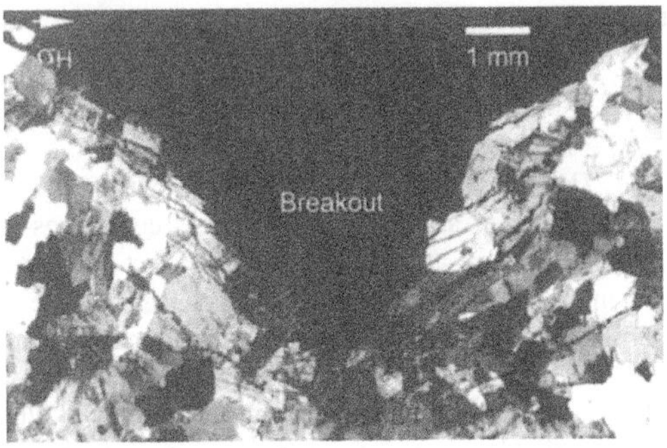

Figure 3. Typical 'V'-shaped breakout along two inclined planes of apparent shear localization (granite).

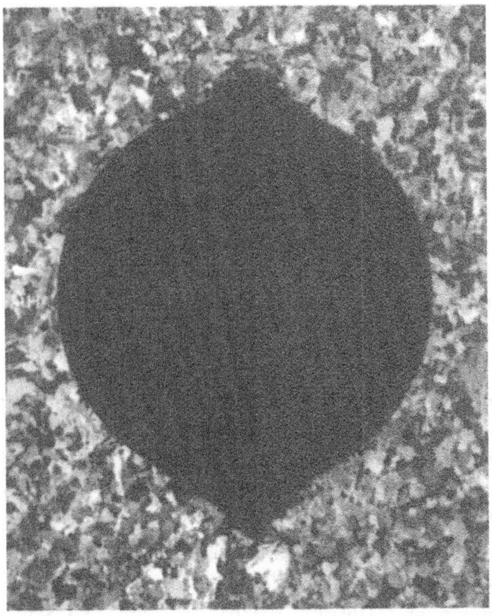

Figure 4. Post-breakout borehole cross section in granite. Breakouts are along the σ_h spring line.

A novel approach to assessing failure mechanism around boreholes is the use of the bifurcation analysis (Vardulakis and Papanastasiou., 1988), which predicts either exfoliation (extensile cracking) or shear failure, depending on the ratio between the tensile and the compressive strengths of the rock and its hardening parameter. The mechanisms of both extensile and shear failure modes have also been explained based on pressure-dependent linear elasticity (Santarelli et al., 1986; Maury, 1987). According to this theory the tangential stress attains its maximum value, not at the borehole wall, but at some distance behind. Failure mechanism will depend on the type of microcracking (intra- or intergranular) which in itself is a function of the difference in toughness between matrix and grains.

In this paper we concentrate on the fracturing mechanisms in granite and sandstones as observed in laboratory experiments of drilling in rock already subjected to a pre-existing state of far-field stress.

2 EXPERIMENTAL SETUP AND PROCEDURE

Two granites (Lac du Bonnet and Westerly) and two Berea sandstones were tested in the reported study. Cubical blocks (10x10x10 cm^3) were prepared and placed in a biaxial apparatus for the application of two unequal horizontal far-field loads; the third (vertical) load was applied through a compression loading frame (Song and Haimson, 1997). After the application of the preset far-field stresses ($\sigma_H > \sigma_v \neq \sigma_h$), core-drilling ($\approx$ 2 cm in dia.) was carried vertically down through the center of the block. Thus a faithful simulation of in situ drilling conditions was achieved.

Under sufficiently high horizontal far-field stresses borehole breakouts were induced along the σ_h spring line. Following drilling, thin epoxy was used to fill the drilled borehole and freeze the breakouts. Thereafter, thin sections were prepared through which to study breakout failure mechanisms. This was mainly done using a petrographic microscope with a polarized light source.

3 FRACTURE MECHANISM IN GRANITE

The two granites tested exhibited similar failure behavior leading to pointed or 'V'- shaped (nicknamed "dog-eared") breakouts. In boreholes drilled under far-field stresses lower than critical, no damage could be observed on the borehole surface. However, thin sections of cross sections of such boreholes reveal considerable microstructural damage immediately behind the borehole wall. In the two zones along the spring line of σ_h, families of extensile microcracks are observed just behind the borehole wall and subparallel to σ_H (Figure 1). These microcracks are both intra- and trans-crystalline. The density of the subparallel microcracks is higher for those closer to the borehole wall, where the compressive tangential stress concentration is larger, and the radial stress approaches zero. These microcracks appear to be extensile, and hence dilatational, similar to those observed parallel to the axial load in uniaxial compression testing. No shear movement along the crack surfaces was detected.

At higher far-field stresses microcracks continue to dilate and grow, coalescing with other fractures, or branching into multiple cracks forming a complex network of discontinuities. These microcracks form thin layers of rock flakes between them which are subparallel to σ_H direction. The rock flakes separate as the microcracks open under increased horizontal stress σ_H. At some critical stress the slender flake between the borehole wall and the nearest microcrack becomes too weak to sustain the load, buckles, and spalls off by shearing off its ends (Figure 2). This is the first stage of the visible-breakout development. The stress concentration is then transferred to the next flake, which in turn fails and the episode repeats itself several times. However, each spalled rock flake is somewhat shorter than the previous one, owing to remnants left at the two ends of the previously failed rock flakes. At some point the flake length becomes subcritical and spalling ceases. This gives rise to the 'V' shape of the breakouts, and the two inclined planes that form this shape can be construed as shear localizations (Figure 3). Lee and Haimson (1993) showed that laboratory induced breakouts in granite (Figure 4) are practically identical in shape to those observed in a 3m dia. circular tunnel, strongly suggesting no size effect biasing our laboratory results.

4 FRACTURE MECHANISM IN WELL-CEMENTED BEREA SANDSTONE

The mechanism of breakout failure in clastic rocks such as sandstones is less obvious, and appears to depend largely on the difference in fracture toughness between the grains and the matrix material. Breakouts in low-porosity ("well-cemented") Berea sandstone (porosity: 17%) were the result of dilatant

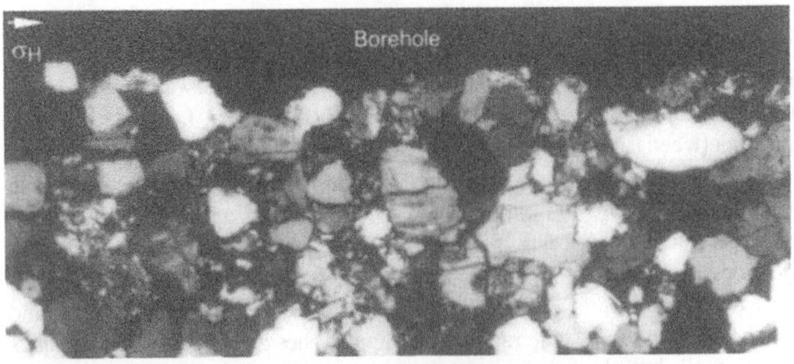

Figure 5. Inter- and intra-granular microcracks in well-cemented sandstone subparallel to borehole wall and to σ_H prior to breakout failure.

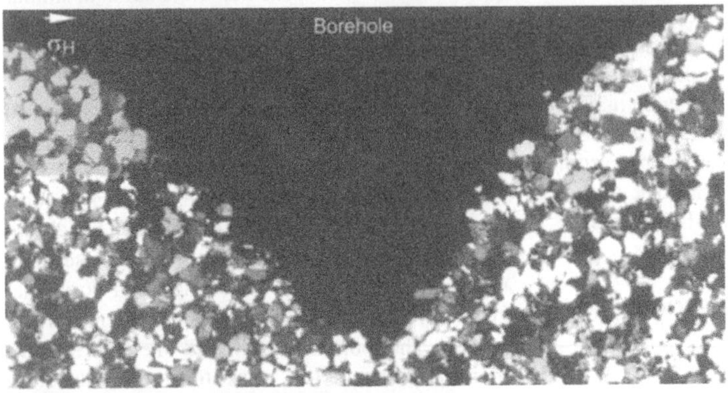

Figure 6. Typical 'V'-shaped breakout in well-cemented sandstone..

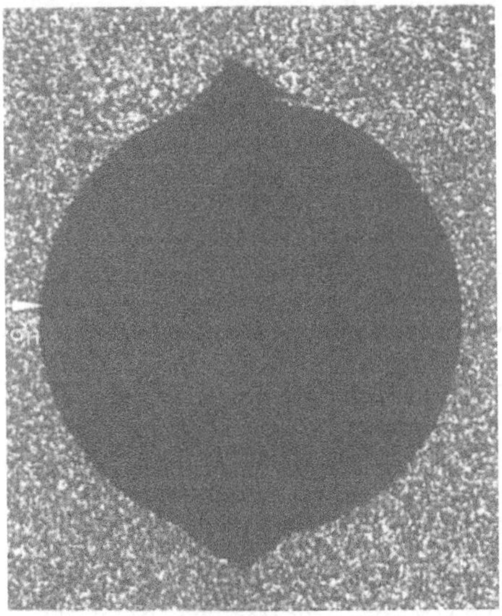

Figure 7. Post-breakout borehole cross section in well-cemented sandstone (compare with granite in Figure 4). Breakouts are along the σ_h spring line.

extensile microcracking subparallel to σ_H direction at two diametrically opposed locations along the σ_h spring line, similar in principle to the observation made in granite. However, the microcracks were mainly intergranular (their propagation made easier by the much lower toughness of the rock matrix) with occasional intragranular grain-splitting segments (Figure 5). Beam-like rock slivers created by the microcracks subparallel to the borehole wall fail under sufficient horizontal far-field-stress, leaving a breakout whose lateral span decreases with depth and forms a 'V' shaped elongation not unlike the

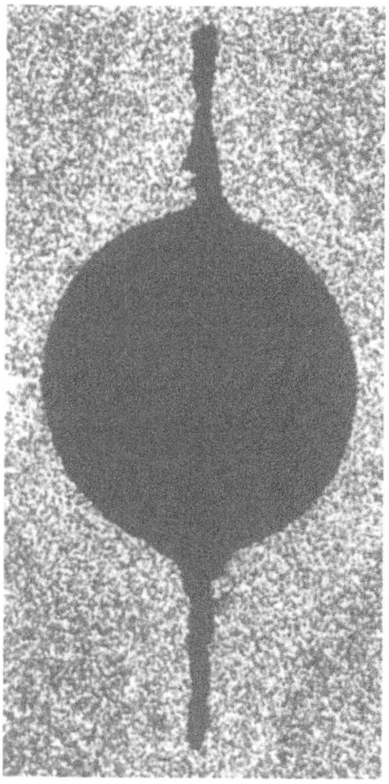

Figure 8. (left) Post-breakout borehole cross section in weakly-cemented sandstone. Slot-breakouts are orthogonal to σ_H direction.

Figure 9. (below) Mechanism of failure in weakly-cemented sandstone is through matrix disintegration leading to grain disaggregation.

Figure 10. (right) Detail of slot- or fracture-breakout, showing intact grains at the free surface (indication of no grain splitting involvement in the failure mechanism).

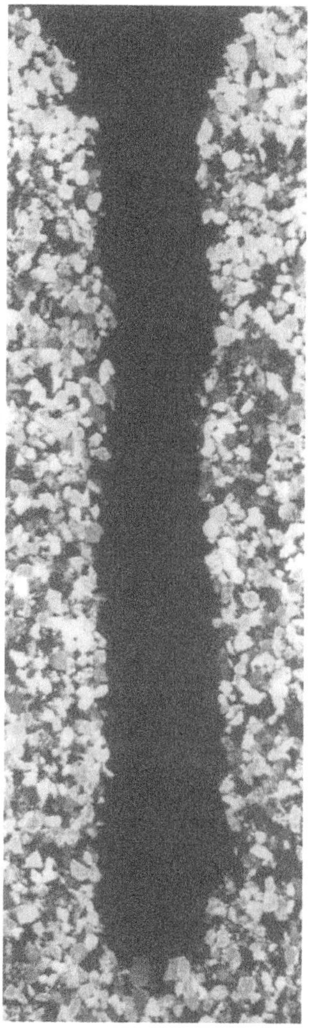

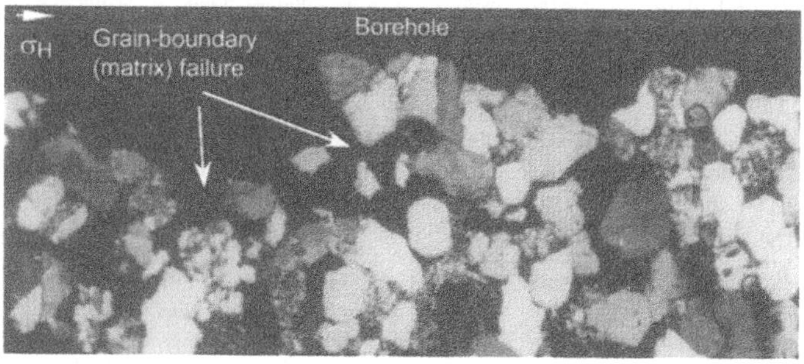

granite case (Figure 6). Post-breakout, a sandstone borehole (Figure 7) looks not unlike the one in granite (Figure 4).

5 FRACTURE MECHANISM IN WEAKLY-CEMENTED BEREA SANDSTONE

Breakouts in the high-porosity ("weakly- cemented") Berea sandstone (porosity: 22.5%) were dramatically different. They did occur at the same locations around the borehole (along the σ_h spring line), where the compressive stress concentration is the highest, but their shape is best described as long linear slots (or fractures) orthogonal to σ_H direction (Figure 8). The occurrence of a linear failure zone normal to σ_H direction is at first counterintuitive. It should be noted, however, that a similar observation was made by Bessinger et al. (EOS, Nov. 1995) in blocks of lightly sintered glass beads.

The mechanism of failure leading to slot-breakouts may be related to the very weak bonding between the quartzitic grains in the higher porosity sandstone. Under sufficient far-field stresses, the rock matrix which is further weakened due to the increase in porosity, disintegrates at the points of highest stress concentration around the borehole wall along the σ_h spring line (Figure 9). This loosens the already fragile bonding and frees the adjacent quartz grains (during drilling this process is aided by drilling fluid washing off loose grains). The disaggregation of the quartz grains creates a higher stress concentration immediately behind them. Inevitably further collapse of the cementing matrix occurs bringing about grain-boundary failure, additional debonding along the σ_h spring line, and subsequent grain removal. The rugged free surface of the slot-like or fracture-like breakout shows whole quartz grains left intact during this failure process (Figure 10).

The surprising results in the high-porosity sandstone suggest that the failure mechanism in some weakly cemented rocks, such as could be encountered in certain fault zones, or in petroleum-rich reservoirs, may be drastically different from the conventional one, and appears not to result from extensile microcracking. This non-dilatational failure mechanism leads to the unexpected and seemingly-paradoxical fracture-like breakout orthogonal to the σ_H direction.

6 CONCLUSIONS

Fracturing mechanism resulting from excessive stress concentration around boreholes can vary with the rock type. Crystalline rocks such as granite undergo dilatant intra- and trans-granular microcracking leading to the well known 'V'-shaped breakouts. In well cemented clastic rocks such as sandstones microcracks are still dilatant but develop mainly intergranularly, culminating in similarly 'V'-shaped breakouts. Most intriguing, however, is the behavior of a weakly cemented Berea sandstone. It developed fracture-like breakouts orthogonal to the maximum horizontal far-field stress direction, as a result of apparent non-dilatational disaggregation of its grains. This type of failure could be a possible source of 'sand production' in oil wells drilled into poorly consolidated sandstones.

ACKNOWLEDGEMENTS: This work was funded by NSF grant EAR-9405836. M. Lee performed the breakout tests in Lac du Bonnet granite.

REFERENCES

Bandis, S.C. , Lindman, J. and N. Barton 1987. Three-dimensional stress state and fracturing around cavities in overstressed weak rock. *Proc. 6th Int. Congr. on Rock Mech.* 769-775, Montreal, A. Balkema, Rotterdam.

Bell J. S. and D. I. Gough 1979. Northeast-southwest compressive stress in Alberta: evidence from oil wells. *Earth and Planetary Sc. Letters.* 45:475-482.

Bessinger, B.A., Liu, Z. and N.G.W. Cook 1995. Laboratory borehole breakout patterns in non-dilatational geologic media: the formation of fracture orthogonal to the maximum compressive stress. *EOS Trans. Am. Geophys. U.* 76: F556.

Brudy, M, Zoback, M.D., Fuchs, K., Rummel, F. and J. Baumgaertner 1997. Estimation of the complete stress tensor to 8 km depth in the KTB scientific drill holes; implications for crustal strength. *J. Geophys Res.* 102:18453-18475.

Haimson B.C. and C.G. Herrick 1986. Borehole breakouts: A new tool for estimating in situ stress?. *Proc. Int. Symp. on Rock Stress and Rock Stress Meas.*. pp. 271-280. Publ. Lulea: Centek.

Haimson, B.C. and I. Song 1993. Laboratory study of borehole breakouts in Cordova Cream: a case of shear failure mechanism. *Int. J. Rock Mech. Min. Sci. & Geomech. Abstr.* 30:1847-1856.

Lee, M.Y. and B.C. Haimson 1993. Laboratory study of borehole breakouts in Lac du Bonnet granite: A case of extensile failure mechanism. *Int. J. Rock Mech. Min. Sci. & Geomech. Abstr.* 30: 1839-1845.

Maury, V. 1987. Observations, researches and recent results about failure mechanisms around single galleries. *Proc. 6th Int. Congr. on Rock Mech.* 741-748. A. Balkema, Rotterdam

Santarelli, F.J., Brown, E.T., and V. Maury 1986. Analysis of borehole stresses using pressure-dependent, linear elasticity. *Int. J. Rock Mech. Min. Sci. & Geomech. Abstr.* 23: 445-449.

Shamir G. and M.D. Zoback 1992. Stress orientation profile to 3.5 km depth near San Andreas fault at Cajon Pass, California. *J. Geophys. Res.* 97:5059-5080.

Song, I. and B.C. Haimson 1997. Polyaxial strength criteria and their use in estimating in situ stress magnitudes from borehole breakout dimensions. *Int. J. Rock Mech. Min. Sci. & Geomech. Abstr.* 34: 498.

Vernik L. and M.D. Zoback 1992. Estimation of maximum horizontal principal stress magnitude from stress-induced well bore breakouts in the Cajon Pass scientific research borehole. *J. Geophys. Res.* 97:5109-5119.

Vardoulakis, I. and P.C. Papanastasiou 1988. Bifurcation analysis of deep boreholes. *Int. J. Num. Anal. Meth. Geomech.* 12: 379-399

Zheng, Z., Kemeny, J., and N.G.W. Cook 1989. Analysis of borehole breakouts. *J. Geophys. Res.* 94:7171-7182.

Mechanics of Jointed and Faulted Rock, Rossmanith (ed.) © *1998 Taylor & Francis, ISBN 90 5410 955 6*

Drilling-induced tensile fractures: Formation and constraints on the full stress tensor

P. Peška
GeoMechanics International, Prague, Czech Republic
M. D. Zoback
Department of Geophysics, Stanford University, USA

ABSTRACT: Advancements in wellbore imaging technologies allow observations of very small-scale features on the borehole wall such as drilling-induced tensile wall fractures which occur only in the rock immediately adjacent to the borehole wall due to the stress concentration. We utilize an interactive software system, SFIB (Stress and Failure of Inclined Boreholes, Peška & Zoback 1997a), to demonstrate how the formation of drilling-induced tensile wall fractures is controlled by the in situ stress and is supported by drilling perturbations related to excess mud weight and borehole cooling. The calculations also illustrate various shapes of drilling-induced tensile fractures under different stress conditions and borehole directions. Relationship between the stress and fractures can be used to constrain the in situ stress which is essential for wellbore stability and hydrocarbon migration. We present a method for stress determination from drilling-induced tensile fractures detected by image data at various depths in multiple variably-oriented wellbores.

1 INTRODUCTION

It is well known that if a vertical wellbore is pressurized a hydraulic fracture will form at the azimuth of the maximum horizontal stress (Hubbert & Willis 1957). Various other authors have theoretically and experimentally demonstrated that the occurrence of hydraulic fractures in inclined boreholes is sensitive both to the state of in situ stress and to the borehole orientation (Daneshy 1973, Bradley 1979, Richardson 1981, Aadnoy & Chenevert 1987, Yew & Li 1988, Roegiers & Detournay 1988, Baumgartner et al. 1989, Kuriyagawa et al. 1989). In this paper we focus on drilling-induced tensile wall fractures which spontaneously form under natural stress conditions, perhaps aided by drilling-related perturbations (Moos & Zoback 1990, Brudy & Zoback 1993, Zoback et al. 1993). Peška & Zoback (1995a) demonstrated that wellbore wall can fail in tension for a wide range of stress states and borehole directions even without any wellbore fluid overpressure. We term these fractures tensile wall fractures because they occur only in the wellbore wall due to the stress concentration. Best conditions for formation of wall fractures are in geothermal fields (Dezayes et al. 1995) where cooling of the rock due to the circulation of relatively cold drilling fluids has a similar effect as fluid pressurization.

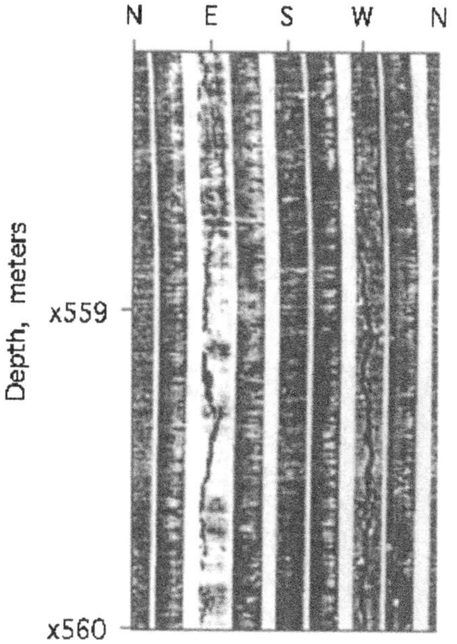

Figure 1. Drilling-induced tensile fractures in a vertical section of a well in the Visund field, North Sea (after Wiprut et al. 1997).

Small-scale features like drilling-induced tensile wall fractures can be detected by modern electrical (FMS/FMI/EMI/STAR) and acoustic (UBI/CBIL/CAST) wellbore imaging devices. Figure 1 shows an example of drilling-induced tensile fractures detected by a Schlumberger FMI log. The cracks occur on both sides of the wellbore at the expected orientation of the maximum horizontal principal stress in the region (~N100°E).

2 FRACTURE FORMATION

In an arbitrarily-oriented borehole, the least tangential stress on the wall is given by

$$\sigma_{t\min} = \frac{1}{2}[\sigma_{zz} + \sigma_{\theta\theta} - \sqrt{(\sigma_{zz} - \sigma_{\theta\theta})^2 + 4\tau_{\theta z}^2}] \quad (1)$$

where σ_{zz}, $\sigma_{\theta\theta}$, $\tau_{\theta z}$ are effective stresses in the borehole coordinate system (Fig. 2) given by formulas originally presented by Hiramatsu & Oka (1962). The least tangential stress, σ_{tmin}, controls fracture formation at the borehole wall: Whenever it exceeds tensile strength of the rock a fracture opens at a position θ where σ_{tmin} is minimum (θ measured clockwise from the bottom side going around the hole; Fig. 2a). At point of fracture initiation, the fracture generally deviates from the borehole axis by an angle ω, where

$$\tan 2\omega = \frac{2\tau_{\theta z}}{\sigma_{zz} - \sigma_{\theta\theta}}. \quad (2)$$

Figure 2a displays a set of tensile wall fractures on opposite sides of a borehole which deviates from vertical by an angle ϕ at an azimuth δ. The orientation of fracture planes is defined by the angles θ and ω that can be directly determined from fracture traces on borehole images (Fig. 2b).

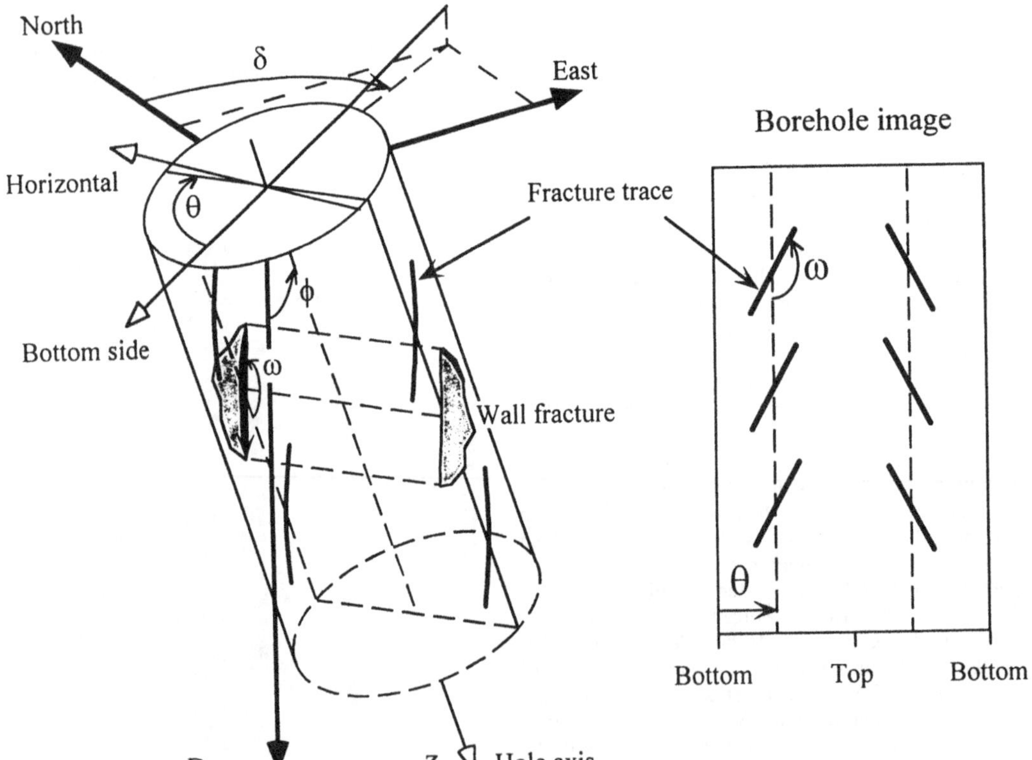

Figure 2. Tensile wall fractures in an inclined borehole. (a) The borehole orientation is defined by angles δ and ϕ, the orientation of fracture planes by the angles θ and ω. (b) Fracture traces form a "chevron" pattern on borehole images.

Stress conditions favorable for fracture formation are illustrated in Figure 3. The figure shows how magnitudes of both horizontal stresses are related to occurrences of drilling-induced tensile fractures in boreholes drilled at three different orientations. The stresses within the polygon correspond to normal, strike-slip and reverse faulting stress states that never exceed a ratio of shear to effective normal stress of 0.7 on any arbitrarily-oriented fault. The stress states at the edges of the polygon correspond to the case of frictional equilibrium where the in situ stress is just equal to the values predicted by Coulomb faulting theory, a condition often observed in the earth (see reviews in Zoback & Healy 1984, 1992). The additional lines subdivide the polygon into regions which correspond to stress conditions favorable for fracture formation in wellbores drilled parallel to S_v , S_{hmin} , and S_{Hmax} , respectively. In a standard near-vertical well, tensile cracks are likely to occur in a strike-slip faulting stress regime, whereas reverse/normal faulting is more favorable for forming cracks in horizontal wellbores drilled parallel to S_{hmin} or S_{Hmax}. A tensile strength of the rock of zero is used, assuming that the cracks will initiate at small pre-existing flaws in the wellbore wall.

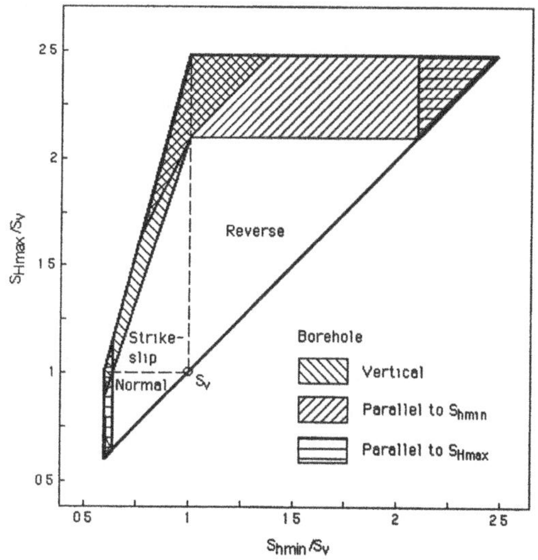

Figure 3. Magnitudes of S_{Hmax} and S_{hmin} required to induce a tensile wall fracture under natural conditions (no excess mud weight and temperature difference) in boreholes of three different orientations. The polygon indicates stress states constrained by the frictional strength of favorably oriented fault planes (frictional coefficient of 0.7). Horizontal stresses are normalized by S_v, weight of the overburden, to make the analysis depth independent.

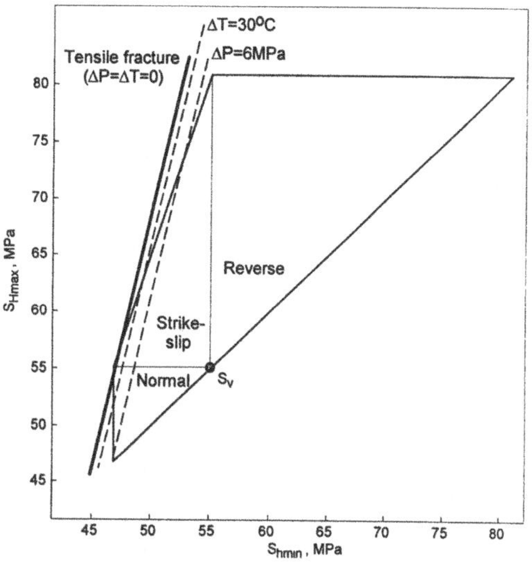

Figure 4. Magnitudes of S_{Hmax} and S_{hmin} (solid line) required to induce a tensile wall fracture in a deviated wellbore. The dashed lines represent the effect of 6 MPa of excess mud weight or 30°C of wellbore cooling (coefficient of thermal expansion of 2.4×10^{-6} °C^{-1}, Young's modulus of 19 GPa, and Poisson's ratio of 0.2).

Excess mud weight and cooling of wellbore fluid can influence the occurrence of drilling-induced tensile wall fractures. Both cause a component of tensile stress to be added to the stresses acting around the wellbore and make the formation of tensile fractures easier as illustrated in Figure 4 for the North Sea case (Fig. 1). The solid line in Figure 4 illustrates the combination of magnitudes of S_{hmin} and S_{Hmax} (azimuth of S_{Hmax} is N100°E) which could lead to the occurrence of drilling-induced tensile wall fractures in deviated section of the well at a depth of ~2750 m (inclined 35° at an azimuth of 280°) if we ignore the excess mud pressure and wellbore cooling. The dashed lines in Figure 4 represent the effect of 6 MPa of excess mud weight (Wiprut et al. 1997), or 30°C of wellbore cooling, on the likelihood of tensile cracks. The lines move into the polygon, indicating that such effects promote tensile failure. While these effects are not negligible, it is clear that the tensile fractures would not occur unless the stress regime was close to strike-slip/normal faulting.

3 FRACTURE IDENTIFICATION

In many cases, it is difficult to distinguish drilling-induced tensile wall fractures from other features on wellbore images. For example, since electrical imaging logs do not sample the entire wellbore circumference it may not be possible to distinguish the wall fractures from steeply dipping natural fractures, especially if data quality is poor. Of course, if a core is available it is easy to say whether fractures on images are natural or induced. Hydraulic fractures and axial drilling-induced tensile fractures may also have similar appearance on wellbore images but hydraulic fractures are often associated with lost circulation as they propagate away from the borehole.

To identify drilling-induced tensile wall fractures on wellbore images, one should use two basic criteria:

1. Induced fractures are discontinuous around the wellbore's circumference (to distinguish them from pre-existing natural fractures which form sinusoids on wellbore images). Also, the fractures should appear 180° apart on a wellbore image. However, this is not a necessary condition because the opposite pad of an image device may be in poor contact with the rock, making it difficult to see the opposite side of the hole. In cases where data quality is poor it is useful to note that continuous natural fractures tend to be visible at their peaks and troughs, whereas induced fractures are steeply dipping.
2. Induced fractures often form a systematic set of fractures with similar features over a depth range in en echelon "chevron" fashion, as it is indicated in Figure 2.

Figure 5 illustrates these criteria. Figure 5a shows the relevant stress components, σ_{tmin} and ω defined by Eqs. (1) and (2), around the wellbore for a stress state determined at a depth of 3213 m in the KTB borehole, Germany (Brudy & Zoback 1993, Peška & Zoback 1995b). It follows from the figure that as the fracture propagates around the wellbore the minimum stress becomes more compressive, which constrains the fracture propagation to a limited range θ_t . Thus, the critical feature that characterizes drilling-induced tensile wall fractures from naturally-occurring ones is that drilling-induced fractures cannot propagate around the entire wellbore. Figures 5b,c model the induced fractures as they would appear on wellbore images. The width of the fracture bands, θ_t , is controlled by σ_{tmin} and the inclination of fractures is governed by ω. The key point illustrated here is that a set of similar, co-planar, en echelon fractures is expected. Although a fracture can initiate at different positions around the wellbore (and different inclinations) where the rock is weak enough, such strength-controlled fractures are unlikely to form a systematic set of fractures as often detected by logging because it is necessary $\sigma_{tmin} \sim 0$ for such features to form in the first place (Fig. 5a). Even with an excess mud pressure of 15 MPa which would theoretically assist in fracture propagation (Fig. 5c), the fracture could not propa-

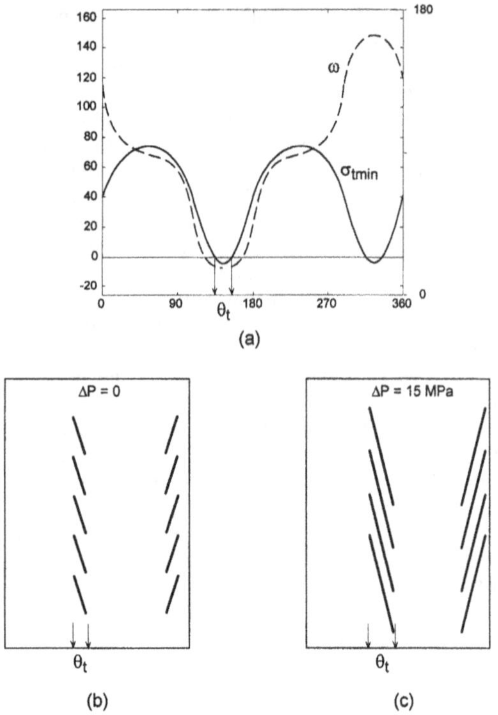

Figure 5. (a) The least tangential stress, σ_{tmin} , and inclination of σ_{tmax} from the hole axis, ω, as a function of position around the borehole and (b) the corresponding fracture pattern on borehole image. (c) The range θ_t increases and fractures tend to be more axial if excess mud weight of 15 MPa is applied.

gate all the way around the wellbore. Note that θ_t increases and the fractures tend to be more axial in this case of wellbore pressurization.

Figure 5 illustrated a case in which the angle ω was nearly constant over the range θ_t and the fracture traces were straight on the simulated wellbore image. However, there are also stress conditions under which the angle ω significantly varies. Figure 6 demonstrates that such curved fractures may even be connected by an axial fracture under favorable stress conditions (also Kuriyagawa et al. 1989). For example, if one takes a stress state determined from axial tensile fractures observed at 2830 m in the discussed Visund field, North Sea (Wiprut et al. 1997) but locally changes the azimuth of S_{Hmax} from N100°E to N130°E, Figure 6 shows the resulting pattern of tensile wall fractures. The fractures are curved and connected by a vertical fracture because ω varies over the range θ_t and reaches 0° or 180° which makes the fracture propagation stop at vertical fractures. Note that the

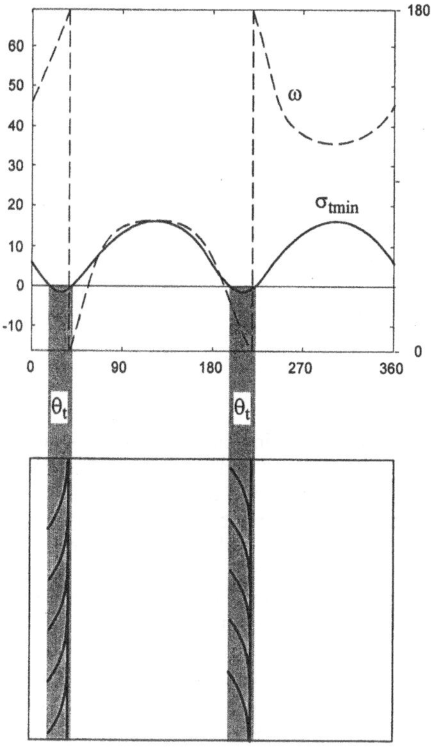

Figure 6. (a) The minimum tangential stress, σ_{tmin}, and the inclination of maximum tangential stress, ω, around the borehole and (b) the corresponding image pattern of tensile wall fractures. The fractures are curved (as ω significantly varies over the range θ_t) and are connected by an axial fracture (as ω reaches 0°/180° within θ_t).

patterns of drilling-induced tensile wall fractures appear 180° apart on the wellbore image, the opposite set of fractures has an opposite curvature. Such an additional criterion for curved fractures helps distinguish drilling-induced fractures from natural fractures which may appear in a "fish hook" shape with opposite sets of the same curvature.

4 STRESS DETERMINATION

The relationship between drilling-induced tensile wall fractures and in situ stress can be used to back-calculate unknown stress components from fracture observations. If one of the principal stresses is assumed to be vertical, four components fully describe the stress state. While the orientation of the horizontal stresses can often be found from breakouts observed in near-vertical wellbores and determination of the vertical and least horizontal principal stress magnitudes is also relatively straightforward (from density logs and leak-off/minifrac tests, respectively), the last unknown component, magnitude of S_{Hmax}, cannot be obtained by direct measurement. However, if a wellbore deviates from vertical, the magnitude of S_{Hmax} can be calculated from wellbore failures such as borehole breakouts (Zoback & Peška 1995) or drilling-induced tensile fractures (Peška & Zoback 1995b, Wiprut et al. 1997).

We developed a method for stress determination that utilizes multiple observations of tensile fractures in a number of deviated wellbores. Similar formulation was addressed by other authors (Hayashi & Abe 1984, Aadnoy 1990a,b, Okabe et al. 1996) who proposed inversion techniques based on a linearization scheme to calculate a best fit stress tensor which most closely matches data on fracture orientation from multiple inclined boreholes. However, instead of the linear inversion techniques our method is based on a grid search in the space of unknown stress components in order to avoid difficulties with strong non-linearity of equations and to keep control of calculations.

Figure 7 illustrates the proposed methodology by using a numerical experiment with drilling-induced fractures observed in five wellbores of different orientation drilled through the horizon of interest (Peška & Zoback 1997b). Poles to the fractures (crosses) and corresponding borehole orientations (open circles) in Figure 7a represent input data displayed in the lower hemisphere projection. The crosses for fracture poles are given by angles θ and ω obtained from wellbore images and the circles for boreholes are defined by angles δ and ϕ (see Fig. 2). For a general stress state, it can be demonstrated that the orientation of tensile wall fractures is a function of five stress parameters (orientation does not depend on the isotropic part of the stress field): three angles, α, β, and γ, describing orientation of the principal stresses and two parameters for relative stress magnitudes, $\Phi = (\sigma_2-\sigma_3)/(\sigma_1-\sigma_3)$ and $\Psi = \sigma_1/\sigma_3$. Since the parameter Ψ can be constrained by the weight of overburden (ratio σ_3/σ_v has to be estimated from leak-off/minifrac tests and density logs), the number of unknown stress parameters reduces to four.

In the first step (Fig. 7a), the interpretation procedure starts with an estimate of the azimuth and inclination of the least principal stress (angles α and β) as they dominate in control of fracture orientation. A reasonable estimate of σ_3 orientation is represented by a stereographic point of intersection of main circles given by fracture poles and corresponding wellbore orientation.

Secondly, Figure 7b shows an average fracture rotation (needed to bring the observed and calculated fracture planes into coincidence) as a function of σ_3 orientation in a region of 9° around the starting estimate. The values displayed for the given α and β are the minimum values taken for any combination of the remaining two parameters γ and Φ (γ is rake of σ_2 - angle between σ_2 and the horizontal measured in a plane normal to σ_3).

Finally, for the orientation of σ_3 determined in the previous steps ($\alpha = 235^\circ$, $\beta = 9^\circ$), Figure 7c shows the average fracture rotation as a function of the stress parameters γ and Φ. Blank regions indicate values inconsistent with observations (insufficient stress concentration), the small circle represents the optimal stress state ($\alpha = 235^\circ$, $\beta = 9^\circ$, $\gamma = 90^\circ$, $\Phi = 0.9$)

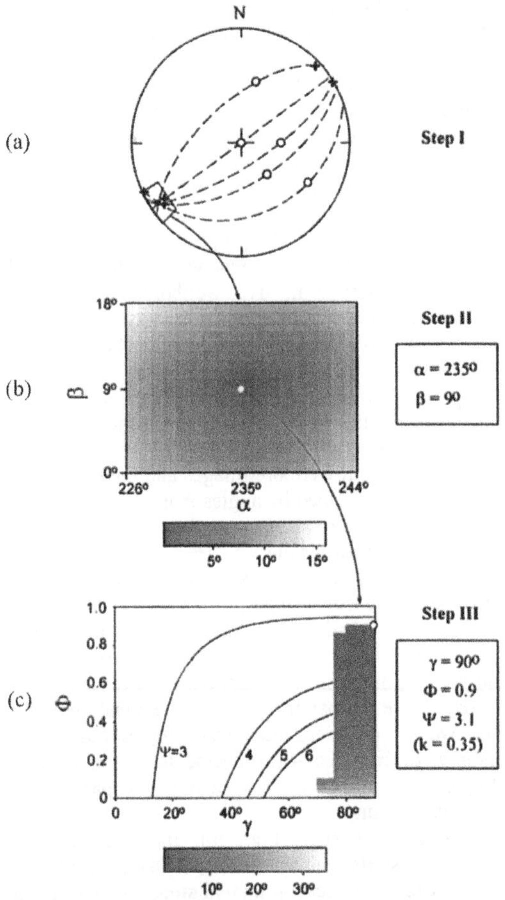

Figure 7. Interpretation of tensile wall fractures. (a) Initial graphic estimate of the orientation of σ_3 obtained from fractures observed in five boreholes. (b) Average fracture rotation as a function of σ_3 orientation. The rotations are calculated for a grid of model parameters ($\Delta\alpha = \Delta\beta = 3^\circ$, $\Delta\gamma = 5^\circ$, $\Delta\Phi = 0.05$). (c) Average fracture rotation as a function of γ and Φ for the orientation of σ_3 determined in Step II ($k = \sigma_3/\sigma_v = 0.35$, Poisson's ratio of 0.25, balanced wellbores, zero tensile strength).

for which the average fracture rotation is minimum. At this point, contours in Figure 7c indicate a value of the parameter Ψ which guarantees the prescribed weight of overburden ($\Psi = 3.1$). Since the stress parameter $\Psi = \sigma_1 / \sigma_3$ is also directly related to the coefficient of friction in the country rock, μ, by the expression

$$\Psi = (\sqrt{\mu^2 + 1} + \mu)^2, \qquad (3)$$

the contours of Ψ provide an independent control of stress models.

5 CONCLUSIONS

Understanding the interaction between in situ stress, engineering practice and wellbore failures is critical for addressing a range of wellbore stability issues and for optimizing reservoir productivity. Using the computer program SFIB we demonstrated a relationship between the stress state, drilling perturbations (excess mud weight or wellbore cooling) and formation and shapes of drilling-induced tensile wall fractures. The main criteria for identification of these fractures are: (1) They occur on opposite sides of the wellbore and do not develop around its entire circumference, (2) they often form a systematic set with similar features over a depth range, (3) they do not propagate away from the borehole. Sensitivity of the drilling-induced tensile fractures to the stress field may help in constraining the full stress tensor. We presented a method based on a grid search in four parameter space of unknown stress components and a simple graphic method to get a starting estimate for the orientation of the least principal stress which governs fracture formation. The proposed method allows to utilize (a) multiple tensile fractures observed at various depths or wellbores, (b) magnitudes of the least principal stress and the vertical stress available from independent leak-off/hydrofrac tests and density logs, and (c) knowledge of the frictional strength of the surrounding rock.

REFERENCES

Aadnoy, B. S. 1990a. Inversion technique to determine the in-situ stress field from fracturing data. *Journal of Petroleum Science and Engineering*, 4: 127-141.

Aadnoy, B. S. 1990b. In-situ stress directions from borehole fracture traces. *Journal of Petroleum Science and Engineering*, 4: 143-153.

Aadnoy, B. S. & M. E. Chenevert 1987. Stability of highly inclined boreholes. *SPE Drill. Eng.*, 2: 364-374.

Baumgärtner, J., J. Carvalho & J. McLennan 1989. Fracturing deviated boreholes: An experimental approach. In V. Maury & D. Fourmaintrauxs (eds), *Rock at Great Depth, Proceedings ISRM-SPE International Symposium.* Elf Aquitaine, Pau: Balkema.

Bradley, W. B. 1979. Failure of inclined boreholes. *J. Energy Resour. Technol.*, 101: 232-239.

Brudy, M. & M. D. Zoback 1993. Compressive and tensile failure of boreholes arbitrarily-inclined to principal stress axes: application to the KTB boreholes, Germany. *International Journal of Rock Mechanics and Mining Sciences and Geomech. Abstr.*, 30: 1035-1038.

Daneshy, A. A. 1973. A study of inclined hydraulic fractures. *Soc. Pet. Eng. J.*, 13: 61-68.

Dezayes, C., T. Villemin, A. Genter, H. Traineau & J. Angelier 1995. Analysis of fractures in boreholes of the Hot Dry Rock project at Soultz-sous-Forest (Rhine graben, France).: *Scientific Drilling,* 5: 31-41.

Hayashi, K. & H. Abe 1984. A new method for the measurement of in situ stress in geothermal fields. *J. Geothermal Res. Soc. Japan,* 6: 203-212.

Hiramastu, Y. & Y. Oka 1962. Stress around a shaft or level excavated in ground with a three dimensional stress state. *Mem. Fac. Eng. Kyoto Univ.,* 24: 56 - 76.

Hubbert, M. K. & D. G. Willis 1957. Mechanics of hydraulic fracturing. *J. Petrol. Tech.*, 9: 153-168.

Kuriyagawa, M., H. Kobayashi, I. Matsunaga, T. Yamaguchi & K. Hibiya 1989. Application of hydraulic fracturing to three-dimensional in situ stress measurement. *Int. J. Rock mech. Min. Sci. & Geomech. Abstr.*, 26: 587-593.

Moos, D. & M. D. Zoback 1990. Utilization of observations of well bore failure to constrain the orientation and magnitude of crustal stresses: Application to continental, Deep Sea Drilling Project, and Ocean Drilling Program boreholes. *Journal of Geophysical Research,* 95: 9305-9325.

Okabe, T., N. Shinohara & S. Takasugi 1996. Earth's crust stress field estimation by using vertical fractures caused by borehole drilling. In *Proceedings VIII International Symposium on the Observation of the Continental Crust Through Drilling, Tsukuba, Japan, Feb.26-28, 1996*: 265-270.

Peška, P. & M. D. Zoback 1995a. Compressive and tensile failure of inclined wellbores and determination of in situ stress and rock strength. *Journal of Geophysical Research,* 100: 12791-12811.

Peška, P. & M. D. Zoback 1995b. Observations of borehole breakouts and tensile wall-fractures in deviated boreholes: A technique to constrain in situ stress and rock strength: In *Proceedings 35th U.S. Rock Mechanics Symposium, Lake Tahoe, California:* 319-325. Balkema.

Peška, P. & M. D. Zoback 1997a. Stress and Failure of Inclines Boreholes - SFIB Version 2.1. User's Manual. Stanford University.

Peška, P. & M. D. Zoback 1997b. Constraining complete stress tensor using drilling-induced tensile fractures in inclined boreholes. In *SRB Project Report*, 63: Paper B1. Stanford University.

Richardson, R. M. 1981. Hydraulic fracture in arbitrarily oriented boreholes: An analytic approach. In *Workshop on Hydraulic Stress Measurements, Dec. 2-5 1981, National Academy Press, Washington D.C.*: 167-175.

Roegiers, J.C. & E. Detournay 1988. Consideration on failure initiation in inclined boreholes. In A. Cundall (ed), *Key Questions in Rock Mechanics*, Brookfield, Vt.: 461-469. Balkema.

Wiprut, D., M. D. Zoback, T. H. Hanssen & P. Peška 1997. Constraining the full stress tensor from observations of drilling-induced tensile fractures and leak-off tests: Application to borehole stability and sand production on the Norwegian margin. *Int. J. Rock Mech. & Min. Sci.*, 34 (3-4).

Yew, C. H. & Y. Li 1988. Fracturing of a deviated well. *SPE Prod. Eng.*, 3: 429-437.

Zoback, M. D. & J. H. Healy 1984. Friction, faulting and in-situ stress. *Annales Geophysicae*, 2: 689-698.

Zoback, M. D.& J. H. Healy 1992. In situ stress measurements to 3.5 km depth in the Cajon Pass scientific research borehole: Implications for the mechanics of crustal faulting. *Journal of Geophysical Research,* 97: 5039-5057.

Zoback, M. D. & P. Peška 1995. In situ stress and rock strength in the GBRN/DOE Pathfinder well, South Eugene Island, Gulf of Mexico. *J. Pet. Technology,* 47:5 82-585.

Zoback, M. D., R. Apel, J. Baumgärtner, M. Brudy, R. Emmermann, B. Engeser, K. Fuchs, W. Kessels, H. Rischmüller, F. Rummel & L. Vernik 1993. Upper-crustal strength inferred from stress measurements to 6 km depth in the KTB borehole. *Nature,* 365: 633-635.

Mechanics of Jointed and Faulted Rock, Rossmanith (ed.) © 1998 Taylor & Francis, ISBN 90 5410 955 6

Fractures in porous Biot medium

A.V. Bakulin
St. Petersburg State University, Geological Faculty, Department of Geophysics, Russia
L.A. Molotkov
St. Petersburg Branch of Steklov Mathematical Institute, Russia

ABSTRACT: We present a calculated seismic signatures of permeable fracture set in initially porous background medium. To address the problem we use approach where plane fractures are modeled as thin and soft Biot layers which in zero-thickness limit are equivalent to so called "linear slip interfaces". Key point is that now fractures possess some permeability and they are put in medium which is itself porous and permeable. By means of matrix averaging technique developed by the authors we construct long-wave equivalent effective model for a medium with a set of parallel fractures. Assuming Biot type of loss in host and fracture materials we analyze the velocity and attenuation behavior of ordinary longitudinal wave. We found that attenuation is most informative parameter. At low seismic frequencies inverse Q has nonlinear behavior with peak and large values along fractures whereas across fractures it has linear behavior and small values.

1 INTRODUCTION

Fractures in elastic medium could be modeled by different ways. Two major approaches suggest that they may be described as finite penny-shaped cracks (Hudson 1981) or as weakness interfaces of infinite extent (Schoenberg 1980). However for practical applications it is useful to have as small number of parameters as it is possible and yet they should have some physical meaning and should be measurable experimentally. In oil industry one more condition should be satisfied: parameters of fracture system must be somehow related to transport properties (permeability) and their anisotropy which is a key parameter in oil exploration. From this point of view penny-shaped cracks are not well suited for this modeling. The case of parallel fractures in host elastic (impermeable) medium is a little better as one can imagine some flow along fractures but still it also leads to some monophase equivalent medium.

In this paper we suggest to improve the second model by considering the host medium as porous one described by Biot model. Moreover single fracture itself is described by special boundary conditions with the jumps of displacements. The fluid can flow now through porous background medium and also along and across fractures. The use of boundary conditions is based on the fact that in elastic medium fractures are well described by so called linear slip interfaces introduced by Schoenberg (1980). This fact have been independently tested on a single fracture (Pyrak-Nolte et al. 1990) and on multiple fractures (Hsu & Schoenberg 1993; Molotkov & Bakulin 1997). Linear slip interface corresponds to the special case of non-welded boundary conditions where all stresses are continuos whereas displacements have a jumps proportional to the stresses. Physically such contact arises from putting a thin elastic layer between elastic half-spaces which is soft compared to the host medium (Schoenberg 1980; Molotkov & Bakulin 1997). If the measuring wavelength is relatively large with respect to layer thickness (thin layer) then it may be replaced by equivalent boundary conditions which are exactly linear slip boundary conditions.

In this paper we use a more straightforward approach rather than deriving these boundary conditions for porous medium which for lossless case may be found in (Bakulin & Molotkov 1997a). First we build an effective (long-wave equivalent) model of thinly-layered porous medium with two constituents and then make a passage to the limit when one type of layers is thin, soft and highly permeable compared to other. After that we discuss seismic signatures of anisotropic attenuation in derived model. Finally it is worth to note that averaging of a thinly-layered porous media was considered by Norris (1993) and Gelinsky &

Shapiro (1995). Elastic Biot moduli derived below are coincident with works of Norris (1993) and Gelinsky & Shapiro (1995). However anisotropic density operators were not obtained by mentioned authors for some reasons given further. Anisotropic densities first appear in paper by Molotkov & Bakulin (1997b) where this problem was solved for lossless case. Here anisotropic frequency-dependent densities are obtained for lossy case.

2 AVERAGING PERIODIC STRATIFIED POROUS MEDIUM

In order to obtain the equations of effective model of layered porous-porous medium we use the method of matrix averaging suggested and developed by Molotkov (1982, 1992). Isotropic Biot model is given by equations

$$\tau_{xx}=P\frac{\partial u_x}{\partial x}+F\frac{\partial u_z}{\partial z}+M\left(\frac{\partial w_x}{\partial x}+\frac{\partial w_z}{\partial z}\right),$$
$$\tau_{zz}=F\frac{\partial u_x}{\partial x}+P\frac{\partial u_z}{\partial z}+M\left(\frac{\partial w_x}{\partial x}+\frac{\partial w_z}{\partial z}\right),$$
$$-p=M\frac{\partial u_x}{\partial x}+M\frac{\partial u_z}{\partial z}+R\left(\frac{\partial w_x}{\partial x}+\frac{\partial w_z}{\partial z}\right),$$
$$\tau_{xz}=L\left(\frac{\partial u_x}{\partial z}+\frac{\partial u_z}{\partial x}\right); \tag{1}$$

$$\frac{\partial\tau_{xx}}{\partial x}+\frac{\partial\tau_{xz}}{\partial z}=\bar{\rho}\frac{\partial^2 u_x}{\partial t^2}+\rho_f\frac{\partial^2 w_x}{\partial t^2},$$
$$\frac{\partial\tau_{xz}}{\partial x}+\frac{\partial\tau_{zz}}{\partial z}=\bar{\rho}\frac{\partial^2 u_z}{\partial t^2}+\rho_f\frac{\partial^2 w_z}{\partial t^2}, \tag{2}$$
$$-\frac{\partial p}{\partial x}=\rho_f\frac{\partial^2 u_x}{\partial t^2}+m\frac{\partial^2 w_x}{\partial t^2}+b\frac{\partial w_x}{\partial t},$$
$$-\frac{\partial p}{\partial z}=\rho_f\frac{\partial^2 u_z}{\partial t^2}+m\frac{\partial^2 w_z}{\partial t^2}+b\frac{\partial w_z}{\partial t},$$

where u_x, u_z - displacements in solid phase, w_x, w_z - displacements of fluid relative to solid times porosity, τ_{xx}, τ_{zz}, τ_{xz} - total bulk stresses in porous medium, p - fluid pressure, $\bar{\rho}$ - volume averaged bulk (global) density, ρ_f - fluid density; $m=\frac{\rho_f\alpha}{\varepsilon}$, α - tortuosity of pore space, ε - porosity; $b=\nu/\kappa$, ν is a fluid viscosity, κ - permeability. For convenience we rewrite last two Eqs. as

$$-\frac{\partial p}{\partial x}=\rho_f\frac{\partial^2 u_x}{\partial t^2}+\mu\frac{\partial^2 w_x}{\partial t^2},$$
$$-\frac{\partial p}{\partial z}=\rho_f\frac{\partial^2 u_z}{\partial t^2}+\mu\frac{\partial^2 w_z}{\partial t^2}, \tag{2,a}$$

where $\mu=m+b/s$ - viscodynamic Biot operator with $s=\partial/\partial t$ which in frequency domain is written in form

$$\mu=m+i\frac{b}{\omega}. \tag{3}$$

Further assume that each period consists of two Biot layers with thicknesses $h_1=\theta_1 h$ and $h_2=\theta_2 h$, where h - thickness of period, θ_1 and θ_2 are the portions of first and second constituent in period. This medium consisting total of n periods occupies a region from 0 to $H=nh$ along vertical z-axis. Boundary conditions between layers are that of welded contact with free flow across interfaces. If we use integral representation for a wave field

$$\begin{pmatrix}u_x\\ w_x\\ \tau_{xz}\end{pmatrix}=-\int_0^\infty\frac{\sin kx}{2\pi i}dk\int_{\sigma-i\infty}^{\sigma+i\infty}\begin{pmatrix}U_x\\ W_x\\ T_{xz}\end{pmatrix}e^{k\tau\eta}d\eta,$$

$$\begin{pmatrix}u_z\\ w_z\\ \tau_{xx}\\ \tau_{zz}\\ p\end{pmatrix}=\int_0^\infty\frac{\cos kx}{2\pi i}dk\int_{\sigma-i\infty}^{\sigma+i\infty}\begin{pmatrix}U_z\\ W_z\\ kT_{xx}\\ kT_{zz}\\ -kP\end{pmatrix}e^{k\tau\eta}d\eta, \tag{4}$$

then we obtain a system of ordinary differential equations with respect to z

$$\frac{d}{dz}\begin{pmatrix}U_x\\ T_{zz}\\ P\\ T_{xz}\\ U_z\\ W_z\end{pmatrix}=k\chi\begin{pmatrix}U_x\\ T_{zz}\\ P\\ T_{xz}\\ U_z\\ W_z\end{pmatrix},\quad \chi=\begin{pmatrix}0 & G\\ B & 0\end{pmatrix},$$

$$G=\begin{pmatrix}\frac{1}{L} & -1 & 0\\ 1 & \bar{\rho}\eta^2 & \rho_f\eta^2\\ 0 & \rho_f\eta^2 & \mu\eta^2\end{pmatrix}, \tag{5}$$

$$B=\begin{pmatrix} \dfrac{\bar{\rho}\mu-\rho_f^2}{\mu}+\dfrac{D}{d} & -\dfrac{FR-M^2}{PR-M^2} & \dfrac{\rho_f}{\mu}-\dfrac{2LM}{PR-M^2} \\ \dfrac{FR-M^2}{PR-M^2} & \dfrac{R}{PR-M^2} & -\dfrac{F}{PR-M^2}^{2} \\ -\dfrac{\rho_f}{\mu}+\dfrac{2LM}{PR-M^2} & \dfrac{F}{PR-M^2} & \dfrac{1}{\mu\eta^2}+\dfrac{P}{PR-M^2} \end{pmatrix}$$

Here χ is so called matrix of layer. Averaging of periodic medium now reduces to simple volume averaging of this matrix (Molotkov 1992)

$$\bar{\chi}=\theta_1\chi_1+\theta_2\chi_2, \tag{6}$$

so all elements of χ should be replaced by their volume averaged analogues in $\bar{\chi}$. After back transformations we obtain the following equations of Biot-like medium

$$\tau_{xx}=\tilde{P}\frac{\partial u_x}{\partial x}+\tilde{F}\frac{\partial u_z}{\partial z}+\tilde{M}\left(\frac{\partial w_x}{\partial x}+\frac{\partial w_z}{\partial z}\right),$$

$$\tau_{zz}=\tilde{F}\frac{\partial u_x}{\partial x}+\tilde{C}\frac{\partial u_z}{\partial z}+\tilde{Q}\left(\frac{\partial w_x}{\partial x}+\frac{\partial w_z}{\partial z}\right),$$

$$-p=\tilde{M}\frac{\partial u_x}{\partial x}+\tilde{Q}\frac{\partial u_z}{\partial z}+\tilde{R}\left(\frac{\partial w_x}{\partial x}+\frac{\partial w_z}{\partial z}\right),$$

$$\tau_{xz}=\tilde{L}\left(\frac{\partial u_x}{\partial z}+\frac{\partial u_z}{\partial x}\right); \tag{7}$$

$$\frac{\partial\tau_{xx}}{\partial x}+\frac{\partial\tau_{xz}}{\partial z}=\tilde{\rho}_x\frac{\partial^2 u_x}{\partial t^2}+\tilde{\rho}_{fx}\frac{\partial^2 w_x}{\partial t^2},$$

$$\frac{\partial\tau_{xz}}{\partial x}+\frac{\partial\tau_{zz}}{\partial z}=\tilde{\rho}_z\frac{\partial^2 u_z}{\partial t^2}+\tilde{\rho}_{fz}\frac{\partial^2 w_z}{\partial t^2},$$

$$-\frac{\partial p}{\partial x}=\tilde{\rho}_{fx}\frac{\partial^2 u_x}{\partial t^2}+\tilde{\mu}_x\frac{\partial^2 w_x}{\partial t^2},$$

$$-\frac{\partial p}{\partial z}=\tilde{\rho}_{fz}\frac{\partial^2 u_z}{\partial t^2}+\tilde{\mu}_z\frac{\partial^2 w_z}{\partial t^2}, \tag{8}$$

where

$$\tilde{\rho}_z=\bar{\bar{\rho}},\ \tilde{\rho}_x=\overline{\left(\frac{\rho\mu-\rho_f^2}{\mu}\right)}+\overline{\left(\frac{\rho_f}{\mu}\right)}^2\overline{\left(\frac{1}{\mu}\right)}^{-1}, \tag{9,a}$$

$$\tilde{\rho}_{fx}=\overline{\left(\frac{\rho_f}{\mu}\right)}\,\overline{\left(\frac{1}{\mu}\right)}^{-1},\ \tilde{\rho}_{fz}=\bar{\rho}_f,\ \tilde{\mu}_x=\overline{\left(\frac{1}{\mu}\right)}^{-1},\ \tilde{\mu}_z=\bar{\mu},$$

$$\tilde{P}=\overline{\left(\frac{FR-M^2}{d}\right)}\tilde{F}+\overline{\left(\frac{2LM}{d}\right)}\tilde{M}+\overline{\left(\frac{D}{d}\right)}, \tag{9,b}$$

$$\tilde{F}=\frac{1}{f}\left\{\overline{\left(\frac{P}{d}\right)}\,\overline{\left(\frac{FR-M^2}{d}\right)}+\overline{\left(\frac{M}{d}\right)}\,\overline{\left(\frac{2LM}{d}\right)}\right\},$$

$$\tilde{M}=\frac{1}{f}\left\{\overline{\left(\frac{M}{d}\right)}\,\overline{\left(\frac{FR-M^2}{d}\right)}+\overline{\left(\frac{R}{d}\right)}\,\overline{\left(\frac{2LM}{d}\right)}\right\},$$

$$\tilde{C}=\frac{1}{f}\overline{\left(\frac{C}{d}\right)},\ \tilde{Q}=\frac{1}{f}\overline{\left(\frac{Q}{d}\right)},\ \tilde{R}=\frac{1}{f}\overline{\left(\frac{R}{d}\right)},\ \tilde{L}=\overline{\left(\frac{1}{L}\right)}^{-1},$$

$$f=\overline{\left(\frac{P}{d}\right)}\,\overline{\left(\frac{R}{d}\right)}-\overline{\left(\frac{M}{d}\right)}^2,\ d=PR-M^2,$$

$$D=\begin{vmatrix} P & F & M \\ F & P & M \\ M & M & R \end{vmatrix}.$$

Obtained formulas are valid for any type of viscodynamic operator. So density operator μ could be defined by any other expression than (3).

This problem was considered first by Norris (1993) and Gelinsky & Shapiro (1995). Elastic moduli from (9,b) are coincident with them however densities expressions (9,a) are different as we did not neglect any terms in layer matrix as did previously mentioned authors.

3 BIOT TYPE OF VISCODYNAMIC OPERATOR

Let us analyze derived model for Biot type of viscodynamic operator (3). Everything below refers to this particular type of loss. Taking into account (3) one can see that densities $\tilde{\rho}_x$, $\tilde{\rho}_{fx}$, $\tilde{\mu}_x$, $\tilde{\mu}_z$ become operators. In frequency domain they appear to be

$$\tilde{\rho}_x=\bar{\rho}_x+\frac{A}{D-i\omega},\ \tilde{\rho}_{fx}=\bar{\rho}_{fx}+\frac{B}{D-i\omega}, \tag{10}$$

$$\tilde{\mu}_z=m_z+i\frac{b_z}{\omega},\ \tilde{\mu}_x=m_x+i\frac{b_x}{\omega}+\frac{C}{D-i\omega},$$

with all the parameters given below

$$\bar{\rho}_x=\overline{\left(\frac{\bar{\rho}m-\rho_f^2}{m}\right)}+\overline{\left(\frac{\rho_f}{m}\right)}^2\overline{\left(\frac{1}{m}\right)}^{-1},\ \bar{\rho}_{fx}=\overline{\left(\frac{\rho_f}{m}\right)}\,\overline{\left(\frac{1}{m}\right)}^{-1},$$

$$m_z=\bar{m},\ b_z=\bar{b},\ m_x=\overline{\left(\frac{1}{m}\right)}^{-1},\ b_x=\overline{\left(\frac{1}{b}\right)}^{-1}, \tag{10,a}$$

$$A=\frac{\theta_1\theta_2(\rho_{f1}-\rho_{f2})^2(\theta_1 b_2+\theta_2 b_1)}{(\theta_1 m_2+\theta_2 m_1)^2},$$

$$B=\frac{\theta_1\theta_2(\rho_{f1}-\rho_{f2})(m_1 b_2-m_2 b_1)}{(\theta_1 m_2+\theta_2 m_1)^2},$$

$$C=\frac{\theta_1\theta_2\,(m_1b_2-m_2b_1)^2}{(\theta_1m_2+\theta_2m_1)^2(\theta_1b_2+\theta_2b_1)},\quad D=\frac{(\theta_1b_2+\theta_2b_1)}{(\theta_1m_2+\theta_2m_1)}.$$

These equations correspond to some homogeneous two-phase medium, which may be called generalized transversely isotropic Biot model. This generalization concerns only equilibrium equations (8) which in time domain are given as

$$\frac{\partial\tau_{xx}}{\partial x}+\frac{\partial\tau_{xz}}{\partial z}=\bar{\rho}_x\frac{\partial^2u_x}{\partial t^2}+A\int_0^t e^{-D(t-\tau)}\frac{\partial^2u_x}{\partial\tau^2}d\tau+$$
$$+\bar{\rho}_{fx}\frac{\partial^2w_x}{\partial t^2}+B\int_0^t e^{-D(t-\tau)}\frac{\partial^2w_x}{\partial\tau^2}d\tau,$$
$$\frac{\partial\tau_{xz}}{\partial x}+\frac{\partial\tau_{zz}}{\partial z}=\bar{\rho}_z\frac{\partial^2u_z}{\partial t^2}+\tilde{\rho}_{fz}\frac{\partial^2w_z}{\partial t^2},\qquad(11)$$
$$-\frac{\partial p}{\partial x}=\bar{\rho}_{fx}\frac{\partial^2u_x}{\partial t^2}+B\int_0^t e^{-D(t-\tau)}\frac{\partial^2u_x}{\partial\tau^2}d\tau+m_x\frac{\partial^2w_x}{\partial t^2}+$$
$$+b_x\frac{\partial\,w_x}{\partial t}+C\int_0^t e^{-D(t-\tau)}\frac{\partial^2w_x}{\partial\tau^2}d\tau,$$
$$-\frac{\partial p}{\partial z}=\tilde{\rho}_{fz}\frac{\partial^2u_z}{\partial t^2}+m_z\frac{\partial^2w_z}{\partial t^2}+b_z\frac{\partial\,w_z}{\partial t}.$$

Three main distinctions should be noted compared to ordinary transversely isotropic Biot model:

1) $\bar{\rho}$ and ρ_f become anisotropic (have different values along and across layers);
2) $\tilde{\rho}_x$, $\tilde{\rho}_{fx}$ become frequency dependent operators;
3) viscodynamic operator $\tilde{\mu}_x$ acquires additional new term.

At low enough frequencies these distinctions do not change phase velocities as they depend only on $\tilde{\rho}_x(0)=\tilde{\rho}_z$. However they do change low-frequency attenuation considerably.

4 PERMEABLE SET OF FRACTURES

It is well known that fractures may be represented by thin elastic layers with small elastic moduli (Hsu & Schoenberg 1993; Molotkov & Bakulin 1997). However in Biot model in order to govern by a fluid flow across interfaces it is necessary to consider a fracture as thin and soft Biot layer. Let us assume that all elastic moduli of second layer in period are proportional to its volume fraction θ_2 which is assumed to be small ($\theta_2\to 0$). For example $P_2=P^*\theta_2$ etc. By this passage to the limit we immediately obtain from (9,b) that

$$\tilde{P}=\frac{F_bR_b-M_b^{\,2}}{d_b}\tilde{F}+\frac{2L_bM_b}{d_b}\tilde{M}+\frac{D_b}{d_b},$$
$$\tilde{F}=\frac{F_bR_b-M_b^{\,2}}{d_b}\tilde{C}+\frac{2L_bM_b}{d_b}\tilde{Q},$$
$$\tilde{M}=\frac{F_bR_b-M_b^{\,2}}{d_b}\tilde{Q}+\frac{2L_bM_b}{d_b}\tilde{R},\qquad(16,\text{a})$$
$$\tilde{C}=\frac{1}{f}\left(\frac{P_b}{d_b}+\frac{P^*}{d^*}\right),\tilde{Q}=\frac{1}{f}\left(\frac{M_b}{d_b}+\frac{M^*}{d^*}\right),$$
$$\tilde{R}=\frac{1}{f}\left(\frac{R_b}{d_b}+\frac{R^*}{d^*}\right),\tilde{L}=\left(\frac{1}{L_b}+\frac{1}{L^*}\right)^{-1},$$
$$f=\left(\frac{P_b}{d_b}+\frac{P^*}{d^*}\right)\left(\frac{R_b}{d_b}+\frac{R^*}{d^*}\right)-\left(\frac{M_b}{d_b}+\frac{M^*}{d^*}\right)^2,$$
$$d^*=P^*R^*-M^{*2},$$

where P_b, F_b, R_b, M_b, L_b are Biot elastic moduli of background medium corresponding to the thick layer material, P^*, R^*, M^*, L^* are coefficients described elastic moduli of thin and soft layer representing a fracture.

More important property of fractures is their high permeability. By permeable fractures we mean much more less flow resistance inside fractures compared to background medium. So for a material of fractures permeability is much larger than in host medium $\kappa_2=\kappa_2^*/\theta_2$ and hence $b_2=b_2^*\theta_2$. Assuming that host medium and fractures are filled with the same fluid we obtain following effective densities in frequency domain

$$\tilde{\rho}_x=\tilde{\rho}_z=\bar{\rho}_b,\ \tilde{\rho}_{fx}=\tilde{\rho}_{fz}=\rho_{bf},$$
$$\mu_z=m_b+i\frac{b_b}{\omega},\ \mu_x=m_b+i\frac{b_x}{\omega}+\frac{C}{D-i\omega},\qquad(16,\text{b})$$
$$b_x=\frac{b_bb_2^*}{(b_b+b_2^*)},C=\frac{b_b^2}{(b_b+b_2^*)},D=\frac{\theta_2(b_b+b_2^*)}{m_2}.$$

Imaginary part of μ_x has two extremes if $C>4b_x$. This causes two instead of one maximums in attenuation of first P-wave along fractures. To analyze the influence of fractures let us consider a concrete example. Parameters of host medium, fracture and effective medium all together are given in Table 1.

Compliant properties of fractures are taken to be close to elastic non-permeable case. Velocities at

Table 1. Properties of constituents and effective model.

	Portion	Porosity	$\bar{\rho}$, kg/m^3	ρ_f, kg/m^3	Tortuosity	$m_x = m_z$, kg/m^3	$\kappa_x \times 10^{-12}$ m^2	$\kappa_z \times 10^{-12}$ m^2	ν, Pa·sec
Host	0.99	0.15	2400	1000	1.8	12000	0.1	0.1	0.001
Fracture	0.01	0.5	1825	1000	1.8	3600	1000	1000	0.001
Effective	-	0.15	2400	1000	1.8	12000	10	0.1	0.001

	P, GPa	C, GPa	F, GPa	M, GPa	Q, GPa	R, GPa	L, GPa
Host	45.6	45.6	5	8.5	8.5	12.5	20.3
Fracture	2.2	2.26	1.06	1.66	1.66	2.37	0.6
Effective	45	35	5	8.18	8.24	12	15.3

	A, kg/sec·m^3	B, kg/sec·m^3	C × 10^9, kg/sec·m^3	D × 10^4, 1/sec	V_{P1X}, m/sec	V_{P1Z}, m/sec	V_{SX}, m/sec	V_{SZ}, m/sec
Host	0	0	0	0	4360	4360	2900	2900
Fracture	0	0	0	0	1110	1110	575	575
Effective	0	0	9.35	2.7	4330	3830	2525	2525

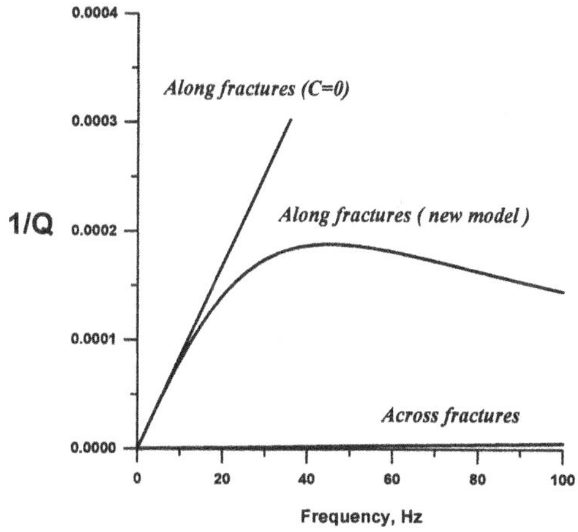

Figure 1. Effect of frequency. Q^{-1} curves along fractures (X) for a first P-wave with fracture permeability 1000 · 10^{-12} m^2 . For comparison shown Q^{-1} curve across fractures. Straight upper line corresponds to the case when additional term in $\tilde{\mu}_x$ is neglected and only Biot terms are retained (C=0).

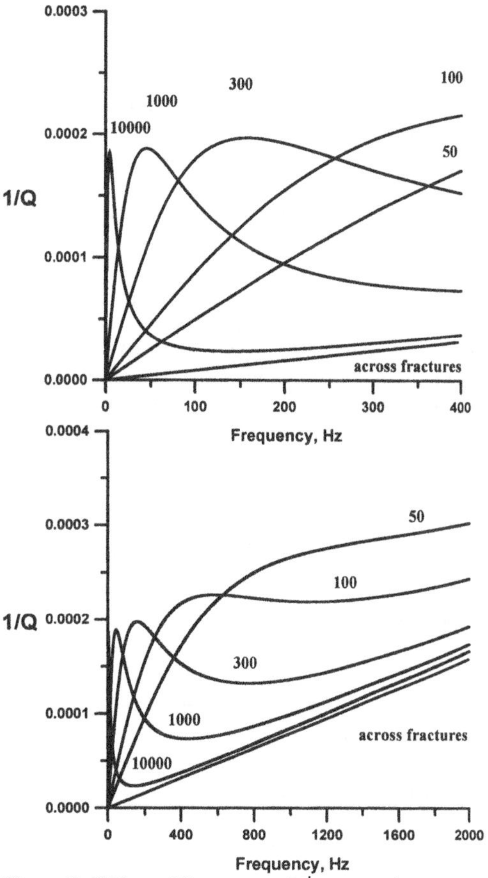

Figure 2. Effect of frequency. Q^{-1} curves along fractures for a first P-wave with fracture permeability 10000, 1000, 300, 100 and 50 in units of 10^{-12} m^2 . All other parameters are given in the table. For comparison shown Q^{-1} curve across fractures for κ_2 =1000·10^{-12} m^2 .

zero frequency are given on Table 1. One can see that P-wave anisotropy is about 12 %. Velocities at low frequency do not depend on A, B, C and D. The most interesting is attenuation behavior.

Attenuation curves for a first P-wave are shown on Figure 1. Three main things should be noted in this graph:

1) Additional maximum occurs at seismic frequency about 40 Hz along fractures (new model);

2) If the additional term $\dfrac{C}{D - i\omega}$ in μ_x is neglected then attenuation does not have a second peak at

seismic frequency and it's value is highly overestimated (straight line with C=0 on Figure 1);
3) Attenuation along and across fractures is different both in behavior (linear – across, nonlinear with peak – along) and in values (much larger along than across).

How does the location of maximum and it's value depend on fracture permeability? One can see on Figure 2 that as fracture permeability increase the maximum goes closer to zero and it's value slightly decreases.

Therefore position of the peak brings the information about fracture set permeability.

5 CONCLUSIONS

Fine layering of poroelastic constituents produces both anisotropic elastic moduli and anisotropic density operators. The expressions for these operators are given which are valid for all frequencies and for any mechanism of loss. If we assume Biot mechanism of loss then at low frequencies moduli anisotropy affects only velocities whereas the density anisotropy affects only attenuation.

Thin and soft Biot layer with high permeability is used as a simple model of fracture in host porous medium. Medium intersected by a set of such highly permeable fractures possesses a very specific attenuation behavior along fractures: inverse Q has additional (to high frequency Biot peak) peak at seismic frequencies (i.e. nonlinear behavior). Across fractures attenuation behavior is linear but it's values are one to two orders of magnitude lower.

Obtained results incorporated only Biot mechanism of loss however any other mechanisms could be easily included to find their influence on seismic properties of finely layered rocks.

REFERENCES

Backus, G.E. 1962. Long-wave elastic anisotropy produced by horizontal layering. *J.Geophys.Res.* 67: 4427-4440.

Bakulin, A.V. & L.A.Molotkov 1998. Effective models of thinly-stratified media containing porous, elastic and fluid layers. *J.Seismic Exploration.*7(1) (in press).

Bakulin, A.V. & L.A.Molotkov 1997a. Poroelastic medium with fractures as limiting case of stratified poroelastic medium with thin and soft Biot layers. *67th Ann.Intern.Mtg. SEG. Exp.Abstr.*:1001-1004.

Bakulin, A.V. & L.A.Molotkov 1997b. Generalized anisotropic Biot model as effective model of stratified poroelastic medium. *59th EAGE Conference and Technical Exhibition: Extended Abstracts.* Geneva, Switzerland. Paper P055.

Gelinsky, S. & S.Shapiro 1995. Poroelastic effective media model for fractured and layered reservoir rock. *65th Ann. Intern.Mtg. SEG. Exp. Abstr.*: 922-955.

Hsu, C.-J. & M.Schoenberg 1993. Elastic wave through a simulated fractured medium. *Geophysics.* 58: 964-977.

Molotkov, L.A. 1982. On equivalence of periodically layered and transversely isotropic media. *Journal of Soviet Mathematics.* 19(4).

Molotkov, L.A. 1992. New method of deriving the equations of effective models of periodic media. *Journal of Soviet Mathematics.* 62(6).

Molotkov, L.A. & A.V.Bakulin 1997. An effective model of a fractured medium with fractures modeled by the surfaces of discontinuity of displacements. *Journal of Mathematical Sciences.* 86(3): 2735-2746.

Norris, A. 1993. Low-frequency dispersion and attenuation in partially saturated rocks. *J.Acoust. Soc.Am.* 94: 359-370.

Pyrak-Nolte L.J., L.R. Myer & N.G.W. Cook 1990. Transmission of seismic waves across single natural fractures. *J.Geophys.Res.*95(B6):8617-8638.

Schoenberg, M. 1980. Elastic wave behavior across linear slip interfaces. *J.Acoust.Soc.Am.* 68: 1516-1521.

Mechanics of Jointed and Faulted Rock, Rossmanith (ed.) © 1998 Taylor & Francis, ISBN 90 5410 955 6

Two technologies of rock samples cutting: Their effect on samples strength properties

P. Konečný Jr & L. Sitek
Institute of Geonics, Academy of Sciences of the Czech Republic, Ostrava, Czech Republic

ABSTRACT: Cubic rock samples of the sandstone and the granite were prepared by two technologies: the technology of high velocity abrasive water jet and the classic cutting with diamond disc cutter. The strength of samples prepared by both technologies was measured and the results were compared.

1 INTRODUCTION

At present intense development as well as modernisation of laboratory methods of measurement of the rock properties always occur. Higher demands are connected with this trend both upon instrumentation of laboratories, computer quality and software outfit as well as upon precise preparing of rock samples tested.

New synthetic materials there are used in geological engineering and it is necessary to know their properties. However, it is problem to cut the samples for laboratory tests, some times. The technology of cutting of materials by high velocity abrasive water jet enables equally shaping of rock samples for needs of laboratory tests equally as cutting by diamond or hard metal disc cutters (Figure 1).

2 CUTTING OF THE SAMPLES USING HIGH VELOCITY ABRASIVE WATER JET

The principle of disintegration by high velocity water jet is based on high energy (stored in the jet) transmission to extremely small area. The material destruction is caused afterwards by complicated physical processes during jet impact (Hood et al. 1990, Hlaváč 1992).

The problems connected to liquid jet involve wide range of up to now realized applications, probably there is an area for new unknown solution as well. It is caused by wide variety of the jets: various liquid pressure upsteam from the nozzle exit, various liquid flow, the use of pure liquid, colloid or non-colloid lotion, the ingredient of abrasive particles, the jet can be generated as continuous, pulsed or cavitating jet etc. In addition the liquid jet technology is "cold technology". It means that the cut surface of material to be cut has no thermal interference.

The use of continuous water jet with add-on abrasive particles (so-called abrasive jet) seems to be the best configuration for our purpose. Abrasive particles are sucked by means of underpressure to mixing chamber. The particles are mixed with the jet here and abrasive jet come into existence after the mixture passing through so-called abrasive tube. Abrasive jet is suitabe for cutting not only conventional engineering and construction materials (metals, glass, rocks and concretes) but new developed materials (ceramics, ceramic babbitt, babbitt with metal matrix, fibreglass etc.) as well.

High pressure pump based on multiplicator FLOW with operating pressure up to 380 MPa was used as the source of high pressure water during cutting. The movement of abrasive jet above rock sample supported six degree of freedom industrial robot or X-Y cutting table, respectively.

The cutting of sandstone samples was performed using following parameters of the abrasive jet: water nozzle dia 0.25 mm, water pressure upstream from the nozzle exit 300 MPa, abrasive tube dia 1.2 mm, stand-off distance 3 mm and cutting velocity 0.6 $mm.s^{-1}$. Czech garnet type GBK (MASH 80) was used as abrasive material, abrasive flow rate was 300 $g.min^{-1}$.

The granite was cut using following parameters of the jet: water nozzle dia 0.33 mm, water pressure 350 MPa, abrasive tube dia 1 mm, stand-off distance 3 mm, cutting velocity 0.5 $mm.s^{-1}$, abrasive material: garnet MASH 120, abrasive flow rate 500 $g.min^{-1}$.

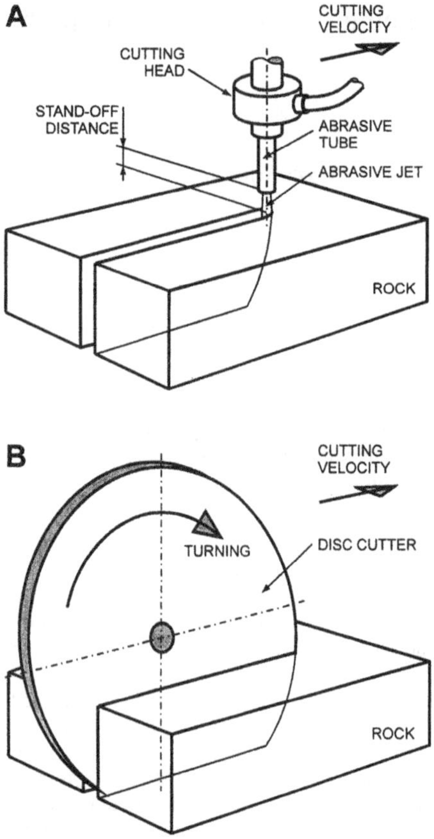

Figure 1. Scheme of the both types of rock samples cutting (A - high velocity abrasive water jet, B - disc cutter)

3 RESULTS

For comparing measurements cubic samples of 8 cm cube edge were applied which were shaped from homogenous Godula sandstone (sample No. 5107) and granite (sample No. 5336) from Žulová (Czech Republic). Density as well as velocity of ultrasonic wave propagation values are mentioned in Table 1. The first four test pieces from both materials were shaped by the above mentioned method of application of high velocity abrasive water jet and the subsequent four specimens of sandstone and granite were cut out of 8 cm high sandstone slabs so that the method of cutting of bottom bearing surfaces (i.e. surfaces perpendicular to force action) was identical in both test series.

Table 1. Basic physical characteristics of the samples.

No.	Density $[kg \cdot m^{-3}]$	Ultrasonic waves velocity $[km \cdot s^{-1}]$
5107/1 (water jet)	2464	3,3
5107/2 (water jet)	2453	3,3
5107/3 (water jet)	2479	3,4
5107/4 (water jet)	2426	3,3
5107/A (dia.-disc)	2474	3,3
5107/B (dia.-disc)	2470	3,2
5107/C (dia.-disc)	2465	3,2
5107/D (dia.-disc)	2458	3,2
5336/1 (water jet)	2583	4,7
5336/2 (water jet)	2602	7,8
5336/3 (water jet)	2592	4,7
5336/4 (water jet)	2586	4,7
5336/A (dia.-disc)	2564	4,8
5336/B (dia.-disc)	2607	4,9
5336/C (dia.-disc)	2589	4,9
5336/D (dia.-disc)	2598	4,8

After the measurement of density and ultrasonic wave velocity simple compression strength of samples as well as Young modulus by mechanical press ZWICK (max. force of 600 kN) were measured on samples. The results mentioned in Table 2 show that no distinct affecting of results has occured during preparing of samples by the one or other methods. Due to the fact that only natural and no synthetic material is concerned, the results are affected more likely by properties of samples then by the method of cutting.

Table 2. Strength & strain properties.

No.	Strength [MPa]	Young modulus [MPa]
5107/1 (water jet)	102	35 000
5107/2 (water jet)	118	37 000
5107/3 (water jet)	129	41 000
5107/4 (water jet)	103	26 000
5107/A (dia.-disc)	117	44 000
5107/B (dia.-disc)	121	43 000
5107/C (dia.-disc)	76	8 000
5107/D (dia.-disc)	115	40 000
5336/1 (water jet)	148	53 000
5336/2 (water jet)	160	60 000
5336/3 (water jet)	131	43 000
5336/4 (water jet)	144	45 000
5336/A (dia.-disc)	129	39 000
5336/B (dia.-disc)	130	42 000
5336/C (dia.-disc)	146	56 000
5336/D (dia.-disc)	145	55 000

4 CONCLUDING REMARKS

The above mentioned results have shown that the rock samples for laboratory experiments prepared by abrasive jet cutting are similar like the rock samples prepared by conventional method of diamond disc cutting.

The importance of preparing samples from rocks and rock like materials by means of high velocity abrasive water jet lies especially in ability to shape materials which are very difficult or impossible to cut by a diamond or hard metal disc cutter. As examples geo-composites can be mentioned - i.e. mixtures of synthetic materials with rock (epoxide resins + sand, polyurethane + gravel etc.), or various kinds of concrete or reinforced concrete, and very hard and abrasive rocks, too.

Just in this area the application of cutting by means of high velocity abrasive water jet appears as the only applicable method which allows how to prepare quickly and high quality way samples for laboratory tests.

ACKNOWLEDGEMENT

The authors wish to thank the Grant Agency of Czech Republic for support of the work presented in this paper by the projects No. 105/96/1531 and RA K1042603 respectively.

REFERENCES

Hood M., R. Nordlund & E. Thimons 1990. A Study of Rock Erosion using High - Pressure Water Jets. *Int. J. Rock Mech. Min. Sci. & Geomech. Abstr.* Vol. 27, No. 2.

Hlaváč L. 1992. Physical description of high energy liquid jet interaction with material. *Geomechanics 91*: pp. 341 - 346, Z. Rakowski (ed.), Balkema, Rotterdam.

Mechanics of Jointed and Faulted Rock, Rossmanith (ed.) © 1998 Taylor & Francis, ISBN 90 5410 955 6

Study of scale effects on the deformability modulus of rock by means of a discontinuous model

F. Re & C. Scavia
Department of Structural Engineering, Politecnico di Torino, Italy

ABSTRACT : The aim of this article is to investigate the effects of scale phenomena on the deformability modulus through an analysis of the mechanical behaviour of specimens of different sizes containing various geometric discontinuity configurations. In order to determine the modulus of deformability of a specimen, a uniaxial compressive test is simulated with the aid of a numerical computation code, which is based on fracture mechanics. In particular, the code was applied to a renormalisation process: as their size increases, the specimens display larger sized cracks but they are seen to be made of a material with a reduced deformability modulus on account of the presence of smaller sized cracks whose value is known from the analysis of smaller specimens. This approach makes it possible to show that the modulus decreases with increasing specimen size with a linear evolution on a semi-logarithmic scale.

1 INTRODUCTION

From both the macroscopic and microscopic point of view, rock masses are essentially non homogeneous and discontinuous media. This aspect is a major cause of the variations that are seen to occur in the mechanical characteristics of rock when shifting from small scale to large scale observations (laboratory rock materials as opposed to rock masses).

In particular, the deformability modulus of rock masses cannot be determined in the laboratory on small sized specimens for use in large scale analyses (e g tunnel excavations), due to the presence in rock masses of discontinuities of different orders of magnitude giving rise to variations in the modulus as a function of specimen size. Some experimental investigations, in fact, have demonstrated that the modulus varies as a function of scale (see, for instance Cunha et al. 1990, Jackson et al. 1990, Gomes 1995). Furthermore, according to some authors, it is possible to define a characteristic volume, referred to as REV (Representative Elementary Volume) above which no appreciable variations take place (Cunha 1990). At present, however, no exhaustive interpretation of scale effects on the deformability modulus has yet been provided and for engineering purposes empirical laws are used linking the quality indices of the rock mass (RMR and Q) with the value of the in-situ deformability modulus (see for instance, Bieniawski 1976, Barton et al. 1974).

The aim of this article is to examine the deformability of large rock volumes and to study scale effects through an analysis of the mechanical behaviour of specimens of different sizes containing various geometric discontinuity configurations. This was done by resorting to a numerical computation code, based on the concepts of fracture mechanics, to analyse rock specimens of increasingly large dimensions. The behaviour of rock masses with a great number of cracks is simulated with reference to a renormalisation process (Creswick et al. 1991)

2. DESCRIPTION OF THE METHODOLOGY

The proposed calculation method is based on the numerical analysis of the mechanical behaviour of rock volumes containing natural discontinuities, simulated as flat cracks.

To this end we have referred to the SOFTEN computation code (Scavia 1995), which is based on the Displacement Discontinuity Method and the concepts of fracture mechanics.

The discontinuities are represented by means of a set of closed displacement discontinuity elements, designed so as to try to simulate crack behaviour in compressive stress fields, whose main characteristics are summarised below :

•displacement discontinuities in the normal direction are nil. This means that no overlapping occurs and that normal stresses are trasmitted through the element without any modification,

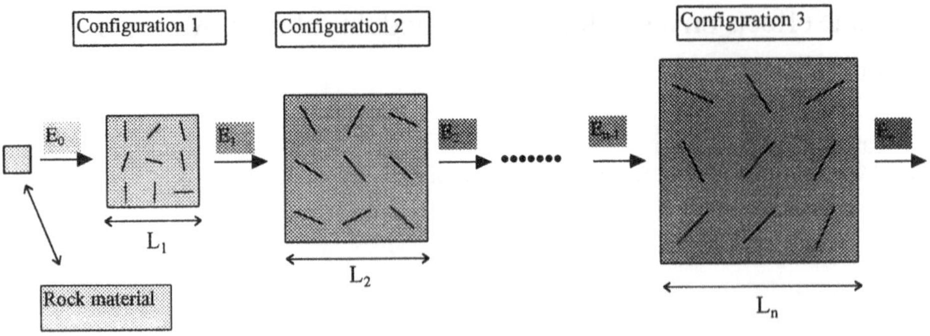

Figure 1, Model of the rock mass obtained through a renormalisation process.

•in the tangential direction, displacements are computed by taking into account the development of frictional resistance on the crack surfaces. With regard to this, an elastic-ideally plastic behaviour of the crack surfaces is assumed.

Thanks to this calculation method it is possible to simulate the behaviour in uniaxial compression of rock volumes containing closed cracks: this yields the stress-strain curves necessary for the determination of the modulus. From the computational standpoint, however, it becomes impossible to simulate the behaviour of large rock volumes containing many cracks of different sizes. To overcome this difficulty, a renormalisation process, as described below, is applied.

With reference to figure 1, in a rock mass it is possible to identify small zones, with sides=L_1, that contain microcracks and where the rock material, among them, is characterised by a deformability modulus, E_0. By expanding the field of observation, we may consider these zones as limited portions of bigger volumes, with side=L_2, where bigger cracks can be discerned. This observation process can be extended to a very large scale (in figure 1. side=L_n) corresponding to the size of the entire rock mass.

Again with reference to figure 1, knowing the value of the deformability modulus of the uncracked material (E_o), it proves possible to work out the modulus for the L_1 sided volume (E_1) by applying the computation code to simulate uniaxial compressive tests. The deformability modulus is determined by taking into account the inclination of the portion of the linear stress-strain curve up to the level triggering crack propagation.

Having determined E_1, it becomes possible to work out the modulus for the greater volume with side=L_2. In this case, the material between the discontinuities is not characterised by a modulus E_o, but rather by modulus E_1, previously determined and allowing for the presence of the smaller cracks that are not represented on this scale. This process is iterated until we reach sizeable volume dimensions, in order to determine the deformability modulus on large rock structures (with side=L_n). To simplify the problem, the following assumptions concerning each of the configurations analysed have been made:

•the rock volumes are invariably assumed to be square shaped;

• the cracks are all the same length,

• the dip of the cracks is generated through a random process in the 0°-180° range;

• to prevent the cracks from intercepting one another, they are arranged so that their centre corresponds to a square mesh whose side is longer than crack length.

Fracture density f is determined according to the formula proposed by Yin & Ehrlacher (1996).

$$f = \frac{N \cdot l^2}{4 \cdot L^2} \qquad (1)$$

where:
- l is crack length;
- N is the total number of cracks;
- L is the side of square specimens

The pattern of cracks created in the specimen is arranged so as to maintain the centre of the edge cracks at a distance from the specimen sides corresponding to half the spacing between cracks Thanks to this arrangement, crack density remains the same regardless of specimen size.

Furthermore, moving from one to another configuration, by increasing specimen scale, the rock specimen side to crack length ratio is also kept constant, and crack density remains the same regardless of crack dimensions.

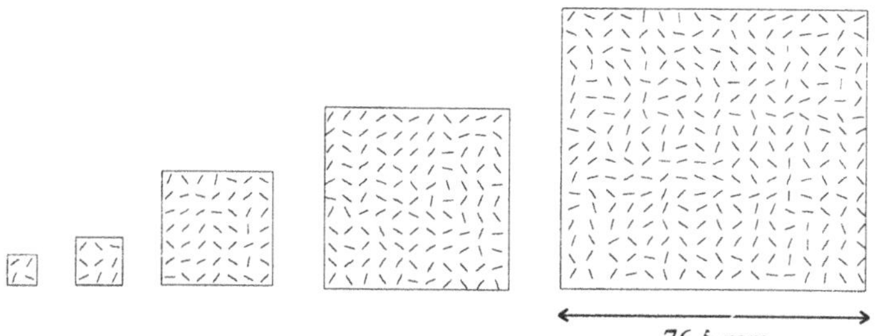

Figure 2. Example of specimens of different sizes and same crack length.

3 NUMERICAL RESULTS

A series of configurations obtained by assuming that crack size in configuration 1 (see figure 1) to be 3 mm, corresponding to the average side of the particles which may be distinguished in granite, have been taken into consideration. Furthermore, a crack density of 0,11 has been taken into account, which, on the basis of equation (1) and the assumptions made, give a spacing between crack centres of 4,5 mm.

Having established these geometrical parameters, in order to perform the renormalisation process it is necessary to establish the value of L_1 (figure 1), i.e., the side of the rock volume containing the smallest cracks Yin & Ehrlacher (1996) point out that, density and crack size being the same, the modulus is seen to increase with increasing volume. An initial series of analyses was conducted in order to confirm this phenomenon for the configuration containing 3 mm long cracks, and to define the most suitable L_1 parameter. With increasing specimen size, the number of cracks increases to a tremendous extent: the specimens used in the analyses had up to 289 cracks

Since the dip of the cracks is a random variable, numerous analyses were performed for each configuration so as to obtain the average modulus values given in table 1. Some of the configurations analysed are exemplified in figure 2.

In these analyses the rock material, assumed to be uncracked, has an elastic modulus of 100000 MPa (E_0 in figure 1), a toughness of 63,2 MPa·m$^{0.5}$ and a Poisson's ratio of 0,2. The cracks, whose shear stiffness is 1 MPa/mm, have been represented with four displacement discontinuity elements.

Each analysis was used to simulate a uniaxial compression test from which it is possible to work out the stress-strain curve up to the onset of crack propagation. In this range, the evolution of the curve is linear (see figure 3) and it is possible to determine the value of the modulus as a function of its inclination.

Table 1. Values of the modulus (E) determined as a function of the side of the rock specimens (L) and the number of cracks (N)

N	L (mm)	E (MPa)
4	9	91852
9	13,5	92373
49	31,5	93561
121	49,5	94443
289	76,5	95745

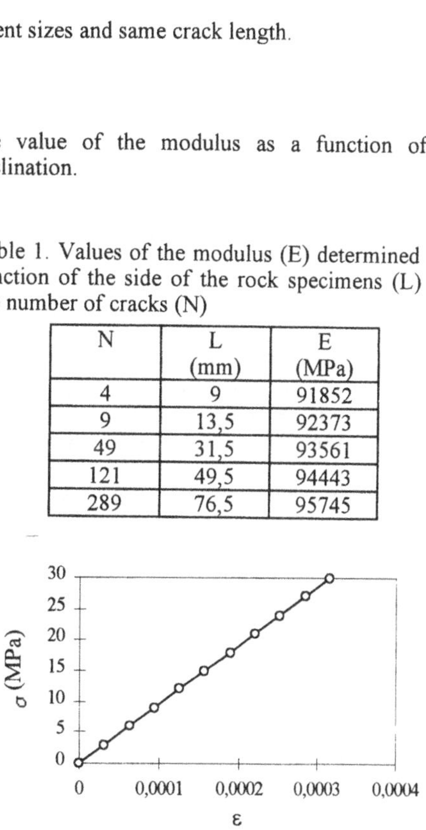

Figure 3. Example of stress-strain curve (numerical analysis).

The values of the deformability modulus determined in this manner are illustrated in figure 4 in relation to specimen size.

As can be seen, the value of the modulus is always lower than in uncracked material (100000 MPa), due to the influence of the cracks on the specimen's mechanical behaviour.

At all events, the scatter in the values is much greater in smaller sized specimens, where the random nature of the inclination of the cracks -

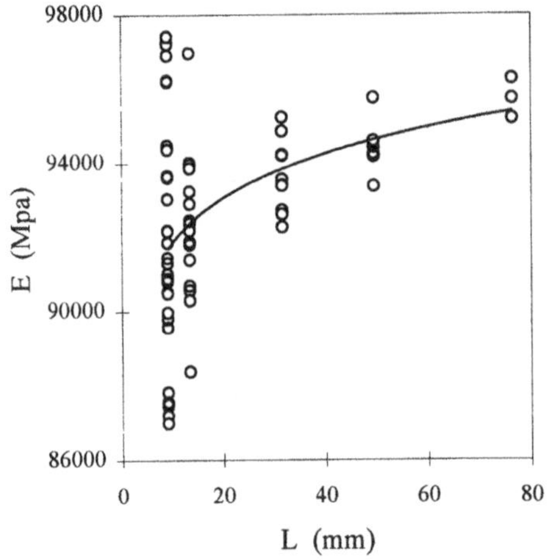

Figure 4. Values of the deformability modulus (E) as a function of specimen size (L).

which are few and of a size comparable to that of the specimen - plays a major role.

On the basis of these analyses, the rock volume containing 121 cracks was deemed to be the most significant for the determination of the E_1 modulus, (figure 1). The L_1 parameter turns out to be 49,5 mm.

Accordingly, configuration 2 was generated:

- the length of the new side L_2 turns out to be 544,5 mm for the volume containing 121 cracks, 33 mm long and spaced 49,5 mm apart.
- the entry value of the deformability modulus is taken to be the mean of the values determined for L_1= 49,5 mm (94443 MPa), thereby admitting the presence of microcracks, 3 mm in length, in the material comprised between bigger cracks.

Table 2. Analyses performed on the basis of the deformability modulus of the material (E_m) and the values calculated (E_c) for the configurations with side L and spacing S between cracks of length l.

L (mm)	l (mm)	S (mm)	E_m (MPa)	E_c (MPa)
49,5	3	4,5	100000	94443
544,5	33	49,5	94443	88712
5989,5	363	544,5	88712	84273
65884,5	3993	5989,5	84273	79515

By the same method, the analysis was extended to specimens of increasingly large size, always assuming the number of cracks to be 121 in each configuration. The results are shown in table 2.

Figure 5 illustrates the ratio between the modulus (E) and the modulus of the uncracked material, E_0, as a function of the length of specimen sides on a semi-logarithmic scale It can be seen that the modulus decreases with increasing scale according to a nearly linear trend on a semi-logarithmic scale.

As it becomes bigger, the rock mass becomes increasingly deformable. For a rock mass of infinite size, no asymptotic modulus value is observed, but this must be ascribed to the assumption of density remaining the same with increasing crack size and scale.

4. CONCLUSIONS AND FURTHER DEVELOPMENTS

A method has been developed to study scale effects on the deformability modulus of cracked rock masses on the basis of the analysis, conducted with

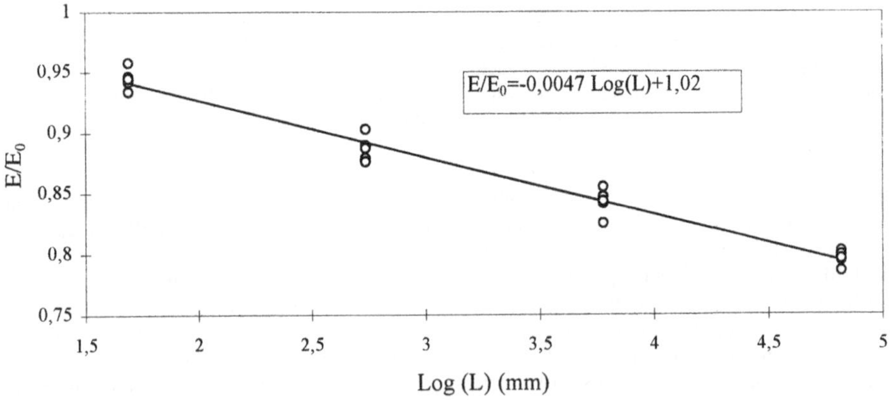

Figure 5. E/E_0 ratio as a function of the length of specimen sides (L) on a semi-logarithmic scale.

the aid of a discontinuous model, of the mechanical behaviour of rock masses with natural discontinuity systems

The numerical results obtained clearly show a decrease in the deformability modulus with increasing rock volume size. The validity of this result is limited to the realm identified by the simplifying assumptions made on joint set geometry and its variations as a function rock specimens. In this connection, further developments of this research project will extend the analysis to different geometrical configurations, so as to enhance the general validity and scope of applicability of the method.

REFERENCES

Barton N., Lien R. & Lunde J. 1974. Engineering classification of rock masses for design of tunnel support. *Rock mechanics* 6 :189-236.

Bieniawski Z. T. 1976. Rock mass classification in rock engineering. In Bieniawski Z. T. (ed), *Exploration for rock engineering, Proc. Symp. Johannesburg* 97-106. Cape Town :Balkema.

Creswick R. J., Farch H.A. & Poole C.P. 1991. *Introduction to renormalization group methods in physics*, John Wiley & Sons.

Cunha, A. P. & Muralha, J. 1990. About lnec experience on scale effects in the deformability of rock masses. In P Cunha (ed), *Scale effects in rock masses* : 219-229. Rotterdam : Balkema.

Cunha A. P. 1990 Scale effect in rock mechanics. In P Cunha (ed), *Scale effects in rock masses* : 219-229. Rotterdam : Balkema.

Gomes M. J L. 1995. Evaluation of characteristics and coefficients determining the scale effect in jointed rock masses. In Rossmanith (ed), *Mechanics of jointed and faulted rock* : 125-130. Rotterdam : Balkema.

Scavia C. 1995. A method for the study of crack propagation in rock structures. *Geotechnique* 45,3: 447-463.

Yin H. P & Ehrlacher A. 1996. Size and density influence on overall moduli of finite media with cracks. *Mechanics of materials* 23 : 287-294.

Jackson R. & Lau J S. 1990 The effect of specimen size on the laboratory mechanical properties of Lac du Bonnet grey granite In P Cunha (ed), *Scale effects in rock masses* : 219-229. Rotterdam Balkema.

Testing

Mechanics of Jointed and Faulted Rock, Rossmanith (ed.) © 1998 Taylor & Francis, ISBN 90 5410 955 6

Crack initiation at a heterogeneity in a rock sample subjected to the Brazilian test

B. Van de Steen, A. Vervoort & J. Jermei
Department of Civil Engineering, KU Leuven, Belgium

ABSTRACT: Exceeding the tensile strength is not a sufficient condition to initiate a fracture at a stress concentrator (e.g., a circular hole). Different theories offering a possible explanation are investigated. Neither Weibull statistics, nor a decline in E modulus as a consequence of damage induced near the stress concentrator can offer a satisfying explanation. A fracture criterion taking the stress gradient into account, and introducing a characteristic distance *c*, allows to cater for a size effect. Applied to a number of Brazilian tests, the characteristic distance seems, especially for small size stress concentrators, to be function of the size of the stress concentrator itself. This is of course inadmissible, and further research will have to reveal if it is not the stress distribution at damage initiation rather than at fracture initiation that should be considered in the yield criterion and secondly if it is not the actual stress distribution rather than an approximation that should be used when averaging the stress along the prospective fracture path.

1 INTRODUCTION

The discussion in this paper forms part of the research on the fracture and damage initiation under mode I conditions in rocks at a stress concentrator. In rocks, both heterogeneities in the material, such as minerals with a modulus of Young significantly different from the surrounding material, asperities at discontinuities, flaws in the material, as well as voids, ranging in size from micro-pores over drill holes to underground excavations can give rise to stress concentrations. To induce a mode I failure, the core samples being used in the study were subjected to the Brazilian test; a hole drilled on the horizontal diameter, parallel to the core axis, acts as stress concentrator (Fig. 1). The tensile stresses induced just above and just below this hole by a diametrical load, acting vertically, were expected to invoke a fracture. This was not always the case, and the fracture often occurred along the vertical diameter. The analysis of the stress distribution at the hole compared to the stress distribution along the vertical diameter together with the discussion of some mechanisms offering a possible explanation for the observations will be presented in this paper.

2 LINEAR ELASTIC ANALYSIS

A total of 56 limestone samples, 15 porphyre samples, 14 sandstone samples and 8 high strength cement samples with diameters ranging from 50 mm to 152 mm were tested. The hole that was drilled in each sample varied in diameter between 4 mm and 38 mm. The average thickness of the 50 mm diameter samples was 21 mm. For the 75 mm diameter samples the average thickness was 30 mm while the average thickness of the 143 mm and 152 mm samples were 33 mm and 36 mm respectively. Being subjected to a Brazilian test, the majority of the samples (approximately 75%), failed along the vertical diame-

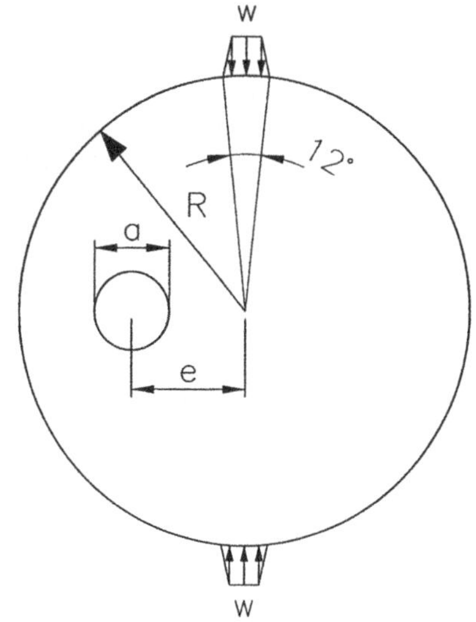

Figure 1: Sample configuration.

ter at tensile stresses close to the tensile strength of the material, as if the hole on the horizontal diameter had no effect. A minority of the samples (approximately 25%) failed at the stress concentrator (hole).

The material was in a first instance assumed to behave linear-elastically and an elastic analysis of the stress distribution, and more in particular of the major principal stress σ_I (tensile) was carried out in the vicinity of the hole and along the loaded diameter. For convenience, tensile stresses are in this paper assigned a possitive value.

The selected sample configuration with its hole on the horizontal diameter has the advantage that the area where the crack is initiated, i.e., at the stress concentrator or along the vertical diameter, can be easily determined. This sample configuration has on the other hand the disadvantage that no analytical solution is available to evaluate the stresses in the vicinity of the hole. The stress analysis has therefore to be carried out by means of a numerical simulation technique. For this study, the finite element code ANSYS was used to carry out the analyses.

The presence of a hole deforms the more or less uniform tensile stress obtained along the vertical diameter in the Brazilian test. The dissimilarity of the stresses along the vertical diameter compared to the stress in the conventional Brazilian test decreases as the eccentricity e decreases. For values of $e \geq 2.5a$, with a being the diameter of the hole that serves as a stress concentrator, the expression valid for the horizontal stress along the loaded diameter (line load W in N/mm) is a good approximation (error < 5%):

$$\sigma_{Braz} = \frac{W}{\pi R} \tag{1}$$

For small holes, an analytical approximation due to Hobbs (1965) can be used as an alternative to the numerical analysis in order to obtain approximate solutions for the linear elastic stress distribution along the boundary of small holes. This approximation involves two steps. In a first step, abstraction is made of the hole and the major and the minor principal stress σ_I and σ_{III} are calculated at the centre of the prospective hole (Andreev, 1995). The stresses σ_I and σ_{III} so obtained, are in a second step used as the far field horizontal and vertical stress with which an infinite plate with a hole in the centre is loaded (Timoshenko & Goodier, 1970). The stresses obtained at the hole boundary in the infinite plate are a fair estimate of the stress distribution at the edge of the hole in the samples being discussed here, provided the diameter of the hole is smaller than 8% of the sample diameter (Jermei, 1997). In this paper, the terms "large hole" and "small hole" refer to this criterion. For the experiments discussed here, the maximal tangential stress σ_θ^{max} becomes (Hobbs, 1965):

$$\sigma_\theta^{max} = \frac{W}{\pi R}\,\frac{2(3R^2-e^2)(R^2-e^2)}{(R^2+e^2)^2} \tag{2}$$

If it is assumed that the laws of linear elasticity are applicable to the undamaged material, the stresses that are calculated above and below the hole are a few times the stress along the vertical diameter. If one accepts that the samples, when they split along the vertical diameter fail as the stress along the vertical diameter reaches the tensile strength, an assumption that was justified by carrying out a series of conventional Brazilian tests to determine the tensile strength, the above means that the material withstands stresses that are locally a few times the tensile strength. An example of a finite element analysis of the major principal stress distribution is given in Figure 2. The analysis is done on a sample, with R = 25 mm, a = 4 mm, e = 10 mm and W= 1 kN/mm. The load has been trapezoidally distributed over an arc of 12° (Fig. 1), corresponding to the presumed load distribution. The influence of this load redistribution remains however limited to the stress distribution in the vicinity of the platens (Jaeger & Cook, 1979).

The ratio of the stress at the opening to the stress along the vertical diameter at failure depends on both the size and the position of the hole, and varies from for example 4.3 for a sample with R = 25 mm, a = 4 mm and e = 10 mm to 3.1 for a sample with R = 71.5 mm, a = 36 mm and e = 36 mm. Whether a crack is

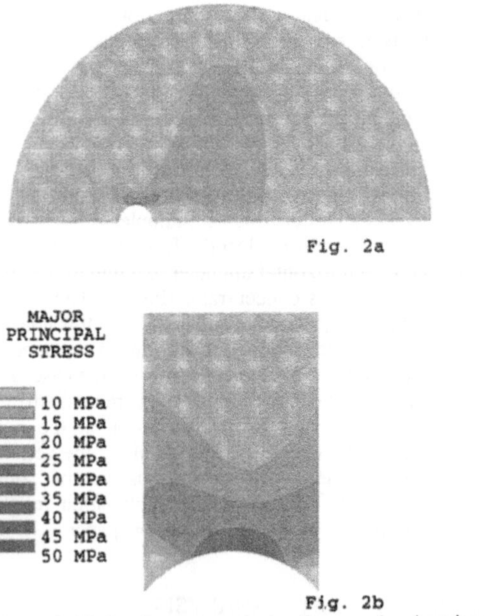

Figure 2: Finite element analysis of a sample, showing the major principal stress distribution σ_I, a) global result (R=25mm); b) detail in the vicinity of the stress concentrator (3 mm x 6 mm).

initiated at the stress concentrator or along the vertical diameter of the sample, depends on the size and on the position of the hole, as well as on the size of the sample itself. This behaviour was observed for the crinoide limestone, the porphyre, the fine-grained mica-rich sandstone as well as for the high-strength cement samples.

As both the size and the position of the hole, as well as the size of the sample seem to be important, the influence of each of these parameters on the maximum tensile stress reached at the edge of the hole will be investigated. These three parameters can be reduced to two by noticing that not the absolute sizes and distances are important, but only the relative distances are relevant. Setting *e/R* equal to *q* in (2), this expression, valid for small holes only, becomes:

$$\sigma_\theta^{max} = \frac{W}{\pi R}\,\frac{2(3-q^2)(1-q^2)}{(1+q^2)^2} \tag{3}$$

An increase in the eccentricity for a constant sample diameter will lead in (3) to an increase in q, and therefore to a decrease in σ_θ^{max}. This finding is confirmed for large holes, based on finite element calculations. In table 1, the stress distribution in a sample with R = 25 mm and one with R = 75 mm are compared. Besides the relative size equivalence (ii and vi), this table also presents the influence of the position of the opening and of the hole size. An increase in the eccentricity (ii and iii or iv and v) results in a decrease of the maximum tensile stress at the hole. For large holes, an increase of the hole size (i and ii) results however in a stress increase. Notice that the expression for small holes (3) is independant of the hole size.

3 FRACTURE CRITERION

In this section a number of possible mechanisms to explain the observed phenomena, will be discussed.

Table 1: Influence of hole size and the relative hole position on the maximum tensile stress.

	R = 25 mm. *W* = 1kN/mm	*R* = 75mm *W* = 3 kN/mm
a = 4 mm *e* = 10 mm	48.3 Mpa (q = 0.4) (i)	-
a = 8 mm *e* = 10 mm	55.9 Mpa (q = 0.4) (ii)	72.6 Mpa (q = 0.13) (iv)
a = 8 mm *e* = 15 mm	31.2 Mpa (q = 0.6) (iii)	67.8 Mpa (q = 0.2) (v)
a = 24 mm *e* = 30 mm	-	56.0 Mpa (q = 0.4) (vi)

3.1 *Weibull statistics*

The question that will be answered in this paragraph is whether the higher than expected stresses calculated at the stress concentrators can be explained by a statistical distribution of the material strength parameters.

The statistical analysis of the test results is focussed on the statistical theory developed by Weibull. Assuming that the tensile strength of the material follows a Weibull distribution (Jaeger & Cook, 1979), the probability that the material withstands the stress levels obtained during the tests was calculated. The Weibull parameters are determined as the parameters of the type III asymptotic distribution of the smallest value (Ang & Tang, 1984) by analyzing the results from a series of Brazilian tests. These parameters are for the experiments equal to 11.23 MPa for the lower limit, 26.5 MPa for the characteristic smallest value and 4.83 for the shape parameter. The probability $Q(\sigma_\theta^{max})$ that the material remains intact when the maximum tensile stress equals σ_θ^{max} is:

$$Q(\sigma_\theta^{max}) = exp[-\int_V (\frac{\sigma_I - 11.23}{26.5-11.23})^{4.83} dv] \tag{4}$$

$$\sigma_I \geq 11.23\ MPa$$

The volume *V* over which the integration is carried out, is defined by the area in which σ_I is greater than 11.23 MPa and by the thickness of the sample. As the stresses are assumed to be homogeneous over the thickness, (4) can be simplified to a surface integral. As pointed out before, no analytical expression is available for σ_I for the sample configurations being considered. An approximation has therefore been used for both the stress distribution and the area of integration: the stress is assumed to decrease linearly, from the point where the maximum value is attained, to the 11.23 MPa contour. This contour is assumed to define a triangular area with the point of maximum tensile stress in the middle of the base. As this stress approximation underestimates the linear-elastically calculated stress with 0 % to 130 %, the result $Q(\sigma_\theta^{max})$ obtained by (4), will be an underestimation as well.

This principle is applied to sample LM1, which failed along the vertical diameter at a load of 20.47 kN. Being characterised by *R* = 25.28 mm, thickness *t* = 20.04 mm, *a* = 4 mm and *e* = 8.2 mm, this results in σ_θ^{max} = 59.25 MPa and one obtains from (4):

$$P(\sigma_\theta^{max}=59.25) = 1 - Q(\sigma_\theta^{max}=59.25) =$$

$$1-\exp[-(t\int_0^{0.5} 2(\int_0^{1.25(0.5-y)} (\frac{\sigma_I(x,y)-11.23}{26.85-11.23})^{4.83} dx)dy)]$$

where

$$\sigma_l\,(x,y)=(59.25-96y)-\frac{(48-96y)}{1.25(1-2y)}$$

$$\Rightarrow\ P(\sigma_\theta^{max}=59.25)=1\text{-}\exp(-7t)\approx 1$$

The initiation and development of a fracture would for the said configuration be a certainty under the prevailing stresses according to the above theory. As the sample clearly did not fail at the stress concentrator, but along the vertical diameter, the size effect as a consequence of the statistical distribution of the strength parameters cannot lie at the basis of the observed behaviour.

3.2 *Damage induced around the hole*

In the models used thus far, a linear-elastic behaviour was assumed in the whole of the sample, and the sample was assumed to be homogeneous. Damage in the vicinity of the opening could however induce elasto-plastic behaviour or render the material inhomogeneous in the region of interest, such that the high stress concentration calculated using linear elasticity concepts are in fact never reached at the edge of the opening. This damage could be induced as a consequence of the drilling of the holes. While the bigger diameter holes were drilled with a core drill, the smaller diameter holes ($a \le$ 12mm) were drilled with an impact drill and are therefore more likely to have been adversely affected by the drilling process. A radius-dependent modulus model for finite hollow cylinders, proposed by Ewy and Cook (1990) to model a mechanically damaged zone near the opening was applied to our test configuration. To explain the observations with the said model, extensive damage is required in the vicinity of the opening (Jermei, 1997). Neither microscopic inspection, nor techniques making use of capillary suction or saturation by means of a fluorescent resin showed evidence of damage at a radial distance in excess of 3 mm from the hole boundary. The proposed mechanism, i.e., that damage induced during sample preparation would lead to a built off of the stress, has therefore to be rejected as explanation for the observed behaviour.

To confirm this conclusion, samples of high strength cement were fabricated. By casting the samples in cylindric moulds, and by placing a removable plug in the cast at the position of the stress concentrators, cylindric samples with a circular hole on the vertical diameter were obtained without any drilling. These cement samples behaved in a similar way to the rock samples: at failure, which happened generally along the vertical diameter, the stresses at the hole were a few times the tensile strength. As no mechanical damage could have been induced at the hole during the sample preparation, these tests confirmed the foregoing conclusion: mechanically induced damage prior to testing cannot be the reason for a built off of stresses.

3.3 *Influence of stress gradient on the fracture criterion*

From observations regarding the size effect, Lajtai (1972) concluded that exceeding the elastically calculated tensile stresses at the edge of the opening was not a sufficient condition to induce fracture, but that it was also necessary to consider the stress gradient along the potential fracture path: "...fracture will be initiated when the maximum tensile stress at a certain distance c beyond the periphery (σ_c), but along the fracture path, reaches the uniaxial tensile strength." Lajtai argues that it is highly unlikely that the high stresses as predicted by the theory of linear elasticity ever develop, but that, also for brittle materials, the stresses will be redistributed over a certain area. This means that the stress peak is actually smoothed out over a certain area. The material will not yield as long as:

$$\sigma_c=\frac{\sigma_\theta^{\max}+\sigma_{2c}}{2}<\sigma_t \qquad (6)$$

Hereby it is tacitly assumed that the stress is uniform over the thickness of the sample, which allows for the stress to be considered along a line instead of over an area. Notice that the stress is averaged over the distance $2c$. When the change in magnitude at the boundary is approximated by means of the stress gradient, (6) becomes:

$$\sigma_c=\sigma_\theta^{\max}+c\left(\frac{\partial\sigma}{\partial r}\right)_{r=a/2}<\sigma_t \qquad (7)$$

where:

σ_c The average stress over the length $2c$; it is at the same time the linear-elastically calculated stress at a distance c from the edge of the opening.

σ_{2c} the elastically calculated tensile stress at a distance $2c$ from the edge of the hole.

$\left(\frac{\partial\sigma}{\partial r}\right)_{r=a/2}$ The stress gradient at the hole boundary (negative value).

Although the stress distribution is non-elastic in detail, only elastically calculated stresses figure in the equation. The distance c at which the critical tensile stress is calculated is considered to be a material constant. The main consequence of introducing this material parameter c, is the introduction of a scale effect. While the maximum tensile stress σ_θ^{max} depends for a homogeneous load at infinity only on the load, σ_c depends on the size of the stress concentrator as well. For the sample configuration being considered

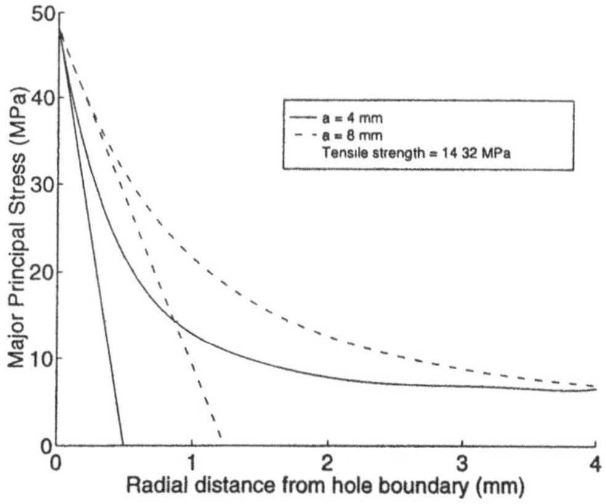

Figure 3: Major principal stress along the potential fracture path for a hole with a = 4 mm and a = 8 mm: calculated solution and gradient approximation.

here, both σ_θ^{max} and σ_c depend on the diameter of the stress concentrator. But where the influence of a on σ_θ^{max} is a consequence of the limited sample size relative to a ((2) and (3) are independent of a), the dependance of σ_c on a does not vanish whatever the sample size may be. To illustrate the influence of the hole size on the stress level in the vicinity of the hole, the main principal stress σ_t along the potential fracture path is plotted in Figure 3 for a = 4 mm and a = 8 mm with R = 25 mm. The load has been adjusted such that σ_θ^{max} is the same in both cases.

The above concept has been applied to determine c for a number of the samples that were tested, assuming the tensile strength σ_t to be equal to 14.32 MPa. The results have been brought together in table 2. The value of the characteristic distance c seems to be function of the hole diameter, with c increasing as a increases. This is not conform with the theory proposed by Lajtai, as c should be a material property. In table 3, the stress σ_c has been calculated for a number of limestone samples that did not fail at the stress concentrator but along the vertical diameter (a = 4, 6 and 8 mm). As distance c, the value obtained from table 2 for the corresponding hole diameter has been used. The criterion proposed by Lajtai is indeed not violated as the value σ_c remains in each case under the tensile strength σ_t = 14.32 MPa.

Table 2: Characteristic distance for a number of samples that failed at the hole obtained with Lajtai's stress gradient analysis.

Name	Diameter 2R (mm)	Hole size a (mm)	2c (mm)
LM2	50.64	4.0	1.06
LM3	74.30	4.0	1.00
LM4	50.62	6.0	1.41
LM5	74.26	6.0	1.32
LM6	50.65	8.0	1.78
LM7	143.13	36.5	4.91

Table 3: Average stress calculated with (7). The value of c is obtained from table 2.

Name	Diam. 2R (mm)	Hole size a (mm)	σ_c (MPa)
LM1	50.57	4.0	12.81
LM8	50.65	4.0	8.92
LM9	74.33	6.0	14.21
LM10	50.70	8.0	7.40

4 DISCUSSION

The problem identified previously, namely that the distance c seems to depend on the hole size could firstly be attributed to the approximation implied by (7), i.e., that the stress at the hole is assumed to decrease linearly. This is clearly not the case, and especially for small diameter holes, this leads to an important underestimation of the reigning principal stress σ_t at some distance from the hole boundary (Fig. 3). One can also notice in Figure 3 that the major principal stress along the prospective fracture path tends to an assymptotic value that is often greater than 50% of the tensile strength. This considerably increases the total distance d over which the major principal stress has to be averaged to obtain an average stress level equal to the tensile strength. This distance d will be up to six times greater than the distance at which the major principal stress gets equal to the tensile strength, and up to 12 times the characteristic distance c obtained with the gradient approximation.

A second reason for the dependance of c on the hole size could be that the yield criterion should be applied to the stress state at damage initiation instead of to the stress state at fracture initiation. The load and the stresses that were so far considered are the elasticly calculated stresses and their associated loads at the moment of fracture initiation; i.e., at the moment a macroscopic fracture is formed. Experiments in which the acoustic emission activity was registered and in which use was made of capillary suction, indicated that fracture initiation at the stress concentrator is preceded by a damage evolution process whereby the material respons ceases to be elastic. The damage evolution process results eventually in a fully developed fracture. The load at which the first evidence of this damage evolution process was found, depended on the actual sample configuration (size and position of the hole). The load at which damage was first noticed ranged between 100% of the load at fracture which means it concerned an unstable fracture not preceded by a distinguishable damage evolution process for openings where e = 0 mm to 90% of the load at fracture for eccentricities in excess of 20 mm. It can furthermore be commented that the first noticeable damage still occurred at stress levels well in excess of the tensile strength of the material.

5 CONCLUSION

Exceeding the tensile strength in a material is not a sufficient condition to initiate a fracture. Neither Weibull statistics nor a non-homogeneous modulus of Young as a consequence of damage initiated during sample preparation can account for the tensile stress at which the material seems to withstand. The introduction of the stress averaging concept of Lajtai, which takes account of the stress gradient and which introduces a characteristic distance, provides for individual samples a fracture criterion that admits the apparent high tensile stresses at the hole. It fails however to explain the full set of observations, as the characteristic distance seems not only to be a function of the material being tested, but of the size of the stress concentrator as well. In the research currently being conducted, it is investigated whether this difficulty can be resolved by taking the actual stress distribution into account and by applying the fracture criterion to the stress distribution at damage initiation rather than to the stress distribution at macro crack initiation.

6 REFERENCES

Andreev G.E. 1995. *Brittle failure of rock materials, test results and constitutive models.* Rotterdam/ Brookfield.

Ang A.H.-S. & W.H. Tang 1984. *Probability Concepts in Engineering Planning Design, Volume II: Decision, Risk and Reliability.* John Wiley & Sons.

Ewy R.T. & N.G.W. Cook 1990. Deformation and Fracture around Cylindrical openings in Rock-I. Observations and Analysis of Deformations. *Int, J. Rock Mech. Min. Sci. & Geomech. Abstr.* Vol 27, No 5, pp 387-407.

Hobbs D.W. 1965. An assessment of a technique for determining the tensile strength of rock. *Brit. J. Appl. Phys.* Vol. 16, p. 259-268.

Jaeger J.C. & N.G.W. Cook 1979. *Fundamentals of rock mechanics.* London.

Jermei J. 1997. *Breukinitiatie tijdens rotsmechanische testen* (In Dutch). M.Sc. Thesis KULeuven, 1997.

Lajtai E.Z. 1972. Effect of Tensile Gradient on Brittle Fracture Initiation. *Int, J. Rock Mech. Min. Sci. & Geomech. Abstr.* Vol 9, No 5, pp 569-578.

Timoshenko S.P. & J.N. Goodier 1970. *Theory of Elasticity.* Third edition. McGraw-Hill.

7 LIST OF SYMBOLS

a	diameter of the hole serving as stress concentrator (mm)
c	characteristic distance, equal to half the distance over which the stress is averaged in the criterion proposed by Lajtai (mm)
d	characteristic distance over which the actual stress is averaged (mm)
e	eccentricity; distance from the sample centre to the centre of the hole (mm)
P	probability that a sample fails
Q	probability that the material withstands the reigning stresses
q	relative eccentricity e/R
r	radial coordinate (origin coincides with centre hole)
R	radius of the sample (mm)
t	sample thickness (mm)
W	external load per unit of thickness (N/mm)
x,y	Cartesian coordinates; origin at the point of maximum tensile strength; y-ax through the centre of the hole
σ_{Braz}	tensile stress along the vertical diameter in a sample subjected to the Brazilian test (MPa)
σ_c	the average stress over the length $2c$; it is at the same time the elastically calculated stress at a distance c from the hole boundary (MPa)
σ_{2c}	the elastically calculated tensile stress at a distance $2c$ from the hole boundary (MPa)
σ_t	tensile strength (MPa)
σ_I	major principal stress (tensile stress positive) (MPa)
σ_{III}	minor principal stress (MPa)
σ_θ^{max}	maximum tensile tangential stress in the vicinity of the hole; origin at the centre of the hole (MPa)

Mechanics of Jointed and Faulted Rock, Rossmanith (ed.) © *1998 Taylor & Francis, ISBN 90 5410 955 6*

Laboratory tests to investigate the effect of presence of discontinuities on rock cutting parameters

M.Cheraghi Seifabad
Department of Civil Engineering, K.N.Toosi University, Iran

ABSTACT : Most of the rock cutting experiments conducted by various investigators dealt with intact materials, despite the importance of discontinuities and their effects on cutting parameters. Furtheremore very limited research have been conducted in cutting jointed rock under saturated conditions. Under in situ conditions, moisture is present in geological layers, therefore there is a need to investigate cutting parameters under such conditions and to compare with dry and ambient moisture content to see the extent of effect. Geological beds often are formed by different layers, for this reason it was decided to investigate the situation where different geological materials are in succession.

1 INTRODUCTION

The performance of drag bit cutting depends on two factors, the rock itself and the cutting parameters. The cutting parameters refers to the tool geometry, and tool speed and the rock characteristics refer amongst others to the presence of discontinuities and the moisture conditions. Fowell and McFeat-Smith (1976) having recognized the important role played by rock mass properties in rock excavation mechanics, decided to seek one rapid and simple method of quantifying the presence of breaks in the rock mass under in situ conditions. They decided to use the break index which was defined as an an average number of breakness planes intersecting horizontal and vertical scan lines per metre for each bed under considration.

Obert and duvall (1967) studied the effect of moisture on different types of rock and found that the effect is less pronounced. The effect of moisture content was investigated by Colback and Widd (1965), who idenfied a 50% reduction of UCS under saturated conditions for shale and sandstone. In general UCS decreases with increase in moisture content. Other researchers Broch (1979) and Vutukuri (1974), have shown that moisture substantially reduces the UCS of rocks. Roxborough and Phillips (1975) carried out cutting tests in both dry and wet states using Bunter sandstone. All forces increased by 20% in wet conditions.

2 MECHANICAL PROPERTIES

2.1 *Uniaxial compressive strength and tensile strength*

A suitable arrangements satisfying ISRM requirements was used for applying and measuring axial load to the specimen (Table 1). The rock cores had nominal dimentions height = 150 mm and diameter =75 mm. LVDTs (linear variable differential transformers) were used for measuring the axial and lateral deformation. Load on the specimens was applied continuously at a constant stress rate of 3 Mpa/min (according to ISRM).

Table 1. Uniaxial compressive strength and tensile strength

Rock	σ_c MPa	σ_t MPa
Springwell sandstone	43.9	3.77
Welton chalk	47.5	4.41

The Brazil disc test was used to measure indirectly the uniaxial tensile of rock specimen.

A cylindrical specimen of diameter 60 mm and thickness 30 mm (ratio 2:1) was compresses diametrical between two platens. A constant loading rate was applied so that failure occured within 20-25 s of loading. The location of Springwell sandstone is Springwell village, SE of Gateshead, North East England and Welton chalk is located in the south shore of river Humber in England.

3 MECHANICAL CUTTING

A core grooving test developed in the University of Newcastle upon Tyne (Roxborough & Phillips 1975) is used extensively in the assessment of the machineability of rock for roadheaders and other drag tool equipped machines. Four cuts are normally made in the rock sample at a constant depth of 5 mm with a tungsten carbide tool 12.7 mm wide, chisel-edged, with a -5 front rake and +5 back clearance angle. This tool is mounted on an instrumented shaping machine. Mean cutting forces (MCF) is the average force acting in the direction of cutting, mean normal force is the average force acting normal to the direction. Mean peak cutting force (PCF) is the average of maximum values of the cutting force, Mean peak normal force (PNF) is the average of the maximum forces generated during cutting to keep the tool in the rock. Yield, (Q) is the mass of rock by the pick per unit length of cut and specific energy, (SE) is defined as the work done to cut a unit volume of rock.

4 CORE GROOVING OF DISCONTINUOUS ROCK

Specimens were water saturated by placing them in a desiccator for three hours. The test were done in different states as follows :

1-intact
2-with a single discontinuity (discontinuity angle 30° to core axis)
3-with a single discontinuity (discontinuity angle 45° to core axis)
4-with a single discontinuity (discontinuity angle 60° to core axis)
5-with a single discontinuity (discontinuity angle 75° to core axis)
6-with a single discontinuity (discontinuity angle 90° to core axis)
7-with a two parallel discontinuities (discontinuities angle 90° to core axis)
8-with three parallel discontinuities (discontinuities angle 90° to core axis)

For all cases the above tests were conducted under dry, ambient, and saturated conditions. For each case a favourable reduction was observed in MCF from dry to ambient and from ambient to saturated condition. In addition the effect of angle was investigated. After observing no relationship between MCF and the relative angle of discontinuity, it was assumed that there may have been an interaction between the influence of angle of discontinuity and spacing of the discontinuity. To investigate the influence of the angle of discontinuity it was decided to conduct some tests with constant discontinuity spacing. However it was not possible to derive any meaningful relationship concerning the investigated parameters.

It was possible to show from the results that there is a reduction in specific energy for the joint spacing:

1-dry specimens:all the forces and SE were reduced proportionally because as a function of the frequency of spacing.
2-ambient : there was no difference for yield and coarseness index between intact specimen and specimen containing a single discontinuity but there was a reduction of specific energy for two discontinuities.
3-saturated : MCF, MNF decreased from intact specimen to specimen containing a single discontinuity and again to two discontinuities, but PCF and PNF increased from intact to single discontinuity and further decreased for double discontinuity. SE increased from intact to single discontinuity then decreased from single discontinuity to double discontinuity.

Comparison of the results of intact specimens with discontinuities show that there was a small increase in SE from intact specimen to spicmen with one discontinuity. However there was a decrease in SE when examined the result of specimen with single discontinuity in comparison with specimen containing two discontinuities . The general conclusions for the three states (dry, ambient, saturated) discontinuity in the rock specimens reduces the strength and the number of discontinuities is not important. This means when spacing is large enough (above 50 mm) rock behaves as intact rock (Cheraghi Seifabad, 1992).

4.1 *Effect of angle of discontinuity for cuttability of rock*

1-dry : if the SE results are examined for the intact specimen and as the angle of the discontinuity ranges from 75° to 30°, it can be seen that the best results, concerning specific energy, correspond to the angle of 75° and the worst to 30°. This is very important because SE was found in this position to be more than the SE of intact rock.
2-ambient : similarly there appears to be a reduction as the angles range from 75° to 30° and the best angle concerning SE is 30°.
3-saturated: with reference to the SE values it can be seen that the best angle is 60° and the worst is 45°.

Because the results showed that there is no trend between angle of discontinuities and cutting forces it was thought that tool spacing might affect the results. For this reason a series of tests were conducted with tool spacing held constant and also with discontinuity held constant. Again for a discontinuity angle between 30°, 45° while the tool spacing was maintained constant, no significant influence on the cutting forces and SE was identified.

4.2 *Effect of spacing of discontinuities*

To investigate the effect of the spacing of the discontinuities angle was kept constant while different spacings were examined. When the discontinuity spacing was chosen to be very small, i.e 10 mm it was found that this spacing is advantageous (in terms of SE). Furthermore it appears that the discontinuum rock behaves as an intact rock if the spacing of discontinuities is

Table 2 Cutting results of 45° angle of discontinuity

Spacing	PNF	PCF	MNF	MCF	Q	SE
	kN	kN	kN	kN	m3/m	MJ/m3
50 mm	2.00	2.19	1.62	1.41	21.4	14.57
60 mm	1.28	2.52	1.41	1.23	25.4	10.70
85 mm	2.21	2.62	1.31	1.48	19.3	16.95
95mm	2.20	2.68	1.55	1.48	20.0	15.97

Table 3 Cutting results of 30° angle of discontinuity

Spacing	PNF	PCF	MNF	MCF	Q	SE
	kN	kN	kN	kN	m3/m	MJ/m3
50 mm	1.95	2.56	1.89	1.33	22.2	13.83
60 mm	1.97	2.63	1.62	1.53	21.8	15.59

greater than 10 mm. In this investigation for an angle of 45° a spacing of 50, 60, 85 and 95 mm was used and also for an angle of 30° a spacing 50 mm and 60 mm was adopted.

Aleman (1983) found that discontinuities spaced every 10 mm will dramatically increase cutting rate of mudstone (Tables 2,3).

4.3 *Effect of type of intercalated geologic materials*

Geological beds often are formed by different layers, for this reason it was decided to investigate the situation where different geological materials are in succession. It was decided to intercalate chalk and clay layers between sandstone (angle of discontinuity was 90° with respect to core axis). When chalk was inserted between sandstone was placed all the forces increased except MCF which remained unchanged, the results are shown in (Table 4).

For the next test China clay was used as filling material between three parts of sandstone with different thickness. The cutting forces decreased with clay intercalations, and also decreased with as the thickness of clay decreased. It appears 10 mm clay thickness results in the lowest forces and 25 mm to the largest forces, the relevant data are shown in Table 5 .

All the forces due to more compacting clay show favourable increase. In another set of experiments where clay was intercalated between

Table 4 Cutting parameters of different types of specimens

Specimen Type	PNF	PCF	MNF	MCF
	kN	kN	kN	kN
intact sandstone	2.70	3.54	1.69	1.73
sandstone with 1 discontinuity	2.12	3.00	1.41	1.31
composite sandstone containing chalk intercalation	2.43	3.17	1.47	1.31

Table 5 Cutting parameters of sandstone with intercalated clay

Specimen Type	MNF	PCF	PNF	MCF
	kN	kN	kN	kN
sandstone without clay, with discontinuity	2.12	3.00	1.41	1.31
composite sandstone with 10mm thickness clay	2.14	2.91	1.31	1.29
composite sandstone with 18mm thickness clay	1.91	2.55	1.10	1.16
composite sandstone with 25mm thickness clay	1.80	2.38	1.02	0.90

chalk the results showed that all the forces were reduced (Table 6):

Table 6 Cutting parameters of chalk with intercalated clay

Specimen Type	PNF	PCF	MNF	MCF
	kN	kN	kN	kN
chalk, one discontinuity	3.31	3.81	1.94	1.60
composite chalk with 10 mm, thickness clay between 3 parts of chalk	2.23	2.82	1.19	1.12

5 CONCLUSIONS

In this work the relationship for various rock properties with different parameters was investigated. Discontinuities angle and spacing are often thought to be important and in geological layeres since most of the time the machine is excavating rock with discontinuities. However it was found for Springwell sandstone that the discontinuity angle is only important for joint angle which is 90°. For other angles; 30°, 45°, 60°, 75°, it was found that the orientation does not influence rock cutting parameters very much. Different spacing more than 50 mm rock behaves as an intact rock. Different sates (dry, ambient, saturated) were chosen to investigate. A specific reduction in the values of the various cutting parameters was found from dry to ambient and from ambient to fully saturated. This suggests that the machine spends less energy to excavate rock in fully saturated condition. In the same way that moisture content reduces the rock cutting parameters. Where in the case of joint filling, it was found that they affect the overall cutting performance more so than the presence of discontinuities.

ACKNOWLEDGMENT

I would like to express my appreciation to Dr. E.K.S. Passaris, senior lecturer in geotechnical engineering department of Newcastle upon Tyne for providing the opportunity to carry out this research.

REFRENENCES

Aleman, V.P. 1981. A starata index for boom roadheaders. *Tunnels and Tunnelling*, 52-55.

Broch, E. 1979. The infuence of water on some rock properties. *4th ISRM Conference, Montreax.*

Cheraghi Seifabad, M. (1992). An investigation into the mechanical cutting of rock materials with particular referefence to fracture mchanics, Univesity of Newcastle upon Tyne.

Colback, P.S.B. & B.L. Widd 1965. The influence of moisture content on the compressive strength of rocks. *Proc 3rd Rock Mech Symposium,* 65-83.

Fowell, R.J. & I. McFeat-Smith 1976. Factors influencing the cutting performance of a selective tunnelling machine. *Tunnelling 76, IMM, Ed. M. J. Jones*, 1-11.

Obert, L. & W.I. Duvall 1967. Rock Mechanics and the design of structures in rock.

Roxborough, F.F. & H.R. 1975. The mechanical properties and cutting characteristics of the Bunter sandstone, report to the Transport and Road Research Laboratory, Univesity of Newcastle upon Tyne.

Vutukuri, V.S. 1974. The influence of liquid on the tensile strength of limestone. *Int. J.Rock Mech. Min. Sci & Geomech Abstr.*11: 22-29.

Mechanics of Jointed and Faulted Rock, Rossmanith (ed.) © 1998 Taylor & Francis, ISBN 90 5410 955 6

Analysis of rock failure after triaxial testing

A. Kožušníková & K. Marečková
Institute of Geonics, Academy of Sciences of the Czech Republic, Ostrava, Czech Republic

ABSTRACT: In paper failure of quartz grains in Carboniferous fine sandstone is analysed in original rock material and in failured test bodies after uniaxial compression test as well as after triaxial stress condition tests. Failure intensity (i.e. sum of crack lengths in quartz grains) increases with increase of confining pressure and also in slip zones in the case of brittle failure. Also character of deformation of test body changes with increase of confining pressure.

1. INTRODUCTION

The process of rock failure is considered to be a continuous one which occurs progressively throughout the brittle behaviour region, in which the rock steadily deteriorates. It happens frequently that sudden failure occurs with complete loss of cohesion across a plane, and this is known as a brittle fracture (Jaeger 1979). The mechanisms of failure change according to the conditions of stress in the specimen or element.

According to Jaeger (1979) the first visible structural damage in the region of elastic behaviour appears as elongated microcracks concentrated in the central zone of the specimen with their long axes oriented parallel to the direction of the maximum principal stress. In the region of ductile behaviour, which usually starts at a value of stress on the order of two-thirds the maximum value of stress, irreversible changes are induced in the rock. There is a remarkable increase in microcracking, which tend to coalesce along a plane in the central zone of the specimen. The fracture grows towards ends of the specimen by the step-wise joining of microcracks. A single fracture plane , inclined at an angle less than 45° to the direction of σ_1 , is typical for a fracture under compressive stresses or for moderate amount of confining pressure, which will be described as a shear fracture. If the confining pressure is increased to the level that the material becomes fully ductile, a network of shear fractures appears, accompanied by plastic deformation of individual crystals.

The transition from brittle to ductile and then to plastic behaviour occurs gradually with increasing stress, and depends on more factors than just stress. Brittle-ductile transition occurs at lower stress levels if the rock is moist or hot, or if it is compressed very slowly. Some rock-forming minerals, such as clays, and halite, become ductile at relatively low temperatures and pressures, whereas quartz and feldspars become ductile only at conditions corresponding to depths of tens of kilometres (Franklin 1989).

In our paper we have analyzed failure of quartz grains from both original sandstone specimen, and samples which have failed in uniaxial compression tests and triaxial stress tests.

2. METHODOLOGY OF MEASUREMENT AND ANALYSIS

2.1. *Characteristics of tested material*

The material selected for tests was Carboniferous fine sandstone. The clastic grains consists predominantly of quartz (42%), rock fragments (22%) and feldspars (10%). The intergranular substance consists of matrix (18 %), cement (4 %), mica fragments (2 %), and traces of coal fragments, altered feldspars, altered rock fragments.

2.2. *Triaxial experiments*

Triaxial tests were performed using triaxial compression equipment GTA 20 - 32 with a maximum confining pressure of 400 MPa. Cylindrical test bodies (with diameter of 36 mm and slenderness ratio of about 2) were used in these tests. Test bodies were protected by a rubber sleeve. In addition to measurement of confining pressure, σ_3 , and axial load, σ_1, longitudinal deformation was measured by sensor of piston movement during the experiment. Triaxial experiments were performed at confining pressures of 50, 150 and 300 MPa. In addition to that, also uniaxial compression test were performed for studied sample.

It is obvious from stress-strain curves (Figure 1) that for experiments with confining pressures $\sigma_3 = 50$ MPa and $\sigma_3 = 150$ MPa, failure of sample occured without distinct plastic deformation while pronounced plastic deformation can be seen at a confining pressure $\sigma_3 = 300$ MPa.

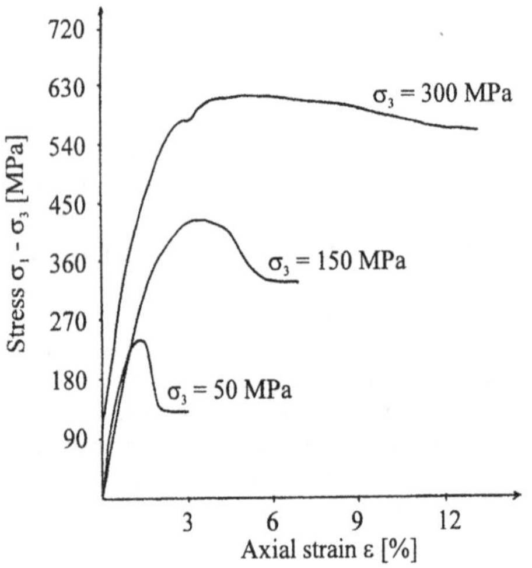

Figure 1. Axial stress-strain curves at different confining pressures.

2.3. *Preparing and system of evaluation of microscopic preparations*

Thin sections were prepared from samples oriented perpendicularly to the surface of failure and intersecting longitudinal axis of test body.

Preparing of microscopic preparations is rather problematic because large thin sections are needed (35 x 55 mm) and also a multiple impregnation is necessary to avoid falling-out of failured grains during grinding and polishing. To ensure proper evaluation the thin section was divided into 6 x 8 squares (5x5 mm each) by means of a slide gauge fixed on the microscope rotary stage. Each square, analysed during microscopy evaluation, included nine viewing fields. Data measured for each square (11,12 ...67,68) was stored in a data file.

Polarisation microscope Amplival (Zeiss) with image analysis Vidas 2.5 was used for analysis of the thin sections. The analysis was performed in polarised light at 80 times magnification.

The thin sections were oriented so that x-axis was parallel to long axis of the sample (and thus parallel to the action of axial force). During measurement cracks longer than 50 μm were scanned manually in all quartz grains within the viewing field. For each crack, its length and orientation with respect to x-axis was recorded in the data file for the corresponding square. The angles are measured in interval of 5,6 degree. The total length of cracks in quartz grains in a square was determined; a histogram of crack angles in relation to x-axis was plotted after evaluation of the nine viewing fields of the square (Figure 2). Based on these histograms, rose diagrams were plotted for each square. The failure intensity is divided into 5 classes based on the total length of cracks in the square.

3. RESULTS OF MICROSCOPIC ANALYSIS

Five thin sections have been processed using microscopic analysis. All five thin sections were prepared from identical material, only loading modes were different:

- original rock material - sample No 3336p
- test body after uniaxial compression test - sample No 3336no
- test body after triaxial test (confining pressure of 50 MPa) - sample No 3336r
- test body after triaxial test (confining pressure of 150 MPa) - sample No 3336zo
- test body after triaxial test (confining pressure of 300 MPa) - sample No 3336h

When comparing Figure 3, Figure 5, and Figure 7, a distinct increase of intensity of quartz grain failure corresponding to increasing confining pressure is obvious. No fields with a failure intensity (i.e. sum of crack lengths in quartz grains within a given field) exceeding 17 mm occured in rock material after uniaxial compression test; whereas no fields with a failure intensity below 13.6 mm were observed in the test body loaded at $\sigma_3 = 300$ MPa. At the same time, a cumulation of most intense failured fields is noticeable in samples 3336no (after uniaxial compression test) as well as 3336r ($\sigma_3 = 50$ MPa), forming a zone corresponding to slip plane.

Contrary to this, the cumulation of most intense failured fields is not so distinct or is missing entirely in samples 3336zo ($\sigma_3 = 150$ MPa) and especially

3336h (σ_3 = 300 MPa). Entire preparation exhibits relatively intense failure at confining pressure of σ_3 = 300 MPa.

The analysis of the predominant crack orientations in quartz grains revealed that the cracks after uniaxial compression test as well as after triaxial tests at lower confining pressures (σ_3 = 50 MPa, σ_3 = 150 MPa) are oriented mostly in one or two directions which correspond to directions of the slip surfaces formed during process of sample deformation, see Figure 4 and Figure 6.

A system of pair cracks oriented at an angle of 45° occurs in 40% of the squares after the uniaxial compression test. Another distinct system of pair cracks was observed in 48% of squares after loading at σ_3 = 50 MPa with an angle of about 37° between the cracks. At a confining pressure of σ_3 = 150 MPa, the angle between the most frequently occurring cracks is 37 ° in average observed in 38% of squares. Cracks in remaining squares of these preparations are either oriented in a single direction corresponding to slip plane and others in varing non repetative direction.

Predominate orientation of cracks (although they are very numerous) in the sample tested at a confining pressure σ_3 = 300 MPa is indistinct. A system of pair cracks forming a 33° angle was observed in only 10% of the squares - see Figure 8 and Table 1.

For comparison a thin section of the original rock material was evaluated with this method. It was evident that no preferable orientation exists of the original cracks in quartz grains and that the intensity of failure of quartz grains is substantially lower than in preparations after strength tests.

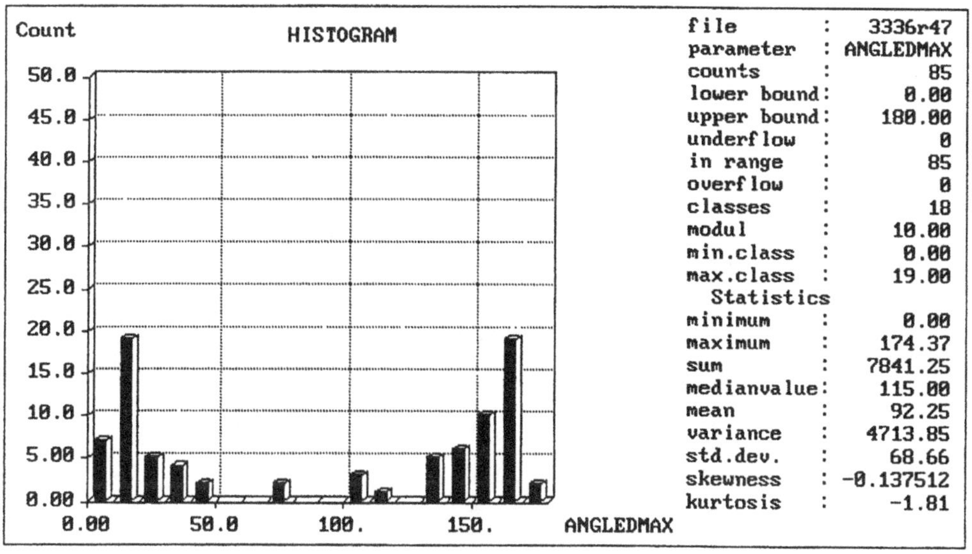

Figure 2. Histogram of crack orientation with respect to x-axis (parallel to long axis of the sample).

Table 1.

Confining pressure σ_3 [MPa]	Σ length - minimum [mm]	Σ length - maximum [mm]	Percentage of squares with pair cracks [%]	Angle between pair cracks [°]
0	3,5	16,9	40	45
50	5,4	22,9	48	37
150	3,8	23,6	38	37
300	13,9	32,3	10	33

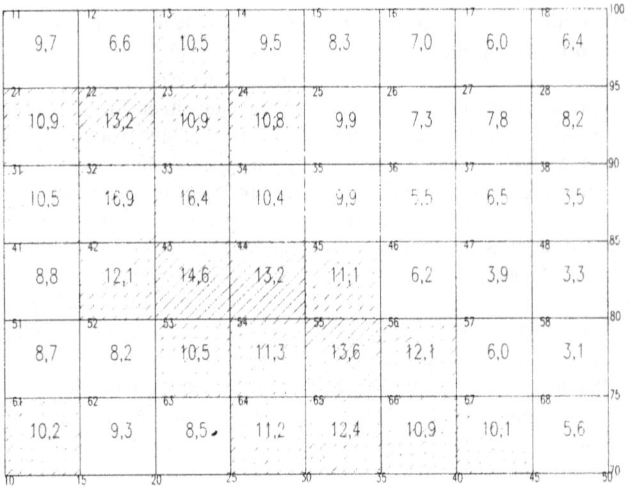

11 9,7	12 6,6	13 10,5	14 9,5	15 8,3	16 7,0	17 6,0	18 6,4	100
21 10,9	22 13,2	23 10,9	24 10,8	25 9,9	26 7,3	27 7,8	28 8,2	95
31 10,5	32 16,9	33 16,4	34 10,4	35 9,9	36 5,5	37 6,5	38 3,5	90
41 8,8	42 12,1	43 14,6	44 13,2	45 11,1	46 6,2	47 3,9	48 3,3	85
51 8,7	52 8,2	53 10,5	54 11,3	55 13,6	56 12,1	57 6,0	58 3,1	80
61 10,2	62 9,3	63 8,5	64 11,2	65 12,4	66 10,9	67 10,1	68 5,6	75

Figure 3. - Failure intensity - sample No. 3336 no (uniaxial strenght).

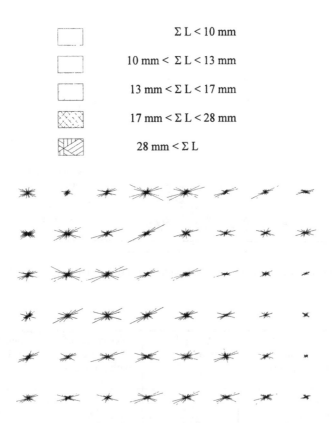

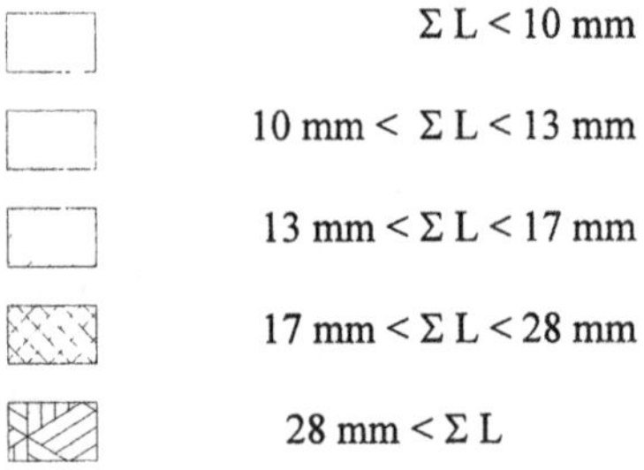

☐ $\Sigma L < 10$ mm

☐ 10 mm $< \Sigma L < 13$ mm

☐ 13 mm $< \Sigma L < 17$ mm

☒ 17 mm $< \Sigma L < 28$ mm

▨ 28 mm $< \Sigma L$

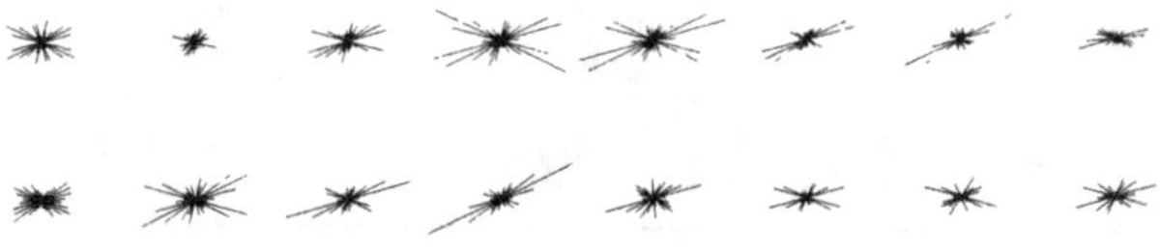

Figure 4. - Rose diagrams of cracks in quartz grains (1 mm ~ 1.4 cracks in this direction) - sample No. 3336 no (uniaxial strenght).

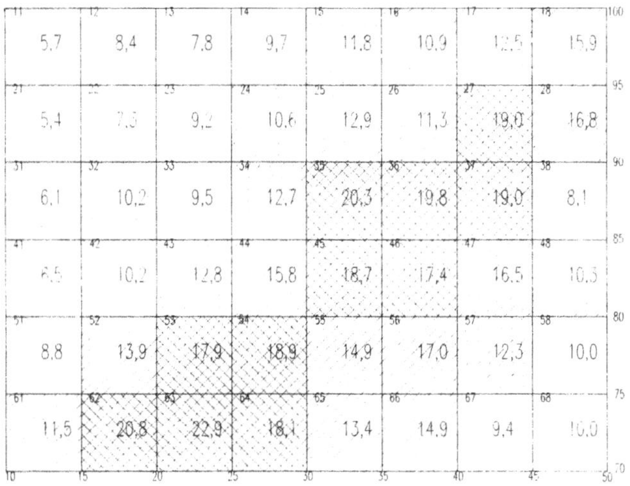

Figure 5. - Failure intensity - sample No. 3336 r (σ_3 = 50 MPa).

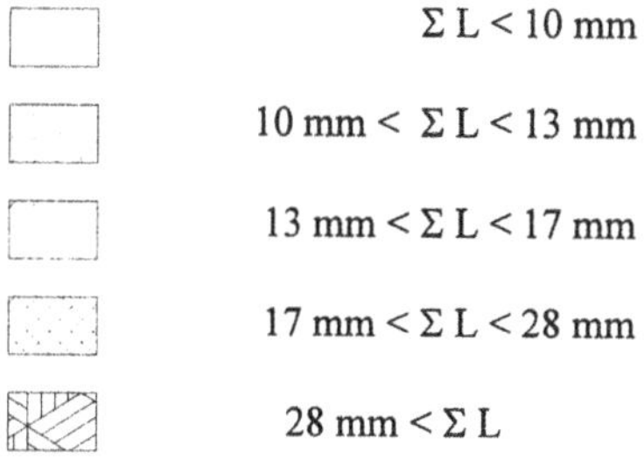

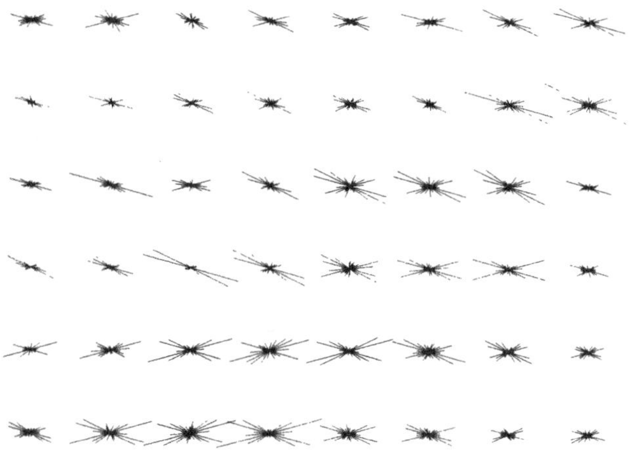

Figure 6. - Rose diagrams of cracks in quartz grains (1 mm ~ 1.4 cracks in this direction) - sample No. 3336 r (σ_3 = 50 MPa).

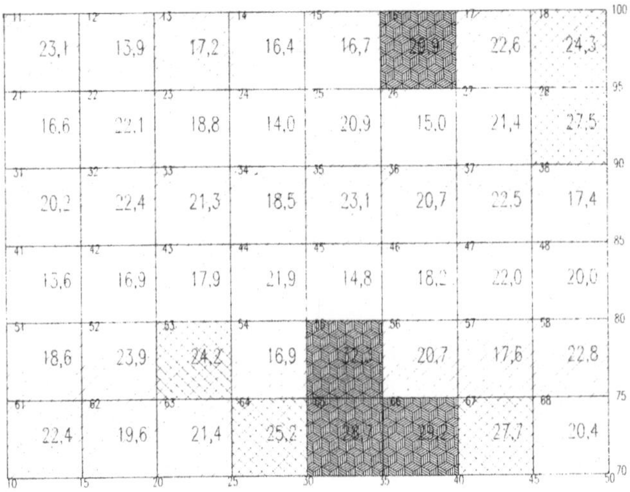

Figure 7. - Failure intensity - sample No. 3336 h (σ_3 = 300 MPa).

Σ L < 10 mm

10 mm < Σ L < 13 mm

13 mm < Σ L < 17 mm

17 mm < Σ L < 28 mm

28 mm < Σ L

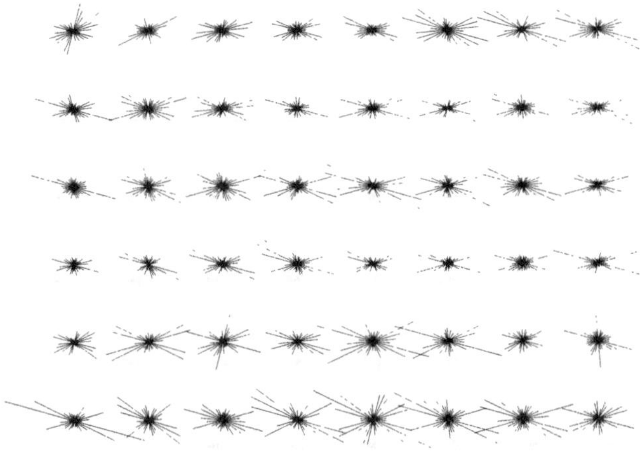

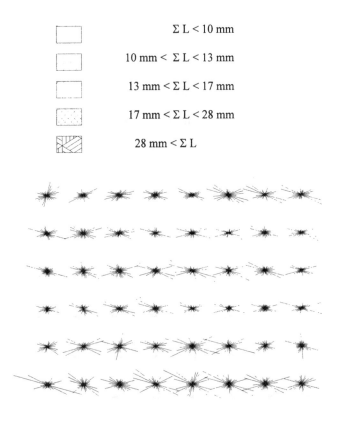

Figure 8. - Rose diagrams of cracks in quartz grains (1 mm ~ 1.5 cracks in this direction) - sample No. 3336 h (σ_3 = 300 MPa).

4. CONCLUSIONS

Failure intensity (i.e. sum of crack lengths in quartz grains) increases with increase of confining pressure and also in slip zones in the case of brittle failure. Also the character of deformation of the test body changes with increase of confining pressure. The deformation of the test body is more homogeneous in the entire volume under higher confining pressures and the failure zone is not close-localised. However, it is obvious from the analysis that the failure is not concentrated to only the slip zone even for samples whose deformation curves correspond to brittle type of failure. An increased number of cracks compared to the original material was also observed outside of the failure zone.

Orientation of cracks is distinct in preparations after both uniaxial compression test and triaxial tests where plastic deformation does not take place distinctly. Cracks in quartz grains are oriented into one or two predominate directions which correspond roughly to directions of slip zones occurring during the deformation process.

This study indicates that stress and strain conditions are reflected also in defects of microscopic size. This finding is in accordance with theories of fracture mechanics.

ACKNOWLEDGEMENTS

Presented work was supported financially by the Grant Agency of the Czech Republic, project No. 105/96/1531. The authors are grateful to Prof. Ove Stephansson and Ph.D. Joanne Fernlund for the review of the manuscript.

REFERENCES

1. Jaeger, C. J. & N.G.W. Cook 1979. *Fundamentals of Rock Mechanics*, 3rd edition. Chapman and Hall, London.
2. Franklin, J. A. & M. B. Dusseault 1989. *Rock Engineering*. McGraw - Hill Publishing Company, New York.

Mechanics of Jointed and Faulted Rock, Rossmanith (ed.) © *1998 Taylor & Francis, ISBN 90 5410 955 6*

Acoustic emission of jointed and intact granite during triaxial compression test

D. Amitrano & D. Hantz
Lirigm, Grenoble, France

ABSTRACT: Acoustic emission (AE) during triaxial compression tests on intact and jointed granite samples was investigated. Artificial rough and smooth joints were tested with confinement pressures from 0 to 80 MPa. The smooth joints were obtained by sawing and rough joints by shear rupture during triaxial compression tests. For the different stages of the sample mechanical behaviour, the following characteristics where analysed : AE count rate, events size distribution and spectral contents corrected for the apparatus response. Our results suggest that AE analysis allows to distinguish different mechanical behaviour : pre-peak, post-peak, shearing on rough or smooth macroscopic surfaces, stick-slip. Moreover the influence of roughness and confining pressure on the b-value is shown.

1 INTRODUCTION

Apart from field observations much studies are dedicated to AE during mechanical tests in laboratory with the aim of better understanding of rock rupture mechanics as well as seismicity. Indeed the similarities of the phenomena physics (wave propagation involved by source fast motion) and the observed statistical distributions have led to consider AE of rock observed in laboratory as a small-scale model for the seismicity in rock masses (rockbursts) or earth crust (earthquakes).

The counting of AE events is the simplest and the most employed parameter for observing the rock fracture or the discontinuity shearing process in field as well as in laboratory. The correlation between AE activity and inelastic strain is obviously shown in a lot of studies (Lockner et al. 1991; Amitrano et al. 1996).

The spectral analysis of AE events has been also investigated but the different studies don't show any concordant results (Ciancara et al. 1991). The use of resonant transducers seems to be the main difficulty for AE spectral analysis.

The observed AE event size distribution are well fitted by power law function :

$$N(>A) = c\,A^{-b} \qquad (1)$$

Where A is the maximum amplitude of AE events, $N(>A)$ is the number of events with maximum amplitude greater than A, a and b being constants. In a log-log representation, this distribution appears linear and b is given by the line slope.

$$\log_{10} N(>A) = a - b \log_{10} A \qquad (2)$$

This distribution exhibits remarkable similarity with the Gutenberg-Richter relation observed for earthquakes.

$$\log_{10} N(>M) = a - bM \qquad (3)$$

where $N(>M)$ is the number of earthquakes with magnitude larger than M in a given area.

Observation of both earthquakes and AE shows variation of the b-value in time and space domains whose explanations are generally given using fracture mechanics theory and/or the self-organised criticality (SOC) concept. Scholz (1968) showed that the b-value decreases before the maximum peak stress of rock specimen and argued for a negative correlation between b-value and stress. Main et al. (1989) observed the same variation but rather invoked a negative correlation between b-value and stress intensity factor K. Following this way Meredith et al. (1990) proposed different patterns of b-value variation driven by the fracture mechanics and the type of rupture (brittle-ductile). Further, authors (Olami et al. 1992; Zapperi et al. 1997) used cellular automata to simulate power-law distribution. The relation between the b-value and the fractal dimension D of AE source location were also investigated (Lockner et al. 1991) and showed decrease of b-value contemporary to strain localisation, i.e. to a decrease in D.

Despite of the fact that discontinuities play a major role in the deformation of rock masses as well as earth crust, few experimental studies have been performed about AE produced during rock joint shearing (Holcomb et al. 1981; Dunning et al. 1985, Li et al. 1990, 1993).

The present paper reports on AE induced in rock samples during the microcracks nucleation and growth before the macro-rupture as well as during the shearing on the joint created by this macro-rupture. This allows us to compare two macroscopically different phenomena: 1) the creation of a macroscopic discontinuity by nucleation, growth and then coalescence of cracks 2) the shearing of this discontinuity. AE produced by this last phenomenon has been compared to that produced by shearing of a smooth sawed surfaces. AE counting, event size distribution and spectral contents of events were analysed.

2 EXPERIMENTAL PROCEDURE

2.1 Rock tested

Triaxial compression tests were performed on Sidobre granite. This rock contains about 71% feldspar, 24.5% quartz, 4% mica and 0.5% chlorite. The density is about 2.65 and the continuity index obtained by sound velocity measurement is about 97%. The sound velocity is about 4 800 m/s. The uniaxial compressive strength is about 190 MPa, the Young modulus 62 GPa and the Poisson coefficient 0.24. The samples were 40 mm in diameter and 80 mm in length. Two series of tests were performed on intact and jointed samples. The smooth joints were obtained by sawing intact samples and were polished to reduce the height of asperities to less than 0.1 mm. The value of the angle between joint surface and the vertical axe was 30° which corresponds to a common value for shearing rupture surfaces obtained during triaxial tests.

2.2 Experimental device

An hydraulical press of 3000 kN capacity has been used. The confining pressure was applied by means of a triaxial cell. The stiffness of the complete loading system (press, piston, sample support) is about 10^9 N/m. The vertical displacement rate was from 1 to 5 μm/s.

Physical Acoustic Corporation (PAC) resonant transducer has been used (peak frequency . 135 kHz effective range frequency · 100 kHz - 1 MHz). The transducer was applied on the outside part of the cell piston which was used as a wave guide. It was connected to a 40 dB preamplifier (PAC 1220A) with adapted filters and, then, to an AE analyser (Dunegan-Endevo 3000 Series) with 40 dB amplification which performed the AE counting. In parallel the signals were digitised after preamplification by means of a fast acquisition board (Imtel T2M50, 8 bits). The sampling frequency was 5 MHz and the duration of the recorded signals was 2048 samples which corresponds to 410 μs. The signal recording trigger was set to 15 mV and the maximal amplitude to 0.5-2 V according to the type of experiment. The board memory segmentation allows to record a few hundred signals per second without dead time.

2.3 Data processing

The AE counting was directly obtained from the analyser. It was very well correlated to AE energy calculated from the digitised signals. The slope of the cumulative curve represents AE activity.

The digitised signals were processed to extract maximal amplitude, energy and spectral contents of each signal. The spectral analysis was performed with correction from the apparatus response using a spectral deconvoluting method proposed by (Guilbert 1995) :

$$E(\nu) = \frac{S(\nu).H^*(\nu)}{H(\nu).H^*(\nu)} \quad (4)$$

Where $E(\nu)$ is the source spectrum, $H(\nu)$ the apparatus transfer function obtained by the Nielsen method (Nielsen 1977) and $S(\nu)$ the spectrum of the digitised signal multiplied by a Hanning function. To avoid numerical amplification a threshold of the term $H(\nu).H^*(\nu)$ was fixed at 1% of the maximum.

The b-value is obtained from the cumulative distribution of events maximum amplitude. The distribution is fitted in a least square sense by a linear function in a log-log representation where the slope gives the b-value. The correlation parameter r^2 is also calculated. The b-value has been calculated for all events recorded during each test. To observe variations of the b-value along the different stages of each test, the b-value has been also calculated for sliding windows of 200 events with an increment of 200 events. According to Pickering et al. (1995) this is an acceptable size of population to calculate the b-value with a good accuracy

3 EXPERIMENTAL RESULTS

3.1 Triaxial tests on initially intact samples

Triaxial tests were performed on initially intact samples with confining pressure from 0 to 80 MPa. The peak states of stress are shown on the Mohr representation in Figure 1.

Four typical stages have been identified in the sample mechanical behaviour. As a typical example the Figure 3 shows the results obtained for a confining pressure of 60 MPa.

Stage 1 is defined by the first linear part of the σ–ε curve. The AE is imperceptible at the chosen amplification level. The b-value is maximum. The stage 2 starts with the beginning of non-linear behaviour. The stress continues to increase, the AE rate increases drastically and continuously and the b-value decreases.

The stage 3 corresponds to the post-peak behaviour. The stress decreases until the macro-rupture which produces a macroscopic discontinuity. The AE activity reaches its maximum value and the b-value is minimum (between 1 and 1.5 regarding to confining pressure).

For high confining pressures (over 40 MPa), the macro-rupture consists in an unique shear surface and is followed by the shearing of this surface. Then, the sample deformation mainly consists in shear along the macro-rupture surface. During this fourth stage, the shear strength is nearly constant and the AE rate decreases slowly. The b-value fluctuates with a mean value somewhat higher than during the stage 3.

The cumulative events size distribution for all the signals recorded during the test can be fitted by a power-law regression with a strong correlation (Figure 2). The study of spectral contents of the signals for the different stages fails to exhibit any significant or systematic pattern.

After dismounting, the joint is filled with powder. The maximal asperity height along lines parallel to the shearing displacement is about 3 mm.

3.2 Triaxial tests on sawed samples

Triaxial tests on samples containing artificial sawed joints were performed with confining pressure from 20 to 80 MPa. The mechanical behaviour exhibits different stages. As an example the Figure 2 presents the results obtained for a confining pressure of 60 MPa. At first the σ–ε curve appears very linear. The stiffness of the sawed sample is equal to than that of an intact one. The AE is imperceptible at the chosen amplification level.

The σ–ε curve appears non-linear after a strain of about 0.005. Then the behaviour is characterised by regular increasing AE activity and a decrease of the σ–ε curve slope until a quasi pure plastic behaviour. In the beginning, the b-value is maximum (near 0.4) and then decreases before the occurring of stick-slip

For the high values of σ_3 (60-80 MPa) typical stick-slip process occurs after a strain of about 0.04. The AE activity is very low after each slip and increases drastically several seconds before the next slip. During the stick-slip process, the b-value stays at a low level.

The events size distribution for the whole tests exhibit very low b-value between 0.25 and 0.7 (Figure 6).

The spectral analysis allows the discrimination between precursor signals occurring few seconds before the slip (type A) and the signals produced during the slip (type B). The type A signals show large amplitudes (between 1 and 2 V) and high frequency contents (Figure 5 A). The type B signals have very large amplitude (full scale saturation) and low frequency (Figure 5 B).

After dismounting, the joint is filled with fine powder and the roughness is larger than before the test: the maximal asperity height reaches 0.5 mm.

3.3 Effect of roughness and confining pressure on the b-value

The b-value observed during the shearing on the rough joints obtained by rupture (range 1-1.6) is much larger than in the case of sawed joints (range 0.25-0.7). This strong effect of the roughness on the b-value is shown in Figure 6.

Moreover, for each behaviour producing AE (before peak, after peak, shearing on rough or smooth surface) the b-value is negatively correlated with the confining pressure (Figure 6). The relations obtained for post-peak and shearing stages are very similar and exhibit the best correlations. On the contrary b-values obtained during pre-peak stages are more variable.

As far as we know, this effect of roughness and confining pressure on the b-value has never been highlighted before.

4 DISCUSSION

Our results are in agreement with the classical observations concerning AE during mechanical laoding of rock samples. AE is perceptible only out of linear behaviour and appears therefore related to

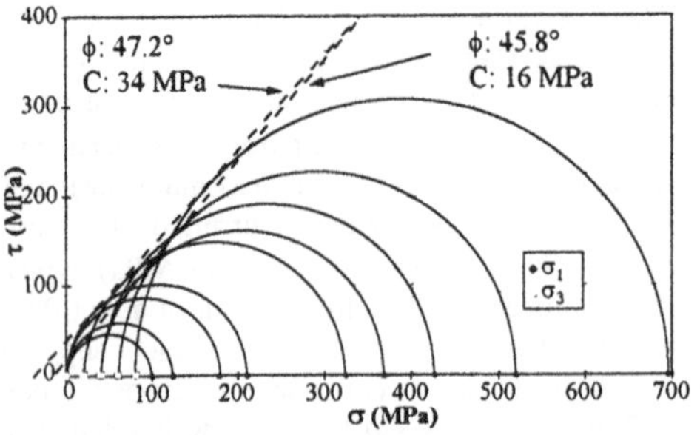

Figure 1 : Mohr representation of the peak state of stress for a triaxial tests series. ϕ and C are given for the inside and outside fitting lines of the Mohr-Coulomb criterion.

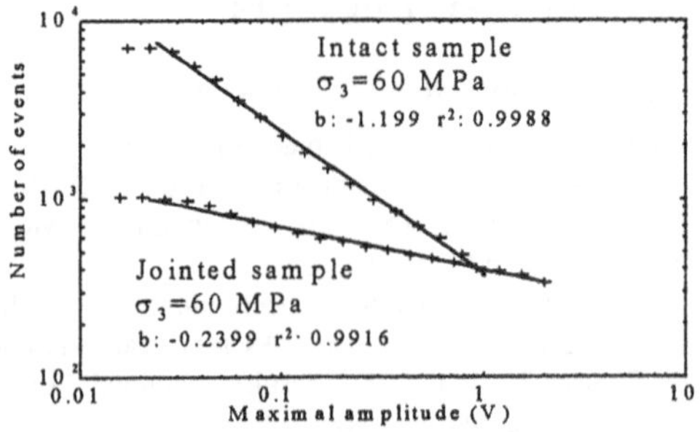

Figure 2 : Cumulative events size distributions for an intact and a jointed sample processed for all the signals recorded during each test.

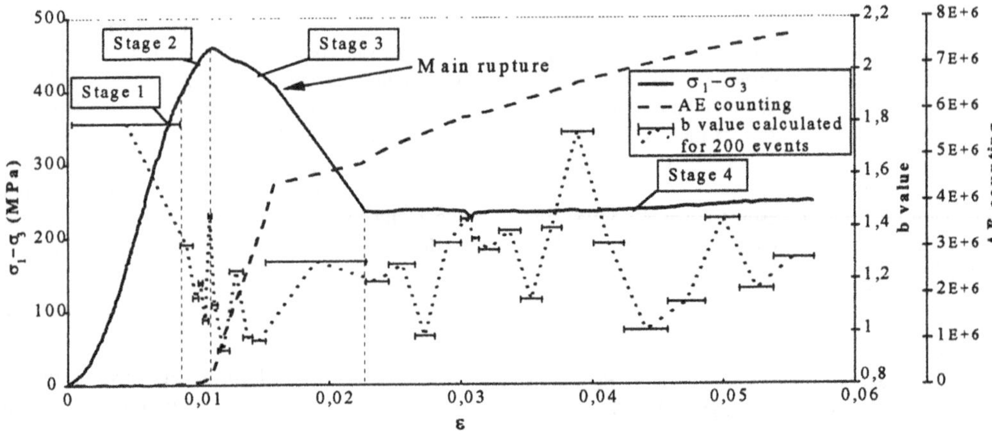

Figure 3 : Triaxial test performed on an initially intact sample with σ_3 = 60 MPa. The difference σ_1-σ_3, the AE counting and the b-value are plotted. The slope of AE counting represents the AE activity. The horizontal bars indicate the duration of 200 events used to calculated each b-value.

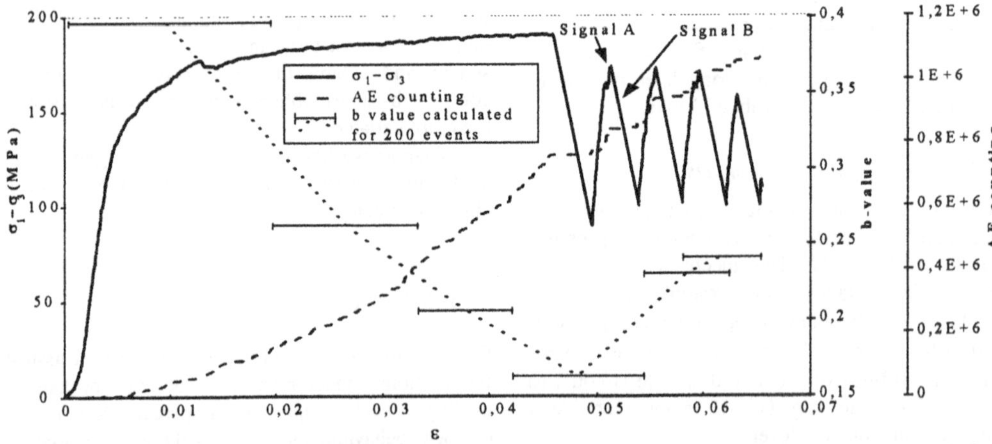

Figure 4 : Triaxial test on a sawed sample with σ_3 = 60 MPa. The two arrows indicate the signal A before a slip and the signal B during the slip whose the spectra are presented in figure 5.

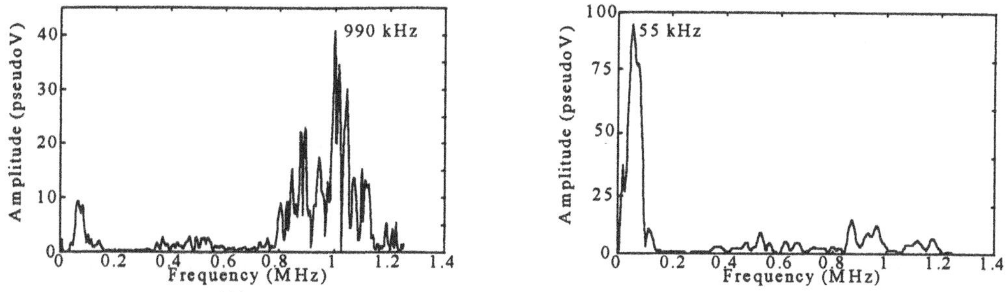

(A) (B)

Figure 5 : Deconvoluted amplitude spectrum of a slip precursor signal (A) and of a signal produced during the slip (B).

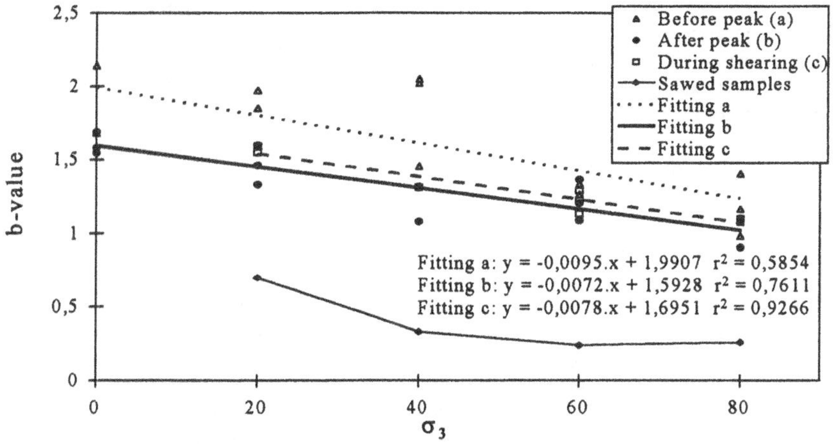

Figure 6 : b-value versus confining pressure for the different stages observed during each test performed on initially intact samples (a-b-c) and sawed samples (d).

inelastic strain, in the case of intact as well as sawed samples .

For an intact sample it increases drastically before the peak stress is reached. This stage is well known to correspond to nucleation and propagation of microcracks. AE activity reaches its maximum value during the post-peak stage which corresponds to the microcracks coalescence leading to the creation of a macroscopic rupture surface. This AE activity increase appears as a simple precursor of the macro-rupture.

During rough surface shearing occurring after the macro-rupture, the AE activity decreases regularly. This decrease may be related to the decrease of roughness which is known to occur as the shear progress.

In the case of sawed sample the AE activity appears to increase as the plastic strain. This may be related to the fact that the roughness is larger after than before the test. This roughness increase indicates that a breaking process occur also for smooth polished surface similarly. This breaking process is similar to the one involved during crack propagation or asperities breaking in the sense that it creates new free surface to release energy.

When the stick-slip process occurs, the AE activity is very low after each slip and increases as a precursor before the next slip.

The spectral analysis is efficient to characterise different behaviours only in the case of stick-slip. Indeed the very low frequency events observed during slip correspond to a large mass motion of all the loading system. On the opposite, the signals generated before the slip correspond to much smaller sources. We assume that the use of resonant transducers allows to distinguish only very different source sizes.

The events size distribution can be well fitted with a power law function whose the pertinent parameter is the b-value. The particularity of our study is that the b-value is obtained by processing digitised signals and not using an AE analogical analyser. This allows us to process the b-value for sliding windows and to verify the good fitting with a power-law for each calculated b-value.

Lockner et al. (1991), using 3D AE events location during triaxial test, showed very well that the b-value decrease corresponds to microcracks localisation. The b-value we obtained for the different behaviours (pre-peak, post-peak, shearing on rough or smooth surfaces) appears negatively correlated with the confining pressure (Figure 6). Therfore we suggest that confining pressure for granite, in the studied range 0-80 MPa, facilitates the plane localisation involving lower b-value during the post-peak stage.

The study of shearing on very rough and smooth joints exhibits different ranges for the b-value : 1-1.6 for rough surfaces 0.25-0.7 for smooth surfaces. We suggest that in the case of smooth surface the lack of obstacles facilitates the occuring of large events, leading to low b-values. On the opposite, in the case of rough surfaces, the asperities restrict large events leading to higher b-values.

The effect of confining pressure on joint shearing can be twice: 1) the better plane localisation during post-peak can result in a less rough rupture surface, 2) a high confining pressure enhances the decrease of roughness during shearing (Cook 1992). These two effects reduce roughness which results in lower b-values.

The studied confining pressure range 0-80 MPa, corresponds roughly to a natural state of stress for a depth range of 0-3200 m. Our results suggest that in this range the b-value variation with depth can reach 0.5.

5 CONCLUSION

AE analysis allows to characterise different mechanical behaviours of intact and jointed samples. The AE activity is closely correlated to non-linear strain and can be use as a precursor of violent ruptures of intact as well as jointed samples. The spectral analysis allow discrimination between precursors and sliding events during the stick-slip process on a smooth surface.

The b-value appears negatively correlated to roughness and to confining pressure.

6 REFERENCES

Amitrano, D., D. Hantz & Y. Orengo (1996). Emission acoustique d'une roche fracturée. *11° Colloque Franco-Polonais en Mécanique des Roches Appliquée*, Gdansk.

Ciancara, B. & K. Takuska-Wegrzyn (1991). Spectral description of seismoacoustic emission. *Vth Conf. AE/MS Geol. Str. & Mat.*, Trans Tech Publication.

Cook, N. G. W. (1992). Natural joints in rock: Mechanical, hydraulical and seismic properties under normal stress. *Int. J. Rock Mech. Min. Sci. & Geomech. Abstr.* 29(3): 198-223.

Dunning, J. D., J. D. Leaird & M. E. Miller (1985). The kaiser effect and frictional deformation. *IVth Conf. AE/MS Geol. Str. & Mat.*, Trans Tech Publication.

Guilbert, J. (1995). Caractérisation des structures lithosphériques sous le Nord-Tibet et sous le Massif Central à partir des données sismologiques du programme lithoscope. Thesis, IRIGM-LGIT. Grenoble.

Holcomb, D. J. & L. W. Teufel (1981). Acoustic emission during deformation of jointed rock. *IIIrd Conf. AE/MS Geol. Str. & Mat.*, Trans Tech Publication.

Li, C. & E. Nordlund (1990). Characteristics of acoustic emission during shearing of rock joints. *Rock Joints*, Balkema, Rotterdam.

Li, C. & E. Nordlund (1993). Acoustic emission / microseismic observations of laboratory shearing tests on rock joints. *Rockbursts an Seismicity in Mines*, Balkema, Rotterdam.

Lockner, D. A. & J. D. Byerlee (1991). Precursory AE patterns leading to rock fracture. *Vth Conf. AE/MS Geol. Str. & Mat.*, Trans Tech Publication.

Main, I. G., P. G. Meredith & C. Jones (1989). A reinterpretation of the precursory seismic b-value anomaly from fracture mechanics. *Geophys. J.* 96: 131-138.

Meredith, P. G., I. G. Main & C. Jones (1990). Temporal variation in seismicity during quasi-static and dynamic failure. *Tectonophysics*. 175: 249-268.

Nielsen, A. (1977). Acoustic emission source based on pencil load breaking. *EWGAE meeting*, Rome.

Olami, Z., H. J. S. Feder & K. Christensen (1992). Self-organised criticality in a continuous, nonconservative cellular automaton modeling earthquake. *Phys. Rev. Lett.* 68(8): 1244-1247.

Pickering, G., J. M. Bull & D. J. Sanderson (1995). Sampling power-law distributions. *Tectonophysics*. 248: 1-20.

Scholz, C. H. (1968). The frequency-magnitude relation of microfracturing in rock and its relation to earthquakes. *Bull. Seismol. Soc. Am.* 58(1): 399-415.

Zapperi, S., A. Vespignani & E. Stanley (1997). Plasticity and avalanche behaviour in microfracturing phenomena. *Nature*. 388(14 august 1997): 658-660.

Mechanics of Jointed and Faulted Rock, Rossmanith (ed.) © 1998 Taylor & Francis, ISBN 90 5410 955 6

Numerical simulation of the Kaiser effect under triaxial stress state

V.L. Shkuratnik & A.V. Lavrov
Moscow State Mining University, Moscow, Russia

ABSTRACT: Three-dimensional numerical simulation of the Kaiser effect in rock samples was carried out using a theoretical model based on the mechanics of wing cracks. Two cycles of axisymmetric triaxial loading were simulated: I-in situ, II-in laboratory. The laboratory loading could be conducted with various values of ratio between axial and confining stress. Dependence of the Kaiser effect on the stress ratio was carefully studied. Based on the simulation results, a new method is proposed for determining values of both in situ principal stresses.

1 INTRODUCTION

The Kaiser effect takes place when rock samples, extracted out of the rock mass, are tested in laboratory. The effect consists in a jump-like increase in acoustic emission activity of the sample, when its stress state nears the rock stress state in situ. This allows us to use the Kaiser effect to estimate geostresses.

The major problem here is that the original in situ stress state can not be completely restored in the laboratory test. This is due to the fact that the laboratory loading is normally carried out as a uniaxial or triaxial axisymmetric compression, while the rock was under triaxial axisymmetric or more complicated stress state in situ.

Some progress has been made to investigate the Kaiser effect under triaxial stress state experimentally (Holcomb 1983, Hughson & Crawford 1987, Holcomb 1993, Li & Nordlund 1993). Numerical simulation of the Kaiser effect with triaxial axisymmetric loading in the 1st cycle ("in situ") and uniaxial loading in the 2nd cycle ("laboratory") has shown a good agreement with experimental results (Shkuratnik & Lavrov 1997, Lavrov 1997).

The purpose of this study is to investigate the Kaiser effect in triaxial axisymmetric laboratory test.

2 THEORETICAL MODEL AND SIMULATION PROCEDURE

Before being loaded in situ (cycle I), the rock is supposed to contain disk-like initial microcracks. The cracks are disposed far enough from each other, so that there is no crack interaction during the loading. Crack planes are oriented chaotically. The angles α, β, and γ between the normal vector to the crack plane and the directions of the principal stresses in the 1st cycle ("in situ") are random values (Fig.1). Each disk-like crack has radius a =0.002m. The coefficient of friction between crack faces μ and the critical stress intensity factor K_{IIc} are equal for all crack (friction and strength microhomogenity).

In the 1st loading cycle ("in situ"), the rock is loaded with the principal stresses σ_1^I, σ_2^I, σ_3^I. Here, the top index (I or II) corresponds to the number of the loading cycle.

In the 1st loading cycle, normal p_n^I and shear p_τ^I stresses act on the disk-like crack plane. The shear stress causes a displacement over the crack plane in the direction which is defined by the unit vector $\vec{e}_\tau^{\,I}$ (Fig.1). If the condition

$$K_{II} > K_{IIc} \tag{1}$$

is fulfilled on the crack contour, two tensile cracks (wing cracks) arise as it is seen in Figure 1 (Adams & Sines 1978, Dyskin et al. 1995). Here, K_{II} is the stress intensity factor given by

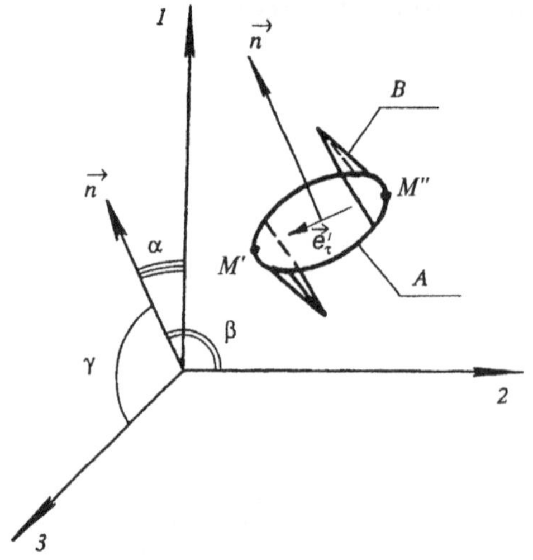

Figure 1. Initial disk-like crack (*A*) with two tensile cracks (*B*): 1, 2, 3 – axes of principal stresses of the first cycle σ_1^I, σ_2^I, σ_3^I

$$K_{II} = 2 p_{eff}^I \sqrt{\frac{a}{\pi}} \tag{2}$$

where $p_{eff}^I = p_\tau^I - \mu p_n^I$ is the effective shear stress over the crack plane in the 1st cycle. In case p_{eff}^I calculated from the above equation is negative, it is supposed to be zero (static friction).

The two wing cracks emerge in two diametrically opposite points M' and M'' on the initial crack contour, so that diameter $M'M''$ is parallel to $\vec{e}_\tau^I$. After completing the 1st loading cycle, the lengths of the wing cracks are strictly determined by the values of the effective shear stresses reached in the 1st cycle.

In the 2nd cycle ("laboratory"), the rock is also loaded axisymmetrically. The stress state in the 2nd cycle is $\sigma_1^{II} > \sigma_2^{II} = \sigma_3^{II}$, and the direction of σ_1^{II} is identical with the direction of σ_1^I in the 1st cycle. In other words, the long axis of the rock sample coincides with the direction of the maximum normal stress in situ.

There are three possibilities for a crack to grow in the 2nd cycle:

1. No wing cracks were born by the disk-like crack in the 1st cycle (condition (1) was not fulfilled). In this case, the disk-like crack will give two wing cracks as soon as condition (1) is fulfilled in the 2nd cycle.

2. Two wing cracks were born by the disk-like crack in the 1st cycle, and projection of the shear stress in the 2nd cycle onto the shear direction in the 1st cycle is negative: $(\vec{e}_\tau^{II} \cdot \vec{e}_\tau^I) < 0$. Here, the unit vector $\vec{e}_\tau^{II}$ points the shear direction over the disk-like crack plane in the 2nd loading cycle.

Wing cracks, which appeared in the 1st cycle, close in the 2nd cycle. The condition for new wings to emerge in the 2nd cycle is (1).

3. Two wing cracks were born by the disk-like crack in the 1st cycle, and projection of the shear stress in the 2nd cycle onto the shear direction in the 1st cycle is positive: $(\vec{e}_\tau^{II} \cdot \vec{e}_\tau^I) > 0$.

In this case, shear stress has a positive component which promotes the present wing cracks to continue growing in the 2nd cycle. The condition for them to restart propagating in the 2nd cycle is given by

$$p_{eff}^{II} \cdot (\vec{e}_\tau^{II} \cdot \vec{e}_\tau^I) \geq p_{eff}^I \tag{3}$$

where p_{eff}^{II} is the effective shear stress in the 2nd cycle and is defined in the same way as p_{eff}^I in the 1st cycle.

The growth of the wing cracks is accompanied with AE. In our model, this is the only source of AE signals.

Note that the cracks of type 3 only are responsible for the Kaiser effect because only for these cracks the start condition is connected with the stresses of the 1st cycle.

In the course of the computer simulation, a rock sample containing 1000 disk-like cracks was supposed to be loaded in two successive cycles. For each crack in the 1st cycle, condition (1) was checked up. The values of p_{eff}^I were memorized for the cracks that gave wings in the 1st cycle (condition (1) fulfilled).

In the 2nd cycle, proportional loading of the rock sample was simulated. The ratio $F^{II} = \sigma_1^{II} / \sigma_2^{II}$ remained constant during the loading. The value of σ_1^{II} was increased from zero to a certain maximum value in equal steps (normally 1 MPa). The value of σ_2^{II} was increased accordingly. The crack type (1, 2, or 3) was determined at each stress level for each crack. The corresponding growth condition was checked up for each crack according to its type. Further, the total number *W* of growing cracks was calculated at each stress level. This value *W* was assumed to be a measure for AE activity at the given stresses

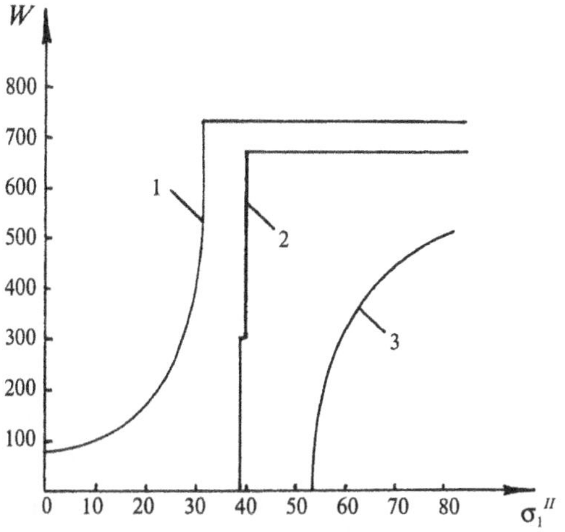

Figure 2. Dependences of AE activity measure on the axial stress in the 2nd cycle for $F^{II}=6$ (curve 1), $F^{II}=4$ (curve 2), $F^{II}=3$ (curve 3); $\sigma_1^I=40$ MPa, $\sigma_2^I=\sigma_3^I=10$ MPa, $\mu=0.3$, $K_{IIc}=0$ for all curves

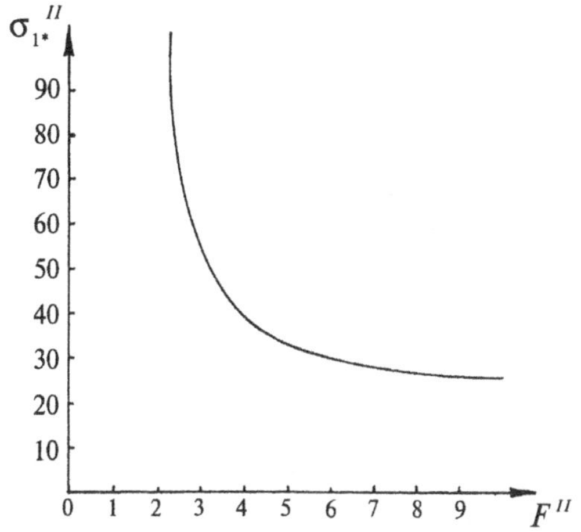

Figure 3. Dependence of the Kaiser effect axial stress on the ratio of the principal stresses in the 2nd cycle for $\sigma_1^I=40$ MPa, $\sigma_2^I=\sigma_3^I=10$ MPa, $\mu=0.3$

σ_1^{II}, σ_2^{II}. As a result, dependences "AE activity measure versus stress σ_1^{II}" were obtained for various combinations of the 1st cycle stresses, for different values of the friction coefficient, and different values of the ratio between the principal stresses in the 2nd cycle.

3 SIMULATION RESULTS

The Kaiser effect is present in the 2nd cycle if the following condition was fulfilled in the 1st cycle:

$$\sigma_1^I > (k+1)\sigma_2^I \tag{4}$$

where

$$k = \frac{2\mu}{(1+\mu^2)^{1/2}-\mu}. \tag{5}$$

If condition (4) was not completed in the 1st cycle, there would be constant high AE activity right from the start of the 2nd cycle loading, independent of the ratio $F^{II}=\sigma_1^{II}/\sigma_2^{II}$.

If condition (4) was fulfilled in the 1st cycle, there will be the Kaiser effect in the 2nd cycle. Some examples of the dependence "AE activity measure versus stress σ_1^{II}" are given in Figure 2. Here, a rock with $\mu=0.3$ was loaded with the stresses $\sigma_1^I=40$ MPa, $\sigma_2^I=\sigma_3^I=10$ MPa in the 1st cycle ("in situ"). Curve 1 corresponds to the laboratory loading with $F^{II}=\sigma_1^{II}/\sigma_2^{II}=6$; curve 2 is for $F^{II}=4$; curve 3 is for $F^{II}=3$. The value of K_{IIc} is zero for all three curves.

Computer experiments have shown that decreasing F^{II} from ∞ (uniaxial laboratory test) over $F^{II}=F^I=\sigma_1^I/\sigma_2^I$ down to $F^{II}=k+1$ leads to increasing σ_{1*}^{II}. Here, σ_{1*}^{II} is the value of the axial stress, at which the Kaiser effect in the 2nd cycle takes place, and is given by

$$\sigma_{1*}^{II} = \frac{F^{II}}{F^{II}-k-1}\left[\sigma_1^I-(k+1)\sigma_2^I\right] \tag{6}$$

The dependence "σ_{1*}^{II} versus F^{II}" is shown in Figure 3.

The shape of the curve $W=f(\sigma_1^{II})$ is also influenced by F^{II}. Decreasing F^{II} from ∞ (uniaxial laboratory test) to $F^{II}=F^I=\sigma_1^I/\sigma_2^I$ leads to decreasing "background" AE activity at lower stress values (at $\sigma_1^{II}<\sigma_{1*}^{II}$). When the stress ratio F^{II} becomes equal to the stress ratio in situ F^I, the curve takes the shape of an ideal step exposed on the zero background (curve 2 in Figure 2). This is the ideal Kaiser effect. Further decrease in F^{II} down to $(k+1)$ changes the curve shape again (curve 3 in Figure 2).

The Kaiser effect is also influenced by the coefficient of friction between crack faces. Rocks with higher value of μ are characterized by higher "background" AE at lower stresses. In rocks with $\mu = 0$, the AE curve for any value of F^{II} is shaped as an ideal rectangular step.

The influence of the critical stress intensity factor on the Kaiser effect was also studied. Increasing K_{IIc} is accompanied with decreasing "background" AE at lower values of σ_1^{II}, but the value of σ_{1*}^{II} remains constant (see also Lavrov 1997). Most of our computer experiments were carried out for the most difficult case $K_{IIc} = 0$. When the 2nd cycle is conducted with $F^{II} = F^I$, varying K_{IIc} leads only to changing AE level at $\sigma_1^{II} > \sigma_{1*}^{II}$. In this case, the emission curve is always shaped as a step, and "background" AE level is zero at any value of K_{IIc}.

4 DISCUSSION

The simulation results have shown that axisymmetric proportional laboratory tests with different values of the stress ratio $\sigma_1^{II}/\sigma_2^{II}$ are to be carried out to estimate geostresses using the Kaiser effect in rock samples extracted from the rock mass. The long axis (σ_1^{II}-axis) of all rock samples should be oriented in the direction of the maximum stress in situ (σ_1^I) (it is supposed that the in situ stress state was axisymmetric). The ideal Kaiser effect, i.e. AE activity jump on the zero "background", would correspond to the loading of the sample with the same stress ratio as it was loaded in situ: $\sigma_1^{II}/\sigma_2^{II} = \sigma_1^I/\sigma_2^I$. In this case, the value of the axial laboratory test stress σ_{1*}^{II}, at which the AE activity abruptly increases, is exactly equal to the stress σ_1^I in situ (complete restoration of the in situ stress state). Both of the in situ stresses can be obtained in this way.

Note that stress measurement based on the Kaiser effect has natural applicability limits. To be estimated from the Kaiser effect, the maximum principal stress in situ must exceed a certain threshold value. If the in situ stress σ_1^I was lower than this threshold, the disk-like cracks did not form wing cracks, therefore the rock sample behaves as a "fresh" one in the laboratory test. The threshold stress is proportional to the confining stress in situ σ_2^I (equation (4)).

Authenticity of the simulation results is confirmed through their stability to varying parameters μ and K_{IIc}. Besides, the quantitative relations obtained correspond with the results of experiments carried out by Holcomb (1983) and Li & Nordlund (1993).

5 CONCLUSIONS

The results of the Kaiser effect simulation with axisymmetric loading in two succesive loading cycles ("in situ" and "in laboratory") allow us to make the following conclusions:

1. For the given confining stress in situ, there is a threshold value of the in situ axial stress to be found out using the Kaiser effect in a laboratory test. If the in situ axial stress did not exceed the threshold value, there is no Kaiser effect in the laboratory test at any principal stress ratio. The threshold value is proportional to the confining stress in situ.

2. In case the in situ axial stress exceeded the threshold, varying ratio of the principal stresses in the laboratory proportional loading results in changing shape of the dependence "AE activity versus stress"; in particular, the background AE at lower stresses changes. If the ratio of the axial stress and the confining stress in the laboratory test is equal to that in situ, the Kaiser effect appears most distinctly. This can be used to estimate the ratio of the in situ principal stresses.

3. Varying critical stress intensity factor does not have any substantial influence on the simulation results.

ACKNOWLEDGEMENT

A.V.Lavrov gratefully acknowledges the support of the International Soros Science Education Program (ISSEP).

REFERENCES

Adams, M. & G.Sines 1978. Crack extension from flaws in a brittle material subjected to compression. *Tectonophysics* 49(1/2):97-118.

Dyskin, A.V., L.N.Germanovich, R.J.Jewell, H.Joer, J.S.Krasinski, K.K.Lee, J.-C.Roegiers, E.Sahouryeh & K.Ustinov 1995. Some experimental results on three-dimensional crack propagation in compression. In H.-P.Rossmanith (Ed.), *Mech. Jointed & Faulted Rock*: 91-96. Rotterdam: A.A.Balkema.

Holcomb, D.J. 1983. Using acoustic emission to determine in-situ stresses: problems and promise. *Proc. Appl.Mech., Bioengng & Fluids Engng Conf.*:11-21.Houston.
Holcomb, D.J. 1993. Observation of the Kaiser effect under multiaxial stress states: implications for its use in determining in situ stress. *Geoph.Res.Letts.* 20(19): 2119-2122.
Hughson, D.R. & A.M.Crawford 1987. Kaiser effect gauging: the influence of confining stress on its response. In G.Herget & S.Vongpaisal (Eds.), *Proc.6th Int.Congr.Rock Mech.*, v.2: 981-985. Rotterdam: A.A.Balkema.
Lavrov, A.V. 1997. Three-dimensional simulation of memory effects in rock samples. In K.Sugawara & Y.Obara (Eds.), *Rock Stress:* 197-202. Rotterdam: A.A.Balkema.
Li, C. & E.Nordlund 1993. Experimental verification of the Kaiser effect in rocks. *Rock Mech.Rock Engng* 26(4): 333-351.
Shkuratnik, V.L. & A.V.Lavrov 1997. Dreidimensionale Computersimulation des Kaiser-Effektes von Gesteinsproben bei triaxialer Belastung. *Glückauf-Forschungshefte* 58(2): 78-81 (in German).

Mechanics of Jointed and Faulted Rock, Rossmanith (ed.) © *1998 Taylor & Francis, ISBN 90 5410 955 6*

Acoustic emission preceding macro crack formation in samples containing a stress concentrator

B.Van de Steen
Department of Civil Engineering, KU Leuven, Belgium

M.Wevers
Department of Metallurgy and Materials Engineering, KU Leuven, Belgium

ABSTRACT: Brazilian tests on samples with a stress concentrator in the form of a circular hole, raised questions with regard to the material behaviour in the vicinity of the stress concentrator. To study the damage evolution and fracture initiation processes in function of the stress history, a number of experiments were carried out whereby the acoustic emission was recorded. After the discussion of the difficulties posed by the set up and the remedies taken, the attention is focussed on the data processing techniques. The test results appertaining to three samples are discussed next, whereby each sample represents a different kind of behaviour as far as fracturing pattern and acoustic behaviour is concerned. This allows to draw some conclusions regarding the micro cracking in the vicinity of the stress concentrators.

1 INTRODUCTION

In tests on rock, carried out in order to study the fracture and damage initiation at stress concentrators, the material either did not yield at the stress concentrator unless the linear-elastically calculated tensile stresses exceeded 3 to 4 times the tensile strength, or the samples did not fail at all at the stress concentrator, although the material at the stress concentrator was subjected to stresses a few times the tensile strength (Van de Steen, unpublished, Van de Steen et al., in press). Microscopic inspection (x 120) of the samples that did not fail at the stress concentrator, did not reveal evidence of damage at the stress concentrators. The attention was therefore turned to the acoustic emission (AE) technique to establish whether any evidence of damage initiation and damage evolution could be picked up prior to fracture. It was further aimed to determine the load level at which the first noticeable activity could be registered, and to determine the approximate position of any possible AE activity prior to fracture. The stresses mentioned in this paper are all calculated by means of a finite element programme, whereby the material is assumed to behave linear-elastically up to the point of fracture or damage initiation.

2 EQUIPMENT AND SAMPLE CONFIGURATION

The samples on which the tests were carried out, consisted of crinoide limestone cores with a diameter of 143 mm and a thickness ranging between 30 mm and 40 mm. They were subjected to a vertical, diametrical load. On the horizontal diameter, a hole was drilled parallel to the core axis. The distance from the centre of the sample to the centre of the hole varied between 20.7 mm and 37.3 mm, while the diameter of the hole was approximately 37.5 mm or 22.5 mm. The tensile strength of the limestone used for the tests is approximately 10 MPa.
The samples were tested with a 5000 kN Dartec press operated in subcell mode (i.e., maximum reach 500 kN). The press was displacement controlled at a rate of 0.005 mm/s. This displacement rate corresponded during the last 20 seconds to a stress increase of approximately 0.06 to 0.12 MPa/s along the vertical diameter. A Mistras 2001 fully digital system from Physical Acoustics Corporation (1995) with two sensors was used to register the AE activity. The sensors were single piezoelectric transducers with a resonance frequency of 300 kHz and a frequency domain of 150 to 325 kHz. Dow Corning high vacuum grease was used as a couplant for the sensors. A 1220A preamplifier with a band pass filter with a bandwidth of 100 to 1200 kHz and an amplification of 40 dB was selected. All measurements were carried out with a fixed threshold value of 40 dB, the signal sampling rate was set at 4 MHz. The peak definition time was set at 30 μs, the hit definition time, determining the end of the event was set at 100 μs; these values were recommended for non-metallic materials and in line with the values obtained by rules of thumb given by the manufacturer of the equipment. These settings were later validated by calibration tests (see further). As rock is a high damping material, and

the chance for late arrivals due to reflections is thus reduced, the hit lockout time was set at 300 μs, the practically attainable minimum value. The registered signal characteristics are time of occurrence, amplitude, energy, number of ring down counts, the signal duration, the rise time, the event position and the load as external parameter.

The sample configuration and the positioning of the sensors is shown in Figure 1. The position of the AE event (further called event) is calculated by comparing the arrival time of an event at one sensor to the arrival time of that same event at the other sensor. If the velocity of the elastic wave and the exact location of the sensors are known, the location of the event can be calculated. The use of only two sensors has however an important limitation as far as event localization is concerned. Strictly spoken, only a linear positioning is possible with two sensors. All events, situated on a hyperbola, which has the sensors as its foci, will give rise to the same arrival time offset (Fig. 1). As the aim of the experiments is to establish the difference in AE activity around the hole compared to the AE activity along the vertical diameter, the sensor position shown in Figure 1, allows to differentiate the events along the vertical diameter from events occurring elsewhere.

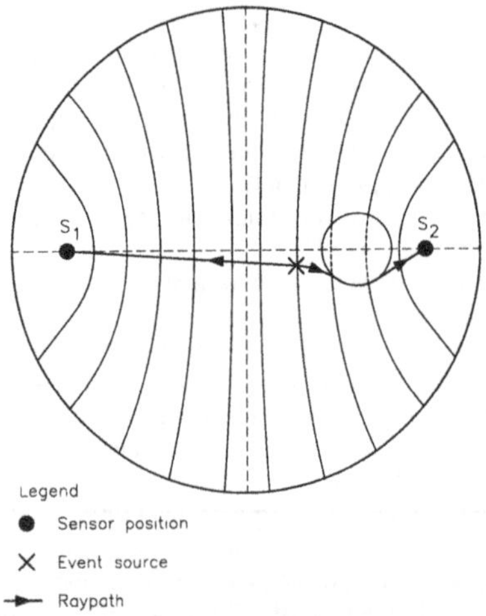

Figure 1: Sample configuration, showing the hyperbola of equal apparent positions, and the raypath approximation used for events shielded by the hole.

Before each experiment, a calibration test was carried out. A pencil lead was broken at a number of predetermined positions, and the AE events associated with the breaking of the pencil leads were recorded (Hsu & Hardy). The aim of this calibration was fivefold. The calibration was firstly used to determine the wave velocity in the limestone, which was approximately 5300 m/s. It secondly allowed to confirm that the selected settings described previously were appropriate for the material and the configuration under consideration. It thirdly provided a means to establish whether the coupling sensor-sample was satisfactory; it fourthly assisted in determining which additional filters had to be devised to get rid of other possible anomalies (see further), and it finally allowed to assess the influence of the hole on the wave propagation.

When the hole is situated in between the acoustic source and one of the sensors, the elastic wave has to cover an additional distance to reach this sensor. This implicates, that the apparent event positions of these events that are shielded from one of the sensors by the hole, will be shifted away from the sensor in question (and therefore apparently away from the hole). The additional distance travelled by the elastic wave, can be estimated by assuming that the raypath of the first arrival follows two straight lines and an arc: the two straight lines both tangent to the hole with the one going through the sensor and the other through the event location, and the arc being formed by the boundary of the hole, in between the two tangent points (Fig. 1).
Notwithstanding this additional difficulty, the two sensor configuration remains valuable in differentiating between events being generated in the vicinity of the hole and events being generated along the vertical diameter.

3 DATA PROCESSING TECHNIQUES

In this paragraph, a short overview will be given of a number of additional filters, used during the data processing.

3.1 *Duration/amplitude filter*

As the high amplitude signal duration is generally longer than the low amplitude signal duration (Pollock, 1995), a plot of the signal duration versus the amplitude, should form a diagonal band across the plot area. Signals under this band, i.e., high amplitude signals of short duration, are usually due to electromagnetic interference (Pollock, 1995). Signals above the band are due to a source which are extended in time, such as friction processes associated with broken pieces of material that slide over each other. As these kinds of processes are not related to the phenomena being investigated, these events were filtered out. The events removed by this filter were all friction type signals, and the number of events removed varied between 0.2 % and 1.1 % of the total number of hits.

3.2 *Energy filter and duration filter*

From the calibration measurements, a number of phantom signals came to the foreground. The "true" signals were all characterized by a ratio of the energy at the sensor hit first to the energy at the sensor hit second of in between 0.5 and 3. The ratio of the signal duration at the sensor hit first to the signal duration at the sensor hit second was in between 0.5 and 2.5. If no hole would be present, these ratios would be expected to be greater than 1, because the signal that has travelled the longest distance can be expected to have been most weakened. This can readily be accepted for the energy, but is also valid for the duration of the signal. While the full signal duration can be extended by a longer travelling time as a consequence of the dispersion of the stress wave (Wevers, 1996), the fact that the signal is dominated by the resonance frequency, and that the dampened tail disappears under the 40 dB threshold, means that a longer travelling distance of the wave also implies a shorter recorded event duration. The presence of the hole, has also its repercussions on the signal duration and its energy content. The hole not only increases the travelling distance, it is also responsible for an additional dampening of the signals that pass it. As a consequence, the signal that reaches the sensor situated behind the hole, is more attenuated than what the travelled distance would normally imply. If the sensor situated behind the hole is first hit, the energy or duration ratio can therefore be less than 1. As the maximum distance a signal can travel without having been reflected is limited, the loss in energy and the decrease of the duration of the signal also have an upper limit.

The number of events removed by the duration filter amount to12 % to 24 % of the total number of hits, while the number of events removed by the energy filter amounts to 10 % to 24 % of the total number of hits.

4 EXPERIMENTAL RESULTS

Whether the sample failed along the vertical diameter or at the hole serving as a stress concentrator depends on both the size and the position of the hole (Van de Steen, unpublished). The samples can be subdivided into three categories: these in which most of the AE activity was concentrated along the vertical diameter; these in which the events mainly originated in the vicinity of the hole; and these in which the main AE activity was situated both along the vertical diameter and in the vicinity of the hole. The samples in which the AE activity was mainly concentrated at the hole, failed at this stress concentrator. The two other categories failed along the vertical diameter like in the conventional Brazilian test. It was the size and the eccentricity of the hole that determined the failure behaviour and hence determined to which of the three categories the sample belonged. The bigger holes combined with the smaller eccentricities favoured fracture and AE activity at the hole, while for the smaller holes, combined with the greater eccentricities the AE activity was concentrated along the vertical diameter.

In further discussions, a sample representing each category will be discussed. Their characteristics are summarised in Table 1.

The histograms showing the number of events in function of the position and the histograms of the number of events in function of the time for each category, as well as the position-time plots are given in Figures 2-4. The distances indicated on the position axes, are the distances from the sensor *S1* (Fig. 1), situated at the left. The samples have been oriented such that the hole is situated at the right-hand side of the sample, closest to the second sensor *S2* (Fig. 1).

Table 1: Characteristics of the 143 mm diameter samples representing each of the three AE activity location categories.

Name	Hole size (mm)	Eccentr. (mm)	Behaviour
LMDg_5	20.6	38.2	Activity mainly along vert. diam. Split along vertical diam.
LMDj_6	37.3	23.5	Activity mainly at hole. Failed at hole. Second. fract. at platen contact.
LMDj_4	38.4	36.4	Activity mainly along vert. diam. and at hole. Split along vertical diam.

5 DISCUSSION

Although the stresses in the vicinity of the hole in sample LMDg_5 were 2.8 times the stresses along the vertical diameter, Fig. 2a clearly shows that no extra AE activity was registered in the vicinity of the hole (centred at 95 mm), but that most of the AE activity was recorded along the vertical diameter. The influence of the hole on the apparent event position is also well illustrated in this figure. While the centre of the sample was situated at 57 mm, and the main fracture was situated at 61 mm, the average of the event histogram, is situated at 52.9 mm. On Figure 2b, one notices two distinct AE activity peaks marked *1* and 2. The second AE activity period, which lasts only 1.5 seconds is clearly associated with the final cracking process. In Figure 2c, which is a combination of the Figures 2a and 2b, one observes that the events that corresponds to peak *1* in Figure 2b have an apparent position situated in between 30 and 70 mm.

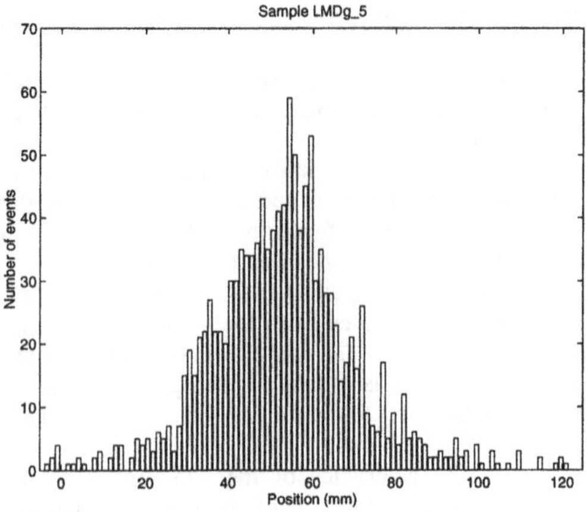

Figure 2a

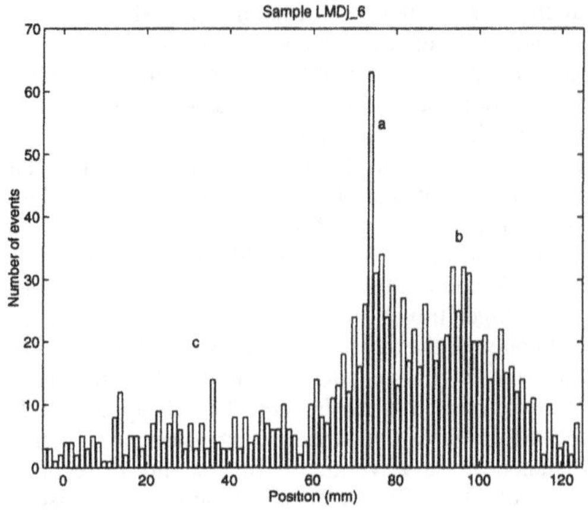

Figure 3a

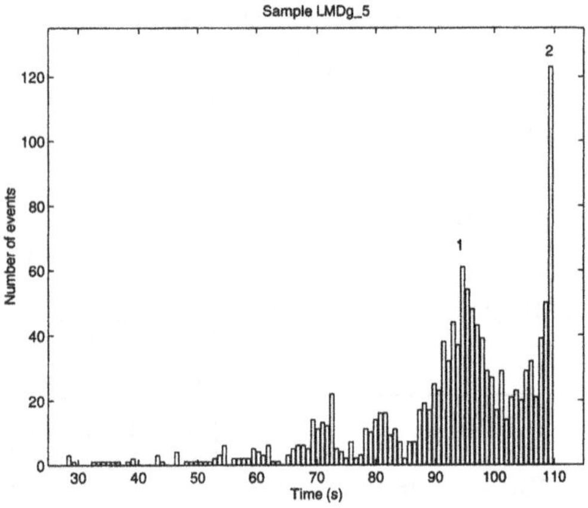

Figure 2b

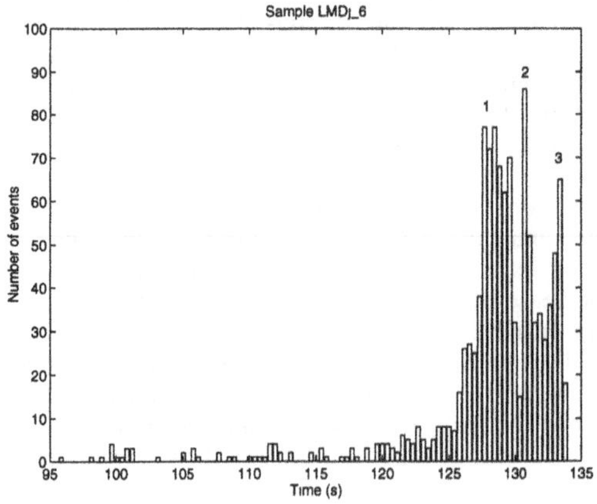

Figure 3b

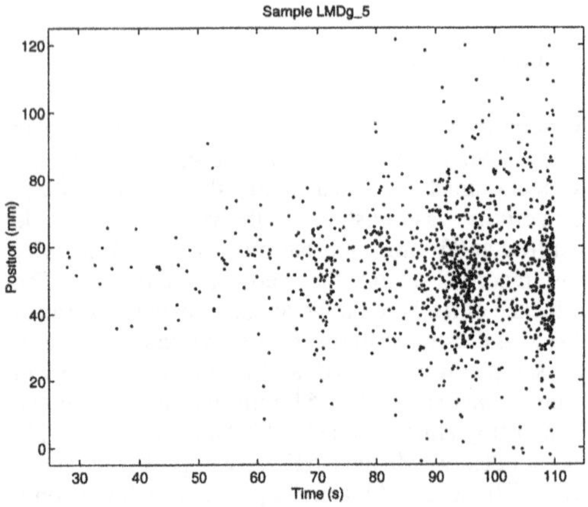

Figure 2c

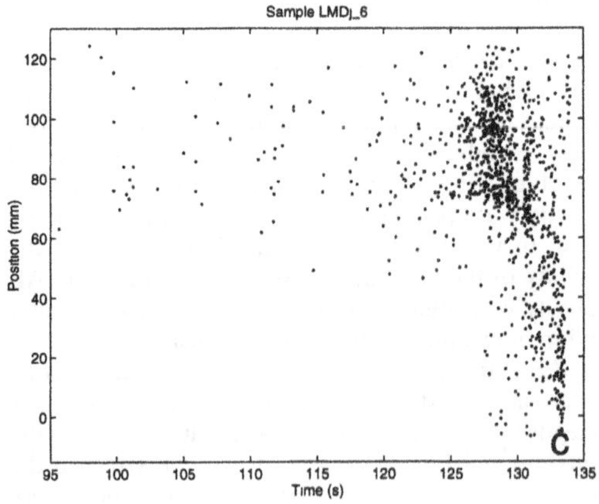

Figure 3c

Figure 2: Histogram of the event localisation (a), and their time distribution (b), supplemented with the time-position point plot (c) of sample LMDg_5.

Figure 3: Histogram of the event localisation (a), and their time distribution (b), supplemented with the time-position point plot (c) of sample LMDj_6.

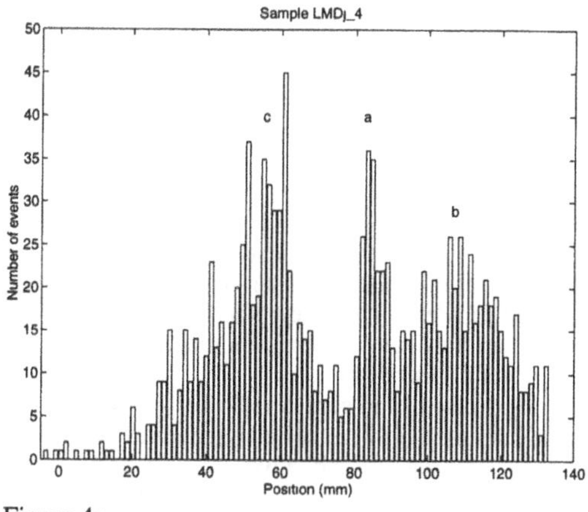

Figure 4a

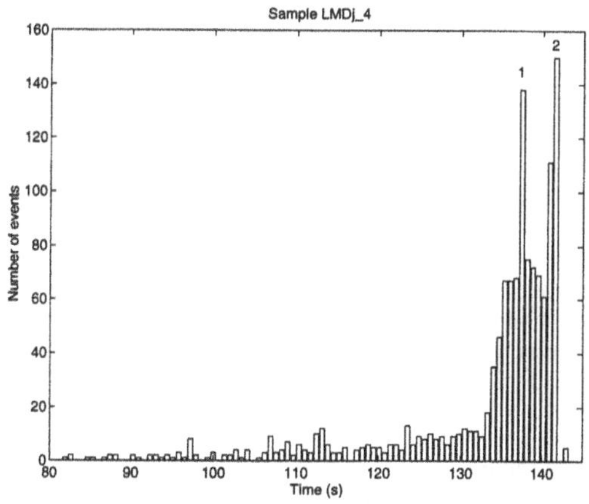

Figure 4b

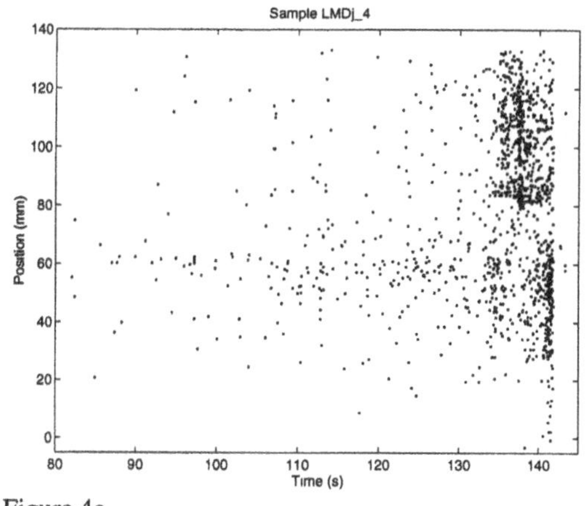

Figure 4c

Figure 4: Histogram of the event localisation (a), and their time distribution (b), supplemented with the time-position point plot (c) of sample LMDj_4.

This could in first instance be interpreted as micro cracking along the vertical diameter. However, when one takes account of the flaking that was visible at the bottom platen, left of the vertical diameter, the cloud of events in Figure 2c in between 90 and 100 s has rather to be explained as a consequence of this flaking. Comparing with the results of the AE behaviour in the other samples and taking the total duration of AE peak *1* into consideration, the flaking together with the accompanying material crushing has to be considered responsible for this first activity peak.

In sample LMDj_6, there is, in contrast to the previous sample, hardly any AE activity in the vicinity of the sample centre (Fig. 3a), which was situated at 59.5 mm. The hole centre is situated at about 83 mm. The apparent position histogram, seems to consist of two subpopulations: one to the left, and one to the right of the stress concentrator, marked *a* and *b* respectively in Figure 3a. The appearance of two subpopulations instead of one is however an artefact induced by the presence of the hole. Physically, all the events originate in the vicinity of the hole, although they find their origin both above and below it. As events occurring nearby the hole can be shielded from either the first sensor or the second sensor, the apparent positions will be shifted either to the left or to the right respectively, away from the sensor from which they are shielded. In Figure 3b, three peaks (*1*, *2* and *3*) are observed. The first as well as the second peak are associated with events originating in the vicinity of the hole: the cloud of points on the position-time diagram (Fig. 3c), corresponding to peak *1* and *2* are situated between 70 and 115 mm, which correlates with the events at and around the peaks *a* and *b* in Figure 3a. They are attributed to damage evolution and fracture formation above and below the hole. The maximum tensile stress at the start of the first activity period is about two times the tensile strength. Peak *3* is associated with a secondary fracture, initiated at the platen contacts, and running along the vertical diameter up to the middle of the sample. The apparent position is offset towards the left-hand side, which is however an artefact (see further). The spatial distribution of this AE activity is somewhat masked on the Figures 3a and 3c, as it is distributed over a relative large area; the peak *3* in Figure 3b corresponds however with the cloud of points marked *c* in Figure 3c and the broad AE activity zone marked *c* in Figure 3a (in between ±10 and 60 mm). The reason for the offset to the left and the wide zone in which the AE activity seems to occur, is that, in order to reach the second sensor, the elastic waves which accompany the secondary cracking, have to travel through a zone already damaged (peaks *1* and *2*, Figure 3b). The presence of additional micro fractures, reduces the elasticity modulus of the solid material, and reduces therefore the speed of the elastic waves (Couvreur, 1997).

The position plot of LMDj_4 (Fig. 4a), shows

features that were encountered in the samples previously discussed. The double peaked distribution *a-b*, associated with the hole centred at 97.5 mm is similar to the histogram obtained in LMDj_6 and characterizes AE activity in the vicinity of the hole. The distribution *c* which shows an offset to the left of the centre (situated at 61 mm), is typical for AE activity along the vertical diameter (see LMDg_5). On the time-event histogram (Fig. 4b), there are only two peaks *1* and *2* visible. The second peak is the actual fracture of the sample along the vertical diameter, and corresponds with the AE activity marked *c* in Figure 4a. The first AE activity period, is rather wide (8 seconds). At the start of the first AE activity period, the maximum tensile stress at the stress concentrator is 2.7 times the stress obtained along the vertical diameter just prior to splitting. When the sample splits, the stress at the hole is 3.1 times the stress along the vertical diameter. In Figure 4c, it can be seen that as the micro cracking in the vicinity of the hole is under way, that, at the same time a dense cloud of points in between 30 and 70 mm indicates that the fracture along the vertical diameter is developing.

6 DAMAGE EVOLUTION

In the three samples under discussion, as well as in the other samples that were tested, exceeding the tensile strength in the vicinity of the hole was never a sufficient condition for micro cracking as evidenced by registerable AE activity. Either no AE activity was recorded in the vicinity of the stress concentrator although the stress had attained a level in excess of the tensile strength or the first signs of micro cracking at the hole, evidenced by measurable AE activity, occurred at stress levels well in excess of the tensile strength. The AE activity associated with a fracture along the vertical diameter directly preceded the fracturing. The AE activity preceding a fracture at the hole started to manifest itself more in advance of the development of a macro fracture. In the sample LMDj_4, the first AE activity was recorded about eight seconds before the sample failed along the vertical diameter. No signs of any damage was visible in the vicinity of the hole upon microscopic inspection (X 120). A radiographic inspection did not reveal any damage in the hole's vicinity either. A confirmation that the AE activity did originate at the stress concentrator and that it was due to damage initiation and evolution, was obtained by a capillary suction technique. Water was sucked up at the top and the bottom of the opening, by micro fractures formed during the test. No water was sucked up at the stress concentrator in sample LMDg_5, which yielded along the vertical diameter as well, but where there had been no AE in the vicinity of the hole.

7 CONCLUSION

A suitable positioning of the sensors, and taking the particularities of the sample configuration into consideration, enabled the determination of the relative position of the source of the events with the two sensor configuration.

The acoustic emission technique proved thereby not only to be an excellent indicator of imminent macro crack initiation in brittle materials, but it also proved to be able to detect the development of damage before there is microscopic or even radiographic evidence of any material degradation.

The experiments confirmed that exceeding the tensile strength is not a sufficient condition to induce damage, and that the AE activity is only triggered at stresses that for a linear-elastic material approach locally attain levels well in excess of the tensile strength. The stress level at which events associated with micro cracking are first identified, depends on the stress distribution in the vicinity of the hole. This stress distribution is determined by both the size and the position of the hole.

8 REFERENCES

Couvreur, J.-F. 1997. Propagations d'ondes ultra-soniques dans des roches sédimentaires, étude de l'endommagement par traitement des signaux. *Ph. D. Thesis UCL*, p30-33.

Hsu, N.N. & Hardy, S.C. Experiments in acoustic emission waveform analysis for characterization of A.E. sources, sensors and structures. *Report of the Centre for Material Science*. p85-106. National Bureau of Standards, Washington, D.C.

Pollock, A.A. 1995. Acoustic Emission Inspection. *Mistras 2001 User's Manual*. Physical Acoustics Corporation, p 278-294.

Physical Acoustics Corporation. 1995. 2001 User's Manual. p. IV 10 - IV 26 and Appendix I. 1995.

Van de Steen, B. 1997. Unpublished report. Fracture initiation at a stress concentrator: experimental results on Limestone of Soignies. *Report submitted to the doctoral advisory committee*. KU Leuven

Van de Steen, B., Vervoort A. & Jermei J. 1998. In press. Crack initiation at a heterogeneity in a rock sample subjected to the Brazilian test. *Paper submitted for the MJFR-3 International Conference on Mechanics of Jointed and Faulted Rock.*

Wevers M. 1996. Fundamentals of acoustic emission. *22nd European conference on acoustic emission testing*. p. 1-10.

Mechanics of Jointed and Faulted Rock, Rossmanith (ed.) © *1998 Taylor & Francis, ISBN 90 5410 955 6*

Effects of discrete memory during rock deformation

Joanna Pinińska
Faculty of Geology, Warsaw University, Poland
Wacław M. Zuberek
Faculty of Earch Sciences, University of Silesia, Sosnowiec, Poland

ABSTRACT: On the basis of acoustic emission and deformation measurements carried on with rock samples from 2000 to 4000 m depth, the memory of paleostress in the sandstone and calcareous series was analysed. Effect of maximum stress memory in the rock induced during the rock deformation in the course of acoustic emission (AE), is called Kaiser effect. The similar effect was observed as a result of cyclic thermal loadings of carboniferous sandstone and mudstone samples. In the course of AE and also in P-wave travel time have been found the changes related to the maximum temperature memory called thermal Kaiser effect.

1.INTRODUCTION

The results obtained hitherto have indicated that during the rock deformation in the course of acoustic emission (AE) there exists the maximum stress memory effect, called Kaiser effect. This effect is closely related to the history of stress acting on the rock and is also connected with the hysteresis of rock deformation (Holcomb, 1981). It has been observed that during the rock sample deformation, under uniaxial continuous increase of compressive loading AE increases abruptly above the maximum stress level applied to the sample (Goodman, 1963; Kurita and Fuji, 1979; Hardy and Shen, 1992). Similar effect has also been found during triaxial compression of sands (Tanimoto and Nakamura, 1981; Chodyń and Zuberek, 1992). Therefore one may state that during deformation, some stress levels are stored in the rock and can be reproduced, so there exists some kind of memory called discrete memory. There have been many attempts undertaken to use the Kaiser effect for in situ stress determinations (Kanagawa et al, 1976; Lord and Koerner, 1979; Yoshikawa and Mogi, 1981) but it has soon turned out that in the triaxial state of stress in hard rocks the effect can be observed only when during the cyclic loading the same stress path is applied because the change in loading conditions can affect the onset of AE during deformation (Holcomb, 1993a; Li, 1998). This very important observation has large practical consequences and causes that, in general, the estimation of in situ stresses by means of Kaiser effect can not be assessed directly. Kaiser effect can be explained by means of a model of rock with uniform distribution of cracks and their unstable enlargement in the triaxial state of stress (Holcomb, 1993b; Li, 1998).

During cyclic thermal loadings of rock samples similar effect has been found. In the course of AE and also in P-wave travel time changes the maximum temperature memory, called thermal Kaiser effect, is observed (Atkinson et al, 1984; Montoto et al, 1989; Carlson et al, 1990; Żogała and Zuberek, 1996). In sedimentary rocks (sandstones, mudstones) the courses of AE and P-wave velocity are clearly related to each other.

If one increases the seasoning time between consecutive heating cycles (up to 1 month), the memory effect gradually disappears. It seems that the discrete memory mechanism and its decay are related to the opening and closure of cracks and microcracks which exist in rocks due to thermal stresses.

The purpose of the paper is to present:

- the AE and deformation measurements carried out on sandstone and limestone samples taken from boreholes from 2 000 to 4 000 m depth with preliminary discussion on the basis of the general theory of the Kaiser effect. We would try to answer the question whether, while using the AE, the estimation of paleostress would be possible,
- the results of cyclic thermal loadings of carboniferous sandstone and mudstone samples. The analysis of AE recordings and P-wave transmission times indicate consistent changes with discrete memory of maximum temperature in the previous

heating cycle, which gradually diminishes with the increase of the seasoning time.

2. ACOUSTIC EMISSION AND DEFORMATION MEASUREMENTS ON PALEOZOIC SEDIMENTARY ROCK CORES

Assuming that Kaiser effect exists, one may expect that the rock cores taken from boreholes, during deformation tests with AE recording will indicate the maximum stress they were subjected to during the geological history, so called paleostress. The laboratory experiments have confirmed the effect in rocks during the primary loading of core samples and during the cyclic compression with increasing maximum load (Łukaszewski, 1996).

Nevertheless, the final explanation of the results, which takes into account the tectonic processes, is difficult because the acoustic emission induced by rock fracturing does not depend only on the state of stress but also on the rock fabric which could be quite different for various rocks (Pinińska, Karska, 1990). This is supported by the results of the deformation and AE measurements carried out on palaeozoic sedimentary rocks - Carboniferous sandstones (from Stężyca) and Devonian limestones (from Mełgiew) from Polish Lowland.

The rocks represent high compressive uniaxial strength (average $R_c > 90$ MPa) with high diagenesis and density ($\rho > 2{,}6$ Mgm^{-3}). They were taken from borehole great depth up to 4,0 km. Rock samples were uniaxially and triaxally compressed with axial, circumferential and volumetric strains and acoustic emission recording. During the tests the anisotropy of P-wave travel has been noted in sandstones.

The analysis of deformation measurements has enabled to estimate the stress limits for the distinguished specified prefailure phases of deformation (Pinińska 1995): compaction ($\sigma_{I/II}$) initial fracturing ($\sigma_{II/III}$), advanced stable fracturing ($\sigma_{III/IV}$) and unstable fracturing ($\sigma_{IV/V}$), up to the ultimate strength (σ_{max}) (fig. 1).

The discrete memory effects have been analysed, taking into account the estimated stress limits for the observed phases of deformation (fig. 2). The course of AE in time units together with loading increase are presented in fig. 3 and 4.

Analysing the data one may notice that in rocks from Stężyca and Mełgiew there is some acoustic emission in the compaction phase. However, in carboniferous sandstones, in that stress range occurs low level of increased emission (fig. 3).

Similar, but more distinct effect occurs in Devonian limestones but it is in the initial fracturing phase (fig. 4) and may be considered as the result of the stresses remembered by rock. Average stress level corresponding to the AE rate increase in Stężyca sandstones is approximately 20 kN and for Mełgiew limestones is approximately 40 kN.

If one assumes that the increased AE rate above that stress level results from Kaiser effect then one may conclude that such stresses have affected the rock mass in the geological history. However, there are two questions to be considered: why are the Kaiser stresses so low because they should be at least twice as high in rocks from those borehole depths under hydrostatic state of stresses; and why does the AE occur mostly in sandstones in

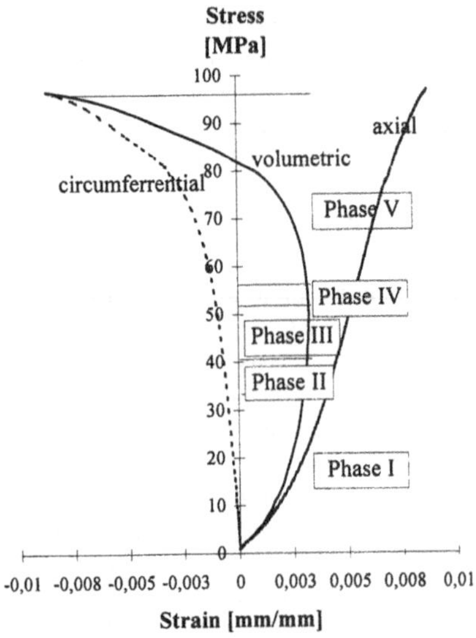

Figure 1. Example of the estimation of prefailure phases of deformation for sandstone sample.

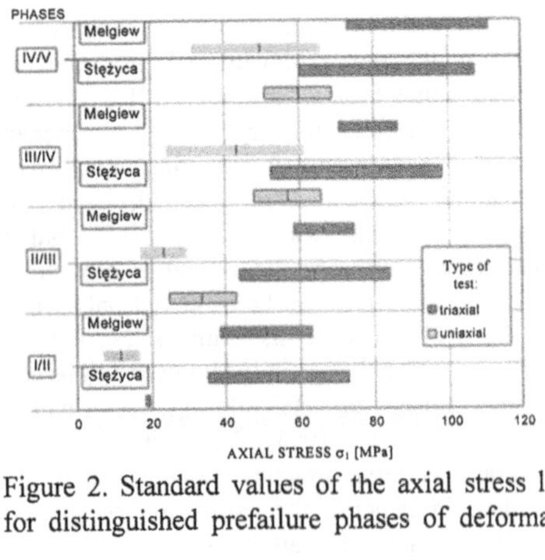

Figure 2. Standard values of the axial stress limits for distinguished prefailure phases of deformation.

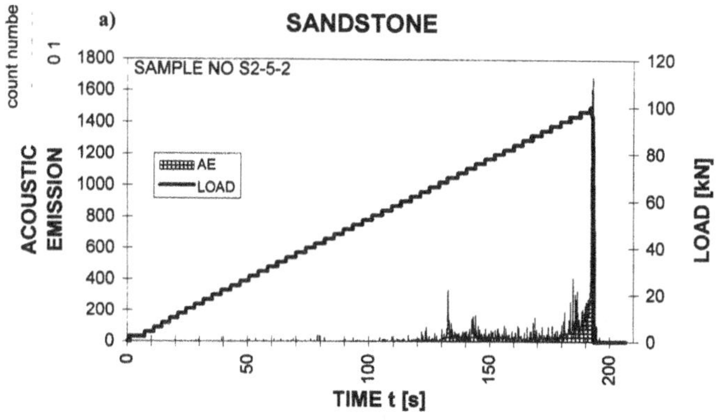

Figure 3. Acoustic Emission (AE) and uniaxial load courses for Stężyca sandstone.

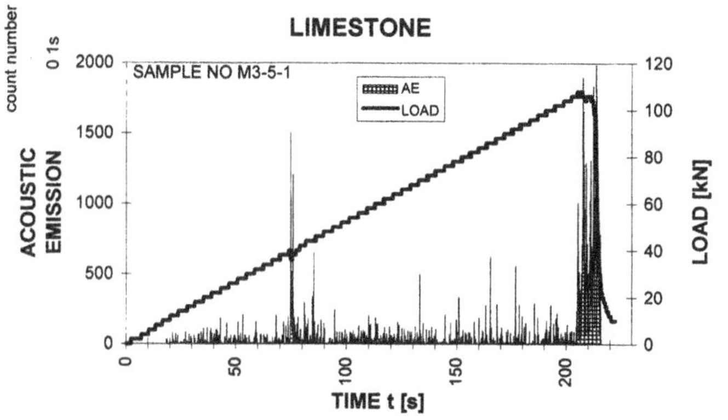

Figure 4. Acoustic Emission (AE) and uniaxial load courses for Mełgiew limestone.

compaction phase while in limestones it occurs in the initial fracturing phase above the stresses connected with elastic strains. It seems that the answer for the first question can be found in the theory of the Kaiser effect (Li, 1998).
In the triaxial compression it has been proved that the differential stress for the Kaiser effect has linear relationship with confining pressure:

$$(\sigma_1 - \sigma_3) = k\sigma_3 + b_{13}$$
$$(\sigma_1 - \sigma_2) = k\sigma_2 + b_{12}$$
$$(\sigma_2 - \sigma_3) = k\sigma_3 + b_{23}$$

where: σ_1, σ_3, σ_3 are principal stress tensor components,

$$k = \frac{2\sin\phi}{1 - \sin\phi}$$

ϕ - friction angle of the rock

b_{13}, b_{12}, b_{23} onset of cumulated AE number increase during corresponding uniaxial compression of taken rock samples (fig. 5).

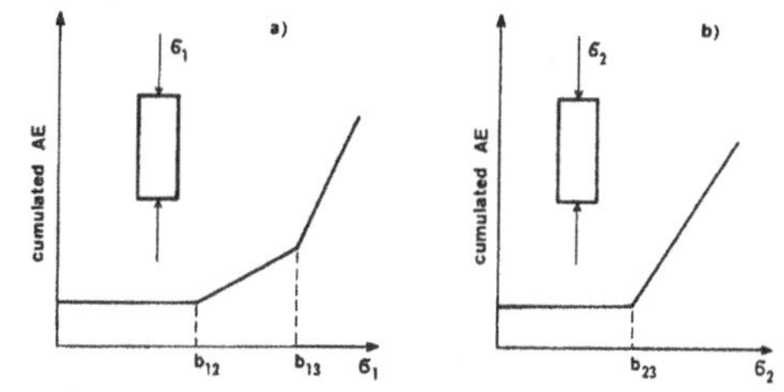

Figure 5. Illustration for the AE onsets during various uniaxial loadings (after Li, 1998).

In that case, as it was concluded by Li (1998), Kaiser effect should be considered to be an indicator of damage in rock rather than a measure of stresses. Then the estimation of in situ stresses can be done only if all three differential stresses have created damage fracturing in rock, which may be very seldom. Thus, the history of regional tectonic stresses can affect the damage pattern in the rock and from simple uniaxial compression tests on relaxed material, with measurements AE we are able to estimate in situ stresses only in case when uniaxial loading follows the same load direction path as a paleostress. In other cases we have to carry out complex compression tests to estimate the proper b_{ij} - AE onset and friction angle values or use oriented boreholes cores, especially for anisotropic rock such as Stężyca sandstone.

In Mełgiew limestones the Kaiser effect occurrence is related to the stable fracturing phase when one can observe significant non-linear circumferential strains which are typical for that phase while the axial strains remain linear.

On the other hand in Stężyca sandstones the initial fracturing phase may occurs very late at stresses 70 - 80 MPa after extensive elastic strain phase and quickly passes into the process of unstable fracturing (after crossing stresses of the order 90 MPa).

In Mełgiew limestones the compaction phase is quickly finished and after limited elastic strains phase

the fracturing phases are occurring. The differences in the deformation development are related to various rock fabric of these rocks and the distinct mechanisms of fracture (Pinińska, 1997).

Intragranular tensional fracturing of singular quartz grains without damage in their cement and visible transgranular dislocations are typical for sandstones. Intergranular fracturing with shearing and significant displacements have been observed in limestones. It leads to the differentiation of the acoustic emission signals energy. Therefore the Kaiser effect in rocks has to be analysed not only in relation to the state of stress in the rock mass but also in connection to the deformation processes, specific for the selected rock.

3. RESULTS OF CYCLIC THERMAL LOADINGS OF SANDSTONES AND MUDSTONES

The rock samples, cylinders of 100 mm diameter and approx. 100 mm height, obtained by cutting the borehole cores were heated in unconfined conditions (Zuberek et al, 1998). All samples were taken from the Upper Silesian Coal Basin and belong to Carboniferous coal bearing strata.

The samples were heated in five cycles sequentially to the temperature of approx. 140^{0}C, 155^{0}C, 170^{0}C, 185^{0}C and 200^{0}C with heating rate of 2,1 - 2,3 ^{0}C/min (fig. 6). In each heating cycle they had been heated twice, meanwhile cooling the sample to the temperature of approx. 30 - 35^{0}C (fig. 5). Between the heating cycles the seasoning time has been changed up to 1 month (720 hours). The P-wave propagation time has been determined before and after each heating using the standard ultrasonic equipment with 40 Kcps transducers.

It has been found that first after cooling the AE increases only at temperatures higher than maximum temperature in the previous heating cycle. However, with the seasoning time increase after cooling (cycles III, IV, V) the growth of AE is observed at lower temperatures. At the same time we have noted that the AE during second heating in the same cycle is in majority of cases evidently lower than in the first one.

So if differences of cumulated AE count numbers between following heatings are determined one can obtain more objective measure of the thermal Kaiser effect although it is not always quite clear.

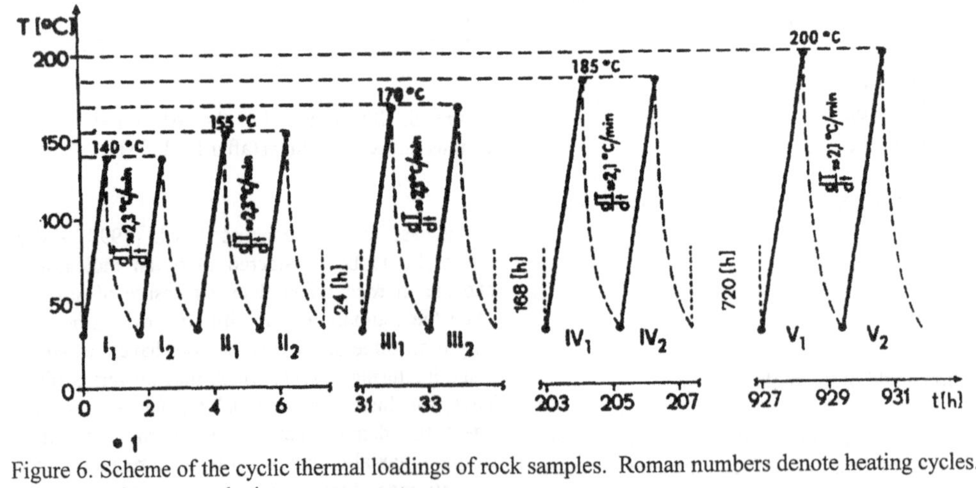

Figure 6. Scheme of the cyclic thermal loadings of rock samples. Roman numbers denote heating cycles. I - points of P-wave velocity measurements.

The thermal Kaiser effect may be quantitatively described by thermal memory coefficient B defined as

$$B = \frac{{}^{n}T_{AE}}{{}^{(n-1)}T_{max}}$$

where: ${}^{n}T_{AE}$ - temperature of the onset of AE in cycle n

${}^{n-1}T_{AE}$ - maximum temperature in the proceeding heating cycle.

Thus, defined B values when close to 1.0 denote that the memory exists. All the deviations form the measure of the maximum temperature memory decay.

In fig. 7a is shown the average B values for sandstones versus the seasoning time t. The results confirm our former research that the memory effect gradually decays with seasoning time increase of the cooled samples (Żogała et al 1995). Considering this, the decrease of the B value is close to linear relation in log-log scale and one may expect that it could be approximated by a power law function.

Comparing the B values to the average values of cumulated AE count number N_Z at 140^{0}C (maximum temperature at the first heating cycle (fig. 7b) one may note that although the standard deviations are very large, they show consistent changes. It means that after the first heating cycle they are lower but with the seasoning time increase they seem to grow.

The P-wave velocities in the samples before heating are approx. 10% higher than after heating (fig. 7c). After the first heating cycle the average P-velocities are the lowest - at the shortest seasoning time but with the seasoning time increase, they grow indicating the return to its previous value in the way somehow similar to the average cumulated AE count number N_Z. The last graph of the fig. 7d illustrates the course of differences of average P-wave velocities before the first and the second heating V_{b2} - V_{b2} with the seasoning time increase t. It occurs that the changes are very similar to those described before. One may conclude that all the described changes are closely related. It may be assumed that P-wave velocity change is related to the corresponding volumetric strain of the sample and during heating and cooling the sample is changing its volume under the thermal stresses. The deformation processes occurring in rocks during the temperature increase from approx. 30^{0}C to 200^{0}C, are probably connected with the development of new microcracks and cracks as well as with the enlargement and opening of already existing ones up to certain sizes typical for the maximum temperature achieved. The microfractures which exist in rock and cracks are going to open mainly due to anisotropic thermal extension of particular mineral grains. As the result, the AE count rate in the following heating cycle will be lower until the maximum temperature from the previous cycle will be reached. The P-wave velocity will be lower, too.

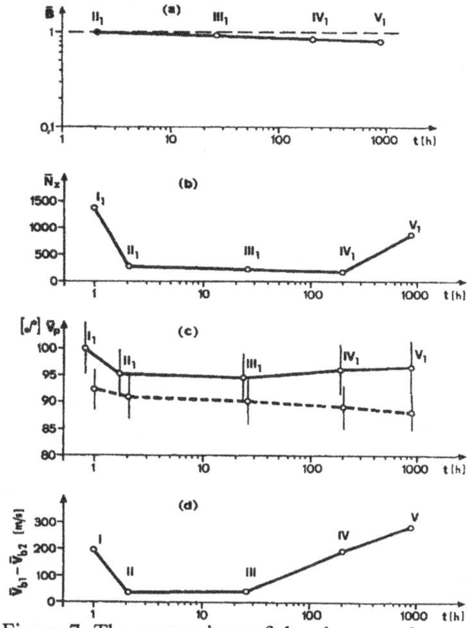

Figure 7. The comparison of the changes of average values of: (after Zuberek et al, 1998)

- the thermal memory coefficient B in AE (log - log scale) (fig. 7a)
- cumulated AE counts N_Z at 140^0 C (fig 7b)
- P-wave velocity v_p before heating (solid line) and after heating (dashed line) (fig.7c); vertical bars - standard deviations
- differential P - wave velocities before the first heating V_{b1}, and before the second one V_{b2} versus the seasoning time t (fig. 7d).

After having cooled the samples, together with the increase of the seasoning time the gradual closure of cracks and adjustment of their inner surfaces occur and therefore the volume of the sample will decrease. This effect is associated with the corresponding increase of the P-wave velocity. If the AE is generated mainly due to shearing between crack surfaces than after closing of previously open cracks, the AE will grow with the seasoning time increase. Therefore it seems that the observed thermal Kaiser effect and its decay can be explained, at least in a qualitative way, by means of the rock fracturing model.

It should be noted that it seems possible to reproduce the maximum temperature of the rock mass with the accuracy of approx. 2-5% providing that the time between taking the samples and laboratory measurements does not exceed tens of hours. However, in the in situ conditions, the existing state of stress can reduce the number of AE counts (Carlson et al, 1990) and may limit the accuracy of maximum temperature determinations. It seems to be reasonable to assume that also in the case of thermal Kaiser effect, it is related rather to damages in rock than to maximum temperatures. So the maximum temperatures can be reproduced only if they have created damage or inelastic deformations in the rock.

4. CONCLUSIONS

1. The Kaiser effect in non oriented rock samples should be considered as an indicator of damage rather than a measure of the paleostress value and the estimation of these stress can be done only if all three differential stresses have created damage in rock or the path a loading follows the paleodamage direction. It also depends on the deformation mechanism, specific for the selected rock.
2. In sedimentary rocks, in similar ways as in magmatic ones, in the course of AE the maximum temperature memory effect exists but it decays with the seasoning time.
3. During heating and cooling the average values of the thermal memory coefficient, cumulated AE counts and P-wave velocities show consistent changes with the increase of the seasoning time increase. These changes can be explained, at least in qualitative way, by means of the model of rock with microfractures and cracks opening and closing due to thermal stresses.

REFERENCES

Atkinson B. K., Mc Donald D., Meredith P. G. 1984. Acoustic response and fracture mechanics of granite subjected to thermal and stress cyclic experiments. *Proc. 3rd Conf. AE/MA in Geologic structures and Materials,* Trans. Tech Publ. Clausthal - Zellerfeld, 5 - 18.

Carlson S. R., Wu M., Wang H. F. 1990. Micromechanical modelling of thermal cracking in granite. *Geoph. Monograph* 56 AGU, 401 - 405.

Chodyń L., Zuberek W. M. 1992. Effect of the discrete stress memory in the acoustic emission in soils. *Acta Geoph. Pol.* v. XL No 2, 139 - 158.

Goodman R. E. 1963. Subaudible noise during compression of rock. *Geol. Soc. Am. Bull* v.74, No. 4, 487 - 490.

Hardy H. R. Jr Shen W. 1992. Recent Kaiser effect studies on rock. *Proc. Progress in Acoustic Emission-VI.* The Japanese Soc. for NDI, 149 - 157.

Hallbauer D. K. , Wagner H. & Cook N.G.W. 1973. Some observations concerning the microscopic and mechanical behaviour of quartzite specimens in stiff, triaxial compression tests. *Int. Journal of Rock Mechanics and Mining Sciences & Geomechanics Abstracts,* vol 10, 713 - 726.

Holcomb D. J. 1981. Memory, relaxation and microfracturing in dilatant rock. *J. Geoph. Res.* v.86, No 137, 6235 - 6248.

Holcomb D. J. 1993a. Observations of the Kaiser effect under multiaxial stress states: implications for its use in determining in situ stress. *Geoph. Res. Letts.* v.20, No 19, 2119 - 2122.

Holcomb D. J. 1993b. General theory of the Kaiser effect. *Int. J. Rock Mech.* v.30, No 7, 929 - 935.

Kanagawa T., Hayashi M., Nakasa H. 1976. Estimation of spatial geostress components in rock samples using Kaiser effect of acoustic emission.

Kurita K., Fuji N. 1979. Stress memory of crystalline rocks in acoustic emission. *Geoph. Res. Lett.* v.6, 9 - 12.

Li C., 1998. A theory for the Kaiser effect in rock and its potential application. *Proc. 6th Conf. AE/MA in Geologic Structures and Materials.* Trans. Tech Publ. Clausthal - Zellerfeld, 171 - 185.

Lord A. E. (Jr.) Koerner R. M. (1979). Acoustic emission in geologic materials. [in:] *Fundamentals of acoustic emission,* Univ. of California, Los Angeles, 261 - 307.

Łukaszewski P. 1996. Acoustic emission monitoring of the crack growth in the Carboniferous rocks. *Ph. D. Thesis* Dept. of Geology, Warsaw University (in Polish).

Montoto M., Ruiz de Argandona V. G., Calleja L., Suarez del Rio L. M. 1989. Kaiser effect in thermocycled rocks. *Proc. 4th Conf. AE/MA in Geologic Structures and Materials.* Trans. Tech Publ. Clausthal - Zellerfeld, 97 - 116.

Pinińska J, Karska Z. (1990). Application of acoustic emission parameters as a criterion of stress-strain behaviour of rocks. *Proc. Int. Con. IAEG,* Amsterdam, 413 - 417.
Pinińska J. 1995. Crack growth in the postcritical path of deformation of sedimentary rocks. *Proc. of Int. Con.* MJFR-2 Balkema, Rotterdam, 113 - 118.
Pinińska J. 1997. Some problems of the stress distribution on structural contacts in natural rocks bodies *Proc. Symp. Mixed problems of the mechanics of nonhomogeneous structures* Lwów, 73 - 79.
Tanimoto K., Nakamura J. 1981. Use of AE technique in field investigation of soil. *Proc. 3rd Conf on AE/MA in Geologic Structures and Materials* Trans. Tech Publ., Clausthal - Zellerfeld, 601 - 612.
Zogała B., Zuberek W.M., Dubiel R. (1995). Studies of the maximum temperature memory effect decay during seismoacoustic emission in sedimentary rocks. *Publ. Inst. Pol. Acad. Sci.* M - 19 (281), 245 - 254 (in Polish).
Zogała B., Zuberek W. M. (1996). Changes of acoustic emission and ultrasonic P-wave velocity in sedimentary rock samples during cyclic heating. *Acta Montana,* Ser A, No 9 (100), 139 - 199.
Zuberek W. M., Zogała B., Dubiel R. (1998). Laboratory investigations of the temperature memory effect in sandstones with measurements of acoustic emission and P-wave velocity. *Proc. Sixth Conf on AE/MA in Geologic Structures and Materials.* Trans. Tech Publ., Clausthal - Zellerfeld, 157 - 168.

Mechanics of Jointed and Faulted Rock, Rossmanith (ed.) © 1998 Taylor & Francis, ISBN 90 5410 955 6

Fracture analysis of loaded rocks by means of acoustic emission

E. Janurová & L. Sosnovec
Institute of Physics, Technical University of Ostrava, Poruba, Czech Republic

ABSTRACT: Dependences of parameters of acoustic emission on longitudinal and transversal deformations on conditions of uniaxial test were measured. Sandstone and coal samples were used. Amplitude of pulses, number of pulses, the mean value of amplitude, duration of pulses were registered. Grain joints failure as the source of acoustic emission is discussed. Comparison of predicted and observed dependences shows quite good agreement in case of number of pulses. Confirmation of grain joints failure model by uniaxial stress cyclic loading test is presented. The type analysis of duration of pulses and amplitude of pulses is discussed, too.

1 INTRODUCTION

Statistics of shape and size distribution of grains in sample of sedimentary rock is similar to statistics of geological objects in sedimentary beds. Therefore, principle of correspondence between behavior of samples of particular rocks and rocks in sedimentary beds could exist. Model of grain joints breakage is used for explanation of acoustic emission in Sandstone samples. Dependence of number of pulses above the level on longitudinal or transversal deformations consists of a few linear parts jointing one another.

2 MODEL OF GRAIN JOINTS BREAKAGE

If joints among the grains are cracked, the grains try to form the most effective layout. Therefore, longitudinal and transversal deformations depend on cracked volume. The following notation is used:

β_n .... the tightest compression porosity
β .... the likelist value of porosity
V .... Volume of cracked domain
V_s .... Volume of sample
δV .... Change of volume of sample
ϵ_{11} .... relative longitudinal deformation
ϵ_{22} .... relative transversal deformation
N_0 .... Volume concentration of grain joints
z .... Number of joints among one grain and its neighbors

Then

$$\delta V = (\beta - \beta_n) \times V. \tag{1}$$

For number of joints among the grains in cracked domain can be written

$$\delta p = N_0 z \frac{V}{2}. \tag{2}$$

From equations (1) and (2) it can be derived

$$\delta p = \frac{N_0 z \delta V}{2(\beta - \beta_n)}. \tag{3}$$

Supposing that deformations are small can be written and fracturing is homogeneous

$$\delta p = \frac{N_0 z V_s(\epsilon_{11} - 2\epsilon_{22})}{2(\beta - \beta_n)}. \tag{4}$$

Eliminating one of deformations final equations occur

$$\delta p = \frac{N_0 z V_s(1 - 2\nu)\epsilon_{11}}{2(\beta - \beta_n)} \tag{5}$$

and

$$\delta p = \frac{N_0 z V_s(\frac{1}{\nu} - 2)\epsilon_{22}}{2(\beta - \beta_n)}. \tag{6}$$

Microcracking is concentrated in domains. Therefore, number of joints depends on volume of domain and type of its spreading. In sample of cylindrical shape with diameter d and height h shear cone close to top and bottom bases exist (see Fig. 1). Former, by means of uniaxial stress shear cones and central domain coalesce, latter, central domain shatter and cracked domain grows.

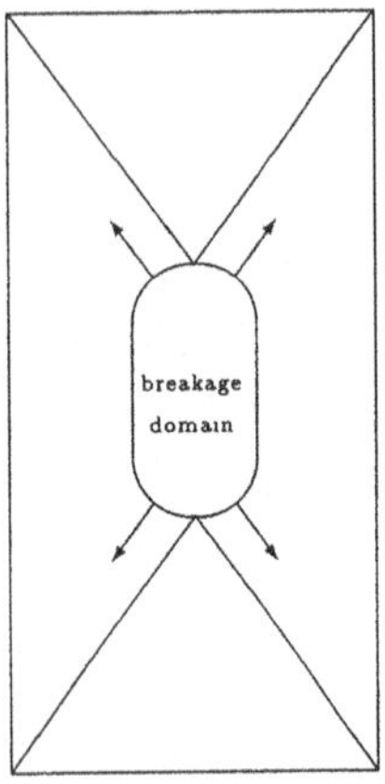

Fig. 1 Shear cones in the sample of cylindrical shape with propagating breakage domain

First, rotating of shear cone is considered. Bases of sample become parallel. Domain of breakage is a thin film with thickness D and surface S as same as mantle of shear cone. D is approximately equal to mean diameter of grain. For volume of this part of breakage domain can be written

$$V = SD \tag{7}$$

or using diameter and height of the sample

$$V = \frac{\pi d D}{6}\sqrt{4h^2 + 9d^2}. \tag{8}$$

With respect to (2) the final eqution has got the form

$$\delta p = N_0 z \frac{\pi d D}{12}\sqrt{4h^2 + 9d^2} \tag{9}$$

Second, central domain of breakage develops. Exchanging of V_s for volume of domain V equations (5) and (6) are valid. Linear dependence of number of breaking joints δp on number of acoustic pulses can be expected. Therefore, dependence of number of acoustic pulses on longitudinal and transversal deformations can be linear too.

Third, shear cone slides on the rest of sample and crowd out it in horizontal direction. Supposing that s is the distance between top of cone and perimeter of the base, surface of cone is equal

$$S = \pi r s \tag{10}$$

Breakage front form breakage cone with distance between top of cone and perimeter of the base s'. Surface of this cone is equal

$$S' = \pi r s' \tag{11}$$

Division (11) by (10) gives

$$\frac{S'}{S} = \frac{s'}{s} = \frac{r'}{r} \tag{12}$$

Marking l_0 as elastic slipping on cone mantle gives for total slipping l_p

$$l_p = l_0 \frac{S'}{S} = \frac{V'}{V} \tag{13}$$

For longitudinal deformation can be written

$$\epsilon_{11} = \frac{l_p}{h}\frac{s}{r} = \frac{l_0 V'}{hSD} \tag{14}$$

By derivation from (2) can be obtained

$$\epsilon_{11} = \frac{l_0}{hSD}\frac{2\delta p}{N_0 z} \tag{15}$$

or finally

$$\delta p = \frac{N_0 z h S D}{2 l_0}\epsilon_{11} \tag{16}$$

3 EXPERIMENTAL OBSERVATIONS

For description of experimental setting (see Knejzlík, Konečný, Rambouský). There are three typical areas in dependence of number of pulses on longitudinal deformation (see Fig. 2). The first one correspond to model of shear cone formation. The bases become parallel when plateau is reached. The second one is transitory area, where homogeneous uniaxial stress in central domain is applied and grains inside are crushed. The third one correspond to sliding of shear cone on remainder of sample as a wedge. Surface of cone mantle become smoother.

Observations of triaxial stress cyclic loading support this model (see Rambouský, Sosnovec). Also dependence of force on longitudinal deformation corresponds this situation.

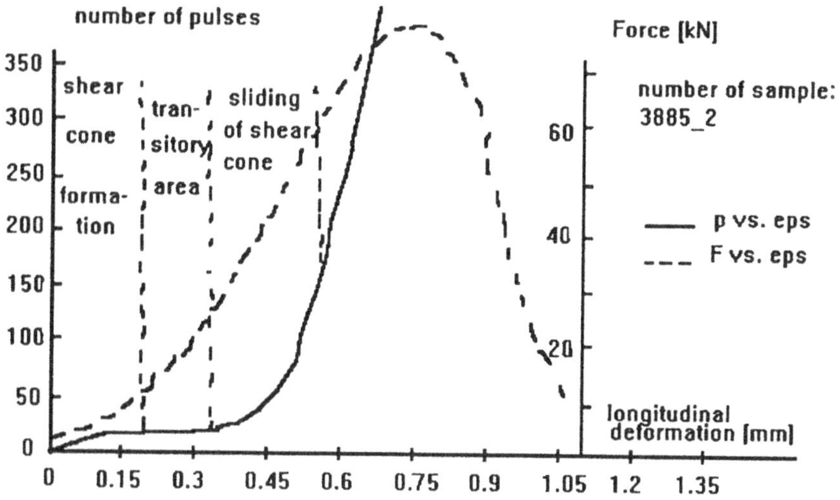

Fig. 2 Dependence of force F and Number of pulses δp on longitudinal deformation

4 CONCLUSIONS

Experimental observations support possibility of using acoustic monitoring in practise. Principle of correspondence between structure of sandstones and real macroscopic sedimentary rock in particles distribution prompts monitoring of underlying rock deformations. This method could be used especially in case of impracticable geometric measurement of deformations and inaccurate tensometric measurement.

ACKNOWLEDGEMENT

The authors would like to thank Mrs. Daria Nováková from the Institute of Geonics of Czech Academy of Sciences for TeX typesetting.

REFERENCES

Rambouský, Z. & Sosnovec, L: *Triaxial Stress Cyclic Loading of Sandstones.* Proceeding of the Second International Conference on the Mechanics of Jointed and Faulted Rock - MJFR-2, Vienna, Austria, 10-14 April 1995, Balkema, Rotterdam, 1995, pp.473-475.

Knejzlík, J. & Konečný, P. & Rambouský, Z: *Scanning of Transforming Properties and Acoustics Emissions in the process of uniaxial stress .* Proceeding of conference Physical Properties of Rocks and their applications in Geophysics, Geology and Ecology, Technical University Ostrava, 1995, (in Czech).

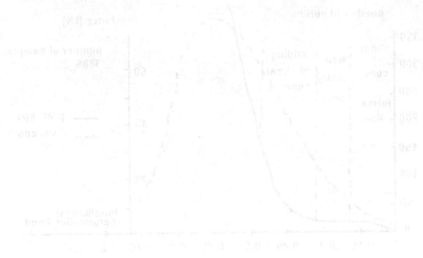

Mechanics of Jointed and Faulted Rock, Rossmanith (ed.) © 1998 Taylor & Francis, ISBN 90 5410 955 6

Crack detection in salt rock and implications for the geomechanical situation

Th. Spies
Federal Institute for Geosciences and Natural Resources (BGR), Hannover, Germany
J. Eisenblätter
Gesellschaft für Materialprüfung und Geophysik, Ober-Mörlen, Germany

ABSTRACT: The different geomechanical behaviour of salt rock is characterized emphasizing the effects of micro- and macrofracturing. Evidence for the possible influence of progressive crack generation on the mechanical stability of underground cavities and the integrity of the rock is summarized. To study current crack generation in salt rock, acoustic emission measurements are applied. Thus regions of high geomechanical loading can be inferred and monitoring can be performed. Results of acoustic emission measurements in a segment of a salt mine are presented in which anhydrite occurs in the hanging wall above cavities in rock salt.

1 INTRODUCTION

Salt rock reveals a large variety of geomechanical behaviour, especially concerning micro- and macrofracturing. Rock salt, which in most salt deposits covers the largest volume, is capable of creep deformation without occurence of fracture in a wide range of the conditions of state. As a consequence, usually no joints resulting from geological influences are found in rock salt strata. In underground mining rooms and drifts tend to close slowly due to creep deformation. Because of its ductile behaviour rock salt is a favourable host rock for the storage of fluids and for the underground disposal of hazardous waste. Nevertheless, in presence of high deviatoric loading microcracks may be generated even in rock salt, for instance in thin zones along the contours of cavities (dilatancy). Compared to rock salt, anhydrite covers less volume in salt deposits but often it is one of the main constituents of the deposits. It exhibits elastic-brittle behaviour and in most cases distinct joint systems are found in anhydrite strata.

The investigation of the micro- and macrofracturing processes is important for the evaluation of the stability of cavities and the hydraulic integrity of the rock. Especially in the case of the underground disposal of hazardous waste in salt rock, in-situ investigations have to be carried out and the long-term safety has to be evaluated (e.g., Langer 1996). Acoustic emission (AE) measurements offer the possibility to detect current crack generation directly as the seismic or acoustic energy radiated during the crack process is recorded and evaluated. The method has been applied to monitor limited volumes in salt mines like single pillars, to determine dilatant zones around galleries and to study thermally induced cracking in rock salt (e.g., Yaramanci 1991, Spies et al. 1997, Eisenblätter et al. 1998). In this paper recent results obtained by operation of a large AE network in a salt mine will be discussed and AE activity in different kind of salt rock will be characterized.

2 MECHANICAL BEHAVIOUR OF SALT ROCK

Figure 1 displays typical results of routine uniaxial testing of salt rock samples in the geotechnical laboratory of the BGR (Schnier & Bleich, pers. comm.). The rate of axial compression is controlled in these experiments at room temperature. The octahedral shear stress is shown as a function of deformation up to the occurence of failure.

In case of anhydrite a linear stress-strain curve and sudden failure is observed implying nearly ideal elastic-brittle behaviour. It is a competent rock having a much higher strength compared to the other types of salt rock. The results for salt clay indicate a small amount of plastic deformation before failure occurs at a low value of strength. In contrast to these rocks, rock salt and sylvinite (hartsalz) show a large portion of plastic deformation (creep).

Because of its competent and brittle properties in a mass of soft rock as rock salt and potash like sylvinite, anhydrite often strongly determines the tectonic form of salt deposits

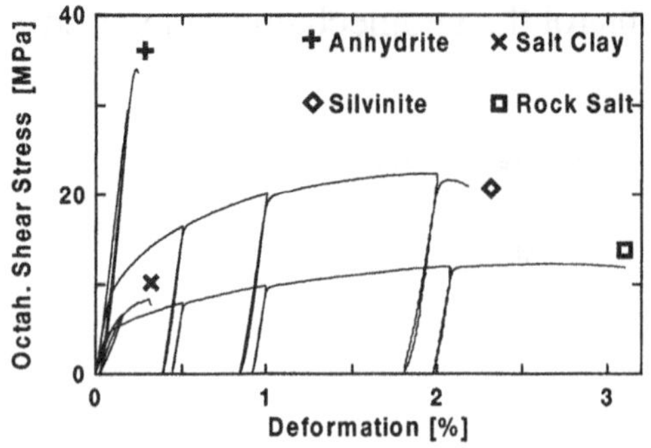

Figure 1. Results of uniaxial compression tests with various salt rock.

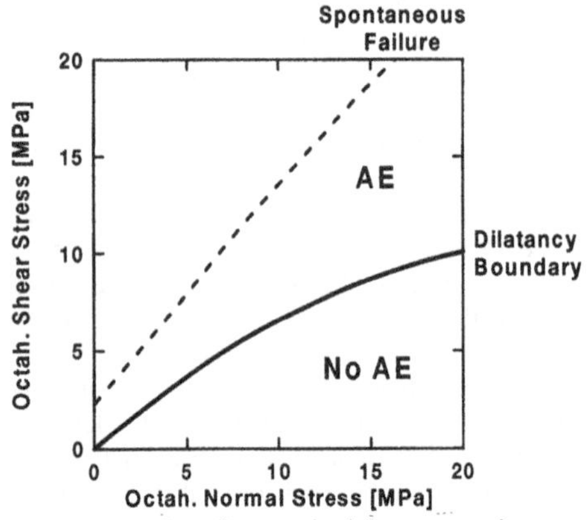

Figure 2. Results of true triaxial compression tests with rock salt.

subjected to salt tectonics and/or halokinesis (e.g., see Richter, 1934). Anhydrite concentrates stresses which in many cases resulted in fracturing and the formation of large blocks. It exhibits distinct joint systems even in case of low tectonic loading in the geological past. Stäubert (1988) investigated joints in anhydrite in eastern Germany and found a range of 0.5 to 1 m for the average joint distance and 1 to 2 m for the average outcrop length of joints. Open joints are found up to a depth range of 400 - 600 m. They represent 20 to 60 % of all investigated joints at single locations. Kamlot (1995) carried out laboratory experiments with jointed and unjointed (compact) anhydrite samples and found that the existence of joints significantly reduces the strength of anhydrite. From analysis of experience in salt mining and from results of geomechanical model calculations Kamlot concluded that mining near or inside the anhydrite can result in opening of joints in the anhydrite leading to higher permeability.

As mentioned above, rock salt usually does not show joints in-situ as a consequence of its ductile behaviour. Figure 2 presents results of laboratory investigations concerning the creep and fracture behaviour of rock salt at room temperature. They have been obtained by true triaxial compressive tests in which the load increase was controlled (Hunsche et al. 1994). Displayed are the stress conditions (octahedral shear stress versus octahedral normal stress or mean normal stress) at the onset of dilatancy and at the occurence of spontaneous failure during the increase of deviatoric load. The dilatancy boundary (DB) is characterized by an increase of sample volume despite compressive stresses. This increase is caused by microfracturing (microcracks in salt grains and along grain boundaries). Below the DB rock salt exhibits creep without generation of microcracks, above the DB creep and generation of microcracks. AE activity which is also recorded in the experiments starts when the DB is crossed and increases strongly during further load increase. This implies that the detection of AE activity in-situ indicates stress conditions above the DB.

Long-term geomechanical consequences of stress conditions above the DB resulting in progressive dilatancy of rock salt are humidity induced creep and creep failure (macrofracturing, see Hunsche et al. 1994). The time after which creep failure occurs is dependent on the distance from the DB in stress space but quantitative results have not been presented in literature up to now. Another consequence of progressive crack generation in rock salt is the increase of permeability as demonstrated by Peach (1991) in the laboratory.

The DB can be regarded as a long-term safety boundary concerning the stability of cavities in rock salt and the integrity of the rock. Using AE measurements zones of current dilatancy in rock salt can be detected.

3 ACOUSTIC EMISSION MEASUREMENTS

Applying this method the seismic or acoustic energy radiated from crack generation in the rock can be recorded (acoustic emission - AE). Figure 3 shows a sketch of the principle and the equipment used. In boreholes piezoelectric transducers are installed which convert the acoustic energy into electric energy. The signals are recorded in the frequency range from 1 to 100 kHz. Figure 3 displays 2 transducers in boreholes as examples but networks up to 24 transducers can be connected to the system used. The central unit of the equipment consists of a transient recorder monitoring the data

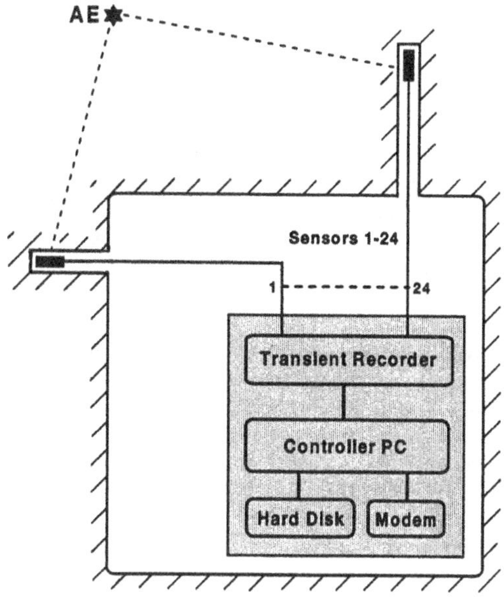

Figure 3. Sketch of acoustic emission measurements and equipment used.

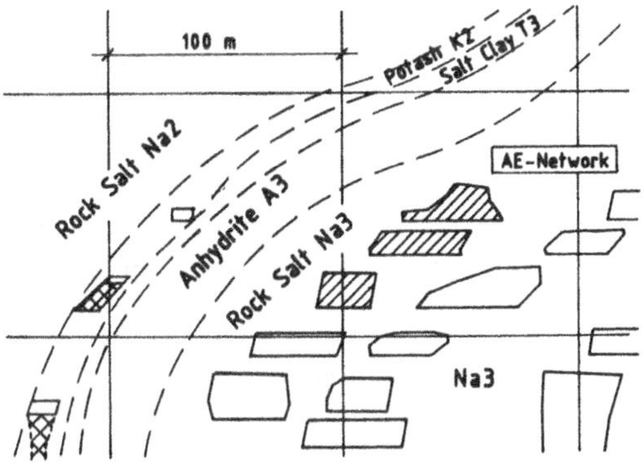

Figure 4. Geology and mining conditions at operational site where the AE network is installed (hatched area at upper three levels).

stream of the 24 transducers, a computer controlling all the actions of data acquisition and evaluation, a large hard disk for storing the data and a modem for telecommunications.

In case of an AE event the computer transfers the signals from the recorder, computes signal parameters and tries to locate the event by automatic picking and inversion of P- and S-traveltimes. A homogeneous velocity model for seismic waves in rock salt is used. The seismic velocities were determined in-situ: 4.5 ms/m for P-waves and 2.5 ms/m for S-waves. From test measurements it was concluded that the error of location can reach a value of 3 m, but in most cases the error was smaller.

The maximum amplitudes A of the 24 transducer signals and the determined distances r of an AE source to the transducers are used to determine a measure of strength analogue to the magnitude in seismology using the relation:

$$A \propto \frac{1}{r} \cdot \exp(-\alpha \cdot r) \qquad (1)$$

where α is the coefficient of damping covering the effects of intrinsic absorption and scattering. The amplitudes are specified in dB.

In a semi-logarithmic plot of the products $A \cdot r$ versus r of all transducers a linear relationship is obtained as expected and a straight line is fitted to the data. The value of this straight line at a reference distance of 50 m is determined and regarded as the magnitude of the AE event. In a next step the frequency distribution of the magnitudes for data of selected spatial domains and temporal intervals are determined and compared.

4 RESULTS

4.1 *Geology and Mining Conditions*

The network of 24 transducers was installed in a segment of a salt mine in northwestern Germany. Figure 4 displays a vertical section as a sketch of the geological conditions and the geometry of rooms in a part of a syncline structure in the mine. The detailed geological structure is more complex. The inner part of the syncline consists of rock salt of the Zechstein Leine series (Na3) separated from rock salt of the Stassfurt series (Na2) by anhydrite (A3), salt clay (T3) and the potash seam (K2). The rooms monitored by the network of 24 transducers are indicated by the hatched signature. The monitored volume amounts to about 100 m × 150 m × 100 m. The average depth of the monitored volume is 400 m. Mining in that part of the mine continued until the 60's, but most of the rooms in the rock salt were mined 60 to 70 years ago.

Figure 4 illustrates a typical geomechanical problem in salt mining: The succession of ductile material (rock salt, potash) and brittle and/or competent material (anhydrite, salt clay). The convergence rates of cavities in this part of the mine are generally low as well as the displacements determined by mine survey, but large mining induced fractures and dripping of brine are observed in rooms at the upper level.

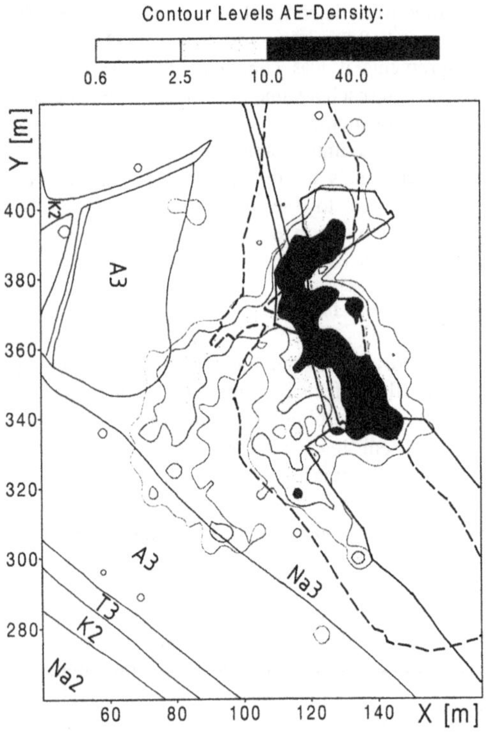

Figure 5. Ground plan of the upper level (continuous lines) and the intermediate of the 3 investigated levels (dashed lines, see Fig. 4), geological boundaries (Na3 - rock salt, A3 - anhydrite) and locations of AE events.

4.2 *Locations of AE Events*

As an example, Figure 5 shows the spatial distribution of events which have been located within a depth interval centered at the upper level in the period of 2 years. The ground plan in the mine coordinate system (Y - North, X - East) displays the contours of the cavities at the upper level as continuous lines, those of the level below (the intermediate level of the three investigated levels, see hatched area in Fig. 4) as dashed lines. The vertical section of Figure 4 intersects the arrangement of rooms in the southern part of Figure 5 perpendicular to the axis of the rooms. The distance of the roof of the intermediate level to the floor of the upper level amounts to about 5 m.

Figure 5 also shows the geological boundaries obtained by intensive geological and geophysical mapping. The abbreviations used for the strata are the same as in Figure 4. Anhydrite and salt clay are not continuous layers but are broken into blocks by salt tectonics.

The 24 AE transducers are installed in 3 to 20 m deep boreholes drilled at the upper level and at the lower level (see Fig. 4). The locations of the transducers are between 300 m and 400 m along the Y-axis of Figure 5. Their average distance is 10 m.

Because of the high number of AE events - in this case nearly 34.000 - the locations in Figure 5 are not displayed as dots for single events but as a density plot. The following procedure has been applied to determine a density presentation of the location data. All locations within the depth interval ranging from 7 m below the upper level to 7 m above the upper level are projected onto the upper level and then summed up in horizontal squares of 1 m × 1 m. Following gridding is performed on a 3 m × 3 m grid and contour lines of AE density are calculated. This spatial smoothing procedure results in a relative measure of AE density in the investigated area. In Figure 5 it ranges from 0.6 to 40 and higher values.

From Figure 5 it can be concluded that most of the events are located near the contours of the rooms in the rock salt (Na3), the dilatant zones extend 5 to 10 m deep into the rock. Especially in pillars between rooms and in locations where contour lines of both mine levels intersect, accumulations of events are found. These are regions where strong stress concentrations are expected. The locations of AE events map these stress concentrations in which local stability problems may arise as spalling from the roof and the walls. Also an increase of permeability has to be expected in these areas. The AE activity along the eastern wall of the rooms cannot be mapped to the south as the room of the upper level prevents the propagation of acoustic waves to the transducers.

Also farer away from the cavities distinct AE activity is found. According to the geological model it is located in the rock salt Na3 near the boundary of rock salt and anhydrite A3. The observation of cracks in that region indicates continuous deformation of the ductile rock salt also farer away from the cavities. At this depth level only few events are found in the anhydrite but at higher depth levels more events are found. Existence of AE events in the anhydrite indicates substantial deviatoric loading in this competent rock which seems to be the consequence of the deformation of the ductile rock salt at its lower boundary.

4.3 *Magnitudes of AE Events*

Figure 6 displays magnitude distributions of events in three subvolumes of the whole monitoring volume. Subvolume A is located in the rock salt near the contours of the rooms including part of the

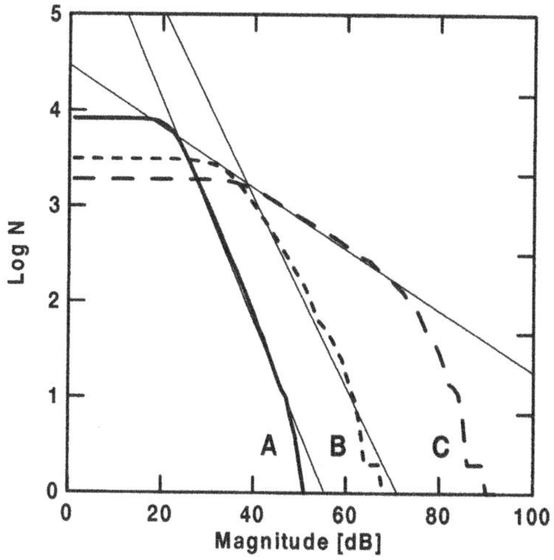

Figure 6. Frequency distributions of AE events in 3 subvolumes: A - rock salt at the cavities, B - rock salt near boundary with anhydrite, C - anhydrite above the upper level.

small pillar between the two rooms at the intermediate level (see Fig. 5). Subvolume B is located west of the cavities in the rock salt and contains the events near the boundary with the anhydrite. Subvolume C covers the whole region west of the cavities (west of X = 100 m in Fig. 5) but extends above the depth interval shown in Figure 5. According to the geological model anhydrite covers the main part of subvolume C. Whereas A and B have nearly the same volume, C is much larger.

In Figure 6 the cumulative frequency distributions of magnitudes M are shown. In this semi-logarithmic plot the distributions are linear as it is found in the case of magnitudes of earthquakes in seismology and in the case of acoustic emission testing in the laboratory. They can be fitted by similar relationships as used there (e.g., see Scholz, 1968):

$$\log N = a - b \cdot M \qquad (2)$$

where N is frequency and a and b are constants. The slope b is dependent on the material and on changes in the stress conditions. In Figure 6 deviations from the linear form of the distributions at values of magnitude below 20 dB and values above 70 dB arise as a consequence of the limited dynamic range of the recording system. In subvolume A low magnitudes down to 20 dB representing weak AE events can be recorded as the distances between events and transducers are small. Subvolumes B and C are farer away from the transducers resulting in a lower magnitude limit of 35 dB.

The curves of A and B represent the variations which are found for many investigated subvolumes in the rock salt. The distributions are narrow, the values of slope b are nearly identical and the largest magnitudes are in the range of 50 to 70 dB. This characterizes the process of dilatancy in rock salt by generation of microcracks. The cracks have small diameters in the order of the grains of the rock salt (10^{-3} to 10^{-2} m) and therefore radiate a limited amount of acoustic energy. Curve C (anhydrite) clearly shows a lower slope and extends to higher values of magnitude. By extrapolation of the linear part of the curve a maximum possible magnitude of 140 dB is obtained. As mentioned above such large magnitudes cannot be recorded by the system but extrapolation is justified by experience from seismology and acoustic emission testing in the laboratory. Two explanations are possible for the large magnitudes in anhydrite which correspond to large release of acoustic energy. As strength of compact anhydrite without joints is much higher than that of rock salt, more elastic energy will be stored until cracks are generated and more acoustic energy is released. Another explanation is the activation of existing joints in form of shear or tensional failure. As stated above, open and healed joints are frequently found in anhydrite. Joints lower the strength of the rock significantly compared to zones which do not contain joints. The last explanation does not require such a high stress level in the rock as in case of compact anhydrite. Such a high level of deviatoric load usually is not expected far away from cavities in the rock. The observed length of joints in anhydrite is in the order of meters. The large AE magnitudes determined in anhydrite are qualitatively in agreement with failure of such joint planes or part of them.

4 CONCLUSIONS

AE measurements offer the possibility to detect zones of current crack generation. In rock salt the progressive generation of microcracks (dilatancy) leads to geomechanical consequences as creep failure, humidity induced creep and permeability increase. In anhydrite shear or tensional failure of joints can be induced by mining which is a different mechanism of crack generation compared to rock salt.

Operation of a large AE network in a salt mine shows that crack generation is limited to zones of 5 to 10 m thickness around cavities in rock salt. Possible failure near the contour of cavities like spalling from the walls or the roof is only a local stability problem. If anhydrite layers are present, AE activity can also be detected farer away from cavities in rock salt. AE events are also found in the competent anhydrite. This indicates a substantial level of deviatoric load resulting from the deformation of the ductile rock salt below the anhydrite. The magnitudes of AE events in rock salt are significantly lower than those of events in anhydrite. This confirms the conclusion that the AE sources in rock salt are microcracks whereas in anhydrite it is most likely failure of existing joints.

REFERENCES

Eisenblätter, J. , Manthei, G., Meister, D., 1998. Monitoring of microcrack formation around galleries in salt rock. *Proc. 6th Conf. Acoustic Emission/Microseism. Activity in Geol. Struct. and Mat.*, Pennsyl. State Univ., 1996. Trans Tech Publications, Series on Rock and Soil Mechanics. Vol. 21: 227-243. Clausthal, FRG.

Hunsche, U., Schulze, O., Langer, M, 1994. Creep and failure behaviour of rock salt around underground cavities. *Proc. 16th World Mining Congress*, Sofia. Vol. 5: 217-230.

Kamlot, P. 1995. Geotechnische Untersuchungen im Salinar zur Ermittlung des gebirgsmechanischen Verhaltens von Anhydrit und Salzton. *Final Report 02E8241A for the Minister of Science, Research and Technology (BMBF, FRG).*

Langer, M. 1996. Underground disposal of wastes requiring special monitoring in salt rock masses: disposal in mines. *Proc. 3. Conf. Mech. Behavior of Salt*, Trans Tech Publications, Series on Rock and Soil Mechanics. Vol. 20: 583-603. Clausthal, FRG.

Peach, C.J., 1991. Influence of deformation on the fluid properties of salt rocks. *Geologica Ultraiectina No. 77*. Univers. of Utrecht.

Richter, G. 1934. Hauptanhydrit und Salzfaltung. *Kali und verwandte Salze und Erdöl*. Vol. 8 - 10: 93-124.

Scholz, C.H. 1968. The frequency-magnitude relation of microfracturing and its relation to earthquakes. *Bull. Seism. Soc. Am.*. Vol. 58: 399–417.

Spies, Th., Meister, D., Eisenblätter, J., 1997. Acoustic emission measurements as a contribution for the evaluation of stability in salt rock. *Proc. 4th Int. Symp. on Rockbursts and Seismicity in Mines*, Krakau: 135-139. Eds. Gibowicz, S.J., Lasocki, S.. Balkema, Rotterdam.

Stäubert, A., 1988. Ergebnisse ingenieurgeologischer Kluftuntersuchungen im Salinar für Abdichtungsinjektionen von Zuflüssen im Kalibergbau der DDR. *Freiberger Forschungshefte*, A753. Freiberg/Sa., FRG.

Yaramanci, U., 1991. Hochfrequenz Mikroseismizität im Steinsalz der Asse um den 945-m-Bereich. *GSF- Forschungszentrum für Umwelt und Gesundheit, GmbH, Report 32/91.* Neuherberg, FRG.

Mechanics of Jointed and Faulted Rock, Rossmanith (ed.) © *1998 Taylor & Francis, ISBN 90 5410 955 6*

Geomechanical behavior of a heated 3m block of fractured tuff

S.C. Blair & W. Lin
Lawrence Livermore National Laboratory, Calif., USA

ABSTRACT: This paper presents an overview of the geomechanical studies conducted at the Large Block Test at Fran Ridge, near Yucca Mountain, Nevada, and results of geomechanical observations made during the first 10 months of heating. This test is being conducted on a block of rock that is 3m x 3m in cross-section and 4.5m high. Heaters have been placed in the rock to simulate a plane heat source, 1.75m above the base of the block. The 3-dimensional geomechanical response of the rock to the heating is being monitored using instrumentation mounted in boreholes and on the surface. Results show that thermal expansion of the block began a few hours after the start of heating. Moreover, expansion in the horizontal direction is consistent with opening of vertical fractures. Opening and sliding of fractures has been correlated with anomalous temperature behavior, indicating that fracture deformation influences the hydrothermal behavior.

1 INTRODUCTION

The Yucca Mountain Site Characterization Project is investigating Yucca Mountain, Nevada, as a potential repository for high-level nuclear waste. When emplaced in a repository, the radioactive decay heat of the waste may cause coupled thermal-mechanical-hydrological-chemical (TMHC) processes in the rock forming the near-field region of a repository. These processes must be understood before model calculations can be performed to confidently predict the performance of a repository over time.

Efforts to understand and characterize coupled processes in a fractured rock mass include the Large Block Test (LBT) currently underway at Fran Ridge, near Yucca Mountain, Nevada. The Large Block Test is being conducted on a rectangular prism of rock that is 3m x 3m in cross-section and 4.5m high. It is a fractured rock mass that was exposed from an outcrop by excavating the surrounding rock, leaving the rectangular prism (see Figure 1). Two sub-vertical sets of fractures and one set of sub-horizontal fractures intersect the block. The sub-vertical fracture sets are approximately orthogonal, with spacing of 0.25 to 1 m and are oriented generally in the NE-SW and NW-SE directions. In addition, the block contains one major sub-horizontal fracture located approximately 0.5m below the top surface. This fracture is visible in Figure 1.

The objective of the LBT is to create, maintain and observe a planar, horizontal region of boiling in the block, so as to observe coupled THMC behavior in a fractured rock mass (Lin et al., 1995). Specifically, the LBT will study the dominant heat transfer mechanism, condensate refluxing, re-wetting of the dry-out zone following the cool-down of the block,

Figure 1: Photograph of Large Block Site. The upper portion of the block is exposed while the lower portion is supported with bracing T shaped grooves on face are locations of fracture monitors.

displacement of fractures, and rock–water interaction. One particular goal of the test is to assess how episodes of opening or shear displacement along fractures are related to changes in the thermal-hydrologic behavior of the rock mass. To this end heaters have been placed in the rock to simulate a plane heat source at a height of 1.75m from the base of the block, and a steel plate fitted with heating/cooling coils has been mounted on the top of the block. This plate is connected to a heat exchanger to allow thermal control of the top surface.

2 INSTRUMENTATION

The 3-dimensional geomechanical response of the rock to the heating is being monitored using borehole extensometers and surface mounted fracture gauges. Multiple-point-borehole extensometers have been deployed in a total of 6 boreholes, 2 of these are horizontal with an E-W orientation, 3 are horizontal with a N-S orientation, and 1 is vertical. The locations of these holes and MPBX anchors in them are shown in Figure 2.

Deformation across fractures that intersect the surface is being monitored using 3-component fracture monitors that have been installed at 17 locations on the surface of the block as shown in Figure 3. A few of the fracture monitor locations are visible as T-shaped grooves in Figure 1.

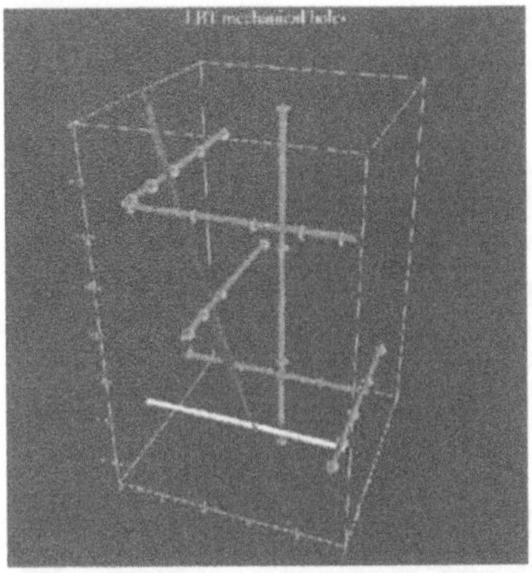

Figure 2: Location of boreholes in Large Block Test used for MPBX measurements. Perspective view is from the Southeast looking Northwest. MPBX anchor locations are shown as disks concentric with the boreholes. Also shown are one observation hole (bright white), and a major fracture plan trending NW-SE.

Fracture deformation at the surface is monitored as follows. One major sub-vertical fracture on each face was chosen for monitoring based on geological mapping studies of the block. Fracture monitors were installed at 3 or 4 locations along the trace of each of the chosen fractures. In addition, motion on the large sub-horizontal fracture near the top of the block is being monitored using one fracture monitor on each face. The fracture monitors measure movement in directions across the fracture, and along the trace of the fracture both parallel and perpendicular to the face. Temperature and fluid movement in the block are also being monitored, and the geomechanical data are being interpreted in conjunction with these other data sets.

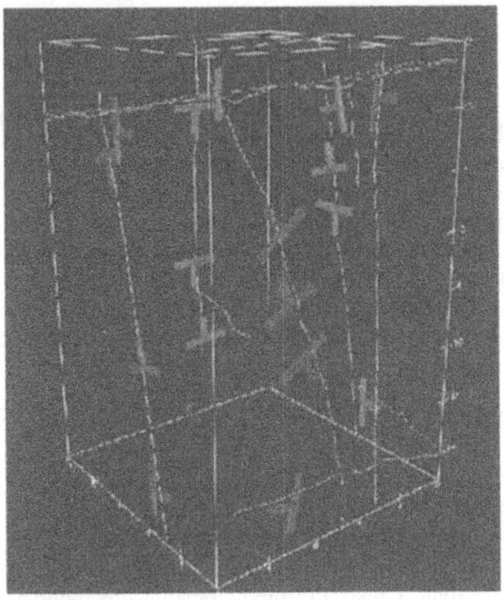

Figure 3. Locations of surface mounted fracture monitors.

3 RESULTS

Heating of the block was started on Feb. 27, 1997, and temperature vs. time measured near the center of the heater plane is shown in Figure 4. This figure shows temperature increased from ambient to 90 °C in the first few hundred hours of heating. A power outage caused a drop in temperature at approximately 600 hours. When power was restored temperatures quickly rose back to pre-outage levels, and near the heater temperature reached approximately 98 °C after 750 hours of heating. After 800 hr. temperature continued to rise more gradually reaching nearly 120 °C, until 2520 hours when a broad thermal disturbance occurred. Figure 4 shows that the temperature dropped sharply to approximately 100 °C. It is important to note that while data for only one temperature sensor are shown in Figure 4, temperatures at several locations in the block jumped to 100 °C at this time. Temperatures remained at this level for about 400 hours and then gradually recovered to levels consistent with those observed prior to 2520 hr. At 4475 hr. temperature suddenly dropped to 100 °C for a second time. However, in this case the temperature recovered after approximately 200 hr. These temperature observations are consistent with the formation of transient heat-pipes within the block. At approximately 5000 hr. the temperature decreases and then becomes steady at approximately 135 °C. This is associated with a change in the power supplied to the heaters.

The overall temperature profile is consistent with conduction dominated heat flow in the block, except for the temperature excursions at 2520 and 4475 hours. These excursions are thought to be breakdowns in the metastable hydrothermal field and formation of transient heat-pipes. While is appears that these thermal anomalies are associated with increased infiltration, thermal-mechanical deformation of fractures, or other changes in the macroscopic flow system, the exact mechanism of their behavior is poorly understood.

Preliminary analysis of deformation has been conducted using data from the MPBX and fracture monitor systems. Results from both these systems show that within a few hours of the heater start-up the block started expanding. This is clearly illustrated in Figure 5 which shows data from one of the MPBX holes that is oriented in the East-West direction. In this figure anchor 0 is at the West face and and anchor 4 is at the bottom of the horizontal hole. Thus, the total deformation observed is approximately 0.01 Overall horizontal deformation of the block after 937 hr. is shown in Table 1. This table shows similar amounts of expansion in both the E-W and N-S directions and that expansion in both horizontal directions is a linear function of height above the base. This is in disagreement with the continuum calculations of Blair et al., 1996 which predicted a non-linear profile of horizontal deformation with height associated with the vertical thermal gradient imposed on the block. Moreover, MPBX data from boreholes in the upper third show that most of the deformation occurs in discrete vertically oriented zones. This may be caused by opening of vertical fractures in this upper region. The observed displacements for the central one third of the block, is in good agreement with predicted values for this zone. Finally, deformation in the lower third of the block is small. This is consistent with the predicted thermal expansion. Data for one of the horizontal holes also indicate that deformation is occurring in a discrete vertically oriented zone. Data also indicate that deformation in the vertical direction is less than that observed in the horizontal direction, and that the region of the block above the heaters is moving upward as a unit.

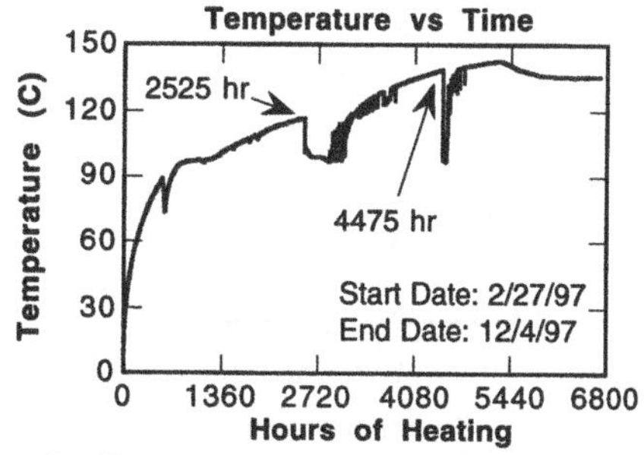

Figure 4. Temperature vs. time for thermocouple located slightly above the heater plane for the Large Block Test.

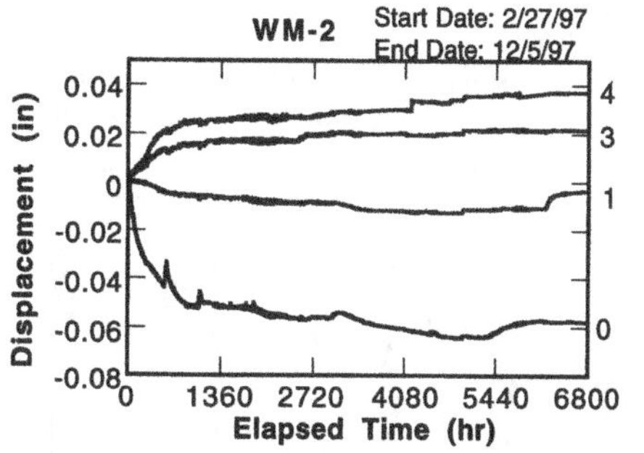

Figure 5. Deformation in E-W direction, recorded using an MPBX system referenced to the center of the block. Numbers at right indicate relative position of anchor.

Table 1. Overall deformation at 937 hours measured using MPBX systems.

Direction	Hole	Height (in)	Expansion (in)
E-W	WM-1	18	0.023
N-S	NM-1	23	0.023
N-S	NM-2	77	0.049
E-W	WM-2	126	0.076
N-S	NM-3	138	0.086

Results from the fracture monitors are consistent with those from the MPBX in that the fractures are generally opening. The overall change in aperture measured at each location is shown in Figure 6. Figure 6a shows that several of the fractures opened between 0.05 and 0.015 in., and that closure of some

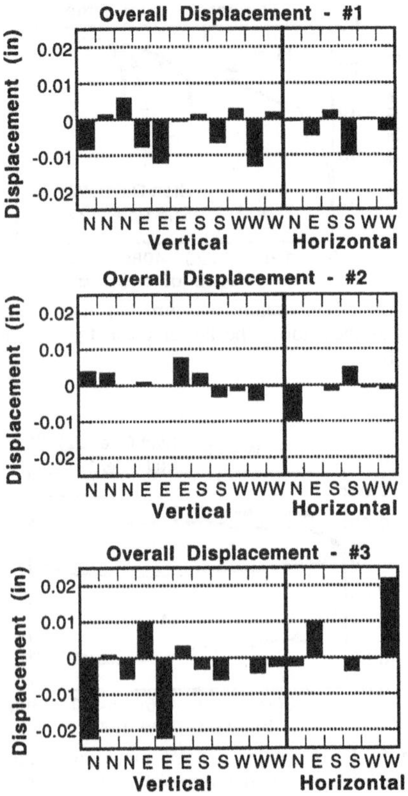

Figure 6. Total displacement observed on selected fractures; (a) changes in aperture (opening is negative), (b) shear parallel to face, (c) shear normal to face.

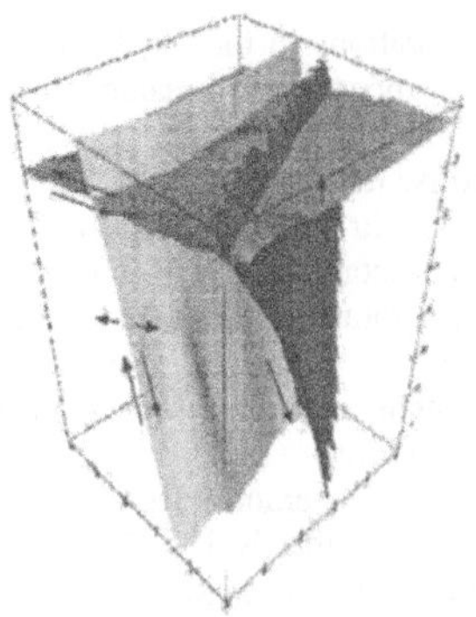

Figure 7. Overall motions observed for Large Block

fractures was observed. Figures 6b and 6c show overall shear movement on the fractures. This figure shows that shear displacements both parallel and perpendicular the faces are both fewer and smaller than normal displacements across the fractures. The large horizontal fracture near the top of the block does show shear offset to the face of the block. This figure shows these to be the largest movements. This is consistent with overall lateral expansion of the block discussed above.

The overall displacements on the block are shown in Figure 7. One of the most prominent fractures observed on the block is a sub-horizontal fracture about 50 cm from the top of the block. Fracture monitor data show that all of the rock above this fracture is moving to the east as a unit. Another major fracture is a North-South trending subvertical fracture that divides the block vertically into eastern and western sub-blocks. Data for this fracture indicate the eastern sub-block is moving upward with respect to the western sub-block.

As mentioned above, one of the goals of the LBT is to study coupled T-H-M processes, and it is interesting to note that the thermal excursion at 2520 hr. was preceded by increased motion on several of the fracture monitors, which indicates that the sudden drop in temperature may be related to a change in the fracture flow network. Displacements measured by the fracture monitors for 2500 - 2550 hr. is show in Figure 6a-c. This Figure shows that significant motion on the large horizontal fracture can be associated with the thermal excursion. More detailed analysis indicates that unlike the thermal response, the motion on this fracture was distributed in time and space. The onset of fracture deformation preceded the thermal event by between 10 and 15 hours.

CONCLUSIONS AND SUMMARY

In summary, the thermomechanical response of the Large Block has been monitored using MPBX and surface mounted fracture monitors. Thermal expansion of the block was evident a few hours after the start of heating. This is verified by data recorded on the fracture monitors and MPBX systems. MPBX data indicate that the block has expanded in the horizontal direction, ant that this expansion is a linear function of the height above the base of the block. Expansion at the top of the block greater than was estimated using continuum assumptions, and much of the deformation has taken place in discrete zones. This is consistent with opening of vertical fractures In the vertical direction, the upper two thirds of the block is extending as a unit.

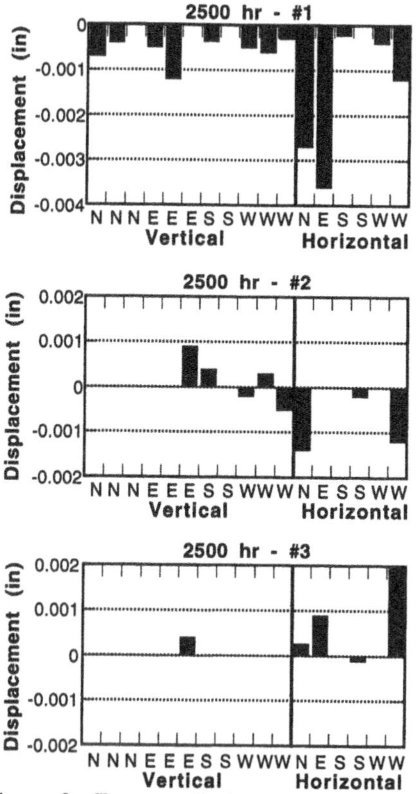

Figure 8. Fracture displacements observed from 2500-2550 hr. (a) changes in aperture (opening is negative), (b) shear displacement parallel to face, (c) shear displacement normal to face.

The hydrothermal response of the block is expected to be dominated by flow in vertical fractures and it is surprising that the major thermal excursion at 2520 hours is associated primarily with movement on a horizontal fracture. It is also important to note that a rainstorm occurred at the site several hours prior to the thermal excursion at 2520 hr. and it is possible that the horizontal fracture served as a primary conduit for focusing water flow within the block.

REFERENCES

Blair, S. C., P. A. Berge & H. F. Wang (1996) "Geomechanical analysis of the Large Block Test", UCRL-ID 122898, Lawrence Livermore National Laboratory.

Lin, W. D.G. Wilder, J. A. Blink, S. C. Blair, T. A. Buscheck, W. E. Glassley, K. Lee, M. W. Owens, & J. J. Roberts (1995) "A heated large block test for high level nuclear waste management", in *Proceedings of the Second International Conference on the Mechanics of Jointed and Faulted Rock*, Vienna, Austria.

This work was supported by the Yucca Mountain Site Characterization Project. Work performed under the auspices of the U.S. Department of Energy by Lawrence Livermore National Laboratory under contract W-7405-ENG-48.

Mechanics of Jointed and Faulted Rock, Rossmanith (ed.) © *1998 Taylor & Francis, ISBN 90 5410 955 6*

Laboratory equipment for the strength and deformability studies of the rock materials under multiaxial stress state

V.E.Levtchouk & E.G.Gaziev
Geotechnical Division, Engineering Institute, National Autonomous University of Mexico, Mexico

ABSTRACT: The special laboratory triaxial compression testing equipment was developed to study the deformational and strength behaviour of the brittle polycrystalline materials in multiaxial stress-strain state. Loading capacities of the equipment permit to realise the experiments with 15 × 15 × 15 cm cubic specimens of rock-like materials such as cement mortar, concrete, sandstone and weak rocks. The concept and design of the equipment permit to realise not only the "classic" triaxial, biaxial or uniaxial experiments, but also to conduct very specific tests for new ideas and theories search, as well as to have detailed studies of pre- and postfailure behaviour. For example, in one of the modes the equipment permits to have the tensor of stresses "sliding" along the strength envelope and the receiving a series of points thus determining the part of the envelope in any state of stress (tri- or biaxial).

The special complex for the experimental data acquisition and processing was developed using personal computer with virtual system of measurements and analysis. The system allows realising the measurements of deformations and loads in multiaxial stress states as well as to register the acoustic emission during the failure progress. The created system manages 20 analogous channels simultaneously with velocity of 1000 measurements per second per channel (the sampling velocity increases when the number of channels in use decreases and can reach up to 10,000 samples per second). Specially developed software and functional basis allows to build relationships between the measuring parameters, "time dependent diagrams", "real time" data saving, and experiment reconstruction with preliminary data processing and analysis (functions fitting, interpolation, extrapolation, signal filtering).

INTRODUCTION

The determination of the strength of brittle polycrystalline materials in a complex stress condition is one of the basic problems of the mechanics, not having until now the satisfactory practical solution. Failure of such materials under combined loading, and especially the brittle failure, was a subject of numerous theoretical and experimental researches in the last years. The increasing application of modern material laws and calculation methods in rock mechanics for underground openings, foundations, slopes, etc. needs the determination of the stress-strain behaviour.

The paper treats with developments for the laboratory equipment and for the stress-strain behaviour studies of the rock materials under multiaxial stress state.

The concept and design of the equipment permit to realise not only the "classic" triaxial, biaxial or uniaxial experiments, but also to conduct very specific tests for new ideas and theories search, as well as to have detailed studies of pre- and postfailure behaviour.

EXPERIMENTAL PROCEDURES

The study of the behaviour of rock-like materials under multiaxial stress state is possible in three different types of investigations.

One type of the rock mass behaviour investigation is so-called "classic" triaxial tests, which are widely spreaded in world practice of the studies of the soils and rocks.

These investigations of the behaviour of the brittle polycristallic (rock) materials in multiaxial stress state are conducted in triaxial confinement machines. In this conditions the lateral confinement is uniform and is being maintained constant during all the experiment ($\sigma_2 = \sigma_3 = \text{const}$).

The other widely spreaded method of the studies is the method of the independent load application along all three axes. This method is usually known as "true" triaxial stress application.

This method permits to manipulate with the values of all three principal stresses and strains during the experiment. It significantly increases the possibilities of studies of the behaviour of material.

Another way to study the material strength and deformational behaviour under multiaxial stress state is to obtain the increment of minor and intermediate stresses σ_3 and σ_2 by means of restriction of a specimen's lateral deformation (dilatation) related with growth of the principal stress σ_1 during the experiment. Rock mass under load, as a rule, has no opportunities of free lateral expansion. Consequently, at increase of one of the acting principal stresses two others grow also. This increment of "lateral" principal stresses goes on after "failure" of a material as well. The increment of lateral confinement results in increment of bearing ability of a rock mass. The degree of the lateral expansion restriction serves for surrounding rock mass stiffness simulation. It permits to simulate the presence of the initial stresses in rock mass as well. The main advantage of this laboratory equipment is in closer simulation of the rock mass behaviour. Having reached the strength envelope the specimen continue to follow it like "sliding" along, and all consequent values of the stress tensor are ultimate as well. Thus, in one experiment it is possible to obtain not just a single point of the strength envelope, but whole part of it. The Figure 1 represents a stress-strain diagram of the concrete specimen obtained in this equipment.

It is seen from the diagram, that having reached the strength envelope at 500×10^{-5} of the strain, the stress-strain curve continued its development "following" the stress envelope even when values of all deformations of the specimen reached significant magnitudes. Besides, it is interesting to note, that the specimen "returns" to it

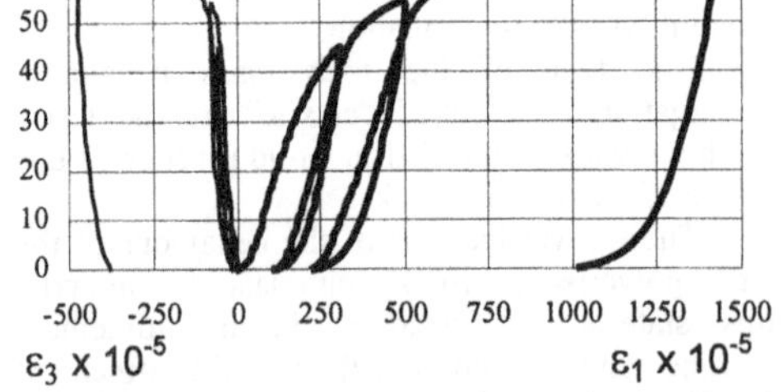

Figure 1. The diagram of ε_1 and $\varepsilon_3 = f(\sigma_1)$. The change of the stresses $\sigma_2 = \sigma_3$ is produced by means of restriction of a specimen's lateral deformation (dilatation) related with growth of the principal stress σ_1 during the experiment.

strength envelope when the loading repeats but withstanding more high level of loads with a greatly high deformations.

The experiments showed that such "self-reinforcement" of the specimen occurs with insignificant increase of lateral stresses (1 – 3 %) while necessary for this stiffness of the lateral confinement system is small enough.

SPECIMENS

There were selected some artificially made polycrystalline materials for equipment adjustment and consequently for some experiments realisation (like a concrete mortar o plaster). There are supposed to be homogeneous, isotropic, with known physical properties. The homogeneity of these materials makes it possible to increase the repeatability of the experiments. The cubes were collated in steel rigid cubic form.

EXPERIMENTAL EQUIPMENT

The special laboratory triaxial compression testing equipment was developed to study the deformational and strength behaviour of the brittle polycrystalline materials in multiaxial stress-strain state. This equipment permits the "true" triaxial stress application in any combinations as well as simulation of the stiffness of surrounding rock mass. Loading capacities of the equipment permit to realise the experiments with 15 × 15 × 15 cm cubic specimens of rock-like materials such as cement mortar, concrete, sandstone and weak rocks.

This laboratory equipment permits to manipulate with the values of all three principal stresses and strains during the experiment. It significantly increases the possibilities of studies of the behaviour of material under constantly changing stress tensor conditions. The experimental equipment developed at the Engineering Institute of the National Autonomous University of Mexico is presented in Figure 2.

The "true" triaxial stress application equipment designed and manufactured at the Engineering Institute contains three mechanically independent axial loads application and deformations control subsystems acting along three mutual perpendicular axis of the cubic specimen. Each of the subsystems consists of a single action hydraulic cylinder and two rigid plates at opposite sides of the sample, which are connected with two rigid tension bars by means of spherical self-adjusting joints.

Figure 2. The "true" triaxial stress application designed and manufactured at the Engineering Institute.

In order to minimise shear stresses on the specimen surfaces, all axial loads application – deformations control subsystems are made mechanically independent and, besides, the plastic film with layer of lubrication is situated between every rigid plate and specimen surface. The spherical self-adjusting joints at the end of each of the rigid tension bars allow small rotations of the rigid plates connected pairwise with the rods. It serves to realise proper loading conditions even in the case when the sample surfaces are not strictly parallel or if the axial loads application and deformations control subsystems are not positioned perfectly.

The single action lateral hydraulic cylinders can apply a maximum stress of approximately 22 MPa while the vertical single action hydraulic cylinder can apply a maximum stress of approximately 130 MPa for cubic specimens with dimensions 15 × 15 × 15 cm.

The loads and deformations on the sample are measured and controlled in a closed loop by a common personal computer equipped with virtual instrumentation system.

The whole system can be divided into two subsystems: control and measurements. The virtual instrumentation system can automatically generate loads or displacements (or both) controlled experimental sequences for each axis. Free switching between the modes is possible.

The control loops of each axial loads application and deformations subsystems are relatively independent one from each other in low level control algorithms (maintaining certain given load or deformation) but they are tightly related at high level control algorithms (maintaining the certain, given by user relationships and ratios).

The oil pressure in each cylinder is measured using special high velocity accurate output transducers. The calculated load is being compared with the measurements of the load cells installed for each direction as well. Then the measured load is being converted to the stress and then compared with required by control algorithm. The virtual instrumentation system generates the control signals for electric valves. The control system with such back loop allows to have necessary stability of the loads, high dynamic range of the loads and accuracy of 22 kPa in stresses.

The measuring subsystem reads the outputs of all sensors independently from load subsystem. For measuring the deformations of the specimen (dimensions changes) in three perpendicular directions 12 high tolerance linear displacement transducers were used (Figure 3).

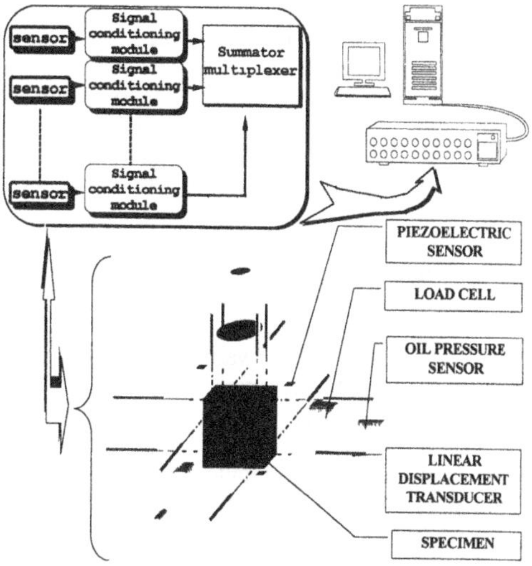

Figure 3. Axial loads application and deformations control subsystems.

The lateral linear displacement transducers have contact directly with the specimen surface through holes in rigid lateral plates and plastic films in order to exclude deformations of the contacts and steel parts. For every lateral direction, four linear displacement transducers permit exact calculation of the deformations. The linear error in length measurements is about 2×10^{-3} mm.

In order to monitor the fracturing of the sample, the measurements system is equipped with piezoelectric transducers mounted in rigid plates. The gauging of frequency and amplitude of acoustic emission is realised by means of a small-sized microphone, established directly on a load plate. The acoustic emission signals are being recorded together with other experiment parameters for further correlation and analysis.

MONITORING AND CONTROLLING THE PROCESS

The special complex for the experimental data acquisition and processing was developed using personal computer with virtual system of control, measurements and analysis. It monitors and controls parameters necessary for maximise operating efficiency. It has modular architecture, quality analog and digital input–output capabilities.

The software provides the operator with a hierarchy of user interface screen displays through which the user can easily navigate the system and experiment. The interface is graphical; it offers a unique combination of flexibility and ease of use to handle the most demanding data acquisition and control.

As the demand increases for measurement and control that are more intelligent and more flexible, the Engineering Institute has applied special combination of technical innovations in experimental studies. The result is "POLICRI" – a state–of–the–art system that is unmatched in terms of usability, reliability and integration into PC-based experimental equipment (Fig.4).

The "POLICRI" system was elaborated at the Instrumental Department of the Engineering Institute.

For registration and analysis of received experimental data, such as: values of applied loads, deformations and acoustic emission at the fracturing process, a special complex of the equipment with virtual system of measurements based on the personal computer was developed in Engineering Institute. The equipment allows to carry out studies of brittle polycrystalline materials behaviour under any stress condition.

Figure 4. The special complex for the experimental data acquisition and processing.

This system has the following characteristics:

- It allows to carry out registration of deformations and loads in a multiaxial stress state, as well as acoustic emission in a sample at the fracturing,

- It allows the simultaneous registration of 20 analog signals (20 channels) with speed of 1000 gauging per second (up to 10,000 samples per second by use only of one channel),

- The program of visualisation of the information and control works in Windows 95 environment.

The program allows: to build on the screen the dependence between measured parameters, and also dependence of these parameters and time, to store the information on a hard disk of the computer in "real time", to reconstruct experiment and to carry out preliminary information processing.

The system is extremely easy to adapt to different experimental conditions. It can operate with different type of sensors (load cells, linear displacement transducers, linear velocity displacement transducers, piezoelectric sensors, microphones, pressure sensors, etc.). POLICRI is developed to deliver significant savings by simplifying installation, configuration, and realisation of the experiment and experimental data postprocessing. Of course, the real benefits of innovative laboratory equipment are not fully realised without a complete software solution.

Through easy-to-use dialog boxes and screens the user can set the sensors and control parameters, signal I/O ranges, watch parameters and experiment states.

In the overall experiment screen the user can view and operate the entire process by clicking the buttons using "mouse". In addition to these "operating screens", the several engineering screens were provided for tune the control and alter the control algorithms running during the experiment. For example, the following experiment paths can be build:

- hydrostatic loading to the certain magnitude of σ_{oct};
- stress loops for the elastic parameters determination;
- a displacements controlled deviatory loading;
- the strength envelope by "sliding" along ultimate values of the stress tensor. Thus, in one experiment it is possible to obtain not just a single point of the strength envelope, but whole part of it;
- the full stress-strain curve (including postpeak part) received at continuous high-speed "real-time" registration of stress, deformations and

acoustic emission during uniaxial loading of a sample.

VIRTUAL INSTRUMENT

Virtual instrumentation is breaking down the barriers of developing and maintaining instrumentation systems that challenge the world of test and measurement. Virtual instrumentation delivers innovative, scalable solutions. They squeeze the most out of the system.

Computer-based instruments use the hardware and software that already is in PC – from the microprocessor and memory to software and firmware – to surpass the performance of stand-alone instruments. Computer-based instruments add user-configurability by expanding functionality and adding flexibility. Signal processing algorithms execute faster on today's PCs than inside traditional instruments.

Unlike a traditional instrument, the designed computer-based virtual instrument can log vast amounts of data; timestamp and annotate the data, communicate directly with databases and information systems automatically generate report and perform high-quality hard copy output. Besides, the user can customise the data analysis and signal processing algorithms with unlimited flexibility. Using standard, off-the-shelf PCs with plug-in instruments saves the time and money by increasing efficiency and ability to solve problems.

The designed instrument permits to share the application experience with the other researchers and to conduct very specific tests for new ideas and theories search, as well as to have detailed studies of pre- and postfailure behaviour. For example by means of the experiments on this equipment, a new constitutive strength criterion was developed (Gaziev 1996). This virtual instrumentation helps in delivery of high-quality products in short experimental cycles. The strategic advantage by using virtual instrumentation is in reduction of the time for the rock mechanic studies, the costs and improvement of the performance, the functionality and flexibility (Gaziev, Levtchouk 1997).

The Figure 5 represents the characteristic stress-strain curve, received at continuous registration of stress, deformations and acoustic emission during loading and failure of a sample. From consideration of bursts of noise, generated in a sample during it microfracturing, it is possible to note, that the first stage of process of microfracturing began at about 50 % from the maximum strength Rc and second stage began, when load has reached 0.8 Rc.

Next, we added a powerful tool that creates

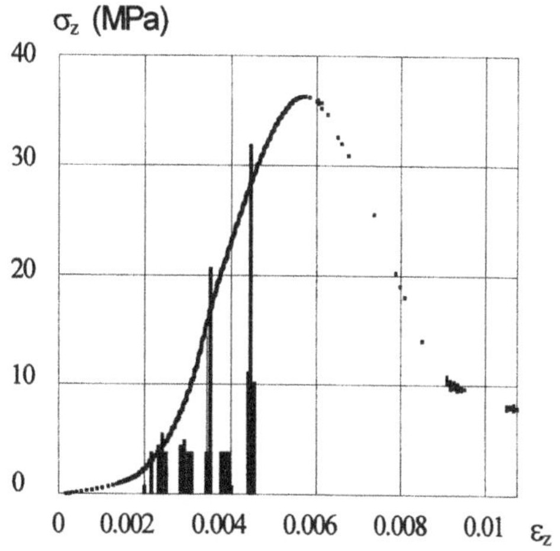

Figure 5. A stress-strain curve, received at continuous registration of stress, deformations and acoustic emission during loading and failure of a sample

complex control strategies and prepares them for download to the experiment directly. The lookout is so intuitive, that no formal training is necessary. By using a plug-in data acquisition board and designed software, it is possible to quickly turn a PC into a powerful tool for the monitoring and control the experiment procedures. Connection to the PC various type of sensors requires selection of the appropriate transducer and logic for signal conditioning hardware. The signal processing capabilities required for signal analysis include digital filtering, peak detection and curves fitting. Digital filtering is used to reject unwanted frequency components in the acquired signal. With peak detection software you can programmatically determine the dominant frequency peaks. The information generated by analysis software can be logged to the hard disk of the PC for later analysis. Specially developed software and functional basis allows to build relationships between the measuring parameters, "time dependent diagrams", "real time" data saving, and experiment reconstruction with preliminary data processing and analysis (functions fitting, interpolation, extrapolation, signal filtering).

Test data is passed from each test as it executes into the Test Engine. The engine uses test data for evaluating the test results, building reports and so on. The module for control strategy creation is a collection of user-modifiable functions and logic compilation module. The user-modifiable functions are defined through a strict interface with the engine, ensuring the rightness and completeness of the generated strategies. While running test sequences, users can place breakpoints to pause the execution. Once the sequence is paused, users can single-step through the sequence or skip ahead to the specific control logic. All operations are accessible through an easy-to-use context menu. The test data is stored in formats compatible with spreadsheet programs (Microsoft EXCEL for example). The program generates three data tables – "raw" experimental data (as it is), processed data and a test table that stores related data (calibration coefficients, offsets, conversion limits, logic of experiment, etc.

CONCLUSIONS

The possibility to control and measure for three independent stresses and strains, as described in this paper, can help in better understanding of the failure mechanisms of rock for complex stress conditions. The developed equipment allows the triaxial stress application in any combinations of the principal stresses as well as simulation of the stiffness of surrounding rock mass.

As an example of results of the tests for new ideas and theories, a new strength criterion was developed. Proposed strength criterion allows more precise estimation of the strength for brittle polycrystalline materials (rock, concrete, plaster, ceramics etc.) in any complex stress state, both biaxial, and triaxial.

The proposed equipment permits to have closer simulation of the rock mass behaviour in the conditions of restriction of an opportunity of free lateral expansion of a sample. Having reached the strength envelope the specimen continue to follow it like "sliding" along, and all consequent values of the stress tensor are ultimate as well.

Complete stress-strain curves of rock under uniaxial and multiaxial compression can be obtained by means of high-speed registration in real time of failure including study of micro and macrofracturing processes and deformations in whole load interval (prefailure and postfailure zones).

The control virtual module allows to have very different experiment paths, for example, to preserve the hierarchy of principal stresses ($\sigma_1 \geq \sigma_2 \geq \sigma_3$).

A powerful computer-based tool permits to create complex control strategies and download it to the experiment directly.

The designed virtual instrument can log vast amounts of data; timestamp and annotate the data, communicate directly with databases and information systems automatically generate report and perform high-quality hard copy output. Besides, the user can customise the data analysis and signal processing algorithms with unlimited flexibility. At any moment it is possible to "reproduce" the recorded experiment with all its peculiarities as many times as it is necessary.

In addition, more detailed input for finite element calculations can be obtained from the tests on such equipment.

The designed laboratory equipment permits to realise practically all variations of the multiaxial studies of the rock specimens. The carefully acquired and processed data are suitable for research work in rock mechanics domain, in tunnelling, foundations, deep open pits and slope stability.

ACKNOWLEDGEMENTS

The authors gratefully acknowledge the support of the presented research work by the researcher of the Engineering Institute of the National Autonomous University of Mexico Prof. Jesús Alberro Aramburu.

REFERENCES

Gaziev, E. 1996. Criterio de resistencia para materiales rocosos policristalinos. *Segunda Conferencia Magistral Profesor "Raúl Marsal"*. México: SMMR.

Gaziev, E. & Levtchouk, V. 1997. The study of behaviour of the brittle materials in post-failure multiaxial stress-strain states. *XI-th Russian conference on rock mechanics with foreign participance,* RUSROCK-97, St-Petersburg, 9 – 11 September 1997: 103-113, St-Petersburg: State Architectural University.

Mining and underground construction

Mechanics of Jointed and Faulted Rock, Rossmanith (ed.) © 1998 Taylor & Francis, ISBN 90 5410 955 6

Fracturing around deep level stopes: Comparison of numerical simulation with underground observations

E.J. Sellers, J. Berlenbach & J. Schweitzer
CSIR Division of Mining Technology, Johannesburg, South Africa

ABSTRACT : The behaviour of the rockmass surrounding gold reef stopes at great depths is controlled by deformation on the fractures and the stability of the hangingwall is influenced by the fracture orientation. Rock type and its associated competency, sedimentary partings, faults, dykes, joints and extension gashes are the major primary and secondary geological features controlling fracture pattern, orientation and frequency. Boundary element techniques for the simulation of the fracture zone are being developed to identify potential hazards, assist in the selection of appropriate support and suggest safer mining methods. An evaluation of some results from the DIGS discrete fracture growth and random tessellation approaches versus selected Witwatersrand environments was undertaken. Predictions of fracture angles, distribution and persistence are similar to underground observations. The models are able to predict the influence of initial stresses and joint patterns on the mining induced fracturing. Further research is required to enable prediction of fracture intensities.

1 INTRODUCTION

The control of geological parameters on the fracturing in the rockmass surrounding deep gold mine stopes in South Africa has been emphasized by many workers (e.g. Jager & Turner 1986; Johnson & Schweitzer 1996; Schweitzer & Johnson 1997). However, numerical modelling studies of mining excavations often neglect the impact of geological features on mining induced fracturing.

The DIGS (Discontinuity and Growth Simulation) boundary element program has been developed (Napier 1990, Napier & Pierce 1995) to be able to model the fracture zone development and include the influence of geological features such as rock competency, reef thickness, bedding planes, dykes or faults. Two approaches for modelling the fracture zone are available. The first (Napier 1990) considers the incremental growth of discrete fractures. The second method (Napier & Pierce 1995) is known as the tessellation approach and is based on the concept that fracture growth can be represented by the activation of selected elements from a random mesh of predefined fracture sites. The tessellation approach was developed to overcome numerical instabilities arising from arbitrary crack intersections.

This study considers a joint approach by rock engineers and geologists to document underground observations in selected mines to assist in the verification of the current DIGS versions. The results of some of these studies are presented, leading to the identification of future activities.

2 DIGS BOUNDARY ELEMENT PROGRAM

The numerical code DIGS is based on the displacement discontinuity formulation of the boundary element method in which a set of fictitious cracks or dislocations in an elastic body, are used to represent the jump in the displacement field across each fracture (Napier 1990).

The displacement discontinuity method can be extended to the modelling of fracture growth as a pseudo-static process if each crack tip is advanced incrementally in a direction that maximises a specified growth criterion (Napier 1990). Two basic modes of fracture in brittle rock are distinguished, namely extension, or cleavage, fracturing and shear fracturing (Napier & Hildyard 1992). The origin of fractures (seed position), fracture growth rule and material properties are defined for each anticipated crack. Growth is accomplished by searching around the seed point or crack tip at a fixed distance with a given angular increment. For extension fractures, applied in this paper, the growth angle is chosen to be the direction of the major (compressive) principal stress. Numerical instabilities may be reduced by preventing crack intersections. This does not alter the fracture pattern.

The problems of pre-selection of the fracture initiation sites and the computational effort required for modelling discrete fracture growth are compounded by the possibility of numerical instabilities arising from the linking of cracks at small angles, and the insertion of very small elements at crack intersection points. These shortcomings led to the development of the tessellation approach in which the region of interest is covered by a random tessellation of inactive elements based on the Delaunay triangulation scheme. The linking of preferentially aligned segments provides an approximation of a discrete fracture path. The segment lengths and angles of intersection are controlled by the choice of tessellation pattern.

A considerable saving in memory can be achieved by the application of the Multipole Method (Peirce & Napier 1995) in which the element influences are transmitted between centre points of an overlain grid of rectangular cells using the Discrete Fourier Transform method. A procedure for selection of the tessellation patterns and the multipole grid size has been developed (Sellers 1997).

Fractures form in the model when induced stresses on crack segments exceed the Mohr-Coulomb failure criterion with given cohesion C_o, friction ϕ and tensile strength T_o. In the incremental activation procedure, applied in this paper, the crack site with the stress state furthest outside the failure criterion is transformed into an active crack by emplacement of a displacement discontinuity element with a specified residual cohesion C_m, friction angle ϕ_m and dilation angle ψ_m. The problem is solved and the activation procedure is repeated until all sites are within the failure envelope.

3 MODELLING OF GEOLOGICAL CONTROLS ON FRACTURE ZONE BEHAVIOUR

Detailed discussions of the relationships between the stability of an excavation or mining layout and geological structures are provided by Adams et al. (1981), Gay & Jager (1986) and others. An overview is given by Schweitzer and Johnson (1997) who divide geological parameters into two classes:

a) Primary features: rock type, sedimentary structures (e.g. planar and cross beds), reef geometry (e.g. rolls), lithological contacts and flow bands.

b) Secondary features: joints, extension gashes, faults, dykes, metamorphism.

Locked in stresses associated with the tectonic history that caused the secondary features may also influence the fracture formation (Gay & Jager 1986).

The primary features can be modelled in the boundary element context by altering the strength of the potential fractures. Layering, bedding and rolls are best modelled with the tessellation approach by altering the strength properties of the potential sites in the relevant regions. In the discrete growth mode, fracture properties are defined by the original seed position and do not change to represent different layers. In both methods, the secondary features are modelled explicitly as displacement discontinuity elements, with a Mohr-Coulomb failure criterion and post-failure properties.

A mining procedure was developed where both the hangingwall and footwall are modelled by displacement discontinuity elements, and are connected by elements representing successive face positions. In each mining step, the boundary conditions on the elements defining the excavation are changed to a state of zero shear and normal stress. The closure of the hangingwall elements is limited to the stope width. Any active or potential elements existing within the new block are removed, to prevent numerical instabilities associated with rigid body motions of separate blocks within the mined out region.

4 UNDERGROUND STUDY SITES

Underground studies concentrated on Carbon Leader and Vaal Reef stopes in the Carletonville and Klerksdorp areas. These two reef horizons are economically important for future mining operations (Schweitzer & Johnson 1997). Detailed stratigraphic descriptions are provided in Berlenbach &

Schweitzer, (1996). Observations were undertaken at Doornfontein, East Driefontein, Vaal Reefs and Western Deep Levels East mines. Site depths varied from 2000m to 3400m below surface.

5 EFFECT OF ROCK TYPE AND BEDDING

The influence of these primary features is best investigated using the Carbon Leader reef. The Carbon Leader is a narrow conglomerate seam, with a maximum thickness of 45 cm, overlying a complex quartzite sequence with alternating incompetent and competent partings. The Carbon Leader is immediately overlain by a competent, medium- to coarse-grained siliceous quartzite, between 1,4 m and 4 m thick. The quartzite is overlain by the Green Bar, a 1 to 2,5 m thick argillaceous unit. The transition between these two lithologies is defined by a pronounced parting plane.

The tessellation model of the Carbon Leader stope consists of a 2 m wide layer of a strong rock (C_0= 18MPa, ϕ = 52° and ϕ_m = 30°) which overlies the reef. A 1 m layer of weak rock (C_0= 18MPa, ϕ = 32° and ϕ_m = 20°) overlies the hangingwall layer and is delimited by two parting planes with a low friction angle (ϕ = 5° and ϕ_m = 2°). The reef and the rest of the hangingwall have the same properties as the hangingwall beam. The footwall is modelled as an intermediate strength material (C_0= 18MPa, ϕ = 43° and ϕ_m = 25°). All other parting planes have ϕ = 10° and ϕ_m = 5°. Studies of the effect of parting plane friction show that low friction angles cause slip and fracturing ahead of the face (Sellers 1997). The values of the cohesion, calculated from laboratory tests (Briggs 1982), are reduced to 30% to account for *in situ* response. All materials have a tensile strength of 5MPa. The predicted fracture pattern at a span of 14m, using the incremental activation rule, is shown in Figure 1.

Fracturing extends ahead of the face and extension and sheared fractures link to form angled shear zones, as observed underground (Adams et al. 1984). Similarities with underground observations include: fractures are more frequent with increasing friction angles of parting planes, fractures are more prominent in the vicinity of the stope, and flatten away from the excavation; shear fractures are more continuous in competent strata; and shear fractures are more widely spaced than the extension fractures in competent footwall or hangingwall rock. However, less intense fracturing is predicted in competent horizons in comparison with weaker, bedded counterparts. This is possibly in disagreement with underground observation. Stress transfer ahead of the face, and periodic bands of fracturing, may be caused by irregular bedding planes (King, et al 1990) and not by low friction on flat parting planes, as in the model.

6 EFFECT OF *IN SITU* STRESS

The underground observations suggested that mining in opposite directions resulted in distinctly different fracture patterns. The best example was found at the Vaal Reefs site. There, they form part of an alternating argillaceous and siliceous quartzite sequence (Berlenbach & Schweitzer 1996). Unconfined strengths range from 180 MPa to 250MPa.

In the hangingwall of the panel that was being mined in a westerly direction, fractures were found to dip between 70° and 80° towards the face. The

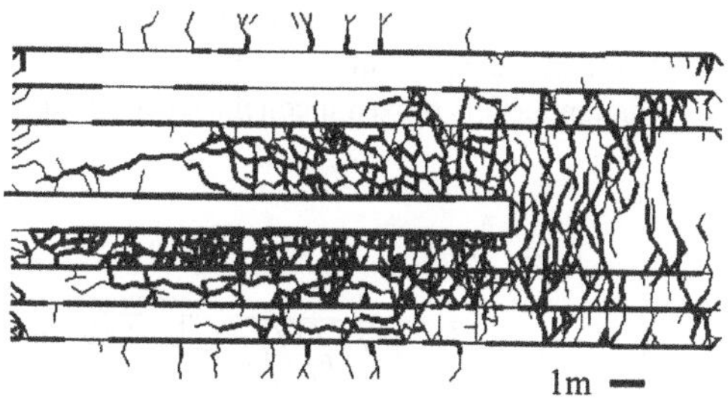

Figure 1: Predicted fracture pattern around Carbon Leader stope after 14 mining steps. (Dark lines indicate open cracks, grey lines are sliding cracks)

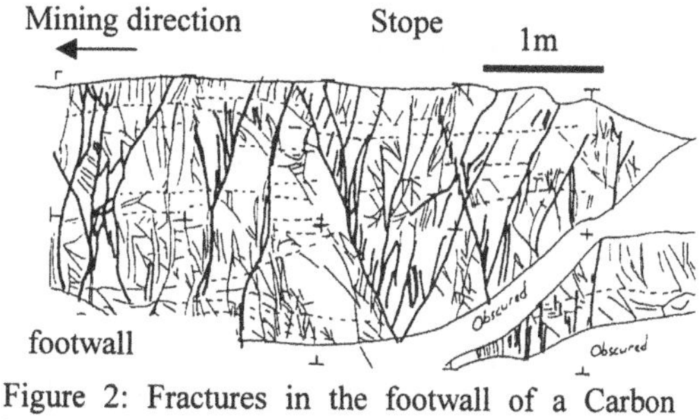

Figure 2: Fractures in the footwall of a Carbon leader stope, mapped from photographs (Brummer & Rorke 1984)

spacing of fractures is of the order of cm-dm (Fig.3a). The fractures are concentrated in a 80 cm thick quartzite beam and, apparently, do not extend across this parting plane. Shear fabrics (s-c fabrics) in argillaceous units within the reef horizon indicate a bedding-parallel movement towards the east (i.e. the hangingwall moved in the direction of the stope). This suggests that the *in situ* stress plunges to the east. Measurements of the stresses in the Klerksdorp district indicate that the major principal stress may plunge either 60° to the east or 30° to the west (Gay et al. 1984). No stress measurements are available near the site itself.

Fractures in the hangingwall of the panel being mined to the east are sub-parallel to the bedding and result in highly unstable hangingwall conditions. However, fractures in the footwall are vertical to steeply dipping towards the face (Fig. 3b). A rockfall in this area, a week before the site investigation, indicated that the large, flat slabs of rock in the hangingwall were difficult to support and were a hazard to miners.

Tessellation and discrete growth models were set up with a major principal stress plunging at 60° to the East. This is achieved by setting the vertical stress σ_v = 60MPa, the horizontal stress to σ_h = 12MPa, and the shear stress σ_{vh} = ± 25 MPa. The tessellation models proved inconclusive as the initial shear stress activated significant regions of the tessellation and obscured the stope fracturing. These fracture zones would correspond to large-scale secondary faulting, which cannot be modelled as discrete discontinuities with the tessellation scheme, alone. In contrast, the discrete models were able to show that the fracture pattern depends on the mining direction. The prevention of crack intersections has permitted stable analyses for spans of more than 14m. Thus, in Figure 4a, most fractures dip at angles between 60° and 75° towards the face, whereas in Figure 4b, the fractures dip at 30° to the face.

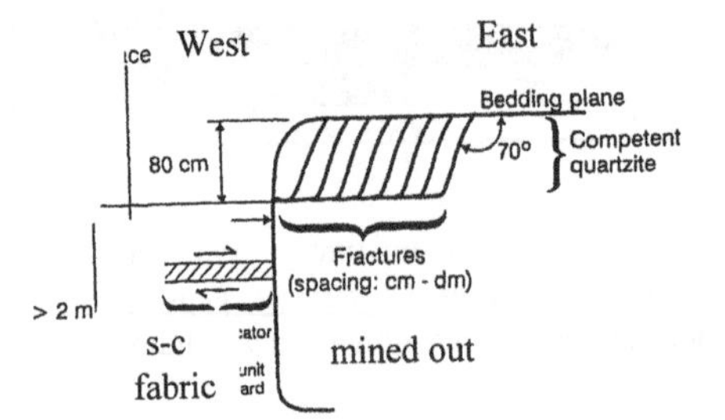

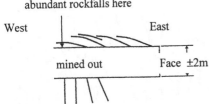

Figure 3: Schematics of fracture patterns observed at Vaal Reef site. A: Steeply dipping fractures when mining westwards b: Flatter dipping fractures when mining eastwards

7 EFFECT OF PRE-EXISTING JOINTS

The dominant geological structures at the

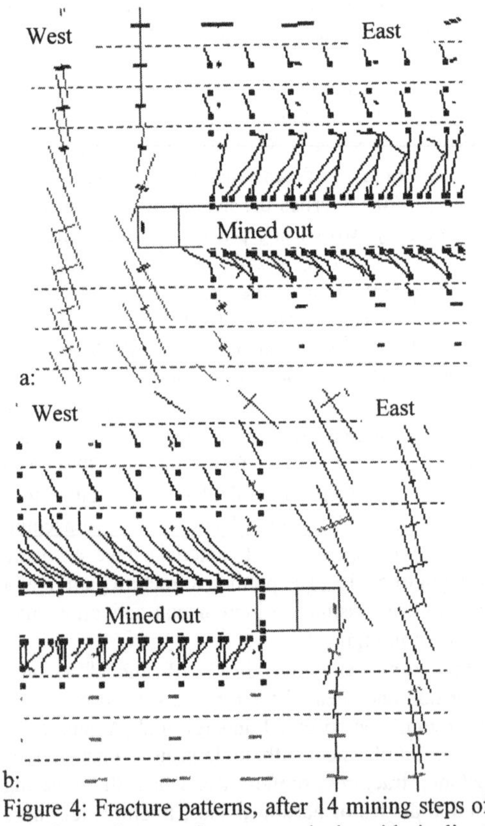

Figure 4: Fracture patterns, after 14 mining steps of 1m, from discrete fracture analysis with inclined major principal stress. A: steep dipping fractures when mining westwards b: shallow dipping fractures mining eastwards (single grey lines are compressive stresses double lines are tensile)

Doornfontein and Western Deep Levels sites were quartz filled extension gashes in the quartzite overlying the Carbon Leader. At Doornfontein, the strike of the extension gashes deviates by approximately 15° from the strike of the face and the average dip is 85° away from the face. Mining induced fractures are consistently parallel to the extension gashes i.e. opposite to conventional concepts of mining induced extension fracturing which suggest that the fractures dip towards the face (Adams et al. 1984). Similarly, fractures are parallel to extension gashes in the footwall quartzites. The relationship between fractures and extension gashes suggests that the gashes, and/or residual stresses, influenced fracture formation. At the Western Deep Levels site, the extension gashes and the associated mining induced fractures dip towards the face.

A model was developed to investigate whether pre-existing joints, or extension gashes, set in the hangingwall, dipping at 80° away from the face with a spacing of 0.5 m could induce extension fractures with the anomalous dip observed at Doornfontein. Due to memory limitations, each joint is assumed to extend only 4m into the hangingwall. The footwall has no jointing for comparison. Each joint is 0.5 m apart and is assumed to have healed (i.e. be quartz filled) and is assigned an initial cohesion of 5 MPa. The friction angle is initially 20° and reduced to 15° upon activation. In this case, the average element size is 0.1 m. The *in situ* stress was 60 MPa in the vertical direction and 30 MPa in the horizontal direction.

As the mining progressed, the joints slightly ahead of the face failed by sliding. Fractures formed ahead of the face and extended into the hangingwall. The mining induced fractures in the hangingwall are observed to dip away from the face, sub-parallel to the jointing, as in the underground situation. Fracturing in the footwall is as expected for a uniform rockmass as the dip decreases with distance from the stope becomes sub-horizontal within 1 m to 2 m of the face.

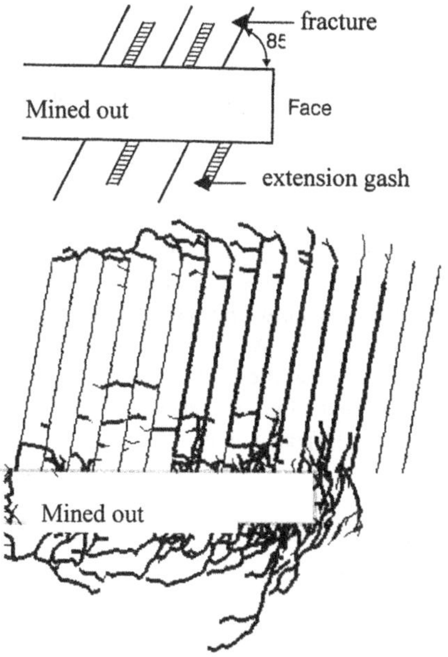

a:

b:

Figure 5: Activated fractures in a tessellation analysis of a stope with a set of steeply dipping, healed, joints/extension gashes in the hangingwall

8 CONCLUSIONS

Two approaches to modelling fracture growth have been incorporated in the DIGS displacement discontinuity element program. The first involves discrete fracture growth and the second considers a pre-defined tessellation of potential sites, which activate sequentially, as the failure criterion is exceeded. The techniques are being developed to model the response of the different types of rockmass occurring around deep level gold mine stopes.

The comparison of modelling predictions with underground observations has shown that the tessellation approach is suitable for the prediction of the distribution and directions of fracturing induced by the interaction of the mining with primary geological structures such as bedding planes and layers of different strength rocks. The fracture intensity is not correctly predicted. The method can also predict the influence of closely spaced joints on the direction of the mining induced fracturing, suggesting that mining induced fracturing may align with mining activated jointing and other secondary geological structures. Methods for distinguishing the induced fractures from the geological features are required to provide unambiguous input parameters for computer modelling. This will lead to a better understanding of the interaction of discontinuities in the stope hangingwall and, therefore, to improvements in support design methodologies.

The discrete growth method is able to model the dependence of the fracturing on the direction of the major principal stress and can be used to investigate ways of overcoming the observations that the rock-fall hazard depends on the mining direction. The tessellation approach can predict a similar response, however, the initial mobilisation of diffuse fracture zones, corresponding to pre-existing faults, is further evidence for the need to incorporate the ability to predict shear localisation and fracture spacing.

Improved modelling of the complex mining environments has been achieved by comparing numerical modelling developments with detailed geological and rock mechanical information. It is, however, recognised that computer modelling will only become a powerful tool in the development, identification and prediction of the most suitable support and mining strategies after extensive verification of the programs with measurements of the rockmass response.

9 ACKNOWLEDGEMENTS

Dr J Napier is especially thanked for his valuable input. The rock mechanics and geology departments of Doornfontein, East Driefontein, Vaal Reef, and Western Deep Levels East Mine are thanked for their assistance. The support of the Safety in Mines Research Advisory Committee (SIMRAC) under project GAP332 is gratefully acknowledged.

REFERENCES

Adams, G.R., A.J. Jager, & C.Roering 1981. Investigations of rock fracture around deep-level gold mine stopes. *Proc. 22'nd U.S. Symp. Rock Mechanics from Research to Application*, Massachusetts: 213-218.

Berlenbach, J.W., & J.K. Schweitzer 1996. Geotechnical environments associated with the auriferous Vaal Reef and Carbon Leader, Witwatersrand Basin, South Africa, *Interim SIMRAC Report GAP 330*, CSIR-Miningtek.

Brummer, R.K. & A.J. Rorke 1984. Mining induced fracturing around deep gold mines stopes, *Report 38/84*, Chamber of Mines of South Africa.

Gay, N.J., D. Spencer, J.J. van Wyk & P.K. van der Heever 1984. The control of geological and mining parameters in the Klerksdorp Gold Mining district. In Gay, N.C., Wainwright, E.H. (eds.) *Proceedings of the 1st International Congress on Rockbursts and Seismicity in Mines*, Johannesburg, 1982, SAIMM: 107-120

Gay, N.C. & A.J. Jager 1986. The influence of geological features on problems of rock mechanics in Witwatersrand mines. In: Anhaeusser, C.R., Maske, S. (eds.) *Mineral Deposits of Southern Africa*, Geol. Soc. S. Afr.: 753-772.

Jager, A.J. & P.A.Turner 1986. The influence of geological features and rock fracturing on mechanised mining systems in South African gold mines. Gold 100, *Proceedings of the International Conference on Gold, Gold Mining Technology*, Johannesburg, SAIMM 1: 89-103.

Johnson, R.A. & J.K. Schweitzer 1996. Mining at ultra-depth: evaluation of alternatives. In: Aubertin, Hassani, Mitri (eds.) *Rock Mechanics*, Balkema, Rotterdam: 359-366.

King, R.G., A.J. Jager, M.K.C. Roberts & P.A. Turner 1989. Rock mechanics aspects of stoping without back-area support. *Research Report 17/89* Chamber of Mines of South Africa.

Napier, J.A.L. 1990. Modelling of fracturing near deep level gold mine excavations using a displacement discontinuity approach, *Mechanics of jointed and faulted rock.* Rossmanith (ed), Balkema: Rotterdam: 709-716,

Napier, J.A.L. & M.W. Hildyard 1992. Simulation of fracture growth around openings in highly stressed, brittle rock, *J. S. Afr. Inst. Min. Metall.* 92:159-168.

Napier, J.A.L. & A.P. Peirce 1995. Simulation of extensive fracture formation and interaction in brittle materials, *Mechanics of jointed and faulted rock - 2*, Rossmanith (ed), Balkema: Rotterdam: 63-75.

Peirce, A. P. & J.A.L Napier 1995. A spectral multipole method for efficient solution of large-scale boundary element models in elastostatics, *Int. J. Num. Methods Eng.* 38: 4009-4034.

Schweitzer, J.K., & R.A. Johnson, 1997. Geotechnical classification of deep and ultra-deep Witwatersrand mining areas, South Africa. Mineralium Deposita, in press.

Sellers, E. & J. Napier 1997. A comparative investigation of micro-flaw models for the simulation of brittle fracture in rock. *Comp. Mechanics*, 20: 164-169.

Sellers, E. J. 1997. A tessellation approach for the simulation of the fracture zone around a stope, *1st SA Rock Engineering Symp.* SANGORM, Johannesburg.

Mechanics of Jointed and Faulted Rock, Rossmanith (ed.) © 1998 Taylor & Francis, ISBN 90 5410 955 6

Mining subsidence in the concealed coalfields of north-east England

N.R.Goulty
Department of Geological Sciences, University of Durham, UK

ABSTRACT: Subsidence profiles can be accurately predicted from empirical experience in the exposed coalfields, provided that the Coal Measures are comprised of weak shale strata. The predicted profiles for both vertical subsidence and horizontal strain are smoothly varying functions of surface position. Where thick beds of strong sandstone or limestone are present in the overburden, however, the subsidence behaviour can be quite different. The outcrop of the Permian Lower Magnesian Limestone in County Durham is deeply fissured above faults which bound multi-seam workings in the underlying Coal Measures. At Selby, in Yorkshire, where the Coal Measures are overlain by about 200 m of Permo-Triassic strata, subsidence due to individual longwall panels does not fully develop until the adjacent panels on both sides have also been worked. These phenomena can be understood in terms of the behaviour of relatively rigid blocks, separated by joints and bedding planes.

1 INTRODUCTION

In the United Kingdom, much of our understanding of mining subsidence has come from observations in the exposed part of the East Pennine Coalfield (Wardell 1954, National Coal Board 1975). Here the Coal Measures strata predominantly comprise shale beds which behave in an incompetent manner during subsidence. The empirical prediction methods developed by the National Coal Board (1975) give very similar results to the methods using influence functions developed in other European countries (Kratzsch 1983). However, in the concealed part of the East Pennine coalfield, beneath the Permo-Triassic cover (Figure 1), the magnesian limestone strata in the Permian have responded in ways which were not anticipated from the general experience in exposed coalfields.

Where the Lower Magnesian Limestone outcrops in County Durham, and little soil or drift cover is present, deep fissures have opened at the ground surface above the edges of old mine workings. Most of the deep mines still operating in England are located in Yorkshire and Nottinghamshire, in the concealed part of the East Pennine Coalfield. The nature of the Permo-Triassic cover in this region, with thick beds of Permian Lower Magnesian Limestone and Triassic Sherwood Sandstone (Buntsandstein equivalent), is such that fissuring is liable to occur here, too, although in many localities it may be obscured by the surficial cover of Quaternary deposits.

The largest coal mine complex currently operating in the UK is at Selby, where the ground surface is close to sea level and so permission has only been granted to mine one seam because of the risk of flooding. Subsidence has been carefully monitored to ensure that it does not exceed 1 m at any point.

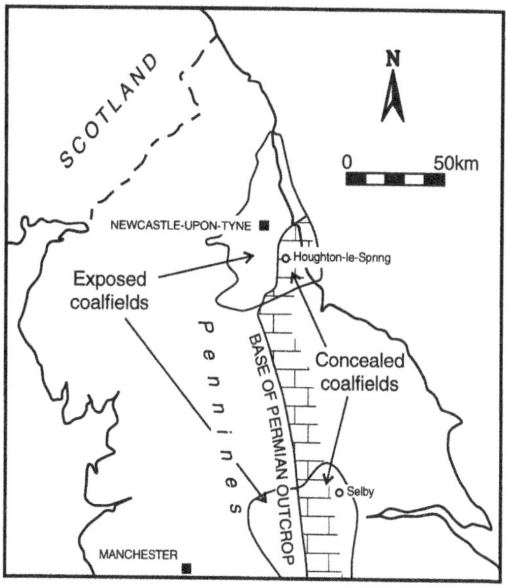

Figure 1. The coalfields of north-east England to the east and north-east of the Pennine hills.

Repeated levelling observations at Selby have shown that the subsidence anticipated above each longwall panel does not fully develop until the panels to either side have also been worked. This implies that some bridging mechanism is taking place in the Permo-Triassic overburden, and tallies with other anomalous observations of reduced subsidence, reported from both exposed and concealed coalfields.

In this paper, I summarise current understanding of subsidence behaviour in exposed and concealed coalfields, distinguishing between active subsidence whilst mining is in progress and residual subsidence which takes place after workings have been halted, and then describe two examples of subsidence behaviour observed in the concealed coalfields of north-east England, concluding with a qualitative explanation.

2 SUBSIDENCE DUE TO LONGWALL MINING

Final subsidence profiles due to longwall coal mining may be predicted using empirical methods (Kratzsch 1983). In the UK, a set of empirical graphs has been published by the National Coal Board (1975). In Germany and Poland, subsidence is calculated by integrating influence functions for the mined-out area. The use of influence functions is also an empirical method, not only because the influence functions are empirical, but also because empirical end corrections need to be applied for the boundaries of the workings. A third alternative is by finite-element calculations using assumed values of rock mechanical properties, and this is also empirical in that the properties themselves are estimated by matching calculations to observations of subsidence.

The horizontal strains and vertical subsidence along the centre line over one end of a longwall panel at depth h are shown in Figure 2. The amplitudes of these profiles scale in direct proportion to the seam thickness worked, and increase non-linearly with increasing width and decreasing depth of the panel. Since the horizontal axis is scaled in units of the depth h, the shape of these profiles is invariant provided that the length of the worked area extends for a distance of at least $0.7h$ to the left of the section shown. The overall process is that the overburden above a panel subsides into the void with a component of horizontal tension pulling the adjacent rock mass inwards. Compressional horizontal strains develop above the panel, with complementary extensional strains to either side. The compressional and extensional strains balance, since the lateral displacement of surface points decays to zero away from the panel. These horizontal strains are principally responsible for any damage which may be caused to surface structures.

A significantly different subsidence profile, known as the 'subsidence development curve', is obtained if it is measured while mining is in progress (Figure 2). When a longwall panel comes to a halt, the amount of residual subsidence which can take place is given by the difference between the subsidence development curve and the final subsidence profile. The residual subsidence is only 7% of the maximum subsidence above the face line, but peaks at more than 20% of maximum subsidence at a distance $0.23h$ behind the face line. The normal expectation has been that residual subsidence should take place within a year of the cessation of mining, although the National Coal Board (1975) did note that, "unusually strong strata in the overburden can delay the residual subsidence (and reduce the amount of total subsidence) but such cases are the exception."

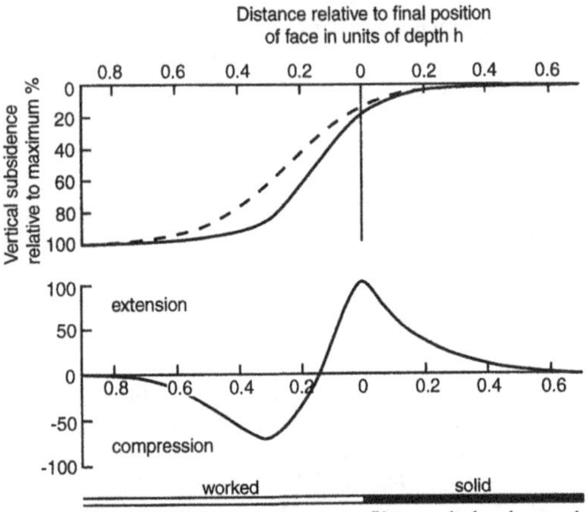

Figure 2. Final subsidence profile and horizontal strains (solid lines) and the subsidence development curve (dashed line) predicted at the ground surface across the end of a longwall panel at depth h.

Wardell (1954) compared subsidence development curves for eleven cases in Yorkshire, and found that all eleven curves were very similar, even though the rates of advance varied by a factor of three and the depths of the workings by a factor of five. Goulty & Al-Rawahy (1996) have recently argued that these well-known and long-established observations demonstrate that subsidence cannot be modelled as a simple viscoelastic process. Instead, they suggested that the subsidence development curve represents the active subsidence which takes place whilst mining is in progress, and responds with a short time constant of the order of days, whereas residual subsidence takes place with a time constant which is generally several months in the exposed coalfields, but could be much longer.

Ferrari (1997) has recently evaluated five case studies from concealed coalfields in England, and concluded that residual subsidence continued for up to eleven years after mining had ceased. His data add significantly to the limited number of observations of delayed subsidence previously reported by Orchard & Allen (1975) and Collins (1978).

As far as the Coal Measures strata are concerned, the mechanical processes involved during active and residual subsidence may be understood qualitatively by reference to Figure 3. Behind a longwall face, the immediate roof breaks off in large pieces and forms the goaf. Typically, there is a transition zone of highly fractured blocks beneath the more rigid main roof, detached from it and constrained to lie on the goaf. The main roof settles in a gentle sigmoidal curve, with bed separation taking place around the points of inflection on the bedding planes. During active subsidence, this zone of intense deformation moves forward with the face, allowing the rocks behind to relax. When the face halts, residual subsidence takes place with a much longer time constant as the rocks in this zone respond to the enhanced stresses by further elastoplastic deformation, allowing bedding planes to close as the strata settle.

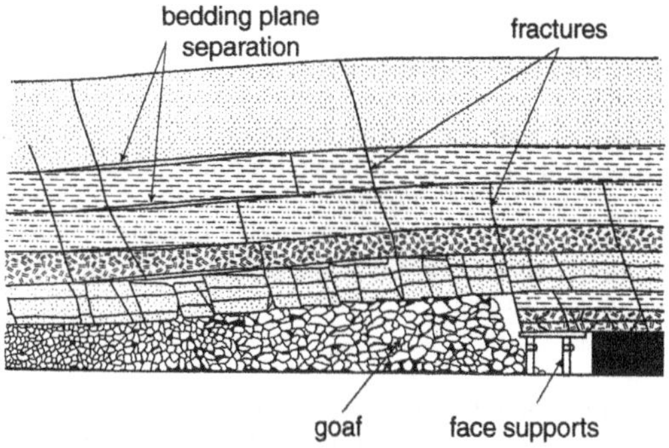

Figure 3. Effect of caving on the beds above a longwall face. After Kratzsch (1983).

As the mechanical processes involved during active and residual subsidence are different, it is not surprising that they exhibit very different relaxation time constants. Furthermore, one would expect residual subsidence to be delayed much longer where thick beds of strong sandstone are present above the workings compared to places where the overburden comprises weak shale strata.

Similar mechanical processes take place within the Permo-Triassic cover in the concealed coalfields, since those strata contain thick beds of strong limestone and sandstone. Indeed, the potential for residual subsidence can remain locked up in the strata almost indefinitely, until triggered by renewed mining activity in the vicinity. Ferrari (1996) reported observations over several panels worked in the concealed coalfields which showed subsidence magnitudes up to 700% greater than predicted. He suggested that potential residual subsidence, associated with subsurface voids remaining from previous workings had been triggered by the later workings and contributed to the observed values.

3 SURFACE FISSURES

All the underground coal mines in County Durham have now been closed. One of the environmental problems resulting from the mining activity has been the development of fissures in the Permian Lower Magnesian Limestone at outcrop above the edges of old mineworkings. Most of the mines in this part of the coalfield were worked by room-and-pillar methods, and where possible the pillars were robbed, or even totally extracted, before the workings were abandoned. Workings halted where they came up against a fault, although commonly seams were worked up to both sides of a fault from opposite directions.

The subsidence profiles over room-and-pillar mine workings may be calculated by the same empirical methods developed for longwall panels. Figure 4 illustrates a hypothetical case where a single seam has been worked up to a fault from both sides. There is a zone of horizontal extensional strain above the fault due to the effects of workings on both sides, with complementary zones of compression to either side. Furthermore, there is a hump in the vertical subsidence profile above the fault.

These profiles may seem counter-intuitive. If the coal seam had been worked progressively across the

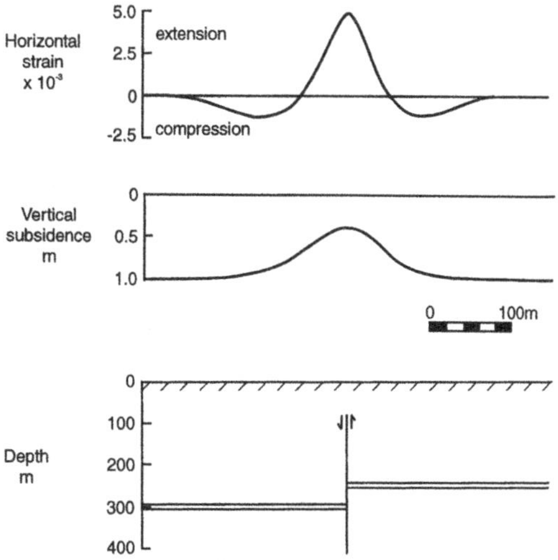

Figure 4. Horizontal strains and vertical subsidence at the ground surface for a single coal seam worked on both sides of a vertical fault.

fault, with the roadways being driven through the fault zone, one would expect uniform subsidence above the fault and zero horizontal strain. However, in practical situations, workings would usually have approached a fault from opposite sides, and at different times, and it is the general experience that final subsidence profiles from successive workings may be calculated independently and superposed as illustrated here (National Coal Board 1975).

A good example is exhibited at Houghton-le-Spring, County Durham, where a zone of fissures in the outcropping Lower Magnesian Limestone was mapped by Turner (1967) and Barrett (1985). There are about 20 m of Permian Basal Sands (Rotliegend equivalent) unconformably overlying the Coal Measures. Individual fissures were up to 2 m wide, and although they have mostly been filled with rubble and colliery spoil, three fissures may still be seen in a road cut (Figure 5).

Figure 5. Fissures in the Lower Magnesian Limestone exposed in a road cut.

Goulty & Kragh (1989) used this site to demonstrate that such fissures could be detected by seismic refraction profiling methods, even where obscured by in-filling or Quaternary drift cover, as they caused a dramatic reduction in the the amplitude of first-arrival seismic head waves refracted through the limestone.

This zone of fissures lies above a near-vertical normal fault in the Coal Measures, with a throw of 100 m down to the north, which does not cut the overlying Permian strata. Seams to either side have been worked from different collieries. A north-south cross-section through the mineworkings, 300 m east of the road cut, is shown in Figure 6 together with the calculated profiles for horizontal strain at the ground surface and vertical subsidence at the base of the Lower Magnesian Limestone. The reason for calculating the subsidence at the base of the limestone is because the limestone behaves as rigid blocks, opening up along joint surfaces, sitting on the incompetent subsided Coal Measures strata below.

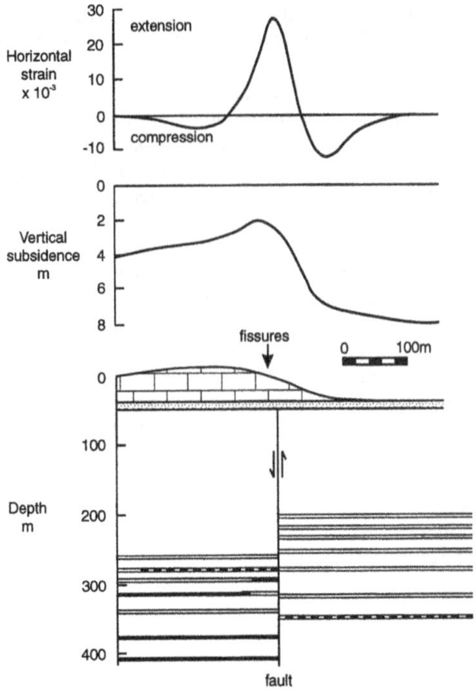

Figure 6. Section through mineworkings beneath Houghton Hill with horizontal strains calculated at the ground surface and vertical subsidence calculated at the base of the Lower Magnesian Limestone. Coal seam thicknesses, given at the right of the figure, are not drawn to scale.

If the horizontal extensional strains at the ground surface were all accommodated by horizontal displacement of rigid blocks of Lower Magnesian Limestone, the opening at the ground surface would be 1.5 m. This is not sufficient to account for the surface observations where the total opening is estimated to be in the range 2-4 m. Furthermore, it would require the compressional strains to either side to be maintained within the Lower Magnesian Limestone, whilst the extensional strains relaxed by shear deformation in the underlying Permian Basal Sands. This would be inconsistent. In fact, the Basal Sands are so soft that they are quarried locally by digging, so presumably they can accommodate shear deformation to decouple all strains in the Coal Measures from the rigid limestone blocks above.

Greater surface displacements of the observed magnitude would be caused by tilting the blocks of Lower Magnesian Limestone according to the subsidence profile at their base (Figure 6). At this horizon, the difference in dip across the fault is 0.1 radian, so an estimated thickness of 30 m for the Lower Magnesian Limestone would account for 3 m of opening at the surface.

4 TRIGGERED RESIDUAL SUBSIDENCE

At the Selby mine complex, in Yorkshire, the stratigraphy is similar to County Durham with the addition of 20 m of surficial Quaternary deposits and a variable thickness of Triassic Sherwood Sandstone conformably overlying the full Permian succession (Figure 7). Only the Barnsley Seam is being mined, even though there are half a dozen seams of economic thickness, to ensure that surface subsidence does not exceed 1 m at any point. The amount of subsidence has been carefully monitored by repeated levelling profiles. The unexpected finding from the levelling has been that the full amount of subsidence predicted above a longwall panel does not develop until the panels to both sides have also been worked.

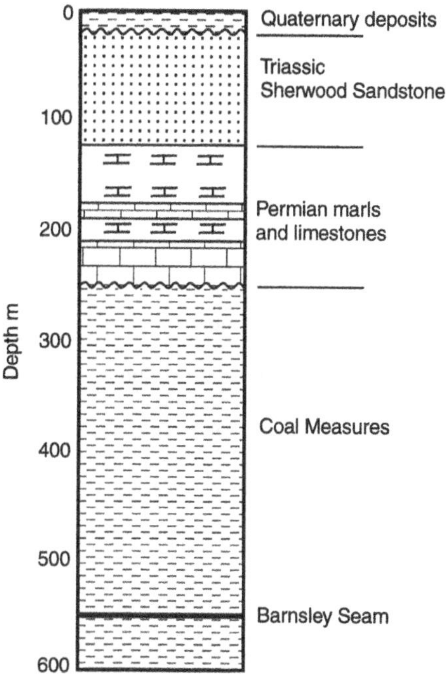

Figure 7. Stratigraphy at Selby.

Pyne & Randon (1986) have suggested that the reduced subsidence is due to bridging in the Permian strata, and reported that wireline log data from a borehole above a worked panel indicated bed separation within the Lower Magnesian Limestone. This idea is consistent with later seismic reflection observations by Al-Rawahy and Goulty (1995), who recorded a 4% reduction in seismic velocity through the Triassic and Upper Permian strata. They attributed the decrease in velocity to a reduction in the vertical effective stress, and the effect of bedding plane separation would be to reduce the vertical effective stress to zero. Again the Permo-Triassic strata are exhibiting the behaviour of rigid blocks, separated by joints and bedding planes. Since the strata would tilt inwards above a worked panel, and the transverse horizontal strains would be compressional, the rigid blocks provide a partial bridging mechanism over the weak shale beds in the Coal Measures which have subsided plastically below.

Figure 8 shows the empirical predictions for the horizontal strain and vertical subsidence profiles at the base of the Permian, 300 m above the Barnsley Seam, due to working a single panel and a full set of adjacent panels separated by roadside pillars of solid coal. The compressional strains due to subsidence over a single panel are reduced only slightly when the adjacent panels are worked, but the extensional strains over the pillars are increased. The subsidence humps over the pillars cause the limestone to tilt inwards over the worked panels and create additional tensile bending stresses. When only a single panel is worked, bridging takes place; but when the adjacent panels have also been worked, the combined effect of extensional strains and bending above the pillars weakens support at the bridging abutments, and may even be sufficient to open vertical fissures. Consequently, full subsidence can develop over the central panel. There is scope for further work to quantify the mechanics of this process.

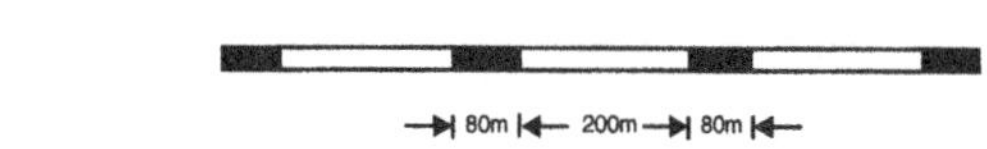

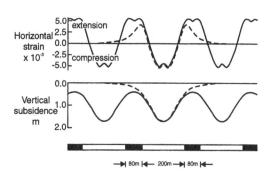

Figure 8. Horizontal strains and vertical subsidence calculated at the base of the Lower Magnesian Limestone over a single panel (dashed curves) and over a set of adjacent panels (solid curves) in the Barnsley Seam at Selby. Extracted thickness is 2.6 m.

5 CONCLUSIONS

Active subsidence whilst mining is in progress takes place with a time constant of the order of several days (Goulty & Al-Rawahy 1996). Residual subsidence after mining has ceased typically takes place in shale-dominated Coal Measures strata with a time constant of several months, but the potential for further residual subsidence can be locked up in the strata where strong beds of sandstone or limestone are present in the overburden (e.g. Ferrari 1997).

The Permian Lower Magnesian Limestone is a strong bed which is present throughout the concealed coalfields of north-east England. It responds to subsidence as rigid blocks, bounded by joints and bedding planes. Consequently, where there is a hump in the subsidence profile at the top of the underlying Coal Measures, deep fissures have opened above the edge of old mine workings. The fissuring is known to be exceptionally severe where several sets of old workings have terminated against a fault. The Triassic Sherwood Sandstone, which forms part of the cover sequence in Yorkshire, the East Midlands and the West Midlands, may respond in the same way. Such fissures could remain undetected where there is a substantial thickness of Quaternary cover, and represent a potential hazard.

Triggered residual subsidence over longwall panels at Selby may also be attributed to rigid-block behaviour in the Lower Magnesian Limestone. Above a single panel, subsidence causes inward tilting and compressional strains, so blocks of limestone can bridge across the subsidence zone in the weak shale-dominated strata beneath. Mining adjacent panels bends and stretches the Lower Magnesian Limestone enough to weaken the bridging abutments, quite probably by opening fissures, thus allowing full subsidence to develop above a panel after both neighbouring panels have been worked.

The clear evidence of bridging in the Permo-Triassic strata at Selby is consistent with a similar mechanism operating in strong sandstone beds within the Coal Measures themselves. A number of cases involving significant amounts of residual subsidence triggered by renewed mining activity have been reported from the concealed coalfields, and the phenomenon seems to be more widespread than had previously been recognised.

The empirical methods of calculation may give valid estimates for the total amount of potential subsidence in the concealed coalfields. Even if this is assumed, the only way of ensuring that subsidence is complete is to check empirical calculations of the total subsidence to be expected against levelling data acquired since before mining started. If a complete history of levelling data is not available, and there are strong beds of sandstone or limestone above the mineworkings, the potential for further residual subsidence should be considered in the design or repair of surface structures.

REFERENCES

Al-Rawahy, S.Y.S. & N.R. Goulty 1995. *Geophysical Prospecting* 43: 191-201.

Barrett, M.A. 1985. Unpublished MSc thesis. University of Durham.

Collins, B.J. 1978. Measurement and analysis of residual mining subsidence movements. In J.D. Geddes (ed.), *Large ground movements and structures*: 3-29. London: Pentech Press.

Ferrari, C.R. 1996. The case for continuing coal mining subsidence research. *Mining Technology* 78: 171-176.

Ferrari, C.R. 1997. Residual coal mining subsidence - some facts. *Mining Technology* 79: 177-183.

Goulty, N.R. & S.Y.S. Al-Rawahy 1996. Reappraisal of time-dependent subsidence due to longwall coal mining. *Quarterly Journal of Engineering Geology* 29: 83-91.

Goulty, N.R. & J.E. Kragh 1989. Seismic delineation of fissures associated with mining subsidence at Houghton-le-Spring, Co. Durham. *Quarterly Journal of Engineering Geology* 22: 185-193.

Kratzsch, H. 1983. *Mining subsidence engineering*. Berlin: Springer.

National Coal Board 1975. *Subsidence engineers' handbook*. London: National Coal Board.

Orchard, R.J. & W.S. Allen 1975. Time-dependence in mining subsidence. In M.J. Jones (ed.), *Minerals and the environment*: 643-659. London: Institution of Mining and Metallurgy.

Pyne, R. & D.V. Randon 1986. Surface environmental aspects of the Selby coalfield. *The Mining Engineer* 146: 77-84.

Turner, M.J. 1967. Unpublished MSc thesis. University of Durham.

Wardell, K. 1954. Some observations on the relationship between time and mining subsidence. *Transactions of the Institution of Mining Engineers* 113: 471-483.

Mechanics of Jointed and Faulted Rock, Rossmanith (ed.) © 1998 Taylor & Francis, ISBN 90 5410 955 6

Fracture development around underground excavations in rock salt

S. Kwon & J.W. Wilson
Department of Mining Engineering, University of Missouri-Rolla, Mo., USA

ABSTRACT: It is generally accepted that rock salt deforms plastically without fracturing. However, underground mine observations of failures show various fracture mechanisms are involved. The most significant fracturing observed at WIPP are bed separation, slabbing of the ribs, and fracturing in the stiff floor. In this study, in situ deformation measurements obtained from the WIPP site were used to understand the fracture behaviors of underground excavations in bedded rock salt.

1. INTRODUCTION

It is generally accepted that rock salt behaves plastically, without fracturing or volume increase. Underground mine observations of failures, however, indicate various fracture mechanisms and suggest that rock salt in situ must be considered as a semi-brittle material. Since the brittle fracture behavior of salt cannot be effectively characterized by laboratory studies, in situ measurements should be considered as the main information source for understanding the failure mechanisms.

The Waste Isolation Pilot Plant (WIPP) is an underground nuclear waste repository in New Mexico. The underground facility at the WIPP site is located 650 m below surface in bedded rock salt deposits. The underground facility has been divided into three areas: (1) the Site and Preliminary Design Validation (SPDV) area; (2) the Experimental area; and (3) the waste storage area. The waste storage area will be made up of eight panels consisting of seven rooms each, where each room is 4 m high, 10 m wide and 100 m long. The four rooms in the SPDV area have similar configurations to the rooms in the storage area. Among the four rooms, SPDV Room 1 and Room 2 collapsed massively in 1991 and 1994. Figure 1 shows a schematic diagram of the underground facilities at the WIPP.

The facility horizon lies within an evaporate sequence consisting of halite, argillaceous halite, and polyhalite. Clay 'G' and Clay 'H' lie about 2 m and 4 m above the roof. A persistent anhydrite and polyhalite layer, identified as Marker Bed (MB) 139, lies about 1.5 m below the floor.

2. FRACTURE MECHANISMS IN ROCK SALT

The fracture patterns frequently observed at the WIPP site were characterized by Cook and Roggenthen (1991) as shown in Figure 2. The following fracture mechanisms related to the development of the fractures. These mechanisms must be considered concurrently to explain the entire fracturing mechanism of the rock salt from its initiation to propagation.

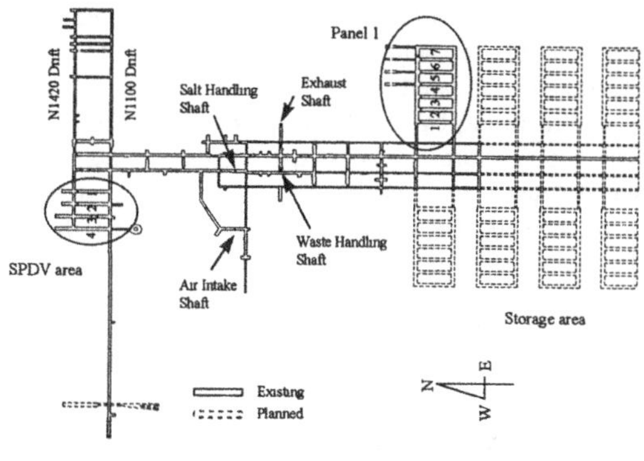

Figure 1. Underground layout of the WIPP facility.

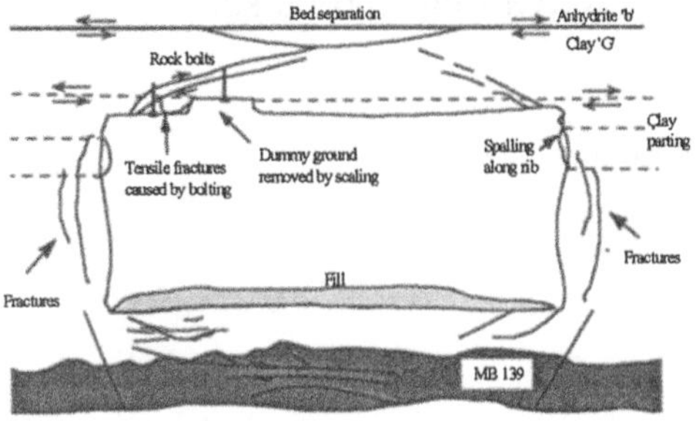

Figure 2. Patterns of fracturing around an excavation (Modified from Cook and Roggenthen, 1991).

a. Fracturing due to shear stress : Immediately after an excavation is created, the initial stress state is redistributed. Because of the stress transfer to the surrounding pillars and abutments, deviatoric stresses develop around the corners of the excavation. The shear stresses around the excavation create fractures when they are higher than the defined limit for the rock.

b. Brittle fracturing under a rapid deformation rate: One unique property of rock salt is that the ultimate strength is highly dependent on the loading rate (Horseman and Passarisi, 1981). With a high loading rate and a corresponding high strain rate, rock salt deforms like a normal elastic rock and shows brittle fracturing (Haupt, 1991; and Lajtai et al., 1994). In contrast, rock salt can exhibit a marked ductility and inelastic deformation with little or no fracture at low loading rates (Aubertin, 1991). This implies that fracturing in rock salt develops only with a sudden stress change. In actual conditions, therefore, the sudden stress change immediately after opening excavation and during the creation of an adjacent excavation should be carefully investigated.

c. Fracturing by tensile stresses: From the fact that the fracture surfaces are rough, show no slickensides or other evidence of shear displacements, Lajtai et al. (1994) concluded that the initiation of the fractures was caused by tensile stresses. The development of fracture zones in rock salt are related to the propagation of a large number of tensile cracks that normally run parallel to the maximum principal stress. The fractures developed almost vertically in the ribs in the SPDV area are typical examples of cracks created by this fracture mechanism.

d. Fracturing in a stiff layer: When a stiff layer is located near an opening surface, fractures develop at that layer because of the stress concentration developed in the layer. At the WIPP site, floor fracturing is observed at many locations. In most cases it is related to the stress concentration in the stiff anhydrite and polyhalite layer MB 139, due to the plastic deformation of the salt layers above and below the layer. Because of the influence of the stiff layer in the floor, fracturing and slabbing in the floor occurs before it develops in the roof.

e. Bed separation across a clay seam: When clay seams are located in the roof or floor, these act as planes of weakness along which shearing and separation can easily take place, because of the negligible tensile and shear strength of the clay (Saeb et al., 1995). Because of the resulting separation across the clay seam, the immediate roof and floor act as "beams" rather than as continuous portions of the rock mass. After the immediate roof or floor beam is isolated, the immediate roof or floor beam continues to be subject to compressive stress due to the increase in horizontal stresses in the beams as a result of the stress redistribution caused by slip (U.S.DOE, 1991).

f. Gas pressure: Many salt and potash deposits contain gas inclusions that suddenly outburst or create a steady influence on the deformational behavior of the rock mass. The gas pressure causes excess tension at newly created surfaces in mine openings which can result in brittle fracturing and an explosion-like gas expansion (Baar, 1977).

3. INVESTIGATION OF THE FRACTURE DEVELOPMENT AT WIPP

3.1 *Calculation of the Bed Separation Across a Roof Clay Seam*

The stability of the excavations are significantly affected by the separations across the clay seams in the roof or floor. Kwon (1996) developed a technique for calculating the separation in the roof layers based on the extension measurements. From Figure 3, the separation across a clay seam, S, can be estimated by the following equation:

$$S = (B-A)\varepsilon_2 - (B-A)\left[\varepsilon_1 - \left(\frac{\varepsilon_1 - \varepsilon_3}{d_3 - d_1}\right)(d_2 - d_1)\right] \quad (3.1)$$

where, d_1, d_2, and d_3 are distances from the roof, ε_1, ε_2, and ε_3 are the bay strains measured

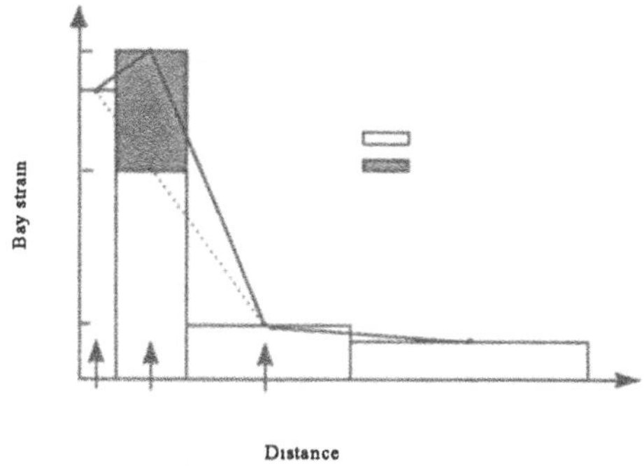

Figure 3. Calculation of the separation of a clay seam between anchors A and B.

between each anchor, and A and B are location of the anchors around the clay seam.

The equation was applied to several measuring locations at the WIPP site and Table 1 lists the results. From the table, the following conclusions could be drawn:

a. Clay 'H' showed a much lower separation than Clay 'G'. The lower separation at Clay 'H' can be explained by the difference in the beam thickness, horizontal displacement of the ribs, and fracturing in the immediate roof beam which decreases the stiffness of the beam.

b. Opening width is an important factor in determining the amount of separation in the roof and floor. The separation rates over the clay seams in openings excavated to less than 7.6 m in width are significantly lower than those openings of 10 m wide. The separation rate increase in the SPDV Room 1 and Room 2 roof after 1000 days can explain the roof falls in the SPDV Room 1 and Room 2.

c. The maximum separation rate was calculated to be about 2.4 cm/year at about 2000 days after excavation. Therefore, a roof support system must be designed to accommodate about 2.4 cm/year of bed separation.

3.2 *Horizontal Displacement Along the Clay Seam in the Roof*

Inclinometers can provide information on rock displacements in a direction perpendicular to the longitudinal axis of the borehole. In the SPDV Rooms, inclinometer measurements have been taken in vertical boreholes up to 15 m deep into the roof and floor. Figure 4 shows the deflection measurements from the inclinometer which is installed in the roof close to the abutment. Because of the influence of the clay seams in the roof, there are sudden changes on the deflection plots. From Figure 4, the slip deformation along each clay seam was calculated and plotted in Figure 5. The slip displacement along Clay 'G' which is located about 2 m above the roof is significantly high compared to the slip along other clay seams.

In order to model the SPDV area, a two-dimensional finite difference simulation program FLAC, was used. The WIPP reference creep model in Equation 3.2 was used for pure halite and argillacious halite and the Mohr-Coulomb model was used for anhydrite and polyhalite.

$$\dot{\varepsilon} = A\exp(-\frac{Q}{RT})(\frac{\sigma_{eff}}{\mu})^n \quad (3.2)$$

Table 1. Calculated separation rates from different sites.

Measuring site				Separation rates for different times (cm/year)				
Name	Location	Width (m)	Layer	1~100	100~500	500~1000	1000~1500	1500~2000
GE263	E300 S400 Roof	4.3	Clay 'G' Clay 'H'			0.05 0.00	0.00 0.00	-0.00 0.00
GE246	E140 S700 Roof	7.6	Clay 'G'	0.64	0.23	0.25	-0.02	
GE249	E140 S3080 Roof	7.6	Clay 'G'	0.13	0.02	0.02	0.00	0.02
GE256	E140 S400 Roof	7.6	Clay 'G'			0.13		
GE257	E140 S400 Roof	7.6	Clay 'G' Clay 'H'			0.18 0.00		
GE259	N1420 TR1 Roof	10	Clay 'G' Clay 'H'			0.33 0.02	0.33 0.05	0.33 0.05
GE214	SPDV Room2 Roof	10	Clay 'G'	0.94	0.51	0.48	0.58	0.84
GE218	SPDV Room1 Roof	10	Clay 'G'	0.99	0.91	0.97	1.22	1.75
GE242	N1100 TR2 Floor	6	MB139	0.28	0.13	0.10	0.18	0.10
GE260	N1420 TR1 Floor	10	MB139		0.84	0.76	0.74	
GE262	N1420 TR2 Floor	10	MB139			0.58	0.33	0.58

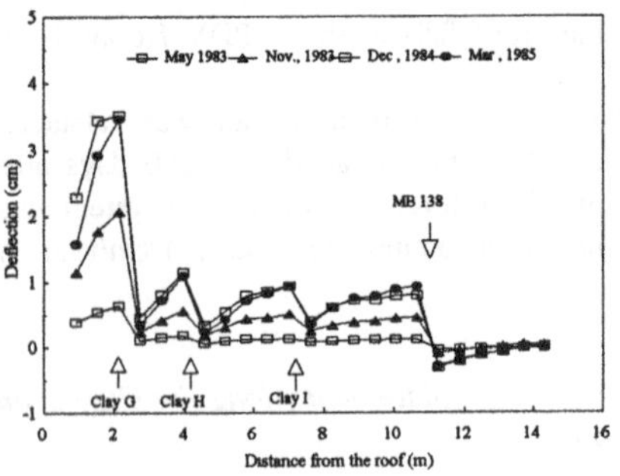

Figure 4. Deflection measurements from the inclinometer in the SPDV Room 1 Roof West.

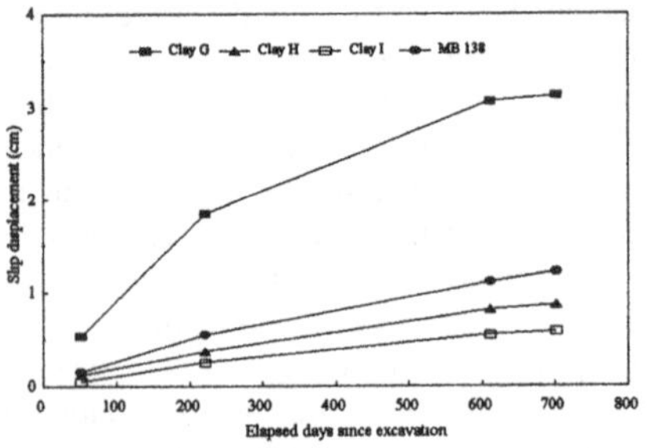

Figure 5. Horizontal slip along the clay seams in the SPDV Room 1 roof.

where, $\dot{\varepsilon}$ is the effective strain rate, σ_{eff} is the effective stress, Q is the activation energy, R is the gas constant, T is the temperature, μ is the shear modulus, and A and n are constants. The clay seams were modeled with interfaces, which allow separation and slip along them.

Figure 6 shows a comparison between the prediction of the slip displacement along Clay 'G' above the rib line from the FLAC simulation, and the actual measurement using an inclinometer. Even though the calculated slip from the deflection plot is normally higher than the prediction from the FLAC simulation, the simulation could predict precisely the slip velocity - especially in the steady state creep stage, which is after 200 days.

3.3 *Distribution of Shear Stress*

If the stress can be back calculated from the strains or strain rates, this would be a powerful alternative to relying on in situ stress measurements, since the deformation measurements are more reliable than direct stress measurements. From the deformation

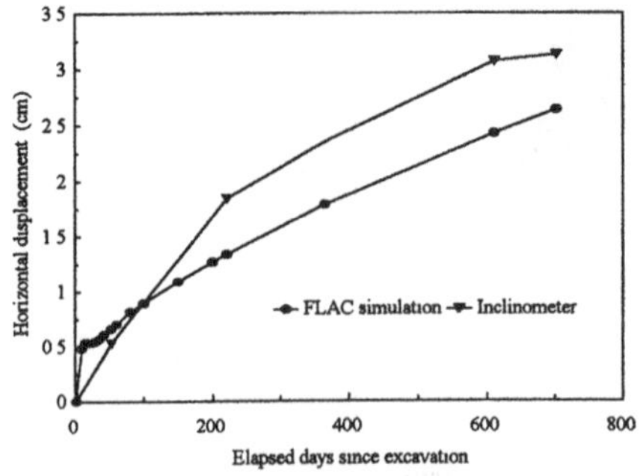

Figure 6. Slip displacements along Clay 'G' above the rib line from the FLAC simulation and inclinometer.

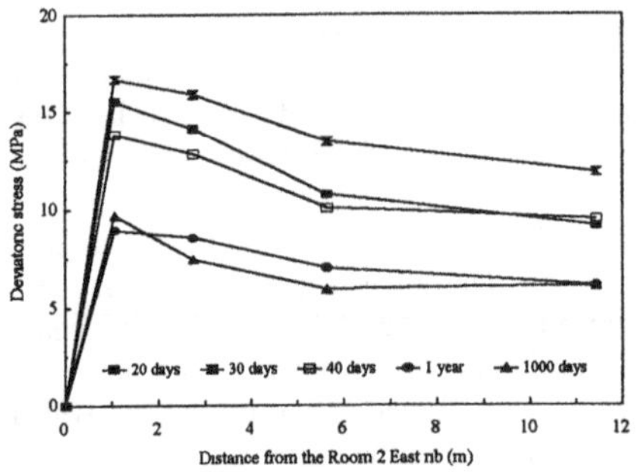

Figure 7. Calculated deviatoric stress distribution in the SPDV Room 2 East rib.

measurements, the deviatoric stress distribution can be calculated by using an adequate constitutive equation. Since the effective strain rate can be calculated from the measured extension, the effective stress can be calculated reversibly from the effective strain rate using Equation 3.3, which is derived from Equation 3.2

$$\sigma_{eff} = \mu\left[\frac{\dot{\varepsilon}}{A\exp(-\frac{Q}{RT})}\right]^{1/n} \qquad (3.3)$$

Equation 3.3 was applied to the SPDV Room 2 East rib and the calculated results are shown in Figure 7. The maximum stress after 20 days was about 16 MPa and decreased with distance from the opening. In the steady state, the maximum deviatoric stress is about 9 MPa. At 30 days after the excavation of SPDV Room 2, the deviatoric stress increased due to the excavation of SPDV Room 1, which was excavated 25 days after SPDV Room 2 excavation at the East side of the SPDV Room 2. The stress increase due to an adjacent excavation is closely related to the distance between the measuring location and the adjacent excavation. The stress increase due to the SPDV Room 1 excavation is about 2.74 MPa at 11 m from the SPDV Room 2, and 2.7 MPa at 5.6 m, 1.77 MPa at 2.7 m, and 1.14 MPa at 1m. The deviatoric stresses in the pillars decreased rapidly and almost uniformly over the whole pillar area from 30 days to 40 days after excavation. The stress variation from one year to 1000 days was small compared to this early stress drop. These results confirm the assumption of a constant stress state in the steady-state creep stage.

3.4 *Influence of Sudden Stress Change*

Because of the brittle fracturing under the rapid deformation rate of rock salt, the deformational behavior of the rock in the early stages should be carefully investigated. The excavation sequence of the opening should also be considered as an important factor which can determine the overall opening stability, since an adjacent excavation changes the stress distribution rapidly.

The seven rooms in the Panel 1 area at the WIPP site were excavated in 1986. The alcoves, TA1, TA2, TA3, and TA4, were excavated from June 1989 to July 1989. The roof conditions of Rooms 4, 5, 6, and 7 are less stable than Rooms 1, 2, and 3 from observation reports. These observations can be related to the excavation of the alcoves, as well as the excavation sequence of the rooms themselves. Figure 8 shows the closure rate contours in the Panel 1 area before and after the excavation of the alcoves.

3.5 *Fracturing in a Stiff Layer*

The in situ deformation measurements made with extensometers can show clearly the development of the fracturing in the floor. Figure 9 and Figure 10 show the relationship between strain and distance into the floor, for SPDV Room 1 and Room 2. For reference, the geology in the floor of SPDV Rooms 1 and 2 have been attached. The SPDV Room 2 shows the strain distribution with a smooth change from the floor surface with depth. This indicates a zone in which the rock deforms in a continuous manner. In SPDV Room 1, however, the influence of the bed of relatively strongly elastic anhydrite as well as polyhalite, can clearly be seen. The

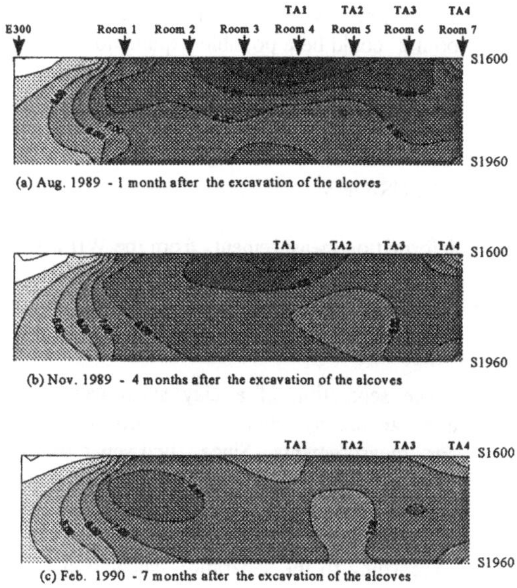

Figure 8. Closure rate contours (cm/year) in the Panel 1 area.

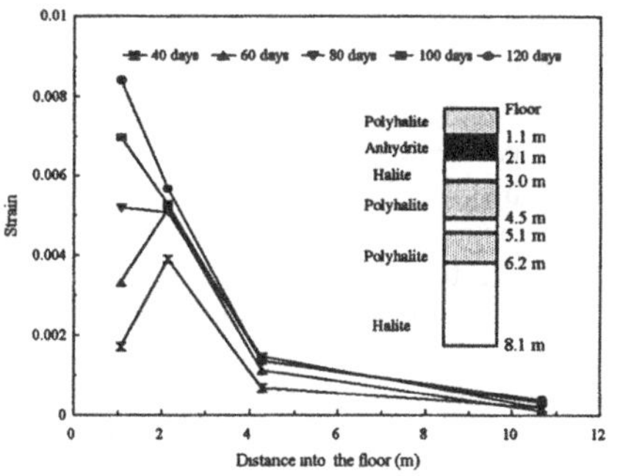

Figure 9. Strain distribution in the SPDV Room 1 floor.

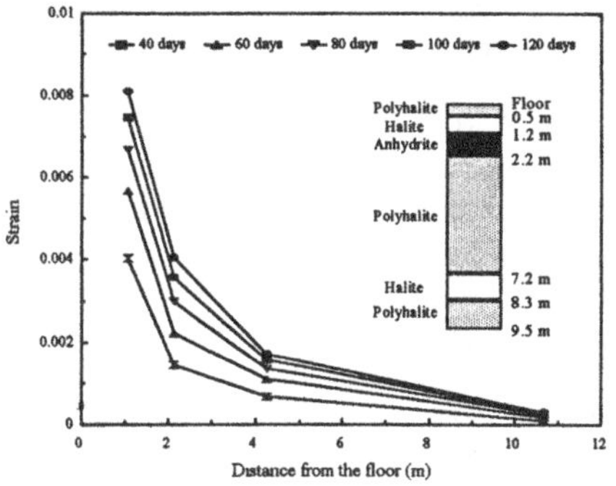

Figure 10. Strain distribution in the SPDV Room 2 floor.

difference in geology in the floor in SPDV Room 1 and Room 2 could be a possible explanation for the three years earlier roof fall in the SPDV Room 1 roof.

4. CONCLUSIONS

Field deformation measurements from the WIPP site were effectively used to investigate the fracture development around underground excavations in rock salt. Important conclusions from this study are:

1. The separation of a clay seam can be estimated accurately and continuously from extension measurements. Since bed separation in the roof can indicate roof conditions more directly, the result from the equation can be used to evaluate the opening stability more accurately.

2. According to the separation calculation, the maximum separation across Clay 'G', which is located about 1.5 m above the roof, occurred in the SPDV Room 1. The maximum separation rate, about 2.4 cm/year at about 2000 days after excavation, can be used in the design of a roof support system.

3. From the calculation of the horizontal displacements along the clay seams in the roof using deflection measurements, it was found that the horizontal slip along Clay 'G' was much larger than other clay seams. The computer simulation using FLAC was able to accurately predict the slip rate along Clay 'G' in the steady state creep stage, as shown in Figure 6.

4. The deformation change after an adjacent opening was excavated was the reaction of the rock mass around an opening to the rapid stress change in the pillars due to the excavation. This situation is similar to the deformational behavior that occurs immediately after excavation. Thus, if the deformation change is measured carefully as adjacent mining takes place, it would be helpful in understanding the deformational behavior before the installation of instruments can take place.

5. The deformation due to the creation of an adjacent excavation can be effectively used to assess the results of a computer simulation. A comparison between the measured deformation after mining of an adjacent excavation and the prediction from a computer simulation can determine the reliability of the assumptions, including rock properties, used in applying the simulation model.

6. From the back-calculation of the deviatoric stress in a pillar, the following conclusions could be drawn: a) the peak deviatoric stress was about 16 MPa at 20 days after excavation and decreased to about 9 MPa in the steady state; b) the deviatoric stress decreased with more or less the same pattern over the whole pillar area; c) the deviatoric stress conditions were almost constant in the steady state; and d) the sudden stress increase due to an adjacent excavation was strongly dependent on the distance between the measuring location and the adjacent excavation.

REFERENCES

Aubertin, M., 1991. An internal variable model for the creep of rocksalt. *Rock Mech. and Rock Eng.* Vol. 24, p. 81-97.

Baar, C.A., 1977. *Applied salt-rock mechanics.* Elasevier Scientific Publishing Co., Amsterdam.

Cook, R.F. and Roggenthen, W.M., 1991. Fracturing around excavations in salt at the WIPP. *Proc. of the 32th U.S. Symp. Rock Mechanics*, University of Oklahoma, p. 889-898.

Haupt, M., 1991. A constitutive law for rock salt based on creep and relaxation tests. *Rock Mech. and Rock Eng.* Vol. 24, p. 179-206.

Horseman, S. and Passaris, E., 1981. Creep tests for storage cavity closure prediction. *First Conf. on the Mechanical Behavior of Salt*, Pennsylvania State University, University park, p. 119-157.

Kelsall, P.C., Case, J.B., and Chabannes, C.R., 1982. A preliminary evaluation of the rock mass disturbance resulting from shaft, tunnel, or borehole excavation, Technical Report, ONWI-411.

Lajtai, E.Z., et al., 1994. En Echelon Crack-Arrays in Potash Salt Rock. *Rock Mech. and Rock Eng.* Vol. 27, p. 89-111.

Kwon,S., 1996. An Investigation of the deformation of underground excavations in salt and potash mines, Ph.D. thesis, Univ. of Missouri-Rolla.

Saeb, S., Francke, C.T., and Patchet, S.J., 1995. Effect of clay seams on the performance of WIPP excavations. *Mechanics of Jointed and Faulted Rock, Vienna*, p. 835-840.

U.S.DOE, 1991. *Report of the geotechnical panel on the effective life of rooms in panel1.* DOE/WIPP 91-023.

Mechanics of Jointed and Faulted Rock, Rossmanith (ed.) © 1998 Taylor & Francis, ISBN 90 5410 955 6

Stability analysis of rock blocks around a tunnel

Chung-In Lee & Jae-Joon Song
School of Civil and Geosystem Engineering, Seoul National University, Korea

ABSTRACT: Using the information of directions and locations of joints on a joint trace map, joints were modeled as three dimensional planes, and the algorithm was coded for finding convex removable blocks in the map and testing the stability of those blocks. This program was used to analyze the stability of rock blocks around a liquified petroleum gas(LPG) storage cavern located at Pyungtaek, Korea, and the detail information of each removable block such as the volume, height and active force were obtained. The analysis using the block theory for the cavern was also carried out.

1 INTRODUCTION

One of the most serious problems in tunnel excavation is the accidents caused by the falling of rock blocks which are made by tunnel surface and discontinuities in the rock mass. The block theory suggested by R.E. Goodman and Gen-hua shi makes it possible to test the stability of rock blocks on a slope or around an underground opening. When this theory is applied to tunnel stability analysis, several joint sets are determined from all of the joints existing in that site and then, key blocks formed with those joint sets are discriminated. In this case, however, only the maximum occurrence ranges and shapes of key blocks show making the support design too conservative. For more precise support design, the information of really existing blocks throughout the tunnel such as volume, shape, size and sliding force are to be known, and for this purpose, locations and lengths of block forming joints must be considered as well as those directions in the stability analysis.

In this study, using the investigations of Gen-hua shi[2], S.F. Hoerger[1], etc., a computer program has been coded, which reconstructs joint traces on a joint map into three dimensional joint net and analyzes the stability of rock blocks after finding them from the net. When this analysis method was applied to an LPG storage cavern, the shape, volume, location and safety factor of each removable block which could be useful information in tunnel support design were obtained. The merits or faults of this approach are easily understood when it is compared with the stability analysis using only joint directions.

2 STABILITY ANALYSIS OF ROCK BLOCKS

Even though a field joint trace map has useful information for joints such as locations, trace lengths and directions, we lose the chance to make the stability analysis more helpful to support design, because of ignoring all the data on the map except joint sets' directions. In this investigation, we take into account directions, locations and trace lengths of joints as well as the shape and size of a cavern, to reconstruct joint traces on the map into 3-d shapes. After finding convex blocks formed by the intersections of joints, removable blocks are separated from infinite or tapered blocks, and mode and stability analysis are carried out. The procedure of this analysis method is illustrated in Fig.1.

2.1 Conversion of joint traces on a map into a 3-D joint net

A 3-d joint net can be made using input data such as dip, dip direction and coordinates of end points of each joint on a field trace map as well as shape and size of a cavern. Fig.2 shows this conversion process with three joint traces on a map.

In Fig.2, the gallery of a circular cavern are

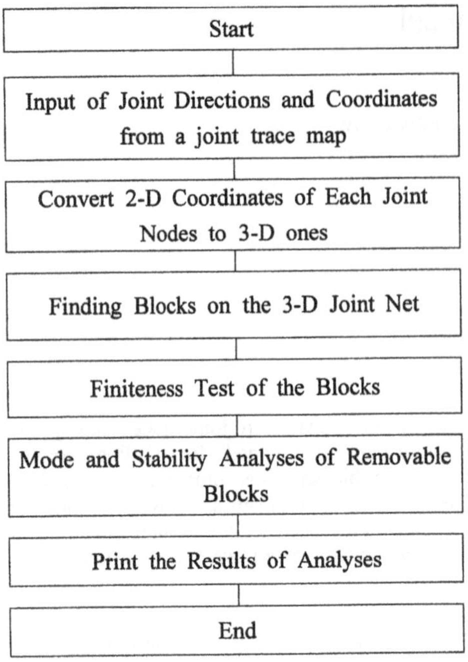

Fig. 1 Flow chart of stability analyses for each block

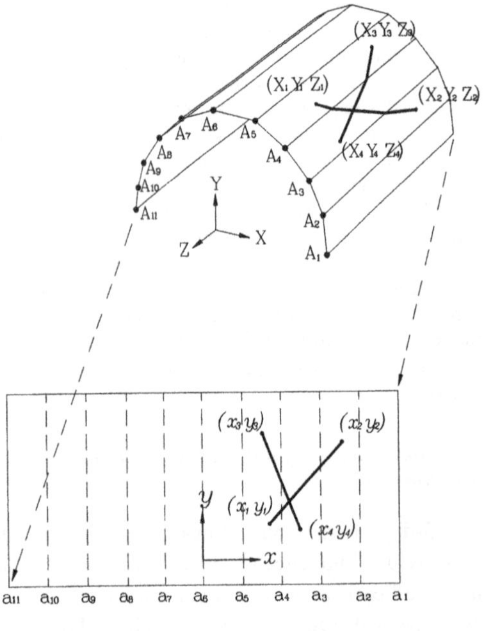

Fig. 2 Conversion of joint traces on a map into a 3-D joint net

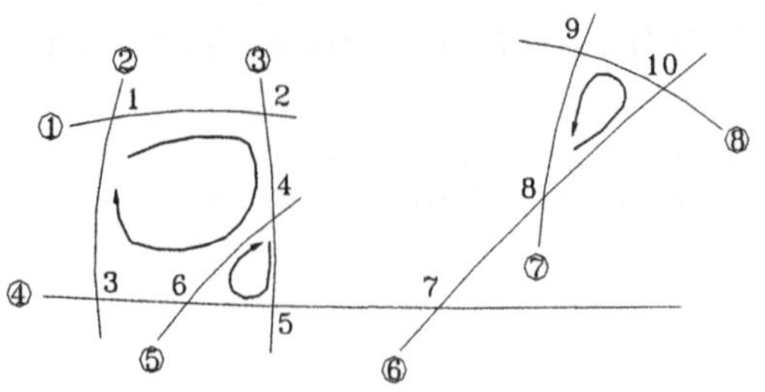

Fig. 3 Finding convex blocks in a joint map including 7 joints

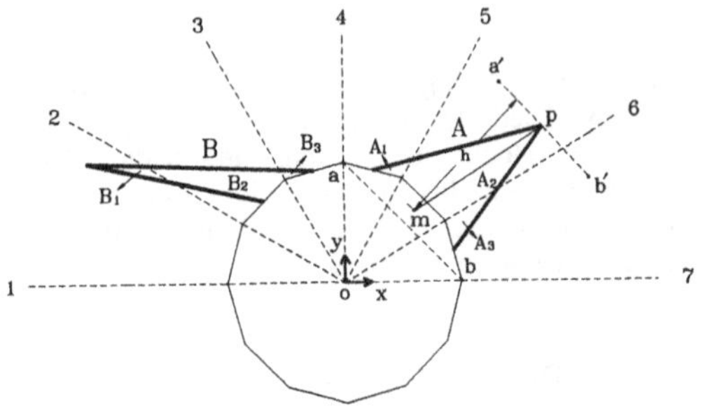

Fig. 4 Subdivision of concave blocks around a circular cavern

simulated with 10 narrow planes having 11 edges parallel to the cavern axis. Point a_1 on the field map corresponds to A_1 in the 3-d gallery, so does (x_{a_1}, y_{a_1}) to $(X_{A_1}, Y_{A_1}, Z_{A_1})$. Because (x_1, y_1) is in a_4-a_5 interval, its corresponding point (X_1,Y_1,Z_1) is in A_4-A_5 in the gallery and (X_1,Y_1,Z_1) is obtained by the following relationships.

$$X_1 = X_{A_4} + \frac{x_1 - x_{a_4}}{x_{a_5} - x_{a_4}} (X_{A_5} - X_{A_4})$$

$$Y_1 = Y_{A_4} + \frac{x_1 - x_{a_4}}{x_{a_5} - x_{a_4}} (Y_{A_5} - Y_{A_4})$$

$$Z_1 = -y_1$$

(x_2, y_2) is converted to (X_2,Y_2,Z_2) by the same method mentioned above.

2.2 *Discrimination of convex blocks*

It is assumed that all of the blocks are convex in the finiteness test or other stability calculations. The following shows a simple procedure of discriminating convex blocks from a joint net.

Four convex blocks can be found in a joint net with eight joints in Fig.3. The basic algorithm for finding a convex block is to complete a loop by repetition of going forward and turning right/left along intersection points until it comes back to a start point. The start point can be chosen in random order but all of the points are to be selected as a start point in turn. Therefore, a block which was previously found can appear later again and in that case, duplicate registering of the block is avoided by checking a list of previously registered convex blocks. When the algorithm is applied to a joint net in Fig.3, four convex loops are discriminated such as '1 2 5 3', '1 2 4 6 3', '4 5 6' and '8 10 9'.

2.3 Calculation of volume and height of a block

Unit weight and volume should be known to calculate the weight of a block. The average unit weight of rock blocks can be determined from those of sample specimens and a volume of a block can be calculated using direction vectors of joints and coordinates of vertex points constructing the block. Because a block around a circular tunnel has a concave form, it is convenient for volume calculation to divide the block into several convex ones, as shown in Fig.4. The volume of a block is the sum of those unit tetrahedrons belonging to it.

The height of a concave block is defined here as a distance from an apex to an average point of vertexes in the bottom, when all of the block edges are projected on a plane of which normal is parallel to the tunnel axis. In Fig.4 the height of block A is the distance h from p to m.

3 CASE STUDY

A stability analysis was carried out for blocks around an LPG storage cavern located at Pyungtaek in the western coastal area of the Korean Peninsula. The cavern, shown in Fig.5, has a span of 18m and a height of 27m. The total length of the cavern is 222m. It was divided into 11 sections for stability tests of rock blocks. The trend and plunge of the cavern axis is N20W and 0° . The rock mass around that cavern is mainly composed of granitic gneiss and is reported to include a lot of nearly vertical steep joints and gentle slope ones. Its joint trace map is on Fig.6. 116 joints of which direction and location had been well recorded, were selected from the trace map. Those joints fall into four sets as illustrated in Table 1. The force acting on a block is delimited to its own weight in this study.

3.1 Applicaton of the block theory

When the block theory was applied to the stability analysis of the LPG storage cavern, 12 removable blocks were found from the stereo-

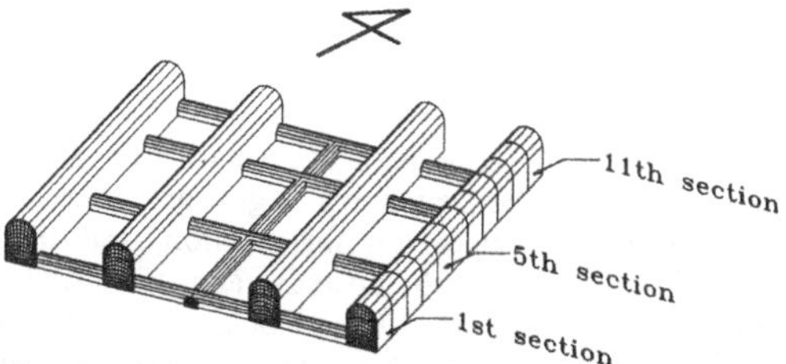

Fig. 5 Bird's-eye view of the LPG storage cavern

Table 1 Directions of four joint sets around the cavern

Joint Set	Dip / Dip direction (degree)
1	20 / 134
2	90 / 27
3	90 / 278
4	24 / 28

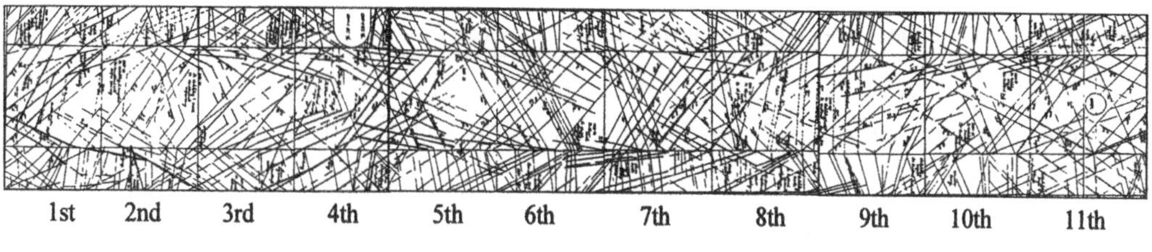

Fig. 6 Joint trace map

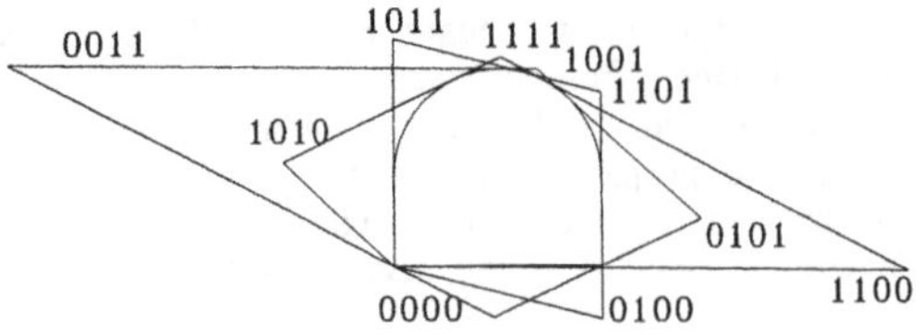

Fig. 7 Cross-sections of the maximum removable blocks

Table 2 Types of blocks found out by the block theory

Block Type	JP code (S.F.)
Key	1 0 0 1 (Lifting)
	1 0 1 1 (Lifting)
	1 1 0 1 (Lifting)
	1 1 1 1 (Lifting)
Potential Key	0 0 1 1 (1.843)
	0 1 0 1 (3.731)
	1 0 1 0 (1.297)
Stable	0 0 0 0
	0 1 0 0
	1 1 0 0
Tapered	0 0 0 1
	1 1 1 0
Infinite	0 1 1 1
	1 0 0 0

Table 3 Specipications of Removable blocks in the 5th section

No.	Joint	Type (S.F.)	Volume (m^3)	Height (m)	Sliding Plane	Force (ton)
1	1 10 5	P (1.00)	0.09	0.44	5	0
2	2 6 9 8	K(F)	1.66	0.79	0	4.48

graphic projection for the four joint sets in Table 1. Assuming all of the joints have a friction angle of 30° , the result of the test is summarized in Table 2. The relations between locations of blocks generated and their types are; removables at the gallery are key blocks falling down by gravity, potential keys exist at the bench and the gallery, and stable blocks at the blottom and the bench. Locations and shapes of the maximum removables are shown in Fig.7. The key blocks and the stable blocks are relatively smaller than the potential keys at the bench. Among the potential keys, it is observed that the smaller they are, the less their safety factors are.

3.2 Results of the stability analysis using a joint trace map

In this analysis the average unit weight of the rock cores was reported to be $2.7 ton/m^3$. The results in the 5th and the 6th section of eleven sections are as follows.

a) The 5th section
Sixteen joints were selected of which eight have dips of more than 60', one had more than 50' and the remaining seven less than 30° . Nine Infinite blocks, one key and one potential key block are generated. The specifications of all removables are shown in Table 3. The small potential key block has a safety factor of 1.0 and its potential sliding plane is the 5th joint. The key block at the gallery is to fall down and its height is less than 1m. An unrolled map and three dimensional view of two removables are in Fig.8 and Fig.9.

b) The 6th section
In this section, six joints have small dips of less than 30° and the remaining eight are steep. In all 106 blocks appear in this section, and there are 37 key blocks, 16 tapereds, and 53 infinites. Safety factors of the key blocks are 0.14 or 0.17. Generally the key blocks have a large volume and weight. The height of the highest

Table 4 Analysis of the block types occurring in all sections

Type \ Section	1	2	3	4	5	6	7	8	9	10	11
Key(falling)	0(0)	0(0)	2(2)	3(3)	1(1)	37(0)	5(3)	0(0)	4(0)	2(0)	1(0)
Potential Key	0	0	0	0	1	0	6	4	0	1	0
Stable	0	0	0	0	0	0	0	0	0	0	1
Tapered	0	0	0	0	0	16	18	0	5	0	7
Infinite	6	8	44	0	9	53	59	22	8	21	44
Total	6	8	46	3	11	106	88	26	17	24	53

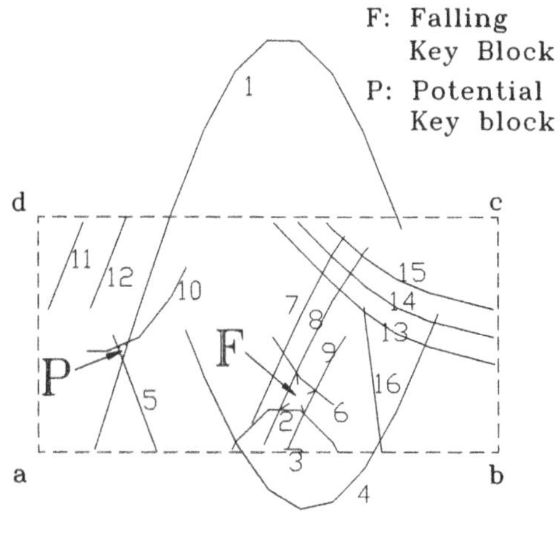

Fig. 8 The unrolled map of the 5th section

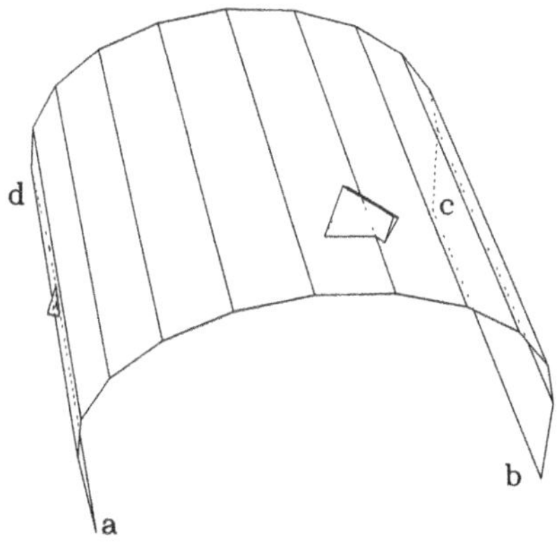

Fig. 9 Cross-section and 3-d view of the blocks in the 5th section

Table 5 Analysis of the moving force of key blocks in all sections

Section No.	Force (ton)			
	0~10	10~100	100~1000	1000~
1	•	•	•	•
2	•	•	•	•
3	•	•	2	•
4	1	2	•	•
5	2	•	•	•
6	•	10	10	17
7	4	1	•	•
8	1	•	•	•
9	•	2	2	•
10	2	•	•	•
11	•	1	•	•
Total/ Average	27.2 / 2.7	684.7 / 42.8	3464.2 / 247.4	65411.8/ 3847.8

key block is 173.5m which seems unreal considering the joint continuity in the rock mass.

All of the blocks generated in eleven sections are arranged according to their types into Table 4. One of the reasons why very few stable blocks throughout the cavern are formed is that selected joints have a tendency to cross each other at the gallery.

If the 'moving force' is defined as the active net force by which a block is to move, the moving force of a falling down block is its weight because no other force except gravity is considered here. The moving forces of key blocks are divided into four groups in Table 5. The largest group has seventeen key blocks which have moving forces of more than 1000 tons and are generated all in the 6th section. These blocks, however, have so large a volume and height that the probability of their appearance in the field is rare. Therefore, the group considered more important is the first one of which moving force is less than 10 tons. This group has ten key blocks of which four are in the 7th section. Five of the ten key blocks are to fall down and two have safety factors of 1.0.

4. DISCUSSION & CONCLUSION

(1) A key block analysis using the block theory shows that four kinds of relatively small key blocks are able to appear at the gallery of the cavern and three kinds of potential key blocks can exist in the bench with safety factors of 1.3~3.7 when the friction angles of joints are assumed to be 30°

(2) From the stability analysis using the computer codes developed in this study, we can find out the shape, volume, height, location and active resultant force of each removable block. A lot of key blocks are shown in the 6th section and most of small key blocks with a moving force of less than 10 tons in the 7th section.

(3) The stability analysis suggested here has several merits for obtaining detail information such as the volume, height, and active resultant force of each removable block, which is much helpful to its support design.

(4) To make the results of the stability analysis of each block more trustworthy, we need to use appropriate information of joint extent in rock mass and joint trace maps showing more precise directions and locations of joints.

REFERENCES

1. S.F. Hoerger, 1988, Probabilistic and deterministic key block analysis for excavation design, Ph.D. Thesis, Michigan Technological Univ., Houghton, MI.
2. G. Shi, R.E. Goodman, 1990, Finding 3-d maximum key blocks on unrolled joint trace maps of tunnel surfaces, *Proc. of the 31st US Symp. on Rock Mechanics*: 219-228.
3. R.E. Goodman, G. Shi, 1985, *Block theory and its application to rock engineering*, Prentice-hall Inc.
4. D. Lin, C. Fairhurst, 1988, Static analysis of the stability of three-dimensional blocky systems around excavations in rock, *Int. J. Rock Mech. Sci. & Geomech. Abstr.* 25:139-147.
5. G. Shi, R.E. Goodman, 1983, Key bolting, *Proc. of Int. Symp. on Rock Bolting*: 143-164.
6. Lap-yan Chan, R.E. Goodman, 1983, Prediction of support requirements for hard rock excavations using key block theory and joint statistics, *Proc. of the 24th US Symp. on Rock Mechanics*: 557-576.
7. J.C. Chern & M.T. Wang, 1993, Computing 3-d key blocks delimited by joint traces on tunnel surfaces, *Int. J. Rock Mech. Sci. & Geomech. Abstr.* 30:1599-1604.

Mechanics of Jointed and Faulted Rock, Rossmanith (ed.) © *1998 Taylor & Francis, ISBN 90 5410 955 6*

Field measurement and model test on tunnel deformation

Yoshiyuki Kojima & Toshihiro Asakura
Railway Technical Research Institute, Tokyo, Japan

Masahiro Nakata & Koji Mitani
Japan Highway Public Corporation, Research Institute, Tokyo, Japan

Toyohiro Ando
East Japan Railway Company, Tokyo, Japan

Kazuyuki Wakana
SHO-BOND Corporation, Tokyo, Japan

ABSTRACT: Filed measurements, model tests and numerical analyses are underway to establish a standard to evaluate soundness of deformed tunnel lining. The present paper reports on the results of (1) case studies of tunnels deformed by squeezing earth pressure; (2) 1/30 scale model lining tests; and (3) crack propagation analysis. Notably it refers to the effects of countermeasures against earth pressure. To sum up the contents; 1) Back-fill grouting can vastly improve the strength of defective tunnel lining with opening behind lining; 2) rock-bolting or inner reinforcement exhibit an ample effect of reinforcement, provided it is preliminarily treated with back-fill grouting; and 3) invert concrete excels in the effect of displacement suppression, but for the purpose of enhancing the structural strength, combination with other lining countermeasure is advocated.

1 INTRODUCTION

At present in Japan there are railway tunnels and road tunnels in practical service, respectively with the following numbers and total lengths; 4600(2900km) and 6500(1800km). Among the railway tunnels, those built before WWII account for about half the total number and they are posing serious problems of how efficiently to inspect, diagnose and repair them to prolong their service life. The road tunnels are equally being plagued with similar problems.

Meanwhile the mountains in Japan are seldom composed of stable hard rocks and more often composed of weakened rocks with complicated folds or faults or of soft rocks such as Neogene tuff or mud rock. Accordingly, there are cases when tunnels are deformed due to earth pressure and call for urgent countermeasures.

In view of such a situation the present authors have been compelled to work out the methods for estimating the soundness of the earth pressure-deformed tunnels and for establishing effective countermeasures in terms of field measurements, lining model tests and numerical analyses.

In the present paper the authors report on the results of their efforts, specifically referring to (1) case studies of tunnels deformed by squeezing earth pressure, (2) 1/30 scale model lining tests and (3) crack propagation analysis. Then they study into the mechanism of tunnel deformation and the effect of countermeasures, and come up with the findings and comments on them.

2 CASE STUDIES OF DEFORMED TUNNELS

Here the discussions are focussed on deformation by squeezing earth pressure. In the following, two features are taken up to investigate the lining structure, deformation behavior and effects of countermeasures.

2.1 *Case I (Tsukayama Tunnel)*

(1) Outline of the tunnel: The tunnel is a double-track railway tunnels, as old as 30 years (opened to service in 1967), length of 1766m. It is located on the Japan Sea side where the sedimentary soft rocks belonging to "Green Tuff region" is widely distributed, specifically situated at the anticline wing partially constituting an active folding. Overburden of the deformed area is about 70m and the surrounding rock is a mud rock liable slaking (uniaxial compression strength 3～6 MPa, natural water content 25～40%, competence factor 2～4). The lining thickness is 50cm, and with no invert provided.

(2) Behavior and countermeasures: Soon after opened to service, the roadbed heaved and the side wall was squeezed, resulting in a bending compressive failure of the arch crown.

To counter this deformation, the sectional opening

was closed with invert concrete in 1970. In consequence, as illustrated in Figure 1, which shows the result of convergence measurement, the maximum convergence rate 36mm/year before provision of invert concrete almost vanished after that.

In 1990, however, with lapse of 20 years, a relatively wide area of the crown which suffered the compressive failure dangled with a lot of shear cracks radiating from the vertical axis. With no time lost, an emergency step was taken to avert a threatening collapse of the crown. This was followed by additional countermeasures, as shown in Figure 2, including back-filling, rock bolting and inner lining (steel fiber reinforced concrete).

(3) Deformation mechanism: Figure 3 illustrates a series of deformation mechanisms which occured. First, a squeezing earth pressure was generated within the surrounding rock mass which had fallen into a secondary stress state when the tunnel was excavated. With lapse of time, the side wall bulged and as a result a strong negative bending moment developed in the crown until a compressive failure occured. When an invert concrete was provided, however, the lining deformation ceased to progress and an axial force built up in the lining steadily, which promoted the compressive failure of the crown. On the contrary, the loosening area of the rock mass gradually expanded with a resultant growth of the loosening vertical earth pressure acting on the arch. At the same time, radiating fissures originated from the failed region of the crown in the arch and ultimately the crown came to droop.

2.2 *Case II (Rokujuri-goe Tunnel)*

(1) Outline of the tunnel: The tunnel is also located in Green Tuff region. Now 27 years old (opened to service in 1970), it is a single-track railway tunnel 6359m in length. Overburden of the deformed area is more than 300m and surrounding rock is Miocene green tuff, competence factor 1.5~2.0 and high smectite content.

The lining section of the deformed area features a side wall with no curvature. The lining thickness is 30 ~45 cm with no invert concrete provided.

(2) Behavior and countermeasures: After the tunnel was opened to service, prominent longitudinal cracks at the side wall to the arch shoulder and compressive failure at the crown locally. The convergence grew approximately in proportion to time lapse, the maximum convergence rate being 30mm/year. The lining of the crown is uniformly thin, with opening behind the lining.

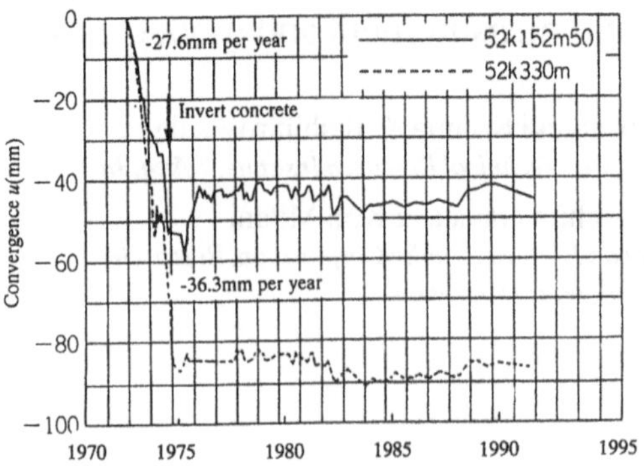

Figure 1. Convergence rate before/after invert concrete

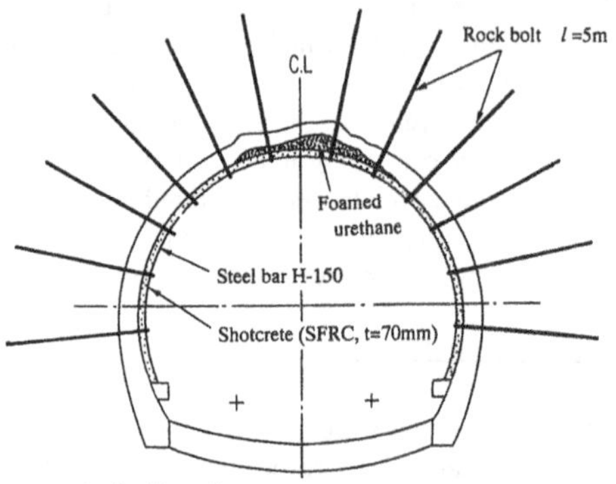

Figure 2. Outline of countermeasure

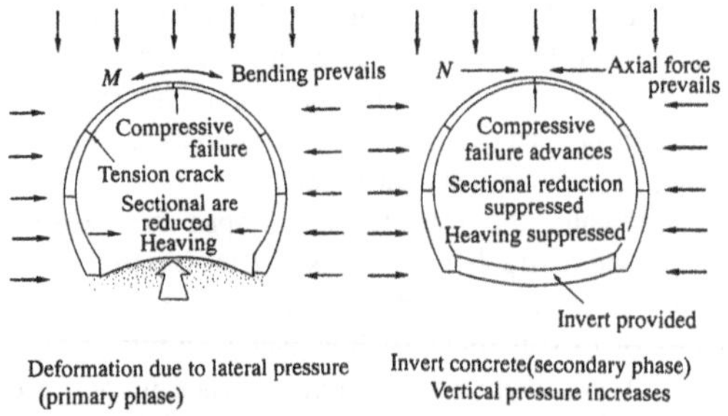

Figure 3. Deformation mechanism

In view of such behavior, lining structure and geological conditions, it can be judged that the deformation has resulted from squeezing earth pressure having acted (mainly lateral pressure) which exceeded the structural strength of the tunnel. The following countermeasures were taken. At first, countermeasure Type-A as illustrated in Figure 4 (back-fill grouting, side wall rock bolting) was executed in 1980. At the spot where this measure proved to be ineffective, in the following year the countermeasure Type-B (arch rock bolting, strutting) was additionally executed. Figure 5 indicates the convergence rate before

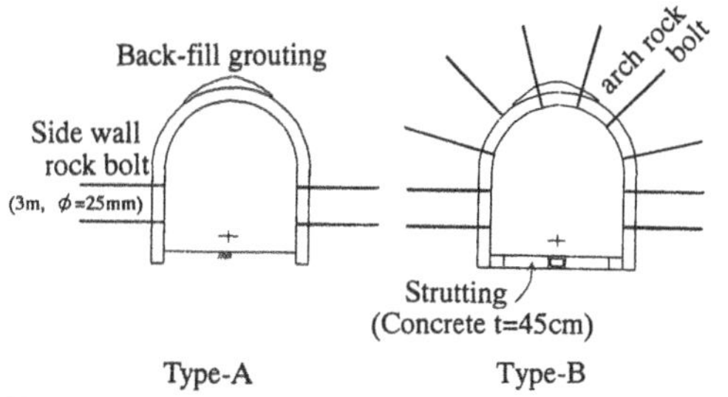

Figure 4. Outline of countermeasures

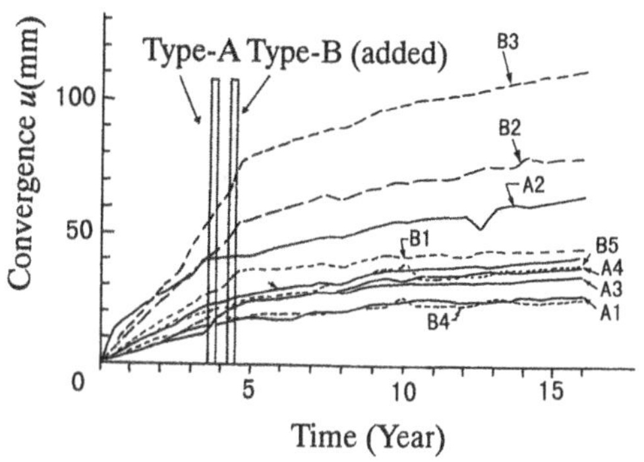

Figure 5. Convergence rate before/after countermeasures

and after the execution of the countermeasure, from which it is observed that the two measures taken successively could effectively suppress the progress of the displacement to about 20%.

(3) Deformed mechanism: The tunnel is designed with a vertical side wall, which is highly vulnerable to lateral pressure (the structural strength of lining, p_c, as calculated by the method described under chapter 4, is; $p_c=6.5\times10^{-2}$MPa). Which means that the earth pressure concerned is not so large-scale; depending on the state of deformation, a relatively minor reinforcement such as Type-A would be enough to address the trouble. At the section where Type-B was added, tension cracks concentrated at arch shoulder, which suggests that the earth pressure mode there must be such that both lateral pressure component and vertical pressure component must have been prevalent. Thus it may be speculated that Type-A alone was short of controlling the deformation and it was addition of Type-B that did realize the intended suppressing effect.

2.3 *Comments*

From these case studies, the following things have been revealed about the behavior of tunnel linings deformed by squeezing earth pressure and the countermeasures to be taken:

①The lining structure and its defects (notably, presence of invert, configuration of side wall, lining thickness of the arch or opening behind the lining) have significant influence on the behavior of deformed lining. Therefore the first thing to be done as a countermeasure should be elimination of these defects or reinforcement to compensate them.

②When the deformation is nothing serious, back-fill grouting or rock bolting will be sufficiently effective. Countermeasures should be designed to match the of deformation behavior.

③Invert concrete has a large effect of suppressing the deformation. When the deformation is something serious, it will be important to execute additional reinforcement of the lining.

3 MODEL EXPERIMENT ON TUNNEL LINING

In this chapter the authors intend to discuss on the results of 1/30 scale model tests device they have developed, specifically referring to ①influence of structural defects, ②effect of back-filling and ③ effect of inner reinforcement work.

3.1 *Test unit*

As shown in Figures 6 and 7, a test unit for a 1/30 scale model of the Shinkansen standard tunnels (equivalent to 2-lane highway tunnels) capable of direct loading (Asakura et al., 1992, 1995) was prepared. The test unit consists mainly of loading/reaction members (consisting of a loading/reaction plate, a cylindrical spring made of hard rubber and double threaded screw bolts), a reaction frame, a bed plate, etc. Within the cross section, a total of 11 sets of the loading/reaction members were set in 11 rows along the tunnel axis to facilitate three-dimensional experiments. At loading points, a steel cylinder was set to directly cause displacement of the lining model. At all points except for the loading points, the hard rubber cylindrical spring was set to induce subgrade reaction.

3.2 *Experimental procedure*

Table 1 shows the materials and their properties used for the experiments. As for similarity relations, only geometric similarity (scale: α) was taken into consideration, assuming the same degree of strength for all materials. Consequently, before cracks occur, deformation and displacement correspond by $1/\alpha$ and

Figure 6. View of tunnel lining test unit

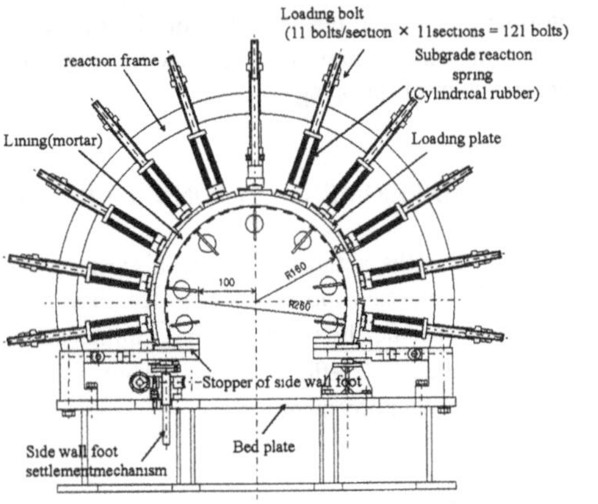

Figure 7. Outline of tunnel lining test unit

Table 1. Experimental material and their properties

Experimental material		Property
Lining	Mortar	E = 1.5×10⁴MPa σc= 30MPa
Subgrade reaction Spring	Hard rubber	Spring constant 78N/mm
Foot reaction spring	Steel plate	It's a condition that there is invert.
Back-fill grouting	Rubber plate	E = 3.0MPa Thickness 0.15,0.30mm
Inner reinforcement	Phosphor bronze plate	E = 1.2 × 105MPa
	Carbon fiber sheet	Fiber area weight 20g/m^2 E = 2.4 × 10^{5}MPa

E: Young's modulus, σc: Unconfined compressive strength

stress corresponds by 1/1 under the same loading pressure. It is considered that, since the lining behavior will be affected by the crack behavior, the similarity will vary after cracks occur.

The experimental procedure is as follows. (1)The lining model on which strain gauges had been set were installed in the test unit together with load cells and displacement meters. (2)Step loading was carried out by displacement control. (3)The experiment was terminated either by (a)ultimate failure of the mode, (b)the stroke limitation of the bolts for loading, or (c)reaching the maximum loading level of the loading unit specified by the design.

3.3 *Experimental results*

(1)Effect of lining deficiency: Figure 8 shows the relationship between load and normal displacement of the lining at the loading point derived from the following two cases where lateral load was applied from both sides. Case 1 shows a sound lining without insufficient lining thickness and openings behind the lining. Case 2 represents a defective lining with openings behind the lining and insufficient lining thickness (1/2 of the regular design thickness) at its crown. As can be seen from the Figure, the sound lining model maintains its durability supported by arch action after cracks occurred. Whilst the defective lining demonstrates brittle failure due to initial cracking.

(2)Effect of back-filling: Figure 8 also shows the effect of back-filling from a case of model experiment where back-filling was applied with soft rubber after cracking had been induced at both side walls and crown by lateral loading from both sides. As is obvious from the Figure, after the application of back-filling, the load increases as the progress of displacement. From this it can be concluded that the lining durability can be restored by the application of back-filling and that back-filling is an effective countermeasure.

(3)Effect of inner reinforcement: Figure 9 shows the effect of the inner reinforcement from cases 4 to 7. In Case 4, force displacement and subsequent cracking were induced vertically in the lining inside its crown and loading was continued without any countermeasures. In Cases 5, 6, and 7, loading was resumed after inner reinforcement had been applied within a range of 60° using carbon fiber sheets of 20g/m^2 in case 5, phosphor bronze plates of t=0.15 mm and t=0.30mm in Cases 6 and 7, respectively. These reinforcing materials are mentioned in reverse order of the magnitude of rigidity. The origin of the coordinate axis represents the point when the countermeasures were applied. In Cases 5 to 7 when the countermeasures were applied, the initial gradient of the curve is steeper than that in the case where no countermeasure was applied, proving that the inner reinforcement has a noticeable effect. Furthermore, the curve gradient becomes steeper in proportion to the rigidity of the materials, confirming that the deformation reducing effect equivalent to the rigidity can be attained.

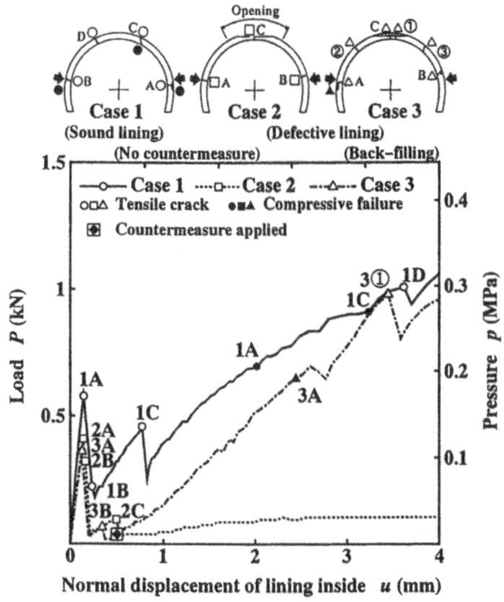

Figure 8. *P- u* curve (case 1 to 3)

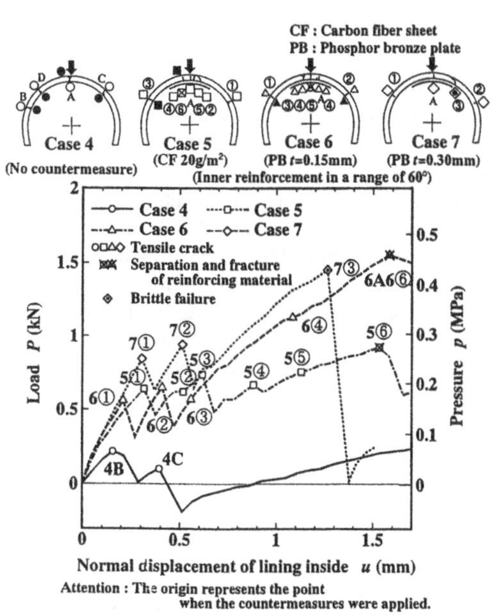

Figure 9. *P- u* curve (case 4 to 7)

As for lining failure mode, when the rigidity of a reinforcing material is low, cracks of the lining are rather dispersed and ductile separation and fracture of the reinforcing material are observed. On the other hand when the rigidity is too high, brittle failures is induced by the concentration of stress near the edges of inner reinforcement.

4 NUMERICAL ANALYSES

In this chapter the authors intend to give an outline of the frame analysis they have developed to take account of the cracking behavior. Further they intend to report on the results of employing their technology in studying the lining strength and comparing the effects of practical countermeasures for deformed linings.

4.1 *Crack propagation analysis method*

In the crack propagation analysis, a structural model (a frame analysis model taking the ground as the spring, the lining as the beam and the crack as the plastic hinge) is formulated for each progress stage of cracking for the sake of calculation and the results of calculation are added up to express the total crack propagation. Thereby the lining element at each stage is assumed as a liner elastic body.

As illustrated in Figure 10, in the crack propagation analysis, the crack generation is evaluated form the stress resultant; the structural model is formulated for each stage of crack generation (analysis step); and then the calculation is repeated until the lining element hits the limit of the compressive strain. The crack is formulated by pin-connection to calculate it on the safe side. As for the propagation of cracking, the calculation is continued with the pin-connected model and the final result is expressed by piling up the stress resultant, the displacement, etc. in each stage.

4.2 *Comparative analysis*

(1)Modeling of tunnel and counter measures: An

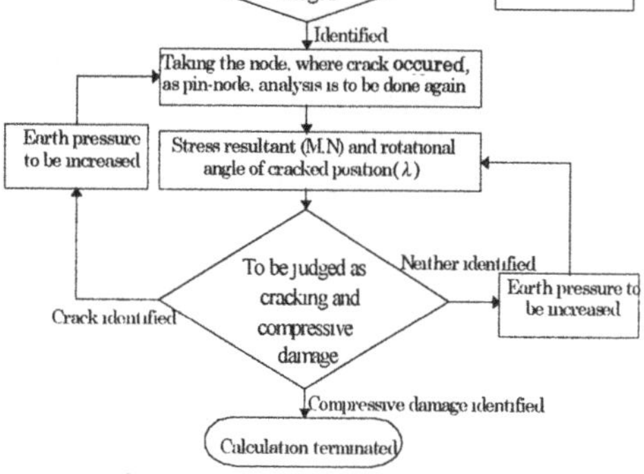

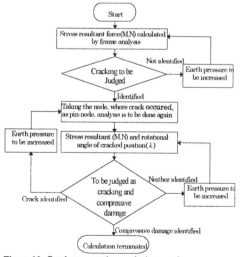

Figure 10. Crack propagation analysis procedure

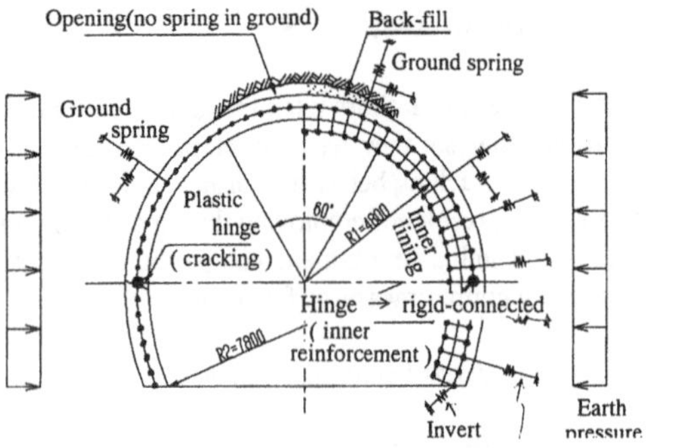

Figure 11. Concept of analysis model

Table 2. Physical properties of lining and ground

Lining	Thickness	70cm
	Young's modulus	2.1×10^4 MPa
	Unit volume	23.5 KN/m^3
	Structural defect	Opening behind lining (arch 60°)
Ground	Young's modulus	5.0×10^2 MPa
	Earth pressure	Horizontal prevalent pressure

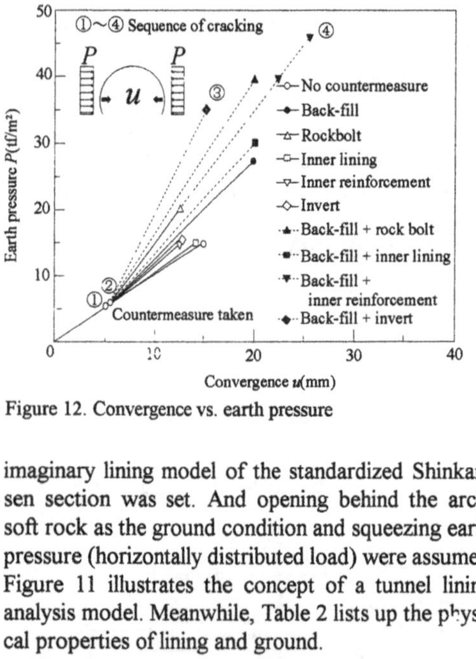

Figure 12. Convergence vs. earth pressure

imaginary lining model of the standardized Shinkansen section was set. And opening behind the arch, soft rock as the ground condition and squeezing earth pressure (horizontally distributed load) were assumed. Figure 11 illustrates the concept of a tunnel lining analysis model. Meanwhile, Table 2 lists up the physical properties of lining and ground.

Five countermeasures which have been often applied for deformed tunnels are selected; back-fill grouting, rock bolting, inner lining, inner reinforcement(carbon fiber sheet) and invert concrete.

(2) Results of analysis: Figure 12 shows the relation between displacement and earth pressure by the type of countermeasures. From Figure 12, it is understood that the best effect of improving the structural strength belongs to back-fill grouting. Any other countermeasure alone cannot be expected to produce good effect. Better effect will come only when it is coupled with back-fill grouting. It is notable that combination of back-fill grouting plus inner reinforcement or rock bolting will bring about a great increase of structural strength. As for the rigidity improvement effect, the invert concrete, which increases the strength in the loaded direction, excels.

5 CONCLUSIONS

Case studies, model testing and numerical analyses of deformed tunnel lining came up with such ample information as follows;

(1) The tunnel lining strength with structural defects can be vastly increased by back-fill grouting.

(2) Given mandatory execution of back-fill grouting, the reinforcing effect of rock bolting or inner reinforcement will be given full justice.

(3) The rigidity enhancing effect of invert concrete is admittedly large, but in order to increase the structural strength, execution of other measures such as rock bolting, inner reinforcement or inner lining will be necessary.

At present, comprising the results of research reported in the present article and based on all the achievements attributable to the present authors, a "Design Manual of Countermeasures for Deformed Tunnels" systematically describing the state-of-the art technology about the deformed tunnel was compiled for publication. (Kojima et al.,1998)

REFERENCES

Asakura, T., et al., 1992. Analysis on the behavior of tunnel lining –Experiments on double track tunnel lining- : QR of RTRI, Vol.33, No.4

Asakura, T., et al., 1995. Countermeasure for deformed tunnel lining by tunnel reinforcement: 8th International Congress on Rock Mechanics, ISRM

Kojima Y., Asakura T., et al., 1998. Design Method of Countermeasures for Deformed Tunnel: QR of RTRI, Vol.39, No.1

Mechanics of Jointed and Faulted Rock, Rossmanith (ed.) © 1998 Taylor & Francis, ISBN 90 5410 955 6

Influence of the rock's joints frequency on the tunnel lining stress state

N.N. Fotieva & K.E. Zalessky
Department of Materials Mechanics, Tula State University, Russia
N.S. Bulychev
Department of Underground Construction, Tula State University, Russia

ABSTRACT: The analytical method of designing circular tunnel linings in the rock weakened by a double-periodic chink set upon the action of the gravitational or tectonic forces, external underground water pressure and internal water head is described in the paper presented. For the analysis of the lining stress state the jointed rock is simulated by a transversely isotropic medium of the equivalent stiffness. On the base of multivariant calculations the dependencies of normal tangential stresses appearing in the linings of different thickness at the every kind of loads on the relative joints frequency are given and analysed.

1 INTRODUCTION

The design method proposed is based on modelling the jointed rock mass weakened by a double-periodic chink set as a transversely isotropic medium of the equivalent stiffness the mechanical characteristics of which are determined by the Erzanov & Kaydarov's method (1970) depending on the relative joints frequency, namely on the relation of the distance between joints centres to the half of the joints length.

For the analysis of the lining stress state caused by the action of the gravitational or tectonic forces, external underground water pressure and internal water head the elasticity theory corresponding plane contact problems for a circular ring supported the opening in a transversely isotropic medium are considered.

2 THE METHOD OF TUNNEL LININGS ANALYSIS

The general design scheme is given in Figure 1.

Here the S_1 isotropic ring of the R_0 external radius and the R_1 internal one the material of which possesses the E_1 deformation modulus and the ν_1 Poisson ratio simulates the lining. The S_0 transversely isotropic medium characterised by the $E_{0,1}$, $E_{0,2}$ deformation modules correspondingly in the plane of isotropy and in direction of normal to that plane, the $\nu_{0,1}$, $\nu_{0,2}$ Poisson ratios and the $G_{0,2}$ shear modulus in planes normal to the isotropy one simulates the jointed rock mass.

The plane of isotropy (in which the Ox axis is located) may be inclined under an arbitrary β angle to the horizontal one.

The deformation characteristics of the transversely isotropic medium simulating the jointed rock mass may be determined on the testing base (Zelensky 1969) or by analytic method (Erzanov & Kaydarov 1970).

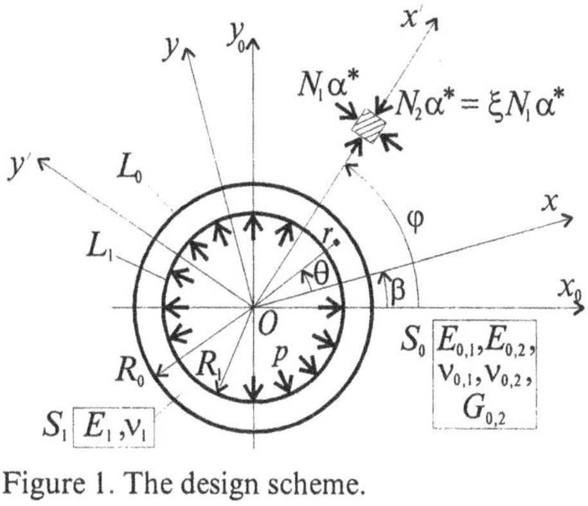

Figure 1. The design scheme.

The action of the rock's own weight or tectonic forces in the rock mass is simulated by a presence of the initial stresses in the S_0 medium the Ox', Oy' main axes of which may be inclined under an arbitrary φ angle to the Ox_0, Oy_0 horizontal and vertical ones. The initial stresses are expressed by formulae

$$\sigma_{x'}^{(0)(0)} = -\xi N_1 \alpha^*, \sigma_{y'}^{(0)(0)} = -N_1 \alpha^*,$$
$$\tau_{x'y'}^{(0)(0)} = 0, \qquad (1)$$

where $\xi = N_2 / N_1$ is the relation between main stresses in an intact rock, α^* = the correcting multiplier introduced for an approximate registration of the influence of the l distance between the lining being constructed and the tunnel face. This multiplier may be determined by empirical formula (Bulychev 1994):

$$\alpha^* = 0{,}64\exp(-1{,}75\ell / R_0). \qquad (2)$$

In case of the action of tectonic forces in the rock mass the N_1, N_2 main stresses and the φ angle being the input data of the lining design are determined on the base of full-scale measurements.

At the gravitational initial stresses field

$$N_{1,2} = \frac{\sigma_{y_0}^{(0)(0)} + \sigma_{x_0}^{(0)(0)}}{2} \pm$$
$$\pm \frac{1}{2}\sqrt{\left(\sigma_{y_0}^{(0)(0)} - \sigma_{x_0}^{(0)(0)}\right)^2 + 4\left(\tau_{x_0,y_0}^{(0)(0)}\right)^2}, \qquad (3)$$
$$\varphi = arctg\left(-\frac{\tau_{x_0,y_0}^{(0)(0)}}{N_1 - \sigma_{x_0}^{(0)(0)}}\right)$$

The initial stresses in an intact transversely isotropic rock mass are determined by formulae (Erzanov, Aitaliev & Masanov 1980):

$$\sigma_{y_0}^{(0)(0)} = -\gamma H, \sigma_{x_0}^{(0)(0)} = \lambda_{x_0}\sigma_{y_0}^{(0)(0)},$$
$$\tau_{x_0 y_0}^{(0)(0)} = -\lambda_{x_0,y_0}\sigma_{y_0}^{(0)(0)}, \qquad (4)$$

where γ = the rock unit weight, H = the depth of the tunnel, $\lambda_{x_0}, \lambda_{x_0,y_0}$ = coefficients depending on deformation characteristics of the transversely isotropic medium and on the β angle of the plane of isotropy.

The action of the pressure of underground water having a high level above the tunnel lining considered as non-penetrated by water is simulated by initial stresses

$$\sigma_x^{(0)(0)} = \sigma_y^{(0)(0)} = -\gamma_w H_w, \qquad (5)$$

where γ_w = the water unit weight, H_w = the static head of underground water in metres counted off from the tunnel axis.

On representing the full stresses in the S_0 medium as sums of the initial stresses and the additional ones caused by a presence of the opening (displacements are considered only as the additional ones) we obtain the boundary problem of the complex variables analytic functions theory for the determination of the two $\Phi_j(z_j)$ $(j = 1,2)$ Lekhnitsky complex potentials characterising the additional stresses and displacements the transversely isotropic S_0 medium and the $\varphi_1(z), \psi_1(z)$ Kolosov-Muskhelishvili complex potentials characterising the S_1 isotropic ring stress-strain state.

The boundary conditions are the following:

$$\varphi_1(t) + t\overline{\varphi_1'(t)} + \overline{\psi_1(t)} = (1 - \beta_1)\Phi_1(t_1) +$$
$$+(1 - \beta_2)\Phi_2(t_2) + (1 + \beta_1)\overline{\Phi_1(t_1)} +$$
$$+(1 + \beta_2)\overline{\Phi_2(t_2)} - q\left(\frac{1+\xi}{2}t + \frac{1-\xi}{2}\bar{t}e^{2i\alpha}\right),$$
$$\text{on } L_0 \qquad (6)$$

$$[x_1\varphi_1(t) - t\overline{\varphi_1'(t)} - \overline{\psi_1(t)}]/2G_1 =$$
$$= (b_{12} - b_{11}\beta_1^2)[\Phi_1(t_1) + \overline{\Phi_1(t_1)}] +$$
$$+(b_{12} - b_{11}\beta_2^2)[\Phi_2(t_2) + \overline{\Phi_2(t_2)}] + \qquad (7)$$
$$+(b_{22}/\beta_1 - b_{12}\beta_1)[\Phi_1(t_1) - \overline{\Phi_1(t_1)}] +$$
$$+(b_{22}/\beta_2 - b_{12}\beta_2)[\Phi_2(t_2) - \overline{\Phi_2(t_2)}],$$
$$\text{on } L_0$$

$$\varphi_1(t) + t\overline{\varphi_1'(t)} + \overline{\psi_1(t)} = -pt \text{ on } L_1 \qquad (8)$$

where

$$x_1 = 3 - 4\nu_1, G_1 = E_1 / 2(1 + \nu_1),$$
$$\alpha = \varphi - \beta, \qquad (9)$$

$t = R_i\sigma$ $(i = 0,1)$ are affixes of points of the

L_i $(i=0,1)$ outlines; t_j $(j=1,2)$ are the affixes of points of the $L_{0,j}$ $(j=1,2)$ outlines restricting the $S_{0,j}$ $(j=1,2)$ infinite areas of the determination of $\Phi_j(z_j)$ $(j=1,2)$ Lekhnitsky complex potentials, obtained from the S_0 area by affine transformations:

$$t_j = 0.5R_0[(1+\beta_j)\sigma_j + (1-\beta_j)/\sigma_j]. \qquad (10)$$

As it is known

$$\sigma = \sigma_1 = \sigma_2 = e^{i\theta}. \qquad (11)$$

The β_j $(j=1,2)$ are determined by formulae:

$$\beta_j = \mu_j / i, \quad (j=1,2) \qquad (12)$$

where μ_j = the roots of the equation

$$b_{11}\mu^4 + (2b_{12} + b_{66})\mu^2 + b_{22} = 0 \qquad (13)$$

The b_{ij} $(i,j=1,2,6)$ are determined by formulae:

$$b_{11} = \frac{1-\nu_{0,1}^2}{E_{0,1}}, \; b_{12} = b_{21} = -\frac{\nu_{0,2}(1+\nu_{0,1})}{E_{0,1}},$$
$$b_{22} = \frac{1}{E_{0,2}}\left(1 - \frac{E_{0,2}}{E_{0,1}}\nu_{0,2}^2\right), \; b_{66} = \frac{1}{G_{0,2}}. \qquad (14)$$

The q and ξ values are determined the following:

$q = N_1\alpha^*$, $\xi = N_2/N_1$ - at the action of gravitational or tectonic forces;

$q = \gamma_w H_e$, $\xi = 1$ - at the action of the underground water pressure;

$q = 0$ - at the action of the internal water head.

The $\Phi_j(z_j)$ $(j=1,2)$ Lekhnitsky complex potentials regular in the $S_{0,j}$ $(j=1,2)$ including the point on the infinity are represented on the L_0 outline in the form:

$$\Phi_j(\sigma) = \sum_{n=1}^{\infty} c_n^{(j)(0)} \sigma^{-n} \quad (j=1,2) \qquad (15)$$

The $\varphi_1(z), \psi_1(z)$ complex potentials regular in the S_1 ring are represented as Loran series:

$$\varphi_1(z) = \sum_{n=1}^{\infty} c_n^{(1)(1)} (z/R_0)^{-n} + \sum_{n=0}^{\infty} c_n^{(3)(1)} (z/R_0)^n, \qquad (16)$$
$$\psi_1(z) = \sum_{n=1}^{\infty} c_n^{(2)(1)} (z/R_0)^{-n} + \sum_{n=0}^{\infty} c_n^{(4)(1)} (z/R_0)^n.$$

Substituting the (15),(16) complex potentials into the (6),(7) boundary conditions and equating the coefficients at the same degrees of the σ variables to each other in the left and right parts of the equations obtained we have the correlations combining the $c_n^{(j)(1)}$ $(j=1,\ldots,4)$ coefficients with the $c_n^{(j)(0)}$ $(j=1,2)$ coefficients. Substituting expressions obtained into the (8) boundary condition we come to an infinite system of linear algebraic equations relative to unknown $c_n^{(j)(0)}$ $(j=1,2)$ coefficients. On solving above system being correspondingly restricted the stress state of the lining may be determined.

The computer program has been developed.

3 EXAMPLES OF THE DESIGN

The results of designing tunnel lining with $R_0 = 3.4m$, $R_1 = 3.0m$ outer and inner radii correspondingly fulfilled from concrete with $E_1 = 30000MPa$, $\nu_1 = 0.2$ deformation characteristics are given below. The tunnel is located in transversely isotropic rock massif with the following characteristics: $E_{0,1} = 10740MPa, \nu_{0,1} = 0.413$, $E_{0,2} = 5230MPa, \nu_{0,2} = 0.198, G_{0,2} = 1200MPa$.

The rock mass is weakened by a double-periodic chink set inclined under the $\beta = 30°$ relatively to the horizontal and having the $\omega / a = 2.5$ relative distance between them (ω = the distance between joints centres, $2a$ = the length of joints).

According to the work by Erzanov, Aitaliev & Masanov (1980) characteristics of the transversely isotropic medium simulating the jointed rock mass in the case considered are the following:

$E_{0,1} = 10740MPa,\ \nu_{0,1} = 0.413,$

$E_{0,2} = 1480MPa,\ \nu_{0,2} = 0.198,$

$G_{0,2} = 450MPa.$

The diagrams of the $\sigma_\theta^{ex} / \gamma H\alpha^*, \sigma_\theta^{in} / \gamma H\alpha^*$ normal tangential stresses appearing on the external and internal outlines of the lining cross-section due to action of the rock own weight are given by solid lines in Figure 2 a, b correspondingly. The same stresses caused by the action of tectonic forces (at $\xi = 2,\ \varphi = 60°$), caused by the underground water pressure and caused by the internal water head are shown by solid lines in Figure 3 a, b, 4 a, b and 5 a, b correspondingly. For comparison the stresses appearing in the lining located in the solid rock mass without joins are given in Figures 2-5 by dotted lines.

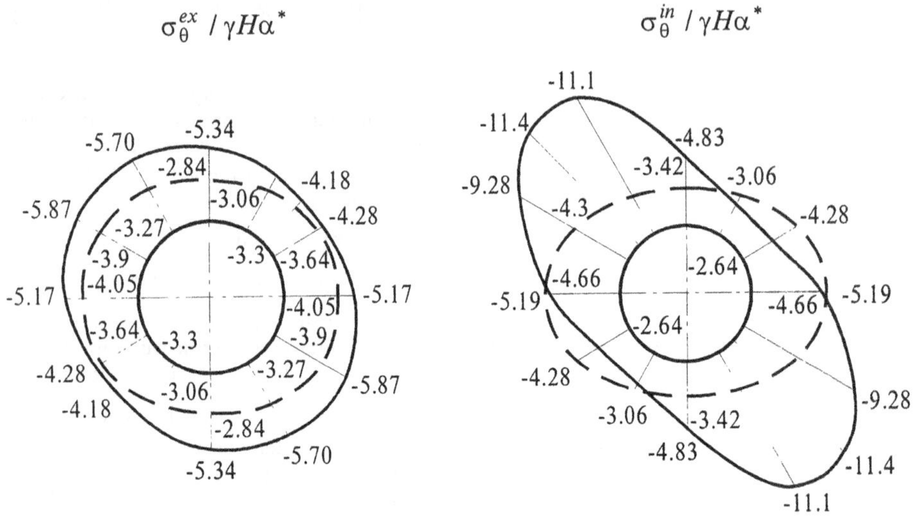

a b

Figure 2. Distribution of the normal tangential stresses in the lining caused by the rock own weight.

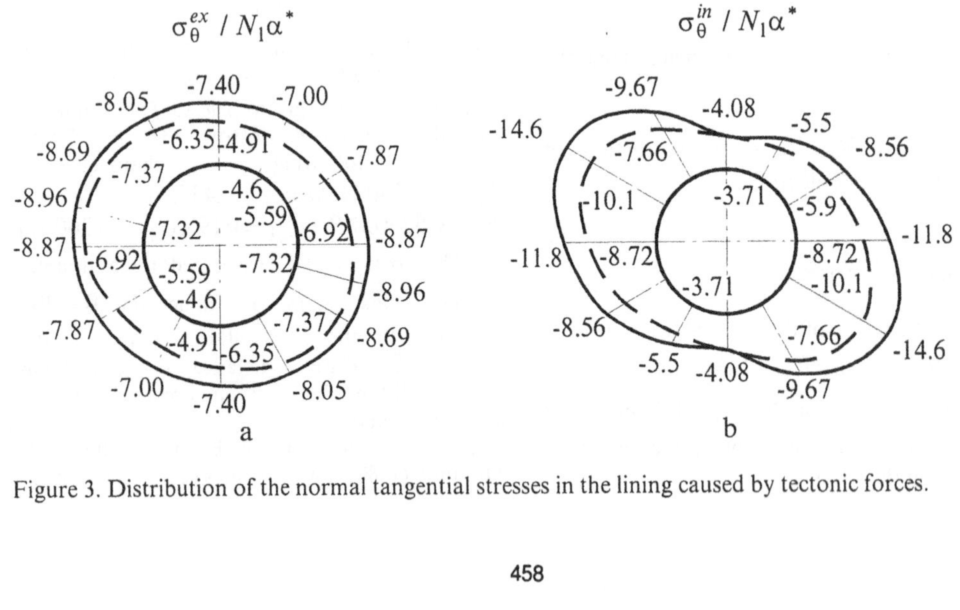

a b

Figure 3. Distribution of the normal tangential stresses in the lining caused by tectonic forces.

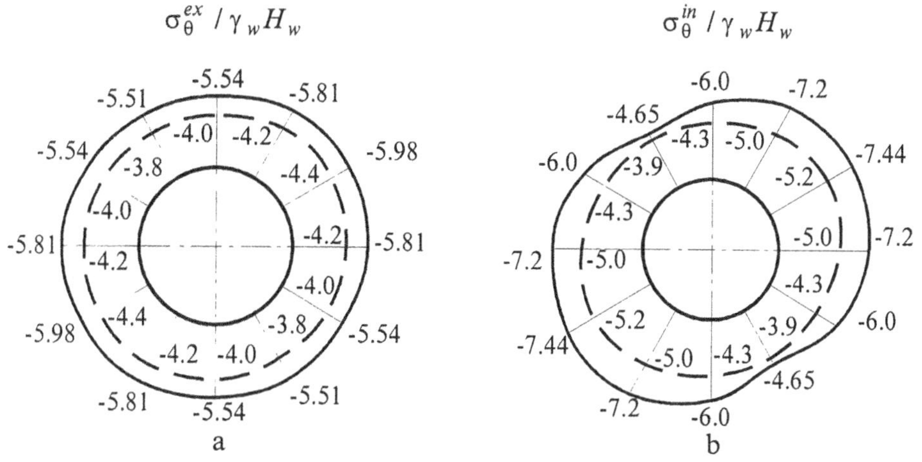

Figure 4. Distribution of the normal tangential stresses in the lining caused by the underground water pressure.

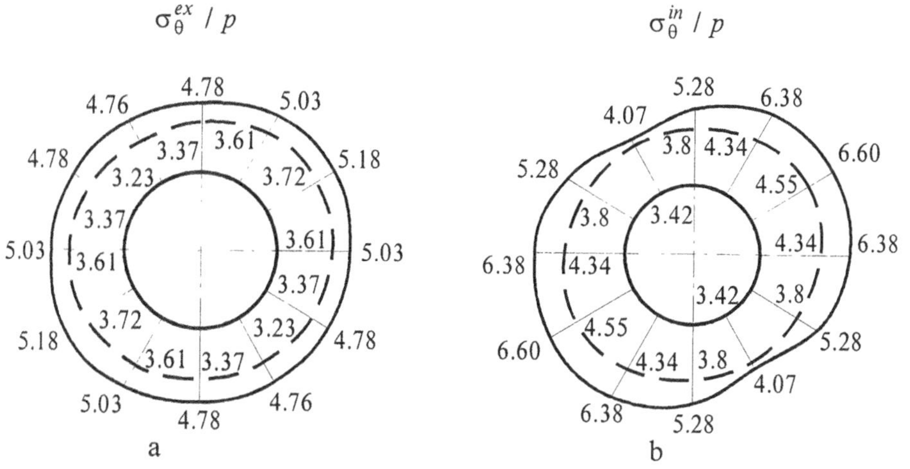

Figure 5. Distribution of the normal tangential stresses in the lining caused by the internal water head.

Dependencies of the $\sigma_\theta^{in} / \gamma H \alpha^*$ appearing in the points of horizontal and vertical diameters due to the rock's own weight on the ω / a joint frequency (at $E_1 = 23000 MPa$, $\beta = 0°$) are given in Figure 6 a, b correspondingly. The curves 1, 2, 3 and 4 correspond to the relative lining thickness $(R_0 - R_1) / R_1 = 0.05,\ 0.1,\ 0.2,\ 0.3$.

The same dependencies for the $\sigma_\theta^{in} / N_1 \alpha^*$ stresses caused by tectonic forces (at $\beta = \varphi$, $\xi = 3$) on the ω / a are given in Figure 7 a, b.

The dependencies of the $\sigma_{\theta\ max}^{in} / \gamma_w H_w$ maximal stresses caused in the lining by the underground water pressure on the ω / a joint frequency are shown in Figure 8. The similar dependencies for the $\sigma_{\theta\ max}^{in} / p$ maximal normal tangential stresses appearing in the internal outline of the lining cross-section due to an internal water head are given in Figure 9 .

For comparison the stresses in case of the solid rock without joints are given in Figures 6-9 in brackets near the corresponding curves.

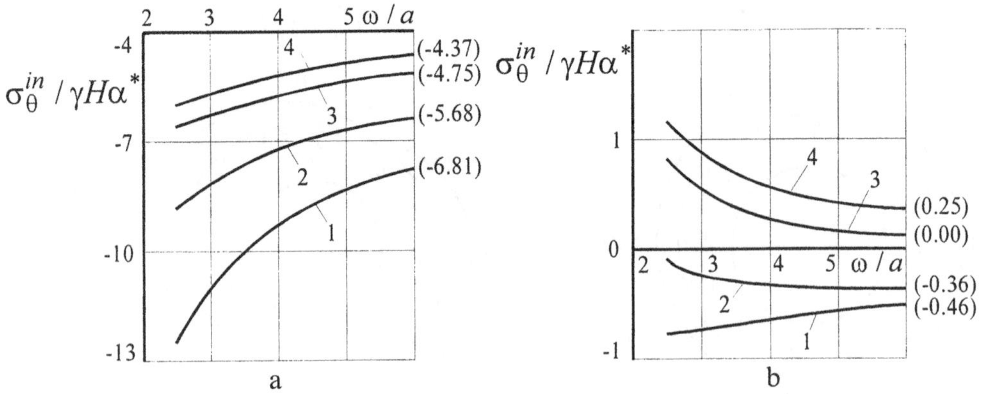

Figure 6. Dependencies of the $\sigma_{\theta}^{in} / \gamma H \alpha^*$ stresses on the ω / a joint frequency in the rock.

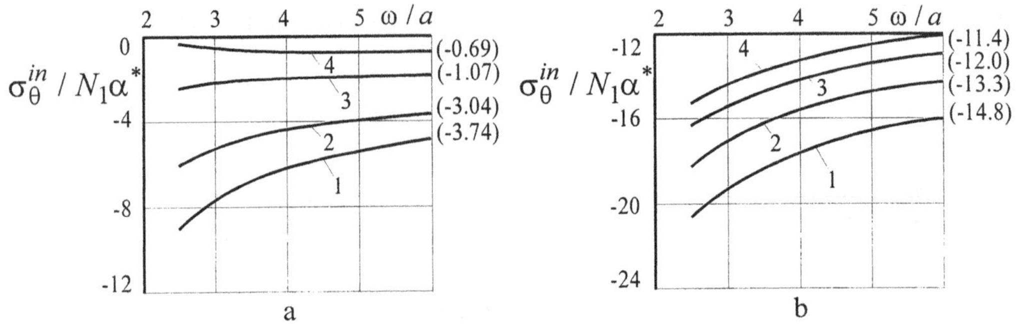

Figure 7. Dependencies of the $\sigma_{\theta}^{in} / N_1 \alpha^*$ stresses on the ω / a joint frequency in the rock.

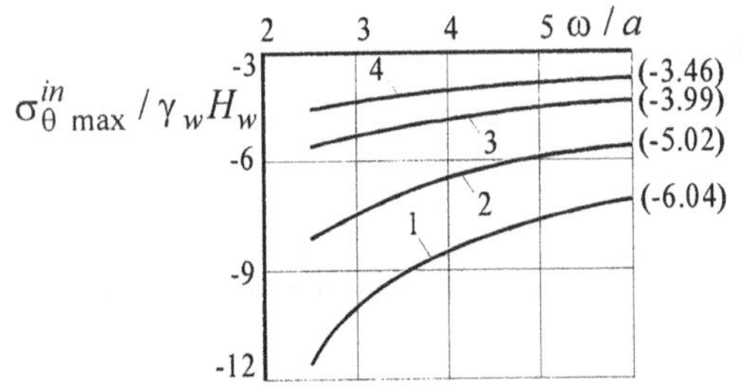

Figure 8. Dependencies of the $\sigma_{\theta\ max}^{in} / \gamma_w H_w$ stresses on the ω / a joint frequency in the rock.

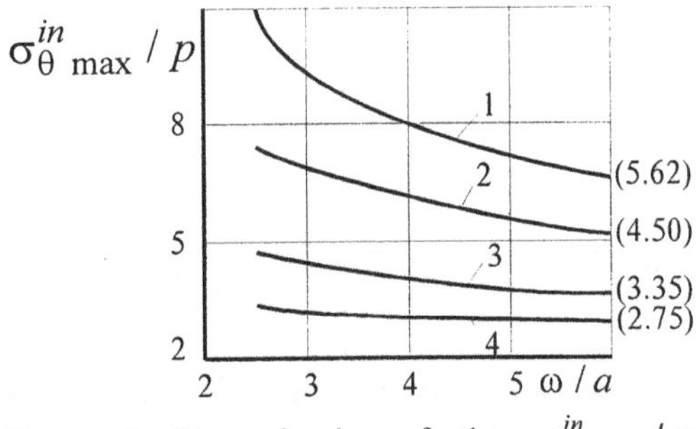

Figure 9. Dependencies of the $\sigma_{\theta\ max}^{in} / p$ stresses on the ω / a joint frequency in the rock.

REFERENCES

Bulychev, N. S. 1994. *Mechanics of underground structure* Moscow: Nedra.

Erzanov, Z. S., Aitaliev, Sh .M. & Z. K.Masanov 1980. *Seismic stress state of underground structures in anisotropic layered massif.* Alma Ata: Nauka.

Erzanov, Z.S. & K.K.Kaydarov 1970. On visco-elastic behaviour of massif with a double periodic chink set. *Proc. of Kaz. Acad.* 3:12-18.

Zelensky, B.D. 1969. Analytic method to evaluate the properties of rock massifs. *Hydrotechnic Construction.* 11:31-35.

Mechanics of Jointed and Faulted Rock, Rossmanith (ed.) © *1998 Taylor & Francis, ISBN 90 5410 955 6*

Numerical analyses of the trial underground exploitation of dimension stone

B. Kovačević-Zelić & S. Vujec
University of Zagreb, Faculty of Mining, Geology and Petroleum Engineering, Croatia

ABSTRACT: Croatia has a long tradition of dimension stone exploitation, mostly by surface extraction. Lately, the dimension stone has been extracted in the world also by underground methods, because of environmental, organisational and economical reasons. In such a case, stability problems require adequate studies in order to avoid expensive and technologically unacceptable artificial support measures. In the Kanfanar quarry near the town of Pazin on the Istrian peninsula an existing active surface exploitation had to be transferred underground because of economical and environmental constraints. The stability analysis was made by numerical modelling with the finite difference code FLAC. These calculations are described in the paper and represent the first phase of the pillar design.

1 INTRODUCTION

Croatia has a long tradition of dimension stone exploitation - both surface and underground. Moreover, it is full of historic monuments, which testify stone application especially in civil engineering and architecture dating back to the Roman period up to these days. Most of them are situated in the coastal regions of Istria and Dalmatia. One of the best preserved amphitheatres in the world is in Pula-Istria (1st Century AD); Diokletian Palace in Split-Dalmatia (4th Century BC) is known as the 'symphony in stone'; one of the most beautiful towns on the Adriatic coast - Dubrovnik was made from local limestone and represents one of the most preserved Middle Ages towns in the world; and throughout Croatia one can find a lot of public buildings, basilicas, cathedrals, temples, houses and palaces made from local stones (Cotman 1995).

Traditionally, dimension stone was extracted mostly by surface methods. Underground exploitation of dimension stone has been spreading lately in the world for three main reasons: economy (heavy overburden), organisation (seasonal weather constraints) and environment (protected areas, important areas for tourism). The latter aspect is the most important in the region of Istrian peninsula in Croatia which is well known as natural protected area and an area of tourism.

The other advantage of underground exploitation is the fact that created subsurface openings can be used later. World-wide they have been already used for many purposes such as for storing of liquids, food, industrial products, strategic materials; for creating protected areas for military and civil use (archives, museums, research centres); for storing of industrial and toxic waste; or as cheese storehouses and wine cellars (Fornaro & Bosticco, 1995).

Underground exploitation of ornamental stone is, from technological point of view, different from surface quarrying only in the first stage, namely in the removal of top slice; descending slices are worked as in conventional quarries. But, in this case stability problems require adequate studies and stability checks in order to avoid expensive artificial support measures (Pelizza et al. 1994, Fornaro & Bosticco 1995).

In the Kanfanar quarry, near the town of Pazin in Istria, an existing active surface exploitation had to be transferred underground because of economical and environmental constraints. The analysis of room stability and pillar design was made by numerical modelling with the 2-dimensional finite difference code FLAC. The results of these analyses are presented in the paper.

2 INPUT DATA FOR NUMERICAL ANALYSES

Croatia has a significant number of quarries producing dimension stones mostly of sedimentary origin. The region of Istria is characterised by

Table 1. Material properties for the numerical analyses.

Material type	Height of layer (m)	Density (kgm^{-3})	Bulk modulus (MPa)	Shear modulus (MPa)	Cohesion (MPa)	Friction angle (°)	Tensile strength (MPa)
Upper roof	5.0	1800	8.5	3.9	0.05	20	0.0
Immediate roof	8.9	2630	7600	4340	17	50	5.3
Exploitable layers	6.8	2635	9345	5878	13	55.5	4.4
Immediate footwall	5.0	2635	9500	5900	15	60	5.0

dimension stone deposits which are situated in limestones and clastites of Upper Jurassic including Lower and Upper Cretaceous till Eocene. In the Kanfanar quarry the limestones of Lower Cretaceous are exploited, which are known under commercial name "Giallo d'Istria" or "Istrian yellow" (Crnković & Jovičić 1993).

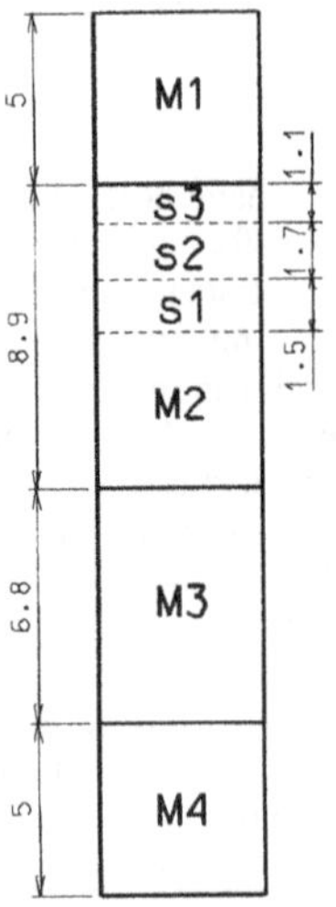

Legend: M1 - Upper roof; M2 - Immediate roof; M3 - Exploitable layers; M4 - Immediate footwall; s1, s2, s3 - Bedding planes.

Figure 1. Schematical representation of geological profile - Kanfanar.

The characteristic geological profile schematically represented in Figure 1. was created according to previously made geological research. The position of the dominant discontinuities, that could affect the results of numerical analyses, was also determined by the in-situ prospection. Three horizontal discontinuities (bedding planes), that are placed in the immediate roof, were obvious at the outcrops. Two sets of vertical discontinuities (fractures of tectonic origin), that are orthogonal to each other, were also observed.

Physical and mechanical properties of intact rock materials (Table 1) were determined by the laboratory tests.

The discontinuities were introduced into the models with the following properties.

cohesion,	$c=0$ MPa,
friction angle,	$\varphi=34°$,
tensile strength,	$\sigma_t=0$ MPa,
normal stiffness,	$k_n=2.7$ GPa/m,
shear stiffness,	$k_s=1.0$ GPa/m

Friction angle was determined by direct shear test. Cohesion and tensile strength are set to zero because of the presumption that the possible joint filling material will be washed out with time. The values of normal and shear stiffness were taken from the literature data for limestones (Bandis et al 1983)

3 NUMERICAL ANALYSES

Regular room and pillar method was chosen for the trial underground extraction of dimension stone Two basic cases with different pillar dimensions had to be examined (Fig. 2, Table 2). These cases were acceptable from the technological point of view (first series of analyses).

It is well known that the mechanical response of rock mass depends far more on the behaviour of discontinuities rather than on the strength of rock material itself. Numerical methods offer a very powerful and effective tool in the analysis of jointed rock structures.

Therefore, the analysis of room stability and pillar design was made by means of numerical modelling with 2-dimensional finite difference code FLAC (Itasca 1991). FLAC is primarily intended for geotechnical engineering applications and has several built-in material models. Mohr-Coulomb plasticity model, ubiquitous joint model, and an interface model are available among others. For the intact rock materials the model of Mohr-Coulomb's plasticity was chosen. Special attention was focused on the influence of discontinuities with different position in the roof-strata.

Basically, there are two different approaches for modelling of jointed rock masses which are available in FLAC. Ubiquitous joint model is one of the implicit methods of the discontinuous rock mass representation. It is an anisotropic plasticity model which assumes a series of parallel weak planes embedded in a Mohr-Coulombs solid. In the explicit methods the discontinuities are directly included into the mesh by the usage of special interface elements. The interfaces are planes upon which slip and/or separation are allowed.

Ubiquitous joint model cannot be used to simulate the behaviour of major discontinuities in mining engineering. Their behaviour has to be modelled directly (Kovačević-Zelić 1994). Because of the nature of discontinuities in the quarry Kanfanar, they were introduced into the models by the usage of interface elements.

Table 2. Types of the analyses.

Dimension	Case 1	Case 2
w (m)	2.8	3.3
l (m)	5.0	5.0
c (m)	5.5	5.5

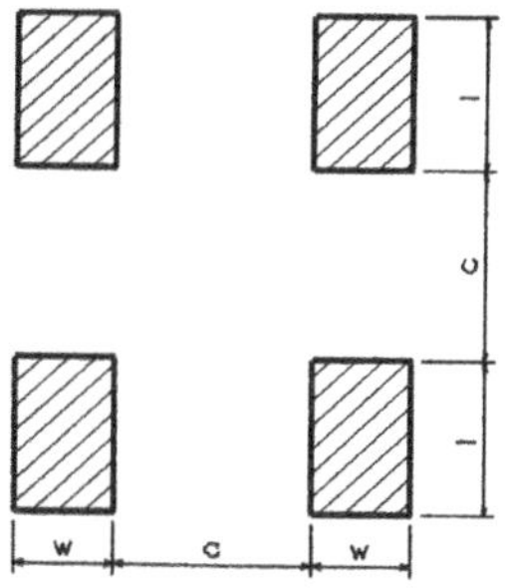

Figure 2. Room and pillar method.

Four different models were created, according to the available geological data:

Model 1 - roof without discontinuities (competent rock),

Model 2 - roof with three parallel horizontal discontinuities,

Model 3 - roof with one vertical discontinuity in the middle of the room,

Model 4 - roof with one vertical discontinuity near the pillar.

It is anticipated that pillar layout could be adjusted with respect to variations in geology i.e. the distribution of the discontinuities. The discontinuities should be mapped continuously with the mining progress. The pillars should be made only in sound rock. Therefore, the numerical models with the discontinuities placed inside pillars were not considered. Instead of that, the models with the enlarged pillar span to 10.0 m were also examined for both cases of pillar dimensions (second series of analyses). These models represent the situation of the necessity of pillar adjustments in the case of discontinuity occurrence in the pillars. Moreover, they represent the crossings too.

4 DISCUSSION OF NUMERICAL RESULTS

The results are presented in terms of stress and strain components. Special attention is focused on the maximum compressive stresses in pillars, the appearance of tensile stresses inside the pillars or in the roof-strata and maximum vertical displacements in rooms.

Tables 3 and 4 show the results of the numerical analyses in terms of maximum compressive stress in pillars σ_{max} and maximum vertical displacement in rooms δ_{max}, for the analyses with the pillar spans of 5.5 m and 10.0 m, respectively.

Three horizontal discontinuities (model 2) did not affect very much the results of numerical analysis in comparison to the model 1, especially if we consider the magnitudes of displacements. It could be explained by the very good quality of intact rock material and the position of discontinuities relatively high in the roof. Therefore, this model was not used for other cases.

Table 3. Results of numerical analyses - c=5 5 m.

	Case 1		Case 2	
	σ_{max} (MPa)	δ_{max} (mm)	σ_{max} (MPa)	δ_{max} (mm)
Model 1	1.211	0.46	1.132	0.420
Model 2	1.223	0.47	-	-
Model 3	1.224	0.50	1.139	0.454
Model 4	1.275	0.65	1.178	0.585

Table 4. Results of numerical analyses - c=10.0 m.

	Case 1		Case 2	
	σ_{max} (MPa)	δ_{max} (mm)	σ_{max} (MPa)	δ_{max} (mm)
Model 1	1.871	0.98	1.676	0 904
Model 2	-	-	-	-
Model 3	1.881	1.00	1.694	0.985
Model 4	2.526	2.40	2.314	2.240

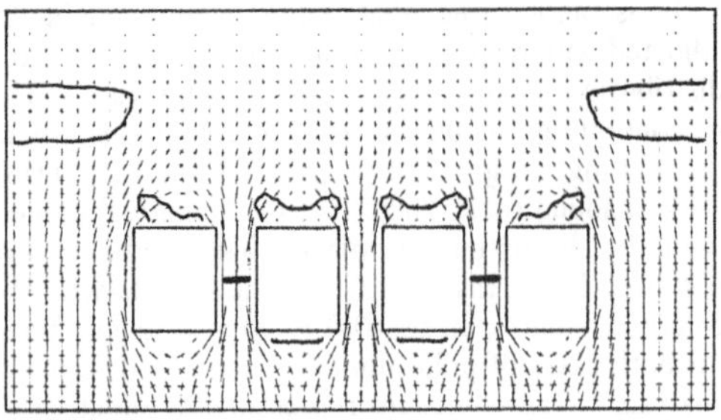

a) Model 1

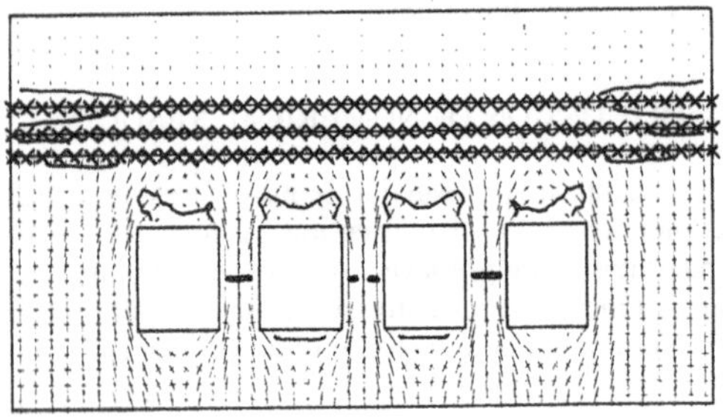

b) Model 2

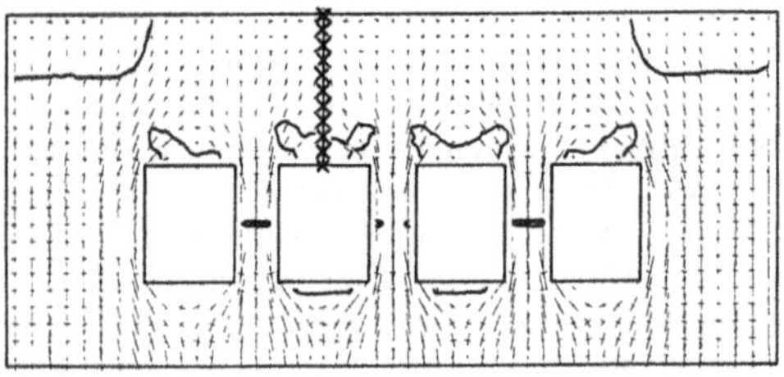

c) Model 3

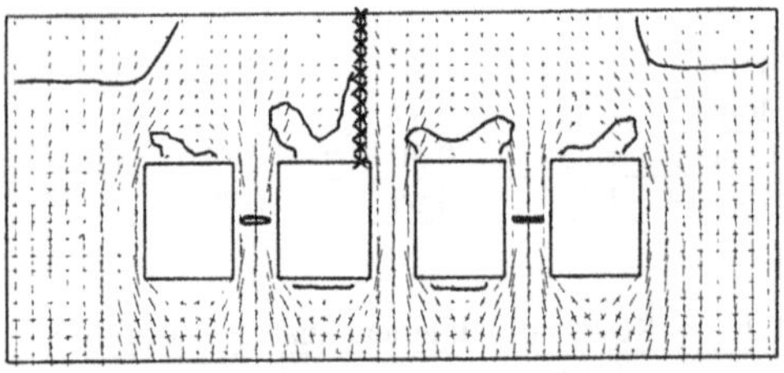

d) Model 4

Legend: xxxx Interface
Tension region contour

Figure 3. Case 1 - Stress distribution for c=5.5 m.

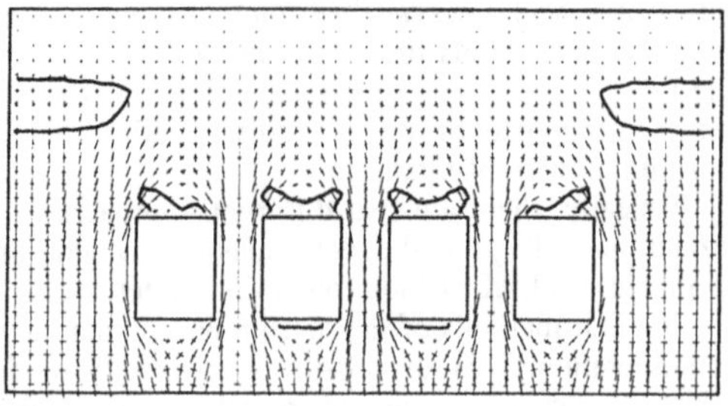

a) Model 1

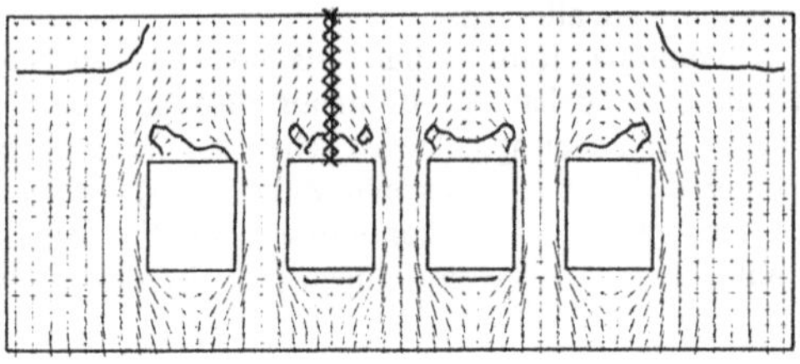

b) Model 3

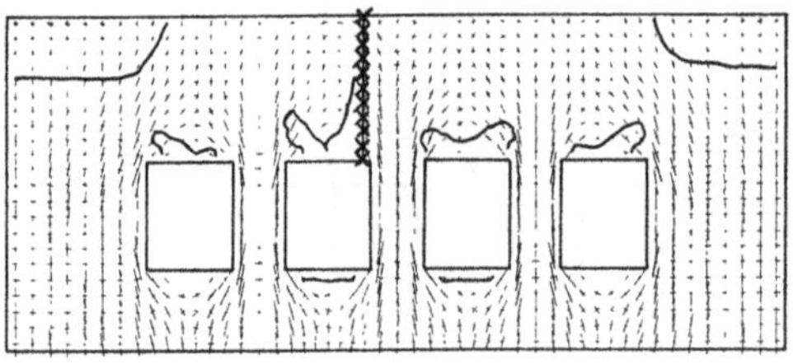

c) Model 4

Legend: xxxx Interface
Tension region contour

Figure 4. Case 2 – Stress distribution for c=5.5 m

If we are looking at the results presented in the tables 3 or 4 separately we can conclude the following: The magnitude of maximum compressive stresses in pillars is almost equal in all analyses as it was expected, except for the models 4. In these models one of the pillars suffers larger loading because of the position of the vertical joint. Moreover, the vertical displacements are also much greater for these models.

It is evident in all models for both pillar spans that the compressive stresses are bigger for pillar width of w=2.8 m (case 1) in comparison to the pillars of w=3.3 m (case 2). The influence of the pillar span enlargement on the magnitude of compressive stresses is even more pronounced. This could be concluded by the comparison of the results for the same models presented in tables 3 and 4.

In all the models for the case 2 (pillar width of 3.3 m), there were only compressive stresses inside the pillars (Figs 4, 6). But, for the case 1 (pillar width of 2.8 m) the zone of mixed stress state (one principal stress is tensile and the other is compressive) was obtained (Figs 3, 5).

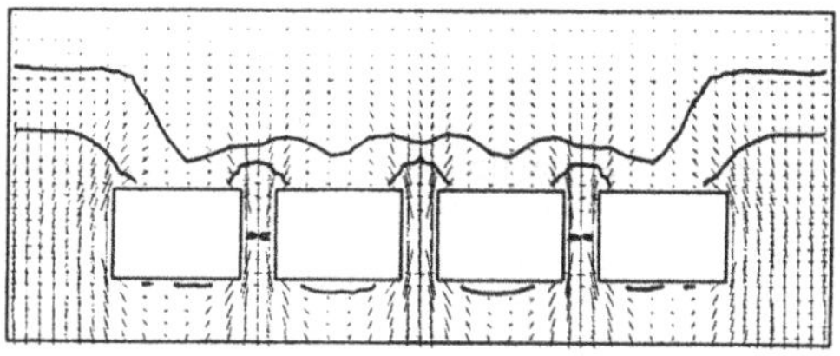

a) Model 1

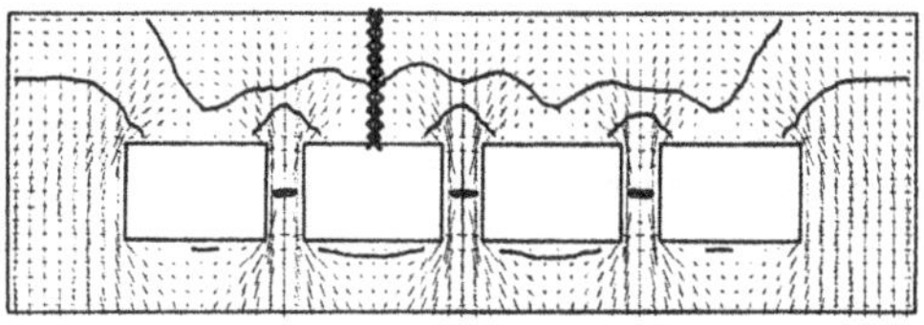

b) Model 3

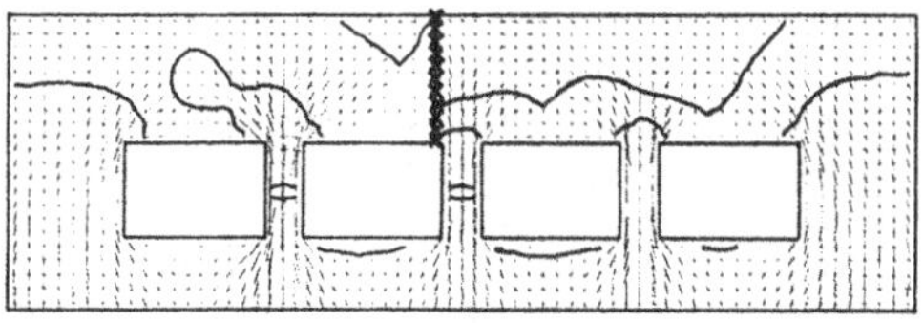

c) Model 4

Legend: xxxx Interface
Tension region contour

Figure 5. Case 1 – Stress distribution for c=10.0 m.

For the cases 1 and 2 with the pillar span c=5.5 m, the zones of mixed stress state were obtained above the opened rooms (Figs 3-4). When the pillar span was enlarged to 10.0 m, the zones of the pure tension were also obtained beside the zones of mixed stress state. The zones of mixed stress state were enlarged in comparison to the same models for the case of c=5.5 m. Moreover, they were connected together for model 4 in case of pillar width of 3.3 m, and for all models in case of pillar width of 2.8 m (Figs 5-6).

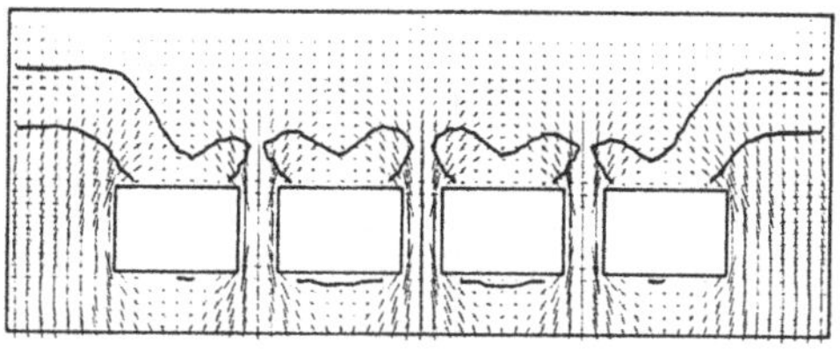

a) Model 1

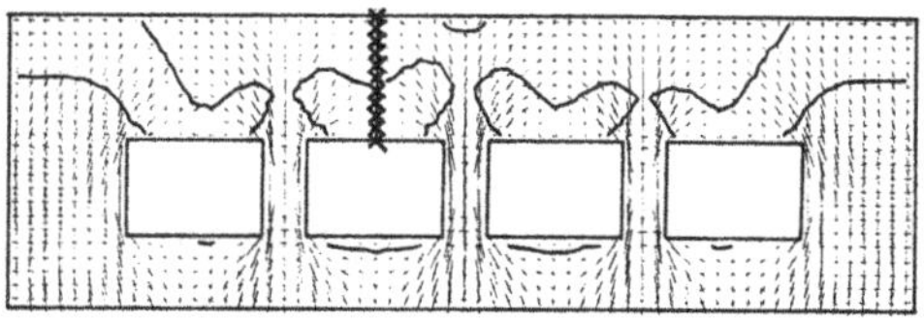

b) Model 3

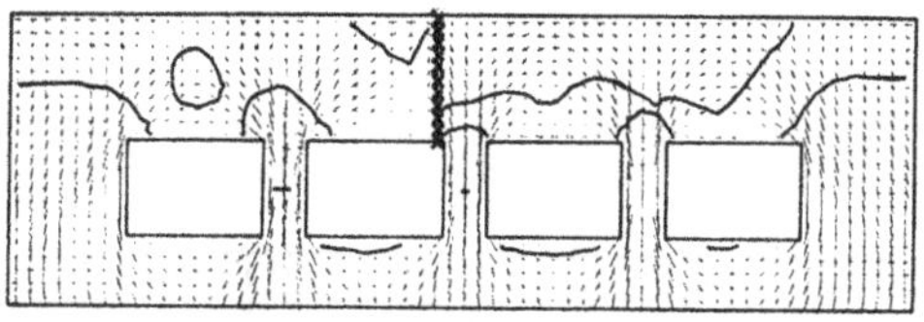

c) Model 4

Legend: xxxx Interface
Tension region contour

Figure 6. Case 2 – Stress distribution for c=10 0 m

However, it should be emphasised that the plasticity indicators were not obtained for the intact rock materials in all models presented in Figures 3-6. In all models with vertical interface elements, the upper roof had to be represented by a constant pressure at the upper boundary. Because of the huge difference in properties of materials M1 and M2 it was not possible otherwise to model interface elements properly.

The influence of the vertical discontinuities, modelled with the usage of interface elements, will be explained hereafter. For the interface element placed in the middle of the room, tension crack was developed. The length of tension crack was approximately 1 m for pillar span of 5.5 m (Fig. 7) and approximately 2 m for pillar span of 10.0 m. It caused only local instabilities and not the general roof failure.

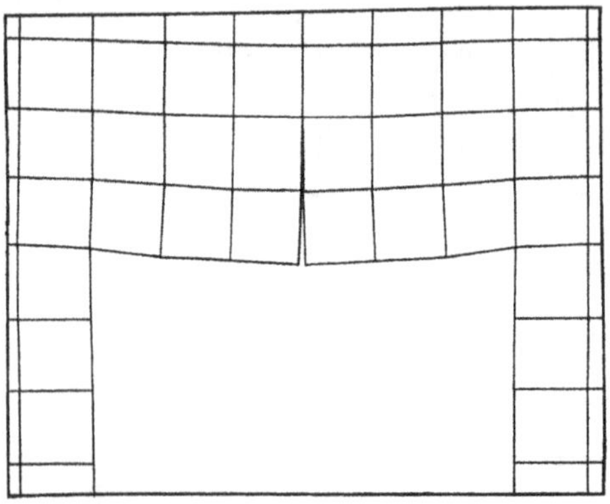

Figure 7. Exaggerated grid distortion for case 1 – model 3.

Interface element near the pillar caused larger vertical displacements (Fig. 8). It did not induce shear failure on interfaces for room width of 5.5 m. But, for the room width of 10.0 m, shear failure and slippage along discontinuity could occur. It could cause a general roof failure. Therefore, the crossings should be designed with special attention and in conformity with measured joint distribution.

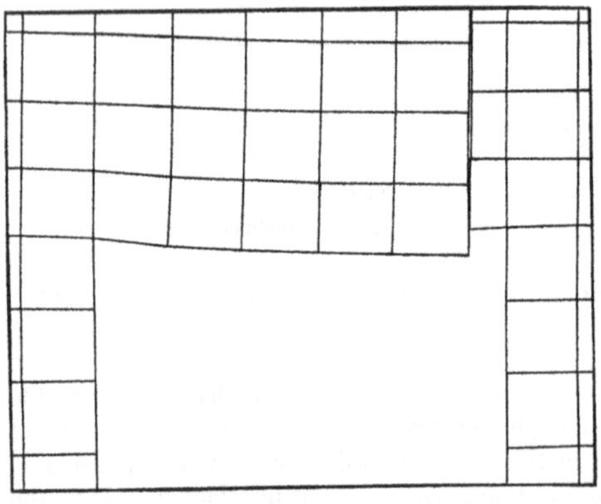

Figure 8. Exaggerated grid distortion for case 1 – model 4.

5 CONCLUSIONS

In the light of previously presented results, the following parameters were chosen for the trial underground exploitation of dimension stone in the quarry Kanfanar. The minimum area of pillar cross section in sound rock must be 16 m^2, minimum pillar width 3.3 m, and maximum room width 5.5 m. It was decided to start the exploitation with the larger pillar widths (case 2), because of the occurrence of the pure tension zones in the roof and the zones of the mixed stress state inside the pillars for the case 1 of pillar dimensions. The pillar dimensions should be adapted to satisfy the condition of minimum pillar area, if they could not be created in sound rock as it is designed because of unfavourable distribution of discontinuities.

It should be mentioned here that all the results were obtained in two-dimensional analyses. In reality, the stability problem of room and pillar method is three-dimensional. Due to the lack of measurements in this phase of the pillar design, presented models are in this respect more qualitative and should be interpreted as such. Adequate in-situ measurements and monitoring is necessary during the trial phase. Special attention should be given to the permanent observation of the distribution of discontinuities. As soon as mining development takes place, refinement of the design should be undertaken. Hopefully, it will result as a reduction of pillar loss. Further research is expected.

REFERENCES:

Bandis, S.C., Lumsden, A.C. & N.R. Barton 1983. Fundamentals of Rock Joint Deformation. *Int. J. Min. Sci. & Geomech. Abstr.* 20(6): 249-268.

Cotman, I. 1995. Stone production in Croatia. *Stone World.* July/95: 50-56.

Crnković, B. & D. Jovičić 1993. Dimension Stone Deposits in Croatia. *The Mining-Geological-Petroleum Engineering Bulletin.* 5: 139-163.

Fornaro, M. & L. Bosticco 1995. Underground Stone Quarrying in Italy-Its Origins, The Present Day and Prospects (Part 4). *Marmomacchine International.* 9. 64-87.

Itasca Consulting Group, Inc. 1991. FLAC Users Manual – Version 3.0. Minneapolis, Minnesota.

Kovačević-Zelić, B. 1994. *Numerical modelling of the rock material behaviour in mining engineering.* M. Sc. Thesis, University of Zagreb. (in Croatian)

Pelizza, S., Mancini, R., Fornaro, M., Peila, D. Cardu, M. & Bosticco, L. 1994. Design Criteria to transfer Underground Ornamental Stone Quarries. *Proceedings of the XVI. World mining congress, Sofia, 12-16 Sept. 1994*: 425-434.

Mechanics of Jointed and Faulted Rock, Rossmanith (ed.) © 1998 Taylor & Francis, ISBN 90 5410 955 6

Load approach on circular tunnels in jointed rock

Ch. N. Marangos
Department of Civil Engineering, University of Thessaloniki, Greece

ABSTRACT: In jointed rock characterised by major joint sets (MJS), the loads acting on the supporting media depend on the joint orientation. For different dips of MJS whose strike is parallel to the tunnel axis, for different initial stress state conditions and for different joint shear parameters, this paper presents series of graphs for the dimensioning of the supporting media.

1 INTRODUCTION

The design of underground opening supporting works is based on the approach of the unstable areas which will be developed around the excavation. In jointed rock the joints have a definite influence on the development of these areas. The shear strength along the joints is considerably smaller than the strength of the intact rock (material which the rock mass is consisted of). Therefore, the joint orientation, their spacing and their shear strength parameters are the sizes which together with the initial stress state (stress state before excavation) and the excavation geometry will determine the position and the extent of the unstable areas.

Taking into consideration the above parameters, the paper is aiming, based on an analysis method of the writer to approach the unstable areas for specific cases of application. Totally one hundred forty cases are examined. For each case the unstable areas are given under the form of graphs aiming to help the engineer to approach the loads acting on the temporary supporting media and to plan the necessary supporting works.

Based on the MJS dip, on its strength parameters and on the value of k factor, the graphs are classified into tables. In the tables, MJS of different dips with angle of friction values according to empirical classification systems and initial stress state conditions characterised by $k \leq 5$ are included.

2 PRINCIPLES OF THE ANALYSIS METHOD

The opening of an underground excavation results in a strong stress redistribution in the narrow area surrounding the excavation; the stress state becomes worse. In hard rock, the problem of setting boundaries to the unstable areas is focused on the determination of the areas in which the shear stress exceeds the joint shear strength.

In Figure 1 the principles of the method are presented. Figure 1a illustrates the static system: A vertical disc located transversally to the tunnel axis is loaded by the geostatic stresses σ_v, σ_h, σ_n, under plane strain conditions. σ_v, $\sigma_h = k\sigma_v$ are the maximum principal stress and the minimum principal stress respectively; σ_n is the intermediate principal stress. In Figure 1b, the applied failure criteria are given: the primary and the secondary stress field (circles 1 and 2 respectively; these circles are referred to the points in the middle of the tunnel walls for $k=0.5$) and the straight lines I and II characterising the shear strength of the intact rock and the joints shear strength of MJS respectively. The only difference between the two diagrams of Figure 1b is the MJS dip β; due to the different dip β, $R<1$ on the left diagram and $R>1$ on the right diagram.

The approach of the unstable areas is carried out as follows: A polar grid (Fig. 1a) is placed on the disc; the grid covers the area which is laying between the tunnel walls (circle 3) and the circle with radius equal to four till six times the tunnel radius (circle 4); this area is the area in which stress redistribution is appeared.

Every MJS is examined separately.

Applying the equations of elastic stresses, the joint shear strength $t=\sigma \cdot tg\varphi$ and the shear stress τ acting on the joint planes of the examined MJS are determined on every grid point. That is at first is assumed that a joint of the examined MJS passes from every grid point. The ratio t/τ determines the partial safety factor, $R=t/\tau$. On grid points where $R \leq 1$, the shear strength will be exceeded. These

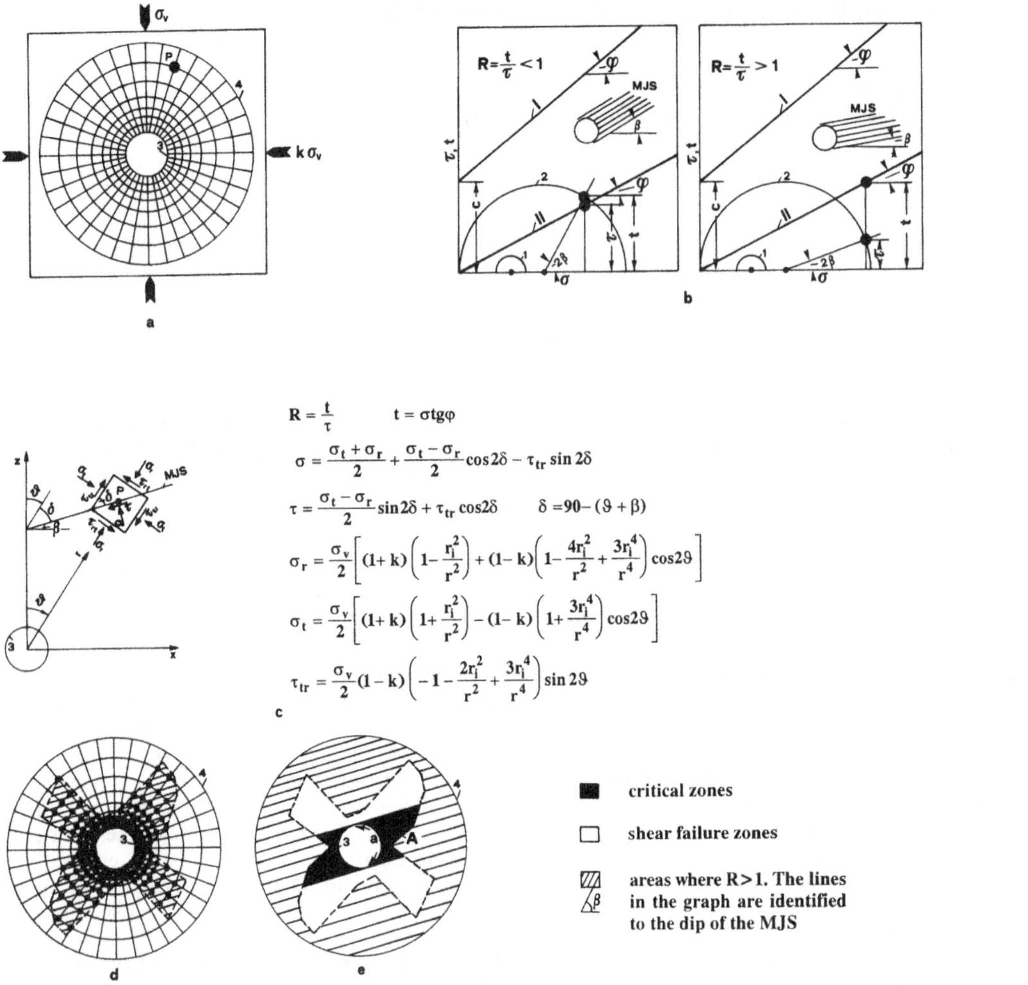

σ_v
P
4
3
$k\sigma_v$
a
$R=\frac{t}{\tau}<1$
$R=\frac{t}{\tau}>1$
MJS
τ, t
b
MJS
P
c
$R = \frac{t}{\tau} \qquad t = \sigma \, tg\varphi$
$\sigma = \frac{\sigma_t + \sigma_r}{2} + \frac{\sigma_t - \sigma_r}{2}\cos 2\delta - \tau_{tr}\sin 2\delta$
$\tau = \frac{\sigma_t - \sigma_r}{2}\sin 2\delta + \tau_{tr}\cos 2\delta \qquad \delta = 90 - (\vartheta + \beta)$
$\sigma_r = \frac{\sigma_v}{2}\left[(1+k)\left(1-\frac{r_i^2}{r^2}\right) + (1-k)\left(1-\frac{4r_i^2}{r^2}+\frac{3r_i^4}{r^4}\right)\cos 2\vartheta\right]$
$\sigma_t = \frac{\sigma_v}{2}\left[(1+k)\left(1+\frac{r_i^2}{r^2}\right) - (1-k)\left(1+\frac{3r_i^4}{r^4}\right)\cos 2\vartheta\right]$
$\tau_{tr} = \frac{\sigma_v}{2}(1-k)\left(-1-\frac{2r_i^2}{r^2}+\frac{3r_i^4}{r^4}\right)\sin 2\vartheta$
d
e
A
critical zones
shear failure zones
areas where R>1. The lines in the graph are identified to the dip of the MJS

Figure 1. Load approach on circular tunnels characterised by one or more continuous MJS whose strike is parallel to the tunnel axis. Principles of the analysis method. a) Static system: vertical disc loaded by the geostatic stresses; polar grid situated on the tunnel surrounding area. 3= The tunnel section. 4= The external limit of the examined surrounding area. b) Failure criteria; left: R<1, right: R>1. c) Equations for the determination of partial safety factors R on a grid point P. d) Partial safety factors R and boundary setting of the failure zones, •=R≤1. e) Boundary setting of the critical zones and approach of p_i acting on the part of the supporting media having width l (case: the unstable area is equal to the critical zone).

points are characterised by small black circles in Figure 1d. Consequently, based on the values of R, it is possible to set the boundaries of zones where the joints will be activated. In Figure 1d the setting of boundaries of these zones (failure zones, R≤1) is illustrated schematically.

Stability problems will appear only in the parts of these failure zones where rock body slidings are possible. These critical zones (characterised in black in Figure 1e) are determined by drawing from the excavation boundaries, the two tangential lines which are parallel to the dip line of the joints (Fig. 1e). The stability of the rock mass included in these critical zones will depend on joint set spacing and on the presence, in these zones, of micro-joints with different orientation or of joints of another set which therefore will permit the sliding towards the excavation of the whole or part of the critical zones. The parts of these critical zones will consist the unstable areas. If more than one MJS exist, each MJS must be examined separately; in this case the critical zones are the sum of the critical zones of all MJS.

Table 1. Critical zones of MJS having dip β=0°.

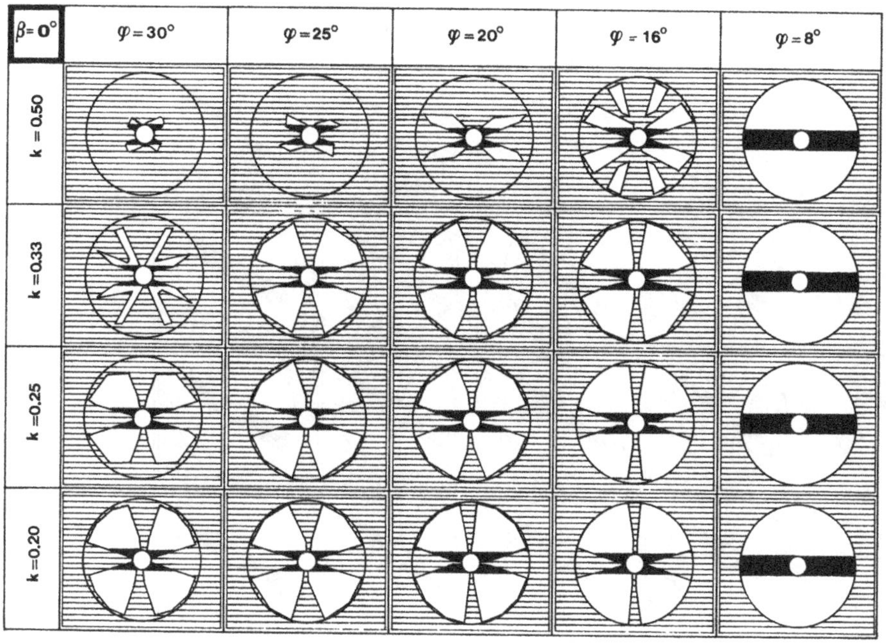

Table 2. Critical zones of MJS having dip β=15°.

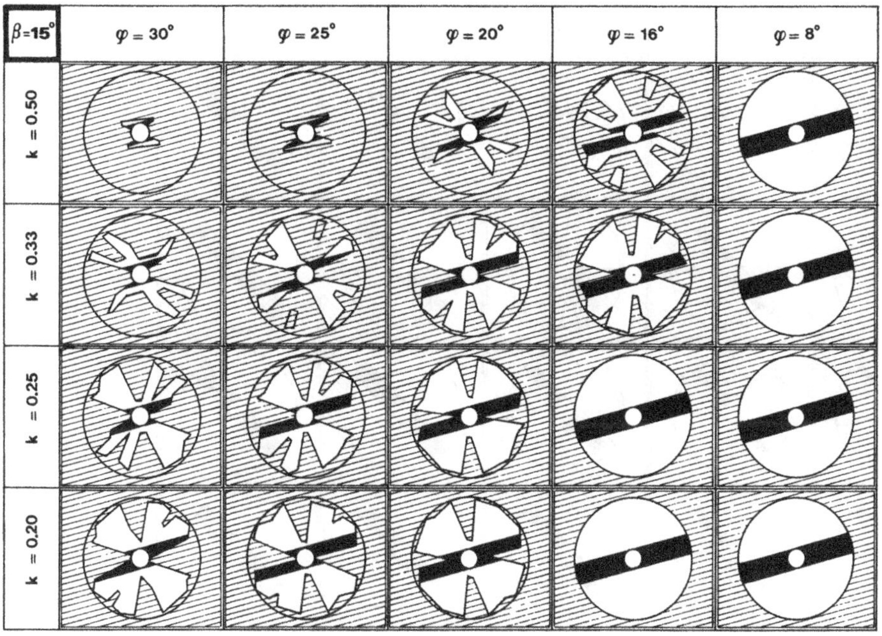

Table 3. Critical zones of MJS having dip $\beta = 30^{\circ}$.

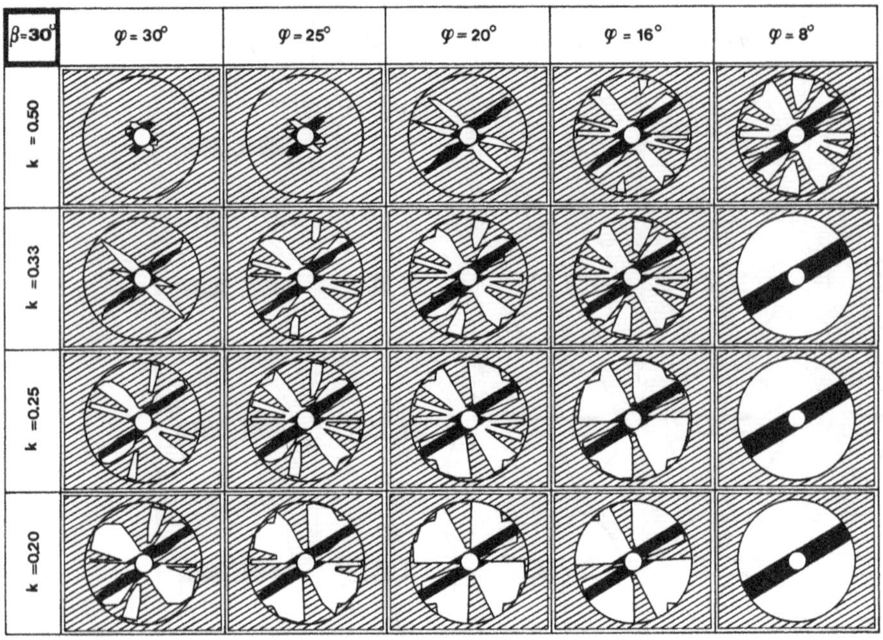

Table 4. Critical zones of MJS having dip $\beta = 45^{\circ}$.

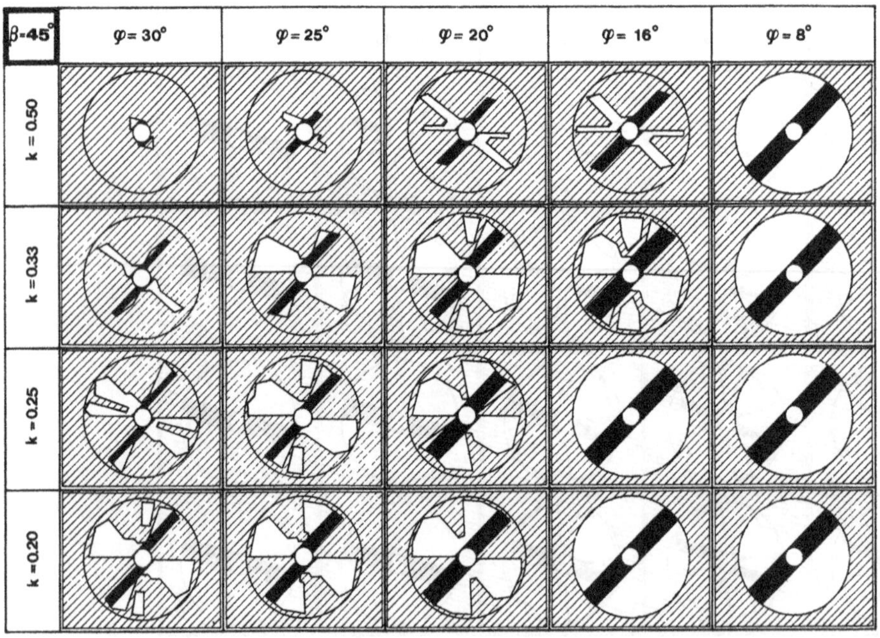

Table 5. Critical zones of MJS having dip β=60°.

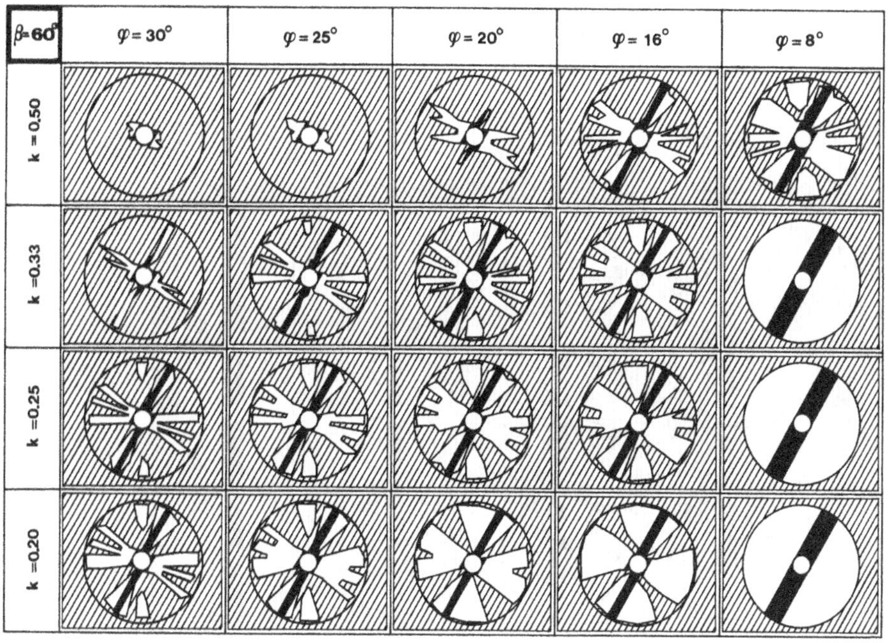

Table 6. Critical zones of MJS having dip β=75°.

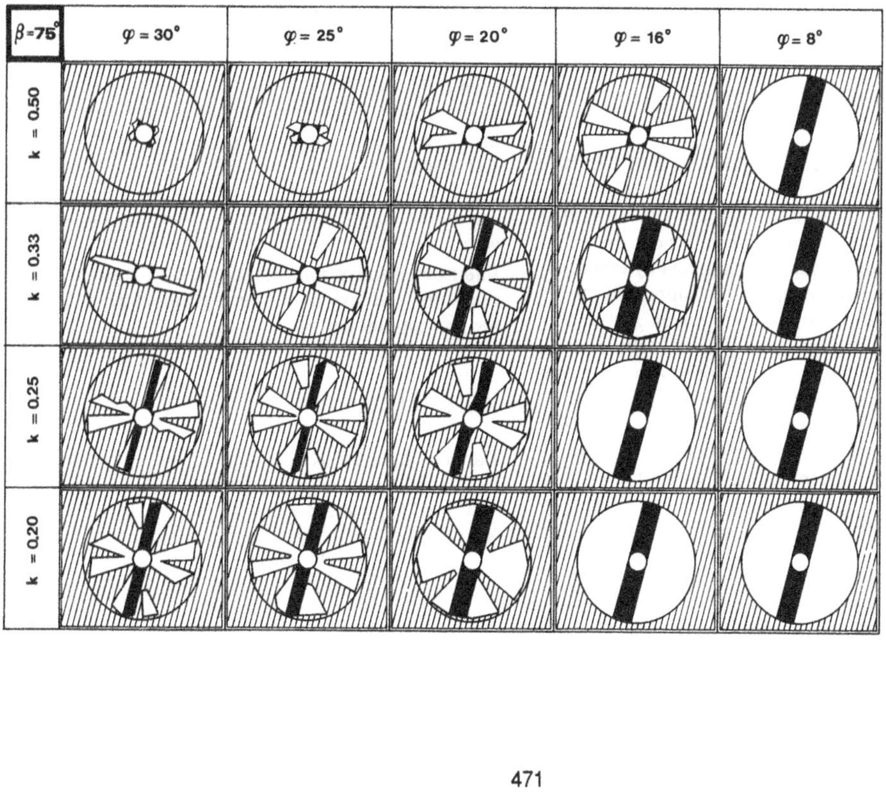

Table 7. Critical zones of MJS having dip β=90°.

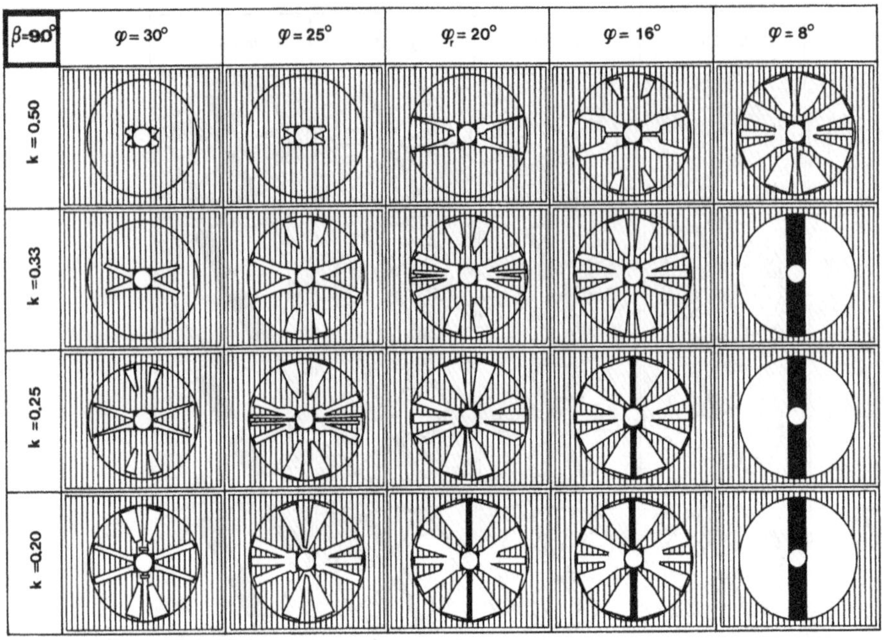

The approach of the loads acting on the temporary supporting media can be carried out as follows: For parts of supporting media which are in contact to the unstable areas whose self-weight forwards their sliding towards the excavation, the following equation can be applied:

$$p_1 = \frac{A\gamma(\sin\beta - \cos\beta\, tg\varphi)}{a} \quad [MPa]$$

p_1= the pressure acting on these parts of the supporting media, A= the unstable area in m^2, γ= the rock unit weight in MN/m^3, a= the unstable area width in m, measured on tunnel walls (Fig. 1e).

The dimensioning of supporting media by this equation presupposes flexible temporary supporting media able to permit failure zone development. Notice that the failure zones are referred to limit static equilibrium conditions and that in the failure zone determination, their self-weight was ignored.

Critical zones whose self-weight contradicts their sliding towards the excavation do not need to be supported; when rock intrusion towards the excavation is observed, the pressure can be determined by Kastner method.

3 TABLE REFERENCES

In the paper, seven Tables concerning cases of MJS with dips $\beta=0^o, 15^o, 30^o, 45^o, 60^o, 75^o, 90^o$ are illustrated. Each Table is referred to a specific dip and includes cases when the angle of friction φ_r of the joints assumes the values $\varphi=30^o, 25^o, 20^o, 16^o, 8^o$ and the k factor the values k=0.5, 0.33, 0.25, 0.20.

For drawing the Tables, a computer program is used. The grid is chosen so that the surrounding area to be investigated till a distance r six times the tunnel radius r_1. The grid consists of polar radii situated every 2° and of homocentric circles of variable step equal to $0.2\ r/r_1$.

The Tables are applied to any tunnel radius and to any tunnel depth. However they can not be applied when the joints are not continuous or when they are filled by cohesive materials ($c\neq0$), because in these cases the extension of the critical zones depends also on tunnel depth.

REFERENCES

Barton N., Lien R., Lunde J. 1974. Engineering Classification of Rock Masses for the design of tunnel support. *Rock Mechanics*. 6. 189-236.

Kastner H. 1971. Statik des Tunnel- und Stollenbaues. *Springer Verlag*. Berlin.

Marangos Ch. 1995. Stability analysis of constructions in jointed rock mass. *Proceedings of the second international Conference on the mechanics of jointed and faulted rock-MJFR-2.* Vienna, Austria.

Mechanics of Jointed and Faulted Rock, Rossmanith (ed.) © 1998 Taylor & Francis, ISBN 90 5410 955 6

Shear creep of discontinuities in hard rock surrounding deep excavations

D.F. Malan, K. Drescher & U.W. Vogler
Division of Mining Technology, CSIR, Johannesburg, South Africa

ABSTRACT: Time-dependent deformations around deep level mines in hard rock occur at rates well in excess of those expected from creep of the solid rock mass. It appears that the time-dependent deformation is controlled by the rheology of the fracture zone. This paper presents the results of a laboratory investigation of the shear creep of discontinuities. It was found that discontinuities with gouge infilling undergo noticeable shear creep when subjected to stress below the shear strength. The steady-state creep rate depends on the shear stress/shear strength ratio, the absolute stress magnitudes and gouge thickness. In comparison, the creep rate of mining induced extensile fractures in hard rock is negligible. Finally, a rheological model to simulate the behaviour of the discontinuities with infilling is proposed.

1 INTRODUCTION

Closure data from the tabular excavations of the deep South African gold mines illustrates significant time-dependent behaviour (Malan 1997, Malan et al. 1997). Laboratory tests indicate that the creep rate of the intact hard rock is low and cannot explain the observed time-dependent deformation. These deformations are the result of the rheology of the fracture zone surrounding these excavations and the time-dependent extension of this zone following a mining increment. Figure 1 illustrates a conceptual diagram of the fracture zone surrounding one of these deep level tabular excavations. Little is known about the role of discontinuity creep in the behaviour of this zone. Legge (1984) measured large displacements on Type II fractures, but it is not clear if these displacements are instantaneous or creep-like. Type I fractures also show some evidence of shearing as the face approaches them. To investigate the creep behaviour of these discontinuities, a laboratory testing program was initiated.

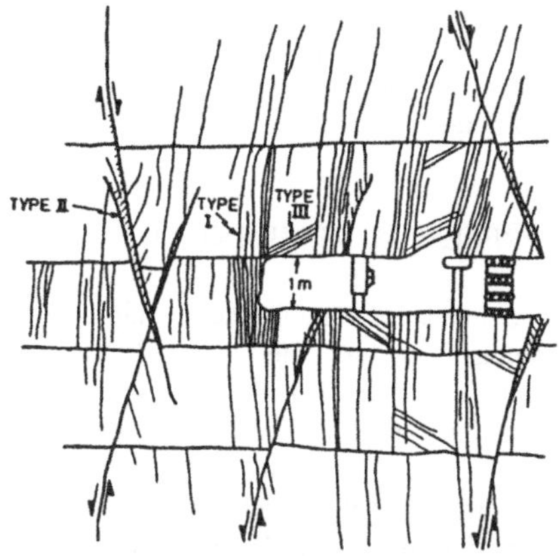

Figure 1. Typical fracture patterns surrounding the tabular excavations in the South African gold mining industry (After Adams et al. 1981). Type I fractures are extensile in nature and form some distance ahead of the face. Type II fractures are mining induced faults which dip away from the stope. Type III are low angle fractures which form close to the face. Also prominent are the bedding planes parallel to plane of the reef.

In the past a number of studies of time-dependent discontinuity behaviour have been related to the investigation of earthquake mechanisms, focusing on time-dependent friction (Dieterich 1978). Other studies (Crawford & Curran, 1981) investigated the rate-dependent behaviour of rock discontinuities to displacement controlled boundary conditions. These tests gave no information about the discontinuity creep behaviour under constant stress conditions. Some workers (Wawersik 1974, Solberg et al. 1978) conducted triaxial creep tests on cylindrical samples of Westerly granite that included discontinuities inclined to the direction of greatest compression.

The axial strain versus time data shows typical creep behaviour with creep rates larger than that of intact rock. If the differential stress was high enough, accelerated creep was observed, leading to sudden violent slip. Schwartz & Kolluru (1984) investigated the influence of stress levels on the creep of unfilled rock discontinuities. Uniaxial creep tests were conducted on samples of gypsum plaster with discontinuities inclined at different angles. This was done to test Amadei's (1979) hypothesis that the ratio of the applied shear stress to the peak shear strength was the critical factor governing discontinuity creep. Schwartz & Kolluru (1984) concluded that the creep appears to depend both on the applied shear stress to shear strength ratio and the absolute shear and normal stress levels across the discontinuity. Bowden & Curran (1984) investigated the creep behaviour of artificially prepared discontinuities in shale. As a direct shear machine was used, it was possible to adjust the shear and normal loads independently. The authors found that the shear creep rate is a non-linear function of the applied shear stress to shear strength ratio. The creep rate is small for small shear stress ratios and very large for large shear stress ratios. For ratios larger than 0.9, the time-dependent displacements are significant and may amount to centimetres in a matter of hours.

2 EXPERIMENTAL METHODOLOGY

The development of suitable shear equipment and a methodology to test the creep behaviour of discontinuities in hard rock is described in an accompanying paper (Vogler et al. 1998). After the appropriate normal stress was applied to the sample, the shear stress was increased in a stepwise fashion with a period of 24 or 48 hours between load increases. This was continued until the shear strength of the discontinuity was exceeded. The pneumatic system ensured a constant air pressure, therefore maintaining a constant shear stress between load increases. The shear stress was increased manually by carefully adjusting a regulator valve to obtain the correct air pressure for the next shear stress level. The duration of this process typically lasted between 60 and 180 seconds. Due to the gradual nature of the shear stress increase, the shear strength could be determined by noting the shear stress when the sample suddenly started to slip in an uncontrolled fashion.

3 DISCONTINUITIES WITH GOUGE INFILLING

Initial tests were conducted on discontinuities containing gouge infilling as these were expected to have a high creep rate and play a prominent role underground. In the underground stopes, Type II fractures and the bedding planes (Figure 1) can contain thick gouge infilling. Artificial gouge was used for the initial tests, as it could be manufactured in sufficient quantities and its properties (such as particle size distribution) were consistent in the different tests. Further tests were conducted by using natural gouge material obtained from underground.

The artificial gouge was obtained by crushing quartzite from Western Deep Levels mine. The uniaxial compressive strength of the intact rock was 237 MPa. From a petrographic analysis, the composition of the rock was found to be 67 % quartz, 32 % muscovite, chlorite and amorphous silica and 1 % pyrite. The broken material was sieved to obtain particle sizes less than 500 μm. The size distribution of the particles is given in Figure 2. The same quartzite was used to prepare two blocks containing a sawcut discontinuity for mounting in the shear box. The sawcut was ground flat to prepare a surface that appeared smooth to the naked eye. The size of the shear area was 5724 mm^2. A typical test would involve mounting the lower quartzite block in the shear box, adding gouge uniformly on the discontinuity to a particular thickness and then adding the top half of the sample. Due to compaction, the thickness of the gouge is reduced slightly when the normal load is applied.

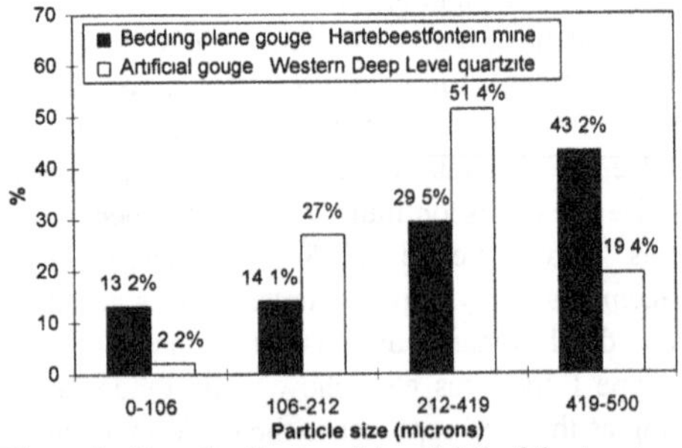

Figure 2. Size distribution (% weight) of the two types of gouge material tested.

The shear creep behaviour of the discontinuity with artificial gouge is illustrated in Figure 3. The last creep phase (for a shear stress of 0.517 MPa) is not plotted on this graph as the small displacement transducers (See Vogler et al. 1998) were out of range. The sample eventually failed at a shear stress of 0.533 MPa giving a friction angle of 28°. Of interest is that a cumulative shear creep displacement of 1.8 mm was recorded for this sample before failure. The normal displacement for this test is given in Figure 4. A convention of dilation being positive and compaction being negative is assumed. The negative displacement in Figure 4 therefore indicates a reduction in the width of the gouge layer. The initial reduction is significant compared to the subsequent steps as this is the compaction due to the application of the normal load. It appears that a further increase in the shear stress leads to further compaction even though the normal stress remains constant.

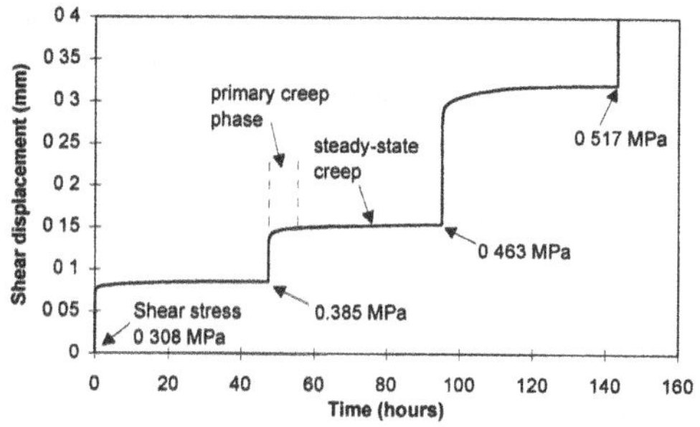

Figure 3. Typical shear creep behaviour of the sawcut discontinuity with gouge infilling and a stepwise increase in shear stress. The normal stress remained constant at 1 MPa. The gouge thickness was 2 mm and the ambient humidity was controlled at 50 %.

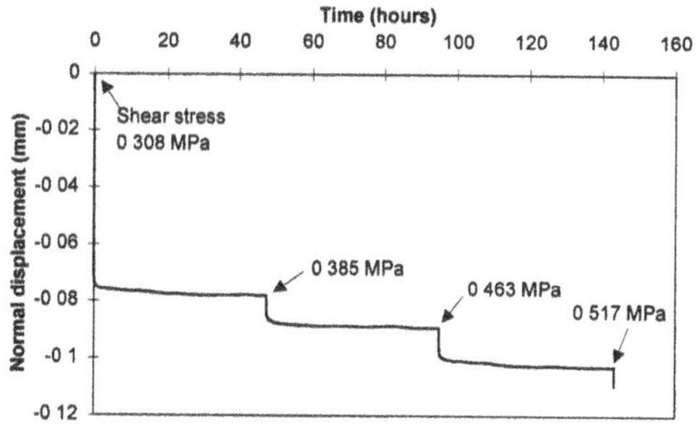

Figure 4. Normal displacement accompanying the shear displacement in Figure 3.

To investigate the effect of the magnitude of the shear and normal stress, different tests were conducted using a normal stress of 0.5 MPa, 1.0 MPa and 1.5 MPa. These tests were also conducted by increasing the shear load in a stepwise fashion and therefore the effect of the shear stress to shear strength (τ/τ_S) ratio could be investigated.

Figure 5 illustrates the incremental shear displacement for the different shear load increments and a normal stress of 0.5 MPa. Note that the magnitude of the instantaneous response is very prominent and increases with the (τ/τ_S) ratio. The steady-state creep rate also increases with the shear stress magnitude. This is illustrated in Figure 6.

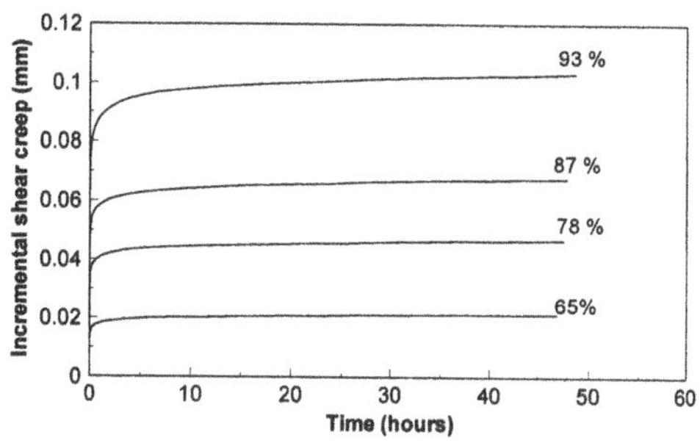

Figure 5. The effect of shear stress magnitude on the creep behaviour. The percentages indicate the magnitude of the shear stress relative to the shear strength (0.276 MPa). The normal stress was 0.5 MPa and the gouge thickness 2 mm.

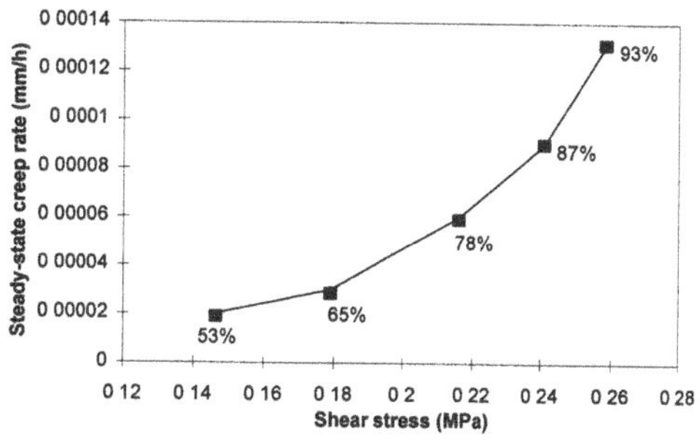

Figure 6. The effect of shear stress magnitude on the steady-state creep rate. The percentages indicate the magnitude of the shear stress relative to the shear strength (0.276 MPa). The normal stress was 0.5 MPa and the gouge thickness 2 mm.

For the tests with normal loads of 1 MPa and 1.5 MPa, behaviour similar to that illustrated in Figures 5 and 6 was observed (Malan 1997). It was, however, noted that the magnitudes of primary phase and steady-state creep rate is not only a function of the (τ/τ_S) ratio, but also of the absolute value of shear and normal stress. For a 1.5 MPa normal load, the steady-state creep rate at a (τ/τ_S) ratio of 0.93 was more than 3 times larger than that of the 0.5 MPa normal load test. This phenomenon has also been observed by Schwartz & Kolluru (1984) for unfilled discontinuities.

The effect of gouge thickness was investigated by performing three tests with a thickness of 0.5 mm,

1 mm and 2 mm. The steady-state creep rate as a function of gouge thickness is illustrated in Figure 7. This data illustrates that, apart from the possible surface creep at the interface between the rock surface and the gouge, there is also a volumetric creep effect in the gouge.

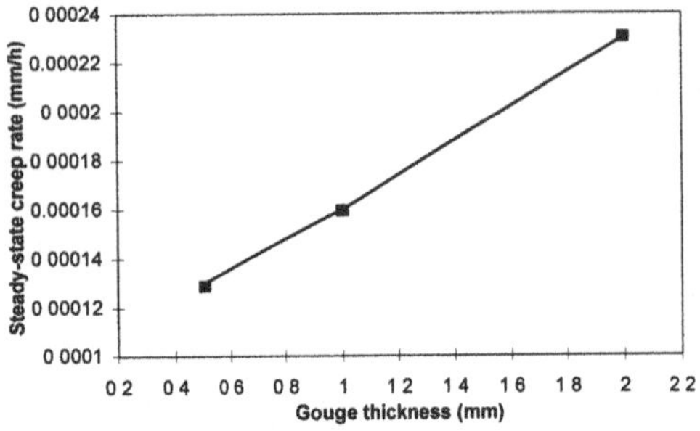

Figure 7. Steady-state creep rate as a function of gouge thickness. For the 1 mm and 2 mm thickness, the (τ/τ_S) ratio was 0.87, while it was 0.86 for the 0.5 mm thickness. The samples were tested at 50 % humidity.

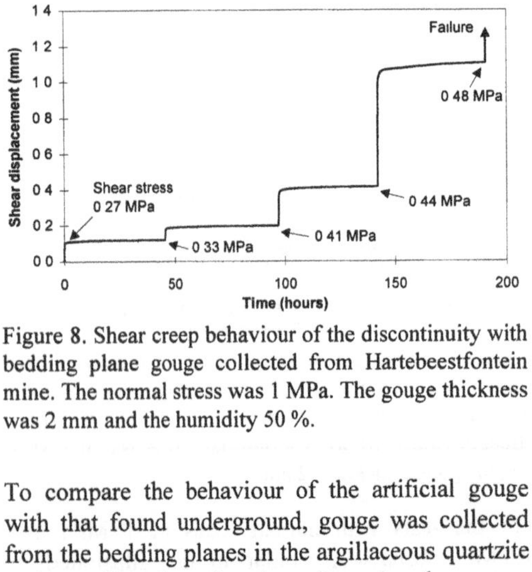

Figure 8. Shear creep behaviour of the discontinuity with bedding plane gouge collected from Hartebeestfontein mine. The normal stress was 1 MPa. The gouge thickness was 2 mm and the humidity 50 %.

To compare the behaviour of the artificial gouge with that found underground, gouge was collected from the bedding planes in the argillaceous quartzite rock at Hartebeestfontein mine. As the gouge contained some large particles which could not be accommodated in the small scale of the experiment, it was sieved to obtain particle sizes less than 500 μm. The size distribution of the remaining material is given in Figure 2. Compared to the artificial material, there is an abundance of fine material with a size of less than 106 μm. In spite of this difference, the experimental behaviour of the natural gouge was generally similar to the results presented above. Figure 8 illustrates the shear creep for a stepwise increase in shear stress.

For the creep data of the artificial and natural gouge presented above, it is clear that the steady-state shear creep rate depends on the applied (τ/τ_S) ratio. This was also found by Amadei (1979) and Schwartz & Kolluru (1984) for other discontinuity types. Malan (1997) proposed a power law to simulate this behaviour namely

$$\dot{D}_{SS} = A\left(\frac{\tau}{\tau_S}\right)^n \tag{1}$$

where $\dot{D}_{SS}$ is the steady-state creep rate, τ is the applied shear stress and τ_S is the shear strength of the discontinuity. From the data above it follows that the parameter A is not a constant, but depends on parameters like gouge thickness and absolute magnitudes of normal and shear stress. The coefficient n and parameter A can be determined from a $\log(\dot{D}_{SS})$ versus $\log(\tau/\tau_S)$ plot. This plot for the data in Figure 6 is illustrated in Figure 9. The linear nature of the curve illustrates the applicability of the model. The parameters for the model in equation (1) were also calculated for the other tests with the higher values of normal stress. The calibrated parameters are given in Table 1.

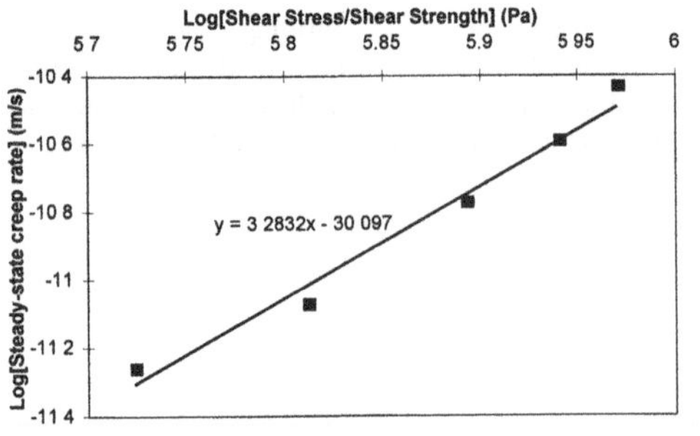

Figure 9. Log (steady-state creep rate) versus Log (τ/τ_S) for the data in Figure 6. The normal stress was 0.5 MPa and the gouge thickness 2 mm. The line was fitted using linear regression.

Table 1. Calibrated values for the power law in equation (1). In all these tests the gouge thickness was 2 mm.

Gouge type	Humidity (%)	Normal stress (MPa)	n	Log(A)
Artificial	50	0.5	3.3	-30.1
Artificial	50	1	2.3*	-23.8*
Artificial	50	1.5	2.9	-27.1
Hartebeest-fontein mine	100	1	2.4*	-24.1*

*These values were calculated from only three data points.

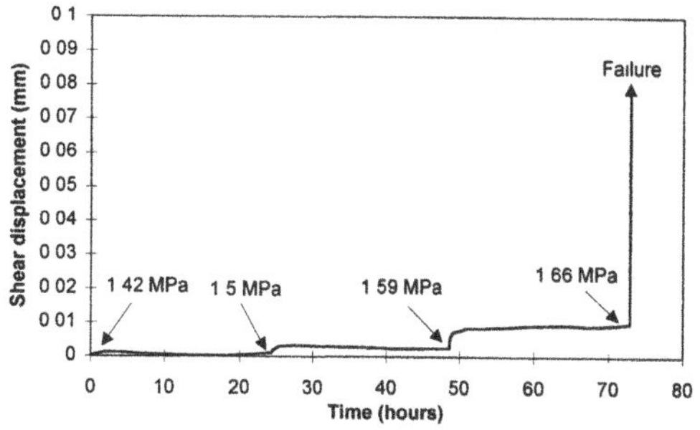

Figure 10. Shear creep behaviour of a mining-induced extensile fracture in lava for a stepwise increase in shear stress. The normal stress was 1 MPa. The sample was tested at the ambient humidity of 50 %.

4 DISCONTINUITIES WITH NO INFILLING

The results of discontinuities with infilling were compared with the creep of mining induced extensile fractures in lava. These fractures belong to the Type I class in Figure 1. The rock is an Alberton Porphyry Formation lava found in the hangingwall of certain stopes at Western Deep Levels mine (Schweitzer & Johnson, 1997). Uniaxial compressive strengths of these lavas can be as high as 436 MPa (Malan 1997). The two opposing surfaces of the fracture were closely matched and were not subjected to any previous shear. The normal load was 1 MPa. Figure 10 illustrates the shear creep behaviour of a typical sample. The sample eventually failed at a shear stress of 1.66 MPa giving an apparent friction angle of 59°. Although some creep behaviour is visible in the last increments before failure, it should be noted that the magnitude of this creep is negligible compared to some of the discontinuities with gouge infilling. From Figure 10 the total creep displacement before failure was 10 μm while for a typical gouge-filled discontinuity (Figure 8), the total creep movement amounted to 1.1 mm. As the creep displacement in Figure 10 was so small, the slight changes in the temperature (±1°C) and the effect of this on the instrumentation affected the recorded data. In the figure the unexpected trends of the creep in the steady-state creep phase of the first two loading stages were a result of these effects.

5 SIMULATING THE CREEP BEHAVIOUR

To enable the simulation of the shear creep behaviour of the discontinuities in numerical codes, Malan (1997) developed a rheological model. This is illustrated in Figure 11. During unloading tests (Malan 1997) it was found that very little of the creep displacement (including the instantaneous response) can be recovered. K_{ds1} and K_{ds2} should therefore only be considered as pseudo-stiffness parameters to simulate the displacement in the forward direction and not for unloading.

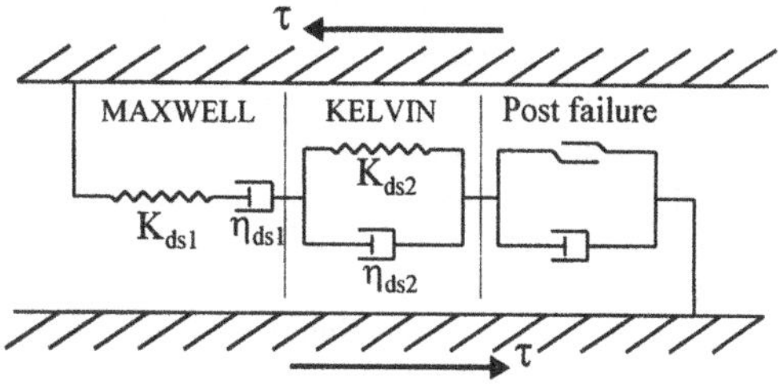

Figure 11. Rheological model to approximate the shear creep of discontinuities with infilling.

Below the yield strength of the discontinuity, the shear displacement can be derived as (Malan 1997)

$$D_{SC} = \tau\left[\frac{1}{K_{ds1}} + \frac{t}{\eta_{ds1}} + \frac{1}{K_{ds2}}\left(1 - e^{\frac{-K_{ds2}t}{\eta_{ds2}}}\right)\right] \tag{2}$$

where K_{ds1} and K_{ds2} are the discontinuity stiffness of the Maxwell and Kelvin components respectively and η_{ds1} and η_{ds2} are interface viscosities. The normal creep D_{NC} is simulated by assuming that

$$D_{NC} = |D_{SC}|\tan\psi \tag{3}$$

where the angle ψ is negative for compaction. This model was successfully fitted to the experimental data sets as illustrated in Figures 12 and 13. The model parameters used were K_{ds1} = 0.675 MPa/mm, K_{ds2} = 7.5 MPa/mm, η_{ds1} = 1100 MPa.h/mm and, η_{ds2} = 14 MPa.h/mm. Note, however, that these model parameters are a function of the (τ/τ_s) ratio, the absolute values of shear and normal stress and other parameters like humidity and gouge thickness. This is further explored by Malan (1997) who used power laws to simulate the stress dependency of the model parameters. The calibrated values given above therefore only apply to the specific experimental parameters given in the caption of Figure 12.

6 CONCLUSIONS

Discontinuities in hard rock with gouge infilling can undergo significant shear creep. This creep behaviour is characterised by a prominent instantaneous response after the shear load increases followed by a primary and steady-state creep phase. The steady-state creep rate increases for increasing shear stress to shear strength ratios and can be modelled with a power law. This creep rate is also affected by the gouge thickness and the absolute values of shear and normal stress. The rheological model discussed in this paper gives a good fit with experimental data and allows for easy implementation in numerical codes. In comparison with gouge-filled discontinuities, the creep behaviour of extensile fractures in hard rock is negligible.

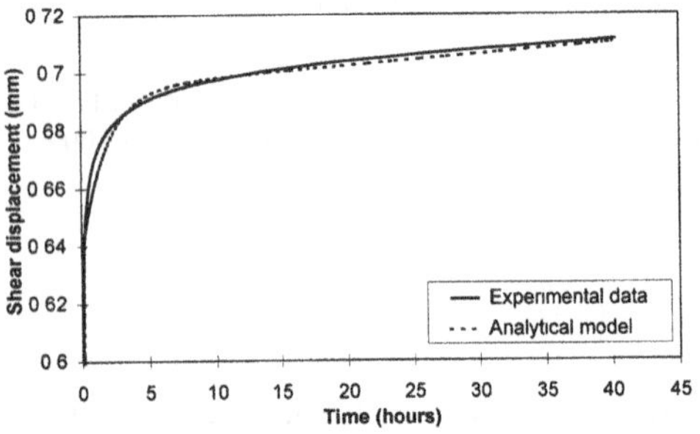

Figure 12. Simulating experimental shear creep using the rheological model in equation (2). The experimental result is for a discontinuity with a gouge thickness (artificial quartzite gouge) of 2 mm subjected to a shear stress of 0.48 MPa (only one loading stage). The normal stress was 1 MPa and the gouge was tested wet.

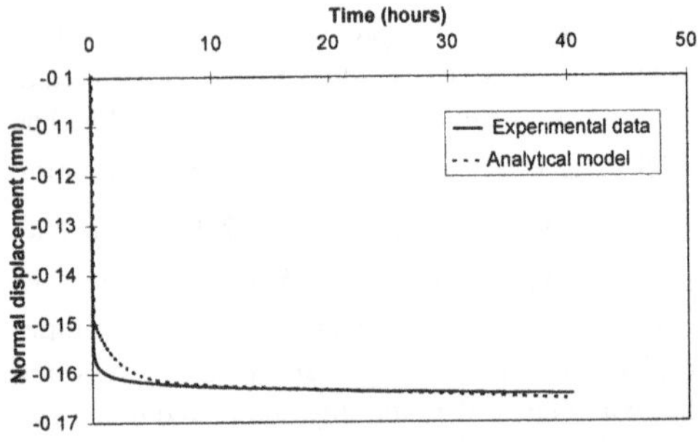

Figure 13. Experimental and simulated compaction (ψ = -13°) accompanying the shear results in Figure 12.

ACKNOWLEDGEMENTS

This work forms part of the research program of Rock Engineering, CSIR Mining Technology and was funded by the Safety in Mines Research Advisory Committee (SIMRAC). D.F. Malan is working towards a Ph.D. degree at the University of the Witwatersrand and the work described here also forms part of that study. The authors would like to thank Dr J. Napier, Dr R. Durrheim and Mr A. Haile for reviewing the manuscript.

REFERENCES

Adams, G.R., Jager, A.J. & Roering, C. 1981. Investigations of rock fracture around deep level gold mine stopes. In: H.H. Einstein (ed.), *Proc. 22nd U.S. Symp. Rock Mech.*: 213-218.

Amadei, B. 1979. Creep behaviour of rock joints. *M.Sc. Thesis*, University of Toronto, Canada.

Bowden, R.K. & Curran, J.H. 1984. Time-dependent behaviour of joints in shale. In: C.H. Dowding & M.M. Singh (eds), *Proc. 25th Symp. Rock Mech.*: 320-327.

Crawford, A.M. & Curran, J.H. 1981. The influence of shear velocity on the frictional resistance of rock discontinuities. *Int. J. Rock Mech Min Sci.* 18:505-515.

Dieterich, J.H. 1978. Time-dependent friction and the mechanics of stick slip. In: J.D. Byerlee & M. Wyss (eds), *Rock friction and earthquake prediction*: 791-806.

Legge, N.B. 1984. Rock deformation in the vicinity of deep gold mine longwall stopes and its relation to fracture. *Ph.D Thesis*, University College, Cardiff.

Malan, D.F. 1997. Identification and modelling of time-dependent behaviour of deep level excavations in hard rock. *Draft PhD thesis*, University of the Witwatersrand, Johannesburg, South Africa.

Malan, D.F., Vogler, U.W. & Drescher, K. 1997. Time-dependent behaviour of hard rock in deep level gold mines. *J. S. Afr. Inst. Min. Metall.* 97:135-147.

Solberg, P.H., Lockner, D.A., Summers, R.S., Weeks, J.D. & Byerlee, J.D. 1978. Experimental fault creep under constant differential stress and high confining pressure. In: Y.S. Kim (ed.), *19th US Symp. Rock Mech.*:118-121.

Schwartz, C.W. & Kolluru, S. 1984. The influence of stress level on the creep of unfilled rock joints. In: C.H. Dowding & M.M. Singh (eds), *Proc. 25th Symp. Rock Mech.*: 333-340.

Schweitzer, J.K. & Johnson, R.A. 1997. Geotechnical classification of deep and ultra-deep Witwatersrand mining areas, South Africa. *Mineralium Deposita* 32: 335-348.

Vogler, U.W., Malan, D.F. & Drescher, K. 1998. Development of shear testing equipment to investigate the creep of discontinuities in hard rock. *Ibid.*

Wawersik, W.R. 1974. Time-dependent behaviour of rock in compression. In: G.B. Wallace (ed.), *Proc. 3rd Congr. Int. Soc. Rock Mech.*: 357-363.

Mechanics of Jointed and Faulted Rock, Rossmanith (ed.) © 1998 Taylor & Francis, ISBN 90 5410 955 6

Improvement of post-pillars behaviour providing its confinement by tailings backfill

M.O.C.H.Costa e Silva
Department of Mining Engineering of I.S.T., Technical University of Lisbon, Portugal

ABSTRACT: This paper presents the experimental work performed in laboratory to provide a better understanding of the improvement of geomechanical behaviour of post-pillars confined by tailings (complex sulphides and marble sawing slimes). The results of the experimental work were analysed according to the Salamon methodology, meaning that mine pillars were assumed to be like rock samples tested in high stiffness loading compressive machines. Some promising conclusions are presented regarding two different aspects, the improvements on peak strength and on the rupture control, being the last one measured by the deformations energies.

1 INTRODUCTION

The environmental protection is one of the major concerns of today's civilisation. Mining industry didn't escape to the environmental policies demands, being forced to minimise the impacts on the environment.

Among these impacts, one of the most important is the one related to the mine tailings superficial deposition. In many cases, the tailing apparent material volume is bigger than the one of the mining itself. Therefore, it sounds logical that, for environmental reasons, the impact created by this tailings deposition should be reduced in a significant way.

The deposition of a large part of the tailings in cavities created by underground mining, seems to be a logical solution for this problem. Nevertheless this procedure can not be performed without a significant mining cost increase.

Regarding this, benefits from the underground deposition of tailings should compensate the mine cost increases. One of these benefits, that can be of various types (reduction of the water pollution problem, reduction of the dusting problem created by the wind on the surface piles of tailings, etc.), is of geomechanical type, resulting on improvements on rock masses global stability and mine works safety. The first referred geomechanical benefit is reflected in the possibility of surface subsidence reduction. It is quite difficult to quantify this kind of benefits, therefore the need of improve research on the matter.

The increase on mine works safety results from the action of backfill as a support factor in the surrounding ground. Nevertheless, the backfilling direct action depends on the reaction that it can offer to the displacements of the rock surfaces limiting the voids. These reaction becomes more significant as the backfill materials compaction increases.

The mine experience shows that the backfill materials direct action as a support element is only valid in peculiar situations, for instance in the mining of vertical or sub-vertical ore deposits, where backfill materials can achieve reasonable consolidations due to its own gravity weight and due to the significant values of horizontal displacements observed in stope walls.

For the majority of the orebodies that gives place to tridimensional type or to horizontal tabular type excavations, the support due to backfill reaction is generally poor as the fill consolidation is low.

The importance in Portugal of the horizontal tabular type of mine excavations justifies investigations on underground backfill benefits, other than the one resulting from simplest tailings deposition. The combination of roof support by pillars with backfill techniques appears with relevant importance in this line of work. As a matter of fact the backfill experience shows that even without tightfill it is possible to obtain good improvements in the pillars behaviour.

This fact becomes even more important when for economical reasons one should go for higher ore recovery and consequently small pillar sections when using room and pillar methods.

Room and pillar, a naturally supported method with abandoned pillars, is in fact a high safety exploitation method, but is not a economic one because the ore recovery ratio depends directly on the pillar area as shows the Tributary Area Theory. In those cases, the ore recovery increase shows economical benefits that can compensate the additional costs of backfill transport and underground deposition.

In mine operations where it is important to analyse the interaction backfill / pillars, the pillars slenderness (height / diameter or width ratio in pillar or sample) are too high to allow the inclusion of the considerations of its resistance in the normal cases of pillars dimensioning studies. Furthermore, in the majority of cases studies of pillars design in underground stoping methods the aim is to ensure with them the global stability of rock masses and not just provide by the pillars a temporary support that gives safe working conditions during mining.

If the global stability is the main goal then the pillar slenderness is always smaller than one; on the contrary, for just a temporary support it is possible to work with slenderness values bigger than one. The pillars with the last characteristics are usually named post-pillars. The behaviour of these pillars is not cited in mining literature in a way that enables a convenient interpretation and the studies of interaction backfill / pillars that are related are not definitively conclusive, so the need of developed experimental work.

2 THE GEOMECHANICAL BEHAVIOUR OF PILLARS CONFINED BY FILL

If the pillars are understood as a temporary roof support excavations, their controlled rupture as the stopes increases its dimensions can be accepted. As a matter of fact the mining backfill experience shows that even without tightfill it is possible to obtain good improvements in the pillars behaviour.

In order to study these improvements, it was realised that it could be interesting to promote a laboratory study of the geomechanical behaviour of post pillars confined by fill that can simulated the pillars that provided a temporary function of roof excavation support.. Therefore the traditional concept of safety factor (ratio of pillar strength by load applied on pillar) has no use here. In order to pursue this aim, the Salamon methodology was adopted and according to this, it can be assumed that mine pillars have a behaviour very similar to the one achieved in laboratory with rock samples tested in high stiffness loading machines. In these situations, the loading system do not transfer its deformation energy to the test specimen, not producing an explosive rupture, and the total curves forces vs. deformations can then be obtained (Fig. 1).

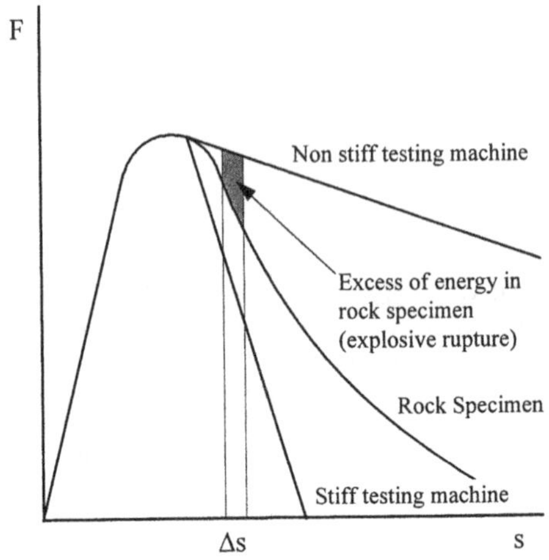

Figure 1. - Total curves forces vs. deformations (PATCHET, 1983).

In this figure, it can be observed that when there is a slow increase deformation Δs there is a variation on deformation energy in loading system and in pillar deformation energy. If the work done by the loading system is minor than the one done on pillar to promote the same deformation Δs, than the stability conditions may be written by this expression

$$\Delta W_p - \Delta W_s > 0 \qquad (1)$$

If the assemblage (stiff loading machine and specimen to be loaded) is represented by one series association of a string and a sample, the load on the string can be find by this equation, $F_s = -R\ \Delta s$ where R is the stiffness constant, and the load on the sample that has the same deformation is $F_p = \Delta s$; where λ is the slope of the curve forces vs deformations. So, the stability condition on rupture is

$$R + \lambda > 0 \qquad (2)$$

Now if the same analogy is used for the pillars in a stope, the slope of the curve, that represents the stiffness characteristics, can be found if the mechanical properties of the roof material and the geometry of the slope are knowed.

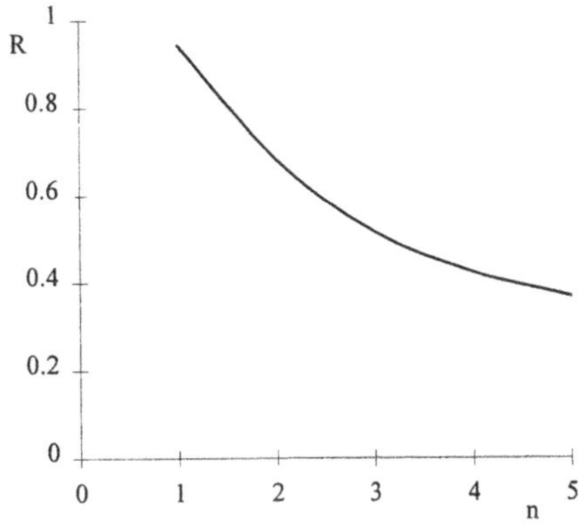

Figure 2. - Geometry effect on stope stiffness. Adapted from SALAMON & ORAVECZ (1976).

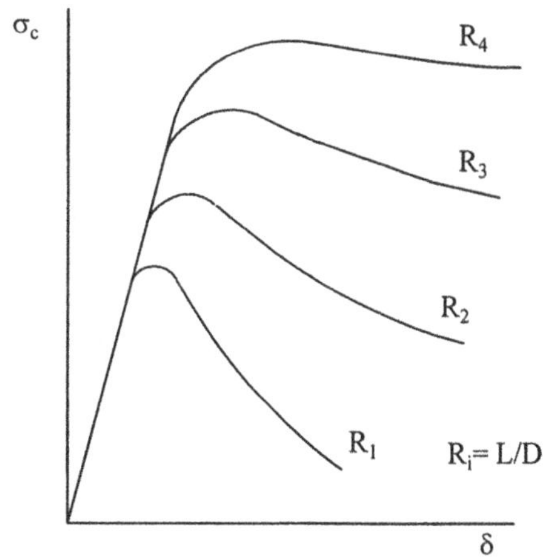

Figure 3. - Pillars slenderness effect on characteristics curves. Adapted from SALAMON & ORAVECZ (1976).

As it can be seen, in Fig. 2 this geometry is very important, as the number of pillars in the stope increase the stiffness decrease.

Notice that, when slenderness decreases (which means that the curves have a smaller slope represented by a more horizontal line), the risk of no controllable ruptures also decreases (Fig. 3).

If it is assumed that the rock masses (material of the roof and floor of the stopes) have an elastic behaviour and it is allowed, in pillars, a different behaviour, not elastic, then these pillars are in a stable equilibrium if there is no additional convergence induced by the increase of works. As matter of fact if the dimension of works increases the stability becomes assured if the deformation

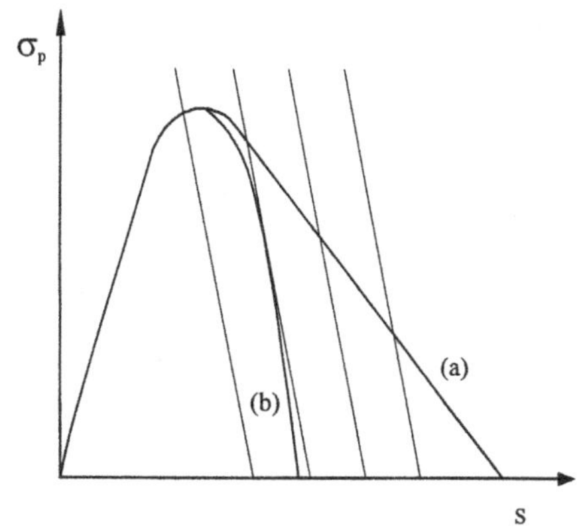

Figure 4. - Pillar behaviour with fill (a) and without (b). Adapted from SALAMON & ORAVECZ (1976).

energy of the rock masses, ΔW_s, during a convergence Δs is inferior than the one necessary, ΔW_p, to compress the pillar of the same Δs. Using the mentioned Salamon methodology this can be translated to the following expression:

$$R+\Lambda > 0 \qquad (3)$$

Where R is the matrix of the stiffness constants of the rock of what the roof and floor are made of and Λ is the matrix of the characteristics of the pillars.

As Salamon the perfect stope stability is assured if $f\,\lambda_m > \lambda_c$, where f is a safety factor, and λ_m is an element of matrix that represents the minimum slope of the curve load / deformation of pillar, and λ_c is the smallest eigenvalue of matrix Λ.

It is possible to increase λ_m, by decreasing the slenderness of pillars, on the other hand λ_c can be decreased, by decreasing the span of the slopes. However these procedures are not in agreement with the need of increase the recovery and so the economy of the method.

So, a logical solution as it has been said in the beginning of this communication may be the improvement of the stability of pillars with high slenderness by confining them with fill.

The more frequent interpretation is the one that explains the improvement in the pillars behaviour, when they are confined by fill, in a similar process as the strengthening effect due to the application of lateral pressions in laboratory triaxial tests. This approach only explains the improvements that sometimes are observed in the peak strengths. However this is an insufficient explication because the increase in peak strengths is not significant in general situations where post pillars offer good contributions for global stability. So, it seems logical to investigate how the pillars confined by fill can contribute for those increases of global stability, more effective in the post peak strength domains.

It is not difficult to understand that if the excess of energy, that can lead to pillars rupture, can be used on the deformation of the backfill that confines the pillars, then its stability increases. This can be seen in Fig. 4. These are the motivations for the experimental work that wishes to contribute to the understanding of the improvement of post pillars behaviour providing its confinement by tailings backfill.

3 EXPERIMENTAL WORK

In order to increase the knowledge on the behaviour of high slenderness (between 2,1 and 5,5) post pillars confined by fill, three rock types models were laboratory tested (two types of marble and one of massive pyrite). As backfill, marble sawing slimes and complex sulphides mine tailings were used (granulometric clay silts). To obtain the models complete load-deformation curves, a load system with a high stiffness was used. Were also used a steel and a copper tubes with known elastics and stiffness characteristics. The tube and the specimen where installed in a parallel association, and it was possible, by using extensometers placed on the wall tubes, to read, in an easy and economical way, the deformations in the tube and therefore in the samples. The problem that was caused by the water present in the backfill was also solved by this method. It should be stated that the option of working with high contents of water in the backfill was a deliberated one (the drainage of the water during a short period of 24 hours was provided, and it was performed with the use of little holes made in the tube for this purpose) in order to test severe conditions that probably will be found in mining practice. It was possible to measure the radial deformations of the samples by using two deflectometers that were in contact with the sample through two holes opened in the tubes for this purpose.

There were performed about a hundred tests with and without confinement by fill. The complete curves were obtained (Fig 5).

The peak strength, the elasticity modulus were calculated as well as, in the post rupture domain, a similar modulus that is the slope of the right side of the curve before reached de peak value. The Poisson ratio and the values of energy deformations after and before reached the peak value were also calculated. These energies, w, were obtained from the areas that were defined for the different curves.

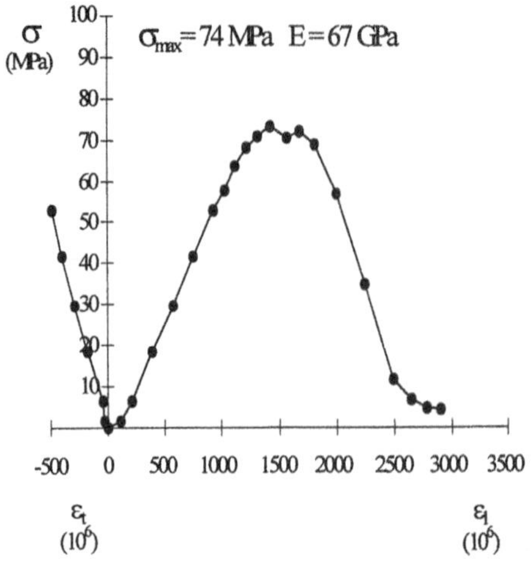

Figure 5. - Experimental complete curve of a marble specimen.

Table 1. Slenderness effects on rupture control in compression tests whit backfill confinement.

Slenderness	Pirite de Aljustrel (Massive Pyrite)	Ruivina Claro (Black Marble)	Estremoz Branco (White Marble)
< 3	Good	Good	Good
> 3	Bad	Bad	Good

Table 2. Slenderness effects on peak strength in compression tests whit backfill confinement.

Slenderness	Pirite de Aljustrel (Massive Pyrite)	Ruivina Claro (Black Marble)	Estremoz Branco (White Marble)
< 3	Good	Medium	Bad
> 3	Good	Medium	Medium

The results of the backfill confinement improvement were analysed regarding two different aspects, the pillar resistance (reflected by the well known triaxial effect on the peak strength value), and its rupture control (reflected on the increase of the values by the more horizontal slopes of the write side of the complete curves). By the analysis of the results it was possible to build tables (Tables 1 and 2) were the improvements were classified as good (>20%), medium (3%-10%) and bad (<3%), for the different slenderness.

4 CONCLUSIONS

It was observed that the two possible improvements by confinement (increase of the peak strength and better rupture control) depends on the pillar material type and its slenderness and on the backfill properties. The increase on pillars peak strength depends clearly of the brittleness of the rock material. If slenderness is inferior than three, the increase reached on rupture control is important for all types of rock; for values of slenderness bigger than three they are only significant for more ductile rocks.

Theses results can lead to some conclusions such as that the triaxial effect resulting from confinement by backfill, as it was done in laboratory and as it was performed in mine practice, is provided almost all by the friction action, that makes difficult the cataclastic flow of the brittle materials more or less fractured and therefore more ductiles.. In the related cases, as the mining practice was simulated, the more direct triaxial effect, as it is commonly obtained in laboratory tests, is of very few importance not only because of the very low values of applied maximum confinement pressures but mostly because of the fact that those values were reached when pillars were completely destroyed. The opposition to the referred cataclastic flow was especially relevant for the samples with slenderness smaller than three, wherein the bottom effects occur more often and therefore the axial splitting was more present. That kind of fractures were relevant for brittle materials. For values of slenderness bigger than three, the backfill confinement effect results eventually from the opposition to bending actions, and its importance can explain the significant increases in peak strength.

Finally it could be admitted that for slenderness bigger than three, in more ductile materials, and consequently, more deformable materials, there are some plastic deformations on broken material that can contribute to the improvement in rupture control.

5 REFERENCES

Costa e Silva, M.O.C.H. 1997 - Contribuition for the Study of Post Pillars Confined by Fill Geomechanical Behaviour. 6th Geotechnical National Congress. Lisbon.

Patchet, S. J. 1983. Mechanical Properties of Rock and Rock Masses. Rock Mechanics in Mining Practice.Budavari.The South African Inst. of Min.and metall.. Johannesburg.

Salamon, M. D. G. 1970. Stability, Instability and Design of Pillar Workings. Int. J. Rock Mech.. Min.Sci.,vol. 7. Pergamon. Oxford.

Salamon, M.D.G. 1983 - The Role of Pillars in Mining. Rock Mechanics in Mining Practice,ed. by Budavari.The South African.Inst. of Min. and Metall. Johannesburg.

Salamon, M. D. G. & Oravecz, K. I. 1976. Rock Mechanics in Coal Mining.Chamber of Mines Of South Africa.

Wawersik, W. R. 1968. Detaild Analysis of Rock Failure in Laboratory Compression Tests. Thesis. University of Minnesota. University Microfilms, Inc.Ann. Arbor, Michigan.

Wawersik, W. R. & Fairhurst, C.1970. A Study of Brittle Rock Fracture in Laboratory Compression Experiments. Int.J. of Rock Mech.and Min.Sci. vol.7.Pergamon. Oxford.

Mechanics of Jointed and Faulted Rock, Rossmanith (ed.) © 1998 Taylor & Francis, ISBN 90 5410 955 6

Extension of concealed fractures: Why did a collapse of the sea-cliff at the western entrance of the Toyohama Tunnel, Hokkaido, Japan occur?

T.Watanabe, N.Minoura, T.Ui, M.Kawamura, Y.Fujiwara & H.Matsueda
Department of Earth and Planetary Sciences, Graduate School of Science, Hokkaido University, Sapporo, Japan

H.Yamagishi
Geological Survey of Hokkaido, Sapporo, Japan

ABSTRACT: A 11,000m^3 rock mass, which was peeled off from the sea-cliff above the western entrance of the Toyohama Tunnel by a deep fracture plane trending NE to EW, slid/fell down and destroyed the tunnel. The deep fracture had been concealed by the collapsed rock mass. The fracture had been partly initiated probably by release of stress in the high and steep cliff; formation of incipient sheeting jont or nectectonoc joint. The extension of the fracture, however, is considered to be accelerated by water supply, which leads to inceases of interstitial water pressure during winter, freezing-thawing process, and rapid change of strain of the rocks in the temperature below zero, make the rocks weaken. The surface structure of the cliff-wall after the collapse suggests that localized, subparallel fractures extended until finally they were connected, forming a continuous deep fracture with a nearly vertical dip.

1 INTRODUCTION

On 10 February 1996 at 8:08a.m, a sea cliff, just above the western entrance of the Toyohama Tunnel on National Highway No.229, Hokkaido, Japan, collapsed and an overburdened rock mass (11,000 m^3) destroyed the tunnel (Fig.1). The rock-mass completely crushed a bus and a car, and small fragmented blocks hit another car that had narrowly escaped the fatal accident. The deaths of the 20 people on the bus driver,passengers and the car-driver were confirmed after 8 days of rescue activities.

Three large blocks (A, B and C) together with numerous smaller blocks slid/fell from the cliff to the shelter against rock-falls and the entrance of the tunnel. Another block (D) had fallen from the cliff many years ago(Fig. 2). The height of the collapsed cliff-wall is ca. 65m (the altitude at the top is 80m asl and at the bottom, the roof of tunnel, it is 15m asl). The volume of block A is ca. 7,200m3, of B ca. 700m3 and of C ca. 100m3 [Investigation Committee on the Rock-Fall Accident at Toyohama Tunnel =ICTT ,1996].

The collapse of the sea-cliff is considered to have occurred by the following process(ICTT,1996) :
1) Nearly vertical discontinuous fractures already existed locally within the sea-cliff.
2)The fractures gradually extended and finally became linked, forming one continuous deep fracture subparallel to the cliff.
3) With the lowering temperature below zero a few days before the tragedy, the cliff surface froze and formed a non-permeable wall for water. The interstitial groundwater pressure in the cliff increased.

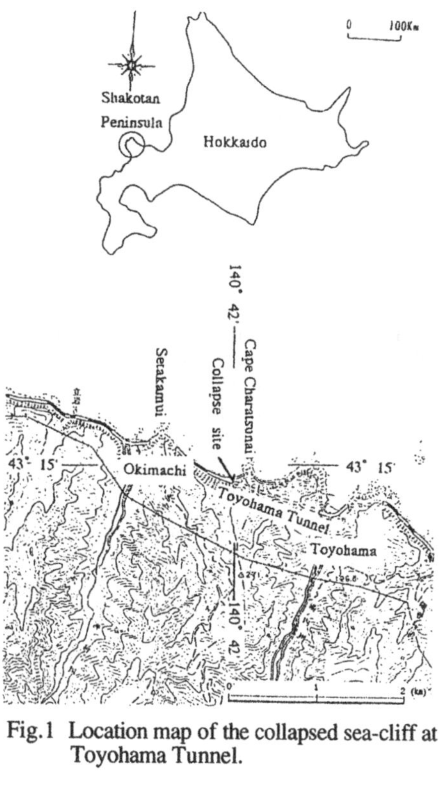

Fig.1 Location map of the collapsed sea-cliff at the Toyohama Tunnel.

Fig.2 Collapse of the sea-cliff(11th of February). Blocks A - C are falling blocks by the collapse on 10th. Block D was fallen over 30 years ago. (from Watanabe et al.,1996, original photo: from ICTT,1996)

4) The increased water-pressure along the fractures caused the wall to peel off, and when the length of fracture from the top of the pelled cliff reached 40-50m, the rock-mass slid down by gravitation onto the tunnel. finally formed a continuous deep fracture. In this paper we dicuss the process that fractures were extended and a continuous fracture was formed.

2. TOPOGRAPHY AROUND THE SEA - CLIFF

The Toyohama Tunnel is located 60km west of Sapporo, Hokkaido. Tthe Toyohama Tunnel runs through the foot of a high, steep cliff composed mainly of Middle Miocene andesitic hyaloclastites.

The average slope angle of the cliff is 70 - 75°, though it is partly and slightly overhung and a hollow topography develops in the left side of the cliff wall. The altitude of the sea-cliff along the road exceeds 150m. The transgression surface of the Jomon Era (ca.6000 B.P.) is considered to be that of ca. 6m asl. This altitude is higher than that of Sapporo (3m asl), suggesting local uplifting in the study area since the Jomon transgression. An abrasion platform also is observed off-shore of the study area and the distance is nearly 200m from the coast line. The formation of this platform suggests an average retreatment rate of ca.0.03 - 0.05m/y. Judging from aerophotographs at the cape of Charatsunai, much rapid retreatment of ca. 0.5-1.1m/year for 18years since 1947 is suggested (Watanabe et al.1996). Fractures cutting through the high and steep cliff rarely develop, but incipient or localized fractures are more common.

3 CLIMATE CONDITIONS

During the winter of 1995-1996 daily average temperature began to record below 0°c since Middle December and further lowered to -120°c until 1st of February with a few exceptional increases above zero during Late December - Early January(Data at the Yoichi automated weather station in Yoichi Town, 10km east of the collapsed site). Then the daily average temperature increased rapidly till 6th up to nearly zero celsius. For three days before the tragedy the temperature again decreased. The temperature increased again on 10th , but in the morning it was still below zero.

Harimaya et al.(1997) reported daily variation in air tmperature at the Bikuni automated weather station,10km west of the collapsed site. Daily variation that the maximum air temperature exceeded 0°c and the minimum was below 0°c, occurred frequently(Fig.3). Effective daily variation for the freezing-thawing process (T_{MAX} >4°c, T_{MIN} <-4°c) recorded only three times during the winter. It is less

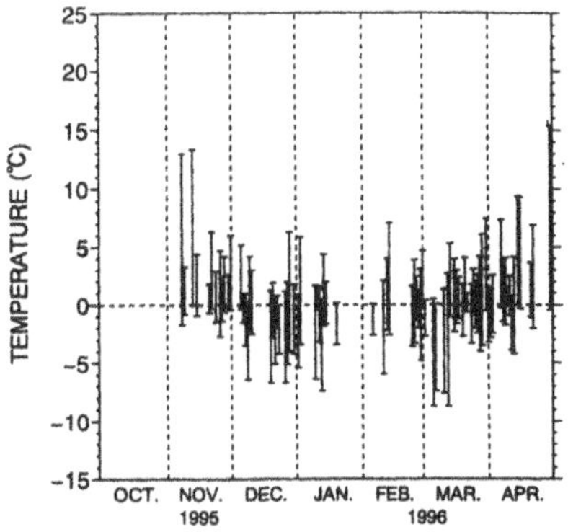

Fig.3 A diagram showing frequency of the daily variation in air temperature ($T_{MAX} > 0°c$, $T_{MIN} < 0°c$) at Bikuni (Harimaya et al.,1997)

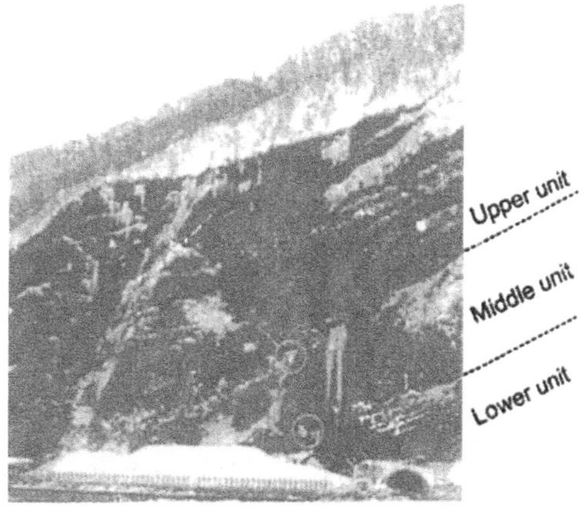

Fig.4 A photo of the sea-cliff including collapse site. Note that icecles in circles and long icecles located in the center of the collapsed part, are produced from groundwater outlets along the unit boundaries. (from Watanabe et al.,1996)

than the average frequency and in early ninties the frequency often recorded 8times. The total depth of accumulated snow-fall in the winter of 1995-1996 is rather deep around the study area and after 5th of February it has recorded the maximum thickness for each day in the last 10 years. Above all, the thickness on 10th of February was the maximum in this winter. These meteorological data suggest that immedeately before the day of the tragedy the surfcace of sea-cliff froze deeply and formed non-permeable wall for water (ICTT, 1996). The depth of frozing wall in mid-winter in the study area is 50-60cm(Fukuda et al., 1997).

4 THE COLLAPSED SEA-CLIFF

Geology, groundwater and surface structure after the collapse on the sea- cliff are described below.

4.1 *Geology*

The rocks comprising the cliff belong to the On'nenai formation of Middle Miocene(Yamagishi, 1981) and a K-Ar age from volcanic blocks from the lower horizon of the formation is 11.2Ma. Alteration of the rocks is generally weak and only smectite is detected. The rocks of collapsed sea-cliff is stratified and lithologically subdivided into three units,i.e., lower, middle and upper units (Fig.4).

The thickness of each unit is 25m+ in the lower unit, 30m in the middle unit and 60m+ in the upper unit at the central part of the collapsed cliff. The strike of the stratified rocks is N70°W and the dip is 30°NE. The dip becomes very gentle to both the western and eastern sides. The lower and upper units are reworked hyaloclastite and the middle unit is mainly dacitic hyaloclastite. The middle unit is further subdivided into at least three parts,i.e.,lower, middle and upper subunits. The lower subunit consists of unsorted volcaniclastic formation. The middle subunit is massive hyaloclastite with isolated lava lobe and coarse grained accessory fragments. The upper subunit is rubble-bearing volcaniclastic rocks gradually changing from the middle subunit. Fine grained volcaniclastic rocks are partly contained in the subunit and less-permeable layers for water develop. Smectite layers or lenses of brownish-yellow colour characteristically occur in the upper subunit and the layers/lenses suggest groundwater penetration for a long term.

The overhanging cliff is composed mostly of the middle unit.

4.2 *Groudwater*

Just after the collapse on 10th of February, groundwater came out on the cliff-surface as shown in Fig.2(Compare with Figs 3 & 4). Therefore, groundwater has been always supplied from the lithologic unit boundaries, mainly between the boundary of the upper and middle units, into fractures and the rocks were satuarated in water. The main outlet of groudwater is located at the center of the collapsed cliff. Water pressure at the two unit boundaries varied from 0.13-1.37kgf/cm^2 in May - June, 1996 in the bore-holes near the collapsed site(ICTT, 1996). It is worthy to note that visible outlets of groundwater shown by hollows or icicles are not olny arranged along the unit boundaries, but also develop in every ca. 20m interval, suggesting pipe-network of groundwater path along the unit boundaries plane.

Fig.5 A photo showning Fr-7 and Fr-8 on the surfce of the collapsed cliff. A curved broken line shows the upper-most margin of the collapsed cliff. (April, 1996)

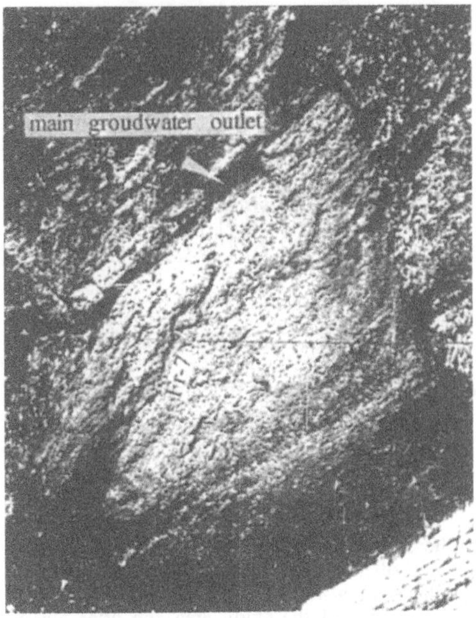

Fig.6 Computer enhanced micro-topograph of the cliff wall appeared after the collapse.

4.3 *Surface structure on the cliffwall after the collapse*

Distinct two fractures were recognized on the cliff wall after the collapse,i.e., Fr-7 and Fr-8 by ICTT (1996). Fr-7 develops only in the left side of the main water outlet shown by long icecles(Fig.5) and disappears in the uppermost part of the middle unit. Fr-7 is not a distinct single fracture (Watanabe et al.,1996). Fr-8 bounds between Surface I and Surface II(see below). Another fracture is recognized on the right side of the main groudwater outlet shown by long icecles.

Judging from the brownish colour, the upper part and left hand side of the cliff-wall after collapse were deeply weathered. Below the weathered part sporadic alteration spots are recognized. The upper most part of the upper-right side(upper unit) was thinly peeled off and it continued to the main surface with Fr-8 and steps. The wall is not a simple smooth-surface and its complicate wavy-surface somehow reflects variation of lithology. Computer enhanced microtopograph of the cliff wall after the collapse reaveals irregular micro-relief structure on the wall surface (Fig.6). Such structure with irregular micro-relief is not in harmony with surface formed by systematic jointing. It is most likely that isolated and discontinuous fractures had been once produced subparallel to the cliff surface, and fractures extended and linked each other. The surface can be classified into two,i.e.,Surface I(S-I, main wall with N60°E trend) and Surface II (S-II, side wall with NW- W trend). On the upper part of S-II, hackle markings, indicating rupture by tensional stress is observed(Watanabe et al.,1996).

No sign indicating shearing is observed on the upper part of the S-I ,except for a few powdered traces caused by the sliding of falling blocks and micro-shearing observed under the microscope.

5 FRACTURE EXTENSION

5.1 *Result of material test*

Block C (thickness is 2m) is considered to have been peeled off from the position of the left side of Fr-7 and it was already a plate peeled off from the wall before the collapse. As described earlier, the upper part of Surface I is also a part of an open fracture, though only the upper-right part was visible from outside. Taking into account that thick blocks A and B have been peeled off from the right side of Fr-7, mechanism on peeling of the right side of Fr-7(the main frontal surface) is most serious. The longest icicles grow from the right side of Fr-7. This strongly suggests the importance of water supply in extension and connection of fractures.

The rocks of the cliff had been saturated in water, and the freezing depth of the rock-wall is 50-60cm within the cliff. An edge of open fracture behind the rock mass will be weakened by the repitition of the freeze and thaw process for a long term(Fukuda et al.,1997). Material tests(Fujii, 1996) reveals that the indirect tensile strength of water-saturated samples is 0.39Mpa which is below the freezing pressure of water and equivalent of interstitial water pressure of 39m of water column. The tests suggest that fracture in rocks can be easily extended by interstitial water pressure in the cliffs or other factors including the freezing pressure of water. The significant effect of expansion and contraction of frozen rocks must examined in detail for the rocks in the study area.

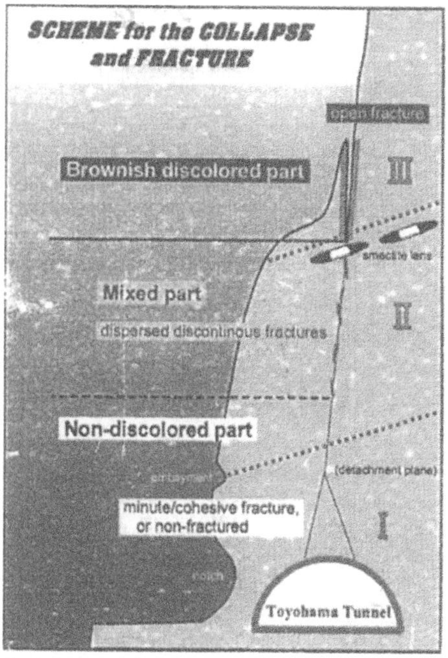

Fig.7 Schematic illustration on fracture-extension and -connection (Kawamura,1997)

5.2 *Incipient Neoectonic Joint*

We now realize that water supply to fractures in poorly fractured hyaloclastite make fractures extend easily. The north-eastern coast zone of Shakotan Peninsula, especially around Toyohama, recorded high rate of regression of cliffs. Gigantic rock-falls historically occurred on adjacent area(ICTT,1996). The coastal areas need to be carefully studied as an active tectonic zone where incipient neotectonic joints may be produced. Fractures develop subparallel to the sea-cliff, but thier direction are not tightly constrained as systematic joints. This may be chacteristics for joints produced in hetrogeneous and less consolidated rocks.

6 CONCLUSIONS

We conclude that the fracture development depends heavily on the watre-supply which leads to increase of interstitial water pressure in mid-winter, effective freezing-thawing process and generation of freezing pressure of water. In the collapsed cliff at the Toyohama Tunnel, pre-existed discontinuos fractures subparallel to the sea-cliff are incipient neotectonic joints which are comparatively irregular joint system. Such joints may be characteristics in hetrogeneous hyaloclastic units.

ACKNOWLEDGEMENTS

We would like to express our thanks to members and supporting staff of the Investigation Committee on the rock-fall accident at Toyohama Tunnel, Members for the Study of Natural Disaster Science in Japan. Co-operative research with Professors M.Fukuda, and T.Harimaya, and Associate Professor Y. Fujii of Hokkaido University is most encouraged our geological study. Technical Staff, Messrs T. Hirama, S. Terada, T. Kuwajima and H. Nomura co-operated in our research.

REFERENCES

Fujii,Y. 1996. Material tests and stress analysis on rock slope collapse at Toyohama Tunnel. *Bulletin of the Natural Disaster Science Center Hokkaido*, 11, 77-89.

Fukuda, M.,T.Harimaya, and K.Harada 1997. The effect of freezing - thawing cycles to rock fall at Toyohama. *Bulletin of the Natural Disaster Science Center,Hokakido, an extra issue*, 51-60.

Investigation Committee on the Rock-Fall Accident at Toyohama Tunnel 1996. *Report on Investigation Committee on the Rock-Fall Accident at The Toyohama Tunnel*, Hokkaido Development Bureau, Sapporo.

Kawamura,M. 1997. Geological background of the collapse accident of the Toyohama Tunnel. *Bulletin of the Natural Disaster Science Center, Hokakido, an extra issue*, 11-19.

Harimaya,T., S.Murai, A.Hashimoto 1997. Freezing - thawing frequency related to the weathering rocks. *Bulletin of the Natural Disaster Science Center Hokkaido, an extra issue*, 41-49.

Watanabe,T., N. Minoura, T. Ui, M. Kawamura, Y. Fijiwara, and H. Matsueda 1996. Geology of a collapse of the sea-cliff at the western entrance of the Toyohama Tunnel, Hokkaido, Japan. *Jour. Nat. Disas, Sci.*, 18, 73-87.

Yamagishi,H. 1981. Geology of the Shakotan Peninsula, Hokkaido, Japan, *Rept. Geol. Surv. Hokkaido*, 52, 1-29.

Mechanics of Jointed and Faulted Rock, Rossmanith (ed.) © 1998 Taylor & Francis, ISBN 90 5410 955 6

Mining induced fractures for roof in logwall stope and their behavior

Hong-Ci Wu
Department of Mining Engineering, Guizhou University of Technology, Guiyang, People's Republic of China

ABSTRACT: The immediate roof, coal seam(foundation)and the main roof(rectangular plate) in longwall stope, for the first time, are drawn separately as elastic, elastoplastic, viscoelastic and viscoplastic foundation plate structures. The boundary element method (BEM) for bent thin plate combines individually with elastic, elasoplastic, viscoelastic and viscoplastic BEM to solve the above mentioned structures. The results show that fractured patterns of the foundation plates are all flat X type or an half of flat X type; Fractured main roof forms a regular or an oblique arch structure with three-gripping; In the transitional process of the foundation from elasticity to viscoplasticity the distance between the fractures of the main roof and the coal wall, and roof convergence rate increase, and the stress concentration factor(k) , the dynamic loading coefficients of primary weighting (q') and of periodic weighting (q″) , and the amount of reverse elastic movement of the main roof and foundation system decrease.

1 INTRODUCTION

State-of-the-art of underground pressure in longwall stope is that many plane models, such as cantilever beam, articulated rock block, preformed crack and voussoir beam models have been constructed using plane methed. The competent roof (or carrying strata) above the coal seam, however, is actually"a plate", a space structure, which is supported by the immediate roof and coal seam(foundatian) . In this paper, overlying competent roof(or main roof) in longwall stope would be regarded as an elastic thin plate, and the immediate roof and coal seam of supporting main roof, according to their rock types and properties, would be respectively classified into elastic, elastoplastic, viscoelastic and viscoplastic foundantions, which means that the immediate roof, coal seam and the carrying strata are drawn as four foundation plate structures and their fractured patterns inducced by minning, formed strutures after fractured and pressure manifestation are analyzed by means of coupling separately between bent thin plate BEM and elastic, elastoplastic, viscoelastic and viscoplastic BEM.

2 FOUNDATION PLATE STRUCTURE IN LONGWALL STOPE

In general, there is 100-200m length of longwall face, 20-50m span of the first weighting, 10-20m span of the periodic weighting and 2-4m layered thickness for main roof in coal face in China, and layered thickness to width (the first and the periodic weighting span) ratio is(1/5-1/13) ,1 4, which meets the condition of elastic thin plate. Therefore, in contile-

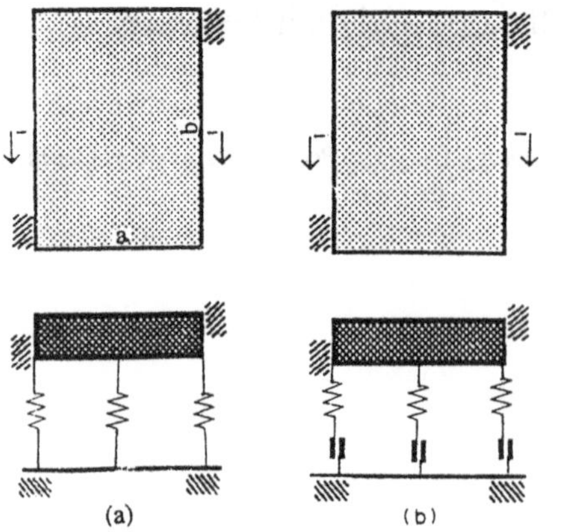

Fig. 1 Elastic (a) and elastoplastic (b) foundation plate structures

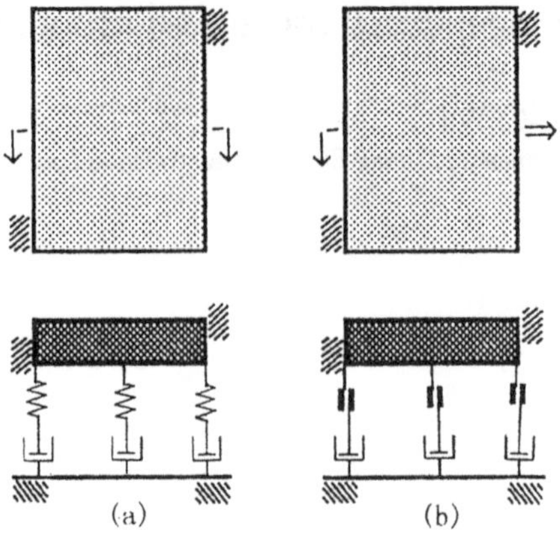

Fig. 2 Viscoelastic (a) and viscoplastic (b) foundation plate structures

vered main roof saging and bed separation, frictional force between layered strata is very small, the main roof can be viewed as an elastic thin plate (or a rectangular plate)(the condition of four sides clamped is only considered in this paper), and the foundations (the immediate roofs and coal seams), according to their types and properties, can be regarded as four classifications, that is, the mechanical systems of the immediate roof, coal seam and the main roof cam be drawn as four foundation plates (Figure 1 and Figur 2) (after Wu, 1997).

Elastic foundation plate: the competent immediate roof(IV classification), such as arenaceous shale of large thickness, greet stone, sandstone, limerock and hard coal seam etc., acted by main roof and its overlying strata, will produce mainly elastic deformation, therefore, they can be described by Hookean substance, which means that they can be drawn as an elastic foundation plate(Fig. 1(a)).

Elastoplastic foundation plate: the stable immediate roof(III classification), such as tight shale, sandy shale and coal seam of medium hardness etc., acted by main roof and its overlying strata, will not only produce elastic deformation, but also plastic deformation, as a result, they can be described by combination Hookean substance with St Venant substance, which means that they can be abstracted as an elastoplastic foundation plate(Fig. 1(b))

Viscoelastic foundation plate: the medium stable immediate roof (II classification), such as untight shale, argillaceous shale, bedding shale and coal seam under medium hardness etc., acted by main roof and its overlying strata, will not only produce elastic deformation, but also creep strain, therefore, they can be described by combination Hookean substance with Neutonian substance, which means that they can be abstracted as a · viscoelastic foundation plate (Fig. 2(a)).

Viscoplastic foundation plate: the unstable immediate roof (I classification), such as clay rock, argillaceous shale, filled sand and soft coal seam etc., acted by main roof and its overlying strata, will produce mainly viscous flow, so they can be described by combination Neutonian substance with St Venant substance, which means that they can be abstracted as viscoplastic foundation plate (Fig. 2(b))(In the Fig., arrow represents advance direction of coal face).

3 NUMERICAL MODELLING FOR FOUNDATION PLATE STRUCTURE

The above four types of foundaticn plate are calculated by means of coupling respectively between bent thin plate BEM (modelling the rectangular plate for the main roof) and elastic (modeling elastic foundation), elastoplastic(modeling elastoplastic foundation), viscoelastic(modelling viscoelastic foundation) and viscoplastic(modelling viscoplastic foundation) BEM.

1. Boundary integral equation for bent thin plate (after Wu, 1997) : Assuming that the rectangular plate for the main roof is composed of domaim Ω and boundary Γ, and that they are continuative, except K corner points(K=8 here). Now a small neighbourhood with radius ρ from the point p(the original point on the plate boundary) in the domain Ω has been removed, a new domain Ω^* is obtained, and boundary of Ω^* is Γ^*, and defining separately N, M, V and $F^{(k)}$ as normal slope, bending moment, equivalent shearing force and discontinuous force of corner point in torque relevant to deflection, and adopting the fundamental solution of u, u_f^*, in the identical equation of work exchangeability, finally boundary integral equation (including the fundamental solution of concetrative force) for bent thin plate can be established.

$$cw\Big|_p+\int_{\Gamma_p}\left\{V_f^*w-M_f^*N+N_f^*M-u_f^*V\right\}ds+\sum_{k=1}^{k}*\left\{F_f^{*(k)}w^{(k)}-F^{(k)}u_f^{*(k)}\right\}=\int_{\Omega}u_f^*qd\Omega \qquad (1)$$

Where $\int_{\Gamma_p}(\bullet)ds=\lim_{\rho\to 0}\int_{\Gamma_\rho}(\bullet)ds$ is Cauchy principal value.

2. Boundary integral equation for an elastic foundation: Based on direct method of BEM, assuming that there is an elastic body V with boundary s, and defining displacement and surface force of infinite body acted by an unit force as U_{ij} and T_{ij}, and moving action point I of unit force from within domain to boundary point p, and considering original stress, the boundary integral equation can be obtained by Betti mutual equalization theorem:

$$c_{ij}^I u_i^I=\int_s t_i U_{ij}ds-\int_s u_i T_{ij}ds+\int_V f_i U_{ij}dV+\int_V \varepsilon_{ij}\sigma_{ij}^0 dV \qquad (2)$$

Where ε_{ij} is induced strain acted by an unit concentrative force, σ_{ij}^0 original stress, f_i body force, and

$$c_{ij}^I=\begin{cases}1 & (I\ in\ V)\\ 1/2 & (I\ on\ s)\end{cases}$$

3. Boundary integral equation for an elastoplastic foundation: Calculating derivate of formula(2), omiting body force, and replacing σ_{ij}^0 by initial stress increment $\dot{\sigma}_{ij}^p$, the boundary integral equation of elastoplastic foundation can be written in the following form:

$$c_{ij}^I \dot{u}_i^I=\int_s U_{ij}\dot{t}_i ds-\int_s T_{ij}\dot{u}_i ds+\int_V \varepsilon_{ij}\dot{\sigma}_{ij}^p dV \qquad (3)$$

4. BEM for viscoelastic foundation: For linear viscoelastic foundation problem, based on E. H. Lee's correspondence principle, fundational governing equation can be written in the following form:

$$\sigma_{ij,j}=0 \qquad (4)$$

$$\varepsilon_{ij}=(U_{i,j}+U_{j,i})/2 \qquad (5)$$

$$p(D)\sigma=Q(D)\varepsilon, \qquad (6)$$

Where D=d/dt is operator, p(D) and Q(D) are polynomials with respect to D.

Constitutive equation of viscoelastic foundation can be obtained by adopting derivative operator method:

$$P_s(D)S_{ij}=Q_s(D)e_{ij}, \qquad (7)$$

$$p_V(D)\sigma_{kk}=Q_V(D)\varepsilon_{kk} \qquad (8)$$

Where S_{ij} and e_{ij} are respectively partial tensors of stress and strain, P_s, Q_s, P_v, and Q_v derivative operators for constant coefficient of real number.

As shown in Fig. 3, viscoelastic foun-

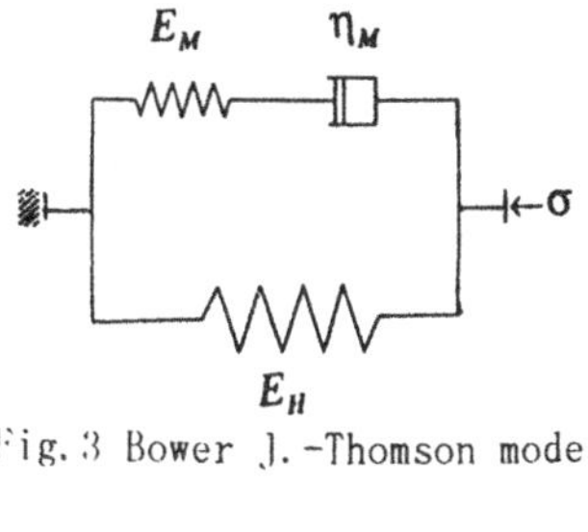

Fig. 3 Bower J.-Thomson model

dation problem can be described by Bower J.-Thomson model, its fundational equation is:

$$\frac{d\sigma}{dt}+\frac{E_M}{\eta_M}\sigma=(E_H+E_M)\frac{d\varepsilon}{dt}+\frac{E_H E_M}{\eta_M}\varepsilon, \quad (9)$$

5. BEM for viscoplastic foundation: Boundary integral equation for viscoplasic foundation can be written in the following form:

$$c_{ij}\dot{u}_j=\int_\Gamma u_{ij}^*\dot{p}_{ij}\,d\Gamma-\int_\Gamma p_{ij}^*\dot{u}_{ij}\,d\Gamma+\int_\Omega u_{ij}^*\dot{b}_j\,d\Omega+\int_\Omega \varepsilon_{jki}^*\dot{\sigma}_{jk}^a\,d\Omega \quad (10)$$

Where Γ is boundary of domain Ω, tensors u_{ij}^*, P_{ij}^*, and ε_{jki}^* respectively represent fundamental solutions of displacement, force and strain induced by unit point load acting along direction i, $\dot{\sigma}_{jk}^a$ is initial stress ratio of viscoplastic foundation problem.

After discretizing formula(10), matrix equation can be obtained.

$$H\dot{u}=Gp+Q\dot{\sigma}^a \quad (11)$$

6. Coupled numerical method: Essential condition of coupled numerical method is compatibility and equilibrium between coupled interfaces (or points). Generally speaking, there on the interfaces between foundations and rectangular plates for main roof exist two conditions: there is force transference between the foundation and the rectangular plate but there are no relative displacements, which means that there are no relative movements or tensile openings, and considering the foundation and the plate as continuous body comprised of different types of materials, it is not difficult for us to evaluate the models by numerical methods; on the other hand, there are relative movements or discontinuities between the foundations and the plates, the discontinuities will be calculated by interface element. In this paper, the former is only considered.

Defining the foundation as F, the rectangular plate as Ω, interface between the foundation and the plate as I, then the condations of compatibility and equilibrium on the interface I are displacement continuity, equal value and opposite direction of surface force at element node, namely

$$\{u_I\}_\Omega=\{u_I\}_F \quad (12)$$

$$\{p_I\}_F=-\{p_I\}_\Omega \quad (13)$$

The formulas (12) and(13) are fundament of coupled numerical method.

7. Numerical model treatment of the foundation plate: Numerical modeling is based on geologic condations of face 1373 (2) at Huachu Colliery, Southwest of China. The length of foundation plate:80 m, the width of foundation plate:40m, the immediate roof in coal face:grey bedded limestone and the main roof: greet stone of grey and dark beddig. Taking into account the boundary condition that the four sides of the rectangular plate are clamped, 16 boundary elements, 272 cells and 8 boundary elements, 144 cells have been respectively used to discretize the long boundary and the short boundary of the plate, therefore, an half of freedom in the models was directly given by the boundary condation, as a result, integral equation number for independent discretization was only an half of total freedom and calculating time was reduced greatly.

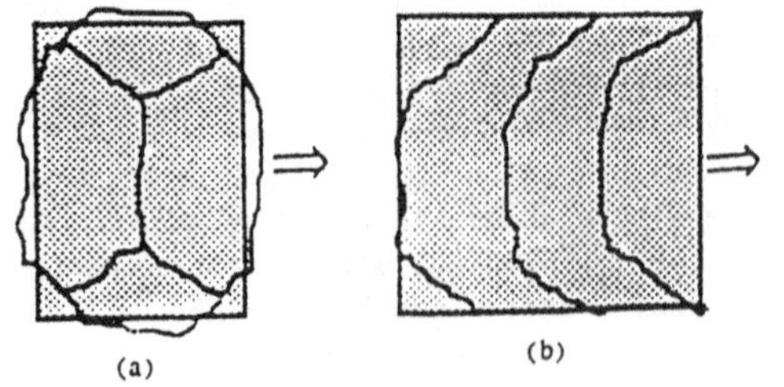

Fig. 4 Fractured patterns for the four foundation plates

4 MINING INDUCED FRACTURES FOR ROOF

Under common circumstance b/a>1 of longwall stope, the first fractured patterns for the elastic, elastoplastic, viscoelastic and viscoplastic foundation plates are all flat X type (after Qian and Zhu, 1986) (Fig. 4(a)), except their fractured position and pressure behavior, and the periodic fractured patterns for the above foundation plates are all an half of flat X type(Fig. 4(b) (In the Figures, arrows indicate advance direction of coal face), which means that, irrespective of the foundation types for the main roofs, the first and the periodic fractured patterns for the main roofs in the common longwall stope are all respectively flat X type and an half of flat X type.

5 FRACTURED ROOF BEHAVIOR

After the foundation plate is broken, its fracture is basically perpendicular to the principal tensile stress locus of cantilever plate for the carrying strata, and there still exists in interlock grip state among the fractured plates in their moving and rotary process, and a regular arch structure with three-gripping after the first fractured is formed (Fig. 5) (circled face is the key grip surface) (bordering often on the key block),

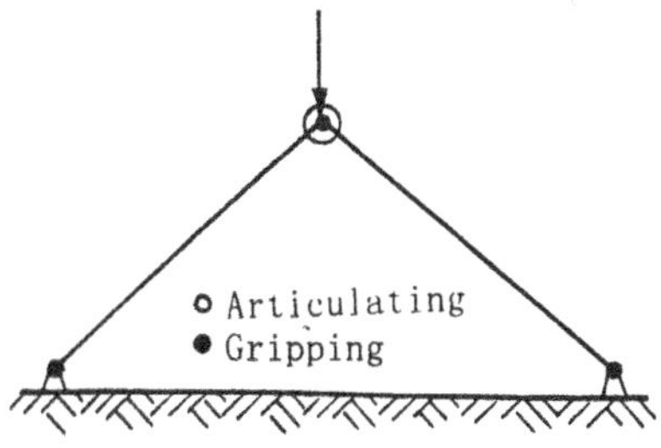

Figure 5 A regular arch structure with three-gripping after the first fractured

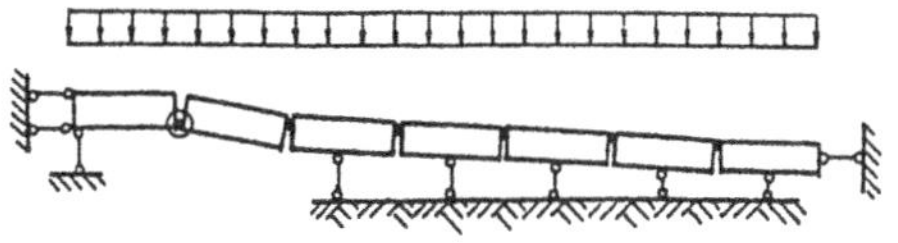

Fig. 6 An oblique arch structure with three-gripping after the periodic fractured

out of balance for the structure will occur the first weighting. And the periodic fractured main roof will form an oblique arch structure with three-gripping (Fig. 6), out of balance for the structure will occur the periodic weighting. For this reason, in the neighbourhood of the key grip face is focal point of artificial support. The purpose of the artificial support is to help to bring about the arch balance with three-gripping to guarantee coal face safety.

6 FOUNDATION EFFECT

The studied results show that the distances between the fractured lines for the main roof and the coal walls, and the roof to floor convergence rates increase, as the foundations(the immediate roofs and coal seams) change from elasticity into viscoplasticity; that as the foundation(the immediate roof and coal seam) changes from elasticity into viscoplasticity, stress concentrative coefficient (k), dynamic loading coefficients of the first weighting(q')and of the periodic weighting (q"), and the recoil quantities for the system of the foundation and main roof decrease.

7 PRESSURE CONTROL IN LONGWALL STOPE

In competent roof condition, adopting boring, grooving, skock blasting and high pressure water flooding etc. enables the foundations (the immediate roofs and coal seams) to change from elasticity into viscoplasticity or plasticity, so that the roof weighting intensity is reduced to guarantee coal face safety.

8 CONCLUSIONS

1. Under common circumstance b/a> 1 in longwall stope, the first fractured patterns induced by mining for elastic, elastoplastic, viscoelastic and viscoplastic foundation plates are all flat X type, except their fractured position and weighting behavior, and the periodic fractured patterns induced by mining for the above four foundatien plates are all an half of flat X type;

2. The fractures of the foundation plates are uneven, and when the foundaion plates are broken, their fractures are basically perpendicuar to the principal tensile stress locus of cantilever plate for the main roof and there still exists in interlocked gripping state among the broken plates in their moving and rotary process. For this reason, the fractured main roof in longwall stope will form the arch structure with three-gripping, out of stability for the structure will occur weighting in coal face;

3. The distance between fractured lines for the main roof and the coal walls, and the roof to floor convergence rates will increase, as the foundations (the immediate roofs and coal seams) change from elasticity into viscoplasticity;

4. Stress concentration coefficients(k), dynamic loading coefficients of the first weighting(q')and of the periodic weighting (q") , and the recoil quantities for the system of the foundation and main roof will reduce, as the foundation alters from elasticity into viscoplasticity;

5. In competent roof condition, adopting boring, grooving, shock blasting and high pressure water flooding etc. enables the foundation to change from elasticity into viscoplasticity, so that the roof weighting intensity is reduced to guarantee coal face safety.

REFERENCES

Qian, M. G. & D. R. Zhu. 1986. Fractured patterns for main roof and their effects on roof pressure in longwall face. J. China Univ. of Min Tech. (2):9-18

Wu, Hong-Ci. 1997. Model for structure of foundation plate in longwall face and rules of its weighting. J. China Coal Society. 22(3):259-264

Wu, Hong-Ci. 1997. Model for structure of gripped rock block and its algorithms. Ground Pressure and Strata Control. (3. 4): 10-12

Wu, Hong-Ci. 1996. Fracturing modeling of rock mass using updated DDA method. Chinese J. Rock Mechanics and Engineering. 15(supp.):556-558

Mechanics of Jointed and Faulted Rock, Rossmanith (ed.) © 1998 Taylor & Francis, ISBN 90 5410 955 6

Research and application of the inside-injection grouting bolts

Yi Gongyou
Shandong Institute of Mining and Technology, People's Republic of China
Zhang Chuanguo
Liangjia Mine, Shandong Province, People's Republic of China

ABSTRACT: By introducing the appliance of combined support system involving bolt, net, spray and grout – injection rockbolt in the extremely difficult stope drift active workings in Liangjia Mine as well as the designing principle of design, basic type and structure of inside – injection rockbolt, this paper illustrates its supporting co – efficience, constructing process, and supporting mechanism together with its economic benefit and future.

KEY WORDS: difficult roadway; inside – grout – injection rockbolt; rational support

It is very difficult to support extremely unstable roadway for it has not only soft roof and floor but soft deposits, and stope drift active workings are affected directly by bearing pressure, sometimes tectonic stress and bulb of pressure. Presently, different types of metal supports are widely used in most countries, but they need rebuilding repeatedly in maintenance even if heavy supports are used, which is uneconomic for coal – yielding countries. It is necessary, therefore, to study new support system.

Our research shows that, besides overlying strata stress, tectonic stress and bulb of pressure, metal supports are usually damaged by two factors: one is the pressure from aged roof whose stress peak is greater than the bearing capability of metal supports thus causing damages to metal supports; the other is the constant vibration of crust and surrounding rock vibration from periodic aged – roof pressure during backstoping which, when obviously, cause unstableness and damages to metal supports. We think that active support is superior to passive support under dynamic pressure. It is generally considered that bolt – spray supports are active while metal supports are passive. But ordinary bolt – spray supports cannot keep stable stope drift active workings in unstable strata for the reason that they cannot provide enough strength required to surrounding rock whether ordinary end fixing bolts or ordinary full – length locking bolts are involved; and it is uneconomic to over – lengthen rockbolts. But if grout is injected into surrounding rock on the basis of ordinary bolt – spray supports, it can not only glue friable rock into a whole, but turn end fixing bolts into full – length locking ones, thus greatly increasing the strength of surrounding rock and its bearing capability. Based on this principle, we design and trial – produce inside – grout – injection rockbolt.

1 TYPE AND STRUCTURE OF INSIDE – GROUT – INJECTION ROCKBOLT

1.1 *Pressure – controlled inside – grout – injection rockbolt*

For roadways which go through broken fault and dripping belt or which need injecting grout for reinforcing, this kind of rockbolt can be used for advanced or pressure – controlled grout – injection. Fig. 1 shows its main structure:

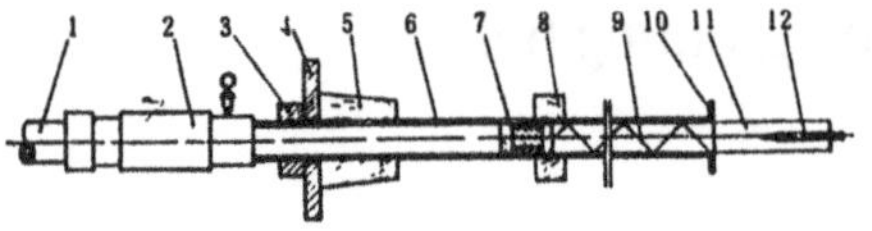

Fig. 1 Structure of pressure - controlled grout - injection rockbolt

1. grout - injection tube 2. back grouting valve
3. rod nut 4. tray 5. grout stop plug
6. rod 7. plug 8. grout control plug
9. spring 10. cement blasting guard
11. rod end 12. inverted wedge

Rod: gapless rail with small holes on it as both grout - injection tube and rockbolt

Rod end: composed of Y - screw - thread steel and iron wedge forming rockbolt together with rapid - hardening cement blasting or resin blasting to ensure the initial bolting strength

Pressure control plug: composed of wood - or rubber - plug and pressure - controlled spring

Grout control plug: made of soft wood, rubber or plastic to control grout and pressure

Grout stop plug: made of soft wood or reclaimed rubber to stop grout from overflowing or plug on spot with vapid - hardening cement blasting, resin blasting cement and sodium silicate binder

Tray: made of metal for supporting rock and enlarging its locking range to increase its locking result

Rod nut: standard nut for fastening the tray and giving prestress to rockbolt and surrounding rock

Grouting valve: Specially for inside - grout - injection rockbolt. It joints directly with the rockbolt in order to stop grout from back - flowing

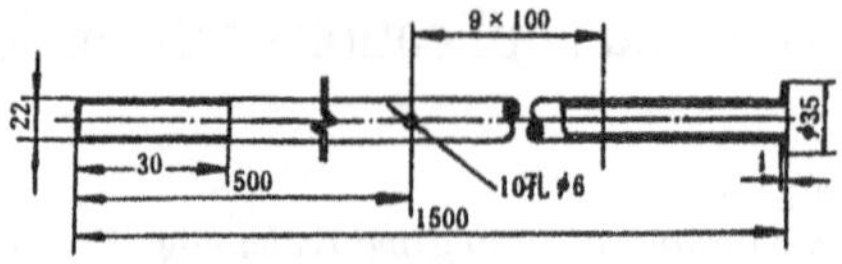

Fig. 2 Illustration to the structure of ordinary inside - grout - injection rockbolt.

1.2 *Ordinary inside - grout - injection rockbolt*

It can be used to the construction, repair and reinforcement of roadways without considering its initial bolting strength or controlling pressure when injecting grout. It consists of rod end, rod, grout stop plug and tray, etc., each having similar function as that of pressure - controlled inside - grout - injection rockbolt. It has been widely used in some mines for its simple structure, easy manufacture and low cost. (See Fig. 2)

Sizes shown in Fig. 2 are trial ones involved Liangjia Mine and should be revised according to different sections and surrounding rock in different mines.

2 APPLIANCE OF INSIDE - GROUT - INJECTION ROCKBOLT

After its performance test and industrialness test in Fenghuangshan Mine, Shanxi Province and Gaokeng Mine, Jiangxi Province, inside - grout - injection rockbolt has been now used in Liangjia Mine and Beizao Mine, Shandong Province, Geding Mine, Guangdong Province, and Linhuan Mine, Anhui Province. The following part deals with its applliance in extremely unstable stope drift active workings of Liangjia Mine, Beizao Mine, Geding Mine.

Liangjia Mine coal series of strata belong to the Third Series of Kainozoic Era. They have the characteristic of low intensity, softness and high brokenness, high water absorption and disintegration. The main roof is oil - bearing, containing over 30% of montmorillonite which belongs to swellen rock. Coal field consists mainly of faults among which 47 are normal faults with drop height over 15m, all the above – mentioned has caused great difficulties to roadway support, especially to stope drift active working, and only 5m of pillars are left along gate road when pillar method is involved. Therefore, neither conventional supports nor metal support will meet the damage. Some have to rebuild before and during backstoping. Most of the conventional supports used in Upper Crossheading No. 2203 had been damaged only at 700m. The roof sank, floor swelt, with the former 3.3m X 3.2m of roadway being reduced to below 2m X 2m. Almost all the I – steel were damaged in Upper Crossheading No. 2203.

Fig. 4 shows the result of our industrial test of combined supports in roadway rebuilding in Upper Crosshending No. 1200 involving bolt, net belt, spray and grout – injection rockbolt. The rockbolts are placed between two rows of ordinary bolts. The convined supports in Fig. 4 is of the same structure as that in Fig. 3 except replacingcurved beams with belts.

To avoid inter – disturbance between driving face and grout – injection, and to let part of the stress release by shapening new roadway under crustal stress and enlarge the gaps, which is advantageous to grout – injection, grout – injection face should lag behind the driving face by 30 – 60m and 15 – 20 days.

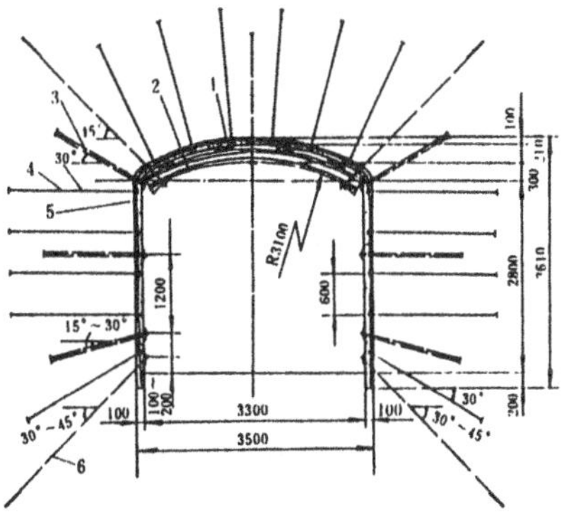

Fig. 3 Illustration to combined supports involving bolt, net, spray, curved beam and grout – injection rockbolt
1. sheeting caps 2. curved beam 3. grout – injection rockbolt 4. ordinary rockbolt 5. blasting cover and metal net 6. full – length rockbolt

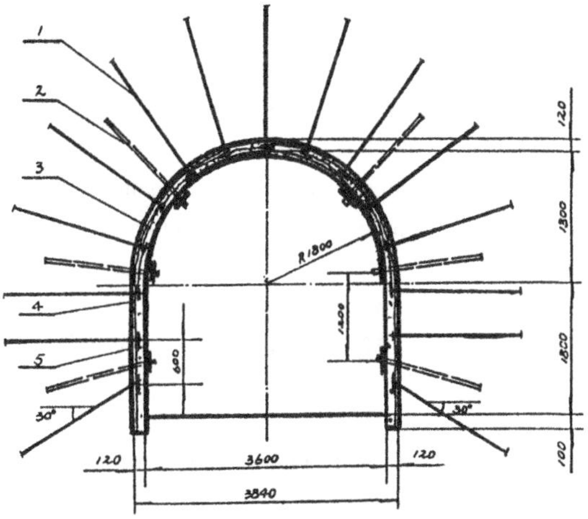

Fig. 4 Illustration to combined supports involving bolt, net, spray, belt and grout – injetion
1. ordinary rockbolt 2. grout – injection rockbolt 3. steel belt 4. metal net 5. spray layer

2.1 *Driving process*

Drilliing holes for short rockbolt on roof – – – – installing rockbolt of 1.6m long – – – – hanging curved beam – – – – drilling holes for long rockbolt on two corners – – – – installing rockbolt of 3.0m long – – – – fixing skid – – – – fastening rockbolt nut and curved beam.

2.2 *Grout injection*

2.2.1 Injecting materials and correlation parameter

Single – staged injection involves 425 # portland cement with the water – cement ration of 0.8:1.0 and sodium silicate of 45Be′ in desity with 4.5% of the weight of cement. Here the grout – injection rockbolts are inside ones of ordinary kind with 1.2m between them. In roadway with good roof, so long as the two walls are steady, so is the roadway;

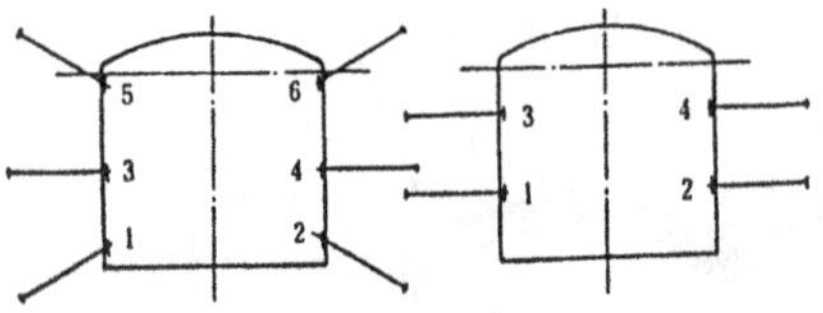

Fig. 5 Illustration to distribution of grout - injection rockbolts

therefore, grout - injection rockbolts will be used only on walls as Fig. 5 shows. To stop grout from backflowing, soft wood or reclaimed rubber plugs are used, or to plug directly on spot with quickhardening cement blasting or cement and sodium silicate binder.

2.2.2 Grout - injection process

Joint injecting pipes in advance. When injecting, do from above and with left and right in turn. Keep pumping pressure within 0 - 1.5MPa and stop when above 1.5MPa. Inject 100 - 200L for each hole. If over 300L has been injected, stop for it is likely to connect with goaves.

Since this technic then still under trial, we used TBW - 50/15 sludger instead, made 100L of iron barrels personally mixed grout by hand But now we have trial - produced corollary equipment of post - injection.

3 BRIEF ANALYSIS TO SUPPORT MECHANISM

Because of insufficient pretightening force and their failure to glue friable rock into a whole, it is difficult for ordinary rockbolts to form a closely build - up arch for bolt and rock. Once they are affected by strong tetonic stress or bearing stress, therefore, rockbolts will become unstable and damaged. But if grout is injected in through grout - injection rockbolts with surrounding rock being glued into a whole, the strength and hardness and bearing capability of surrounding rock will be greatly increased. Furthermore, grout can turn end - fixing bolts into full - length bolts, which doubles and redoubles its locking force and increases greatly the bearing stress of surrounding rock. This support system can be considered as a unit involving three groups: one is a bearing arch (beam) formed by bolt, net and spray; another is a combined arch (beam) formed by rockbolt locking bolt and rock; and the other is a rock arch (beam) glued with grout in its extension range. These three groups of arch (beams) are glued into a whole by rockbolts and give full scope together, which will increase greatly the bearing capability of support thus keeping roadway stable. Fig. 6 shows the structure of combined arch (beam) of support mechanism.

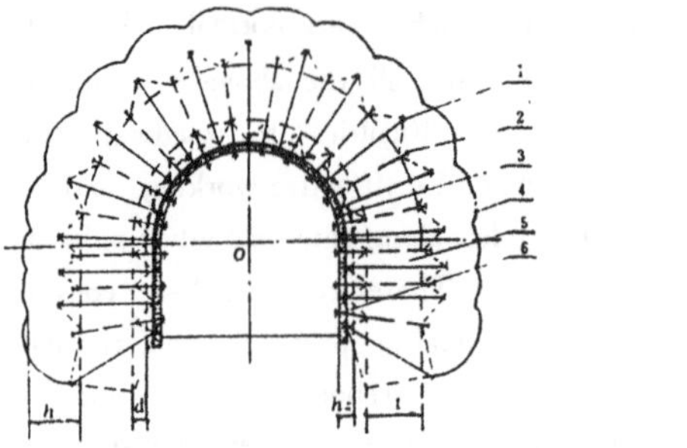

Fig. 6 Illustration to the structure of combined arch (beam) of support mechanism
1. ordinary rockbolt 2. grout - injection rockbolt 3. spray layer and steel belt 4. extension range of grout 5. combined arch formed by bolt and rock 6. combined arch formed by bolt, net and spray

4 TECHNICAL & ECONOMIC ANALYSIS AND ITS FUTURE OF APPLICATION & DISSEMINATION

Practice proves that working on stope face can be kept normal only after rebuilding once or twice between construction and backstoping whether bolt - spray supports or metal supports are involved in extremely unstable stope drift active workings; whereas we can have normal working without any rebuilding if bolt - spray supports and grout - injection rockbolts are used together. Compared with metal supports or conventional supports, 30% of direct expense can be saved with combined supports. And furthermore, most surrounding rock cracks will be blocked if grout – injection rockbolts are used, which not only tightens the roadways, but also prevents fire or even puts out fire. This tech has brought safety to operation in the pit.

Extremely unstable roadway is the most difficult to support at present in stope drift active workings. Except a few mines involving bolt – spray supports, most mines in the world use various metal supports that are necessary to rebuild before and after back – stoping, which greatly affects normal operation in the pit, increases ton – coal cost and also brings danger to operation. If conbined supports are beneficial to comprehensive mechanized coal mining method, and reduce maintenance thus increasing working efficiency. This tech, therefore, has fairly good economic and social benefit. Besides, it can be used in extremenly unstable roadways and blockhouses of such underground engineerings as railway tunnelings, highway tunnelings and so on, thus having a bright future for application and dissemination.

MAJOR REFERENCES

[1] 岩土工程中的锚固技术．中国岩土锚固工程协会编著．地震出版社, 1992 年

[2] 矿山压力及其控制．钱鸣高著．煤炭工业出版社, 1991 年

[3] Разработка и обоснование технологии упрочнения тонкотрещиноватых горных пород цементными растворами. А. В. Угляница, В. В. ПЕРМИН, Кемерово – 1997

[4] Coal Mine Ground Control. SYD S. PenG John Wiley & Sons, Inc New York, 1978

[5] 煤矿注浆技术．煤科总院建井所注浆室编．煤炭工业出版社, 1978

[6] 锚杆支护手册 . А. П. 希罗科夫著．煤炭工业出版社, 1992

Mechanics of Jointed and Faulted Rock, Rossmanith (ed.) © *1998 Taylor & Francis, ISBN 90 5410 955 6*

Research on an IDSS for rock support design in Jinchuan mine

Liu Tongyou & Zhou Chengpu
Jinchuan Non-ferrous Metal Corporation Jinchang, Gansu, People's Republic of China
Gao Qian
University of Science and Technology of Beijing, People's Republic of China
Ma Nianjie
China University of Mining and Technology, Beijing, People's Republic of China

ABSTRACT: Considering the complex geological conditions in No.2 Mine of Jinchuan non-ferrous metal Corporation, an Intelligent Decision Support System for Rock Support Design (IDSSRSD) has been studied and developed. This paper states the research ideas of the system, the structure of the system and the key technologies of the system; Finally the advances and problems which is being studied of the system are included.

1 INTRODUCTION

Jinchuan nickel mine lies in Jinchang city, which is situated in Gansu province, China. The mine is now the largest mineral deposit of nickel and copper in China and in the top rows of the world in ore reserves. Total quantity of ore is more than three hundred and twenty four million tons .The quantity of nickel metal is up to four thousand and one hundred thousand tons, and there are copper, cobalt, gold, silver, platinum, palladium etc. all together 18 kinds of metals. The designed produce capacity of the first project is nine hundred and ninety thousand tons per year. The second is two thousand six hundred and four thousand tons for No.2 mine. The mine began capital construction in 1986 and tried to produce in 1995 as well as full plough into operation in 1996. The same year produced ore was up to 2,000,000 tons. There are many kinds of import equipment about 135 suits in the mine and the ratio of mechanized production is 90% at present .At present the mine is the largest in produced capability and the highest level in modernization of production in China.

The field of mine underwent more one time geological construct actions so the engineering geological conditions in the mine are very complex. Faults and joints in the rock masses are excellent development and the horizontal stress (original stress) is high. More than 75% of the development openings are located in fractured or jointed rock masses, so it is very difficult to drive and maintain the stability of tunnels (Liu Tongyou 1996 , Liu. Tongyou, Tian Yongsui & Guo Shugao 1996).

In order to heighten the optimization support design level of entries in the fractured rock masses, an intelligent decision support system for rock support design has been studied and developed.

2 DESIGN IDEAS OF IDSSRSD

It is one of the most difficult problems for us to make decision of support design for tunnels in fractured rock. This is because support design contains a lot of uncertain and unknown factors. The theoretical analyses and numerical simulation are able to aid decision but these methods are not capable of simulating accurately the interaction between rock and support .Generally speaking, these models are only used to analyze a special procedure at a special time. So designers have to make design decisions with the help of practical experiences as well as engineering judgement. In fact the stability researches and support designs belong to an half-structure problem (E. Hoek & E. T. Brown 1980).

Intelligent Decision Support Systems (IDSS) were studied and developed just for the solution of no structure or half-structure problems. It is attracting for us to study and develop an Intelligent Decision Support System for Rock Support Design (IDSSRSD) for exaction and support design of entries in jointed and fractured rock in Jinchuan mine based on the theory and technology of IDSS.

DSS has been developed rapidly since its conception came into being in the 1970's. Through

*** This research was partly supported by coal fund of China**

continual probing, studying, applying, developing and perfecting of many experts and scholars in the academic world for more than 20 years, the connotation of DSS's conception and its theoretical basis and its relation with other technology have already been obvious and are getting ripe (Yao Qingda & Yang Wu 1990, Xu Jiepan 1993, Xu. Jiepan & Zhang Zhongmou 1990).

With the advance of theoretical research of DSS, many applying probes of DSS have been conducted . Chen Shifu developed an intelligent decision support system, NUIDSS (Chen Shifu et al 1994), which is used in economic forecast and management of Jiangsu province. Co-operating with Nanjing Forestry University, Chen Wenwei exploited an IDSS, PCFES, which can forecast the insect disaster of forestry (Chen Wenwei 1994) in China. Gao Hongchen (Gao Hongshen 1996) also developed an IDSS, IPDSS (Integrated oil refinery Production Decision Support System) (Gao Hongshen 1996). A DSS of tunnel engineering has been given for aided exaction (Kalamaras, G. S. 1997).

Because the complex geological conditions and difficulties in mining, co-operation with institutes and universities researches in home and abroad, researches have been performed in engineering geology, mining technology as well as support design of rock (Liu Tongyou, Gao Qian & Zhao Qianli in press). The researches

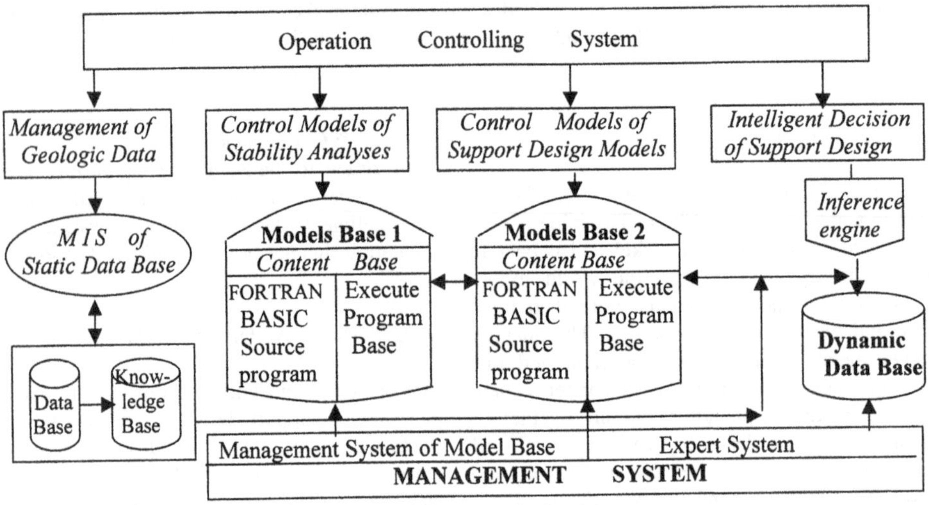

Figure 1 The structure of IDSSRSD

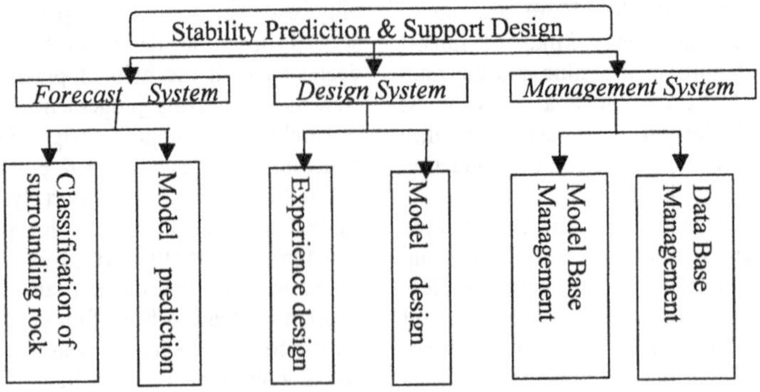

Figure 2 Functions of IDSSRSD

not only solved many problems of mining production but also promoted the advance of rock mechanics. The study results also established a foundation for development of IDSSRSD.

3 STUDY AND DEVELOPMENT OF IDSSRSD

3.1 The structure of the IDSSRSD

IDSSRSD was exploited by co-operation with University of Science & Technology of Beijing. The system contains Human Computer Interface, Data base Management System, Model base Management System and Expert System. The structure of IDSSRSD is shown in Figure 1 .

3.2 The functions of IDSSRSD

IDSSRSD has capability of stability prediction of the surrounding rock of entries, support design and information management. The functions of the system is seen in Figure.2.

3.3 The management information system of IDSSRSD

The Management Information System is a subsystem of IDSSRSD. The information of management deals with data of engineering geology including faults, shear zones and mainly rock, mining information related with mining methods and technologies. The subsystem contains 5 databases. They are YSWX.dbf, YSQD.dbf, YTQD.dbf, DLGZ.dbf and ZHSJ.dbf respectively. YSWX.dbf contains the parameters of physical quality of rock and ore in the No.2 mine . YSQD.dbf and YTQD.dbf are Data bases of parameters of different kinds of rock or ore strength and rock or ore masses respectively. DLGZ.dbf deals with the data of orientations and mechanical quality of faults and shear zones. ZHSJ.dbf is a database that contains experience parameters of support design of entries for all kinds of surrounding rock .

The subsystem was developed using the FoxPro management information software, which was developed by Microsoft Corporation. The subsystem possesses the functions of adding, deleting, revising, static analyses and printing output of the information. The information also is used in drawing some knowledge by using neural network expert system.

Forecast System deals with prediction of stability of surrounding rock based on classification of rock masses or forecasting models. The aim of the function is offering information for support design. Design system contains some models and knowledge which provide information for support design decision .

3.4 Forecast system

(1) Forecast by knowledge reasoning

The system contains a knowledge base and an inference engine. Based on the system the class of surrounding rock can be forecasted by receiving information from the dialog of human-computer. The system is composed of five parts, that is, knowledge base, dynamic data base, inference engine and user interface.

① Knowledge base

The expert system of IDSSRSD was developed with the help of the exploiting tool of expert system <Tian Ma>, which was developed by 6 studying units, that is , Institute of Mathematics, ACADEMIA SINICA, Zhejiang University etc. . The representation of knowledge uses "form + rule". The form knowledge and rule knowledge are saved in different files respectively .

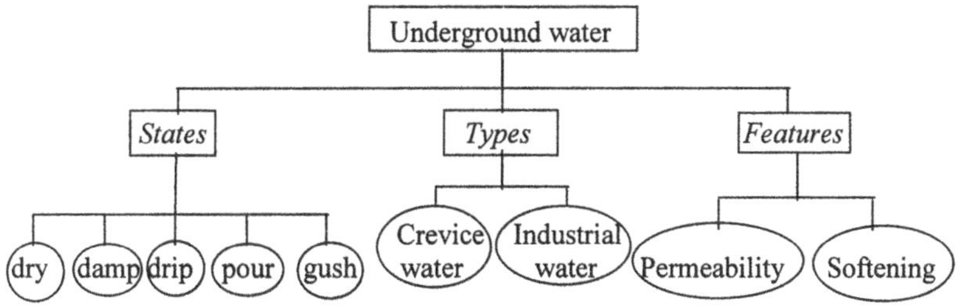

Figure 3 A tree structure figure of underground water factor

A form knowledge consists of environment head and one or several form definition bodies, that is,

$$\begin{cases} \#environ\{ \\ \quad \} \end{cases} \quad \text{Environment head}$$

$$\begin{cases} \#definecfm \\ \#endcfm \end{cases} \quad \text{Form definition body No.1}$$

...

$$\begin{cases} \#definecfm \\ \#endcfm \end{cases} \quad \text{Form definition body No.n}$$

In order to get a form knowledge, tree structure figures are used. Figure 3 shows a tree structure figure of underground water factor.

The rules in rule base are divided into five classes: that is, General Rule; Ante Rule; Trig Rule; Confirm Rule; and Deep Rule. General rules are production rules. The formation of a rule is as follows:

if condition 1 and condition 2 and ¡- condition n then conclusion

Based on knowledge base the class of surrounding rock can be determined and parameters of physical mechanical quality of rock masses are estimated as well as the types of structure of rock mass are recognized.

② Dynamic Data Base (DDB)

The dynamic base is used as saving original facts and obtained results from inference engine as well as the facts of answer of users, known facts and inferential facts .In other words, the system will save these needed facts for solving current problem. The contents of dynamic database come from the facts of user's answer through computer's questions by human computer interface.

③ User interface

The user interface is a bridge of users and expert system. In the system users can select the needed forecast methods according to the menu shown on computer's screen only by pressing Y or N and inputting some data.

(2) Forecast models of stability of surrounding rock

Forecast models are based on theoretical analyses or numerical simulation to estimate the stability of a surrounding rock. Considering the advance of rock mechanics , the system contains theoretical models, numerical models and key block models(Fig. 4).

3.5 Design system

Similarly, the forecast system of stability of surrounding rock design system involves theoretical models and numerical models. Selection of design models depends on designer's experiences, also refers to computer's hints that come from knowledge reasoning with the help of the knowledge base of the system and user's answer. The knowledge base of the design system can not only help selection of design models but also aid support design experience decision based on knowledge reasoning . The knowledge of knowledge base come from

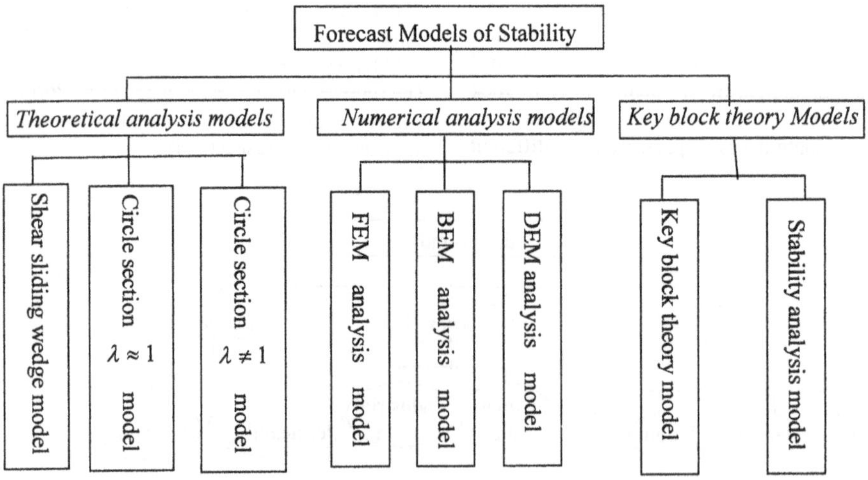

Figure 4 Forecast models of stability of surrounding rock

research results of Jinchuan mine field. Considering excavation and optimum design of some larger span and important tunnels, the system contains reliability analyses of stability of surrounding rock and optimum design models.

4 THE KEY TECHNOLOGIES OF THE IDSSRSD

In order to shorten studying time, Distributed Multimedia-Intelligence Decision Support System Platform (DM-IDSSP) which was studied and developed by National Defense University of Science & technology (Chen Wenwei 1994) was used to develop IDSSRSD .

DM-IDSSP possesses powerful functions that contain a development tool of expert system, machine learning, artificial neural network and model base management system etc. . Considering the features of IDSSRSD, the following key technologies were studied for IDSSRSD.

(1)Interface of expert system and model base

Model analyses need calculated parameters. The data file is used to offer needed parameters for traditional model analyses. But in IDSSRSD parameters come from knowledge reasoning based on knowledge base. Transition of the parameters has to be solved. Here a temporary data file *.TMP is set up .When carrying out knowledge reasoning , the temporary file automatically is formed . Model calculation first gets the parameters from the file *.TMP, then performs computation and the results are saved in the another file *.RUT. The calculated results is aided decision information or needed data of other model . So the transited ways by files can realize the aim of multi-models aided decision .

(2) Interface of expert system and database system

Data base has two types. One is static database which saves original data. The other is dynamic database which saves both reasoning knowledge and model computed results. With the help of database management system, information of data in the databases can be added, deled, revised and printed . Expert system also can obtain data from databases by reading of inference and aid decision .

(3) Knowledge acquisition from examples

Knowledge acquisition is the most important in IDSSRSD. In order to acquire knowledge from engineering cases, a method using Fuzzy Neural Network (FNN) model to extract fuzzy logic rules from examples is used (Zeng Shuqing Yan Jianjun & He Yongbao 1995). Knowledge acquiring algorithm there are five steps:

① Acquiring initial rule base;

② Tuning initial rules into FNN network, and performing learning by BP algorithm;

③ Clustering each hidden nodes in FNN network, forming simplified FNN network;

④ Continue to train the simplified FNN network using BP algorithm;

⑤ Tuning the network into rules, obtaining revised rules.

5 THE ADVANTAGES OF IDSSRSD

Compared with traditional design method IDSSRSD has the following advantages :

1 Combine models of stability forecast and design ,database and expert system into an entirety . Before IDSSRSD stability evaluated, selecting parameters and support design are carried out separately. The reliability depends on to some extent the experiences and theoretical capability of designers. IDSSRSD provides a tool that aids users enhancing analyzing ability and design level.

2 Posses capability of multi-models forecast and optimum decision.

The system provides a function of multi-models analyses and optimum decision. It is necessary for designers to solve complex mining design problems. Results of multi-models analyses can be displayed in table and printed when needed With the help of knowledge reasoning optimum design can be determined.

3 Models in the system can access data form database.

In the past needing parameters of a model analyses are provided by form a data file in advance . So it is not easy for users who do not understand the calculating method or do not possess operating experiences .The Model computed in IDSSRSD is not set up data file and models can access data directly from databases .

4 Building a larger database and database management system for stability forecast of surrounding rock and support design for mining entries in fractured and jointed rock masses.

In the past research results were saved by studying reports. So it is not convenient for us to manage, exchange and apply. The system has collected all data and research reports . The data were saved in disc with the help of database. Then data can be easily managed ,exchanged and applied.

5 The system posses the function of knowledge reasoning based on static and dynamic databases formed by initial information and inference respectively.

Due to complex geologic conditions and various influence factors, support designs have to consider special conditions, mining environment and technology . So design not only need model decision but also experience knowledge .The system can perform knowledge reasoning acquiring expert knowledge.

6 Set up a synthetic system with multi-functions IDSSRSD contains forecast of parameters of rock masses, stability analyses, support designs functions. etc. Making use of the system, designer can obtain experiential knowledge and deep information from model analyses.

7 The system operation under Chinese way
For convenience of application to users, development of IDSSRSD is under a Chinese system .

There are some shortcomings in the system , such as model management system is not perfect etc. The research of perfecting IDSSRSD still is continuing, by aid finance of Jinchuan Non-ferrous Metal Corporation and coal science fund. of China.

REFERENCES

Chen Shifu, Pan Jingui & Xu Dianxiang 1994. NUIDSS: An intelligent decision support system. *Journal of software, Vol.5, No.6, June, 32-37.*

Chen Wenwei 1994. Decision support system and development technique. Tsinghua University Press, Beijing .

E. Hoek & E. T. Brown 1980. Underground excavation in rock.

Gao Hongshen 1996. Decision support system - theory method case. Tsinghua University Press, Beijing .

Kalamaras, G. S. 1997. A computer-based system for supporting decisions for tunneling in rock under conditions of uncertainty. *Int. Journal Rock Mechanics Mining Sci. Vol.34, No. 3/4, 588, no.147.*

Li Haiquan 1995. Intelligence decision support system. *Computer System Application, No.2 12-14.*

Liu Tongyou 1996. On rock mechanics and engineering geology in Jinchuan. *Chinese J. of Rock Mechanics and Engineering, Vol.15 No.2 97-101.*

Liu Tongyou, Tian Yongsui & Guo Shugao 1996. Application and development of the calculated methods of rock stability for Jinchuan mine. *Research about Engineering Geology and Rock Mechanics for Mining of Jinchuan Nickel Mine, (1), 158-166.*

Liu Tongyou 1996. Problems of engineering geology and rock mechanics for Jinchuan mine. *Research about Engineering Geology and Rock Mechanics for Mining of Jinchuan Nickel Mine, (1), 171- 175,.*

Liu Tongyou, Gao Qian & Zhao Qianli in press. System analysis and integrated management for underground mining. *Lanzhou University Press.*

Liu Tongyou, Ma Nianjie & Gao Qian in press. stability reliability for roadway project. *China University of Mining & Technology Press.*

Tian Yongsui 1996. Analyzing methods of stability for rock masses and application in Jinchuan mine. *Research about Engineering Geology and Rock Mechanics for Mining of Jinchuan Nickel Mine, (1), 282-289.*

Xu Jiepan & Zhang Zhongmou 1990. The Design and Implementation of an intelligent decision support system generator. *Computer Research and Development, No.10 34-42.*

Xu Jiepan 1993. Some discussion of a knowledge model for intelligent DSS. *Journal of Software, Vol.4 No.2, 9-14.*

Yao Qingda & Yang Wu 1990. New a generation decision support System, *Institute of Software Zhongshan University .*

Zeng Shuqing, Yan Jianjun & He Yongbao 1995. Fuzzy neural network to extract rules from examples. *PR & AI, Vol.8 NO.1, 82-85.*

Zhang Hongen 1996. Research about blasting and bolt support for fractured and jointed rock. *Research about Engineering Geology and Rock Mechanics for Mining of Jinchuan Nickel Mine,(1), 526-536.*

Mechanics of Jointed and Faulted Rock, Rossmanith (ed.) © *1998 Taylor & Francis, ISBN 90 5410 955 6*

A control system for tunnel boring machines in jointed rock

A.H.Zettler, R.Poisel & D.Lakovits
Institute for Geology, Technical University of Vienna, Austria
W.Kastner
Department of Automation, Technical University of Vienna, Austria

ABSTRACT: More and more tunnels are excavated using tunnel boring machines (TBM). The possibilities to use the informations available during the excavation of a tunnel in order to determine the required support or evaluate a rock mass class (RMC) are discussed in the paper. Two different examples are used to show the power of a fuzzy based evaluation system. The same procedure can be used to control the tunnel boring machine during the excavation. The advantages and disadvantages of the suggested system are discussed.

1. INTRODUCTION

More and more tunnels are excavated using tunnel boring machines (TBM). The geological, hydrological and geomechanical conditions of rock masses where tunnels have to be driven through are getting more and more difficult. During the excavation of a tunnel the driver of the machine wants to increase his knowledge of the surrounding rock and of the interaction between the rock and the machine in order to increase the excavation rate and in order to minimise the risk of a collapse of the tunnel. A concept which is to help him in the future to optimise the performance of the machine and to avoid dangerous situations will be discussed in the paper.

Two examples are used to show the power of a fuzzy based system for evaluation of the data gathered during the excavation of a TBM driven tunnel. Before starting the excavation little is known about the rock mass and about the hydrogeology but knowledge increases while excavating. The data of the machine such as the thrust of the cutter, the required power consumption and the boring meters per day provide increasing information about the interaction between the rock and the machine. Both sets of data gathered during excavation and evaluated by a fuzzy system may be used for the classification of the rock mass and for controlling the boring machine.

2. SCHWARZACH PILOT TUNNEL (EXAMPLE 1)

The Schönberg tunnel is the main building to by-pass the town of Schwarzach St. Veit in Salzburg, Austria. The geological, hydrological, and geomechanical conditions were investigated by a 3km long pilot tunnel excavated with a 3.6 m diameter tunnel boring machine (TBM). The minimal and maximal overburden was 20m and 140m, respectively. Geological investigations showed that the tunnel was situated mainly in phyllite with a dominant joint set K1 and with dominant heavy jointed crushed zones.

2.1 Parameters

The required support for the Schwarzach pilot tunnel were evaluated taking three parameters into account (Fig. 1). The first parameter was the angle α between the tunnel axis and the strike of the foliation. The second parameter was the angle β between the tunnel axis and the dip direction of the intersection of foliation and joint set K1. The third parameter was the joint spacing. The investigations showed that the whole tunnel was not influenced by water. Water was not a parameter in this back calculation. The parameters were determined every 50 m in 57 cross sections in right angles to the tunnel axis. The measured parameters were evaluated using a fuzzy

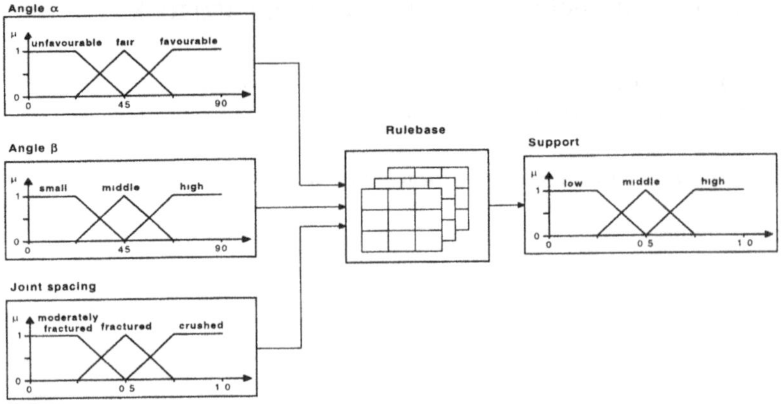

Figure 1: Fuzzy Set – Schwarzach; angle α : between the tunnel axis and the strike of the foliation ;angle β : between the tunnel axis and the dip direction of the intersection of foliation and joint set K1

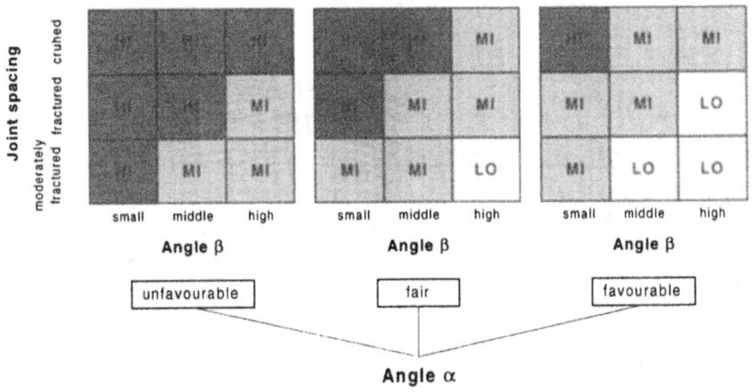

Figure 2: Rule base 1 - emphasis on rock structure orientation

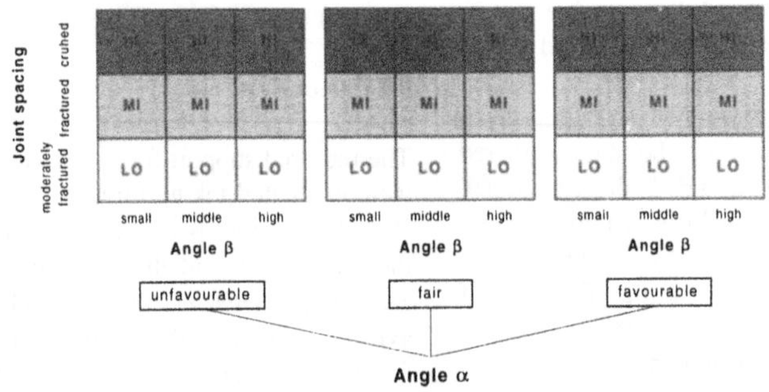

Figure 3: Rule base 2 - emphasis on joint spacing

set. The membership functions of the three input parameters and the output parameter are given in Fig. 1. The classes for the angle α between the tunnel axis and the strike of the foliation were called: unfavourable, fair, and favourable, and the angle β between the tunnel axis and the dip direction of the intersection of foliation and joint set K1 were called: small, middle, and large. The joint spacing classes were called: moderately fractured, fractured, and crushed. The membership function for the output variable support was divided into three classes called: low, middle, and high .

2.2 Rule bases

Different rule bases were used for the investigations. The first rule base (Fig. 2) emphasis the structure; which means, it puts more attention on the angle α between the tunnel axis and the strike of the foliation and the angle β between the tunnel axis and the dip direction of the intersection of foliation and joint set K1. This system was based on theoretical considerations: e. g. for an unfavourable angle α and a small angle β the value of support measures should be high even if the joint spacing is in the range: moderate fractured, because the geometrical conditions are unfavourable; if the angle α is favourable and the angle β is high the value of support measures should be low if the joint spacing is in the range: not fractured (see Fig. 2).
The second rule base emphasised the joint spacing more than the geometrical conditions expressed by the angle α and the angle β (Fig. 3). This rule base was a strict ordered system where not much attention was paid to the circumstance that the geometrical condition of the system would be unfavourable if the angle α was unfavourable and the angle β was small.

2.3 Results

In the fuzzy logic calculation the minimum maximum inference procedure (Cox, 1994) and the centre of gravity method (COG) without a weighting function was used.
Results for rule base 1:
A comparison of the actual support and calculated support based on the fuzzy rule base 1 (the structure was emphasised) showed that in some parts the results coincide quite good but in most cross sections the calculated support is in disagreement with the field data. This means that either the rule base or the system to describe the rock mass behaviour had to be changed (Lakovits, 1998).
Results for rule base 2:
Fig. 4 compares the calculated support for fuzzy rule base 2, in which the joint spacing was emphasised, with the required support. The same field data for the required support which were used to evaluate rule base 1 were now in good agreement compared to the to the fuzzy based calculated results using rule base2. The coincidence of both the actual support and the calculated support was not only for many cross sections in a row but also changed if the required support changed. This back calculation shows that for the special geological conditions (phyllite), with the low overburden and only a small influence of water the joint spacing dominants the required support (Lakovits, 1998).

3. EVINOS WATER SUPPLY TUNNEL (EXAMPLE 2)

The Evinos tunnel is a water supply tunnel in Greece to supply Athens (Jäger and Rudigier, 1994) . The 30km long tunnel has been excavated by 4 tunnel boring machines (TBM's); two open type TBM's erecting New Austrian Tunnelling Method (NATM) support and two Double Shield TBM's erecting a pre-cast honeycomb final lining. The excavation diameters varied from 4.0 m to 4.20m. The minimal and maximal overburden was 600m and 1300m, respectively. The main geological formations are fine grained flysch-facies (intercalations with banded clay-stones thin bedded, fine grained sand and silt-stones), irregular, chaotic structure of sandstone clay and silt-stone without any bedding, Triassic and Jurassic limestone (medium to thin bedded limestone with milimiter thin clay layers) and radiolarian chert (thin bedded radiolarian with intercalations of thin bedded clay-stone; the percentage of clay stone varies from 10% to 50 %) (Fig. 5). In the investigated areas the water has no big influence as can be seen in Fig. 6. In sections where the water is of greater importance it has to be taken into account. Including the water as parameter would enlarge the rule base but not change the whole system.
The data used in the paper were taken from an open type TBM excavation.

3.1 Parameters

The main target of the investigations was to evaluate the machine data in order to use them in the future to control a TBM. Those easy available machine data were the thrust of the cutter, the power consumption to maintain the machine, and the boring meters per day (Fig. 5). Those parameters were evaluated using a fuzzy based system (Fig. 6). All input parameters were divided in three classes. The classes for the thrust of the cutter were called: small, middle, and great. The classes for the power consumption to maintain the machine were called: low, middle, and high. The classes for the boring meters per day were called: few, middle, and many. The output variable rock mass class (RMC) was divided in six classes

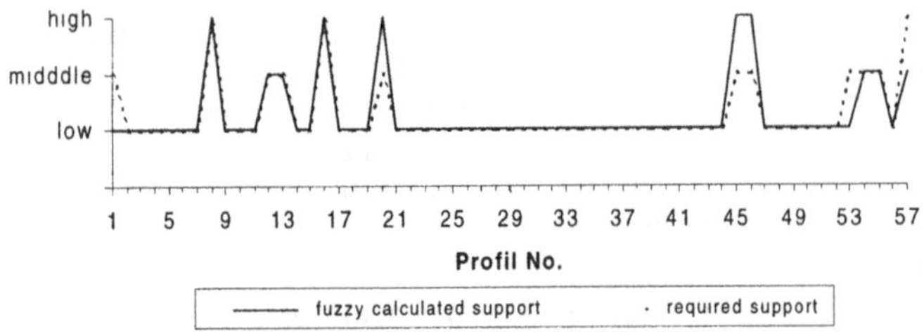

Figure 4: Required support – fuzzy calculated support for Rule base 2

(Fig. 6), they were enumerated from class I to class V$^+$, where class I indicate the best class and class V$^+$ stands for weak rock. The output RMC were enumerated from number 1 to number 6 (number 1 without support and increasing support up to number 6).

3.2 Rule base

As a very first approximation one rule base for the whole tunnel was used, not taking the influence of the changing geological conditions into account. Fig 8 shows the fuzzy logic rule base to evaluate the rock mass class (RMC) of the Evinos tunnel. The rules were based on experiences, theoretical considerations and tuned by an adaptive algorithm in order to get a good approximation of the actual RMC: e. g. if the thrust of the cutter is small, and the power consumption to maintain the machine is low, and the boring meters per day are few, than the RMC is high, and needs more support.

3.3 Results

The calculated RMC and the actual RMC are compared in order to evaluate the fuzzy system. The tuning of the rule base showed that if the thrust of the cutter, and the boring meters per day increase, and the power consumption remains constant at a low range, the RMC increases, which means that the required thrust must be a function of the RMC (the confinement of the rock) also, and not only a function of the drillability of the rock mass. The rule base shows further that a small value of the thrust of the cutter combined with a middle range power consumption leads to a RMC IV no matter if few or many boring meters per day are excavated.

The tuned rule base allows also the prediction that if the power consumption is middle, and the boring meters per day are few the RMC is in the upper range depending on the thrust of the cutter. A small thrust of the cutter leads to a better RMC than a middle or a great thrust. The RMC are constant at class IV if the boring meters per day are increased to middle range assuming the same middle range power consumption. If the boring meters per day are further increased to the range many the RMC decreases further (from RMC IV to RMC III; more stable rock) assuming the same middle range power consumption. If the power consumption is high the rock becomes better (from RMC IV to RMC II). For this range of power consumption and an excavation rate few boring meters per day the thrust of the cutter indicate whether the RMC III or the RMC IV has to be taken into account. For a middle range thrust of the cutter the RMC is higher (RMC IV) than for a small or for a great thrust of the cutter (RMC III).

This behaviour changes if the increasing rate of boring meters per day indicate a better RMC. In this case the middle range thrust of the cutter leads to the lowest range of the RMC (RMC II). Fig. 8 compares the fuzzy evaluated and the actual RMC. The results of both evaluation systems are in good coincidence. The results from tunnel meter 2000 to tunnel meter 3000 are in one geological formation in chert. The same procedure was used to evaluate the behaviour of a different petrographical formation and the results do not coincide to the same degree (Fig. 8). This leads to the assumption that for different geological formation different rules are required.

4. CONCLUSIONS

Both cases show that it is to some extend possible to use a fuzzy logic based system to evaluate a tunnel excavated by a TBM. If the conditions are clear enough it is possible to use a very small rule base with only a few rules. In both examples (Schwarzach and Evinos) it was possible to get reasonable close

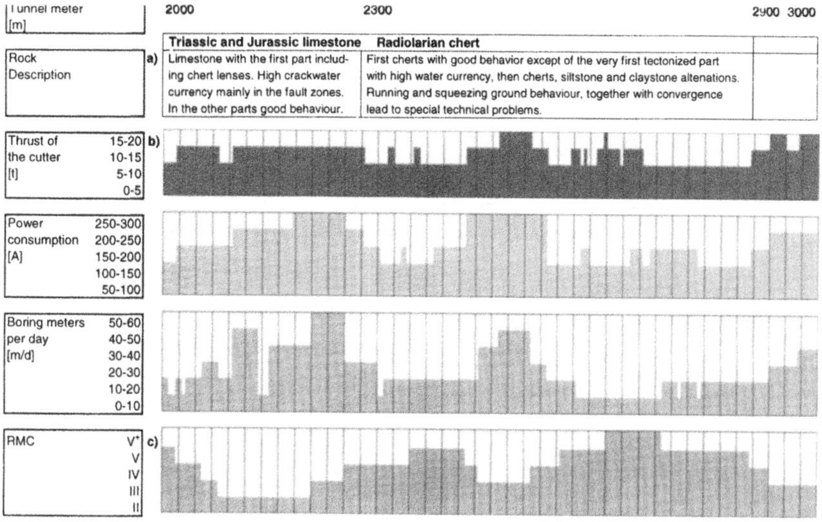

Figure 5: Evinos water supply tunnel: a) Geology , b) Machine data (thrust of the cutter, boring meters per day, and required power consumption to maintain the machine), c) Rock mass class (RMC)

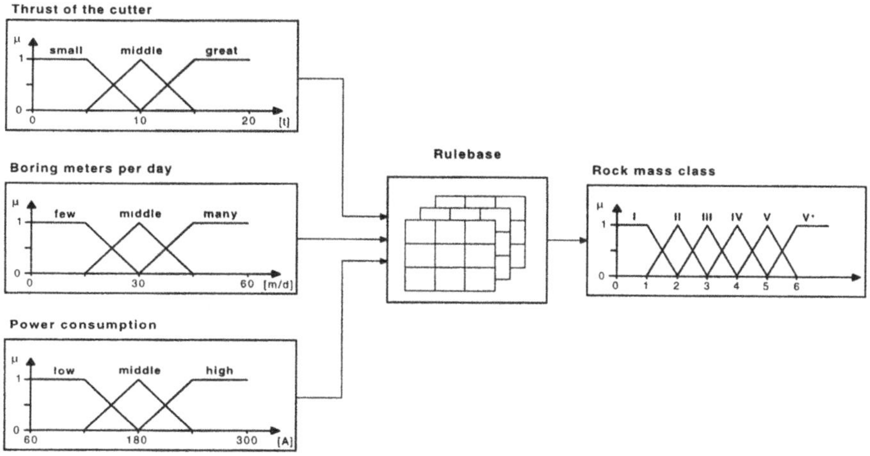

Figure 6: Fuzzy Set for Evinos

results compared to the measured data in the field. In the first example the joint spacing was the leading parameter. If this parameter is controlled during the excavation all the other observations are additional informations (Zettler, Poisel & Stadler; 1996). In both cases different systems were used to evaluate the measures in the field, that means that for different cases different leading parameters are necessary. That shows further that it is very difficult and to some extend not possible to decide in advance what the leading parameters are in the field. On the basis of geological, hydrological, and geomechanical investigations it should be possible to detect the most important and therefore leading parameters which determine the e. g. required support or the RMC.

The second example showed that it is possible to calculate e. g. a RMC by including parameters like thrust of the cutter, power consumption to maintain the machine and boring meters per day. The relations which lead to a quite good relationship between

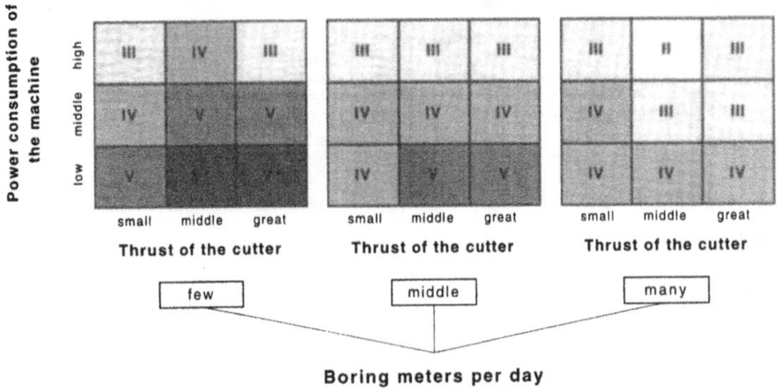

Figure 7: Rule base for Evinos

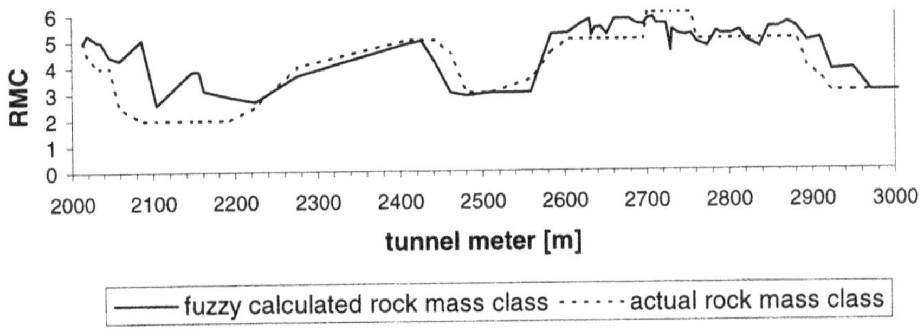

Figure 8: Actual rock mass class – fuzzy calculated rock mass class

actual RMC and calculated RMC gives hope that it is possible to use a more complex interrelation matrix to control a TBM and/or to optimise the required thrust, the necessary cutting tool change, etc. The dependencies of petrography, water, geomechanical conditions, tectonical circumstances etc. are too complicated to investigate one overall rule base. Based on geological, hydrological and geomechanical investigations the introduced system could be able to guide the driver during excavating a tunnel using a TBM. The great advantage of this fuzzy system is that it is possible to evaluate the fuzzy rules.

ACKNOWLEDGEMENTS

The authors gratefully acknowledge the support of the provincial government of Salzburg (Austria) and of JAEGER limited.

REFERENCES

Cox, Earl, (1994): „The Fuzzy Systems Handbook" Academic Press, Inc.

Fairhurst, C. and Lin, D. (1985): „Fuzzy methodology in tunnel support design" Proc. 26th U. S. Rock Mechanics Symposium, pp. 269 - 278.

Jäger, M and G. Rudigier (1994): „Baudurchführung des 30 km langen Trinkwasserstollens Evinos-Mornos in Griechenland" Felsbau 12 (1994) # 6 pp. 500 - 504 (in German).

Lakovits D. (1998): „Gebirgsklassifizierung von mechanisch vorgetriebenen Tunnelbauten mit Hilfe von Fuzzy Logik Konzepten". Master Thesis Technical University of Vienna (in German).

Zettler, Alfred H. R. Poisel and G. Stadler. (1996): „Bewertung geologisch-geotechnischer Risiken mit Hilfe von Fuzzy Logic und Expertensystemen" Felsbau 14 (1996) # 6 pp. 352 - 357 (in German).

Mechanics of Jointed and Faulted Rock, Rossmanith (ed.) © 1998 Taylor & Francis, ISBN 90 5410 955 6

Voussoir beam analysis of an underground marble quarry

A. I. Sofianos
National Technical University, Athens, Greece

A. P. Kapenis
Edafomichaniki Ltd, Maroussi, Greece

C. Rogakis
Dionyssomarble Co., S.A., Ambelokipi, Athens, Greece

ABSTRACT: The exploitation of marble in Greece is extended lastly underground. Its productivity is largely dependant on the ability to excavate large stable openings. The prediction of the mechanical response of such an exploitation is endeavoured with the observational method. Numerical models and closed form solutions provided by the voussoir beam theory are used to analyse the structural behaviour of the excavation. The necessary initial estimates for the deformability data of the in-situ surrounding rock are evaluated by back-analysing the response of a pilot tunnel. These data are used then to calculate at a first stage the stability of the enlarged opening. This observational method may be repeated then contiguously at the various subsequent stages of the excavation enlargement, which is performed in slices, by using the updated data evaluated at each stage of the underground structure, to the analysis of the following ones.

1 INTRODUCTION

Dionyssos Marble Co., which is the second largest marble producer in Greece, is recently forced by environmental restrictions to adopt a strategy of extending its operations underground. Such quarries are only few all over the world yielding only 2% of the world marble production, and this is the first one attempted in Greece. It is located at the Pentelikon mountain, 20 km north-east of Athens at an average elevation of 430 m. Of the two main marble horizons existing in the area, the lower one is the main source of the well known white marble which has been exploited since the ancient times. The thickness of the lower horizon varies between 500 and 650 m and underlies the 600m thick Kaissariani schist horizon, which becomes outcropping there.

Currently, a single experimental exploitation is operated by the company, where the extraction is based on a modified room and pillar method. Due to insufficient experience of the underground rock mass behaviour, the maximum room size was limited initially to a height of 15m and a width of 15 m with 10 m wide square pillars. The experience gathered during this pilot quarrying gives encouraging results for the feasibility of wider and higher underground exploitations. This will not only increase recovery but also will facilitate more efficient utilisation of the machinery used for the extraction, such as diamond saw and diamond wire machines. Lastly, it is decided to extend the existing underground exploitation to the north-east boundary starting with a new entrance. The aim is to allow there for working conditions similar to those of the surface exploitation. Thus, a preliminary target is set to increase the size of the underground rooms up to 25 m or more wide and 30 m high, supported by square pillars 10 m wide.

Productivity of the marble and hence profitability is intimately connected with reliable design of the subsurface excavations. Important is the evaluation of the limiting dimensions of the rooms in conjunction with their pertinent safety margins against a stability failure. The observational method (Peck, R.B., 1969) is used for the prediction of the stability of these openings. Their analysis is accomplished with design charts and closed form solutions provided by voussoir beam theory, and numerical simulation of the excavation procedures. The initial mechanical properties of the surrounding rock are based on available data and on the back-analysis of the measured response of a pilot tunnel.

2 PILOT OPENING

In order to study the behaviour of the rock in-situ, the pilot tunnel shown in Figure 1, starting at the new entrance, with a 6 m wide 3 m high room was excavated and finally widened to 9m. In this region the roof of the tunnel consists of marble which is overlaid by a schist horizon.

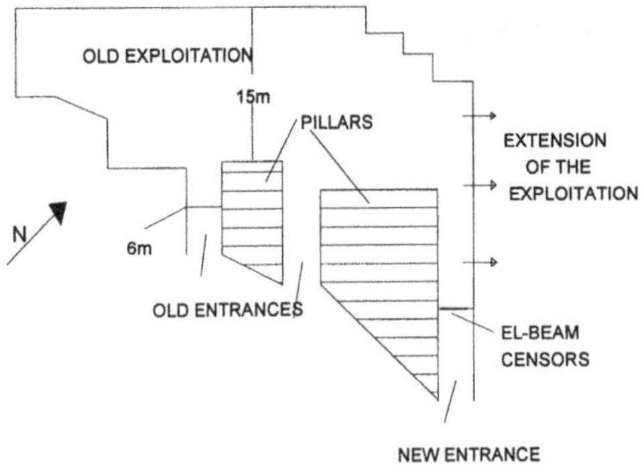

Figure 1. Plan view sketch of the pilot tunnel

The intact rock deformation modulus E, the Poisson's ratio ν, the unconfined rock strength q_u, and the unit weight γ of these rocks are estimated, according to laboratory tests (e.g. Tsoutrelis et. al 1997; Livadaros 1994) and literature review data (e.g. Lama & Vutukuri 1978) to be as given in Table 1.

Table 1. Laboratory mechanical parameters

Horizon	E[GPa]	ν	q_u[MPa]	γ[kN/m^3]
Schist	<4	0.34	>5	23
Marble	<47	~0.25	>42	26

In order to evaluate the in-situ deformation modulus of the rock mass the deflections of the roof were measured during the excavation of the roof using a multiple EL Beam (Slope Indicator 1994) sensor system.

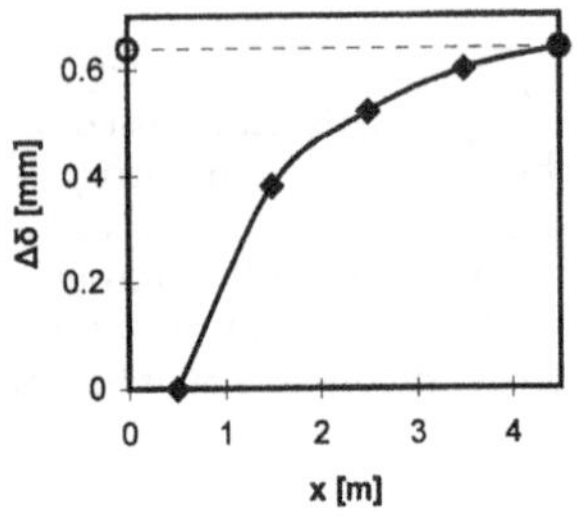

Figure 2. Measured deflections across the span

When the exploitation of the entrance started, a set of five El-beam sensors, 1m long each, was installed into the roof to monitor the roof deflections across the span during the exploitation progress. Each EL-beam set consists of an electrolytic tilt sensor attached to a rigid metal beam one meter long, which is mounted on anchor bolts that are set into the roof. The measuring accuracy of the sensor is 0.005 mm per meter of beam. All five beam sensors were then linked end-to-end and were installed at the crown, almost at the face of the excavation; laterally, they started 0.5 m from the pillar. Changes to the tilt angle of each beam, caused by the roof displacements, were measured at regular time intervals and were then translated into vertical displacement. The distribution of the measured deflections Δδ across the span at the completion of the exploitation are illustrated in Figure 2. These values do not contain any deflections prior to the installation of the instruments at the crown. However, it is anticipated that these initial deflections which are reacted by column action of the rock face are insignificant compared to the deflections undertaken by bending action of the roof beam.

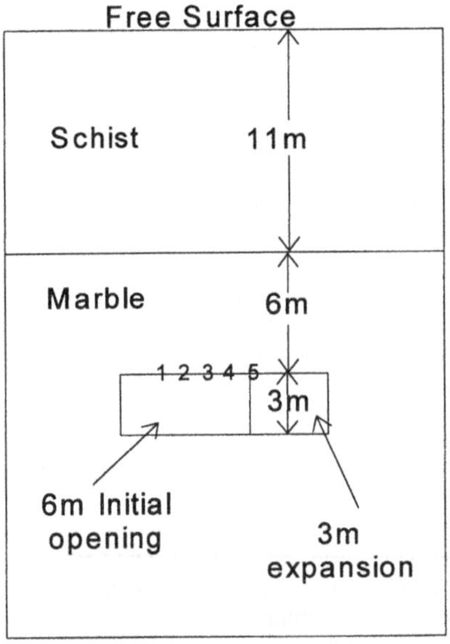

Figure 3. Model of the pilot tunnel excavation

The analysis of the deflections of the tunnel is performed with an elastic finite element computer code. The conceived model, which is shown in Figure 3, assumes a marble stratum 6m thick immediately above the roof of the tunnel overlaid by a schist stratum 11m thick. The deformability of the roof strata is defined by a range of values for E below the upper limits given in Table 1 and the values of ν given in the same Table. Loading of the strata is due to their own unit weight given also in the Table. Output of the model is the deflection at midspan of the tunnel roof and at a distance 0.5m from the abutment, which is the starting point for the EL-beam sensor set. Of interest is the differential deflection $\Delta\delta$, which is defined as the deflection δ at midspan minus the deflection δ_1 at the starting point of the EL-beam sensor set. In Figure 4 the differential deflection $\Delta\delta$, as evaluated by the numerical model is related to the modulus of deformation of the marble of the tunnel roof for three values of the deformability modulus of the schist. It may be observed that the three curves almost coincide. Thus, by using the graph of the figure, the in-situ deformability modulus of the surrounding rock may be evaluated from the measured deflection to be ~4GPa.

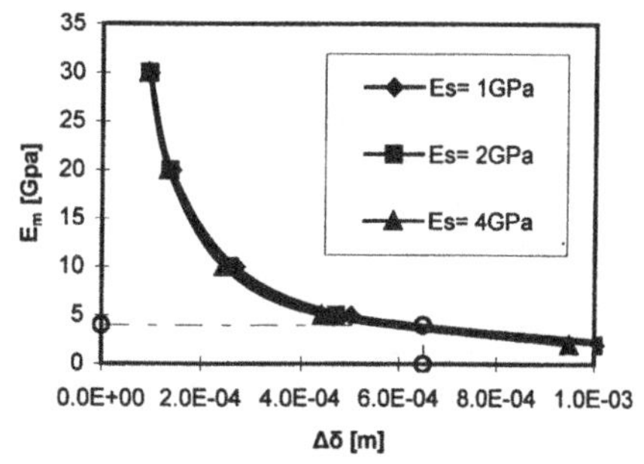

Figure 4. Pilot tunnel deflection sensitivity study

3 FINAL OPENING

The particular scheduled to be enlarged with a step by step excavation of vertical and horizontal slices, until it reaches a maximum width of 30 m and a height of 18 m. For this span the roof is anticipated either to contain a crack at midspan or a crack may be induced. The behaviour of such a cracked rock roof is analysed using numerical and analytical procedures.

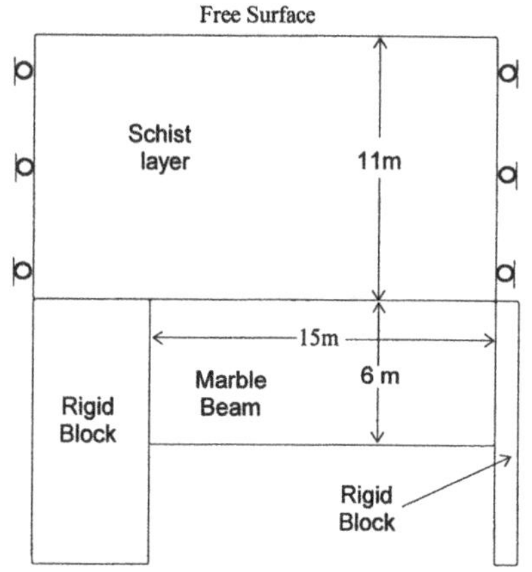

Figure 5. Model of the enlarged opening

3.1 Numerical simulation

The numerical analysis of such an opening necessitates the use of a discontinuum code; UDEC™ (1993) is considered as a most appropriate one. The model used is shown in Figure 5. It consists of two rigid blocks, one at the abutment and one at the midspan, and two deformable ones which simulate the schist and marble layers. The left rigid block simulates the pillar, whereas the right one imposes the appropriate boundary conditions at the midspan. The marble rock is assumed to be elastic, whereas the schist is considered to follow an elasto-plastic behaviour with strength parameters c=1MPa and $\varphi=30^{\circ}$ of the Mohr-Coulomb failure criterion. The values of Table 1 are assumed for ν and γ. For the modulus of deformability of the schist and the marble the same value ranges used for the pilot tunnel modelling are used. The boundaries between the rigid blocks and the marble beam are considered as joints of which the left does not allow vertical movement. The interface between the two rock horizons is modelled as a joint too. Loading is assumed to be only gravitational.

Elastic behaviour is observed for all models within the schist horizon. Output of the computations are the deflections at midspan and the mean normal stress σ_{nj} acting on top of the marble horizon. In Figure 6 the calculated deflections are

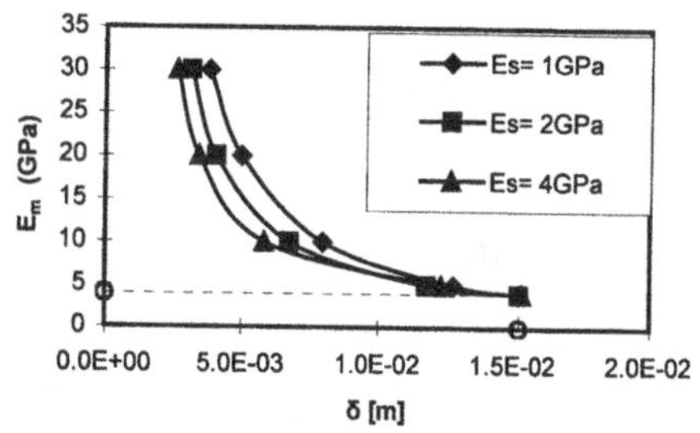

Figure 6. Enlarged opening deflection sensitivity study.

related to the in-situ deformability of the marble for three values of the deformability of the overlying schist. It may be observed that for the back-analysed, in Figure 4, value of E for the marble, the anticipated deflection is 15.2÷15.3mm.

In Figure 7 the mean normal stress σ_{nj} acting immediately over the span on the interlayer surface between the marble and the schist is related to the

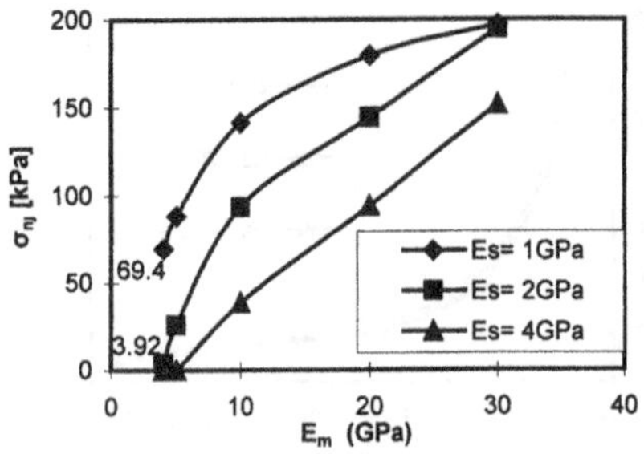

Figure 7. Enlarged opening interlayer stress sensitivity study

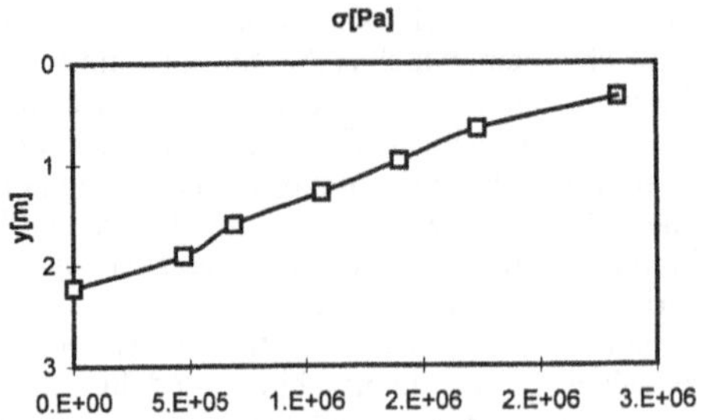

Figure 8. Stress distribution at the midspan joint of the enlarged opening for E_s=1GPa, E_m=4GPa

deformability E_m of the marble for the same three values of the deformability E_s of the schist.

In Table 2 the evaluated deflections δ at midspan, the maximum (extreme) stresses σ_{xm} at midspan, the interlayer normal mean stresses σ_{nj} and the surcharge factor k_q are given for the estimated, in Figure 4, value for the deformability modulus of the marble. There, it may be observed that although the mean stress σ_{nj} on the interlayer is significantly increased with the reduction of the deformability of the schist, the deflection is not affected by it. This is because this surcharging pressure acts close over the pillars, whereas it becomes zero for most part over the span.

$$k_q = 1 + \frac{\sigma_{nj}}{\gamma \cdot t_m} \tag{1}$$

The latter values, indicate that full arching develops for E_s>4GPa, whereas only partial arching develops for lower values.

In Figure 8 the stress distribution across a joint at the midspan is drawn, from the top of the marble stratum, for the back-calculated value of E_m.

3.2 Analytical methods

If cracks occur in the marble roof stratum only at the abutments, the deflection of this non surcharged stratum may be approximated by the elastic solution (Timoshenko & Goodier, 1970) for a free rotating high wall beam, given by :

$$\delta_n = \frac{5}{32} \cdot Q_n \cdot s_n^3 \cdot \left[1 + \frac{2.4}{s_n^2} \cdot \left(0.8 + \frac{\nu_m}{2}\right)\right] \approx 2mm \tag{2}$$

A fifth of this value is obtained if cracks are not anticipated at the abutment. However, these values are much smaller from those corresponding to a cracked beam.

Table 2. Response of final opening; E_m=4GPa

E_s[GPa]	1	2	4
δ[mm]	15.2	15.2	15.3
σ_{xm}[MPa]	2.3	2.6	2.7
σ_{nj}[kPa]	69.4	3.92	0
k_q	1.44	1.03	1

The analytical procedure developed by Sofianos (1996) allows for the evaluation of the maximum deflection and strain in a voussoir beam, if full arching, i.e k_q=1, is developed by the overlying strata. This may be evaluated as follows :

$$\begin{aligned} s_n &= \frac{s}{t_m} = 5 \\ n &= \frac{1}{0.22 \cdot s_n + 2.7} = 0.26 \\ Q_n &= \frac{\gamma_m \cdot s}{E_m} = 0.000195 \\ z_{on} &= 1 - \frac{2}{3} \cdot n = 0.82 \\ s_z &= \frac{s_n}{z_{on}} = 6.06 \\ \delta_{zo} &= \delta_z \cdot (1-\delta_z) \cdot (1-\delta_z/2) = \\ &= Q_n \cdot s_z \cdot \left[\frac{1}{16} \cdot s_z^2 + \frac{1}{3}\right] = 0.0031 \approx \delta_z \\ \delta &= t \cdot z_{on} \cdot \delta_z = 15.4mm \end{aligned} \tag{3}$$

$$\begin{aligned} \varepsilon_x &= \frac{Q_n \cdot s_z}{4 \cdot n \cdot (1-\delta_z)} = 0.00113 \\ \sigma_x &= E_m \cdot \varepsilon_x = 4.5MPa \end{aligned} \tag{4}$$

where,

γ_m : unit weight of the marble, i.e. 26kN/m^3

E_m : modulus of deformability of the marble, i.e. 4GPa

t_m : thickness of the marble stratum above the roof, i.e. 6m

s : span of the opening, i.e. 30m

δ : deflection at midspan

v_m : Poisson's ratio of the marble assumed 0

k_q : Surcharge factor, assumed 1.

An almost complete agreement may be seen between the deflection evaluated by eq. 3 and the calculated numerically value given in Table 2. The extreme stress σ_x evaluated by eq. 4 is a mean value of a smaller one at midspan and a larger one at the abutment. This may be compared with the numerically calculated extreme stress at the midspan given in Table 2.

The small values of this deflection and of the equivalent maximum stress ensure stability against buckling failure or crushing of the rock respectively. This may also be illustrated graphically in Figure 9, for the above calculated pair of values of s_n and Q_n. The state corresponding to this pair of values lies well to the left of the buckling curve (FSb=1) and lies almost at the curve with ε_x=0.001.

However, for such thick beams, a danger against sliding failure exists. In the graph of this figure it may observed that the geometry of the beam defined by s_n requires, in order to ensure against sliding failure, a friction angle larger than $\theta_{max}=35^o$. Thus, a persistent vertical crack at the abutment, pre-existing or induced, with a friction angle less than 35° might cause sliding failure to the roof.

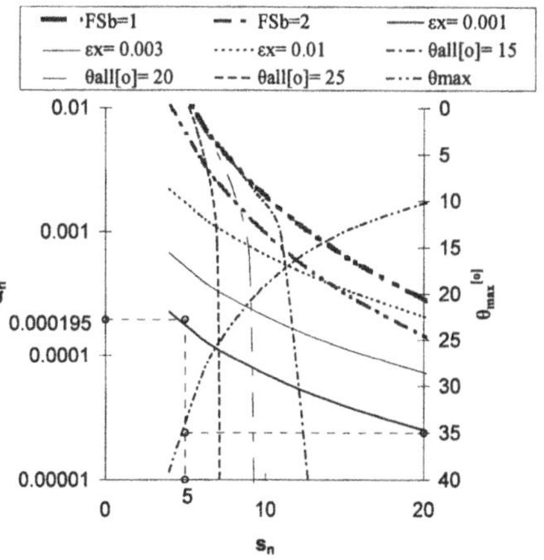

Figure 9. Roof stability chart

By monitoring with further measurements during the exploitation of the next wall slices and by back-calculating the in-situ properties of the surrounding rock, the confidence on the predicted in-situ behaviour of the rock strata will be increased, and a close control may be established on the response of the rock mass to the process of excavation.

4 CONCLUSIONS

The observational method has been used in order to evaluate, at first stage, the stability of a large unsupported roof span in an opening of an underground exploitation. There, the existence of pre-existing or induced vertical cracks is assumed. Major unknowns to any analysis of the stability of such an opening are the deformability properties of the surrounding rock. In order to evaluate these properties, measurements were obtained during the opening of a pilot tunnel, which were then back-analysed numerically. Forward numerical analysis was then performed to evaluate the stability of the final exploitation. Further, forward analysis is performed by using established stability charts and closed form solutions based on voussoir beam theory. However, the latter may provide immediate answer only if any uniform surcharge on the marble roof beam is known, e.g. if complete arching is developed by the overlying strata.

ACKNOWLEDGEMENTS

The authors are grateful to Dionyssomarble Co. for their help during the site investigation and for permitting publication of collected information.

REFERENCES

Lama R.D. and Vutukuri V.S., 1978. *Handbook on mechanical properties of rocks*, vol. 2, Trans Tech publications.

Livadaros R. 1994. *Investigation of the consequences from the use of the detonating cord in marble block extraction.* Ph.D. thesis presented at the NTUA, in Greek.

Peck R.B. 1969. Advantages and limitations of the observational method in applied soil mechanics. *Geotechnique* 19, No.2, 171-187.

Slope Indicator, 1994, *Applications Guide.* Geotechnical, Environmental and Structural Instrumentation, Second Edition.

Sofianos A.I. 1996. Analysis and design of an underground hard rock voussoir beam roof. *Int. J. Rock Mech. Min. Sci. & Geomech. Abstr.* Vol.33, No. 2, pp. 153-166.

Timoshenko S.P. and Goodier J.N., 1970. *Theory of elasticity.* McGraw Hill, Kogakusha.

Tsoutrelis Ch.E., N. Gikas, and Nomikos P. 1997. *Improvement of dimension stone quarrying techniques by using detonating cord.* PENED 91ED652, in Greek.
UDEC, 1993. Itasca consulting group, Inc.

Slope stability

Mechanics of Jointed and Faulted Rock, Rossmanith (ed.) © 1998 Taylor & Francis, ISBN 90 5410 955 6

Stability analysis of jointed rock slope

S.Ohtsuka
Department of Civil and Environmental Engineering, Nagaoka University of Technology, Japan
M.Doi
Department of Civil Engineering, Takamatsu National College of Technology, Japan

ABSTRACT : Stability analysis for jointed rock structure is developed by taking account of contact condition at cracks based on the lower bound theorem in plasticity. The applicability of the proposed method is examined through the comparisons with the upper bound calculations conducted by Tamura(1990). The analysis is applied to the stability assessment of jointed rock slopes where typical inherent failure modes might take place. The non-linear shear strength of crack against the confined stress is also introduced into the analysis.

1 INTRODUCTION

The stability of rock structure is largely affected by macroscopic cracks distributed in the rock. It is very important to take account of discontinuity of cracks in the assessment of stability. Many researches have been accumulated on the stability assessment, however, the structure has been usually modeled into two extreme conditions of (1)continuity model and (2)fully discrete block model. However, the behavior of rock structure has both aspects of continuity of intact rock and discontinuity of cracks in fact. The failure mode of jointed rock structure might be determined by the relationship of shear strengths between intact rock and cracks.

This study employs the numerical procedure for the stability assessment of jointed rock structures based on the lower bound theorem in plasticity. The contact condition of crack is directly introduced into the analysis. With the use of joint element to estimate the traction along cracks, the system of lower bound analysis gets fallen into the continuum mechanics framework. It can be easily solved with finite element discretization technique (Ohtsuka, 1997a, 1997b).

This study discusses the applicability of the analysis with the ultimate bearing capacity of footing on the ground which includes discontinuous line inside. Tamura(1990) gave the upper bound solution on the ultimate bearing capacity with the assumption on dissipation energy along discontinuous line. It gives good case studies and the comparisons are conducted between lower and upper bound analyses.

Rock slopes shows some typical failure modes of jointed rock structures as sliding, toppling and buckling depending on the conditions of geometry and shear strength characteristics of macroscopic cracks. This study investigates the effects of both geometry and shear strength of cracks on the stability of jointed rock slope. The shear strength of crack is well known to show the non-linearity against confined stress (Jaeger, 1971). This study develops the stability analysis which takes account of non-linear strength of cracks. The applicability of developed method is also discussed.

2 STABILITY ANALYSIS OF JOINTED ROCK

This study employs the lower bound theorem in plasticity (Koiter, 1960) for the stability assessment of rock structures. The rocks is modeled as the elastic perfectly plastic material.

2.1 *Lower Bound Theorem*

The residual stress which yields due to plastic deformation is defined here as the difference between the true stress $\boldsymbol{\sigma}$ and the elastic stress $\boldsymbol{\sigma}^e$ such as

$$\boldsymbol{\sigma}^r = \boldsymbol{\sigma} - \boldsymbol{\sigma}^e. \tag{2.1}$$

$\boldsymbol{\sigma}$ and $\boldsymbol{\sigma}^e$ satisfy the equilibrium equation, respectively so that $\boldsymbol{\sigma}^r$ is self-equilibrate.

The lower bound theorem assures that a rock structure is safe against the external force $\boldsymbol{F}(t)$ if any time independent residual stress $\overline{\boldsymbol{\sigma}}^r$ can be found everywhere in the structure satisfying

$$\boldsymbol{\sigma}^e(t) + \overline{\boldsymbol{\sigma}}^r = \boldsymbol{\sigma}^s(t), \quad f(\boldsymbol{\sigma}^s(t)) < 0 \tag{2.2}$$

where f is a yield function of rock. If a structure is safe for the applied load, the behavior of it is proven to shakedown to be elastic against any repeat of load.

2.2 *Analytical Procedure*

The lower bound theorem gives a lower solution for the exact stability so that it is formulated as a maximization problem. With the use of linear yield functions such as

$$N^T\boldsymbol{\sigma} = N^T\left(\boldsymbol{\sigma}^e + \overline{\boldsymbol{\sigma}}^r + \boldsymbol{\sigma}_o\right) \leq \boldsymbol{K} \tag{2.3}$$

where the suffixes "o" denotes the initial stress, the lower bound analysis can be formulated as a linear programming problem with finite element discretization technique (Maier,1969).

The bearing capacity analysis against the external force $\boldsymbol{F}$ can be formulated with a load factor α against $\boldsymbol{F}$ such as

$$s = \max\left\{\alpha \left| \begin{array}{c} N^T\left(\alpha\boldsymbol{\sigma}^e + \overline{\boldsymbol{\sigma}}^r + \boldsymbol{\sigma}_o\right) \leq \boldsymbol{K} \\ B^T\boldsymbol{\sigma}^e = \boldsymbol{F} \\ B^T\overline{\boldsymbol{\sigma}}^r = \mathbf{o} \end{array}\right.\right\} \tag{2.4}$$

where s indicates the ultimate bearing capacity. B is a matrix correlating the stress vector with the external force vector. The second and third equations describe the equilibrium equations on elastic and residual stresses, respectively. It is noted that the redistribution of stress is considered with the residual stress $\overline{\boldsymbol{\sigma}}^r$ which is determined by solving the boundary value problem.

2.3 *Contact Condition of Joint*

The traction $\boldsymbol{q}$ at cracks has two components of normal stress q_n and shear stress q_s. On the normal stress working on cracks, the extension in stress is usually not permitted as $q_n \leq 0$ since the extension stress is defined as positive. Even if there is a certain cohesion c_n at cracks, the contact condition on q_n is followed by $q_n \leq c_n$.

On the contrary, the contact condition on the shear stress at cracks can be defined by introducing shear models. The contact condition of friction model is described as $q_n tan\phi_d - c_d \leq q_s \leq -q_n tan\phi_d + c_d$, where c_d, ϕ_d are the cohesion and frictional angle at cracks. These conditions are expressed in the following equation:

$$\begin{bmatrix} 1 & \tan\phi_d \\ -1 & \tan\phi_d \\ 0 & 1 \end{bmatrix}\left\{\begin{array}{c} q_s \\ q_n \end{array}\right\} - \left\{\begin{array}{c} c_d \\ c_d \\ c_n \end{array}\right\} \leq \mathbf{o}. \tag{2.5}$$

The contact conditions at cracks indicate the constraint conditions on possible traction at cracks so that it can be expressed with Eq.(2.3) in linear equation such as

$$\overline{N}^T\left(\left\{\begin{array}{c} \boldsymbol{\sigma}^e \\ \boldsymbol{q}^e \end{array}\right\} + \left\{\begin{array}{c} \overline{\boldsymbol{\sigma}}^r \\ \overline{\boldsymbol{q}}^r \end{array}\right\} + \left\{\begin{array}{c} \boldsymbol{\sigma}_o \\ \boldsymbol{q}_o \end{array}\right\}\right) \leq \overline{\boldsymbol{K}}. \tag{2.6}$$

2.4 *Stability Analysis of Rock Structure*

By employing the joint element proposed by Goodman *et al.*(1968), the equilibrium equation can be expressed in terms of stress $\boldsymbol{\sigma}$ in rocks and the traction $\boldsymbol{q}$ at cracks. The bearing capacity analysis for rock structure including cracks is formulated in the same way of Eq.(2.4) as the following:

$$s = \max\left\{\alpha \left| \begin{array}{c} \overline{N}^T\left(\alpha\left\{\begin{array}{c} \boldsymbol{\sigma}^e \\ \boldsymbol{q}^e \end{array}\right\} + \left\{\begin{array}{c} \overline{\boldsymbol{\sigma}}^r \\ \overline{\boldsymbol{q}}^r \end{array}\right\}\right. \\ \left. + \left\{\begin{array}{c} \boldsymbol{\sigma}_o \\ \boldsymbol{q}_o \end{array}\right\}\right) \leq \overline{\boldsymbol{K}} \\ \overline{\boldsymbol{B}}^T\left\{\begin{array}{c} \boldsymbol{\sigma}^e \\ \boldsymbol{q}^e \end{array}\right\} = \boldsymbol{F} \\ \overline{\boldsymbol{B}}^T\left\{\begin{array}{c} \overline{\boldsymbol{\sigma}}^r \\ \overline{\boldsymbol{q}}^r \end{array}\right\} = \mathbf{o} \end{array}\right.\right\} \tag{2.7}$$

It is noted that the redistribution of stress is taken into account with the residual stress $\overline{\boldsymbol{\sigma}}^r$ in rock and the residual traction $\overline{\boldsymbol{q}}^r$ at cracks which are determined by solving the boundary value problem.

The joint element introduces two springs to crack as the normal stiffness k_n and shear stiffness, k_s. The physical meaning of these springs is not clear, but the joint element is introduced in this study to express the equilibrium equation in terms of traction at crack. In joint elements the contact conditions are considered as the yield function of the elements. The detachment and sliding at cracks are taken as the plastic deformation of joint elements. Therefore, the stiffness coefficients of k_n, k_s are taken large enough not to take account of joint element effects.

3 CONSIDERATION ON LOWER BOUND SOLUTION

3.1 *Material and Boundary Conditions*

Tamura (1990) developed a stability analysis with the constitutive equation of stress-strain rate relationship based on the rigid perfectly plastic assumption taking account of discontinuous velocity field. He showed some numerical analyses on the ultimate bearing capacity. In order to compare with the Tamura's results, the ultimate bearing capacity analysis is conducted against the boundary condition of Fig.1 under the plane strain condition.

The rocks are modeled as the Mises and Drucker-Prager type materials. The yield function is approximated by the linear inequality equation of Eq.(2.6). The elastic moduli of rock are taken as $E=100MPa$ and $\nu=0.33333$. The macroscopic crack is arranged at level 1 and 2 as shown in Fig.1. The shear strength of crack is set as a cohesion model. The effect of crack shear strength on the ultimate bearing capacity is investigated. The upper and lower bound solutions are directly compared in each case.

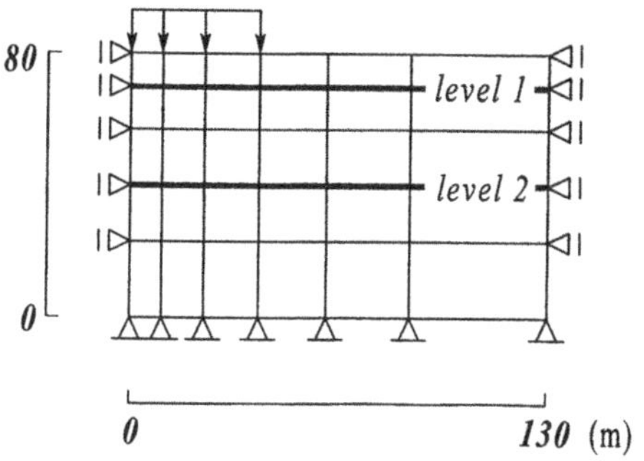

Fig.1 Boundary condition and finite element mesh.

3.2 *Comparison of Upper and Lower Solutions*

Tamura gave the upper bound solution on the ultimate bearing capacity in Fig.2 in the case of Mises type material. The bearing capacity is normalized by the strength of rock as q_u/c and plotted against the shear strength of crack, c_d. Two cases are conducted for installed levels of macroscopic crack. In each case the ultimate bearing capacity increases with the cohesion at crack and closes to the Hill's theoretical solution $(\pi+2)c$. It indicates that the failure mechanism is firstly affected by the crack, however, the effect gets faded with the increase of crack shear strength. The same problem is solved by the proposed lower bound analysis. The additional parameters for analysis are set as k_n, $k_s=10^{10}kPa/m$ and $c_n=0kPa$. Fig.3 shows the obtained results. It shows the fairly good agreement with the Tamura's result of Fig.2.

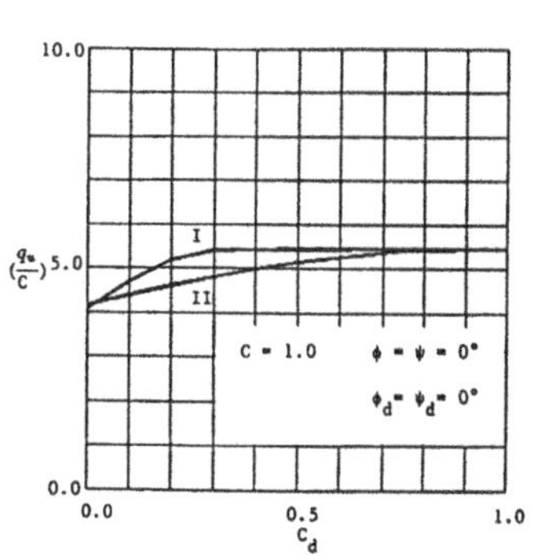

Fig.2 Crack cohesion effect on ultimate bearing capacity (Mises case, (Tamura, 1990)).

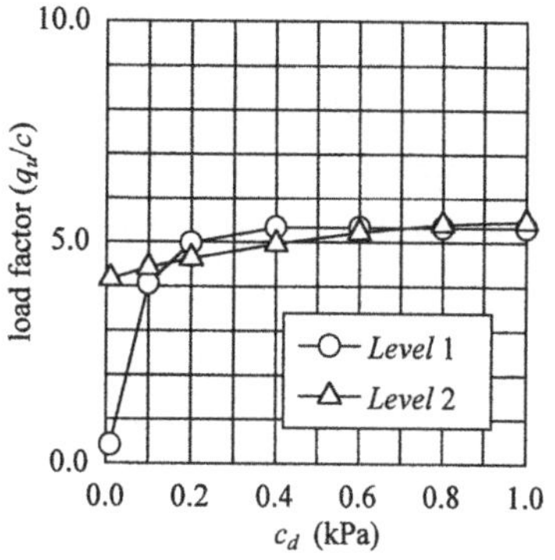

Fig.3 Numerical results on crack cohesion effect on ultimate bearing capacity (Mises case).

Secondary, the rock ground is modeled as the Drucker-Prager material. Fig.4 shows the result by Tamura where the macroscopic crack is located at the level 2 in Fig.1. The material constants employed for rock are the cohesion $c=1kPa$ and the internal frictional angles of $\phi=20$, 30°. In the figure ψ indicates the dilation angle so that the associated flow rule is employed when $\phi=\psi$. There found that the ultimate bearing capacity increases with the cohesion of crack, but it settles to certain magnitude depending on the internal frictional angle of rock.

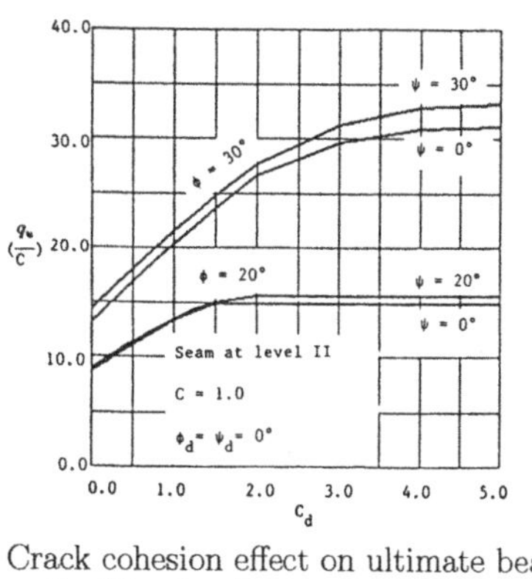

Fig.4 Crack cohesion effect on ultimate bearing capacity (Drucker-Prager case, (Tamura, 1990)).

Fig.5 indicates the numerical results of the proposed method corresponding to Tamura's result of Fig.4. There is a little difference in a load factor, but it shows a good agreement with Tamura's computation. From the view point of lower and upper bounds, the coincidence of these results indicates the proposed method could give a rational solution on rocks including macroscopic cracks inside.

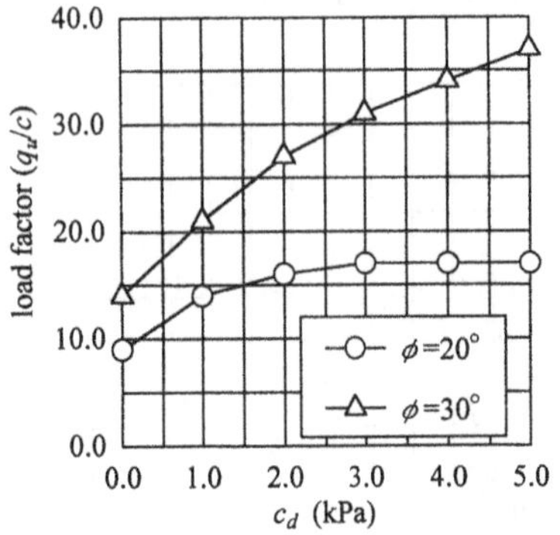

Fig.5 Numerical results on crack cohesion effect on ultimate bearing capacity (Drucker- Prager case).

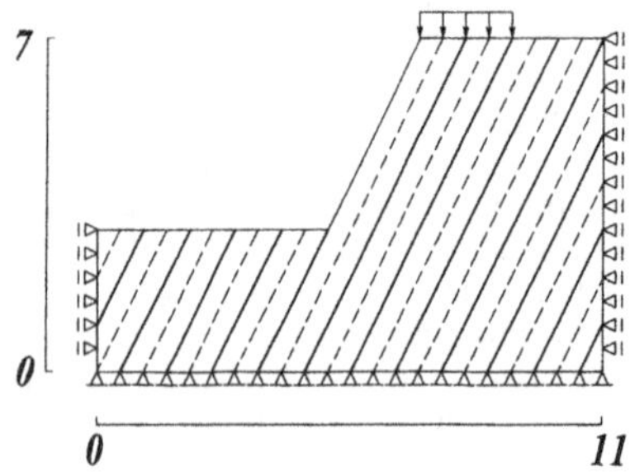

Fig.6 Geometry of jointed rock slope (type1).

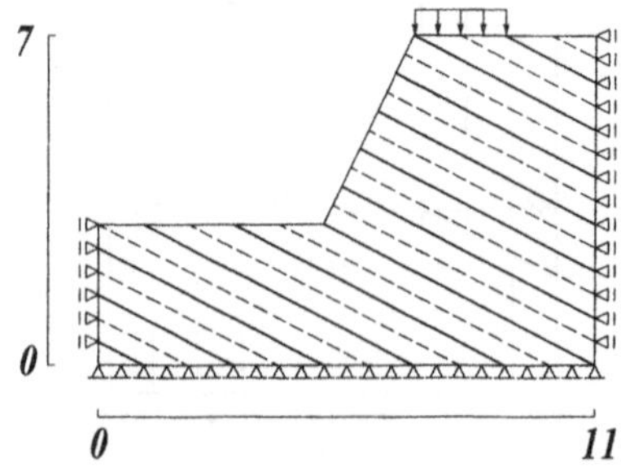

Fig.7 Geometry of jointed rock slope (type2).

4 ULTIMATE BEARING CAPACITY OF JOINTED ROCK SLOPE

4.1 *Boundary Conditions*

Figs.6 and 7 show the rock slopes which include inclined macroscopic cracks (type 1 and 2). The inclination angles are different in these two figures. In each figure, two spaces of installation are considered. One is a fine installation case and the spacing is $0.5m$ and the other, a coarse installation case of spacing $1m$.

Figs.8 and 9 also show the rock slopes (type 3 and 4), which constitute of rock blocks. The block pattern is different in these two figures. The effect of geometry and strength characteristics of cracks on the ultimate bearing capacity is investigated on the types 1 to 4. The rock and crack constants employed in analyses are exhibited in Table.1.

Table 1. Rock and crack constants.

E	$1000\ MPa$	ν	0.33333
γ_t	$23\ kPa/m$		
c	$1000\ kPa$	ϕ	0°
c_n	$10\ kPa$	ϕ_d	0°
k_n	$10^{12}\ kPa/m$	k_s	$10^{12}\ kPa/m$

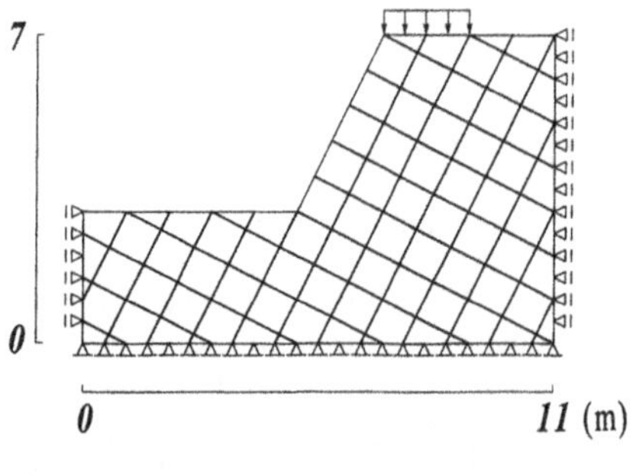

Fig.8 Geometry of jointed rock slope (type3).

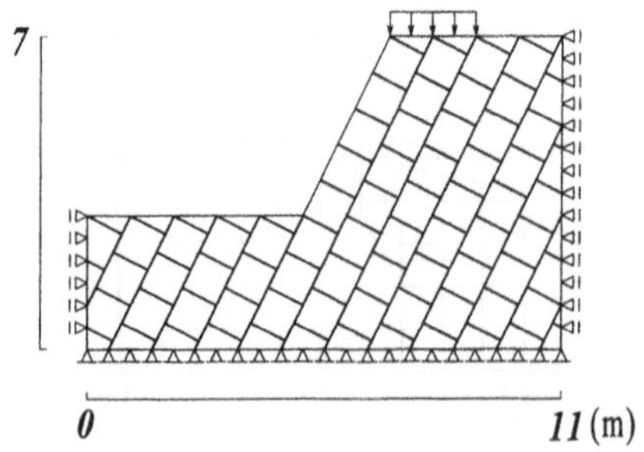

Fig.9 Geometry of jointed rock slope (type4).

4.2 *Case Studies*

Firstly, the ultimate bearing capacity of types 1 to 4 are investigated for the coarse installation of cracks. The rock slope is taken as the Mises material as shown in Table.1 and the shear strength of crack is set as variable. The change in ultimate bearing capacity with crack shear strength is demonstrated in Fig.10. In the figure, each bearing capacity increases with crack shear strength and settles to the bearing capacity without cracks. So that the rock failure mode might take place when the crack shear strength is large. On the contrary, the mixed failure of rock and crack takes place in the case of low crack shear strength. There is not so much difference in bearing capacity between types 1 and 2. The reason of this behavior might be due to the low density of crack installation. It will be discussed later with the fine installation case. The bearing capacity of types 3 and 4 are obtained smaller than those of types 1 and 2. It is natural because the installed cracks of types 3 and 4 cover all cracks of types 1 and 2., but it clearly indicates that the stability of block type slope is lower. In types 3 and 4, the space of crack installation is same, but the configuration is different. The figure shows the difference in bearing capacity due to crack configuration clearly. It suggests this method properly takes account of the mixed failure of rock and crack.

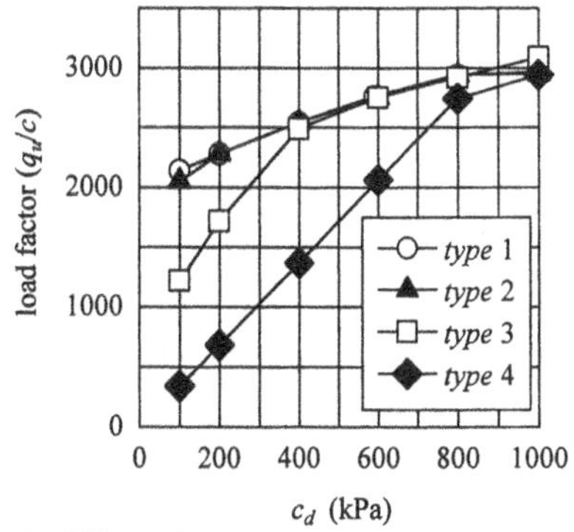

Fig.10 Effect of crack shear strength on bearing capacity (coarse case).

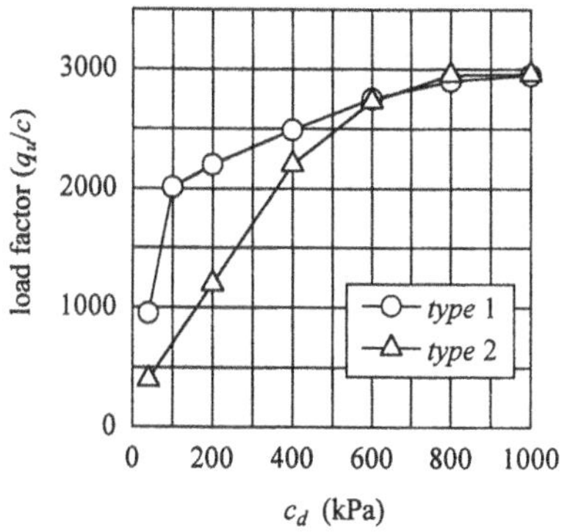

Fig.11 Effect of crack shear strength on bearing capacity (fine case).

Fig.11 illustrates the ultimate bearing capacity in the case of fine installation of cracks. It shows bearing capacity of types 1 and 2. The general relationship between bearing capacity and crack strength is similar with the coarse installation case. However, apparent difference in bearing capacity between types 1 and 2 is observed when the crack shear strength is low.

5 NON-LINEAR PROPERTY OF CRACK STRENGTH

5.1 *Non-linearity of Crack Strength*

The shear strength of macroscopic crack is well known to show the non-linearity against the confined stress as shown in Fig.12 (Jegaer, 1971). The contact condition of crack is, therefore, described by the bi-linear function as the following:

$$\begin{bmatrix} 1 & \tan\phi_{dL} \\ -1 & \tan\phi_{dL} \\ 1 & \tan\phi_{dH} \\ -1 & \tan\phi_{dH} \\ 0 & 1 \end{bmatrix} \left\{ \begin{array}{c} q_s \\ q_n \end{array} \right\} - \left\{ \begin{array}{c} c_{dL} \\ c_{dL} \\ c_{dH} \\ c_{dH} \\ c_n \end{array} \right\} \leq \mathbf{o}. \qquad (5.1)$$

In the above equation, c_{dL} and ϕ_{dL} denote the cohesion and friction angle in low confined stress and c_{dH} and ϕ_{dH} denote those in high confined stress. The stability analysis of jointed rock structure is easily conducted by employing Eq.(5.1) instead of Eq.(2.5).

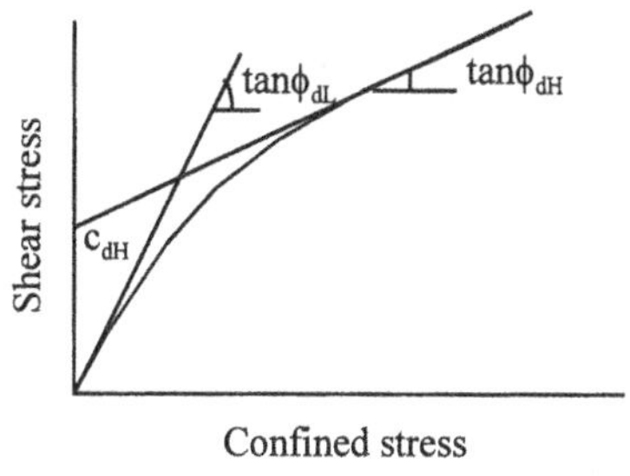

Fig.12 Non-linear property of crack strength.

5.2 *Ultimate Bearing Capacity of Footing*

The ultimate bearing capacity of Fig.1 is estimated for the non-linear shear strength property of crack. The discontinuous line is set as level

2. The ground is modeled as the Drucker-Prager material as c=1.0kPa and ϕ=30°. The shear strength of crack in low confined stress is modeled as c_{dL}=0kPa and ϕ_{dL}=30° and the magnitude of c_{dH} and ϕ_{dH} are changed to investigate the effect of non-linearity of crack shear strength on bearing capacity. Fig.13 indicates the computation result. ϕ_{dH} is set as o° in case 1 and 10° in case 2. The bearing capacity is obtained depending on c_{dH} and ϕ_{dH}, which reflects the non-linear property of crack shear strength. In whole region of c_{dH} the bearing capacity of case 2 is larger than that of case 1. It is because the internal friction angle ϕ_{dH} of case 2 is larger than that of case 1.

Secondary, c_{dH} and ϕ_{dH} are kept constant and the effect of ϕ_{dL} is investigated. c_{dL} is set as 0kPa and c_{dH} is taken as 0.5 to 2.5kPa for case study. Fig.14 shows the effect of non-linear property of crack shear strength on the ultimate bearing capacity. The bearing capacity is obtained to increase with the internal friction angle of ϕ_{dL}, which reflects the non-linear characteristics of crack strength. The range of ϕ_{dL} affecting the bearing capacity shifts depending on c_{dH}.

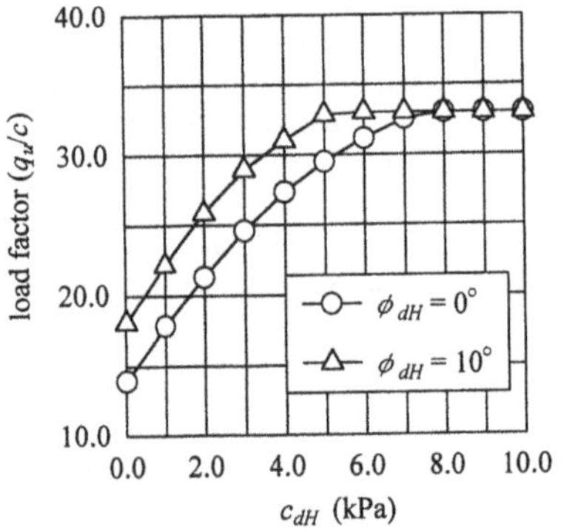

Fig.13 Effect of non-linear crack strength (c_{dH} is variable).

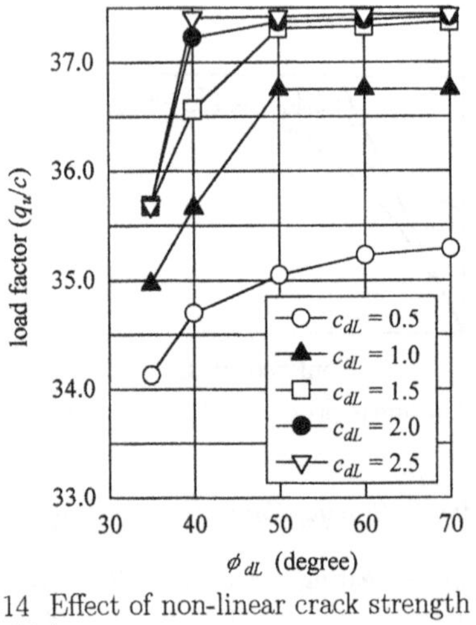

Fig.14 Effect of non-linear crack strength (ϕ_{dL} is variable).

6 CONCLUSION

Stability analysis for jointed rock structure is developed by taking account of contact condition at cracks based on the lower bound theorem in plasticity. The comparisons with the upper bound calculations conducted by Tamura(1990) showed good coincidence and it assured that the proposed method gave rational solutions. Through some numerical analyses on jointed rock slopes, it was examined that the proposed method could estimate the stability taking account of the inherent failure modes of (1)crack failure, (2)mixed failure of crack and rock, and (3)rock failure. The non-linear shear strength of crack was also fairly considered in the stability analysis.

ACKNOWLEDEMENT

The authors wish to thank Mr. Yamaji, T., Nagaoka University of Technology, for his assistance and valuable comments to conduct this research.

REFERENCES

Goodman,R.E., Taylor,R.L. and Brekke,T. (1968). A model for the mechanics of jointed rock, *Proc. of ASCE*, 94, SM3, pp.637–659.

Jaeger,J.C. (1971). Friction of rocks and the stability of rock slopes, *Geotechnique*, Rankine Lecture, Vol.21.

Koiter,W.T. (1960). General theorems for elastic plastic solids, *Progress of Solid Mechanics*, Chap.6, Vol.2, North Holland Press.

Maier,G. (1969). Shakedown theory in perfect elastoplasticity with associated and nonassociated flow-laws: a finite element linear programming approach, *Meccanica*, Vol.4, No.3, pp.1–11.

Ohtsuka, S. (1997a). Effect of Microstructure on Crack Shear Strength in Rock, *Proc. of Int. Symp. on Deformation and Progressive Failure in Geomech.*, pp.75–80.

Ohtsuka, S. (1997b). Bearing Capacity Analysis of Rock Structures Including Cracks, *Proc. of 9th Int. Conf. of Int. Assoc. for Comp. Meth. and Adv. on Geomech.*, Vol.1, pp.739–744.

Tamura,T. (1990). Rigid-Plastic Finite Element Method in Geotechnical Engineering, *Computational Plasticity, Current Japanese Material Research*, Vol.7, Elsevier, pp.135–164.

Mechanics of Jointed and Faulted Rock, Rossmanith (ed.) © 1998 Taylor & Francis, ISBN 90 5410 955 6

Stability analysis of wedges in jointed rock masses using finite element analysis

A. Fahimifar
Department of Civil Engineering, Amirkabir University of Technology, Tehran, Iran

ABSTRACT : Limit equilibrium analysis of wedges and blocks is not thorough enough to evaluate all aspects of the problem . As an alternative and complementary approach , finite element model was used to analyze stability of wedges in rock slopes , and the performance of the model was compared with limit equilibrium method .
A rock mass containing a wedge was analyzed in two and three dimensions , and pre-determined joints were modeled by interface elements , and Lusas Finite Element Program was used for analysis . Using the program for rock slopes with different geometries and joint sets, various aspects of the slopes were evaluated in two and three dimensions and the factor of safety for each case was determined, and also variations of significant parameters such as friction angle and cohesive strength were examined .

1 INTRODUCTION

There are various analytical methods to evaluate rock slopes, among them graphical, limit equilibrum and kinematic methods are of more significant . However, non of them is not thorough enough to examine all aspects of a rock slope, particularly in a jointed rock mass .

It has been proved that numerical methods, especially finite element analysis is a powerful model to analyze complicated engineering problmes. Examination and evaluation of various methods in analysis of wedges in jointed rock masses and comparison with finite element model would reveal performance of each method, and would be resulted in a more comprehensive and valuable solution for a rock slope analysis .

2 STRUCTURAL DATA COLLECTION

The rock slope stability is generally controlled by discontinuities existing through out the rock mass . In order to characterize potential of failures and deriving appropriate model, the site of Irankouh Mine (in South-East of Iran) was selected for investigation . Data collection and measurment of discontinuity properties were performed using scanline survey according to ISRM suggested methods (Brown, 1986) . Geomechanical properties of joints such as orientation , spacing , JCS (joint compressive strength) , persistence , JRC , aperture and fillings, and condition of water and weathering of the rock mass were investigated .

For determination of strength parameters of intact rock and joints, appropriate specimens from various
locations were prepared , then uniaxial and shear tests were carried out in the laboratory , thereby corresponding Mohr-Coulomb envelops were derived .

3 ANALYSIS OF DATA

On the basis of data collected and using graphical method the most critical joint sets were detemined .

Figure 1 shows that there are four major joint sets controlling the rock mass stability . Sliding potentials resulting from these joint sets are clearly observed . Limit equilibrium method is not able to characterize sliding types and potentials . For this purpose graphical method was used for data analysis .

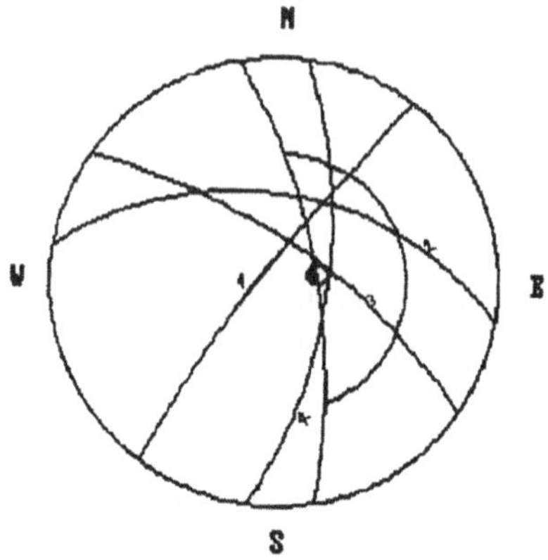

Figure1:Major joint sets produeing in sliding potentials.

Using geometrical and geomechnical data , limit equilibrium and kinematic analysis were performed, and safety factor for wedge sliding, under various conditions, were examined . Sensitivity analysis was also carried out in order to show variation of safety factor to strength parameters and water condition in tension crack .

Figures 2 and 3 show two typical graphs obtained through the analysis . Figure 2 illustrates variation of safety factor for a rock slope in dry and saturated rock conditions for different magnitudes of friction angle and cohesive strength with no tension crack . However , figure 3 illustrates variation of safety factor for the same rock slope with consideration of tension crack . As is observed, reduction in safety factor in water filled tension crack , in comparisoin with figure 2 , is very clear .

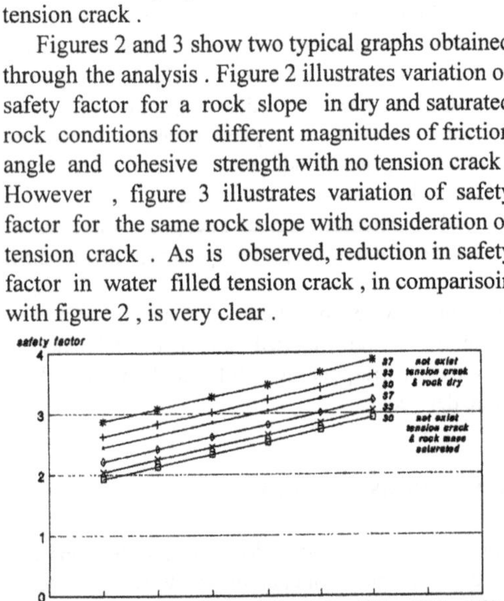

Figure 2 : Sensitivity analysis of a slope without tension crack .

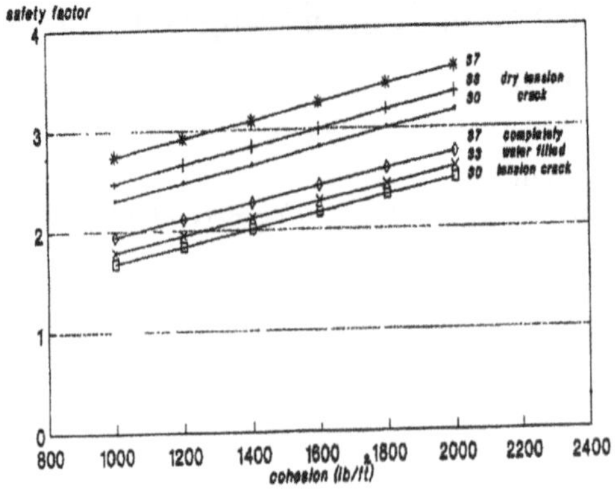

Figure 3 : Sensitivity analysis of a slope with a tension crack .

4 FINITE ELEMENT ANALYSIS OF ROCK SLOPES

Finite element model has been extensively used in various engineering problems . This method is easily used in continuum media . However , a jointed rock mass is a discontinous media because of various structural features throughout the rock mass . Therefore , a linear behavior model for the rock mass and one type of element for finite element mesh cannot be applied in the analysis . Different workers have proposed appropriate methods to model discontinuities in rock masses (Goodman , 1976 , Saeb et al , 1992) . Joint and interface elements have been used extensively to model discontinuities in rock masses . In the present work interface element was used in the analysis .

Due to non-linear behaviour of interface element the rock mass surrounding the wedeg dose not behave linearly . Therefore, the relationship between stress and strain in the rock mass is not linear . In this work, Lusas Finite Element Program (1990) which is a powerful non-linear program , was used for analysis .

5 DISCUSSION OF RESULTS

Two different rock slopes containing wedges, in two and three dimensions, were analyzed . Figure 4 illustrates a rock slope in three dimension . The slope has a height of 40 m, slope face angle of 65^0 , the first joint set inclination of 58^0 and second joint set inclination of 45^0 . This slope was analyzed for several cases and conditions . Mohr-Coulomb yield criterion was used in the analysis and magnitudes of cohesions and friction angles were examined . Figure 5 shows the slope after deformation . As is observed wedge sliding along the intersection line of the two joint sets is very clear . This figure also clearly shows that the upper part of the wedge is in tension . Figure 6 illustrates shear stress contours in the slope . As is observed shear stresses near toe of the slope and on the discontinuity surfaces are high . This may be attributed to the high normal stress in this region .

Direction of major principal stresses near toe of the slope is also inclined which may be used for identification of sliding potentials . Shear stresses over the discontinuity surfaces have various magnitudes which may be due to the variations of normal stress on the discontinuity surface .

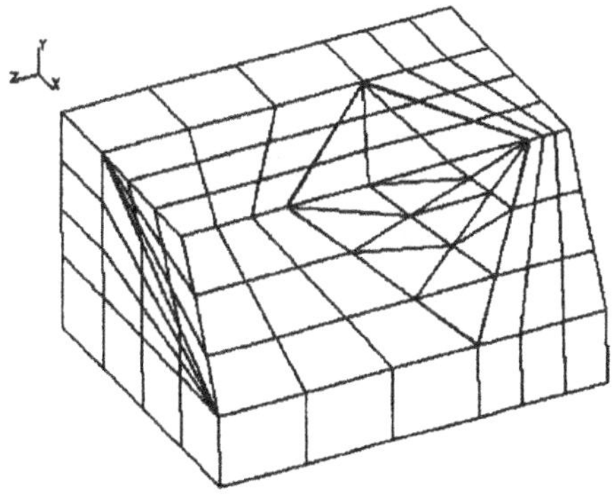

Figure 4 : Finite element mesh for a rock slope containing two major joint sets .

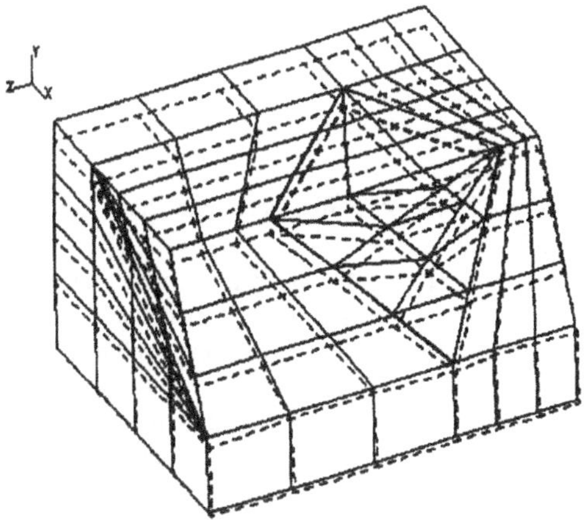

Figure 5 : Finite element mesh for a deformed rock slope with two major joint sets .

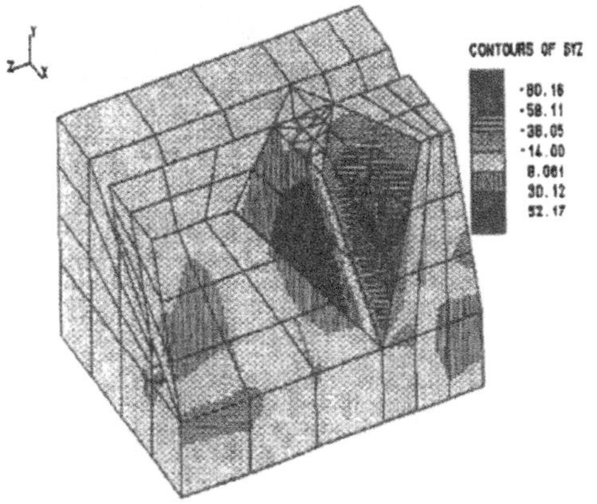

Figure 6 : Shear stress contours for a rock slope with two major joint sets .

Figure 7 shows a slope in two dimension with 70 meter height, 65^0 face inclination and a joint set with 40^0 inclination . Figure 8 shows the deformed slope after running the program using interface element as the discontinuity model . Sliding movement along the discontinuity is clearly observed . Yield points over the sliding surface is also observed in figure 9 . Comparison of the displacements through the slope and along the discontinuity reveals that displacement through the slope is not considerable, however, sling movement along the discontinuity is predominant . Shear stress contours in figure 10 illustrates the yield zone and proves the nature of deformability in figure 8 . Because of the presence

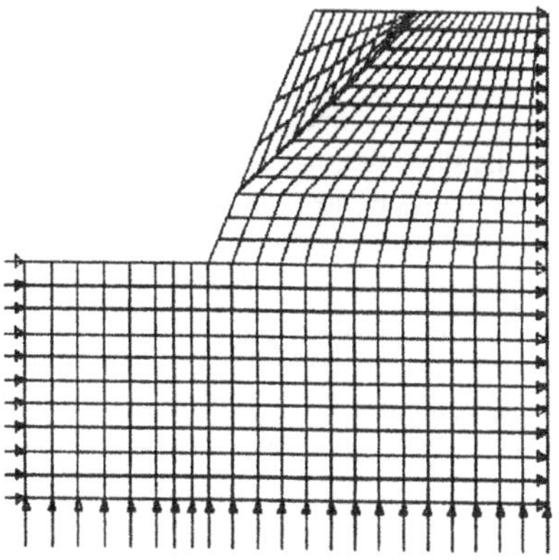

Figure 7 : Finite element mesh for a slope with a major joint set .

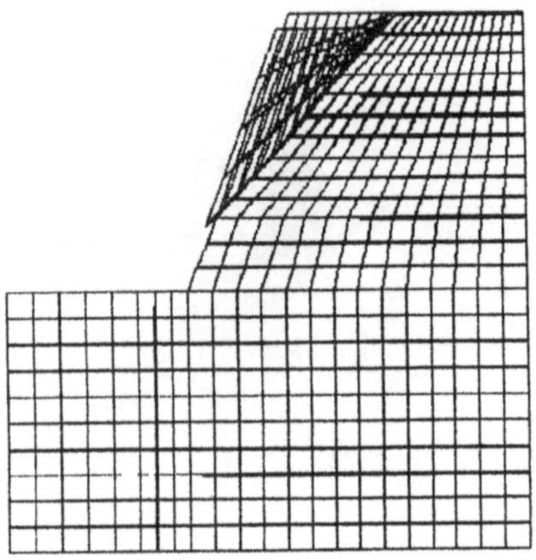

Figure 8 : Finite element mesh for a deformed slope with a joint set .

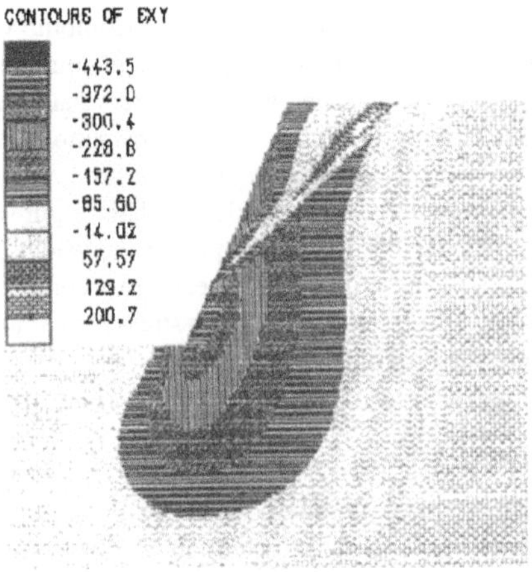

Figure 10 : Shear stress contours in a rock slope .

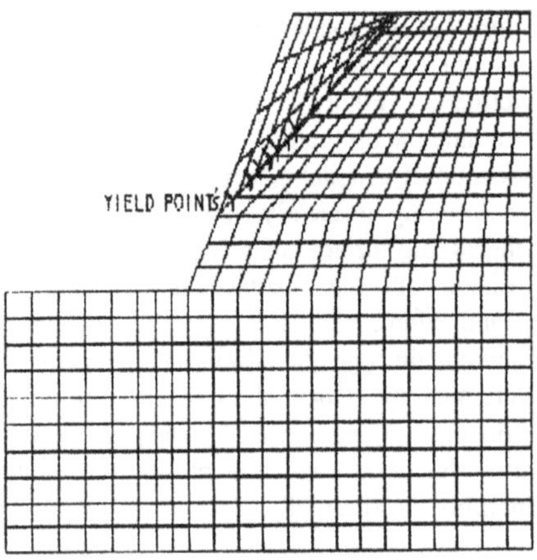

Figure 9 : Presentation of yield points in the rock slope .

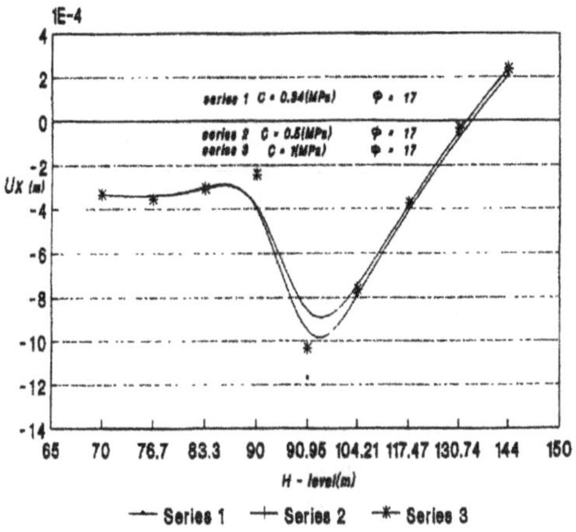

Figure 11 : Horizontal displacement-slope height with constant friction angle and variable conhesion.

of discontinuity in the slope , the stresses in the elements near the slope face have been released , and therefore , the wedge is prone to instability.

In order to show the effects of discontinuity strength parameters on the wedge stability , the program was run for various magnitudes of friction angles and conhesions . Figure 11 illustrates horizontal displacement relative to slope height with constant friction angle and variable conhesion . The minimum point on the curves shows the location of discontinuity with a maximum displacement . Increasing cohesion from 0.34 MPa to 1 MPa the amount of displacement has reduced $0.8 \times 10^{-4}\ m$. Figure 12 shows the effect of friction angle in controlling displacement . Increasing friction angle from 17^0 to 45^0, displacements have reduced significantly .

Figures 13 , 14 and 15 illustrate variations of shear stress , normal and shear displacements along a rock layer respectively , for linear and non-linear analyses using Mohr-Coulomb yield criterion . These results have been obtained in the nodal points near the rock layer towards the slope face . As is observed in figure 13 variations of shear stress for linear and non-linear analyses is no significant . However , variations of normal and shear displacements as in figures 14 and 15 are considerable .

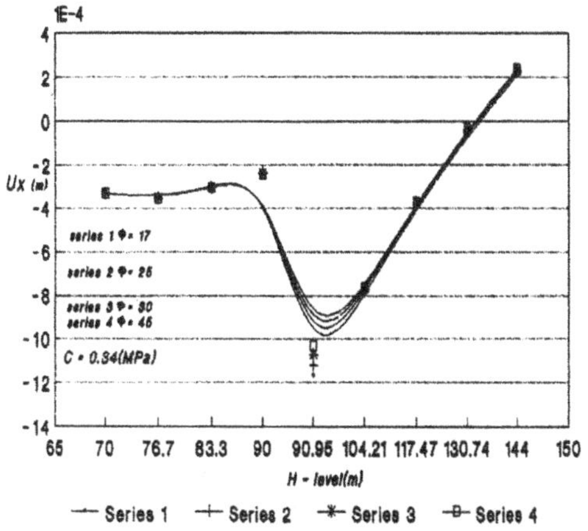

Figure 12 : Horizontal displacement-slope height with variable friction angle and constant cohesion .

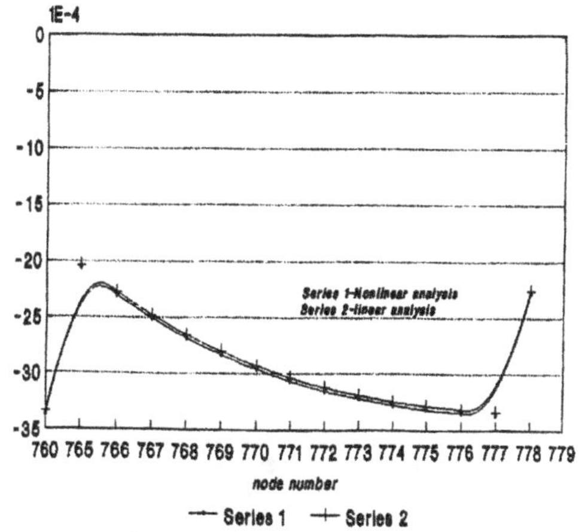

Figure 15 : Vertical displacements in a rock layer for elastic and plastic analyses .

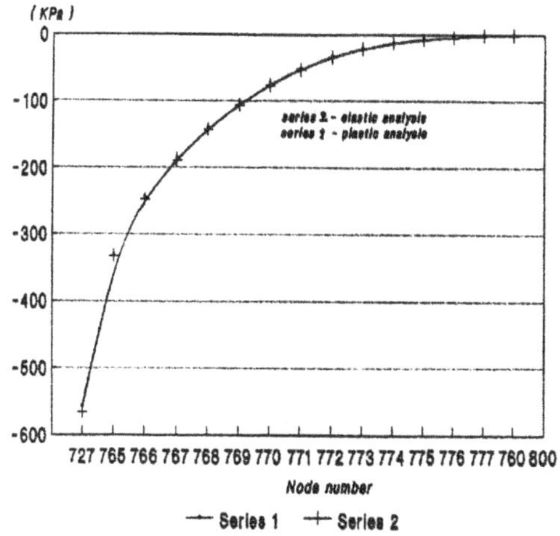

Figure 13 : Shear stress in a rock layen for elastic and plastic analyses .

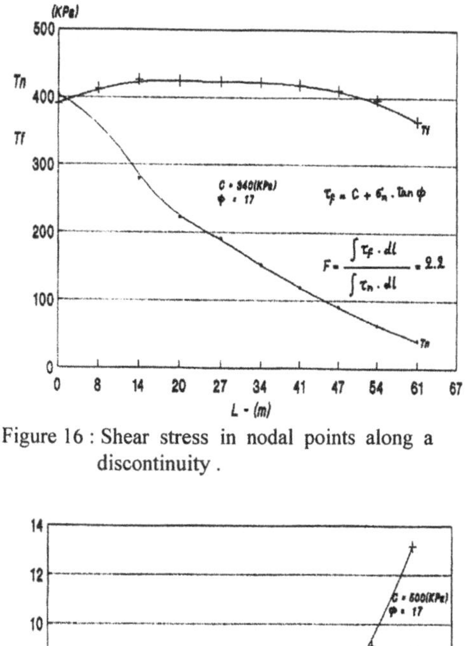

Figure 16 : Shear stress in nodal points along a discontinuity .

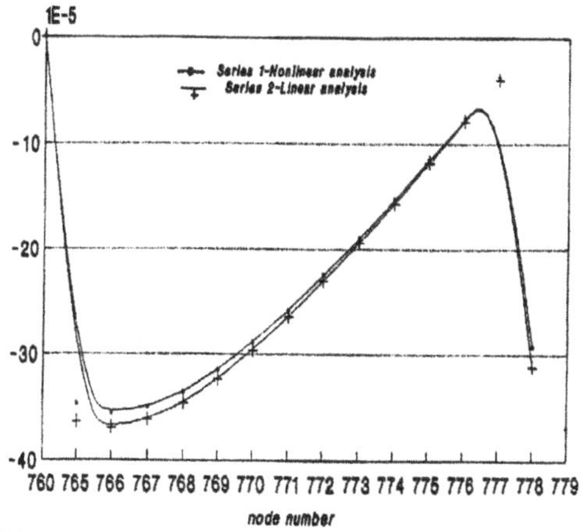

Figure 14 : Horizontal displacements in a rock layer for elastic and plastic analyses .

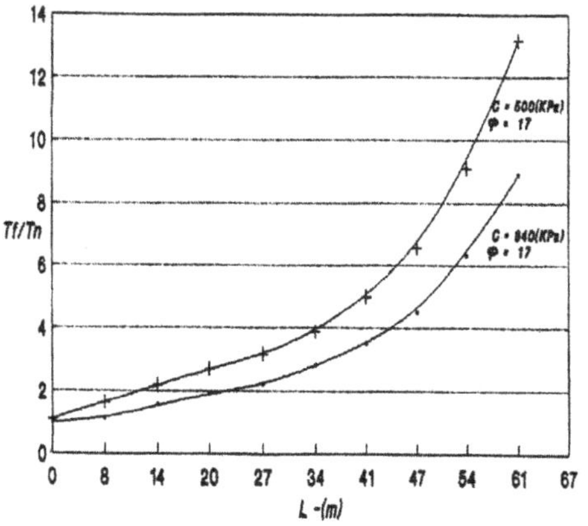

Figure 17 : Variations of τ_f / τ_n along a discontinuity for a constant friction angle .

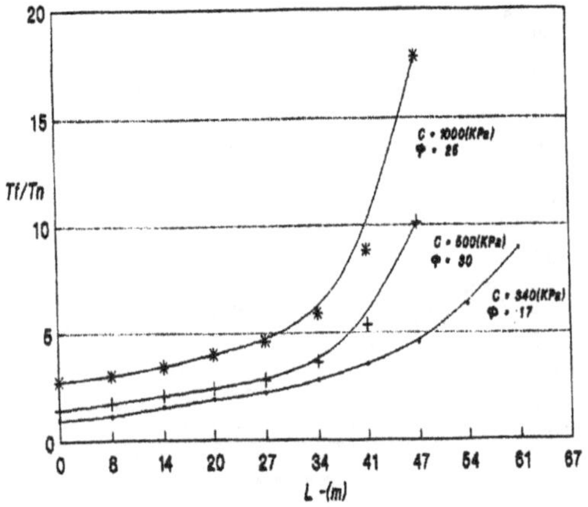

Figure 18 : Variaton of τ_f / τ_n along a discontinuity for variable friction angles and cohesions .

Figure 16 illustrates the magnitude of safety factor at various points along a discontinuity . As is observed the amount of safety factor has become 2.2, however , for the same discontinuity and the same conditions safety factor in limit equilibrium method has become 1.94 . This difference may be attributed to the fact that in finite element model shear stress along the discontinuity is not the same at various points . Figures 17 and 18 show the non-linearity of safety factor along a discontinuity for different friction angles and cohesive strength (Framarzi , 1996) .

6 SUMMARY AND CONCLUSIONS

Stability analysis of rock slopes with particular reference to wedge analysis was investigated . Limit equilibrium , graphical and finite element methods were examined . On the basis of finite element model rock slopes containing wedges were analyzed.

The most important conclusions may be summarized as follows :

a) non of the methods applying for analysis of rock slopes is not thorough enough to evaluates all aspects of a rock slope .

b) Finite element analysis is able to characterize the zones of tension through the slope and therefore , to identify potentials of failures .

c) Finite element model is not singly able to assess various aspects of a slope .

REFERENCES

Brown , E. T. 1986. Rock characterization testing and monitoring. ISRM suggested methods. pergamon press.

Goodman , R. E. 1976. Methods of geological engineering in discontinuous rock . West publishing company .

Saeb, S.& B.Amadei 1992. Modeling of rock joint under shear and normal loading. Int.J. rock mech.& minig sci. Vol 29, No3 PP267-278.

Lusas Finite Element Program , 1990 ., Finite Element Analysis LTD , UK .

Framarzi, L.1996 . Finite element analysis of rock slopes. MSc Dissertation , Amirkabir University of Technology , Tehran; Iran.

Mechanics of Jointed and Faulted Rock, Rossmanith (ed.) © *1998 Taylor & Francis, ISBN 90 5410 955 6*

Analysis of stability factor of slope in jointed rock masses

P.A. Fonaryov
Moscow Automobile and Roads Institute, Russia

ABSTRACT: This article deals with grapho-analytical method of the analysis of stability factor in jointed rock masses on the basis of calculating scheme of rock masses stability estimation depending on their tensile strength developed by the author.

1 INTRODUCTION

The analysis of rock slope stability factor depending on such parameters as slope steepness, strata dip of the slope and an inclination of the sliding slope to the horizon is of greet interest and importance during the designing and construction of engineering structures in jointed masses. This problem is the most actual on designing and construction automobile highways and railroads in mountains where engineering-geological conditions of the route may be very variable on short parts. The steepness of the natural slope (ω) and dugout slope (δ), conditions of rock deposition (α), strata azimuth scattering and axes of the highway change in such cases. Structure-texture specifications of jointed rock masses inner structures (ρ, m, l_0), consisting mainly of structure blocks acting on each other (Maslov & Fonaryov 1987, Fonaryov 1980) have great influence on the rock slope stability.

2 MATHEMATICAL ANALYSIS

The solution of the above mentioned problem is presented at the definite example of calculation scheme, suggested by the author for the estimation of slope stability factor in jointed block rock masses with taking into consideration their possibility to extension resistance.

According to this calculation scheme (Fig.1) parameter L_Z (the distance from the edge of the slope to the tension fissure (crack) determines the mass of the sliding block (P), the length of the slickenside and extension resistance (E) via the depth of the tension fissure (Z_{CR}). The given parameter L_Z influences greatly the relationship between shear and holding forces, hence the equation

$$F_S = \frac{P\cos\beta . tg\varphi + cL + E[\cos(\beta-\alpha) + tg\varphi\sin(\beta-\alpha)]}{P\sin\beta} \quad (1)$$

for continuation of the analysis it would be necessary to simplify it, expressing all the parameters via L_Z and carrying out all the necessary conversions. As a result we got

$P\cos\beta . tg\varphi = A_1 + A_2 L_Z + A_3 L_Z^2$;

$cL = B_1 + B_2 L_Z$;

$E[\cos(\beta-\alpha) + tg\varphi\sin(\beta-\alpha) = C_1 + C_2 L_Z + C_3 L_Z^2$;

$P\sin\beta = D_1 + D_2 L_Z + D_3 L_Z^2$;

where A_1 ... D_3 - some coefficients, P - sliding block mass, φ and C - parameters of shear strength along the rock fissure, the meaning of the rest parameters is clear from the calculation scheme (Fig.1)

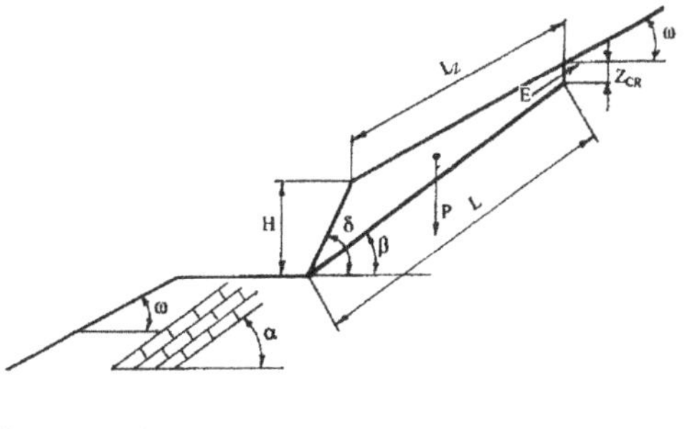

Fig.1 Calculation scheme

after that equation (1) can be presented in the following way

$$F_S = \frac{M_1 + M_2 L_Z + M_3 L_Z^2}{D_1 + D_2 L_Z + D_3 L_Z^2}, \qquad (2)$$

where
$M_1 = A_1 + B_1 + C_1; M_2 = A_2 + B_2 + C_2; M_3 = A_3 + C_3$.

In this form equation (1) can be better studied and it is easier to find the parameter L_Z, at which function $F_S(L_Z)$ has its minimum as we are interested only in minimal magnitude of the rock mass stability factor.

As a result of the multifactor analysis (Fonaryov 1986) the presence of parabolic relationship and the magnitude of the reserve coefficient depending on the position of the tension fissure (Z_{CR}) relatively to the slope edge (L_Z) and the sliding slope inclination to the horizon (β) were stated. Both parameters in general determine the relationship between shear and holding forces at the jointed rock mass slope.

For preliminary analysis we will use analytical geometry methods permitting to determine the character of stability factor changes depending on one of the two above-mentioned parameters characterizing structure and properties of the rock mass slope (Fonaryov 1994). In this case, according to the theory, vertical asymptotes of the function $F_S(L_Z)$ are found from the condition that the equation denominator is equal to zero (2), that is possible in the points equal

$$L_{Z12} = \frac{-D_1 \pm \sqrt{D_2^2 - 4D_1D_3}}{2D_3}. \qquad (3)$$

The function $F_S(L_Z)$ breaks in these points the limiting magnitudes of the function $F_S(L_Z)$ at $L_Z \to \infty$ is found from the expression

$$\lim F_S(L_Z) = \frac{M_3}{D_3}, \qquad (4)$$

which is the function's horizontal asymptote.

The magnitudes L_Z, in which the plot crosses the horizontal axis, are equal to

$$L_{Z34} = \frac{-M_2 \pm \sqrt{M_2^2 - 4M_1M_3}}{2M_3}. \qquad (5)$$

The point, in which the plot crosses the vertical axis, is found from the condition $L_Z = 0$, i.e.

$$F_S(L_Z) = \frac{M_1}{D_1}. \qquad (6)$$

As we see, the problem of the minimum function $F_S(L_Z)$ determination is simplified as its change character is already known and we can state whether this function has a minimum or not, or it is necessary to find the limit to which it will tend, especially in the field of positive argument magnitudes, i.e. in the range $0 \le L_Z \le L_Z$ max, where

$$L_{ZMAX} = \frac{H(ctg\delta . tg\beta - 1)}{\sin\omega(1 - ctg\omega . tg\beta)}. \qquad (7)$$

In function minimum is present in the given range, definite magnitudes L_{Zcr} and $F_{S\,min}$ are found by differential calculation methods.

According to the theory, if the function derivative $F_S(L_Z)$ only once changes to zero, $F_S = f(L_Z)$ in this point is minimal. Hence, the first derivative from along L_Z will change finally to

$$F_S'(L_Z) = \frac{G_1 L_Z^2 + 2G_2 L_Z + G_3}{(D_1 + D_2 L_Z + D_3 L_Z^2)^2}, \qquad (8)$$

where $G_1 = M_3D_2 - M_2D_3; G_2 = M_3D_1 - M_1D_3;$
$G_3 = M_2D_1 - M_1D_2$.

Expression (7) will be equal to zero, when $L_Z = L_{Zcr}$

$$L_{ZCR} = \frac{-G_2 + \sqrt{G_2^2 - G_1G_3}}{G_1}, \qquad (9)$$

where L_{ZCR} - critical magnitude of argument L_Z, at which $F_S(L_Z)$ has minimal magnitude.

3 DEFINITE EXAMPLE

Let is consider the possibilities of the suggested method at the definite example of the slope recommended by SNiP 2.05.02-85 (1986) p.6.35: H = 16 m; $\delta = 79^0$ (1 : 0.2); $\alpha = \beta = 30^0$ (slickenside coincides with the bedding); $\omega = 25^0$; $\gamma = 2.5$ t/m^3; $\varphi = 25^0$; $\rho = 0.5$ (coefficient of the structure at block bandage in the layers correspondingly); m = l_0 = 0.2 (thickness of layers and the magnitude of block bandage in the layers correspondingly); C = 5 t/m^2 (50 kPa).

3.1 *Variant 1*

The magnitudes of coefficients M ... D as a result of finished calculations are as follows, M_1 = 214.56; M_2 = 16.27; M_3 = -0.041; D_1 = 27.61; D_2 = 16.09; D_3 = -0.057.

Furthermore, according to equations (3 ... 6) we'll define vertical asymptotes: L_{Z1} = -1.71; L_{Z2} = 283.99: horizontal asymptote - $\lim F_S(L_Z) = 0.72$; points of crossing with horizontal axis: L_{Z3} = -12.77: L_{Z4} = 410.36: point of crossing with vertical axis - $F_S(L_Z = 0) = 7.77$, and make a plot $F_S = f(L_Z)$ (Fig.2).

As it was mentioned above, the range of function $F_S(L_Z)$ determination is the sum of positive magnitudes L_Z , at which the range of function changes will involve only positive magnitudes of $F_S(L_Z)$. So the field of the function study is reduced greatly.

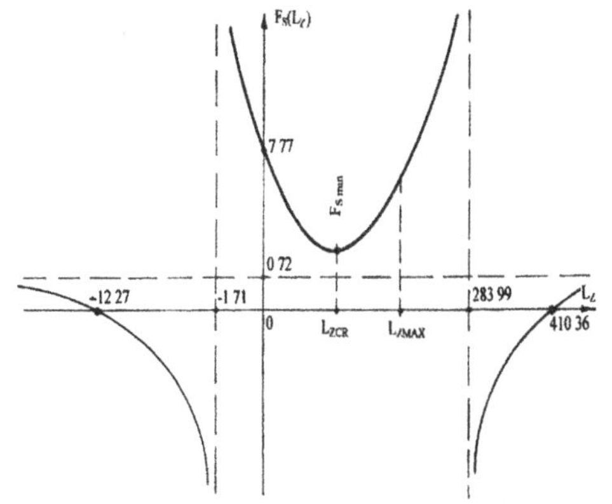

Fig.2 The character of the relationship $F_S = f(L_Z)$ at $\beta > \omega$

The magnitudes $F_{S\ min}$ and L_{Zcr} are determined in future according to the equations (2 and 9). As a result we find, that L_{Zcr} = 72.18 m; $F_{S\ min}$ = 1.32.

We can see that the magnitude $F_{S\ min}$ = 1.32 differs from the limiting magnitude equal to 0.72. It is explained by the fact that in the problems with definite conditions the independent variable (for example, L_Z) can usually change only in limited range ($0 \leq L_Z \leq L_Z$ max). So the minimal magnitude of the function $F_S(L_Z)$ is found not for the whole curve, but only for the given interval. In this case one magnitude $F_S(L_Z) = 7.77$ at $L_Z = 0$, and the second - 1.46 at L_Z max = 141.46 m. So the magnitude $F_S(L_Z = L_{Zcr}) = 1.32$ corresponds to the minimal magnitude of the function $F_S(L_Z)$.

3.2 *Results analysis*

However, from practical $F_{S\ min}$ experience, slickenside does not always coincide with the bedding and more often is much steeper to the horizon, i.e. cuts the bedding. Besides, the previously made factor analysis (Fonaryov 1986) showed that the relationship $F_S(\beta)$ has also the parabolic character. But taking into consideration the complexity of determination of the angles β_{min} by analytical geometry method, corresponding to the minimal magnitude, i.e. F_S the minimal magnitude of angle β in the range between α and δ , with the corresponding determination L_{Zcr} and $F_S(L_{Zcr}, \beta)$. The minimal magnitude of the function $F_S(L_{Zcr}, \beta)$ and will be absolute minimum of the given function, i.e. $F_{S\ min\ min}$ of the rock mass slope and angle $\beta = \beta_{min}$. The example of the graphic solution of the given problem is shown at the Fig.3 for the cases of rock mass tension resistance and without rock mass tension resistance (E = 0). In this case the angles of slickenside inclination β_{min} practically do not differ and are equal to 51.5^0.

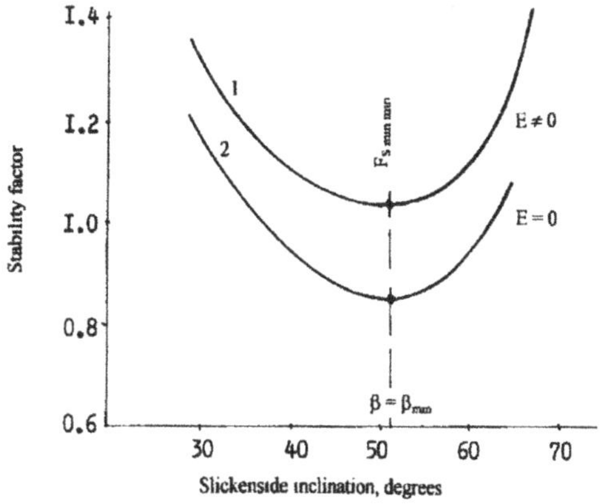

Fig. 3 Relationship $F_{S\ min}$ from the slickenside inclination

The calculated magnitude $F_{S\ min\ min}$ = 1.03 corresponds to the tension fissure position L_{Zcr} = 15m from the slope edge. In the absence of tension resistance (E = 0 at l_0 = 0 - in the absence of block bandage in layers) $F_{S\ min\ min}$ = 0.85 at L_{Zcr} = 4.91m from the slope edge. As we see, taking into consideration of tension resistance of the block rock massive increases $F_{S\ min\ min}$ from 0.85 to 1.03, i.e. so though unstable slope turns out to be stable though close to the limiting state.

So, the most possible slickenside corresponding to the minimal slope stability lies under the angle 51.5^0 to the horizon and does not coincide with bedding $\alpha = 30^0$. At this = 1.32 turns out to be highly overestimated, as at $\beta = \beta_{min}$ $F_{S\ min}$ = $F_{S\ min\ min}$ = 1.03 and rock mass slope is in this case close to the limited state. Without taking into consideration tension resistance the slope practically must crash, as it is in overlimited state $F_{S\ min\ min}$ = 0.85.

Determination of the stable slope steepness with the planned magnitude F_S (for example > 1.25, but < 1.5) may be done by selecting magnitudes of angle δ to the direction of its decreasing or increasing with periodical determination the most undesirable falling of slickenside, i.e. angle β_{min}. The example of such solution of the problem is shown on Fig.4, according to which the planned range of F_S magnitude changes will correspond to the slopes with steepness in the range 68^0 - 58^0 at cohesion equal 50 kPa. The line $F_{S\ min\ min}$ is shown in Fig. 4 in dash line.

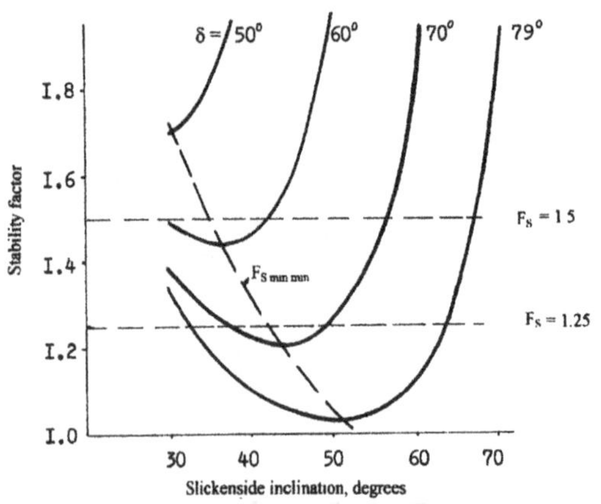

Fig. 4. The dependence of $F_{S\ min}$ on the slickenside inclination for slopes with different slopes.

It is known that slickenside may coincide with the bedding or have steeper inclination to the horizon. The preliminary analysis allowed us to state that at the slope steepness to 50^0 slickenside coincides with the rock bedding and at the larger steepness has the larger angle of inclination to the horizon. In this case β_{min}., as analysis showed, is linked with linear function with angle δ. And the change of internal friction angle affects the magnitude of β_{min}. angle to less extent, than cohesion magnitude change.

3.3 *Variant 2*

Let us consider one more frequent case when angles $\alpha = \beta = \omega$, that is quite possible, when steep falling rock mass slopes armor the slope surface the slope surface. In this case equation (2) turns into

$$F_S(L_Z) = \frac{M_1 + M_2 L_Z}{D_1 + D_2 L_Z}, \qquad (10)$$

that is characteristic for fraction linear function which can be expressed on the plot in the form of equalsided hyperbole, asymptotes of which are parallel to the coordinate axis.

According to analytical geometry methods we will analyze the plot of the fraction-linear function concerning the solution of the planned problem at the same initial data, and the magnitudes of the angles α, β and ω will be 30^0. The conducted calculations showed that coefficients magnitudes turned out to be equal $M_1 = 214.56$; $M_2 = 17.49$; $M_3 = 0$; $D_1 = 27.61$; $D_2 = 15.38$; $D_3 = 0$. So vertical asymptote of hyperbole will be equal

$$L_{Z1} = \frac{-D_1}{D_2} = -1.79; \qquad (11)$$

horizontal asymptote at

$$L_Z \to \infty - \lim F_S(L_Z) = \frac{M_2}{D_2} = 1.13; \qquad (12)$$

the crossing point of hyperbole with the vertical axis

$$F_S(L_Z = 0) = \frac{M_1}{D_1} = 7.77; \qquad (13)$$

and with horizontal axis

$$L_{Z2} = \frac{-M_1}{M_2} = -12.31. \qquad (14)$$

The plot of the given function is shown in Fig.5. The absence of function minimum in this case does not allow us to use the first derivative for the determination of the minimal magnitude of rock mass stability factor of rock slope. But it is possible to find this magnitude in this case too. So, according to (12) the minimal possible magnitude will be 1.13. The real magnitude $F_S(L_Z)$ will be determined by L_Z found on the definite section and may be in the range of 7.77 at $L_Z = 0$ to 1.13 at $L_Z = \infty$.

3.4 *Results analysis*

The analysis of angles α, β and ω influence on the stability of slopes with different steepness showed that in the case of their equality (at the armouring of the slope surface) the most dangerous conditions ($F_S < 1$), in this case, take place of slope steepness changes in the range from 34^0 to 70^0 (Fig.6).

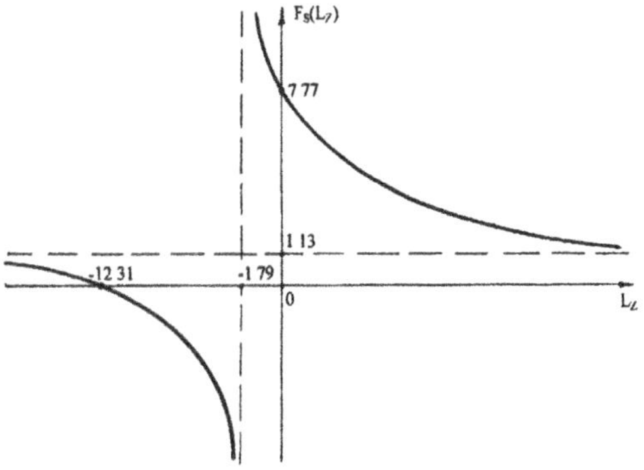

Fig.5 Relationship character F_S at $\alpha = \beta = \omega$.

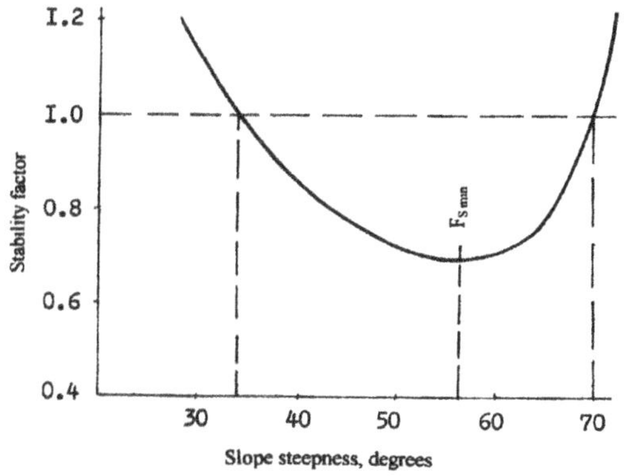

Fig 6 The Change of F_S in the slopes of different steepness at $\alpha = \beta = \omega$

On the basis of expression (2) analysis it is stated that parabolic relationship in the field of magnitudes L_Z that are interesting for us does not take place in all cases. The given statement is justified only at the slickenside inclination that is larger than slope steepness (ω), that is confirmed by the results of the field tests, when in the back side of the landslide we can see vertical or steepfalling break-off wall, i. e. There is $Z_{CR} > 0$. So the case $\beta = \omega$ should be considered as particular.

4 CONCLUSIONS

On the analysis of the conducted analysis we may make a conclusion that slickensides are not always the potential slickensides of landslides and the most dangerous from the point of view of rock slope stability. The magnitude of stability factor, defined from the assumption of coincidence of slickenside and bedding turns out overestimated and, according to the experience, is often the reason of rock slope destruction in future.

The analysis of another case, at which slickenside inclination coincides with the bedding and rock strata armour the slope, surface, showed, that the relationship $F_S(L_Z)$ may have hyperbolic character. This may be the reason of landslide occurence in water interface sections of the slope with the output of the displacement surface to the opposite side of slope water interface. But magnitude of stability factor in this case (at the same initial parameters) turns out to be higher, than in the steeper case related to rock mass layers dip, slickenside inclination to the horizon.

The given method of rock mass stability criterion determination and the results obtained allow us to be more objective in the stability analysis of the existing and newly designed slopes in jointed rock masses.

REFERENCES

Maslov N.N., Fonaryov P.A. 1987. Modelirovanie treshinovatyh skaljnyh massivov.//Ingenernaya geologia, N.1, 23-28.

Fonaryov P.A. 1986. Ustojchivostj skaljnyh sklonov, podrezannyh pri stroitelstve avtomobilnyh dorog.- Voprosy geotehnicheskogo obespechenia dorognogo stroitelstva.//Trudy MADI, , 45-47.

Fonaryov P.A. 1980 Strukturno-teksturnye osobennosti stroeniya skalnyh massivov i harakter narusheniya ih ustojchivosti.- Materialy nauchno-tehnicheskoy konferentsii molodyh uchenyh MADI // Deponirovano v TSINIS v 1980 godu. Bibliotechny ukazatelj deponirovannyh rukopisej, vypusk.5, N.1993.

Fonaryov P.A. 1994 Graphoanaliticheskij metod opredeleniya ustoychivosti skaljnyh otkosov//Geoekologiya, N.3, , 99-103.

SNiP 2.05.02.-85 1986 Avtomobiljnye dorogi. Moskwa,. 52 s.

Mechanics of Jointed and Faulted Rock, Rossmanith (ed.) © *1998 Taylor & Francis, ISBN 90 5410 955 6*

The influence granular structure of rockfalls on their spreading along mountain slopes

V.V.Adushkin & V.G.Spungin
Institute for Dynamics of the Geopheres Russian Academy of Sciences, Russia

ABSTRACT: The authors studied two rock avalanches that were triggered on the northwest slope of Moiseev Mountain, Novaya Zemlya, by seismic shaking after a strong underground nuclear explosion. Through investigation of motion characteristics and distribution of the rock avalanches, analysis of granular structure and mineralogical composition of rock deposits we have established that the runout length of the avalanches is influenced by the granular structure of the constituent rock debris.

1 INTRODUCTION

Landslides, rock falls, and mass wasting are the characteristic phenomena found on mountain slopes. In the case of a major rock fall, rock masses can flow as a liquid and their runout length can be much larger than their fall height. This phenomenon is known as a rock avalanche. Catastrophic rockfalls in the Alps, the Cordilleras, the Pamirs and other mountain systems are well known. Triggering mechanisms of these rockfalls may be either artificial or natural, for example earthquakes, explosions, or tectonic activity. After the initial loos of stability, the motion of the rock mass is governed only by the forces of gravity and friction and becomes independent of the triggering mechanism (Scheidegger1975).

Investigations of many avalanches demonstrate that the apparent coefficient of friction H/L (where H is the rockfall height, L is the runout length) is not constant ranging from 0.1 to 1.0. Furthermore, the avalanche volume influences this coefficient. It is well known (Scheidegger 1973) that H/L decreases with increasing slide mass volume for volumes larger than about 10^5-$10^6 m^3$ due to their high potential energy. It is necessary to stress, however, that the dispersion in the experimental data is very large. There is some evidence that many other factors effect the process of distribution of rock avalanches and their runout length. The review by O. Hungr (1990) contains descriptions of nine hypotheses that attempt to explain the unexpected long runout and, consequently small apparent friction of these slides. It has been proposed that avalanche deposits may attain ther small apparent friction because of stream fluidization under the influence of air, water, dust, soil and so on. But even dry avalanches of large volume, without water saturation, can have a large runout length. Recently, Campbell (1989) proposed that the low friction may be a result of simple granular mechanics; the low friction is due to the bulk of the slide riding nearly as a solid block, on top of an active layer within which all of the energy dissipation is confined. The most elaborate theoretical model developed to explain this phenomenon is the theory of acoustic fluidization by Melosh (1979).

All the theories presented so far do not consider the influence of the average size and structure of the granules that make up the avalanche debris on the runout length. Also, the influence of initial mineralogical composition on the runout length has not been properly investigated. These factors are worthy of consideration because different types of rocks have different strengths, densitys, and other characteristics that can influence the process of the energy exchange between the particles in the debris. The reason is that natural rock avalanches are dispersed in space and time, and usually the investigator can look only at avalanche deposits, which are often disrupted by more recent geological events. In this connection, underground nuclear tests on Novaya Zemlya make it possible to obtain unique data. The time and place of the rockfall are known in advance, allowing the opportunity to record the whole process in time, as well as to determine the geometric sizes of rockfall areas and avalanches deposits, undisturbed by erosion.

2 DESCRIPTION OF THE EXPERIMENT

The rockfall on the northwest slope of Moiseev Mountain, Novaya Zemlya, were triggered by seismic shaking after a large underground nuclear explosion (experiment A-10, Michailov 1992). For all practical purposes, the explosion can be described as totally contained. The rockfall, induced by the dynamic force of the explosion, took place at the steepest section of the slope. Broken rock masses swelled up over the slope surface as a dome and then rolled downhill. The rock masses were accelerated to high velocity over the steeper part of the slope by gravity and then spread out over the gentler part of the slope as the flow, passed over the surfase topography.

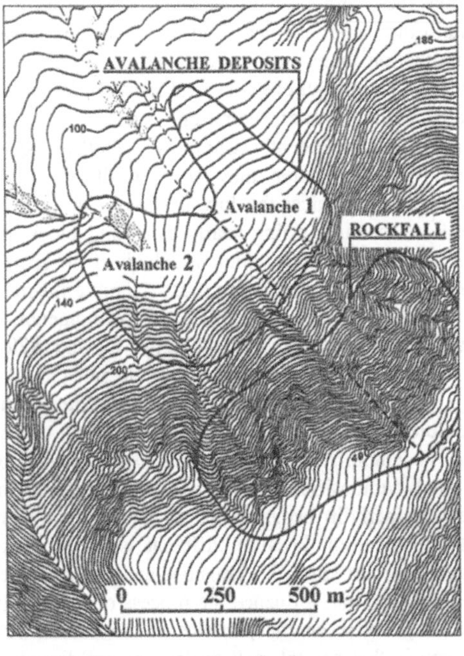

Figure 1. Map showing the configuration of the rockfall area and the deposit area of Avalanche 1 and 2, Novaya Zemlya, 23 August 1975. The contour inter val is 5 m.

The topographic sketch map in Figure 1 shows the initial position of the rockfall region on the slope

Figure 2. Oblique aerial view of the area of Avalanche 1.

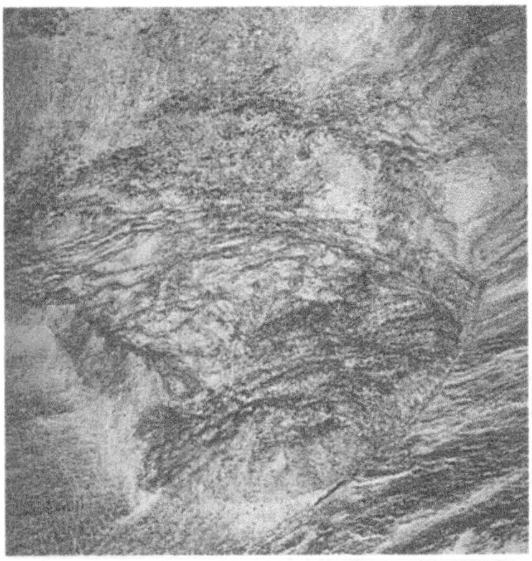

Figure 3. Oblique aerial view of the area of Avalanche 2.

and the final position of the avalanche deposits. We identify the two lobes as Avalanche 1 and Avalanche 2. General views of the avalanche deposits are shown in Figure 2,3.

2.1 *Experimental procedure*

The rockfall and avalanche motion was captured by high speed film photography. The vertical acceleration ofthe slope surface was measured using light points and velosimeters. The volumes and areas of the avalanches were calculated using surface mapping and aerial photographs with an accuracy of about 10%-20%. The granular structure of the avalanche's rock mass was studied on surface of the avalanche deposits by profile measurements of the size of rock fragments on the avalanche surfase. The mineralogical composition of the rocks was studied by microscopic analysis.

2.2 *Description of rock avalanches*

The total volume of the rockfall is about $5\times10^6 m^3$. From the film, it was established that initially, both avalanches moved as a single block which then subdivided into two parts with a volume ratio of approximately 1:4. Figure 4, derived from the film photography, shows profiles of the Avalanche 1 rock mass, as it moved down the side of the mountain. At the initiation of motion, the landslide is obscured by

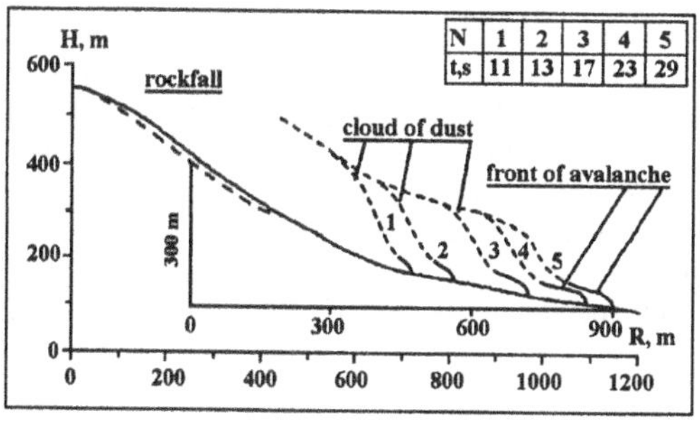

Figure 4. Dynamics of Avalanche 1.

a cloud of dust which follows the slide; the contour labeled (1) in the Figure is the first in which the toe of the landslide becomes visible as it passes through the front of the dust cloud. Successively numbered contours represent later stages in the evolution of the slide. The initial acceleration of the slide occurs over the first 500 meters of the runout. It begins on the slope of 35^0 and continues until the slope falls to 15^0, after which the slide decelerates. Most of the subsequent motions occurs on slopes that vary from 7^0 to 10^0. The maximum velocity of the landslide toe is about 45 m/s and occurs about 12 seconds into the slide. The most important geometrical characteristics of the avalanches are presented in Table 1.

Table 1. Geometric parameters of Avalanches 1 and 2.

Geometric parameters	Avalanche 1	Avalanche 2
Volume of avalanche deposits (V)................................	$1\times10^6 m^3$	$4\times10^6 m^3$
Area covered by avalanche deposits (S)...........................	$0.97\times10^5 m^2$	$1.6\times10^5 m^2$
Averaged thickness of avalanche deposits (h).....................	10 m	25 m
Gravity center height of rockfall with respect to distal boundary of avalanche deposits (H).............	310 m	300 m
Horizontal runout length for avalanche tip deposits from gravity center of rockfall (L)..........	900 m	770 m
Length of avalanche deposits on horizontal (l)..............................	540 m	485 m
Effective coefficient of friction (H / L)................................	0.34	0.39

2.3 *Paradox of the smaller energy dissipation of Avalanche 1 relative to Avalanche 2*

Analysis of the rock avalanches' geometric parameters show a considerable comparative discrepancy between potential energies and runout lengths of the two avalanches. Table 1 indicates that initial potential energy of Avalanche 1 equals:

$$E_p^1 = 1\times10^6 m^3 \times 2700\ kg/m^3 \times 9.8\ m/s^2 \times 310\ m = 8.2\times10^{12}\ J$$

The potential energy calculation for Avalanche 2 shows:

$$E_p^2 = 4\times10^6 m^3 \times 2700\ kg/m^3 \times 9.8\ m/s^2 \times 300\ m = 31.7\times10^{12}\ J$$

As both avalanches start at nearly the same elevation (300 - 310 m), the difference in their potential energies is largely due to the difference in the avalanche volumes. Furthermore, the conditions that govern the motion of both avalanches were approximately identical. For example, the average slope angle and seismic intensity (characterized by the maximum particle velocity) are very close for both avalanches and have values of about 30^0-35^0 and 10-25 m/s, respectively. Both avalanches were

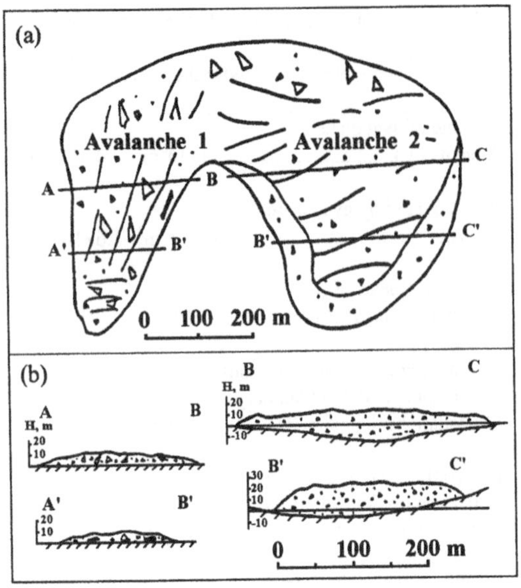

Figure 5. Surface structure and cross-section of Avalanche 1 and 2.

unconfined and concluded their runouts on a valley surface of about 7^0-9^0.

A larger runout for Avalanche 2 would be expected because of its initially larger potential energy. But in the case we see a reverse situation: the runout length of Avalanche 1, with a smaller volume, is approximately 1.2 times longer than the runout length of Avalanche 2, which has a larger volume. Similarly, the value H_1/L_2 is smaller than the value H_2/L_2 by a factor of approximately 1.2.

The square area covered by deposits from Avalanche 2 is also relatively smaller. Thus the S value for Avalanche 2 is only 1.6 times greater than the

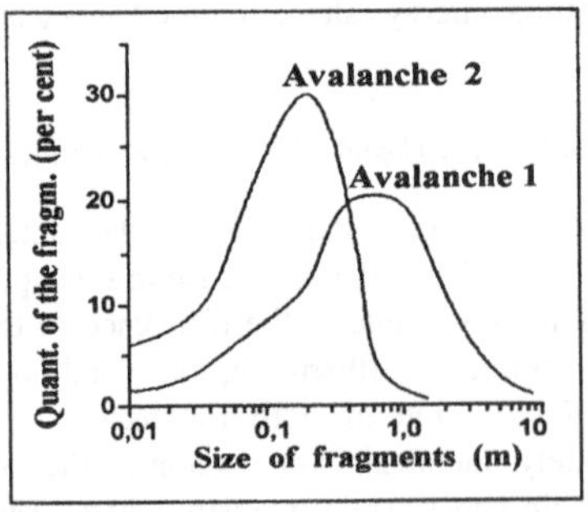

Figure 6. Size frequency distribution of rock fragments of the avalanche deposits.

area covered by Avalanche 1 deposits. This is due to the lower height of the Avalanche 1 deposit. The mean height of the Avalanche 1 deposit is 2.5 times smaller than the mean height of the Avalanche 2 deposit.

3 THE INFLUENCE OF THE GRANULAR STRUCTURE AND AVERAGED SIZE OF GRANULES OF ROCK AVALANCHES ON THEIR RUNOUT LENGTH

3.1 *Established differences between Avalanche 1 and Avalanche 2*

Through field analysis, we conclude that, in this case the lenth of runout depends upon the granular size distribution in the avalanche body, which is very different for these two avalanches. The particle size distributions for the two are shown in Figure 6. This figure shows that the maximum size of the granules in Avalanche 2 deposits is less than 1 m, and that the granular size is narrowly distributed with the most abun dant granular size being about 15-30 cm (Figure 7). On the other hand, deposits of Avalanche 1 consist of debris with widely distributed granular sizes, varying from a few centimeters to few meters (Figure 8). The granular size distribution in this case is very similar to the one of described by Sadovsky (1983).

In the rockfall region the rock massif is characterized

Figure 7. View of part of the Avalanche 2 deposit.

Figure 8. View of part of the Avalanche 1 deposit.

by monoclinic bedding (at an angle of 40^0- 60^0), consisting of Silurian-aged metamorphic terrigenous rocks. At the point of origin of the rockfall, the rock mass is composed of mica schist, crystalline schist and mica-crystalline schist, consisting of sericite (30-70%), feldspar (30-75%), quartz (1-15%) and some chlorite, tremolite and dark-colored minerals (5-15%). The initial quantity and structure of cracks in the rockfall region are approximately identical for the source areas of both avalanches. Using microscopic analysis, we have established that the differences between the debris of Avalanches 1 and 2 can be attributed to the differences in the mineral compositions and the rock's texture, probably due to local variations in the metamorphic processes.

3.2 *The differences between types of motion in the rock avalanches*

The larger runout length and smaller energy dissipation of Avalanche 1 relative to Avalanche 2, can probably be ascribed to the fact that the two avalanches had different types of motion due to their different granular structure, especially while moving over the more gently sloping portions of the surface topography (less than 10^0-15^0). On the steeper parts of the slope (30^0-40^0), velocities and directions of the motion of both avalanches were approximately the same. Collisions between granules are insignificant and the granular mass accelerates freely, under the influence of gravity, converting the potential energy of the deposit into kinetic energy. But on the more gentle parts of the slope, collisions between granules become more important. These collisions lead to an increase in local heating and plastic deformation that changes the material properties as well as the directions of motion of the constituent granules. Consequently, on this part of the slope, the dissipation of the kinetic energy in Avalanche 2 became several times larger than the dissipation of the kinetic energy in Avalanche 1.

It is known from theory (Goldsmith 1965) that energy dissipation during impact depends upon elastic properties, masses, shapes, and relative velocities of the impacting particles. In the case of simple central collision when the collision occurs at the point situated on the straight line connecting gravity centers, the dissipation of kinetic energy for two particles equals:

$$\Delta E_k = \frac{m_1 \times m_2}{2(m_1 + m_2)} \times (1 - k^2) [(V_1 - V_2) - (U_1 - U_2)]^2$$

where m_1 and m_2 are the masses of the particles, V_1 and V_2 are the velocities before the collision, U_1 and U_2 are the velocities after collision and $k = (U_1 - U_2) / (V_1 - V_2)$ is the coefficient of restitution.

In the limiting case $\Delta E_k = 0$ if both bodies are absolutely elastic ($k = 1$), or

$$\Delta E_k = \frac{m_1 \times m_2}{2(m_1 + m_2)} \times [(V_1 - V_2) - (U_1 - U_2)]^2$$

if both bodies are completely inelastic ($k = 0$).

Experimental investigations (Goldsmith 1952) demonstrate that for natural rocks and soils the coefficient of restitution lies between 0.1 - 0.2 (clay) and 0.75 - 0.9 (marble). The average value of this coefficient for rocks of Moiseev mountain area is between 0.4 and 0.5.

The influence of the impactor's mass difference on the energy dissipation can be derived from the coefficient $m_1 \times m_2 / 2(m_1 + m_2)$. It is easy to see that increasing m_2 relative to the value m_2, i.e. increasing the mass difference, increases the coefficient $m_1 \times m_2 / 2(m_1 + m_2)$ and consequently, the energy dissipation ΔE_k, that is seemingly opposite of what we have observed. However, it is necessary to bear in mind that, together with a reduction in mass, the difference of impacting granules number ("n") of the granules increases and, accordingly, the number of granule collisions increases too. This is correct under the condition that the total volume of rock mass is constant. It is can be shown that the total loss of kinetic energy by "n" collisions of two granules with unit masses are vastly greater than it is by one collision of two granules with masses 1 and "n"x1.

In addition, as C. Campbell noted in the discussion of these data, after a collision, the resulting change in the impact particles' velocities will be more-or-less randomly distributed. In this way, a portion of the forward-directed kinetic energy of the landslide is converted into randomly-directed kinetic energy of its constituent particles. As far as the bulk motion of the slide is concerned, the energy is lost at the moment of conversion from forward to randomly directed motion, rather than when it is eventually dissipated to heat by collisional inelasticity. Furthermore, the random particle motions will drive particles together, increasing the collision rate, thus increasing the rate of conversion.

The relative presence or absence of this process may be used to explain the differences in the behavior of Avallanches 1 and 2. We assume that the motion of the avalanches with different granular sizes, on the less steep part of the slope proceeds as follows: For Avalanche 1, consisting of granules of different sizes (up to few meters in diameter), large blocks are uniformly distributed in the avalanche body and are surrounded by blocks of smaller sizes. These large blocks carry a significant part of total kinetic energy of the avalanche (about 20%-30%). Collisions with smaller particles will do little to change the motion of these large blocks and only collisions with blocks of equal or larger size will cause the conversion of forward-directed kinetic energy into randomly-directed energy. But due to this intervention of small and medium blocks, large blocks can "flow" in the stream without colliding with one another. Thus, little of the forward-directed kinetic energy of the large blocks will be converted to randomly-directed energy and will persist to carry the slide forward.

Furthermore, Figures 2 and 5 show transverse hummocky relief and transverse ridges on the surface of Avalanche 2. A possible explanation for this relief may be found in the lateral motion of the avalanche granules, which is driven by the redirection of kinetic energy into traverse directions. On the surface of Avalanche 1, we can see ridges elongated in the direction of the avalanche motion and the absence of transverse ridges, perhaps indicating the relative absence of random particle motions (Figure 1 and 5).

4 CONCLUSIONS

We find that the rock avalanche with various sizes of granules (ranging from centimeters to 5 - 10 meters) has a longer runout in comparison to the rock avalanche with a more uniform size distribution (commonly 20 - 30 centimeters, maximum up to 1 meter). The smaller energy dissipation of the avalanche with various sizes of granule can be ascribed to different types of motion of the avalanches, especially while moving over the more gently sloping portions of the surface topography.

It is found that granular structure and the average size of granules in a rockfall can be depended on the local characteristics of the geological structure of the rockfall region. Furthermore, the small differences in the mineral composition, texture and structure between rocks of one type and time of genesis can influence the granular structure and average size of rock avalanche deposits.

The peculiarities of rock avalanche propagation allows for predicting the runout nature or artificial rock avalanches.

ACKNOWLEDGMENTS

We thank Charles Campbell for useful discussion the data. We are very grateful to Rodney Matzko for his helpful editting the English version of the paper.

REFERENCES

Campbell C.S. 1989. Self-lubrication for long runout landslides, J. Geology, v. 97, p. 653 - 665.

Goldsmith W. 1952. The coefficient of restitution, Bull., Appl. Mech. Div., Amer. Soc. Eng. Educ., 2.

Goldsmith W., 1965. Impact the theory and physical behavior of colliding solids, Berkeley - London - Mosca.

Hungr O., 1990. Mobility of Rock avalanches. Submitted to the Bulletin of National Research. Centre for Disaster Prevention, Tsukuba, Japan. July.

Melosh H.J. 1979. Acoustic fluidization: a new geological 7513-7520.

Michailov V.N. and other, 1992. Nuclear explosions in USSR., part 1 North test site, reference book, Moscow, 195 p. (in Russian)

Sadovskii M.A. On the size distribution of solid jointing, Dokladw AS USSR 1983, t 269, 1, p. 69-72, (in Russian).

Scheidegger A.E. 1973. Physical aspects of natural catastrophes, Elsevier Scientific Publishing Company, Amsterdam - Oxford - New-York,

Mechanics of Jointed and Faulted Rock, Rossmanith (ed.) © 1998 Taylor & Francis, ISBN 90 5410 955 6

The new method for modelling of slopes stability at changeable temperature field

A. M. Demine
The Comprehensive Institute for Mineral Resources Exploitation of Russian Academy of Sciences, Russia
K. A. Gulakyan
Institute for Mechanics, Moscow State University, Russia

ABSTRACT: The paper deals with the results of modelling the breakdown of benches with vertical slopes. Material properties were purposively changed during the test. Critical values of the model strength properties, that were equivalent to natural rock; value of relative buckling during the failure; critical values of some deformation parameters, corresponding to failure moment, were defined during the study.

1. INTRODUCTION

It is known, that modelling is an effective method of geomechanical research. It allows to understand the mechanism of slopes deformation development, as well as to predict them in a good time. Glesen was the first, who began to model benches and slopes stability, purposively changing the model materials (sand) strength properties by wetting; as a result a slope was failed. Further investigations were carried out in another way: the model was broken down by external load increase. In this case, strength and deformation parameters were constant, that didn` t give the real modelling pattern, in comparison with the natural processes. Also, centrifugal modelling method requires expensive equipment and special rooms. The method of equivalent model materials can`t be used for breaking-down. It doesn`t allow to keep boundary conditions for slopes. Both methods mentioned above don`t provide strength and deformation properties change of model material.

2. OBSERVATION

Moisture content change of rock, namely clay rock (open pit slides as a rule) significantly influences open working and dump stability under specific geologic conditions. Identity of clay rock strength parameters change depending on moisture content and the one for equivalent model materials depending on temperature, was confirmed by laboratory tests. Control of model temperature conditions, corresponding to natural rock moissture content change may assist in changing the model strength and deformation properties during test in a good time. A series of model tests with temperature changing was carried out at the Department of Natural Processes Mechanics of the Institute for Mechanics, Moscow State University named after Lomonosov. The aim of investigations was studing of 2 versions of stability loss of benches with vertical slopes in uniform massif and with weakening surface (area). Maximum bench altitude was selected for quantitative characteristic, determining the degree of slope stability under different types of deformations.

3. CALCULATIONS

The modelling conditions were as follows;
planar-stressed state
scale - = 200
mechanical and physical properties:
1. Rock: density: γ - 2,53 gr/cm^3; coherence C - 0,3 MPa, internal frictions angle φ - 33^0.
2. Weakening area C_0' - 0,01 MPa, φ' - 8°.
According to similarity theory created by M.V. Kirpichev, L.I. Sedov and G.N. Kuznetsov, the main similarity conditions of modelling the rock deformation and failure are as follows:

$$\frac{\tau_H}{\tau_M} = \frac{C_H}{C_M} = \frac{\gamma_H L_H}{\gamma_M L_M}, \quad (1)$$

$$\varphi_H = \varphi_M, \quad (2)$$

$$\frac{t_H}{t_M} = \left(\frac{L_H}{L_M}\right)^{\frac{1}{2}}, \quad (3)$$

$$\frac{\eta_H}{\eta_M} = \frac{\gamma_H (L_H)^{\frac{3}{2}}}{\gamma_M (L_M)^{\frac{3}{2}}}. \quad (4)$$

where τ - shearing strength; η - coefficient of viscosity , L - linear dimension; t - time. Indices correspond to the values under natural and modelling conditions. Geometric similarity during deformation will be maintained, when relative deformation of the model and natural materials at appropriate points are equal and displacement scale corresponds to geometric scale.

$$\frac{\delta_H}{\delta_M} = \frac{L_H}{L_M} = \lambda. \quad (5)$$

Under these conditions the disturbance and fracture areas of the model and natural materials have similar arrangement. To observe boundary conditions during physical modelling, the model height shouldn't exceed it's width by more than 2-3 times.

Multipurpose roll-over stand and devices for axial compression and single-plane shearing, providing short-and-long- term tests of thermoplastic materials under controlled temperature conditions, were used in modelling.

Special research on the selection of composite thermoplastic optimum composition were carried out. This material have properties depending on temperature change as well as higth stability to repeated tests. As a results, the mixture consisting of quarts sand (80%), bentonite (10%) - as inert fillers and lubricant grease (10%) as binding component, was selected as composition for obtaining materials, equivalent to modelling rock-aleurolite. The wide range of strength and deformation properties variations with the temperature change caused the selection of this composition. Figures 1-3 show the results of short-and-long- term tests on model material shearing. Corresponding parameters are given in these plots on the base of therock taking into account the modelling scale.

Test temperature conditions were set from the research results. The model temperature was gradually varied, from 20° C to 75° C; as a results it's strength and rheological properties were hanged and model material was deformed. Tests were

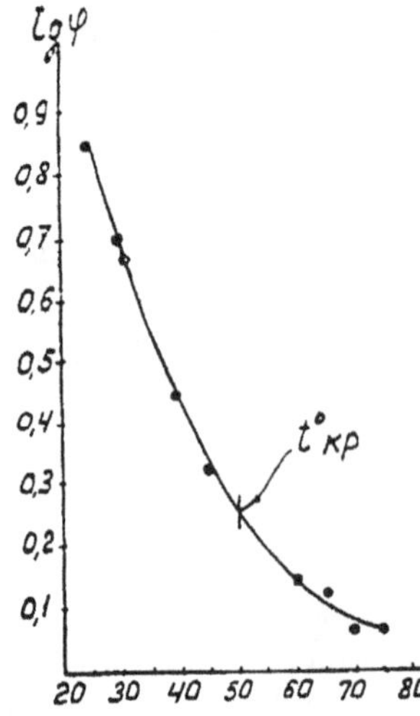

Fig.1

Equivalent-material adhesion versus temperature. t^0_{kr}-temperature, providing stability of the model with uniform structure.

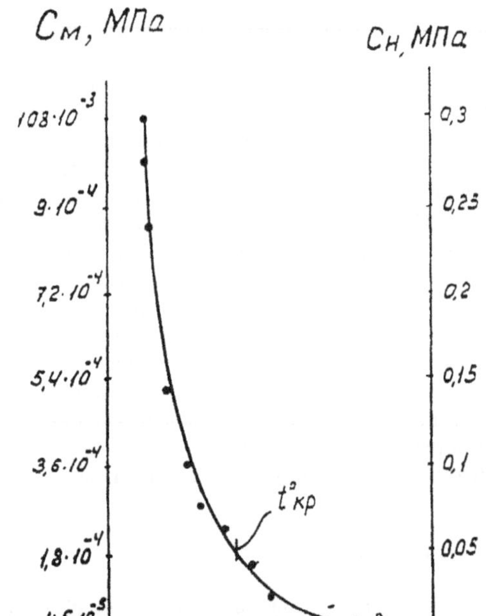

Fig. 2

Equivalent-material coefficient of friction versus temperature.

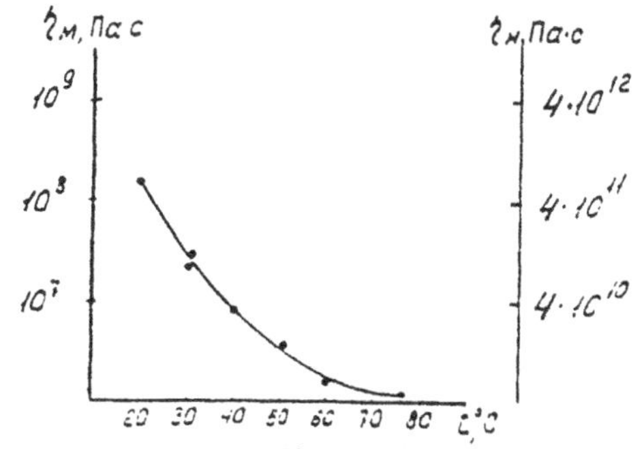

Fig. 3
Equivalent-material coefficient of viscosity versus temperature.

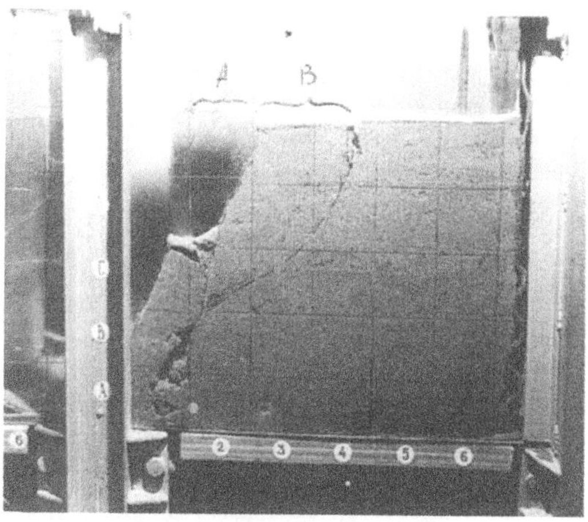

Fig.4
Uniform model failure

considered to be completed when deformation and caving of the part of model massif followed by the new equilib state, was achieved. Models of vertical slope benches having uniform structure and poor contact, including towards slope at the angle 30°, 45°, were studied.

At the first, the uniform structure model in the form of right-angled massif of equivalent -material, having the following characterictics: γ=1,8 g/cm^3, C (at 30°) = 10^{-3} MPa, (at 30°C) = 33°, was studed. The model dimensions are: 0,5 m - height (H), 0,5 m - length; 0,4 m - width.

Heating of the model - using special heating devices, placed at the model`s base and it`s rare part. Temperature control was carried out by means of contact thermometers and deformation measurement - by indicators (accuracy - 0,1 mm). Model tests started under the temperature 30° C, with removal of the model protecting front wall. The tests were carried out using photography. There was no deformation during 24 hours. Then, model`s temperature was graduallyincreased with an 1-hour inrterval in every stage.

Model deformation occured at 52° C temperature as massif border subsidence along the fracture breaking-off-area A, Fig. 4.

The failure prism width under the natural conditions - 10 m. After reaching ultimate relative strain value h_1/H - 0,08, the above mentioned massif was separated and failed along steep inclined surface ($\alpha_{cr.}$ = 70), where h_1 - vertical subsidence of the first block. Deformation processes with new fracture forming - area B, Fig, 4, were observed immediately after massif border breaking-off. The failure prism width - 15 cm (under the natural conditions - 30 m). The second block was separated and shifted along circular surface. Block shift was incompleted - it`s movement stopped near the shifted block. The second block ultimate strain value (h_2/H) was 0,04.

The process active stage proceeded 3 hours 25 min. Taking into account the time scale (3), the duration of the equivalent period under natural conditions is as follows:

$$t_{nat\ total} = t_M \cdot \lambda = 205_{min} \cdot 14 = 48 hrs$$

incl. 12 hrs. = 1 stage, 36 hrs. = 2 stage. Critical values of the "C" and "φ" parameters that caused the model active deformation and breakedown, were 1,8*10^{-4} MPa and 14° respectively. Models of the massif with specified weakening surface were made of the same equivalent material as in the first case. The presence of weakening surface (area) being formed when the model concentrated heating by the integrated heating devices.

Two versions of the problem were studied: weakening surface angle a) 45° and b) 30°. Test procedure was as follows: at the first, the model massif was heated to 30°C by peripheral heaters, arrannnnged along the model outline. Thereby, the model was introduced into a selected modelling scale (λ=200) and necessary values C=1* 10^{-3} MPa, φ=35°, for material-equivalent were obtained. Then, the heating device, forming-weaakening area was switched on.

Heating was carried out gradually with the temperature raising by 2 degrees every hour. In the

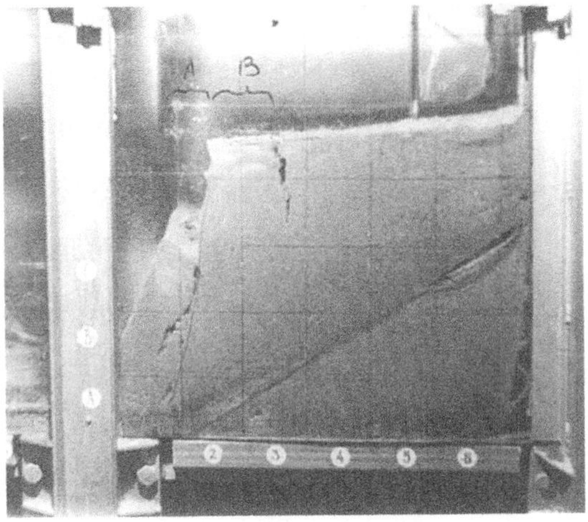

Fig. 5
The failure of model having weakening surface, inclined at 30^0 angle to slope.

first case (weakening area angle is 45°) the upper part o-f the model failed along given shifting area when reaching 34° C ($C = 7*10^{-4}$ MPa, φ = 29°) In the second case (weakening area angle is 30°) the deformation occured at 45° C temperature. Here, the deformation was observed as sliding of the upper part of massif, followed by the entire model surface subsidence - h/H = 0,1. After 36 min. (8 hr. 30 min. under natural conditions), massif border shearing to 7 cm (14 m under natural conditions) broke-off and failed along steep (≈ 75°) fracture in the form of slab. The declined block width is 6 cm (12 m).
Afterwards, the second formed fracture separated area of 12 cm width (24 m). The second shifting was incompleted and the model completely stabilized (Fig. 5).
Then, the deformation wasn't observed despite the temperature raising to 60° C ($C \approx 6*10^{-5}$MPa, $\varphi = 8°$) in weakening area.

4. CONCLUSION

Duration of the second stage was 50 min.12 hrs. under natural conditions). The study shown, that stability loss of the benches with vertical slopes depends on model massif tension and components strength. Also, critical values of modelling rock strength properties causing irreversible deformation processes starting and developing, incl, adhesion (C) in the range of 0,16 - 0,023 MPa (under natural conditions) and internal friction angle (φ) - 14-24°, were determined.
Benches deformation started with subsidence of the massif marginal part (h/H) in the range of 0,1 - 0,08, followed by the slab's (25 - 30 m width) breaking-off and failing. This stage duration is 8 - 12 hrs. Then, the deformation developes after the same pattern-subsidence of the massif marginal part (h/H) in the range of 0,06 - 0,04, it's breaking-off and shearing. The width of the broken-off block is compared with that of the former slab. The model width is 0,1 - 0,3 H per unit values, that corresponds to 0,14 - 0,33 H for open pit landslides.
In all cases brocken-off blocks sheared along steeply dipping fractutres (≈70°) that sometimes (in uniform massives) transform into cylinrical surface. Duration of the second stage of deformation process was 12 - 30 hrs. under natural conditions. Overall duration of deformation processes active stage varied from 24 - 48 hrs. Rock massif structure was of great importance for slope stability. Weakening areas in two test versions lied to stability loss and bench failure even in case of higher strength properties compared with that of uniform structure massives.

5. ACKNOWLEDGMENTS

The present paper should be considered as the first stage of research intended for definition of mechanism and dynamics of the bench stability loss and deformation processes under standart and specified geomachanical conditions at quarries. Volumetric model tests providing three - dimentional features reproduction, are planned. Thus, a new research method for evaluation of slopes and benches stability in uniform and cross-linked massives under natural conditions is developed.

REFERENCES

1. Glesen, J. Uber die Standfestigkeit vou Tagebaukippen //Zeitschrift fur das Berg. - Hutten und Salinenwesen, 1931. - N 1. - 3, 1- 47.
2. Gulakayn, K.A., Kyuntsel, V.V., Postoev, G.P. Landslide processes prediction. - M.: Nedra, 1977. - 135 p.

3. Demine, A.M. Gulakyan, K.A et al. Method for modelling landslides deformations of dump and natural slopes: A.S. 941579 USSR MKU E 21 C 41/00/.

4. Demine , A.M. Classification of slopes deformation in quarries and dumps according to their manifestation mechanism //Development of method for efficient exploitation of mineral resources by open-pit mining. - M: Offprint duplicator IPKON , 1991.

5. Demine, A.M. Regularity of slopes deformation manifestation in quarries. - M: Nauka, 1981, 144 p.

6. Model study of rock pressure manifestations. M.: Ugletechizdat, 1959. - 284 p.

7. Pokrovskiy, G.I. Centrifugal modelling . M.: ONTI, 1935, 135 p.

Hydromechanics

Mechanics of Jointed and Faulted Rock, Rossmanith (ed.) © *1998 Taylor & Francis, ISBN 90 5410 955 6*

The behavior of induced pore fluid pressure in undrained triaxial shear tests on fractured porous analog rock material specimens

Guy Archambault, Stéphane Poirier & Alain Rouleau
Centre d'Etudes sur les Ressources Minérales, Université du Québec à Chicoutimi, Qué., Canada

Sylvie Gentier
BRGM, Direction de la Recherche, Orléans, France

Joëlle Riss
Centre de Développement des Géosciences Appliquées, Université de Bordeaux 1, Talence, France

ABSTRACT: Six series of triaxial tests were performed on intact and fractured porous rock analog material specimens for various confining and pore pressures and drainage (drained and undrained) conditions. It was found that physico-mechanical properties are affected by water and Terzaghi's effective stress law holds for this analog material. Under undrained test conditions, pore water pressure increases as the specimens volume is reduced, and vice versa. The maximum induced pore pressure is a function of the confining pressure, initial pore pressure and volumetric strain. Similarly to the intact material, the fractured samples showed that the behavior of pore pressure is closely related to the volume change. The application of a deviatoric stress initially produced an increasing phase of pore pressure during the friction mobilization phase. After the friction mobilization phase, the roughness is mobilized and this corresponds to the beginning of a progressive decrease of pore pressure. The increase and decrease in pore pressure is different from one sample to the other mainly due to the influence exerted by the roughness morphology different for each fracture.

1 INTRODUCTION

The important role of pore-pressure in hydromechanical stability of engineering works in rock masses (intact or jointed) was recognized since several decades (Lane 1970). However, few experimental studies of pore-pressure effect on rock behavior are available in the literature (Aldrich 1969, Bruhn 1972, Goodman & Ohnishi 1973, Mesri et al. 1976, Ismail & Murrell 1976). Pore fluids have significant influences on the physico-mechanical behavior of rocks attributed to the physico-chemical interaction between the fluids and rock constituents (Terzaghi 1945, Serdengeti & Boozer 1961, Boozer et al. 1963, Colback & Wiid 1965, Aldrich 1969, Parate 1973) or to the pore fluid pressure (u). In this latter case, if was shown that for many porous rocks the Terzaghi's effective stress law ($\sigma'=\sigma-u$) was valid (Robinson 1959, Serdengecti & Boozer 1961, Handin et al. 1963, Murrell 1965, Aldrich 1969, Dropek et al. 1978).

Few experimental works were dedicated to pore pressure behavior and effects, under undrained conditions, in rocks during their deformation, failure and post-failure phases (Aldrich 1969, Ismail & Murrell 1976); while Mesri et al. (1976) studied the pore-pressure response in rock to undrained change in all-round or isotropic stress. Fewer studies were devoted to the same problem in jointed rock (Lane 1969, Goodman & Ohnishi 1973) but for smooth saw cut joints. No such studies were performed on rough irregular joints submitted to shear under undrained conditions. In this case the stress-dilatancy behavior and asperities morphology on the joint surfaces play a major role on the induced pore pressure behavior.

2 PORE PRESSURE BEHAVIOR IN UNDRAINED TESTS ON INTACT SPECIMENS

Under undrained condition, interstitial water is trapped within the void spaces (pores and micro-cracks) in the specimen. Dilation of the sample volume decreases the pore pressure, while contraction increases it. For this condition, it is assumed that water can move freely within the sample but not out of it. As a modification of the pore pressure changes the state of applied effective stress and consequently, the rock behavior; it is important to define the relation between the pore pressure variation resulting from a change in the state or applied stress. Skempton (1954) has developed such an empirical equation for soils in the following form :

$$\begin{aligned} u_i &= B\cdot[\Delta\sigma_3 + A\cdot(\Delta\sigma_1 - \Delta\sigma_3)] \\ &= B\cdot\Delta\sigma_3 + \bar{A}\cdot(\Delta\sigma_1 - \Delta\sigma_3) \end{aligned}$$

where $\bar{A} = A\cdot B$. The parameters B and $\bar{A}$ are respectively the induced pore pressure coefficients for a variation of the isotropic stress ($\Delta\sigma_3$) and of the stress deviator ($\Delta\sigma_1 - \Delta\sigma_3$).

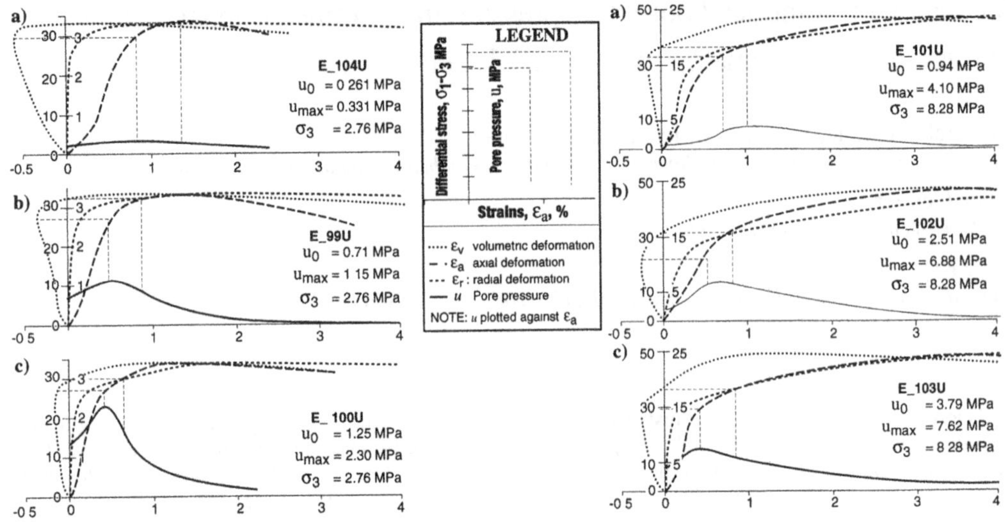

Figure 1. Material deformability under undrained conditions

Mesri et al. (1976) studies the pore pressure response in rock to undrained change in all-round or isotropic stress and demonstrate that the B coefficient rapidly decreased with increasing effective isotropic stress to values ranging from 0.33 to 0.69 depending on the type of rock and the isotropic stress level. As the confining pressure stays constant during an undrained triaxial test ($\Delta\sigma_3$=0), this study will evaluate the coefficient $\bar{A}$ only related to the variation of the stress deviator ($\Delta\sigma_1$ - $\Delta\sigma_3$).

2.1 *Pore pressure behavior in relation with stress-strain variation*

A first series of tests was performed at a confining pressure (σ_3) equal to 2.76 MPa while for a second series, σ_3 was set at 8.28 MPa. For both series, the initial pore pressures (u_0) were approximately set to $0.1\sigma_3$, $0.3\sigma_3$ and $0.5\sigma_3$. A third series was also realized with σ_3 set at 13.8 MPa and this series, not shown in this paper, show a quasiplastic behavior (Poirier et al. 1994, Poirier 1996).

In the first phase of the stress-volumetric strain curve (Fig. 1), a contraction of the specimen was observed (ε_v decrease) and pore pressure increased. This phenomenon result from cracks and pores closure. At the maximum contractancy (ε_v min), the pore pressure was approximately at its maximum value (u_{max}). Thereafter the specimens began to dilate, new cracks were developed and pore pressure decreased. When the specimens reached their initial void ratio (ε_{v0}) the pore pressure became equivalent to their initial value (u_0). Similar results were first observed by Lane (1969) (Fig. 2) in relation with Bieniawski fracture concept (Bieniawski 1967). It was observed that in both series of tests, the maximum pore pressure (u_{max}) was reached before failure. This is easily explained by the fact that ε_v min occurs always before failure. Similar observations can be found in the literature (Aldrich 1969, Lane 1969).

The deformability behavior of the analog material is affected by pore pressure (Fig. 1). The whole volumetric contraction is caused by the axial strain ε_a, showing differences in the non linear pre-elastic domain but a similar behavior in the elastic domain. Increasing the initial pore water pressure (u_0) leads to a decrease in the pre-elastic axial compressibility. This may be caused by differences in pore water pressure inside the cracks compared with the hydrostatic pore water pressure present in the samples. This reduces the effective stress tending to close the cracks. This occur if the normal axis to the crack's plane is nearly parallel to the axial stress (σ_1) direction.

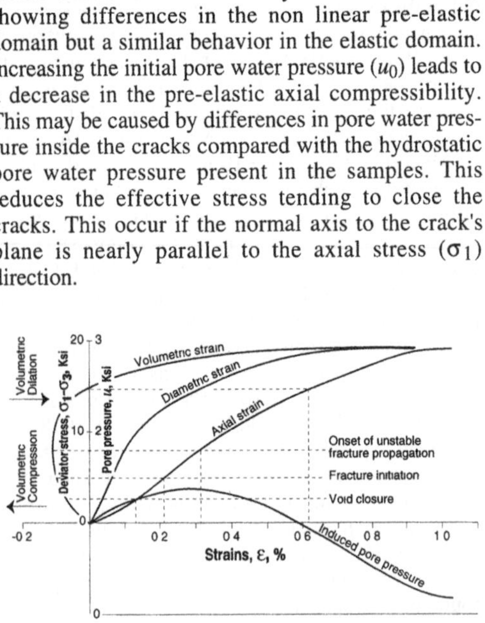

Figure 2. Undrained triaxial compression test of Berea sandstone, σ_3 = 13.8 MPa, u_0= 6.9 MPa. (after Lane, 1969).

Moreover, the influence of the initial pore pressure on the axial deformability seems to decrease with increasing initial effective confining pressure (σ'_3). The first series were influence more by u_0 than the second. This was with approximately the same three ratios u_0/σ_3. All these observations had direct applications on the effective stress law for discontinuities.

2.2 *Induced pore pressure variation*

The first series of undrained tests (σ_3 = 2.76 MPa) results demonstrates (Fig. 1A) that the higher the initial pore pressure (u_0), the higher the maximum induced pore pressure (u_{max} - u_0) will be. Figure 1B shows that this is also valid for the second series of tests (σ_3 = 8.28 MPa) except for E103U. Aldrich (1969) postulated that the maximum induced pore pressure is only function of the initial effective confining pressure $(\sigma'_3)_0$. This study shows that the maximum induced pore pressure is function of the confining pressure (σ_3), the initial pore pressure (u_0) and the volumetric strain (ε_v).

Aldrich's hypothesis is restricted to rock materials characterized by a short initial inelastic crack closure domain. The assumption that changes of pore water volume are essentially equal to variation in bulk volume, must be taken with caution, especially in the pre-elastic non linear domain. This explain the observation that for greater maximum contractancy, ε_{vmin}, (E104U compared to E99U and E99U compared to E100U), the observed maximum induced pore pressure (u_{max}-u_0) is lower (Fig. 1). It is shown in fig. 1 that most of the volumetric contractancy is in fact created in the pre-elastic non linear domain. This limits the previous hypothesis on the proportionality between volume variation and pore pressure behavior.

2.3 *Induced pore pressure and $\bar{A}$ coefficient in relation with the stress deviator*

The induced pore pressure (u_i) response to an increase of the stress deviator (σ_1-σ_3) follows the evolution of the voids volume: (u_{max}-u_0) (where u is the instant pore pressure and u_0 is the initial pore pressure) increases non-linearly until unstable fracture propagates then u decreases to reach zero (Fig. 3). But for high σ_3 (13.8 MPa), the decrease of u stops before zero because of the material plasticity condition reached at this state of stress. Higher values of u_{imax} are attained for higher values of initial effective confining pressure $(\sigma'_3)_0$ caused by the increasing of fracturing strength and the decreasing dilatancy. But, the decreasing rate of pore pressure, from the unstable fracturing point, seems independent from the confining pressure (Fig. 3). These observations agreed with those in the works done by Aldrich (1969), Bruhn (1972) and Ohnishi (1973) and like them, a linear relationship between u_{imax} and $(\sigma'_3)_0$ was observed.

Figure 3 shows that within the same series of tests (same σ_3) the linear portion of the curves to u_{imax} are characterized by the same slopes, $\bar{A}$ = $u_i/\Delta(\sigma_1$-$\sigma_3)$, and all the series of tests show similar slopes between u_{imax} and the failure point. It can also be observed that the higher the confining pressure (σ_3), the higher the pore pressure increasing rate will be, caused by the reduction of dilatancy with increasing σ_3. Induced pore pressure at failure is either negative or positive, depending on the initial effective confining pressure $(\sigma'_3)_0$ and it increases with increasing $(\sigma'_3)_0$ so that a pressure threshold can be defined around $(\sigma'_3)_0 \cong 9$ MPa. Similar observations were done by Bruhn (1972) and Ohnishi (1973) for various rocks giving different threshold values.

Figure 4 shows the evolution of the Skempton (1954) pore pressure coefficient $\bar{A}$ ($\Delta u_i/\Delta\sigma_1$ because $\Delta\sigma_3$=0, the confining pressure, was maintained constant during testing) with the variation of the stress deviator (σ_1-σ_3) and various initial effective confining pressure $(\sigma'_3)_0$ which influences the rate of variation of u for an increase of (σ_1-σ_3). During the contraction phase, in the specimen under testing, u increases and the value of $\bar{A}$ varies from 0.015 to 0.04 for stress deviators between 4 to 28 MPa in the first series of tests; while in the second series of tests $\bar{A}$ show a greater variability and is varying from values of 0.01 to 0.30 for stress deviators between 5 and 35 MPa; and finally the third series show a

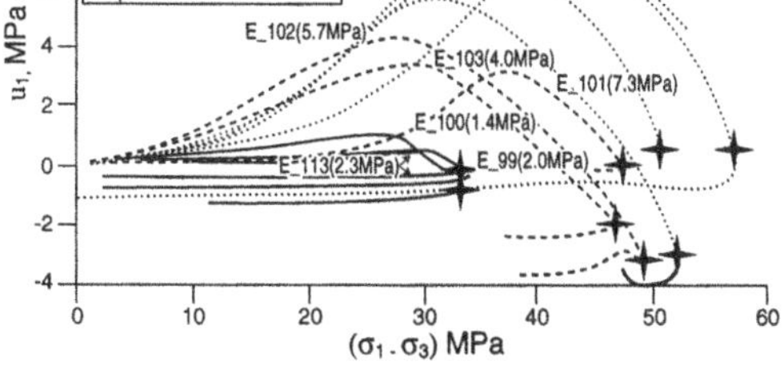

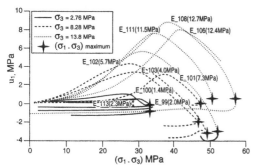

Figure 3. Induced pore pressure evolution with applied stress deviator on intact samples.

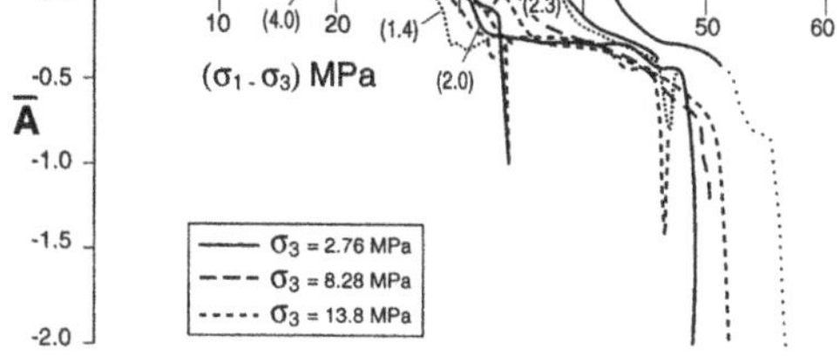

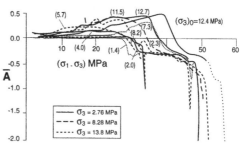

Figure 4. Evolution of the coefficient $\bar{\mathbf{A}}$ (Δu_i / $\Delta(\sigma_1$ - $\sigma_3)$ with the stress deviator for the series of undrained triaxial tests on intact samples.

progressive increasing of $\bar{A}$ from 0.03 to 0.42 for stress deviators varying from 6 to 40 MPa. From these test results, it may be observed that for an initial effective confining pressure $(\sigma'_3)_0 < 7$MPa, the $\bar{A}$ coefficient is approximately constant from the beginning of loading to the initiation of dilatancy or u_{imax} while $\bar{A}$ increases gradually for $(\sigma'_3)_0 > 7$ MPa between the same limits. From loading starting point to the dilatancy initiation phase, the $\bar{A}$ coefficient behavior is directly related to $(\sigma'_3)_0$ and it may be stated that the higher the initial effective confining pressure $(\sigma'_3)_0$ the higher the stress deviator $(\sigma_1-\sigma_3)$ must be to initiate dilatancy (Figs. 3 et 4) or to reach u_{imax} (or $\bar{A} = 0$).

After u_{imax}, the test results show that the pore pressure coefficient $\bar{A}$ decreases rapidly, following an almost identical slope for all the tests (Fig. 4) to reach a plateau at an approximative negative value $\bar{A}$ = -0.25 to -0.35. It may be noticed that the decrease of $\bar{A}$ correspond to the beginning of stable fracturing or dilatancy phase. $\bar{A}$ becomes nul when unstable fracturing begins corresponding to u_{imax}. From this point, induced pore pressure decreases and the sign of $\bar{A}$ changes (positive to negative) and continue to decrease after this plateau (around $\bar{A} = -0.3$). During this dilation phase, it seems that $(\sigma'_3)_0$ does not influence the behavior of $\bar{A}$ coefficient. Figure 5 illustrates the schematic behavior and relationships between the various parameter under undrained triaxial tests on intact specimens : the parameters u_i, ε_v, $\bar{A}$ in relation with $(\sigma_1-\sigma_3)$. Regarding the progressive increment of the coefficient with the increment $\Delta(\sigma_1-\sigma_3)$ within the same test for $(\sigma'_3)_0 > 7$ MPa, the only plausible explanation for this behavior is that for this confining pressure the rock analog material passed from a brittle to a brittle-ductile behavior (near plasticity) in which ε_v, u_i and $\bar{A}$ show a non-linear mobilization of the three parameters with increasing $(\sigma_1-\sigma_3)$ and this is confirmed by the u behavior in the second and third series of tests.

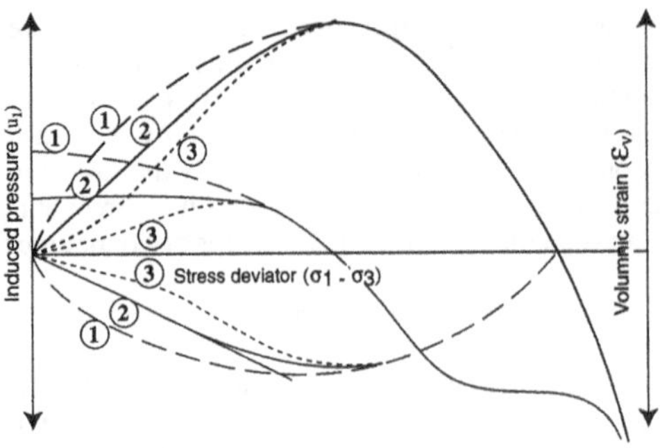

Figure 5. Schematic illustration of the relationships between the parameters u_i, $\bar{A}$ and ε_v with the stress deviator $(\sigma_1 - \sigma_3)$ evolution.

3 PORE PRESSURE BEHAVIOR IN UNDRAINED TESTS ON JOINTED SPECIMENS

The triaxial tests were performed on the same material jointed specimens with the fracture planes at around 30° with applied σ_1 under three confining pressures (σ_3 = 2.76, 5.53 and 8.28 MPa) and three initial pore pressure for each of them. Two series of tests are shown in Figure 6. The stress path (τ and σ_n) followed in the triaxial shear test on the fracture under shearing is different from the stress path in direct shear test under constant normal stress. In the triaxial shear test, each shear stress increment on the fracture shear plane is accompanied by an increment in normal stress. Pore pressure increases is controlled both by the variation in porosity of the material itself and of the fracture. But the reduction in pore pressure results mainly from dilatancy on the fracture because the effective confining pressure was maintained under the intact material dilatancy initiation threshold.

Profiles were taken on each fracture surface, around 10 for each surface tested, to characterize the fracture surfaces roughnesses. On the basis of the relationship between the z_2 statistical parameter of roughness and the JRC coefficient developed by Tse and Cruden (1979), JRC values were computed for the nine fractures. JRC values between 7.21 and 8.9 with a mean around 8 were computed for the fractures and this agreed with other methods of estimation.

3.1 *Pore pressure and stress deviator behavior with strain variation*

Like in the direct shear test on joints under constant or variable normal stiffness, the stress-strain (or displacement) relationships obtained from the triaxial shear test results (Fig. 6) show a peak with dilatancy mobilization and asperities failure followed by a stress drop and as the normal stress on the joint increases progressively so for the shear stress and there is a progressive increment of $(\sigma_1-\sigma_3)$. In the case of undrained shear tests, dilatancy mobilization caused a reduction in pore pressure and an increase in effective normal stress on the joint. In the triaxial shear test on fracture, if there is no rotation mechanism integrated in the specimen set-up within the triaxial cell, compatibility necessitates the development of a conjugated fracture and this happened in the present tests with the rising of $(\sigma_1-\sigma_3)$ after the stress drop when the stress level was sufficient to provoke a new failure (indicated by an arrow on the stress-strain curves in Fig. 6). The development of the new fracture correspond to a decreasing in the pore pressure after a stabilization plateau corresponding to friction and degradation on the fracture under shearing and the reduction of dilatancy by the progressive increments of the normal stress acting on it. Analysis of the pore pressure variation is restricted to the phase before the development of the new fracture.

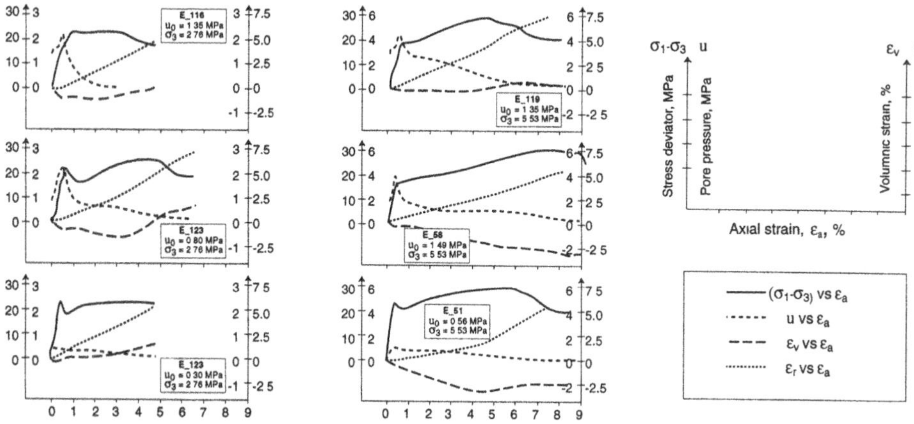

Figure 6. Joint deformability under undrained conditions

Similarly to the intact rock analog material specimens under pore pressure and undrained conditions, the fractured samples submitted to the same loading conditions showed that the pore pressure (u) behavior is closely related to the volume variation. For similar effective stress state, the joint specimens induced slightly higher pore pressure than the intact specimens because of the greater deformability of joints caused by the closure between the surfaces inducing the increase in pore pressure. So, as observed in figure 6, the application of a deviatoric stress cause initially an increasing pore pressure, during the friction mobilization phase, resulting from an added closure between the joint surfaces as noticed previously (Archambault et al., 1996). The increase of the pores pressure to u_{imax} shows fluctuations and local variations caused by the irregular morphology of the fracture surfaces on which small slips and dilatancy occurred but of very low intensity. Pore pressure decreases thereafter with the phase of roughness and dilatancy mobilization before peak shear-strength and the effective normal stress increased with the reduction of pore pressure on the fracture under shearing (dilatancy hardening).

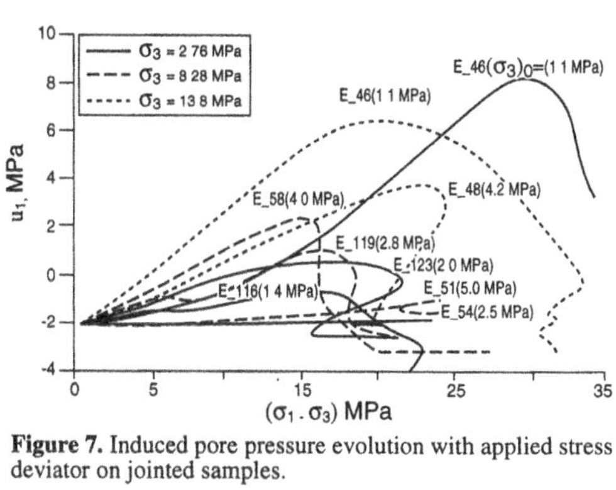

Figure 7. Induced pore pressure evolution with applied stress deviator on jointed samples.

After peak shear strength, asperities failed on the fracture surface with shear displacement and the dilatancy rate decline progressively as well as the stress deviator (σ_1-σ_3); while the dilatancy rate tends towards stabilization, the stress deviator re-increased with the increasing normal stress on the fracture. Pore pressure decreases also and its decreasing rate follows the declining rate of dilatancy. Also, the pore pressure maximum decreasing rate correspond to the maximum rate of dilatancy. Pore pressure stabilize itself finally at the end of asperities degradation phase indicating that contractancy begins at this point of shear displacement (Fig. 6).

3.2 *Induced pore pressure and $\bar{A}$ coefficient variation with (σ_1-σ_3)*

Induced pore pressure (u_i) variation differs from one test to the other (Fig. 7). While certain tests (E.48 and E.58) show a constant increasing of induced pore pressure to u_{imax}, others demonstrate irregular increment (E-46, E-116 and E.119) or parabolic increasing (E.49, E.123) or almost no variation (E.51, E.54) in induced pore pressure to u_{imax} with the increasing stress deviator (σ_1-σ_3). After u_{imax} value and the increasing rate of induced pore pressure ($\Delta u_i/(\Delta\sigma_1$-$\sigma_3) = \bar{A}$) are proportional to $(\sigma_3)_0$ like in the intact samples. This can be explained by the fact that the higher the confining pressure, the higher the normal stress under shearing and the lower the dilatancy will be on the fracture, inducing higher pore pressure. Beyond u_{imax}, the decrease in induced pore pressure is so irregular in each test and so different from one test to the other, that it is difficult to evaluate the influence exerted by $(\sigma'_3)_0$. The increase and decrease in pore pressure (u) differs for each sample mainly due to the influence exerted by the roughness morphology different in each fracture.

$\bar{A}$ coefficient behaves quite irregularly in each test and from test to test (Fig. 8) with an almost

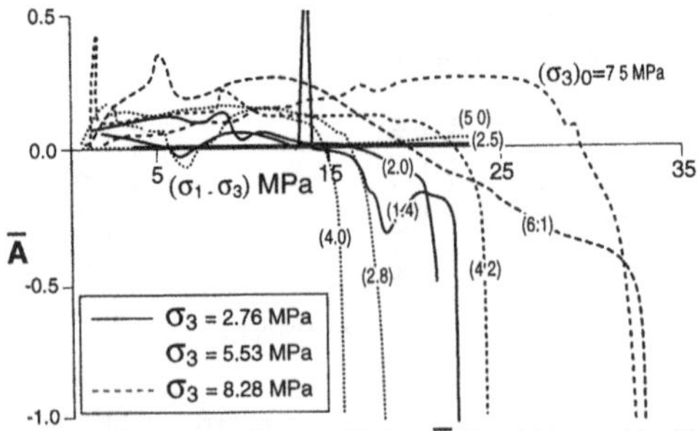

Figure 8. Evolution of the coefficient $\bar{A}$ ($\Delta u_i / \Delta(\sigma_1 - \sigma_3)$) with the stress deviator for the series of undrained triaxial tests on joint samples.

unpredictable evolution of pore pressure changing continually with the irregular dilatancy-contractancy behavior on the fractures submitted to shear used by the variable roughness morphology characterizing each fracture. Globally, the values of $\bar{A}$ varies from 0.01 to 0.3 and are proportional to the values of $(\sigma'_3)_0$ varying from 0.3 MPa or 4.1 MPa and correspond to the increasing pore pressure during the friction mobilization phase with local $\bar{A}$ variations caused by the roughness morphology of the fractures. Mobilization of roughness and dilatancy with shear displacement being dependent on roughness morphology on each fracture show specific decreasing rate of pore pressure of the $\bar{A}$ coefficient in relation with the dilatancy rate depending on the slip-shear processus of asperities on the fracture surfaces. Larger shear displacement after peak shear strength show larger variations in the $\bar{A}$ coefficient passing from positive to negative values with varying $(\sigma_1\text{-}\sigma_3)$. But mainly $\bar{A}$ trends towards 0 with shear displacement and the progressive degradation of asperities on the fracture. It seems that there is no relation, in this phase, between the $\bar{A}$ coefficient and the confining pressure (σ_3) beyond peak shear strength.

REFERENCES

Aldrich, M.J. 1969. Pore pressure effects on Berea sanstone subjected in experimental deformation. Geo. Soc. Am. Bull., 80: 1577-1586.

Archambault, G., Gentier, S., Riss, J., Flamand, R. & Sirieix, C. 1996. Rock joint shear mechanical behavior with 3D surfaces morphology and degradation during shear displacement. In: prediction and performance in rock mechanics and Rock engineering, EUROCK'96, Barla (ed.), Balkema, 247-254.

Bieniawski, Z.T. 1967. Mechanism of brittle fracture of rock. Int. J. Rock Mech. and Min. Sci., 4(4): 395 parts 1-2.

Boozer, G.D., Hiller, K.H. & Serdengecti, S. 1963. Effects of pore fluids on the deformation behavior of rocks subjeted to triaxial compression. 5th Symp. Rock Mech., 579-624.

Bruhn, R.W. 1972. A study on the effects of pore pressure on the strength and deformability of Berea Sanstone in triaxial compression. US Dept. of the Army technical report, Eng. Study No 552.

Colback, P.S.B. & Wiid, B.L. 1965. The influence of moisture content on the compressive strength of rocks. Proc. Rock Mech. Symp., Toronto, 65-83.

Dropek, R.K. & Johnson, J.N. 1978. The influence of pore pressure on the mechanical properties of Kayanta sandstone. J.G.R. 83(B6): 2817-2824.

Goodman, R.E. & Ohnishi, Y. 1973. Undrained shear testing of jointed rock. Rock Mechanics, 5: 129-149.

Handin, J., Hager, R.V., Friedman, M. & Feather, J.N. 1963. Experimental deformation of sedimentary rock under confining pressure: pore pressure tests. Am. Assoc. Petrol. Geol. Bull., 47: 717-755.

Ismail, I.A.H. & S.A.F. Murrell. 1976. Dilatancy and the strength of rocks containing pore water under undrained conditions. Geophys. J.R. asr. Soc., 44: 107-134.

Lane, K.S. 1969. Pore pressure on Berea sandstone subjected to experimental deformation: Discussion. Geo. Soc. Am. Bull., 80: 1587-1590.

Lane, K.S. 1970. Engineering Problems Due to Fluid Pressure in Rock. Rock Mechanics - Theory and Practice. W.H. Somerton (Ed.). American Insitute of Mining, Metallurgical and Petroleum Engineers, New-York, pp. 501-540.

Mesri, G., Adachi, K. & Ullrich, C.R. 1976. Pore pressure response in rock to undrained change in all-round stress. Geotechnique 26(2): 317-330.

Murrell, S.A.F. 1965. The effect of triaxial stress systems on the strength of rocks at atmospheeric temperature. JGR Astron., 10: 231-281.

Ohnishi, Y. 1973. Laboratory measurement of induced water pressures in joined rocks. Doct. thesis, Univ. of California, Berkeley, CA.

Parate, N.S. 1973. Influence of water on the strength of limestone. Trans. A.I.M.E., Soc. Min. Eng., 254: 127-131.

Poirier, S. 1996. Étude expérimentale du comportement de la pression interstitielle et de son influence sur le comportement physico-mécanique d'un matériau poreux intact ou fracturé par essais triaxiaux non drainés. Unpubl. M.Sc. Thesis, Université du Québec à Chicoutimi, 150 p.

Poirier, S., Archambault, G. & Rouleau, A. 1994. Experimental testing of pore water influences on physico-mechanical properties of a porous rock analog material. In Proc. of 47th Canadian Geotechnic Conf., Halifax, NS, pp. 418-427.

Robinson, L.H. 1959. The effect of pore and confining pressure on the failure process in sedimentary rock. 3rd Symp. Rock Mech. Quart. Colo. School Min., 54: 177-199.

Serdengecti, S. & Boozer, G.D. 1961. The effects of strain rate and temperature on the behavior or rocks subjected to triaxial compression. Proc. 4th Symp. Rock Mech., 83-97.

Skempton, A.W. 1954. The pore pressure coefficient A and B. Géotechnique, 4: 143-147.

Terzaghi, K. 1945. Stress conditions for the failure of saturated concrete and rock. Proc. of ASTM, 45: 777-792.

Tse, R. & Cruden, D.M. 1979. Estimating joint roughness coefficients. Int. J. Rock mechanics and mining sciences and geomechanical, Abstracts, 16: 303-307.

Mechanics of Jointed and Faulted Rock, Rossmanith (ed.) © 1998 Taylor & Francis, ISBN 90 5410 955 6

Evaluation of jointed rock permeability using a high pressure triaxial apparatus

B. Indraratna & A. Haque
Department of Civil and Mining Engineering, University of Wollongong, N.S.W., Australia

W. Gale
Strata Control Technology (SCT), Wollongong, N.S.W., Australia

ABSTRACT: The permeability characteristics of jointed coal and sandstone are investigated in the laboratory under triaxial conditions. A high pressure triaxial apparatus was designed for measuring the triaxial strength and permeability of both intact and fractured rocks. In this equipment, air or water can be made to flow through the rock specimens depending upon the porosity of the rock in order to estimate the air or water permeability. Test results verify that the permeability of the fractured rock decreases with the increasing confining pressure. It is also observed that the air permeability is greater than the water permeability for a fractured coal specimen. The laboratory measurements agree with the field permeability data obtained from hydro-fracturing test conducted on coal.

1 INTRODUCTION

The rock mass in nature remains in a triaxial stressed state, and the strength or flow characteristics under triaxial conditions represent accurately the field situation. There are a few types of triaxial cells for rocks which have been commonly used for strength testing of soft and hard rocks in the laboratory, such as the one designed by Mesri et al., 1976. In order to test weak rocks, Bro (1996) introduced a new type of triaxial cell, which had no facility of measuring the permeability. Recently, Singh (1997) has conducted drained triaxial tests on sandstone specimens to measure the permeability and volumetric strains as a function of differential stresses. This equipment, however, cannot be used for undrained testing and effective stress measurement, because, no facilities are provided for pore pressure monitoring. In conventional triaxial cells, essential features such as the use of flexible sample sizes, measurement of pore pressure and volume change under low to high confining pressure are not uncommon. Nevertheless, the provisions for measuring the permeability of a fractured rock under triaxial conditions are rarely included. Therefore, the requirement of triaxial testing apparatus having all the essential features for determining both the triaxial strength and permeability of intact and fractured rocks has provoked the authors to design a unique high pressure triaxial apparatus. In addition to determining the strength and permeability of intact and fractured rocks, this apparatus can also measure the volume changes and pore pressures under triaxial conditions. The wide range of possible applications include drained and undrained tests of intact and fractured rocks with volume changes and pore pressure measurements, permeability measurements etc. Only the results in relation to the permeability of jointed rocks are presented within the scope of this paper.

2. HIGH PRESSURE TRIAXIAL APPARATUS

The high pressure triaxial apparatus comprises of five different parts namely, (i) high pressure cell assembly, (ii) volume change device, (iii) entrance and exit pressure measurement system, (iv) axial loading device and (v) the digital display unit, details of which are shown in a schematic diagram in Figure 1. Further descriptions of these items are given in the following sections:

2.1 *High pressure cell assembly*

The cell is made from high strength steel having a 100 mm internal diameter and a 120 mm height. The cell can withstand a pressure of up to 150 MPa. Specimen sizes up to a maximum of 54 mm in diameter and 110 mm in height can be tested. The cell is confined at the top and bottom by a thick stepped steel plate which is firmly held in place by six high tensile strength bolts. The air or water inlet and outlets as well as the strain meter connections are made through the bottom plate. The outlet at the

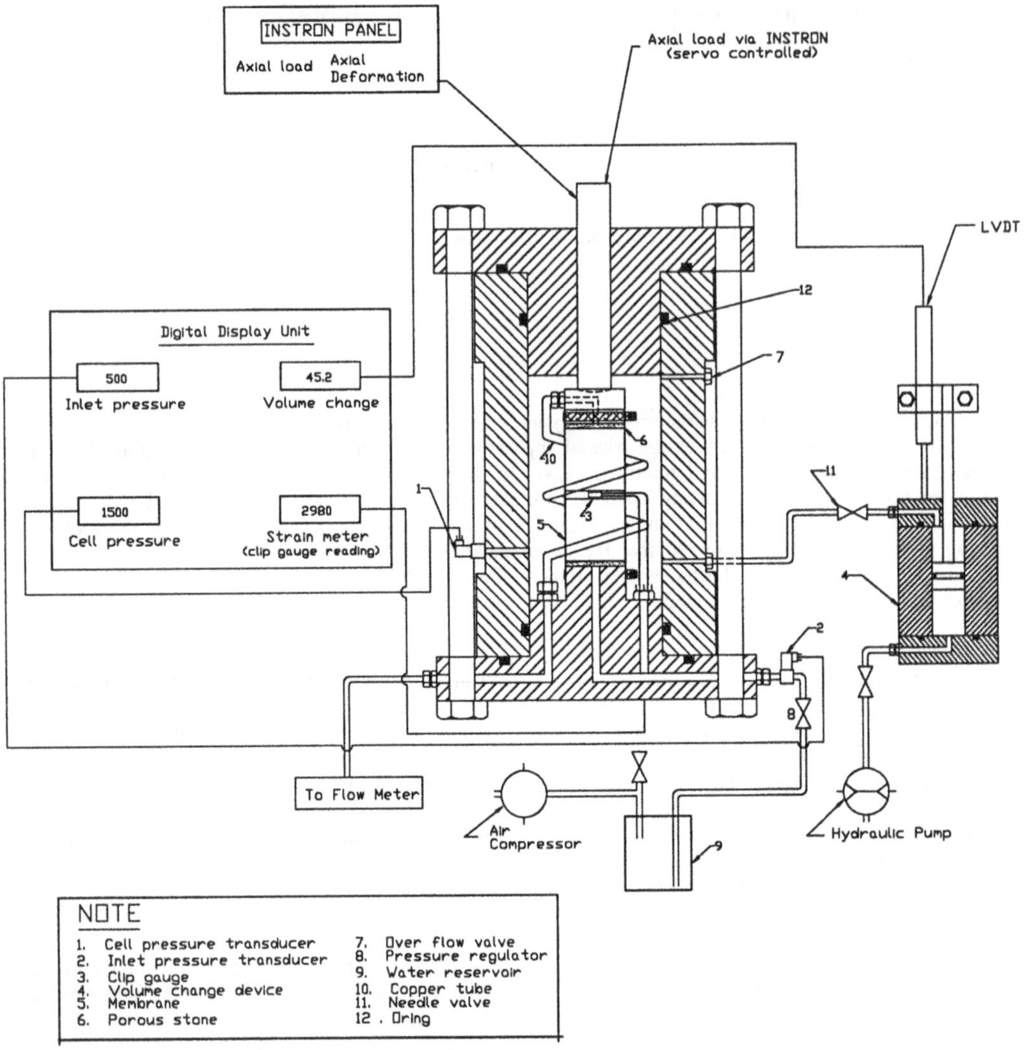

Figure 1. Schematic diagram of the high pressure triaxial apparatus.

bottom of the cell wall is used to apply the cell pressure, and the top outlet carries the overflow. Two transducers, one at the inlet end and the other at the outlet end are provided to measure the pressures at a given confining pressure.

2.2 *Volume change device*

The volume change device consists of a cylindrical chamber having an internal diameter of 25 mm and a height 90 mm. A piston is attached co-axially to the cylindrical chamber, in which the piston moves up or down depending on the volume increase or decrease of the specimen, and a LVDT is used to monitor the piston movement. The top chamber of the volume change device is connected to the cell and the bottom is connected to a hydraulic jack. Once the cell is filled with oil, the volume change device is connected to the cell, and subsequently, the required cell pressure is applied by a hydraulic jack (Fig. 1). The opening or closure of a regular rock fracture can be measured by a specially designed and sensitive clip gauge, which is fitted circumferentially to the mid-height of specimen.

2.3 *Air or water flow device*

Compressed air is used to force water or air to flow through the fractured rock specimens for the determination of permeability. The flow of air or water is recorded via a flow meter at the outlet.

2.4 *Axial loading device*

Once the triaxial cell is placed under the servo-controlled compression test machine (INSTRON), an axial load equivalent to the confining pressure is applied to the top of the specimen.

3 EXPERIMENTAL STUDY

3.1 *Preparation of specimen*

Coal and sandstone specimens were prepared by coring a block of rock to obtain a height to diameter ratio of about two. Subsequently, the specimen was wrapped in a specially designed membrane (2 mm thick, Polyurethane TU800) and placed inside the cell. Two hose clamps were used to tighten the top and bottom end caps. After setting up the cell within the loading frame, the predetermined confining pressure was applied using a hydraulic jack. Depending upon the porosity of the specimen, either compressed air or water was circulated to obtain a steady flow. In particular, water was forced fed through saturated specimens while air was circulated through dry specimens.

3.2 *Testing procedure*

After setting up the specimen inside the cell, the required cell pressure can be applied in several stages. Once the cell pressure is set to the predetermined value, air or water pressure is applied to the inlet end. It is of important to note that the inlet pressure must always be maintained by at least 300 kPa below the level of confining pressure, in order to prevent the flow of air or water through the membrane and specimen. The flow of air or water should be measured at the outlet end at constant interval of time until a steady state condition is achieved.

4 PERMEABILITY MEASUREMENT

Once the flow rate reaches a steady state condition, the permeability of the rock for a given confining pressure is determined using the following methods.

4.1 *Water flow permeability, k_w*

The permeability (k_w) due to the flow of water through the fractured rock can be determined using the Darcy's law as given below:

$$k_w = \frac{q\mu}{A}\frac{\Delta L}{\Delta P} \qquad (1)$$

where, q = flow rate at steady state condition, A = specimen area, μ = viscosity of water at test temperature, ΔP = inlet and exit pressure difference and ΔL = specimen height.

4.2 *Air flow permeability, k_a*

The permeability (k_a) due to the flow of air through the fractured rock can be calculated on the basis of the measured air flow rate and inlet driving pressure, as represented by the following equation (ASTM, D4525-90):

$$k_a = \frac{2Q\mu LP_{out}}{A(P_{in}^2 - P_{out}^2)} \qquad (2)$$

where, Q = flow rate, μ = viscosity of air at test temperature, A = cross-section area of specimen, L = height of the specimen, P_{in} = inlet pressure and P_{out} = outlet pressure (usually atmospheric).

5 RESULTS AND DISCUSSIONS

5.1 *Critical pressure difference (σ_3 - P_{in})*

Several tests were conducted on a coal specimen having micro-fissures (CS1). For a given confining pressure (500 kPa), the inlet air pressure was varied. The air flow rate was measured at constant time intervals until a steady state condition was reached. Subsequently, the inlet pressure was increased, and the corresponding flow rate was measured. The measured air flow rate (Q) is divided by the mean air pressure [(P_{in} + P_{out})/2] to obtain the normalised flow rate (Q') which is plotted against the air pressure difference (P_{in} - P_{out}) in Figure 2. It is observed that the normalised flow rate varies linearly up to a pressure difference (P_{in}-P_{out}) of 200 kPa for a given confining pressure of 500 kPa. A bi-linear variation in Q' is observed for a pressure difference beyond 200 kPa, which is attributed to the flow taking place between the membrane and the specimen. Therefore, it is essential to keep the difference in confining pressure and the inlet air or water pressure (σ_3 - P_{in}) by at least 300 kPa to avoid flow through the interface of the membrane and specimen. The corresponding permeability of the specimen for various driving pressure is plotted in Figure 3. It is evident that the permeability of the specimen for a given confining pressure is almost constant within the critical pressure difference zone. However, it varies significantly, as the driving pressure is increased further beyond the critical point.

For the same specimen, tests were conducted to determine the change in permeability for various confining pressures ranging from 0.5 MPa to 2 MPa. The observed values of the air permeability are given in Table 1. This shows that the permeability significantly decreases with the increase in confining pressure (P_c).

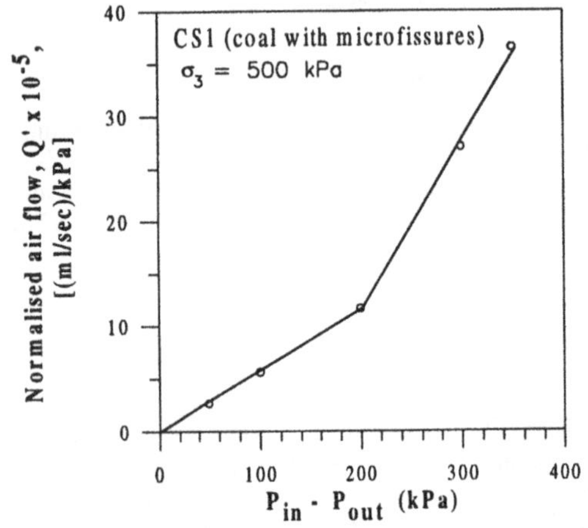

Figure 2. Variation of normalised flow (Q') with pressure difference for coal specimen (CS1).

Table 1 Air permeability, k_a for CS1 specimen.

Confining pressure, σ_3 (kPa)	Air permeability, k_a x 10^{-2} (milli-Darcy)
500	6
1000	0.25
2000	0.035

5.2 *Permeability of fractured specimen*

Several tests were conducted on a fractured coal specimen (CS2) for a range of confining pressure under increasing driving pressure. For a given confining pressure, the steady state flow rates were recorded for various driving pressures and the air permeability was calculated using Eqn. 1. The measured normalised air flow rate (Q') is plotted against the air pressure difference (P_{in} - P_{out}) in Figure 4. As expected, the variation in the normalised flow rate (Q') against the pressure difference (P_{in} - P_{out}) is approximately linear, as the pressure difference is maintained below the critical point for all the tests.

In order to determine the water flow permeability (k_w), water is forced to flow through the same fractured specimen under various confining pressures until a steady state flow rate was recorded. The

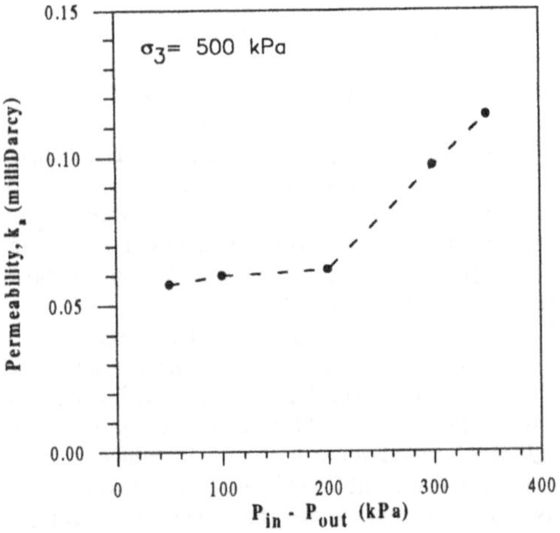

Figure 3. Variation of permeability with pressure difference for the coal specimen (CS1).

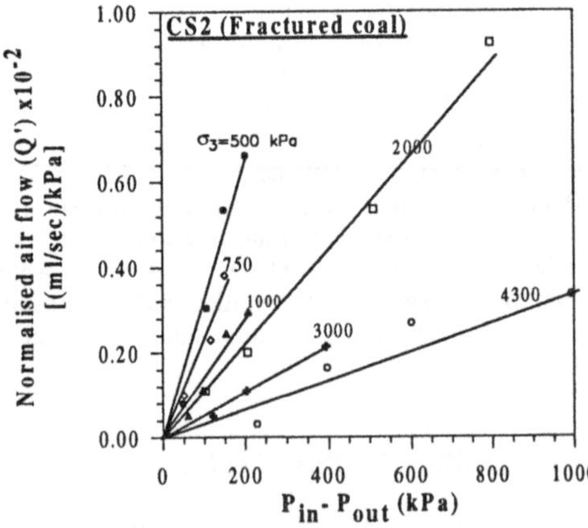

Figure 4. Variation in normalised flow rate (Q') against pressure difference for various confining pressures.

water permeability (k_w) was then calculated using Eqn. 2. The changes in both air and water permeability of the fractured specimens are plotted in Figure 5. As expected, the results verify that the permeability of fractured specimens is significantly higher than that of relatively intact specimens (Table 1) tested under the same confining pressures. It is of relevance to note that the permeability of relatively intact and fractured specimens decreases considerably as the cell pressure is increased. However, as indicated in Figure 5, the reduction in permeability seems to be marginal when the cell pressure is increased beyond say, 3.5 MPa. Figure 5 also shows that the water permeability (k_w) is considerably smaller than the air permeability (k_a), except at high levels of confining pressure (>3.5 MPa), at which the maximum closure of joints makes the difference between k_a and k_w values very small or negligible. Available field permeability results for coal based on hydro-fracturing (Edgoose et al., 1996) are also plotted for comparison. This data indicate a slightly higher k_w than the laboratory values especially at small confining pressures. This is not surprising, because in the field, the larger 'sample size' is expected to intersect a greater number of fissures, thereby making the apparent permeability greater (Hoek & Brown, 1980).

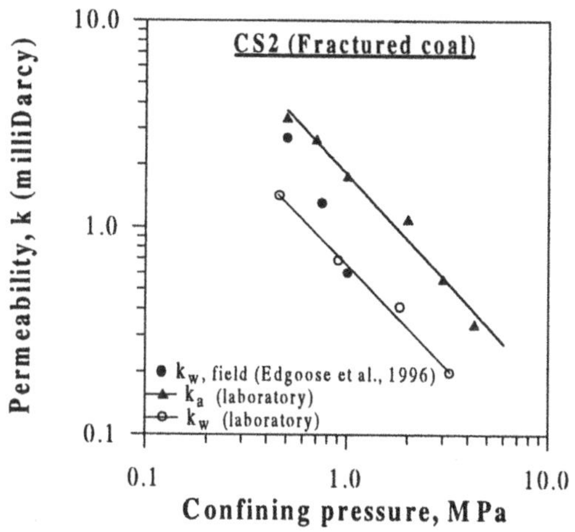

Figure 5. Permeability of fractured coal specimen (CS2) under various confining pressures.

5.3 *Permeability of split sandstone specimen*

Under Brazilian test condition, an intact sandstone specimen was split into almost two equal halves. Using a LVDT, the surface profile of the split specimen was constructed and plotted in Figure 6. The flow characteristics were determined by circulating water through the fracture under a constant driving pressure of 300 kPa. The variation of flow rate per unit hydraulic head (Q/ΔH) against the increased confining pressure of 0.5 MPa to 1.0 MPa is plotted in Figure 7. It is verified that the changes in flow rate becomes negligible as the confining pressure is increased beyond 2 MPa, indicating the closing of joint aperture at higher stress levels.

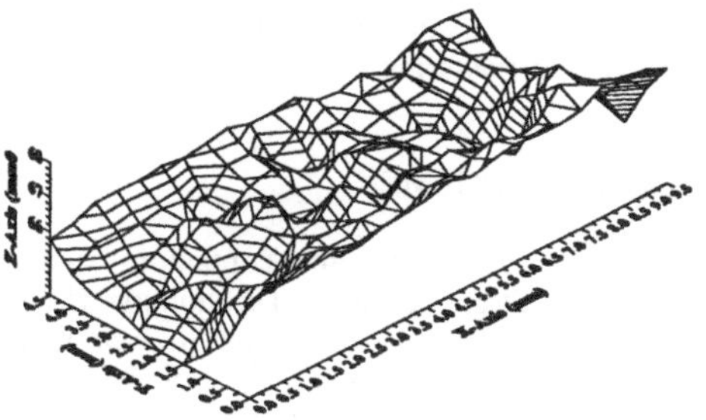

Figure 6. Surface profile of the split specimen under Brazilian test condition.

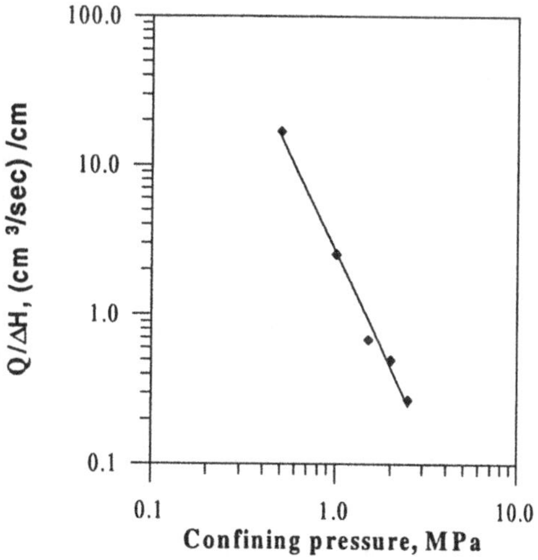

Figure 7. Variation of flow rate per unit gradient (Q/ΔH) against confining pressure.

6 CONCLUSIONS

The newly developed triaxial apparatus can determine the permeability of the intact and fractured rocks under triaxial stress states, which is useful in the design of underground mines through jointed rocks, in particular, where there is a potential risk of mine inundation. This apparatus can also be used successfully to conduct drained and undrained strength tests of soft to medium rocks with the facility of measuring pore pressure and volume changes. The laboratory measured values of permeability for jointed coal and sandstone specimen is observed to drop quickly with the increase in confining pressure. Furthermore, this drop in permeability becomes marginal at considerably high confining pressure as the joint openings start to close at relatively high pressures. The laboratory measured permeability agrees well with the field measurements based on hydro-fracturing.

7 ACKNOWLEDGEMENT

The authors acknowledge the continuous support provided by Strata Control Technology (SCT), Wollongong. Sincere thanks are due to Alan Grant (Senior technical staff, Dept. of Civil & Mining Engineering) for his assistance during the construction of the equipment.

8 REFERENCES

ASTM Standards. (1990). D 4525: Standard test method for permeability of rocks by flowing air. *Annual Book of ASTM Standards*, Vol. 04.08, 825-828.

Bro, A. (1996). A Weak Rock Triaxial Cell. Technical Note. *Int. J. Rock Mech. Min. Sci. & Geomech. Abstr.*, Vol. 33, No.1, 71-74.

Edgoose, J., Casey, D.A. & Enever, J.R. (1996). An integrated testing capability for in-situ stress measurement and determination of reservoir parameters. *Proc. 7th ANZ Conference on Geomechanics*, Adelaide, South Australia, (Jaska, M.B., Kaggwa, W.S. & Cameron, D.A. eds.), 879-884.

Hoek, E. & Brown, E. T. (1980). *Underground excavation in rock.* IMM, London, 527p.

Mesri, G., Adachi, K. & Ullrich, C.R. (1976), Pore pressure response in rock to undrained change in all round stress. *Geotechnique*, 26, No.2, 317-330.

Singh, A.B. (1997). Study of rock fracture permeability method. *J. Geotech & Geoenvironmental Engineering*, ASCE, Vol. 123, No.7, 601-608.

Mechanics of Jointed and Faulted Rock, Rossmanith (ed.) © *1998 Taylor & Francis, ISBN 90 5410 955 6*

Channelling effects in the hydraulic behaviour of rock-joints

M. Kharchafi, T. Jacob, P. Egger & F. Descoeudres
Laboratoire de Mécanique des Roches, Ecole Polytechnique Fédérale de Lausanne, Switzerland

ABSTRACT: The cubic law is commonly used to predict the hydraulic behaviour of rock joints. Yet, test results show that this model can constitute a valuable approximation only in specific conditions of geometry and aperture of the joints. This paper outlines the importance of channelling effects with respect to the relative aperture/roughness size of the joints. Thanks to a set of hydraulic tests carried out on artificial joints with regular and well defined geometry, the validity of the cubic law is discussed. An evaluation of the channel network size and orientation that is established in a real rock joint (granite) after a slight shearing is presented, for different degrees of closure.

1 INTRODUCTION

Depending on the size, the density and the persistence of the fractures, the rock mass can be considered as a continuous medium (homogenisation) for the fissured rock or as a discontinuous medium (isolated behaviour of the joint) for the rock mass containing large fractures.
It is well known that the three aspects: morphology of joints, mechanical behaviour and hydraulic behaviour are interconnected. At first, our approach was to separate these aspects for understanding their specific contributions. The investigations presented in this article concerne the behaviour of artificial or real joints whose shape has been well determined. The morphological reconnaissance and digitalisation of the real rock joints have been carried out using an automatic profilometer developed at our laboratory.

2 STATE OF THE ART.

The majority of rock joint hydraulic behaviour models that are referred to in literature and used in engineering are based on the hypothesis of laminar flow between two smooth planes. However the real joints are neither smooth nor parallel and they have number of contact points that implicate a tortuosity flow that passes through a channel network which depend on the geometry of the joint.

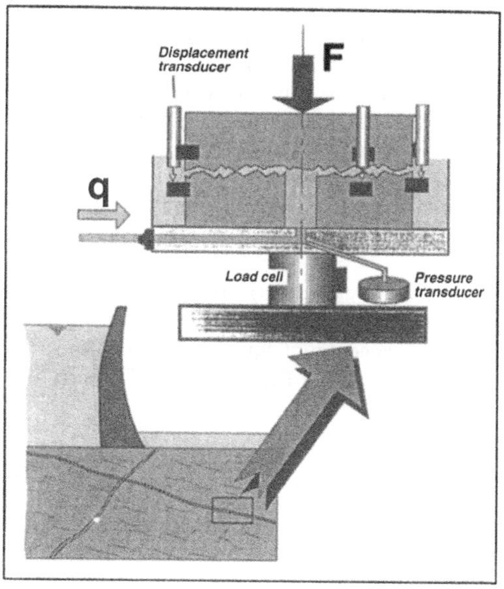

Fig.1: Principle of the hydromechanical tests

A number of formulations have been proposed for adapting the cubic law to the case of flow between two rock surfaces (Louis 1974, Barton et al 1990). Most of them propose a correction of the aperture e by an equivalent hydraulic aperture e_h. Thus the cubic equation becomes

$$q = -\frac{g}{12 \cdot v} \cdot J \cdot (e_h)^3 \qquad \text{Eq. 1}$$

where e_h is a function of the surface roughness, ν is the kinematic viscosity, J is the pressure gradient, q is the flow rate. That means, this equation always assumes a flow between two smooth parallel planes.

During the last decade a number of experimental studies have been made contradicting the approach based on the flow between two smooth planes or at least limiting its application.

Gentier (1987), by injecting coloured water during a permeability test on rock joints, has noted that the fluid comes out only through a limited number of points. This shows that the flow does not concern all the joint surface, but only a limited number of channels. This number rapidly decreases when the normal stress increases on the joint (closure of the joint) and it eventually stabilises.

Raven and Gale (1985), by carrying out tests on 5 granite samples containing natural fractures, showed that the results became far from the cubic law when the normal stress increased. They explained this by the fact of an increasing contact surface area, which increases the tortuosity of the flow. The cubic law in this case underestimates the conductivity of the joint.

Iwano and Einstein (1995) have carried out permeability tests on joints generated by three different ways: natural joints, joints generated by Brazilian tests or by direct traction, in the same natural rock (granodiorite). According to the type of the joint (i.e. according to its initial geometry), the results obtained compared to the cubic law were different.

Durham et al (1995) have published some experimental results on granite joints which showed that the exponent n of equation 2 varies from 1 to 8 with the aperture e. Pyrak-Nolte (1987) also found that the exponent can go up to 9.8.

$$q = -\frac{g}{12 \cdot v} \cdot J \cdot e^n \qquad \text{Eq. 2}$$

Boulon et al (1993) have tried to take into account the tortuosity of the flow by decomposing the joint into an assembly of independent parallelepipedal pipes with different heights. The flow rate in the joint was estimated by the application of the cubic law to each channel.

Brown (1987) published a study in which he calculated the flow between two numerically generated surfaces by the integration of Reynold's equations using the finite difference method. He compared the results thus obtained with those calculated by different approaches based on the cubic law. He concluded that the cubic law can constitute an acceptable approximation only when the joints are relatively open compared to their roughness.

3 SCOPE OF THE STUDY

As described above, it seems important to consider the tortuosity flow for modelling the hydraulic behaviour of rock joints. Especially in the case of joints under high normal stress (small aperture of the joint), the tortuosity (flow through a finite number of channels) becomes a major factor. With the help of some tests on well determined joint shape, we studied the effects of relative aperture/asperities size on the hydraulic permeability of rock joints on the one hand, of the shape and size of channel networks on the other hand.

4 EXPERIMENTAL SET-UP

The experimental set up used is schematised in figure 1. This computer controlled equipment allows the application of normal stress on a cylindrical sample (Φ15 cm) containing a horizontal joint and allows radial flow through the joint under controlled pressure or controlled flow rate. During the test the following parameters are measured: the closure of the joint at 3 points by inductive transducers, the axial force on the joint, the injection pressure of the fluid and the flow rate.

The complete procedure of a typical test is described in the following section.

5 TESTS ON ARTIFICIAL JOINTS

Joint-like models were made by regularly arranging one layer of steel balls (simulating the asperities) between two lead plates (simulating the constitutive rock).The analogy with the flow in real rock joint is shown in figure 2.

Three steel ball diameters (1, 2 and 3mm) and two different fluids were used, water and oil of 130 mPa.s viscosity

At different stages of loading under the force F, the steel balls penetrate the lead plate simulating the different apertures e of the joint. For each one of these apertures, a diverging radial flow test was carried out under different imposed pressures or flow rates.

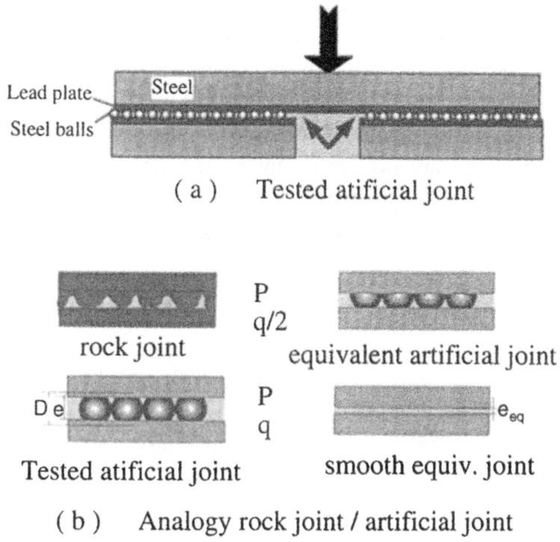

Fig.2: Principle of the tests on artificial joints

the flow rate as a function of the injection pressure (figure 3b). Thus the q/p ratio is representative of the joint permeability tests.

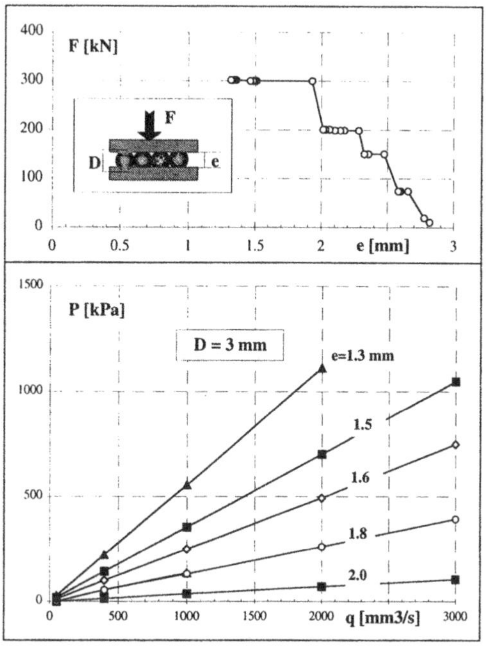

Fig.3a-b: *Typical test results on an artificial joint*

5.1 Typical results

A complete test result on an artificial joint carried out with steel balls of 3mm diameter and oil as fluid is presented in figures 3. The initial aperture after the application of a small normal force was e = 2.82 mm. The permeability was considerable for this particular aperture and the measured pressures were small. The figure 3b shows the results obtained for smaller apertures, e = 2mm to 1.3mm. A permeability test was carried out for each one of these apertures with the flow varying between 400 and 3000 mm^3/s.

5.2 Discussion

To study the influence of asperity size on the joint permeability, three different diameter steel balls were used for setting the artificial joints. On each one of these joints, tests similar to the one described above were carried out.

In all these cases (different asperities, different apertures) the tests have shown a linear variation of

Figure 4 shows the variation of the ratio q/p as a function of the mechanical aperture e for the different joints tested (D =1, 2 and 3mm). By comparing with the behaviour of a plane joint (cubic law), we can define, for each diameter (size of the asperities) and for each aperture of the joint, an equivalent plane joint aperture e_{eq} that has the same permeability. For example, the figure 4 shows that an equivalent plane joint aperture e_{eq} = 0.2mm corresponds to an aperture e of 0.7mm for a joint asperity of D = 1mm, 1.3mm for D = 2mm and 1.9 mm for D = 3mm.

In figure 5 the hydraulic response of these rough joints are compared in logarithmic scale with the cubic law that governs the flow between two smooth parallel planes. One can see that for D = 1 (small asperities) the exponent n of equation 2 is 3.8 which is close to 3 (cubic law). In case of bigger asperities D =2 and D = 3mm, n is in the order of 8.2 and 8.4 respectively. It appears that the exponent n increases non linearly with the size of asperities.

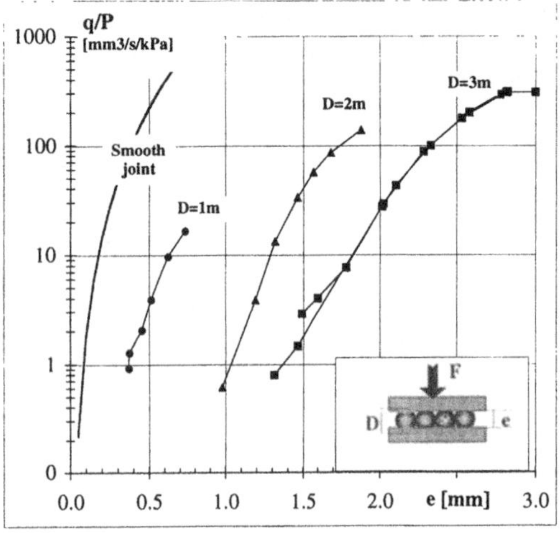

Fig.4: Flow rate vs joint aperture

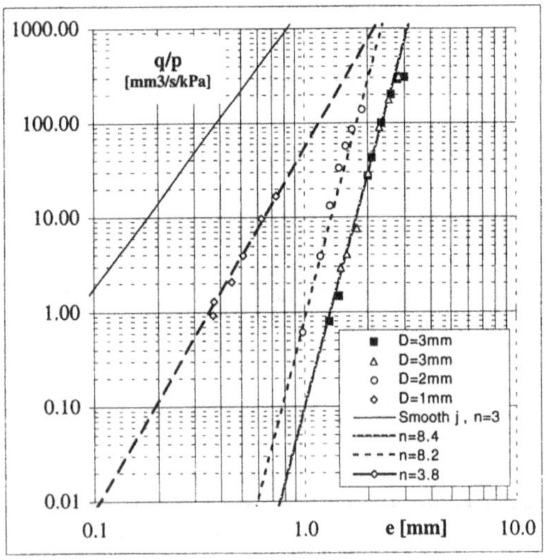

Fig.5: Flow rate vs joint aperture

5.3 Conclusions

From the background of the tests on artificial joints whose form is well determined and thus allows us to estimate the form and the volume of the contributing flow space for each joint aperture, we can deduce the following points:

- The approximation by a flow between two smooth planes is only valid for particular conditions such as large aperture of the joint (small normal stress) and reduced size of asperities(fig. 5).
- In the case of a nearly closed joint with wide asperities or which is sheared (staggering of surfaces), the number of contact points becomes very large and the most of the established flow passes through a network of pipes determined by the form and roughness of the joint.
- Especially under high normal stresses, the parameters relative to the volume (fig 6) like mechanical aperture or average aperture or porosity of the joint are not representing the flow through a real rock joint. Quantities relative to the section and to the number of flow channels should be more appropriate.

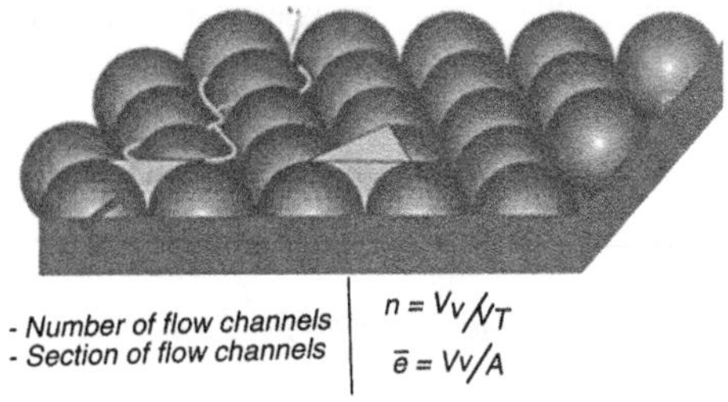

Fig.6: Consideration of type of flow in an artificial joint.

6 FLOW NETWORK IN SHEARED ROCK JOINT

The influence of the imbrication of the two rock joint surfaces on its permeability is more important than its form and its roughness. Under high normal stress in spite of high roughness the permeability can become almost zero if the two surfaces correspond perfectly.

To study this aspect, we tried to evaluate the channel flow network that is established in a rock joint after a slight shearing and according to its degree of closure.

The test was carried out on a rock joint obtained by splitting a sample of granite gneiss. The two circular surfaces (diameter 140mm) were therefore identical and corresponded perfectly. The shape of these surfaces, obtained with a meshing of 1x1mm is shown in figure 7.

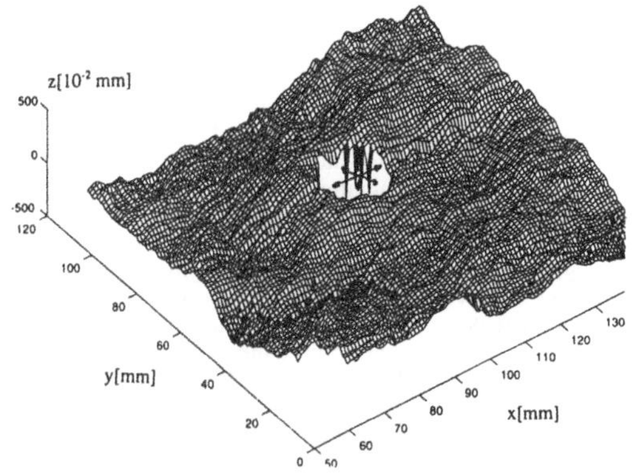

Fig.7: A digitised rock joint surface. 1x1 mm mesh.

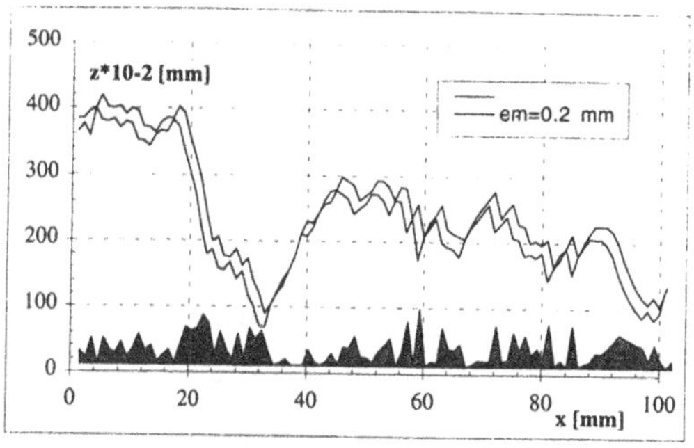

Fig.8: *Relative position of a sheared joint profile*

To simulate a shearing, we displaced (numerically) one of the two surfaces by 1mm in the x direction. A number of vertical displacements in the Z direction simulating the effect of joint closure under normal stresses were tested.

Figure 8 shows, for example, the relative positions of a randomly chosen profile of the joint for an aperture $e_m = 0.2$mm. The variation of the corresponding hydraulic aperture e_h estimated at each point of the profile is also represented.

We observed that for $e_m = 0.3$ mm, the hydraulic aperture varies between 0 and 1 mm. The small number of contact points (5% of the total number of joint points discritized) and the homogeneous distribution of the hydraulic aperture tends to assume the flow between two parallel planes.

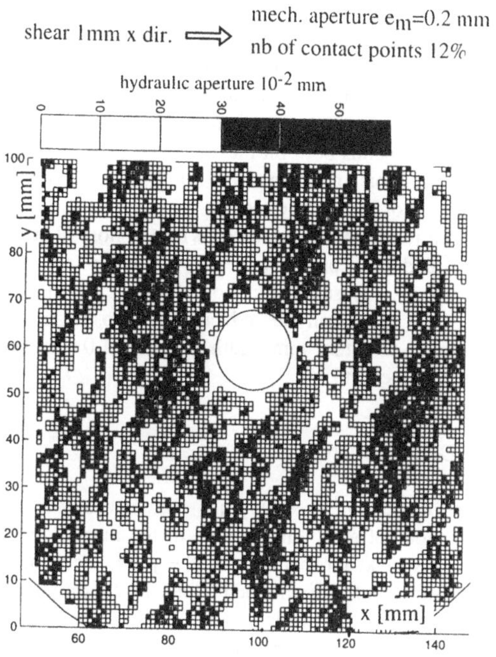

Fig.9: *Hydraulic aperture and flow channels in a sheared rock joint.*

For a greater mechanical closure $e_m = 0.2$ mm (fig.†8), the maximum e_h value reduces proportionally and the contact point zone increases to 12% of the total surface. According to the imposed contact areas, the hydraulic flow is then going to be modified to a preferential flow through a channel network.

The representation of the entire hydraulic apertures of the joint (fig. 9) confirms this aspect and gives an idea of the orientation and size of the flow network.

Only around 5% of the two surfaces are in contact for the mechanical aperture e_m =0.3mm. According to the normal distribution law the hydraulic aperture has a mean value of 0.33 mm and a standard deviation of 0.15mm. At this stage, the distribution of the apertures on the whole surface allows to suppose that the hydraulic flow established between the two surfaces makes the hypothesis of flow between two planes and the application of the cubic law possible, even though the aperture e varies.

For the joint comparatively more closed $e_m = 0.2$ mm, the contact surface is only 12% of the total surface. The figure 9 shows clearly the establishment of preferential channel flow networks. The mean aperture is 0.25 mm with the shown distribution. The flow channels shows a preferential orientation. In this case it is around 45° from the X-axis. It depends both on the shearing direction (here the direction of increasing X) and the oriented structure of the surface rock.

As the joint closure increases with the normal stress the trend to a preferential flow network increases and keeps the same direction. When $e_m = 0.1$ mm, the joint flow can be reduced to 3 to 4 parallel channels of 1 to 2 mm width and 0.4 to 0.6 mm of effective height.

7 CONCLUSIONS

The difficulty in studying the hydromechanic behaviour of rock joints is due to the interaction of three aspects: morphology of the joint, mechanical behaviour and hydraulic behaviour. Furthermore, the problem of describing and characterising the morphology and the roughness of rock surfaces is also very challenging.

The tests presented here on artificial joints allowed us to see the importance of considering the tortuosity of flow passing through preferential channels. Tortuosity flow is important when the joint is mechanically closed and when the asperities are big. Thus the characterisation of the flow through the form and sections of the channels seems primarily important.

Studies carried out on a rock joint show the type of channels that are established in a real case as a function of its compression and shearing solicitations. The size of the channels and the preferential orientation can be shown in a significant way. Our final objective is to generalise this flow channel identification for any joint as a function of its morphology.

Acknowledgements

The works presented here and presently carried out at the Rock Mechanics Laboratory of Swiss Federal Institute of Technology, Lausanne, on the identification and on the characterisation of the rock joint roughness and its hydraulic behaviour are made possible in great part by the financial support of the Swiss Federal Department of Water Economy, as well as of the Swiss National Science Foundation.

REFERENCES

Louis C. , 1974. Introduction à l'hydraulique des roches. Bulletin du B.R.G.M., No.4, pp. 283-356

Kharchafi M., Egger P., Descoeudres F. 1996. Description de la rugosité et perméabilité des joints rocheux. Proc. Séminaire sur la recherche dans le domaine des barrages, Lausanne.

Barton N., Bandis S. 1990. Review of predective capabilities of JRC-JCS model in engineering practice Proc. Int. Symp. on Rock Joints, Loen, pp. 603-610.

Gentier S. 1987. Comportement hydromécanique d'une fracture naturelle sous contrainte normale. Proc. Int. Cong. on Rock Mech., Monreal, Vol.1 , pp. 105-108.

Raven K., Gale J. 1985. Water flow in natural rock fracture as a function of stress and sample size. Int. J. Rock Mech. Min. Sci. & Geomech. Abst., Vol.22 , No.4 , pp. 251-261.

Iwano M., Eistein H. 1995. Laboratory experiments on geometric and hydromechanical characteristics of three different fractures in granodiorite. Proc. 8th Int. Cong. on Rock Mech., Tokyo, Vol.2 , pp. 743-750.

Durham W., Bonner B. 1995. Closure and fluid flow in discrete fractures. Proc. Fractured and jointed rock masses, Lake Tahoe, pp. 441-445.

Boulon M., Selvadurai A., Benjelloun H., Feuga. 1993. Influence of rock joint degradation on hydraulic conductivity. Int. J. Rock Mech. Min. Sci. & Geomech. Abst., Vol.30 , No.7 , pp. 1311-1317.

Brown S. 1987. Fluid flow through rock joints: the effect of surface roughness J. of Geophysical research, Vol.92 , No.B2, Feb. , pp1337-1347.

Hakami E., Larsson E. 1996. Aperture measurement and flow experiments on a single natural fracture. Int. J. Rock Mech. Min. Sci. & Geomech. Abst., pp 395-404.

Wei Z., Egger P., Descoeudres F. 1995. Permeability predictions for jointed rock masses Int. J. Rock Mech. Min. Sci. & Geomech. Abst., Vol.32 , No.3 , pp. 251-261.

Mechanics of Jointed and Faulted Rock, Rossmanith (ed.) © 1998 Taylor & Francis, ISBN 90 5410 955 6

Modelling of hydro-mechanical coupling in rock joints

X. Dunat, M. Vinches & J.-P. Henry
Laboratoire GERM, Ecole des Mines, Alès, France

M. Sibai
Laboratoire de Mécanique de Lille, Villeneuve d'Ascq, France

ABSTRACT : Hydro-mechanical coupling in rock joints in generally expressed as a relation between the flow rate and the applied stresses. The present paper presents an experimental procedure based on equilibrium similar to the coupling test in porous media. An example on a very closed granite joint is presented : the « strain » parameter in the continuum approach is the joint displacement. A brief description of the coupling is presented by describing the joint as a set of small pins which are gradually put in contact under the external loading. The comparisons between the experiments and this modelling on several loading paths and also on the flow rate indicates the potential of this approach with regards to the simplicity of the model.

1. INTRODUCTION - REFERENCES

The notion of hydro-mechanical coupling in rock joints was introduced a long time ago by hydrologists who had noticed a strong relation between the flow rate and the applied stresses (Lomize 1951, Louis 1969). The interpretation of the « coupling » was first refined with the determination of the nature of the flow (laminar or turbulent regime, Witherspoon and al. 1980) then by the modelling of the joint by two parallel planes with the opening being expressed as a function of the applied stresses (Detournay 1979). The interpretation was then refined by the introduction of models describing the contacts inside the fracture (ex : Gentier 1986) or numerical models taking into account the progressive closure of the joint and the channel effect.

All these approaches mainly concentrate on the study of the flow not on the stress. The stress is secondarily considered to « globally disturb » the flow. Apart from the numerical models (Lin 1994) which introduce an elastic behaviour of the rock (unable to predict the hysteresis often observed by experimentalists), the coupling is not a real coupling. In the porous media, the coupling is described locally by the action of the stresses tensor on the fluid pressure thanks to the strain tensor and of course, the rheological law (Coussy 1991).

A model of local behaviour for rock joint, based on this approach has been introduced (Henry and Sibai 1997). It did not allow the study of the influence of the joint on the surrounding matrix. We propose to consider it back, and modify it in order to study this influence.

These models define locally the action of the normal stress and of the fluid pressure on the joint displacement. The definition being local, the fitting tests will be performed in a uniform stress field. Thus, the first part of this paper is dedicated to the presentation of these tests, and the results on a granite joint are given as an example.

2. EXPERIMENTAL DESCRIPTION OF COUPLING

2.1 *Experimental mechanism*

The experimental mechanism devised at the Laboratoire de Mécanique de Lille within the framework of the Soutoudeh PhD thesis (1995) allow to test cylindrical sleeved samples, with a 65mm diameter, and a 120mm height, split by a joint in the diametrical plane.

The mechanism is composed of a high pressure hydrostatic unit (20MPa), two pumps for the controlled confining and fluid pressures, an acquisition unit for the measurements of the displacements, calibrated capillaries of different diameters for measurement of the injected or expelled volumes of fluid.

The figure 1 represents, in detail, the sample put on its support and its equipment. The fluid pressure is measured in four points P_1, P_2, P_3, and P_4 by pressure sensors. P_2 and P_3 carry out a direct measure of the pressure in the joint. The displacement at the joint level is obtained by a global measure at the joint level (three LVDT captors in opposition - precision 0,1μm) and the measure of the rock matrix strain determined by extensometry gages. This experimental equipment allows the very accurate measurement of the joint displacements under any uniform loading conditions, the realisation of the pressure profile, the relationship between the flow rate and the loading during the percolation tests.

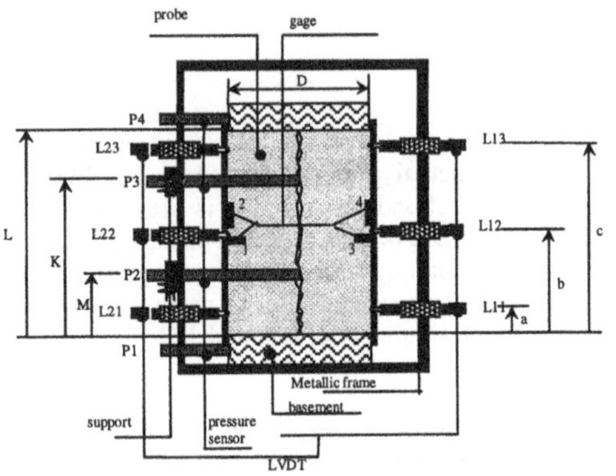

Figure 1: Measurement equipment set-up around the rock joint

2.2 *Tests types definition*

The coupling in porous media requires, in a first step, the characterisation of the skeleton behaviour (Elasticity : Young modulus E, plasticity : behaviour low) and in a second step coupling tests in a uniform stress field : study of the strain field versus the increase of the stress level (ex. : confining pressure) or/and the interstitial fluid pressure.
For the coupling in rock joints, we are following a similar process in defining two fundamental tests :

type 1 test or joint characterisation test
type 2 test or coupling test

In the second type of test, the applied load Pc is perfectly uniform (confining pressure) and fluid pressure P_i in joint is kept uniform (controlled by pressure sensors). These conditions correspond to drained coupling tests in porous media ; the study of the joint displacements for any path in the graph P_c - P_i is the translation of the coupling.

The percolation test, called type 3 test will be used mainly for the validation of the proposed model.

2.3 *Results*

The results presented below are obtained on a granite joint generated by Brazilian tests. The figure 2 gives the results of type 1 tests . One can notice that the three levels of sensors give very similar results for the joint displacement. The figure gives the mean value of the previous displacements and shows the existence of an important hysteresis during the loading - unloading cycle.

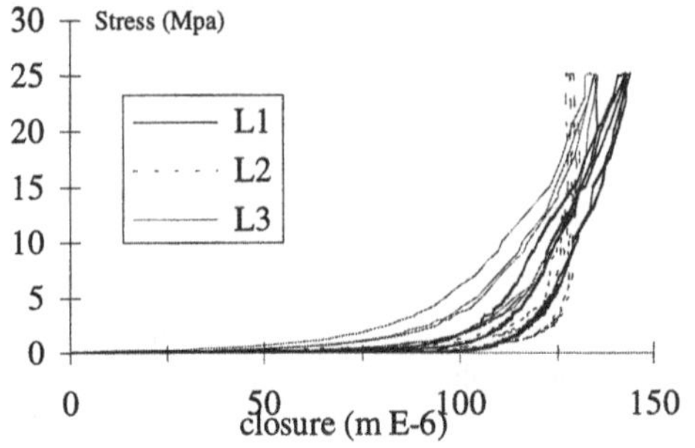

Figure 2 : Loading test of the joint (P_i=0)
Measurements on the sensors during the loading-unloading cycle.

3. MODELLING

3.1 *Presentation of the first model, hypothesis*

The presented model is in one dimension , the width of the joint is supposed equal to one, and is the length unit of an area being representative of the joint area. The length (in the flow direction) will be equal to the unit too. On this representative area, we propose to describe the joint as an assembly of n pins of relative height Ch(i) and of width S(i) (i=1,n) represented on figure 3 who will successively be put in contact under the normal external loading. We suppose the pins do not interact.

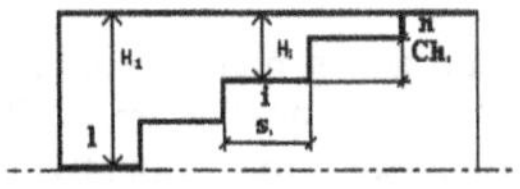

Figure 3 : Joint modelling

If we write down $2e_0$, the maximum crushing of joint, we can norm the thickness, and the pins height h(i) varying to 1 at 0 with h(1)=1 and h(n)=0. We suppose that, in the first model, the pins strain have no influence in matrix itself.

During the closure ε of the joint, the pin strain ε(i) will be equal to (1-ε+h(i)). Then we make the following hypothesis :

The width is in geometric progression of ratio q_1,

The relative height is in geometric progression of ratio q_2,

any pin behaviour is of elastic type (elastic modulus E) with plasticity with strain hardening of slope M. The elastic limit is written σ_0. This behaviour law will allow to take into account the observed irreversibility, due to the friction at the joint level.

3.2 *Parameters determination*

We have 7 model parameters. There are 4 geometric parameters : n the number of pins, q_1 and q_2 the geometric ratios, e_0 the opening, and 3 behaviour law parameters : E (elasticity), M (plasticity) and σ_0 the threshold.

Let us write ε_r the residual strain, it corresponds to the strain of first pin that allows us to write :

$$M=E[1-(\sigma_0/E)-\varepsilon_r]/[1-(\sigma_0/E)] \quad (1)$$

Please note that S(i) and Ch(i) are only defined by the geometric ratios :

$$S(1)=[q_1-1]/[q_1^n-1] \quad (2)$$

If we suppose what the joint can be totality closed by a σ_M stress, and write down n_e the number of pins then over the elastic limit, the equilibrium allows us to write :

$$\sigma_M=E\sum_{i=1}^{n} S(i)\varepsilon(i)-[E\varepsilon_r/(1-\sigma_0/E)]\sum_{i=1}^{n_e} S(i)[\varepsilon(i)-\sigma_0/E] \quad (3)$$

This relation (3) allows the explicit calculation of E and M as functions of σ_M. We must notice that, if σ_0 is known, n_e is easily computed as a function of this parameter. It turns out that it will be enough to calculate q_1, q_2 and σ_0, after having arbitrarily chosen a n value, in order to simulate the joint behaviour.

The calculation of this last parameters are not explicit. We have used the inversion methods(Shao, Henry 1997) for their determination. As a result, for n=7 and for 7 experimental values, a very good correspondence was observed between the model and the experimental data used to fit the model .

The figure 4 gives the comparison between the model and the experimental data.

3.3 *Second model*

If we now suppose what the pins have an influence on the matrix on a height h_{inf}, we could consider that the height of the pins, we write down now as ht, vary from $h_{inf}+1$ to h_{inf}, with $ht(1)=h_{inf}$ and $ht(n)=h_{inf}$. If we write down $h(i)=ht(i)-h_{inf}$, h(i) remains equal to its value in the first model. We suppose, keeping the previous hypothesis, that the pin elastic modulus E is equal to the modulus of the matrix. In this case, h_{inf} represents the height up to which the joint affects the matrix.

This new hypothesis allows us to write the strain in pin i, according to the following equation :

$$\varepsilon(i)=(1+h(i)-\varepsilon)/ht(i) \quad (4)$$

The relation between E and M becomes :

$$M=E*[1-\varepsilon_r-(h_{inf}+1)*(\sigma_0/E)]/[1-(h_{inf}+1)*(\sigma_0/E)] \quad (5)$$

and equilibrium allows us to write the total stress according to the following equation :

$$\sigma_M=E\sum_{i=1}^{n} S(i)\varepsilon(i)-[E\varepsilon_r/(1-h(1)*\sigma_0/E)]\sum_{i=1}^{n_e} S(i)[\varepsilon(i)-\sigma_0/E] \quad (6)$$

the modulus E being now imposed, this relation allows us to calculate $\varepsilon(i)$ as a function of σ_M.

We obtain an excellent correspondence with experimental values for a new model with 7 pins, calculated from 5 values from the experimental curve.

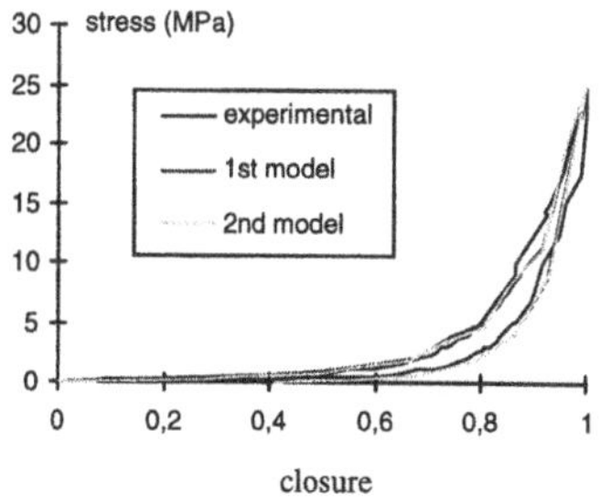

Figure 4 : Comparison between experimental data and modelling for n=7

We have given, as an example in the table 1, the h_{inf} values obtained for the granite and for two other materials : marble and schist.

We can now compare the prediction of this second model with experimental data on the figure 4. We can see a rather good agreement between the model an the experimental values. H_{inf} corresponds to the ratio of the first pin height to the opening e_0. It thus represents the ratio of the matrix height whose mechanical behaviour is affected by the joint, and the opening. These values seem to us physically acceptable for slightly opened joints.

Tableau 1 : h_{inf} values for different materials

	Granite	marble	schist
H_{inf}	140,7	123,9	257,0

4. COMPARISON EXPERIMENT - MODELS

Let us now compare the prediction of different models for type 2 and type 3 tests with the experimental data. The ε value for $\sigma = \sigma n$ is the only parameter used to fit the model.

4.1 Type 2 test

The type 2 test corresponds a pressure loading of the joint up to a constant confining pressure Pc, then we increase the fluid pressure starting from zero. For the models, when the fluid pressure increases, the pins go successively unstuck, one after the other. As soon as there is no more contact, the fluid pressure covers a more important area. The calculations from two models are not difficult and the results are presented in figure 5 $P_c = P_n$. The quality of the predictions shows the good aptitude of the model to describe the hydro-mechanical coupling. It is then possible to deduce the notion of effective stress by the relation :

$$\sigma_{eff} = \sigma_n - (1 - S_c) P_i \qquad (7)$$

with S_c the pins area in contact and P_i the fluid pressure.

4.2 Type 3 tests.

The models that we have presented are local models. We can now integrate them in non constant normal stresses fields or non constant fluid pressure fields P_i, for example when there is a flow (one dimension case only).

Assuming that, for the couple (σ_n, P_i) applied in a point of the joint, we have j pins in contact, and that the joint closure is equal to ε, the surface area available for the flow A_e is given by the relation :

$$A_e(\sigma_n, P_i) = A_e(\varepsilon) = \sum [1 - h(i) - \varepsilon] S(i) \qquad (8)$$

This allows us to calculate the flow rate of fluid Q, by unit of width, assuming a laminar flow between the different pins :

$$Q(\sigma_n, P_i) = (gJ/12\upsilon)\left[\sum [1 - h(i) - \varepsilon]^3 S(i)\right] e_0^3 \qquad (9)$$

with J the hydraulical gradient, ν the cinematic fluid viscosity, e_0 the initial joint opening. We thus obtain the local average speed :

$$V_{moy} = Q(\sigma_n, P_i) / e_0 A_e \qquad (10)$$

The parameter e_0 will be calculated from only one test, interpreted according to the present method. In order to calculate the flow rate in a percolation test, one needs to perform a discretisation in the flow direction, using ΔPi = constant and apply the previous procedure. The figure 6 and 7 give a comparison example between the models predictions and the experimental results, e_0 has been fitted for two cases : σ_n= 15 MPa, P_1= 12,896 MPa and P_n= 0 MPa.

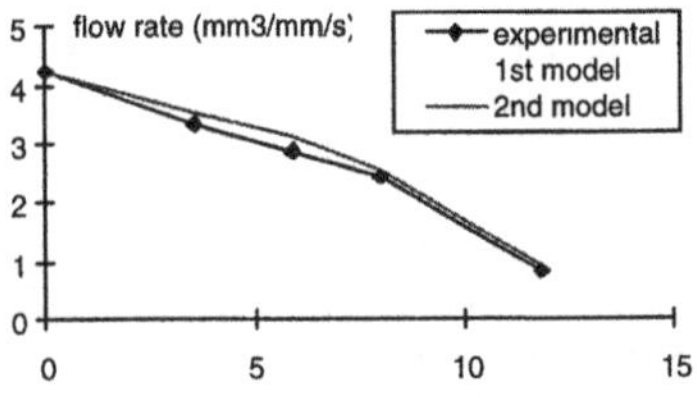

Figure 6 : Evolution of the flow rate with for σ_n=15MPa, P_1=12,896MPa

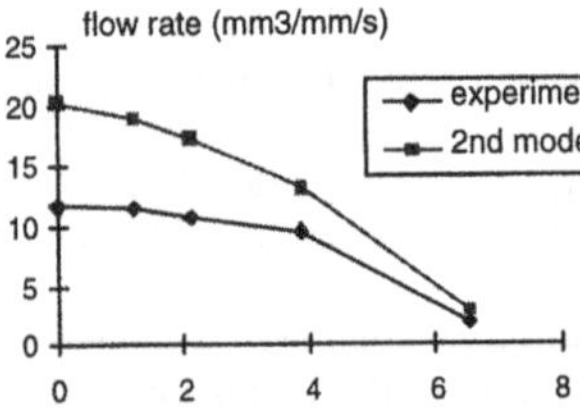

Figure 7: Evolution of the flow rate versus the draining off pressure σ_n=7,5MPa

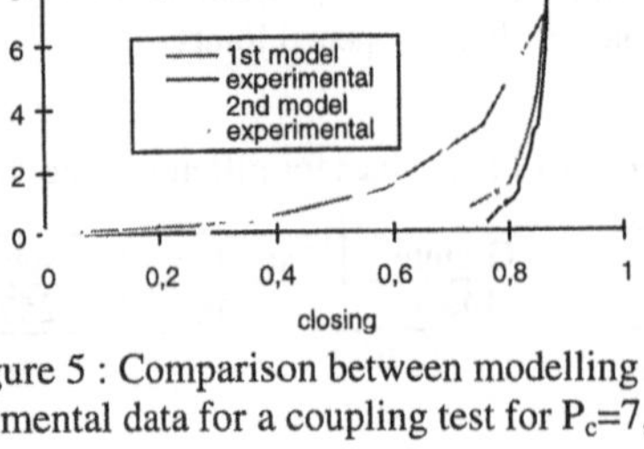

Figure 5 : Comparison between modelling and experimental data for a coupling test for P_c=7,5MPa.

5. INFLUENCE OF h_{inf}

The study of the influence of h_{inf} was performed on a 10 pins model.

5.1 Influence on the mechanical behaviour.

We have simulated the mechanical behaviour with h_{inf} values ranging from 100 to 220, where the value calculated by inversion is about 180. For superior values we obtain aberrant values for the plasticity module M.

The simulation of the mechanical behaviour is coherent, for values between 130 and 200 (Figure 8). For inferior values, we obtain a poor prediction notably for the loading cycle, with predicted stress values inferior to the experimental results. Over 220, we have a unloading path very similar to the loading path due to unrealistic values of M.

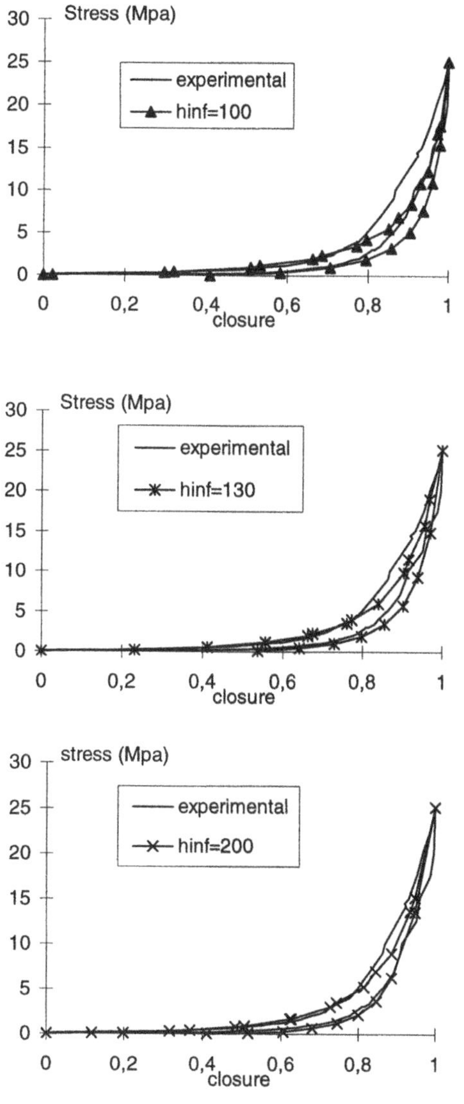

Figure 8 : Influence of h_{inf} on mechanical behaviour

5.2 *Influence on M*

As we can observe it on figure 9, the relation M- h_{inf} looks like an hyperbolic function. At first, M is stable, then decreases very quickly, and can reach negative values (not represented on the figure).

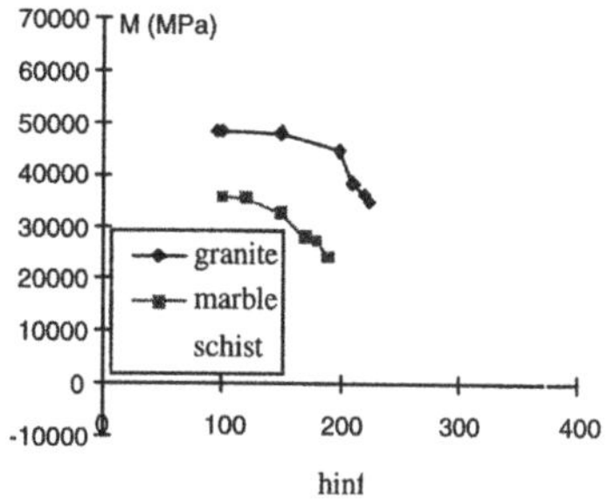

Figure 9 : influence of h_{inf} on module M

This phenomenon can be explained by the fact that if h_{inf} increases, the strain of a pin $\varepsilon(i)$ for a given closure ε is smaller. As the plasticity module M is smaller than the elasticity module E, to obtain the same total stress, the elasticity limit σ_0 must increase. According to equation (5), if σ_0 increases, M decreases following an hyperbolic function. The increase of σ_0 is confirmed by calculation and is represented of the figure 10.

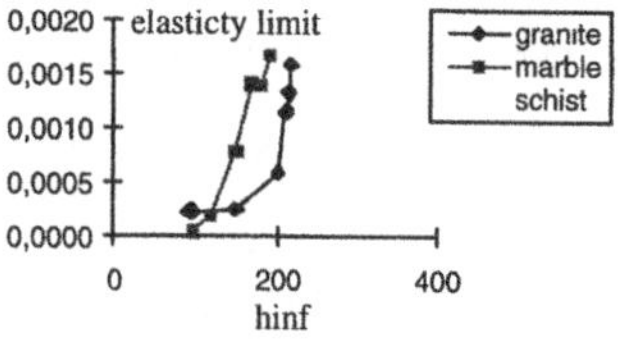

Figure 10 : influence of h_{inf} on σ_0/E

6. CONCLUSION

We have presented local, very simple, physical models using deformable pins allowing to carry out hydro-mechanical coupling and flow rate calculation. These models allow us to obtain good coupling prediction. They present some particularities. The equations used in the percolation study allow us to extract the pressure distribution in the joint. An example is given in figure 11 which shows that the pressure loss is absolutely not linear in the joint, a result that was difficult to interpret up to now, and was observed by Soutoudeh (1995).

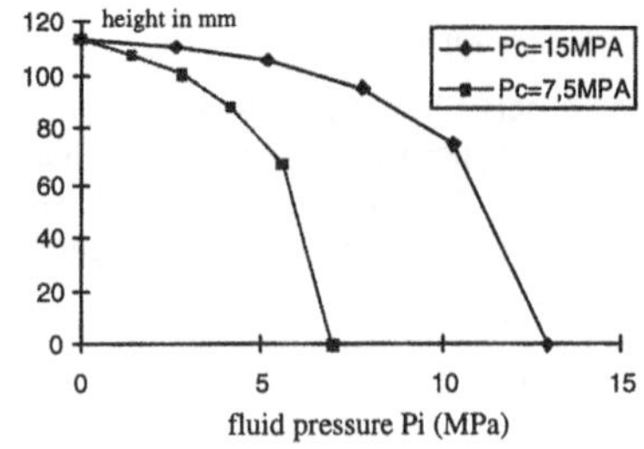

Figure 11 : Modelling of the fluid pressure along the rock joint

REFERENCES

COUSSY 0. (1991) - Mécanique des milieux poreux. Ed. Technip

DETOURNAY E. (1979) - The interaction of déformation and hydraulic conductivity in rock fracture : an experimental and analytical study. Improved Stress determination procedures by hydraulic fracturing. *Final report. Mineapolis,* University of Minnesota, Vol.2

GENTIER S. (1986) - Morphologie et comportement hydromécanique d'une fracture naturelle dans un granite sous contrainte normale. *Thèse.* Université d'Orléans

LIN J. (1994) - Etude du comportement hydromécanique d'une fracture sous contrainte normale. *Thèse.* INPL

LOMIZE G.M. (1951) - Water flow through jointed rock. *Gosenergoizdat,* Moscou (en russe)

LOUIS C. (1969) - A study of groundwater flow in jointed rock and its influence on the stability of rock masses. *Rock mechanics research report Nr. 10,* Imperial College, University of London

SHAO J.F. DAHOU A; et HENRY J.P. (1991) - Application de la théorie des problèmes inverses à l'estimation des paramètres rhéologiques. *Revue Française de Géotechnique,* n°57, PP.75-80

SOUTOUDEH H. (1995) - Etude expérimentale et modélisation du couplage hydromécanique de joints rocheux. *Thèse.* Université de Lille

WHITERSPOON O.A., WANG J.S.Y, IWAI K. et GALLE J.E. (1980) - Validity of cubic law for fluid flow in deformable rock fracture. *Water Ressources Research* Vol.10, n°16, p.1016.

Mechanics of Jointed and Faulted Rock, Rossmanith (ed.) © 1998 Taylor & Francis, ISBN 90 5410 955 6

On the experimental study of mass transport in a fracture model

J.H.Lee & J.M.Kang
School of Urban, Civil and Geosystem Engineering, Seoul National University, Korea

ABSTRACT: In this experimental study, we attempted to investigate how the aperture's statistical and spatial characteristics affect flow behavior in fractured rock . We conducted water flooding and tracer test using a transparent single fracture model. This study indicates that fracture flow can be represented by the anisotropic ratio of aperture correlation(r[ϒ]). Results show that larger dispersion and smaller recovery of oil occur as r[ϒ] increases.

1 INTRODUCTION

Flow characteristics in fracture have been studied by using the parallel plate model. In these approaches, the spatial distribution for dissolved species in solute transport and phases in water flooding are assumed to be uniform. However, the spatial distribution should be determined when the mass transport between fracture and rock matrix is considered.

In past decades, considerable progresses have been made to describe variable apertures of fracture more realistically. Tsang & Tsang[1)] suggested that the mean aperture, standard deviation and correlation length can be useful parameters to represent the hydrological characteristics of a fracture in their numerical works. The spatial correlations have a great effect on multiphase flow in fractures, as shown in numerical work of Pruess & Tsang[2)]. However, it is difficult to determine a single value of correlation length from semivariogram constructed by laboratory or field data.

To overcome these defficiencies, we suggested a new variable r[ϒ], which is the ratio of noramlized sill value of transverse direction to that of longitudinal direction, and investigated experimentally its sensitivity to solute transport and water flooding in a single fracture model.

2 EXPERIMENTAL SETUP & PROCEDURES

2.1 Fracture models and holder

Five transparent fracture models made of epoxy were molded from silicone rubber castings of fractured sandstones. This method was originally developed and used by Gentier[3)]. More details on the procedures of making fracture models may be found in the work by Hakami et al[4)]. All fracture models were mated and carefully machined approximately 3.75cm thick, 5cm wide and 29cm long. Intersections of the fracture with the sides of the sample were sealed with silicone sealant.

We made a transparent cylindrical sample holder which can accomodate the fracture model in order to maintain stable confining pressure up to 150psi. Two endpieces of the holder were installed on the ends of sample with silicone sealant. For complete sealing, leaking tests were performed before conducting flow tests.

2.2 Monitoring system

Fig.1 shows a schematic diagram of the experimental apparutus in this study. We used two reciprocating pumps for dyed water and pure water or oil to keep constant concentrations of dyed water. Image data from flowing tests were captured with CCD camera and stored in a PC. Effluent concentration and pressure data were obtained using inline spectrophotometer and pressure transducer, respectively.

2.3 Procedures

After placing a fracture model into the coreholder, we filled the annulus between the holder and the fracture model with water for better visualization. Constant confining pressure of gas was applied to the holder through a fluid

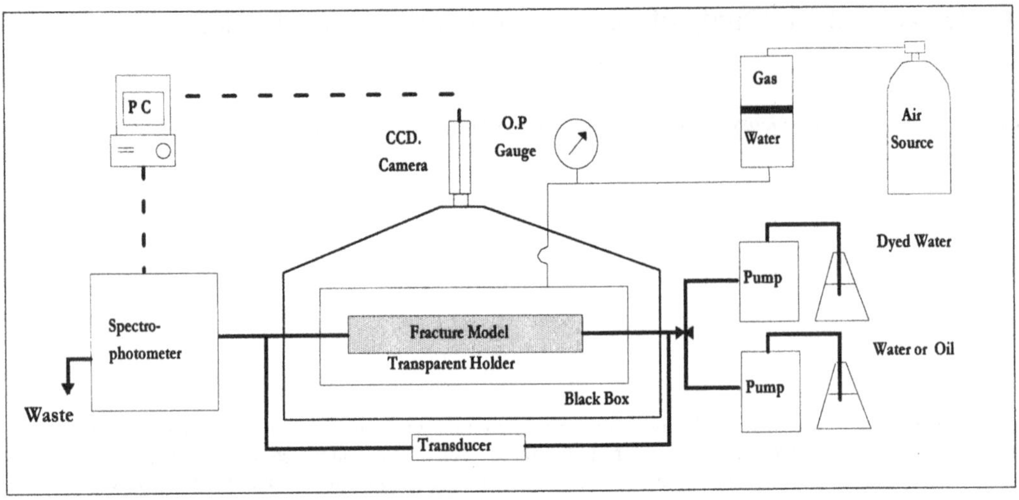

Fig. 1 The schematic diagram of experimental apparatus in this study

accumulator. The void volume of the fracture model was measured by resaturation method and corrected for system dead volume. With water injected at a constant rate, we measured the hydraulic apeture of the fracture model.

The aperture distributions of fracture models have been measured with image analyzer adopting a light attenuation theory. We acquired two images of each sample, one filled with water and the other filled with dyed water. We subtracted the latter from the former to remove the material and light source effects. All samples have 130×620 pixels approximately, each pixel covers about 0.39mm wide×0.45mm long. Image values at each pixel were converted to aperture values using a similar method used in the work of Persoff et al[5].

Step type tracer tests were conducted by displacing water with dyed water when the effluent concentration and images were monitored. Pure water flushing was then performed for next runs.

After finishing the tracer test, the model was saturated with Mineral oil(13cp at 20℃) for water flooding. The image data and total oil recovery were obtained.

The pattern of solute transport is discussed with breakthrough curves from tracer tests, and the results of water flooding are shown with the total amount of oil recovery.

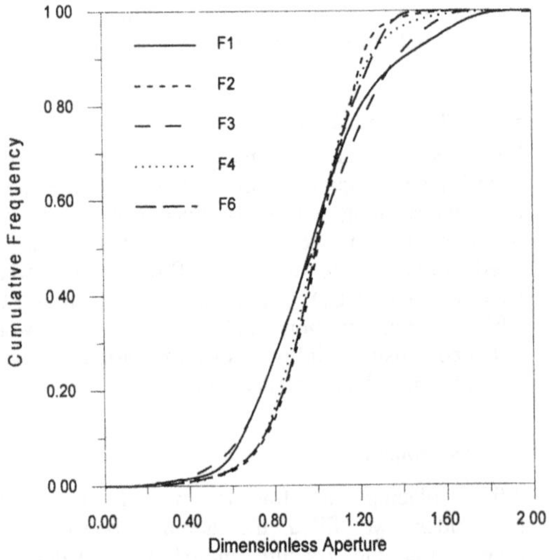

Fig. 2 The cumulative distributions of samples

3 RESULTS AND DISCUSSION

3.1 Morphologic characteristics of fractures

The cumulative aperture distrbutions for each sample are similar to the form of nomal distribution, as illustrated in Figure 2. For the purpose of comparison between samples, apertures are divided by average values.

The semivariograms depicted by executing GAM2 (from GSLIB, programed by Deutsch) were shown in Figure 3 and Figure 4 in terms of normalized variogram(Υ_n). As shown in the figures, it is difficult to determine a single value of correlation length(λ). However normalized sill value(Υ_{nmax}) chosen at a maximum value of Υ_n can be determined conveniently. Υ_{nmax} implies the variance of apertures in the corresponding direction. In other words, if the ratio of $\Upsilon_{nmax,T}$ to $\Upsilon_{nmax,L}$($r[\Upsilon]$) is greater than 1, there is a flatter path along longitudinal direction than that along transverse direction. Therefore, $r[\Upsilon]$ can be useful parameter for correlation length when the spatial correlation shows anisotropy.

Two parameters, which is coefficient of variation(C_v) and $r[\Upsilon]$, are employed to describe the fracture's morphology. All the statistical values are summarized in Table 1.

3.2 Tracer tests

The results of the tracer tests are shown in Figure 5 in terms of normalized dye concentration versus (injection void volume - void volume when C is 0.5). Injection void volume is calculated to divide the cumulative volume of dyed water injected by void volume of fracture. To explain the solute transport pattern in fracture, $\triangle t$ and $t_{0.5-}$ were calculated and shown in Table 2 with other parameters. $\triangle t$ and $t_{0.5-}$ are as followed.

$$\triangle t = t_{0.95} - t_{0.05}$$
$$t_{0.5-} = (t_{0.5} - t_{0.05})/(t_{0.95} - t_{0.05})$$

Here, t_x is the injection void volume when normalized effluent concentration has reached x.
$\triangle t$ is a modified measure of dispersion and the reciprocal of $t_{0.5-}$ represents the portion of large flow rate channel.

The value of $\triangle t$ is decreased with $r[\Upsilon]$ almost irrespective of C_v when the spatial correlation shows anisotropy . In case of the small $r[\Upsilon]$, the flow path tends to be built up in transverse direction and the front of the solute is well spreaded along the transverse direction compared with that along the longitudinal direction. When the spatial correlation shows isotropy for the case of F4, $\triangle t$ is as large as for the F3 or F6. We note that $\triangle t$ is controlled by C_v as much as $r[\Upsilon]$ in isotropic condition. The values of t_{50-} seems to be the function of C_v only. This indicates that the portion of large flow rate channel is more sensitive to C_v than $r[\Upsilon]$.

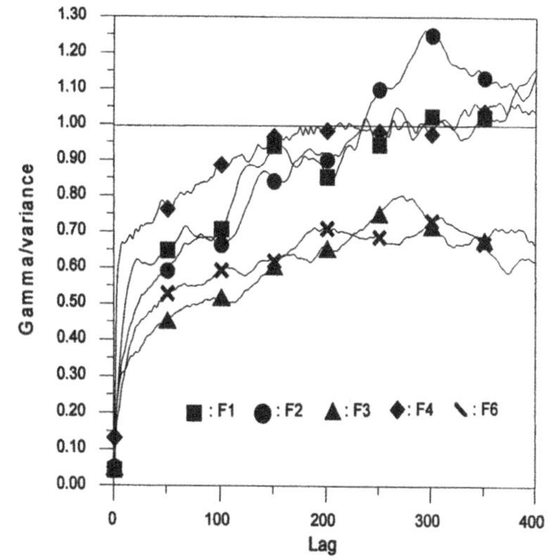

Fig. 3 The semivariogram for longitudinal direction

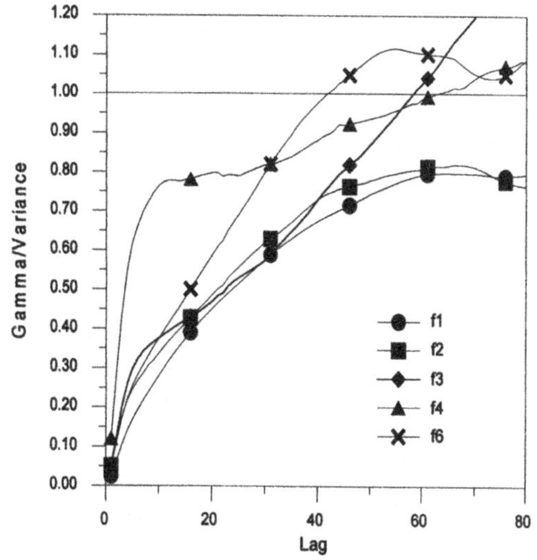

Fig. 4 The semivariogram for transverse direction

Table 1. Basic parameters of aperture distribution

Sample	μ (μm)	σ (μm)	C_v	$\Upsilon_{nmax,T}$	$\Upsilon_{nmax,L}$	$r[\Upsilon]$	h (μm)	h/μ
F1	424	127	0.30	0.80	1.17	0.68	60	0.14
F2	467	93	0.20	0.83	1.26	0.66	135	0.29
F3	234	68	0.29	1.20	0.81	1.48	76	0.33
F4	195	42	0.22	1.06	1.05	1.01	52	0.27
F6	225	46	0.20	1.12	0.74	1.51	109	0.48

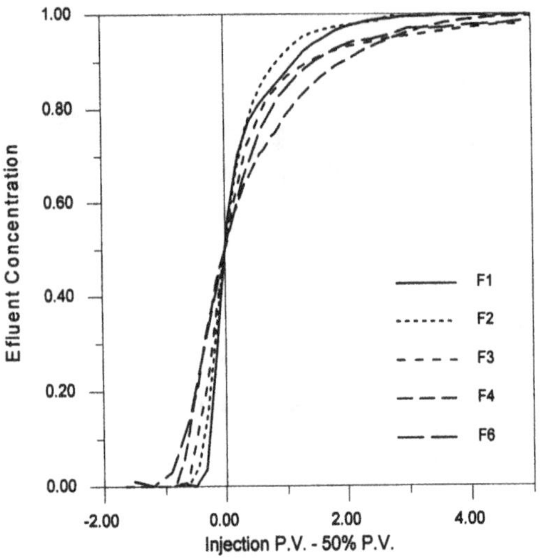

Fig. 5 The breakthrough curves for tracer tests

Table 2. Parameters for tracer test interpretation

Sample	Cv	r[Υ]	△t	$t_{0.5-}$
F1	0.30	0.68	1.96	0.16
F2	0.20	0.66	1.71	0.26
F3	0.29	1.48	3.34	0.17
F6	0.20	1.51	3.46	0.25
F4	0.22	1.01	3.49	0.21

Table 3. Total oil recovery of water flooding

Sample	Cv	r[Υ]	Total oil recovery
F1	0.30	0.68	0.54
F2	0.20	0.66	0.77
F3	0.29	1.48	0.43
F6	0.20	1.51	0.70
F4	0.22	1.01	0.55

3.3 Water flooding

With visual inspection of water and oil drop on the epoxy plate, we observed similar wetting for both phases. We carefully assumed that the wetting condition of epoxy used in this study is slightly water wet.

With decreasing C_v and r[Υ], the trend of increasing total oil recovery becomes clear. And the reduction amount of oil recovery due to increases of C_v is greater than that due to increasing r[Υ]. This result shows that C_v value has a great effect on total recovery of oil as well as r[Υ].

Total recovery of oil was rather small than expected, as presented in Table 3. This relatively large residual oil saturation suggests that variable apertures should be included when imbibition is considered between the rock matrix and fracture.

4. CONCLUSION

(1) The ratio of the normalized sill value of transverse direction to that of longitudinal direction can be used to represent spatial characteristics instead of correlation length when the spatial correlation shows anisotropy.

(2) The dispersion of solute transport is dominantly contolled by r[Υ] when r[Υ] is less than 1.

(3) The reduction amount of oil recovery with increasing C_v is greater than that with increasing r[Υ].

REFERENCES

1. Tsang, Y.W. & Tsang, C.F. 1990, Hydrological characterization of variable-aperture fractures, *Proc. of the Int. Symp. on Rock Joints,* edited by N. Barton & O. Stephansson, 383-390. Rotterdam : Balkema.

2. Pruess, K. & Tsang, Y.W. 1990. On two-phase relative permeability and capillary pressure of rough-walled rock fractures, *Water Resources Research* 26(5):1915-1926.

3. Gentier, S., Billaux, D. and van Vliet, L. 1989. Laboratory testing of the voids of a fracture, *Rock Mechanics and Rock Engineering* 22(5):149-157.

4. Hakami, E. & Barton, N. 1990. Aperture measurement and flow experiments using transparent replicas of rock joints, *Proc. of the Int. Symp. on Rock Joints,* edited by N. Barton & O. Stephansson, 383-390. Rotterdam:Balkema.

5. Persoff, P. & Pruess, K. 1995. Two-phase flow visualization and relative permeability measurement in natural rough-walled rock fractures, *Water Resources Research* 31(5):1175-1186.

Mechanics of Jointed and Faulted Rock, Rossmanith (ed.) © *1998 Taylor & Francis, ISBN 90 5410 955 6*

Hydromechanical behavior of a fracture: How to understand the flow paths

S.Gentier
BRGM, Orléans, France

J.Riss
CDGA, Université Bordeaux 1, Talence, France

D.Hopkins
LBNL, MS:46A-1123, Berkeley, Calif., USA

E.Lamontagne
CERM, UQAC, Chicoutimi, Qué., Canada

ABSTRACT: A combination of experimental and modeling techniques have been developed to identify and monitor the evolution of flow paths in natural fractures under normal and shear stresses. A casting method allows the morphology of the fracture to be analyzed in conjunction with the results of traditional hydromechanical tests. Modeling indicates that deformation of the fracture surfaces that occurs with increasing stress results in changes in void space geometry that could have a substantial affect on the hydromechanical response of the fracture. For fractures under a shear stress, the spatial distribution of flow is shown to be related to the damage zones that evolve with increasing tangential displacement. Results of laboratory experiments, during which fluid recovery is monitored, show that flow is progressively re-oriented during shearing in a direction subperpendicular to the shear direction.

1 INTRODUCTION

Changes in either the normal or shear stresses acting on a fracture induce notable changes in the flow paths that can substantially change the hydraulic behavior of the fracture. The modifications of the paths can be observed using specially designed instrumentation during laboratory experiments, and can be understood by simultaneous analysis of the morphology of the fracture along with the results of traditional hydromechanical tests. The aim of the work presented here is to synthesize the approaches and techniques developed to date that help us understand the evolution of flow paths in natural fractures under stress.

2 HYDROMECHANICAL BEHAVIOR UNDER NORMAL STRESS

2.1 *Experimental results*

Under normal stress, the hydromechanical behavior of a fracture can be summarized as indicated in Figure 1. The intrinsic transmissivity of the fracture decreases rapidly at first with increasing normal stress, for stresses at the lower end of the range, then decreases less rapidly until a critical stress is reached, at which point the intrinsic transmissivity tends to become stable. Simultaneously, the closure of the fracture decreases, then stabilizes. This relatively well known behavior can be observed by analyzing both the number and location of the outlets around the fractured sample at various stress levels. The results of such a study, used in parallel with an analysis of the morphology of the sides of the fracture, clearly shows the relationship between the mean distance between the outlets and the size of the different structures that define the topography of the fracture walls and the void space (Gentier 1987). Effectively, the mean distance between the outlets at the lowest normal stress corresponds to the smallest range determined from the variogram of asperity heights and the range determined from the variogram of apertures.

2.2 *Modeling of hydromechanical behavior under a normal stress*

As the results described above indicate, the morphology of the void space partially controls the evolution of the flow paths, and modeling is based on a map of the fracture apertures. Once a cast of the fracture has been obtained, a pixel map is recorded and a map of apertures is calculated (Gentier et al., 1989). Figure 2 shows the variation in thickness of a fracture cast in shades of gray. On the image, dark gray corresponds to the thicker fracture voids and the brighter gray to the thinner fracture voids.

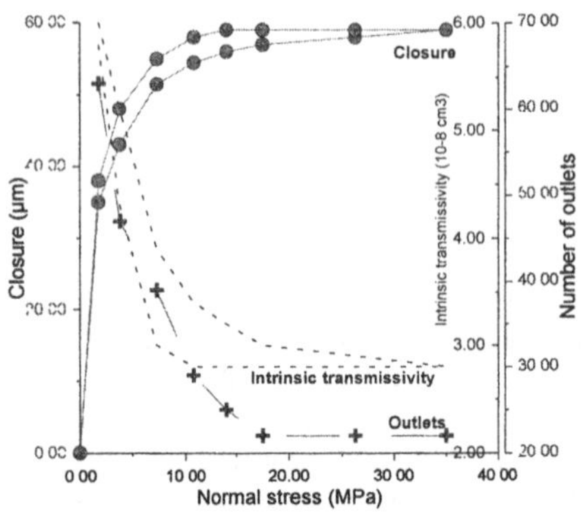

Figure 1. Evolution of the intrinsic transmissivity, fracture closure and number of outlets as a function of normal stress.

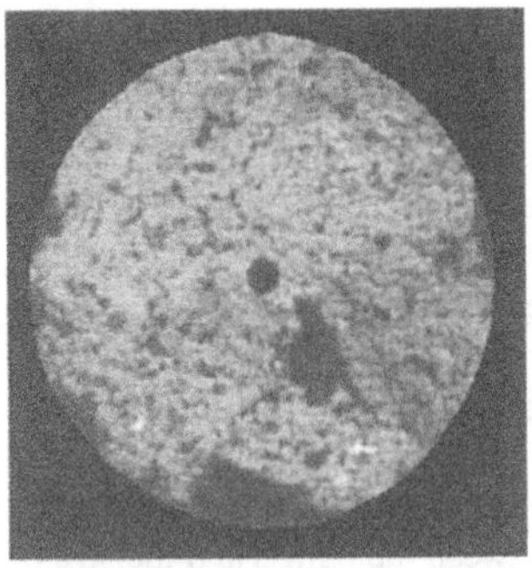

Figure 2. Gray level image of a fracture cast.

The first approach to modeling, which is the simplest one, is based on the assumption that under a normal load, there is uniform closure across the fracture meaning that any two points on opposite sides come together by the same amount. This can be modeled by specifying successive thresholds corresponding to the progressive average closure of the fracture with increasing normal load (Fig. 3a). At each increment of closure, it is assumed that the aperture has been reduced equally across the fracture, so one can easily get an estimate of the number and size of contacts from the original aperture map.

The second approach is based on a mechanical model that accounts for deformation of the fracture surfaces, asperity deformation, and mechanical interaction between contacting asperities (Hopkins, 1990). From the gray-level image of the fracture cast, the model can be used to calculate changes in aperture with increasing normal load. The gray-level image is transformed into a regular grid of elements and the average aperture over each element is calculated (Gentier & Hopkins 1997). These data, along with the mechanical properties of the material, are the input to the model. The output is the deformation of the fracture surfaces and asperities, the aperture at any point on the fracture, and the force acting at each contacting asperity. The result, expressed as (x,y,force) is transformed back into pixels. These new images (Fig. 3b) show the changes in aperture that occur under increasing load. The gray levels on these images are proportional to the applied force acting at each contact point, as calculated by the model.

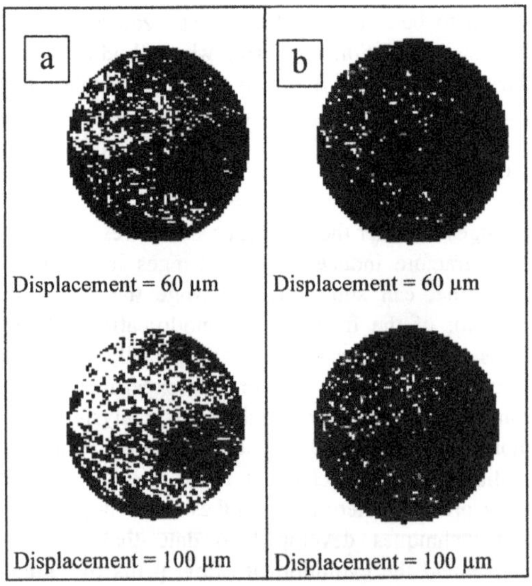

Figure 3. Results of the two models:
a- uniform closure over the fracture
b- Hopkins model

Comparing the images that result from application of the two methods helps in understanding the mechanical behavior of the fracture. The mechanical model indicates that deformation of the fracture surfaces results in less contact area than is predicted by the first model. It is also important to note that unlike the first approach, the mechanical model predicts changes in void geometry with increasing normal stress that result from the roughness of the fracture surfaces. These results have important implications for fluid flow through the fracture.

One approach to modeling flow in fractures is based on the observation that, in most fractures, flow occurs in channels, particularly when the normal load is high. It is, therefore, unnecessary to consider the non-conducting paths. Skeletonization, a discretization technique derived from a classical mathematical morphology algorithm (Gentier et al. 1989), is used and is an appealing technique because it accounts in a natural way for preferential flow paths. The image is first binarized based on a specified threshold (Fig. 4a). Flow is assumed to occur inside a pixel only when its gray-level, corresponding to aperture, is above the threshold. The black and white image obtained is then simplified and reduced to its skeleton (Fig. 4b). This new picture represents all the connections existing in the binarized image for one-pixel-wide channels lying at the geometrical center of these connections. The two-dimensional image is thus reduced to a mesh of one-dimensional channels. Dead-ends and small loops are then discarded (Fig. 5a). For every channel that is part of the skeleton, the importance of the flow path is determined using a modification of the skeletonization algorithm. For each skeleton path, the gray levels along the path are integrated such that the gray level assigned to a final skeleton pixel is the integral of the gray levels of the pixels in the flow section it represents. At this stage, a new simplified pixel map showing preferential flow paths is obtained. Figure 5 illustrates the evolution of a skeleton as the closure is increased. To simulate flow, a mesh consisting of one-dimensional elements, each element representing a portion of the channel, bounded by two nodes representing channel intersections, is deduced from the skeleton. The aperture, which is used to calculate the transmissivity, is given by the harmonic mean of the series of pixels along the line element it represents. The flow is then computed in the mesh using a classical Galerkin finite-element formalism (Billaux & Gentier 1990).

3 HYDROMECHANICAL BEHAVIOR UNDER SHEARING

3.1 *Experimental results*

The experimental set-up consists of a classical shear machine, to which instrumentation for injection and a system of water recovery have been added. The general details are outlined in Gentier et al. (1997).

Fluid is recovered in eight sectors located at the periphery of the joint and weighed continuously during the injections. The conducted flow is radially divergent injection. Samples used are mortar replicas of a granite core sample containing a natural fracture at its midpoint. These samples are cylinders, 120 mm in diameter, and are cast from replicas of the natural fracture analyzed in previous work (Gentier et al. 1997). In those studies, experiments were performed for three directions of shearing (90°, 180° and 270°) for a normal stress held constant at 7 MPa. Only the experimental data corresponding to the 270° direction of shearing are presented here (Figure 6).

A closure phase, whose magnitude does not exceed twenty microns, appears at the beginning of

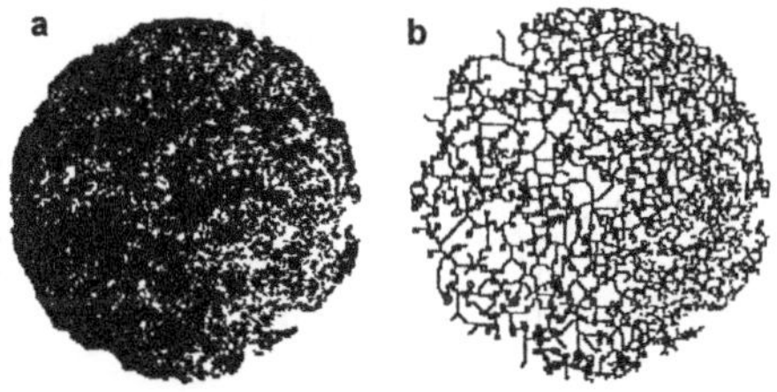

Figure 4. Binary image (a) and skeleton (b) obtained for a specified level of fracture closure.

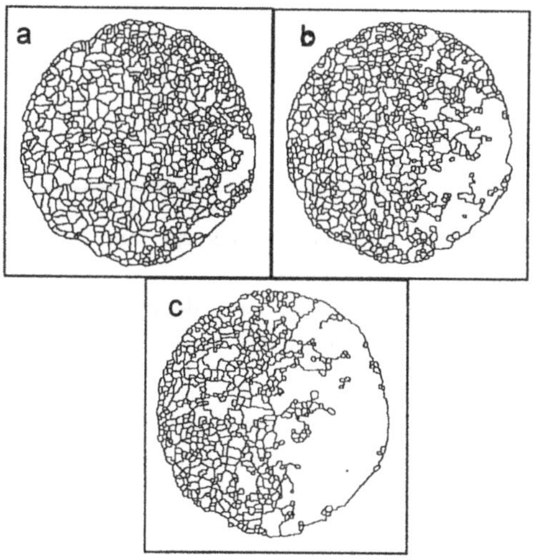

Figure 5. Evolution of the fracture skeleton for increasing levels of fracture closure: 50 μm (a), 71 μm (b) and 88 μm (c).

shearing until a tangential displacement of 0.250 mm is reached. Thereafter, dilatancy is initiated and dominates (α_{max} = 14°). The peak shear stress (8.5 MPa) appears at a displacement of 0.393 mm for this shearing direction. A phase of residual shear behavior begins at approximately 2 mm of tangential displacement. However, the dilatancy is stable at approximately 8°. In this direction, the dilatancy for large shear displacements can be simply understood as an inclination of the mean shear plane. The influence of the shear direction is presented in detail in Gentier et al. (1997). In summary, the anisotropy in shear behavior with respect to the three directions tested appears mainly in the peak shear strength (6.3 MPa in the 90° direction for 0.246 mm of tangential displacement, and 8.5 MPa in the 180° direction for 0.263 mm of tangential displacement) and pre-peak phase (closure phase between 0.180 and 0.250 mm). In contrast, the dilatancy mobilized at the very beginning of shearing does not really differ from one direction to the other (α_{max} = 14 to 16°).

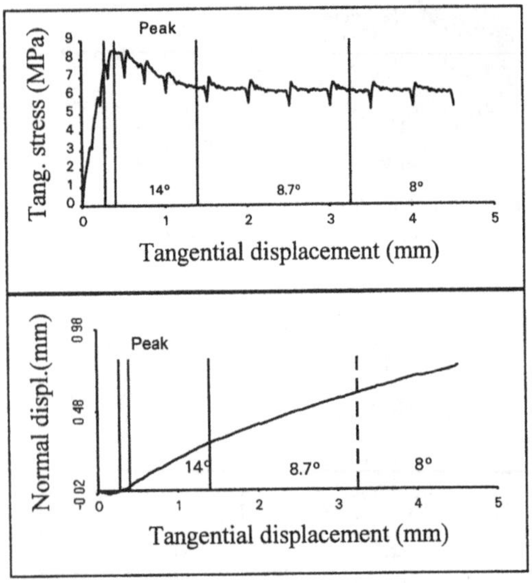

Figure 6. Mechanical behavior of the fracture (shear direction : 270° - normal stress :7 MPa).

Global intrinsic transmissivity (Fig. 7), decreases (by an order of magnitude) in the first phase of closure of the joint. After the closure phase, dilatancy is mobilized and it induces a small increase in intrinsic transmissivity until the peak is reached. After the peak, during the softening phase that is attributed to the degradation of asperities, intrinsic transmissivity increases substantially (approximately two to three orders of magnitude) and then stabilizes. For the two other directions of shearing, the global behavior is the same. For this fracture, the anisotropy does not seem to play an important role in determining the magnitude of the intrinsic transmissivity corresponding to each phase; it seems to be more closely related to the tangential displacement corresponding to variations in transmissivity.

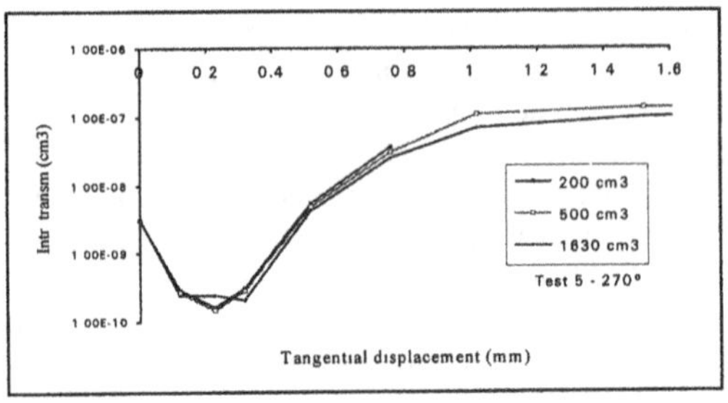

Figure 7. Evolution of global intrinsic transmissivity as a function of tangential displacement.

Beyond the global hydraulical behavior, a precise analysis of the recovery pattern and its relation to tangential displacement is possible due to the compartmental system of recovery. Figure 8 illustrates the percentage of recovery per each sector and the various recovery patterns associated with each test. The details of the results are complex, but general overall behavior can be postulated. It appears that:

- The pre-peak phase shows fluctuations in the recovery pattern linked to the initial closure phase followed by the beginning of dilatancy.
- The recovery pattern is systematically modified as the shear stress passes through the peak as compared to the recovery pattern during the corresponding pre-peak phase.
- The softening phase is characterized by a specific recovery pattern; during this phase, the fracture is more open, and the flow occurs more easily through large channels that bypass contact-damaged zones.
- The reduction in dilatancy rate associated with the beginning of the residual shear behavior is accompanied by a systematic modification in the recovery pattern. Generally, passing through peak shear strength causes a reorientation of the main recovery direction sub-perpendicular to the shear direction. The gradual passage to residual shear behavior tends to modify recovery patterns more-or-less progressively towards greatest isotropy.

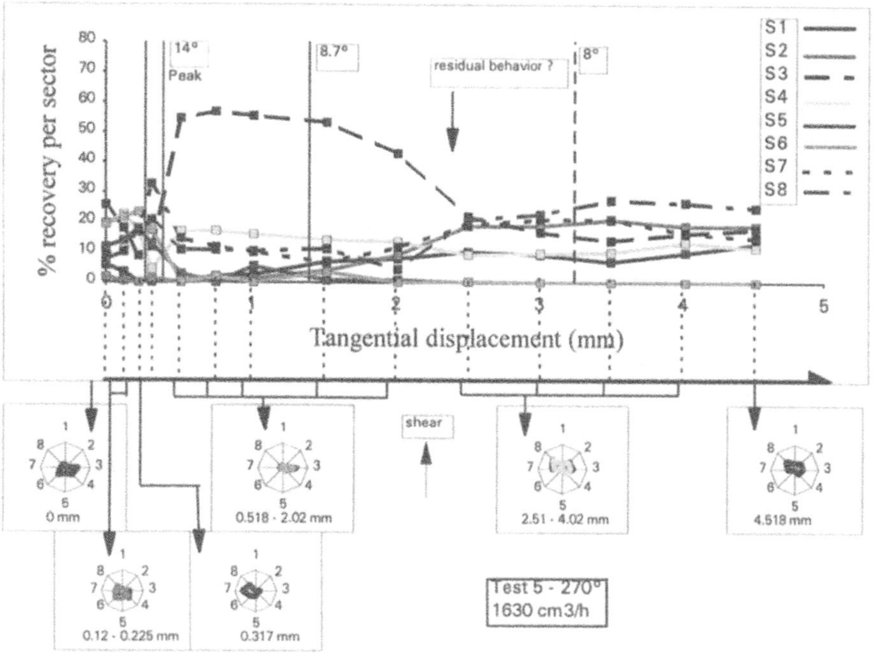

Figure 8. Evolution of the recovery percentage per sector and recovery patterns associated with each test.

3.2 *Degradation of the walls*

The characterization of damage zones occurring during shear tests with various normal stresses, shear directions and tangential displacements, is based on the acquisition of gray-level images of the joint surfaces, followed by segmentation which allows identification of the damaged zones and generation of binary images (Fig. 9). The damaged areas are identified in such a way that the damage contours can be superposed on a topographic map of the corresponding joint wall (Fig. 10). The spatial features of the fracture are studied using geostatistical methods. The experimentally derived variograms of heights and first derivatives have been fitted with three nested models. The surface is reconstructed by kriging.

For this direction of shearing, damaged areas are created in a direction globally perpendicular to the shear direction and this arrangement becomes more and more pronounced as the normal stress is increased.

3.3 *Casting of the voids*

The final step is to obtain a cast of the void space at the end of the shear test by injecting dyed silicon resin into the fracture before removing the sample from the shear box. The sample is removed when the resin has cured. An example is shown in Figure 11. The rough cast includes three kinds of information: contact areas (blank areas), aperture map of the void space (gray levels) and gouges that are included in the resin (black areas). A special image processing technique allows identification of the three different phases.

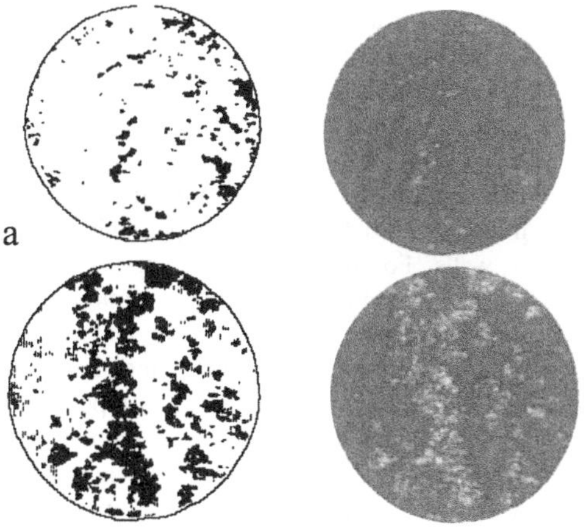

a

b → shear

Figure 9. Gray-level and binary images of the wall (shear direction : 270°, tangential displacement : 5 mm, normal stresses : 7 MPa (a) and 21 MPa (b)).

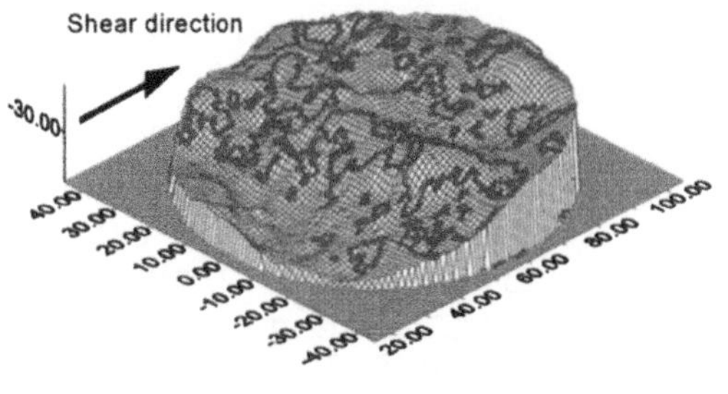

Figure 10. Damaged areas superposed on a map of the topography of the wall of the fracture reconstructed by kriging (shear direction : 270°, tangential displacement : 5 mm, normal stress : 21 MPa).

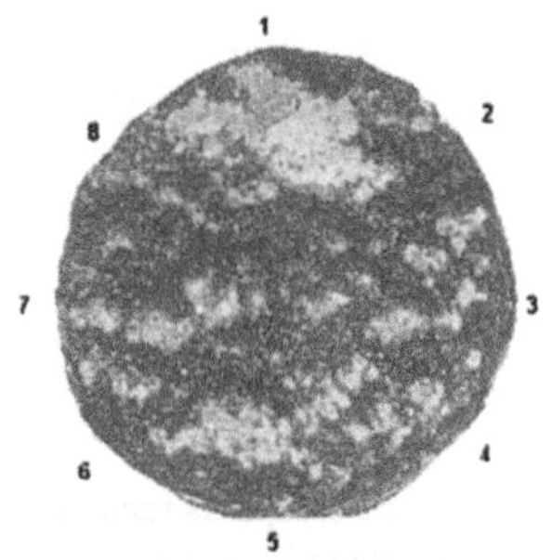

Figure 11. Example of a fracture cast made at the end of a shear test ((shear direction : 270°, tangential displacement : 5 mm, normal stress : 7 MPa).

CONCLUSIONS

The experimental and modeling methods described in this paper show how we can expect to develop an understanding of the evolution of the flow paths in a fracture as a function of the stresses it is subjected to. First, we are able to obtain a map of the apertures corresponding to zero applied normal stress, and from that we can determine the topography of the fracture walls. For the case of an applied normal stress, we are able to calculate the evolution of the contact area and deformation of the void space from the initial aperture map. Then, using the algorithms described, we are able to postulate a topology of the possible channels and their evolution with increasing normal stress. Finally, we are able to calculate flow through the resulting network. For the case of fractures under a shear stress, we are able to estimate the evolution of the damage zones as a function of the tangential displacement and the normal stress. The final casting of the voids at the end of the shear test allows us to see how the void space changes during shearing.

In future work, the evolution of the damage zones and changes in void space observed experimentally will be compared to results from the Hopkins model which will be extended to analyze shear behavior. Using the same procedure described above, it will be possible to estimate flow. The ultimate objective is to compare modeling and experimental results to validate and improve the models and help develop our understanding of the hydromechanical behavior of natural fractures.

ACKNOWLEDGMENT

This research was carried out as part of a project funded by BRGM, Orléans, France.

REFERENCES

Billaux, D. & S. Gentier 1990. Numerical and laboratory studies of flow in a fracture. In N. barton and O. Stephansson (eds.), *Proc. International Symposium on Rock Joints, Loen, Norway*. Balkema, Rotterdam, 369-373.

Gentier, S., E. Lamontagne, G. Archambault & J. Riss 1997. Anisotropy of flow in a fracture undergoing shear and its relationship to the direction of shearing and injection pressure. *Int. J. Rock Mech. & Min. Sci.*, 34:3-4, Paper No. 258.

Gentier, S. & D. L. Hopkins 1997. Mapping fracture aperture as a function of normal stress using a combination of casting, image analysis and modeling techniques. *Int. J. Rock Mech. & Min. Sci.*, 34:3-4, Paper No. 132.

Gentier, S., D. Billaux & L. Van Vliet 1989. Laboratory testing of the voids of a fracture. technical Note. *Rock Mechanics and Rock Engineering*, 22, 149-157.

Gentier, S. 1987. Morphologie et comportement hydromécanique d'une fracture naturelle dans un granite sous contrainte normale - Approche expérimentale et théorique. Document BRGM n°167, 637p.

Hopkins, D. L., N. G. Cook & L. R. Myer 1990. Normal joint stiffness as a function of spatial geometry and surface roughness. In N. Barton and O. Stephansson (eds.), *Proc. International Symposium on Rock Joints, Loen, Norway*. Balkema, Rotterdam, 203-210.

Riss, J., S. Gentier, G. Archambault & R. Flamand 1997. Sheared rock joints : Dependence of damage zones on morphological anisotropy. *Int. J. Rock Mech. & Min. Sci.*, 34:3-4, Paper No 94.

Mechanics of Jointed and Faulted Rock, Rossmanith (ed.) © 1998 Taylor & Francis, ISBN 90 5410 955 6

Experiments on hydro-thermo-mechanical and chemical parameters in a fracture

S.Gentier & P.Baranger
BRGM, Orléans, France

L.Bertrand & L.Rouvreau
ANTEA, Orléans, France

J.Riss
CDGA, Université Bordeaux 1, Talence, France

ABSTRACT: The work presented here consists of a detailed study of the relationship between fracture morphology, flow, stress and fluid-rock chemical reactions in a fracture in a granite. The objectives of this work were the design, development and experiments of percolation to prove the feasibility of such studies for the understanding of coupled phenomena. From a hydromechanical point of view, the intrinsic transmissivity of the fracture remained constant during the two percolations considered even if the flow pattern was not stable in time. All the results of the chemical analysis showed that the chemical reactions were very sensitive to the hydromechanical conditions imposed to the fracture. The analysis of the morphology of the fracture walls showed that local changes were induced by the percolation of the fluid.

1 INTRODUCTION

Tests to validate coupled computational codes, carried out as part of the CEC Chemval/Mirage program, have shown the difficulty in finding experimental results that can be used to this end. Very few experiments have been conducted, and these were not generally aimed at validating coupled computational codes. More often, only some aspects of the coupling are studied (Drew et al. 1990, Serra et al. 1997). As a result, we do not have sufficient data with which to properly constrain the models. Nevertheless, it has been through modeling that we have been able to clearly improve our capabilities for predicting the response of a rock medium to the prospective storage of high-level radioactive waste. The rise in temperature due to the presence of radioactive waste is likely to provoke dissolution-precipitation phenomena that can in turn result in either plugging or an opening of the discontinuities. The use of modeling, therefore, needs to be preceded by a phase of verification and validation, and it is precisely within this specific validation context that we conceived our experiment.

This paper consists of a detailed presentation of the laboratory experimental device designed for that specific objective and the results obtained during the first percolations tests (Gentier et al. 1996). The overall objective is to perform an experiment allowing analysis of the relationships between morphology, flow, stress and chemical water/rock reactions and by which we can expect to validate chemical - thermo - hydro - mechanical models including models explicitly integrating the morphology of the fracture void space and its variation on the scale of a single fracture. All the work has been carried out on a 12-cm-diameter core sample collected across a fracture in a granite.

2 EXPERIMENTAL DEVICE

The specially designed test equipment enables a normal stress to be applied to the mean plane of the fracture, a divergent radial injection of the fluid into the fracture plane with compartmental recovery of the fluid for a hydraulic balance and chemical analysis of the exiting percolate, and allows the fractured sample to be maintained at a given temperature. A schematic diagram of the device is shown in Figure 1.

The test equipment includes a loading frame equipped with a hydraulic jack for applying axial loads and a hydraulic device for controlling the jack pressure. Normal stress can be applied in steps up to 15-20 MPa. The injection circuit is composed of an injection head located at the base of the sample and a HPLC pump that provides flow rates ranging from a few cm^3/h up to 600 cm^3/h. A thermostatic bath is inserted between the pump and the injection head to adjust the fluid to the desired temperature before it enters the fracture. The test cell ensures a controlled temperature throughout the percolation tests and a slight over-pressure around the sample to avoid evaporation of the fluid. The maximum possible temperature is 120°C, but the presently available over-pressure (<10 kPa) restricts the maximum usable temperature to 80°C to avoid boiling of the fluid. A fluid recovery device incorporates a fluid reception tank divided into four 90° compartments (or sectors) around the fractured sample. Each compartment overflows into the recovery circuit, itself connected to a weighing system. This allows a directional study of the flow and allows us to monitor any changes in direction over the course of the experiment.

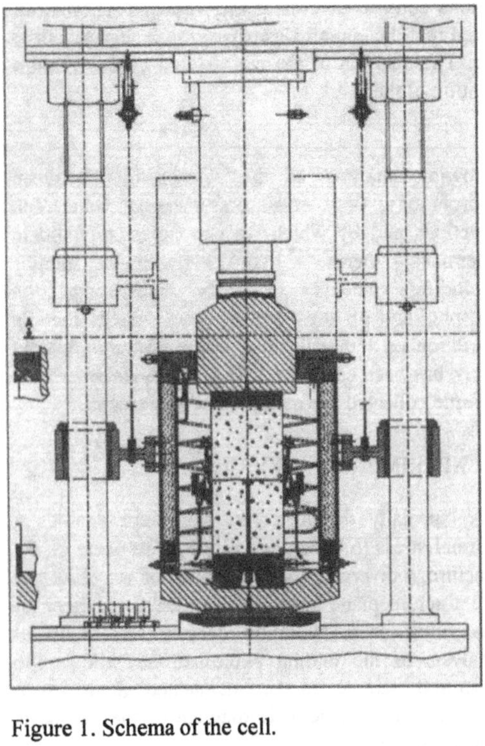

Figure 1. Schema of the cell.

Data are recorded continuously to monitor the mechanical and hydraulic response of the test specimen. The axial force applied to the sample is recorded, along with the closure of the fracture as measured by four linear displacement transducers (LVDT) placed around the fracture. Hydraulic measurements include the injection flow rate, the injection pressure, and the fluid mass recovered in each compartment (or sector). The fluid temperature on injection, the fluid temperature at the central injection point in the fracture, the exiting fluid temperature in each of the four recovery compartments, and the cell temperature (at the bottom and top) are also monitored. Samples of the exiting fluid are collected regularly from the recovery system for chemical analysis (major elements).

3 INITIAL CHARACTERIZATION OF THE FRACTURE

The main objective of the initial characterization is to establish a reference base for the fracture morphology, making it possible to estimate the effects of the chemical interactions between the percolating fluid and the walls of the fracture.

3.1 Microroughness analysis

With respect to the objective of analyzing the effect of chemicals, the scale of characterization of the morphology had to be much finer than is customary in rock mechanics. To be able to record modifications of the topography due to dissolution or precipitation, characterization at the micron scale was imperative. Few industrial devices are available for this case because of the size and weight of each half sample. A device designed specifically to carry the sample and to reassemble it very precisely in the same position before and after percolations has been built. The repeatability of the profile is approximately 5 µm. Twenty eight profiles have been recorded (Fig. 2).

3.2 Petrographic map

Subsequent to the morphological analysis, the minerals on the surface of the wall of the fracture were mapped as shown in Figure. 2 (Wu 1995).

4 CONDITIONS IMPOSED DURING THE PERCOLATIONS

The goal of the two percolations described here was to test the influence of the flow rate, the temperature and the chemistry of the fluid injected (acidified water : HF 10^{-3} M and 10^{-2} M). We have used such an acidified fluid to enhance the chemical reactions between fluid and fracture wall.

During the first percolation, the fluid injected was alternatively demineralized water and acidified water (pH = 3). The normal stress was maintained constant at 6 MPa during the percolation. The flow rate was 0.28 ml/mn (17ml/h), and the temperature was the room temperature which varied between 21 and 23°C. The duration of the percolation was eight days.

During the second percolation, acidified water (pH = 2) was injected continuously. The normal stress applied was the same as for the previous experiment (6 MPa) and two flow rates were imposed (0.67 ml/mn and 0.2 ml/mn). For each of these flow rates, two temperatures were imposed (room temperature and a temperature of 35°C). The duration of the percolation was fifteen days.

The analyses performed were for six major elements : sodium, potassium, magnesium, calcium, aluminum and silica. These analyses were made by atomic absorption spectrometry, except for silica which is determined by colorimetry.

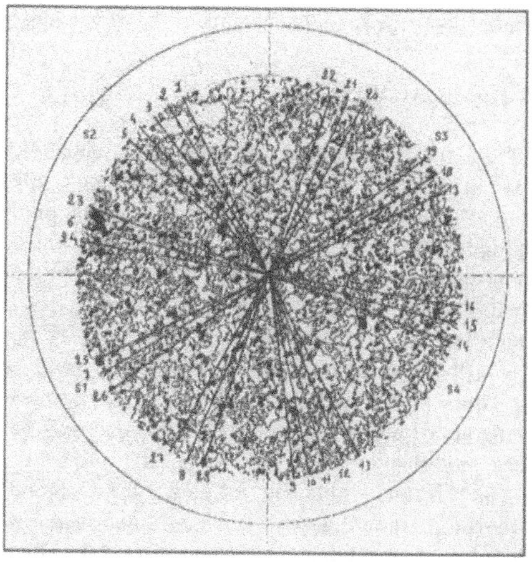

Figure 2. Mineralogical map of a side of the fracture and location of the profiles recorded for the micrometric analysis of the fracture sides.

5 HYDROMECHANICAL BEHAVIOR

Before applying the imposed constant normal stress and beginning percolation, a series of load cycles are applied to ensure that the fracture is well mated. This consists of the application of cycles of loading and unloading with increasing loads. During the last loading cycle an injection is performed to quantify the hydromechanical behavior of the fracture and to provide a reference for comparison from one percolation to the other (Fig. 3). The curves corresponding to the intrinsic transmissivity as a function of the normal stress all exhibit the same behavior. At first, in the lower range of stresses, the intrinsic transmissivity decreases quickly with increasing normal stress, then decreases less and less quickly reaching a quasi stabilized value for normal stresses in the upper range of applied stresses. However, the curve obtained after the first percolation and just before the second percolation show higher intrinsic transmissivities for normal stresses lower than 5 MPa.

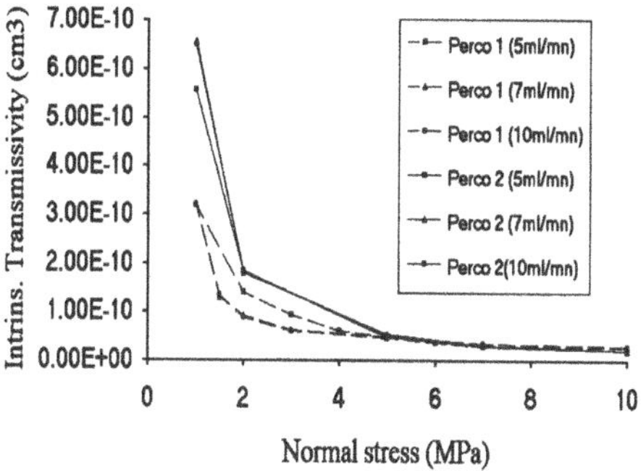

Figure 3. Evolution of the intrinsic transmissivity as a function of normal stress and of flow rate before each percolation.

After having matched the fracture surfaces, the normal stress is applied and maintained constant throughout each percolation. The mechanical behavior of the fracture was consistent through the two percolations. The curves of fracture closure as a function of the normal stress are reproducible.

During the first percolation which consists of a succession of injections of acidified water followed by injections of demineralized water to clean the fracture, the results show a spatial instability as well as pressure. After reaching a steady state, the injection pressure oscillates. Simultaneously, variations occur in the recovery of the various sectors. However, no geometrical variation is observed in the global closure of the fracture. The precipitation of amorphous chemical compounds (gels of aluminum or silica) due to local variations in pH could be responsible for temporary plugging of channels that could cause the observed fluctuations in the injection pressure.

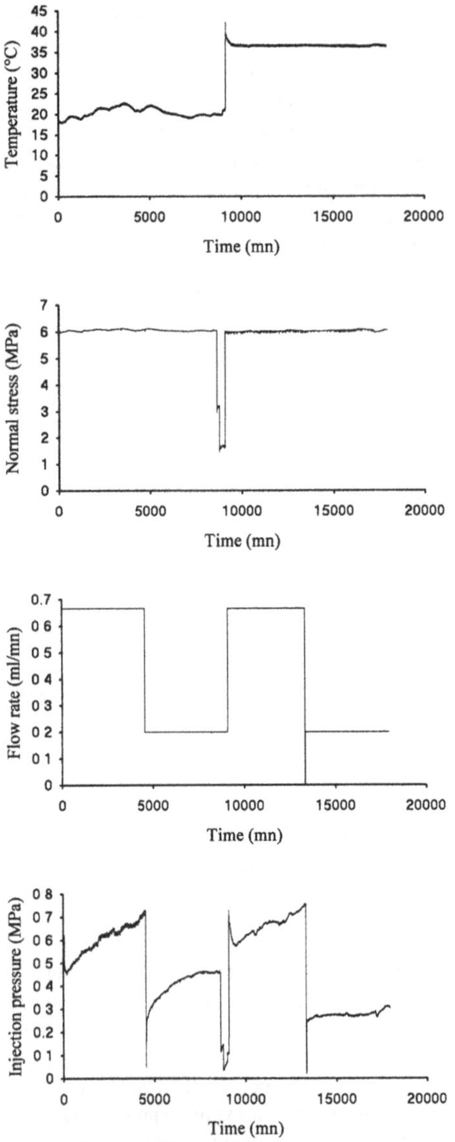

Figure 4. Hydro-thermo-mechanical conditions during the second percolation.

The various phases of the second percolation are summarized in Figure 4. During the second percolation, performed with acidified water (HF 10^{-2} M), the hydraulic behavior was more stable (as indicated by an absence of precipitation of amorphous chemical compounds due to a lower pH). For a flow rate of 0.67 ml/mn, the injection pressure did not stabilize definitively. After having reached a slight plateau, pressure increased continuously during the two corresponding phases. The temperature did not seem to influence significantly the range of observed pressures. For a flow rate of 0.2 ml/mn, the injection pressure stabilized after a relatively long time and remained constant during the two phases of injection. However, the injection pressure measured during the second phase of injection at this flow rate is much lower than that measured during the first phase of injection. This may be due to the influence of the temperature. The exact role of the temperature is difficult to estimate because the parameter acts simultaneously on the viscosity of the fluid, on the opening of the fracture and on the chemical dissolution of minerals which can cause an increase in permeability. The effect of the decreasing of the viscosity with temperature can be estimated to approximately 60%. But this effect was not observed for the other flow rate.

However, even if the range of the intrinsic transmissivity is similar, the recovery of the fluid in the various sectors is not constant over a given percolation or from one percolation to another. This demonstrates the spatial instability of the flow paths.

6 CHEMICAL RESULTS

For the first percolation where the main objective was to specify the experimental protocol, the analyses of the major elements in the exiting fluid highlights the major chemical phenomena which affect the fracture. The influence of cleaning of the fracture with demineralized water is not fully quantified but it seems that its presence may be responsible for the increase in pH that is favorable for the precipitation of chemical amorphous compounds (gels of aluminum or silica) that may be cause modification of the flow channels .

The results obtained during the second percolation show clearly that any modifications (flow rate, stress, temperature) that disrupt the flow inside the fracture find expression immediately in the chemical response. This phenomenon is illustrated in Figure 5 by the evolution of the concentration of aluminum and potassium. The influence of the flow rate is particularly clear as is the influence of the temperature. As expected, the concentrations increase when the flow rate decreases and when the temperature increases. These modifications have repercussions directly on the rate of dissolution of the minerals in contact with the fluid, and the flow velocity of the fuid. The ratio

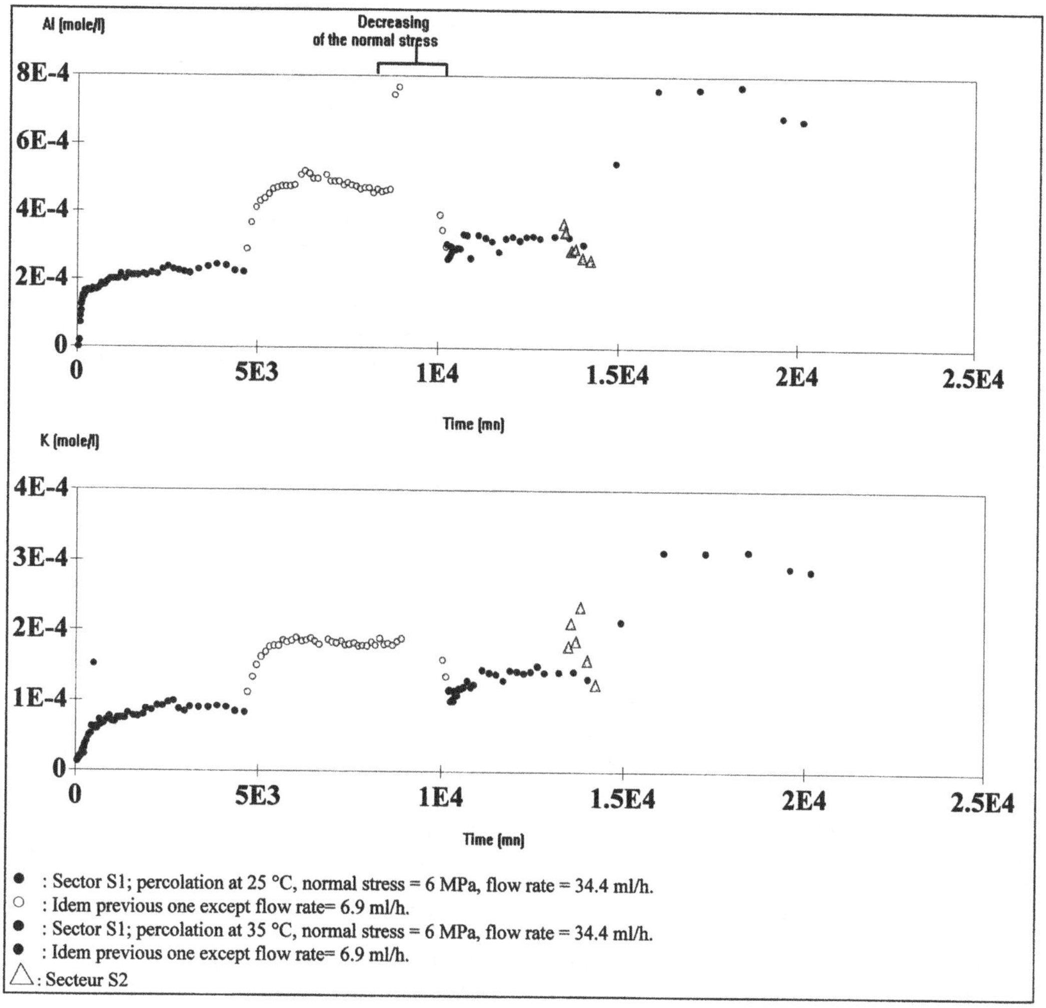

Figure 5. Evolution of the concentration of Aluminum and Potassium during the second percolation.

between these two parameters is essential for the system because it can cause opening or closing of the fracture by chemical dissolution/precipitation processes. Local fluctuations of hydro-thermo-mechanical and chemical conditions, not recordable at the global scale, can be observed thanks to variations in the chemical composition of the fluid analyzed at the exit points of the fracture. The measurement of these variations constitutes an efficient means of estimating these local fluctuations. The access of the fluid to areas richer in some minerals (e.g. feldspars, ...) could explain the main variations observed during the different phases.

7 MORPHOLOGY OF THE FRACTURE SUBSEQUENT TO PERCOLATIONS

Following the percolations, the profiles defined in Figure 2 were re-recorded. The comparison between the profiles recorded before and after the percolations show that fluid circulation seems to have locally modified the topography of the walls of the fracture. An example of these profiles are shown in Figure 6. The changes in height are due to dissolution for most cases. The magnitude of the differences between the initial profiles and the final profiles is on the order of tens of microns for the mean difference.

Superposition of the microroughness profiles recorded before and after percolation with the petrographic map confirms the high reactivity of HF towards quartz and feldspars.

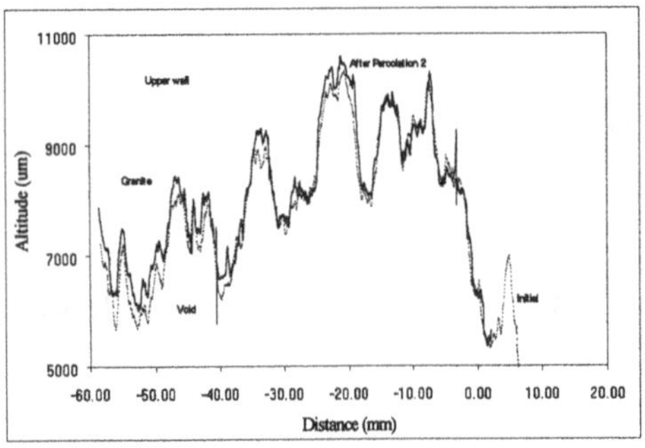

Figure 6. Example of a profile recorded before and after the two percolations.

CONCLUSION

The main conclusions derived from the two percolations are the following :

- the chemical reactions are extremely sensitive to the hydrodynamic conditions existing in the fracture -- the velocity of the fluid flow directly influences the degree of dissolution of the minerals that are in contact with the fluid;
- the percolations of acidified fluid induce local modifications of the topography of the walls of the fracture. The preferential dissolutions of quartz and feldspars by HF are observed and can be identified by the microroughness analysis ;
- the role of the temperature is difficult to estimate because the temperature has a mixed influence on mechanics and hydraulics, and chemical parameters of the system -- specific tests must be done to better determine the role of temperature.
- the observed variations in the hydraulic behavior do not modify the mechanical behavior.

In conclusion, these experiments show that strong coupling exists between the hydraulics and the chemistry when the equilibrium between the percolating fluid and the percolated medium is modified. Hydraulic behavior is strongly coupled with the mechanics which modifies the geometry and topology of the flow paths. The coupling of these phenomena are complex. The experiments performed to date indicate that the various phenomena (mechanical, hydraulic, thermal and chemical) can be discriminated and analyzed. The device used to perform these experiments seems reliable and allows measurements of data needed to validate models. However, the procedure can be improved, in particular, to better characterize the influence of the chemical composition of the fluid injected. Regarding the chemical analysis, we are not yet able to link all the observations done on the global chemical composition of exiting fluid to the chemical reactions occurring in the fracture. One way to improve the analysis would be to identify the main flow paths, as deduced from the map of the void space, and superposed on the map of the minerals to estimate the surface area of each mineral phase in contact with the fluid in various directions of flow.

ACKNOWLEDGMENT

This research was carried out in the framework of a project funded by BRGM and the Commission of the European Communities (DGXII).

REFERENCES

Drew, D. W., D. M. Grondin & T. T. Vandergraaf 1990. The large block radionucleide migration facility. Atomic Energy of Canada Limited, TR-519.

Gentier, S., P. Baranger, L. Bertrand, L. Rouvreau & J. Riss 1996. Expérience d'écoulement dans une fracture pour la validation des modèles couplés chimie-hydro-thermo-mécanique en milieu fracturé. Collection Sciences et techniques nucléaires, Commission Européenne, EUR 17123 FR.

Serra, H., C. Guy & J. Schott 1997. Etude de la percolation de fluide à travers de l'Hawaite de Mururoa fracturée artificiellement. Expérimentation et modélisation de l'interaction eau/roche. Séance spécialisée de la Société Géologique de France : Hydrodynamique et interaction fluides-roches dans les roches poreuses et fracturées, Montpellier, pp 153-154.

Wu, J. 1995. Description quantitative et modélisation de la texture d'un granite: granite de Guéret (France). Thèse de l'université Bordeaux 1. 255 p.

Mechanics of Jointed and Faulted Rock, Rossmanith (ed.) © *1998 Taylor & Francis, ISBN 90 5410 955 6*

What is the usefulness of Fluid Inclusion Planes (FIP)?

M. Lespinasse
UMR 7566 G2R, University HP Nancy 1, Vandœuvre, France

ABSTRACT: The Fluid Inclusion Planes (**FIP**) are mode I cracks which occurs in sets with a predominant orientation perpendicular to the minimal stress axis σ3. They results from the healing of open crack and appear to be fossilized fluid pathways. One can use the physico-chemical differences in the included fluids to separate different sets of FIP. On the other hand, it should be possible to use FIP geometry to relate the differente stages of fluid percolation to a regional succession of deformational events. Therefore, the systematic measurements of microstructural marker orientations together with detailed fluid inclusion and host mineral characterization may results in the determination of fluid pathways as a function of the changes occuring in the stress field.

1. INTRODUCTION.

Most fluid migrations in rocks are favoured by fissure permeability which forms during brittle deformation. The deformation is in some instances related to fluid pressure and movement of faults. Evidences of paleofluid migration through the fractured rock may be very scarce, whatever the observation scale, when little change occurs in the mineral assemblages resulting from fluid-rock interactions (dissolution, alteration, new cristallization). The best record of formed fluid percolation are paleofluids trapped as fluid inclusions in healed microcracks of the rock forming minerals or within the infilling of microstructures (the Fluid Inclusion Planes, **FIP**). However, the repeated microfracturing and healing of the rock forming minerals yield complex superimposed patterns of healed microcracks. Such patterns are often difficult to interpret due to the lack of suitable chronological criteria. These problems have been recently documented and solved by coupling deformation studies, detailed examination at all scales of the relationships between trapped fluids and their host structures, and studies of fluid inclusions.

2. WHAT ARE THE FIP?

The FIP result from the healing of former open cracks and appear to be fossilized fluid pathways (review in Roedder, 1984), (**Figure 1**). Microcracks should provide valuable information about the local stress in rocks and can be assimiled as (σ1 - σ2) planes (Tuttle, 1949, Lespinasse and Pêcher, 1986; Kowallis et al., 1987; Boullier and Robert, 1991).

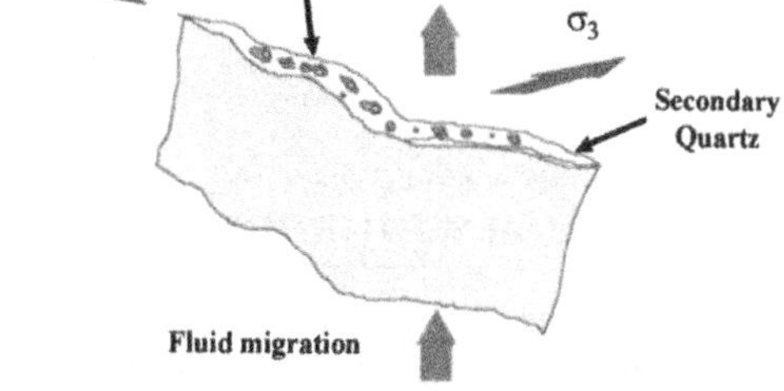

Figure 1: Description of a Fluid Inclusion Plane (FIP). After the formation of a mode I crack, a fluid is percolating through the crack and is trapped as secondary Fluid Inclusion and form a FIP

The FIP are mode I cracks which occurs in sets with a predominant orientation perpendicular to the minimal stress axis σ3. These mode I cracks are propagating in the direction which favors the maximal decrease of total energy of the system (Gueguen and Palciauskas, 1992). They do not disrupt the mechanical continuity of mineral grains and do not exhibit evidences of displacement contrarly to the mode II and III. The FIP are mainly characterized in minerals from which cristals may crack according to the regional stress field, independently of their cristallographic properties (as demonstrated for quartz by Lespinasse and Cathelineau, 1990), and may easily trap fluids as fluid inclusions when healing. In other minerals (carbonates, feldspars), the fluids are not always preserved from further disturbances and cracks display more complex patterns resulting from the presence of easy cleavages, subgrain boundaries or twin planes. The rate of healing is short in quartz (compared to geological times) as shown by Brantley (1992).

Frequently FIP form well defined networks which permitt an elaboration of a chronology (Figure 2). After a first generation of FIP formation, another crack family can be formed with trapping of a second fluid. This second generation of FIP crosscut generaly the first one. Thus, one can admitt that the FIP are good records of successive episods of crack initiation and fluid migration (Pecher et al., 1985).

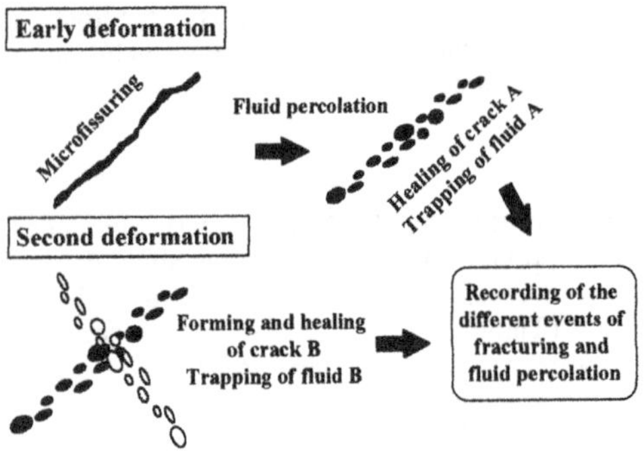

Figure 2: FIP are good records of successive episods of crack initiation and fluid migration

3. HOW TO ANALYSE THE FIP?

The methodology used here links paleostress data obtained from field studies and the microfracture network features (Lespinasse, 1991). Systematic three-dimensional (3 D) measurements of microfractures or FIP are carried out on orientated wafers with a Universal stage, and an image analyzer (Lapique et al., 1988). In this analysis, we determined the FIP features (orientations, dips and length of each healed crack family) (Figure 3). Data are generally collected in quartz grains. Results are presented on stereograms. Microthermometric properties of each FI can be related to a specific geometry and finally to a $\sigma 3$ direction (FIP are mode I cracks). Fluid inclusions are systematically studied in healed microcracks of known orientation from the granite quartz grains using oriented fluid inclusion wafers (200 µm thick). Microthermometric characterization of the inclusion fluids are carried out using a heating-freezing Chaixmeca stage (Poty et al., 1976).

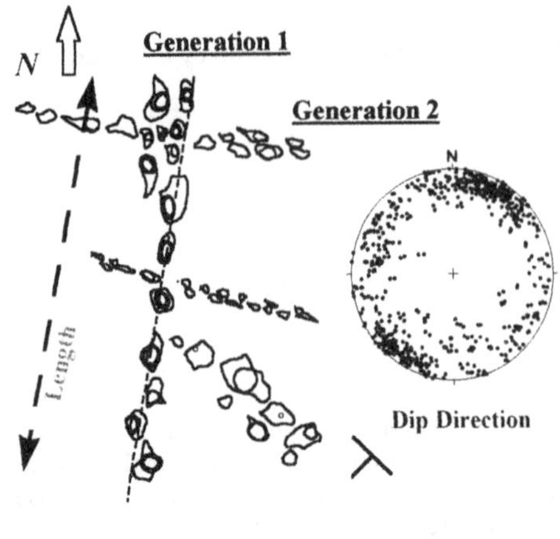

Figure 3: Presentation of the different parameters collected for each FIP family. Geometrical parameters (length, thickness, Dip Direction) are collected by using an image analyser and presented on stereographic projections.

4. WHY FIP ARE USEFULLNESS?

4.1. Withness of stress orientation.

A study of the brittle deformation of the La Marche granite (NW french Massif Central) indicates a relationship between paleostress field and the geometry of the FIP (Lespinasse and Pecher, 1986) (Figure 4).

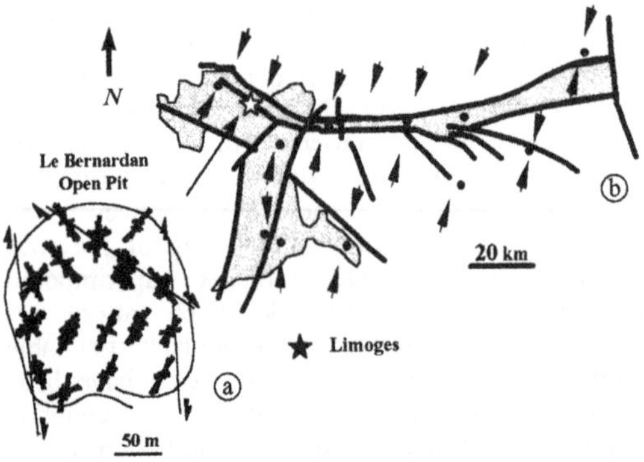

Figure 4: Relationships between FIP orientations and regional paleostress field. (a) Sketch map of FIP orientations (rose diagrams) in the Le Bernardan open pit. The main orientation is NNE-SSW. (b) Regional NNE-SSW compression in the La Marche zone (NW french Massif Central) Arrows show the direction of the maximum principal stress $\sigma 1$ at each sampling site. $\sigma 1$ was determined by striated fault planes analysis (Lespinasse and Pecher, 1986)

A statistical study of the distribution of the FIP in the Le Bernardan open pit leads to the following conclusions. FIP exhibits several distinct preferred orientations on the scale of a grain, which may be observed in many samples; the orientations of the FIP are similar to those of micro- and mesoscale fractures in the granite; the dominant FIP direction is parallel to the main direction of regional shortening (NNE-SSW). Thus, FIP can be used as microstructural markers of paleostress fields like tension gashes.

More complex patterns are usually found within folded metamorphic rocks where the mechanical discontinuities (fold, for instance) induce local stress reorientations (Cathelineau et al., 1990). In addition, early microfissuring in metamorphic quartz is in part erased by quartz recristallization, as hidden by late microfissuring associated with retrograde metamorphism (Alvarenga et al., 1990).

4.2 Fluid pressure in fault systems and Paleostress quantification

One can use the physico-chemical differences among the inclusion fluids to separate different sets of FIP ; in the other hand, it should be possible to use the FIP geometry to relate the different stages of fluid percolation to a regional succession of deformational events. Applications are important in the reconstruction of paleofluid pressure and stress quantification.

Paleostress magnitudes can be estimated by using fault slip data, rupture and friction laws for dry conditions (Angelier, 1989). However, their estimation is difficult if fluids are present during deformation, the lithostatic load and the fluid pressure being usually unknown. Lespinasse and Cathelineau (1995) have shown that the quantitative estimation of the lithostatic load and the fluid pressure during a tectonic event can be derived from paleofluid analysis in fluid inclusion planes (FIP). As the FIP are healed mode I cracks (Lespinasse et Pêcher, 1986; Lespinasse et Cathelineau, 1990; Cathelineau et al. , 1993) , oriented in a consistant manner relative to regional or local structures, stress and fluid features may be obtained for a given deformation event (Pecher et al. 1985).

This approach has been applied to a fault system which affects an Hercynian granite of the NW French Massif Central (Leroy, 1978).

A NW-SE compression has been defined from a population of 51 faults characterized by orientations around N60°E to N110°E for dextral strike slips and N135°E to N175°E for sinistral movements. The stress ratio has been determined with fault slip data around 0.52 ± 0.08 (Etchecopar et al., 1981) and by using Angelier's method, the ratio Y (σ_3 / σ_1) was estimated around 0.27 ± 0.03 and the friction coefficient (μ) around 0.58 ± 0.1. The dominant FIP trend is NW-SE, vertical or dipping toward the SW (Figure 5). The fluid inclusions from NW-SE FIP are characterized by homogenization temperatures with a mode around 300°C and ice melting temperatures with a mode around -1.0°C.

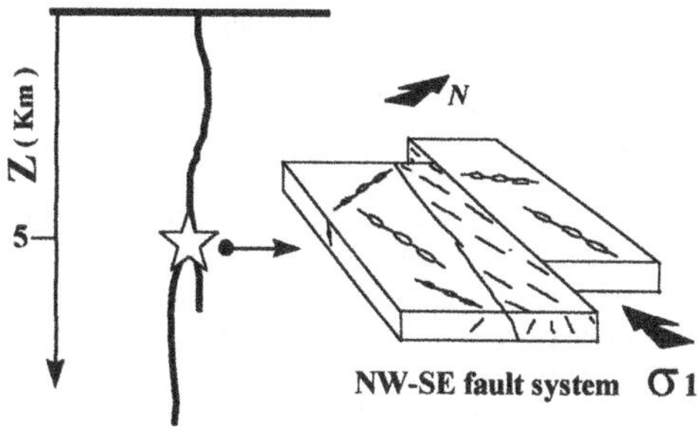

Figure 5: Schematization of the Margnac 805 sinistral fault zone. FIP are essentially oriented NW-SE parralel to the fault They have been formed during the senestral shearing of the fault.

From these data, a reconstruction of the P-T conditions of the fluid migration has been carried out. Since fluid composition (H_2O - x NaCl) and minimum trapping temperatures are known, the fluid density can be determined, and representative isochores for the studied fluid can be drawn (Figure 6).

The possible P-T pairs of the trapping conditions may be estimated making the following assumptions:

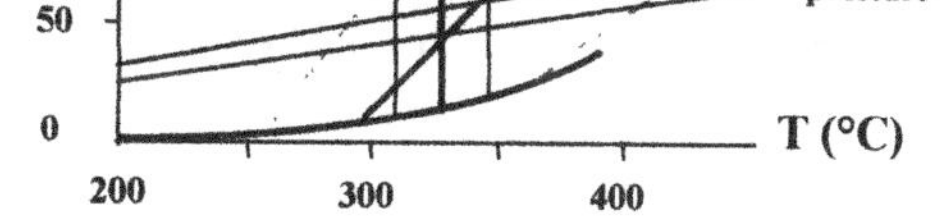

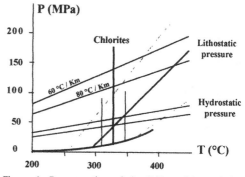

Figure 6: Reconstruction of the P-T conditions of fluid migration, with the representative isochores (isochore for average fluid features (solid lines), and two extrema (dashed lines), the fossil geothermal gradients is in the range (b) 60° to (a) 80°km^{-1} for hydrostatic and lithostatic conditions Trapping pressure is determined in considering, hydrostatic fluid pressures and chlorite (Chl) formation temperature estimated at 330° ± 20°C

1. Fluids from the Margnac area are considered as hydrothermal solutions from a fossil geothermal system characterized by geothermal gradients in the range 60°-80°C/km (Leroy, 1978).

2. Pressures may be considered as hydrostatic because the fluids are migrating in open spaces (faults and quartz-depleted granites of high permeability) in this hydrothermal system characterized by high water/rock ratios (Turpin, 1984). In addition, early migration of vapors confirms the relatively low fluid pressures typical of geothermal systems and therefore the hydrostatic regime of this hydrothermal activity.

3. The use of chlorite geothermometry (Cathelineau and Nieva, 1985) and crystal-chemical data obtained on chlorite associated with the alteration process yields a temperature estimate of 330° ± 20°C (Cathelineau, 1987).

These assumptions yield estimates of the most probable range of pressures during fluid migration. Figure 8b shows that fluids in NW cracks are trapped under pressures which can be estimated at about 50 ± 10 MPa. Considering that fluids are trapped in faults under hydrostatic conditions (water density of 1000 kg /m^3, neglecting the density changes with increasing temperature), the trapping depth of the major fluid migration (also present in NW-SE cracks) responsible for quartz leaching can be estimated at about 5 km. These data are in agreement with previous reconstructions (Leroy, 1978). Thus, assuming a vertical column of granite (density of 2700 kg/m^3), $\sigma_v = 132 \pm 10$ MPa.

The presence of fluid in rocks during deformation can drastically change the conditions of rupture in fault systems and also the stress magnitude. In a fluid-saturated rock, the effective stress is given by $\sigma'_n = (\sigma_n - P_f)$ (Hubbert and Rubey, 1959). The fluid pressure at a depth z in a rock mass of average density ρ can be defined in relation to the overburden pressure (vertical stress) σ_v by means of the pore fluid ratio: $\lambda_v = P_f/\sigma_v$ (Sibson, 1981, 1989). Thus, the effective overburden can be written:

$$\sigma'_n = (\sigma_n - P_f) = \rho g z\,(1 - \lambda_v). \qquad (1)$$

where g is the acceleration of gravity and z is the depth. The magnitude of each effective stress axes can be expressed as follows:

$$\sigma'_1 = \sigma_1 - P_f > \sigma'_2 = \sigma_2 - P_f > \sigma'_3 = \sigma_3 - P_f \qquad (2)$$

These expressions describe the relationships between fluid pressure and stresses. When pore spaces are interconnected to the surface, $P_f = \rho g z$, and a state of hydrostatic fluid pressure prevails with $\lambda_v = 0.4$. However, if P_f is lithostatic, λ_v approaches 1 and the vertical effective stress (σ'_v) is zero (σ'_v= 0). In an "Andersonian stress state" (Anderson, 1951), during a strike slip regime of faulting, the relationships between stress axes can be expressed as follows (Yin and Ranalli, 1992):

$$\sigma_1 - \sigma_3 = [2\mu\rho g z(1-\lambda) + 2C]/[(\mu^2+1)^{1/2} + \mu(2R-1)] \qquad (3)$$

where μ is the static coefficient of rock friction and C is cohesion. μ is related to the angle θ between the plane of failure and σ_1 by:

$$\theta = 1/2 \tan^{-1}(1/\mu). \qquad (4)$$

In the case of fault reactivation, one can consider that the cohesion of the rock is very poor and tends to zero (C = 0); then equation (3) becomes

$$\sigma_1 - \sigma_3 = 2\mu\rho g z(1-\lambda)/[(\mu^2+1)^{1/2} + \mu(2R-1)] \qquad (5)$$

These relations show that stress intensities inferred from this type of failure analysis depend strongly on the fluid pressure. In this study, we determine the fluid pressure and, consequently, the vertical stress magnitude, by the analysis of paleofluids trapped as fluid inclusions in minerals during the brittle deformation event.

In spite of the uncertainities in the values of parameters such as Pf, σ_v, R and μ, a quantification of stress magnitudes has been attempted. Values of σ_v, R, and μ considered in the calculations are 132 ± 10 MPa, 0.52 ± 0.08, and 0.58 ± 0.08, respectively. The vertical effective stress magnitude depends strongly on the fluid pressure. Therefore further calculations have been carried out taking into account the data derived from the FIP P-T reconstruction.

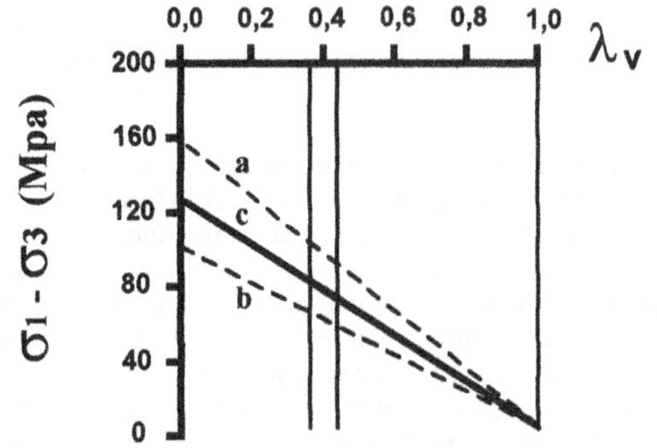

Figure 7: Determination of the stress magnitude differences (σ_1-σ_3) on a function of the pore pressure factor (λ_v). Each line (a, b, c) is calculated with different values of R and μ. Line a, R = 0 52, μ = 0.58. Line b, R = 0.46 and μ = 0.66 Line c, R = 0.64 and μ = 0 50. The hydrostatic domain correspond to λ_v within the 0.37-0.43 range assuming that rock density in the upper crust may vary from 2300 to 2700 kg/m^3.

Relations between the fluid pressure and the stress differences ($\sigma_1 - \sigma_3$) are based on equation (4). Uncertainities in the values of R and μ yield to the uncertainity domain included within two extreme lines.

Schematic representation of the relations (Figure 4) suggests the following (1) for a lithostatic fluid pressure ($\lambda_v = 1$), the effective stresses are equal to zero; (2) for $\lambda_v = 0$ (dry conditions), the values of $\sigma_1 - \sigma_3$ are in the range of the results obtained from fault slip inversion; (3) for hydrostatic conditions ($\lambda_v = 0.4$), $\sigma_1 - \sigma_3$ values are in the 70-105 MPa range.

5. CONCLUSION

The following conclusions can be drawn from this study:

1. The tested method links data sets generally obtained and interpreted separately. It clearly shows that the use of the FIP data linked to paleostress analysis may give important constraints (fluid pressure and the vertical load) on stress magnitude reconstruction for specific tectonic events.
2. Fluid inclusion planes are excellent microstructural markers and may give quantitative data on the geometry of the open permeability and on the geometry of the migration of fluids during specific tectonic events. In addition, the interpretation of FIP data may yield a pore pressure and vertical load estimates for a specific tectonic event.
3. The application of the method to a study site has given a rough estimate of stress magnitudes under wet conditions, a data extremely difficult to obtain from standard studies, especially when it concerns paleotectonic events. Different parameters such as the stress ratio R and the friction coefficient μ are not well constrained because they depend on the rock properties, on the nature of the faults, and on the computer solution (especially for the ratio R). However, their extreme values can be taken into consideration in the calculations to bracket the estimates. In addition, it is assumed in the calculations that faults are of "Andersonian type".

REFERENCES

Alvarenga, C., Cathelineau, M. & J. Dubessy 1990. Chronology and orientation of N2-CH4, CO2-H2O H2O rich fluid inclusion trails in intrametamorphic quartz veins from the Cuiaba gold district, Brazil. *Min Mag*, 54: 245-255.

Anderson, E. M., *The dynamics of faulting*, Oliver & Boyd, Edinburgh, 1951.

Angelier, J. 1989. From orientation to magnitudes in paleostress determinations using fault slip data, *J. Struct. Geol.*, 11: 37-50.

Boullier, A. M., & F. Robert. 1991. Paleoseismic events recorded in Archean gold-quartz vein networks, Val d'Or, Abitibi, Quebec, Canada, *J. Struct. Geol.*, 14: 161-179.

Brantley, S. 1992. The effect of fluid chemestry on microcracks lifetimes. *Earth Planet. Sc. Let.*, 113: 145-156.

Cathelineau M., MC. Boiron, S. Essarraj, J. Dubessy, M. Lespinasse , & B. Poty. 1993. Fluid pressure variations in relation to multistage deformation and uplift: A fluid inclusion study of Au-quartz veins, *Eur. J. Mineral*, 5: 107-121.

Cathelineau, M., Lespinasse, M., Bastoul, A., Bernard, Ch., & J. Leroy. 1990. Fluid migration during contact metamorphism: the use of oriented Fluid Inclusion Trails for a time/space reconstruction. *Minl Mag.* 54: 169-182.

Cathelineau, M.. 1987. Les interactions entre fluides et roches: thermométrie et modélisation. Exemple d'un système géothermique actif (Los Azufres, Mexique) et d'altérations fossiles dans la chaine varisque), Doct. thesis, 503 pp., Inst. Nat. Polyt. Lor., Nancy, France.

Cathelineau., M., & D. Nieva. 1985. A chlorite solid solution geothermometer. The Los Azufres (Mexico) geothermal system, *Contrib. Mineral. Petrol.*, 91: 235-244, 1985.

Etchecopar, A., G. Vasseur, & M. Daignières. 1981. An inverse problem in microtectonic for determination of stress tensor from fault striation analysis, *J. Struct. Geol.*, 3: 51-65.

Guéguen Y. & V. Palciauskas. 1992. *Introduction à la Physique des Roches*. Hermann, Paris, 299 p.

Hubbert, M. K., & W. W. Rubey. 1959. Role of fluid pressure in the mechanics of overthrust faulting, *Geol. Soc. Am Bull.*, 70: 115-205.

Kowallis, B. J., H. F. Wang, & B. A. Jang. 1987. Healed microcrack orientations in granite from Illinois borehole UPH-3 and their relationship to the rock's stress history, *Tectonophysics*, 135: 297-306.

Lapique, F., M. Champenois, & A. Cheilletz. 1988. Un analyseur vidéographique interactif description et applications, *Bull. Mineral.*, 6: 258-263.

Leroy. J. 1978. The Margnac and Fanay uranium uranium deposits of the la Crouzille District (Western Massif Central, France): Geologic and fluid inclusions studies, *Econ. Geol.*, 73: 1611-1634.

Lespinasse, M. & A. Pecher. 1986. Microfracturing and regional stress field: a study of preferred orientations of fluid inclusion planes in a granite from the Massif Central, France, *J. Struct Geol.*, 8: 169-180.

Lespinasse, M. & M. Cathelineau. 1990. Fluid percolations in a fault zone: A study of fluid inclusion planes (FIP) in the St Sylvestre granite (NW French Massif Central), *Tectonophysics*, 184: 173-187.

Lespinasse, M. & M. Cathelineau. 1995. Paleostress magnitudes determination by using fault slip and fluid inclusions planes (FIP) data. *Journal of Geophys Res,* 100: 3895-3904.

Pecher. A., M. Lespinasse, & J. Leroy. 1985. Relations between fluid inclusion trails and regional stress field: A tool for fluid chronology. An example of an intragranitic uranium ore deposit (northwest Massif Central, France), *Lithos*, 18: 229-237.

Poty, B., J. Leroy, & L. Jachimowitz. 1976. Un nouvel appareil pour la mesure des temperatures sous le microscope, l'installation de microthermométrie Chaix-Meca, *Bull. Soc. Fr. Mineral._Cristallogr.*, 99: 182-186.

Roedder E. 1984. *Fluid Inclusions*, Rev. Mineral., Mineralogical Society of America, Washington, D.C. 12: 644p. Sibson, R. H.. 1981. Fluid flow accampanying faulting: field evidence and models, *Earthquake Prediction: An International Review,* edited by D.W. Simpson & P.G. Richards, pp 593-600, Maurice Ewing Ser., vol4.

Sibson, R.H., 1989. High-angle reverse faulting in northern New Brunswick, Canada, and its implications for fluid pressure levels. *J. Struct. Geol.,* 11, 873-877.

Turpin, L. 1984. Alterations hydrothermales et caracterisation isotopique (O. H. C)des mineraux et des fluides dans le massif uranifere de St Sylvestre. Extension a d'autres gisements intragranitiques francais, *Mem.6*, 190 p., Geol. Geochim. Uranium, Nancy, France.

Yin, Z. M., & G. Ranalli. 1992. Critical stress difference, fault orientation and slip direction in anisotropic rocks under non Andersonian stress systems, *J. Struct. Geol*, 14, 237-244.

Mechanics of Jointed and Faulted Rock, Rossmanith (ed.) © *1998 Taylor & Francis, ISBN 90 5410 955 6*

Quantification and simulation of fissural permeability at different scales

J. Sausse, M. Lespinasse & J. L. Leroy
UMR 7566, Géologie et Gestion des Ressources Minérales et Energétiques, UHP Nancy 1, Vandœuvre, France

A. Genter
BRGM, DR/GIG, Orléans, France

ABSTRACT: Two different scales of fractures collected in Hercynian granites were investigated. At the borehole scale, macrofractures properties were sampled continuously. The resulting fissural calculated permeability was evaluated versus depth and compared with the geometry (azimuth, dip) of fractured zones as well as with the fracture filling. Results shown that a global approach of borehole data are not representative of that on a local scale and that "efficient" fluid pathways can be decribed. At microscopic scale, cracks are modelized by finite shapes (discs). Water-rock interactions are therefore quantified in terms of precipitation or dissolution processes thanks to the knowledge of fissural permeability, porosity and surface exchange. The main results of the simulation indicate that a granitic rock can strongly and quicly be altered by a meteoric fluid.

1. INTRODUCTION

Discontinuities such as fractures and fissures are potential sites for fluid circulation and have important implications for the hydraulic properties of the rock. It is therefore important to quantify fluid flow in these discontinuities, in order to characterize and understand fluid transfers. A major problem facing theoretical modeling of fluid flow is that the fracture porosity of a rock is often poorly known. The precise geometrical description of the fractured medium therefore requires high quality datasets of fractures.

2. THEORITICAL APPROACHES

After Scheidegger (1974), the different models of permeability calculation can be grouped in two families: models based on the hydraulic radius notion and defining the fluid flow structures as finite shapes (Guéguen and Dienes, 1989, Guéguen and Palciauskas, 1992; Long and al, 1985; Billaux, 1990) and models based on a an equivalent porous media definition (Walsh and Brace, 1984). Dullien (1979) qualifies the first model as "statistical" and the second as "geometrical" model.

2.1. Geometrical model: Macrofracturation.

Fissural permeability studies based on borehole sampling data are difficult because of a major problem: the tridimensionnal extension of fractures in space is unknown. An extrapolation of data even in the near-well volume, is basically very difficult. The characterization of the rock hydraulic properties needs therefore the definition of some equivalent geometrical properties (equivalent media). Several approaches allow to link these equivalent properties (porosity and permeability) to the geometrical parameters of the fractured media. The model of Snow (1965) proposes that a volume of rock can be characterized by infinite fracture planes grouped in different and independant systems (Figure 1). Each group of fractures (similar orientations and geometrical properties) produces an equivalent directionnal conductivity which is directly related to the cubic apertures of the fractures and to their normal spacing. Turpault (1989) has shown that the width of an alteration zone around a fracture is proportional to the quantity of fluids passing through the walls. The fracture width is therefore related to the importance of paleocirculations through the rock.

The equivalent directionnal conductivity for fractures m is:

$$K_m = \frac{g}{12 \cdot \eta} \cdot \frac{A_m^3}{S_m} \tag{2.1}$$

$$S_m = \frac{1}{N_m} \cdot \sum_{k=1}^{N_m} s_{m,k} \tag{2.2}$$

$$A_m = \left(\frac{1}{N_m} \cdot \sum_{k=1}^{N_m} a_{m,k}^3 \right) \tag{2.3}$$

where g=acceleration due to gravity ($m.s^{-2}$), η=kinematic viscosity of the fluid ($m^2.s^{-1}$), A_m=average width characteristic of the fractures system m (m), am,k=aperture of the fracture k owing to the fracture system m (m), S_m=average normal spacing of the fracture system m (m), sm,k=spacing between the fractures $k-1$ and $k+1$ owing to the fracture system m (m), Nm=number of fractures in the fractures system m.

Several hypothesis must be done in the Snow (1965) model:

1. The Poiseuille law is used as an approximation of the Navier-Stokes general equation for an individual fracture.
2. The effective pressure gradient for each fractures is equal to a constant global gradient pressure characteristic of the entire fractured volume (homogenization).

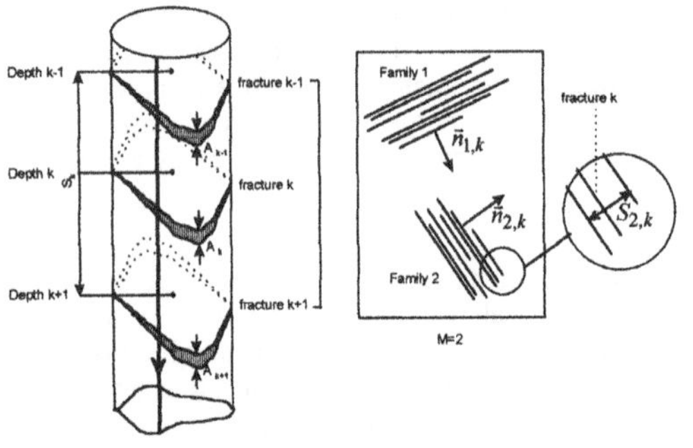

Figure 1: Available data collected on cores and geometrical fracture parameters useful for equivalent hydraulic properties calculations (Ak and Sk are the width and fracture k spacing respectively)

2.2. Statistical Models: Microfracturation.

In this section, we consider only fluid flows taking place through the microfissural system (sample scale). Microcracks are important markers in geology. In the context of fluid behaviour during deformation, fluid inclusion trails appear to be fossilized pathways of hydrothermal solution migration. Microcracks and fluid inclusion planes may therefore be used as reliable structural markers to reconstruct the geometry of fluid migration (Lespinasse and Cathelineau, 1990).

The knowledge of the microcrack geometrical parameters is enabled by microscopic scale of observation. Studies of rock thin sections enable to quantify microcrack extensions which can't be investigated at the scale of the quarry or on cores. The fissural permeability is directly related to the fracture lengths and therefore to their connectivity. Several works based on the Percolation Theory (Broadbent and Hammersley, 1957; Stauffer, 1985) lead to a possible estimation of the permeability tensor (Ayt Ougougdal, 1994; Canals and Ayt Ougougdal, 1997). Its determination needs a complete description of the three-dimensional geometry of the microcracks network, including definitions of crack family orientations, average lengths, apertures and volumic densities.

This approach needs to assume that a microcrack is a disc of constant radius and aperture. This modeling leads to the description of networks composed by geometrical oriented discs (Guéguen and Dienes, 1992).

The permeability tensor (2.4) can be calculated thanks to the computer code PERMEA developed by Canals (1997). This program uses the percolation theory and necessitates two assumptions:

1. Microcracks show finite radius.
2. Centers of fissures are distributed in agreement with a Poisson law.

$$\overline{\overline{K}} = P_f \cdot K_f \cdot \overline{\overline{W}}_f \qquad (2.4)$$

$$\text{with } K_f = \frac{\varepsilon.\pi^3 \cdot N_f \cdot c_f^2 \cdot e_f^3}{32} \qquad (2.5)$$

for an isotropic system of cracks and equation (2.6) for an anisotropic one:

$$K_f = \frac{\varepsilon \cdot \pi \cdot N_f \cdot c_f^2 \cdot e_f^3}{12} \qquad (2.6)$$

where K_f=global permeability tensor expressed in the crack plane (geographic reference), K_f=permeability characteristic of the the fracture system f, e_f=average aperture characteristic of the fracture system f, N_f=volumic density of the fracture system f, c_f=average radius characteristic of the fracture system f, ε=roughness of cracks planes, W_f=rotationnal matrix allowing the orientation in the crack plane reference of the tensor axis.

The definition of a finite volume (volume of disc) is therefore essential for the determination of crack permeabilities, porosities and surface exchanges. In other way, this model allows to define the main parameters of a fluid-rock interaction quantification.

3. APPLICATIONS

The different methods of fissural permeability quantification were applied at different scales of observation: the macroscopic scale (borehole sampling data, Soultz-sous-Forêts granite, Bas-Rhin, France) and the microscopic scale (thin sections data, Brézouard granite, Vosges, France) of fracturation. Two different approaches of permeability quantification were done and related to the fluid-rock interaction phenomena.

3.1. Macroscopic scale.

Soultz-sous-Forêt, located in the Upper Rhine Graben, is the European deep geothermal 'Hot Dry Rocks' test site. Three boreholes have been drilled: GPK1 (1987, 1992), GPK2 (1994) which represent the geothermal doublet (production and injection wells) and a reference well EPS1 (1990) which has been fully cored. A large database for EPS1 has been collected (Genter and Traineau, 1992) using cores as well as BHTV borehole wall imagery logs. These detailed datasets present an excellent opportunity to study the structural and mineralogical properties of the Soultz granite over a depth interval of 810 m between 1420 and 2230m depth. This granite has been strongly altered by fluid percolations which have completely sealed the majority of fractures. The purpose of this study is to describe and understand the mechanisms of fluid circulation that were responsible for the propagation of alteration through fractures (vein alterations). Fractures have been characterized by the following parameters: depth of intersection with the borehole, orientation, mineralogy of fracture fillings, type of discontinuity, fracture thickness.

Correlations between these variables have been statistically studied, and the nature of fracture filling is chosen as a key parameter for the descriptive study of the fracture hydraulic properties.

The main results are these:

1. Fracturing, although generally occurring in two main groups (N10°E, 70°W and N170°E, 70°E) is not homogeneously distributed with depth. Ten secondary families of fractures have been described throughout the borehole.

2. Fractures of EPS1 well are systematically sealed by hydrothermal products. Some of them are partly sealed. Twelve different fillings minerals have been observed as fracture fillings and grouped into 5 major categories (quartz, calcite, illite, hematite, chlorite).

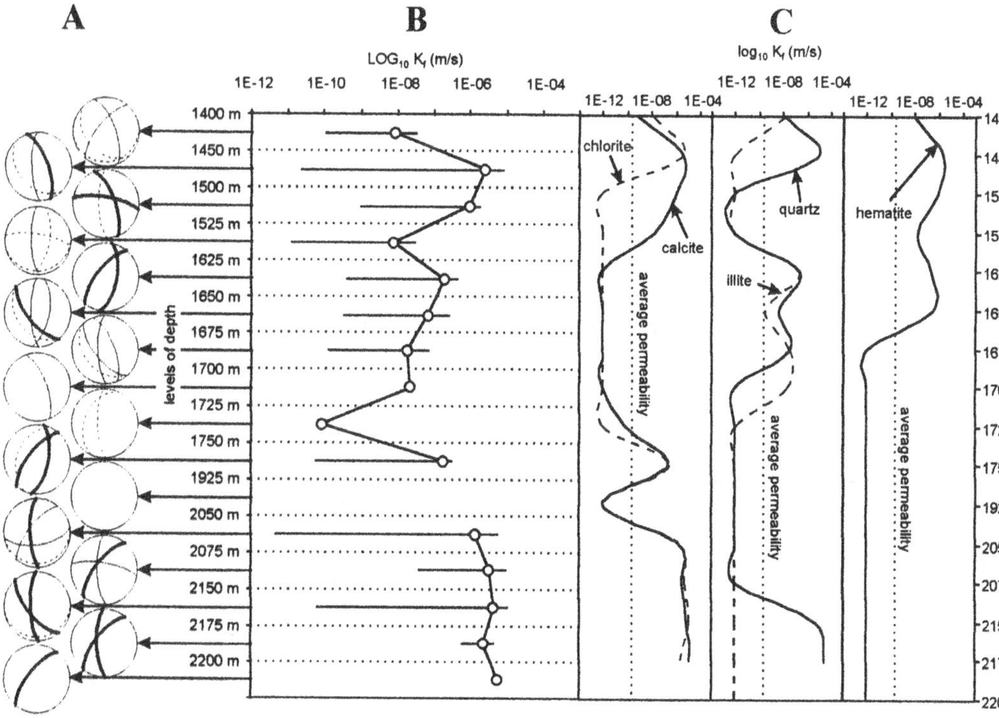

Figure 2· Equivalent hydraulic fractures properties of EPS1 drilling.

A) Stereographic projections (Schmidt plot, lower hemisphere) representing cyclographic traces of average fracture planes present in each zone. The trace's size is directly function of the fracture conductivities: dashed line (1E-12 to 1E-9 m/s), simple line (1E-9 to 1E-7 m/s), thick line (1E-7 and more m/s). B) Logarithm of the average directional conductivities in m/s calculated for each zone. The bars represent the amplitude of equivalent conductivities between their minima and maxima. The average conductivity is marked with a white circle. C) Logarithm of the directional conductivities (m/s) calculated for each level and for each type of fracture filling: "efficient" fractures systems for hematite, calcite, chlorite, quartz and illite precipitations

3. Relationships between the filling mineralogy, depth and orientation of fractures can be analysed.

A study of the correlations between these variables shows that some trends emerged, inferring that each depth zone is characterized by "efficient" fracture systems.

The occurrence of these systems as a function of depth is the basis of the fracture paleopermeabilities study. The assumption that each zone shows homogeneous mineralogical and structural properties has been made. Each fractured depth zone will be therefore considered as a homogeneous medium with constant physical properties.

Calculations of equivalent permeability [Snow, 1965] have been performed in two steps (Figure 2):

1. All the fractures present in a given zone were included in the calculation (Figure 2B).

This study demonstrates that paleopermeabilities are variable both within a single zone (weak or strong permeability of the different fracture families) as well as across different zones. The average permeability (0,83 m/year) is close to those given by Le Carlier and al [1994] and is clearly not representative of the whole EPS1 well. Three main zones are observed with an alternance of high permeability values (top of the granite and basement of EPS1 well) and lower ones (1525-1750 m). The maximal permeability (10^{-6} m/s) corresponds to the deeper depths where a permeable zone was observed.

The presence of main conductive systems consisting principally of NNW-SSE fractures with E-W dips for the top of granite, and NNE-SSW fractures with W dips for the basement of EPS1 has been defined in Figure 2A. Despite real structural heterogeneities (depth zones characterized by different fracture families), paleo- fluid flows seem to have been controlled by similar orientation planes characterized by large thicknesses, oriented globally along the axis of the three drill sites EPS1, GPK1 and GPK2: N150°E.

2. Calculations have been performed as a function of the presence or absence of certain fillings (Figure 2C).

Each zone is defined by the presence or absence of different fracture systems corresponding to a precise sealing mode: hematite (paleosurface alteration), quartz-clay (vein alteration) or calcite-chlorite fractures (pervasive alteration) constitute three different systems. This assumption has allowed to show different evolutions of directional conductivities concerning the three sealing events. Systems through which fluid percolated during the quartz-illite deposits are characterized by different fracture associations, thus different thicknesses and different permeabilities, than the other vein alterations. The main results are illustrated in Figure 2C.

Fractures sealed by hematite are characteristic of the top of the granite and present high permeability values which do not affect the deeper part of EPS1.

The association quartz-illite in fractures occurs in the high permeability zones.

Fractures containing chlorite are not very important in the permeability estimation but must be taken into account for the 2050-2300m zones.

Fractures sealed by calcite are characterized by an heterogeneous behaviour. This kind of fracture sealing is absent in certain zones or has a major effect on permeability in other zones.

The study of vein alterations allows the visualization of the fracture behaviour as a function of the different filling mineralogies. The aim of this approach using borehole data is not for deciphering chronological alteration succession (order of crystallization of minerals on vein walls or indications on their period of precipitation are not taken into account) but it is more an attempt for detecting "efficient" fracture systems corresponding to certain alteration event. The study of vein alteration in the Soultz EPS1 core shows that different fracture networks are present throughout the core, implying that estimation of the average fluid flow is not representative of that on a local scale.

3.2. Microscopic scale.

This work proposes a simulation which quantifies the percolation of a meteoric fluid through cracks in a granitic rock.

The fissural permeabilities are generally quantified by statistic methods based on the analysis of the fissural connectivity. However, difficulties occur by taking into account the fluid rock interactions. A fluid can react with a rock and modifies mineralogical assemblages by dissolving a pre-existing mineral or precipitating a new mineral assemblage. The formation of new mineral phases depends on the rock chemistry, the nature of the fluid and the time duration of fluid-rock interactions.

Thus, this study is an attempt to quantify the evolution of the physical parameters of the flow (permeability, connectivity of microcracks, porosity) during a fluid-rock interaction.

Fields data have been acquiried in the so-called "granite du Brézouard" which outcrops in the Northern french Vosges massif. The interaction between a meteoric fluid and this granite has been simulated considering a vertical column in which the fluid is percolating through cracks (Figure 3).

The fissural dataset is initially defined by studying oriented thin sections with image analysis. The different fracture sets observed are characterized by their average aperture, radius and volumic density. These initial geometrical parameters allow to calculate the initial rock permeability (0,180 mD for the maximal component Kz), porosity (4,5%) and surface exchange (110 cm^2 per cm^3 of rock).They are the basis of the fluid-rock interaction quantification and are characteristic of the initial, non altered, homogeneous granite.

The first phase of fluid-rock simulation shows 8 alteration events (Figure 3) and 8 zones in the granite. Each zone corresponds to a particular degree of alteration and therefore to a particular mineralogy. The water-rock equilibrium is reached after 23 years and for a fluid penetration of 13 m. The distance of penetration of the fluid has been estimated with a capillarity model.

The rock is therefore segmented in elementary volumes (cylinders) allowing a discretization of the chemical and physical modifications of the rock.

Each zone is characterized by a particular dissolution (primary minerals) or precipitation (secondary minerals) rate (Figure 3). The alteration is most intense at the top of the granite because of the fluid initial under saturation compared to the rock. The dissolution is the main phenomena and the dissolution/precipitation ratio is therefore considered as infinite. Then, zones 2 to 8 are characterized by the deposition of new mineral phases and the dissolution/cristallisation ratio decreases: reactions become less and less marqued until the water-rock equilibrium.

The modification of the fissural parameters is taking into account with three hypothesis:

1. Crack density, azimuth, and dip, are constant data during alteration.

2. Secondary mineral precipitations affect the crack extremities (generally characterized by thinner apertures). Each crack set is therefore defined by a radius negative variation and a new characteristic diameter.

3. Dissolution of primary minerals is homogeneous and affects the whole crack. Dissolution implies therefore an increasing of apertures and fissural radius.

After the fluid-rock interaction simulation, new fissural parameters are defined and each zone is characterized by a new permeability, porosity and surface exchange (Figure 3).

The main results are these:

The different components of the permeability tensor (Kx, Ky, Kz) have a similar evolution during the fluid-rock interaction. The top of granite (zone 1) is characterized by a strong increase of permeability (+40%) and porosity (11%) compared to the initial non altered rock. These values can be explained by the importance of dissolution in the first steps of the alteration. The following zones (2 to 7) are marqued by a less marqued opening of the crack system because of the balance between precipitation of new mineral phases and dissolution. Zone 8 is practically not affect by the alteration.

Surface exchanges are not characterized by the same evolution. This parameter is directly fonction of the crack radius and is therefore less modify by the dissolution phenomena. A decrease is observed for the zones 2-7 where precipitation occurs.

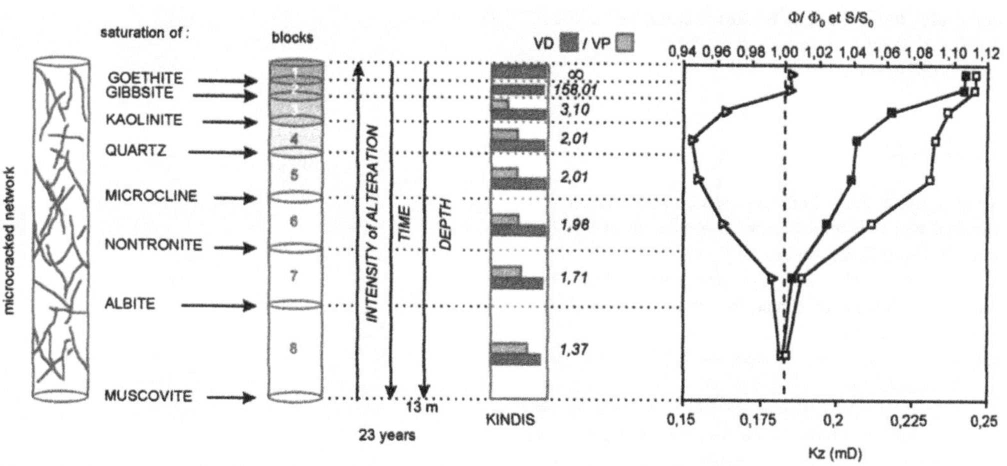

Figure 3. Results concerning the evolution of the fissural parameters: porosity (ϕ/ϕ_0, ϕ_0 and ϕ are respectively the intact rock and the altered rock porosities), surface exchange (S/S_0, S_0 and S: intact rock and altered rock surfaces) and permeability (Kz. maximal component of the permeability tensor) after the simulation of the fluid rock-interaction. The relative variation of ϕ and S are compared to the absolute variation of the permeability (in milliDarcy). The rock profile of alteration is composed by 8 events of alteration The water-rock equilibrium is nearly obtained in 23 years (indicated time directly given by KINDIS) and corresponds to a depth of fluid penetration of 13 m. The differents rates between the volume of the dissolved minerals (VD) and volume of precipitated phases (VP) show that dissolution is the main phenomena of alteration. A global opening of the crack network is therefore observed and well marqued at the top of the granite

This simulation shows that under meteoric conditions a granite can be strongly altered in a short time. During the first step of alteration, the fissural permeability increases. This work, which is based on a simple model shows that the propagation of the alteration can be extremely rapid. This approach shows that a time-space evolution of alteration phenomena and their consequences on fluid flows can be quantified.

The simulations have been carried out through a cooperation between CGS-CNRS (Strasbourg, France) and CREGU (Vandoeuvre, France). The fluid-rock interactions evolution with time were quantified with the "KINDIS" software (Madé et al, 1990, 1991).

DISCUSSION

The study of paleo fluid flows in the Soultz granite shows that several stages of alteration can be identified by the way of statistical studies of fracture macroscopic properties. A real complexity of the alteration phenomena is present at this scale. The rock matrix porosity and its importance in the fluid-rock interactions were not taken into account. However, the second part of this paper shows that the microscopic scale provides excellent pathways for fluid percolation. A fluid which reacts with opened microcracks can strongly altered the rock matrix in a short time.

These two approaches concerning the quantification of present or paleo fluid percolations demonstrate therefore that fractures must be studied at different scales in order to describe the whole alteration phenomena.

REFERENCES

Ayt Ougougdal, M. 1994. Controles magmatiques, structuraux et hydrothermaux de la formation des épisyénites de la Marche Occidentale. *Thesis INPL, Nancy.*

Billaux, D. 1990 Hydrogéologie des milieux fracturés. Géométrie, connectivité et comportement hydraulique. *Doc. Bur. Rech. geol. min.* 186.

Broadbent, S. E. & J. M. Hammersley 1957. Cristals and mazes. *Proc.-Camb. Phil. Soc.* 63: 629-641.

Canals, M. & M. Ayt Ougougdal 1997. Percolation on anisotropic media, the Bethe lattice revisited. Application to fracture networks *Nonlinear Processes in Geophysics (European Geophysical Society)*. 4: 11-18.

Dullien, F. A. C. 1979. Porous Media, Fluid Transport and Pore Structure. *Academic Press, N. Y.*

Genter A. & H. Traineau 1992. Borehole EPS1, Alsace, France: preliminary geological results from granite core analysis for Hot Dry Rock research. *Scientific Drilling*. 3: 205-214.

Guéguen Y & V Palciauskas V. 1992. Introduction à la Physique des Roches. *Hermann, Paris.*

Guéguen Y & J. Dienes 1989. Transport properties of rocks from statistics and percolation. *Mathematical Geology*. 21: 1-13.

Le Carlier de Veslud C., J.J. Royer & L. Florès 1994 Convective heat transfer around the Soultz-sous-Forets geothermal site: implication to oil potential. *First Break*, 12 (11): 553-560.

Lespinasse M. & M. Cathelineau 1990. Fluid percolations in a fault zone: a study of fluid inclusion planes in the St Sylvestre granite, northwest Massif Central, France *Tectonophysics*. 184: 173-187.

Long J. C S., P. Gilmour & P. Whiterspoon 1985. A model for steady fluid flow in random tridimensional networks of disk-shaped fractures *Water Ressour. Res.* 21(8): 105-115.

Madé B. 1991. Modélisation thermodynamique et cinétique des réactions géochimiques dans les interactions eau-roche *Thesis, Université Louis Pasteur de Strasbourg*

Madé B., A. Clément & B. Fritz 1990. Modélisation cinétique et thermodynamique de l'altération le modèle géochimique KINDIS. *C. R. Acad. Sci. Paris* 310 (II): 31-36.

Scheidegger A. E 1974. The physics of flow through porous media. *Univ. of Toronto Press, Toronto.*

Snow, D. T. 1965. A parallel Plate model of fractured Permeable Media, *PhD. Dissertation, University of California.*

Stauffer D. 1985. Introduction to percolation theory. *Taylor and Francis, Londres.*

Turpault M. P. 1989 Etude des mécanismes des altérations hydrothermales dans les granites fracturés. *Thesis, Université de Poitiers.*

Walsh J. B & W. F. Brace 1984. The effect of pressure on porosity and the transport properties of rock. *J. Geophys. Res.* 89: 9425-9431.

Mechanics of Jointed and Faulted Rock, Rossmanith (ed.) © *1998 Taylor & Francis, ISBN 90 5410 955 6*

Investigating fractured-porous systems – The aquifer analogue approach

C.I. McDermott, M. Sauter & R. Liedl
Applied Geology, University of Tübingen, Germany

ABSTRACT: The investigation concept of the aquifer analogue approach is being used to characterise fractured porous media. New laboratory and sample recovery techniques are presented enabling the determination of the three dimensional physical parameters of undisturbed fractured porous sandstone bench scale samples, (30cm diameter x 40 cm height). Gas tomographical flow and transport experiments with helium as a tracer in a specially designed experimental cell provide different scaled multi-dimensional information on the hydraulically important units and their respective hydraulic and transport parameters. This information is combined to generate an analogue of the fractured aquifer. Analysis of the results to date using a multiple shelled model have shown that the dimensionality of the flow field as well as the structure of the sample plays an important role in the understanding of the system.

1 THE AQUIFER ANALOGUE APPROACH

The detailed investigation of flow through fractured porous aquifer systems is generally very difficult, mainly due to the high contrast in the hydraulic properties of highly permeable fractures and the usually much less permeable rock matrix. The investigation of the individual flowpaths with generally available techniques is almost impossible. Pumping tests or tracer tests only provide integral information, i.e. averaged flow and transport properties. In contrast it is known from numerical modelling and from laboratory investigations that the characterization of the fracture geometry (length, aperture, orientation, connectivity etc.) as well as the interaction between fracture and matrix is a prerequisite for the understanding of transport in these systems.

The basic idea of the aquifer analogue approach is to investigate an outcrop area whose lithological and sedimentological properties can be considered to be analogous to a more or less inaccessible aquifer. The concept comprises examining in detail multiple sec-

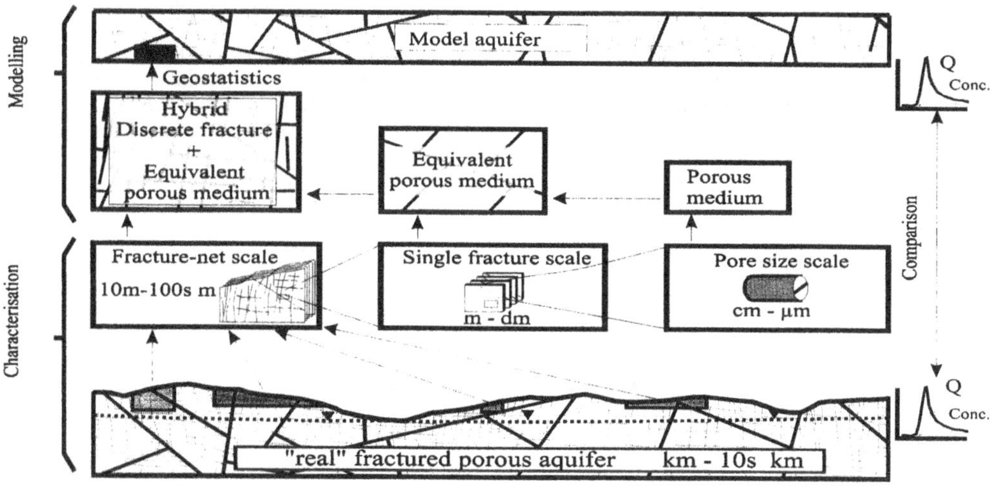

Figure 1: The aquifer analogue concept

tions of a natural outcrop at a small scale (sub meter) to middle scale (several meters) and transfering the results to a real fractured rock aquifer. Appropriate controlled flow and transport experiments (laboratory and field) can be conducted providing information on the spatial distribution and variability of hydraulic and transport parameters. After completion of the experiments, the laboratory samples will be cut into sections allowing detailed geometrical characterisation, thus a comprehensive data set on the fracture inventory and the matrix properties will be available supplementing the results from integral measurements. The advantage of this type of approach is the detailed investigation of real systems as well as the availability of results from controlled experiments. The comprehensive data sets are used for calibration and to a certain extent also validation of discrete fracture and equivalent models, which is usually not possible for field sites to such a detail due to the lack of information.

Analogue studies have mainly been conducted in the field of petroleum exploration in order to characterise the properties of petroleum reservoirs (e.g. Flint & Bryant, 1993). Jussel (1992) used the concept in order to identify the geometric and hydraulic characteristics of various sedimentological units in quatenary fluvial deposits. In the field of fractured rock aquifers analogue studies have relatively rarely been employed and been limited to two-dimensional outcrop situations (e.g. Kiraly, 1969; Billaux et al., 1989) frequently within the context of the investigation of future nuclear waste repositories.

This paper presents the general investigation concept of the aquifer analogue approach (Figure 1). At the bottom of this figure natural exposures of the aquifer formation are shown where field measurements and samples for laboratory investigations can be taken. In the centre of Figure 1 the different investigation scales are presented along with the applicable model type. The model which is derived from this procedure is shown at the top.

In the field it is difficult to provide fully controlled conditions and there is always uncertainty as to the volume affected by the test. It was therefore decided to first develop a tomographical experimentation technique under controlled conditions on laboratory bench scale samples enabling three-dimensional parameter identification. At a later stage, these techniques will be applied in the field at a larger scale, hence providing multi-scaled information on aquifer properties.

2 SAMPLING

Three types of cylindrical bench scale samples (30cm diameter, 40cm height), classified according to the nature of the fractures, were selected for the investigation of the fractured porous system. Firstly

Step 1

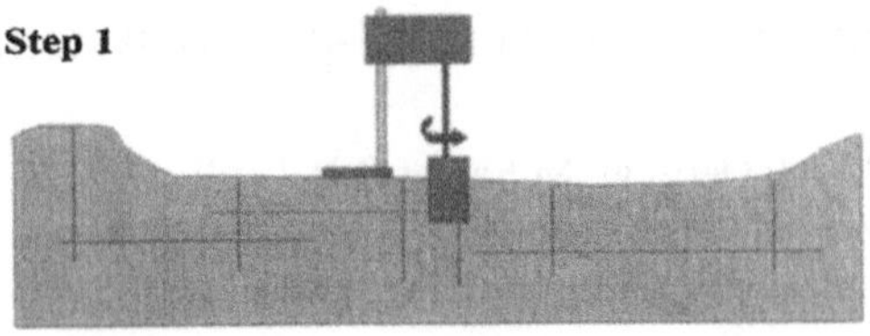

A drilling machine is anchored to insitu fractured porous rock and a core is drilled using a water flush.

Step 2

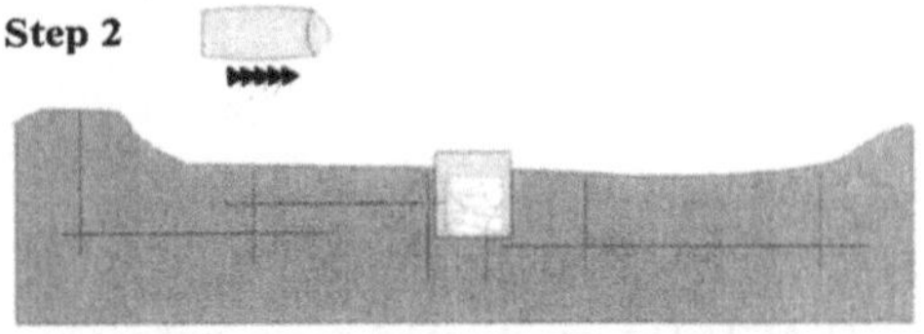

A prewarmed membrane is placed in the anulus and allowed to cool shrinking to the size of the sample and exerting a stabilising pressure on the sample

Step 3

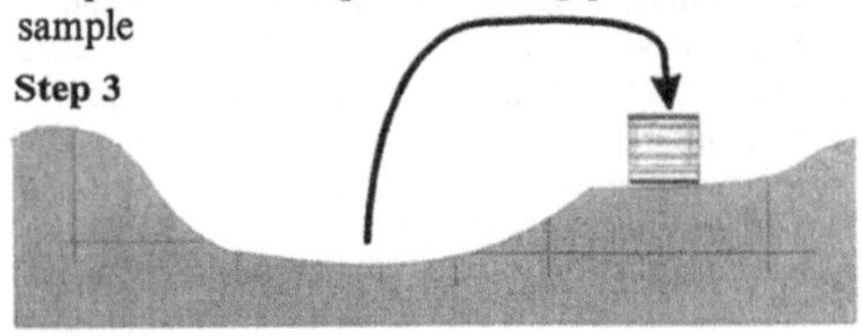

The surrounding rock material is removed, the sample is further supported and removed to the laboratory.

Figure 2: Sample Recovery

samples with open unfilled fractures, secondly samples with sand filled fractures and thirdly samples with clay filled fractures.

There is no established technique for the recovery of undisturbed highly fractured samples at a bench laboratory scale. The greatest problem with such material is that during sampling the fractures, natural plains of weakness, open up and the sample disintegrates, even before reaching the laboratory (Alexander et al., 1996). However, for the study of the fractured system, highly fractured bench scale samples are required which can then be placed under controlled laboratory conditions. The sample recovery procedure is described in Figure 2.

3 LABORATORY EXPERIMENTS

Flow and transport experiments were conducted using gas flow and gas tracer transport techniques in order to obtain integral measurements in the water unsaturated sandstone samples. The resulting gas parameters are then converted into hydraulic parameters. Gas flow techniques allow a relatively rapid and accurate determination of the parameter field. When considering three dimensional parameter identification, comprising several hundred multiple tests, an efficient measuring technique is a prerequisite.

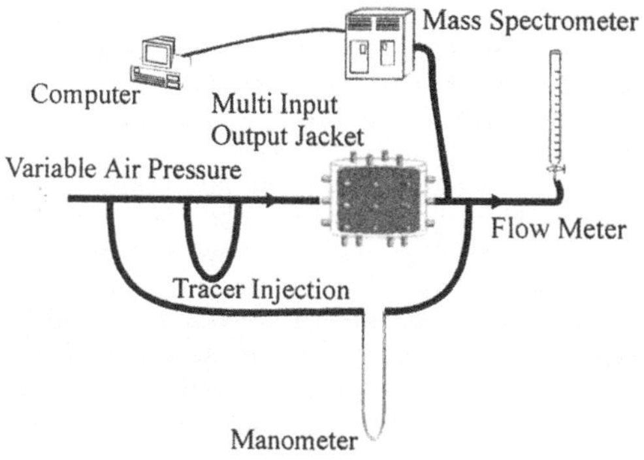

Figure 3: The experimental set up.

The experimental set-up is shown in Figure 3. A stable linear flow field is established across the sample. The flow rate across the sample is recorded using a bubble meter. Gas tracer is injected via a flow-through loop and the breakthrough of the gas tracer is recorded using a mass spectrometer. The exact mass of the tracer placed into the system is known as well as the time of input. Hydraulic conductivities were calculated using a one dimensional flow field model (Darcy's law).

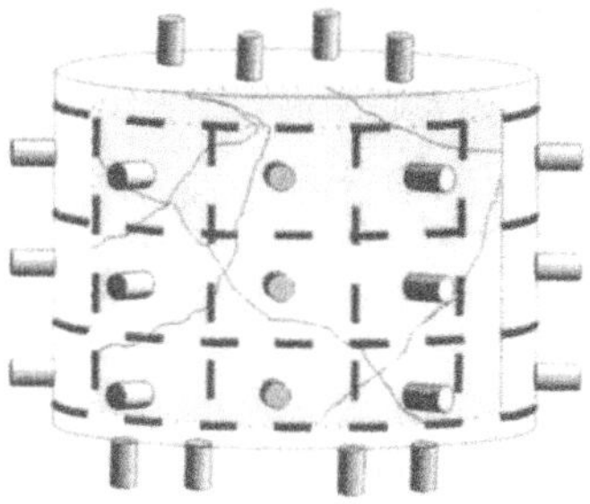

Figure 4: The Multiple Input Output Jacket, MIOJ

A key element in this system is the design of the experimental cell, named by ourselves as the multiple input output jacket (MIOJ) enabling the three dimensional investigation of the samples. The cell consists of a sealing membrane, with open spaces in order to allow selective access to the sample and a supporting brace. The MIOJ is shown in detail in Figure 4. Depending on the experimental set-up, different surface areas can be connected to a predefined pressure system thus allowing a variety of adjustable boundary conditions in an enclosed system. By varying the flow direction across the sample the anisotropy of flow and transport parameters can be determined.

4 EXPERIMENTAL RESULTS AND MODELLING APPROACH

The dependency of hydraulic conductivity on flow direction is shown in Figure 5 for a sample with clay filled fractures. Because the clay fill acts as a barrier to flow, highest hydraulic conductivities are associated with the flow direction parallel to the fractures (0°-90°; 180°-270°). Figure 6 presents the conductivity tensor for a sample with sand filled fractures. It is apparent that highest hydraulic conductivities occur in N-S (0°-180°) direction.

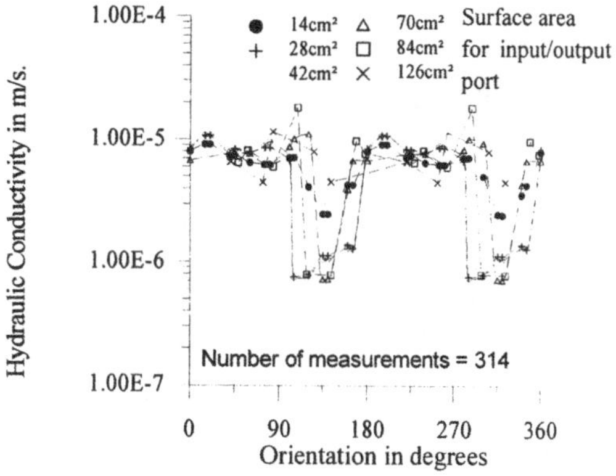

Figure 5: Hydraulic conductivity for a sample with clay filled fractures, all measurements over a flow path length 290mm.

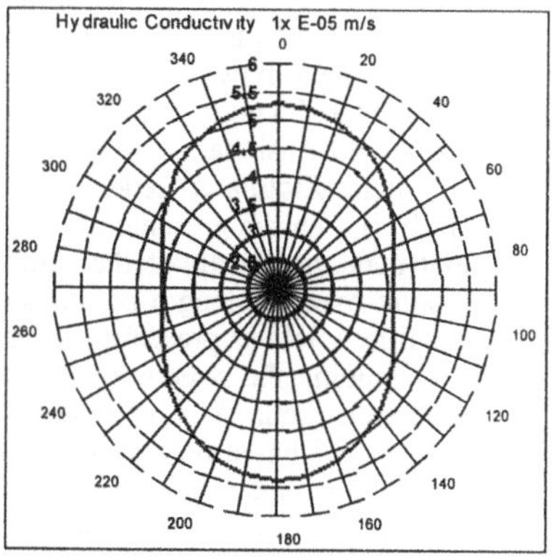

Figure 6: Hydraulic conductivity tensor for a sample with sand filled fractures. Only measurements over a flow path length of 290mm are presented.

Using the MIOJ it is also possible to gain an insight into the scale dependency of physical parameters as considered by a number of authors, e.g. (Tidwell & Wilson, 1997.

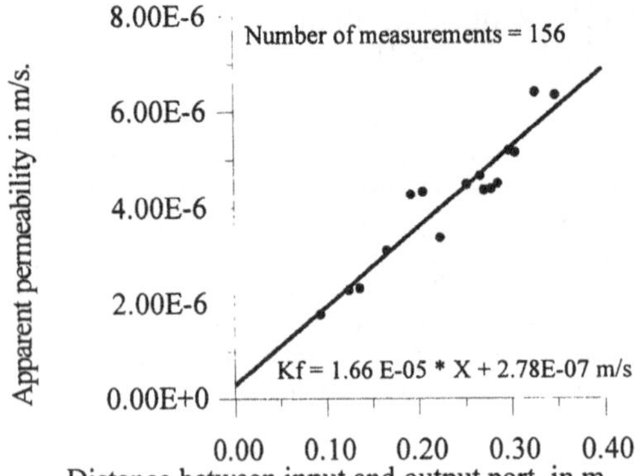

Figure 7: Relationship between path length and one dimensional apparent permeability for matrix dominated flow paths.

Figure 7 presents the relationship found between path length of measurement (distance between input and output port) and the apparent one dimensional hydraulic conductivity for matrix dominated flow. All known fracture influenced results, i.e. measurements where a fracture was directly connected to a input/output port, have been removed. It is apparent that with increasing input port/output port distance there is an increase in apparent hydraulic conductivity. The authors propose two explanations for this trend. Firstly, with an increase in the input/output port distance there is an increased probability that a fracture in the sample has had a positive influence on the flow, leading to a higher hydraulic conductivity. Secondly, and considered to be the dominant influence, is the effect of applying a one-dimensional model to derive apparent hydraulic conductivity in a two- or three-dimensional flow field.

With a point input and a point output it is possible to generate a three dimensional flow field as in the case of a dipole field (Zlotnik & Ledder 1996; Heer & Hadermann, 1994). This is presented in Figure 8b. Given an input over the length of the sample and likewise an output over the length of the sample a two dimensional flow field as presented in Figure 8a develops. On examining the structure of these flow fields as presented in Figure 8, it is apparent that a one-dimensional model based on the area of the input and output window underestimates the available cross-sectional flow area as well as the average flow path length.

Considering the one-dimensional flow model, assuming the cross-sectional flow area and pressure difference to be constant then an increase in the real path length produces an erroneously lower apparent hydraulic conductivity when Darcy's law is applied. Likewise assuming the pressure difference and the flow path length to be constant then an increase in the real cross-sectional flow area produces an erroneously higher apparent hydraulic conductivity. When considering the geometry of the flow system it is apparent that for 3D flow fields, the actual flow path length is a linear function and the cross-sectional area a quadratic function of the distance between input and output port thus leading to an increase in apparent hydraulic conductivity with increasing input port/output port distance using a one-dimensional model for interpretation.

With the additional transport information from tracer breakthrough curves, it can be shown that the dimensionality of the flow path plays a determining role in the response of the system to a certain flow field and tracer experiment. Effective porosities, determined from independent measurements of the matrix porosity are not consistent with those evaluated from transport experiments using a one-dimensional model. The only way to reconcile the results from flow and transport experiments was by assuming a larger effective cross-sectional area between input and output port as suggested by a multi-dimensional model.

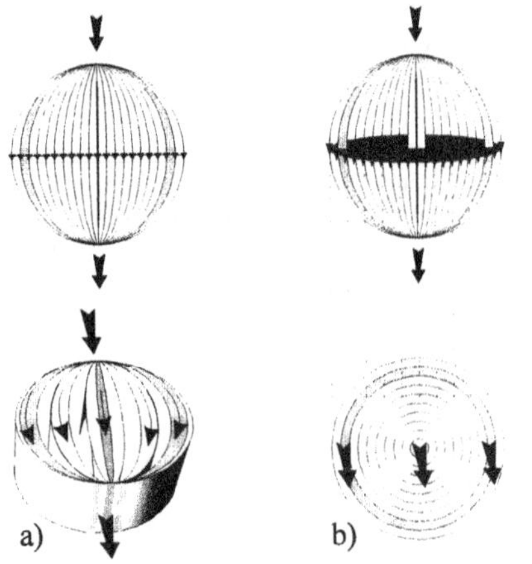

Figure 8: Two and three dimensional flow fields. a) Two dimensional flow field (slit input – slit output), b) Three dimensional flow field (point input – point output)

Applying this theory to experimentation in the experimental cell, the system can be set up to develop one, two or three dimensional flow fields as shown in Figure 8. This obviously has an effect on the flow characteristics, on the shape of the breakthrough curve, as well as on the interpretation method. To model tracer breakthrough in this system in detail using a finite element model is time consuming. Therefore a simple analytical model has been developed to determine model parameters by fitting measured and modelled breakthrough curves.

The principle behind this model is apparent from Figures 8a and 8b. The flow field is divided up into a series of shells around the centre flow line, i.e. the direct connection between input and output port. When the flow is three dimensional then the shells have a ball or onion form as shown in Figure 8b and when the flow is two dimensional then the shells have a curved form as shown in Figure 8a. Each shell has a calculable volume and a central path length, from which the average cross sectional flow area can be calculated by dividing the volume of the shell by the central path length. Each shell is then represented mathematically as a one dimensional pipe of length equivalent to the central path length for that shell and of cross sectional area equivalent to the average cross sectional area. The pressure difference from one end of the pipe or shell to the other is constant and known for all pipes. In such a manner the contribution of each shell to the flow and transport in the system can be calculated and combined to provide a simple representative model of the system. The diffusion within the system is considered to be one dimensional along the length of the shells and the diffusion coefficient of the tracer gas in air is used to describe the diffusion of the gas in the system.

Darcy's law for compressible flow is used to calculate the flow within the fields but incompressible flow is assumed for simplicity when considering diffusion. Experimentally low pressure gradients (150mbar/flow path length) were applied allowing such assumptions. Geometrically considered, there is an interplay between two parameters affecting the flow within the system, that is firstly the path length of flow and secondly the cross sectional flow area of the system. Figure 9 presents the increase of path length with increasing distance from the flow centre for two and three dimensional flow, and Figure 10 the increase in cross-sectional area available to the flow field with increasing distance from the flow

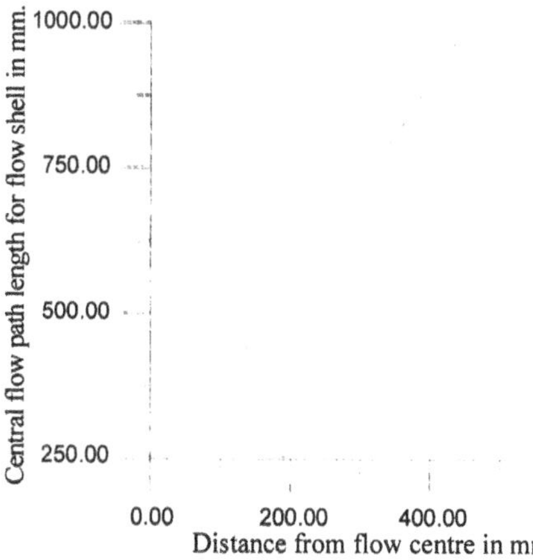

Figure 9: Relationship between flow path length of individual shells against maximum distance of shell from the centre of the shell model.

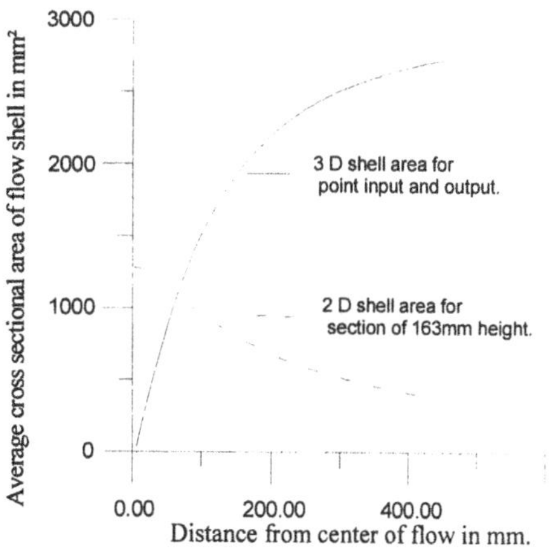

Figure 10: Relationship between equivalent cross sectional area of individual shells against distance of shell from the centre of the model.

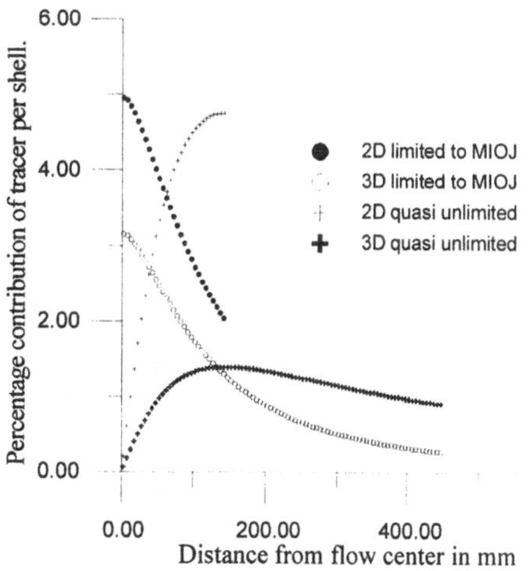

Figure 11: Percentage arrival of tracer per model shell

centre, again for two and three dimensional flow fields.

The mass of tracer carried in each shell is directly related to the proportion of flow carried in each shell and this is presented in Figure 11 as a percentage contribution of tracer mass per shell against the maximum distance of the shell from the flow centre. It is found that for 3D flow configurations (Figure 8b) the main portion of tracer mass is advected within the outer shell, while for 2D flow patterns (Figure 8a) tracer transport occurs predominantly in the centre shell. Converting this data into mass flux per second, produces advection only curves in Figure 12. The inflection point in the arrival of tracer from the three dimensional flow field is due to the fact that although the outer shells carry more of the tracer

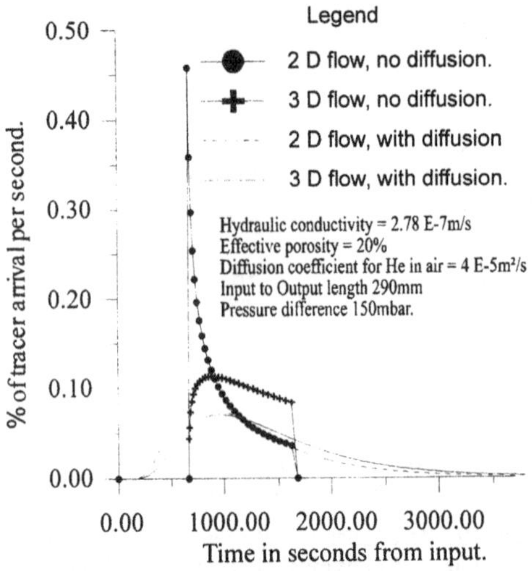

Figure 12: Advective transport only compared to advective and diffusive transport.

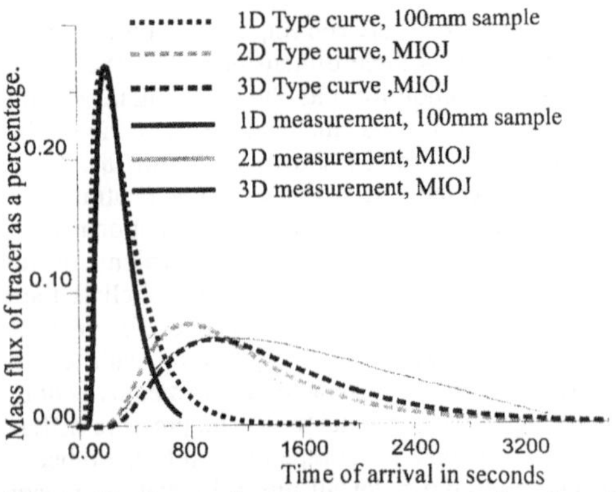

Figure 13: Comparison between measured and modelled breakthrough curves for one-, two- and three-dimensional flow fields (1D results from Jaritz, 1998).

than the inner shells the difference in arrival times for the outer shells are much larger than for the inner shells. When diffusion is included in this model, the breakthrough curves, presented in Figure 12 (with diffusion) are derived.

Using the above model to fit the experimental breakthrough curves allows accurate and rapid determination of the system parameters. The flow within the system can be forced into a certain dimensionality and then modelled accordingly. Figure 13 presents just such a case where two and three dimensional flow scenarios were induced. The measured tracer breakthrough curves were modelled using the homogeneous multiple shell approach. The agreement between the modelled curves and the measured curves is quite clear.

5 CONCLUSIONS

A new experimental method has been presented, which, although in its early stages of development, has already provided interesting results with respect to the three dimensional characterisation of fractured porous media. What is of most importance in the determination of the relevant parameters for use in the construction of an aquifer analogue is the dimensionality of the flow field. This factor not only affects the evaluation of flow measurements but also determines to a major degree the shape of the breakthrough curves. In any study using an induced flow field with a higher dimensionality this is of critical importance for the later accurate determination of the aquifer parameters. These findings are of particular importance when dealing with the commonly encountered complex dipole flow field between two wells. The "onion-shell" model, or multi-shell model provides a quick and accurate tool for the derivation of the important aquifer parameters and the assessment of relative flow contribution of the different parts of the flow system.

REFERENCES

Alexander W. R., Frieg B., Ota K., Bossart P. (1996): Untersuchung der Retardation von Radionukliden im Wirtgestein, *Nagra informiert* 27, Pp 43-55.

Billaux, D., Chiles, J.P., Hestir, K. & J. Long, (1989): Three-dimensional statistical modelling of a fractured rock mass – an example from the Fanay-Augères mine. *Int. J. Mechanics Mining Sci. Geomech. Abstr.*, 26, 281-299.

Flint, S.S. & I.D. Bryant, (1993): The geological modelling of hydrocarbon reservoirs and outcrop analogues. *Specs. Publ. Int Ass. Sedimentology*, 15, 269p.

Heer W., Hadermann J., (1994): Grimsel Test Site Modelling Radionuclide Migration Field Experiments, *PSI-Bericht Nr 94*-13

Jaritz R., (1998): *Quantifizierung der Heterogenität einer Sandsteinmatrix am Beispiel des Stubensandsteins,* Ph.D.-Thesis, University of Tübingen, Geowissenschaftliche Fakultät, in prep.

Jussel, P., (1992): Modellierung des Transports gelöster Stoffe in inhomogenen Grundwasserleitern. Ph.D. Thesis, Inst. für Hydromechanik und Wasserwirtschaft, ETH-Zürich, 323p.

Kiraly, L., (1969): Statistical analysis of fractures (orientation and density). *Geol. Rundschau*, 59.

Tidwell V. C. & Wilson J.L., (1997): Laboratory method for investigating permeability upscaling, *Water Resources Research*, 33, No.7, p1607-1616.

Zlotnik V. & Ledder G., (1996): Theory of dipole flow in uniform anisotropic aquifers, *Water Resources Research,* 32, No. 4, p1119-1128.

Mechanics of Jointed and Faulted Rock, Rossmanith (ed.) © 1998 Taylor & Francis, ISBN 90 5410 955 6

Damage mechanics approach for hydromechanical coupling in rock joints under normal stress

M. Bart & J.F. Shao
EUDIL, LML, University of Lille, France

ABSTRACT : In this paper, hydromechanical model is proposed for a fracture submitted to normal stress. The progressive joint closure is considered as a modification process of the fluid flow space, this process is characterised by internal compaction variable and the evolution of this variable is expressed as a function of joint closure. Furthermore, in order to describe hydromechanical coupling, the Biot's effective stress concept for porous media is generalised to rock joints. The proposed coupled model, in association with fluid flow laws, provides a complete set of equations for modelling hydromechanical behavior of rock joints (pressure distribution in joint, flow rate prediction and joint closure). The numerical predictions from our model give good agreements with experimental data and with the results from other models on two granites.

1 INTRODUCTION

The fractured rock massif is concerned by numerous designs of civil engineering (dams, underground radioactive waste repositories). Their investigation requires to take fractured aspect into consideration and the safety of these structures strongly depends on the mechanical and hydromechanical behavior of rock fractures and joints. Many laboratory mechanical investigations have been performed on the mechanical behavior of rock joints. However, further works are needed for a better modelling of hydromechanical coupling in rock joints.

The mechanical behaviour of a joint is investigated from simple compression tests carried out on intact and fractured rock samples. The characteristics of a joint are achieved by subtraction of displacements due to the elasticity of intact rock to global displacements of fractured samples (Fig. 1). The no linear character of the normal stress - displacement curve for a fracture and the presence of a maximal closure V_{max} showing up asymptotical behaviour have been confirmed by many authors (Goodman 1974, Bandis et al 1983, Gentier 1986, Benjelloun 1991,...). Different models are developed in order to describe the mechanical behaviour of the joint. There exists two classes of models. On the one hand, some authors have proposed more or less simple analytical relations expressing normal stress σ_n versus closure V, the parameters used in these models are calculated by adjusting numerical results to experimental data (Goodman 1976, Bandis et al 1983). On the other hand, some authors have established models which, from mechanical properties of intact rock and from the geometry of the fracture, provide for the mechanical behaviour of the rock joint. There exist two ways to consider the geometry of the fracture : either it is assimilated to a set of asperities (Gentier 1986) or it is considered as a set of voids (Tsang & Witherspoon 1981), (Fig. 2, 3). Our mechanical model is developed using the second approach and will be presented in the second part of this paper.

The previous works on fluid flow in rock joints have shown that for a joint with smooth walls, the flow can be assimilated to a flow through a void with parallel planes and therefore can be described by Navier and Stockes equations. The average velocity checks the following equation :

$$\vec{v} = -\frac{\rho g}{12\mu} e^2 \nabla H \qquad (1)$$

where μ is the dynamic fluid viscosity and ∇H is the hydraulic head. And, the flow rate Q is found by the relation generally called the cubic law :

$$Q/\Delta H = Ce^3 \quad (2)$$

where C is a constant which depends on the properties of the fluid and the system geometry and e is the distance between the two planes.

For rock fractures, the wall surfaces are irregular and the equations (1) and (2) can not be applied directly and this, for two reasons. On the one hand, the rock joint is composed of two walls separated by an aperture $e(x,y)$. Because of joint roughness, this aperture varies according to the localisation (x,y). On the other hand, there exist some points where $e(x,y) = 0$ and these contact points involve a tortuous and bidimensional flow (Tsang 1991). Furthermore, these two parameters, roughness and channeling , which control the flow in the fracture, vary according to normal stress and flow pressure. Therefore, for each stress, the state of the fracture is completely modified and consequently, the flow rate varies. The modelling of these phenomena is very difficult. The hydromechanical models are based on the cubic law but, for the most part, they correspond to simple hydraulic models which take the variation of aperture under normal stress into consideration but not the effect of flow pressure. Our hydromechanical model, by using the effective stress concept, takes into account the aperture variation due to normal stress and flow pressure. It is presented in the third part.

The proposed model will be tested through some experiments which are presented in the first part of this paper.

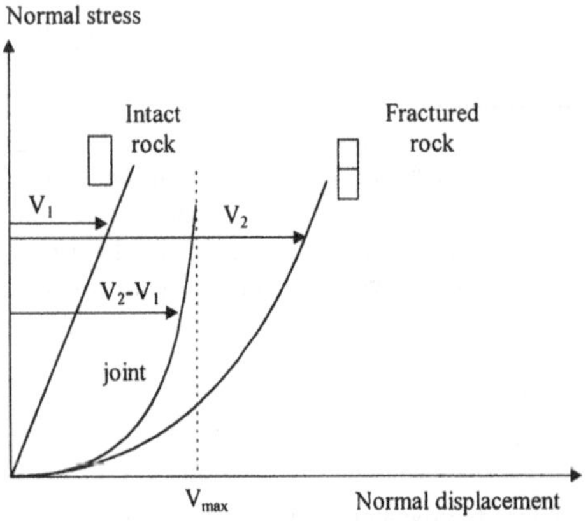

Figure 1. Mechanical behaviour of intact rock and fractured rock samples submitted to normal stress (Diagram after Benjelloun 1991).

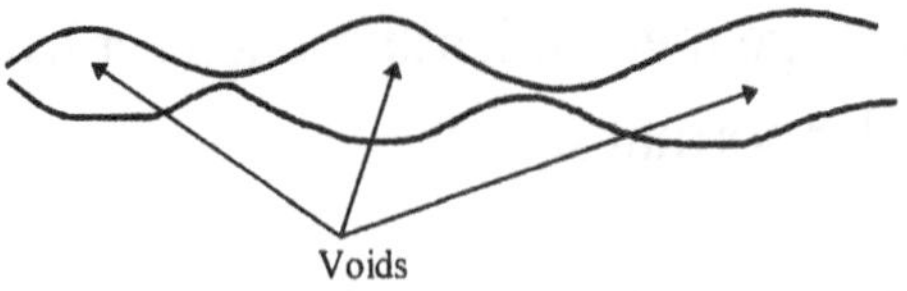

Figure 2. Schematic presentation of a fracture by the model of voids (after Tsang & Witherspoon 1981).

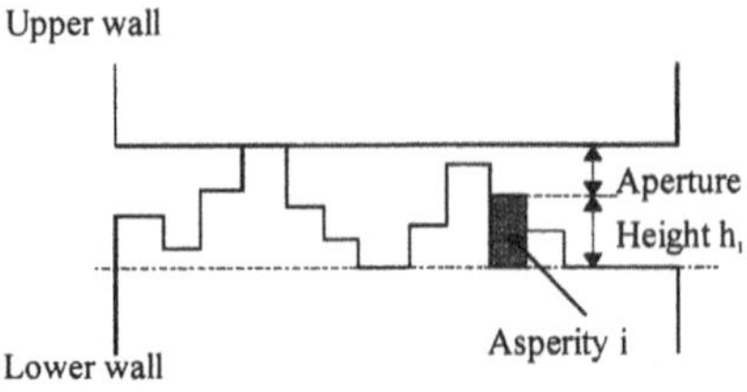

Figure 3. Schematic presentation of a fracture by the model of asperities.

2 SUMMARY OF EXPERIMENTAL TESTS

In this section, a short summary of experimental results on Lanhelin granite by Benjelloun's experiments (1991) and on Tennelles granite by Sotoudeh 's (1995) is given. These tests were performed on artificial rock joints. The results obtained will be used to check the hydromechanical model proposed in this paper.

Benjelloun's experiments (1991) consist in some hydraulic conductivity tests under constant normal stress applied by a machine. The flow is realised by hydraulic gradient between the ends of the samples and a central spout (Fig. 4).

Sotoudeh (1995) has realised three kinds of tests. At first, some mechanical tests, without flow pressure, have been realised. In order to have a uniform normal stress, the normal load is applied by a confinement pressure. Then, coupling tests are also performed : during these tests, the confinement pressure is constant and fluid pressure varies but remains uniform in the joint. Finally, flow tests have been conducted under different confinement pressure and fluid pressure. The flow is built up with the help of a hydraulic gradient between the top and bottom of the sample. The experimental device allows the measure of fluid pressure at four joint levels and then gives the fluid pressure distribution (Fig. 5).

3 MECHANICAL BEHAVIOR

In the present model, we consider rock fracture as a set of voids (Tsang & Witherspoon 1981) and accordingly, the progressive joint closure under normal stress is seen as a modification process of void space.

The principal point of the model is to describe the process through a compaction variable. For the mechanical behavior of a joint submitted to normal stress, this variable is defined by a scalar d^*. Its evolution is expressed as a function of normalised joint closure and checks the following asymptotical conditions : in the initial state $(\sigma_n = 0)$, d^* is taken equal to zero and for the final state when the joint is completely closed, d^* equals to 1. The complete joint closure corresponds to the state when the behaviour of fractured rock can be compared with that of the intact rock.

The second important point is the definition of this complete closure state. Numerous authors show that this phenomenon occurs for a stress level about 15 - 20 MPa. However, more detailed studies reveal that for this level, the fracture is not completely closed. Iwai (1976) found experimentally that the contact surface ratio ω for simple fracture under 20 MPa is included between 0.1 - 0.2 for granite and 0.25 - 0.35 for basalt. Futhermore, Tsang & Witherspoon (1981) precise that even if in the mechanical point of view, fractured rock and intact rock have the same behaviour , in the hydraulic point of view, the fracture is still opened and the permeability of the fractured sample is different from that of intact sample. Besides, Kranz & al (1979) found that the flow rates for a fractured rock and for intact rock are similar for a normal stress of about 200 - 300 MPa. Consequently, such a level stress has been chosen for the present model and is supposed to correspond to a complete joint closure.

Therefore, the mechanical behaviour is controlled by the following equations. The normalised joint closure is expressed by :

$$V^* = V/V_{max} \tag{3}$$

and the relation normal stress - displacement V^* is given by :

$$\sigma_n = E^m d^* V^* \tag{4}$$

From experimental data, the compaction variable is expressed by the following relation :

$$d^* = 1-(1-V^*)^{a_0} \tag{5}$$

The modulus E^m corresponds to the stress level when the fracture is completely closed i. e. $V^* = 1$ and $d^* = 1$. The value of E^m may depend on the rock studied. The parameter a_0 is a parameter fitted from experimental data and for some cases, it will be chosen as a function of normalised joint closure.

The model is now tested from Haji Sotoudeh's mechanical tests and will be compared with two classical models : Goodman (1976) and Bandis et al (1983) (Fig. 6). The parameters used have been listed in tables 1 and 2, σ_i is the pre-loading stress and V_i is the closure for this stress. We can note that K_{ni}, A and t have been calculated by numerical fitting of experimental. The Goodman's model generally gives better results but, this model includes two fitting parameters. So, we can conclude that the present model gives similar results as the two models.

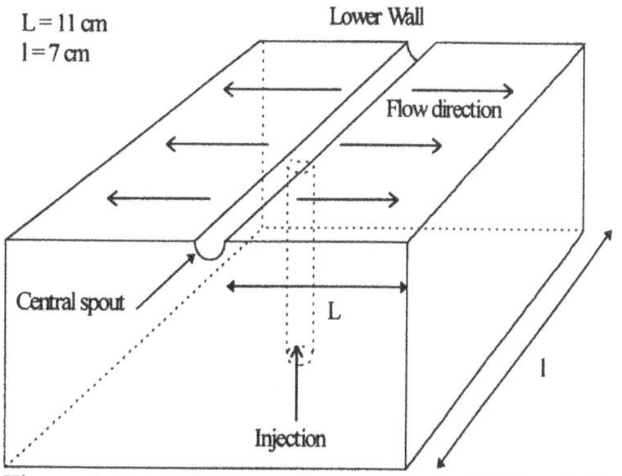

Figure 4. Benjelloun 's tests on sample L2 (Lanhelin granite).

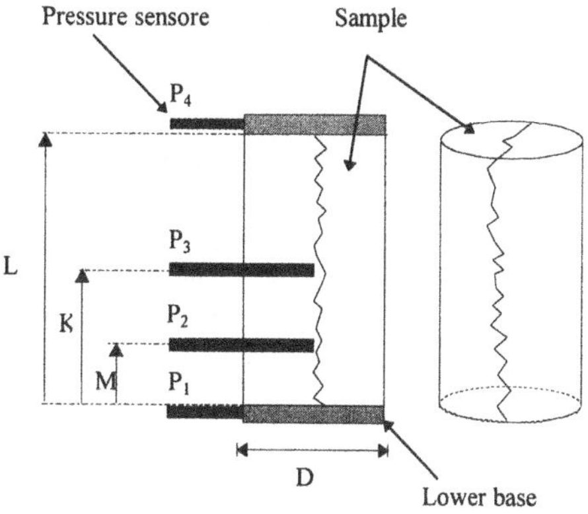

Figure 5. Haji Sotoudeh tests on Tennelles granite (D = 64 mm, M = 35 mm, K = 69.6 mm, L =113.5 mm).

Table 1. Model mechanical parameters.

Paramètres	Haji Sotoudeh cycle 1	Haji Sotoudeh cycle 4	Benjelloun
V_i (μm)	0	104	108
σ_i (MPa)	0.26	0.39	0.151
V_m (μm)	96	202	160
E^m (MPa)	200	200	200
a_0	0.0262	0.0347	0.008

Table 2. Bandis and Goodman Parameters.

Bandis	Goodman
$V_m = 96$ μm	$V_m = 96$ μm
$\sigma_{ni} = 0.26$ MPa	$\sigma_{ni} = 0.26$ MPa
$K_{ni} = 0.0103$ MPa/μm	$A = 5.511$ MPa/μm
	$t = 0.993$

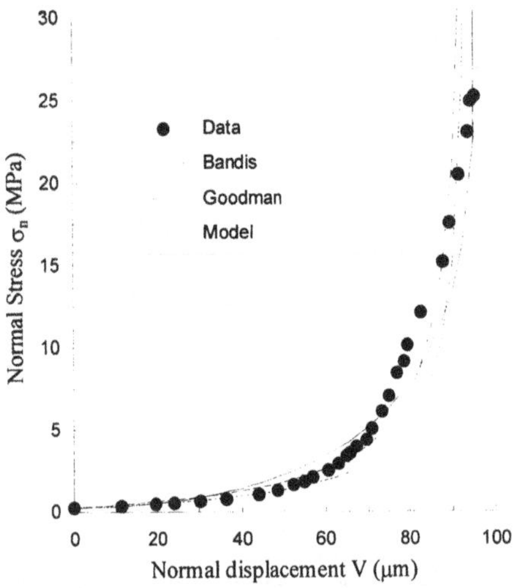

Figure 6. Simple compression test (1st cycle) on Tennelles granite (HAJI SOTOUDEH 1995) - Experiments and numerical simulations.

4 HYDROMECHANICAL BEHAVIOR

In order to describe the hydromechanical behaviour, it is necessary to consider the coupled effects of normal stress and fluid pressure in the fracture. This will be made by extending the Biot's effective stress concept.

4.1 *Hydromechanical coupling law - Effective stress concept*

In our model, the Biot's effective stress concept for porous media is generalised to rock joints and the effective stress is expressed by the following equations :

$$\sigma_{eff} = \sigma_n - \beta \cdot P_f \tag{6}$$

where β is the generalised Biot's coefficient which allows to account for the hydromechanical coupling. Indeed, it describes the effect of fluid pressure on the mechanical behaviour of rock joints. But, on the other hand, this coefficient itself depends on the compaction variable :

$$\beta = b(1 - d^*) \tag{7}$$

where b is a material constant determined from experimental data.

The equations (4) and (6) constitute the coupled hydromechanical model and they can be written in the following form :

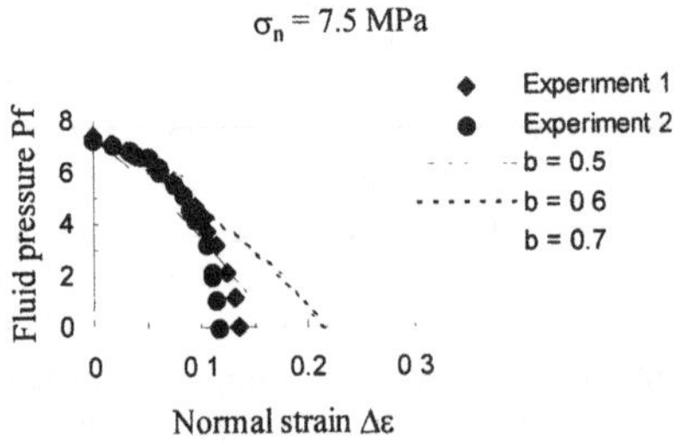

Figure 7a. Haji Sotoudeh coupling tests - Experiments and numerical simulations for $\sigma_n = 7.5$MPa.

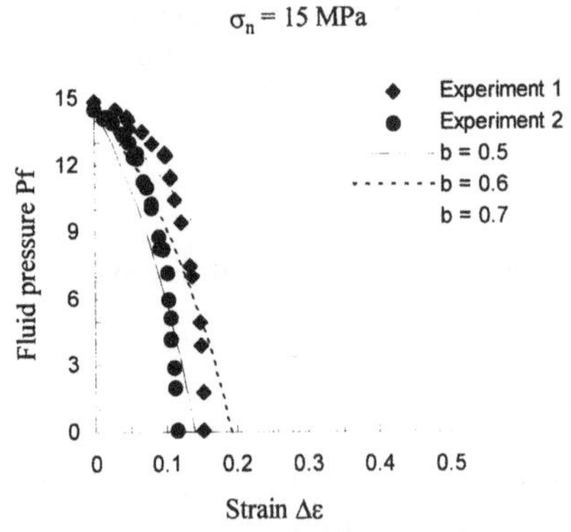

Figure 7b. Haji Sotoudeh coupling tests - Experiments and numerical simulations for $\sigma_n = 15$MPa.

$$\sigma_n - \beta \cdot P_f = E^m d^* V^*$$

i. e.

$$\sigma_n = E^m d^* V^* + \beta \cdot P_f \quad (8)$$

In order to justify the used effective stress concept and to calculate the parameter b, Haji Sotoudeh's coupling tests (1995) are simulated. The tests have been realised on the fourth cycle of the mechanical loading and the mechanical parameters have been listed in table 1. The results are presented on Figures 7a, b for different values of b and the choice of b = 0.6 seems reasonnable for the set of tests.

4.2 *Hydromechanical behaviour modelling :*

The aims of this work is to predict the flow rate and fluid pressure distribution in joint. A numerical solution using finite differences method is considered. The principle is to divide the hydraulic charge between injection pressure P_0 and drainage pressure P_n in n sections (Fig. 8). For each section i, the fluid pressure is constant and the following hypothese are used :

◊ the hydromechanical coupling law (8) is valid,
◊ the flow equations (1) and (2) are locally valide,
◊ the open surface ratio is equal to β_i .

The distance between two sections is noted Δx_i . Each section is defined by its aperture e_i , its open surface ratio β_i and its fluid pressure P_{fi} which is expressed by the following relation :

$$P_{fi} = P_0 - i \cdot \Delta P \quad (9)$$

with

$$\Delta P = \frac{P_0 - P_n}{n} \quad (10)$$

From the equations (1) and (2), the average velocity in the flow direction and the flow rate for a section i are expressed :

$$v_i = \frac{e_i^2}{12\mu} \cdot \frac{\Delta P}{\Delta x_i} \quad (11)$$

$$Q_i = \frac{e_i^3 \beta_i}{12\mu} \cdot \frac{\Delta P}{\Delta x_i} \quad (12)$$

The aperture e_i is defined by :

$$e_i = e_0 - V_i \quad (13)$$

where e_0 is the initial aperture of the joint and V_i the joint closure for the section i calculated from the relation (8) due to normal stress σ_n and the hydraulic pressure P_{fi} .

As there is fluid mass conservation, by noting L the joint length, we have :

$$Q = Q_1 = Q_2 = ... = Q_n$$

that is leads to

$$Q = \frac{\sum_{i=1}^{n} e_i^3 \beta_i \Delta P}{12\mu L} \quad (14)$$

and therefore, the length Δx_i is given by :

$$\Delta x_i = L \cdot \frac{e_i^3 \beta_i}{\sum_{i=1}^{n} e_i^3 \beta_i} \quad (15)$$

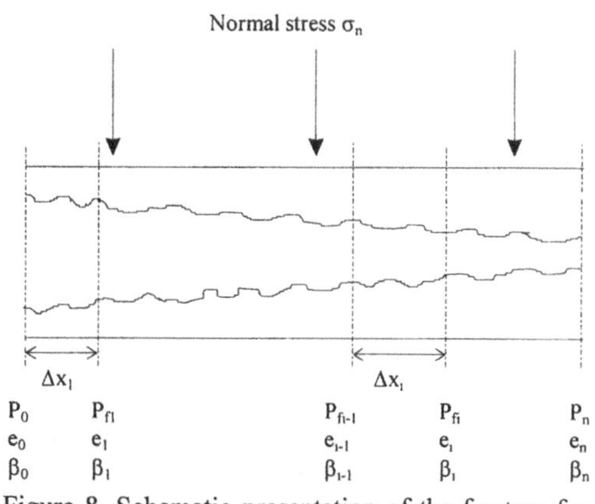

Figure 8. Schematic presentation of the fracture for the hydromechanical model.

The hydromechanical coupling model so defined is validated through Benjelloun's tests (1991) and Haji Sotoudeh's tests (1995). Benjelloun used an injection pressure lower than 5.10^{-2} MPa, and this, in order to consider normal stress as effective stress. So, we have used pressure fluid of the same order of magnitude and b was taken equal to 0.7 but it had no real influence in this case. The numerical results have been compared with the Tsang & Witherspoon

Table 3. Hydromechanical models parameters.

Tsang & Witherspoon	
β =	26.66
ω =	0.10 for σ = 20 MPa
b_0 =	176.5 μm
Model	
P_0 =	0.05 MPa
P_n =	0 MPa
b =	0.7
n =	10
e_0 =	176 μm

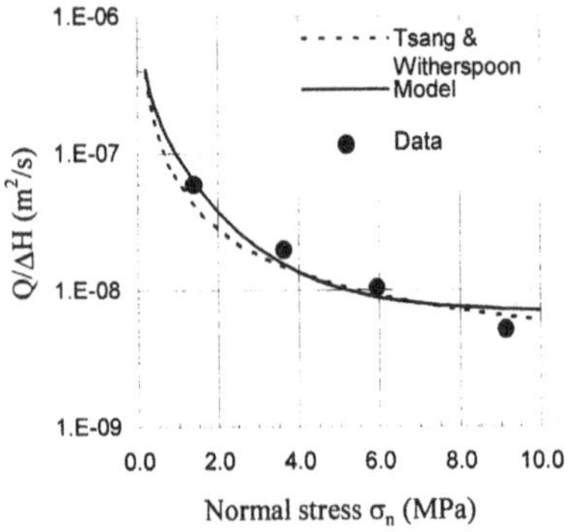

Figure 9. Benjelloun hydraulic conductivity tests - Experiments and numerical simulations.

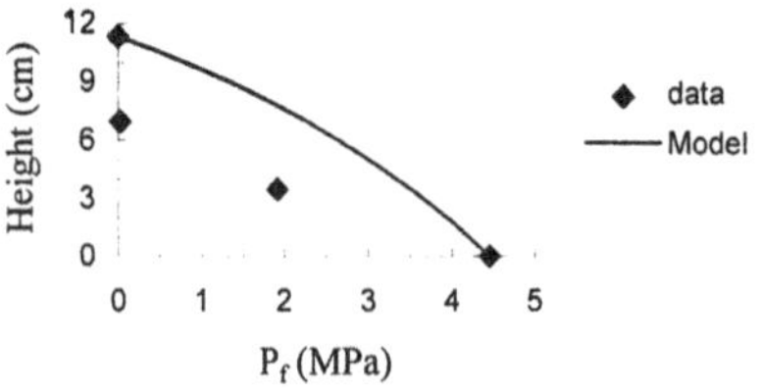

Figure 10a. Haji Sotoudeh flow tests - Experiments and numerical simulations for $\sigma_n = 5\,\text{MPa}$ and $P_{inj} = 4.46\,\text{MPa}$.

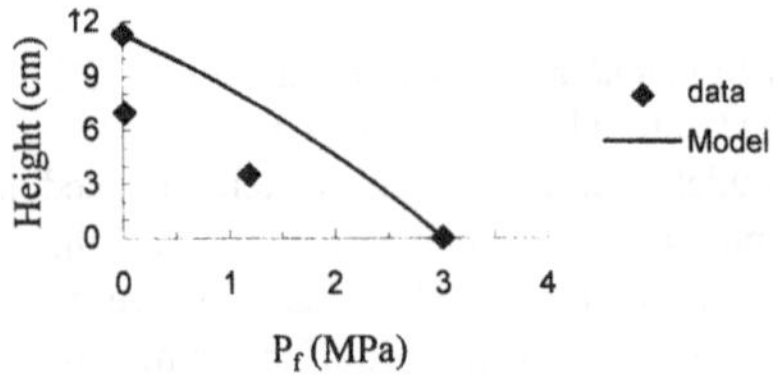

Figure 10b. Haji Sotoudeh flow tests - Experiments and numerical simulations for $\sigma_n = 5\,\text{MPa}$ and $P_{inj} = 3\,\text{MPa}$

model (1981) in figure 9. The set of the model parameters has been listed in table 1 and table 3. Both the two models give good agreements with data. However, as the flow pressure in this case is quite small, the good agreements obtained allows only to confirm the performance of the mechanical model to predict the closure variation.

Then, Haji Sotoudeh's flow tests in order are simulated to predict the pressure distribution in the joint. The results are presented in the figures 10a, b and are satisfactory except for the pressure at the point n°3. Further validation is still necessary to have a better understanding.

5 CONCLUSION

In this paper, a complete hydromechanical model for joints submitted to normal stress is presented and tested. The progressive closure is considered as a modification process of voids and described by a compacity variable. Furthermore, in order to describe the coupling process, the Biot's effective stress concept is generalised to rock joints. The coupled model, with fluid flow laws, provides a set of equations for complete modelling of hydromechanical behaviour. The numerical predictions are in good agreement with experimental data and with other classical models.

However, flow rate predictions with high fluid pressure would be necessary in order to complete the model validation. Furthermore, the mechanical behaviour modelling will be extended to the study of joint responses under shear stress. Finally, it would be interesting to establish possible correlation between the model's parameters and joint roughness.

REFERENCES

BANDIS S., LUMSDEN A.C., BARTON N. (1983) - Fundamentals of rock joint deformation, *Int. J. Rock. Mech. Min. Sci.*, vol 20, n° 6, p 249-268.

BENJELLOUN Z. H. (1991) - Contribution à l'étude expérimentale et numérique du comportement mécanique et hydraulique des joints rocheux, doctoral thesis (in french), University of Grenoble.

BOUSSINESQ J. (1868) - Mémoire sur l'influence des frottements dans les mouvements réguliers des fluides, J. Liouville, 13, p 377-424.

GENTIER S. (1986) - Morphologie et comportement hydromécanique d'une fracture naturelle dans le granite sous contrainte normale - Etude

expérimentale et théorique, doctoral thesis (in french), University of Orléans.
GOODMAN R. (1974) - Les propriétés mécaniques des joints, *Progrès en mécanique des joints*, vol 1, t. A (3ème congrès de la Soc. Int. Mech. Roches, Denver).
GOODMAN R.E. (1976) - Method of geological engineering in discontinuous rocks, *West Publishing Company*.
HAJI SOTOUDEH M. (1995) - Etude expérimentale et modélisation du couplage hydromécanique des joints rocheux, doctoral thesis (in french), University of Lille.
IWAI K. (1976) - Fundamental studies of the fluid flow through a single fracture, Ph.D. thesis, Univ. of Calif., Berkeley.
KRANZ R. L., FRANKEL A. D., ENGELDER T., SCHOLZ C. H. (1979) - The permeability of whole and jointed Barre granite, *Int. J. Rock. Mech. Min. Sci.*, vol 16, p 225-234.
TSANG C.F. (1991) - Coupled hydromechanical - thermochemical processes in rock fractures, Reviews of Geophysics, vol 29, n° 4, p 537-551.
TSANG Y.W., WITHERSPOON P.A. (1981) - Hydromechanical Behavior of a Deformable Rock Fracture Subject to Normal Stress, Journal of Geophysical Research, vol 86, n° B10, p 9287-9298.

Mechanics of Jointed and Faulted Rock, Rossmanith (ed.) © 1998 Taylor & Francis, ISBN 90 5410 955 6

Hydromechanics of fractured rock masses: Results from an experimental site in limestone

Y. Guglielmi
University of Franche-Comté, Besançon, France

ABSTRACT: This paper presents the first results of the hydromechanical stimulation of a jointed rock mass that was submitted to a controlled hydrostatic pressure. Under water pressures of about 0.7 bars, the behaviour of the rock mass is elastic and the deformation is localized along joints. For a pressure lower than the lithostatic one, deformation correlates with the permeability of joints : the most permeable joints open and this opening induces the closing of the least permeable joints. For a pressure in the range of the lithostatic pressure, the tilting of blocks amplifies the deformation.

1 INTRODUCTION

The deformation of a jointed rock mass under natural piezometric pressures or rainfall infiltration can cause a local increase by a factor 2 or 3 of the mass joints apertures (DURAND 1992 ; CROCHET et al. 1983 ; CROCHET 1983). Such an increase in the joint aperture can slightly modify hydrodynamics of the reservoir. If the relationships are experimentally well established for one single fracture (WITHERSPOON et al. 1980) or for meter-scale joint networks (JOUANNA 1972), large and permeable jointed rock masses remain poorly studied « in situ » due to the general difficulty in determining boundary conditions in such large sites. For this reason, we have developped an experimental site in a karstified limestone rock mass where we can follow natural filling and emptying of the reservoir by opening and closing the main spring which is equiped with a water gate. We have studied the effects of fluid pressure and lithostatic stress on the deformation which is normal to the joints.

2 DESCRIPTION OF THE SITE

The chosen reservoir is made of a 17 m-thick pile of lower Cretaceous limestone limited to the top and to the bottom by a two-meter thick impervious glauconious marl layer (Figure 1 A). We worked on the lower part of the aquifer very near the reservoir main spring of annual average yield of 12 $l.s^{-1}$. This main spring appears right on a fault zone which plays the role of a relative natural dam to groundwater by a vertical contact between permeable calcareous and rather impervious calcareous marls. Seven meters higher along the same boundary, a temporary spring T1 overflows during experiments of rock mass saturation. Up stream in the river besides the site (30 m far from the main spring) another secondary spring T2 also temporary overflows along a N80-70N fault, 6 m higher than the main spring.

The rock mass is characterized by three directions of discontinuities (Figure 1 B):

- the bedding surface S_0 underlined by tiny marly laminae (0.1 mm thick) with a N40-45E attitude and of low average hydraulic aperture (estimated to 0.7 mm). Some S_0 joints are karstified with micro-channels of 1.6 mm in average apertures;
- the N140-75NE faults contain a 4 to 5 cm-thick breccia zone often karstified with channels of 2.3 mm in average hydraulic aperture. Many contact zones (bridges) made of calcitic breccia remain between the two fracture planes ;
- the N90 to N60 trending and 70 to 80 SE to NW dipping faults present the largest karstic channels of a 4-5 mm average aperture.

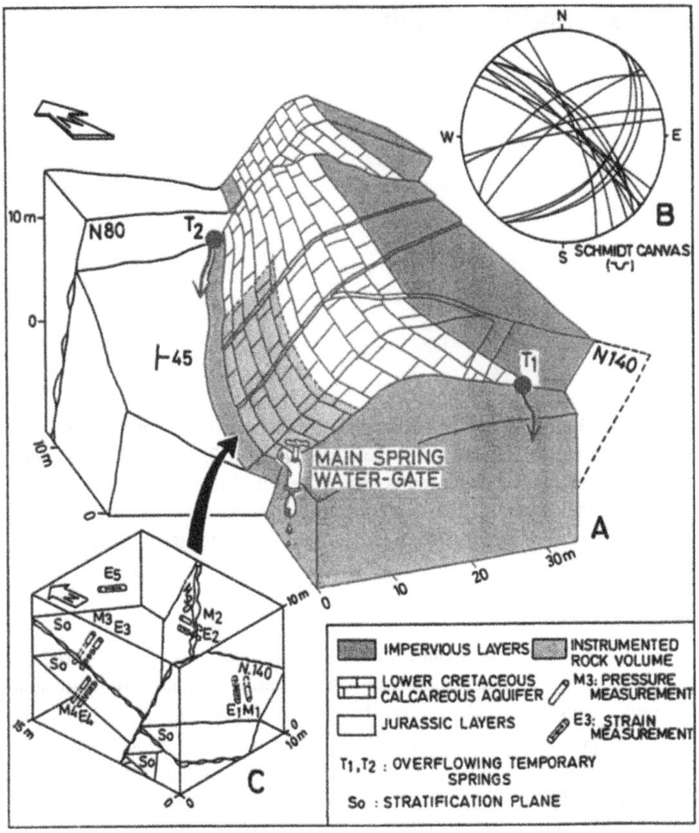

Figure 1 : Geology of the experimental site ; A-General geometry ; B-Schmidt network of the three directions of discontinuities ; C- Measurement points location in the joint network of the instrumented rock volume.

3 EXPERIMENT PROTOCOLE

On the main spring of the aquifer a water-gate has been settled in to induce artificial hydrostatic pressure elevations in the rock mass. The pressure is monitored along the three joint directions (Figure 1 C). At each monitored point two holes were drilled normally to the joint 20 cm from each other. In the first hole, a chamber allows hydrostatic pressure monitoring in the joint, in the second one, a vibrating-wire extensometer is centered on the joint in order to monitor the deformation normal to the joint. The extensometer is included into concrete in order that the deformation measurement is that of a concrete cylinder (7 cm diameter) which long axis coincides with the extensometer. The temperature of the extensometer is also recorded. All measurements occur in a 1500 m^3 rock volume. Yields of the different springs were also registered all along the experiment.

The presented experiment corresponds to one hour closing of the water-gate (Figure 2). Repeated over ten times it yielded the same results each time. Pressure measurements of this experiment have been made with mechanical manometers with a 0.01 bar accuracy. Strain measurements are given with an instrument accuracy of 0.5 μm/m and a temperature accuracy of 0.1 °C. During the first 30 mn of this experiment measurements were recorded every 30 s. Afterwards pressures were followed every 1 to 5 mn. When the water-gate was reopened at the end of the experiment, all the measurements were recorded every 30 s.

During the experiment we followed 5 pressure-deformation points (Figure 1 C):

- two points on S_0 joints ; one on a very permeable joint (E3, M3) and the other on a low permeable joint (E4, M4) ;
- one point on the N90 (E2, M2) and one on the N140 (E1, M1) faults ;

- one point situated in a non-fractured block (E5) to monitor the rock matrix strain. This extensometer is orientated N130-45SE, that is parallel to the bedding surface.

4 RESULTS

Before the closing of the water-gate, there was no pressure in the reservoir and the yield of the main spring was 10 $l.s^{-1}$. When the water-gate was closed and during the first 11 mn, water filled the rock mass: (transitional state). After 11 mn the yields of springs T1 and T2 increased (on Figure 2 we represented the cumulative variations of T1 and T2 yields). A stabilization occured 20 mn later. From the 11th to the 20th mn, the flow in the joints was in a transitional state influenced by the hydraulic boundaries of the site (springs T1 and T2). After 20 mn, there was a steady state characterized by the constant overflow of springs T1 and T2. When the water-gate was reopened, water pressure fell down to zero in less than 30 s and up to 5 mn depending on the joints permeability. Overflowing springs T1 and T2 and the main spring yield variations versus time look like a typical natural karstic spring. The hydrograms present a rapid decrease followed by a slow dry up and a final stabilization at the initial yield of 10 $l.s^{-1}$.

The deformation-versus-time curves are all very noisy (Figure 2). We reduced this phenomenon by a sliding mean on a 5 mn period. We can also see a slow deviation of the signal due to the thermomechanical effect of sun radiations on the rock outcrop. When the water-gate was closed, the signal was clearly characterized by deformations ranging from 1 μm/m to 200 μm/m depending on the point. When the water-gate was opened a residual deformation remained while water pressure in the joints equalled to zero. On extensometer E3 the extension state gave way to a shortening state. A return to the initial mechanical state occurred slowly after 30 to 40 mn (except for E3 where it is much longer). So the mechanical behaviour of rock mass appeared roughly elastic for the low hydrostatic pressure of the experiment. The extensometer located in the non-fractured rock did not record any deformation which proves that strain is localized along joints.

Three deformation-versus-time curves could be observed (note that scales of strains values are different from one point to the other):

- along the least permeable bedding surface (Figure 2 E4, M4), there is a 1 μm/m closure while hydrostatic pressure in the joint increases up to 0.44 bars ;
- along the permeable N140 fault (Figure 2 E1, M1), the aperture occurs as soon as the water-gate is closed. Deformation-versus-time and pressure-versus-time curves are similar ;
- along the most permeable bedding surface (Figure 2 E3, M3) and N90 fault (Figure 2 E2, M2), the aperture occurs also for low water pressures (0.15 to 0.2 bars). However between 0.4 and 0.44 bars an important aperture of the joint takes place (about 100 μm/m) while water pressure slightly decreases of about 0.05 bars.

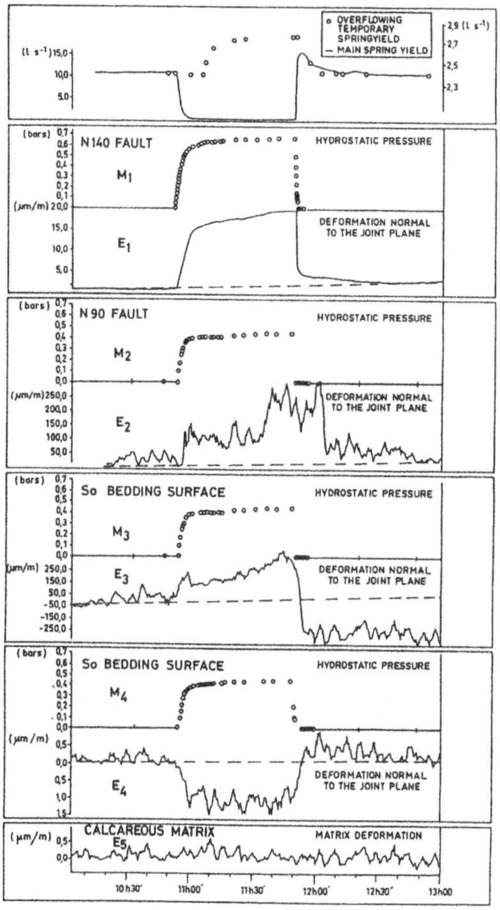

Figure 2 : Hydromechanical variations during a one hour closing of the water-gate.

5 INTERPRETATION

Even if we can only deal with relative normal values (due to the measurement method), the very different ranges of deformations cannot be explained by different ranges of water pressures. Indeed, consequent deformations take place for very weak hydrostatic pressure variations. The orientation of the joint plane relative to the principal lithostatic stress axis explains that the largest deformations take place along the N90 steeply dipping fault. This is due to the weak influence of the lithostatic stress on such a subvertical plane. Nethertheless a similar range of deformation is recorded along the permeable S_0 joint (Figure 2 E3, M3) whose angle between the main stress axis and the one normal to the plane is more important. Similarly the deformation range is much lower along the N140 fault (Figure 2 E1, M1) which is vertical. This reveals the indirect effect of the hydraulic aperture of joints (straightened linked to their geological history) on deformation. Therefore, the more joints are opened, the less abundant the bridges are between the two planes, and as a result, the less rigid they are. On this site S_0 joints are underlined by marl films which eases deformation.

If we consider the three deformation types observed, only the deformation occurring normally to the N140 fault plane (Figure 2 E1) can be explained by the water pressure increase in this joint. Concerning the very permeable S_0 joint and the N90 fault, deformation only follows water pressure up to 0.4-0.44 bars. This value is about the same as the lithostatic stress one estimated at these points between 0.35 to 0.7 bars. Over 0.4 to 0.44 bars, deformation is amplified by a phenomenon that we attribute to the tilting of non fractured blocks. These blocks remain unstrained as measured with extensometer E5 (note that non fractured rock deformations are only measured in one arbitrary direction). On the site, the tilting is mainly guided by N70-90 faults and S_0 joints. The closure of the low permeable S_0 joint (Figure 2 E4) is linked to the opening of the most permeable one (Figure 2 E3) which happened at the beginning of the experiment. In that case, we must also admit the displacement of an undeformed matrix block.

6 CONCLUSIONS

Our experimental site allows the qualitative measurement of hydromechanical coupling in a near-surface jointed rock mass. Our main result is to show that in any case the hydromechanical behavior of a jointed rock mass cannot only be modelized by combination of elementary pressure-strain relations valid for a unique joint. Our experiment reveals two distinct hydromechanical behaviors of the rock mass. First, for an hydrostatic pressure lower than the maximum local stress, the permeability regulates joints deformation. The most permeable ones open with an hydrostatic pressure increase to the detriment of less permeable joints which close. Second, when the hydrostatic pressure reaches principal stress value, the deformation is amplified independently from pressure. That deformation is caused by the tilting of blocks following the joints network geometry. This phenomenon has already been described in more important rock masses (Durand 1992). In both cases, we can see deformations independant from hydrostatic pressure increase. Indeed, coupling measurements of natural deformation along the different joint directions of a rock mass with measurements of piezometric or spring yield can lead to the determination of the permeability tensor orientation of that rock mass.

REFERENCES

Crochet, P. 1983. Fracturation profonde en milieu cristallin. Etude in situ de la déformation des milieux karstiques. Thèse de Docteur ingénieur, USTL, Montpellier, 240 p.

Crochet, P.,Lesage, P., Blum, P.A. & M. Vadell 1983. Extensometric deformations linked with rainfalls. *Annales Geophysicae*, 1, 4-5, 329-334.

Durand, V. 1992. Structure d'un massif karstique. Relations entre déformations et facteurs hydro-météorologiques. Thèse de Doctorat en Géologie, USTL, Montpellier, 207 p.

Jouanna, P. 1972. Effet des sollicitations mécaniques sur les écoulements dans certains milieux fissurés. Thèse de Doctorat, Université Paul Sabatier de Toulouse, 263 p.

Witherspoon, P.A., Wang, J.S.Y., Iwai, K. & J.E. Gale 1980. Validity of cubic law for fluid flow in a deformable rock fracture, *Water Resource Research*, Vol. 16, n° 6, p. 1016-1024.

Dam

Mechanics of Jointed and Faulted Rock, Rossmanith (ed.) © 1998 Taylor & Francis, ISBN 90 5410 955 6

Contribution of a 3D model to the analysis of the hydromechanical behaviour of a jointed rock mass dam foundation

H. Bargui
TOTAL Scientific and Technical Center, Saint-Remy-Les-Chevreuse, France (Formerly: Centre de Géologie de l'Ingénieur)

J. Billiotte & R. Cojean
Centre de Géologie de l'Ingénieur, Ecole des Mines de Paris & Ecole National des Ponts et Chaussées, Paris, France

ABSTRACT: The paper describes numerical analyses of hydromechanical behaviour of a fissured rock mass dam foundation. It shows that the dam load induces a highly compressed area, in which hydraulic conductivity and flow lines are modified. This modification may affect the stability of the dam foundation. The numerical analyses are carried out by using a 3D coupled mechanical-hydraulic model, called BRIG3D.

INTRODUCTION

It is well known that a dam loading may seriously affect the distribution of the natural stress within the rock mass surrounding the dam. Some highly compressed zones can appear and modify the in situ hydraulic parameters. These modifications may endanger the stability of the dam foundation and abutments.

The location of these highly compressed areas depends on the geological site conditions. It is mainly related to the direction of the applied dam force with respect to the orientation of existing sets of discontinuities such as faults, bedding planes, foliations and other planes of weakness. These discontinuities introduce a further anisotropic behaviour of a rock mass which imposes severe complications for the design of the dam foundations. Numerical models based on the distinct element method (Cundall, 1971, Kawai *et al.*, 1981, Belytschko *et al.*, 1984) seem to be an adequate tool to analyse the hydromechanical response of such a medium.

This paper describes analyses of hydromechanical behaviour of a jointed rock mass dam foundation. These analyses are performed by using a computational model, called BRIG3D (Tahiri, 1992, Bargui, 1997), which is a coupling of a 3D static distinct element model and a hydraulic model. The latter involves flow through connected planes of discontinuity which is solved by mean of a boundary element approach.

PRINCIPLE OF THE MECHANICAL MODEL

The developed distinct element model simulates a fissured rock mass as a set of rigid blocks interacting mechanically at their interfaces. These blocks are assumed to be rigid and only interface deformation can occur. The movement of each block i is characterised by a translation vector $\vec{U}_i$ and a rotation vector $\vec{W}_i$, with respect to its centroid.

For analysing the deformation of a jointed rock mass, the mechanical approach introduced in this model uses the principle of the minimum potential energy. This energy, denoted $\mathcal{E}$, is given by the sum of the joint deformation energy and the external load work, since it is assumed that blocks are rigid. $\mathcal{E}$ is expressed as a function of blocks centroid displacement. In order to take the non-linearities due to blocks movement and joints behaviour laws into consideration, the energy minimisation is performed by using incremental procedures.

For any given calculation step, the deformation energy of any interface or joint J common to the blocks i and j by way of their respective faces F_i and F_j, is given by:

$$E(\mathrm{J}) = \int_S (\frac{1}{2}\delta\vec{\epsilon}.\delta\vec{\sigma} + \delta\vec{\epsilon}.\vec{\sigma}_0)ds$$

where S is the joint surface, $\delta\vec{\epsilon}$ is the incremental deformation of the joint J, $\delta\vec{\sigma}$ is the incremental stress and $\vec{\sigma}_0$ is a cumulative stress vector due to previous steps.

The incremental deformation $\delta\vec{\epsilon}$ is given by the incremental relative displacement between the faces F_i and F_j. To be evaluated, first the displacement $\delta\vec{u}_p$ of any material point within the face F_p (p=i or j) is approximated by:

$$\delta\vec{u}_p = \delta\vec{U}'_p + B_p\,\delta\vec{W}'_p \qquad (1)$$

where $\delta\vec{U}'_p$ is the incremental displacement of the block p and $\delta\vec{W}'_p$ its incremental rotation vector. These two vectors are expressed with respect to a local cartesian frame $(\vec{t}_x, \vec{t}_y, \vec{n})$ where $\vec{n}$ is the unit normal to the joint plane pointing towards the block j.
In the equation (1), B_p is a matrix given by:

$$B_p = \begin{bmatrix} 0 & z - z_p & -(y - y_p) \\ -(z - z_p) & 0 & x - x_p \\ y - y_p & -(x - x_p) & 0 \end{bmatrix}$$

where (x, y, z) and (x_p, y_p, z_p) are respectively the co-ordinates of the considered material point and the block centroid, with respect to the local frame. The deformation $\delta\vec{\epsilon}$, equal to $(\delta\vec{u}_j - \delta\vec{u}_i)$, can therefore be expressed in a matrix form as:

$$\delta\vec{\epsilon} = [B_{ij}]\,\delta\vec{U}'_{ij} \qquad (2)$$

where

$$[B_{ij}] = [-[B_i]\ ,\ [B_j]] = [B]$$

and

$$\delta\vec{U}'_{ij} = \begin{bmatrix} \delta\vec{U}'_i \\ \delta\vec{U}'_j \end{bmatrix}$$

By introducing the joint normal and shear stiffness K_n and K_t, the incremental stress vector can be expressed as:

$$\delta\vec{\sigma} = [K]\,\delta\vec{\epsilon} \qquad (3)$$

where

$$[K] = \begin{bmatrix} K_t & 0 & 0 \\ 0 & K_t & 0 \\ 0 & 0 & K_n \end{bmatrix}$$

is the joint stiffness matrix.
By using the transformation matrix $[T]$, from the global to the local co-ordinate frames, the displacement vector $\delta\vec{U}'_{ij}$ can be calculated as follows:

$$\delta\vec{U}'_{ij} = [T]\,\delta\vec{U}_{ij} \qquad (4)$$

where $\delta\vec{U}_{ij}$ is the displacement vector with respect to the fixed global frame.

Substituting equation (4) into equations (2) and (3) leads to:

$$\delta\vec{\epsilon} = [B]\,[T]\,\delta\vec{U}_{ij}$$

and

$$\delta\vec{\sigma} = [K]\,[B]\,[T]\,\delta\vec{U}_{ij}$$

The joint deformation energy can be written as:

$$E(\mathrm{J}) = \frac{1}{2}\,\delta\vec{U}^t_{ij}\,[K_{\mathrm{J}}]\,\delta\vec{U}_{ij} + \delta\vec{U}^t_{ij}\vec{f}_0$$

where

$$[K_{\mathrm{J}}] = \int_S [T]^t\,[B]^t\,[K]\,[B]\,[T]\,ds$$

$$\vec{f}_0 = \int_S [T]^t\,[B]^t\,\vec{\sigma}_0\,ds$$

With regard to all existing joints, the potential energy of the studied field can be expressed as follows:

$$\mathcal{E} = \frac{1}{2}\,\delta\vec{U}^t\,[\mathcal{K}]\,\delta\vec{U} + \delta\vec{U}^t\vec{\mathcal{F}}_0 - \delta\vec{U}^t\vec{F}_{ext}$$

where $[\mathcal{K}]$ is the global stiffness matrix, $\delta\vec{U}$ is the blocks displacement vector, $\vec{F}_{ext}$ is the external load and $\vec{\mathcal{F}}_0$ is the initial contact effort. The external and the contact efforts ($\vec{F}_{ext}$ and $\vec{\mathcal{F}}_0$) are expressed with respect to the blocks centroid.
The principle of the minimum potential energy leads to a system of linear equations:

$$[\mathcal{K}]\,\delta\vec{U} = \vec{F}_{ext} - \vec{\mathcal{F}}_0$$

The displacement vector $\delta\vec{U}$ constitutes the main unknown variable and must satisfy both blocks static equilibrium and boundary conditions. To include the non-linearities, an iterative procedure is used until small residual efforts (non-equilibrated forces and moments at blocks centroid) are obtained.

PRINCIPLE OF THE HYDRAULIC MODEL

This model simulates fluid flow through the blocks interface network. Flow through each interface is assumed to be steady and planar with a uniform hydraulic conductivity. It is therefore governed by the Laplace equation which is solved here by using a boundary element method (Huyakorn & Pinder, 1983). To ensure the hydraulic connection between all interfaces, fluid mass equilibrium is written at their intersection lines.
The model uses a cubic law to relate the hydraulic conductivity of any interface to its hydraulic aperture. Along the interface, this aperture is consid-

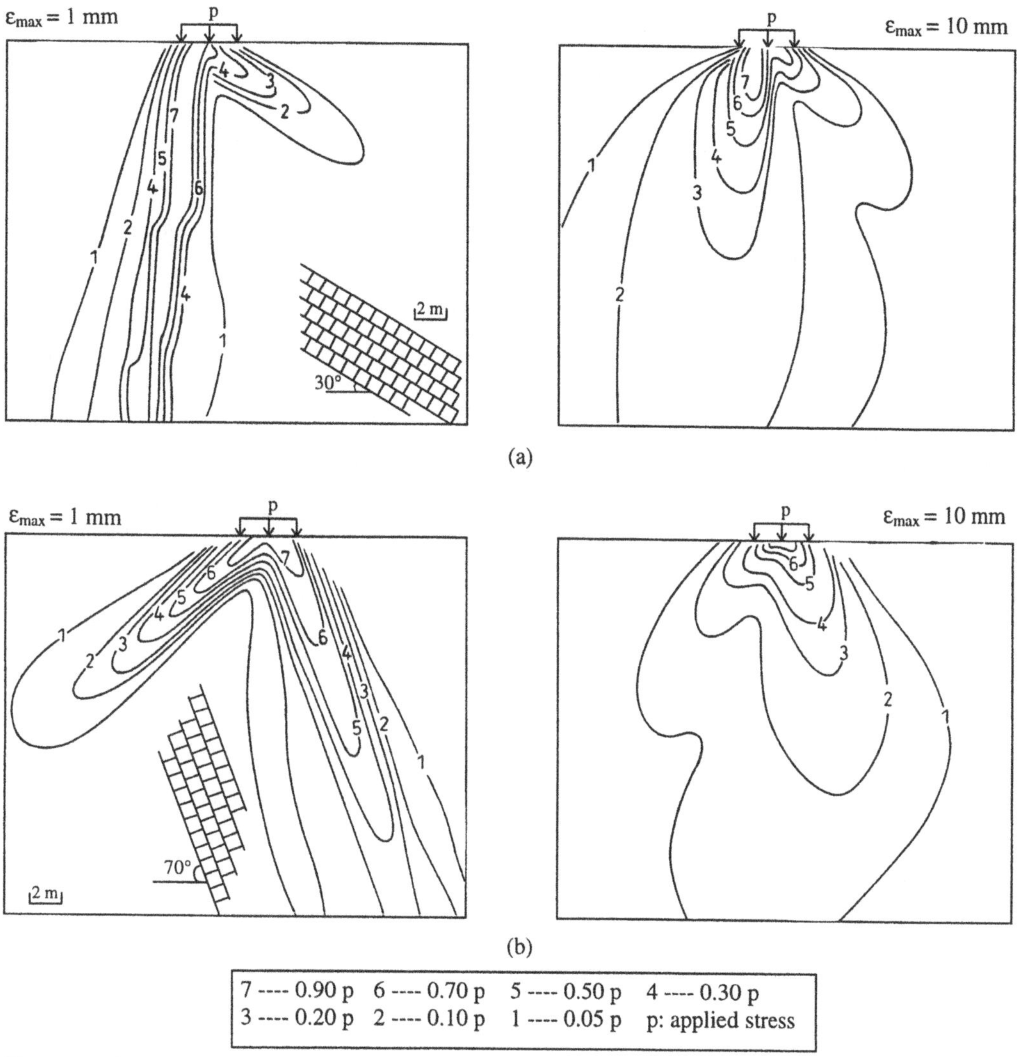

Figure 1: Normal stress contour (bulbs of pressure): (a) Infinite joint inclination of 30 degrees (b) Infinite joint inclination of 70 degrees (no gravity).

ered to be uniform but depending on the blocks relative movement, which closes or opens the interface.

The hydraulic aperture of an interface, being subject to an average normal deformation $\bar{\epsilon}_n$, is assumed to be equal to:

$$e_h = e_0\,(1 - \frac{\bar{\epsilon}_n}{\epsilon_{max}}) + e_{res} \qquad (5)$$

where e_0 and e_{res} denote respectively the initial and the residual hydraulic aperture of the interface (e_{res} may be equal to zero) and ϵ_{max} is its maximum mechanical closure.

Hence the hydraulic conductivity can be expressed by:

$$K_h = \frac{\rho\,g}{12\,\nu}\,e_h^3 \qquad (6)$$

where ρ is the fluid density, ν is the fluid viscosity and g is the gravity.

The equations (5) and (6) are used to couple this hydraulic model to the mechanical model introduced above. The reverse coupling is established by evaluating first the flow pressure resultant relative to blocks centroid, and then adding it to the external blocks loading.

NUMERICAL APPLICATION

The numerical model BRIG3D is applied to examine the effect of a dam, founded on a jointed rock mass, on the behaviour of this rock mass medium. Two orthogonal sets of joints are assumed to exist at the rock foundation (Fig. 1), one being finite, the other infinite. Normal behaviour of joints is controlled by the hyperbolic law introduced by Bandis *et al.* (1983) with a maximum mechanical closure ϵ_{max} and an initial stiffness K_{ni}. The normal stiffness K_n of each joint depends therefore on its normal deformation. This stiffness increases when the assigned maximum closure decreases. Shear behaviour of these joints is described by an elastic perfectly plastic law with a Mohr-Coulomb slip criteria. The shear stiffness is denoted K_t and the friction angle is denoted ϕ (cohesion is equal to zero).
To locate the area which is highly compressed due to the dam loading, first only a surface loading is considered. Two infinite fracture dippings are used (30 and 70 degrees). For each case, joints are supposed to all have the same mechanical properties (see table 1). The distributions of resulting compressive stress are presented in figure 1. In accordance with existing experimental and theoretical models (Gaziev & Erlikhman, 1971, Goodman, 1980), these results confirm the anisotropic character of the fractured rock media in relation with their structural characteristics. In fact, the induced compressive stress in the discontinua studied here is mainly concentrated in two directions. As shown by the equal normal stress contour (Fig. 1), the influenced zones become deeper and narrower when the maximum closure decreases. As confirmed by Bargui (1997), this is due to the increase of the joints stiffness ratio (normal stiffness over shear stiffness) rather than to the increase of only normal stiffness.
The coupled hydromechanical behaviour of the foundation is conducted by using a downstream dip angle of 70 degrees of the infinite fracture. This configuration is in accordance with the structural setting of the rock mass foundation of the Malpasset dam in France.
The mechanical and hydraulic conditions are given in figure 2. To prevent high pressure downstream, and thus blocks uplifting, a low water level is considered (10 m). In fact, the relatively low fracture spacing leads to small blocks unable to resist by their weight to any high pressure level.
The joints have the same maximum mechanical closure (0.001 m) and, initially, they all have the same conductivity which corresponds to a hydraulic aperture of 0.002 m. A residual aperture of 0.001 m is assigned to each joint.
At hydromechanical equilibrium, the joints conductivity is presented in figure 3 by a joint line width proportional to its new hydraulic conductivity. In this figure, the highly compressed joints are not drawn. Hence, the finite joints, highly compressed at great depth under blocks weight, are not plotted. The downstream fracture located at zone a_1 corresponds to the highly compressed area due to reservoir pressure. This downstream area behaves as a quasi impermeable screen which is subject to the total hydraulic head as shown by the equipotential lines presented in figure 4. The upstream zone a_2 corresponds to a decompressed area (fracture opening) due to the dam thrust. In addition, as illustrated in figure 4, high hydraulic gradients, and thus high water pressures, are produced downstream and lead to an opening of some fractures located near the surface (zone a_3 in Fig. 3). This is a critical zone because blocks could be lifted up due to the highly upward gradient.

Table 1: Joint input parameters in both studied cases (30 and 70 degrees inclination angle). For each case, two maximum closure values are considered.

Shear stiffness K_t (MPa)	100
Initial normal stiffness K_{ni} (MPa)	0.5
Maximum closure ϵ_{max} (mm)	10 and 1
Friction angle ϕ (degrees)	25
Cohesion (MPa)	0

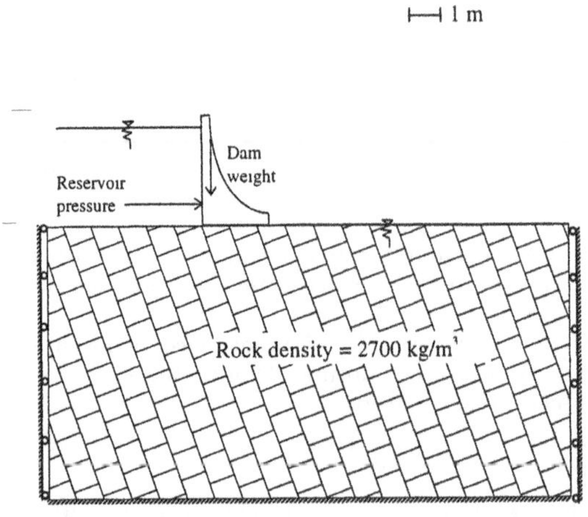

Figure 2: Mechanical and hydraulic boundary conditions. The boundary limits are supposed to be impermeable.

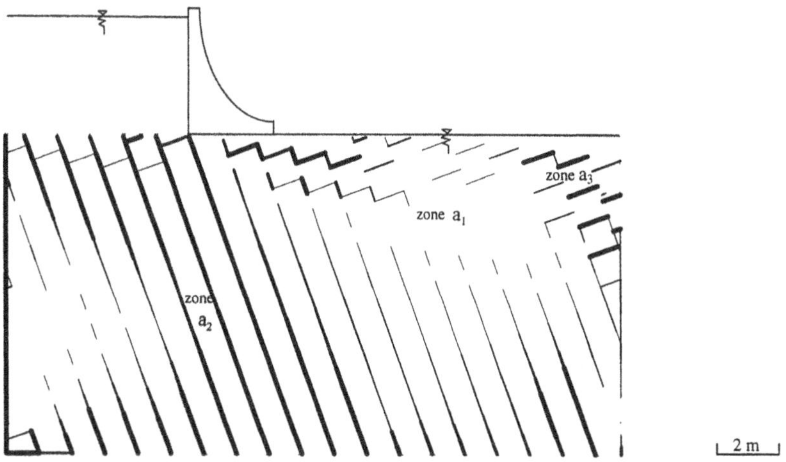

Figure 3: Distribution of the joints hydraulic conductivity at the hydromechanical equilibrium (for the 10 meters water level). The hydraulic conductivity is given by the joint line width (the minimum width corresponds to a conductivity equal to $2K_{res}$, where K_{res} is the residual conductivity).

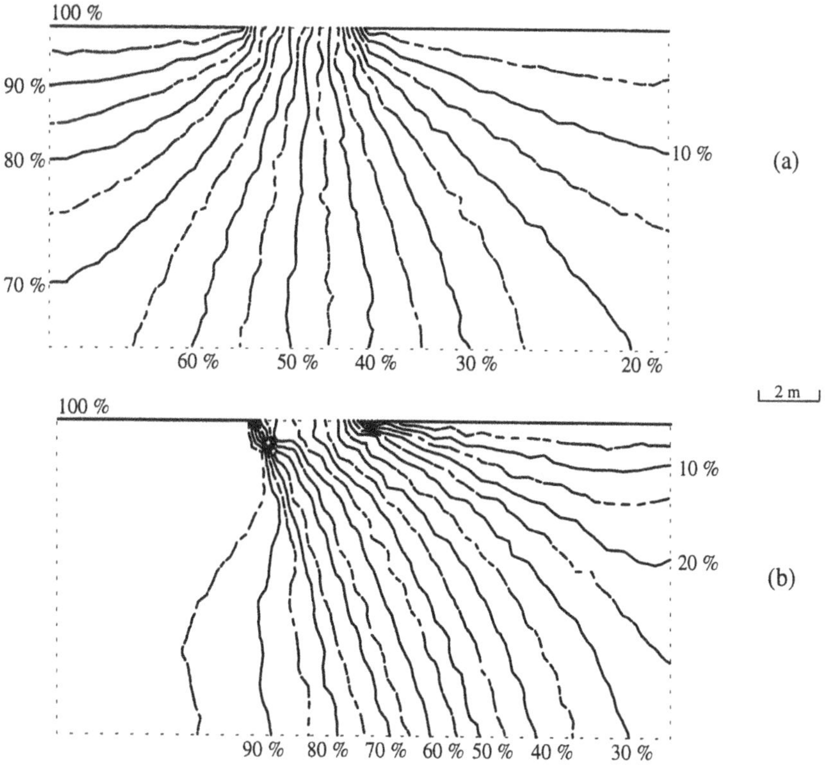

Figure 4: Equipotential contours: (a) The existing joints have all the same hydraulic conductivity (intial state) (b) At hydromechanical equilibrium (joint conductivity is modified under blocks movement) for the 10 meters water level.

CONCLUSION

The fissured rock mass has a complex behaviour strongly related to its discontinuities distribution as well as their mechanical properties. The model BRIG3D allows a complete 3D analysis of such a medium including a fully hydromechanical coupling. Through the application presented here, it is demonstrated that BRIG3D simulates well the influence of a dam loading on its rock mass foundation. In particular, it shows that the dam thrust compresses the downstream area and decompresses the upstream area under the dam. This modifies dangerously the flow lines and pressure as suggested by Bernaix (1967) when studying the Malpasset dam failure.

REFERENCES

Bandis S.C., Lumsden A.C. & Barton N.R. 1983. Fundamentals of the rock joint deformation. *Int. J. Rock Mech. Min. Sci. & Geomech. Abstr.*, Vol. 20(6), pp 249-268.

Bargui H. 1997. *Modélisation des comportements mécaniques et hydrauliques de massifs rocheux simulés par des assemblages de blocs rigides. Introduction d'un couplage hydro-mécanique*, Ph.D thesis, Ecole Nationale des Ponts et Chaussées, France.

Belytschko T., Plesha M. & Dowding C.H. 1984. A computer method for stability analysis of caverns in jointed rock, *International Journal for Numerical and Analytical Methods in Geomechanics*, Vol. 8, pp 473-492.

Bernaix J.B. 1967. *Etude géotechnique de la roche de Malpasset*, Dunod, Paris.

Cundall P.A. 1971. A computer model for simulating progressive, large-scale movements in blocky rock systems. *Proceedings of the Symposium of the International Society of Rock Mechanics.* Nancy, France, Vol. 1, paper II-8.

Gaziev E.G. & Erlikhman S.A. 1971. Stresses and strains in anisotropic rock foundation, *Symposium International de Mécanique des Roches*, Nancy, France.

Goodman R.E. 1980. *Introduction to rock mechanics*, John Wiley & Sons, USA.

Huyakorn P.S. & Pinder G.F. 1983. *Computational methods in subsurface flow*, Academic Press, USA.

Kawai T., Takeuchi N. & Kumeta T. 1981. New discrete models and their application to rock mechanics, *Proceedings of the International Symposium on Weak Rock*, Tokyo, pp 725-730.

Tahiri A. 1992. *Modélisation des massifs rocheux fissurés par la méthode des éléments distincts*, Ph.D thesis, Ecole Nationale des Ponts et Chaussées, France.

Mechanics of Jointed and Faulted Rock, Rossmanith (ed.) © 1998 Taylor & Francis, ISBN 90 5410 955 6

Behavior of jointed rock masses during construction of embankment dams

Yoshikazu Yamaguchi, Masaki Kawasaki, Hiroyuki Ishikawa & Hitoshi Yoshida
Public Works Research Institute, Ministry of Construction, Tsukuba, Japan

Norihisa Matsumoto
Japan Dam Engineering Center, Tokyo, japan

ABSTRACT: Many sites for embankment dams in Japan are composed of soft, weathered, or jointed rocks. Because these rocks often have many fine joints, it is difficult to reduce the permeability in these rock foundations, particularly in shallow areas, by cement grouting. However, because of the large deformability of these foundations, they will be considerably deformed due to fill placement. If the deformation takes place in a compressive direction and fine joints become narrower, the permeability of the foundation will decrease. The deformation and the change in permeability of foundation due to the fill placement of embankment dams were measured at two dam sites composed of jointed rocks. This paper reports the measurements, and discusses effective and rationalized grouting specifications in consideration of the consolidation influence on the reduction of permeability due to the fill placement in embankment dams.

1 INTRODUCTION

Recently in Japan, because most of the sites that have satisfactory geological conditions for the construction of dams have already been developed, dams tend to be constructed on sites that do not always have good geological conditions. In selecting the type of dams, an embankment dam is often chosen for sites that are low in strength for concrete dams. Therefore, many sites for embankment dams under planning or construction are composed of soft, weathered, or jointed rocks. Since the higher grouting pressure causes fracturing or lifting of the surface of these rock foundations, the maximum pressure should be kept low. In addition, these rocks often have many fine joints, which cannot be penetrated by cement particles. Therefore, it is often difficult to reduce the permeability in these rock foundations, particularly in shallow areas, by cement grouting, even if microfine cement is used.

On the other hand, because of the large deformability of these foundations, especially in shallow areas, they will be considerably deformed due to fill placement. If the deformation takes place in a compressive direction and fine joints become narrower, the permeability of the foundation will decrease.

In our previous research (Matsumoto & Yamaguchi 1987), the deformation and the change in permeability of the foundation due to the fill placement of embankment dams were measured at three dam sites. The observation areas of two dam sites were composed of soft rocks with a deformation modulus ranging from 200 to 300 MPa. Another site consisted of jointed rocks with a deformation modulus of 660 MPa. The results measured at the three sites revealed that the dam foundation was consolidated due to fill placement with a height in the range of 15 to 60 meters, and the permeability of the foundation decreased with the increase of fill placement. Based on these results, we proposed an effective design method of grouting for soft or jointed rock foundations.

In this research, the same measurement was made in riverbeds and abutments at two other embankment dam sites by installing extensometers and permeability test holes. These two dam sites are composed of jointed rocks with a deformation modulus of more than approximately 1000 MPa. This paper reports the measurements, and discusses effective and rationalized grouting specifications in consideration of the consolidation influence on the reduction of permeability due to fill placement in embankment dams.

2 PROFILES OF INVESTIGATED DAMS

The deformation and the change in permeability of jointed rock masses due to the fill placement of embankment dams were measured at Okuno Dam and Suetakegawa Dam. Table 1 summarizes the profiles of the two dams and the geological outlines of the foundations surrounding the measuring areas, while Figure 1 shows the locations of both dams on a map of Japan. This map also shows the locations of the three dams introduced in our previous study (Matsumoto & Yamaguchi 1987).

Table 1. Profiles and geological outlines of investigated dams.

Name of dam	Height (m)	Crest Length (m)	Volume ($\times 10^3 m^3$)	Year of completion	Geological age	Measuring area	Foundation rock	Rock mass classification
Okuno	63.0	323.0	1,804	1989	Neogene period, Cenozoic era	Riverbed	Andesite	C_M
Suetakegawa	89.5	275.0	2,560	1991	Palaeozoic era	Riverbed & Left abutment	Hornfels	C_M

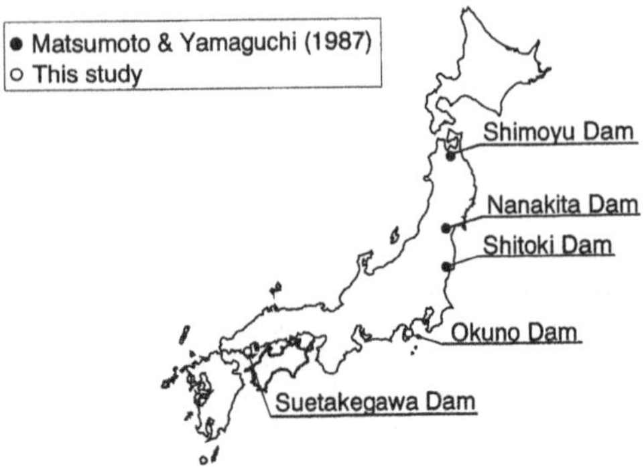

Figure 1. Location of investigated dams.

3 CASE STUDY AT OKUNO DAM

3.1 *Geology and measurement*

The foundation for Okuno Dam consists mainly of andesite with intercalated pyroclastic rocks such as tuff breccia and tuff. Many large and small faults also exist at the dam site. Figure 2 shows a geological map along the dam axis.

The earth pressure, rock mass displacement, and change in permeability of the foundation at the riverbed were measured during the fill placement of the embankment dam. Figure 3 shows the arrangements of the earth pressure gauges, the extensometers, and the permeability test holes. The initial measurement lengths of the three extensometers R-1, R-2, and R-3 were 60 m, 30 m, and 10 m respectively. The lengths of the permeability test holes WT-1 and WT-2 were both 25 m, and each permeability test was performed in 5 stages with each test section length set at 5 m. The permeability test method conformed to the technical guidelines for Lugeon water tests (River Bureau, Ministry of Construction 1984). The permeability was evaluated with the Lugeon value. The rock mass measured is composed of relatively hard andesite, and it is classified as C_M class (Japan Society of Engineering Geology 1992).

3.2 *Measured results*

Figure 4 illustrates the change with time in height of fill placement, total earth pressure and rock mass displacement. Figure 5 shows the relationship between the height of fill placement and total earth pressure. The symbol α in the figure represents the ratio of the measured total earth pressure to the product of the wet unit weight of the impervious material and the height of fill placement. When the straight lines in the figure are drawn, 19.6 kN/m^3 was substituted for the unit weight of the impervious material based on the results of construction control testing. Figure 6 shows the relationship of the total earth pressure with the compressive strain of the rock mass obtained by dividing the rock mass displacement by the initial measurement length of the extensometer. The figure also shows the deformation moduli calculated from the straight-line gradients of the relationship between the total earth pressure and rock mass strain. These figures reveal the following:

1. There is a relatively good correspondence between the height of fill placement and rock mass displacement / total earth pressure.

2. While α ranged from 0.6 to 0.7 during the early stage of fill placement, it declined as fill placement advanced, so that at a fill placement height over 30 m, it had converged to about 0.45. This arching in the impervious zone is caused by stress concentration in the filter zone resulting from the higher rigidity of the filter zone adjoining the impervious zone (Sakamoto *et al.* 1994).

3. The longer the measurement length of the extensometer, the larger the measured rock mass displacement and the calculated deformation modulus. This is believed to be a result of the fact that the rigidity of the rock mass increases with depth, in addition to the dispersion with depth of the stress generated in the rock mass by the embankment load. In either case, the measured rock has a deformation modulus of at least 1000 MPa.

The relationship between the height of fill placement and the Lugeon value is shown in Figure 7. Figure 8 shows the ratio of the Lugeon value at the fill placement height of 48 m to the Lugeon value immediately after the permeability test hole was drilled. These figures reveal the following:

1. Although not as conspicuously as it appeared in previous measured results in rock masses with a deformation modulus less than 660 MPa (Matsumoto & Yamaguchi 1987), a tendency for the permeability to decline due to fill placement was noted. This tendency was not very remarkable because the stages where the

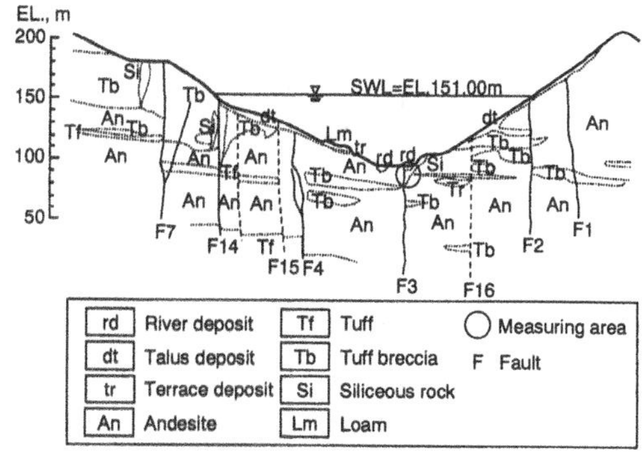

Figure 2. Geological map along dam axis.

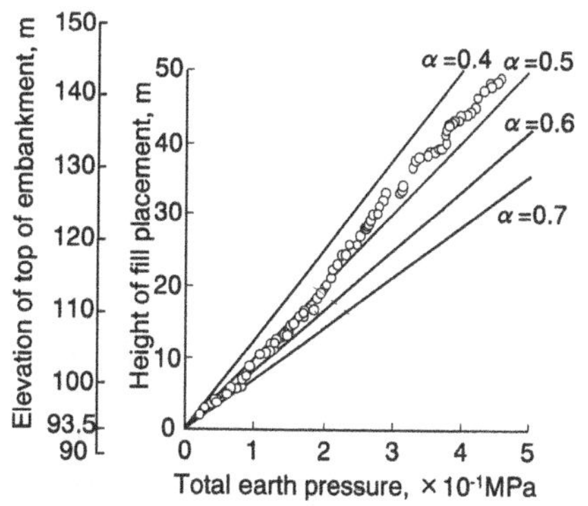

Figure 5. Relationship between height of fill placement and total earth pressure.

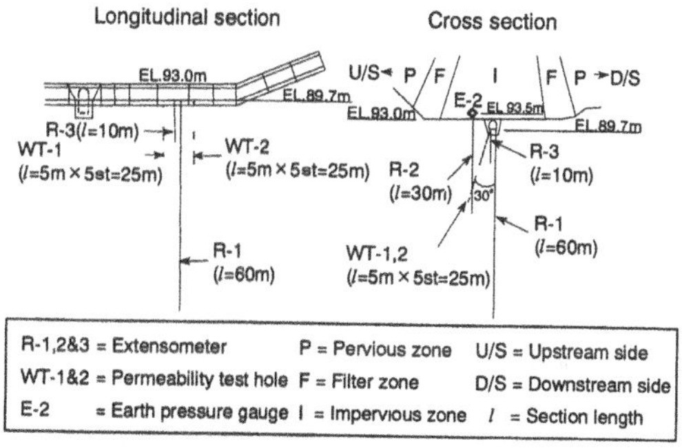

Figure 3. Layout of measuring devices.

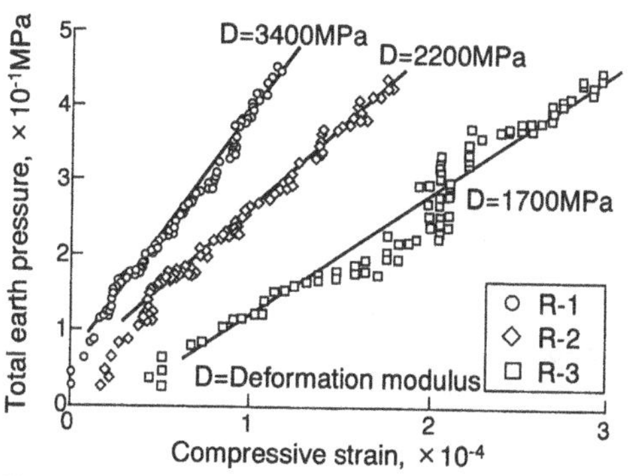

Figure 6. Relationship between total earth pressure and compressive strain.

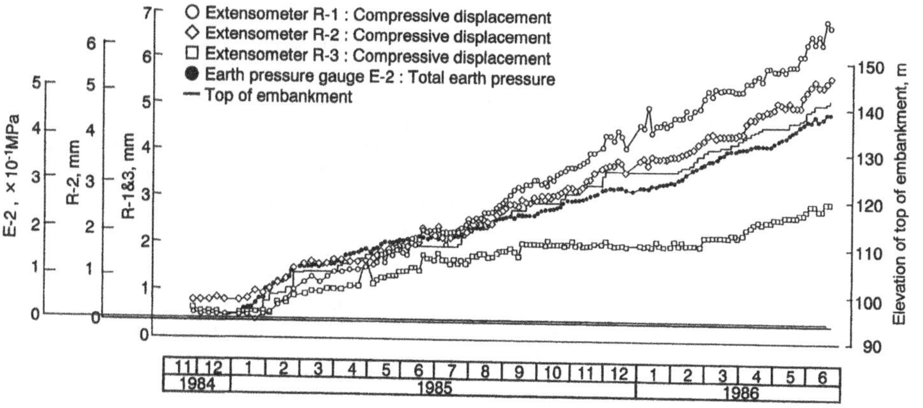

Figure 4. Change of measured results with time.

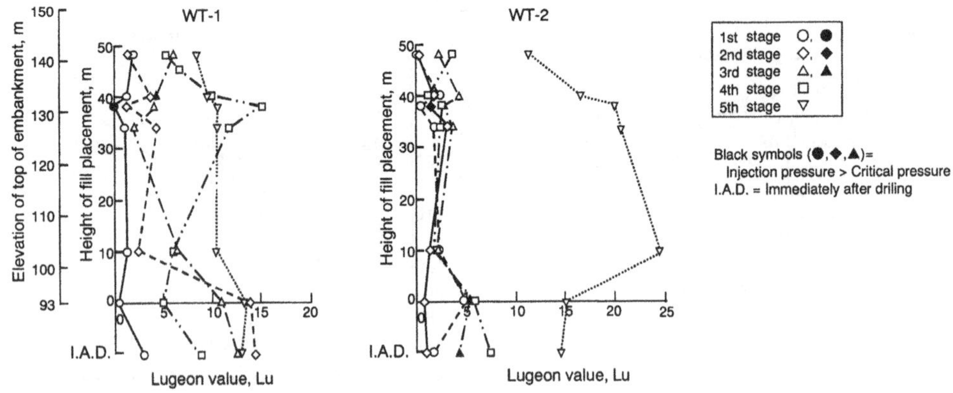

Figure 7. Relationship between height of fill placement and permeability.

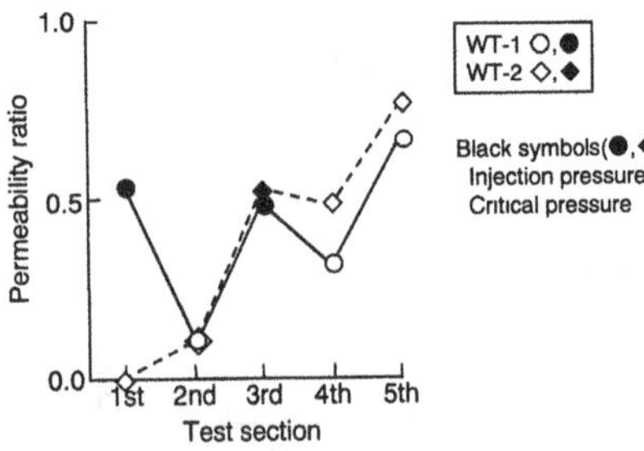

Figure 8. Permeability ratio.

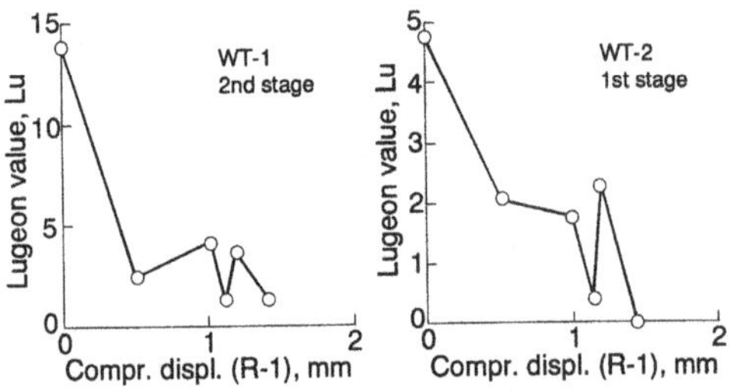

Figure 9. Relationship between compressive displacement and permeability.

initial permeability was relatively high at more than 10 Lu were not very numerous at only four out of ten stages and even the maximum initial permeability was about 15 Lu, and because the compressive displacement due to the fill placement was small because of the comparatively large deformation modulus of the rock mass of more than 1000 MPa. However, because there is insufficient information about the aperture of the joints, their directional distribution, continuity and so on, this explanation is speculative.

2. The ratio of the Lugeon value at the fill placement height of 48 m to the Lugeon value immediately after the permeability test holes were drilled was smaller at the shallower stages than at the deeper stages. In other words, the decline in permeability due to fill placement is more conspicuous at the shallower stages. Because it is generally more difficult to improve imperviousness by grouting in the shallower foundation with small overburden loading, it is expected that grouting accounting for the decline in permeability due to fill placement would rationalize dam foundation treatment.

The decline in permeability due to fill placement is thought to have been caused by compressive displacement of rock mass. Figure 9 presents the relationship between the rock compressive displacement measured with the extensometer R-1 and the Lugeon value obtained at the second stage for hole WT-1 and at the first stage for hole WT-2. In these two stages, the decline in permeability due to fill placement was particularly remarkable and the Lugeon water test was performed with an injection pressure lower than the critical pressure. When this figure was prepared, no consideration was given to the incline from the vertical direction of the permeability test hole. The trend of the relationship between the Lugeon value and the compressive displacement was unchanged even when the results from the other extensometers R-2 and 3 were used. These figures reveal that the decline in permeability due to fill placement is undoubtedly caused by the compressive displacement of the rock mass.

4 CASE STUDY AT SUETAKEGAWA DAM

4.1 *Geology and measurement*

The foundation for Suetakegawa Dam is mainly

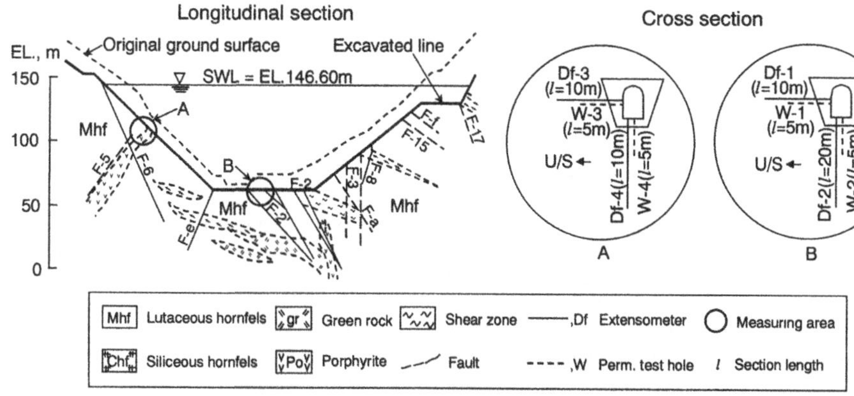

Figure 10. Geological map along dam axis and layout of measuring devices.

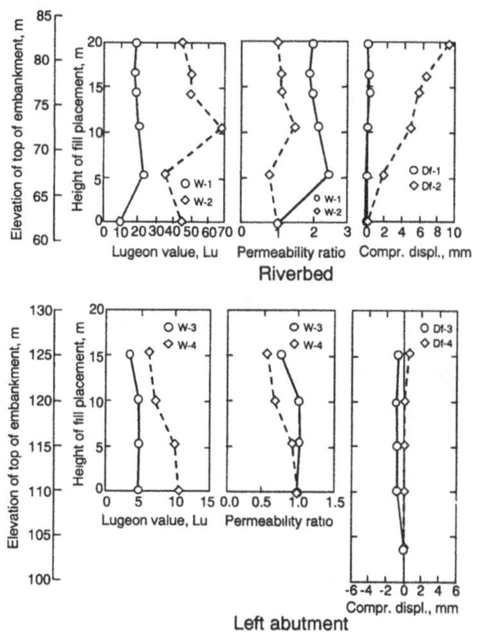

Figure 11. Relationship between height of fill placement and permeability/compressive displacement.

composed of lutaceous hornfels with subordinate amounts of siliceous hornfels, green rock, and porphyrite. Many faults and shear zones are also distributed. A geological map along the dam axis is shown in Figure 10.

Fragments of the unweathered lutaceous hornfels have a specific gravity of 2.7 and an absorption of 0.5%, and are extremely hard with an unconfined compressive strength in excess of 100 MPa. But because drill cores recovered from the measurement points are considerably brecciated, the rock mass is classified as C_M class (Japan Society of Geological Engineering 1992).

As illustrated in Figure 10, extensometers and permeability test holes were installed in the riverbed and at an intermediate elevation of the left abutment to measure the deformation and change in permeability of the foundation due to the fill placement of the embankment dam. The extensometers and permeability test holes were installed in the upstream/downstream (horizontal) direction as well as in the vertical direction. The permeability test was performed for only one stage at each hole in the test section of 5 m. The testing method conformed to the technical guidelines for Lugeon water tests (River Bureau, Ministry of Construction 1984).

4.2 *Measured results*

Figure 11 shows the relationship of the fill placement height with the Lugeon value and the rock mass displacement. All permeability tests were performed at an injection pressure below the critical pressure.

First the measured results in the riverbed are analyzed.

1. The initial Lugeon values obtained at holes WT-1 and WT-2 were approximately 10 Lu and 45 Lu respectively. Although the Lugeon values from both test holes increased temporarily at the time of a certain fill placement height, they tended to decrease due to later fill placement. But even the Lugeon values obtained at the placement height of 20 m were larger than the initial values.

2. The compressive displacement in the vertical direction measured by extensometer Df-2 increased almost proportionally with the height of fill placement, but the displacement in the horizontal direction measured by Df-1 remained almost unchanged from its value at the beginning of the embankment work. The deformation modulus of the rock mass calculated from the

relationship of the compressive strain found by dividing the compressive displacement from Df-2 by its initial measurement length of 20 m with the total earth pressure found as the product of the wet unit weight of the impervious material and the fill placement height was about 800 MPa. Based on the results of the construction control testing, 2.13 kN/m^3 was substituted for the wet unit weight of the impervious material. The value 1.0 was substituted for α because it was a measurement up to the relatively low fill placement height of 20 m.

3. It is impossible to explain the increase in permeability at the time of a certain fill placement height in terms of its correspondence with the tensile displacement of the rock mass. An injection pressure lower than the critical pressure was applied in all of the permeability tests. It is suspected that because the rock mass measured was considerably jointed, looseness around the test holes caused an increase in the permeability. As in the case of Okuno Dam, it is essential to prepare a rock mass model based on considerable joint information in order to quantitatively evaluate the displacement-permeability relationship in jointed rock mass due to fill placement. In either case, there was no conspicuous decline in permeability in the riverbed due to fill placement at Suetakegawa Dam.

The next step is the analysis of the measured results in the abutment.

1. The initial Lugeon values were about 5 Lu at the vertical hole WT-4 and about 10 Lu at the horizontal hole WT-3. Measurements were performed up to a maximum fill placement height of 15 m, but the Lugeon value obtained at both holes tended to decline as fill placement advanced.

2. The horizontal displacement measured at Df-3 appeared on the tensile side when measurements commenced, but remained almost constant after the permeability test started. The vertical displacement measured by Df-4 tended to gradually increase toward the compressive side as fill placement continued, but its absolute value was about one order of magnitude smaller than the value measured by Df-2 installed in the riverbed. This indicates that the rock mass at the measurement area has a deformation modulus of several thousand MPa.

3. The measurements in the vertical direction particularly indicate relatively good correspondence between the compressive displacement and the decline in permeability of the rock mass. But the decline in the permeability was not as conspicuous as that observed in the rock masses with a deformation modulus less than 660 MPa (Matsumoto & Yamaguchi 1987).

5 CONCLUSIONS

At two embankment dams constructed on jointed rock masses, the deformation and the change in permeability of rock masses due to fill placement were measured, and the following information was obtained.

1. In jointed rock masses, the measurements revealed that as the embankment work advanced, compressive deformation of the rock mass occurred, causing a decline in its permeability. But this decreasing trend was not as conspicuous as that observed in our previous research (Matsumoto & Yamaguchi 1987).

2. The decrease in permeability due to fill placement was more remarkable in a shallow foundation. Because it is more difficult to improve imperviousness by grouting in a shallow foundation with small overburden loading, it is expected that grouting accounting for the decline in permeability due to fill placement would rationalize dam foundation treatment.

3. These measurements did not reveal a very marked decrease in permeability due to the fill placement because the quantity of compressive deformation was small as a result of the large deformation modulus in excess of 1000 MPa of the rock mass investigated.

4. In order to conduct a thorough analysis of the quantitative relationship of the compressive deformation and the decline in permeability of jointed rock masses due to the fill placement of embankment dams and execute effective grouting accordingly, it is necessary to obtain detailed information about joints such as aperture of the joints, their directional distribution, continuity, and so on.

REFERENCES

Japan Society of Engineering Geology 1992. *Rock mass classification in Japan.*

Matsumoto, N. & Y. Yamaguchi 1987. Deformation of foundation and change of permeability due to fill placement in embankment dams. *Proc. 6th Int. Congr. Rock Mech.*: 177-180.

River Bureau, Ministry of Construction, Japanese Government 1984. *Technical guidelines for Lugeon water tests.* (in Japanese)

Sakamoto, T. *et al.* 1994. Safety assessment based on the observed behavior of zoned rockfill dams. *Proc. 18th Int. Congr. Large Dams, Q.68*: 925-953.

Mechanics of Jointed and Faulted Rock, Rossmanith (ed.) © 1998 Taylor & Francis, ISBN 90 5410 955 6

Ultimate bearing capacity of the foundation of a gravity dam, on an isotropic transverse rock media

A. Serrano
ETSICCP, Università Politécnica, Madrid, Spain

C. Olalla
Laboratorio Geotecnia-CEDEX, Madrid, Spain

ABSTRACT: The ultimate bearing capacity of the foundation of a rollcrete gravity dam is determined, under the assumption that the behaviour of the rock mass is non linear and that the transverse isotropy is due to the presence of a single familiy of discontinuities. The calculation procedure is based on previous papers published by Serrano & Olalla. A complete sensitivity analysis is conducted to find the dependence of the Safety Coefficient on the different parameters involved: Hoek & Brown model, linear strength of the planes of weakness and geometrical configuration. It is obtained that a sufficient safety margin exists with respect to the failure mechanism studied.

1 INTRODUCTION

This study belongs to the consulting activity that the Geotechnics Laboratory at the Public Works Studies and Research Centre, has been carrying out for the Hydraulic Works Administration (Ministry of Public Works). It has been conducted in order to determine the suitability of the Atance Dam foundations (Spain). The Atance Dam is a rollcrete dam located in province of Guadalajara.

2 CALCULATION HYPOTHESIS

2.1 *General hypotheses*

The calculation procedure used is based on Sokolovskii's characteristics method for linear media, (1965), but applied to rock masses using non-linear failure theories (Serrano & Olalla (1994; 96; 98). These theories set out a procedure for calculating the ultimate bearing capacity for a homogeneous medium with isotropic or anisotropic behaviour.

Basically, it is assumed that;

1. The rock mass fails according to the non-linear failure criterion defined by Hoek & Brown (1980).
2. Where appropriate, the failure surface is completed by sliding throughout the length of the "planes of weakness". The failure criterion throughout the planes of weakness is Coulomb's linear criterion.
3. It is assumed that the weight of the rock mass itself does not contribute to the quantification of the ultimate bearing capacity. In this case, the hypothesis is conservative, in the sense that the contribution of the ground weight itself is kinematically opposed to the movement.
4. A two-dimensional analysis is conducted.
5. In overall terms the rock mass is considered to behave isotropically but superimposed to it there is a family of a) parallel planes of weakness, b) being very closely spaced, having a theoretically infinite frecuency, c) with absolute persistence throughout the foundation and d) running in the same direction as the foundations and their corresponding actions.
6. There is no possibility of generation of three-dimensional wedges.
7. The behaviour of the rock mass, is taken to be ductile in nature.

2.2 *Failure mechanism*

This chapter is a synopsis of the Serrano & Olalla (1998) paper, focused on the analysis of the failure mechanism produced in this case. This failure mechanism begins through the rock mass under the foundation and ends along the planes of weakness merging at the ground surface.

Considering the dip of the planes of weakness, the mechanism produced under those boundary conditions is conditioned,

- by the rupture through the rock mass underlying the foundation. In this case the strength is governed by the Hoek & Brown failure criterion, and

- in the surrounding area of the foundation by the rupture along the planes of weakness. (Fig. 1).

For its theoretical solution the first need is to solve the global equilibrium of the wedge. The forces acting at the outcropping surface are known. Consequently, analysing the Mohr's circle, bearing in mind that the forces at the dip direction must be on the strength line of the planes of weakness, then, this case is transformed into a homogeneous and isotropic one, with a new slope inclination value.

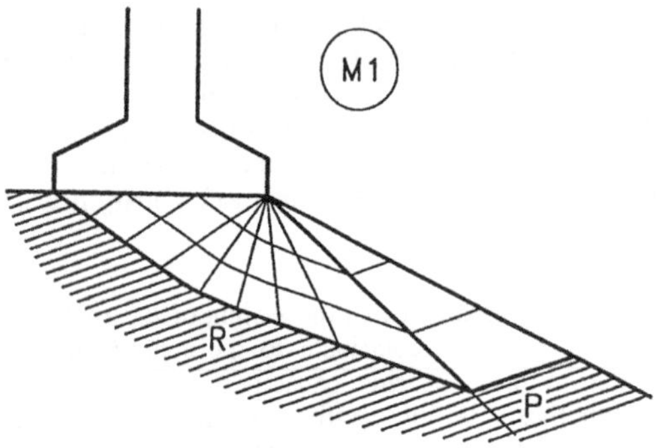

Fig. 1. Failure mechanism

2.3 *Specific hypotheses*

For the analysis of the dam foundation, in addition to the above hypotheses, the following assumptions have also been made:

1. Downstream from the dam, the adjacent foundation surface is horizontal.
2. Water pressure is the only external force exerted upon the dam, acting at the upstream face and at the foundation.
3. The effect of the uplift pressure acting on the foundation surface, (U), is considered to range from 100% of its theoretical intensity, defined by the water head at the upstream face zero at the downstream's toe. Figure 2 shows a diagram of the forces and pressures exerted. It can be seen, that the angle of inclination respect to vertical, being U = 100%, is about 30°. When there is no uplift pressure the angle of inclination of the loads is approximately 20°. (Lower percentages of uplift pressure acting at the foundation surface means, obviously, that the pore water pressure has been drained).

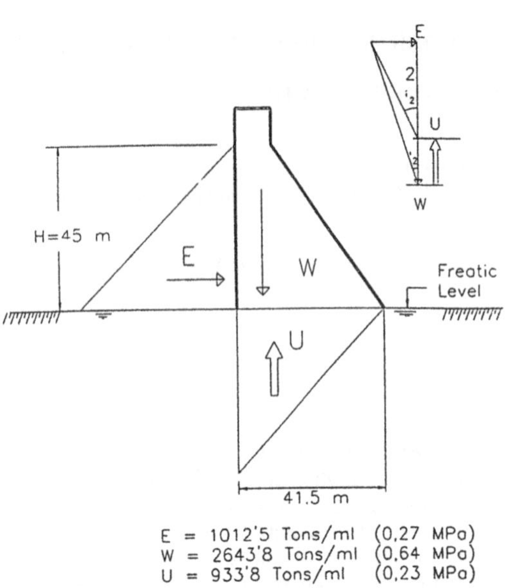

Fig. 2. Diagram of the external loads exerted on the ground surface

3 PARAMETERS USED

3.1 *For the rock mass*

The failure criterion of Hoek & Brown (1980), needs three parameters, m, s, and UCS (Unconfined Compressive Strength). The first two ones can be evaluated from the value of m_0 and Bieniawski's (RMR) index (1974), according to Hoek & Brown (1988). According to the CEDEX studies the rocks concerned are "mica-schists and paragneisses", (CEDEX, 1997), whose grain-size can range from medium to coarse. The range of values for the different parameters involved are presented in the table below:

VALUE	m_o	RMR	UCS (MPa)
mean	25	55	35
maximum	33	60	50
minimum	10	50	25; 10*

(*) Theoretical; extremely low

3.2 *For the planes of weakness*

An extensive series of seven field tests was conducted in order to find out the shear strength of the planes of weakness underlying the dam foundations (CEDEX, 1997).

The residual strength behaviour that have been considered in the calculations were ranging from 0 to 0.2 MPa for cohesion and between 30° and 50° for the internal friction angle: 0.2 MPa and 50° represents the mean values, respectively.

The mean value of the dips of the planes of weakness was 10°. Acording to Serrano & Olalla (1998) the value of the angle that must be introduced in the calculations is 90° + 10° = 100°. However, in view of the fact that the least favourable value possible is closer to 90°, (as can be deduced from the geological and geotechnical studies performed), the calculations have been made for one single value of 95° (considering the nomograms included in the 1998's papers already mentioned).

4 RESULTS OBTAINED

The sensitivity of the safety coefficient was studied for a considerable variation in the parameters. A total of 41 cases were calculated, using a computer program coded to solve the different situations. The value of the Safety Coefficient (SC) is obtained by dividing the vertical component that produces the rock mass failure, (Ultimate Bearing Capacity = UBC), by the real vertical working load (W-U), where W = 26.4 MN/ml. and U, can range from 0 to 9.3 MN/ml. (Fig 2).

4.1 *Dependence of the angle of inclination of loads*

A parametric analysis of the incidence of the angle of inclination of the loads (i_2) has been developed. However, in fact, (i_2) represents the effect of the uplift pressure.

When the reservoir is empty, the angle of inclination is always 0°, and when the reservoir is full, the angle of inclination can range from 20° to 30°.

The reservoir would have to be partially empty or empty for the load to have an inclination below 20°. Table below, shows the values of the vertical loads (VL) that are exerted in each case.

i_2 (°)	VL (MN/ml)	VL (MPa)
30°	17.5	0.42
27°	19.9	0.48
25°	21.7	0.52
20° (*)	26.4	0.64
< 20°	26.4	0.64

(*) In fact, this situation really takes place when the angle is 21°

Figure 3 shows the incidence of the angle (i_2).

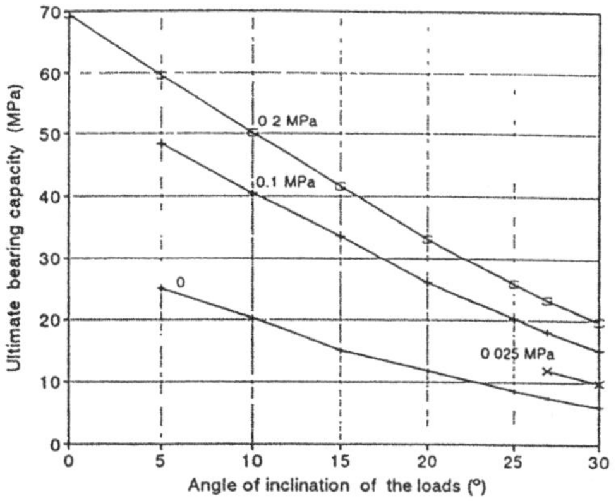

Fig. 3. Effect of the angle of inclination of the loads on the Ultimate Bearing Capacity, for different cohesion values (m_o = 25; RCS = 3500 t/m^2; RMR = 55; φ = 40°)

4.2 *Function of discontinuity strength parameters*

A parametric study was also undertaken, with the purpose of deducting the sensitivity of the method to the strength parameters of the planes of weakness. The results obtained, can be seen in Figure 4. It shows that the Safety Coefficient is extremely dependent on the cohesion and the internal friction angle of the planes of weakness.

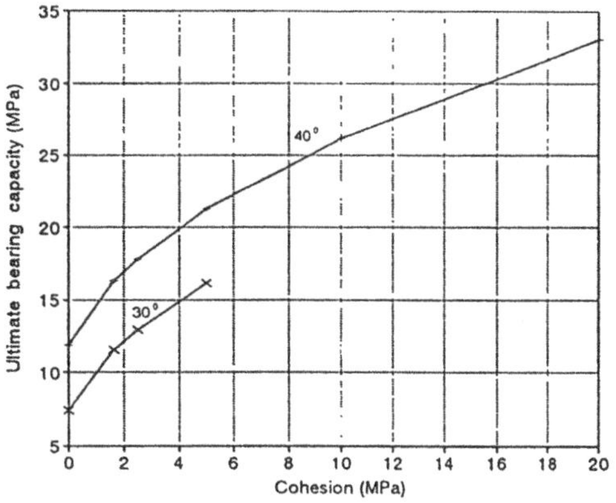

Fig. 4. Effect of the "shear strength" of the planes of weakness on the Ultimate Bearing Capacity (m_o = 25; UCS = 35 MPa; RMR = 55; φ = 40°; i_2 = 20°)

4.3 *Rock mass strength*

It was considered advisable to make calculations using the lowest set of rock mass strength parameters, within reasonable limits:

a) For the rock mass: m_o = 10; UCS = 10 and 25 MPa; RMR = 50.

b) For the discontinuities: c = 0; and φ = 30°.

c) For the geometrical data: A dip angle of 5°, (that is to say χ = 95°) and (i_2) ranging from 20° to 30°.

Figure 5 shows the results obtained as a function of i_2 and UCS.

4.4 *Network of characteristic lines*

The failure mechanism which takes place under the "average" hypothesis (I) and under the "least favourable" hypothesis (II) can be seen in Figure 6. The angles that define the failure mechanism have also been indicated. A 27° angle of inclination has been assumed for the loads, and this corresponds to an uplift pressure reduction of roughly 30%. The main data are in table below.

It can be seen, that given the extensive development of the failure surfaces throughout the discontinuities, the continuity hypothesis for the planes of weakness is very much on the safe side, according to the in situ investigations.

Figure 7 shows Mohr's Circle at failure, for the very extreme case II. It depicts the normal and shear stresses at the foundation surface, in contact with the dam base.

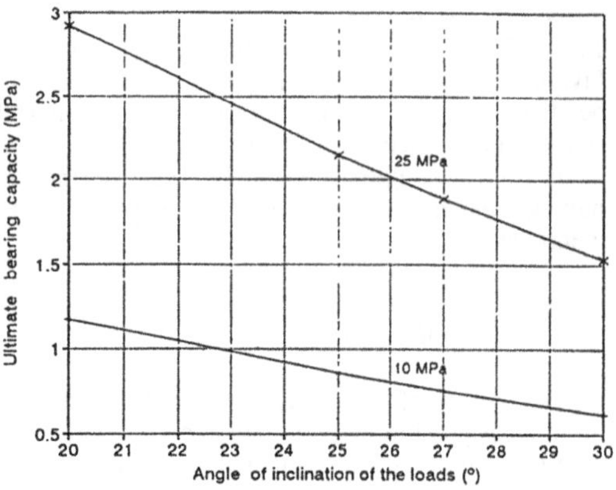

Fig. 5. Effect of the Unconfined Compressive Strength and the inclination of the loads on the Ultimate Bearing Capacity value (m_o = 10; RMR = 50; c = 0; φ = 30°)

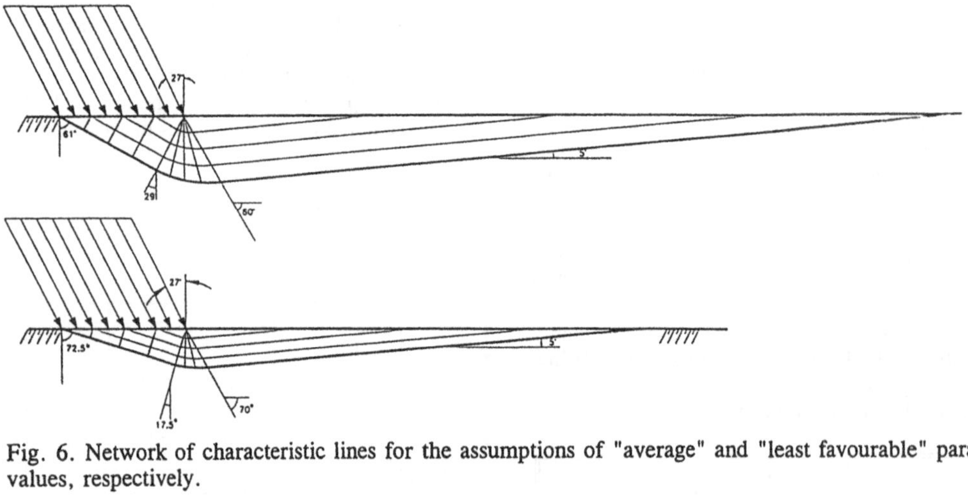

Fig. 6. Network of characteristic lines for the assumptions of "average" and "least favourable" parameter values, respectively.

HYPHOTHESES	m_o	UCS (MPa)	RMR	c (MPa)	φ	χ	i_2	UBC (MPa)	SC
I	25	35	55	0.10	40°	95°	27°	18.13	36
II	10	10	50	0	30°	95°	27°	0.75	1.6

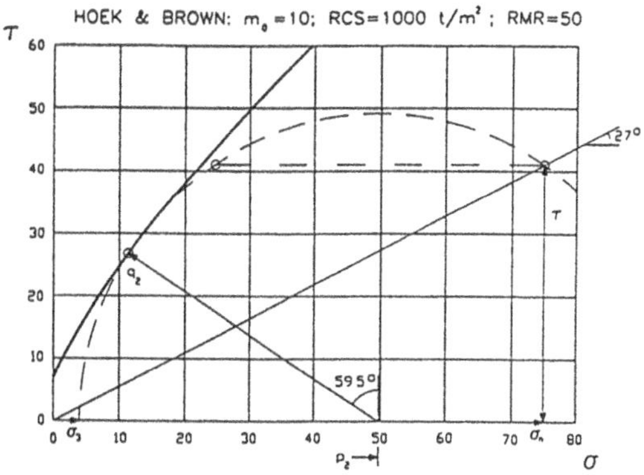

Fig. 7. Mohr's Circle at the foundation plane failure for the assumption of "average" parameter value

5. CONCLUSIONS

1. The Serrano & Olalla model (1994, 1996 and 1998), has been applied to calculate the ultimate bearing capacity in an anisotropic rock medium, when the failure is plastic in nature.
2. The rupture surface is a) through the rock mass according to the Hoek & Brown criterion, and b) along the planes of weakness according to the Mohr-Coulomb criterion.
3. The two-dimensional medium has been analysed. The weight of the rock mass itself has not been considered in this case.
4. It is assumed that one single family of planes of weakness exists. These discontinuities are assumed to be parallel, highly persistent, close together, etc.
5. A complete sensitivity analysis was conducted, to find out how the results can be affected by variations of the parameters that define the rock mass strength and the strength of the planes of weakness.
6. With respect to the hypothesis for parameters described as "average" (through the rock mass $m_0 = 25$; UCS = 35 MPa; RMR = 55 and at the planes of weakness c = 0.10 MPa and $\varphi = 40°$) and with 30° inclination of load, a safety coefficient is obtained that is close to 36.
7. Where the hypothesis for parameters described as "very much on the safe side" are concerned (through the rock mass $m_0 = 10$; UCS = 10 MPa; RMR = 50 and at the planes of weakness c = 0 MPa and $\varphi = 30°$) and a 30° inclination of load, a safety coefficient is obtained that is close to 1.6. These are the most unfavourable hypotheses that can be imagined.
8. In the light of those results, it is considered that a sufficient margin of safety exists with respect to the failure mechanism studied in this work.

ACKNOWLEDGEMENTS

The authors want to thank Eng. Salvador Madrigal of the Confederación Hidrográfica del Tajo for his permission to publish the data contained herein.

6. REFERENCES

SOKOLOVSKII, V.V. (1965). *"Statics of Granular Media"*. Pergamon Press, London, U.K.

HOEK, E. & BROWN, E.T. (1980). *"Empirical Strength Criterion for Rock Masses"*. Journal of Geotechnical Engineering Division, American Society of Civil Engineers, Vol. 106, GT9, pp. 1013-1035.

HOEK, E. & BROWN, E.T. (1988). *"The Hoek-Brown Failure Criterion, a 1988 Update"*. Proc. 15th Canadian Rock Mechanics Symposium. Univ. of Toronto. October.

SERRANO, A. & OLALLA, C. (1994). *"Ultimate Bearing Capacity of Rock Masses"*. Int. Journal of Rock Mechanics and Mining Science. Vol. 31. nº 2, pp. 93-106.

SERRANO, A. & OLALLA, C. (1996). *"Allowable Bearing Capacity in Rock Foundations, Based on a Non Linear Criterion"*. Int. Journal of Rock Mechanics and Mining Science. Vol. 33, nº 4, pp. 327-345.

SERRANO, A. & OLALLA, C. (1998a). *"Ultimate Bearing Capacity of an Anisotropic Rock Mass, Part I: Basic Modes of Failure"*. International Journal of Rock Mechanics and Mining Science. In Press.

SERRANO, A. & OLALLA, C. (1998b). *"Ultimate Bearing Capacity of an Anisotropic Rock Mass, Part II: Procedure for its determination"*. International Journal of Rock Mechanics and Mining Science. In Press.

CEDEX (1997). *"Estudio de la cimentación de la Presa del Atance (Guadalajara)"*. Tomos I y II. Informe Técnico del Laboratorio de Geotecnia del CEDEX, para el Ministerio de Obras Públicas. Junio.

Mechanics of Jointed and Faulted Rock, Rossmanith (ed.) © 1998 Taylor & Francis, ISBN 90 5410 955 6

Mechanism and assessment of the interstratal and intrastratal fractures

Su Shengrui
Department of Engineering Geology, Xi'an Engineering University, People's Republic of China

ABSTRACT: In the area under study gently dipping intrastratal and interstratal fractures are main discontinuities that influence the stability of engineering site. This paper, on the basis of field observation, presents the basic characteristics of the two kinds of fractures, discusses their mechanism and assesses their influence on the engineering. It is thought that the majority of both the intrastratal and interstratal fractures was formed by shearing during flexural folding and developed further under the action of shearing caused by the subsequent tectonic stress field. According to the geometric relations between the fractures and orientations of the slopes and constructions, they have little influence on the stability of the rock mass at the dam site.

1 INTRODUCTION

Rock discontinuities provide parameters for the design of various engineering structures involving rock foundations. In the design of rock quarries, open pit mines, cut slopes, building foundations, tunnels and other underground openings and dam abutment the influence of discontinuities must be taken into account (Doyuran et al. 1993). There are often reports on the gently dipping discontinuities in magmatic rocks concerning engineering. Their mechanism is still in argument (Xu et al. 1993). A large – scale dam has been planned to be built in the area studied in this paper. Geological surveying indicated that the engineering geology conditions there are relatively good except gently dipping intrastratal and interstratal fractures. This paper, on the basis of field observation, presents the basic characteristics of the two kinds of fractures, discusses their mechanism and influence on the engineering.

2 GEOLOGICAL BACKGROUND

The dam area under studying is located on the Jinshajiang River in southwest China. It is in the west part of the Yangtz platform and the transitional belt from the Qianghai – Tibet Plateau to the Sichuan basin. The neotectonic movement in the region is characterized by integrated and intermittent uplifting. Historical earthquake studying shows that the modern microseismicity has a lower activity level. Numerical simulation suggests that the region has a lower density of strain energy without apparent accumulation. It also suggests that after 200 years of rheologic process, the maximum principal stress and maximum shear stress are stabilized in the primary state, displaying high stability (Yan et al. 1996). The recent tectonic stress field has inherited the general features of the one of late Himalayan period, and is predominated by NWW – SEE horizontal compression stress. The strata there include limestone of Yangxin Fm. of Lower Permian and the Emeishan basaltic rocks of Upper Permian. The dam is mainly concerned with the basaltic sequence that consists of 14 subsequences and each of them is made up of lower fine basalt and upper welded breccia lava and tuff. Tectonically the area is situated in a triangular block surrounded by three faults, that is, the NE trending Lianfeng – Huayinshan fault, the NS trending Erbian – Jingyang fault and the NW trending Mabian – Jingyang fault. The area is situated in the Leibo – Yongsheng anticline structure basin and the western limb of the Yongshen anticline, making up a southeast – dipped monoclinal structure consisting of bedded basaltic sequences. On the profile along the river, the dip angle of the strata changes from low angle (10° – 15°) to lower angle (5° – 10°) then low angle (10° – 15°) from the upper reach to lower reach and thus the strata show a steep – gentle – steep shape. There is not large fault in the area except three small – scaled faults with a displacement less than 6m. The main fractures are interstratal and intratratal ones that often distribute in the gentle part of the strata with steep – gentle – steep shape.

Table 1. Tectonic evolution and resulting structures

Tectonic stages	Orientation of σ_1	Resulting structuress
1	NWW	NE – trending gentle folds and low dip – angle interstratal fractures, NE – NNE trending low dip – angle intrastratal fractures, NEE and NWW trending high dip – angle lateral – sliding fractures
2	NEE	secondary superimposed folds, NNW – NW trending low dip – angle fractures, EW and NNE trending high dip – angle fractures, NW trending interstratal fractures
3	NNE	NWW trending low dip – angle fractures, NNW – NW trending high dip – angle fractures
4	NNW	NEE trending low dip – angle fractures, NEE trending compression fractures with high dip – angle, NW trending high dip – angle fractures
5	NWW	reactivity of NNE – NE rending low – dip angle fractures, NNW and NEE trending high dip – angle fractures

Table 2. Preferred orientation of the intrastratal fractures

Preferred orientation	percentage
NEE	30.0
NNE	22.0
NWW	21.1
NNE	17.5
NS	6.3
EW	3.1

3 BASIC FEATURES OF THE INTRASTRATAL AND INTERSTRATAL FRACTURES

The interstratal fractures are those that develop along the interfaces between two basaltic subsequences or in the basaltic subsequence above and near the bedding surface. Their occurrence is controlled by the undulating and connection characteristics of the interfaces and the lithology. There are dislocations along the 13 interfaces of the 14 subsequences. Field observation demonstrates that the interstratal fractures between the ninth and the tenth, the twelfth and the thirteenth, the first and second, the eighth and the ninth , the seventh and eighth subsequences are more continuous and apparent than others. They may be classified into dislocating –, fissuring – and welded – type. The first type, with apparent displacement, is composed of fractured and crush breccia and rock debris with a grain size of 1 – 3cm and fissures beside it. The second type is characterized by low angle fractures without apparent displacement. The third type is cohesive. According to the displacement and scale, the interstratal fractures may be grouped into strongly dislocated and slightly dislocated types. The former, with a fractured zone thickness of 20 – 30cm, extends over 200m, has flat surface and is displayed by breccia, epigenetic mud, debris and striations. The later, with a fractured zone thickness of 5 – 20cm, discontinuously extends less than 200m and makes up the majorities of the interstratal fractures.

The intrastratal fractures refer to those that develcp within basaltic subsequences. They often distribute discontinuously within the lower subsequence, that is, the fine basalts, and extend for 50 – 100m, but not into the upper welded breccia lava and tuff and across the lower bedding surface of the subsequence. Their altitude may be grouped into N30° – 60°E∠SE or NW∠10° – 25°, N10° – 45°W ∠NE or SW∠8° – 25° and EW∠S or N∠6° – 20° among which the predominant one is N30° – 60°E ∠SE. Among the 14 basaltic subsequences, the sixth and eighth subsequences that have the larger thickness of fine basalt and lie in the lower elevation, have the higher percentage of fractures than others. The surfaces are rough and the fractured zone is composed of crush and fractured breccia with a thickness of 10 – 20cm.

In terms of the bifurcation, tranvertion and restriction relations between the intrastratal fractures and their relations with the evolution of the regional tectonic stress field, the fractures were formed by five tectonic stages (Table 1).

The statistic result of the intrastratal fractures in the area is shown in Table 2. The former three sets constitute the majority of the intrastratal fractures and had controlled the weathering of rock mass at the slope and foundation, extended long and dislocated apparently.

4 MECHANISM OF THE INTRASTRATAL AND INTERSTRATAL FRACTURES

Folding is a kind of bending deformation. The bending of rockmass near earth surface inhabits certain plastic and ductile nature with brittle fracture. According to mechanism, folding may be classified into cross bending and flexural bending (Ramsay 1967). Observation and theoretical computation demonstrate that the mechanism of deformation and fracture for a fold differ greatly between the axis and the limbs. At the limbs, the deformation and fracture are produced by shearing, which results in flexural slip or discrete slipping (Roth et al. 1982).

Regional structural studies suggest that the fold in the area was formed by flexural bending. Generally, at the earlier stage of folding, i.e. when the rock strata were still horizontal, two sets of conjunctive shearing joints which has higher dip angle were formed. With the continuous action of tectonic stress, the fold was produced and the rock strata became inclined. At this stage, two kinds of deformation and fracturing associated with the folding would occur subsequently. One is the relative shear sliding between upper and lower strata along the bedding surface, which produced the above mentioned interstratal fractures. The another one is the intrastratal shearing deformation and fracturing between two bedding surfaces induced by the relative sliding along the bedding surfaces, which leads to the formation of above mentioned intrastratal fractures. As a result, the intrastratal fractures do not extends from one subsequence across the bedding surface to another one and display parallel and echelon – like combination of fractures. The fractures derived from the folding make up the main part of interstratal and intrastratal fractures of the area.

The fractures, especially the intrastratal ones, developed further due to subsequent tectonic movement in the area. Roeder shearing (Mandl 1988) may be the leading mechanism for the development of intrastratal fractures. Its direct results are manifested by the fracturing of the bridge structure between the earlier intrastratal fractures (Gammond 1987), forming wavy or bifurcating combination in which the main fractures are breccia – or debris – type, and furthermore, the formation of compound intrastratal fractures with braided or netted combination in which the main fractures are breccia – type bearing debris or debris – type with breccia.

The difference in the occurrence of intrastratal fractures between the upper tuff and welded breccia lava and the lower fine basalt is attributed to their difference in mechanical properties. The fine basalt has higher compression strength (average 265Mpa) than the tuff and breccia lava (average 143.7MPa).

5 THE ENGINEERING GEOLOGICAL ASSESSMENT

The effect of discontinuities on the stability of rock mass depends on the geometric relations of the discontinuities with orientation of slopes, adts and the important constructions. According to the extending distance, the intrastratal and interstratal fractures may be unable to form potential continuous sliding borders between different blocks. Since their dipping direction is perpendicular to the orientation of slopes, it is clear that their existence does not affect stability of the slopes apparently. Furthermore, the spacing of the two kinds of fractures is also large. However, it has been observed that the unloading induced by local stress concentration under the valley had prompted the weathering of the rock mass at the dam foundation.

Except the intrastratal and interstratal fractures, there still exist some steep dipping joints which orient NWW, NEE and NNW. The geometrical relations between the dam and joints show that at the northeast side of the river the detrimental joint should orient NNW and at the southwest side it should orient NEE. Fortunately, both of them are less developed.

Therefore, the intrastratal and interstratal fractures and steep dipping joints have little influence on the stability of the rock mass at the dam site. As a result, from the viewpoint of rock mass structure, the geological condition at the dam site is excellent.

6 CONCLUSIONS

The majority of both the intrastratal and interstratal fractures was formed by shearing during flexural folding and developed further under the action of shearing caused by the subsequent tectonic stress. The intrastratal and interstratal fractures and steep dipping joints have little influence on the stability of the rock mass at the dam site.

The author thanks Prof. Huang Rungqiu, Wang Shitian and Dr. Zhou Zhidong for useful discussion during various stages of completing this work. I gratefully acknowledge funding by the Opening Laboratory of Geotechnical Research under the Ministry of Geology and Mineral Resources of China.

REFERENCES

Doyuran, V., Ayday, C. & Karahanoglu N. 1993. Statistical analyses of discontinuity parameters. Eskisehir Marble and Porsuk dam peridotite in Turkey. Bulletin of the international Association of Engineering Geology. 48: 15 – 31

Gammond, J.F. 1987. Bridge structures as sense of displacement criteria in brittle fault zones. J. Struct. Geol. 9 (5/6): 609 – 620

Mandl, G. 1988. Mechanics of tectonic faulting – models

and basic concepts. Developments in Structural Geology. 1. (Series editor: Zwart, H.J.). Elservier
Ramsay, J.G. 1967. Folding and fracturing of rocks. McGraw – Hill New York
Roth, W.H., Sweet, J. & Goodman, R.E. 1982. Numerical and physical modeling of flexural slip phenomena and potential for fault movement. Rock Mech. Suppl. 12. Springer – Verlog
Xu Bin, Xu Weiya, Yao Jiajian & Jin Shuyan. 1993. Comprehensive genetic study of gently dipping structural plane in granite rock mass at three – gorge damsite on the Changjiang River. Journal of Engineering Geology. 2, (2): 1 – 13
Yan Ming, Wang Shitian & Li Yusheng. 1996. Back analysis of the regional tectonic stress field in the Xiluodu area by FEM. Journal of Chengdu Institute of Technology. 23(1):26 – 31

Supplement

Mechanics of Jointed and Faulted Rock, Rossmanith (ed.) © 1998 Taylor & Francis, ISBN 90 5410 955 6

The influences of perforation parameters to wellbore instability and sand production

A. Samsuri & S. K. Subbiah
Department of Petroleum Engineering, University of Technology, Malaysia

ABSTRACT: This paper presents a series of laboratory experiment had been performed to see the influences of perforation parameter to wellbore instability and sand production. Shot density and perforation pattern has been varied for entire laboratory work and the stability test conducted to see the effects on the wellbore instability of the sandstone wellbore model. The results show that as the shot density increases the wellbore stability decreases. Whereas as the perforation pattern changes from spiral to inplane and inline the wellbore stability decreases. The sand particles produced were found by sieve analysis to be oversized 500 micron. Generally, stable perforated wellbore produced less sand particle, therefore minimizing the sand production problem.

1 INTRODUCTION

In petroleum production, the formation fluid have to go through the wellbore before collected at the surface. Therefore designing the wellbore would be an important factor that should be considered during the field development period. Knowning that a reservoir is in the equilibrium state between overburden and pore pressure in triaxially stressed state.

When perforation was created in the production zone, the stresses intended to redistribute to the surrounding rock. Hence the surrounding rock has to carry the redistributed stress. In other word the formation will be in fully stressed condition with maximum on the perforation tunnel correspond to Jaeger's conclusion on a hollow cylinder (Jaeger et.al (1979) and Obert et.al (1967)). This situation will lead to formation failure or collapse as the production continue. In other word the pore pressure had been reduced which will increase the effective stress as suggested by Tegazhi where $\sigma_{eff} = \sigma - cp$. This phenomenon also will lead to a situation where crushed material will be produced on the perforation face which will be produced as sand production. As the results it may cause the total casing collapse, increased in the rig time and not cost effective. Therefore maintaining the wellbore stability is primarily important and depends on the controllable and uncontrollable factors. The uncontrollable factor are such as overburden pressure, pore pressure. The controllable factor are such as the shot density, perforation pattern, borehole inclination angle and flow rate. In this paper the perforation parameter has been varied to see the effects of shot density and perforation pattern to the wellbore instability and sand particle produced under the static condition.

2 METHODOLOGY

Laboratory work has been conducted in determining the basic rock mechanical properties such as porosity, permeability, compressive strength, tensile strength, Young Modulus, Poisson ratio, triaxial shear strength and angle of friction. All tests were conducted correspondence to American Standard Testing Method (ASTM - Suggested Method by Brown).

Sandstone samples were cored for 2 inches diameter and cut for 5 inches length (for the compressive strength and triaxial shear strength). As for the tensile strength the 2 inches diameter core sample was cut for 1 inch length. All the specimen then tested under a Servo Controller machine at constant loading rate of 0.7 $MN/m^2/sec$ and the Brazilian testing method was used for the tensile strength test where the specimen was loaded with constant rate of 0.2 $MN/m^2/sec$ until failure.

2.1 *Wellbore model preparation*

The 6 inches diameter coring bit was used to core the cylindrical wellbore model which was then cut into 6 inches length. In addition the 2 inch borehole was cored at the center of the core after which the model then has dried in an oven to remove the moisture. The model then saturated with glycerol which represent formation fluid. The steel pipe of 1 inch OD was then cut into correct length to represent the casing.

The correct casing dimension was then placed in the middle of the borehole and the G Class Cement has been squeezed into the annulus as the bonding agent between casing and formation. The G Class Cement slurry was prepared according to API Spec.

10 with the cement water ratio was 2 to 1. The system then left over in the ambient condition for 24 hour in order to allow the cement to set. As final procedure in preparing wellbore model, the sandstone were perforated with different shot density and perforation pattern i.e. 6 SPF, 8 SPF and 16 SPF of spiral, inplane and inline.

2.2 *Special platen preparation*

For the stability test, a special platen has to be designed. Cylindrical steel bar with 6 inches diameter were cut into 5 inches length. Then a hole of 2 inches where drilled at the center of the steel bar. Since the steel bar not been treated yet, the hardening process should be done before the stability test in order to avoid the error during the stability test due to the platens. The two platens were prepared weretop and bottom platens..

The platens were fired in a kiln at temperature of 500 °C then quenched in water. After cooling, again the platens then were put in the kiln for second heat cycle about 2 hours and left to cool down to room temperature. The platens surface were cleaned and smoothed including the center hole.

A temperature of 550 °C is necessary because, the platens must get the maximum tensile strength, ductility and toughness due to increasing coalescence of carbides. The 2 hours period for second heating (tempering) is also necessary in order to aid in restoring the toughness as recommended by Garmo et. al.

2.3 *Stability test*

Servo Controller Compression Machine was used for the stability test. The machine is fully computerized and a program has been created in order to run the correct stability test. The machine software will be able to detect the failure by itself correspondence to the program.

The wellbore model was then put at the centre of the servo machine. The fabricated platens were then placed carefully on the top and bottom of the wellbore model. The wellbore model was then loaded at rate of 0.7 $MN/m^2/sec$ until the wellbore model failed. In other part a plot of axial load versus displacement was plotted by the software on top of the raw data. Thereafter the model failed, any sand particles produced were carefully collected for the sieve analysis.

2.4 *Sieve analysis*

The sand particles produced after the wellbore model failure were collected and weighed before being sieved. This weight will be recorded as recovered weight. The laboratory disc of the sieve machine was then carefully emptied and clean on the nested sieves. As precaution and minimizing the experiment error the sieve column was closed off with cover to avoid contamination by dust or loss of sand particles.

The nested sieve column was then shaken for 10 minutes by using electric sieve shaker (Vicker (1978), Craigh (1978) and Wills (1979)). After 10 minutes, the content of each sieve was carefully brushed off and then weighed. Consequently, the weighed for each particle size range as defined by the mesh size range of successive sieves was calculated. The amount of oversized and undersized sand particles were then calculated. Cumulative percentage oversized is defined as cumulative percentage of particle retained on the sieves. The cumulative percentage undersized is the cumulative percentage of particle passing through all the sieves and collected at bottom of the sieve shaker.

3 RESULTS AND DISCUSSION

Table 1 shows details of the sandstone basic mechanical properties.

In general, all the perforated wellbore models failed and sand particles were produced. Basically, the perforated wellbore stability, sand particles produced and the particles size distribution depends on the shot density and perforation pattern. Table 2 shows the summarize results of laboratory results.

Table 1 - Mechanical properties of sandstone

PARAMETER	AVERAGE VALUE
Density, ρ	2.03 g/cc
Porosity, ϕ	11 %
Permeability, K	8.63 mD
Compressive Strength, C_o	32 MN/m^2
Tensile Strength, T_O	1.82 MN/m^2
Young Modulus, E	6.95 GN/m^2
Poisson Ratio, ν	0.3
Triaxial Shear Strength, S_O	12.90 MN/m^2
Angle of Friction, Φ	39.39°

Table 2 - Summary of the results

Factor Increased	Aspect		
	Perforated Wellbore Stability	Sand Particles Produced	Size Distribution (> 500 μm)
Shot Density	Decreases	Increases	Increases
Spiral Pattern	Strongest	Lowest	Lowest
Inplane Pattern	Intermediate	Intermediate	Intermediate
Inline Pattern	Weakest	Highest	Highest

3.1 Effects of shot density to wellbore stability

From observation found that all models with various shot density failed under compression and shear stress. In general, it can be said that the wellbore stability decreases as the shot density increases.

Figure 1 shows the effect of shot density to wellbore instability. Found that, for 6 SPF the model failed at 24.14 MN/m^2 for spiral pattern, 15.83 MN/m^2 for inplane pattern and 14.08 MN/m^2 for inline pattern. As for the 8 SPF the spiral pattern failed at 19.55 MN/m^2 followed with 13.61 MN/m^2 for inplane pattern and the inline pattern failed at 10.97 MN/m^2. Finally the for 16 SPF model exhibit the weakest model which failed at 16.85 MN/m^2, 10.15 MN/m^2 and 7.69 MN/m^2 for spiral, inplane and inline pattern respectively.

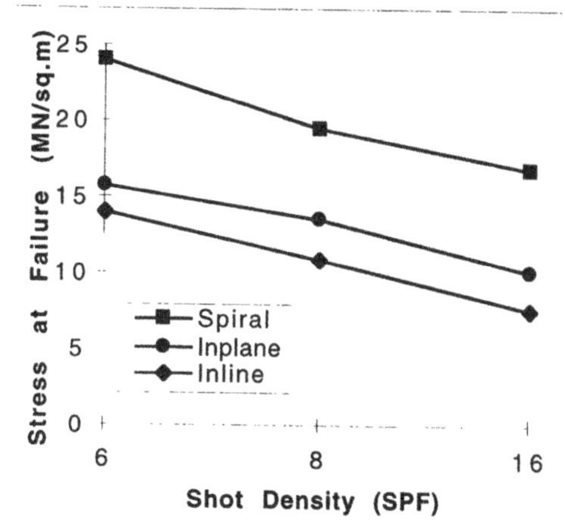

Figure 1 - Relationship between shot density and wellbore stability

The shot density increases can be defined as reducing the amount of the rock mass. Thus more stress have to be redistributed. As the results, the rock mass strength being reduced since have to carry the rearrange stress due to perforation process. Greater stress concentration in the rock between perforation tunnel can be experienced as the number of shot density increases.

Therefore, 16 SPF shot density will be exposed to more stress concentration compared to 8 SPF and 6 SPF. Thus the 16 SPF wellbore model failed at lower axial stress since the wellbore couldn't hold much higher stress and the yield point has been achieved earlier compare to 8 SPF and 16 SPF models.

3.2 Effect of perforation pattern to wellbore stability

Perforation pattern has a great effect on the wellbore stability. As can be seen in Figure 2, the stresses at failure are changing correspond to the perforation patterns depending upon the shot density. These results show that the spiral pattern gives the most stable wellbore. Then followed by inplane pattern and the inline pattern gives the least stable welllbore.

The above phenomenon understandable since, the perforation tunnels in the inline pattern are in one vertical line which is parallel to the applied load. Thus, resulting in a rock mass stress to the applied vertical stress which is lower than for an inplane

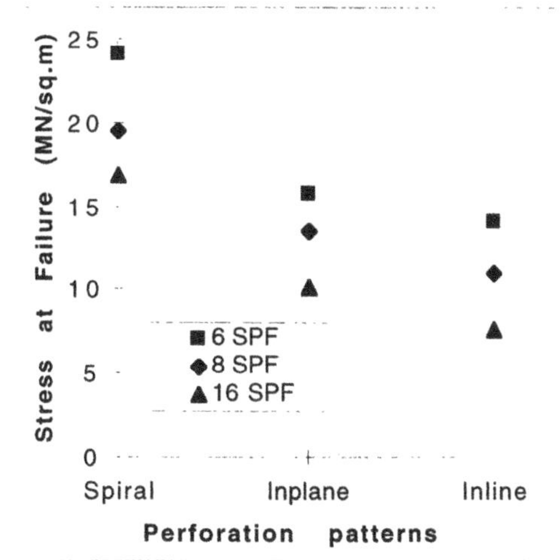

Figure 2 - Relationship between perforation pattern and wellbore stability

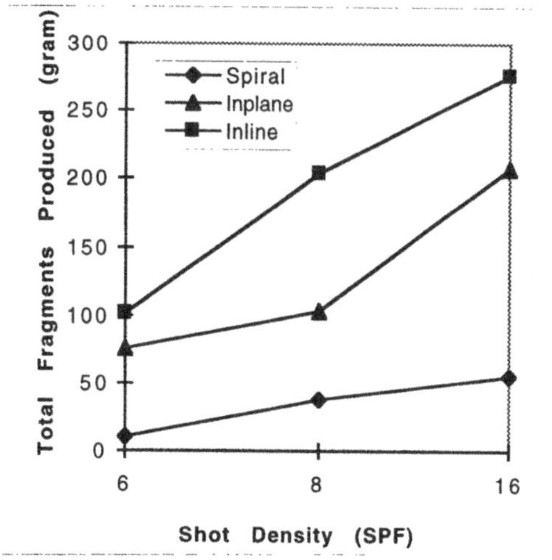

Figure 3 - Relationship between shot density and sand fragments produced

pattern where the perforation tunnels are in one horizontal line perpendicular to applied vertical/axial load. Therefore. the rock mass stress to the applied vertical stress is higher in inplane pattern than inline pattern.

As for the spiral pattern, the perforation tunnels are in a plane inclined to the applied vertical stress, resulting in higher rock mass stress to the applied vertical stress than for the other two previous patterns. The spiral pattern also produces the greatest distance between each successive perforation and therefore a stronger perforated structure.

3.3 Effect of shot density to sand fragment produced.

Figure 3 shows that increasing the shot density from 6 to 16 SPF will increases sand fragment produced from 11.08 gram to 55.28 for spiral pattern. Whereas for inplane pattern the increment was from 74.67 gram to 208.25 gram. Sand fragment produced for inline pattern was from 101.15 to 275.91 gram.

The results show that the amount of sand fragment produced by collapse perforated wellbore

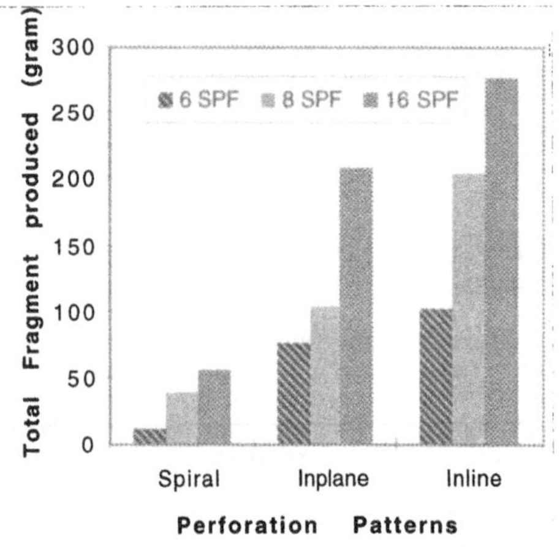

Figure 4 - Relationship between perforation pattern and sand fragments produced

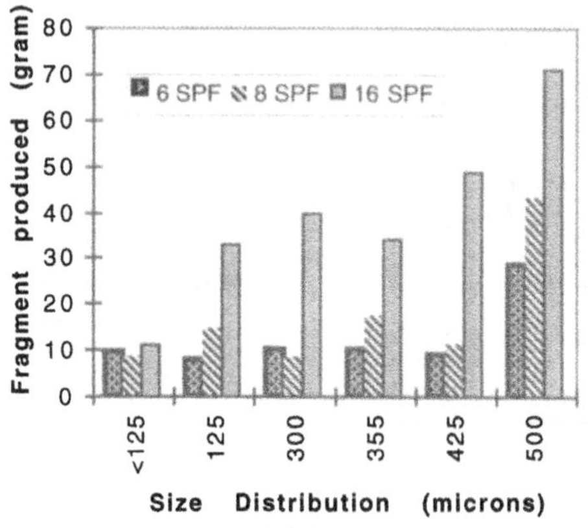

Figure 6 - Size distribution for inplane pattern

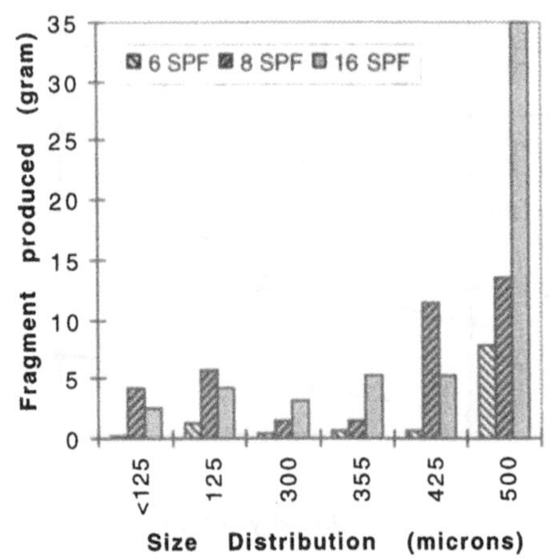

Figure 5 - Size distribution for spiral pattern

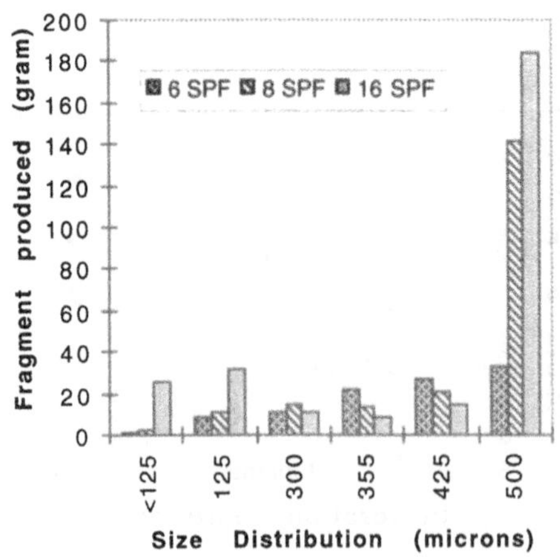

Figure 7 - Size distribution for inline pattern

increases as the shot density increases. The overall results also shows that the sand fragments produced depends on perforation pattern and phasing angle.

3.4 Effect of perforation pattern to sand fragment produced.

Changing the phasing angle of perforation is actually changing the perforation pattern. For instant phasing angle changes from 0° to 90° also means changing the perforation pattern from inline to inplane or spiral.

Figure 4 shows that as the perforation changes from inline to inplane and spiral (which also changes of phasing angle from 0° to 90°) the amount of sand fragments produced at failure decreases, since the perforated wellbore stability increases. The effect of perforation pattern/phasing angle also depends on the shot density. The effect becomes greater as the shot density increases.

As the results it can be said that the sand fragment produced has close relationship with wellbore instability. Where as the perforated wellbore stability increases the amount of sand fragment produced decreases.

3.5 Size distribution of sand fragments produced

Understanding the sand fragment size is important for further studies and application or controlling of the sand production. The size distribution of the sand fragments produced are as shown in Figure 5, 6 and 7 for spiral inplane and inline, respectively.

For spiral (Figure 5) pattern, the oversized 500 micron of sand fragments produced increases as the shot density increases. The increment is 10.63% to 33.58%. The inplane pattern also exhibit (Figure 6) the same phenomenon where the 500 microns size sand particles increased from 6.43 % to 15.9 % as the shot density increases. As for the inline pattern (as shown in Figure 7) the oversized fragments increases from 5.57 % to 31.73 % as the shot density increases from 6 SPF and 16 SPF.

Generally, less 500 microns sand fragments were produced by decreasing the shot density and by changing the perforation pattern from inline to inplane and spiral. Thus, it can be concluded that the spiral perforation pattern with 6 SPF shot density produced the least large sand particles.

4 CONCLUSIONS

It can be concluded that all perforated wellbore failed as the in-situ stress increases and produced sand fragments depending on shot density, perforation pattern and phasing angle. The wellbore stability decreases as the shot density increases. The spiral pattern appears to be most stable pattern followed by inplane and inline pattern.

The sand fragments produced increases as the wellbore stability decreases i.e. shot density increases. The amount of sand fragments produced also increases as the perforation pattern changes from spiral to inplane and inline. Whereas the same phenomenon occured as the phasing angle changes from 90° to 0°.

Big portion of the sand fragment produced were found by sieve analysis to be oversized 500 microns. The sand fragments created within the perforated wellbore or the rock adjacent to the perforation tunnel surface will contribute to any sand production problems.

Generally, stable perforated wellbore can minimizing the sand fragments production in the wellbore, therefore minimizing the sand production problems.

Understanding the effect of the perforation parameters i.e. shot density and perforation pattern to the wellbore stability and sand production, optimization of production and minimizing the sand production problem can be done. Consideration of the wellbore stability effects in designing phase of petroleum field development can be accomplished the optimizations program.

REFERENCES

Craigh, R.F," Soil Mechanics ", 2nd. edition, Van Nostrand Reinhold Co., 1978

Jaeger, J.C, Cook & N.G.W., " Fundamental of Rock Mechanics", 3rd edition, Chapman & Hall London, 1979.

Obert, L & Duvall, W.I., " Rock Mechanics & The Design of Structure in Rock", John Wiley & Sons Inc., 1967.

Vickers,B , " Laboratory Work in Civil Engineering", Granada Pub., London, 1978.

Wills, B.B, " Mineral Processing Technology " , Pergaman Press, Oxford, 1979

Samsuri. A. A study of Perforation Stability by Physical and Numerical Modelling. PhD Thesis, University of Straytlyde. 1990

Subbiah,S.K. Wellbore Instability Studies by Physical Modelling. MEng. Thesis, University Technology Malaysia.1997.

Author index

Printed and bound by CPI Group (UK) Ltd, Croydon, CR0 4YY
23/10/2024
01777679-0018